AF694034

Nijkamp and Parnham's Principles of Immunopharmacology

Michael J. Parnham
Frans P. Nijkamp • Adriano G. Rossi
Editors

Nijkamp and Parnham's Principles of Immunopharmacology

Fourth revised and extended edition

Springer

Editors
Michael J. Parnham
Translational Drug Validation
Fraunhofer IME-TMP and
Goethe University Frankfurt,
Frankfurt am Main, Germany

Frans P. Nijkamp
Utrecht University
Utrecht, The Netherlands

Adriano G. Rossi
The Queen's Medical Research Institute
MRC Centre for Inflammation Research
University of Edinburgh
Edinburgh, UK

ISBN 978-3-030-10809-0 ISBN 978-3-030-10811-3 (eBook)
https://doi.org/10.1007/978-3-030-10811-3

This Springer imprint is published by the registered company Springer Nature Switzerland AG
The registered company address is: Gewerbestrasse 11, 6330 Cham, Switzerland

Preface to the First Edition

The rapid developments in immunology in recent years have dramatically expanded our knowledge of mammalian host defence mechanisms. The molecular mechanisms of cellular interactions during immune responses have been unravelled, the intracellular responses involved in signal transduction delineated and an ever-increasing number of soluble mediators of immune and inflammatory responses have been discovered.

The initial result of this explosion of knowledge has been to provide the researcher and the clinician with an arsenal of diagnostic tools with which the immunological bases of disease processes can be investigated. This has made disease diagnosis much more precise, enabling the physician to tailor therapy much more closely to the individual patient's needs. However, better understanding of disease processes only provides a gradual improvement in therapy. This is because the new molecular targets that have been uncovered must first be tested as potential bases for immunomodulatory drug actions and then the new compounds must be subject to extensive development studies. As a result of the molecular unravelling of the immune system, we now understand more precisely the mechanisms of action of some established therapies, such as anti-allergic and anti-asthma agents, including the corticosteroids. New rational treatments based on molecular mechanisms also are now entering clinical practice and are making their mark on cancer, infectious and autoimmune disease therapy.

Concomitantly with these advances in understanding of molecular mechanisms of immunity, immunodiagnosis and immunotherapy, it has become possible to test more accurately the way in which a variety of drug classes interact with the immune system. It is of particular importance to regulatory authorities that the toxic side-effects of immunomodulatory drugs can be distinguished from their beneficial therapeutic effects.

Currently, it is only possible to obtain an overview of these various aspects of immunopharmacology by reading a range of immunological, pharmacological, diagnostic and toxicological literature. Good immunological textbooks are available, while immunopharmacology is covered mainly in terms of the inflammatory response. *Principles of Immunopharmacology* is intended to provide for the first time in a single volume a basic understanding of immunological mechanisms, a review of important immunodiagnostic tools and a description of the main pharmacological agents which modify the immune response, together with an introduction to immunotoxicology. As such we hope that it will be useful as a reference text for physicians, researchers and students with a rudimentary knowledge of immunology.

We, the editors, are grateful to all the authors who have invested their time and effort into this volume. We have received continuous help and encouragement from Petra Gerlach and Katrin Series of Birkhäuser Verlag and particular thanks are due to Dinij van der Pal for administrative assistance.

Utrecht, The Netherlands
Frankfurt, Hessen, Germany
March, 1999

Frans P. Nijkamp
Michael J. Parnham

Preface to the Second Edition

Our knowledge of immunological processes and their modulation has progressed considerably since the first edition of *Principles of Immunopharmacology*. Molecular mechanisms have been elucidated so that we are now in a position to understand many of the complex pathways leading from surface stimulation to cellular responses. We now appreciate much better that the innate immune response is also regulated by far more external and internal stimuli than was previously realised and are starting to understand the role of memory and regulatory cells. Advances in genomics and proteomics have enabled the identification of many genes and new proteins that are intimately involved in the responses of the immune system. We have sought to include the most important of these advances in the first part of this second edition of *Principles of Immunopharmacology*. In addition to including new mechanisms in the section on the immune response, we have also included the new techniques of the genomic and proteomic revolution in the diagnostics section, as methods such as microarrays have now become an essential aspect of cellular analyses.

Inevitably, our increased understanding of immune mechanisms has opened opportunities for the development of novel drugs to treat inflammation and disorders of the immune system. Biologicals are now commanding worldwide interest, both in research and development and in clinical practice. The section on therapy has now been expanded to accommodate these new therapeutic approaches, as well as describing our improved understanding of the mechanisms of action of established agents. The final section on immunotoxicology has also been updated, particularly with regard to new regulatory changes.

While maintaining the unique approach of providing sections on immunology, immunodiagnostics, therapy and immunotoxicology in a single volume, we have also introduced a new double-column format to provide easier access to the text. Important statements are highlighted and instead of giving annotations in the margins, key terms are now indicated in the text and presented in a glossary[1] at the end of the book. A new appendix summarises important characteristics of commercially available therapeutic agents.

[1]Words included in the glossary are highlighted in the text with CAPITAL LETTERS.

We are very grateful to many of the contributors to the first edition, who have kindly revised and modified their chapters, as well as to the additional authors who have added totally new information. The preparation of this second edition has been the result of close collaboration with Dr. Hans-Detlef Klueber and his colleagues at Birkhäuser Verlag. Thanks for all your help, advice and hard work. We hope you, the reader, will find the new edition useful and informative.

Utrecht, The Netherlands
Frankfurt, Hessen, Germany
March, 2005

Frans P. Nijkamp
Michael J. Parnham

Preface to the Third Edition

The last 5–6 years since the publication of the second edition of *Principles of Immunopharmacology* have seen several notable changes in our understanding of the immune system. Not least of these has been the expansion of the number of defined T cell subsets with the rapid appreciation of their roles in infectious and autoimmune diseases. In parallel, the spotlight has fallen increasingly on the contribution of sub-populations of dendritic cells and the recognition that these cells represent potential targets for drug therapy. These areas are given detailed attention in this third edition, together with a new chapter on pre- and probiotics and a greater emphasis on therapeutic biologicals, which are providing the major thrust in new immunomodulatory drug development. All other chapters have been updated by the authors, bringing the book in line with the latest progress in each field. The format of the textbook, with sections on immune mechanisms, diagnostics, therapeutic agents and immunotoxicology, a glossary of keywords and the collection of appendices, has been retained and illustrations have been improved. An introduction has been added, providing an overview of inflammation, the immune response and its pharmacological modification as a framework for the subsequent details given in each chapter. Most of these improvements have been made in response to comments from authors and readers, which we much appreciate.

Once again, we are very grateful to all the contributors for taking the time to revise or contribute new chapters. We are particularly grateful to Els Tange-Bijl for her organisational talent and expert co-ordination of the contacts with authors and her persistence in pressing them to deliver the manuscripts. We also thank Anke Brosius and her colleagues at Springer Basel for the professional desk editing of the final text and Dr. Hans-Detlef Klueber for sustaining the publisher's support and guidance for the project. We hope you, the reader, find the final product informative and useful.

Utrecht, The Netherlands
Frankfurt, Hessen, Germany
October, 2010

Frans P. Nijkamp
Michael J. Parnham

Preface to the Fourth Edition

With the very rapid and extensive advances in the therapy of immunological disorders, it is now barely possible with a multi-authored book to keep up with all the most recent changes in the field. The variety of approaches to both diagnosis and therapy has diversified to cover genetic, cellular, biological and small molecular agents. We have sought to cover the most important advances, retaining the format of the previous volumes with sections on immunity, immunodiagnostics, immunotherapy and immunotoxicology and an emphasis on explanatory illustrations. In this new edition, all chapters have been thoroughly revised, incorporating suggestions provided by previous contributors and academic teachers and in several cases rewritten by new authors. New chapters have been added to cover epigenetic regulation of immunity, imaging of inflammation, treatment of sepsis, therapy of soft tissue autoimmunity and the use in autoimmune disease of stem cell transplantation.

We are again very grateful to all the hard work of the authors for their contributions, many for the third or fourth time. We also give a special word of thanks to Garry G. Graham who kindly checked some of the manuscripts. This volume has taken some time to complete as the publisher went through radical changes and we are grateful to the continued support provided at Springer by Sajeni Ravindranatha Das, her predecessor Sowmya Ramalingam, Sushil Kumar Sharma, C. Dulcy Nirmala, Ingrid Fischer, Dr. Martina Himberger and Dr. Hans-Detlef Klueber who has been with the project since the second edition. We hope this volume remains as useful and informative to the reader as the previous editions.

Frankfurt, Hessen, Germany	Michael J. Parnham
Utrecht, The Netherlands	Frans P. Nijkamp
Edinburgh, UK	Adriano G. Rossi
November, 2018	

Contents

Abbreviations

α	Alpha
β	Beta
γ	Gamma
Ω-3	Omega 3
ω-3	Omega 3
3-HK	3-Hydroxykynurenine
5-FU	5-Fluorouracail
5-HT	Serotonin
A23187	Calcium ionophores
AA	Arachidonic acid
Ab	Antibody
ACE	Angiotensin converting enzyme
ACh	Acetylcholine
ACTH	Adrenocorticotropic hormone
ADA	Antidrug antibody
ADCC	Antibody-dependent cellular cytotoxicity
ADCP	Antibody-dependent cell PHAGOCYTOSIS
ADP	Adenosine diphosphate
ADRs	Adverse drug reactions
AE	Adverse event
AFC assay	Antibody forming cell assay
Ag	Antigen
AGM	Aorta-gonad-mesonephros
AgRP	Agouti-related protein
AICD	Activation-induced cell death
AID	Activation-induced cytidine deaminase
AIDS	Acquired immune deficiency syndrome
ALG	Antilymphocyte globulin
ALL	Acute lymphocytic leukemia
ALXR	Affinity G-protein-coupled lipoxin receptors
AMD	Age-related macular degeneration
AMP	Adenosine monophosphate
ANC	Absolute neutrophil count
ANLL	Acute non-lymphocytic leukemia
AOL	Amine oxidase-like domain
APC	Allophycocyanin
APC	Antigen-presenting cell

APRIL	A proliferation-inducing ligand
ARDS	Acute respiratory distress syndrome
Asn	Asparagine
ATG	Antithymocyte globulin
ATL	Aspirin-triggered lipoxins
ATP	Adenosine triphosphate
ATRA	All-trans retinoic acid
AZA	Azathioprine
BAFF	B cell-activating factor
BALT	Bronchus-associated lymphoid tissue
BBB	Brain-blood barrier
BCG	Bacillus Calmette-Guérin
Bcl-2	B-Cell lymphoma 2
BCMA	B cell maturation antigen
BCNU	1,3-bis(2-chloroethyl)-1-nitrosourea
BCR	B cell receptor
BDCA	Blood dendritic cell antigen
BET	Bromodomain and extraterminal
BH_4	Tetrahydrobiopterin
BLNK	B cell linker protein
BM	Bone marrow
BMD	Bone mineral density
BMT	Bone marrow transplantation
BRM	Biological response modifier
BT	Buehler test
Btk	Bruton's tyrosine kinase
BU	Busulfan
C/EBP	CCAAT/enhancer binding proteins
C1q	Complement component 1q
C3a	Complement component 3a
C5a	Complement component 5a
Ca^{2+}	Calcium ions
$[Ca^{2+}]i$	Intracellular calcium ions
CAF	Cancer-associated fibroblast
CAP	Community-acquired pneumonia
CAPS	Cryopyrin associated periodic syndromes
CBP	CREBBP; CREB binding protein
CCR7	C-C chemokine receptor type 7
CD	Cluster of differentiation, e.g., CD3, CD8, etc.
CD40L	CD40 ligand
CDC	Complement-dependent cytotoxicity
cDC	Conventional dendritic cell
cDNA	Complementary deoxyribonucleic acid
CDP	Common dendritic progenitor
CDR	Complementarity determining region
CFA	Complete Freund's adjuvant
CGD	Chronic granulomatous disease
cGMP	Cyclic guanosine monophosphate

CGRP	Calcitonin gene-related peptide
CHD1	Chromodomain helicase DNA binding protein 1
CHMP	EU Committee on Human Medicinal Products
C_HN	Constant domain number N of the heavy chain
CIAS1	Cold-induced Autoinflammatory Syndrome 1
CIDP	Chronic inflammatory demyelinating polyneuropathy
CILP	Common ILC progenitor
CLC	Cardiotrophin-like cytokine
CLR	C-type lectin receptor
CML	Chronic myelogenous leukemia
CMV	Cytomegalovirus
CNS	Central nervous system
CNTF	Ciliary neurotrophic factor
COPD	Chronic obstructive pulmonary disease
COX	Cyclooxygenase
CP	Choroid plexus
CpG	Cytosine-phosphate-guanine
CR	Complete response
CRH	Corticotrophin-releasing hormone
CSA	Cyclosporine A or Ciclosporin
CSF	Cerebrospinal fluid
CSF	Colony-stimulating factor
CT-1	Cardiotrophin-1
CTA	Cancer/testis antigen
CTL	Cytotoxic T lymphocyte
CTLA	Cytotoxic T-lymphocyte antigen
CVID	Common variable immunodeficiency
CX3CL1	Fractalkine
CY	Cyclophosphamide
CYP	Cyclophilin
cysLT	Cysteinyl leukotrienes
Da	Dalton
DA	Dopamine
DAG	Diacylglycerol
DAMP	Damage-associated molecular pattern
DC	Dendritic cell
DC-SIGN	Dendritic cell-specific intracellular adhesion molecule 3-grabbing non-integrin
DFS	Disease-free survival
DHODH	Dihydroorotate dehydrogenase
DIT	Developmental immune toxicity
DLI	Donor leukocyte infusion
DLN	Draining lymph nodes
DMARD	Disease-modifying antirheumatic drug
DMR	Differently methylated regions
DNA	Deoxyribonucleic acid
DNMT	DNA methyltransferases
DP	Double-positive

DP	PGD receptor
DTH	Delayed-type hypersensitivity
DTP	Diphtheria-tetanus-polio
E2A	TCF3; transcription factor 3
EBF1	Early B-cell factor 1
EBV	Epstein-Barr virus
ECM	Extracelular matrix
EDTA	Ethylenediaminetetraacetic acid
EGCG	(−)-Epigallocatechin-3-gallate
eGPx	Extracellular glutathione peroxidase
EGR2	Early growth response 2
ELAM	Endothelial-leukocyte adhesion molecule
ELISA	Enzyme-linked immunosorbent assay
ELISpot	Enzyme-Linked ImmunoSpot
ELISPOT	Enzyme-linked immunospot assay
EMA	European Medicines Agency, formerly EMEA
eNOS	Endothelial nitric oxide synthase
EOGRTS	Extended One-Generation Reproduction Toxicity Study
EP	PGE receptor
EPO	Erythropoietin
ERK	Extracellular signal-regulated kinases
EU	European Union
EZH	Enhancer of zeste 1 polycomb repressive complex 2 subunit
$F(ab')_2$	Divalent antigen-binding fragment of immunoglobulins
Fab	Fragment antigen-binding, antigen-binding portion of immunoglobulin
FACS	fluorescence-activated cell sorting
FasL	Fas ligand
Fc	Fragment crystallisable, constant portion of immunoglobulin, mediates binding to Fc receptors
FcR	Fc receptor
FcRn	Neonatal Fc receptor for IgG
FcαR	Fc-alpha receptor
FcγR	Fc-gamma receptor
FcγRI	Fc gamma receptor I
FcδR	Fc-delta receptor
FcεR	Fc-epsilon receptor
FDA	Food and Drug Administration
FGF	Fibroblast growth factor
FKBP	FK506-binding protein
FLAP	Five-lipoxygenase-activating protein
Flt3/FLT3	Fms-like tyrosine kinase 3
FOS	Fructo-oligosaccharides
FOXO1	Forkhead box O1
FOXP3	Forkhead box P3
FP	PGF receptor
FXIa	Coagulation factor XIa

GALT	Gut-associated lymphoid tissue
GATA	GATA binding protein
GBS	Guillain-Barré Syndrome
GCs	Glucocorticoids
G-CSF	Granulocyte colony-stimulating factor
GI	Gastrointestinal
GIST	Gastrointestinal stromal tumors
GM-CSF	Granulocyte macrophage-colony stimulating factor
GOS	Galacto-oligosaccharides
GPMT	Guinea pig maximization test
GPx	Glutathione peroxidase
GRP	Gastrin-releasing peptide
GRs	Glucocorticoid receptors
GVHD	Graft-versus-host disease
GVT	Graft-versus-tumor
H_2O_2	Hydrogen peroxide
H3K27ac	Histone 3 lysine 27 acetylation
H3K27me3	Histone 3 lysine 27 trimethylation
H3K36me	Histone 3 lysine 36 trimethylation
H3K4me1	Histone 3 lysine 4 monomethylation
H3K79me	Histone 3 lysine 79 methylation
H3K9me3	Histone 3 lysine 9 trimethylation
H4K20	Histone 4 lysine 20
HAART	Highly active antiretroviral therapy
HAT	Histone acetyltransferases
HBsAg	Hepatitis B surface antigen
HCMV	Human cytomegalovirus
HD	Hodgkin's disease
HDAC	Histone acetyltransferases
HDT	High-dose chemotherapy
HGPRT	Hypoxanthine-guanine phosphoribosyltransferase
Hib	*Haemophilus influenzae* type b
HIV	Human immunodeficiency virus
HLA	Human leukocyte antigen
HMGB	High-mobility group box
HMW	High molecular weight
HPA	Hypothalamic-pituitary-adrenal
HPETE	Hydroperoxyeicosatetraenoic acids
HPV	Human papillomavirus
HSC	Hematopoietic stem cell
HSCT	Haematopoietic stem cell transplantation
HSP	Heat shock protein
i.m.	Intramuscular
i.v.	Intravenous
IAC	Immunoaffinity chromatography
IBD	Inflammatory bowel disease
ICAM	Intercellular adhesion molecule
ICE	IL-1α-converting enzyme

ICH	International Council for Harmonisation of Technical Requirements for Pharmaceuticals for Human Use
ICOS-ligand	Inducible costimulator ligand
ID2	Inhibitor of DNA binding 2
IDO	Indolamine 2,3 dioxygenase
IEC	Intestinal epithelial cells
IFI44L	Interferon induced protein 44 like
IFN	Interferon
IFN-I	Type I interferon
IFNα	Interferon alpha
IFNγ	Interferon gamma
Ig	Immunoglobulin
IgA	Immunoglobulin of class A
IgD	Immunoglobulin of class D
IgE	Immunoglobulin of class E
IGF	Insulin-like growth factor
IgG Ab	Immunoglobulin G antibodies
IgG	Immunoglobulin of class G
IgH	Ig heavy chain
IgM	Immunoglobulin of class M
IIC	Infiltrating immune cell
IKAROS	IKAROS family zinc finger 1
IL	Interleukin
IL-1	Interleukin 1
IL-1RA	IL-1 receptor antagonist
IL-1RAcP	IL-1 receptor accessory protein
IL-2	Interleukin 2
IL-4	Interleukin 4
IL-6	Interleukin 6
IL-10	Interleukin 10
IL-12	Interleukin 12
ILC	Innate lymphoid cell
ILSI	International Life Sciences Institute
IMIG	Intramuscular immunoglobulin
IMPDH	Inosine monophosphate dehydrogenase
iNOS	Inducible nitric oxide synthase
IP	Inducible protein
IP	PGI_2 receptor (Chap. 9)
IP-1	Inflammatory protein-1
IP-10	Interferon c-induced protein 10 (CXCL10)
IP_3	Inositol trisphosphate
IPV	Inactivated polio vaccine
IQPP	International Quality Plasma Program
IRF	Interferon regulatory factor
IRS	Insulin receptor substrates
IRSE	IFN-stimulated response elements
ISP	Immature single-positive
ITAM	Immunoreceptor tyrosine-based activation motif

ITIM	Immunoreceptor tyrosine-based inhibitory receptor
ITP	Immune thrombocytopenia purpura
iTregs	Induced regulatory T cells
IU	International unit
IV	Intravenous
IVIG	Intravenous immunoglobulin
JAK	Janus kinase
JmjC	Jumonji C
JNK	c-Jun-N-terminal kinases
KA	Kynurenic acid
KDM	Histone lysine demethylases
KIR	Killer immunoglobulin-like receptor
KLH	Keyhole limpet haemocyanin
KMT	Lysine methyltransferase
KO	Knockout
L	Ligand
LAF	Lymphocyte activation factor
LAG	Lymphocyte-activation gene
LAK	Lymphokine-activated killer
LBP	LPS binding protein
LC	Langerhans cells
LDL	Low-density lipoproteins
LEMS	Lambert-Eaton myasthenic syndrome
LIF	Leukemia inhibitory factor
LLNA	Local lymph node assay
LMW	Low molecular weight
LN	Lymph node
L-NAME	L-N^G-nitro-arginine methyl ester
L-NMMA	L-N^G-monomethyl arginine
LOXs	Lipoxygenases
LPS	Lipopolysaccharide
LRP4	Lipoprotein receptor-related protein 4
LSK	Lineage marker negative, Sca-1+ and C-kit+
LT	Lymphotoxin
LTi	Lymphoid tissue inducer cells
LTs	Leukotrienes
MAA	Melanoma-associated antigens
mAb	Monoclonal antibody
MALT	Mucosa-associated lymphoid tissue
MAPK	Mitogen-activated protein kinase
MBL	Mannan-binding lectin
MCD	Multicentric Castleman's diesease
MCP	Monocyte chemoattractant protein
M-CSF	Macrophage colony-stimulating factor
MDP	Macrophage/monocyte dendritic progenitor (Chap. 5)
MDP	Muramyl dipeptides
MDR	Multidrug-resistant
MDSC	Myeloid-derived suppressor cell

MG	Myasthenia gravis
MHC	Major histocompatibility complex
MHLW	Ministry of Health, Labour and Welfare in Japan
mIg	Membrane-bound immunoglobulin
MIP-1	Macrophage Inflammatory Protein-1
MLL	Mixed lineage leukemia
MLV	Multilamellar vesicles
MM	Malignant melanoma
MMF	Mycophenolate mofetil
MMN	Multifocal motor neuropathy
MMP	Matrix metalloproteinase
MMR	Measles-mumps-rubella
MoA	Mechanism of action
MODS	Multiple organ dysfunction syndrome
MOX	Monooxygenase
MPA	Mycophenolate acid
MPL	Monophosphoryl lipid A
mRNA	Messenger ribonucleic acid
MS	Multiple sclerosis
MSC	Mesenchymal stem cell
MT	Metallothionein
MTD	Maximum tolerated dose
mTOR	Mammalian target of rapamycin
MTP-PE	Muramyl tripeptide phosphatidylethanolamide
MTX	Methotrexate
MuSK	Muscle-specific kinase
NAb	Natural antibodies
NADPH	Nicotinamide adenine dinucleotide phosphate
NALP	NACHT, LRR and PYD containing proteins
NCI	National Cancer Institute
NE	Norepinephrine
NF-κB	Nuclear factor kappa-B
NIEHS	US National Institute of Environmental Health Sciences
NIOSH	US National Institute for Occupational Safety and Health
NK	Natural killer
NKA	Neurokinin A
NKB	Neurokinin B
NKR	Natural killer receptors
NKT	NK T lymphocyte
NLR	Nod-like receptor
NMDA	*N*-methyl-D-aspartate
nNOS	Neuronal nitric oxide synthase
NO	Nitric oxide
NOD	Nucleotide-binding oligomerization domain
NOXs	Non-phagocytic oxidases
NPY	Neuropeptide Y
NSCLC	Non-small cell lung cancer
NSIADs	Non-steroidal anti-inflammatory drugs

NTP	US National Toxicology Program
NuR77	nr4a1; Nuclear receptor subfamily 4, group A, member 1
O_2^-	Superoxide anion
ODN	Oligonucleotides
OECD	Organization for Economic Co-operation and Development
OID	Optimal immunomodulatory dose
$ONOO^-$	Peroxynitrite
OPV	Oral polio vaccine
ORR	Objective Response Rate
OS	Overall survival
OSM	Oncostatin M
p/gpXXX	Protein/Glycoprotein with the mass XXX
P300	E1A binding protein p300
PAF	Platelet-activating factor
PAMP	Pathogen-associated molecular pattern
PAX	Paired box 5
PBL	Peripheral blood leukocyte
PBMCs	Peripheral blood mononuclear cells
PCR	Polymerase chain reaction
PCR2	Polycomb repressive complex 2
PD	Pharmacodynamics
PD1/PD-1	Programmed cell death protein-1
pDC	Plasmacytoid dendritic cell
PDE5	Phosphodiesterase type 5 inhibitor
PDGF	Platelet-derived growth factor
PDGFR	Platelet derived growth factor receptor
PD-L1	Programmed cell death ligand-1
PE	Phycoerythrin
PEG	Polyethylene glycol
PerCP	Peridinin chlorophyll protein
PFS	Progression-free survival
PG	Prostaglandin
PGI_2	Prostacyclin
PI3-kinase	Phosphoinositide 3-kinase
PID	Primary immunodeficiency
pIgR	Polymeric immunoglobulin receptor
PIP2	Phosphatidylinositol 4,5-bisphosphate
PIP3	Phosphatidylinositol (3,4,5)-trisphosphate
PK	Pharmacokinetics
PKA	Prekallikrein activator
PKB	Protein kinase B
PKC	Protein kinase C
PKC-α	Protein kinase C-α
PLA_2	Phospholipase A_2
PLC	Phospholipase C
PLNA	Popliteal lymph node assay
PML	Progressive multifocal leukoencephalopathy
PMNL	Polymorphonuclear leukocytes

PPARγ	Peroxisome proliferator-activated receptor-gamma
PRA	Panel-reactive antibody
PRM	Pathogen recognition molecules
PRMT	Protein arginine methyltransferases
PrP^{Sc}	Prion protein scrapie
PRR	Pattern recognition receptor
PSCT	Peripheral stem cell transplant
PTK	Protein tyrosine kinase
PU.1	SPI1; Spi-1 proto-oncogene
PUFA	Polyunsaturated fatty acid
Px	Peroxidase
QA	Quinolinic acid
Qdot	Quantum dot
QOL	Quality of life
QSAR	Quantitative structure-activity relationship
QSEAL	Quality Standards of Excellence, Assurance and Leadership
RA	Reporter antigen
RA	Rheumatoid arthritis
RAGE	Receptor for advanced glycation end products
RANTES	Regulated on activation, normal T cell expressed and secreted
RA-PLNA	Reporter antigen popliteal lymph node assay
RBC	Red blood cells
RCC	Renal cell carcinoma
RCT	Randomized clinical trial
RDA	Recommended daily allowance
REACH	Registration, Evaluation and Authorisation of Chemicals
RFS	Recurrence-free survival
RhD	Rhesus D
rhIL	Recombinant human interleukin
RIG	Retinoid acid-inducible gene
RIVM	The National Institute for Public Health and the Environment in the Netherlands
RLR	RIG-like receptor
ROS	Reactive oxygen species
rRNA	Ribosomal RNA
Rrp	Receptor-related protein
RTK	Receptor tyrosine kinase
RUNX	Run-related transcription factor
s.c.	Subcutaneous
SAM	*S*-adenosyl methionine
SAR	Structure-activity relationship
SC	Secretory component
SCC	Squamous cell carcinoma
SCF	KIT ligand
SCID	Severe combined immunodeficiency
SCIG	Subcutaneous immunoglobulin
SCT	Stem cell transplantation

Sel P	Selenoprotein P
SEPT9	Septin 9
SETD2	SET domain-containing 2
sFcγRIIb	Soluble Fc-gamma receptor IIb
SH2	Src homology 2
SIgA	Secretory immunoglobulin class A
SIgM	Secretory immunoglobulin class M
SIRS	Systemic inflammatory response syndrome
SLE	Systemic lupus erythematosus
SLN	Sentinel lymph node
SNS	Sympathetic nervous system
SOD	Superoxide dismutase
SP	Single-positive (Chap. 2)
SP	Substance P
SPI-C	Spi-C transcription factor
SPM	Specialized pro-resolving mediator
SRBC	Sheep red blood cells
SRS-A	Slow-reacting substance of anaphylaxis
STAT	Signal transducer and activator of transcription
TAA	Tumor-associated antigen
TAC	Total antioxidant capacity
TACI	Transmembrane activator and calcium-modulating cyclophilin ligand interactor
TAM	Tumor-associated macrophage
TAMP	Tumor-associated molecular pattern
TAP	Transporter associated with antigen processing
TB	Tuberculosis
T-bet	TBX21; T-box 21
TCD	T-cell-depleted
TCGF	T cell growth factor
TCR	T cell receptor
TDO	Tryptophan dioxygenase
TdT	Terminal deoxynucleotidyl transferase
TEE	Thromboembolic events
TF	Transcription factor
Tfh	T follicular helper cell
TGF-ß	Transforming growth factor beta
T_H	T-helper cell
TIL	T-cell-infiltrating lymphocyte
TK	Protein tyrosine kinase
TLR	Toll-like receptor
TLRL	Toll-like receptor ligand
TM	Thrombomodulin
TME	Tumor microenvironment
TNF	Tumor necrosis factor
TNFR	TNF receptor
TNFSF	TNF super family
TNFα	Tumour necrosis factor alpha

TP	TXA_2 receptor
TPEN	*N*,*N*,*N*,*N*′-tetrakis (2-pyridylmethyl)ethylenediamine
TPO	Thrombopoietin
TR	Thioredoxin reductase
TRAIL	TNF-related apoptosis-inducing ligand
Treg	Regulatory T cell
TREM	Triggering receptor expressed on myeloid cells
TSA	Tumor-specific antigen
TUR	Transurethral resection
TXs	Thromboxanes
UNG	Uracil-N-glycosylase
VAP	Ventilator-associated pneumonia
VCAM	Vascular cell adhesion molecule
VEGF	Vascular endothelial cell growth factor
VIP	Vasoactive intestinal peptide
VLP	Virus-like particle
WHO	World Health Organization
XLA	X-linked agammaglobulinaemia
XSCID	X-linked severe combined immunodeficiency
ZA	Zoledronic acid

1 Initiation, Propagation and Resolution of Inflammation

Michael J. Parnham, Frans P. Nijkamp, and Adriano G. Rossi

Defence against invading organisms (such as bacteria, fungi and parasites) is an essential requirement for survival for all living organisms. Several protective mechanisms are shared across vertebrates and invertebrates and form the basic components of innate immunity. These include phagocytosis or engulfment of foreign particles (particularly by the leukocytes or white blood cells, the infantry of the host defence system) and also recognition of microbial components by pattern response molecules, such as Toll-like receptors. Such responses to invasion muster cellular and humoral defences in a co-ordinated attack strategy which we recognize as inflammation.

Final manuscript submitted on May 01, 2018.

M. J. Parnham (✉)
Translational Drug Validation, Fraunhofer IME-TMP and Goethe University Frankfurt, Frankfurt am Main, Germany
e-mail: parnham@med.uni-frankfurt.de; michael.parnham@ime.fraunhofer.de

F. P. Nijkamp
Utrecht University, Utrecht, The Netherlands

A. G. Rossi
Centre for Inflammation Research, University of Edinburgh, Edinburgh, UK

1.1 The Acute Inflammatory Response

The cardinal signs of inflammation are redness, heat, swelling, pain and sometimes loss of function, which provide unmistakable evidence of the battle taking place. This process represents the first line of defence. Initially, when a tissue is injured or invaded by an infectious pathogen, BLOOD COAGULATION and THROMBOSIS regulate bleeding and trap leukocytes. Local hormones or mediators, such as prostaglandin E (PGE) or histamine, cause the local blood vessels to dilate and increase the flow of blood as well as enhance the supply of "reinforcements". Under the influence of pro-inflammatory mediators (e.g. histamine, cytokines), the ENDOTHELIUM of the vessel wall becomes leaky, gaps being formed between the cells, providing the opportunity for activated blood proteins and cells to invade the damaged tissue. Mediators, such as cytokines and chemokines, then activate the leukocytes, guiding them along concentration gradients and adhesion molecules expressed on the surface of the cells act as "stepping stones" to guide the cells into the battle zone. Here the various mediators stimulate the cells to release microbe-killing constituents, such as reactive oxygen species or lysozyme and specific mediators which orchestrate the specific populations of leukocytes

M. J. Parnham et al. (eds.), *Nijkamp and Parnham's Principles of Immunopharmacology*,
https://doi.org/10.1007/978-3-030-10811-3_1

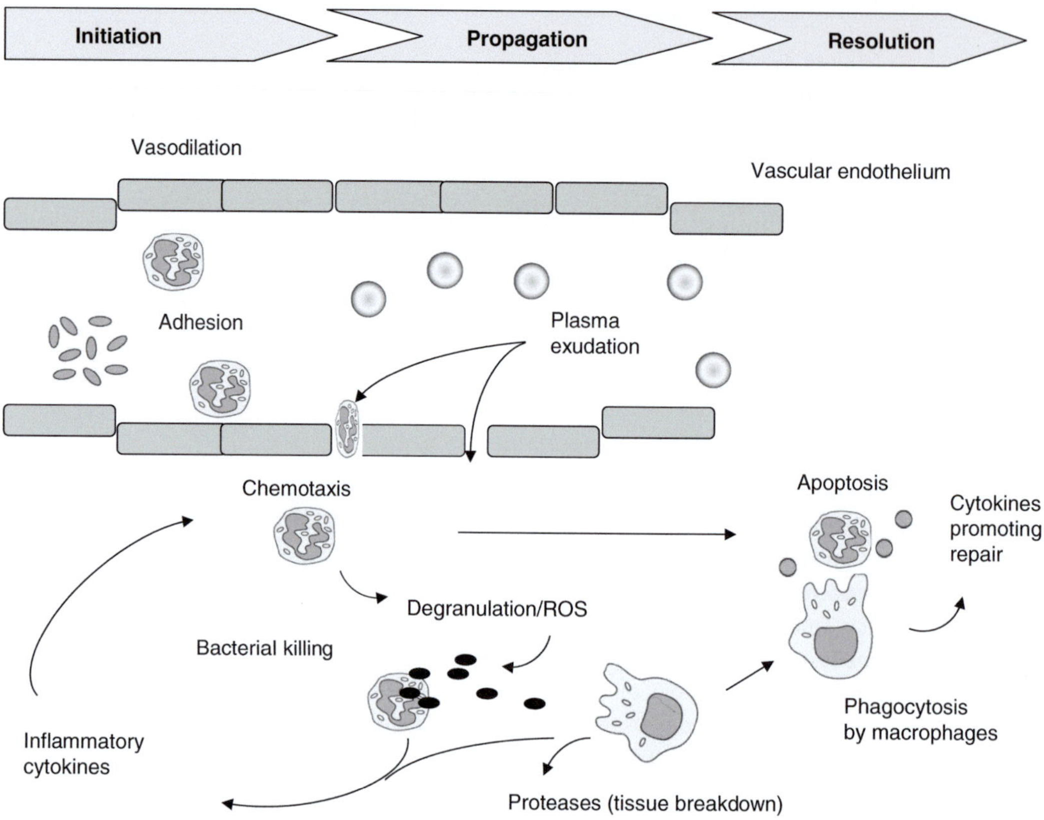

Fig. 1.1 Simplified presentation of acute inflammation. In the presence of an inflammatory stimulus, such as bacteria or tissue injury, inflammation is initiated by damage to microvessels causing blood coagulation and platelet aggregation. Local inflammatory mediators, including eicosanoids, amines and peptide cytokines and chemokines, are generated (particularly by tissue macrophages) and act on the vascular endothelium and circulating leukocytes to cause the leukocytes (initially neutrophils and potentially other populations at later time points) to adhere to the endothelium. Vasodilation also occurs. This facilitates already ongoing plasma exudation. The process is further propagated by diapedesis (transendothelial migration) of leukocytes along a concentration gradient of chemokines. Within the tissue, in contact with the bacteria or injurious stimulus, the leukocytes release bactericidal proteases and reactive oxygen species (ROS), which also cause bystander tissue injury and neutrophil death by necrosis. Tissue macrophages are activated and joined by monocytes entering from the circulation, which become macrophages. These cells then initiate the resolution phase, when neutrophils die by apoptosis. Apoptotic cells are phagocytosed by macrophages which start to generate repair-inducing products and cytokines which stimulate regrowth of surrounding tissue

needed to deal with the inflammatory stimulus (Fig. 1.1). The COMPLEMENT SYSTEM in the blood may also be activated to generate proteins which attract leukocytes to the inflamed site while immunoglobulins coat foreign particles to make them more easily subject to phagocytosis by binding to Fc receptors for immunoglobulins on the surface of the phagocytic cells. A primary bacterial infection elicits a defence reaction that is a typical example of this type of acute inflammatory response.

1.2 Resolution of Inflammation

In preparation for the successful defence of the tissue, already within a few hours, the "cleaning-up operation" is initiated. This is heralded by the suicide of the initial foot soldiers (such as NEUTROPHILIC LEUKOCYTES) by APOPTOSIS or programmed cell death. These cell carcasses and other debris are rapidly engulfed by MONONUCLEAR PHAGOCYTES, including macrophages, using specific receptors and molecules that result in dampening of the inflammatory response. During this active process, macrophages change their phenotype from pro-inflammatory to anti-inflammatory/tissue repair and release lipids termed SPECIALIZED PRO-RESOLVING MEDIATORS (SPMs) (e.g. lipoxins, resolvins, protectins, maresins) and other mediators (e.g. IL-10 and TGF-β) which also may directly or indirectly stimulate tissue repair and the regrowth of healthy tissue. Once again, local mediators direct this active resolution of the inflammatory process.

1.3 Adaptive Immunity

Adaptive immunity represents a sophisticated and finely tuned approach to defence, akin to the complex guided weaponry of modern military arsenals. The generation of specific antibodies by activated B lymphocyte cells or cellular immune responses, mediated by effector T lymphocyte cells and antigen-presenting phagocytes, allows the organism to neutralize invading microbes or foreign particles without necessarily causing discomfort to the host. This requires prior exposure to a foreign protein, the antigen, and its uptake by leukocytes, for instance, during a previous infection.

The intracellular processing of antigens by ANTIGEN-PRESENTING CELLS (APCs), in particular by the "professionals"—the DENDRITIC CELLS (DCs) in primary lymphoid organs—is followed by presentation of the antigen to the awaiting T lymphocytes. This occurs, in association with the self-restrictive MAJOR HISTOCOMPATIBILITY COMPLEX (MHC) MOLECULES, on the surface of the APCs. The T LYMPHOCYTES, crucial cells for the adaptive immune response, have already undergone a very rigorous selection process, based on genetic factors, antigenic stimuli and cellular interactions in the thymus, long before their recruitment to the defence force. Depending on the battle zone (antigen, pathogen, cytokines produced, etc.), a variety of subtypes of T cells are produced which determine the type of defence reaction to be mounted. T helper lymphocytes then turn their attention to the B LYMPHOCYTES, which have been recruited from the bone marrow. In concert with the DCs and other APCs, the T cells (within special areas of lymphoid organs) help B cells to respond to the specific antigen and to change into plasma cells which generate specific immunoglobulin antibodies. Now when a new exposure to the antigenic protein occurs, as in a renewed episode of infection, the antibodies bind antigen, forming immune complexes which activate complement and stimulate phagocytes to engulf the immune complexes and destroy the antigen (including infected cells which express the antigen on their surface).

1.4 Hypersensitivity and Chronic Inflammation

However, these complex responses, as with military weapons, can go awry and contribute to HYPERSENSITIVITY, involving cross-reactivity between an external molecule and a component of the host tissue, resulting in attack by the immune system on its own host tissues. In this way, immune reactions can occur to otherwise innocuous environmental particles, such as dust or pollen or even drug molecules, and lead to injurious hypersensitivity reactions. Sustained chronic inflammatory or autoimmune responses may ensue which can maim or kill the very host organism that should be protected. The inability to kill infectious pathogens, such as *Mycobacteria tuberculosis*, or the persistence of an irritant, together with the inadequacy of resolution of acute inflammation, can result in prolongation of the inflammatory process. Often an antigenic stimulus may persist, possibly the result of genetic susceptibility to an imbalance in adaptive immune responses and/or viral infection. Alternatively, cross-reactivity between a foreign antigen and an endogenous structural molecule may result in the development of autoimmunity, such as rheumatoid arthritis or multiple sclerosis, in which the immune cells attack host tissues. Low-level persistent inflammation is thought to be one of the underlying causes of cancer, in which cells are transformed by genetic mutations, escape surveillance by cells of the innate and adaptive

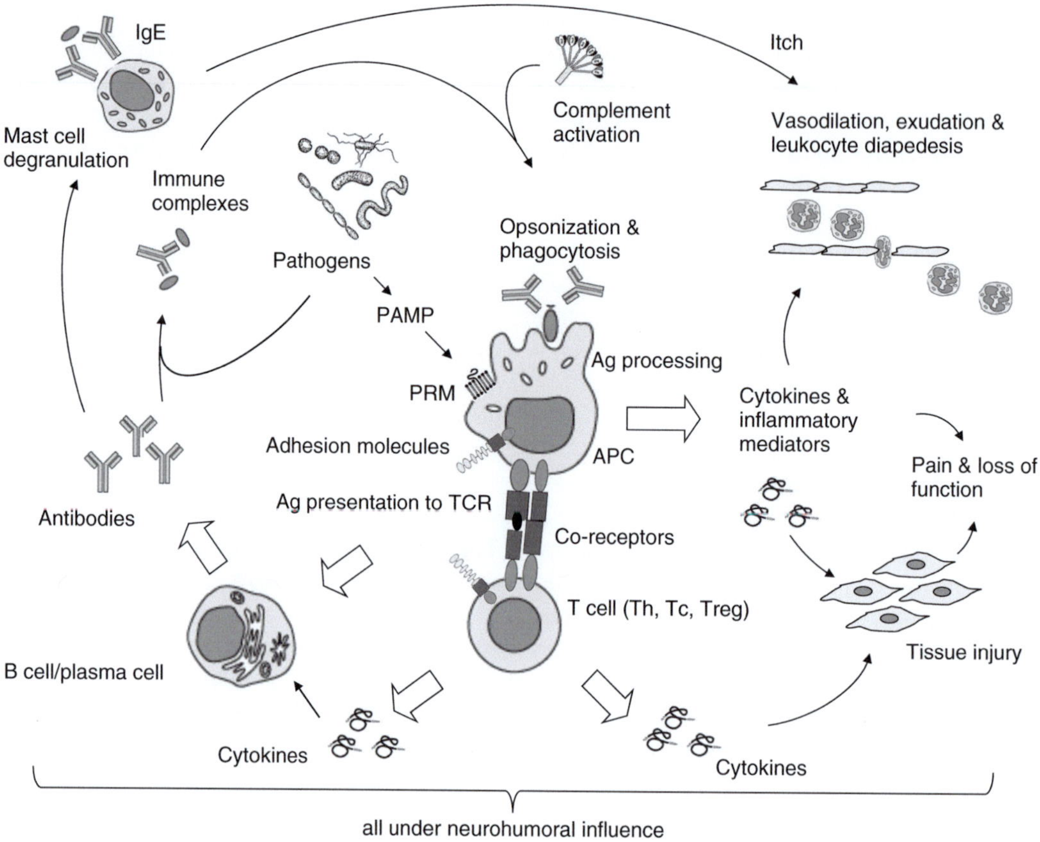

Fig. 1.2 Summary of some of the major processes in immune defence and sites of drug action. *APC* antigen-presenting cell (e.g. dendritic cell or macrophage), *PAMP* pathogen-associated molecular patterns, *PRM* pathogen recognition molecules

immune system and proliferate to form destructive tumours. The immune system does not act in isolation and other systems in the organism, including the endocrine and central and peripheral nervous systems, and the microbiome mainly located in the gut are able to modify host defence. Figure 1.2 summarizes the processes described above.

1.5 Measuring Host Defence Reactions

A variety of techniques are used in the diagnosis and evaluation of immune status. In addition to microbiological assessment of pathogen involvement and the standard clinical biochemical determinations in blood and urine, the analysis of different subsets of white blood cells has become an essential tool. This is achieved by detecting cell surface molecules by FLOW CYTOMETRY. Antibody determination allows for an assessment of prior exposure to antigen or infectious agent and the harnessing of the immune response to produce radio- and enzyme-linked immunoassays (ELISAs) has revolutionized the detection of a wide variety of molecules, well beyond those involved in host defence reactions. Today, methods for analysis of tissue samples, cell populations or even single cells using protein and gene arrays and polymerase chain reaction enhancement of small quantities of material have improved immensely the specific diagnosis of disease states. Importantly, novel imaging techniques in cells, experimental animal models and humans have made significant advances in the understanding of the mechanisms and processes governing inflammation and have improved diagnosis and treatment of inflammatory disease in patients.

These analytical methods are also important in determining the type of therapy that needs to be administered when host defence reactions are deranged.

1.6 Pharmacological Modulation of Immune Responses

The goal of pharmacological intervention is to modify the chain of command in host defence reactions, to:

- Facilitate rapid resolution of inflammation and avoid unnecessary tissue damage.
- Enhance inadequate, beneficial defence reactions.
- Dampen overactive immune responses.
- Restore balance to disturbed immune homeostasis.

The inhibition of acute, non-infectious inflammation and local pain or itch is frequently achieved with non-steroidal anti-inflammatory or anti-allergic drugs. In inflammatory airway diseases, in which allergic reactions are often involved, a combination of anti-inflammatory and bronchodilating drugs is required. In bacterial infections, additional support to endogenous host defence can be gained through the immunomodulatory properties of some antibiotics, while immunostimulatory agents can strengthen the immune system in its defence against infections.

The most effective means to overcome attack by an infectious agent is to actively immunize the organism against the specific pathogen, using vaccines, or else to administer immunoglobulins to provide temporary passive immunization. Similar enhancement, in this case against inadequate immune defence, is offered by immunomodulators and IMMUNE CHECKPOINT INHIBITORS, which turn off endogenous immune feedback inhibition, in cancer therapy.

When the immune response has become excessive and pathological, a number of immunosuppressives and antirheumatic drugs are indicated, as in rheumatoid arthritis or transplant rejection, while repeated antigen administration can rebalance the immune response, as in allergy immunotherapy. Older, less-specific drugs, such as CYTOTOXIC AGENTS, are being used much less commonly to suppress excessive immune reactions, but some long-established drugs, including acetylsalicylic acid or corticosteroids, have never lost their therapeutic usefulness.

Equally, the scientific basis for therapy with dietary and plant-derived agents—many with their roots in folklore—has become stronger in recent years. At the same time, the increasing use of highly specific biological agents, including monoclonal antibodies and recombinant cytokines, is revolutionizing immunopharmacology, particularly in cancer and the treatment of autoimmune diseases. Such biologicals, although needing to be administered by injection because of their protein nature, have the considerable advantage that they target highly specific molecules. As a result, not only are discreet pathological mechanisms attacked, but also the likelihood of adverse effects of the drugs is reduced. Undesirable effects were common among the older cytotoxic, antirheumatic and anti-inflammatory agents, but the science of toxicology has also progressed, and regulatory authorities governing the registration of drugs now provide clear guidance on the testing that is needed to improve the safety of new immunopharmacological agents. Immunopharmacologists are developing specific agents that actively promote the resolution of inflammation without rendering the host susceptible to infection.

This textbook provides the reader with an overview of all these principles of immunopharmacology.

Selected Readings

Alessandri AL, Sousa LP, Lucas CD, Rossi AG, Pinho V, Teixeira MM. Resolution of inflammation: mechanisms and opportunity for drug development. Pharmacol Ther. 2013;139(2):189–212.

Beck A, Wurch T, Bailly C, Corvaia N. Strategies and challenges for the next generation of therapeutic antibodies. Nat Rev Immunol. 2010;10:345–52.

Buckley CD, Gilroy DW, Serhan CN, Stockinger B, Tak PP. The resolution of inflammation. Nat Rev Immunol. 2013;13(1):59–66.

Buckley CD, Gilroy DW, Serhan CN. Proresolving lipid mediators and mechanisms in the resolution of acute inflammation. Immunity. 2014;40(3):315–27. https://doi.org/10.1016/j.immuni.2014.02.009.

Cash JL, Norling LV, Perretti M. Resolution of inflammation: targeting GPCRs that interact with lipids and peptides. Drug Discov Today. 2014;19(8):1186–92.

Dalli J, Serhan CN. Pro-resolving mediators in regulating and conferring macrophage function. Front Immunol. 2017;8:1400.

Dorward DA, Lucas CD, Rossi AG, Haslett C, Dhaliwal K. Imaging inflammation: molecular strategies to visualize key components of the inflammatory cascade, from initiation to resolution. Pharmacol Ther. 2012;135(2):182–99.

Duffin R, Leitch AE, Fox S, Haslett C, Rossi AG. Targeting granulocyte apoptosis: mechanisms, models, and therapies. Immunol Rev. 2010;236:28–40.

Fox S, Leitch AE, Duffin R, Haslett C, Rossi AG. Neutrophil apoptosis: relevance to the innate immune response and inflammatory disease. J Innate Immun. 2010;2(3):216–27.

Fullerton JN, Gilroy DW. Resolution of inflammation: a new therapeutic frontier. Nat Rev Drug Discov. 2016;15(8):551–67.

Galli SJ, Tsai M, Piliponsky AM. The development of allergic inflammation. Nature. 2008;454:445–54.

Gilroy DW, Lawrence T, Perretti M, Rossi AG. Inflammatory resolution: new opportunities for drug discovery. Nat Rev Drug Discov. 2004;3(5):401–16.

Hallett JM, Leitch AE, Riley NA, Duffin R, Haslett C, Rossi AG. Novel pharmacological strategies for driving inflammatory cell apoptosis and enhancing the resolution of inflammation. Trends Pharmacol Sci. 2008;29(5):250–7.

Iwasaki A, Medzhitov R. Control of adaptive immunity by the innate immune system. Nat Immunol. 2015;16(4):343–53.

Jones HR, Robb CT, Perretti M, Rossi AG. The role of neutrophils in inflammation resolution. Semin Immunol. 2016;28(2):137–45.

Kaufmann SHE. Immunology's foundation: the 100-year anniversary of the Nobel Prize to Paul Ehrlich and Elie Metchnikoff. Nat Immunol. 2008;9:705–12.

Mantovani A, Marchesi F, Malesci A, Laghi L, Allavena P. Tumour-associated macrophages as treatment targets in oncology. Nat Rev Clin Oncol. 2017;14(7):399–416.

Mantovani A. Wandering pathways in the regulation of innate immunity and inflammation. J Autoimmun. 2017;85:1–5.

Medzhitov R. Origin and physiological roles of inflammation. Nature. 2008;454:428–35.

Nathan C, Ding A. Nonresolving inflammation. Cell. 2010;140(6):871–82.

Nathan C. Neutrophils and immunity: challenges and opportunities. Nat Rev Immunol. 2006;6(3):173–82.

Nourshargh S, Alon R. Leukocyte migration into inflamed tissues. Immunity. 2014;41(5):694–707.

Perretti M, Leroy X, Bland EJ, Montero-Melendez T. Resolution pharmacology: opportunities for therapeutic innovation in inflammation. Trends Pharmacol Sci. 2015;36(11):737–55.

Poon IK, Lucas CD, Rossi AG, Ravichandran KS. Apoptotic cell clearance: basic biology and therapeutic potential. Nat Rev Immunol. 2014;14(3):166–80.

Robb CT, Regan KH, Dorward DA, Rossi AG. Key mechanisms governing resolution of lung inflammation. Semin Immunopathol. 2016;38(4):425–48.

Serhan CN, Brain SD, Buckley CD, Gilroy DW, Haslett C, O'Neill LA, Perretti M, Rossi AG, Wallace JL. Resolution of inflammation: state of the art, definitions and terms. FASEB J. 2007;21(2):325–32.

Serhan CN, Savill J. Resolution of inflammation: the beginning programs the end. Nat Immunol. 2005;6(12):1191–7.

Serhan CN. Pro-resolving lipid mediators are leads for resolution physiology. Nature. 2014;510(7503):92–101.

Udalova IA, Mantovani A, Feldmann M. Macrophage heterogeneity in the context of rheumatoid arthritis. Nat Rev Rheumatol. 2016;12(8):472–85.

Part I

Immunity

Hematopoiesis and Lymphocyte Development: An Introduction

2

Menno C. van Zelm, Magdalena B. Rother, and Frank J. T. Staal

2.1 Introduction: Blood Cell Development and Immunity

The production of blood and immune cells is a continuous process throughout life and essential for the existence of any individual human being. This process is referred to as HEMATOPOIESIS (from the Greek hematos, blood, and poiein, to generate) and includes the generation of three major cell lineages: (1) ERYTHROCYTES (red cells) that transport oxygen, (2) THROMBOCYTES (platelets) that mediate blood clotting, and (3) LEUKOCYTES (white blood cells) that function in host defense and consist of various types of GRANULOCYTES, LYMPHOCYTES, and MONOCYTES. The major site of HEMATOPOIESIS is the BONE MARROW, but it can occur at other (extramedullary) locations as well.

Final manuscript submitted on February 26, 2018.

M. C. van Zelm (✉)
Department of Immunology and Pathology, Central Clinical School, Monash University, Melbourne, VIC, Australia

The Alfred Hospital, Melbourne, VIC, Australia
e-mail: menno.vanzelm@monash.edu

M. B. Rother
Department of Immunology, Erasmus MC, University Medical Center, Rotterdam, The Netherlands

F. J. T. Staal
Department of Immunohematology and Blood Transfusion, Leiden University Medical Center, Leiden, The Netherlands

The human body has several ways of combating bacteria, viruses, and other pathogens, as we are exposed to millions of microorganisms on a daily basis. Collectively, these potentially pathogenic microorganisms or substances in the external environment are referred to as ANTIGENS. The first line of defense is formed by physical barriers, such as our epithelial layers [1], and physiological defense mechanisms, such as the low pH in the stomach and a slightly acidic layer on the skin [2]. When pathogens succeed in passing these physical and physiological barriers, cells of our IMMUNE SYSTEM play an important role in attacking them [3]. These initially concern cells that are capable of engulfing and digesting microorganisms, local resident MACROPHAGES, as well as GRANULOCYTES and MONOCYTES that enter from the blood. In addition, NATURAL KILLER (NK) CELLS are recruited from blood and are capable of killing infected cells. All these cell types are part of the INNATE IMMUNE RESPONSE (see Chap. 8). In addition to being responsible for the early defense against microbes, these innate immune cells play an important role in activating the ADAPTIVE IMMUNE RESPONSE. The adaptive immune response consists of T and B LYMPHOCYTES and ANTIBODIES that respond based upon specific recognition of an antigen. Each B or T cell carries a unique antigen receptor that can recognize an ANTIGEN, i.e., a stretch of amino acids,

M. J. Parnham et al. (eds.), *Nijkamp and Parnham's Principles of Immunopharmacology*,
https://doi.org/10.1007/978-3-030-10811-3_2

polysaccharides, or lipids present on or derived from pathogens. B LYMPHOCYTES produce ANTIBODIES, which are soluble proteins that can neutralize antigens as part of the humoral immune response. T LYMPHOCYTES either function as "stimulators" in immune responses or mediate cellular immunity through direct killing of transformed or virus-infected cells. In addition to antigen specificity, adaptive immune responses result in the generation of immunological memory, i.e., long-lived memory B and T cells that are specific for the ANTIGEN they have encountered and neutralized. Compromised adaptive immunity can lead to severe illness and death, underscoring its importance. Basic cellular processes are similar for HEMATOPOIESIS and reactions of mature immune cells. Thus, pharmacological agents affecting mature LYMPHOCYTE proliferation and maturation can also affect the differentiation of blood cells. For example, immunosuppressants with an antiproliferative mode of action can show BONE MARROW depression as an adverse side effect.

2.2 Hematopoietic Stem Cells

The STEM CELLS that give rise to all blood lineages mainly reside in the bone marrow and are known as HEMATOPOIETIC STEM CELLS (HSC). Figure 2.1 shows a schematic overview of hematopoiesis. It is estimated that about 1 in 100,000 cells in the BONE MARROW is a true HSC [4]. Despite their low frequency, HSC have the ability to produce high numbers of new blood cells each day. The potential to sustain the supply of blood throughout an individual's life-span can be attributed to two features that characterize HSC: self-renewal and multipotency. Self-renewal of HSC is defined as the ability to divide while retaining undifferentiated features. The multipotency of HSC refers to their capacity to differentiate into multiple different cell types, in this case all types of blood cells. The multilineage differentiation capacity of HSC was initially demonstrated by the prevention, through injection of bone marrow cells, from hematopoietic failure after total-body irradiation of mice [4]. HSC are cells with a slow division rate. Up to 20% of all divisions are asymmetrical, resulting in one daughter cell that retains stem cell features and other daughter cells that differentiate [5–7]. It is difficult to identify HSC by their size and shape, since their morphology and behavior in culture resemble that of white blood cells. The development of MONOCLONAL ANTIBODIES recognizing cell surface markers, as well as fluorescence-activated cell sorting (FACS; see Chap. 16), has facilitated the purification of small subsets of cells. In combination with in vivo and in vitro assays, populations that contain cells with HSC capacities have been identified. Besides being present in the BONE MARROW, human HSC can also be isolated from umbilical cord blood and, when mobilized by GRANULOCYTE-COLONY STIMULATING FACTOR (G-CSF), from peripheral blood [8, 9]. This is an important pharmacological means to manipulate the numbers of STEM CELLS that can be used for clinical transplantation applications. Murine HSC are characterized by the surface expression of Sca-1, C-kit, and CD38, low expression levels of the Thy-1, low to absent CD34, and the lack of lineage markers (B220, Mac-1, Gr-1, CD3, CD4, CD8, and Ter119). The most widely used HSC population in the mouse is the so-called LSK population: lineage marker negative, Sca-1+ and C-kit+. Within this population two subsets can be distinguished, namely, those with long-term and short-term repopulating capacities [10–13]. Human HSCs are currently defined as CD34+ CD38− CD90+ CD49f+ [14]. For both mouse and man, other markers are being continuously evaluated and added to diagnostic methods in an attempt to more precisely define true HSC. Of note are the so-called SLAM markers, CD150 and CD48, which further subdivide the LSK population into cells enriched for long-term or short-term repopulating stem cells and multipotent progenitors [15, 16]. Although primitive hematopoiesis in the embryo starts in the yolk sac, current evidence strongly suggests that the AORTA-GONAD-MESONEPHROS (AGM) region is the first source of true HSC [17]. These HSC subsequently colonize the liver and then the BONE MARROW. Upon birth, hematopoiesis is almost completely restricted to the BONE MARROW [18]. Because they can give sustained

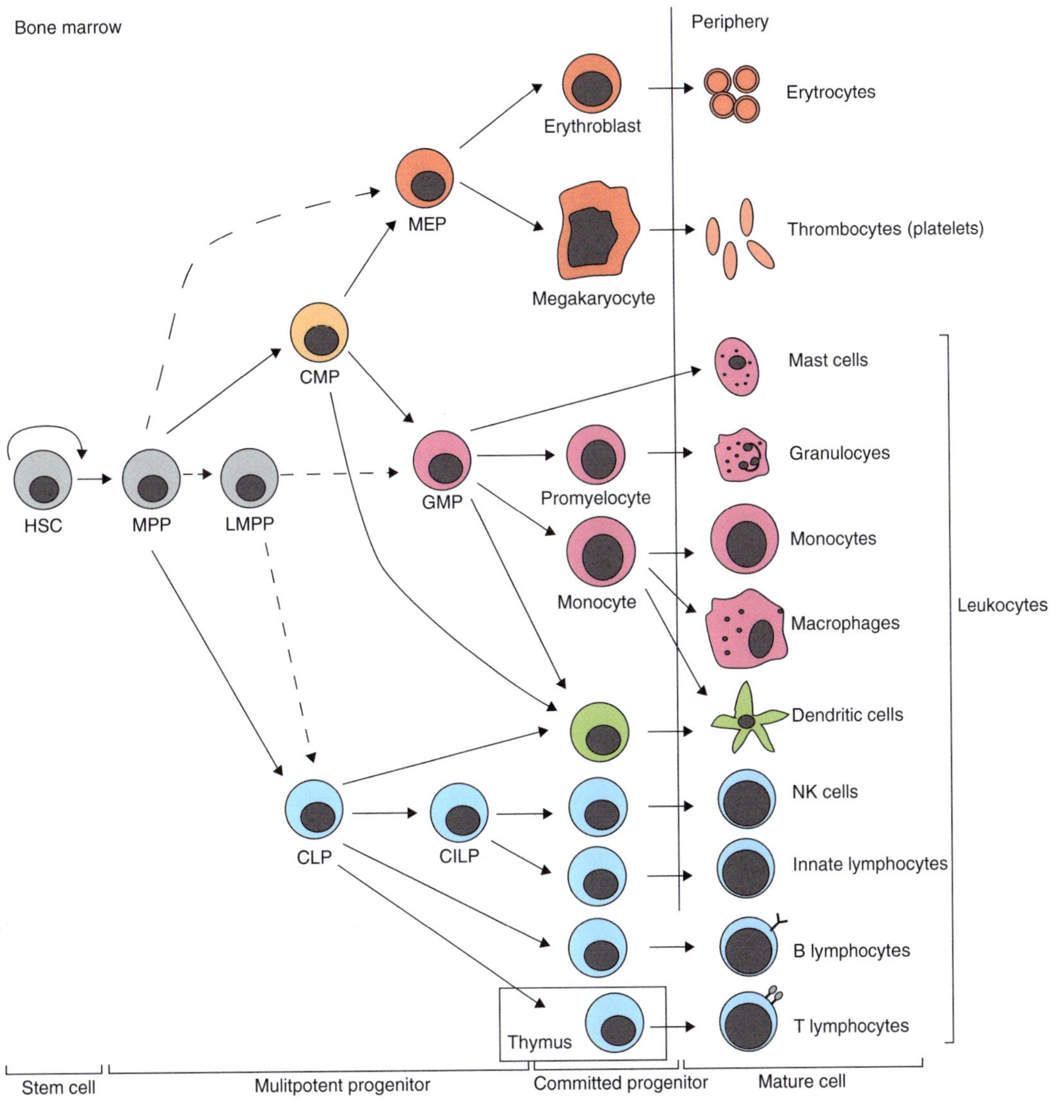

Fig. 2.1 Schematic overview of hematopoiesis. Hematopoietic stem cells (HSC) are responsible for blood cell production throughout the lifetime of an individual. Pluripotent HSC can give rise to several different hematopoietic lineages via progressively restricted progenitor stages while retaining the capacity for self-renewal. Lineage-committed progenitor cells produce progeny destined to differentiate into red cells, platelets, granulocytes, monocytes, and lymphoid cells (solid arrows) [79]. The dotted arrows represent an alternative model of hematopoiesis [80]. *MPP* multipotent progenitor, *LMPP* lymphoid-primed multipotent progenitor, *CMP* common myeloid progenitor, *GMP* granulocyte/monocyte progenitor, *MEP* megakaryocyte/erythrocyte progenitor, *CLP* common lymphoid progenitor, *CILP* common innate lymphocyte progenitor

reconstitution of all blood lineages, transfer of HSC has been used in numerous therapeutic protocols. Conditions that are regularly treated by HSC transfer include leukemia, lymphoma, various types of inherited anemia, inborn metabolic disorders, Wiskott-Aldrich syndrome, and severe combined immunodeficiency (SCID) [19–24].

2.3 Lymphocyte Development

T cells, B cells, and NK cells are the three major types of cells that can be distinguished within the LYMPHOCYTE lineage. NK cells are innate immune cells and play a key role in the host defense against virally infected cells as well as tumors. Upon activation of NK cells, granules that contain cytotoxic proteins are released from their cytoplasm, resulting in the destruction of the target cell. T and B LYMPHOCYTES make up the adaptive arm of the IMMUNE SYSTEM and are each able to specifically recognize a specific ANTIGEN. T cells bear T CELL RECEPTORS (TCR) on their cell surface that are capable of recognizing unique ANTIGENS when presented in the context of MAJOR HISTOCOMPATIBILITY COMPLEX (MHC) molecules by innate immune cells or by B cells (see Chap. 4). Similar to T cells, B LYMPHOCYTES express specialized antigen RECEPTORS with a single antigenic specificity, called B cell ANTIGEN receptors or IMMUNOGLOBULINS (Igs). These Igs are membrane bound but can also be secreted when B cells mature into plasma cells. The secreted forms of B cell RECEPTORS are called ANTIBODIES and contribute to humoral immunity (see Chap. 4). B, T, and NK lymphocytes develop from HSC through a common progenitor cell in a process called lymphopoiesis [25, 26]. Key decisions made during the multistep process of lymphopoiesis depend on a large number of signals conveyed by cell-cell interactions, soluble factors, and the extracellular matrix within stromal microenvironments at specialized sites of maturation.

2.4 Specificity: Rearrangement of Genes Encoding Antigen Receptors

For HSC to become specialized B and T LYMPHOCYTES, the cells have to undergo multiple developmental steps. This is more extensive than for the other leukocytes, because B and T cells need to generate a unique antigen receptor through rearrangement of genomic DNA in the Ig and TCR loci, respectively. If each unique antigen receptor were coded by a separate gene, the size of the human genome would not suffice. Therefore, these loci contain exons that encode the constant domains, which are equipped to provide structure, effector function, and signaling of antigen receptors [27, 28], but not a germline-encoded antigen-binding domain. Instead, this is a variable domain that is formed through recombination of variable (V), diversity (D), and joining (J) genes. V(D)J RECOMBINATION is a highly specialized process of cutting and pasting of genetic elements and assures the generation of an antigen RECEPTOR with unique specificity in each developing cell, thereby providing an organism with an extensive antigen-receptor repertoire [29].

In loci containing V, D, and J segments, the D to J rearrangement takes place first, followed by V to DJ rearrangements. An example of V(D)J RECOMBINATION is shown in Fig. 2.2. The initiation of RECOMBINATION is directed by recombination signal sequences (RSSs) that flank the coding gene segments. RSSs are built up of 2 conserved sequences, a heptamer (conserved 7 base pair sequence) and a nonamer (conserved 9 base pair sequence), separated by a non-conserved spacer sequence of 12 or 23 base pairs [30, 31]. The consensus heptamer sequence is CACAGTG, and the nonamer consensus sequence is ACAAAAACC. Generally, RECOMBINATION occurs between an RSS with a 12 base pair spacer and an RSS with a 23 base pair spacer, the so-called 12/23 rule [31]. The recombinase-activating gene (Rag-)1 and Rag-2 proteins bind to an RSS as a heterodimer and capture another RSS [32]. Subsequently, Rag1 and Rag2 initiate double-stranded DNA breaks on both sites on the borders of the RSSs and the gene segments. The intervening DNA is excised and forms an episomal circle through ligation of both RSSs in a signal joint [33]. Prior to their ligation, the coding ends on the chromosome are modified through excision and addition of nucleotides, creating ultimately a coding joint with a unique DNA sequence. Random nucleotides are introduced by Terminal deoxynucleotidyl transferase (TdT). This enzyme is specifically expressed in progenitor B and T lymphocytes and at least in part regulated by IL-7 [34].

Ig molecules are composed of two identical Ig heavy chains (IgH) and two identical Ig light chains. The *IGH* locus contains V, D, and J genes, whereas the Ig kappa and Ig lambda light chain loci only have V and J segments (Fig. 2.2). T cells either carry a receptor composed of one TCRα

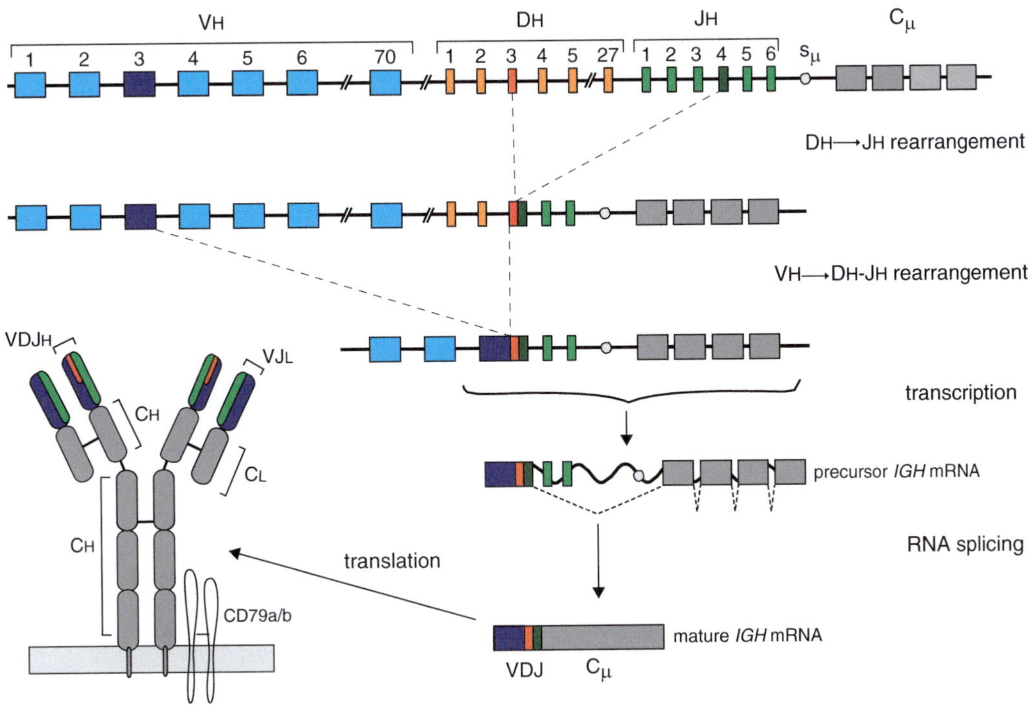

Fig. 2.2 Schematic diagram of sequential rearrangement steps, transcription, and translation of the *IGH* gene during B cell differentiation. V(D)J recombination in the IGH locus starts with a DH to JH gene rearrangement, followed by the coupling of a VH gene to the DJH joint. Subsequently, the VDJH exon is transcribed and spliced to the constant μ exons followed by translation to Igμ protein. Similar recombination events occur between V and J genes in the Ig kappa or Ig lambda loci of a functional IgL protein. Finally, two identical copies of the IgH and IgL chains complex into the B cell antigen receptor, which is expressed on the B cell membrane with CD79a and CD79b proteins

and one TCRβ chain (αβ-T cells) or one TCRγ and one TCRδ chain (γδ T cells). The *TCRA* and *TCRG* loci are composed of V and J segments, while the *TCRB* and *TCRD* loci contain V, D, and J segments. Thus, each antigen receptor is composed of protein chains encoded by two loci, one containing V, D, and J genes and one only V and J genes, and the combinations as well as the junctional region processing together result in the high diversity within the total lymphocyte compartment.

2.5 T Cell Development

T cell development is a highly regulated, multistep process aimed at generating mature, functional T cells bearing TCR that are capable of recognizing a broad range of antigens in the context of self-MHC. In contrast to all other hematopoietic lineages that develop in the specialized microenvironment of the BONE MARROW, development of T cells from pluripotent HSC takes place in the thymus. Throughout life, lymphoid progenitors from the bone marrow seed the THYMUS and differentiate into T cells. Mature T cells express a heterodimeric TCR that is either composed of one TCRα chain and one TCRβ chain or one TCRγ chain and one TCRδ chain. When progressing through T cell development, cells undergo lineage commitment, TCR gene rearrangements, proliferation, and selection. Using cell surface markers, several T cell developmental stages can be distinguished (Fig. 2.3). Primarily, thymocytes are subdivided into double-negative (DN), double-positive (DP), and single-positive (SP) populations, referring to the expression of the co-receptors CD4 and CD8. The most immature thymocytes lack expression of both CD4 and CD8 and are therefore called

DN. In mouse and humans, additional but different surface markers are used to further subdivide the DN stage. For mouse, the markers CD25 and CD44 are used: CD44+ CD25− cells are called DN1 cells, CD44+ CD25+ cells are referred to as DN2, DN3 cells express CD25 but no CD44, and DN4 cells express neither CD25 nor CD44. The most immature human thymocyte population is characterized by the expression of CD34, but lacks CD1a and CD38 expression, and resembles the murine DN1 population [35–41]. The next stage of differentiation is marked by the expression of both CD34 and CD38 and resembles the murine DN2 stage. The most mature human DN stage that can be discerned is made up of cells expressing CD34 and CD38 as well as CD1a. Thymocytes undergo a substantial number of cell divisions, six to ten, in the first DN stages [42]. The first round of gene rearrangement takes place in DN thymocytes, and rearrangements are initially detected at the *TCRD* locus, mainly during DN2 and DN3 stages of development [27, 43]. Subsequently, *TCRG* rearrangements occur, mostly at the DN3 stage. When a TCRγδ is successfully formed, the developmental path of γδ T cells diverges from αβ-T cell development, most likely at the DN3 stage [44]. If rearrangements at the *TCRD* locus are nonfunctional, rearrangement of the *TCRB* locus will proceed. Successful rearrangement of *TCRB* gene segments is tested by expression of the TCRβ chain on the cell surface paired with the invariant pre-Tα receptor in the pre-TCR complex. This process is also referred to as β-selection, a major checkpoint during T cell development. Signaling through the pre-TCR will result in entry of the cell cycle. During proliferation cells go through immunophenotypic changes: immature single-positive (ISP) cells arise when thymocytes express a co-receptor in the absence of high levels of CD3. In humans, ISP cells express co-receptor CD4, whereas in most strains of mice, they express CD8. Subsequently, both CD4 and CD8 are expressed, and therefore these cells are referred to as double positive. In the DP stage, *TCRA* gene rearrangements are initiated. The DP stage makes up approximately 85% of all thymocytes. After *TCRA* rearrangement, a TCRαβ heterodimer is expressed on the cell surface. This unique TCR is then tested for the recognition of

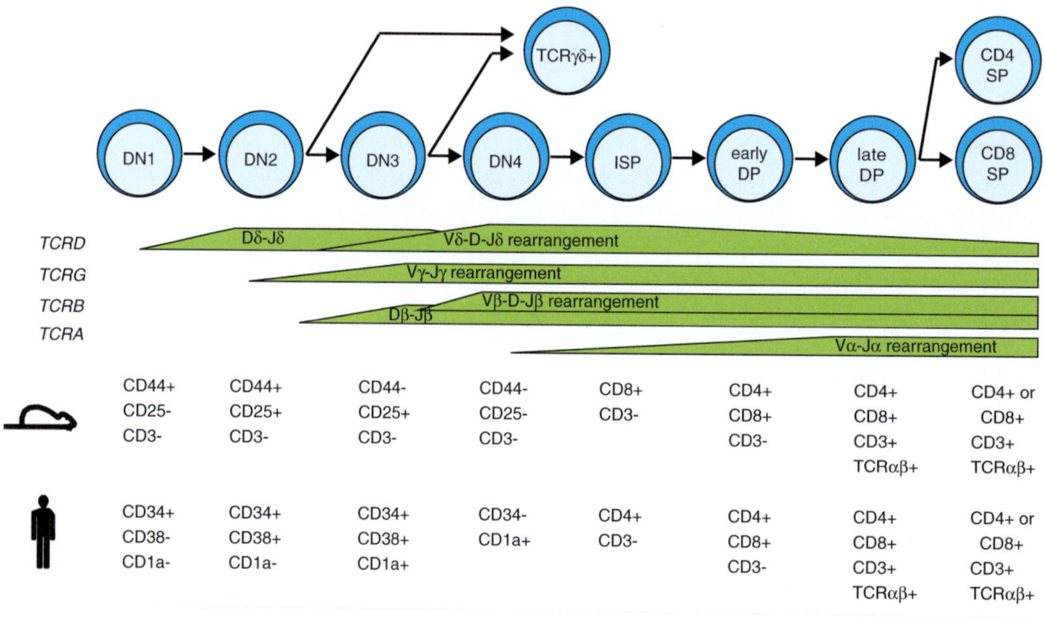

Fig. 2.3 T cell development. Consecutive stages of development are described for mouse and human according to the surface markers that are expressed. Based on a previously presented model, stages of development that are comparable between mouse and man are grouped together. The rearrangement status of the T cell receptor loci is depicted

self-MHC molecules in a process called positive selection, while negative selection tests for the absence of self-reactivity. After failing the selection criteria, approximately 95% of thymocytes die through the induction of APOPTOSIS [45]. After successfully undergoing positive and negative selection processes, thymocytes that express a functional TCR commit to either the CD4+ T helper lineage or the CD8+ CYTOTOXIC T lineage, ready to migrate to the periphery.

2.6 B Cell Development

Precursor B cells undergo stepwise differentiation in the bone marrow, during which each developing B cell creates a unique antigen receptor by V(D)J recombination in a similar fashion as to precursor T cells in the thymus. The B cell antigen RECEPTOR (BCR) consists of two identical copies of the Ig heavy chain (IgH) and two identical copies of the Ig light chain (Igκ or Igλ) and is expressed on the membrane with CD79a and CD79b. In contrast to the TCR, the BCR can recognize complete, unprocessed ANTIGEN without the need for presentation by other immune cells. During precursor B cell differentiation, five functionally different stages can be identified based on the stepwise rearrangement of IgH and Ig light chains. The five precursor B cell subsets are defined by cytoplasmic and membrane Ig expression but can also be identified by unique combinations of cell surface markers (Fig. 2.4) [46, 47]. Prior to commitment to the B cell lineage, a stem cell is first restricted to the lymphoid lineages. This step is succeeded by specification to the B cell lineage. At this stage, the cells are identified as pro-B cells and have started V(D)J RECOMBINATION of the *IGH* loci with incomplete DH–JH gene rearrangements. E2A and EBF are important transcription factors that promote B cell specification, inhibit other cell fates, and activate RAG gene transcription [48]. Subsequently, E2A and EBF induce transcription of Pax-5, which commits the precursor cell to the B cell lineage [49]. Pax-5

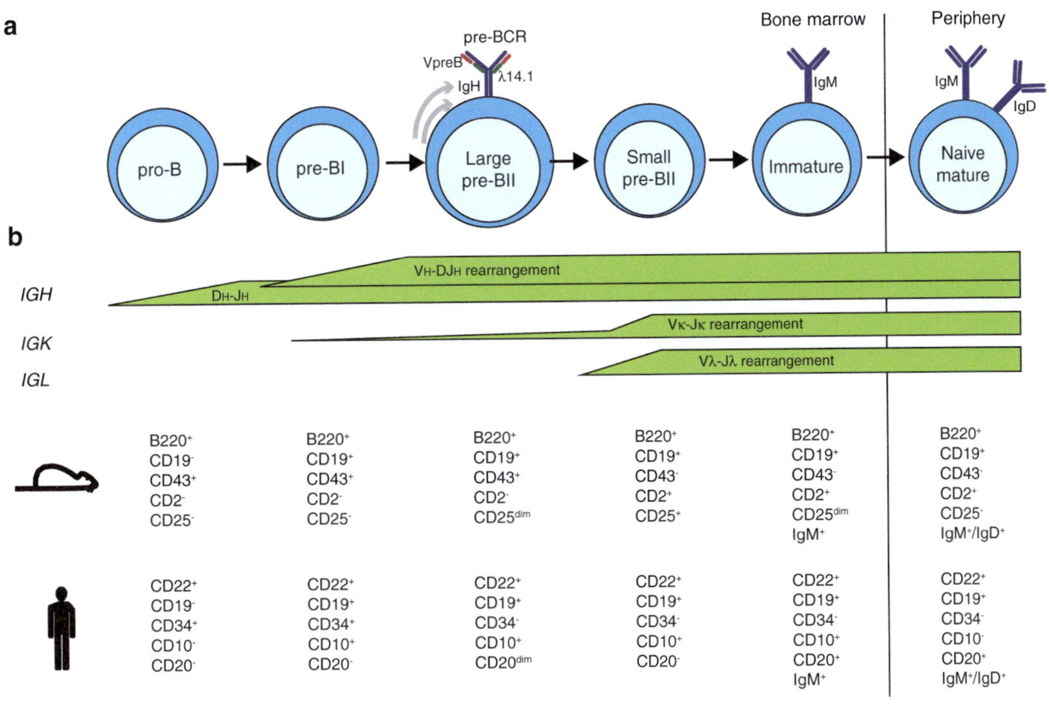

Fig. 2.4 B cell development in the bone marrow. Consecutive stages of development are described for mouse and human according to the surface markers that are expressed. Stages of development that are comparable between mouse and man are grouped together. The rearrangement activity of the immunoglobulin loci is depicted

induces complete VH–DJH gene rearrangements and the transcription of multiple B cell-specific proteins, such as CD79a, BLNK, and CD19 [50]. These early committed precursor B cells are identified by membrane CD19 expression and named pre-B-I. Upon successful recombination of one of the *IGH* loci, the VDJH exon is spliced to the μ constant exons, producing Igμ heavy chain protein. In the absence of a functional Ig light chain, Igμ is expressed on the membrane with CD79a, CD79b, and two proteins that form a SURROGATE light chain: λ14.1 (named λ5 in mice) and VpreB. This complex is called the pre-BCR, and although it is expressed on the cell membrane, it is difficult to detect. Still, Igμ expression can be detected intracellularly in all precursor B cells that have generated a functional Ig heavy chain. Cytoplasmic Igμ + cells are defined as pre-B-II. Directly upon membrane expression, the pre-BCR initiates multiple processes via signaling cascades [51]. Despite some controversy about the need for a LIGAND, it appears that pre-BCR activation is dependent on stromal cell-derived galectin-1 and adhesion proteins (INTEGRINS) [52]. Upon activation, the kinase Lyn is recruited and signals for induction of proliferation (large pre-B-II cells). Lyn signaling also leads to the downregulation of the Rag proteins to prevent potential further *IGH* gene rearrangements on the second allele (allelic exclusion). Finally, after several rounds of proliferation, the pre-BCR is downregulated to enable further differentiation and the induction of Ig light chain gene rearrangements. The latter two processes depend on additional signaling proteins, especially BLNK and BTK. In the now small pre-B-II cells, Ig light chain Vκ–Jκ gene rearrangements are initiated first. If this leads to a functional protein that can pair with the Igμ heavy chain, the cell will express an Igκ + BCR. Unsuccessful Vκ–Jκ gene rearrangements are followed by Vλ–Jλ gene rearrangements. If these yield functional protein, the B cell will express an Igλ + BCR. Membrane expression of the BCR is readily detectible, and these IgM+ cells are named immature B cells [53]. At this stage, the majority of all B cells express autoreactive BCRs [54]. Autoreactive B cells are removed from the REPERTOIRE by APOPTOSIS or by generation of a new BCR by additional rearrangements that replace the Ig light chain (RECEPTOR editing). Once the cell fulfills the right criteria, it will transcribe two splice variants of the Ig heavy chain, thus co-expressing IgM and IgD BCRs, and migrate to peripheral LYMPHOID ORGANS. These recent BONE MARROW emigrants are transitional cells, and once they develop into naive mature B cells, they are ready to encounter and respond to their cognate antigen.

2.7 NK and Innate Lymphocyte Development

The third member of the lymphoid lineage, the NK cell, plays an important role in defense against virally infected cells and tumors as well as activation of adaptive immune responses [55]. NK cells are large granular cells that make up 10–15% of all circulating lymphocytes and are immunophenotypically characterized as CD3–CD56+ in humans [56]. For murine NK cells, other markers are used, namely, DX5, NKG2D, and NK1.1 [57]. In comparison to B and T cell development, relatively little is known about the development of NK cells. Unlike B and T cells, development of NK cells does not involve the process of V(D)J recombination. A significant part of NK cell development takes place in the BONE MARROW, and IL-15 has been identified as a critical mediator of NK cell development, since mice that lack either IL-15 or IL-15R are NK cell deficient [58, 59]. In early phases of NK cell development, factors produced by stromal cells in the bone marrow mediate the generation of NK precursor cells. These precursors are receptive for IL-15 and are able to develop in mature NK cells. The final NK cell differentiation step can occur in the bone marrow, but other sites, such as lymph nodes, could also provide factors necessary for terminal NK cell differentiation. CYTOKINES other than IL-15 have also been described to contribute to NK cell development. The most important contributors are IL-2 and IL-7. The only CYTOKINES that can support the development of NK cells in vitro are IL-2, IL-15, and IL-7, although resulting NK cells are functionally and phenotypically immature [60–62]. The mature NK cell population can be subdivided into $CD3^{-}CD56^{dim}$ and $CD3^{-}CD56^{bright}$ cells [63]. $CD3^{-}CD56^{dim}$ cells express high levels of CD16 and killer IMMUNOGLOBULIN-like RECEPTORS (KIR). The highly CYTOTOXIC nature of these cells is underscored by their ability to mediate direct cytotoxicity, antibody-dependent cellular cytotoxicity (ADCC), as well as lymphokine-activated killing. In contrast, the expression of CD16 and KIR is low or absent on $CD3^{-}CD56^{bright}$ cells, and these cells have a low toxicity potential. $CD3^{-}CD56^{bright}$ cells act as immunoregulatory cells and are capable of producing large amounts of interferon-γ (IFNγ) and IL-10. A distinct feature of mainly $CD56^{dim}$ NK cells is their ability to quickly mediate cellular toxicity. NK cell responses are regulated by a balance of signals from activating and inhibiting receptors. In the absence of activating RECEPTOR ligation, effector function is inhibited as long as the KIR molecules are bound to HLA class I molecules on the membrane of a TARGET cell. As soon as the activating receptor-ligand interaction overrides the inhibitory KIR signals, NK cells are activated. Once NK cells are activated, granules that hold several types of CYTOTOXIC proteins are released from their cytoplasm, which results in the destruction of the TARGET cell.

For many decades, it was thought that NK cells were the only lymphocytes without a specific antigen receptor. However, in recent years, several other innate lymphocyte subsets have been identified that are formally named INNATE LYMPHOID CELLS (ILC) [64, 65]. ILC, including NK cells, differentiate in the BONE MARROW from the CLP through a common ILC progenitor (CILP; Fig. 2.1). Subsequently, an NK cell progenitor gives rise to cytotoxic NK cells, while three types of non-cytotoxic ILC derive from a “helper-like” progenitor. In fact, three distinct groups of ILC can be identified that show parallels with helper T cell subsets. ILC1 parallel TH1 in that they can produce IFNγ and tumor necrosis factor (TNF) and function in responses to intracellular bacteria and parasites. ILC2 show Th2 properties with the production of IL-4, IL-5, IL-9, and IL-13 and their function in anti-helminth infection. Finally, ILC3s can produce diverse CYTOKINES including IL-17A and IL-17F, and they function in antibacterial immunity. Importantly, ILC3s also encompass lymphoid tissue inducer cells (LTi), which were already discovered in 1997 and are essential for lymph node formation during embryogenesis and postnatal lymphoid follicle formation [66]. In contrast to B, T, and NK cells, ILC are very infrequent in human blood [67] and are typically located in the lymphoid tissue, skin, and mucosa.

2.8 Immunopharmacology and Hematopoiesis

Immunopharmacology mainly deals with the effects of agents on mature cells of the IMMUNE SYSTEM to treat autoimmunity, allergies, and acquired immunodeficiencies. Commonly used immunosuppressive and other drugs that are aimed at targeting such aberrant immune responses can affect hematopoiesis and therefore cause unwanted side effects. Examples include the DNA-damaging or DNA synthesis blockers cyclophosphamide, mycophenolate mofetil, and AZATHIOPRINE, which are DNA-damaging agents ([68–72]; see Chaps. 30 and 31). There are also drugs aimed at mobilizing HSC from the bone marrow into the peripheral blood to harvest stem cells for transplantation purposes or to strengthen the number of blood cell under cytostatic cancer treatment. These are formulations of G-CSF, a GROWTH FACTOR for GRANULOCYTES, but other GROWTH FACTORS such as stem cell factor, IL-8, and GM-CSF have mobilizing properties as well but are not commonly used due to side effects. The majority of drugs that are toxic to the bone marrow are anticancer agents, but there is considerable variability in the severity of bone marrow depression that they induce. The anticancer agents 1,3-bis(2-chloroethyl)-1-nitrosourea (BCNU) and busulfan (BU) are DNA alkylating agents used for CNS tumors and leukemias, respectively. They can severely compromise the bone marrow and resulting HEMATOPOIESIS. Cyclophosphamide is also used in hematological tumors and is known to deplete the bone marrow of all but the most primitive precursors. Doxorubicin (Adriamycin) is active against tumor cells through several mechanisms including inhibition of mitochondrial cytochrome C oxidase, intercalation of DNA, and generation of toxic free radicals. Clinical use of doxorubicin is limited by myelosuppression and cardiotoxicity [73, 74]. Rapamycin is an immunosuppressive drug that blocks GROWTH FACTOR-induced cell proliferation [75, 76]. This drug does not affect the bone marrow very much but has profound effects on T cell development in the thymus. The immunosuppressive drugs CICLOSPORIN and FK-506 have a more restricted mechanism of action (see Chap. C13), i.e., inhibition of calcineurin A activity with subsequent blockade of intracellular signal transduction, leading to synthesis of CYTOKINES such as IL-2, an important GROWTH FACTOR for peripheral T cells [77]. While preventing T cell activation and thereby graft rejection in transplantation settings, these drugs have an interesting activity on T cell development as they appear to block APOPTOSIS during the negative selection of T cells. This could potentially increase the incidence of autoimmune reactions.

The introduction of monoclonal antibody therapies for the treatment of immunological disease has allowed more specific targeting of immune cells to ameliorate disease. Still, most antibodies target both normal and abnormal cells and can therefore affect a complete cell lineage for a long period of time. An example is anti-CD20 therapy with, e.g., rituximab, which has been successful in the treatment of both B cell malignancies and autoimmune disease [78]. As CD20 is expressed on B cell progenitors and mature B cells (Fig. 2.4), anti-CD20 treatment results in a near-complete depletion of all circulating B cells. B cell recovery usually takes 6–9 months but can take up to 2 years with still lower numbers than prior to treatment. This is potentially due to the fact that progenitor B cells are targeted. Similar effects are to be expected from anti-CD19 and anti-CD22 therapies, while targeting of survival and activation factors, such as BAFF or APRIL, will not result in long-term depletion as these do not affect progenitor cells.

Selected Readings

Artis D, Spits H. The biology of innate lymphoid cells. Nature. 2015;517(7534):293–301.

Berkowska MA, van der Burg M, van Dongen JJ, van Zelm MC. Checkpoints of B cell differentiation: visualizing Ig-centric processes. Ann N Y Acad Sci. 2011;1246:11–25.

Dharmarajan TS, Widjaja D. Erythropoiesis-stimulating agents in anemia: use and misuse. J Am Med Dir Assoc. 2009;10(9):607–16.

Kaufman DS. Toward clinical therapies using hematopoietic cells derived from human pluripotent stem cells. Blood. 2009;114(17):3513–23.

Manz MG, Di Santo JP. Renaissance for mouse models of human hematopoiesis and immunobiology. Nat Immunol. 2009;10(10):1039–42.

Sandy AR, Maillard I. Notch signaling in the hematopoietic system. Expert Opin Biol Ther. 2009;9(11):1383–98.

Schaniel C, Moore KA. Genetic models to study quiescent stem cells and their niches. Ann N Y Acad Sci. 2009; 1176:26–35.

Sigvardsson M. New light on the biology and developmental potential of haematopoietic stem cells and progenitor cells. J Intern Med. 2009;266(4):311–24.

Staal FJ, Luis TC. Wnt signaling in hematopoiesis: crucial factors for self-renewal, proliferation, and cell fate decisions. J Cell Biochem. 2010;109(5):844–9.

Important Websites

http://www.aai.org/
http://www.ehaweb.org/
http://www.efis.org/
http://www.hematology.org/
http://www.hcdm.org/

References

1. Marchiando AM, Graham WV, Turner JR. Epithelial barriers in homeostasis and disease. Annu Rev Pathol. 2010;5:119–44.
2. Turner JR. Intestinal mucosal barrier function in health and disease. Nat Rev Immunol. 2009;9:799–809.
3. Janeway CA Jr, Medzhitov R. Innate immune recognition. Annu Rev Immunol. 2002;20:197–216.

4. Lorenz E, Uphoff D, Reid TR, Shelton E. Modification of irradiation injury in mice and guinea pigs by bone marrow injections. J Natl Cancer Inst. 1951; 12:197–201.
5. Giebel B, Zhang T, Beckmann J, Spanholtz J, Wernet P, Ho AD, et al. Primitive human hematopoietic cells give rise to differentially specified daughter cells upon their initial cell division. Blood. 2006;107:2146–52.
6. Leary AG, Strauss LC, Civin CI, Ogawa M. Disparate differentiation in hemopoietic colonies derived from human paired progenitors. Blood. 1985;66:327–32.
7. Mayani H, Dragowska W, Lansdorp PM. Lineage commitment in human hemopoiesis involves asymmetric cell division of multipotent progenitors and does not appear to be influenced by cytokines. J Cell Physiol. 1993;157:579–86.
8. Gluckman E, Broxmeyer HA, Auerbach AD, Friedman HS, Douglas GW, Devergie A, et al. Hematopoietic reconstitution in a patient with Fanconi's anemia by means of umbilical-cord blood from an HLA-identical sibling. N Engl J Med. 1989;321:1174–8.
9. Matsunaga T, Sakamaki S, Kohgo Y, Ohi S, Hirayama Y, Niitsu Y. Recombinant human granulocyte colony-stimulating factor can mobilize sufficient amounts of peripheral blood stem cells in healthy volunteers for allogeneic transplantation. Bone Marrow Transplant. 1993;11:103–8.
10. Morrison SJ, Weissman IL. The long-term repopulating subset of hematopoietic stem cells is deterministic and isolatable by phenotype. Immunity. 1994;1:661–73.
11. Osawa M, Hanada K, Hamada H, Nakauchi H. Long-term lymphohematopoietic reconstitution by a single CD34-low/negative hematopoietic stem cell. Science. 1996;273:242–5.
12. Smith LG, Weissman IL, Heimfeld S. Clonal analysis of hematopoietic stem-cell differentiation in vivo. Proc Natl Acad Sci U S A. 1991;88:2788–92.
13. Uchida N, Weissman IL. Searching for hematopoietic stem cells: evidence that Thy-1.1lo Lin- Sca-1+ cells are the only stem cells in C57BL/Ka-Thy-1.1 bone marrow. J Exp Med. 1992;175:175–84.
14. Notta F, Doulatov S, Laurenti E, Poeppl A, Jurisica I, Dick JE. Isolation of single human hematopoietic stem cells capable of long-term multilineage engraftment. Science. 2011;333:218–21.
15. Kim I, He S, Yilmaz OH, Kiel MJ, Morrison SJ. Enhanced purification of fetal liver hematopoietic stem cells using SLAM family receptors. Blood. 2006;108:737–44.
16. Sintes J, Romero X, Marin P, Terhorst C, Engel P. Differential expression of CD150 (SLAM) family receptors by human hematopoietic stem and progenitor cells. Exp Hematol. 2008;36:1199–204.
17. Tavian M, Biasch K, Sinka L, Vallet J, Peault B. Embryonic origin of human hematopoiesis. Int J Dev Biol. 2010;54:1061–5.
18. Kumaravelu P, Hook L, Morrison AM, Ure J, Zhao S, Zuyev S, et al. Quantitative developmental anatomy of definitive haematopoietic stem cells/long-term repopulating units (HSC/RUs): role of the aorta-gonad-mesonephros (AGM) region and the yolk sac in colonisation of the mouse embryonic liver. Development. 2002;129:4891–9.
19. Filipovich AH, Stone JV, Tomany SC, Ireland M, Kollman C, Pelz CJ, et al. Impact of donor type on outcome of bone marrow transplantation for Wiskott-Aldrich syndrome: collaborative study of the International Bone Marrow Transplant Registry and the National Marrow Donor Program. Blood. 2001;97:1598–603.
20. Grunebaum E, Mazzolari E, Porta F, Dallera D, Atkinson A, Reid B, et al. Bone marrow transplantation for severe combined immune deficiency. JAMA. 2006;295:508–18.
21. Haddad E, Landais P, Friedrich W, Gerritsen B, Cavazzana-Calvo M, Morgan G, et al. Long-term immune reconstitution and outcome after HLA-nonidentical T-cell-depleted bone marrow transplantation for severe combined immunodeficiency: a European retrospective study of 116 patients. Blood. 1998;91:3646–53.
22. Krivit W. Allogeneic stem cell transplantation for the treatment of lysosomal and peroxisomal metabolic diseases. Springer Semin Immunopathol. 2004;26:119–32.
23. Steward CG, Jarisch A. Haemopoietic stem cell transplantation for genetic disorders. Arch Dis Child. 2005;90:1259–63.
24. Thomas ED, Lochte HL Jr, Lu WC, Ferrebee JW. Intravenous infusion of bone marrow in patients receiving radiation and chemotherapy. N Engl J Med. 1957;257:491–6.
25. Hermiston ML, Xu Z, Weiss A. CD45: a critical regulator of signaling thresholds in immune cells. Annu Rev Immunol. 2003;21:107–37.
26. Landree MA, Wibbenmeyer JA, Roth DB. Mutational analysis of RAG1 and RAG2 identifies three catalytic amino acids in RAG1 critical for both cleavage steps of V(D)J recombination. Genes Dev. 1999;13:3059–69.
27. Borst J, Brouns GS, de Vries E, Verschuren MC, Mason DY, van Dongen JJ. Antigen receptors on T and B lymphocytes: parallels in organization and function. Immunol Rev. 1993;132:49–84.
28. Davis MM, Bjorkman PJ. T-cell antigen receptor genes and T-cell recognition. Nature. 1988;334:395–402.
29. Akamatsu Y, Oettinger MA. Distinct roles of RAG1 and RAG2 in binding the V(D)J recombination signal sequences. Mol Cell Biol. 1998;18:4670–8.
30. Lieber MR. The mechanism of V(D)J recombination: a balance of diversity, specificity, and stability. Cell. 1992;70:873–6.
31. van Gent DC, Ramsden DA, Gellert M. The RAG1 and RAG2 proteins establish the 12/23 rule in V(D)J recombination. Cell. 1996;85:107–13.
32. Curry JD, Geier JK, Schlissel MS. Single-strand recombination signal sequence nicks in vivo: evidence for a capture model of synapsis. Nat Immunol. 2005;6:1272–9.

33. Difilippantonio MJ, McMahan CJ, Eastman QM, Spanopoulou E, Schatz DG. RAG1 mediates signal sequence recognition and recruitment of RAG2 in V(D)J recombination. Cell. 1996;87:253–62.
34. Rother MB, Jensen K, van der Burg M, van de Bovenkamp FS, Kroek R, van IWF, et al. Decreased IL7Ralpha and TdT expression underlie the skewed immunoglobulin repertoire of human B-cell precursors from fetal origin. Sci Rep. 2016;6:33924.
35. Dik WA, Pike-Overzet K, Weerkamp F, de Ridder D, de Haas EF, Baert MR, et al. New insights on human T cell development by quantitative T cell receptor gene rearrangement studies and gene expression profiling. J Exp Med. 2005;201:1715–23.
36. Luis TC, Staal FJ. WNT proteins: environmental factors regulating HSC fate in the niche. Ann N Y Acad Sci. 2009;1176:70–6.
37. Luis TC, Weerkamp F, Naber BA, Baert MR, de Haas EF, Nikolic T, et al. Wnt3a deficiency irreversibly impairs hematopoietic stem cell self-renewal and leads to defects in progenitor cell differentiation. Blood. 2009;113:546–54.
38. Pike-Overzet K, de Ridder D, Weerkamp F, Baert MR, Verstegen MM, Brugman MH, et al. Gene therapy: is IL2RG oncogenic in T-cell development? Nature. 2006;443:E5; discussion E6-7.
39. Staal FJ, Luis TC, Tiemessen MM. WNT signalling in the immune system: WNT is spreading its wings. Nat Rev Immunol. 2008;8:581–93.
40. Staal FJ, Weerkamp F, Langerak AW, Hendriks RW, Clevers HC. Transcriptional control of T lymphocyte differentiation. Stem Cells. 2001;19:165–79.
41. Weerkamp F, Pike-Overzet K, Staal FJ. T-sing progenitors to commit. Trends Immunol. 2006;27:125–31.
42. Penit C, Lucas B, Vasseur F. Cell expansion and growth arrest phases during the transition from precursor (CD4-8-) to immature (CD4+8+) thymocytes in normal and genetically modified mice. J Immunol. 1995;154:5103–13.
43. van Dongen JJ, Comans-Bitter WM, Wolvers-Tettero IL, Borst J. Development of human T lymphocytes and their thymus-dependency. Thymus. 1990;16:207–34.
44. Ciofani M, Knowles GC, Wiest DL, von Boehmer H, Zuniga-Pflucker JC. Stage-specific and differential notch dependency at the alphabeta and gammadelta T lineage bifurcation. Immunity. 2006;25:105–16.
45. Surh CD, Sprent J. T-cell apoptosis detected in situ during positive and negative selection in the thymus. Nature. 1994;372:100–3.
46. Ghia P, ten Boekel E, Rolink AG, Melchers F. B-cell development: a comparison between mouse and man. Immunol Today. 1998;19:480–5.
47. van Zelm MC, van der Burg M, de Ridder D, Barendregt BH, de Haas EF, Reinders MJ, et al. Ig gene rearrangement steps are initiated in early human precursor B cell subsets and correlate with specific transcription factor expression. J Immunol. 2005;175:5912–22.
48. Nutt SL, Kee BL. The transcriptional regulation of B cell lineage commitment. Immunity. 2007;26:715–25.
49. Nutt SL, Heavey B, Rolink AG, Busslinger M. Commitment to the B-lymphoid lineage depends on the transcription factor Pax5. Nature. 1999;401:556–62.
50. Sato H, Saito-Ohara F, Inazawa J, Kudo A. Pax-5 is essential for kappa sterile transcription during Ig kappa chain gene rearrangement. J Immunol. 2004;172:4858–65.
51. Hendriks RW, Middendorp S. The pre-BCR checkpoint as a cell-autonomous proliferation switch. Trends Immunol. 2004;25:249–56.
52. Espeli M, Mancini SJ, Breton C, Poirier F, Schiff C. Impaired B-cell development at the pre-BII-cell stage in galectin-1-deficient mice due to inefficient pre-BII/stromal cell interactions. Blood. 2009;113:5878–86.
53. van Zelm MC, Szczepanski T, van der Burg M, van Dongen JJ. Replication history of B lymphocytes reveals homeostatic proliferation and extensive antigen-induced B cell expansion. J Exp Med. 2007;204:645–55.
54. Wardemann H, Yurasov S, Schaefer A, Young JW, Meffre E, Nussenzweig MC. Predominant autoantibody production by early human B cell precursors. Science. 2003;301:1374–7.
55. Cichocki F, Sitnicka E, Bryceson YT. NK cell development and function—plasticity and redundancy unleashed. Semin Immunol. 2014;26:114–26.
56. Caligiuri MA. Human natural killer cells. Blood. 2008;112:461–9.
57. Di Santo JP, Vosshenrich CA. Bone marrow versus thymic pathways of natural killer cell development. Immunol Rev. 2006;214:35–46.
58. Kennedy MK, Glaccum M, Brown SN, Butz EA, Viney JL, Embers M, et al. Reversible defects in natural killer and memory CD8 T cell lineages in interleukin 15-deficient mice. J Exp Med. 2000;191:771–80.
59. Lodolce JP, Boone DL, Chai S, Swain RE, Dassopoulos T, Trettin S, et al. IL-15 receptor maintains lymphoid homeostasis by supporting lymphocyte homing and proliferation. Immunity. 1998;9:669–76.
60. Miller JS, Alley KA, McGlave P. Differentiation of natural killer (NK) cells from human primitive marrow progenitors in a stroma-based long-term culture system: identification of a CD34+7+ NK progenitor. Blood. 1994;83:2594–601.
61. Mrozek E, Anderson P, Caligiuri MA. Role of interleukin-15 in the development of human CD56+ natural killer cells from CD34+ hematopoietic progenitor cells. Blood. 1996;87:2632–40.
62. Shibuya A, Nagayoshi K, Nakamura K, Nakauchi H. Lymphokine requirement for the generation of natural killer cells from CD34+ hematopoietic progenitor cells. Blood. 1995;85:3538–46.
63. Jacobs R, Hintzen G, Kemper A, Beul K, Kempf S, Behrens G, et al. CD56bright cells differ in their KIR

repertoire and cytotoxic features from CD56dim NK cells. Eur J Immunol. 2001;31:3121–7.

64. Artis D, Spits H. The biology of innate lymphoid cells. Nature. 2015;517:293–301.
65. Gasteiger G, Rudensky AY. Interactions between innate and adaptive lymphocytes. Nat Rev Immunol. 2014;14:631–9.
66. Mebius RE, Rennert P, Weissman IL. Developing lymph nodes collect CD4+CD3- LTbeta+ cells that can differentiate to APC, NK cells, and follicular cells but not T or B cells. Immunity. 1997;7:493–504.
67. Hazenberg MD, Spits H. Human innate lymphoid cells. Blood. 2014;124:700–9.
68. Eisenbrand G, Berger MR, Brix HP, Fischer JE, Muhlbauer K, Nowrousian MR, et al. Nitrosoureas. Modes of action and perspectives in the use of hormone receptor affinity carrier molecules. Acta Oncol. 1989;28:203–11.
69. Jendrossek V, Handrick R. Membrane targeted anticancer drugs: potent inducers of apoptosis and putative radiosensitisers. Curr Med Chem Anticancer Agents. 2003;3:343–53.
70. Reed E, Kohn EC, Sarosy G, Dabholkar M, Davis P, Jacob J, et al. Paclitaxel, cisplatin, and cyclophosphamide in human ovarian cancer: molecular rationale and early clinical results. Semin Oncol. 1995;22:90–6.
71. Schilsky RL, Ratain MJ, Vokes EE, Vogelzang NJ, Anderson J, Peterson BA. Laboratory and clinical studies of biochemical modulation by hydroxyurea. Semin Oncol. 1992;19:84–9.
72. Senderowicz AM. Novel direct and indirect cyclin-dependent kinase modulators for the prevention and treatment of human neoplasms. Cancer Chemother Pharmacol. 2003;52(Suppl 1):S61–73.
73. Case DC Jr. Long-term results of patients with advanced diffuse, non-Hodgkin's lymphoma treated with cyclophosphamide, doxorubicin, vincristine, prednisone and bleomycin (CHOP-Bleo). Oncology. 1983;40:186–91.
74. Loehrer PJ Sr, Chen M, Kim K, Aisner SC, Einhorn LH, Livingston R, et al. Cisplatin, doxorubicin, and cyclophosphamide plus thoracic radiation therapy for limited-stage unresectable thymoma: an intergroup trial. J Clin Oncol. 1997;15:3093–9.
75. Webster AC, Lee VW, Chapman JR, Craig JC. Target of rapamycin inhibitors (sirolimus and everolimus) for primary immunosuppression of kidney transplant recipients: a systematic review and meta-analysis of randomized trials. Transplantation. 2006;81:1234–48.
76. Zangari M, Cavallo F, Tricot G. Farnesyltransferase inhibitors and rapamycin in the treatment of multiple myeloma. Curr Pharm Biotechnol. 2006;7:449–53.
77. Bierer BE, Hollander G, Fruman D, Burakoff SJ. Cyclosporin A and FK506: molecular mechanisms of immunosuppression and probes for transplantation biology. Curr Opin Immunol. 1993;5:763–73.
78. Dorner T, Lipsky PE. B cells: depletion or functional modulation in rheumatic diseases. Curr Opin Rheumatol. 2014;26:228–36.
79. Bryder D, Rossi DJ, Weissman IL. Hematopoietic stem cells: the paradigmatic tissue-specific stem cell. Am J Pathol. 2006;169:338–46.
80. Adolfsson J, Mansson R, Buza-Vidas N, Hultquist A, Liuba K, Jensen CT, et al. Identification of Flt3+ lympho-myeloid stem cells lacking erythro-megakaryocytic potential a revised road map for adult blood lineage commitment. Cell. 2005;121:295–306.

T Cell Subsets and T Cell-Mediated Immunity

3

Femke Broere and Willem van Eden

3.1 Introduction

T cell-mediated immunity is an adaptive process of developing antigen (Ag)-specific T LYMPHOCYTES to eliminate viral, bacterial, or parasitic infections or malignant cells. T cell-mediated immunity can also involve aberrant recognition of self-antigens leading to autoimmune inflammatory diseases. Ag specificity of T LYMPHOCYTES is based on recognition through the TcR of unique antigenic peptides presented by MHC molecules on antigen-presenting cells. T cell-mediated immunity is the central element of the adaptive immune system and includes a primary response by naïve T cells, effector functions by activated T cells, and persistence of Ag-specific memory T cells. T cell-mediated immunity is part of a complex and coordinated immune response that includes other effector cells such as MACROPHAGES, NATURAL KILLER CELLS, MAST CELLS, BASOPHILS, EOSINOPHILS, and NEUTROPHILS.

Final manuscript submitted on October 27, 2015.

F. Broere · W. van Eden (✉)
Division of Immunology, Faculty of Veterinary Medicine, Utrecht University, Utrecht, The Netherlands
e-mail: f.broere@uu.nl; w.vaneden@uu.nl

3.2 Biology of the T Lymphocyte Immune Response

Each T LYMPHOCYTE expresses a unique T CELL RECEPTOR (TCR) on the surface as the result of developmental selection upon maturation in the thymus (see Chap. 2 on Hematopoiesis). Mature T LYMPHOCYTES, known as naïve T cells, circulate through blood and the lymphatic system and reside in secondary LYMPHOID ORGANS (Fig. 3.1). Naïve T cells are those that have not yet encountered foreign Ag and have not yet been activated. Antigenic peptides are presented to the naïve T LYMPHOCYTE in secondary LYMPHOID ORGANS by DENDRITIC CELLS (DC), which are the most efficient "professional" Ag-presenting cells (APC) since they also provide co-stimulatory signals for effective T cell activation. DC acquire Ag in non-lymphoid tissues throughout the body and migrate into secondary LYMPHOID ORGANS guided by inflammatory stimuli and CHEMOKINES. APC generate antigenic peptides from a pathogenic agent or a self-antigen by ANTIGEN PROCESSING and display them on the cell surface in the context of MHC molecules. The recombinant variability of individual αβTCR, on the other hand, ensures that at least a few naïve T cells will have high-affinity binding to an antigenic peptide derived from virtually any pathogen. TCR engagement triggers a cascade of intracellular signalling events resulting in activation of the naïve T cell.

M. J. Parnham et al. (eds.), *Nijkamp and Parnham's Principles of Immunopharmacology*,
https://doi.org/10.1007/978-3-030-10811-3_3

The activated T cells rapidly proliferate (clonal expansion), migrate through the tissues to the sites of Ag presence, and perform effector functions such as cell-mediated cytotoxicity and production of various CYTOKINES (soluble mediators of the immune response). Cytotoxic CD8+ T cells are very effective in direct lysis of infected or malignant cells bearing the Ag, while CD4+ T helper cells produce CYTOKINES that can be directly toxic to the target cells or can stimulate other T cell effector functions and B cell ANTIBODY production, as well as mobilize powerful inflammatory mechanisms (Fig. 3.1) (see Chap. 6 for cytokine review).

Most effector T cells will disappear after the antigenic agent is eliminated, although others will remain and form memory T cells. Unlike naïve T cells that live for a few months or effector cells that disappear at the end of the immune response, memory T cells may survive for years in LYMPHOID ORGANS and peripheral tissues. The easily activated memory T cells can perform immediate effector functions in peripheral tissues or undergo activation and clonal expansion in LYMPHOID ORGANS to mount a secondary immune response if the same Ag appears again. Memory T cells respond much faster to the Ag than naïve T cells. Thus, in the case of infection, they help to eliminate pathogens at an early stage, thereby effectively preventing the spreading of disease.

3.2.1 The Exhausted T Cell Phenotype

In the case of chronic infections or cancer, T cells may become terminally differentiated and display a phenotype which is characterized by a reduced ability to produce inflammatory cytokines and the presence of the inhibitory receptors

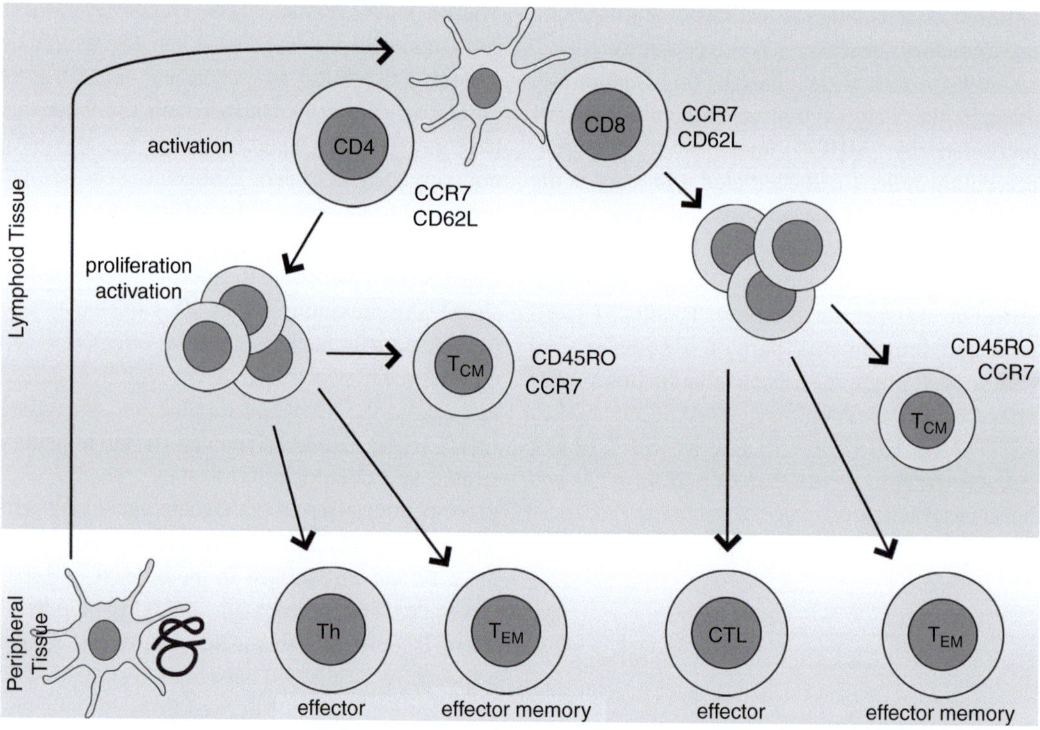

Fig. 3.1 Development of T cell-mediated responses is a sequential process. Antigen-presenting cells can take up antigen in peripheral tissues and migrate to secondary lymphoid tissues. Naïve T cells will be activated by recognition of MHC/peptide complexes on the APC, proliferate, and differentiate into effector or memory T cells. Both CD8 (CTL) and CD4 (Th) effector T cells will migrate to peripheral tissues to exert their function. In addition, memory T cells can develop into CCR7− effector memory cells (T_{EM}) that will migrate to peripheral tissues or CCD7+ central memory T cells (T_{CM}). These, in turn, can recirculate through lymphoid tissues. CCR7 is a chemokine receptor involved in T cell homing into lymphoid tissues

PD1 (programmed cell death protein 1), LAG3, TIM3, and CD160. This may be a functionally impaired or exhausted phenotype that is compatible with a progressive weakening or shutdown of the T cell response. Alternatively, it is possible that this phenotype represents a functional adaptation to a stable phenotype that balances control of immunity with avoidance of immune-mediated pathology. In this latter case, this status of the T cell would reflect "a finely tuned effector population that is optimized to fulfil a certain level of effector function and pathogen control without causing overwhelming immunopathology" [1].

3.3 Composition of the T Cell Network

3.3.1 Lymphoid Organs

The primary LYMPHOID ORGANS—the BONE MARROW and thymus—are sites of HAEMATOPOIESIS and clonal selection of T cells. The T cell-mediated immune response begins in the secondary LYMPHOID ORGANS: the spleen, lymph nodes, and organized lymphoid tissues associated with mucosal surfaces including Peyer's patches, tonsils, and bronchial, nasal, and gut-associated lymphoid tissues. The secondary LYMPHOID ORGANS have specialized T cell-rich zones where naïve T LYMPHOCYTES are concentrated; these include the PERIARTERIOLAR LYMPHOID SHEATH of spleen (PALS) and the PARACORTEX of lymph nodes. Naïve T cells reside in the spleen for just a few hours and in the lymph nodes for about 1 day before they leave via splenic veins or via efferent lymphatic vessels, respectively. Migrating naïve T cells eventually reach the bloodstream and soon after enter new LYMPHOID ORGANS, repeating the cycle until they become activated by antigenic peptides or die by neglect.

3.3.2 T Cell Subsets

Thymic selection results in the appearance of T cells with two types of TCR. The majority express Ag-binding αβ-chains in the TCR, which are disulphide-linked heterodimers of Ig superfamily proteins (Fig. 3.2) forming unique structures on each T cell. αβTCR T cells have a very diverse REPERTOIRE of Ag recognition receptors and represent mature T cells that circulate through the secondary lymph organs and develop adaptive immune responses. A small fraction of the T cells expresses γδ-chains in TCR appears to be much less heterogenic than αβTCR T cells, resides in skin and certain mucosal surfaces, and may play a role in the initial response to microbial invasion. Although the functions of γδ-TCR T cells are not fully understood, they are considered to be a relatively primitive part of the innate T cell response and will not be reviewed in this chapter.

αβTCR T cells are subdivided into several groups on the basis of lineage markers and functional activities. Two major surface co-receptor

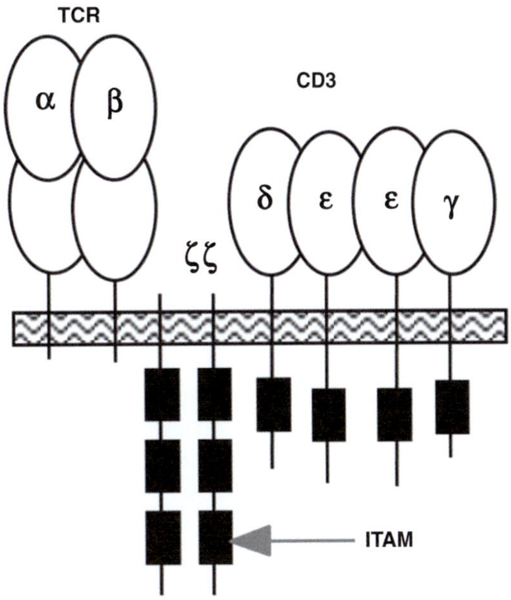

Fig. 3.2 T cell receptor complex consists of αβ-heterodimers responsible for antigen recognition and CD3 molecules involved in intracellular signalling. Immunoglobulin-like αβ-chains are formed upon gene rearrangement and have high variability among individual T cells. Non-polymorphic CD3 chains (ζ, δ, ε, γ) contain intracellular immuno-receptor tyrosine-based activation motifs (ITAMs) initiating cascades of signal transduction. ZAP70 recruitment and phosphorylation result in LAT (Linker for Activation of T cells) activation, which will induce activation of MAPK, NF-kB, and Ca signalling pathways

molecules, CD4 and CD8, define two separate T cell lineages with different functions. CD4+ cells recognize Ag in the context of MHC class II molecules (only expressed on so-called professional APC such as B cells, macrophages, and DC) and produce CYTOKINES when developed into effector T helper cells. CD8+ LYMPHOCYTES are activated by Ag peptides presented by MHC class I molecules (expressed on all nucleated cells) and form effector cytotoxic T LYMPHOCYTES (CTL).

On the other hand, the functional status of the T cells allows us to distinguish naïve, effector, and memory cells, as each of these displays extensive diversity in terms of phenotype, function, and anatomic distribution. Naïve T cells are the most homogenous representatives of CD4+ and CD8+ subsets. Upon activation, however, they can be further distinguished by their cytokine profiles. Thus, activated CD4+ T helper cells can be subdivided into Th1, Th2, Th17, and Treg subsets based on production of signature cytokines. In the case of the Th1/Th2 dichotomy, the characteristic cytokines are IFN-γ (Th1) versus IL-4 and IL-5 (Th2) [2]. CD8+ LYMPHOCYTES also can be assigned to Tc1 or Tc2 subsets according to their cytokine profile [3], although they do not produce the same quantities of CYTOKINEs as CD4+ helpers and are not efficient in B cell activation (see Chap. A3). Theoretically, both effector and memory LYMPHOCYTES of CD4+ and CD8+ lineage can be divided into subsets based on the above criteria. In addition, there are subsets of regulatory T cells that make T cell heterogeneity even more complex. Treg cells can be subdivided into naturally arising cells (nTreg) that are generated in the thymus and inducible Treg (iTreg) that are converted into Treg upon activation in the periphery [4]. Many of the specific cell surface markers representing various T cell subsets can be very useful in the design of drugs for selective manipulation of the immune response (Table 3.1).

Many proteins are upregulated or downregulated rapidly after T cell activation, like adhesion molecules or molecules involved in effector functions.

Table 3.1 Phenotypic markers associated with naïve, effector, or memory T cells

	Naïve	Effector	T_{EM}	T_{CM}
T cell subset markers				
CCR7	+++	–	+/–	+++
CD62L	+++	–	+/–	+++
CD45RO	+	+++	+++	+
CD45RA	+++	–	+	++
CD95	+/–	+++	++	+/–
Granzyme B	–	+++	+/–	–
CD25	–	+	–	–
CD127	++	+/–	+	+++
CD28	++	–	+	++

3.3.3 Naïve T Cell Markers

Naïve T cells circulating in the blood express L-selectin (CD62L), CC chemokine receptor 7 (CCR7), and leukocyte function antigen-1 (the αLβ2 integrin LFA-1). These mediate the rolling, adhesion, and extravasation of the cells through the high endothelial venules (specialized venules found in lymphoid tissues) in peripheral lymph nodes and mucosal LYMPHOID ORGANS.

Survival of naïve cells is maintained by low-affinity TCR/self-antigen interaction and signalling as well as by the presence of IL-7. These signals are normally sufficient to maintain homeostasis of naïve T cells for several months.

3.3.4 Effector T Cell Markers

High-affinity interactions of TCR with foreign Ag peptide-MHC on mature APC following activation are reflected in phenotype changes. Activated T cells express CD69 (a very early activation antigen) and CD25 (IL-2Ra). Other important surface receptors of activated T cells are CD40 ligand, which stimulates APC through binding to CD40, leading to the upregulation of CD80 (B7-1) and CD86 (B7-2) on APC, and CD28, which binds to CD80 and CD86 and propagates a co-stimulatory signal, thereby enhancing growth factor (IL-2) production and increasing T cell activation.

TNF receptor family molecules OX-40, CD27, and 4-1BB also can be found on primary activated T cells. These receptors were found to sustain T cell proliferation and survival of activated T LYMPHOCYTES upon their binding to the corresponding ligands on the APC. At the peak of their proliferation, CD4+ effector cells were also found to change the pattern of adhesion receptors such as CD62L and sPSGL-1 (sialylated form of p-selectin glycoprotein ligand 1) and chemokine receptor CXCR5. CD8+ CTL could also be characterized by expression of perforin and granzymes, proteins required for cytolytic functions. A particular set of surface markers may predict the homing capacity of effector T cells. For example, CXCR5 receptor helps CD4+ CD62L−, sPSGL-1−, and CXCR5+ T cells to migrate into B cell-rich FOLLICLES of the lymph nodes and support ANTIBODY production. In contrast, the absence of CCR7 and CD62L on CTL allows them to migrate into inflamed nonlymphoid tissues such as the lung or gut and to clear pathogenic agents in these tissues.

3.3.5 Memory T Cell Markers

Memory T cells, unlike effector T cells, are not blasts nor do they enter the cell cycle. However, they are capable of circulating in lymphoid and nonlymphoid compartments. According to the location, memory T cells are divided into central and effector memory cells and express corresponding surface markers. For example, among three phenotypes of CD8+ memory cells that have been identified (CD45RA−, CCR7+; CD45RA−, CCR7−; CD45RA+, CCR7−), the CCR7+ T cells are non-cytotoxic central memory T cells, while CCR7− are effector memory T cells [5]. Upon contact with the appropriate Ag, effector memory cells can execute effector functions instantly, whereas central or lymphoid memory cells can rapidly proliferate, expand, and acquire effector functions. CD4+ memory T cells also appear to be heterogenic. At least two subsets of CD45RA− CD4+ memory cells have been identified in humans. The central memory cells express CCR7 and CD62L and reside in LYMPHOID ORGANS, producing IL-2 upon stimulation. Some of these have been found to migrate into certain inflammation sites depending on the expression of chemokine receptors such as CCR4, CCR6, and CXCR3. The other CCR7 subset with low CD62L expression produces IFN-γ and IL-4 upon stimulation and apparently represents effector memory cells.

3.3.6 Detection of T Cells with Peptide-MHC Multimers

It is possible to detect antigen-specific T cells with the use of peptide-MHC multimers, in most cases tetramers. This technology started with the successful detection of CD8+ T cells with avidin-biotin-based MHC I tetramers [6]. More recently it became also possible to detect CD4+ T cells with MHC II tetramers. pMHC multimers are now mostly linked to fluorochromes, which enable T cell detection through conventional flow cytometry. One of the drawbacks of tetramers for T detection is that the affinity required for pMHC binding exceeds that of T cell activation. This means that sensitivity is critical and that only TcRs with relatively high binding affinities are detected. Recent advances have shown better sensitivity with the use of higher-order multimers or by inclusion of antibodies against the pMHC multimer [7]. With this technology, monitoring of therapeutic interventions at the level of antigen-specific T cells has become a more common practice.

3.3.7 T Cells with Conserved (Invariant) T Cell Receptors

Besides peptides, T cells were shown to possess the capability to recognize antigens of a nonproteinaceous nature as well. For example, mycobacterial lipids were shown to become visible for T cells when presented in the context of a so-called nonclassical MHC molecule: CD1. Recent studies done with CD1 tetramers have shown the presence of, for example, germline-encoded mycolyl-reactive (GEM) T cell populations in the blood of TB patients. In contrast to the polymorphic classical MHC molecules, all humans do express nearly identical CD1 proteins. In correspondence with this, the CD1-restricted T cell populations were found to bear conserved T cell receptors. Therefore, while the concept of conserved or invariant T cells initially was formed on the basis of the discovery of the CD1d-restricted NKT cells, it now appeared that besides NKT cells and the so-called mucosal-associated invariant T (MAIT) cells, additional subsets of human T cells with invariant T cell receptors, such as GEM T cells, exist [8].

3.4 Effectors of T Cell-Mediated Immunity

3.4.1 CD4+ Helpers

Two major functional T helper subpopulations are distinguished by their cytokine profiles (Fig. 3.3). Th1 cells produce mainly IFN-γ but also IL-2, TNF-α, and lymphotoxin. Th1 cells enhance pro-inflammatory cell-mediated immunity and were shown to induce delayed-type hypersensitivity (DTH) and B cell production of opsonizing ISOTYPES of IgG and mediate the response to some protozoa like *Leishmania* and *Trypanosoma*. Th2 cells secrete IL4, IL-5, IL-6, IL-10, and IL-13, promote non-inflammatory immediate immune responses, and were shown to be essential in B cell production of IgG, IgA, and IgE. Th1 and Th2 development routes appear to be mutually antagonistic. This has given rise to the model of polarization of immune response in accordance with the nature of the Ag and the surrounding CYTOKINE milieu. For example, IFN-γ and IL-12 are known to support Th1 cells, while IL-4 and IL-10 assist Th2 development. Although the evidence for the polarized cytokine secretion profiles of Th1 and Th2 is indisputable, several recent studies have shown more complex

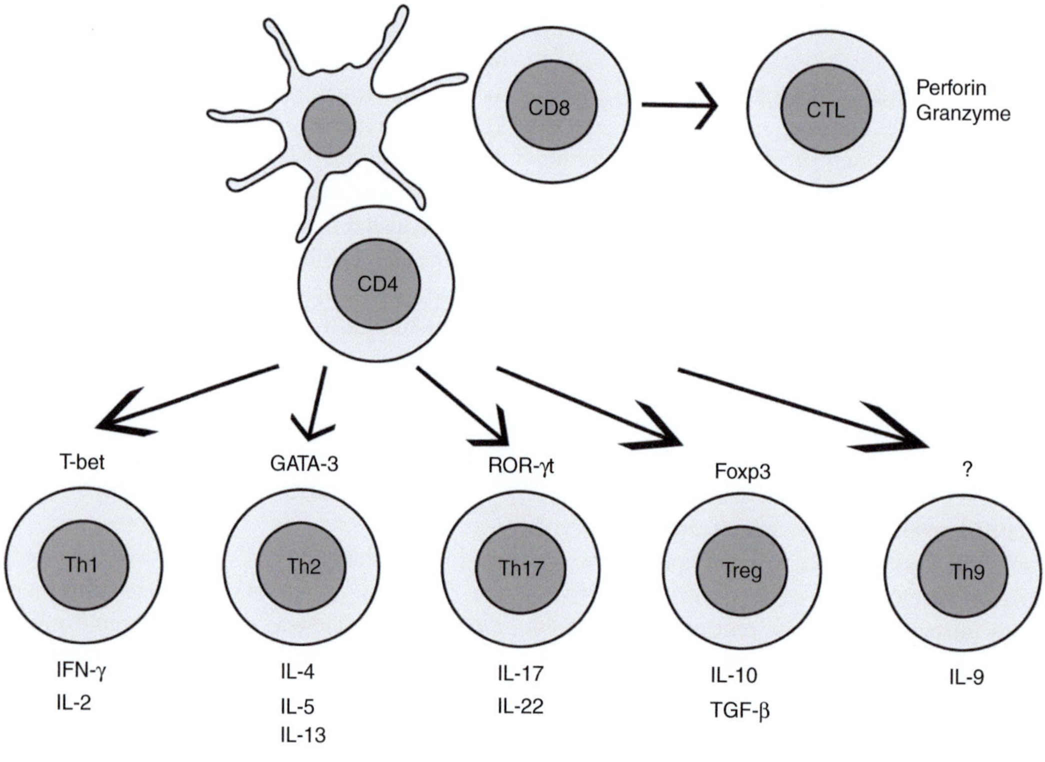

Fig. 3.3 Differentiation of effector T cells: antigen-activated T cells will differentiate into different phenotypes depending on the cytokines in the local environment and can be characterized by their cytokine profile and by transcription factors. Th1 cells produce IFN-γ and IL-2 and express T-bet. Th2 cells produce IL-4, IL-5, and IL-13 and express GATA3. Th17 cells produce IL-17 and IL-22 and express ROR γt. Treg can be divided into different subsets based on the expression of FoxP3 and/or the production of IL-10, TGF-β, and IL-35

patterns of CYTOKINE interaction in different models of immune response, including autoimmune models that are inconsistent with the simple dichotomy paradigm.

Since CD4+ T cells are central in the origin and regulation of autoimmunity, emphasis has been placed on the characterization of Th subsets and their possible roles in the inflammatory process. With the discovery that the p40 subunit of the pro-inflammatory cytokine IL-12 can not only dimerize with the p35 subunit to form IL-12 but also with p19 to create IL-23, the former dogma that IL-12-driven Th1 responses were the critical contributors to inflammation had to be revised [9]. It was found that IL-23 induced production of CD4+ T cells that secrete pro-inflammatory cytokine IL-17A. Subsequently, these cells were characterized as a separate Th subset, called Th17. Th17 cells are regarded as a major effector lineage with pro-inflammatory actions in diseases like rheumatoid arthritis, psoriasis, and Crohn's disease. Contribution of Th1 cells to inflammatory diseases is still possible, although complex, given the additional regulatory contributions of IL-12 and IFN-γ in inflammation.

Th17 cells also play a prominent role in infection. In fact, Th17 is the first subset that is generated during infection. The IL-17 receptor is expressed on fibroblasts, epithelial cells, and keratinocytes. Contact with IL-17 leads to production by the latter cell types of IL-6 and chemokines like CXCL8 and CXCL2 and GM-GSF (granulocyte-macrophage colony-stimulating factors). Altogether, this leads to recruitment of neutrophils and macrophages into the site of infection and enhances the bone marrow production of these cells. IL-22 produced by Th17 cells co-operates with IL-17 in the induction of antimicrobial peptides, such as β-defensins in epidermal keratinocytes, thereby enhancing the innate acute inflammatory response in infection.

It is anticipated that a growing spectrum of Th subset lineages will be discovered and defined by the external stimuli they respond to and the transcription factors they can induce (see Fig. 3.3). IL-12, IFN-γ, and transcription factors STAT1, STAT4, and T-bet lead to the production of Th1 cells. IL4 in combination with STAT6 and GATA-3 generates Th2 cells. Follicular T helper cells (T_{FH}) were recently defined to develop under the influence of IL-6 and transcription factor Bcl-6. Th17 cells develop in the presence of TGF-β, IL-6, and IL-23 and are characterized by the transcription factors RORγt, RORα, and STAT3. Recently, Th9 cells also were proposed, a subset that develops under the influence of IL-4 and TGF-β and that produces IL-9 [10]. Their function is associated with allergy, skin inflammatory conditions, and the control of extracellular pathogens.

There are now several subsets which may have potential to produce immunological disease. Adoptive transfer of Th1 or Th17 cells produces EAE and uveitis. Colitis in mice is produced by Th1, Th2, Th17, and Th9 cells. T_{FH} can mediate the pathogenic antibody response in experimental lupus models [11]. Th22 cells were also proposed to exist and to exert a role in inflammatory skin diseases.

3.4.2 CD8+ Cytotoxic T Lymphocytes

The CTL are derived from activated naïve CD8+ cells, proliferate in the presence of IL-2, and can expand their number many thousandfold at the peak of a primary immune response. The dramatic clonal expansion of CD8+ CTL in comparison to CD4+ cells most likely can be attributed to the relatively easy activation by the Ag/MHC class I complex and better survival in the circulation. Rapid expansion and the ability of single CD8+ CTL to destroy more than one target cell while sparing "innocent" bystanders make CTL very efficient Ag-specific effector cells. Destruction of selected cells by CTL requires the establishment of cell contact with the target cell and Ag recognition, thus initiating the release of cytolytic granules into the IMMUNOLOGICAL SYNAPSE. CTL, unlike naïve T cells, do not require co-stimulatory signals upon Ag recognition in order to kill. Therefore, they can destroy a variety of target cells bearing "foreign" Ag.

3.4.3 Mechanisms of Cell-Mediated Cytotoxicity

Two major pathways of cytotoxicity have been described in CTL: Ca^{2+}-dependent perforin/granzyme-mediated APOPTOSIS and Ca^{2+}-independent Fas ligand/Fas-mediated APOPTOSIS (Fig. 3.4). Both pathways are initiated via TCR signalling. Lytic granules (secretory lysosomes containing granzymes, perforin (PFN), and the proteoglycan serglycin (SG)) [12] appear to be transported into target cells as one complex. Granzymes are effector molecules capable of inducing APOPTOSIS in target cells via caspase-dependent and caspase-independent mechanisms. Granzymes enter into the target cell directly via plasma membrane pores formed by PFN or via receptor-mediated endocytosis. In the latter case, PFN mediates the translocation of granzymes from endocytic vesicles into the cytosol. Proteoglycan SG presumably serves as a chaperone of PFN until the complex reaches the plasma membrane of the target cells. Lytic granules represent a very efficient natural drug delivery system.

Fas-mediated APOPTOSIS is initiated by binding of Fas molecules to the target cell via Fas ligand on the CTL. The Fas molecule is a member of the TNF receptor superfamily with an intracellular "death" domain initiating caspase-dependent APOPTOSIS upon binding to Fas ligand. TCR cross-linking was shown to induce upregulation of Fas ligand expression on the cell surface of CTL and in cytolytic granules. Fas-mediated APOPTOSIS appears to be a general phenomenon not restricted to CTL. It was found to be involved in the control of cell proliferation and homeostasis among other cells.

3.4.4 Regulatory T Cells

Regulatory T cells (Treg) include more than one cell type that are critical in the maintenance of peripheral tolerance, down-modulate the amplitude of an immune response, and prevent autoimmune diseases. There is enough evidence at present to conclude that regulatory T cells participate in all cell-mediated immune responses, directly affecting Th1, Th2, Th17, CTL, and B cell reactions against "self" and "foreign" Ag. The mechanisms by which Treg exert their function are still not completely clear, but immunosuppressive cytokines such as TGF-β, IL-10, and IL-35 play an important role.

Although the majority of Treg appears within the CD4+ T cell set, suppressor activity was also reported among CD8+ T cells. In the last few years, most attention, however, was focused on CD4+ regulatory cells and particularly the nTreg

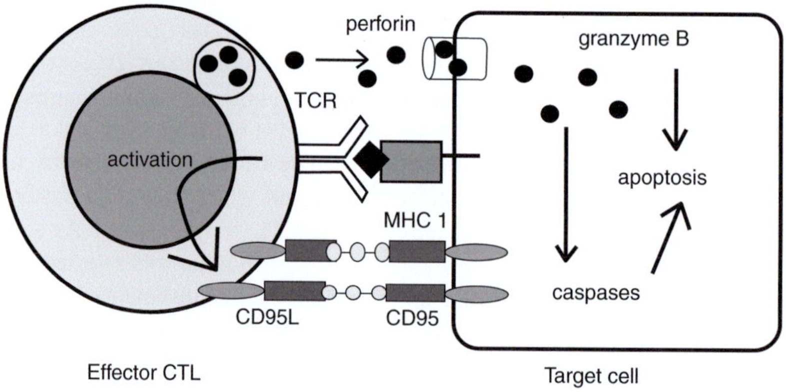

Fig. 3.4 CTL cytotoxicity can be mediated by two distinct pathways. One mechanism is via secretion of perforin and granzyme B from cytolytic granules. Perforin creates pores in the membrane of the target cell to enable granzyme B entry into the cell. Granzyme activates caspases that induce apoptosis. The second mechanism is via interaction between CD95 (Fas) and CD95L (FasL). TCR-mediated activation induces CD95L expression on the CTL. Binding of CD95 on the target cells will induce sequential caspase activation leading to apoptosis

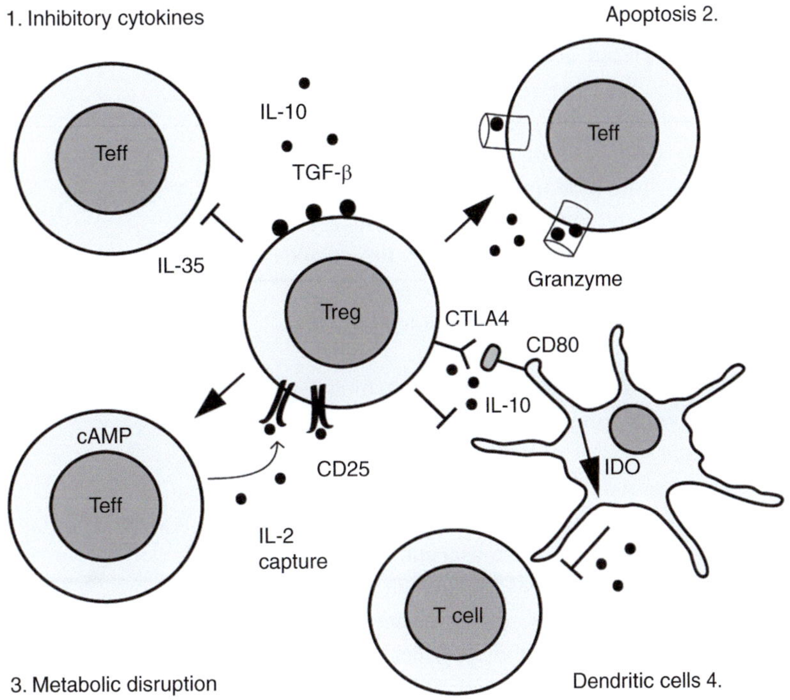

Fig. 3.5 Several mechanisms mediate Treg cell function. (1) Inhibitory cytokines such as IL-10 TGF-B and IL-35 can suppress T cell activation. (2) In some cases, cytolysis has been described as a potential suppressive mechanism, killing effector cells in a granzyme A- and B-dependent fashion. (3) Cytokine deprivation, through binding of IL-2, leads to metabolic disruption of target cells or direct cAMP-mediated inhibition. (4) DCs are targeted via direct cell-cell interactions, via CTLA4, for example, or via suppressive cytokines such as IL-10

that are characterized by constitutive expression of the α-chain of the IL-2 receptor (CD25) and the transcription factor Foxp3 [13]. nTreg arise from the thymus and represent about 10% of the total CD4 population.

Foxp3 is essential in the development and function of nTreg. The absence of functional Foxp3 results in severe systemic autoimmune diseases in mice and man. Foxp3 inhibits IL-2 transcription and induces upregulation of Treg-associated molecules, such as CD25, CTLA-4, and GITR [14], that can downregulate the immune response of adjacent cells.

In addition to nTreg, induced Treg develop in the periphery from naïve CD4+ T cells in the presence of TGF-β and IL-10 or in the absence of co-stimulation, especially in mucosal tissues. Within the population of iTreg, the heterogeneity is even more complex. Tr1 cells [15] depend on IL-10 for their induction and their suppressive action, whereas Th3 cells [16] depend on TGF-β for their suppressive action.

The inhibitory effect of all Treg primarily requires stimulation of the TCR. Upon activation, cells may mediate their function via direct cell contact through inhibitory molecules such as CTLA4, but they may also function via secretion of IL-10 and TGF-β. IL-10 can suppress differentiation of Th1 and Th2 cells directly by reducing IL-2, TNF-α, and IL-5 production but also indirectly by downregulating MHC and co-stimulatory molecules on antigen-presenting cells, thereby reducing T cell activation. The mechanism of suppression will most likely depend on the type of Treg, the nature of the immune response, the antigen, and the site of inflammation (Fig. 3.5) [17].

3.5 Mechanisms of T Cell Activation

3.5.1 Antigen Presentation

Antigenic peptides are derived by different molecular mechanisms of Ag processing, from pathogens residing either in the cytosol or in vesicular compartments of the infected cell. MHC class I molecules bind to the antigenic peptides, which originate in the cytosol of APC as a result of a multimolecular complex of proteases (proteasomes) and are transported to the endoplasmic reticulum by TAP-1 and TAP-2 (transporter associated with Ag processing-1 and -2). The newly assembled MHC/peptide complexes in the endoplasmic reticulum are then translocated through the Golgi to the cell surface. Virtually all cells of the body express MHC class I molecules at different levels and thus present antigenic peptides to CD8+ CTL and become potential targets of destruction, depending on the Ag.

MHC class II molecules, in contrast, bind peptides deriving from pathogens that appear in intracellular vesicles of the cell or from extracellular proteins internalized by endocytosis. MHC class II molecules are transported from the Golgi to endosomes and lysosomes as a complex bound to the non-polymorphic invariant chain instead of a peptide. Subsequently, the invariant chain is degraded and replaced with peptides generated by vesicular acid proteases at acid pH in the endosomal compartments. MHC II/peptide complexes appear on the surface of only a few types of immune cells, including MACROPHAGES, B cells, and DC [18].

Another important mechanism is CROSS-PRESENTATION of Ag, a process in which "professional" APC may present an Ag transferred from other cells. This enables extracellular antigens to be presented by MHCI and to activate CTL. Several studies have shown that DC can actually initiate a T cell response against MHC class I-restricted antigens by cross-presentation. CROSS-PRESENTATION also may serve as a mechanism for T cell tolerance to self-ANTIGENS in the periphery [19].

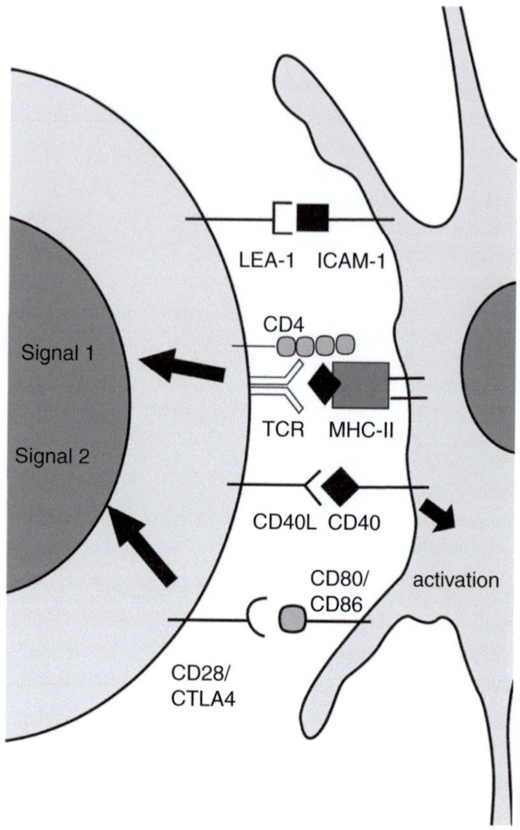

Fig. 3.6 Effective T cell activation requires interaction with multiple surface receptors on both T cells and APC. Binding of MHC class II peptide complex to the TCR and CD4 induces signal 1 in the T cell. Positive co-stimulation (signal 2) is provided by binding of CD80 or CD86 to CD28, whereas binding to CTLA4 will inhibit T cell activation. Other interactions, such as binding of LFA-1 and ICAM-1, will ensure further intensified cell-cell interactions. Binding of CD40 and CD40L will induce an activating signal in the APC, enhancing the expression of MHC molecules and co-stimulatory receptors

3.5.2 Molecular Mechanisms of T Lymphocyte Activation

Activation of naïve T cells is the most critical step in developing immunity and requires a complex interaction of TCR, co-receptors, and accessory molecules on the surface of the T cell with corresponding ligands on the APC (Fig. 3.6). TCR-Ag/MHC interaction provides an Ag recognition step and initiates intracellular signalling. Co-receptors such as CD4 and CD8 assist the TcR signal. Co-stimulatory molecules such as CD28 and CTLA-4 initiate their own intracellular signals that enhance or modulate the TCR signal. Accessory molecules such as LFA-1 or CD2 provide adhesion at the cell contact site, strengthening the interaction between the T cell and APC and allowing sustained signal transductions. The αβ-chains of TCR are non-covalently associated with invariant chains of the CD3 complex (ζ, δ, ε, and γ) (Fig. 3.2). Intracellular parts of CD3 chains include one or multiple ITAMs (immunoreceptor tyrosine-based activation motifs). ITAMs provide sites of interaction with protein tyrosine kinases (PTK) that propagate the signalling events [20].

Src family protein tyrosine kinases *Fyn* and *Lck* phosphorylate ITAMs upon TCR cross-linking by Ag/MHC, and fully phosphorylated ITAMs recruit PTK ZAP-70 to the complex via their SH2 domains. This allows LCK to trans-phosphorylate and to activate ZAP-70. The activated ZAP-70 interacts and phosphorylates SLP-76 and LAT (Linker for Activation of T cells). SLP-76 appears to be involved in actin cytoskeleton changes, while LAT is a membrane-associated protein that upon phosphorylation provides binding sites for a number of critical signalling proteins, including Grb2, Ras, and PLC-γ. PLC-γ plays a critical role in regulation of Ca^{2+} flux as it cleaves 4,5-biphosphate (PIP2) to diacylglycerol (DAG) and inositol 1,4,5-triphosphate (IP_3) upon activation by PI3 kinase. DAG stimulates PKC, while accumulation of IP_3 is the initial trigger for release of intracellular Ca^{2+} that, in turn, triggers the opening of the plasma membrane Ca^{2+} release-activated Ca^{2+} (CRAC) channels. Cascade of the signalling actions eventually results in activation of transcription factors including NF-AT, ELK-1, Jun, and ATF-2 and immune gene expression.

Although the first phosphorylation events occur within a few seconds of TCR cross-linking, the sustained contact and interaction of T cells with APC is required for full T LYMPHOCYTE activation. Earlier studies of TCR engagement have focused on IMMUNOLOGICAL SYNAPSE (IS)-dynamic clustering of different surface molecules at the contact point between T cell and APC involving TCR/CD3, co-receptors, and accessory molecules [21]. Original studies reported the formation of a ring-type structure of TCR/pMHC in the centre and the formation of spatially segregated regions of supramolecular activation complexes (SMAC). These SMAC were held to initiate signal transduction. However, the current idea is that TCR signalling is initiated in microclusters of TCR, LAT, and ZAP70-mediated phosphorylation of LAT. However, for full T cell activation, prolonged TCR activation via TCR-peptide MHC interaction remains essential. IS that contains central SMAC formed on the cell surface may provide prolonged cellular interaction and sustained signalling leading to the Ca^{2+} flux, actin cytoskeleton reorganization, and full-blown T cell activation.

3.5.3 Tolerance

An essential part of T cell-mediated immunity is the development of non-responsiveness towards naturally occurring self-ANTIGENS while mounting effective immune responses against "foreign" antigens [22]. Breakdown of self-tolerance will result in the development of autoimmune diseases. Self-reactive T cells, both CD4+ and CD8+, have been shown to be responsible for initiating and mediating tissue damage in many experimental animal models of organ-specific autoimmunity as well as in human studies.

Immunological tolerance is achieved by different mechanisms at different stages. Initially, potential self-reactive T LYMPHOCYTES are deleted during T cell development in the thymus. High-affinity interaction of TCR on immature thymocytes with self-Ag on thymic stromal cells results in apoptosis and elimination of such T cells in the process known as negative selection. T cells with TCR of low to moderate affinity to self-ANTIGENS escape from the thymus and migrate to the periphery. These T cells are normally "ignorant" to self-Ags or develop tolerance after initial activation.

Although the Ag-specific TCRs of T cells do not possess an intrinsic mechanism to distinguish self from non-self peptides, the activation by self-Ag is different to that by "foreign" Ag, mainly due to the absence of co-stimulatory signals from non-activated APC. This is in contrast to activated APC that upregulate co-stimulatory molecules during inflammation, infections, or other pathological conditions. Partial activation of T cells in the absence of co-stimulatory signals leads, instead of activation, to the state of T cell unresponsiveness towards further stimulation, also known as ANERGY [23].

In most cases, co-stimulatory molecules will direct T cell response towards either activation or tolerance. Simple absence of co-stimulatory signals was shown to induce ANERGY in effector T cells in vivo and in vitro, while naïve T cells may require a negative signal of CTLA-4 engagement to develop anergy and become tolerant.

Self-reactive cycling T cells may also undergo programmed cell death after re-exposure to the same Ag in the process called activation-induced cell death (AICD). AICD is mediated by death receptors (FAS/FAS ligand interaction of CD4+ T cells and by TNFRII/TNF interaction of CD8+ T cells) that involve interaction of caspase-dependent, death-inducing signalling complexes (DISC).

Peripheral tolerance can be also controlled by immune cytokine divergence and by regulatory T cells. Both natural and adaptive CD4+ regulatory cells have been implicated in the regulation of the autoimmune response. Thymus-derived CD25+ nTreg cells suppress other types of cell activation by largely unknown mechanisms. They require strong co-stimulatory signals for induction and maintenance, with Foxp3 expression. Adaptive (antigen-induced) regulatory T cells are generated in the periphery by suboptimal antigenic signals and rely on CYTOKINES such as IL-10 and TGF-β for suppression. These cells of varying phenotype appear often under special conditions such as chronic viral infections. Regulatory T cells present new possibilities for the treatment of autoimmune disorders and for the graft survival of transplanted organs.

3.6 Summary

T cell-mediated immunity includes priming of naïve T cells, effector functions of activated T helpers and CTL, and long-term persistence of memory T cells. Development of an effective immune response requires proper activation of T LYMPHOCYTES by APC in secondary LYMPHOID ORGANS and migration of the responding T cells to the sites of Ag presence in the body. The efficiency of T cell activation in LYMPHOID ORGANS depends on the concentration of an antigenic peptide and affinity of TCR towards the Ag/MHC complex and is facilitated by inflammatory stimuli, co-stimulatory signals, and CYTOKINES. CD8+ naïve T cells develop into effector CTL after interaction with APC, while CD4+ naïve T cells differentiate into T helper cells of major T helper types: Th1 (producing IL-2, IFN-γ, TNF-α, and LT-α) or Th2 (IL-4, IL-5, IL-6, IL-10, and IL-13).

Absence of inflammatory stimuli may induce insufficient activation of DENDRITIC CELLS resulting in induction of ANERGY and apoptosis among T cells instead of activation and productive response. This may serve as a mechanism of tolerance to self-Ags. Circulation and extravasation of T LYMPHOCYTES are orchestrated by multiple adhesion receptors whose expression and avidity are modulated by CYTOKINES and chemokines. In the process of mediating effector functions, some activated T cells undergo activation-induced cell death (AICD), while others undergo activated T cell autonomous death after the inflammation wanes, thus terminating the immune response. Only a small population of Ag-specific memory cells remains in LYMPHOID ORGANS and throughout the tissues for a long time after the immune response is over. When exposed to the Ag a second time, memory cells rapidly acquire and mediate effector functions, thereby preventing spread of pathogenic infection.

Acknowledgements We are indebted to Sergei G. Apasov and Michail V. Sitkovsky who created the basis of this chapter and contributed to the early editions of this book.

References

1. Speiser DE, et al. T cell differentiation in chronic infection and cancer: functional adaptation or exhaustion? Nat Rev Immunol. 2014;14:768–74.
2. Mosmann TR, Cherwinski H, Bond MW, Giedlin MA, Coffman RL. Two types of murine helper T cell clone. I. Definition according to profiles of lymphokine activities and secreted proteins. J Immunol. 1986;136:2348–57.
3. Croft M, Carter L, Swain SL, Dutton RW. Generation of polarized antigen-specific CD8 effector populations: reciprocal action of interleukin (IL)-4 and IL-12 in promoting type 2 versus type 1 cytokine profiles. J Exp Med. 1994;180:1715–28.

4. Wing K, Sakaguchi S. Regulatory T cells exert checks and balances on self tolerance and autoimmunity. Nat Immunol. 2010;11:7–13.
5. Sallusto F, Lanzavecchia A. Heterogeneity of CD4+ memory T cells: functional modules for tailored immunity. Eur J Immunol. 2009;39:2076–82.
6. Altman JD, et al. Phenotypic analysis of antigen-specific T lymphocytes. Science. 1996;274:94–6.
7. Dolton G, et al. More tricks with tetramers: a practical guide to staining T cells with peptide-MHC multimers. Immunology. 2015;146:11–22.
8. Van Rhijn I, Moody DB. Donor unrestricted T cells: a shared human T cell response. J Immunol. 2015;195:1927–32.
9. Palmer MT, Weaver CT. Autoimmunity: increasing suspects in the CD4+ T cell lineup. Nat Immunol. 2010;11:36–40.
10. Veldhoen M, et al. Transforming growth factor-beta 'reprograms' the differentiation of T helper 2 cells and promotes an interleukin 9-producing subset. Nat Immunol. 2008;9:1341–6.
11. Linterman MA, et al. Follicular helper T cells are required for systemic autoimmunity. J Exp Med. 2009;206:561–76.
12. MacDermott RP, et al. Proteoglycans in cell-mediated cytotoxicity. Identification, localization, and exocytosis of a chondroitin sulfate proteoglycan from human cloned natural killer cells during target cell lysis. J Exp Med. 1985;162:1771–87.
13. Fehervari Z, Sakaguchi S. Peacekeepers of the immune system. Sci Am. 2006;295:56–63.
14. Shevach EM, Stephens GL. The GITR-GITRL interaction: co-stimulation or contrasuppression of regulatory activity? Nat Rev Immunol. 2006;6:613–8.
15. Battaglia M, Gregori S, Bacchetta R, Roncarolo MG. Tr1 cells: from discovery to their clinical application. Semin Immunol. 2006;18:120–7.
16. Weiner HL. Induction and mechanism of action of transforming growth factor-beta-secreting Th3 regulatory cells. Immunol Rev. 2001;182:207–14.
17. Vignali DA, Collison LW, Workman CJ. How regulatory T cells work. Nat Rev Immunol. 2008;8:523–32.
18. Germain RN. MHC-dependent antigen processing and peptide presentation: providing ligands for T lymphocyte activation. Cell. 1994;76:287–99.
19. Luckashenak N, et al. Constitutive crosspresentation of tissue antigens by dendritic cells controls CD8+ T cell tolerance in vivo. Immunity. 2008;28:521–32.
20. Weiss A. The right team at the right time to go for a home run: tyrosine kinase activation by the TCR. Nat Immunol. 2010;11:101–4.
21. Grakoui A, et al. The immunological synapse: a molecular machine controlling T cell activation. Science. 1999;285:221–7.
22. Matzinger P. The danger model: a renewed sense of self. Science. 2002;296:301–5.
23. Schwartz RH. T cell anergy. Annu Rev Immunol. 2003;21:305–34.

4 Antibody Diversity and B Lymphocyte-Mediated Immunity

Ger T. Rijkers and Bob Meek

4.1 Antibodies and Immunoglobulins

Host defense against infections with microorganisms depends on the complex interplay between cells and proteins of the immune system, which together are capable to recognize and specifically interact with molecular structures of the microorganism. The proteins involved in this process include TOLL-LIKE RECEPTORS, defensins, collectins such as mannose binding lectin, surfactant proteins, and immunoglobulins. Among these proteins, the immunoglobulins are special because of their extreme diversity in primary structure. It is estimated that up to 10^{12} different immunoglobulin molecules can be formed (see also below). This endows the host with a large spectrum of defense molecules that can bind specifically to virtually any given microorganism. The immunoglobulins in serum comprise approximately 10–20% of total protein; upon electrophoresis they end up in the gamma region of the spectrum, hence the alternative name of gamma globulins for immunoglobulins.

Antibody molecules are immunoglobulins, and immunoglobulins are antibodies. Still, it can be confusing to intermix these terms. The term antibody should only be used for immunoglobulins with known specificity for antigen, such as anti-blood group A antibodies or anti-measles antibodies. Furthermore, the term antibodies is also used when describing the interaction of an immunoglobulin with antigen. The term immunoglobulin is used when dealing with molecular or biochemical characteristics of these proteins.

4.2 Structure of Immunoglobulins

Immunoglobulins are glycoproteins consisting of four polypeptide chains, comprising two identical heavy chains of 400–500 amino acids and two identical light chains of ca. 200 amino acids (Fig. 4.1). These chains are held together by disulfide bridges and noncovalent protein-protein interactions. On the basis of the primary structure of the heavy chains, the

Final manuscript submitted on November 28, 2016.

G. T. Rijkers (✉)
Department of Sciences, University College Roosevelt, Middelburg, The Netherlands

Laboratory for Medical Microbiology and Immunology, St. Elisabeth Hospital, Tilburg, The Netherlands
e-mail: g.rijkers@ucr.nl; g.rijkers@etz.nl

B. Meek
Laboratory for Medical Microbiology and Immunology, St. Antonius Hospital, Nieuwegein, The Netherlands

M. J. Parnham et al. (eds.), *Nijkamp and Parnham's Principles of Immunopharmacology*,
https://doi.org/10.1007/978-3-030-10811-3_4

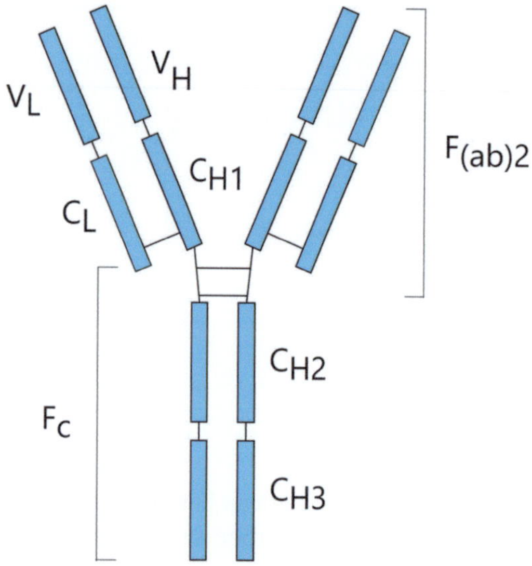

Fig. 4.1 Schematic structure of an IgG immunoglobulin molecule. The immunoglobulin molecule has two identical heavy chains, each composed of three constant domains (C_{H1}, C_{H2}, C_{H3}) and a variable domain (V_H). Two identical light chains (composed of a constant (C_L) and a variable domain (V_L)) are connected to the heavy chains by disulfide bonds. Fragments of the molecule that can be obtained after treatment with proteolytic enzymes are indicated with $F(ab)_2$, the antigen-binding fragment, and Fc, the fragment that can be obtained in crystallizable form

immunoglobulins are divided into five classes or ISOTYPES, namely, IgG, IgA, IgM, IgD, and IgE. IgG is further subdivided into four subclasses IgG1, IgG2, IgG3, and IgG4, while IgA is subdivided into two subclasses IgA1 and IgA2. There are two types of light chains, kappa (κ) and lambda (λ) chains. Heavy chains and light chains each have a specific domain structure. Heavy-chain molecules consist of four domains, three of them are termed constant domains (C_{H1}, C_{H2}, C_{H3}; see Fig. 4.1) because of the limited variation in amino acid sequence, while the fourth domain shows considerable sequence variation (V_H). Light chains have a two-domain structure (C_L and V_L). The combination of the variable domains of one heavy and one light chain determines the specific recognition of an antigen and the binding to it. The constant domains determine the biological activity of the formed antibody-antigen complex (see Sect. 4.9). In contrast to IgG, the heavy chains of IgM and IgE consist of four constant domains and one variable domain. IgG is shown in Fig. 4.1 in its monomeric configuration. Immunoglobulins can also appear as multimeric structures. IgM consists of five monomers (pentamer), and IgA appears in serum predominantly as a monomer but can also appear as a dimer. In secretion fluids and on mucosal surfaces of the respiratory and gastrointestinal tract, IgA is present as secretory IgA. Secretory IgA is dimeric IgA coupled to a J chain and a so-called secretory component. The secretory component is important for the transport of IgA through the epithelial cells to external secretions and the protection of the IgA molecule from proteolytic digestion.

4.3 The Generation of Antibody Diversity

The combination of the variable domains of the heavy and the light chain constitutes the binding site for antigen [1]. Immunoglobulins are capable of recognizing a wide variety of different antigens because of the large variation in amino acid sequences within the variable domains. The genes encoding the V_H domain are organized in clusters, each of which encodes parts of the variable domain (gene segments). During the process of gene rearrangement in precursor B lymphocytes, three gene segments are joined together, one so-called variable (V) segment, one diversity (D) segment, and one joining (J) segment. In humans, there are more than 60 V gene segments, about 30 D and 6 J gene segments. The three segments are joined together at random combination, allowing for more than 10,000 different combinations (Fig. 4.2a). This process takes place in the bone marrow and requires the activity of two enzymes, the recombinase activating genes RAG-1 and RAG-2. The rearrangements take place in a defined order; first a given D gene segment joins with a given J gene segment, and subsequently a given V gene segment is joined to the combined DJ sequence. In addition, nucleotides at the ends of the gene

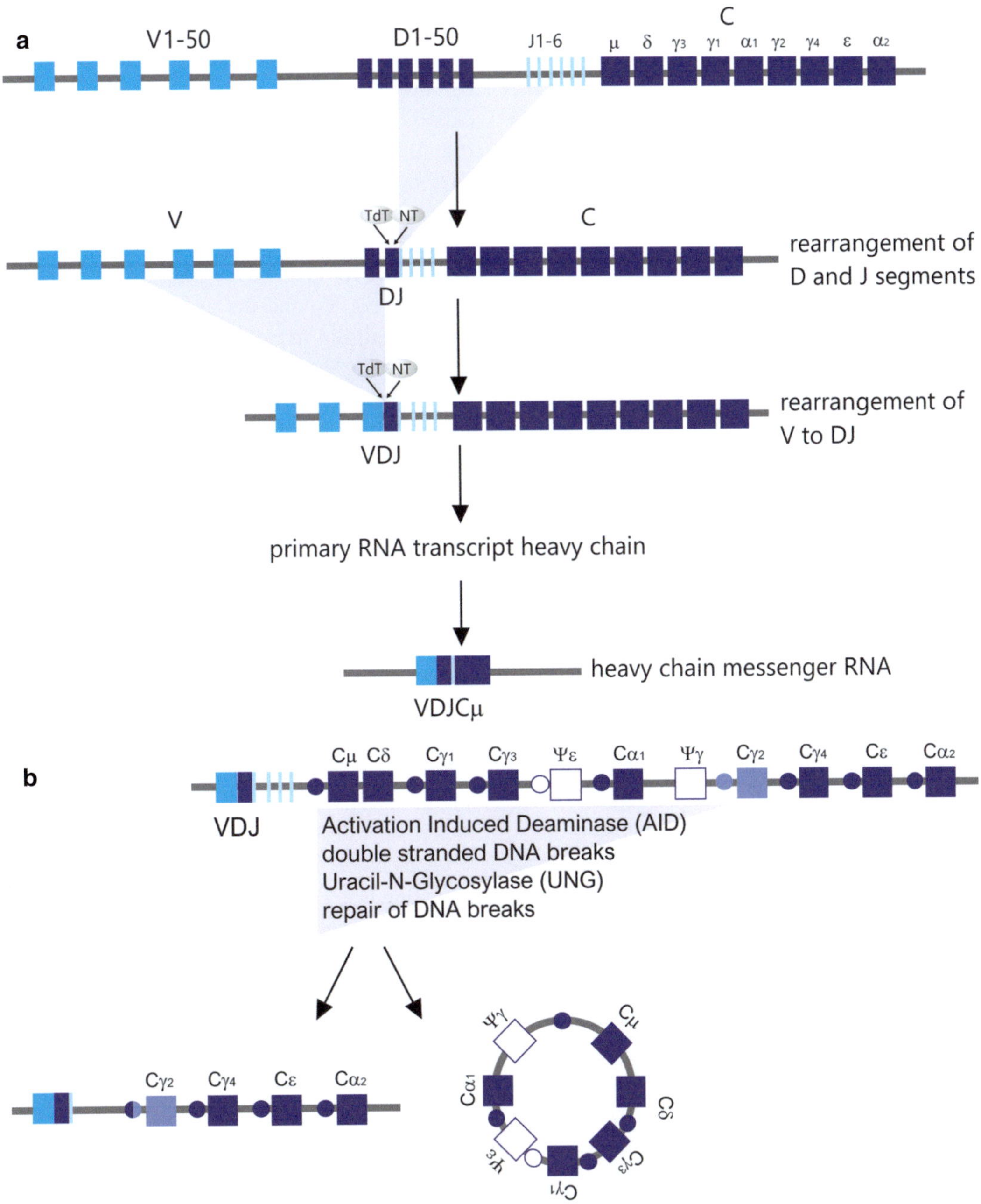

Fig. 4.2 Molecular mechanisms of rearrangement of the immunoglobulin gene segments (panel **a**) and class switch recombination (panel **b**). (Panel **a**) Early during B lymphocyte development, rearrangement of one of the D gene segments to one of the J segments takes place regulated by the recombinase-activating genes RAG-1 and RAG-2. Subsequently, rearrangement of one of the V gene segments to the DJ segment occurs. Before D is joined to J and DJ to V, the enzyme terminals deoxynucleotidyl transferase (TdT) and endonucleotidase (NT) can add (TdT) or remove (NT) extra nucleotides to or from the open ends of the DNA strands. The primary RNA transcript is being processed by splicing the VDJ segment to a constant region gene segment (C), in this example Cμ. (Panel **b**) Every C_H gene segment is preceded by a so-called switch-region (indicated by bullets in the figure). During class switch recombination, double-stranded DNA breaks do occur in the switch regions of Cμ and the C_H gene segment of the isotype to which the switch occurs (γ_2 in this example). The enzyme activation-induced deaminase and uracil-N-glycosylase play a (yet undefined) role in this process

segments can be enzymatically removed by nucleotidases (NT), while extra nucleotides can be added by the enzyme terminal deoxynucleotidyl transferase (TdT). These processes of deletion or addition of nucleotides change the germline nucleotide sequences and thus contribute to greater diversity. Next, the gene segments encoding the variable domain and the segments encoding the constant part are joined and transcribed into mRNA, spliced together, and translated into a complete heavy chain. Also during this joining process, several nucleotides can be added or deleted. A similar process occurs for the immunoglobulin light chain, except that the variable region of the light chain is composed of V and J segments only. The amino acid sequence variability in the V regions is especially pronounced in three hypervariable regions; these are at the positions of the V, D, and J segment juncture. The hypervariable regions of the H and L chains combined form the antigen-binding sites (i.e., the COMPLEMENTARITY DETERMINING REGIONS, CDR1, CDR2, and CDR3).

Early during B lymphocyte development, the rearrangements of the coding segments of the variable regions take place, and the final product, the immunoglobulin, is expressed on the surface as the membrane-bound immunoglobulin (mIg). Following a functional check of the mIg complex, B lymphocytes either egress from the bone marrow or go through a second round of gene rearrangements. This process, receptor editing, is initiated when the mIg binds too strongly to (self-)proteins presented by stromal cells in the bone marrow and mainly involves the light chain. When the immature B lymphocyte is unable to functionally rearrange a new set of a V and a J segment, it will be forced into apoptosis. This way, self-reactivity is prevented.

Upon egress from the bone marrow, daughter cells derived from such a B lymphocyte form a clone of B lymphocytes, with all cells expressing identical immunoglobulins, and therefore all have identical antigen specificity and antigen-binding capacity. However, during an immune response, additional diversity can be generated by SOMATIC MUTATIONS and by altering the heavy-chain constant region, called CLASS SWITCHING (illustrated in Fig. 4.2b). Class switching is regulated by two enzymes: activation-induced cytidine deaminase (AID) and uracil-N-glycosylase (UNG). The first transcript to be produced as a B lymphocyte develops, after VDJ joining, contains the exon for V_H and the exon for C_H of the μ chain, resulting in expression of IgM and production of IgM when stimulated by binding of an antigen. Upon prolonged exposure to the antigen, the B lymphocyte will switch the class of antibody being produced. DNA encoding a different constant domain gene segment is joined to the original V_H exon, while intervening DNA encoding, the μ heavy-chain gene segment is eliminated, resulting in the so-called excision circle (see also Fig. 4.2b). Thus, the B lymphocyte will produce another immunoglobulin molecule with different functional characteristics but identical specificity for the antigen [2]. At the same time, but also independent from the process of class switching, mutations in the variable region may arise. This results in B lymphocytes with (small) differences in the AFFINITY of the antibody molecule that is expressed on the membrane, either lower or higher affinity to the antigen. When antigen concentration becomes low, only B lymphocytes with high affinity for that particular antigen will be activated/selected. This increase in the affinity of antibodies during a humoral immune response is called affinity maturation. Both class switching and AFFINITY MATURATION are facilitated in lymph nodes and the spleen.

In the absence of specific antigen, mature B lymphocytes survive in the peripheral circulation for only a few days. Cells that do not encounter antigen within this period of time undergo apoptosis (see Box 4.1). This is necessary in order to maintain an optimal and more or less constant number of B lymphocytes in the peripheral circulation.

Box 4.1: Apoptosis

There are two different ways in which cells can die. Cells can die in an uncontrolled manner called necrosis because they are damaged by injurious agents such as toxic agents leading to cell lysis, or they undergo programmed cell death or apoptosis. Apoptosis is an active, closely regulated process which is initiated by triggering of the so-called FAS molecule on the cell surface. Alternatively, the influx of granzymes through perforin openings in the cell membrane also induces apoptosis. Cells that undergo apoptosis, also called programmed cell death, display a characteristic series of changes. The cells shrink and form bubble-like blebs on their surface, and the chromatin (DNA + proteins) in the nucleus are degraded forming a vacuolar nucleus. There are different reasons why cells undergo apoptosis. One is because of proper development, such as formation of fingers and toes of a fetus by apoptosis of the tissue between them. The other reason is for the benefit of the organism. Cells with DNA damage, tumor cells, or cells of the immune system which attack the own body tissues need to be cleared from the organism. Defects in the apoptotic machinery are associated with autoimmune diseases such as rheumatoid arthritis and lupus erythematosus.

4.4 B Cell Receptor and Signal Transduction

B lymphocyte activation is initiated by specific recognition of antigen by the antigen receptor, i.e., the mIg. Resting, primary B lymphocytes express two isotypes of mIg: mIgM and mIgD. Both mIgM and mIgD (as well as other mIg isotypes; see below) are expressed on the cell surface in association with Igα and Igβ molecules; collectively such a complex is called the B cell receptor (BCR) complex. Igα (CD79a) and Igβ (CD79b) are the protein products of the MB-1 and B29 genes, respectively, and both belong to the Ig superfamily [3]. Igα and Igβ fulfill at least three different functions: they are required for expression of membrane Ig on the surface of the B lymphocyte, they act as transducer elements coupling the antigen receptor to intracellular signaling molecules by virtue of the ITAM motif (see below), and they contain sequences for efficient internalization of antigen when it is bound to mIg.

One of the first signs of cellular activation after antigen-induced ligation of the BCR is the increase in the activity of protein tyrosine kinases (PTKs; see Box 4.2). Because the cytoplasmic domains of mIgM and mIgD consist of only three amino acids, it could be assumed that Igα (cytoplasmic domain of 61 amino acids) and Igβ (48 amino acids) serve a role in signal transduction. Of

Box 4.2: Protein Tyrosine Kinases (PTKs)

Protein tyrosine kinases are enzymes that catalyze the phosphorylation of tyrosine residues in proteins with adenosine triphosphate (ATP) or other nucleotides as phosphate donors. The definition of protein tyrosine kinase activity is catalysis of the reaction: ATP + a protein tyrosine = adenosine diphosphate (ADP) + protein tyrosine phosphate. Cell-to-cell signals concerning growth, differentiation, adhesion, motility, and death are frequently transmitted through tyrosine kinases. PTKs represent a diverse, large superfamily of proteins, including both transmembrane receptor tyrosine kinases and soluble cytoplasmic enzymes. Activation of the PTK domain of either class of PTK enzymes results in interaction of the protein with other signal-transducing molecules and propagation of the signal along a specific signal transduction pathway.

```
CD3γ      GQDGVRQSRASDKQTLLPNDQLYQPLKDREDDQYSHLQGNQLRRN
CD3δ      GHETGRLSGAADTQALLRNDQVYQPLRDRDDAQYSHLGGNWARNN
CD79a     RKRWQNEKLGLDAGDEYEDENLYEGLNLDDCSMYEDISRGLQGTYQDVGSLNIGDVQLEKP
CD79b             DKDDSKAGMEEDHTYEGLDIDQTATYEDIVTLRTGEVKWSVGEHPGQE
FcεR1-γ    RLKIQVRKQAITSYEKSDGVYTGLSTRNQETYETLKHEKPPQ
LMP2              FGQDGEANIQEERLYELLSDPRSALGLDGLIAEN

ITAM                 D.......D..Y..L.......Y..L
                     E       E                I
```

Fig. 4.3 The immunoreceptor tyrosine-based activation motif (ITAM). The amino acid sequence (given in the single letter code) of the cytoplasmic domains of human CD3γ, CD3δ, Igα (CD79a), Igβ (CD79b), the γ chain of the type I Fcε receptor (FcεRI-γ) and of the EBV encoded LMP2 protein are shown. The cytoplasmic domains are aligned to indicate the ITAM sequence motifs. See text for further explanation

crucial importance for signal transduction is the ITAM motif (*i*mmunoreceptor *t*yrosine-based *a*ctivation *m*otif), present in the cytoplasmic domain of Igα and Igβ (Fig. 4.3). This amino acid motif resides in a 26-amino acid sequence and consists of a tyrosine (Y) followed two residues later by a leucine (L) or isoleucine (I), a sub-motif that is repeated once after six to seven variable residues. The complete ITAM motif also contains two aspartate (D) or glutamate (E) residues at characteristic positions (Fig. 4.3). The ITAM motif is found in Igα and Igβ, in the CD3γ and CD3δ chains of the T cell receptor (TCR) complex, and in the γ chain of the Fcε receptor type I. More or less truncated forms of ITAM are present in CD22 and Fcγ receptor type II. Also a viral encoded protein, LMP2 from Epstein-Barr virus, contains an ITAM. The central role of ITAM in cellular signaling through the BCR complex (as well as through the TCR complex for that matter) has become apparent from studies in which single amino acid substitution receptor mutants and CHIMERICAL RECEPTOR molecules have been used.

Upon triggering of mIg, a number of cytoplasmic PTKs become associated with the BCR. These include kinases of the src family, such as lyn, fyn, and blk, as well as the syk tyrosine kinase and Bruton's tyrosine kinase (Btk). The binding is mediated by the interaction of the src homology 2 (SH2) domain within the tyrosine kinase with phosphorylated tyrosine residues within the ITAMs of Igα and Igβ. Note that this model suffers from a "chicken-and-egg" problem: if binding of SH2 domains is on phosphorylated ITAM tyrosines, how do ITAM tyrosines become phosphorylated initially? It has been found, however, that an alternative interaction is possible, not depending on phosphotyrosine: the ten N-terminal residues of src kinases can interact with a specific sequence within the ITAM of Igα (DCSM motif).

Following phosphorylation of Igα and Igβ by the src family PTKs, syk is recruited and activated [4]. Binding of src and syk kinases to (phosphorylated) ITAMs triggers a series of downstream signaling events, in which adapter proteins are involved. B cell adapter molecules, such as B cell linker (BLNK) and Bam32, function as conduits to effectively channel upstream signals to specific downstream branches. These include activation of phospholipase Cγ2, GTPase activating protein, MAP kinase (all through the N-terminal regions of lyn, fyn, and blk), of phospholipase Cγl (through syk), of the guanine nucleotide releasing factor Vav, and of p85 phosphoinositide 3-kinase (PI 3-kinase, through the SH3 domains of fyn and lyn). Activated PI-3 kinase in turn results in the phosphatidylinositol (3,4,5)-trisphosphate (PIP3)-mediated recruitment of Btk to the plasma membrane where it is involved in activation of phospholipase Cγ. Thus, originating from the BCR, several cytosolic PTKs are activated, resulting in the initiation of several distinct cell signaling pathways (Ras, phospholipase C, PI-3 kinase) via various adapter molecules. These signaling pathways result in the activation of a set of protein kinases, which in turn phosphorylate cytoplasmic and nuclear substrates and ultimately activate transcription (Fig. 4.4).

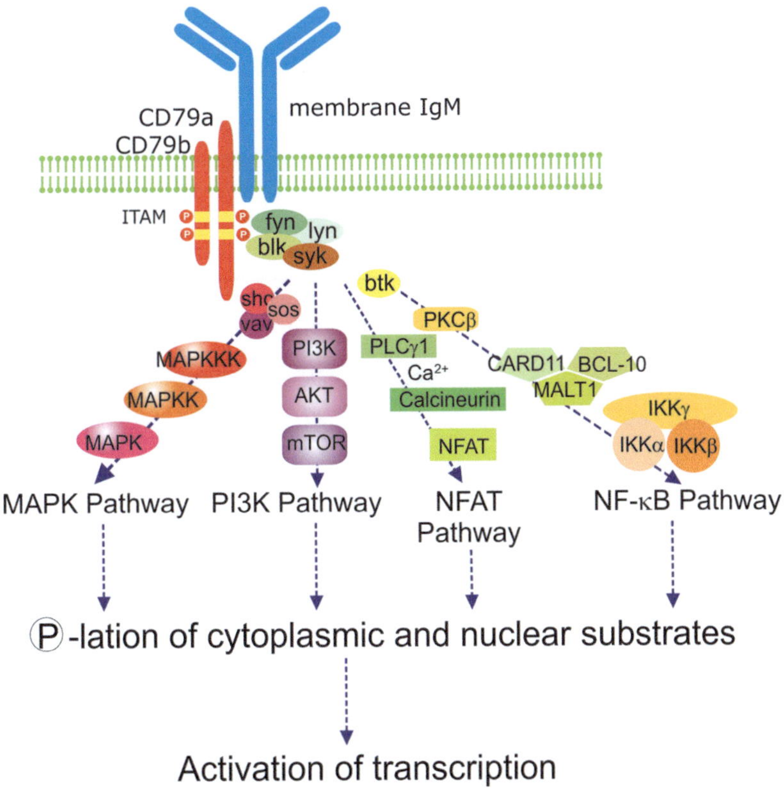

Fig. 4.4 Signaling through the BCR complex. The membrane immunoglobulin is composed of disulfide-linked heavy- and light-chain molecules flanked by a noncovalently associated dimer of Igα and Igβ. The cytoplasmic domains of Igα and Igβ contain the ITAM. Phosphorylation (P) of the tyrosine residues in the ITAMs allows src kinases (lyn, fyn, blk) and syk to associate with the BCR. Activated kinases lead to further phosphorylation of ITAMs, autophosphorylation, as well as phosphorylation of a number of cell signaling molecules. The latter include Shc and Vav (activating Ras and thus also leading to activation of MAP kinase), phospholipase Cγ1 which activate the NFAT pathway, PI-3 kinase, and finally the NF-κB pathway. All these pathways result in phosphorylation of cytoplasmic and nuclear substrates (including transcription factors) leading to activation of transcription and thus B lymphocyte activation

4.5 B Lymphocyte Costimulation

Whereas the events described above are causally linked to B lymphocyte proliferation and differentiation into antibody-secreting plasma cells, in only a few cases is triggering of the BCR by specific antigen sufficient to ensure subsequent B lymphocyte activation and differentiation. In all other instances, involving the vast majority of naturally occurring antigens, the process of B lymphocyte activation and differentiation depends on activation of additional receptors on the B lymphocyte. A number of these receptors interact with counter-receptors on T lymphocytes, thus providing the structural basis for the interaction between these two cell types in the process of antibody formation. Other receptors on B lymphocytes have ligands that are expressed on other cell types (such as monocytes, endothelial cells, etc.) or have soluble ligands.

A major coreceptor on B lymphocytes is CD40 [5]. This 50-kDa glycoprotein is a member of the so-called TNF receptor superfamily and is expressed on all mature B lymphocytes. The counterreceptor for CD40 is the CD40 ligand (CD40L or CD154), a 39-kDa cell surface glycoprotein which is expressed on activated T helper cells. The role of CD40 and CD40L for the process of B lymphocyte activation is depicted schematically in Fig. 4.5. Antigen that is bound to the BCR is internalized and processed, and peptide fragments derived from the antigen are subsequently presented in major histocompatibility complex (MHC; see Box 4.3) class II molecules expressed on the B lymphocyte surface. These peptides can be recognized by specific T lymphocytes, leading to T lymphocyte activation and thereby expression of CD40L. The interaction of CD40L with CD40 results in progression of the B cell activation process, including acquisition of the capacity to proliferate in response to cytokines produced by the activated T lympho-

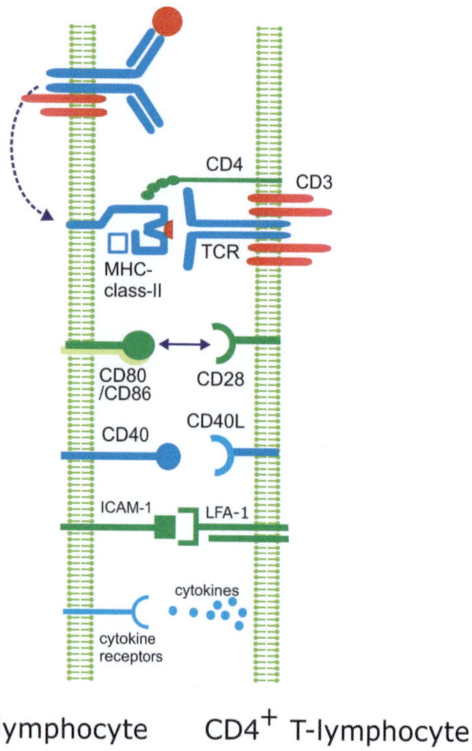

Fig. 4.5 Interaction between B and T lymphocytes. A peptide/MHC class II complex expressed on a B lymphocyte can be specifically recognized by a T cell receptor expressed on a CD4+ T cell, considered to be signal 1. To induce an effective and sustained immune response, an additional signal is needed. CD40L interacts with the CD40 molecule on the B cell, resulting in an increase of expression of CD80/CD86. Ligation of CD28 by CD80/CD86 provides a costimulatory signal (signal 2) needed for T cell activation and production of IL-2. Subsequently, several types of cytokines can be produced, and B cell differentiation into antibody-secreting plasma cells is stimulated

> **Box 4.3: Major Histocompatibility Complex (MHC)**
>
> MHC molecules, also known as human leukocyte antigens or HLA, are the products of a cluster of genes in the human DNA known as the major histocompatibility complex (MHC). There are two types of MHC molecules: MHC class I and MHC class II. MHC class I molecules designated as HLA-A, HLA-B, and HLA-C are expressed on all nucleated cells, while MHC class II molecules (HLA-DR, HLA-DQ, and HLA-DP) are primarily expressed on antigen-presenting cells (APC) like macrophages, dendritic cells, and B cells. MHC molecules bind small protein fragments, called peptides, and form MHC/peptide complexes at the cell surface. Recognition of these MHC/peptide complexes by the T cell receptor is required for T cell activation. B cells, however, are also able to directly recognize antigens via their B cell receptor.

cyte. The signal received through CD40 is also important for the process of class switching, the mechanism through which antibodies of immunoglobulin classes other than IgM are produced. The biological significance of the interaction between CD40 and CD40L is illustrated in a human immunodeficiency disease, the so-called X-linked hyper-IgM syndrome. Affected patients carry mutations in the CD40L gene resulting in the inability to produce antibodies other than IgM because the switch to IgG or IgA is not possible. This disease thus is an example of a defect in T lymphocytes that is manifested in impaired function of B lymphocytes.

Additional costimulatory receptor-counterreceptor pairs contribute to successful interaction between B lymphocytes and T lymphocytes, such as CD28 and the inducible costimulator (ICOS) on T lymphocytes and B7.1/B7.2 (CD80/CD86) and ICOS ligand (B7h) on B lymphocytes, respectively [6]. As a result of CD40-CD40L interaction, the expression of B7 on the B lymphocyte is upregulated. Ligation of CD28 by B7 provides a costimulatory signal that is required for T cell activation, proliferation, production of IL-2, and cell survival. Both MHC/peptide complex-T cell receptor interaction (signal 1) and CD28-ligand interaction (signal 2) are needed for the induction of a sustained immune response.

4.6 Cytokine Regulation and Isotype Switching

The full stimulatory effect of the interaction between T lymphocytes and B lymphocytes depends not only on binding of cell surface receptors and counter-receptors but also on production of T cell cytokines that promote various stages of B lymphocyte proliferation and terminal differentiation into plasma cells. The soluble cytokines should not be considered merely as endocrine hormones, because they are secreted at the sites of direct cell-cell contact, and therefore the particular B lymphocyte engaged in cellular interaction with the relevant T lymphocyte benefits most from these growth and differentiation factors. The interaction between T and B lymphocytes takes place in the follicles of the spleen and lymph nodes. A specialized subset of CD4⁺ T lymphocytes, T follicular helper cells, Tfh cells, provide T cell help to the B lymphocyte [7].

The cytokines that regulate B lymphocyte growth and differentiation predominantly include interleukin (IL)-4, IL-6, IL-10, and IL-21 [8, 9]. IL-4 acts as a costimulator for signals received through the BCR and CD40 in promoting B lymphocyte growth. IL-6 and IL-21 primarily regulate the differentiation of an activated B lymphocyte into an antibody-secreting cell. IFN-γ and IL-21 promote switching to IgG, while IL-10 and TGF-β drive activated B lymphocytes to IgA class antibodies. Both IL-4 and the related cytokine IL-13 can cause switching to IgE and IgG4 production (Fig. 4.6).

The molecular mechanisms that govern the changes in the isotype of the immunoglobulin heavy chain used in the B lymphocyte have been described above. Cytokines play a role in this

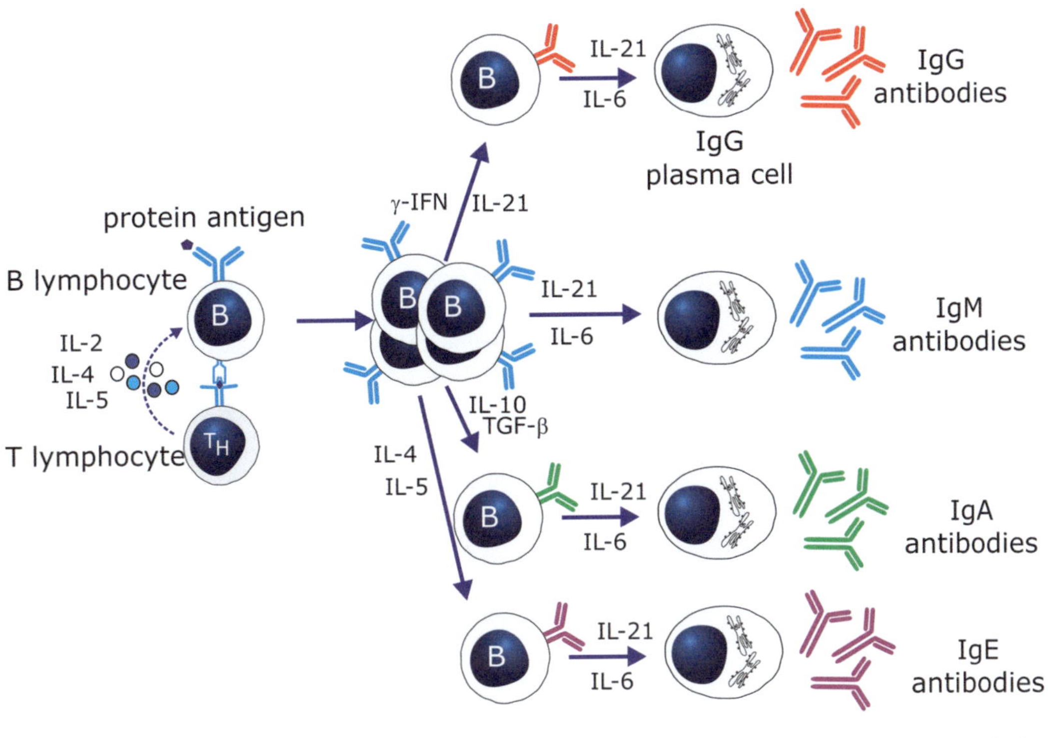

Fig. 4.6 Cytokine regulation of B lymphocyte activation and class switching. The various cytokines which regulate the initial phase of B lymphocyte activation and the subsequent class switching and differentiation into antibody-secreting cell are indicated

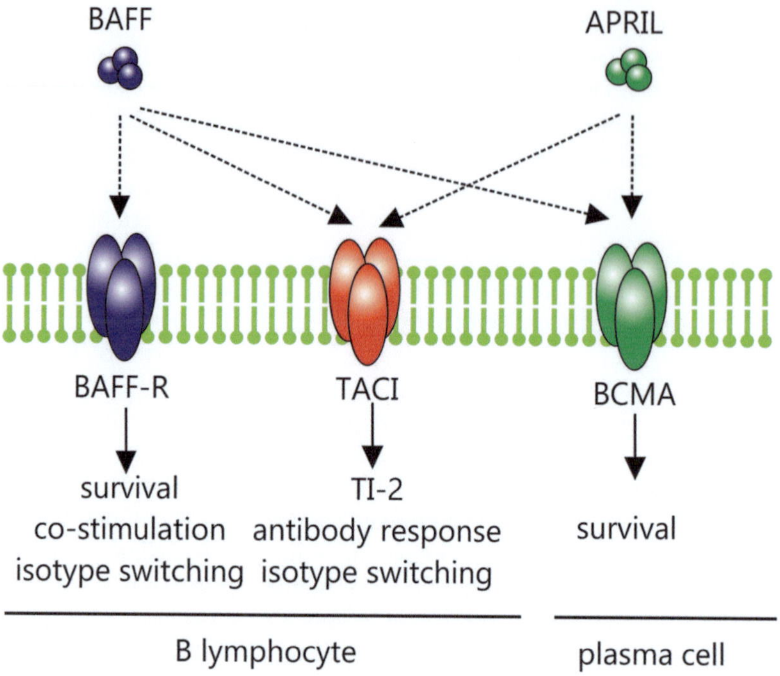

Fig. 4.7 Regulation of B lymphocyte activation and function by BAFF and APRIL. BAFF and APRIL are cytokines belonging to the TNF-family. BAFF can bind to three different receptors: the BAFF receptor (BAFF-R), TACI, and (weak) BCMA (B cell maturation antigen). APRIL binds to TACI and to BCMA. The most important cellular effects of binding of BAFF and APRIL to the various receptors are indicated

process by making the DNA in switch regions accessible for the enzymes AID and UNG. The different switch regions are more or less cytokine-specific activated: IL-21 promotes switching to IgG1 and IgG3 and TGF-β and IL-10 to IgA1. IFN-γ regulates switching to IgG1 and IL-4 and IL-13 to IgE. Two other cytokines (BAFF and APRIL), produced by macrophages, dendritic cells, and epithelial cells, drive T lymphocyte-independent isotype switching of B lymphocytes [10]. BAFF (B cell-activating factor) can bind to the BAFF receptor and to TACI (transmembrane activator and calcium-modulating cyclophilin ligand interactor) on B lymphocytes or to BCMA (B cell maturation antigen) on plasma cells (Fig. 4.7). APRIL (a proliferation-inducing ligand) binds to both TACI and BCMA.

4.7 T Cell-Independent B Lymphocyte Activation

In the section above, the important role of interaction with T lymphocytes for the process of B lymphocyte activation has been emphasized. However, there is a category of antigens that is unable to activate T lymphocytes, whereas B lymphocyte responses and induction of antibodies can be readily demonstrated. These types of antigens are called T cell-independent antigens, and major representatives are capsular polysaccharides from encapsulated bacteria such as *Streptococcus pneumoniae*, *Haemophilus influenzae* type b, and *Neisseria meningitidis*. Neither processing and antigen presentation in the context of MHC class II molecules nor specific T lymphocyte activation has been demonstrated for polysaccharide molecules. There are several unique aspects about how the immune system has organized the response to T cell-independent (TI) antigens. The spleen is very important in the immunity against encapsulated bacteria, and within the spleen there resides a dedicated B cell subset called marginal zone (Mz) B cells. These cells are considered to be innate-like B lymphocytes as their immunoglobulin genes are less diversified and TLR signaling is very important for their activation.

The marginal zone of the spleen is the area functionally specialized for filtration of blood. The Mz B cells are surrounded by macrophages, dendritic cells, neutrophils, and innate lymphoid cells (ILCs) (see also Chap. 8) [11]. Especially ILCs are important for immune responses in the Mz in general and activation of Mz B cells in particular. In the Mz their function is similar to antigen-specific T cells

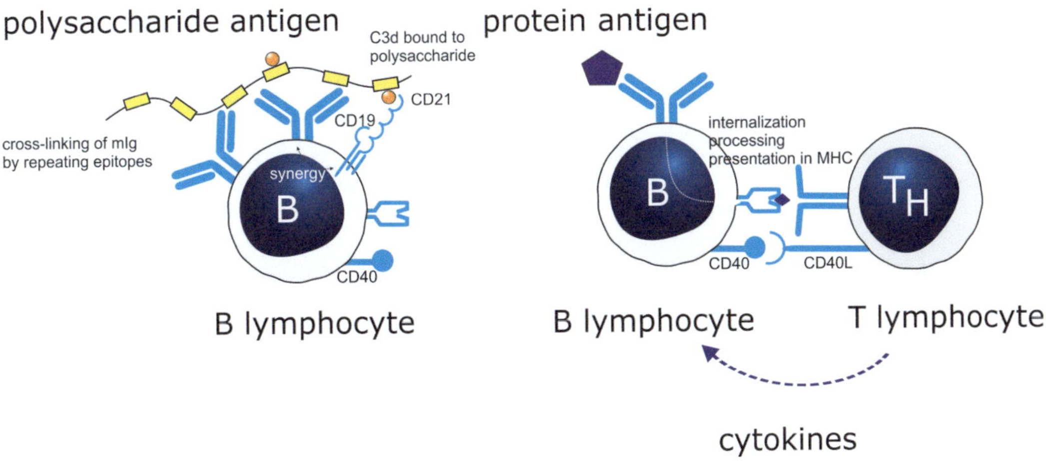

Fig. 4.8 B lymphocyte activation by T cell-independent and by T cell-dependent antigens. Left panel: A polysaccharide antigen with repeating epitopes cross-links mIg on the surface of the B lymphocyte. Deposited C3d is bound by CD21, which provides a synergistic signal for B lymphocyte activation. Right panel: A protein antigen is bound and internalized by mIg; following intracellular processing, peptide fragments are expressed in MHC class II molecules. Specific T lymphocytes recognize these peptides in the context of MHC, become activated, and express CD40L. Upon interaction of CD40L with CD40, cytokine production is initiated (see Fig. 4.5 for more details)

as they provide costimulatory signals such as BAFF production, expression of CD40L and Delta-like 1 important for the function of Mz B cells, their differentiation toward plasma cells, and activation of neutrophils in the response against TI antigens.

Next to ILCs, indirect evidence points toward a role for the CD19/CD21 receptor complex on B cells in the response against TI antigens [12]. CD19 is a 95-kDa glycoprotein of the immunoglobulin superfamily that is expressed throughout B lymphocyte development. A specific ligand for this molecule has not been identified, although purified CD19 protein does bind to bone marrow stromal cells. Activation of CD19 by specific antibodies provides a costimulatory signal for B lymphocyte activation through the BCR. Indeed, the cytoplasmic domain of the CD19 molecule contains cell signaling motifs. On mature B lymphocytes, CD19 is expressed in a molecular complex that includes CD21, the TAPA-1 protein (CD89), and the Leu-13 molecule. The prevailing model is that in this complex CD19 acts as the signal-transducing moiety for CD21. CD21 is a 145-kDa glycoprotein of the complement receptor family, which is expressed on mature B lymphocytes (and also on follicular dendritic cells and at a low level on a subpopulation of T cells). CD21 is the receptor for the complement component C3 split products iC3b, C3dg, and C3d. CD21 also serves as the cellular receptor for the Epstein-Barr virus and as an interferon α receptor. Furthermore, CD21 can interact with CD23. The (chemical) coupling of C3d to protein antigens lowers the threshold for antibody induction 100- to 1000-fold [11]. Bacterial polysaccharides, through the alternative pathway of complement activation, can generate C3 split products, which become deposited on the polysaccharide. Natural complexes of polysaccharide and C3d, thus formed, can cross-link mIg and CD21 on polysaccharide-specific B lymphocytes. This mechanism may be required in parallel to engagement of CD40/CD40L for B lymphocyte activation (Fig. 4.8). Compatible with this mechanism is the finding that children up to the age of 2 years who are unable to respond to polysaccharide antigens have a reduced expression of CD21 on B lymphocytes [13].

4.8 Primary and Secondary Antibody Response

The first contact of the immune system with a given antigen will induce what is called a primary (antibody) response. B lymphocytes

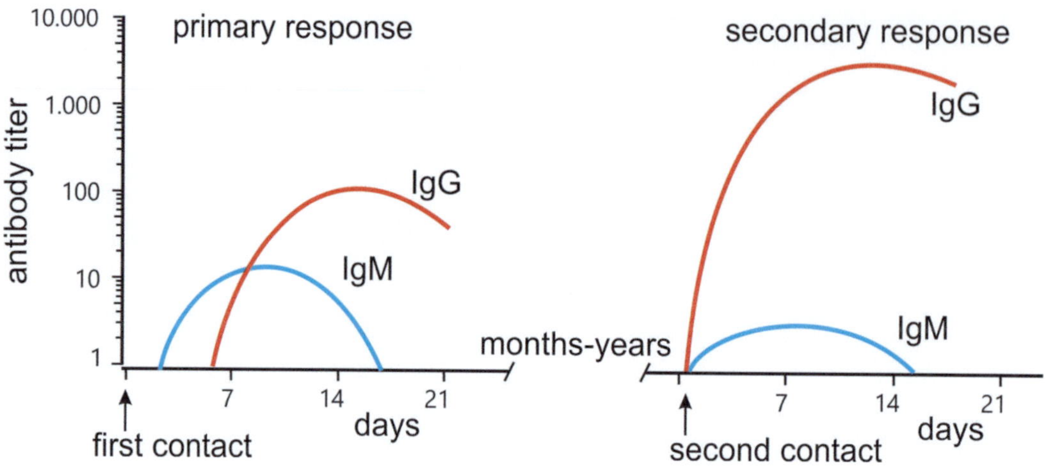

Fig. 4.9 The response triggered by the first encounter with a given antigen is called the primary antibody response. During this response IgM appears first, followed by IgG. The most prevalent class of antibody, IgG, is produced when the same antigen is encountered again. This response is called the secondary antibody response. It is faster and results in higher antibody titers than the primary antibody response

become activated and differentiate into plasma cells (along routes as described above). Plasma cells are highly differentiated cells, which maximally produce 10^4 antibody molecules per second, equaling 40% of the total protein-synthesizing capacity of the cell. The lifetime of a plasma cell is 3–4 days in the initial phase of the antibody response. These plasma cells are called short-lived plasma cells. A second contact with the same (protein) antigen elicits a secondary antibody response, which is produced by long-lived plasma cells and differs in a number of aspects from a primary response [14]. The latency period (time between contact with antigen and start of antibody production) is shorter in a secondary response (Fig. 4.9), and antibody levels attained are much higher (1–2 orders of magnitude). While during a primary antibody response predominantly IgM, and to a lesser extent IgG, antibodies are produced, IgG, IgA, and IgE antibodies are the major classes during a secondary response. The affinity of antibodies produced increases during the response; there may be a 100- to 1000-fold difference in affinity between antibodies produced by the short-lived plasma cells at the start of a primary response and at the end of a secondary response. This process (affinity maturation) is the combined effect of somatic hypermutation of CDR1 and CDR2 regions during B lymphocyte proliferation and the selection of the B lymphocytes with the highest affinity (as described earlier).

A primary antibody response takes place in follicles and the marginal zone of spleen and lymph nodes. During a secondary response, bone marrow is the major site of antibody production. Long-lived plasma cells in the bone marrow can survive for at least 90 days in the absence of cell division. Antibodies which are secreted in mucosal tissue of the respiratory and gastrointestinal tract are produced locally by the bronchus-associated lymphoid tissue (BALT) and gut-associated lymphoid tissue (GALT), respectively (Fig. 4.10).

During the primary immune response, a fraction of antigen-specific B lymphocytes do not differentiate into plasma cells but into so-called memory B lymphocytes. Whether an activated B lymphocyte differentiates into a plasma cell or into a memory B lymphocyte is largely determined by the relative expression of transcription factors: Blimp-1 is the master regulator for plasma cell generation, while Bcl-6 promotes memory B lymphocyte formation. Bcl-6 suppresses apoptosis genes, which contributes to the longevity of memory B lymphocytes [15]. Naive and memory B lymphocytes

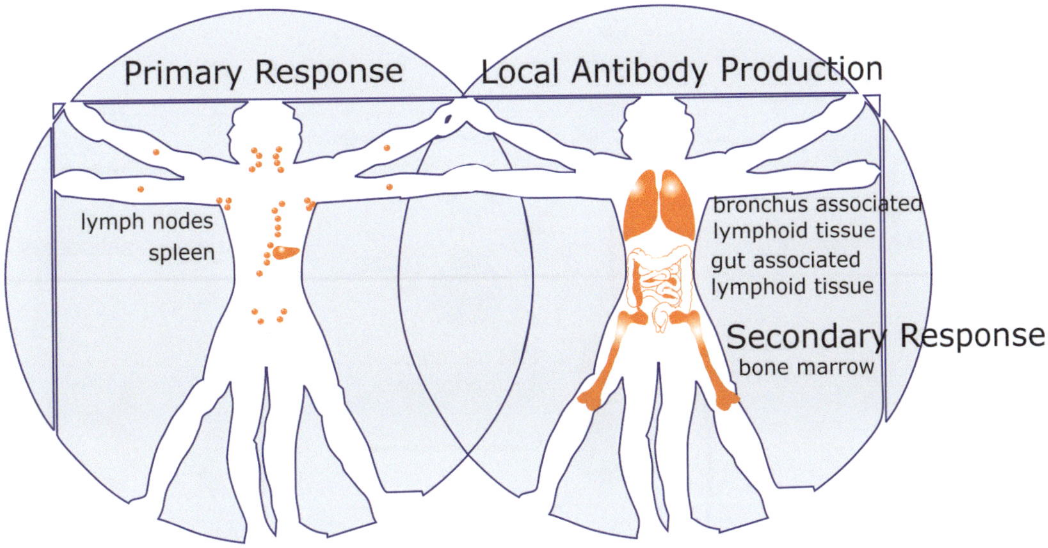

Fig. 4.10 Sites of antibody production

differ in expression of the isotypes of mIg and affinity of mIg. Naive primary B lymphocytes express mIgM and mIgD; memory B lymphocytes have lost mIgD and express mIgG or mIgA, with or without mIgM. Adult circulating B lymphocytes can be separated into three subpopulations on the basis of CD27 and mIgD expression: $IgD^{+}CD27^{-}$ represent naive B lymphocytes, $IgD^{+}CD27^{+}$ non-switched memory B lymphocytes, and $IgD^{-}CD27^{+}$ memory B lymphocytes ([16]; see also Fig. 4.11). Because of the above-described affinity maturation, the affinity of mIg for antigen on memory B lymphocytes is higher than on primary B lymphocytes.

All characteristics of a primary and secondary antibody response as described above hold true only for protein antigens. For polysaccharide antigens, a second contact with antigen induces an antibody response that is similar in terms of kinetics and magnitude as a primary response. Affinity maturation does not occur, and isotype distribution of antibodies does not change. Moreover, anti-polysaccharide antibodies use a restricted number of V_H and V_L genes, whereas anti-protein antibodies are more heterogeneous. Finally, IgG anti-protein antibodies to the vast majority of antigens are of the IgG1 subclass; IgG anti-polysaccharide antibodies in adult individuals are predominantly IgG2.

4.8.1 Turnover of Antibodies

Following production by plasma cells, individual IgG antibodies have a half-life ranging from 120 to 180 h in the blood. In general, antibodies are one of the most abundant proteins in the blood. When an antibody does not encounter its antigen, it gradually decays, mainly due to oxidation and activity of proteases encountered upon extravasation into tissues. Important for extension of the half-live of IgG are the FcRn receptors (neonatal Fc receptors) expressed on endothelial and various hematopoietic cells, such as monocytes and dendritic cells. Endothelial cells transport material across membranes in a process called transcytosis, while hematopoietic cells continuously sample material from their environment in a process called endocytosis. Soluble proteins such as antibodies are included during transport. FcRn receptors take part in the transcytotic and endocytotic machinery, among others, to salvage IgG antibodies. FcRn receptors do not bind IgG antibodies at neutral pH. During transport, the pH is lowered upon which FcRn receptors bind the Fc part of IgG. This prevents IgG from being broken down

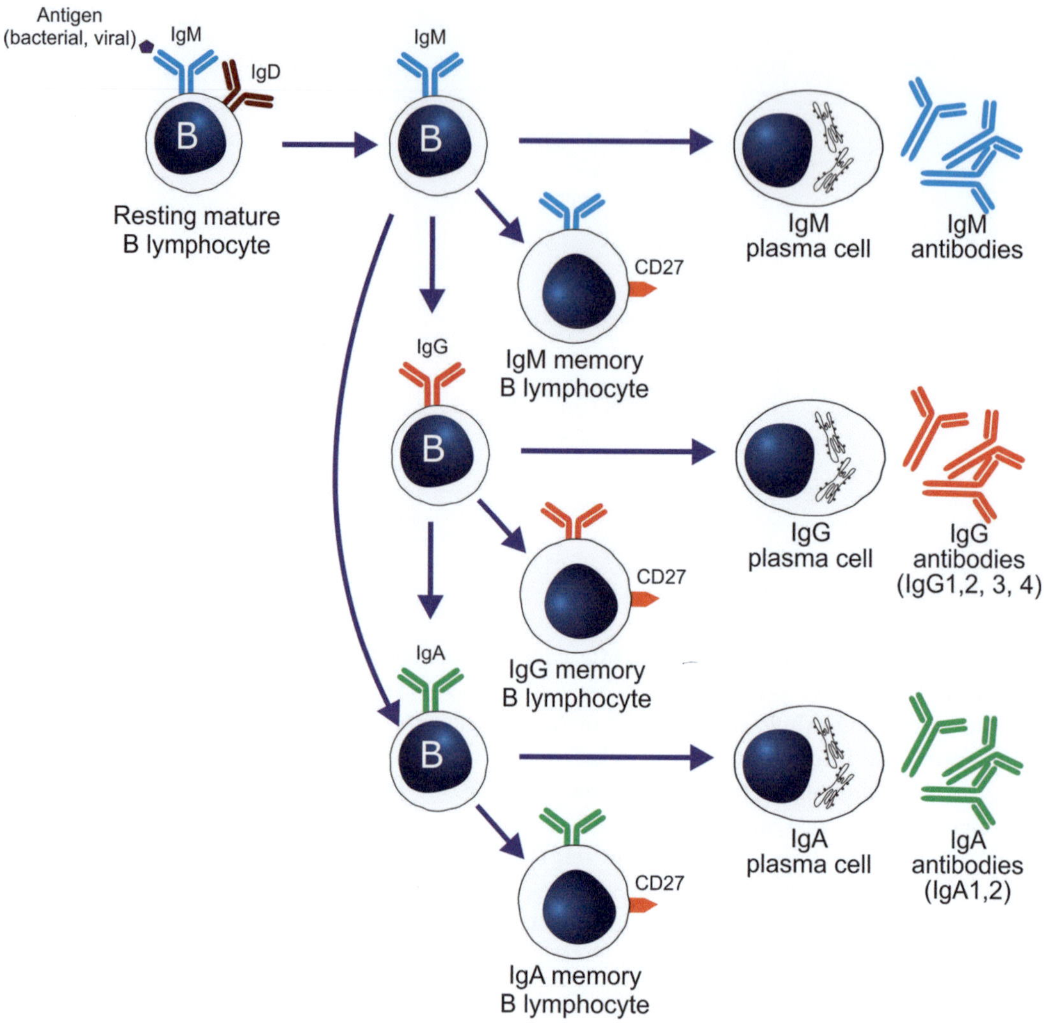

Fig. 4.11 Generation of memory B lymphocytes

in lysosomes. Subsequently, the complex of FcRn with IgG is re-routed toward the extracellular space (blood or tissue), after which IgG is released [17].

FcRn receptors are also expressed in epithelial cells lining the lung, intestines, and placenta and important for transport of IgG antibodies across membranes into the lumen and transplacentally from mother to fetus.

4.9 Biological Functions of Antibodies

The biological functions of antibodies are exerted by the different parts of the molecule: binding of antigen is carried out by variable V_H and V_L domains, while effector functions are mediated by the constant domains (in particular C_{H2} and/or C_{H3} domains). In a few cases, antibody molecules can have a direct biological effect by binding to antigen. The bacterial toxins mentioned at the beginning of this chapter are neutralized when bound by antibodies. Antiviral antibodies can limit the spread of virus particles through the body. For all other functions of antibodies, interaction of the antibody molecule with other effector mechanisms is required. Clearance of immune complexes and antibody-opsonized microorganisms depends on the functional integrity of the Fc part of the antibody molecule. The Fc part can either interact with soluble biologically active molecules, such as the complement system (see Chap. 8) or bind to Fc receptors which are expressed on a variety of cells of the immune system [18]. Fc receptors expressed on monocytes, macrophages, and granulocytes are essential for phagocytosis of immune complexes and opsonized microorganisms. Fc receptors for IgG (Fcγ receptors) are expressed on monocytes and macrophages; neutrophilic granulocytes also express receptors for IgA (Fcα receptors). Depending on the class of antibodics in an immune complex, the complement system becomes more or less efficiently activated. This will enhance phagocytosis by monocytes, macrophages, and granulocytes, since these cells also express complement receptors in addition to Fc receptors.

Fc receptors for IgE (Fcε receptor) are primarily expressed by mast cells. In allergic individuals, Fcε receptors have constitutively bound IgE; exposure to allergens causes cross-linking of Fcε receptors, resulting in mast cell degranulation and histamine release. Apart from phagocytosis and degranulation, Fc receptors also mediate cytotoxicity in a process called ADCC (antibody-dependent cellular cytotoxicity). Target cells (e.g., tumor cells) to which antibodies are bound can be recognized by Fc receptors expressed on cells with cytotoxic potential. The killing process itself is complement-independent. Monocytes, neutrophilic and eosinophilic granulocytes, and natural killer (NK) cells display ADCC activity. ADCC can be a mechanism for removal of tumor cells and has been implicated in tissue damage that occurs in autoimmune diseases.

The Fcγ receptor expressed on B lymphocytes (FcγIIb) plays a role in downregulation of B lymphocyte activation. When high IgG antibody concentrations are reached during an immune response, antigen-IgG complexes will be formed which can cross-link the BCR and FcγIIb on the surface of the B lymphocyte (see also Fig. 4.12). The cytoplasmic domain of FcγIIb contains a YSLL motif, which has been termed ITIM for immunoreceptor tyrosine-based inhibitory motif. Tyrosine phosphorylation of this motif causes the association of a protein tyrosine phosphatase. When brought in close proximity to ITAMs, this enzyme causes tyrosine dephosphorylation and therefore inhibits BCR signaling. This mechanism, by which IgG antibodies interact with FcγIIb and co-crosslink with the BCR, is an example of active downregulation of the antibody response. There are other examples of cell surface receptors with either intrinsic (CD45) or associated protein tyrosine phosphatase activity (CD22). CD22 is a member of the sialic acid-recognizing Ig superfamily of lectins (Siglecs) [19]. The precise role of this receptor-ligand pair in regulation of B lymphocyte activation is currently being delineated.

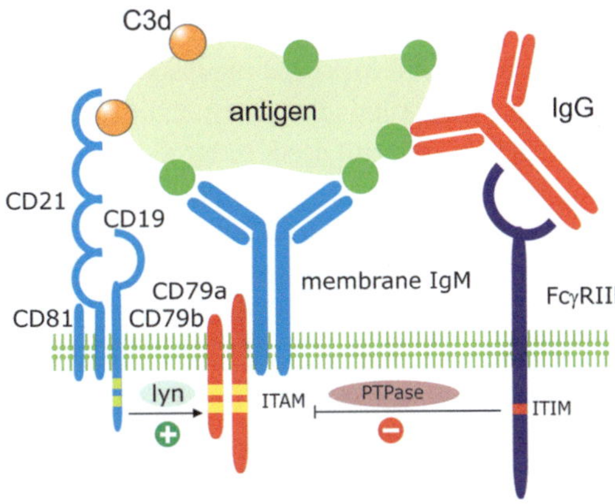

Fig. 4.12 Regulation of BCR activation by coreceptors. The cytoplasmic domains of Igα and Igβ contain the ITAM. The Fcγ receptor IIb binds the Fc part of IgG antibodies in immune complexes. The cytoplasmic domain of FcγRIIb contains an immunoreceptor tyrosine-based inhibitory motif (ITIM) composed of the amino acids YSLL. ITIM can activate a protein tyrosine phosphatase (PTPase) which dephosphorylates tyrosine residues of ITAM. CD21 (type 2 complement receptor) binds C3d deposited on the antigen or on IgG antibodies. The cytoplasmic domain of CD19 has a positive regulatory effect on signaling through mIg (see text for further explanation)

4.10 Regulatory B Cells

Next to being responsible for humoral immunity, a fraction within the B cell population is capable of inhibiting innate and adaptive cellular immune responses via production of IL-10. As yet there are no specific surface markers for these so-called B10 cells, except that they usually can be found within the memory (CD27+) subset [20]. In the blood, B10 cells constitute a minor B10pro subset that can mature into IL-10+ B10 cells in vitro following activation via CD40L (via T cells) and TLR ligand CpG [21]. In animal disease models, administration of in vitro expanded regulatory B cells is capable of inhibiting disease. In humans, their importance in vivo still needs to be confirmed.

4.11 Clinical Relevance and Future Prospects

The integrity of the humoral immune system is crucial for host defense against bacterial and certain viral infections. Inborn or acquired deficiencies in humoral immunity result in increased susceptibility to potentially life-threatening infections. A dysregulated humoral immune system may result in conditions such as allergy. These and other clinical aspects are discussed elsewhere in this volume (see Chaps. 10, 11, and 12).

Most of what is known about the cellular and molecular aspects of B lymphocyte activation and regulation has been gathered from experiments performed during the last decades. Detailed knowledge of the mechanisms that govern the regulation of expression of cell surface receptors and signaling mechanisms allows pharmacological intervention in antibody-mediated immunity. Intervention is possible at three levels: outside the B lymphocyte, at the cell surface, and intracellularly. The importance of Mz B cells in response to encapsulated bacteria has been revealed in patients suffering from autoimmune lymphoproliferative syndrome (ALPS). An impairment of FAS-mediated apoptosis of lymphocytes results in displacement of B cells from the marginal zone, causing patients to suffer from infection by encapsulated bacteria and poor IgM (but not IgG) response against T cell-independent antigens such as polysaccharides [22].

The coupling of polysaccharides to protein carriers changes the nature of the anti-polysaccharide antibody response from T cell independent to T cell dependent. These polysaccharide-protein conjugate vaccines bypass the selective unresponsiveness to T cell-independent antigens early in life and thus constitute novel and effective tools in prevention of infectious diseases. Conjugate vaccines for *Haemophilus influenzae* type b and *Streptococcus pneumoniae* now have demonstrated to be able to prevent invasive diseases in otherwise susceptible populations [23]. In case of uncontrolled proliferation of B lymphocytes (B cell lymphoma) or unwanted autoantibody production (autoimmune diseases), B lymphocytes can be eliminated by monoclonal antibodies directed to the CD20, CD22, CD30, or CD52 molecules. These reagents (biologicals) are powerful tools in managing these abnormalities of the humoral immune system [24, 25]. Care must be taken for the negative side effects associated with complete depletion of B lymphocytes; when anti-CD20 or anti-CD22 is used, these patients essentially become immunodeficient. The relevance of regulatory B10 cells needs to be established, but there is already interest to activate and/or expand these cells in autoimmune diseases, such as SLE and RA. Interestingly, B10 cells are functionally impaired in SLE patients possibly due to defective CD40 signaling. Furthermore, B10pro cells are expanded in patients suffering from chronic B cell lymphocytic leukemia (B-CLL).

Selected Readings

Burger JA. B-cell receptor signaling in chronic lymphocytic leukemia and other B-cell malignancies. Clin Adv Hematol Oncol. 2016;14:55–65.

Candando KM, Lykken JM, Tedder TF. B10 cell regulation of health and disease. Immunol Rev. 2014;259:259–72.

Eibel H, Kraus H, Sic H, Kienzler AK, Rizzi M. B cell biology: an overview. Curr Allergy Asthma Rep. 2014;14:434–40.

Hardy RR, Hayakawa K. B cell development pathways. Ann Rev Immunol. 2001;19:595–621.

Leavy O. Immune memory: Sequential evolution of B cell memory. Nat Rev Immunol. 2016;16:72–3.

Siegrist CA, Aspinall R. B-cell responses to vaccination at the extremes of age. Nat Rev Immunol. 2009;9:185–94.

Stavnezer J, Guikema JE, Schrader CE. Mechanism and regulation of class switch recombination. Annu Rev Immunol. 2008;26:261–92.

Vickery JD, Michael CF, Lew DB. Evaluation of B lymphocyte deficiencies. Cardiovasc Hematol Disord Drug Targets. 2013;13:133–43.

Wienands J, Engels N. Control of memory B cell responses by extrinsic and intrinsic mechanisms. Immunol Lett. 2016;178:27–30. https://doi.org/10.1016/j.imlet.2016.05.010. pii: S0165-2478(16)30086-4

Zubler RH. Naive and memory B cells in T-cell-dependent and T-independent responses. Springer Semin Immunopathol. 2001;23:405–19.

Recommended Websites

http://www.antibodyresource.com/educational.html (portal with many links to relevant websites)

http://primaryimmune.org (many links to primary immunodeficiencies, including B lymphocyte deficiencies)

References

1. Li Z, Woo CJ, Iglesias-Ussel MD, Ronai D, Scharff MD. The generation of antibody diversity through somatic hypermutation and class switch recombination. Genes Dev. 2004;18:1–11.
2. Edry E, Melamed D. Class switch recombination: a friend and a foe. Clin Immunol. 2007;123:244–51.
3. Brezski RJ, Monroe JG. B-cell receptor. Adv Exp Med Biol. 2008;640:12–21.
4. Gauld SB, Cambier JC. Src-family kinases in B-cell development and signaling. Oncogene. 2004;23:8001–6.
5. Elgueta R, Benson MJ, de Vries VC, Wasiuk A, Guo Y, Noelle RJ. Molecular mechanism and function of CD40/CD40L engagement in the immune system. Immunol Rev. 2009;229:152–72.
6. Sharpe AH, Freeman GJ. The B7-CD28 superfamily. Nat Rev Immunol. 2002;2:116–26.
7. Wali S, Sahoo A, Puri S, Alekseev A, Nurieva R. Insights into the development and regulation of T follicular helper cells. Cytokine. 2016; 87:9–19. https://doi.org/10.1016/j.cyto.2016.06.010. S1043-4666(16)30138-7
8. Takatsu K, Kouro T, Nagai Y. Interleukin 5 in the link between the innate and acquired immune response. Adv Immunol. 2009;101:191–236.
9. Awasthi A, Kuchroo VK. The yin and yang of follicular helper T cells. Science. 2009;325:953–5.
10. Mackay F, Silveira PA, Brink R. B cells and the BAFF/APRIL axis: fast-forward on autoimmunity and signaling. Curr Opin Immunol. 2007;19:327–36.
11. Magri G, Miyajima M, Bascones S, Mortha A, Puga I, Cassis L, et al. Innate lymphoid cells integrate stromal and immunological signals to enhance antibody production by splenic marginal zone B cells. Nat Immunol. 2014;15:354–64.
12. Fearon DT, Carroll MC. Regulation of B lymphocyte responses to foreign and self-antigens by the CD19/CD21 complex. Annu Rev Immunol. 2000;18:393–422.
13. Rijkers GT, Sanders EA, Breukels MA, Zegers BJM. Infant B cell responses to polysaccharide determinants. Vaccine. 1998;16:1396–400.
14. Tarlinton D. B-cell memory: are subsets necessary? Nat Rev Immunol. 2006;6:785–90.
15. Crotty S, Johnston RJ, Schoenberger SP. Effectors and memories: Bcl-6 and Blimp-1 in T and B lymphocyte differentiation. Nat Immunol. 2010;11:114–20.
16. Shivarov V, Shinkura R, Doi T, Begum NA, Nagaoka H, Okazaki IM, Ito S, Nonaka T, Kinoshita K, Honjo T. Molecular mechanism for generation of antibody memory. Philos Trans R Soc Lond B Biol Sci. 2009;364:569–75.
17. Martins JP, Kennedy PJ, Santos HA, Barrias C, Sarmento B. A comprehensive review of the neonatal Fc receptor and its application in drug delivery. Pharmacol Ther. 2016;161:22–39. https://doi.org/10.1016/j.pharmthera.2016.03.007.
18. Nimmerjahn F, Ravetch JV. Fc-receptors as regulators of immunity. Adv Immunol. 2007;96:179–204.
19. Mahajan VS, Pillai S. Sialic acids and autoimmune disease. Immunol Rev. 2016;269(1):145–61. https://doi.org/10.1111/imr.12344.
20. Lykken JM, Candando KM, Tedder TF. Regulatory B10 cell development and function. Int Immunol. 2015;27:471–7. https://doi.org/10.1093/intimm/dxv046.
21. Yoshizaki A, Miyagaki T, DiLillo DJ, Matsushita T, Horikawa M, Kountikov EI, et al. Regulatory B cells control T-cell autoimmunity through IL-21-dependent cognate interactions. Nature. 2012;491:264–8.
22. Neven B, Bruneau J, Stolzenberg MC, Meyts I, Magerus-Chatinet A, Moens L, et al. Defective anti-polysaccharide response and splenic marginal zone disorganization in ALPS patients. Blood. 2014;124:1597–609.
23. Snapper CM. Differential regulation of polysaccharide-specific antibody responses to isolated polysaccharides, conjugate vaccines, and intact Gram-positive versus Gram-negative extracellular bacteria. Vaccine. 2016;34:3542–8. https://doi.org/10.1016/j.vaccine.2015.12.077.
24. Dörner T, Radbruch A, Burmester GR. B-cell-directed therapies for autoimmune disease. Nat Rev Rheumatol. 2009;5:433–41.
25. Zappasodi R, de Braud F, Di Nicola M. Lymphoma immunotherapy: current status. Front Immunol. 2015;1(6):448. https://doi.org/10.3389/fimmu.2015.00448.

Dendritic Cells

5

Christina Dillmann, Geert R. Van Pottelberge, Ken R. Bracke, and Guy G. Brusselle

5.1 Introduction

Maintaining immune tolerance against the body's own tissue and trillions of commensal microorganisms and food antigens, while launching a robust adaptive immune response against foreign pathogens and tumors, is a mammoth challenge for the immune system. DENDRITIC CELLS (DCs) are powerful and highly versatile antigen-sensing and antigen-presenting cells and are able to preserve the balance between tolerance and immunity. They were first described by Paul Langerhans in 1868, but their function remained unknown until 1973 when it was shown that DENDRITIC CELLS [1] have a unique immunological key function: they are capable of presenting antigens to lymphocytes in order to induce and regulate adaptive immune response. Hence, DCs are called professional antigen-presenting cells and link innate to adaptive immunity. Acknowledging importance of the discovery that DCs are key players in the immune system and his part in investigating DC biology, Ralph Steinman was awarded the Noble Prize for Physiology and Medicine in 2011 [2].

DCs stand guard in the skin, the mucosal surfaces (respiratory, gastrointestinal, and urogenital MUCOSA), and the organs where they continuously sense for danger signals and capture antigens (Fig. 5.1). DCs then migrate toward draining LYMPH NODES (LN) while integrating the information of danger signals with the processed antigen. In the LYMPH NODES, DCs present the processed antigen on major histocompatibility complex (MHC) molecules and make contact with naïve lymphocytes. At this point, DCs are capable of selecting T cells with an antigen-matched T cell receptor (TCR) for clonal expansion. In addition to stimulation of clonal proliferation of T cells, DCs can direct and control the nature and extent of the T cell response. DCs thus play a central role in the initiation of primary adaptive immune responses and the enhancement of secondary immune responses. In this way, DCs form the crucial link between innate and adaptive immunity [3].

5.2 Biology and Ontogeny of the Dendritic Cell

DENDRITIC CELLS form a complex and heterogeneous immune cell network, traditionally defined by characteristic anatomical location and phenotype. Currently, the DC family is clustered into different subpopulations, each possessing distinct phenotypic and functional properties. The two

Final manuscript submitted on April 09, 2018.

C. Dillmann (✉)
Faculty of Medicine, Institute of Biochemistry I, Goethe-University Frankfurt, Frankfurt, Germany
e-mail: Dillmann@biochem.uni-frankfurt.de

G. R. Van Pottelberge · K. R. Bracke · G. G. Brusselle
Laboratory for Translational Research in Obstructive Pulmonary Diseases, Department of Respiratory Medicine, Ghent University Hospital, Ghent, Belgium

M. J. Parnham et al. (eds.), *Nijkamp and Parnham's Principles of Immunopharmacology*, https://doi.org/10.1007/978-3-030-10811-3_5

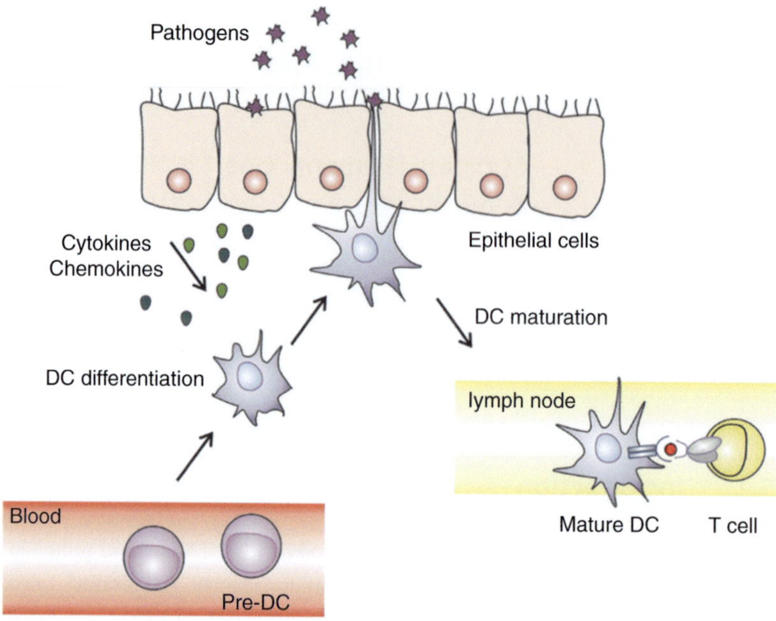

Fig. 5.1 The dendritic cell sentinel paradigm. Dendritic cell precursors (pre-DCs) migrate from the blood stream into border regions like the mucosa where they differentiate to immature dendritic cells under the influence of cytokines and growth factors, mainly released by epithelial cells. DCs then capture antigens and sense for danger signals. They then mature and migrate toward draining lymph nodes where they interact with naïve T cells

main DC lineages are conventional DCs (cDCs) and PLASMACYTOID DCs (pDCs) [4]. Murine cDCs are separated into two discrete groups, the first characterized by the expression of CD8α+ and CD103+ and are called cDC1 and the second expressing CD11b+ and CD172+, named cDC2. In humans cDC1 express CD141 and cDC2 express CD1c. PLASMACYTOID DCs represent a small and specialized subset of DCs expressing a mixture of markers (mouse, Siglec-h, CD11c, B220; humans, CD123, CD303, ILT7) (Table 5.1) [5].

Conventional DC lineages as well as pDCs are short-lived hematopoietic cells that are constantly substituted by blood-derived precursors. Hematopoietic stem cells (HSC) differentiate in the bone marrow via several intermediates into the macrophage/monocyte DC progenitor (MDP), which expresses fms-like tyrosine kinase 3 (Flt3) and high levels of the stem cell growth factor receptor c-kit (Fig. 5.2) [6]. On the one hand, MDPs give rise to monocytes/macrophages and on the other hand to common DC progenitors (CDP) which exclusively differentiate into DC subsets in an Flt3-ligand-dependent manner. CDP finally differentiate into either pDC or pre-conventional DCs (pre-cDCs), which migrate into the circulation to populate distinct organs. After emigration into the tissue, pre-cDCs in mouse and man differentiate into the major cDC subsets [7, 8]. Due to DC turnover in the tissue, replacement of pre-DCs is vital to maintain homeostasis.

5.2.1 Regulation of DC Differentiation

DC differentiation is highly regulated through different cytokines and requires the concerted action of several lineage-determining transcription factors. The most prominent cytokine in DC generation is Flt3L. Mice lacking either Flt3L or the corresponding receptor Flt3 (CD135) have reduced numbers of DC progenitors, tissue cDCs, and pDCs. Furthermore Flt3L injection in human volunteers resulted in massive increased amounts of DCs in the circulation. On the transcriptional level, cDC1 development is driven by IRF8, Batf3, and Id2, whereas terminal differentiation of tissue-specific cDC2 subsets requires IRF4 and Notch2. PLASMACYTOID dendritic cell development is strictly dependent on the basic helix-loop-helix protein E2-2, which is repressed by Id2 during conventional DC development. E2-2 further induces the expression of several pDC-specific genes like ILT7 in humans and the constitutive expression of IRF7 which is indispensable for pDCs to produce large amounts of

Table 5.1 Murine and human DC subsets [4–8, 11, 12, 40–43]

	Mouse				Human			
	Phenotype	Localization (subset)	Origin (BM → circulation)	TFs and soluble factors/receptors involved in development	Phenotype	Localization (subset)	Origin	TFs and soluble factors/receptors involved in development
cDC1	CD8α CD103	Lymphoid tissue (CD8), e.g. spleen or LN Non-lymphoid tissue (CD103) Migratory fraction of LN-DCs (CD103)	HSC CDP into pre-cDC	TF: Irf8, Batf3, Id2 Soluble Factor: Flt3L Receptor: Flt3, CSF-R2	CD141	Blood Tonsils LNs Non-lymphoid tissue (e.g. lung, skin, liver) Migratory CD141 DCs (from tissue to draining LNs)	HSC hCDP (CD34+) into hpre-cDC	TF: Irf8, Batf3, Id2 Soluble factor: Flt3L Receptor: Flt3 Limitations: CD141 also expressed on pDCs, dermal CD14+ DCs, CD1c+DCs (but human cDC1 have no CD14, CD11b, CD11c)
cDC2	CD4 CD11b CD172	Lymphoid (CD4) e.g. spleen Non-lymphoid organs e.g. lung, intestine, dermis Further subtypes: e.g. tissue-resident DCs	HSC CDP into pre-cDC	TF: Notch2, Relb, Irf4, Zeb2 (subset-dependent) Soluble factor: Flt3L Receptor: Flt3, CSF-R2	CD1c CD11b CD11c Sirpα	Blood LN Spleen Tonsils Non-lymphoid tissue, e.g. skin, lung	HSC hCPD (CD34+) into hpre-cDC	TF: Id2, Irf4 Soluble factor: Flt3L, GM-CSF Receptor: Flt3
pDC	Siglec-h BST-2 CD11c MHC II	Lymphoid tissue Blood	HSC CDP into pDC	TFs E2-2, Runx2, Zeb2 Cytokines: Flt3L	CD123 CD303 CD304 ILT-7 BST-2	Lymphoid tissue Blood (in non-lymphoid tissue only under inflammatory conditions)	HSC hCDP (CD34+) into pDC	TF: E2-2, Irf4, Irf8 Soluble factor: Flt3L

DC subsets were clustered based on DC ontogeny due to recent [8] nomenclature
Abbreviations: *BM* bone marrow, *CDP* common dendritic progenitor, *HSC* hematopoietic stem cell, *hpre-DC* human progenitor DC, *TF* transcription factor

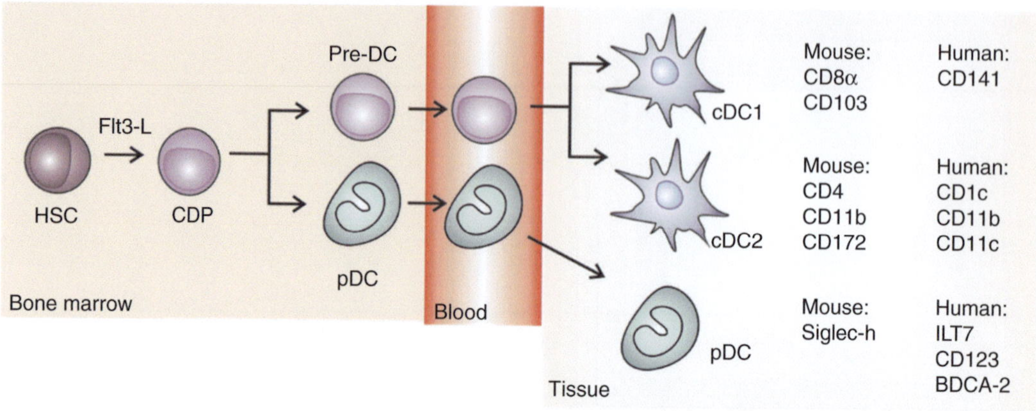

Fig. 5.2 The ontogeny of the DC network. The dendritic cell network is characterized by its complexity and heterogeneity. But DC subsets can be clustered due to their origin and characterized by molecules expressed in the surface of the cell. Conventional DCs (cDCs) and plasmacytoid DCs (pDCs) develop from hematopoietic stem cells (HSC) in the bone marrow into common dendritic progenitors (CMP) into either pDCs or pre-DC, which then migrate via the blood into the tissue. The progenitor population of pre-DC can further develop into cDC1 and cDC2 members

type I interferon (IFN-I) due to viral stimuli. In contrast to cDCs, there is strong evidence for a lymphoid bias of pDC generation since mice lacking IL-7-Rα show reduced total pDC numbers. In addition to the cytokine milieu and the interaction of TFs, the microenvironment of the developmental niche in the bone marrow (BM) shapes the fate of progenitor cells. For instance, CDPs in the bone marrow might migrate to regions near sinusoids where high levels of oxygens and the sphingolipid S1P are detectable, which in turn favors pDC generation from CDPs via suppressing the transcription factor HIF1-α and stimulating the S1P-receptor 4, both involved in pDC development [9, 10].

5.2.2 Tissue Distribution of DCs

Compared to cDCs which mature at the site of infection, pDCs are fully developed in the BM [11] and migrate through the blood directly to the LYMPH NODES via high endothelial venules (HEVs), where they can be found in close contact to T cells, iNKT cells, B cells, and NK cells. Furthermore pDCs are found in nonlymphoid tissue like the skin and are also found within tumors.

A common link for all cDC subtypes is made by the fact that all cDCs develop from CDPs into pre-DCs in the BM. Pre-DCs are the migratory precursors, expressing CD11c and leaving the BM via the blood into lymphoid and nonlymphoid tissue where they differentiate toward either the cDC1 or cDC2 lineage. This final development step which includes phenotype and functional properties is then influenced by the local microenvironment. However scenarios in which the fate of the cDC subset is already decided in the BM are also discussed [6].

Besides DCs which develop in a Flt3/Flt3L-dependent manner via CDPs, DC-like subsets with functional similarities but distinct development pathways exist. Langerhans cells (LCs) expressing CD207, high levels of MHC-II in the steady state, and CD11b are a specialized form of DCs [12] with monocyte/macrophage ontogeny and DC-like functions seeding the epidermis of the skin prenatally. LCs, therefore, exhibit an embryonic origin, a long lifespan, and self-renewable capacities. Development, proliferation, and migration of LCs are dependent on TGF-β since tgfb$^{-/-}$ mice lack LCs and a conditional ablation of Tgfbr1 negatively affects LC migration in vivo [13]. Moreover, reports indicate that under inflammatory conditions, LCs leave the dermis and are spontaneously replaced by short-lived monocyte-derived LC-like cells. Next to Langerin (CD207), the expression of

Birbeck granule (tennis racket-shaped organelles) is a key marker of LCs. Analogue langerin-positive, Birbeck granule-positive LCs have also been reported in the respiratory and genital mucosa. These LCs are mainly located in the epidermis and the epithelium of the MUCOSA. LCs reside at the outer layer of the skin and represent the first line of immunological defense. Like all antigen-presenting cells (APCs), LCs migrate to the draining LYMPH NODES after they capture antigens, and recent studies show that LCs are clearly involved in Th_{17} and T_{FH} cell differentiation and are able to dampen cytotoxic CD8+ T cell responses [14].

Another DC subpopulation termed inflammatory or monocyte-derived DCs (mo-DCs) develops from $Ly6C^{high}$ expressing monocytes during acute inflammation. However, the factors involved in the generation of this DC subpopulation are poorly understood.

5.3 Mechanisms and Pathways

5.3.1 Innate Immune Functions and Pathogen Recognition

All DC lineages are equipped with a broad set of pattern recognition receptors (PRRs) which allow the recognition of certain pathogen-associated molecular patterns (PAMPs) not found in the host, danger signals released by dying or damaged cells (danger-associated molecular patterns, DAMPs), and tumor-associated molecular patterns (TAMPs) from altered or transformed cells (Fig. 5.3).

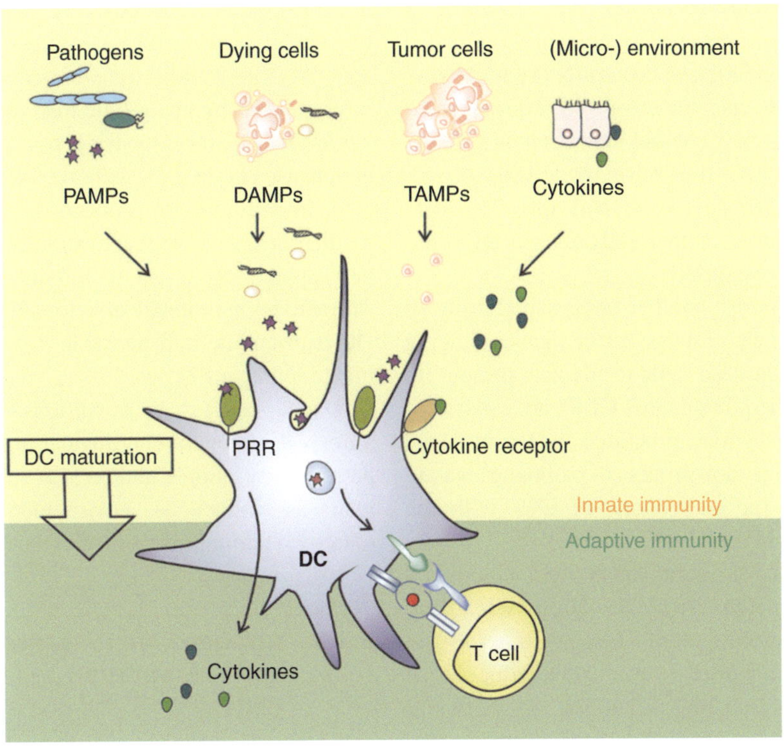

Fig. 5.3 Innate and adaptive immunity. Dendritic cells (DCs) participate in the innate immune response by pattern recognition of microbial signals and recognition of cellular stress and damage. Upon activation by these signals, DCs mature and activate the adaptive immune response. In addition, these DCs are also capable of producing antimicrobial peptides, contributing to first-line antimicrobial defense. Finally, DCs produce mediators that attract other immune cells, enhancing the innate immune response. Abbreviations: *DAMP* damage-associated molecular pattern, *PAMP* pathogen-associated molecular pattern, *PRR* pattern recognition receptor, *TAMP* tumor-associated molecular pattern

Most PRRs belong to one of the five families classified by their protein domain homology. These families consist of toll-like receptors (TLRs), C-type lectin receptors (CLRs), nucleotide-binding and oligomerization domain (NOD)-like receptors (NLRs), retinoid acid inducible gene (RIG-I)-like receptors (RLRs), and AIM2-like receptors (ALRs). TLRs and CLRs are basically membrane-bound, whereas NLRs, RLRs, and ALRs are cytosolic and only able to recognize patterns derived from intracellular pathogens.

TLRs are the most prominent PRRs and expressed at the surface of DCs or in the endosomal compartment. After TLR activation all DC subsets undergo a maturation process triggered by intracellular signaling events which induce a maturation program. This includes a transient loss of endocytic/phagocytic receptors and diminished antigen capture, changes in cellular morphology with development of large cytoplasmatic protrusions (dendrites), translocation of MHC-II antigen-presenting molecules to the cell surface, upregulation of costimulatory molecules like CD80/86, and an enhanced expression of lymphoid homing chemokine receptor CCR-7. These mature DCs then travel toward the secondary lymphoid organs (LNs and spleen), where they make contact with T cells to steer the adaptive immune response.

CLRs comprise another large superfamily of PRRs recognizing a large set of ligands ranging from fungi via dead cells to glycans present on tumor cells [15]. In general, CLRs are capable of binding carbohydrate structures and are involved in the internalization process of pathogens but are also crucial for interactions of DCs with the extracellular matrix.

To ensure an anticipatory and fine-tuned immune response, each DC subset selectively expresses a specific set of TLRs and other PRRs, which drive downstream events that will polarize the immune response depending on particular subsets of pathogens and stress signals. For instance, pDCs are specialized in recognition of viral PAMPs through endosomal TLR-7 and TLR-9, triggering the production of large amounts of IFN-I, especially interferon-α, which is the cornerstone of the early antiviral defense mechanisms. In contrast, cDC subsets are capable of detecting both bacterial and viral PAMPs through different TLRs and a specific set of NOD-like receptors and RIG-I -like receptors. CLRs (like DEC-205 or dectin-1) are highly expressed on immature DC subsets and are internalized after pathogen uptake.

Next to DC maturation after PRR activation, the subsequent release of cytokines and soluble mediators, such as interleukins (IL), like IL-1, IL-6, and IL-12, or members of type I and type II IFNs directly influences and regulates effector cells of the innate and the adaptive immunity. In addition, DCs release a series of innate immune molecules such as pentraxins and neuropeptides contributing to the first line of defense.

Innate immune cells like natural killer cells (NK cells) and neutrophils can be directly targeted by the APC to induce effector functions. For instance, DCs can prime NK cells through soluble factors like cytokines (IL-12 or IFN-I) or direct cell-cell contact (CD70-CD24 interaction). Primed NK cells in turn are potent effector cells and are able to kill transformed or pathogen-infected cells by using lytic mediators like granzyme B or receptor-triggered induction of apoptosis via, e.g., Fas/FasL or Trail/Trail-R [16]. In addition, the production of chemotactic mediators after TLR-9 activation of DCs leads to neutrophil recruitment to inflamed tissue [17]. The activation of innate effector cells by different DC subsets is indispensable for an adequate innate immune response.

Under steady-state conditions, DCs are not unresponsive or immunological quiescent. They permanently sample and present self-antigens as well as innocuous environmental proteins, thereby retaining immune tolerance [18].

5.3.2 Antigen Sampling, Processing, and Presentation by Dendritic Cells

The term antigen includes any substance evoking an adaptive immune response. To do so, antigens must be taken up, processed, and loaded onto MHC molecules and can then be recognized in the draining LNs by T cells via their T cell receptor (TCR). The uptake can be triggered via receptors like the CLRs DEC-205 or the Mannose receptor (MR), whereas antigen processing and loading of peptides to MHC molecules take place within DCs and are mediated by a complex machinery (Fig. 5.4). Next to exogenous antigens, which have to be internalized, DCs can be infected or transformed, thereby being exposed to endogenous antigens, which are generated within the DCs. In addition to peptides, lipids of self and foreign sources can act as antigens and are loaded on CD1 molecules.

5.3.2.1 MHC-II Antigen Presentation

MHC-II molecules generally bind peptides generated by lysosomal proteolysis in the endocytic or phagocytic compartments. Immature DCs take up antigens, like pathogens or dying cells, via receptor-mediated endocytosis, micropinocytosis, or phagocytosis. Those antigens then enter the vesicular pathway including early and late endosomes or phagosomes ending in proteolytic cleavage of proteins into peptide fragments in the lysosomes. MHC-II expression is

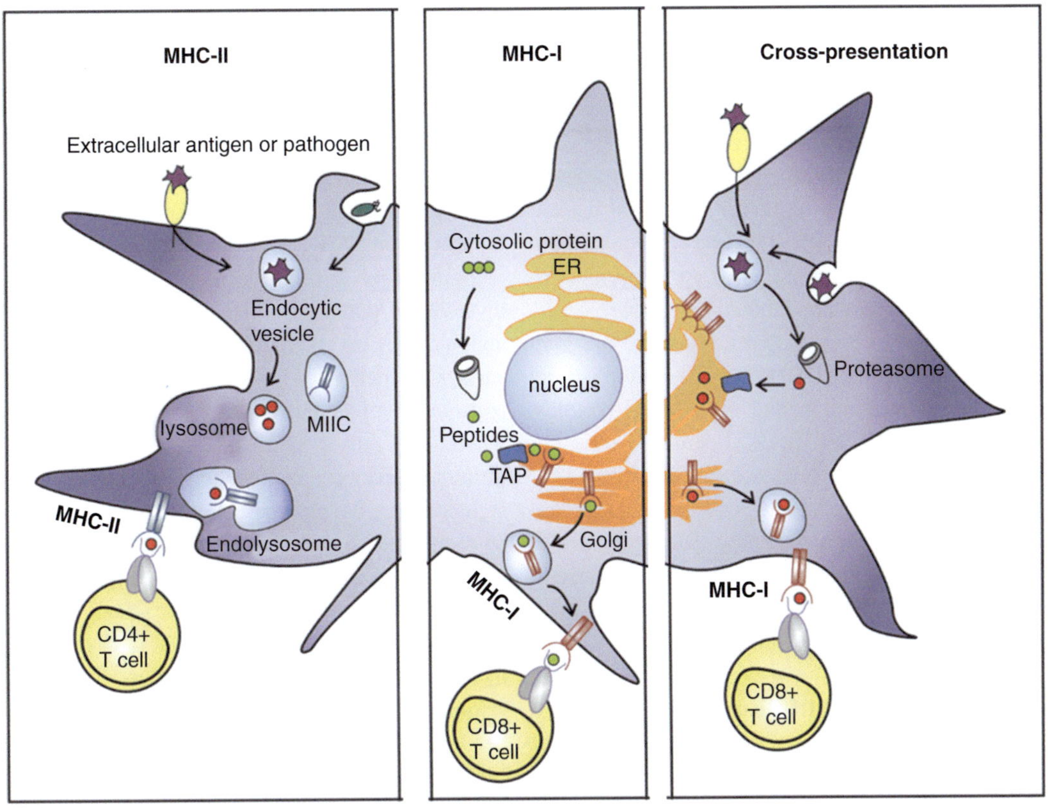

Fig. 5.4 Antigen processing and presenting mechanisms. Extracellular antigens are presented on major histocompatibility class (MHC)-II molecules. These antigens are taken up by DCs through different mechanisms like receptor-mediated endocytosis or phagocytosis. The antigens are packed into endosomes or phagosomes and were digested and then degraded into peptide fragments. The MHC-II molecules are assembled in the endoplasmic reticulum (ER) shuttled via MHC-II-enriched compartments (MICCs) to the endocytic system where they were loaded with the antigenic peptides. These antigen-MHC complexes are then transported to the cell membrane and induce CD4+ T cell responses. Intracellular cytoplasmatic antigens are presented by MHC-I molecules after processing by the proteasome in the cytosol and uptake in the ER through TAP (transporter associated with antigen processing). The MHC-I molecules loaded with processed antigen are then transported to the cell membrane and induce CD8+ T cell responses. Cross-presentation is the presentation of extracellular antigens by the MHC-I pathway to induce CD8+ T cell responses. Antigens presumably leak out of the phagosome into the cytosol where they are further processed by the proteasome

more restricted compared to MHC-I, limiting the MHC-II pathway to APCs, B cells, and endothelial cells [19]. The goal of this pathway is to sample the extracellular compartment and present the antigens to CD4+ T cells. MHC-II molecules are synthesized and assembled in the endoplasmic reticulum (ER) and transported either directly or via the plasma membrane to the *trans*-Golgi network into the endosomal system. Some specific intracellular lysosome-related compartments are enriched with MHC-II molecules. Captured antigen is shuttled to the so-called MIICs (MHC-II-rich compartments) and loaded onto MHC-II [19]. Peptide-loaded MHC-II molecules are then transported to the cell surface where antigens are presented to CD4+ T cells.

5.3.2.2 MHC-I Antigen Presentation

Next to exogenous antigens, uncommon events such as an infection or cellular transformation occurring within the DCs must be communicated to the adaptive immune system. Foreign or misfolded proteins are labeled for proteasome-driven degradation, transported via the antigen peptide transporter (TAP) into the endoplasmic reticulum (ER), and loaded onto MHC-I molecules. Peptide-loaded MHC-I shuttles through the *trans*-Golgi network, is exported into vesicles, and transported to the plasma membrane where the complex can be recognized by CD8+ T cells [20]. Since MHC-I is expressed on a variety of immune cells, MHC-I-triggered antigen presentation is not restricted to APCs.

5.3.2.3 Cross Presentation

DCs have the capability to take up antigens from the extracellular environment and present them through MHC-I molecules, inducing a CD8+ T cell response. This process is known as "cross presentation" or "cross priming" and is crucial for the initiation of immune responses to viruses that do not directly infect APCs, as in the case of an influenza infection. It is well established that all human DC subsets are at least partially able to cross-present antigens to CD8+ T cells and that the ability to do so depends on the nature of the antigen, the internalization receptor, and intracellular trafficking properties and kinetics [21].

5.3.2.4 CD1 Lipid Presentation

Next to proteins, lipids can also act as antigens. The MHC-I-related CD1 molecules (CD1a, b, c, d, and e) are a family of glycoproteins expressed on DCs and other APCs that present foreign and self-lipids to a specialized T cell population, called natural killer T cells (NKTs). Extracellular, foreign lipids from pathogens must be internalized, whereas self-lipids are commonly generated within the APC and presented to NKTs.

5.3.3 DC-Lymphocyte Interaction: The Formation of the Immunological Synapse

Antigen-bearing DCs migrate from the nearby tissue through afferent lymphatics into the LYMPH NODES where they contact naïve T cells. Theoretically, there are billions of antigens presented on MHC molecules which should be recognized by T cells via their TCR. Hence the diversity of potential antigens for T cells is high and requires an expanded repertoire of T cells, each with their own assembled TCR. T cells are recruited from the blood into the LYMPH NODES where they meet DCs. The formation of the immunological synapse in general requires three signals. First, the antigen bound to the MHC must be recognized by the TCR; second, costimulatory molecules (CD40, CD80, CD86) expressed by DCs must be present to activate the CD28 receptor on T cells; and the third signal is the soluble mediator released from DCs which induces T cell priming and activation (Fig. 5.5a). The interplay between all signals induces the generation of T cell effector cells, which rapidly proliferate and traffic to inflamed or damaged tissue, where they release cytokines and fulfill their specific effector functions. By contrast, DCs delivering signal one in the absence of signal two induce T cell anergy instead of CTL or Th_1 effector responses. Furthermore, DCs are able to directly skew T cell effector functions into immunosuppressive responses by releasing anti-inflammatory cytokines (e.g., IL-10) or expressing immune modulatory molecules on their surface, which then drive CD4+ T cells into regulatory effector T cells (T_{regs}).

a

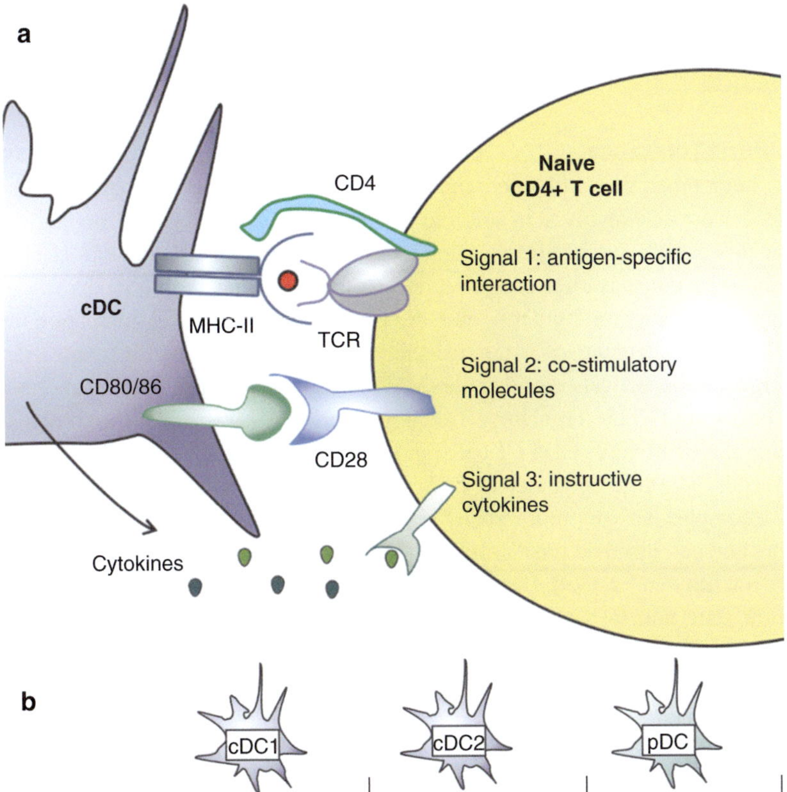

b

	cDC1 Mouse	cDC1 Human	cDC2 Mouse	cDC2 Human	pDC Mouse	pDC Human
TLR pattern	3,4,11,13	3,8	5,6,7,9,13	1-8	7,9	7,9
T cell response	Th1, CTL	Th1, CTL	Th2, Th17	Th1, Th2 Th17, Treg	Th1, Th2 Treg	Th1, Th2 Treg
Cross-presentation	Yes	Yes			Yes	Yes

Fig. 5.5 The immunological synapse. (**a**) Dendritic cells interact with naïve T cells, steering the proliferation and differentiation of antigen-specific T cell clones. The processed antigen, presented on MHC-II, is recognized by the matched T cell receptor together with the CD4 receptor on the T cell. A costimulatory signal from mature DCs is provided through CD80 and CD86, each interacting with a CD28 molecule on the T cell. Finally, a third signal is provided by the production of cytokines by the DCs. Depending on the third signal produced by DCs, CD4+ T cell differentiation is skewed toward specific T cell subpopulations (T helper 1, T helper 2, T helper 17, or regulatory T cells). (**b**) Different DC subpopulations induce different T cell effector responses. Cells of the cDC1 population are specialized in cross-priming antigens to CD8+ T cells in mouse and man; however they are able to induce Th_1 effector cells by presenting exogenous antigens via MHC-II to CD4+ T cells. Members of the cDC2 population are more specialized in Th_2 and Th_{17} effector cell priming and not able to cross-prime antigens. Whereas pDCs are able in cross-priming, their potential of inducing CTL responses in vivo is currently discussed. All DC subpopulations express different sets of TLRs and are therefore not only specialized in the effector T cell response they induce but also in the pathogen or antigen they recognize and capture

The outcome of the T cell response depends on the MHC class molecule that presents the antigen. On the one hand, MHC-I-loaded antigens are recognized by CD8+ T cells. After correct binding and priming, CD8+ T cells develop cytotoxic effector functions and are capable of killing infected or transformed cells. On the other hand, antigens presented via MHC-I are recognized by CD4+ T cells, leading to the development of T helper effector functions. CD4+ T helper cells then fulfil various functions depending on the different subpopulation (like Th_1, Th_2, or regulatory

T cells). The fine-tuning of T cell priming into the different subsets is mediated via the cytokines released from DCs.

5.3.3.1 Immune Tolerance

DCs initiate both protective and tolerogenic T cell responses but are also involved in autoimmunity [22]. The process of tolerance implies that no adaptive immune response is elicited against the antigen. Tolerance induction by DCs includes several mechanisms: silencing of differentiated antigen-specific T cells, activation and expansion of naturally occurring CD25+ regulatory T cells, and the differentiation of naïve CD4+ T cells into regulatory T cells. The tolerogenicity of DCs is mediated by a complex network of microenvironmental factors and cell intrinsic mechanisms and until recently not fully understood. Depending on their immature state and localization, each DC subset is able to induce tolerogenic immune responses. The induction of central tolerance within the thymus occurs during T cell development and leads to the elimination of autoreactive lymphocytes. Different DC subsets, including migratory CD8+ DCs or CD11b+ DCs which migrate through the blood into the thymus, are involved in this process. Although lymphocyte selection is efficient, not all autoreactive T cells are eliminated and must be kept under control within the periphery. Under noninflammatory conditions, DCs which exhibit an immature or unactivated phenotype patrol through the system and present self-antigens on their MHCs, which leads, in the case of a match between the DC and the T cell, to T cell anergy and clonal depletion. Central and peripheral immune tolerance are active processes in which several subpopulations of DCs are involved.

5.3.3.2 Functional Specialization of DC Subsets

In contrast to immature DCs, mature DC subsets induce strong T cell effector functions. Cells of the cDC1 population comprising in mouse of lymphoid tissue-resident CD8+ DCs and nonlymphoid tissue-resident CD103+ DCs, and in humans including CD141+ DCs, are strong inducers of CD8+ T cell responses during microbial infections, due to their superior capability to cross-prime antigens [4, 23]. Furthermore, splenic CD8+ cDCs produce IL-12 which promotes the polarization of Th_1 effector cells. However, the role of cDC1s in the activation of CD4+ T cells needs further investigation, since in vivo models which discriminate between CD8+ and CD103+ DCs are currently unavailable.

Cells of the cDC2 population induce Th_2, Th_{17}, and regulatory T cell responses in mice and man depending upon the pathogen or antigen stimulus. Murine CD11b+ DCs express high levels of MHC-II compared to CD8+ DCs and secrete large amounts of IL-23p19 which induces the generation of Th_{17} effector cells [24]. They, therefore, play a pivotal role in the induction of specific CD4+ T cell responses to extracellular pathogens.

Each cDC subset has a preference for a T cell subset, cDC1 or cDC2 populations inducing either CD4+ or CD8+ T cells (Fig. 5.5b).

PLASMACYTOID DENDRITIC CELLS produce large amounts of IFN-I, especially IFN-α in a TLR-7- or TLR-9-mediated manner. Although they do not show a DC morphology in their mature state, pDCs are capable of presenting endogenous and exogenous antigens to T cells. Murine studies in vivo in which pDCs are depleted and studies using TLR9-activated human pDCs point to a role for pDC-mediated induction of T_{regs}. However, pDCs also suppress antigen-specific CD4+T cell responses elicited by cDC subsets. Moreover, their impact on induction of CTL responses is controversial, since BST2 mAB-directed depletion of pDCs in vivo blocked CTL effector functions. However, the depletion is not specific, and a contribution by cDCs cannot be ruled out [25]. Taken together, pDC are able to cross-prime antigens but also induce regulatory T cell responses or suppress T cell effector cells to maintain immune tolerance or dampen overshooting immune responses. One possibility by which pDCs can suppress T cell proliferation is via the expression of indoleamine-2,3-dioxygenase (IDO). IDO metabolizes the amino acid tryptophan, which is essential for T cell metabolism, into kynurenine, causing T cell suppression and favoring immune tolerance [26].

5.3.4 Pathophysiological Relevance

The complex dendritic cell network in which all different DC subpopulations have their own specialized functions and biology plays a pivotal role in several immunological-driven diseases. Several studies in mice and man point to a dependence on DC localization, maturation status, and polarization in pathologies, like allergic asthma, cancer, and autoimmune and infectious diseases.

During *infections*, DCs are indispensable for the initiation of the adaptive immune response against the invading microorganism. However, various microorganisms try to evade the immunological response by altering DC function. *Coxiella burnetii*, *Salmonella typhi*, HIV, and herpes simplex virus achieve this deception by blocking DC maturation. Certain pathogens (e.g., *Yersinia pestis* and *Salmonella typhi*) selectively inject toxins into phagocytes such as DCs to destroy them, whereas some viruses can induce apoptosis of DCs. Finally, microbes can also interfere with the controlling function of DCs, switching the T cell response away from the protective Th_1 response toward a non-protective Th_2 response.

The role of DC subsets during every stage of *HIV infection* has been well investigated [27]. Although nearly every DC subpopulation expresses the receptors needed for HIV entry, they are poorly infected by the virus compared to T cells. DCs are one of the first cells to come in contact with HIV during early infection. pDCs then produce, in a TLR-dependent manner, large amounts of IFN-I, leading to the expression of various antiviral genes like myxovirus resistance 2 protein (Mx2), which was recently shown to suppress cellular infection from multiple HIV strains. Nevertheless, through the expression of receptors that bind virions (viral particles which have not infected a cell yet and reside outside of the cells) and the interaction of DCs with CD4+ T cells, DCs participate in viral spread. In addition, the chronic activation of DCs in later stages of the disease and their influence on T cell immunity confer on DCs an essential role in HIV pathology.

Furthermore, the role of DCs in *allergic diseases* such as asthma has been well described. Here, different lung DC subsets are capable of inducing adaptive Th_2-skewed T cell proliferation directed against an intrinsically harmless antigen. Most allergens contain motifs that trigger PRRs on DCs leading to DC maturation and induction of Th_2 and/or Th_{17} responses. Furthermore, close communication between lung epithelial cells and DCs exists, triggering the release of cytokines needed for Th_2 immunity [28].

Inflammatory bowel disease (IBD) is a heterogeneous inflammatory disorder consisting of two main forms, ulcerative colitis and Crohn's disease. In both types of IBD, patients develop severe inflammation within the colon and the small intestine. It is well established that genetic factors favoring IBD disease initiation could be linked to the ability of DCs to respond to pathogens [29]. Although currently there are still many open questions about the specific DC subsets that are involved in pathogenesis, new approaches using tolerogenic DCs to treat IBDs are under investigation.

Heterogeneity of DC populations also exists within *solid tumors* [30]. However, for an adequate antitumor immune response, the nature of the DC population is relevant. In general, mature DCs are considered to be immunostimulatory, but they must recognize neo-antigens on tumor cells to yield an anticancer cytotoxic effector T cell response rather than generating regulatory T cells. The cDC1 subpopulation is the most likely group since they capture both exogenous and endogenous antigens and present them on MHC-I and MHC-II. They favor the generation of cytotoxic and protective T cells which lyse the tumor. However, a number of events within the tumor are able to suppress DC activation and maturation. Anti-inflammatory cytokines (e.g., IL-10) and other mediators (e.g., PGE2, S1P) released from tumor-associated immune cells or dying tumor cells are able to induce tolerogenic DCs which suppress antitumor immunity [31].

DCs perform various complex functions during inflammation and maintain immune tolerance. Any kind of imbalance in DC function or DC subset composition can lead to defective or exaggerated immune responses and tissue damage. Therefore, it is not surprising that DCs are linked to several *autoimmune diseases*. It is clear that not just changes in numbers of DCs can induce autoimmunity but alterations in DC functionality might induce inflammatory responses against self-antigens and favor autoimmune reactions [32]. Some common pathogenic patterns in the roles of DCs during autoimmune diseases are becoming clear. One of these patterns is the pathogenic overexpression of IFN-I, called the interferon signature, during systemic lupus erythematosus (SLE), rheumatoid arthritis (RA), or psoriasis, which can be linked to defective pDC activation. Immune complexes (ICs) containing self-DNA, peptides, and autoantibodies are able to activate pDCs via TLR-7 and TLR-9, thereby favoring an excessive immune response. However, IFN-I was shown to have a protective role in multiple sclerosis (MS) [33]. Another pattern common to several autoimmune diseases, manifested in different murine models, is the abnormal maturation state and maturation time point of DC subsets. But these mouse findings have not yet been translated into human pathogenesis. Moreover, the extraordinary specificity and diversification of DC subpopulations remain to be applied to the tissue-specific initiation and suppression of human autoimmunity.

5.4 Pharmacological Implications

Based on their strategic position in the immune system, DCs are attractive target cells for the development of new preventive and therapeutic strategies. In view of the multiple DC phenotypes (with different immunological functions), modulation of a specific DC subset is difficult to achieve in vivo.

Glucocorticoids (GCs) have anti-inflammatory and immunosuppressive properties and are widely used to treat inflammatory, autoimmune-driven, and allergic diseases (see Chaps. 23 and 34) [34]. In addition to their direct impact on T cell biology, through inhibition of T cell proliferation or induction of T cell apoptosis, GCs alter TLR-induced DC maturation, therefore favoring tolerogenic DCs which dampen inflammatory responses [35]. In general, GCs bind to the GC-receptor which induces the downregulation of several pro-inflammatory cytokines (e.g., IL-1β or IL-12) but also affects neutrophil recruitment to the site of infection.

The ultimate goal of *vaccination* is the induction of an antigen-specific immune response with long-lasting immunologic memory to protect against subsequent disease (see Chap. 19). Vaccines are defined as any biological preparation inducing adaptive immune responses to particular diseases. Adjuvants co-administered directly or indirectly with the antigen stimulate the maturation of DCs and their presentation of antigens. Apart from promoting antigen multimerization and internalization, particulate adjuvants (including mineral salts such as alum) directly target the INFLAMMASOME in the DCs, inducing the production of several pro-inflammatory cytokines [36].

Based on the increasing understanding of DC biology, strategies are being established to use *DCs as vaccines* for the treatment of cancer and infectious diseases like HIV [37, 38]. In principle, monocyte-derived DCs (mo-DCs) or natural DCs generated ex vivo from patients are loaded with disease-specific antigens and then administered back to the patient, where they should induce an adaptive immune response against infected or transformed cells. The first clinical trial with DC VACCINES for *anticancer vaccination* was published in 1996. Now *next-generation DC vaccines* are under investigation and are already in early clinical trials [38]. However, despite the promising in vitro and preclinical studies showing an efficient priming of DCs on antitumor effector T cells and the large number of clinical trials and tumor patients, cancer vaccination to date has shown only modest effects in the clinic. There are many reasons for this. For instance, the quality of the tumor antigen loaded onto the DCs and the antigen-delivery system must be optimized. Furthermore, the impact of the tumor microenvironment (TME) which has powerful immunosuppressive characteristics strongly influences antitumor immunity. The impact of the TME on tumor-related inflammation was dramatically shown by the use of antibodies against immune checkpoints (CTLA-4 PD-1, PD-1L) (see Chap. 12 and 25). New approaches linking DC subset specificity and tumor immunity indicate that next-generation DC vaccines could exhibit promising clinical efficacy. Using pDCs loaded with tumor-associated antigens (TAAs), researchers recently showed that this vaccine induces strong CD4+/CD8+ effector T cells and a measurable IFN-I response in melanoma patients [39]. Taken together, the efficiency of DC vaccination still depends on a variety of confounding issues. The heterogeneity of the antigen, the composition of the TME, and the cancer type are only a few them.

5.5 Summary

DENDRITIC CELLS are professional antigen-presenting cells of hematopoietic origin. They form a surveillance network in the skin, at the mucosal surfaces, and in the internal organs. In these locations, DCs continuously sense for danger signals and sample antigens. DCs migrate toward the draining LYMPH NODES, process the sampled antigens, and present them on MHC molecules to naïve T cells. Depending on the expression of costimulatory molecules and the production of cytokines, mature DCs steer the T cell differentiation toward a Th_1, Th_2, Th_{17}, or regulatory T cell response. Furthermore, DCs are indispensable for the maintenance of immune tolerance. In this way, DCs are indispensable to induce and modulate the adaptive immune response.

Selected Readings

Blum JS, Wearsch PA, Cresswell P. Pathways of antigen processing. Annu Rev Immunol. 2013;31:443–73.

Garg AD, Coulie CG, Van den Eyden BJ, Agostinis P. Integrating next-generation dendritic cell vaccines into the current cancer immunotherapy landscape. Trends Immunol. 2017;38(8):577–90.

References

1. Steinman RM, Cohn Z. Identification of a novel cell type in peripheral lymphoid organs of mice. J Exp Med. 1973;137:1142–62.
2. Mellman I, Nussenzweig M. Retrospective. Ralph M. Steinman (1943-2011). Science. 2011;334(6055):466.
3. Bancherau J, Steinman RM. Dendritic cells and the control of immunity. Nature. 1998;392:245–52.
4. Guilliams M, et al. Dendritic cells, monocytes and macrophages: a unified nomenclature based on ontogeny. Nat Rev Immunol. 2014;14(8):571–8.
5. Collin M, Bigley V. Human dendritic cell subsets: an update. Immunology. 2018;154(1):3–20.
6. Schlitzer A, McGovern N, Ginhoux F. Dendritic cells and monocyte-derived cells: two complementary and integrated functional systems. Semin Cell Dev Biol. 2015;41:9–22.
7. Lee J, et al. Restricted dendritic cell and monocyte progenitors in human cord blood and bone marrow. J Exp Med. 2015;212(3):385–99.
8. Merad M, et al. The dendritic cell lineage: ontogeny and function of dendritic cells and their subsets in the steady state and the inflamed setting. Annu Rev Immunol. 2013;31:563–604.
9. Dillmann C, et al. S1PR4 is required for plasmacytoid dendritic cell differentiation. Biol Chem. 2015;396(6-7):775–82.
10. Weigert A, et al. HIF-1alpha is a negative regulator of plasmacytoid DC development in vitro and in vivo. Blood. 2012;120(15):3001–6.
11. Breton G, et al. Circulating precursors of human CD1c+ and CD141+ dendritic cells. J Exp Med. 2015;212(3):401–13.
12. Kaplan DH. Ontogeny and function of murine epidermal Langerhans cells. Nat Immunol. 2017;18(10):1068–75.
13. Kel JM, et al. TGF-beta is required to maintain the pool of immature Langerhans cells in the epidermis. J Immunol. 2010;185(6):3248–55.
14. Flacher V, et al. Murine Langerin+ dermal dendritic cells prime CD8+ T cells while Langerhans cells induce cross-tolerance. EMBO Mol Med. 2014;6(9):1191–204.
15. Dambuza IM, Brown GD. C-type lectins in immunity: recent developments. Curr Opin Immunol. 2015;32:21–7.
16. Barreira da Silva R, Munz C. Natural killer cell activation by dendritic cells: balancing inhibitory and activating signals. Cell Mol Life Sci. 2011;68(21):3505–18.
17. Sacramento L, et al. Toll-like receptor 9 signaling in dendritic cells regulates neutrophil recruitment to inflammatory foci following Leishmania infantum infection. Infect Immun. 2015;83(12):4604–16.
18. Geijtenbeek TB, et al. Self- and nonself-recognition by C-type lectins on dendritic cells. Annu Rev Immunol. 2004;22:33–54.

19. ten Broeke T, Wubbolts R, Stoorvogel W. MHC class II antigen presentation by dendritic cells regulated through endosomal sorting. Cold Spring Harb Perspect Biol. 2013;5(12):a016873.
20. Blum JS, Wearsch PA, Cresswell P. Pathways of antigen processing. Annu Rev Immunol. 2013;31:443–73.
21. Chiang MC, et al. Differential uptake and cross-presentation of soluble and necrotic cell antigen by human DC subsets. Eur J Immunol. 2016;46(2):329–39.
22. Audiger C, et al. The importance of dendritic cells in maintaining immune tolerance. J Immunol. 2017;198(6):2223–31.
23. Theisen D, Murphy K. The role of cDC1s in vivo: CD8 T cell priming through cross-presentation. F1000Res. 2017;6:98.
24. Schlitzer A, et al. IRF4 transcription factor-dependent CD11b+ dendritic cells in human and mouse control mucosal IL-17 cytokine responses. Immunity. 2013;38(5):970–83.
25. Kohli K, Janssen A, Forster R. Plasmacytoid dendritic cells induce tolerance predominantly by cargoing antigen to lymph nodes. Eur J Immunol. 2016;46(11):2659–68.
26. Mellor AL, Lemos H, Huang L. Indoleamine 2,3-dioxygenase and tolerance: where are we now? Front Immunol. 2017;8:1360.
27. Manches O, Frleta D, Bhardwaj N. Dendritic cells in progression and pathology of HIV infection. Trends Immunol. 2014;35(3):114–22.
28. Gaurav R, Agrawal DK. Clinical view on the importance of dendritic cells in asthma. Expert Rev Clin Immunol. 2013;9(10):899–919.
29. Bates J, Diehl L. Dendritic cells in IBD pathogenesis: an area of therapeutic opportunity? J Pathol. 2014;232(2):112–20.
30. Laoui D, et al. The tumour microenvironment harbours ontogenically distinct dendritic cell populations with opposing effects on tumour immunity. Nat Commun. 2016;7:13720.
31. Gardner A, Ruffell B. Dendritic cells and cancer immunity. Trends Immunol. 2016;37(12):855–65.
32. Ganguly D, et al. The role of dendritic cells in autoimmunity. Nat Rev Immunol. 2013;13(8):566–77.
33. Sozzani S, Del Prete A, Bosisio D. Dendritic cell recruitment and activation in autoimmunity. J Autoimmun. 2017;85:126–40.
34. Hackstein H, Thomson AW. Dendritic cells: emerging pharmacological targets of immunosuppressive drugs. Nat Rev Immunol. 2004;4(1):24–34.
35. Larange A, et al. Glucocorticoids inhibit dendritic cell maturation induced by Toll-like receptor 7 and Toll-like receptor 8. J Leukoc Biol. 2012;91(1):105–17.
36. De Gregorio E, D'Oro U, Wack A. Immunology of TLR-independent vaccine adjuvants. Curr Opin Immunol. 2009;21(3):339–45.
37. García F, et al. Dendritic cell based vaccines for HIV infection. Hum Vaccin Immunother. 2014;9(11):2445–52.
38. Garg AD, et al. Integrating next-generation dendritic cell vaccines into the current cancer immunotherapy landscape. Trends Immunol. 2017;38(8):577–93.
39. Tel J, et al. Natural human plasmacytoid dendritic cells induce antigen-specific T-cell responses in melanoma patients. Cancer Res. 2013;73(3):1063–75.
40. Schlitzer A, et al. Identification of cDC1- and cDC2-committed DC progenitors reveals early lineage priming at the common DC progenitor stage in the bone marrow. Nat Immunol. 2015;16(7):718–28.
41. Atmatzidis DH, Lambert WC, Lambert MW. Langerhans cell: exciting developments in health and disease. J Eur Acad Dermatol Venereol. 2017;31(11):1817–24.
42. van der Aar AMG, et al. Cutting edge: loss of TLR2, TLR4, and TLR5 on Langerhans cells abolishes bacterial recognition. J Immunol. 2007;178(4):1986–90.
43. Scott CL, et al. The transcription factor Zeb2 regulates development of conventional and plasmacytoid DCs by repressing Id2. J Exp Med. 2016;213(6):897–911.

6 Cytokines

Bastian Schirmer and Detlef Neumann

6.1 Introduction

CYTOKINES designate a broad category of factors that are produced by immune cells and/or act on immune cells. They are small proteins or peptides, mostly glycosylated, that regulate cellular growth, differentiation, and/or activity in a para- or autocrine manner. Those peptides, which are involved in growth and differentiation of target cells, are also named GROWTH FACTORS. Others regulate immune or inflammatory reactions and, in order to differentiate them from other factors, can be called MEDIATORS. However, there are examples for cytokines able to do both, and also factors other than cytokines can be assorted into these groups. The majority of cytokines are not exclusively synthesized by a single specific cell type. Moreover, certain cytokines are produced by both immune and nonimmune cells. On the other hand, a given cytokine can act on several cell types. And again, both immune and nonimmune cells may be affected by the same cytokine. Generally, cytokines are most important in regulatory processes of the immune system and can be classified into five subgroups: INTERLEUKINS, TUMOR NECROSIS FACTORS, INTERFERONS, CHEMOKINES, and COLONY-STIMULATING FACTORS. This, however, is not an unambiguous and broadly accepted definition or classification of cytokines, but it is rather used to give this chapter a structure.

In the 1960s it was demonstrated that supernatants of stimulated immune cells could regulate function and growth of leukocytes. Out of those supernatants, factors were partially purified and designated by function-related names such as lymphocyte activation factor (LAF) or T-cell growth factor (TCGF). These were grouped according to their major producing cells: monokines or lymphokines. In the 1970s Cohen suggested [1] the use of the term cytokines as a more general denomination. Since it became apparent that identical molecules had been described by different groups giving them different names, the term INTERLEUKINS (IL) was proposed at the second International Lymphokine Workshop held in 1979 "as a system of nomenclature ... based on the ability to act as communication signals between different populations of leukocytes." Concomitantly, the names IL-1 for LAF and IL-2 for TCGF were introduced.

Also in the 1960s, a cytotoxic activity produced by lymphocytes in response to mitogen or antigen was described and called lymphotoxin (LT). About 10 years later, another cytotoxic activity able to kill certain transplantable tumor cells was detected. The respective proposed factor was named TUMOR NECROSIS FACTOR (TNF) [2]. Again 10 years later, the two, indeed

Final manuscript submitted on July 04, 2016.

B. Schirmer · D. Neumann (✉)
Institute of Pharmacology, Hannover Medical School, Hannover, Germany
e-mail: schirmer.bastian@mh-hannover.de; neumann.detlef@mh-hannover.de

M. J. Parnham et al. (eds.), *Nijkamp and Parnham's Principles of Immunopharmacology*,
https://doi.org/10.1007/978-3-030-10811-3_6

different, corresponding proteins were identified: TNF from macrophages and LT from lymphocytes [3]. Confusingly, TNF was renamed TNF-α and LT TNFβ. When in 1993 LTβ (and the LTαβ complex) was identified, along with the realization that its biologic effect differs from that of TNF-α, the nomenclature was again revised, and the former TNFβ is now called LTα, while there is no further need for indexing TNF by an α.

INTERFERONS (IFNs) have been originally described in 1957 by the virologists A. Isaacs and J. Lindemann. IFNs are classified as factors produced by virus-infected cells capable of conferring resistance to infection with homologous or heterologous viruses [4]. However, soon it was recognized that they also regulate immune responses, and, thus, they became a subgroup of the cytokines. The regulation of immune and inflammatory reactions is the predominant function of IFNγ, also called type 2 or immune interferon. Structurally different from IFNγ are members belonging to the subgroups IFNα and IFNβ, which, both composed of series of members, make up the large family of type 1 interferons.

In 1987 a gene encoding a neutrophil-attractive protein, named IL-8, was cloned. The structure of this protein, today designated CXCL8, indicated that other structurally similar chemoattractive proteins have been identified earlier. This family of structurally related cytokines involved in migration and activation of immune cells was designated CHEMOKINES at the Third International Symposium of Chemotactic Cytokines in 1992 [5]. Chemokines, small molecular weight proteins ranging in size from approx. 8 to 15 kDa, are characterized by the position of the first two of four cysteine residues in highly conserved positions within their amino-terminal protein sequence [6]. Chemokines were divided into four major groups, designated CXC, CC, C, and CX_3C, where X denotes amino acids positioned between the two characterizing cysteine residues. The addition of L followed by a number to this nomenclature indicates a specific chemokine ligand, while the addition of R and a number indicates a chemokine receptor.

COLONY-STIMULATING FACTORS (CSFs) were first described in 1966 by Metcalf, reflecting the observation that they upon addition to bone marrow cells in semisolid medium promote the formation of granulocyte or monocyte colonies [7]. CSFs predominantly function as inducers of growth and differentiation of hematopoietic stem cells but can also activate fully differentiated immune cells, thus belonging to the group of cytokines.

Since hematopoiesis, immune cell development and maturation (Fig. 6.1), and activation of an immune response, including inflammation, are covered by other chapters of this textbook, here we will just very briefly introduce these issues and focus on the functions of cytokines in the diverse biological processes and how they can be targeted to obtain a therapeutic benefit.

To this end, cytokines themselves can be exploited as drugs. Examples for this are CSFs and IFNs as described in the following. However, from a quantitative point of view, much more fruitful proved to be the possibility of developing inhibitors of cytokine activities. Their therapeutic potential is largely demonstrated in the groups of interleukins, TNFs, and chemokines.

A major breakthrough in the understanding of an immune reaction was the recognition that different immune mechanisms are executed by different cells, which are, in turn, regulated by again different cell populations. In this regard, an important finding was the identification of several subsets of $CD4^+$ T helper cells (Th cells), which develop, i.e., polarize, from naïve Th cells upon activation with the cognate antigen [8]. These Th cell subsets are characterized by the specific cytokine profile they secrete. As such, Th1 cells mainly produce IFNγ; Th2 cells IL-4, IL-5, and IL-13; Th17 cells IL-17; and regulatory T cells (Treg) IL-10 and TGF-β. These cytokines are produced exclusively by the respective subset and not only promote specific types of immune effector functions but also cross-inhibit the polarization of other Th cell subsets.

More recently, an additional family of lymphoid cells was identified, which, however, differ from "conventional" lymphoid cells, T cells, and B cells, inasmuch as they do not rearrange DNA segments to develop a huge repertoire of receptors with different specificities [9]. Thus, the

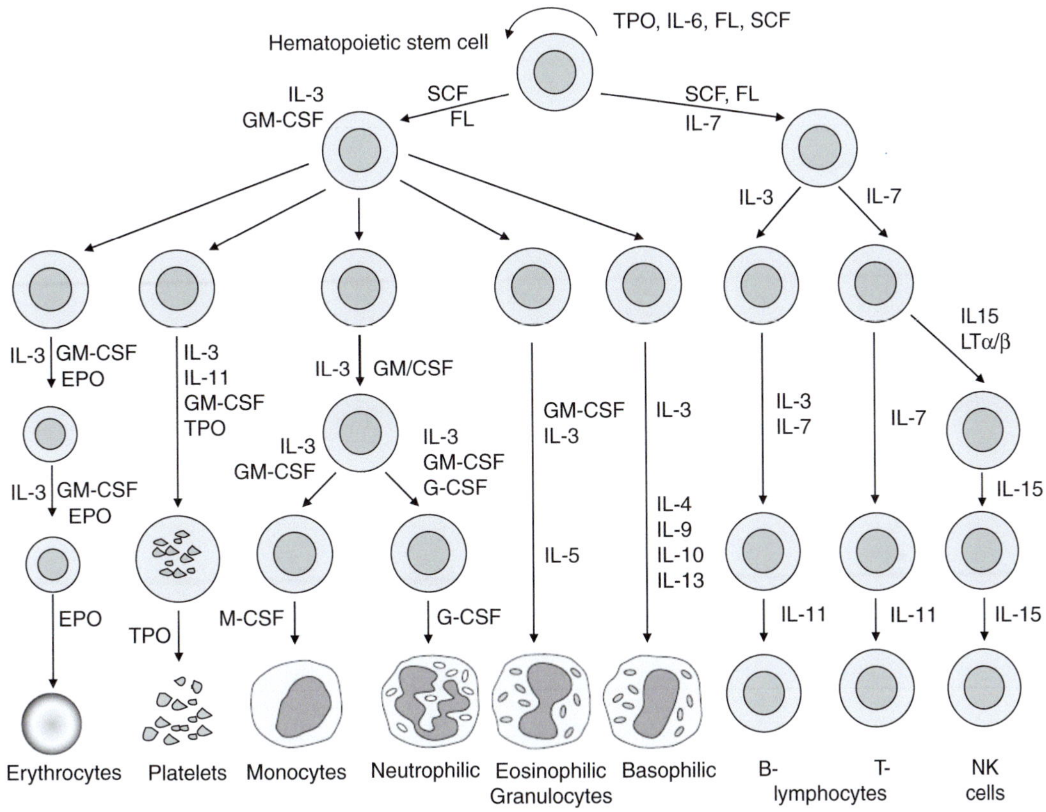

Fig. 6.1 Cytokines involved in the differentiation of cells of the immune system. *CSF* colony-stimulating factor, *EPO* erythropoietin, *G* granulocyte, *IL* interleukin, *M* monocyte, *SCF* stem cell factor, *TPO* thrombopoietin

newly discovered lymphoid cells were named innate lymphoid cells (ILCs). In analogy to the Th cell subsets, also ILCs are subgrouped according to specific cytokines they exclusively produce. Thus, group 1 ILC (ILC1) are identified by production of IFNγ, while ILC2 produce IL-5 and IL-13, and ILC3 release their key cytokine IL-22 (Fig. 6.2).

6.2 Interleukins

Interleukins mediate signals from one leukocyte to the other, from any other cell to a leukocyte, or vice versa. Interleukins have been numbered consecutively upon their characterization/identification. Thus, this numbering does not contribute to their systematic classification. Today (April 2016), the list of interleukins comprises 39 entries (Table 6.1); however, several of them appear with a number of isoforms, indicated by an index (Greek or Arabic letter) after the number.

6.2.1 The IL-1 Family

Already the first interleukin in that list, **IL-1**, exists in two isoforms, IL-1α and IL-1β, and is eponymous for a large family of cytokines acting on structurally related receptors [10]. The family of IL-1 cytokines consists of IL-1, IL-18, IL-33, IL-36, IL-37, and IL-38. IL-36 exists in different isoforms, too: IL-36α, IL-36β, and IL-36γ. All these IL-1 family ligands are agonistic at their respective receptors but exert different functions: while IL-1, IL-18, IL-33, and IL-36 are pro-inflammatory, IL-37 delivers an anti-inflammatory signal at the IL-18 receptor. The mechanism how IL-38 exerts its anti-inflammatory function

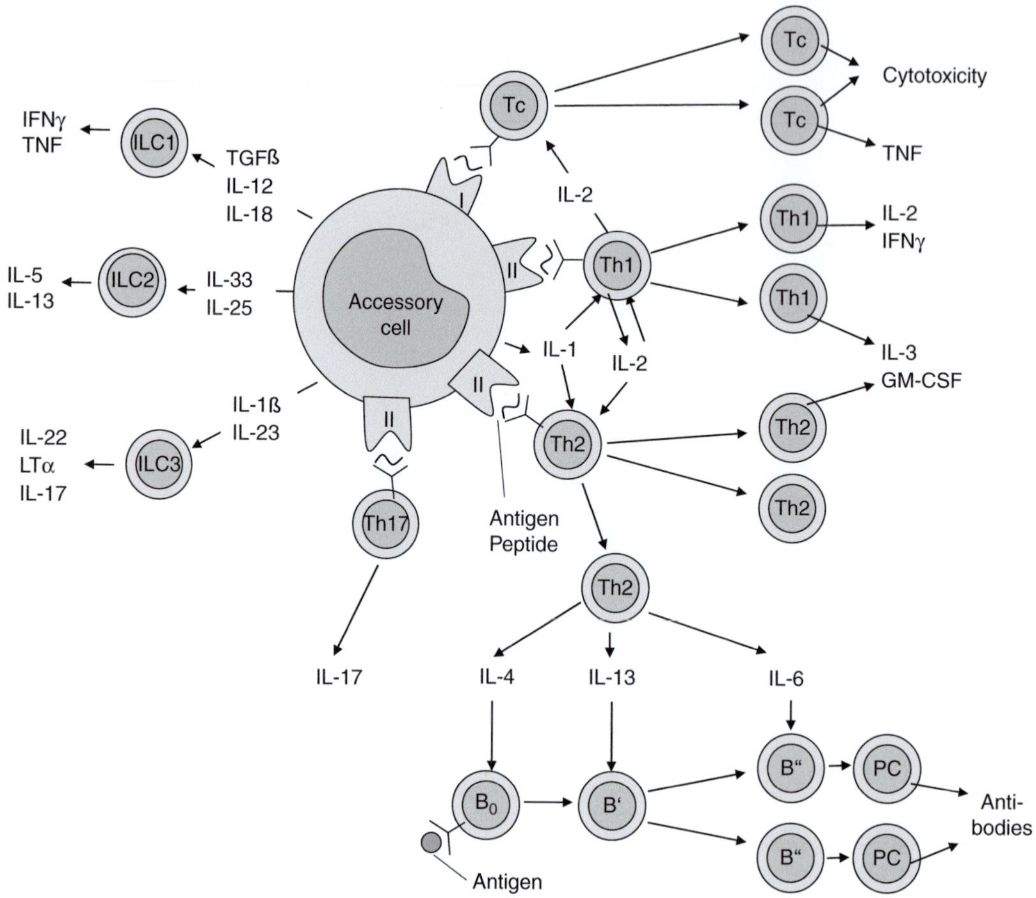

Fig. 6.2 Activation of T and B lymphocytes. *B* B lymphocyte, B_0 virgin B lymphocyte, *B′*, *B″* activated B lymphocytes, *GM CSF* granulocyte/monocyte colony-stimulating factor, *IFN* interferon, *IL* interleukin, *PC* plasma cell, *Th* T helper lymphocyte, *Tc* cytotoxic T lymphocyte, *TNF* tumor necrosis factor, *I* MHC class I molecule, *II* MHC class II molecule

is not well elaborated yet. Moreover, for IL-1 and IL-36, isoforms with antagonistic, thus competitive binding activity at the respective receptor exist, called IL-1RA and IL-36RA, where RA denotes "receptor antagonist" [10–12].

6.2.2 IL-1

IL-1 is produced upon infection mainly by macrophages, endothelial cells, and fibroblasts and acts on virtually all cells and each organ. Generally, IL-1 together with TNF serves as the major mediator of inflammatory diseases, including autoimmune, infectious, and degenerative ones. In the central nervous system (CNS), it acts as endogenous pyrogen inducing fever, and systemically, it induces the acute phase response, either directly or indirectly via the induction of IL-6 production. Both central and systemic effects are executed in order to initially promote and then orchestrate the innate and the adaptive immune reaction.

IL-1α and IL-1β are distinct gene products but bind to the same receptor, the type I IL-1 receptor (IL-1RI), and therefore bear similar, but not identical, biological activities. IL-1RA as well binds IL-1RI, but with a higher affinity than IL-1α or IL-1β do. Thus, the agonists IL-1α/IL-1β and the antagonist IL-1RA compete for IL-1RI binding. The agonist complex composed of IL-1α or IL-1β and the IL-1RI recruits the IL-1 receptor accessory

Table 6.1 Interleukins

Cytokine	Molecular mass (kD)	Predominant producer cells	Major functions
IL-1α IL-1β	17	Monocytes	Activation of T lymphocytes and inflammatory cells
IL-1RA	17	Macrophages	Complete IL-1 receptor antagonist
IL-2	15	T lymphocytes	Proliferation of T lymphocytes, promonocytes, NK cells
IL-3	14–28	T lymphocytes	Differentiation and propagation of early myeloid progenitor cells
IL-4	15–20	T lymphocytes	Differentiation and proliferation of Th2 cells and B lymphocytes, inhibition of macrophage activation
IL-5	45–60	T lymphocytes	Maturation of eosinophils
IL-6	26	T lymphocytes, many other cells	Activation of B and T lymphocytes and other cells, stem cell expansion
IL-7	25	Stromal cells	Maturation of T and B lymphocytes
IL-8	10	Monocytes/macrophages	Chemotaxis and activation of granulocytes, chemotaxis of T lymphocytes
IL-9	37–40	T lymphocytes	Propagation of mast cells, megakaryocytes
IL-10	17–21	T lymphocytes	Inhibition of cellular immune and inflammatory reactions, propagation of mast cells
IL-11	23	Stromal cells	Maturation of lymphocytes, proliferation of myeloid and megakaryocytic progenitor cells
IL-12	70	Monocytes/macrophages, dendritic cells	Differentiation and activation of Th1 lymphocytes, NK cells
IL-13	17	T lymphocytes	Activation and proliferation of B lymphocytes, inhibition of cellular immune reactions
IL-14	60	T lymphocytes	Activation of B lymphocytes
IL-15	14–15	Epithelial cells	Activation and proliferation of T lymphocytes, NK cells
L-16	14	T lymphocytes	Chemotaxis of T lymphocytes, macrophages
IL-17	35	T lymphocytes	Induction of pro-inflammatory cytokines
IL-18	24	Monocytes/macrophages, dendritic cells	Differentiation and activation of Th1 lymphocytes
IL-19	18	Monocytes	Differentiation of Th2 lymphocytes
IL-20	18	Monocytes, keratinocytes	Skin and kidney inflammation, downregulation of T-cell responses
IL-21	14	T helper lymphocytes, esp. Th2 and Th17 cells	Regulation of B-cell differentiation and activation, differentiation of NK cells
IL-22	17)	Th17 lymphocytes, mucosal NK cells	Augmentation of innate immune reactions in the skin and mucosa
IL-23	60	Monocytes, dendritic cells	Th17 differentiation, activation of monocytes
IL-24	33	Monocytes, Th2 cells	Induction of inflammation in the skin and lung
IL-25	18	Mast cells, eosinophils, Th2 cells	Induction of Th2 cytokines (IL-4, IL-5, IL-9, IL-13)
IL-26	18	T lymphocytes, NK cells	Local immune mechanisms in the skin or lung, induction of IL-8 and IL-10
IL-27	60	Dendritic cells, placentary trophoblasts	Regulation of differentiation of Th1 cells, suppression of Th17 development
IL-28A, IL-28B	24, 22	Lymphocytes, dendritic cells, virus-infected cells	Induction of type I interferons, development of tolerogenic DC and Treg
IL-29	26–35	Like IL-28	Like IL-28
IL-30			Reported as p28 chain of Il-27
IL-31	24	Th2 cells	Hematopoiesis, modulation of cell proliferation of many cell types
IL-32	25	T lymphocytes, NK cells	Differentiation and activation of monocytes

(continued)

Table 6.1 (continued)

Cytokine	Molecular mass (kD)	Predominant producer cells	Major functions
IL-33	32	Endothelial cells, smooth muscle cells, keratinocytes	Increase in Th2 cytokine release, activation of mast cells, basophils, and eosinophils
IL-34	39	Many cells	Proliferation of monocytes and monocytic progenitors
IL-35	70	Treg, dendritic cells (unknown, speculative), placentary trophoblasts	Control of effector T cells, immune modulator at feto-maternal barrier in the placenta
IL-36α, IL-36β, IL-36γ	17	Monocytes, lymphocytes, other cells	Activation of T lymphocytes and inflammatory cells
IL-36RA	17	Monocytes, lymphocytes, other cells	Reduction of inflammation
IL-37	19	Monocytes	Antagonistic to IL-18
IL-38	18	Unknown, speculative	Antagonistic to IL-36α, IL-36β, IL-36γ
IL-39	54	B lymphocytes (unknown, speculative)	Promotes inflammation (unknown, speculative)

protein (IL-1RAcP), resulting in a ternary complex able to induce signal transduction, thus to elucidate a biological effect, i.e., inflammation. The complex of IL-1RA and IL-1RI, in contrast, due to steric hindrance is unable to recruit the IL-1RAcP and, therefore, to generate a biological response. Thus, IL-1RA is an endogenous regulator of IL-1 activity, essentially necessary to limit the acute inflammatory response, as demonstrated by patients born with inactive IL1RA, who suffer from severe systemic and local inflammation.

IL-36 proteins are produced by epithelial cells and immune cells and induce the production of cytokines promoting Th1 and Th17 cell polarization. IL-36 acts via the receptor called IL-1R-related protein 2 (IL-1Rrp2), which, when ligand-bound, also recruits IL-1RAcP. Biologically, IL-36 is highly similar to the IL-1 and thus reflects most of the aspects depicted there, including the existence of an endogenous receptor antagonist, IL-36RA. This, however, gave no rise for effective therapeutic developments so far [13].

6.2.3 Pharmacological Implications for IL-1

Human IL-1RA has been cloned and is now being produced commercially as recombinant, N-terminally methionylated, and non-glycosylated therapeutic protein (ANAKINRA) in an *E. coli* expression system. ANAKINRA is approved by the European Medicines Agency (EMA) for the therapy of rheumatoid arthritis (RA) and cryopyrin-associated periodic syndromes (CAPS). CAPS are rare diseases resulting from a gain-of-function mutation in the cryopyrin gene (CIAS1/NALP3) which is part of the inflammasome. The activated inflammasome induces the proteolytic maturation and the secretion of IL-1β [14]. Thus, CAPS patients show elevated systemic concentrations of IL-1β and then subsequently produced IL-6, as well as some other acute phase (AP) proteins. In synovial fluid and plasma of patients with RA, elevated concentrations of IL-1β can be found, too, and IL-1β is thought to be a central mediator of the synovitis characteristic for RA. ANAKINRA reduces the biological activity of IL-1β by competitive antagonism at the IL-1RI and decreases the systemic concentrations of IL-6 and AP proteins. Consequently, RA or CAPS patients treated with ANAKINRA exhibit significantly ameliorated inflammatory symptoms [15].

Another strategy for blockade of IL-1 signaling is the use of monoclonal antibodies against the cytokine itself. Indeed, the fully human $IgG_{1,\kappa}$ monoclonal IL-1β antibody CANAKINUMAB, which is produced in murine hybridoma cells, has been approved for the treatment of systemic juvenile idiopathic arthritis (SJIA), gout arthritis, and CAPS. CANAKINUMAB significantly reduces IL-1β activity by establishing high-affinity IL-1β/IL-1β antibody complexes, thus leading to less IL-1RI receptor activation and less inflammatory symptoms in these diseases [16].

A third way to block IL-1 signaling is the neutralization of the circulating interleukin by binding to soluble IL-1R. Therefore, a dimeric fusion protein of the ligand binding part of the IL-1RI and the IL-1RAcP to the human Fc-domain of IgG1 (RILONACEPT) has been constructed and was approved in Europe as orphan drug for the treatment of CAPS (see above), but the marketing authorization has been voluntarily withdrawn by the producing company due to commercial reasons.

6.2.4 IL-18 and IL-33

IL-18 and IL-33 are mainly involved in Th1- and Th2-directed adaptive immune reactions, respectively, such as autoimmune diseases, metabolic syndrome, or inflammatory bowel disease and response to parasites, lung inflammation, or fibrosis. Both interleukins are produced by a variety of cell types, including epithelial cells and monocytic cells. While IL-18 promotes the production of IFNγ (originally IL-18 was named IFNγ-inducing factor) in polarized Th1 cells, IL-33 drives the synthesis of Th2-type cytokines, IL-4, IL-5, and IL-13, in Th2 cells and ILC2. IL-18 and IL-33 are recognized by receptor complexes similar to the IL-1 receptor composed of a binding protein and an accessory protein, called, in the case of IL-18, IL-18Rα and IL-18Rβ [17] and, for IL-33, ST2, which associates upon ligand binding with the IL-1RAcP [18]. Interestingly, IL-18Rα can also bind IL-37, which, however, does not act as a classical competitive antagonist. It seems that the complex of IL-37 and IL-18Rα recruits another membrane molecule, TIR8, which together induce anti-inflammatory signaling. Consequently, mice, upon transgenic overexpression of IL-37, are protected against experimentally induced colitis and ischemic diseases. IL-37, since it induces dendritic cells (DC) to promote regulatory T-cell (Treg) activation, seems to be a regulator of the adaptive immune response.

6.2.5 Pharmacological Implications for IL-18

Because IL-18 has shown antitumor effects in preclinical animal models, the therapeutic potential of recombinant IL-18 in treating solid tumors, metastatic melanoma, lymphomas, or ovarian cancer is currently evaluated in clinical trials, and preliminary results point to no clear-cut but more complex role of IL-18 in several cancers [19, 20]. The benefit of ANTAGONIZING IL-18 in treating chronic inflammatory conditions is the subject of ongoing trials with monoclonal IL-18 antibodies (targeting inflammation in type II diabetes mellitus) or with recombinant IL-18BP (targeting inflammation in Still's disease). To date, no final results of these studies have been published.

6.2.6 The IL-2 or Common Cytokine Receptor γ-Chain Family

The pleiotropic **IL-2** mainly induces proliferation of both CD4 (helper) and CD8 (cytotoxic) T-cell and natural killer (NK)-cell activities. IL-2 is primarily produced by activated T cells themselves and thus provides a prototypic auto- or paracrine growth factor. It is necessary for the clonal expansion of activated T cells, a process by which a single, antigen-specific T cell gives rise to up to 10^7 descendants. Since this is a central step in the activation of adaptive immunity, nature has created a backup system, IL-15, which can restore inadequate IL-2 activity [21, 22]. The receptors of these two cytokines use an identical component, the common cytokine receptor γ-chain (γ_c), a feature they share also with IL-4, IL-7, IL-9, and IL-21. Therefore, they are summarized to the γ_c family of cytokines. γ_c is essential for the receptor functions, and, thus, mutations in this protein affect the biological activity of all IL-2 family cytokines, which are essential for the normal development and activity of lymphoid cells. Indeed, such mutations are found in humans with X-linked severe combined immunodeficiency (XSCID).

Besides its main biological targets, T and NK cells, IL-2 also act on B cells, ILC, and neutrophils, in which it promotes proliferation and augments cytokine production. Resting T cells express the IL-2 receptor (IL-2R) β-chain together with γ_c forming a receptor dimer with intermediate affinity for IL-2. Upon activation of T cells, expression of IL-2 and of the IL-2α chain is induced, the latter giving rise to a trimeric high-affinity IL-2R, composed of IL-2Rα, IL-2Rβ, and IL-2Rγ_c. Thus, activated T cells are maximally responsive to auto- or paracrine stimulation by IL-2 [23].

During a primary immune response, antigen-specific naïve T cells are activated, expand clonally due to the activity of IL-2, and differentiate into effector T cells. The orientation of effector T-cell differentiation, called polarization, is instructed by means of key cytokines and transcription factors, such as IL-4 and GATA3, which promote the polarization of Th2 cells. Th2 cells control humoral immunity and are mainly involved in the reaction against extracellular antigens but also in allergic inflammation. IL-21, together with IL-23, and RORγt induce polarization of Th17 cells. Th17-type cytokines, IL-17A, L-17F, and IL-22, are involved in the immune response to bacterial infections and also in autoimmune diseases. Finally, IL-2 in combination with TGF-β promotes the differentiation of Treg, necessary for maintaining self-tolerance.

6.2.7 Pharmacological Implications for IL-2

This responsiveness of T cells can impede the success of organ transplantation, where it can ultimately lead to transplant rejection. The murine/human chimeric monoclonal $IgG_{1,\kappa}$ antibody BASILIXIMAB binds with high affinity to the IL-2Rα chain, thus preventing the binding of the key T-cell activator IL-2. BASILIXIMAB is approved in conjunction with other immunosuppressants for the prophylaxis of acute rejection of renal transplants, where it dampens the inadequate cellular immune response.

The immunoregulatory function of IL-2 is harnessed in cancer treatment, too. A modified form of the human IL-2 gene, recombinantly expressed in *E. coli* (ALDESLEUKIN), shows immunostimulatory and antitumorous activity in vivo. ALDESLEUKIN is approved for the treatment of metastatic renal cell carcinoma [24]. Together with histamine dihydrochloride, ALDESLEUKIN is used to treat acute myeloid leukemia (AML). It is hypothesized that histamine via the H_2-receptor inhibits generation of reactive oxygen species (ROS) by neutrophilic granulocytes. ROS are increased in the tumor environment, thereby inhibiting cytokine action on NK and T cells. Thus, IL-2 stimulation of NK and T cells is more effective with co-administered histamine [25, 26].

6.2.8 Th2-Type Cytokines

Structurally, IL-4 is a member of the above discussed IL-2 cytokine family. However, functionally it is grouped together with **IL-5** and IL-13 into the Th2-type cytokines cells. Of these, IL-5 is the key factor regulating activation, recruitment, differentiation, and proliferation of eosinophils. The receptor for IL-5 (IL-5R), which is composed of two chains, α and β, belongs to the group of type 1 cytokine receptors. The IL-5Rα chain which virtually is exclusively expressed on eosinophils specifically binds IL-5 but with low affinity. The IL-5-bound IL-5Rα chain does not generate a cellular signal, but upon polymerization with the β-chain, resulting in a high-affinity IL-5R complex, the cell is adequately activated. The β-chain, also called common β-chain, is not an exclusive component of the IL-5R but is also shared by the receptors for IL-3 and GM-CSF. This chain, essential for the signal transduction, associates with the protein tyrosine kinases JAK2 and Lyn. Of note, by alternative splicing a soluble IL-5Rα chain can be formed. This soluble receptor chain binds IL-5 as well as does the membrane-inserted chain, but is unable to form a signal-transducing complex. Thus, it sequesters free IL-5 and functionally acts as antagonist.

In lung biopsies of a subgroup of asthmatic patients, IL-5 mRNA expression was found to be enhanced as compared to controls. Moreover, the quantity of IL-5 mRNA in those samples correlated with the clinical severity of the disease.

In mouse models of asthma, the inhalative provocation enhanced IL-5 expression and blockade of IL-5 activity resulted in the reduction of eosinophil numbers in the lung.

Also IL-31 can be regarded a Th2-type cytokine, an atypical one, since it is not exclusively produced by Th2 cells but also by mast cells, macrophages/monocytes, and DC. Its synthesis, however, seems to be regulated by IL-4, and functionally it is clearly associated with allergic diseases, and it is a major factor in the generation of pruritus. Therefore, in addition to IL-5, it provides a highly attractive drug target.

6.2.9 Pharmacological Implications for IL-5 and IL-13

Since 2016, patients with severe, refractory eosinophilic asthma can be treated with the humanized monoclonal $IgG_{1,\kappa}$ IL-5-antibody MEPOLIZUMAB. This antibody binds IL-5 with high affinity and specificity and thereby decreases the IL-5-initiated signal transduction leading to growth, differentiation, recruitment, and survival of eosinophilic granulocytes. MEPOLIZUMAB thereby significantly lowers the rate of asthmatic exacerbations and the dosage of oral glucocorticoids needed for the effective control of asthma activity [27, 28]. Other possible indications for MEPOLIZUMAB currently under investigation are COPD with eosinophilic bronchitis, hypereosinophilic syndrome, and eosinophilic esophagitis.

Two phase III clinical trials have investigated the clinical efficacy of the IL-13 antibody LEBRIKIZUMAB in patients with asthma (LAVOLTA I/II). However, the results of the studies are inconsistent so far [29, 30].

6.2.10 The IL-6 Family

IL-6 is eponymous for a family of cytokines that originally were characterized by their helical structure and by sharing a common receptor component, the glycoprotein 130 (gp130). Besides IL-6, the family includes IL-11, leukemia inhibitory factor (LIF), oncostatin M (OSM), cardiotrophin-1 (CT-1), ciliary neurotrophic factor (CNTF), and cardiotrophin-like cytokine (CLC). As for its structure, the above discussed IL-31 is also a member of the IL-6 family of cytokines. It, however, does not signal via the gp130, but via a complex composed of the gp130-like IL-31 receptor A (IL-31RA) chain and the OSM receptor ß subunit.

IL-6 is one of the key mediators regulating inflammation. Upon activation by infectious agents, IL-6 is produced by macrophages, endothelial cells, and T cells and activates many tissue cells including endothelial cells and parenchymal cells to produce effector molecules of inflammation. In the liver, IL-6 together with IL-1 and TNF induced the secretion of acute phase proteins, and in the CNS, they are pyrogenic. The regulation of IL-6 activity is complex. The IL-6-specififc IL-6 receptor (IL-6R) is expressed in its membrane-bound form only on liver cells and leukocytes. After ligand binding the receptor associates with gp130 and activates target cells. gp130 is expressed on a large variety of cells, which do not express the IL-6-specific receptor chain but nevertheless can respond to IL-6. This responsiveness is due to the generation of a soluble form of the IL-6R (sIL-6R), able to bind IL-6 as well and then to associate with membranous gp130 (mgp130). This complex of IL-6, sIL-6R, and mgp130 is able to generate signal transduction, called IL-6 trans-signaling. IL-6 trans-signaling is inhibited by a soluble form of gp130 (sgp130) that sequesters the IL-6/sIL-6R complex, thereby avoiding the association with mgp130. Notably, the interaction between IL-6 and the membrane-bound IL-6R is not affected by the latter mechanism. Thus, sgp130 focuses IL-6 activity on cells expressing the IL-6R.

6.2.11 Pharmacological Implications for IL-6 Family Members

Soluble and membrane-bound IL-6 receptors can be blocked pharmacologically by monoclonal antibodies, thereby preventing pro-inflammatory IL-6 signal transduction. For the treatment of the chronic inflammatory rheumatoid arthritis, the recombinant humanized monoclonal IgG1 IL-6 antibody TOCILIZUMAB is approved in Europe [31]. Alternatively, not the IL-6 receptor but the circulating cytokine itself can be bound by antibodies. SILTUXIMAB is a chimeric (human/murine) $IgG_{1,k}$ antibody against IL-6, interfering with the binding of IL-6 to both soluble and membranous IL-6R. However, although SILTUXIMAB blocks the very same signal-transduction pathway as TOCILIZUMAB does, it is currently approved only for the treatment of multicentric Castleman's disease (MCD). MCD is a rare, aggressive lymphoproliferative disease, which is partly caused by an overproduction of systemic IL-6, leading to the pathologic expansion of lymphoid cells, especially B cells [32].

The IL-6 protein has been fused recombinantly to the extracellular domain of its cognate receptor IL-6R creating the superagonistic protein "hyper-IL-6" capable of activating even cells containing only gp130. Thus, hyper-IL-6 is more than 100-fold as potent as IL-6 alone, and the fusion protein can be efficiently used for in vitro expansion of hematopoietic stem cells [33]. In principle, the resultant stem cells could be used for transplantation, but currently there are no convincing data proving a superiority of ex vivo expanded stem cells over unmanipulated stem cells.

Since IL-11 acts as platelet growth factor, inducing megakaryopoiesis in vivo, the therapeutic effect of recombinant human IL-11 (rhIL-11) in thrombocytopenias is currently being investigated [34].

CNTF has been proven effective in rescuing retinal ganglion cells in preclinical animal models of, e.g., retinitis pigmentosa [35]. Thus, CNTF-secreting eye implants for the treatment of retinitis pigmentosa are now being tested in clinical trials for their clinical efficacy.

6.2.12 The IL-10 Family

The **IL-10** family of cytokines consists of nine members: IL-10, IL-19, IL-20, IL-22, IL-24, IL-26, IL-28A, IL-28B, and IL-29 [36]. Based on their biological functions, three subgroups can be defined. The first consists of only IL-10 itself, which mainly reduces excessive inflammatory responses. The second subgroup is composed of IL-19, IL-20, IL-22, IL-24, and IL-26. These cytokines primarily protect epithelial cells from damage by extracellular pathogens and promote tissue remodeling and wound healing. Members of the third subgroup, IL-28A, IL-28B, and IL-29, are also called type III IFNs (IFN λs). Similar to type I and type II IFNs, type III IFNs induce antiviral responses, however, primarily on epithelial cells.

IL-10 itself is mainly produced by monocytes and T cells, i.e., Th2 cells and induced Treg. It reduces inflammatory processes and thus limits collateral inflammation-induced tissue damage. In fact, together with TGF-β, which is produced by iTreg as well, IL-10 is the most potent anti-inflammatory cytokine.

The receptor for IL-10 is a tetrameric complex, composed of the two IL-10 receptor(R)1 chains and IL-10R2 chains, the latter being structurally similar to the IFN receptor. While IL-10R1 mediates specific ligand binding, IL10R2 is essential for initiating signal transduction, which involves Tyk2- and JAK1-mediated STAT3 phosphorylation. Phosphorylated STAT3 molecules form homodimers and translocate into the nucleus to promote expression of gene products that limit inflammation.

The most prominent phenotype of mice deficient in IL-10 or IL-10R expression is the spontaneous development of chronic enterocolitis associated with an increased prevalence for colorectal carcinomas. Thus, these mice are a relevant model for human inflammatory bowel diseases (IBD), and indeed, in genome-wide association studies, polymorphisms in the genes encoding IL-10 or the IL-10R have been identified as susceptibility loci with association to IBD [37, 38].

6.2.13 Pharmacological Implications for IL-10

The role of IL-10 in immune responses is largely an anti-inflammatory one. Therefore, recombinant human IL-10 (rhIL-10) has been tested for the treatment of Wegener granulomatosis, psoriasis, IBD, and RA [28]. Most interestingly, there are innovative approaches for targeted delivery of IL-10 to inflamed tissues, e.g., by fusing the cytokine to the antibody fragment F8 (targeting vessels in inflamed tissue) or by using genetically modified organisms producing rhIL-10 for local gastrointestinal delivery [39, 40]. On the other hand, since IL-10 is involved in the development of pathogenic B cells in patients suffering from systemic lupus erythematosus (SLE) [41], an IL-10 antibody (BT-063) is currently being tested for clinical efficacy in treating SLE.

6.2.14 The IL-12 Family

A unique feature of the family of **IL-12** cytokines, which comprises besides IL-12 itself cytokines IL-23, IL-27, and IL-35, is their composition as heterodimeric proteins. Individual pairs of an α-chain (p19, p28, or p35) with a β-chain (p40 or Ebi3) constitute the individual cytokine. Thus, IL-12 is composed of p40 and p35, while IL-23 consists of p40 combined with p19. The β chain of IL-27 and IL-35 is Ebi3, which pairs with p28 or p35, respectively. The remaining two combinations, p28/p40 and p19/Ebi3, are poorly characterized so far [42, 43].

The main function of IL-12, produced mainly by microbe-activated DCs, macrophages, and B cells, is the induction of IFNγ expression by T cells and NK cells, thereby promoting Th1 polarization. IL-23 has pro-inflammatory properties as well but plays a role in the Th17-type immune response. The role of IL-27 and IL-35 has not been ultimately defined; however, they seem to be rather anti-inflammatory.

The receptors for IL-12 cytokine family members are dimeric protein complexes, which, common with the cytokines, share subunits. Pro-inflammatory IL-12 and IL-23 share the IL-12Rβ1 chain that pairs with the IL-12Rβ2 chain and the IL-23R chain, respectively [44]. The probably anti-inflammatory IL-27 and IL-35, in contrast, share with each other and with IL-6 family members the gp130 chain, which associates with WSX-3 (IL-27 receptor) or IL-12Rβ2 (IL-35).

Moreover, some of the IL-12 cytokine chains are functional as monomers or homodimers. Homodimers of p40 antagonize IL-12 by competing for receptor binding. Interestingly, they still are able to activate DCs. Monomers of p28, also referred to as IL-30, probably can inhibit IL-6- and IL-27-induced signaling by sequestering gp130.

6.2.15 Pharmacological Implications for IL-12 Family Members

A dysregulation of IL-12/IL-23 signaling has been associated with the pathology of psoriasis/psoriasis arthritis and Th1/Th17 inflammatory diseases like rheumatoid arthritis and Crohn's disease. Since IL-12 and IL-23 share the p40 subunit, both cytokines can be targeted by a single monoclonal p40 antibody. Indeed, the recombinantly produced human monoclonal $IgG_{1,\kappa}$ antibody USTEKINUMAB reduces the bioactivity of IL-12/IL-23 in vivo. It is efficient and approved for the treatment of psoriasis/psoriasis arthritis and is evaluated in clinical trials for its efficacy in other Th1-type inflammatory diseases such as multiple sclerosis, RA, and Crohn's disease [45].

An alternative approach to inhibit IL-12/IL-23 signaling is the use of inhibitors of cytokine synthesis. The small molecule STA-5326 mesylate, which can be administered orally, downregulates IL-12p35 and IL-12/IL-23p40 synthesis on the transcriptional level and is currently being tested in clinical trials for its use in the treatment of rheumatoid arthritis and Crohn's disease [46].

Monoclonal antibodies targeting solely IL-23p19 are being evaluated for their efficacy in the treatment of RA, CD, and psoriasis/psoriasis arthritis at the moment. The IL-23p19 antibody TILDRAKIZUMAB, for example, has proven effective and superior to placebo in a phase IIb clinical trial [47].

Instead of inhibiting IL-12 bioactivity, clinical trials regarding antitumor effects of IL-12 focus on the use of recombinant human IL-12 (rhIL-12) or plasmids encoding IL-12 (pIL-12). Subcutaneously administered rhIL-12 promotes activation of NK cells and of the cytotoxic T-cell response and is thus thought to boost the antitumor activity of the immune system. pIL-12 can be directly injected into tumorous tissue and subsequently be electroporated into the local cells. This leads to a targeted expression of IL-12 in the tumor. Both approaches are evaluated in the treatment of some cancers like lymphomas, prostate cancer, and melanoma. Very recently, pIL-12 has been started to be tested as "boosting agent" for DNA vaccination, e.g., against HIV-1 or some cancers [48].

6.2.16 The IL-17 Family

Human **IL-17A** was originally described in 1993 (initially named CTLA8) and gave rise to the identification of five homologous proteins (IL-17B–IL-17F), which together are grouped to the IL-17 family of cytokines. Notably, the IL-17 proteins occur as homodimers, and only one heterodimer, composed of IL-17A and IL-17F (IL-17A/IL-17F), has been identified [49]. The main IL-17-producing cell type is a subpopulation of effector T cells, referred to as Th17 cells, which are generated from naïve $CD4^+$ T cells due to the activity of a series of cytokines including TGF-β, IL-21, IL-1β, IL-6, and IL-23. Besides this, IL-17 is also produced by $CD8^+$ T cells, γδ-T cells, NK cells, lymphoid tissue inducer cells, macrophages, neutrophils, and group 3 innate lymphoid cells.

So far, five IL-17 receptor (IL-17R) subunits, called IL-17RA–IL-17RE, have been identified, which pair to generate IL-17 isoform-specific heterodimeric receptors. Thus, IL-17A, IL-17F, and IL-17A/IL-17F, the most abundant and best described IL-17 isoforms, are recognized by the IL-17RA/IL-17RC dimer, while IL-17RA/IL-17RB binds IL-17E, and IL-17RA/IL-17RE binds IL-17C. IL-17B and IL-17D bind to IL-17RB and IL-17RD, respectively, both paired with a so far not identified second receptor chain.

IL-17 is a pleiotropic cytokine; thus, its receptors are expressed on a wide variety of cell types. The lack of IL-17 or IL-17R expression in mice enhances susceptibility to infections with extracellular bacteria and fungi, indicating a central role of IL-17 in host defense. This effect is mediated, at least in part, by the IL-17-induced expression of other mediators by, e.g., epithelial cells or macrophages. These mediators include growth factors such as G-CSF and chemokines such as CXCL8 (IL-8) or CCL20, which activate granulopoiesis and attract neutrophils or lymphocytes to the site of infection, respectively. IL-17-induced mediators also include pro-inflammatory cytokines (IL-1, IL-6, TNF), linking the activity of IL-17 with inflammatory diseases. Indeed, in patients suffering from rheumatoid arthritis or psoriasis enhanced IL-17 concentrations or Th17 cell numbers have been detected, indicating a beneficial effect of drugs targeting IL-17 in these diseases.

6.2.17 Pharmacological Implications for IL-17

Currently, two recombinant monoclonal anti-IL-17A antibodies (SECUKINUMAB and IXEKIZUMAB) are approved for the therapy of severe plaque psoriasis. In clinical studies, patients receiving SECUKINUMAB or IXEKIZUMAB had significantly reduced "psoriasis area and severity indices" (PASI) as compared to patients receiving placebo. Even in comparison to those patients receiving the comparator anti-psoriatic therapy ETANERCEPT (see below, TNF), the IL-17A antibodies showed a significantly better improvement of PASI scores. SECUKINUMAB has been proven effective in the treatment of arthritic manifestations of psoriasis and Morbus Bechterew, too, and is therefore also approved for these indications. Both SECUKINUMAB and IXEKIZUMAB are currently being evaluated for their efficacy in the treatment of rheumatoid arthritis, and preliminary results point to a possible approval for treating RA, too [50–52]. Interestingly, a clinical trial investigating the efficacy of SECUKINUMAB in the treatment of active Crohn's disease had to be terminated, because the placebo arm of the study exhibited higher reductions of disease activity as compared to the patients treated with IL-17 antibody [53].

6.3 Tumor Necrosis Factors (TNF)

TNF, together with IL-1 and IL-6, are the main mediators of induction and orchestration of an inflammatory reaction, the so-called master cytokines of inflammation (Fig. 6.3). As such, TNF is involved in virtually all diseases with a contribution of inflammation. TNF, a homo-trimer in its active form, is synthesized mainly by macrophages either in a membrane-bound or in a secreted form. As suggested by its name, it possesses antitumor activity but is also involved in the induction of fever and acute phase protein; in cell proliferation, differentiation, migration, and survival; and in the induction of apoptosis. TNF was assigned to a large superfamily of distantly related proteins, the TNF superfamily (TNFSF) with more than 20 members (Table 6.2) [54]. Most of them are type 2 transmembrane proteins and only TNFSFL1 (LTβ) and TNFSFL2 (TNF) represent true cytokines. TNF effectively activates many tissue cells, including endothelial cells and parenchymal cells, to secrete effector molecules of inflammation, thereby contributing to the inflammatory process. In addition, TNF also enhances the activities of

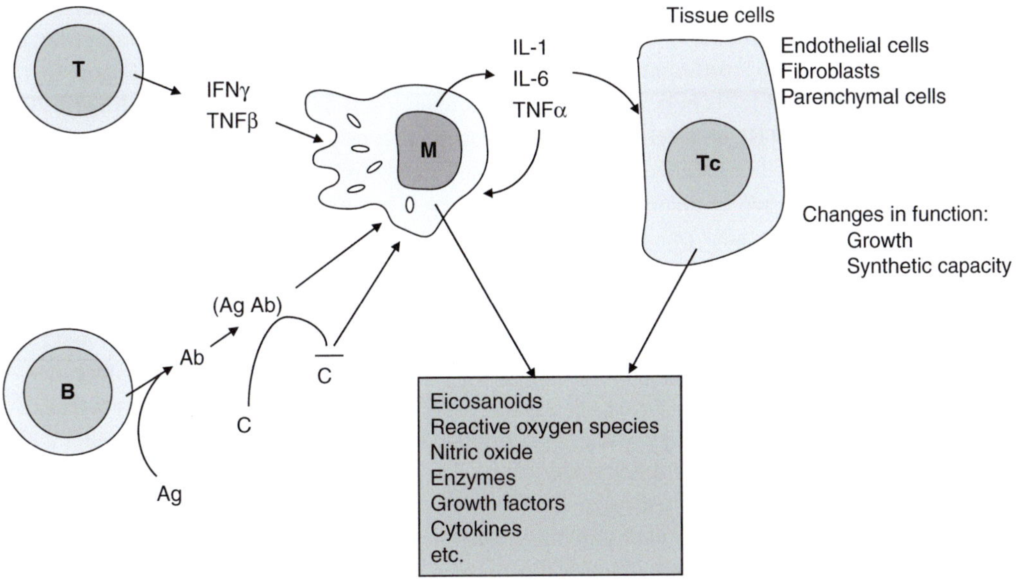

Fig. 6.3 Immune reactions and inflammation. *Ag* antigen, *Ab* antibody, *B* B lymphocyte, *C* complement, *C* activated complement components, *IFN* interferon, *IL* interleukin, *M* macrophage, *TC* tissue cells, *TNF* tumor necrosis factor

Table 6.2 Tumor necrosis factor superfamily (TNFSF) (selected members)

Cytokine	Molecular mass (monomers)[a] (kD)	Predominant producer cells	Major functions
TNF (TNFSF2)	17 (s)[b]	Monocytes and many others	Defense against pathogens, inflammation, induction of apoptosis
LTα (TNFSF1)	25 (m)[c] 17(s)	T lymphocytes	cachexia, shock
LTβ (TNFSF3)	33 (m)	T Lymphocytes	Development, homeostasis, and structure of lymphoid organs, virus defense
CD40L (TNFSF5)	29 (m)	T lymphocytes	B-lymphocyte activation
FasL (TNFSF6)	31 (m)	T lymphocytes	Induction of apoptosis
TRAIL (TNFSF10)	30 (m)	T lymphocytes	Induction of apoptosis

[a]Generally forming multimers such as homotrimers
[b]s = soluble protein
[c]m = membrane-bound protein

mononuclear phagocytes and other leukocytes, in an autocrine and paracrine manner, thereby amplifying the inflammatory reaction in an autoregulatory loop.

Two types of TNF receptors (TNFR) have been identified: TNFR1 (p55) and TNFR2 (p75). In analogy to their corresponding ligands, their active form is trimeric, and together with related receptors, they are grouped into a family, the TNFR superfamily (TNFRSF). TNFR1, but not TNFR2, contains a so-called death domain (DD), able to activate caspase cascades resulting in apoptosis. Both TNFR1 and TNFR2 are able to recruit TRAF proteins leading to activation of NF-κB (nuclear factor κB), a central transcription factor associated with inflammation, i.e., cell death/survival, activation, differentiation, and cytokine production. TNFR1 is believed to transmit the majority of TNF-induced biological responses; however, whether TNF leads to NF-κB activation and inflammation or to caspase activation and apoptosis depends on the cellular context.

6.3.1 Pharmacological Implications for TNF

A chimeric (mouse/human) antibody directed to TNF (INFLIXIMAB) represented the first example of a specific cytokine-blocking antibody and has since proven efficacious in the treatment of psoriasis, psoriasis-arthritis/rheumatoid arthritis, or inflammatory bowel diseases (IBD). INFLIXIMAB—in combination with METHOTREXATE— significantly reduced the clinical severity of arthritis as well as the progression of tissue destructions as apparent by X-ray measurements in clinical studies. Regarding IBD patients, INFLIXIMAB has proven effective for the induction and perpetuation of clinical remission—a subgroup of patients even showed persistent, steroid-free remission under INFLIXIMAB therapy. The successful drug INFLIXIMAB was followed up by further antibodies including completely human antibodies with similar properties and indications (ADALIMUMAB, GOLIMUMAB) and a construct of the F_{ab} fragment of a humanized antibody coupled to polyethylene glycol (PEG), which has been approved for the therapy of rheumatoid/psoriasis arthritis or axial spondyloarthritis (CERTOLIZUMAB-PEGOL) [55, 56]. The coupling of peptides or proteins to PEG is called pegylation and serves the purpose of altering the pharmacokinetics of the primary substance: Pegylated proteins are taken up more slowly from subcutaneous depots and show delayed excretion, too. Illustrating the principle of self-limitation, the extracellular domains of TNFR can be released during immune or inflammatory reactions. As these soluble receptor fragments contain the full cytokine binding site, they bind their cognate cytokine TNF and thereby dampen its biological effect. A dimeric TNF-receptor construct (ETANERCEPT)—in which the ligand binding domain of the human TNFR2/p75 was fused to the Fc-domain of human IgG1 to increase affinity and half-life in vivo—has been approved with similar indications, effectiveness, and side effects as INFLIXIMAB [57]. The major side effect of the therapy with anti-TNF drugs is an increased risk of infections—as would be expected!—including recurrence of tuberculosis and occasionally septic shock [58].

6.4 Interferons (IFNs)

IFNγ is the lead cytokine produced by Th1 cells, which develop due to the activity of IL-12. During polarization of the Th cell immune reaction, Th1-derived IFNγ suppresses the generation of Th2 cells, which are involved in the defense against extracellular pathogens and in allergic reactions. Other cells producing IFNγ include $CD8^+$ cytotoxic T cells, NK cells, and type 1 innate lymphoid cells. The main function of IFNγ is to activate macrophages, e.g., in case of bacterial infection. Indeed, together with LTβ, IFNγ constitutes the most important macrophage-activating factors, which promote the immune reaction by enhancing the expression of antigen-presenting MHCII molecules as well as the production of cytokines and the execution of macrophage effector mechanisms. In addition, IFNγ participates in NK-cell activation and affects B-cell differentiation. Thus, IFNγ mediates adaptive immunity-driven effector mechanisms against bacterial infection and, likewise, may provide also antiviral and antitumor activities.

Type 1 IFNs, IFNα and IFNβ, can be secreted by almost all cell types; however, some cells likely are specialized for this function. Thus, a dendritic cell (DC) subset, plasmacytoid DC, and fibroblasts produce abundant amounts of IFNα and IFNβ, respectively, upon stimulation by, e.g., viral nucleic acids or proteins. Thirteen isoforms of IFNα are known so far (Table 6.3), each encoded by a single gene. These genes, *Ifna1*, *Ifna2*, *Ifna4*, *Ifna5*, *Ifna6*, *Ifna7*, *Ifna8*, *Ifna10*, *Ifna13*, *Ifna14*, *Ifna16*, *Ifna17*, and *Ifna21*, are located together in a single gene cluster. Two genes encode the IFNβ isoforms IFNβ1 and IFNβ3. The major function of type 1 IFNs is to induce an antiviral status in cells neighboring the virus-infected cells and to activate innate immune mechanisms, i.e., macrophages and granulocytes, combatting viral infections [59].

All type 1 IFNs bind to the same receptor, type 1 IFN receptor (IFN1R). Consequently, mice devoid of IFN1R expression are highly susceptible to virus infection. Ligand-activated IFN1R is a heterodimer consisting of the chain IFNαR1 and IFNαR2, which initiate signaling via a Tyk-JAK-STAT pathway [60]. Activated STAT molecules eventually bind to specific DNA sequences, IFN-stimulated response elements (IRSE), which regulate the expression of a wide range of genes involved in the antiviral response.

6.4.1 Pharmacological Implications for IFNs

Recombinant human IFNγ is used therapeutically for the prevention of severe infections in patients with septic granulomatosis or malignant osteopetrosis. The exact mechanism of how IFNγ exactly improves the immune competence of these patients is unknown, but it is hypothesized that IFNγ induces a respiratory burst and an upregulation of HLA-DR/Fc receptors on macrophages, thereby increasing antibody-/cell-dependent cytotoxicity [61].

IFNα (IFNα-2a and IFNα-2b) is indicated in chronic hepatitis B and hepatitis C as well as in the treatment of some malignant tumors, including hairy cell leukemia or chronic myeloid leukemia, non-Hodgkin lymphoma, cutaneous T-cell leukemia, malignant melanoma, hypernephroma, or bladder carcinoma [62]. Unfortunately, it is ineffective against the majority of carcinomas. Covalent conjugates of IFNα-2a or IFNα-2b with monomethoxy-polyethylene glycol (Peginterferon α-2a or α-2b) have been introduced for the treatment of hepatitis B and C.

Approved indications for recombinant IFNβ (IFNβ-1a and IFNβ-1b) are relapsing remitting forms of multiple sclerosis and first manifestations of an inflammatory demyelinating process [63]. Antitumor effects of IFNβ in humans are uncertain.

Clinical side effects of IFNs include the common "flu-like" syndrome (fever, fatigue, shivering, muscle and joint pain), paresthesias, disturbances of the central nervous system, major depression, gastrointestinal disturbances, cardiac symptoms, granulocytopenia, thrombocytopenia, or anemia.

Table 6.3 Interferons

Cytokine	Molecular mass (kD)	Predominant producer cells	Major functions
IFNα1 IFNα2 ⋮ IFNα15	19–26	Dendritic cells	Induction of antiviral activity, inhibition of tumor cell growth, activation of cells
IFNβ	23	Fibroblasts	Induction of antiviral activity
IFNγ	17–25	T-lymphocytes	Induction of antiviral activity, activation of macrophages, immunoregulation

6.5 Chemokines

Inflammatory—including allergic—diseases usually are confined to certain organs. This implies that all cells of the immune system which participate in the underlying pathomechanisms must emigrate from the blood vessels and invade into the perspective tissue. On the other hand, antigenic material of infective agents penetrating into the body must be taken up by antigen-presenting dendritic cells in order to be carried to the adjacent lymphoid organs such as the regional lymph nodes to initiate an effective immune response. The very complex migration of leukocytes which proceeds in several subsequent defined steps and, similarly, the migration of dendritic cells are controlled by various cytokines. Among these the family of chemokines plays a pivotal role [64].

More than 40 chemokines have been cloned in human beings (Table 6.4). According to the number and position of cysteine residues in the N-terminus, four groups can be distinguished, which are also represented at the genomic level by gene clusters. Those chemokines, in which the two characteristic cytokines C are separated by an arbitrary amino acid X which are denominated as CXCL1 to CXCL15, are predominantly chemotactic for neutrophils and some—forming a small separate subcluster—for T-lymphocyte subsets. Chemokines with directly adjacent cysteins, CCL1 to CCL27, mostly attract monocytes/macrophages (and again a subset of lymphocytes) and XC chemokines and XCL1 and XCL2 (or CL1 and CL2) lymphocytes. The CX_3CL family contains only one member.

Chemokine receptors all belong to the G-protein-coupled receptors. The 18 receptors are subclassified into 4 groups according to their binding specifications. Thus, CXC chemokine binds to CXC receptors (CXCR), CC chemokines to CCR, XC chemokines to XCR, and CX3C chemokines to CX_3CR. So far, six CXCR, ten CCR, one XCR, and one CX_3CR have been identified [65]. Within these groups, selectivity of single ligands/receptors is overlapping, but not between the groups.

Chemokines have fundamental roles in development, homeostasis, and function of the immune system. Within the immune system, they can be divided into the homeostatic and the inflammatory category. The homeostatic chemokines are constitutively expressed regulating the structural organization and cellular composition of peripheral lymphoid organs such as lymph nodes. They also govern the recirculation of lymphocytes.

Table 6.4 Chemokines (some selected)

Chemokine family	Chemokine	Predominant producer cells	Major function
CXC	CXCL1 (KC, Groα)	Macrophages, neutrophils, epithelial cells	Chemotaxis of neutrophils
	CXCL4 (PF4)	Platelets	Blood coagulation, wound healing
	CXCL8 (IL-8)	Endothelial cells, epithelial cells, monocytes, fibroblasts	Chemotaxis of neutrophils
	CXCL10 (IP-10)	Endothelial cells, monocytes, fibroblasts	Chemotaxis of monocytes, lymphocytes
	CXCL12 (SDF-1α/SDF-1β)	Bone marrow cells	Recruitment of hematopoietic stem cells
CC	CCL2 (MCP-1)	Monocytes, macrophages, dendritic cells	Chemotaxis of monocytes, memory T cells, dendritic cells
	CCL3 (MIP-1α)	Macrophages	Chemotaxis of neutrophils, eosinophils
	CCL5 (RANTES)	T cells	Chemotaxis of T cells, eosinophils, basophils
	CCL11 (eotaxin)	Epithelial cells	Chemotaxis of eosinophils
XC	XCL1 (lymphotactin)	T cells, NK cells, NKT cells	Chemotaxis of dendritic cells
CX3C	CX3CL1 (fractalkine)	Endothelial cells	Chemotaxis of T cells, monocytes

The expression of inflammatory chemokines is strongly induced by pro-inflammatory stimuli, predominantly in cells of the immune system such as macrophages and T lymphocytes but also in fibroblasts. Inflammatory chemokines participate in the development of inflammatory and immune reactions not only by their chemotactic properties but also as potent activators of their target cells.

6.5.1 Pharmacological Implications for Chemokines

The chemokine receptors CCR5 (and to a lesser extent CXCR4) have raised great interest as co-receptors (in addition to CD4) for the entry of human immunodeficiency virus (HIV) into macrophages and T lymphocytes. Selective blockade of the human CCR5 leads to an inhibition of the entry of HIV-1 with CCR5 tropism. Thus, the low molecular weight CCR5 antagonist MARAVIROC is approved for the treatment of HIV infections and AIDS [66].

During a screening for HIV-inhibiting drugs in the early 1990s, the selective CXCR4 antagonist PLERIXAFOR was discovered. One could speculate that this drug would be another possible inhibitor of viral entry, but since PLERIXAFOR has very limited oral bioavailability, the drug has never been explored and marketed for HIV therapy. Instead, it has gained approval for mobilization of hematopoietic stem cells into the peripheral blood prior to autologous transplantation (in combination with G-CSF). It is hypothesized that PLERIXAFOR leads to leukocytosis and release of hematopoietic stem cells by preventing the cognate ligand CXCL12 from binding to CXCR4 [67].

6.6 Colony-Stimulating Factors (CSFs)

The major differentiation factors of the myelomonocytic cell lineage are well known. Several of them are named colony-stimulating factors (CSF) according to the observation which led to their discovery, to stimulate outgrowth of colonies from bone marrow cell cultures. Some of these factors—e.g., stem cell factor (SCF) and multi-CSF (synonymous with IL-3)—regulate early differentiation steps. Others, such as granulocyte/monocyte (GM) CSF, control intermediate steps or selectively induce the final differentiation into mature (neutrophilic) granulocytes (G-CSF) or monocytes (M-CSF). Similarly, erythropoietin, synthesized in the kidney, promotes generation of erythrocytes, and thrombopoietin, which is synthesized in the liver and spleen, promotes formation of platelets (Table 6.5).

Table 6.5 Myeloid differentiation factors, erythropoietin, and thrombopoietin

Cytokine	Molecular mass (kD)	Predominant producer cells	Major functions
Stem cell factor	36	Bone marrow stromal cells	Self-renewal and differentiation of stem cells
Flt3 ligand	25 (m)[a]	Bone marrow stromal cells	Self-renewal and differentiation of stem cells
	16 (s)[b]		
IL-3	14–28	T lymphocytes	Differentiation and propagation of myeloid progenitor cells
Granulocyte-macrophage CSF	14–35	T lymphocytes, monocytes, endothelial cells, fibroblasts	Differentiation and propagation of myeloid progenitor cells
Granulocyte CSF	18–22	Monocytes	Propagation and maturation of granulocytes
Macrophage CSF	35–45	Endothelial cells	Propagation and maturation of monocots
	18–26	Fibroblasts, monocytes	
Erythropoietin	30–32	Peritubular renal capillary cells	Maturation of erythrocytes
Thrombopoietin	31	Liver, kidney	Maturation of platelets, self-renewal of stem cells

[a]Membrane-bound protein
[b]Soluble protein

6.6.1 Pharmacological Implications for CSFs

As gene technology has facilitated the production of sufficient amounts, CSFs were the first cytokines exploited as drugs. Erythropoietin has become established as the drug of choice for the treatment of severe anemias during terminal renal diseases or due to cytostatic therapy, and thrombopoietin has been applied successfully in clinical trials for the treatment of thrombocytopenias [68]. A relatively new class of agents is the group of TPO-receptor mimetic agents. ROMIPLOSTIM is a fusion protein between a human IgG1-Fc domain and a TPO-receptor binding domain ("peptibody"), whereas ELTROMBOPAG is a non-peptidic small molecule targeting the TPO receptor. Both agents act as agonists and are approved for the treatment of severe thrombocytopenias [69]. Filgrastim (human recombinant G-CSF with an additional methionine, generated from bacteria) was the first CSF approved for the treatment of granulocytopenias. Similar to lenograstim (human recombinant G-CSF from eukaryotic cells), it promptly and selectively increases up to 100-fold the number of functionally active neutrophils, for instance, in patients with cytotoxic drug-induced neutropenias [68]. Treatment with G-CSF markedly reduced the incidence and severity of infections leading to hospital admissions in patients who have received chemotherapy because of malignant tumors. In tumor patients, however, the therapy has not led to an increase in life expectancy. All other CSFs have been evaluated in clinical trials, and some are approved in countries outside Europe.

Therapeutically administered CSFs are intended to substitute for the loss of a patient's own differentiation factors. To increase their half-lives in vivo, CSFs can be pegylated, too, sometimes additionally via a carbohydrate linker. Despite of the close relationship between these "biologicals" and the endogenous CSFs, these drugs can cause side effects, too. For the CSFs the most prominent side effect is bone and muscle pain, since the administered CFS dose largely exceeds the endogenous concentration and thus induces massive proliferative bone marrow cell activity. Other side effects include dysuria, sometimes elevation of liver enzymes and uric acid, and rarely, a drop in blood pressure, eosinophilia, or allergic reactions [70].

6.7 Other Drugs Targeting Cytokine Signaling

Suppression of immune and inflammatory reactions can be achieved by inhibition of CYTOKINES in several ways: (1) the inhibition of CYTOKINE synthesis, (2) the decrease of CYTOKINES in free active form, (3) the blocking of the interaction with their receptor, or (4) the inhibition of CYTOKINE-dependent signaling (Fig. 6.4). For each mechanism, at least one clinically relevant example exists.

The by far most predominant target cell for immunosuppressants is the T helper lymphocyte. Cytostatic drugs such as AZATHIOPRINE decrease the number of (T) lymphocytes and thereby also of cytokine-producing cells. Monoclonal antibodies directed against T-cell epitopes in part also work by this mechanism; antibodies against CD3 or the INTERLEUKIN-2 receptor mainly prevent activation or proliferation, respectively. They are effective in preventing transplant rejection and also in the treatment of autoimmune diseases (an example being lupus erythematosus). The immunosuppressants CYCLOSPORINE A and TACROLIMUS very selectively block the synthesis of T-lymphocyte cytokines, predominantly of their growth factor IL-2 [71]. Thus, without being cytotoxic, these drugs interfere with T-cell receptor signaling and are approved for prophylaxis and treatment of transplant rejection. GLUCOCORTICOIDS (e.g., prednisone) are similarly immunosuppressive by interfering with the gene expression of IL-2 and other cytokines synthesized by T lymphocytes [72]. They also represent the most efficient anti-inflammatory drugs available at present. Although they affect multiple pro-inflammatory mechanisms, their efficacy largely relies on the capacity to block gene expression of most pro-inflammatory cytokines including IL-1 to IL-8, TNF, or IFNγ.

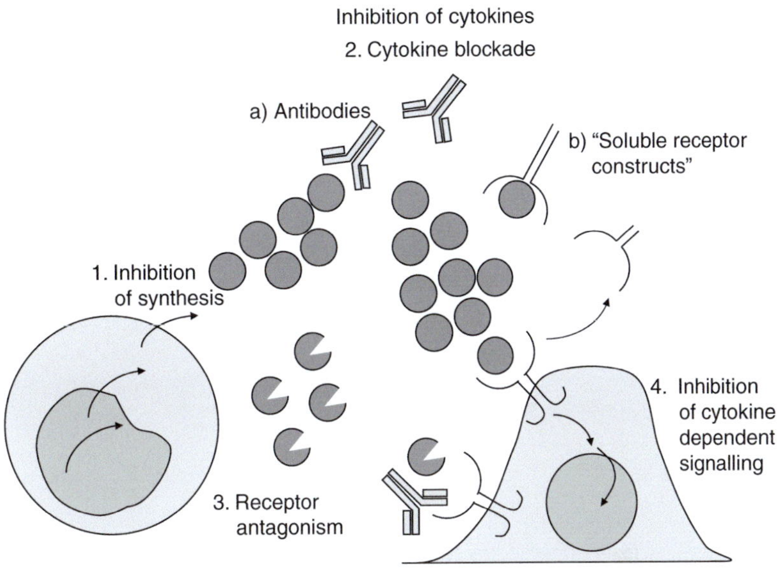

Fig. 6.4 Inhibition of cytokines. : cytokine, : cytokine antagonist, : functionally active cytokine receptors, : extracellular ("soluble") domains of receptors

In the last decades, the signal-transduction pathways engaged by cytokine receptors have been investigated extensively. A core finding was that there exist several phosphorylation cascades in which key protein kinases offer targets for pharmacological interventions. Although quite a number of selective protein kinase inhibitors have been found, only very few of them so far have been approved. The inhibitors SIROLIMUS and EVEROLIMUS, both blocking the serine/threonine protein kinase mTOR (mechanistic target of rapamycin, a synonym for sirolimus), are approved for the prevention of transplant rejection. They prevent mTOR from being activated by IL-2 signaling, normally leading to proliferation of activated T lymphocytes [73].

Regarding the treatment of autoimmune or inflammatory diseases, components of the JAK/STAT signaling pathway seem to be a promising drug target, since over 50 different cytokines exert their effect by activating this common pathway. Janus kinases (JAK) selectively bind to cytokine receptors and, upon activation, transmit the signal by phosphorylating either themselves or other signal proteins, such as the transcription factor STAT (signal transducer and activator of transcription). The JAK1/JAK2 inhibitor RUXOLITINIB is approved for the therapy of myeloproliferative disorders (primary myelofibrosis and polycythemia vera), with a gain-of-function mutation of JAK2 being the most common cause of the disease. RUXOLITINIB is currently under evaluation for the treatment of RA and psoriasis. The US Food and Drug Administration (FDA) has granted approval for the JAK1/JAK3 inhibitor TOFACITINIB for the treatment of RA, but the European Medicines Agency (EMA) refused approval of the drug because of unresolved drug safety concerns. TOFACITINIB also showed beneficial effects in preclinical model of psoriasis and ulcerative colitis and is now being evaluated in clinical trials for its efficacy in these diseases [74].

6.8 Summary and Outlook

It was not long after their discovery and subsequent molecular characterization that cytokines were tested for their therapeutic potential. This was only made possible by gene technology, which allowed sufficient amounts to be produced in good quality. Some of them—interferons or the colony-stimulating factors were subsequently established as drugs with great medical and even economic importance (Tables 6.6 and 6.7; see Appendices 1 and 2). Not all high-flying hopes, however, have been fulfilled, especially with

Table 6.6 Therapeutically relevant cytokines

Cytokine	Target cells	Indication
IFNα	Virus-infected cells, tumor cells	Viral infections, malignant tumors
IFNβ	Virus-infected cells	Viral infections
IFNβ	Virus-infected cells	IFNβ
TNF/LT	Monocytes/macrophages, tumor cells	TNF/LT
TNF/LT	Monocytes/macrophages, tumor cells	TNF/LT
TNF/LT	Monocytes/macrophages, tumor cells	Malignant tumors
IL-2	T lymphocytes, monocytes, NK cells	Malignant tumors, viral infections (AIDS)

Table 6.7 Therapeutically relevant cytokine inhibitors

Mode of action	Example drugs	Indication
Inhibition of cytokine synthesis		
Reduction of the number of cytokine-producing cells		
Cytostatic immunosuppressants	Azathioprine, leflunomide	Organ transplantation, autoimmune diseases, inflammatory diseases
Monoclonal antibodies to cells	Muromonab CD3	Organ transplantation
Regulation of cell activity		
Regulatory cytokines	(Interleukin-4, interleukin-10)*	
Calcineurin inhibitors	Cyclosporine, tacrolimus	Organ transplantation, severe autoimmune diseases, severe inflammatory diseases
Regulation of cytokine gene expression		
Glucocorticoids	Prednisone	Autoimmune diseases, inflammatory diseases
Decrease of the concentration of cytokines in active (free) form		
Monoclonal antibodies against cytokines		
TNF antibodies	Infliximab, adalimumab	Psoriasis, psoriasis/rheumatoid arthritis, inflammatory bowel diseases
IL-1 antibodies	Canakinumab	Systemic juvenile arthritis, gout arthritis, cryopyrin-associated periodic syndromes (CAPS)
IL-12/IL-23 antibodies	Ustekinumab	
Soluble cytokine receptors		
Soluble TNF receptor constructs	Etanercept	Psoriasis, psoriasis/rheumatoid arthritis, inflammatory bowel diseases
Soluble IL-1 receptor constructs	Rilonacept	Cryopyrin-associated periodic syndromes (CAPS)
Receptor blockade		
Monoclonal antibodies against cytokine receptors		
IL-2 receptor antibodies	Basiliximab	Organ transplantation
IL-6 receptor antibodies	Tocilizumab	Rheumatoid arthritis
Cytokine antagonist		
Antagonist at the IL-1 receptor	Anakinra	Rheumatoid arthritis, cryopyrin-associated periodic syndromes (CAPS)
Inhibition of cytokine-dependent signaling		
Protein kinase inhibitors	Sirolimus	Organ transplantation

regard to the treatment of malignant tumors. Thus, after a period of drawbacks, new strategies have begun to evolve, which allow high local concentrations to be selectively generated, the most sophisticated approach involving the use of genetically altered cells.

On the other hand, cytokines are now known to be crucial participants in the pathogenesis of many diseases. The realization that long-known and valuable drugs, such as the glucocorticosteroids, act predominantly by suppressing the synthesis of certain cytokines has prompted a search for mechanisms by which the synthesis or function of individual cytokines can be blocked more selectively. Even though cytokines or their inhibitors such as cytokine-specific antibodies have developed into indispensable drugs in important indications, it is certain that this is only the beginning. This assumption is based on the growing evidence that these molecules contribute to many more diseases than those anticipated originally; important examples are atherosclerosis, congestive heart failure, or neurodegenerative diseases.

Selected Readings

Dembic Z. The cytokines of the immune system: the role of cytokines in disease related to immune response. 1st ed. London: Academic Press; 2015.

Thomson AW, Lotze MT, editors. The cytokine handbook. 4th ed. London: Academic Press; 2003.

Yoshimoto T, Yoshimoto T, editors. Cytokine frontiers: regulation of immune responses in health and disease. 1st ed. Berlin: Springer Verlag; 2013.

Recommended Websites

Registry/results database of clinical trials around the world. http://www.clinicaltrials.gov. Accessed June 2016.

Data base of drugs approved in the European Union. http://www.ema.europa.eu/ema/

Horst Ibelgauft's Cells|Cell communication|Cancer cells|Cytokines online pathfinder encyclopedia (COPE). http://www.cells-talk.com. Accessed June 2016.

References

1. Cohen S. Cell mediated immunity and the inflammatory system. Hum Pathol. 1976;7:249–64.
2. Carswell EA, Old LJ, Kassel RL, Green S, Fiore N, Williamson B. An endotoxin-induced serum factor that causes necrosis of tumors. Proc Natl Acad Sci U S A. 1975;72:3666–70.
3. Kelker HC, Oppenheim JD, Stone-Wolff D, Henriksen-DeStefano D, Aggarwal BB, Stevenson HC, et al. Characterization of human tumor necrosis factor produced by peripheral blood monocytes and its separation from lymphotoxin. Int J Cancer. 1985;36:69–73.
4. Isaacs A, Lindemann J. Virus interference. I. The interferon. Proc R Soc Lond B Biol Sci. 1957;147:258–67.
5. Taub DD, Oppenheim JJ. Review of the chemokine meeting the third international symposium of chemotactic cytokines. Cytokine. 1993;5:175–9.
6. Rollins BJ. Chemokines. Blood. 1997;90:909–28.
7. Metcalf D. The Florey lecture, 1991. The colony-stimulating factors: discovery to clinical use. Philos Trans R Soc Lond Ser B Biol Sci. 1991;333:147–73.
8. Mosmann TR, Sad S. The expanding universe of T-cell subsets: Th1, Th2 and more. Immunol Today. 1996;17:138–46.
9. Klose CS, Artis D. Innate lymphoid cells as regulators of immunity, inflammation and tissue homeostasis. Nat Immunol. 2016;17:765–74.
10. Garlanda C, Dinarello CA, Mantovani A. The interleukin-1 family: back to the future. Immunity. 2013;39:1003–18.
11. Palomo J, Dietrich D, Martin P, Palmer G, Gabay C. The interleukin (IL)-1 cytokine family—balance between agonists and antagonists in inflammatory diseases. Cytokine. 2015;76:25–37.
12. Boraschi D, Lucchesi D, Hainzl S, Leitner M, Maier E, Mangelberger D, et al. IL-37: a new anti-inflammatory cytokine of the IL-1 family. Eur Cytokine Netw. 2011;22:127–47.
13. Gabay C, Towne JE. Regulation and function of interleukin-36 cytokines in homeostasis and pathological conditions. J Leukoc Biol. 2015;97:645–52.
14. Campbell L, Raheem I, Malemud CJ, Askari AD. The relationship between NALP3 and autoinflammatory syndromes. Int J Mol Sci. 2016;17.
15. Furst DE. Anakinra: review of recombinant human interleukin-I receptor antagonist in the treatment of rheumatoid arthritis. Clin Ther. 2004;26:1960–75.
16. Gram H. Preclinical characterization and clinical development of ILARIS(®) (canakinumab) for the treatment of autoinflammatory diseases. Curr Opin Chem Biol. 2016;32:1–9.
17. Bossu P, Del Giudice E, Ciaramella A, Neumann D, Boraschi D. IL-18 and IL-18 receptors in the development of autoimmunity. Eur Cytokine Netw. 2000;11:515–6.
18. Martin NT, Martin MU. Interleukin 33 is a guardian of barriers and a local alarmin. Nat Immunol. 2016;17:122–31.
19. Palma G, Barbieri A, Bimonte S, Palla M, Zappavigna S, Caraglia M, et al. Interleukin 18: friend or foe in cancer. Biochim Biophys Acta. 2013;1836:296–303.
20. Srivastava S, Salim N, Robertson MJ. Interleukin-18: biology and role in the immunotherapy of cancer. Curr Med Chem. 2010;17:3353–7.
21. Liao W, Lin JX, Leonard WJ. IL-2 family cytokines: new insights into the complex roles of IL-2 as a broad regulator of T helper cell differentiation. Curr Opin Immunol. 2011;23:598–604.
22. Waickman AT, Park JY, Park JH. The common γ-chain cytokine receptor: tricks-and-treats for T cells. Cell Mol Life Sci. 2016;73:253–69.
23. Malek TR. The biology of interleukin-2. Annu Rev Immunol. 2008;26:453–79.

24. McDermott DF, Cheng SC, Signoretti S, Margolin KA, Clark JI, Sosman JA, et al. The high-dose aldesleukin "select" trial: a trial to prospectively validate predictive models of response to treatment in patients with metastatic renal cell carcinoma. Clin Cancer Res. 2015;21:561–8.
25. Monczor F, Copsel S, Fernandez N, Davio C, Shayo C. Histamine H2 receptor in blood cells: a suitable target for the treatment of acute myeloid leukemia. Handb Exp Pharmacol. 2017;241:141–60.
26. Romero AI, Thorén FB, Aurelius J, Askarieh G, Brune M, Hellstrand K. Post-consolidation immunotherapy with histamine dihydrochloride and interleukin-2 in AML. Scand J Immunol. 2009;70:194–205.
27. Deeks ED. Mepolizumab: a review in eosinophilic asthma. BioDrugs. 2016;30(4):361–70.
28. Ortega HG, Liu MC, Pavord ID, Brusselle GG, FitzGerald JM, Chetta A, et al. Mepolizumab treatment in patients with severe eosinophilic asthma. N Engl J Med. 2014;371:1198–207.
29. Hanania NA, Noonan M, Corren J, Korenblat P, Zheng Y, Fischer SK, et al. Lebrikizumab in moderate-to-severe asthma: pooled data from two randomised placebo-controlled studies. Thorax. 2015;70:748–56.
30. Corren J, Lemanske RF, Hanania NA, Korenblat PE, Parsey MV, Arron JR, et al. Lebrikizumab treatment in adults with asthma. N Engl J Med. 2011;365:1088–98.
31. Navarro-Millán I, Singh JA, Curtis JR. Systematic review of tocilizumab for rheumatoid arthritis: a new biologic agent targeting the interleukin-6 receptor. Clin Ther. 2012;34:788–802.e3.
32. Fajgenbaum DC, Kurzrock R. Siltuximab: a targeted therapy for idiopathic multicentric Castleman disease. Immunotherapy. 2016;8:17–26.
33. Fischer M, Goldschmitt J, Peschel C, Brakenhoff JP, Kallen KJ, Wollmer A, et al. I. A bioactive designer cytokine for human hematopoietic progenitor cell expansion. Nat Biotechnol. 1997;15:142–5.
34. Negahdaripour M, Nezafat N, Ghasemi Y. A panoramic review and in silico analysis of IL-11 structure and function. Cytokine Growth Factor Rev. 2016;32:41–61.
35. Wen R, Tao W, Li Y, Sieving PA. CNTF and retina. Prog Retin Eye Res. 2012;31:136–51.
36. Commins S, Steinke JW, Borish L. The extended IL-10 superfamily: IL-10, IL-19, IL-20, IL-22, IL-24, IL-26, IL-28, and IL-29. J Allergy Clin Immunol. 2008;121:1108–11.
37. Engelhardt KR, Grimbacher B. IL-10 in humans: lessons from the gut, IL-10/IL-10 receptor deficiencies, and IL-10 polymorphisms. Curr Top Microbiol Immunol. 2014;380:1–18.
38. Shouval DS, Biswas A, Goettel JA, McCann K, Conaway E, Redhu NS, et al. Interleukin-10 receptor signaling in innate immune cells regulates mucosal immune tolerance and anti-inflammatory macrophage function. Immunity. 2014;40:706–19.
39. Galeazzi M, Bazzichi L, Sebastiani GD, Neri D, Garcia E, Ravenni N, et al. A phase IB clinical trial with Dekavil (F8-IL10), an immunoregulatory 'armed antibody' for the treatment of rheumatoid arthritis, used in combination with methotrexate. Isr Med Assoc J. 2014;16:666.
40. Martín R, Chain F, Miquel S, Natividad JM, Sokol H, Verdu EF, et al. Effects in the use of a genetically engineered strain of Lactococcus lactis delivering in situ IL-10 as a therapy to treat low-grade colon inflammation. Hum Vaccin Immunother. 2014;10:1611–21.
41. Vadasz Z, Peri R, Eiza N, Slobodin G, Balbir-Gurman A, Toubi E. The expansion of CD25 high IL-10 high FoxP3 high B regulatory cells is in association with SLE disease activity. J Immunol Res. 2015;2015:254245.
42. Wang RX, Yu CR, Mahdi RM, Egwuagu CE. Novel IL27p28/IL12p40 cytokine suppressed experimental autoimmune uveitis by inhibiting autoreactive Th1/Th17 cells and promoting expansion of regulatory T cells. J Biol Chem. 2012;287:36012–21.
43. Ramnath D, Tunny K, Hohenhaus DM, Pitts CM, Bergot AS, Hogarth PM, et al. TLR3 drives IRF6-dependent IL-23p19 expression and p19/EBI3 heterodimer formation in keratinocytes. Immunol Cell Biol. 2015;93:771–9.
44. Langrish CL, McKenzie BS, Wilson NJ, de Waal Malefyt R, Kastelein RA, Cua DJ. IL-12 and IL-23: master regulators of innate and adaptive immunity. Immunol Rev. 2004;202:96–105.
45. Teng MW, Bowman EP, McElwee JJ, Smyth MJ, Casanova JL, Cooper AM, et al. IL-12 and IL-23 cytokines: from discovery to targeted therapies for immune-mediated inflammatory diseases. Nat Med. 2015;21:719–29.
46. Wada Y, Lu R, Zhou D, Chu J, Przewloka T, Zhang S, et al. Selective abrogation of Th1 response by STA-5326, a potent IL-12/IL-23 inhibitor. Blood. 2007;109:1156–64.
47. Papp K, Thaçi D, Reich K, Riedl E, Langley RG, Krueger JG, et al. Tildrakizumab (MK-3222), an anti-interleukin-23p19 monoclonal antibody, improves psoriasis in a phase IIb randomized placebo-controlled trial. Br J Dermatol. 2015;173:930–9.
48. Tugues S, Burkhard SH, Ohs I, Vrohlings M, Nussbaum K, Vom Berg J, et al. New insights into IL-12-mediated tumor suppression. Cell Death Differ. 2015;22:237–46.
49. Beringer A, Noack M, Miossec P. IL-17 in chronic inflammation: from discovery to targeting. Trends Mol Med. 2016;22:230–41.
50. Patel DD, Lee DM, Kolbinger F, Antoni C. Effect of IL-17A blockade with secukinumab in autoimmune diseases. Ann Rheum Dis. 2013;72(Suppl 2):ii116–23.
51. Garnock-Jones KP. Secukinumab: a review in moderate to severe plaque psoriasis. Am J Clin Dermatol. 2015;16:323–30.
52. Farahnik B, Beroukhim K, Zhu TH, Abrouk M, Nakamura M, Singh R, et al. Ixekizumab for the treatment of psoriasis: a review of phase III trials. Dermatol Ther (Heidelb). 2016;6:25–37.

53. Hueber W, Sands BE, Lewitzky S, Vandemeulebroecke M, Reinisch W, Higgins PD, et al. Secukinumab, a human anti-IL-17A monoclonal antibody, for moderate to severe Crohn's disease: unexpected results of a randomised, double-blind placebo-controlled trial. Gut. 2012;61:1693–700.
54. Ward-Kavanagh LK, Lin WW, Šedý JR, Ware CF. The TNF receptor superfamily in co-stimulating and co-inhibitory responses. Immunity. 2016;44: 1005–19.
55. Magro F, Portela F. Management of inflammatory bowel disease with infliximab and other anti-tumor necrosis factor alpha therapies. BioDrugs. 2010;24(Suppl 1):3–14.
56. Selmi C, Generali E, Massarotti M, Bianchi G, Sciré CA. New treatments for inflammatory rheumatic disease. Immunol Res. 2014;60:277–88.
57. Scott LJ. Etanercept: a review of its use in autoimmune inflammatory diseases. Drugs. 2014;74:1379–410.
58. Downey C. Serious infection during etanercept, infliximab and adalimumab therapy for rheumatoid arthritis: a literature review. Int J Rheum Dis. 2016;19:536–50.
59. Sadler AJ, Williams BR. Interferon-inducible antiviral effectors. Nat Rev Immunol. 2008;8:559–68.
60. Borden EC, Sen GC, Uze G, Silverman RH, Ransohoff RM, Foster GR, et al. Interferons at age 50: past, current and future impact on biomedicine. Nat Rev Drug Discov. 2007;6:975–90.
61. Gautam S, Fatehchand K, Elavazhagan S, Reader BF, Ren L, Mo X, et al. Reprogramming nurse-like cells with Interferon-γ to interrupt chronic lymphocytic leukemia cell survival. J Biol Chem. 2016;291(27):14356–62.
62. Kiladjian JJ, Mesa RA, Hoffman R. The renaissance of interferon therapy for the treatment of myeloid malignancies. Blood. 2011;117:4706–15.
63. Marziniak M, Meuth S. Current perspectives on interferon beta-1b for the treatment of multiple sclerosis. Adv Ther. 2014;31:915–31.
64. Sallusto F, Baggiolini M. Chemokines and leukocyte traffic. Nat Immunol. 2008;9:949–52.
65. Thiele S, Rosenkilde MM. Interaction of chemokines with their receptors—from initial chemokine binding to receptor activating steps. Curr Med Chem. 2014;21:3594–614.
66. Van Der Ryst E. Maraviroc—a CCR5 antagonist for the treatment of HIV-1 infection. Front Immunol. 2015;6:277.
67. Burger JA, Peled A. CXCR4 antagonists: targeting the microenvironment in leukemia and other cancers. Leukemia. 2009;23:43–52.
68. Steensma DP. Hematopoietic growth factors in myelodysplastic syndromes. Semin Oncol. 2011;38:635–47.
69. Kuter DJ. The biology of thrombopoietin and thrombopoietin receptor agonists. Int J Hematol. 2013;98:10–23.
70. Smith TJ, Khatcheressian J, Lyman GH, Ozer H, Armitage JO, Balducci L, et al. 2006 update of recommendations for the use of white blood cell growth factors: an evidence-based clinical practice guideline. J Clin Oncol. 2006;24:3187–205.
71. Choudhary NS, Saigal S, Shukla R, Kotecha H, Saraf N, Soin AS. Current status of immunosuppression in liver transplantation. J Clin Exp Hepatol. 2013;3:150–8.
72. Liberman AC, Druker J, Perone MJ, Arzt E. Glucocorticoids in the regulation of transcription factors that control cytokine synthesis. Cytokine Growth Factor Rev. 2007;18:45–56.
73. Klawitter J, Nashan B, Christians U. Everolimus and sirolimus in transplantation-related but different. Expert Opin Drug Saf. 2015;14:1055–70.
74. Aittomäki S, Pesu M. Therapeutic targeting of the Jak/STAT pathway. Basic Clin Pharmacol Toxicol. 2014;114:18–23.

7 Epigenetic Regulation of Immunity

Shahzad N. Syed, Eduard Resch, and Nathalie Dehne

7.1 Introduction

Immunity is a collective response to a foreign pathogen, insult, or injury by specialized cells of the body. The hallmarks of immune responses are specificity, speed, and self-regulation. During inflammation, immune cells are in a dynamic microenvironment that interacts with and complements the plasticity of immune cell phenotypes [1]. Gene regulation is the underlying key process in activation and differentiation of immune cells. For decades, work concentrated on the discovery of factors binding to specific DNA sequences and signaling networks activating those factors while overlooking the impact of chromatin structure on the regulation of transcription. Today it is accepted that chromatin structure as well as the DNA sequence significantly defines transcription, a phenomenon often referred to as epigenetic regulation. The term "epigenetics" was first used by C. H. Waddington in 1942. Although at that time he had no knowledge of the mechanisms of gene regulation, he used the term to describe the interaction of genes with their environment to create a phenotype. Robin Holliday suggested that DNA METHYLATION is a mechanism to control gene regulation, thus defining the first mechanism of epigenetic regulation. In the 1990s, the term was loosely used until 2002 when Sir Adrian Bird proposed a unifying definition. He defined epigenetics as "the study of mitotically and/or meiotically heritable changes in gene function that cannot be explained by changes in DNA sequence" [2]. In 2008, a consensus definition of the concept of *epigenetic trait* as "stably heritable phenotype resulting from changes in a chromosome without alterations in the DNA sequence" was formulated at a Cold Spring Harbor meeting [3], which is widely used today.

Epigenetic chromatin marks are regulated in response to environmental cues and persist after the original stimulus has resolved, providing a mechanism for extending transient short-lived signals into a more stable and sustained cellular response lasting several hours or days. Epigenetic studies have been mostly descriptive, but during the last decade, a clearer understanding of the underlying mechanisms has emerged mostly due

Final manuscript submitted on November 10, 2016.

S. N. Syed
Faculty of Medicine–Institute of Biochemistry I, Goethe-University Frankfurt, Frankfurt am Main, Germany
e-mail: syed@biochem.uni-frankfurt.de

E. Resch
Branch for Translational Medicine and Pharmacology TMP, Fraunhofer Institute for Molecular Biology and Applied Ecology IME, Frankfurt am Main, Germany
e-mail: eduard.resch@ime.fraunhofer.de

N. Dehne (✉)
Faculty of Health Sciences, Joint Faculty of the Brandenburg University of Technology Cottbus – Senftenberg, Brandenburg Medical School Theodor Fontane and the University of Potsdam, Potsdam, Germany
e-mail: Nathalie.dehne@fgw-brandenburg.de

M. J. Parnham et al. (eds.), *Nijkamp and Parnham's Principles of Immunopharmacology*, https://doi.org/10.1007/978-3-030-10811-3_7

to extensive use of next-generation sequencing techniques (see Box 7.1). Events that are involved in epigenetic gene regulation include:

- Enzyme-catalyzed covalent modifications of DNA and/or histones, as well as nonhistone proteins that do not alter the DNA sequence.
- Selective binding of the transcriptional machinery to recognition elements on modified chromatin.
- ATP hydrolysis-driven changes in DNA-histone topography.
- Modulation of chromatin structure by interactions with other proteins or nucleic acids.

We shall discuss histone and DNA modifications that are related to transcriptional regulation and alterations in chromatin structures, but epigenetic modifications specific for DNA damage responses or DNA replication are beyond the scope of this chapter.

Box 7.1: Next-Generation Sequencing (NGS)
Sequencing methods based on the Sanger sequencing reaction require a significant amount of pure DNA of a single sequence. In contrast, NGS techniques allow analysis of a large number of different DNA molecules in parallel. The development of these methods in the 1990s enabled sequencing of the human genome much faster than originally estimated. Today a number of companies exist that have developed different techniques, the most prominent being the pyrosequencing method developed by 454 Life Sciences, the ion semiconductor from Ion Torrent, sequencing by ligation or SOLID sequencing, and especially the sequencing by the synthesis method invented by Illumina. These techniques and also sample preparation methods differ vastly between the platforms but they all have in common that a sample can be analyzed that contains a large number of DNA strands with different sequences in parallel without isolation and amplification of each strand. This allows, for example, analysis of the complete human genome (mechanically fragmented) in one run. The sequences obtained are usually between 36 and 400 bp and need to be aligned or organized by bioinformatic methods. As prices for sequencing have declined over the last 10 years, NGS-based analyses have increased, and therefore, the amount of data has increased. This, in turn, has increased the requirement for new bioinformatic tools and experts.

7.2 Cell Differentiation as a Perfect Example of Epigenetic Regulation

7.2.1 Macrophage Differentiation

The process of immune cell differentiation from hematopoietic stem cells (HSC; see also Chap. 2) provides a good example of epigenetic regulation. Cells gradually differentiate from pluripotent stem cells to the different lineages of immune cells. This process is controlled by a fine-tuned expression of so-called lineage-defining transcription factors, which ultimately create a cell-type-specific chromatin structure through epigenetic signals [4, 5]. Epigenetic regulation in general follows a simple scheme that is composed of three categories of signals (Fig. 7.1) [3]. First an “Epigenator,” a signal from the environment that activates the process, is needed. This “Epigenator” will subsequently activate the “Initiator,” which could be a DNA binding protein that initiates alterations of the chromatin structure. To sustain this new structure, the signals need the “Maintainer,” which can be a protein or complex that alters DNA METHYLATION or histone modifications (Fig. 7.1). One example is the differentiation of macrophages from HSCs, which is initiated by M-CSF (or CSF1; colony stimulation factor 1). This cytokine is the “Epigenator” in this concept and induces via its receptor (CSF1R) upregulation of the transcription factor PU.1 (or SPI1; Spi-1 proto-oncogene). PU.1, the “Initiator,” binds to the DNA and is associated with the histone mark H3K4me1 (histone 3 lysine 4 monomethylation). The enzyme/

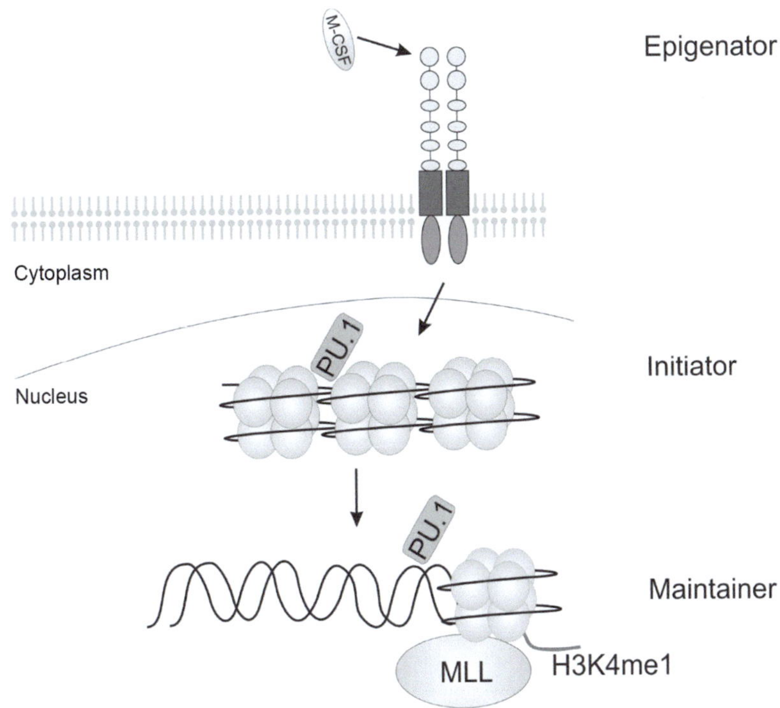

Fig. 7.1 Epigenetic regulation follows a simple Scheme. A signal from the environment, the Epigenator, is translated within the cell to activate a DNA binding protein, the Initiator, to initiate a histone or DNA modification that is maintained by the Maintainer to establish the new state of the chromatin. In the context of differentiation of hematopoietic stem cells to macrophages, M-CSF (the Epigenator) activates its receptor to induce PU.1 (the Initiator) expression. DNA binding of PU.1 is followed by H3K4 methylation (H3K4me1), a signal that is catalyzed by the histone lysine methyltransferases KMT2 (the Maintainer) also called mixed lineage leukemias (MLL)

complex that catalyzes and maintains H3K4me1 at histones in the proximity of PU.1 binding sites has not been identified but most likely is a KMT2C (or MLL3; lysine methyltransferase 2C) or KMT2B (MLL4) histone methyltransferase-containing complex [6, 7]. Nevertheless, it is clear that binding of PU.1 to the DNA is followed by H3K4 methylation. Thus, the unknown MLL complex that maintains H3K4me1 is the "Maintainer." Therefore, by upregulation of PU.1 and its DNA binding, M-CSF initiates epigenetic signaling that results in the formation of a chromatin structure that is typical for myeloid cells. How is this accomplished? Metazoan DNA is organized in NUCLEOSOMEs (for details see Sect. 7.4"). The number and density of NUCLEOSOMEs defines the accessibility of a segment of DNA for transcription factors and the RNA polymerase complexes to initiate transcription of the coded genes. A high number of NUCLEOSOMEs result in a closed, inaccessible conformation and are found in regions with genes that are not transcribed, while a low number of NUCLEOSOMEs allow transcription and gene expression. The distinction between accessible euchromatin and inaccessible heterochromatin is facilitated by histone tail marks, specifically methylation or acetylation of histone 3 lysine 27 (H3K27), lysine 9 (H3K9), and H3K4 (Fig. 7.2).

These marks are recognized by chromatin remodeling complexes that either introduce new NUCLEOSOMEs to the DNA or remove them by eviction or displacement. In this context, H3K4 methylation leads to the removal of NUCLEOSOMEs and, thus, is an important histone mark commonly found in enhancers or promoters of actively transcribed genes [6, 7]. For the example of epigenetic regulation of macrophage differentiation, upregulation and DNA binding of PU.1 and introduction of the H3K4 methylation

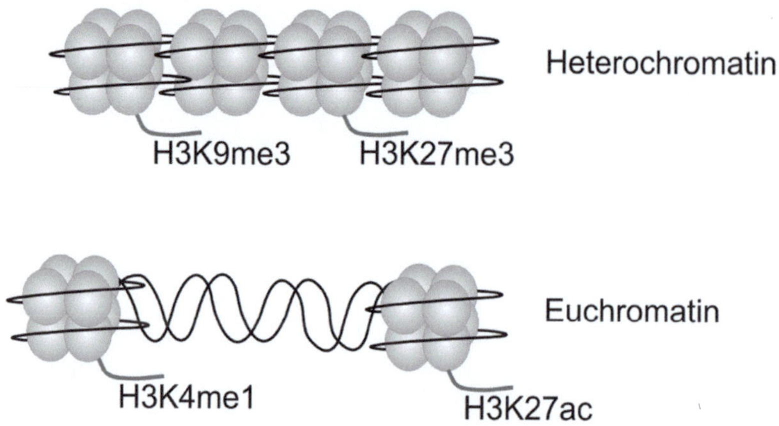

Fig. 7.2 Histone modifications found in distinct areas of the genome are associated with dense heterochromatin or open euchromatin. Heterochromatin is characterized by a high density of nucleosomes, and the tails of histone 3 (H3), lysine 9, and lysine 27 are frequently trimethylated (H3K9me3 and H3K27me3). In contrast, euchromatin contains a low number of nucleosomes, and H3 tails are methylated at lysine 4 (H3K4me1) and often acetylated at lysine 27 (H3K27ac)

provoke the formation of the macrophage-specific enhancer/promoter landscape of open/accessible chromatin. Moreover, once H3K4 is methylated, the modification blocks DNA METHYLATION and thus epigenetic inactivation at this site.

7.2.2 Interaction of Lineage-Defining Transcription Factors Is Required for Immune Cell Differentiation

Although PU.1 is the major lineage-defining transcription factor for macrophages, it is also expressed in other immune cells, including B cells, indicating that expression of PU.1 alone does not define a macrophage. Lineage-defining transcription factors act in concert with each other, and in the case of macrophage differentiation, C/EBPs (CCAAT/enhancer binding proteins), EGR2 (early growth response 2), IRF (interferon regulatory factor), and RUNX (run-related transcription factor) are required and found in the proximity of PU.1 DNA binding sites [6, 7]. The combination of these factors defines an enhancer landscape that is specific for macrophages. Differentiation of hematopoietic cells in general depends on the interaction of different transcription factors to establish lineage fate by creating a specific chromatin structure (see also Chap. 2, Table 7.1 [4, 8, 9]). B-cell fate is

Table 7.1 Epigenetic control of immune cell differentiation

Signal (Epigenator)	Transcription factor (Initiator)	Resulting cell fate
IL-3 EPO TPO	GATA2	Megakaryote/ erythroid lineage
IL-3 IL-7	E2A IKAROS	Lymphoid lineage
IL-4 IL-7	EBF1 FOXO1 PAX5	B cells
IL-7 IL-2	GATA3 Notch1	T cells
SCF IL-2	ID2 IKAROS PU.1	NK cells
GM-CSF M-CSF FLT-3 ligand IL-4	RelB PU.1 IKAROS ID2	Dendritic cells
IL-3/GM-CSF	PU.1 C/EBPs	Myeloid lineage
M-CSF GM-CSF	PU.1 EGR2 C/EBPs	Monocytes/ macrophages
G-CSF GM-CSF	Gfi-1 C/EBPα	Granulocytes

Cytokine signaling coordinates the differentiation of immune cells. From the epigenetic point of view, the cytokine is the “Epigenator” that induces expression of the “Initiator,” a transcription factor. Upon DNA binding of these transcription factors, histone 3 lysine 4 methylation is induced, which leads to the opening of the respective DNA area and, thus, allows transcriptional upregulation of the encoded gene. Thereby, these lineage-defining transcription factors specify the cell-type-specific gene expression pattern established by the expression of EBF1 (early B-cell factor 1), FOXO1 (forkhead box O1), and PAX5 (paired box 5). The expression of ID2 (inhibitor of DNA binding 2) leads to the differentiation of NK cells and innate lymphoid cells and is repressed by the B-cell fate factor EBF1. Similarly, T-cell fate is established by GATA3 (GATA binding protein 3), partially by repressing factors required for the B-cell fate. Thus, by creating a cell-type-specific chromatin landscape, repression of pioneer factors that drive differentiation toward a different lineage is a common trait of these transcription factors. Returning to our scheme (Fig. 7.1), the cytokines that initiate the hematopoietic cell fate are the "Epigenators" in this system, while the lineage-defining transcription factors represent the "Initiators." The epigenetic signal is always H3K4me1, and thus, the recruitment of the histone LYSINE METHYLTRANSFERASE complex is the "Maintainer" of the achieved chromatin state of the differentiating cell.

7.2.3 Epigenetic Regulation Explains Macrophage and T-Cell Subtypes

The mechanisms outlined above only explain the differentiation of a prototype of macrophages but do not explain the observation that each tissue contains a specific (tissue-specific) macrophage type that harbors its own individual enhancer landscape. In this respect, myeloid cells are a special case as their differentiation does not lead to a terminally defined cell type but rather to a cell that still allows for high plasticity. During embryonic development, stem cells from the yolk sac provide the first macrophage-like immune cells that populate the developing tissues in the body [10]. Later, hematopoietic stem cells residing first in the liver and later in the bone marrow provide monocytes that invade tissues and are able to replace most of the first yolk sac-derived macrophages. It was believed that the origin, either from the yolk sac, fetal liver, or bone marrow, defines the function of these tissue-resident macrophages. In the last few years, however, data has emerged showing that specific signals in the tissues are recognized by monocytes/macrophages and induce expression of a new set of transcription factors that, as a consequence, alters the chromatin structure to subsequently form a new enhancer landscape typical for the individual macrophage subtype (Table 7.2). Transplantation and ex vivo stimulation experiments have shown that indeed the appropriate stimulus is able to recreate the enhancer landscape of these macrophages and, thus, recreates the tissue-specific macrophage type [11, 12].

In contrast to tissue-resident macrophages, which still show a high plasticity toward homeostatic or stress stimuli from the environment, T-helper cell activation leads to a terminal differentiation of the cell (see Chap. 3). Nevertheless, as a signal from the environment, in the case of T-cell activation/differentiation, the coactivation of the T-cell receptor and a specific cytokine is required (see Fig. 7.3). This cytokine signal, the "Epigenator," upregulates a specific transcription factor, the "Initiator," that is able to alter chromatin states to allow expression of subset-specific genes like IFNγ in Th1 or IL-4 in Th2 cells by initiating the introduction of active histone marks, while gene loci of the genes specific for other subsets gain repressive histone marks. This illustrates that epigenetic gene regulation is an underlying process of activation and differentiation of diverse immune cell types.

Table 7.2 Epigenetic regulation of tissue-specific macrophages

Epigenator	Initiator	Resulting macrophage type	References
Retinoic acid	GATA6	Peritoneal macrophages	[77]
TGF-β	Smad2/3	Microglia	[78, 79]
Unknown	SPI-C	Splenic red pulp macrophages	[80]
Unknown	PPARγ	Alveolar macrophages	[81]
Unknown	Nur77	Patrolling monocytes	[82]

Tissue-specific signals like retinoic acid or TGF-β initiate the activation and upregulation of transcription factors that, in context with the lineage-defining factors, create a landscape of active enhancers that defines the tissue-specific responses of macrophages (incomplete list; contains only identified signals)

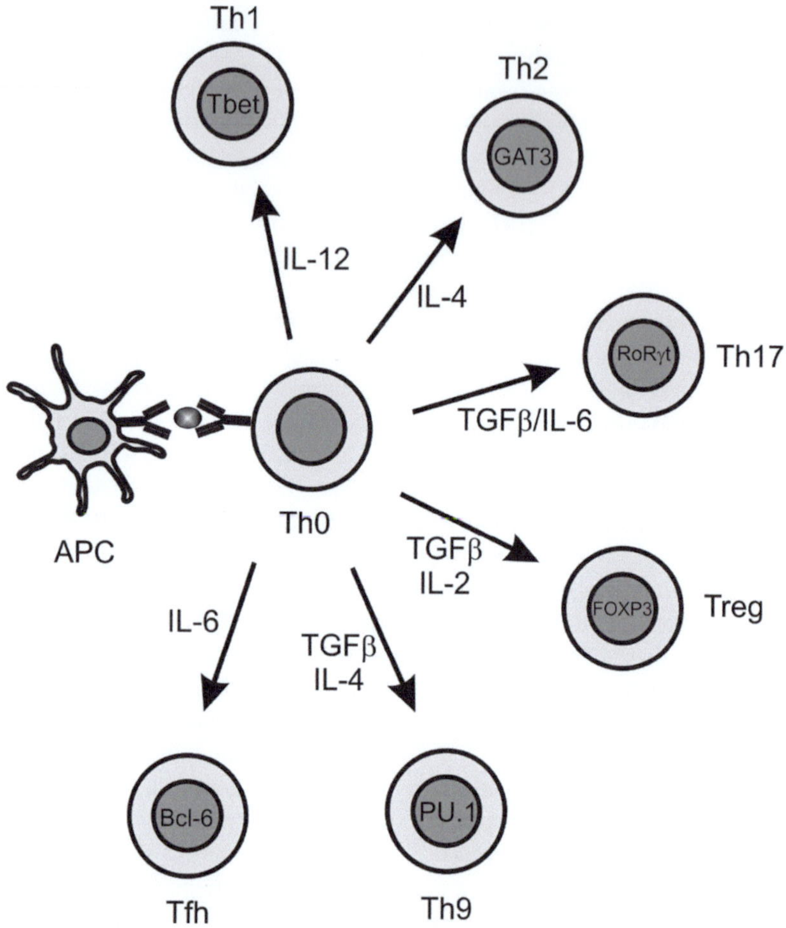

Fig. 7.3 T-helper cell differentiation. T-helper cell differentiation is driven by a specific cytokine produced by the antigen-presenting cell (APC). Stimulation of a naïve T-helper cell (Th0) leads to the upregulation of a transcription factor, which defines the T-helper cell subtype, for example, stimulation with IL-12 leads to the upregulation of T-bet, which defines the Th1 cell

The established enhancer landscape defines the repertoire of binding sites that is used by stimulus-specific transcription factors like NF-κB and thereby defines the cell-type-specific gene signature observed after activation with the same stimulus in diverse cells. This means that transcription factors not only differ in their cooperation with transcriptional coactivator or repressor complexes but also in their ability to bind DNA wrapped around NUCLEOSOMEs. Lineage-defining transcription factors have the ability to recognize DNA sequences that are bound to histones, while stimulus-specific factors require NUCLEOSOME-free DNA interaction [13, 14]. Thus, only regions with histone marks that induce NUCLEOSOME removal will be accessible to those factors that respond to a signal such as NF-κB activated by bacterial infections, and those regions differ between the different cell types. Interestingly, genome-wide analysis of single nucleotide polymorphisms (GWAS) that are associated with diseases frequently describes loci that are located in regulatory regions rather than in the coding sequences, underlining the central role of these regulatory elements in health and disease.

7.3 Role of DNA Methylation During Differentiation

In addition to HISTONE METHYLATION, DNA METHYLATION is an important signal to establish immune cell differentiation. DNA METHYLTRANSFERASES (DNMTs) are enzymes that catalyze DNA METHYLATION in a cytosine at the 5′-carbon position of the

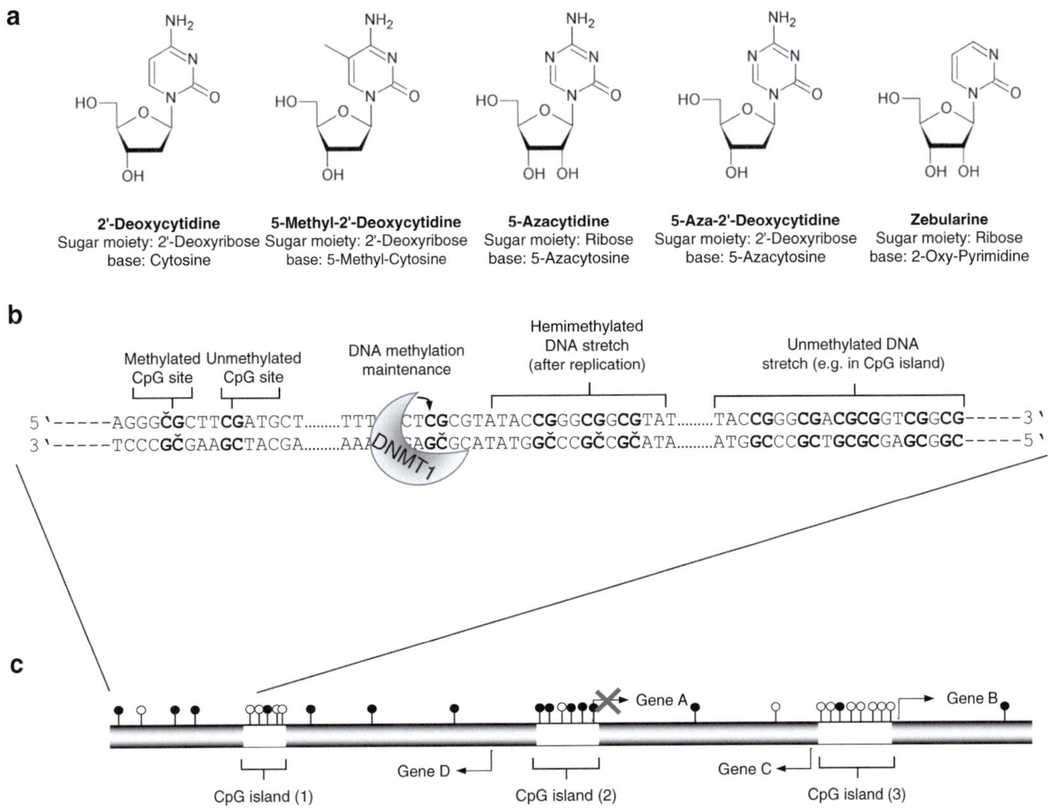

Fig. 7.4 Mechanisms of DNA methylation. (**a**) Chemical structures of nucleosides and DNA methyltransferase inhibitors. 2′-Deoxycytidine is the unmodified nucleoside incorporated into the DNA strand, abbreviated with "C." 5-Methyl-2′-Deoxycytidine is the methylated nucleoside, abbreviated as "Č." 5-azacytidine, 5-Aza-2′-deoxycytidine, and zebularine are examples of nucleoside analog inhibitors. (**b**) Schematic drawing of a snapshot of a DNA double strand during replication. Hemimethylated DNA stretches are substrates of the "maintenance" DNA methyltransferase 1 (DNMT1), copying the methylation mark of the mother strand to the daughter strand, as each CpG site is paired with another CpG site on the complementary strand. Unmethylated CpG sites are predominantly found at so-called CpG islands. (**c**) CpG islands are stretches of DNA (up to 1000 base pairs) exhibiting a high density of CpG sites (illustrated by lollipop symbols). On the genome-wide scale, most CpG sites are methylated (filled lollipop). 60% of gene promoters overlap with a CpG island (here: gene A, B, C). These methylated genes are poised for transcription, whereas the majority is not methylated (open lollipop, CpG island 1 and 3). Promoters whose CpG island is methylated (here: gene A) are usually silenced. Genes not associated with a CpG island (here: gene D) are regulated differently. For more details, please refer to the main text

pyrimidine ring, resulting in 5-methyl-cytosine (Fig. 7.4). In mammals, three active DNMTs have been identified: DNMT1, DNMT3A, and DNMT3B (Table 7.3). Cytosine methylation occurs primarily at so-called CpG sites, which refers to the DNA sequence of a cytidine (C) linked by a phosphate (p) group to a guanosine (G) in the 5′ to 3′ direction (not to be confused with the "CG" pairing in a DNA double strand). Due to the antiparallel, double strand nature of DNA, each CpG site on a single DNA strand is paired with another CpG site on the opposite DNA strand. During DNA replication, the "maintenance" DNMT1 is recruited to the hemimethylated DNA (in which one of the two complementary strands is methylated), transferring the methylation marks of the mother strand to the newly synthesized daughter strand (Fig. 7.4b). By this mechanism, methylation marks are inherited to the daughter cells. In contrast, DNMT3A and DNMT3B are able to methylate CpG sites de novo [15].

Table 7.3 Function of mammalian DNA methyltransferases

Enzyme	Activity	Functions, loss-of-function phenotypes	References
DNMT1 (in mouse: Dnmt1)	Maintenance methylation	Knockout: genome-wide loss of DNA methylation, aberrant expression of imprinted genes, reactivation of silent retrotransposons; lethal at embryonic day 9.5 (E9.5)	[83]
DNMT3A (in mouse: Dnmt3a)	De novo methylation	Knockout: failed establishment of methylation imprinting in germ cells, male mice are sterile; lethal at weeks 4–8, human patients with acute myeloid leukemia (AML) frequently exhibit a somatic mutation in the DNMT3A gene (in 25% of patients)	[84, 85]
DNMT3B (in mouse: Dnmt3b)	De novo methylation	Knockout: loss of methylation of centromeric minor satellite repeats; lethal at embryonic day 14.5 (E14.5) Double knockouts (Dnmt3a and Dnmt3b): no de novo methylation in embryonic stem cells and early embryos; lethal before embryonic day 11.5 (E11.5) In humans: DNMT3B mutations associated with centromeric region instability and facial anomalies syndrome (ICF)	[84, 86]
DNMT3L (in mouse: Dnmt3l)	No activity	Expression is limited to germ cells and early developmental stages; DNMT3L interacts with DNMT3A and DNMT3B facilitating de novo DNA methylation, knockout: viable, male mice exhibits severe defects in spermatogenesis	[87]

Humans contain ~28 million CpG sites throughout their genome, and roughly three-fourths are methylated in cells of somatic tissue [2]. Unmethylated CpG is predominantly found in so-called CpG islands, which are DNA stretches of up to 1000 base pairs with a high frequency of CpG sites, which often overlap with gene promoter regions (Fig. 7.4c). Methylation of specific promoters is observed during cell differentiation and usually results in the silencing of the expression of the corresponding gene. Aberrant promoter methylation, as well as demethylation, accompanied by aberrant gene silencing or expression, respectively, has been shown in malignant cancer cells [16]. About 20% of CpG sites exhibit a dynamic methylation state. These so-called DIFFERENTLY METHYLATED REGIONS (DMRs) are predominantly found remote to promoter regions and overlap with gene regulatory elements, like enhancers [17, 18].

DNA METHYLATION is often associated with a decrease in gene expression. Methylation of transcription factor binding sites in the DNA sequence impedes transcription factor binding which is crucial for gene expression. Alternatively, methylated DNA recruits repressor complexes containing a methyl-CpG binding protein, which also impedes gene expression [19]. DNA METHYLATION is highly interconnected with histone modifications (see below). Knockout in mice of specific LYSINE METHYLTRANSFERASES responsible for the methylation of histone H3 at lysine position 9 (H3K9) revealed that DNA METHYLATION depends on this histone modification, as the knockout mutants exhibited severe defects in DNA METHYLATION [20–22]. Further experiments showed that DNMTs are recruited to DNA via H3K9 methylation. Hence, H3K9 methylation and DNA METHYLATION are hallmarks of heterochromatin formation and gene silencing. In contrast, methylation of H3K4, a hallmark of active promoters and enhancers, blocks DNA METHYLATION. DNMTs (shown for DNMT3A) need unmethylated H3K4 to be able to dock to the DNA and cause activation [23].

During early embryonic development, cells pass through a wave of genome-wide DNA demethylation, designed to reestablish the DNA METHYLATION pattern. However, differences in the pattern evolve during cell differentiation and tissue development. The overall methylation pattern of a differentiated cell is quite stable and resembles that of cells from the same tissue but differs from the pattern of cells of another tissue [24]. This is also the case during the differentiation of immune cells (monocytes,

T cells, B cells, granulocytes, etc.). Hence, it is even possible to discriminate between them by their specific DNA METHYLATION fingerprint [25]. As a cell, for instance, a T cell, differentiates, the process is accompanied by ongoing changes of this fingerprint. More than 100 DMRs related to genes have been identified, which are pivotal for T-cell lineage commitment from conventional CD4+ T cells to CD4+ CD25+ regulatory T cells (Tregs) [26]. One noted example is FOXP3 (forkhead box P3), a master transcriptional regulator of Treg development and function (Fig. 7.3). A DMR in the FOXP3 locus is hypermethylated in conventional T cells and hypomethylated in Tregs [27]. Pharmacological targeting of DNA METHYLATION using DNMT inhibitors (which are described in more detail in the paragraph below) resulted in increased FOXP3 expression in conventional T cells.

Changes in DNA METHYLATION are also involved in the V(D)J recombination (cf. Chap. 2) of the T-cell receptor (TCR) and immunoglobulin (Ig) during T- and B-cell development, respectively. V(D)J recombination depends on the chromatin state of the TCR and the Ig coding region [28], which regulates the accessibility of the V(D)J recombinases (RAG1, RAG2) [29]. Accessible sites for recombination are characterized by HISTONE ACETYLATION and H3K4 methylation, whereas inaccessible sites exhibit markers in the form of heterochromatin formed by H3K9 and DNA METHYLATION [30, 31].

Whereas DNA METHYLATION is conducted by DNMTs, there is no enzyme that performs DNA demethylation. One mechanism to remove methylated DNA is to block the maintenance of DNA METHYLATION during replication (e.g., by downregulation of DNMT1). The other is the repetitive oxidation of the methyl group of cytosine by the TET dioxygenases, resulting in hydroxymethyl, formyl, and finally carboxyl cytosine. The oxidized forms are not recognized as methylation marks anymore and are either not renewed during DNA replication or are actively replaced by unmodified cytosine by a base excision and repair mechanism [32].

7.4 Histone Code

DNA in eukaryotic cells is organized in NUCLEOSOMEs to fit meters of DNA into a nucleus with a diameter measured in microns, but this organization also presents an obstacle to transcription. As such, the positioning of NUCLEOSOMEs along the DNA must be tightly regulated to ensure that lineage-appropriate genes are accessible for transcription and inappropriate genes are silenced. NUCLEOSOMEs are composed of four core histone proteins each wrapped by approximately 147 base pairs of DNA. Histone 2A, 2B, 3, and 4 proteins, the components of the NUCLEOSOME, are alkaline, globular proteins with tails that protrude from the NUCLEOSOME complex and are targets for multiple posttranslational modifications (Fig. 7.5) [33]. An appreciable number of known posttranslational modifications are found on histone tails, the most prominent being methylation, acetylation, and phosphorylation (Fig. 7.5). Phosphorylation of histone tails is a common signal during the process of DNA repair and mitosis. We shall, therefore, concentrate on HISTONE ACETYLATION and METHYLATION and discuss their impact on chromatin structures and on transcription.

Posttranslational modifications can direct the positioning of NUCLEOSOMEs by altering the affinity of the histone for the DNA or may serve as substrates for chromatin remodeling complexes that deposit, evict, and reposition NUCLEOSOMEs [34, 35]. HISTONE ACETYLATION is believed to destabilize the dense chromatin structure by reducing the net positive charge of the histone as well as serving as a substrate for chromatin remodeling complexes, such as ATP-dependent SWI/SNF complexes, which rearrange chromatin to increase DNA accessibility. Therefore, HISTONE ACETYLATION is found in regulatory regions of actively transcribed genes. In contrast, HISTONE METHYLATION is associated with both active and repressed chromatin states, depending on the precise residue modified and the degree of modification. HISTONE

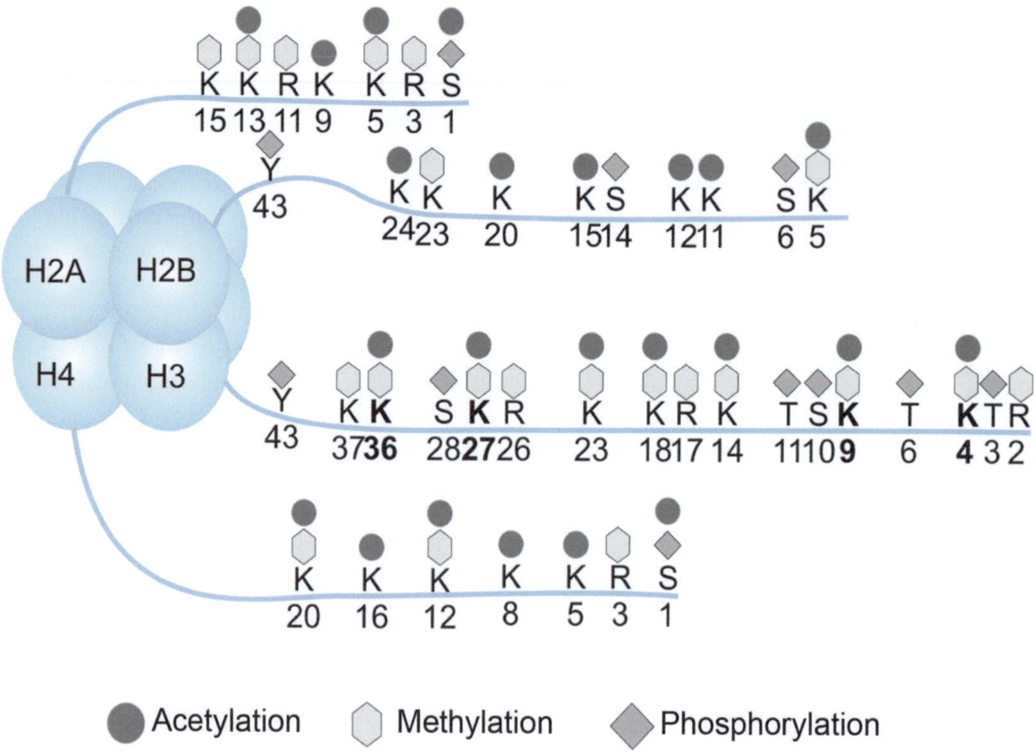

Fig. 7.5 Histone tail residues show a great variety of modifications, the most common ones being acetylation, methylation, and phosphorylation. Here the N-terminal tails of histones 2A, 2B, 3, and 4 are shown that protrude out of the core nucleosome complex. Amino acid residues such as lysine (K), arginine (R), threonine (T), serine (S), or tyrosine (Y) in these tails are posttranslationally modified. The combination of these modifications is read by protein complexes, determines chromatin structure, and regulates transcription and also replication and DNA repair. This is summarized as the histone code

METHYLATION conveys effects on transcription solely through indirect means, acting as a substrate for NUCLEOSOME remodeling complexes such as CHD1 (chromodomain-helicase-DNA binding protein 1) that binds to H3K4me1.

7.4.1 Histone Acetylation in Immune Responses

Integral to the process of immune cell activation and, thus, to alteration of the transcriptional program are epigenetic modifiers, including HISTONE ACETYLTRANSFERASES (HATs) and HISTONE DEACETYLASES (HDACs). These modifiers bring about posttranslational modifications to histone proteins that result in changes to chromatin structure and function [36]. HATs acetylate the lysine residues in core histones leading to a less compact and more transcriptionally active chromatin. HDACs have the opposite effect and repress transcription through tightening the chromatin structure and, thus, reducing the accessibility to binding of transcription factors and other regulatory proteins to the DNA and, thereby, influence gene expression (Fig. 7.6). Reversible modification of the terminal tails of core histones constitutes the major epigenetic mechanism for remodeling higher-order chromatin structure and controlling gene expression. Blocking this action can result in hyperacetylation of histones, thereby affecting gene expression.

Mammalian HDAC enzymes are classified into four major classes in line with their homology to yeast [37] (see Table 7.4). Class I, II, and IV HDACs are zinc-dependent enzymes, while class III HDACs enzymes are dependent on nicotinamide adenine dinucleotide (NAD+) [38]. Since these proteins differ greatly with respect to

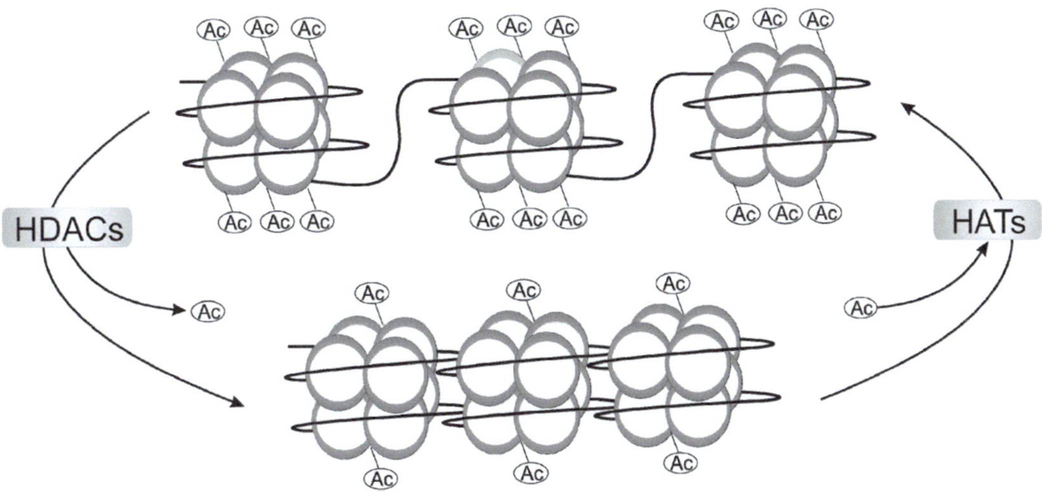

Fig. 7.6 Histone acetylation. Histone deactylases (HDACs) are enzymes involved in removing the acetyl group (Ac) from histones. Histone acetyltransferases (HATs) reversibly transfer the acetyl groups to the core histones by neutralizing the positive charge of the lysine residues in histone N-terminals. This results in an open chromatin structure accessible to transcriptional factors and leads to activation of gene transcription, while histone deacetylation by HDACs usually leads to gene transcriptional repression

Table 7.4 HDACs/HATs

Group	Class/type	Members	Location	Inhibitors
HDACs zinc-dependent	Class I	HDAC1	Nucleus	FK-228, MS-275
	Class I	HDAC2	Nucleus	FK-228, Apician
	Class I	HDAC3	Nucleus, cytoplasm	Apician, RGFP136
	Class I	HDAC8	Nucleus, cytoplasm	
HDACs zinc-dependent	Class IIa	HDAC4	Nucleus, cytoplasm	TSA, phenyl butyrate
	Class IIa	HDAC5	Nucleus, cytoplasm	
	Class IIa	HDAC7	Nucleus, cytoplasm	
	Class IIa	HDAC9	Nucleus, cytoplasm	
HDACs zinc-dependent	Class IIb	HDAC6	Cytoplasm	Tubacin, TSA, SAHA
	Class IIb	HDAC10	Nucleus, cytoplasm	
HDACs NAD^+-dependent	Class III	SirT1	Nucleus, cytoplasm	
	Class III	SirT2	Cytoplasm	
	Class III	SirT3	Mitochondria	
	Class III	SirT4	Mitochondria	
	Class III	SirT5	Mitochondria	
	Class III	SirT6	Nucleus	
	Class III	SirT7	Nucleus	
HDAC zinc-dependent	Class IV	HDAC11	Nucleus	
HATs	Type A	Gcn5, p300/CBP, and $TAF_{II}250$	Nucleus	Anacardic acid, MG149
HATs	Type B	Hat1	Cytoplasm	

HDACs and HAT members, location, and inhibitors. Class III HDACs are referred to as sirtuins and share homologous sequences with the yeast Sir2. Type A HATs contain a bromodomain, which helps them recognize and bind to acetylated lysine residues on histone substrates. Type B HATs lack a bromodomain, as their task is to recognize newly synthesized core histones, which are unacetylated. Only selected examples of HATs are shown in the table; for more details refer Ref. [88]

their subcellular localization and functions, pan-HDAC inhibitors have limited effectiveness and produce several side effects. In contrast, HATs are subdivided into two subclasses based on their subcellular localization. Type A HATs are located in the nucleus and are involved in the regulation of gene expression through acetylation of nucleosomal histones, whereas type B HATs have a cytoplasmic location and are responsible for acetylating newly synthesized histones prior to their assembly into NUCLEOSOMEs. Expression of these proteins in cells of the immune system varies greatly, and understanding the level of expression would assist the development of refined immunomodulatory therapeutic strategies. Furthermore, the multitude of isoforms provides further control over the complex immune response dynamics during effector cell functioning.

Apart from targeting histone acetyl groups, HDACs and HAT isoforms also regulate gene expression through various mechanisms including the targeting of nonhistone substrates such as transcription factors and coregulators, chaperones, signaling and motility mediators, as well as DNA repair proteins [39]. Transcription factors such as NF-κB, FOXP3, T-bet (also called TBX21; T-box 21), or GATA3 are integral to the development of immunity. HDAC enzymes can bind NF-κB subunits p50 and p65 to suppress activity, but upon stimulation HDACs are displaced from the subunits, which results in NF-κB target gene transcription. HDAC enzymes are therefore critical to the modulation of inflammatory responses mediated by IL-1β, IL-6, IL-8, and GM-CSF through regulation of the histone and NF-κB and AP-1 acetylation status. Oxidative stress in the context of cigarette smoke, chronic obstructive pulmonary disease, and air pollutants are all known to inhibit HDAC activity, thereby enhancing transcription of pro-inflammatory genes via NF-κB and exacerbating these activities [40].

Repression of the acetylation of nonhistone proteins such as TFs (e.g., STAT-3, GATA1, FOXP3) is also a very prominent mechanism by which HDACs regulate immune responses. These TFs are involved in differentiation, effector functions, and plasticity of almost all immune cell types, but by acting as substrates for HDACs, they provide an extra layer of regulation of immunity. A noteworthy example is dendritic cells (DC) whose function is inhibited by HDAC acting on histones and the non-histone target STAT-3. In this case, pharmacological inhibition of HDAC leads to acetylation and activation of STAT-3, which is critical for the induction of indoleamine 2,3-dioxygenase (essential for DC-mediated tolerance induction in T cells) and the regulation of DC function [41]. This negative regulation of DC function has an important implication in the induction of effector immunity. Similarly, epigenetic mechanisms have been associated with the transcription and expression of *Foxp3*. Histone H3 and H4 acetylation, just like DNA METHYLATION, is associated with a conserved region within the FOXP3 locus for $CD4^{+}CD25^{+}$ Treg but not with conventional $CD4^{+}CD25^{-}$ T cells suggesting an important role for epigenetic regulation in maintaining FOXP3 expression. Dysregulation of this epigenetic response may alter T-cell homeostasis and lead to exacerbated immunopathologies such as multiple sclerosis, rheumatoid arthritis, etc. [42].

Polarization of immune cells such as macrophages and $CD4^{+}$ T-helper (Th) cells plays a crucial role in both innate and adaptive immunity. In T cells, HDACs control the differentiation of naïve $CD4^{+}$ T cells into Th1 and Th2 by reversing the hyperacetylation of H3 and H4 at the IFN-γ promoter, which is stably inherited by fully differentiated Th2 effector cells. GATA3 helps maintain repression of Th1-associated gene expression by binding to HDAC enzymes, which then interact with the IFN-γ gene to allow stable Th2 differentiation [43]. The Th1 cytokine IFN-γ or Th2 cytokines, IL-4 and IL-13, polarize macrophages into classical (M1) and alternative (M2) phenotypes, respectively, and are characterized by a unique set of marker genes. These marker genes (e.g., iNOS in M1 and FIZZ1 and ARG-1 in M2 macrophages) are significantly induced by the dominance of either Th1 or Th2 cytokines.

HDACs also regulate macrophage polarization, first by regulating the expression of polarizing Th cytokines for CD4+ T cells and secondly by regulating the expression of polarized macrophage markers genes. HDAC11 (class IV HDAC) serves as an epigenetic silencer of IL-10 expression in macrophages. HDAC11 interacts with the distal promoter region of *Il10*, leading to the development of an hypoacetylated and condensed chromatin structure. These structural changes inhibit the binding of STAT-3 on the distal promoter of the *Il10* gene resulting in the silencing of *Il10* expression [44].

7.4.2 Histone Methylation in Immune Responses

As outlined above, HISTONE METHYLATION is an important signal regulating chromatin structure and transcription. HISTONE METHYLATION is found at lysine and arginine residues and is one of the most stable histone marks. Lysine can be mono-, di-, or trimethylated, while arginine residues can undergo mono- and dimethylation in a symmetric or asymmetric manner, referring to methylation of separate or the same nitrogens, respectively (Fig. 7.7). Lysine METHYLTRANSFERASES (KMTs) and protein arginine methyltransferases (PRMTs) catalyze the reaction using *S*-adenosyl methionine (SAM) as a methyl donor (Fig. 7.7). KMTs are highly conserved, and dysregulation is associated with cancer and other diseases [45, 46]. The enzymes are ubiquitously expressed but show varying expression levels in immune cells. KMT4, KMT3c, and KMT1F show low expression, while KMT2A, KMT2B, and KMT3B show high expression in immune cells, especially in T and B cells. The enzyme group consists of 52 proteins that contain the 130 amino acid long catalytic SET domain, with the exception of KMT4 (also known as Dot1L). KTMs show a high degree of enzymatic specificity for the lysine residue substrate and also for the degree of methylation (Table 7.5). KMT2A (also known as MLL1) catalyzes the methylation of H3K4 to H3K4me2; however, when associated with its endogenous interacting proteins, KMT2A can trimethylate H3K4 (for more details compare [45, 46]). This example also underlines the fact that most histone modifying enzymes are arranged as multiple protein complexes that are specifically recruited to either promote or repress transcription of a gene.

HISTONE LYSINE DEMETHYLASES (KDM) are required to remove the methylation on the lysine and arginine residues. There are 33 members catalyzing removal of one or more specific modifications, which are all expressed in immune cells. Expression of KDM4B and KDM4B is low in human and mouse immune cells, while KDM3B is extremely highly expressed. Nevertheless, these enzymes also are specific for their lysine residue and methylation degree of choice (see Table 7.5) [45, 47]. The first KDM was discovered in 2003, and although many enzymes have been identified through homology search of the catalytic domains amine oxidase-like/AOL (only LSD1 and LSD2) or Jumonji C (JmjC), no enzyme has currently been shown to demethylate H3K79me1-H3K79me3 and H4K20me3. KDMs are oxygen-dependent dioxygenases that use either FAD (AOL domain) or 2-oxoglutarate and iron (JmjC domain) as cofactors (Fig. 7.7). In both cases, the oxidative removal of the methyl group produces formaldehyde as a by-product.

HISTONE LYSINE METHYLATION marks are the major determinants for the definition of active and inactive regions of the genome; however, the role of arginine methylation is not well understood. The best described histone lysine methylations are H3K4 methylation, H3K79me3, HeK36me3, linked to active transcription, and H3K9 and H3K27 trimethylation associated with gene repression. These marks, as well as the KMTs and KDMs that catalyze them, are not equally distributed throughout the genome [45]. Genome-wide studies and bioinformatic analysis have identified up to 51 different chromatin states in the human genome coupled to a specific pattern of histone modifications.

The combination of specific marks allows the identification of several states of enhancer and promoter activity which correlate well with the

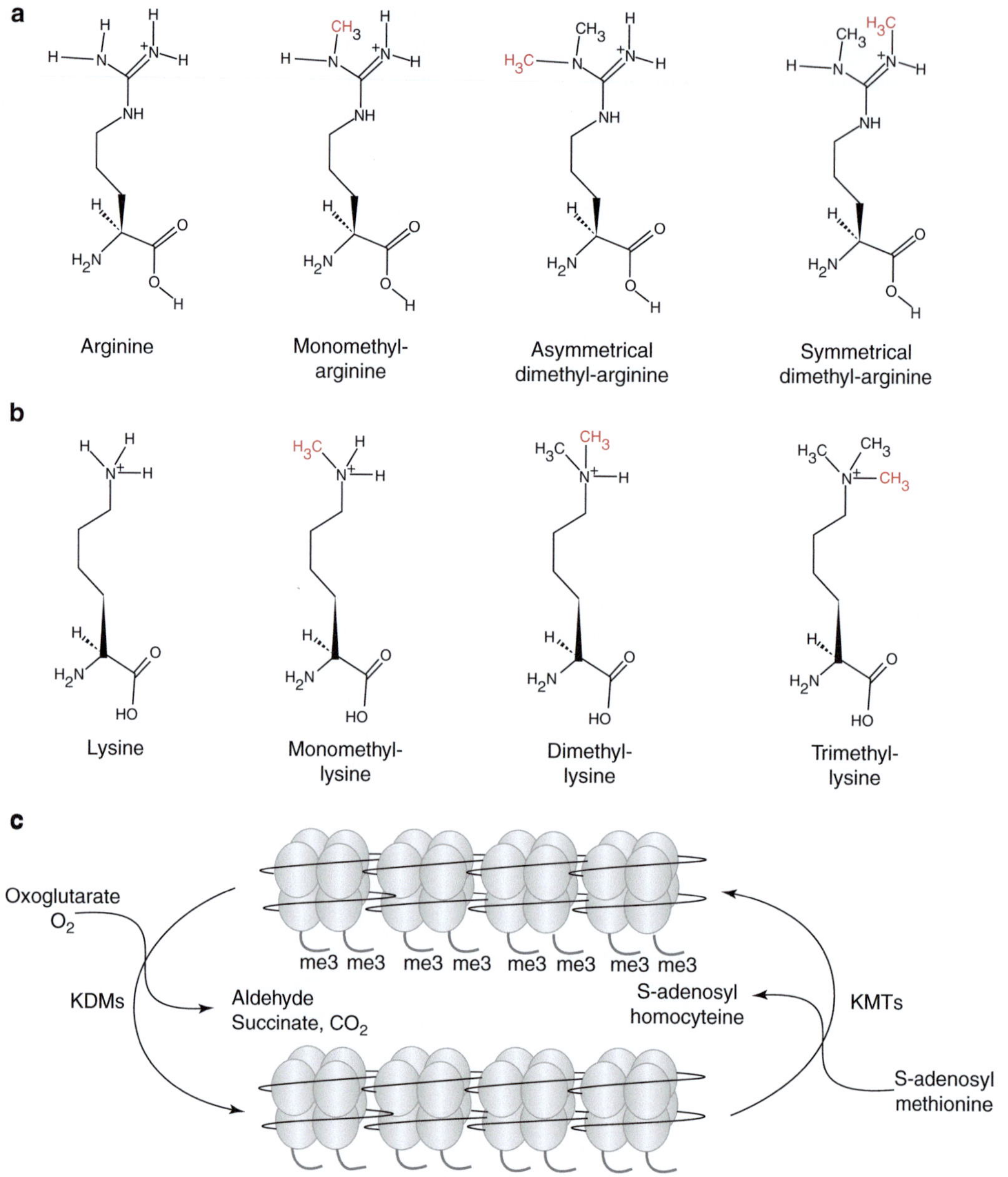

Fig. 7.7 Histone tail methylation. (**a**) Arginine residues are either mono- or dimethylated. When the second methyl group is added to the same amino group as the first, this is referred to as asymmetric, and when the other amino group is methylated, this is called symmetric. (**b**) Lysine residues can be mono-, di-, or trimethylated. (**c**) Lysine methyltransferases (KMTs) catalyze the transfer of a methyl group from *S*-adenosyl methionine (SAM) to the lysine group by producing *S*-adenosyl homocysteine. Histone demethylases (KDMs) are able to remove most methyl groups in a reaction that requires oxoglutarate and oxygen and produces aldehyde, succinate, and CO_2

expression of the corresponding gene. Figure 7.8 shows the marks present at the most extreme states, an active gene locus and an inactive/repressed locus. Looking at the promoter site of active genes, the H3K4me3 mark is enriched and indicates that these genes are either expressed in a cell or are able to be induced upon activation. In the regions flanking H3K4me3, NUCLEOSOMEs frequently have H3K4me1/2, while H3K4me1 is found also in enhancer regions of the genes that are expressed or inducible [6, 7, 45]. Upon binding of transcription factors to enhancer or

Table 7.5 Histone methyltransferase and demethylases

Histone methyl transferase (KMT)	Histone lysine residue	Histone demethylase (KDM)
KMT2A (MLL1) KMT2B (MLL2) KMT2C (MLL3) KMT2D (MLL4) KMT2e (MLL5) KMT2F (hSET1A) KMT2G (hSET1B) KMT2H (ASH1) KMT7 (SET7–9)	H3K4	KDM1 (LSD1) KDM1B (LSD2) KDM5A (JARID1A) KDM5B (JARID1B) KDM5C (JARID1C) KDM5D (JARID1D) KDM2B (FBXL10)
KMT1A (SUV39H1) KMT1B (SUV39H2) KMT1C (G9a) KMT1D (GLP) KMT1E (SETDB1) KMT1F (CLL8) KMT8D KMT8E KMT8F (RIZ1)	H3K9	KDM1 (LSD1) KDM3A (JMJ1A) (JMJD1C) KDM3B (JMJ2B) KDM3C (JMJ2A) KDM4A (JMJD2A) KDM4B (JMJD2B) KDM4C (JMJ2C) KDM4D (JMJD2D)
KMT6A (EZH2) KTM6B (EZH1) KMT3F (WHSC1L1) KMT3G (WHSC1)	H2K27	KDM6A (UTX) KDM6B (JMJD3) KDM6AL (UTY)
KTM3A (SET2) KTM3B (NSD1) KTM3C (SMYD2)	H3K36	KDM2A (FBXL11) KDM2B (FBXL10) KDM4A (JMJD2A) KDM4B (JMJD2B) KDM4C (JMJD2C)
KTM4 (Dot1L)	H3K79	Unknown
KTM5A (PR-SET7–8 KTM5B (SUV4-20H1) KTM5C (SUV4-20H2)	H4K20	Unknown

Histone lysine methyltransferases (KMTs) specifically methylate lysine residues of histone tails. The histone lysine demethylases (KDMs) are able to remove these modifications. They also show a high specificity for the lysine residues, but several enzymes recognize more than one lysine group

promoter regions and subsequent recruitment of the common cofactors p300 (E1A binding protein p300) or CBP (also called CREBBP; CREB binding protein), H3K27 is acetylated. Therefore, H3K27ac marks active enhancer/promoters.

H3K36me3 and H3K79me3 are enriched in gene bodies of actively transcribed genes, H3K79me3 being most abundant at the transcriptional start site, while H3K36me3 is more prominent toward the transcriptional stop site (Fig. 7.8) [45]. H3K36 methylation is catalyzed by SETD2 (SET domain-containing 2), a yeast homologue, which interacts with hyperphosphorylated RNA polymerase II to establish H3K36me3 along the gene bodies during transcriptional elongation. The presence of H3K27me3 within a gene body correlates with transcriptional repression and needs to be removed to allow active transcription. KDM6B can relieve this inhibition by demethylation of H3K27 and is, therefore, commonly found at the transcriptional start site and throughout the gene body of active genes.

The repressive marks, H3K27me3 and H3K9me, are frequently found in inactive genes and are associated with a dense chromatin structure called heterochromatin (Fig. 7.8). H3K27 trimethylation is catalyzed by KMTs EZH1 or EZH2 (enhancer of zeste 1 polycomb repressive complex 2 subunit), which require additional proteins such as the polycomb complex (PRC2) to exert catalytic activity, thereby, restricting their activity to defined areas of the genome. In the mammalian genome, PCR2 complexes are recruited by long noncoding RNAs to specifically silence genes. Such examples include X chromosome inactivation or silencing of the human HOXC locus during embryonic development [46]. Likewise, T-cell differentiation is accompanied by methylation of H3K27 of histones in loci of genes that are repressed in the specific T-cell subtype. The IFNγ locus, for example, is methylated at H3K27 during Th2 cell differentiation [48] H3K27me3 is thought to be one of the earliest histone modifications that appeared during eukaryotic evolution and is important to specifically restrict gene expression during development and differentiation.

As explained above, H3K9me3 is linked to DNA METHYLATION and gene silencing. In contrast to H3K4 and H3K27, KMTs catalyzing H3K9 methylation form more transient interactions and, thus, are found in a number of different complexes with different roles in regulating chromatin structure and transcription [46]. H3K9me3 is highly enriched in heterochromatin regions such as pericentric chromatin and other regions containing repetitive DNA elements and is

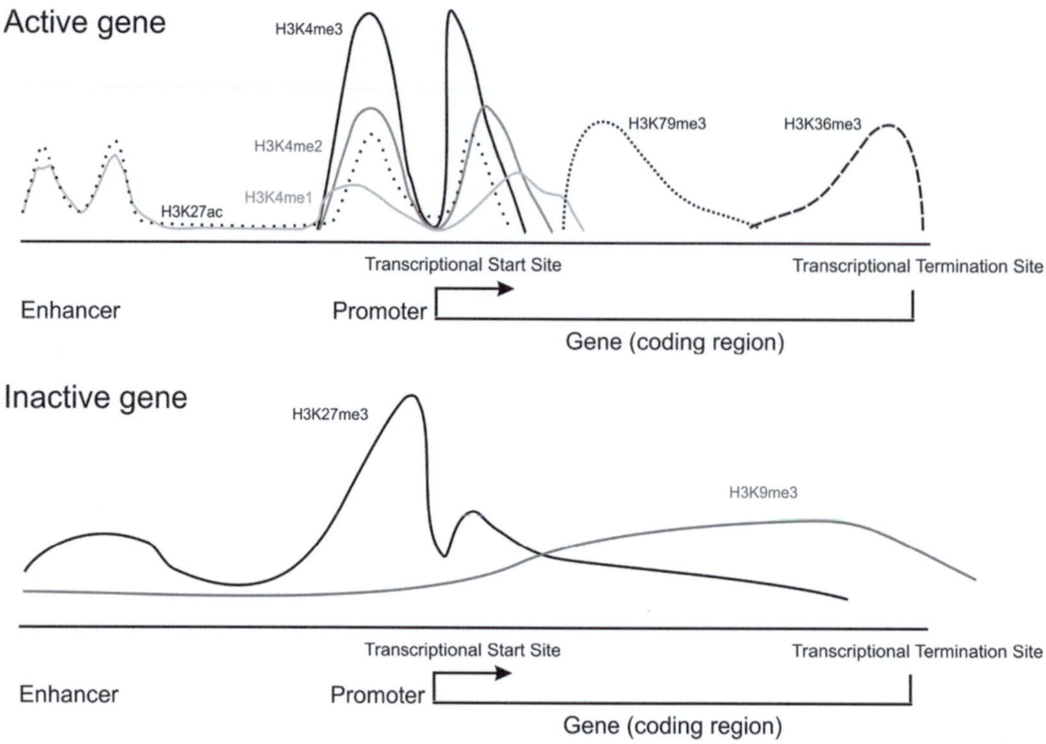

Fig. 7.8 Distribution of histone tail modifications throughout a gene. The histone marks, histone 3 lysine 4 (H3K4) mono-, di-, or trimethylation (me1, me2, or me3), are found in the promoter or enhancer of an actively transcribed gene. These areas also contain the H3K27 acetylation (ac) mark. H3K79me3 and H3K36me3 are found in the so-called gene body, which refers to the coding regions of a gene. In inactive genes, H3K27me3 and H3K9me3 are enriched

important to maintain chromatin stability. Although it is less frequent in coding regions, it is a key player in repressing expression of lineage-inappropriate genes and is implicated in the regulation of cytokine expression in immune cells. In T cells exposed to TGF-β, for example, IL-12 expression is repressed by increased H3K9 trimethylation [49].

7.5 Epigenetics in Immune Responses and Pharmacological Implications

7.5.1 DNA Methylation

The recent awareness of the role of chromatin remodeling for transcriptional control has paved the way for the development of novel compounds that specifically interfere with epigenetic reactions and thus have therapeutic potential. This has been explored primarily for cancer therapy, but as several of the enzymes targeted by these new inhibitors are involved in immune reactions as well, they provide promising drugs for the treatment of inflammatory-driven diseases such as bacterial or viral infection, autoimmune diseases, or chronic inflammatory disorders including psoriasis and atherosclerosis.

As previously discussed, immune cell differentiation, especially of T and B cells, is connected to profound changes in the DNA METHYLATION landscape associated with the regulation of genes, which are crucial for the development and lineage commitment. Moreover, this process can be manipulated by external factors. Autoimmune diseases, for example, systemic lupus erythematosus [50], rheumatoid arthritis [51], and multiple sclerosis [52], are characterized by a genetic predisposition. However, because of the high discordance in the onset of these complex diseases, the influence of environmental factors has been suggested to play a crucial role, too, manifesting itself in the epigenetic landscape, for example, by DNA METHYLATION marks. Therefore, current studies seek to identify biomarkers in the DNA METHYLATION pattern related to these diseases. For example, the analysis of cell-free, circulating DNA in the patient's blood plasma for the methylation of the septin (SEPT)9 gene is an FDA-approved in vitro diagnostic screening test for colorectal cancer. SEPT9 promoter is not methylated in normal colon mucosa, but methylated in malignant cells, which release DNA into the blood flow [53]. For systemic lupus erythematosus, the methylation level of the interferon-induced protein (IFI44L) promoter in peripheral blood cells is suggested to be a biomarker [54]. In rheumatoid arthritis, 207 DMRs were identified in fibroblast-like synoviocytes in contrast to fibroblast-like synoviocytes from osteoarthritis patients [55]. However, further research is needed to identify, validate, and develop DNA METHYLATION marks for the use as prognostic biomarkers, as several requirements need to be met (diagnostic sensitivity and specificity, positive and negative predictive values) [56].

7.5.1.1 Methylation Inhibitors

The fact that DNA METHYLATION is a reversible modification with an impact on gene expression, influenced by external factors and disturbed in several diseases, makes it an appealing target for pharmacological interventions. To date, the only FDA- and EMA-approved DNMT Inhibitors (5-azacytidine and 5-aza-2′-deoxycytidine) are used to treat specific forms of leukemia (myelodysplastic syndrome, acute myeloid leukemia, chronic myelomonocytic leukemia, Fig. 7.4a) [56]. Additional DNMT inhibitors (e.g., 5-fluoro-2′-deoxycytidine, 5,6,-dihydro-5-azacytidine, zebularine, SGI-110), also belonging to the group of nucleoside analogs, were developed over the last few years to improve the pharmacological profile, chemical stability, and bioavailability. Several of these reached clinical trials for the treatment of hematological malignancies as well as cervical, ovarian, and non-small cell lung cancer and hepatocellular carcinoma [57].

The mechanism of action of these inhibitors is not completely understood. For example, 5-azacytidine, 5-aza-2′-deoxycytidine, and zebularine are incorporated into the DNA during replication, replacing the endogenous cytidine nucleotide (Fig. 7.4). The analogs are not methylated and trap the DNMTs by covalent binding, thereby triggering the proteasomal degradation of the DNMTs and reducing DNA METHYLATION levels. It is reported that DNMT inhibitors induce demethylation of tumor suppressor and cell differentiating genes to reestablish their expression. Other studies have shown that treatment with DNMT inhibitors augments the expression of cancer antigens in the cancer cells, resulting in a more efficient immunogenic reaction. Moreover, 5-azacytidine is primarily incorporated into RNA instead of DNA, leading to the inhibition of ribonucleotide reductase, thereby, impeding DNA replication and DNA repair of the cells.

In patients with acute myeloid leukemia, treatment with 5-azacytidine increased Treg-cell populations after allogeneic stem cell transplantation [58]. Treg cells are important in the development and perpetuation of tolerance after transplantation as they reduce the risk of graft-versus-host disease (GVHD) without impairing the graft-versus-leukemia (GVL) effect [59]. The immunomodulatory effects of DNMT inhibitors have also been tested in autoimmune diseases including multiple sclerosis. In the murine experimental autoimmune encephalomyelitis (EAE) model, low-dose pretreatment with 5-aza-2′-deoxycytidine increased Treg cell number, inhibited effector T-cell activity, and prevented the development of EAE [60].

7.5.1.2 Recent Limitations and Prospects

As 5-azacytidine and 5-aza-2′-deoxycytidine are approved drugs used in patients, a large amount of data has accumulated through clinical studies on their efficacy, pharmacological benefit, and also detrimental side effects. Hematological toxicities, nausea, and constipation are side effects often observed in treated patients [61, 62]. The immunomodulating capacities of DNMT inhibitors, being exploited in allogeneic stem cell transplantation, also exhibit adverse effects, as several DNMT inhibitors induce a lupus-like autoimmunity, rendering CD4+ T-cells autoreactive [63]. Non-nucleoside-based inhibitors (e.g., RG108) which do not have to be incorporated into the DNA have also been developed [64]. Moreover, with several approved drugs (procainamide, an antiarrhythmic agent, and hydralazine, a vasodilator), additional targeting of DNMTs was observed, but the underlying mechanisms are not clear.

To reduce side effects, drugs with a higher specificity for targeting of DNA METHYLATION or demethylation are needed. For this to happen, a deeper understanding of the regulation of DNA METHYLATION and its interconnection with histone modifications in different cellular settings is required.

7.5.2 Histone Acetylation

7.5.2.1 T-Cells

The molecular and clinical implications of the HDAC inhibitors (HDACi) were initially identified in cancer therapy. HDACi are a potential source of novel immunomodulatory drugs aimed at treating a wide range of diseases. In particular, prime targets are immune disorders mediated by a T-cell anergic state, which is associated with repression of the IL-2 gene promoter region [65]. This occurs as a result of epigenetic imprinting that can be inherited over multiple cell division cycles. $CD4^+$ T-cell activation requires antigenic and costimulatory signals and depends on HISTONE ACETYLATION and chromatin remodeling at the IL-2 promoter. However, anergy results from the lack of costimulation, as well as histone hypoacetylation of both IL-2 and IFN-γ promoters at which HDAC activity is thought to maintain the anergic phenotype. Inhibition of HDAC activity, in this context, restores HISTONE ACETYLATION in conjunction with reduced IKAROS expression (which interacts with corepressor complexes) and relieves this anergic state. Nevertheless, most of the HDAC inhibitors are nonselective and target not only nuclear histones but also cytoplasmic nonhistone proteins. Treatment of mice with trichostatin A (TSA; class I/II HDACi) augments natural $Foxp3^+$ Treg cells as well as Treg gene expression and the suppressive function of these cells. TSA treatment of these Treg cells also increased mRNA levels of Foxp3, CTLA4, GITR, and IL-10, which are important mediators of the response. The TSA activity was associated with acetylation of Foxp3 as well as histones contained within Treg-specific genes, providing further evidence for the fact that nonhistone proteins are targets for HAT/HDAC enzymes [66]. The most commonly used HDAC inhibitors target multiple HDACs, which makes it difficult to determine whether the biological consequences of HDAC inhibition (including clinical toxicities) are due to inhibition of a specific HDAC, the combined effect of inhibiting multiple HDACs, and/or effects on one or more multiprotein complexes that incorporate specific HDACs as key enzymatic components.

7.5.2.2 Macrophages

Activation of Toll-like receptor (TLR) signaling in macrophages induces HISTONE LYSINE ACETYLATION in genes of inflammatory cytokines. Acetylated lysine residues in the histone tails are recognized by the bromodomain and extraterminal (BET) proteins that are composed of Brd2, Brd3, Brd4, and Brtd. Among them, Brd4 is well known to contribute to the transcription of inflammatory cytokine genes by forming a complex with positive transcription elongation factor-b and RNA polymerase II at the transcription start site. A synthetic compound, I-BET, suppresses the expression of inflammatory cytokine genes in lipopolysaccharide (LPS)-stimulated macrophages by interfering with binding of BET proteins to acetylated histones. Administration of I-BET protected mice from endotoxin shock induced by LPS treatment or heat-killed *Salmonella typhimurium* [67]. This finding demonstrated the therapeutic potential of drugs that interfere with the binding of BET to acetylated histones in the prevention of inflammatory disorders. Furthermore, I-BET and related compounds JQ1 and I-BET151 strongly suppress Myc expression and also show great promise for the treatment of Myc-driven cancers such as acute myeloid leukemia (AML) and multiple myeloma [68].

Another area in which HDACs are established targets is cancer immunotherapy. The discovery that HDAC function is dysregulated in cancer and has a pivotal role in tumorigenesis revealed the molecular trigger that directs pathological manifestation. Altered HDAC function and recruitment could result in hypoacetylation and repression of genes required for normal growth and development and disruption of the HDAC/HAT balance, favoring abnormal acetylation and inappropriate protein expression. HDACi are used in modulating antitumor immunity by enhancing the tumor antigenicity, thereby preventing tumor escape. HDACi such as sodium butyrate, TSA, and trapoxin A among others upregulate MHC class I and class II proteins; CD40, CD80, and CD86 antigens necessary for costimulation; as well as adhesion molecules such as intercellular adhesion molecule-1 (ICAM-1) on acute myeloid leukemic (AML) cells, human neuroblastoma tumor cells, and mouse plasmacytoma cells in vitro [69]. Diverse mechanisms are involved in the antitumor efficacy of these compounds by modulating activating receptors of antitumor cells such as NK cells, $CD8^+$ T-cells, etc.

7.5.3 Histone Methylation

Currently, no compounds targeting HISTONE METHYLATION are in clinical use, although inhibitors of H3K9 methyltransferase Dot1L (EPZ004777) are in phase 1 trials as treatment for AML and ALL [70]. An inhibitor (ORY-1001) targeting the H3K9 demethylase LSD1 is also in phase 1 trial for AML and SCLC treatment. H3K9 methyltransferases are important in mediating the silencing of specific genes during endotoxic shock. Additionally, LSD2 directly demethylates NF-κB and is, therefore, involved in a regulatory circuit that controls the expression of pro-inflammatory genes in DCs, suggesting the use of the new inhibitors in sepsis or severe infections as well.

Another KMT inhibitor currently in clinical trial is the EZH2 inhibitor GSK126, for the treatment of various tumors including B-cell lymphomas. EZH2 catalyzes H3K27 methylation and is implicated in the differentiation of natural killer cells, indicating a possible application in viral infections [70]. The polycomb repressor complex (PCR), which includes the subunit EZH2, is involved in HIV latency. Knockdown of PCR components leads to a reactivation in latency-infected cells. An inhibitor targeting the antagonistic KDM4 enzymes was used in a mouse model of herpes simplex virus reactions and reduced virus reactivation significantly [71].

Recently, GSK-J1 a JMJD3 subfamily inhibitor was discovered that binds competitively to the 2-oxoglutarate cofactor for the KDMs [72]. LPS responses in macrophages require JMJD3-dependent demethylation to induce a majority of LPS-inducible genes, including TNF-α [73]. Furthermore, JMJD3 is involved in the upregulation of IL-4-dependent genes in alternative macrophage activation [74, 75]. JMJD3 is also critical for T-cell differentiation, and ablation increases Th2 and Th17 differentiation [76]. Although none of these inhibitors are currently in clinical use for inflammatory diseases, the development of small compounds targeting epigenetic modulators has just started and continued focus on these mechanisms will lead to the development of new classes of potent agents for the treatment of these diseases.

7.6 Summary

Epigenetic regulation of gene expression describes mechanisms, which induce modifications, mostly to histones or DNA, that regulate chromatin structures and, thereby, transcription, without altering the DNA sequence. This additional layer of gene expression regulation is controlled by signals from the environment and often lasts longer than the signal itself. This allows cells to adapt to changes in their environment, which is fundamental for immune cells as they are the cells that detect insults or infections and coordinate removal of pathogens and promote regeneration. Two main mechanisms control epigenetic reactions within the cell.

First, posttranslational modifications to the protruding tails of histone proteins control the density of NUCLEOSOMEs in a DNA region. Signals like HISTONE TAIL ACETYLATION and METHYLATION of H3K4 or H3K79 and H3K36 result in the removal of NUCLEOSOMEs and in an open euchromatin conformation of the DNA section. This allows DNA binding of transcription factors, such as NF-κB and RNA polymerase complexes, to upregulate expression of the gene encoded within this region. In contrast, methylation of H3K27 or H3K9 leads to an incorporation of NUCLEOSOMEs and to a dense, heterochromatin conformation, which prevents DNA binding and, thus, expression of the gene.

The second mechanism acts in concert with HISTONE METHYLATION on H3K27 or H3K9. METHYLATION of cytosine bases in the DNA supports the heterochromatin structure of the DNA and also inhibits binding of transcription factors and RNA polymerase complexes. Both mechanisms in concert create a chromatin structure that allows expression of certain genes and prevents expression of others, coordinated by the expression of lineage-defining transcription factors. Lineage-defining transcription factors, such as PU.1 for macrophages or EBF1 for B cells, act in concert with other factors to create this cell-type-specific landscape of open and closed chromatin to allow the cell-type-specific reaction to occur, as reflected by gene expression observed after stimulation. Furthermore, HISTONE ACETYLATION and METHYLATION are also involved in the finely tuned activation of transcription. Thus, epigenetic mechanisms control all steps of transcriptional regulation of cell activation and differentiation making them promising targets for the development of new drugs to control inflammatory reactions in several diseases.

Selected Readings

Amit I, Winter DR, Jung S. The role of the local environment and epigenetics in shaping macrophage identity and their effect on tissue homeostasis. Nat Immunol. 2016;17(1):18–25. https://doi.org/10.1038/ni.3325.

Arrowsmith CH, Bountra C, Fish PV, Lee K, Schapira M. Epigenetic protein families: a new frontier for drug discovery. Nat Rev. Drug Discov. 2012;11(5):384–400. https://doi.org/10.1038/nrd3674.

Black JC, Van Rechem C, Whetstine JR. Histone lysine methylation dynamics: establishment, regulation, and biological impact. Mol Cell. 2012;48(4):491–507. https://doi.org/10.1016/j.molcel.2012.11.006.

Bolden JE, Peart MJ, Johnstone RW. Anticancer activities of histone deacetylase inhibitors. Nat Rev Drug Discov. 2006;5(9):769–84. https://doi.org/10.1038/nrd2133.

Du J, Johnson LM, Jacobsen SE, Patel DJ. DNA methylation pathways and their crosstalk with histone methylation. Nat Rev. Mol Cell Biol. 2015 Sep;16(9):519–32. https://doi.org/10.1038/nrm4043.

Heinz S, Romanoski CE, Benner C, Glass CK. The selection and function of cell type-specific enhancers. Nat Rev. Mol Cell Biol. 2015;16(3):144–54. https://doi.org/10.1038/nrm3949.

Li E, Zhang Y. DNA methylation in mammals. Cold Spring Harb Perspect Biol. 2014 May 1;6(5):a019133. https://doi.org/10.1101/cshperspect.a019133.

Rogatsky I, Adelman K. Preparing the first responders: building the inflammatory transcriptome from the ground up. Mol Cell. 2014;54(2):245–54. https://doi.org/10.1016/j.molcel.2014.03.038.

Smale ST, Tarakhovsky A, Natoli G. Chromatin contributions to the regulation of innate immunity. Annu Rev. Immunol. 2014;32:489–511. https://doi.org/10.1146/annurev-immunol-031210-101303.

Tessarz P, Kouzarides T. Histone core modifications regulating nucleosome structure and dynamics. Nat Rev. Mol Cell Biol. 2014;15(11):703–8. https://doi.org/10.1038/nrm3890.

Recommended Websites

http://fantom.gsc.riken.jp/5/
https://genome.ucsc.edu/ENCODE/
https://www.immgen.org
http://immunet.princeton.edu
http://www.immuneprofiling.org/hipc/page/show

References

1. Ivashkiv LB. Epigenetic regulation of macrophage polarization and function. Trends Immunol. 2013;34(5):216–23. https://doi.org/10.1016/j.it.2012.11.001.
2. Bird A. DNA methylation patterns and epigenetic memory. Genes Dev. 2002;16(1):6–21. https://doi.org/10.1101/gad.947102.
3. Berger SL, Kouzarides T, Shiekhattar R, Shilatifard A. An operational definition of epigenetics. Genes Develop. 2009;23(7):781–3. https://doi.org/10.1101/gad.1787609.

4. Cvejic A. Mechanisms of fate decision and lineage commitment during haematopoiesis. Immunol Cell Biol. 2016;94(3):230–5. https://doi.org/10.1038/icb.2015.96.
5. Rogatsky I, Adelman K. Preparing the first responders: building the inflammatory transcriptome from the ground up. Mol Cell. 2014;54(2):245–54. https://doi.org/10.1016/j.molcel.2014.03.038.
6. Heinz S, Romanoski CE, Benner C, Glass CK. The selection and function of cell type-specific enhancers. Nat Rev Mol Cell Biol. 2015;16(3):144–54. https://doi.org/10.1038/nrm3949.
7. Smale ST, Tarakhovsky A, Natoli G. Chromatin contributions to the regulation of innate immunity. Annu Rev Immunol. 2014;32:489–511. https://doi.org/10.1146/annurev-immunol-031210-101303.
8. Wolff L, Humeniuk R. Concise review: erythroid versus myeloid lineage commitment: regulating the master regulators. Stem Cells. 2013;31(7):1237–44. https://doi.org/10.1002/stem.1379.
9. Miyazaki K, Miyazaki M, Murre C. The establishment of B versus T cell identity. Trends Immunol. 2014;35(5):205–10. https://doi.org/10.1016/j.it.2014.02.009.
10. Amit I, Winter DR, Jung S. The role of the local environment and epigenetics in shaping macrophage identity and their effect on tissue homeostasis. Nat Immunol. 2016;17(1):18–25. https://doi.org/10.1038/ni.3325.
11. Gosselin D, Link VM, Romanoski CE, Fonseca GJ, Eichenfield DZ, Spann NJ, et al. Environment drives selection and function of enhancers controlling tissue-specific macrophage identities. Cell. 2014;159(6):1327–40. https://doi.org/10.1016/j.cell.2014.11.023.
12. Lavin Y, Winter D, Blecher-Gonen R, David E, Keren-Shaul H, Merad M, et al. Tissue-resident macrophage enhancer landscapes are shaped by the local microenvironment. Cell. 2014;159(6):1312–26. https://doi.org/10.1016/j.cell.2014.11.018.
13. Buenrostro JD, Giresi PG, Zaba LC, Chang HY, Greenleaf WJ. Transposition of native chromatin for fast and sensitive epigenomic profiling of open chromatin, DNA-binding proteins and nucleosome position. Nat Methods. 2013;10(12):1213–8. https://doi.org/10.1038/nmeth.2688.
14. Soufi A, Garcia MF, Jaroszewicz A, Osman N, Pellegrini M, Zaret KS. Pioneer transcription factors target partial DNA motifs on nucleosomes to initiate reprogramming. Cell. 2015;161(3):555–68. https://doi.org/10.1016/j.cell.2015.03.017.
15. Foulks JM, Parnell KM, Nix RN, Chau S, Swierczek K, Saunders M, et al. Epigenetic drug discovery: targeting DNA methyltransferases. J Biomol Screen. 2012;17(1):2–17. https://doi.org/10.1177/1087057111421212.
16. Honda T, Tamura G, Waki T, Kawata S, Terashima M, Nishizuka S, et al. Demethylation of MAGE promoters during gastric cancer progression. Br J Cancer. 2004;90(4):838–43. https://doi.org/10.1038/sj.bjc.6601600.
17. Yagi S, Hirabayashi K, Sato S, Li W, Takahashi Y, Hirakawa T, et al. DNA methylation profile of tissue-dependent and differentially methylated regions (T-DMRs) in mouse promoter regions demonstrating tissue-specific gene expression. Genome Res. 2008;18(12):1969–78. https://doi.org/10.1101/gr.074070.107.
18. Polansky JK, Kretschmer K, Freyer J, Floess S, Garbe A, Baron U, et al. DNA methylation controls Foxp3 gene expression. Eur J Immunol. 2008;38(6):1654–63. https://doi.org/10.1002/eji.200838105.
19. Nan X, Ng HH, Johnson CA, Laherty CD, Turner BM, Eisenman RN, et al. Transcriptional repression by the methyl-CpG-binding protein MeCP2 involves a histone deacetylase complex. Nature. 1998;393(6683):386–9. https://doi.org/10.1038/30764.
20. Dong KB, Maksakova IA, Mohn F, Leung D, Appanah R, Lee S, et al. DNA methylation in ES cells requires the lysine methyltransferase G9a but not its catalytic activity. EMBO J. 2008;27(20):2691–701. https://doi.org/10.1038/emboj.2008.193.
21. Lehnertz B, Ueda Y, Derijck AA, Braunschweig U, Perez-Burgos L, Kubicek S, et al. Suv39h-mediated histone H3 lysine 9 methylation directs DNA methylation to major satellite repeats at pericentric heterochromatin. Curr Biol. 2003;13(14):1192–200.
22. Leung D, Du T, Wagner U, Xie W, Lee AY, Goyal P, et al. Regulation of DNA methylation turnover at LTR retrotransposons and imprinted loci by the histone methyltransferase Setdb1. Proc Natl Acad Sci U S A. 2014;111(18):6690–5. https://doi.org/10.1073/pnas.1322273111.
23. Guo X, Wang L, Li J, Ding Z, Xiao J, Yin X, et al. Structural insight into autoinhibition and histone H3-induced activation of DNMT3A. Nature. 2015;517(7536):640–4. https://doi.org/10.1038/nature13899.
24. Ziller MJ, Gu H, Muller F, Donaghey J, Tsai LT, Kohlbacher O, et al. Charting a dynamic DNA methylation landscape of the human genome. Nature. 2013;500(7463):477–81. https://doi.org/10.1038/nature12433.
25. Reinius LE, Acevedo N, Joerink M, Pershagen G, Dahlen SE, Greco D, et al. Differential DNA methylation in purified human blood cells: implications for cell lineage and studies on disease susceptibility. PLoS One. 2012;7(7):e41361. https://doi.org/10.1371/journal.pone.0041361.
26. Schmidl C, Klug M, Boeld TJ, Andreesen R, Hoffmann P, Edinger M, et al. Lineage-specific DNA methylation in T cells correlates with histone methylation and enhancer activity. Genome Res. 2009;19(7):1165–74. https://doi.org/10.1101/gr.091470.109.
27. Lal G, Bromberg JS. Epigenetic mechanisms of regulation of Foxp3 expression. Blood. 2009;114(18):3727–35. https://doi.org/10.1182/blood-2009-05-219584.
28. Slotkin RK, Martienssen R. Transposable elements and the epigenetic regulation of the genome. Nat Rev Genet. 2007;8(4):272–85. https://doi.org/10.1038/nrg2072.

29. Sen R, Oltz E. Genetic and epigenetic regulation of IgH gene assembly. Curr Opin Immunol. 2006;18(3):237–42. https://doi.org/10.1016/j.coi.2006.03.008.
30. Cobb RM, Oestreich KJ, Osipovich OA, Oltz EM. Accessibility control of V(D)J recombination. Adv Immunol. 2006;91:45–109. https://doi.org/10.1016/S0065-2776(06)91002-5.
31. Goldmit M, Ji Y, Skok J, Roldan E, Jung S, Cedar H, et al. Epigenetic ontogeny of the Igk locus during B cell development. Nat Immunol. 2005;6(2):198–203. https://doi.org/10.1038/ni1154.
32. Wu H, Zhang Y. Reversing DNA methylation: mechanisms, genomics, and biological functions. Cell. 2014;156(1–2):45–68. https://doi.org/10.1016/j.cell.2013.12.019.
33. Xu YM, Du JY, Lau AT. Posttranslational modifications of human histone H3: an update. Proteomics. 2014;14(17–18):2047–60. https://doi.org/10.1002/pmic.201300435.
34. Petty E, Pillus L. Balancing chromatin remodeling and histone modifications in transcription. Trends Genet. 2013;29(11):621–9. https://doi.org/10.1016/j.tig.2013.06.006.
35. Tessarz P, Kouzarides T. Histone core modifications regulating nucleosome structure and dynamics. Nat Rev Mol Cell Biol. 2014;15(11):703–8. https://doi.org/10.1038/nrm3890.
36. Bolden JE, Peart MJ, Johnstone RW. Anticancer activities of histone deacetylase inhibitors. Nat Rev Drug Discov. 2006;5(9):769–84. https://doi.org/10.1038/nrd2133.
37. Chuang DM, Leng Y, Marinova Z, Kim HJ, Chiu CT. Multiple roles of HDAC inhibition in neurodegenerative conditions. Trends Neurosci. 2009;32(11):591–601. https://doi.org/10.1016/j.tins.2009.06.002.
38. Gregoretti IV, Lee YM, Goodson HV. Molecular evolution of the histone deacetylase family: functional implications of phylogenetic analysis. J Mol Biol. 2004;338(1):17–31. https://doi.org/10.1016/j.jmb.2004.02.006.
39. Marks PA. Histone deacetylase inhibitors: a chemical genetics approach to understanding cellular functions. Biochim Biophys Acta. 2010;1799(10–12):717–25. https://doi.org/10.1016/j.bbagrm.2010.05.008.
40. Rahman I, Marwick J, Kirkham P. Redox modulation of chromatin remodeling: impact on histone acetylation and deacetylation, NF-kappaB and pro-inflammatory gene expression. Biochem Pharmacol. 2004;68(6):1255–67. https://doi.org/10.1016/j.bcp.2004.05.042.
41. Mellor AL, Munn DH. IDO expression by dendritic cells: tolerance and tryptophan catabolism. Nat Rev Immunol. 2004;4(10):762–74. https://doi.org/10.1038/nri1457.
42. Shen Z, Chen L, Hao F, Wu J. Transcriptional regulation of Foxp3 gene: multiple signal pathways on the road. Med Res Rev. 2009;29(5):742–66. https://doi.org/10.1002/med.20152.
43. Usui T, Preiss JC, Kanno Y, Yao ZJ, Bream JH, O'Shea JJ, et al. T-bet regulates Th1 responses through essential effects on GATA-3 function rather than on IFNG gene acetylation and transcription. J Exp Med. 2006;203(3):755–66. https://doi.org/10.1084/jem.20052165.
44. Villagra A, Cheng F, Wang HW, Suarez I, Glozak M, Maurin M, et al. The histone deacetylase HDAC11 regulates the expression of interleukin 10 and immune tolerance. Nat Immunol. 2009;10(1):92–100. https://doi.org/10.1038/ni.1673.
45. Black JC, Van Rechem C, Whetstine JR. Histone lysine methylation dynamics: establishment, regulation, and biological impact. Mol Cell. 2012;48(4):491–507. https://doi.org/10.1016/j.molcel.2012.11.006.
46. Herz HM, Garruss A, Shilatifard A. SET for life: biochemical activities and biological functions of SET domain-containing proteins. Trends Biochem Sci. 2013;38(12):621–39. https://doi.org/10.1016/j.tibs.2013.09.004.
47. Kooistra SM, Helin K. Molecular mechanisms and potential functions of histone demethylases. Nat Rev Mol Cell Biol. 2012;13(5):297–311. https://doi.org/10.1038/nrm3327.
48. Chang S, Aune TM. Dynamic changes in histone-methylation 'marks' across the locus encoding interferon-gamma during the differentiation of T helper type 2 cells. Nat Immunol. 2007;8(7):723–31. https://doi.org/10.1038/ni1473.
49. Wakabayashi Y, Tamiya T, Takada I, Fukaya T, Sugiyama Y, Inoue N, et al. Histone 3 lysine 9 (H3K9) methyltransferase recruitment to the interleukin-2 (IL-2) promoter is a mechanism of suppression of IL-2 transcription by the transforming growth factor-beta-Smad pathway. J Biol Chem. 2011;286(41):35456–65. https://doi.org/10.1074/jbc.M111.236794.
50. Miao CG, Yang JT, Yang YY, Du CL, Huang C, Huang Y, et al. Critical role of DNA methylation in the pathogenesis of systemic lupus erythematosus: new advances and future challenges. Lupus. 2014;23(8):730–42. https://doi.org/10.1177/0961203314527365.
51. Liu Y, Aryee MJ, Padyukov L, Fallin MD, Hesselberg E, Runarsson A, et al. Epigenome-wide association data implicate DNA methylation as an intermediary of genetic risk in rheumatoid arthritis. Nat Biotechnol. 2013;31(2):142–7. https://doi.org/10.1038/nbt.2487.
52. Huynh JL, Casaccia P. Epigenetic mechanisms in multiple sclerosis: implications for pathogenesis and treatment. Lancet Neurol. 2013;12(2):195–206. https://doi.org/10.1016/S1474-4422(12)70309-5.
53. Song L, Li Y. SEPT9: a specific circulating biomarker for colorectal cancer. Adv Clin Chem. 2015;72:171–204. https://doi.org/10.1016/bs.acc.2015.07.004.
54. Zhao M, Zhou Y, Zhu B, Wan M, Jiang T, Tan Q, et al. IFI44L promoter methylation as a blood biomarker for systemic lupus erythematosus. Ann Rheum Dis. 2016. https://doi.org/10.1136/annrheumdis-2015-208410.
55. Nakano K, Whitaker JW, Boyle DL, Wang W, Firestein GS. DNA methylome signature in rheumatoid arthritis. Ann Rheum Dis. 2013;72(1):110–7. https://doi.org/10.1136/annrheumdis-2012-201526.

56. Treppendahl MB, Kristensen LS, Gronbaek K. Predicting response to epigenetic therapy. J Clin Invest. 2014;124(1):47–55. https://doi.org/10.1172/JCI69737.
57. Subramaniam D, Thombre R, Dhar A, Anant S. DNA methyltransferases: a novel target for prevention and therapy. Front Oncol. 2014;4:80. https://doi.org/10.3389/fonc.2014.00080.
58. Goodyear OC, Dennis M, Jilani NY, Loke J, Siddique S, Ryan G, et al. Azacitidine augments expansion of regulatory T cells after allogeneic stem cell transplantation in patients with acute myeloid leukemia (AML). Blood. 2012;119(14):3361–9. https://doi.org/10.1182/blood-2011-09-377044.
59. Sanchez-Abarca LI, Gutierrez-Cosio S, Santamaria C, Caballero-Velazquez T, Blanco B, Herrero-Sanchez C, et al. Immunomodulatory effect of 5-azacytidine (5-azaC): potential role in the transplantation setting. Blood. 2010;115(1):107–21. https://doi.org/10.1182/blood-2009-03-210393.
60. Chan MW, Chang CB, Tung CH, Sun J, Suen JL, Wu SF. Low-dose 5-aza-2′-deoxycytidine pretreatment inhibits experimental autoimmune encephalomyelitis by induction of regulatory T cells. Mol Med. 2014;20:248–56. https://doi.org/10.2119/molmed.2013.00159.
61. Martinez-Frances A. Adverse effects of azacitidine: onset, duration, and treatment. Adv Ther. 2011;28(Suppl 4):1–5. https://doi.org/10.1007/s12325-011-0021-5.
62. Wijermans PW, Lubbert M, Verhoef G, Klimek V, Bosly A. An epigenetic approach to the treatment of advanced MDS; the experience with the DNA demethylating agent 5-aza-2′-deoxycytidine (decitabine) in 177 patients. Ann Hematol. 2005;84(Suppl 1):9–17. https://doi.org/10.1007/s00277-005-0012-1.
63. Quddus J, Johnson KJ, Gavalchin J, Amento EP, Chrisp CE, Yung RL, et al. Treating activated CD4+ T cells with either of two distinct DNA methyltransferase inhibitors, 5-azacytidine or procainamide, is sufficient to cause a lupus-like disease in syngeneic mice. J Clin Invest. 1993;92(1):38–53. https://doi.org/10.1172/JCI116576.
64. Gros C, Fahy J, Halby L, Dufau I, Erdmann A, Gregoire JM, et al. DNA methylation inhibitors in cancer: recent and future approaches. Biochimie. 2012;94(11):2280–96. https://doi.org/10.1016/j.biochi.2012.07.025.
65. Telander DG, Malvey EN, Mueller DL. Evidence for repression of IL-2 gene activation in anergic T cells. J Immunol. 1999;162(3):1460–5.
66. Minucci S, Pelicci PG. Histone deacetylase inhibitors and the promise of epigenetic (and more) treatments for cancer. Nat Rev Cancer. 2006;6(1):38–51. https://doi.org/10.1038/nrc1779.
67. Nicodeme E, Jeffrey KL, Schaefer U, Beinke S, Dewell S, Chung CW, et al. Suppression of inflammation by a synthetic histone mimic. Nature. 2010;468(7327):1119–23. https://doi.org/10.1038/nature09589.
68. Delmore JE, Issa GC, Lemieux ME, Rahl PB, Shi J, Jacobs HM, et al. BET bromodomain inhibition as a therapeutic strategy to target c-Myc. Cell. 2011;146(6):904–17. https://doi.org/10.1016/j.cell.2011.08.017.
69. Maeda T, Towatari M, Kosugi H, Saito H. Up-regulation of costimulatory/adhesion molecules by histone deacetylase inhibitors in acute myeloid leukemia cells. Blood. 2000;96(12):3847–56.
70. Arrowsmith CH, Bountra C, Fish PV, Lee K, Schapira M. Epigenetic protein families: a new frontier for drug discovery. Nat Rev Drug Discov. 2012;11(5):384–400. https://doi.org/10.1038/nrd3674.
71. Liang Y, Vogel JL, Arbuckle JH, Rai G, Jadhav A, Simeonov A, et al. Targeting the JMJD2 histone demethylases to epigenetically control herpesvirus infection and reactivation from latency. Sci Transl Med. 2013;5(167):167ra5. https://doi.org/10.1126/scitranslmed.3005145.
72. Hojfeldt JW, Agger K, Helin K. Histone lysine demethylases as targets for anticancer therapy. Nat Rev Drug Discov. 2013;12(12):917–30. https://doi.org/10.1038/nrd4154.
73. De Santa F, Narang V, Yap ZH, Tusi BK, Burgold T, Austenaa L, et al. Jmjd3 contributes to the control of gene expression in LPS-activated macrophages. EMBO J. 2009;28(21):3341–52. https://doi.org/10.1038/emboj.2009.271.
74. Ishii M, Wen H, Corsa CA, Liu T, Coelho AL, Allen RM, et al. Epigenetic regulation of the alternatively activated macrophage phenotype. Blood. 2009;114(15):3244–54. https://doi.org/10.1182/blood-2009-04-217620. [pii] blood-2009-04-217620
75. Satoh T, Takeuchi O, Vandenbon A, Yasuda K, Tanaka Y, Kumagai Y, et al. The Jmjd3-Irf4 axis regulates M2 macrophage polarization and host responses against helminth infection. Nat Immunol. 2010;11(10):936–44. https://doi.org/10.1038/ni.1920. [pii] blood-2009-04-217620
76. Li Q, Zou J, Wang M, Ding X, Chepelev I, Zhou X, et al. Critical role of histone demethylase Jmjd3 in the regulation of CD4+ T-cell differentiation. Nat Commun. 2014;5:5780. https://doi.org/10.1038/ncomms6780.
77. Okabe Y, Medzhitov R. Tissue-specific signals control reversible program of localization and functional polarization of macrophages. Cell. 2014;157(4):832–44. https://doi.org/10.1016/j.cell.2014.04.016.
78. Butovsky O, Jedrychowski MP, Moore CS, Cialic R, Lanser AJ, Gabriely G, et al. Identification of a unique TGF-beta-dependent molecular and functional signature in microglia. Nat Neurosci. 2014;17(1):131–43. https://doi.org/10.1038/nn.3599.
79. Abutbul S, Shapiro J, Szaingurten-Solodkin I, Levy N, Carmy Y, Baron R, et al. TGF-beta signaling through SMAD2/3 induces the quiescent microglial phenotype within the CNS environment. Glia. 2012;60(7):1160–71. https://doi.org/10.1002/glia.22343.
80. Kohyama M, Ise W, Edelson BT, Wilker PR, Hildner K, Mejia C, et al. Role for Spi-C in the development of red pulp macrophages and splenic iron homeostasis. Nature. 2009;457(7227):318–21. https://doi.org/10.1038/nature07472.

81. Schneider C, Nobs SP, Kurrer M, Rehrauer H, Thiele C, Kopf M. Induction of the nuclear receptor PPAR-gamma by the cytokine GM-CSF is critical for the differentiation of fetal monocytes into alveolar macrophages. Nat Immunol. 2014;15(11):1026–37. https://doi.org/10.1038/ni.3005.
82. Hanna RN, Carlin LM, Hubbeling HG, Nackiewicz D, Green AM, Punt JA, et al. The transcription factor NR4A1 (Nur77) controls bone marrow differentiation and the survival of Ly6C-monocytes. Nat Immunol. 2011;12(8):778–85. https://doi.org/10.1038/ni.2063.
83. Li E, Zhang Y. DNA methylation in mammals. Cold Spring Harb Perspect Biol. 2014;6(5):a019133. https://doi.org/10.1101/cshperspect.a019133.
84. Okano M, Bell DW, Haber DA, Li E. DNA methyltransferases Dnmt3a and Dnmt3b are essential for de novo methylation and mammalian development. Cell. 1999;99(3):247–57.
85. Ley TJ, Ding L, Walter MJ, McLellan MD, Lamprecht T, Larson DE, et al. DNMT3A mutations in acute myeloid leukemia. N Engl J Med. 2010;363(25):2424–33. https://doi.org/10.1056/NEJMoa1005143.
86. Ehrlich M. The ICF syndrome, a DNA methyltransferase 3B deficiency and immunodeficiency disease. Clin Immunol. 2003;109(1):17–28.
87. Liao HF, Tai KY, Chen WS, Cheng LC, Ho HN, Lin SP. Functions of DNA methyltransferase 3-like in germ cells and beyond. Biol Cell. 2012;104(10):571–87. https://doi.org/10.1111/boc.201100109.
88. Roth SY, Denu JM, Allis CD. Histone acetyltransferases. Annu Rev Biochem. 2001;70:81–120. https://doi.org/10.1146/annurev.biochem.70.1.81.

Innate Immunity: Phagocytes, Natural Killer Cells, and the Complement System

8

Michael J. Parnham and Adriano G. Rossi

8.1 Introduction

In the second half of the nineteenth century, Eli Metchnikoff discovered that bacteria can be ingested (phagocytosed) by LEUKOCYTES present in the blood of a variety of animals. At about the same time, Paul Ehrlich found that certain agents dissolved in blood had bactericidal potential. The scientific discussion on the importance of cellular versus humoral factors in our defense against bacteria that followed came to an end when it was recognized that both components enforce each other's effect within the context of the IMMUNE SYSTEM. In 1908, both scientists shared the Nobel Prize for Physiology and Medicine. Further investigations on the nature and the working mechanism of the cells and the proteins that constitute our immunological defense system showed that each of these components is made up of several different constituents. In its turn, this led to the view that a functional distinction exists between the adaptive and the innate branch of the IMMUNE SYSTEM.

While an INNATE IMMUNE SYSTEM of some form at least appears to be present in all multicellular organisms, the adaptive IMMUNE SYSTEM represents a more recent adaptation that emerged some 500 million years ago and is present only in jawed vertebrates. As indicated previously (Chaps. 2–4), the adaptive branch is executed by LYMPHOCYTES, i.e., white blood cells capable of generating ANTIBODIES or performing CYTOTOXIC T-cell responses against foreign structures (ANTIGENS) to the body and killing virus-infected cells. LYMPHOCYTES are able to differentiate between ANTIGENS that belong to the body and those that are foreign. Importantly, LYMPHOCYTES can display immunological memory: once they have encountered foreign material, they will recognize and respond faster and much more efficiently upon subsequent encounters. The INNATE IMMUNE SYSTEM is largely responsible for the actual elimination of the source of infection, which may either be viral, bacterial, fungal, or parasitic in nature. The major components of the INNATE IMMUNE SYSTEM include PHAGOCYTES, INNATE LYMPHOID CELLS, NATURAL KILLER CELLS, and the COMPLEMENT SYSTEM. PHAGOCYTES are white blood cells capable of uptake (PHAGOCYTOSIS) and intracellular killing of microbes. The recognition of microbes by phagocytes may occur

Final manuscript submitted on May 3, 2017.

M. J. Parnham (✉)
Translational Drug Validation, Fraunhofer IME-TMP and Goethe University Frankfurt, Frankfurt am Main, Germany
e-mail: parnham@med.uni-frankfurt.de; michael.parnham@ime.fraunhofer.de

A. G. Rossi
Centre for Inflammation Research, University of Edinburgh, Edinburgh, UK
e-mail: a.g.rossi@ed.ac.uk

M. J. Parnham et al. (eds.), *Nijkamp and Parnham's Principles of Immunopharmacology*,
https://doi.org/10.1007/978-3-030-10811-3_8

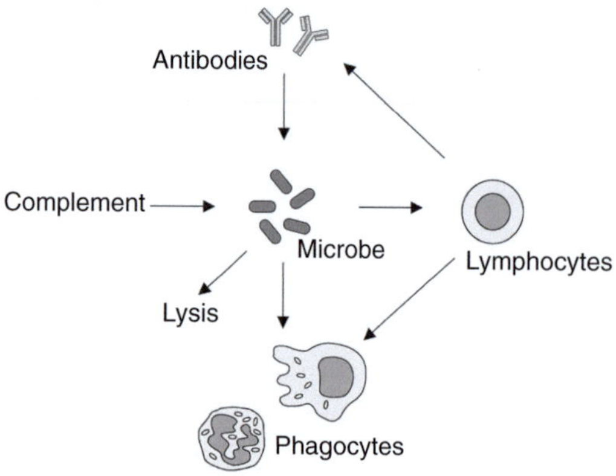

Fig. 8.1 Cooperative action of innate and adaptive immune systems in host defense. Microbes can bind antibodies and complement fragments. This may result in direct complement-mediated lysis, phagocytosis by (recruited) macrophages and granulocytes, and/or antigen presentation by dendritic cells to trigger adaptive immune responses that results in the generation of (more) antibodies that may further help to eliminate the pathogen. Innate lymphocytes (ILCs) are also present in tissues and may form a first line of defense for the adaptive immune response, as they mirror some functions (e.g. cytokine release) of antigen-responsive lymphocytes.

either directly or after binding of ANTIBODIES and complement proteins to the surface of the microbes (Fig. 8.1). NATURAL killer (NK) cells are lymphocyte-like cells with CYTOTOXIC potential against certain virus-infected and tumor cells. They are a sub-group of INNATE IMMUNE CELLS (ILCs) which are tissue-associated lymphocytes that do not express antigen receptors, but can form memory cells. The ILCs include ILC1, ILC2 and ILC3 cells (see Chap. 1). ILC1s produce several cytokines generated by Th1 cells and contribute to defence against intracellular bacteria. ILC2s produce a number of cytokines generated by Th2 cells and contributing both to defence against helminths and to allergic and asthmatic reactions. ILC3s release a variety of cytokines, including IL-17, IL-22 and GM-CSF and regulate intestinal tissue homeostasis and inflammation. Innate B cells also occur in tissues and produce natural antibodies. ILCs are the subjects of intense research but as yet have not been implicated in effects of immunomodulatory drugs. The COMPLEMENT SYSTEM consists of a series of proteolytic enzymes in blood and other bodily fluids, capable of lysing microorganisms or marking them for elimination by PHAGOCYTES, often in a fashion that can be further enhanced by specific ANTIBODIES. The activities of these innate systems have to be tightly regulated, because in principle they cannot only be harmful to the pathogen but also to the host itself. This chapter will give a short description of each of these innate systems, except for ILCs, their clinical relevance, and the potential for therapy in case of failure.

8.2 Phagocytes

Many cell types are capable—to some extent—of internalizing and killing microorganisms, which may sometimes help to eliminate infection. However, MACROPHAGES and GRANULOCYTES are the actual "professional" PHAGOCYTES, because these cells are equipped with a motile apparatus for actively moving to sites of infection with surface RECEPTORS to bind microorganisms, granules filled with CYTOTOXIC proteins, and with an enzyme that can generate toxic oxygen radicals. MACROPHAGES and GRANULOCYTES are formed in the BONE MARROW from pluripotent hematopoietic precursor cells that differentiate along the myeloid lineage under the influence of growth and differentiation hormones (see Chap. 2).

MACROPHAGE progenitors are released into the blood as immature MONOCYTES. These may subsequently traffic to the various tissues and organs where they further differentiate into MACROPHAGES. In contrast, some subpopulations of MACROPHAGES develop from local self-replenishing precursors. MACROPHAGES constitute a very HETEROGENEOUS population of cells, and this heterogeneity reflects their functional versatility. MACROPHAGES not only are important for host defense but also play a role in normal tissue development and homeostasis, e.g., by clearing (apoptotic) dead cells.

GRANULOCYTES take about 14 days to develop and are released into the blood from the bone marrow as mature cells. Most GRANULOCYTES become so-called "neutrophilic" GRANULOCYTES, which exhibit a high antimicrobial potential. Other granulocyte types are "eosinophilic" GRANULOCYTES, involved in anti-parasite defense, and "basophilic" GRANULOCYTES, which lack the ability to phagocytose but can release HISTAMINE that plays a role in inflammatory reactions. In the tissues, MACROPHAGES have an estimated life span in the order of weeks to months, depending on the subpopulation. Instead, neutrophilic GRANULOCYTES survive only several days after release from the BONE MARROW. Thus, neutrophilic GRANULOCYTES (NEUTROPHILS) need to be formed in much larger numbers than MACROPHAGES for efficient surveillance against microorganisms. Indeed, in healthy adults about 10^{11} NEUTROPHILS are newly generated from the BONE MARROW each day, and this can increase a further tenfold during infection. MACROPHAGES are formed at not more than 10^{9} per day. PHAGOCYTES end their life either through necrosis as a result of PHAGOCYTOSIS and the release of toxic mediators, which becomes apparent as pus formation, or through mainly APOPTOSIS (programmed cell death) and subsequent removal by MACROPHAGES.

The critical importance of PHAGOCYTES for host defense against microorganisms is illustrated by the recurrent, life-threatening infections that occur in patients with a genetic or acquired shortage or deficiency of these cells. Patients with a shortage of NEUTROPHILS, e.g., in a condition known as NEUTROPENIA, may be treated with relevant GROWTH FACTORS, such as granulocyte colony-stimulating factor (G-CSF), to stimulate granulocyte production and thereby to compensate for their apparent lack. Complete cure may be achieved by BONE MARROW transplantation.

8.3 Phagocyte Mobility

NEUTROPHILS and MONOCYTES have the ability to move actively to the site of an infection. This is caused by the release in these areas of so-called chemotactic substances (lipids and proteins), small molecules of bacterial or host origin that diffuse into the surroundings and can bind to specific RECEPTORS on the PHAGOCYTES. PHAGOCYTES are able to "sense" concentration gradients of these chemotactic agents and to move toward the source of these agents at the site of infection. This process is called CHEMOTAXIS. However, PHAGOCYTES in the blood must first pass the blood vessel wall before moving into the tissues [1] (see Chap. 1). This multistep process of extravasation is initiated by reversible interaction of L-selectin, an ADHESION MOLECULE on the surface of LEUKOCYTES that interacts with carbohydrate structures on endothelial blood vessel cells and by similar interaction between E-SELECTIN on the endothelial cells with carbohydrate structures on LEUKOCYTES. Under normal conditions, this ROLLING of PHAGOCYTES over the endothelium leads to stable adhesion and spreading of the PHAGOCYTES on the vessel wall, a process in which surface receptors like INTEGRINS play a decisive role (Fig. 8.2). Finally, DIAPEDESIS (transendothelial migration) and movement of the PHAGOCYTES into the tissues take place. In infected or inflamed areas, these processes are strongly increased by the formation of complement fragment C5a, by formylated bacterial peptides, by the LEUKOTRIENE, LTB4, and by other CHEMOTAXINS. These agents, which are detected by distinct G protein-coupled seven-transmembrane (7TM)-spanning RECEPTOR family members, which amongst others by raising the intracellular Ca^{2+} concentration trigger a variety of cellular changes, such as cytoskeletal rearrangements, that lead to CHEMOTAXIS. This also involves an increase in expression of adhesion proteins on the surface of the PHAGOCYTES, such as INTEGRINS. Moreover, the INTEGRINS are also "activated" by means of a change in their configuration, which leads to a higher AFFINITY for their endothelial counter-structures and supports their recruitment to inflammatory sites. In addition, microbial components, like bacterial ENDOTOXIN, induce local MACROPHAGES to produce INTERLEUKIN-1β (IL-1β) and TUMOR NECROSIS FACTOR-α (TNF-α). These CYTOKINES, as well as ENDOTOXIN itself, activate the local endothelial cells to upregulate the expression of intercellular ADHESION

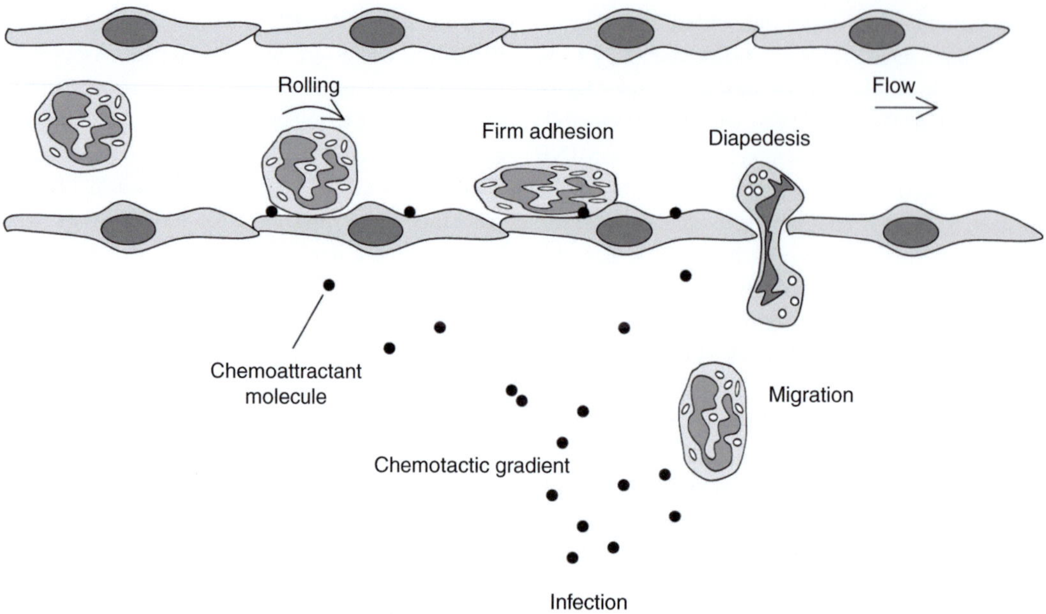

Fig. 8.2 Phagocyte recruitment during inflammation. The process of phagocyte extravasation involves a series of events. The initial interaction with the vessel wall is termed 'rolling' and this is mediated by interactions between lectins and their carbohydrate ligands. In particular, (i) L-selectin is expressed on the phagocyte, and sialyl Lewis-X carbohydrates are expressed on a variety of surface protein carrier molecules on endothelial cells, and (ii) E-selectin is expressed on endothelial cells and sialyl Lewis-X on phagocytes. Chemoattractant molecules (black dots) produced by the inflamed tissue and immobilized by the endothelium then induce activation of integrins on the phagocytes. This results in firm adhesion of these molecules to ICAM-1, ICAM-2, VCAM-1, that are in turn up-regulated on the endothelium by inflammatory cytokines. This is followed by the diapedesis of the phagocytes through the endothelial layer and their migration to the source of inflammation.

MOLECULE-1 (ICAM-1), E-SELECTIN, and vascular ADHESION MOLECULE-1 (VCAM-1), which strongly enhances phagocyte adhesion. Endothelial cells also produce PLATELET-ACTIVATING FACTOR (PAF) and IL-8 under these conditions, which remain bound to the endothelial cells and stimulate phagocyte recruitment.

In the process of DIAPEDESIS, the endothelial cells that form the lining of the blood vessel wall participate actively. Binding of LEUKOCYTES to ICAM-1 or VCAM-1 on the endothelial cells induces signaling in the endothelial cells that leads to relaxation of endothelial cell-cell junctions, especially by disrupting the intercellular interactions mediated by, for example, cadherins. This enables the LEUKOCYTES to squeeze themselves between two adjacent endothelial cells. The migratory process itself is governed by multiple interactions between the LEUKOCYTES and the endothelial cells in the intercellular cleft between two endothelial cells (Fig. 8.2).

The formation of chemotactic CYTOKINES, termed CHEMOKINES, by resident tissue cells or already immigrated leukocytes is also critical for the influx of PHAGOCYTES into inflamed tissues. CHEMOKINES of the C-C chemokine family, of which monocyte chemotactic protein-1 (MCP-1, CCL2) is the prototype, act in a PHAGOCYTE-selective fashion and primarily attract MONOCYTES and MACROPHAGES, whereas CHEMOKINES of the C-X-C family, such as IL-8 (CXCL8), mediate the recruitment of NEUTROPHILS and EOSINOPHILS. In tissues, PHAGOCYTES migrate by a co-ordinated and dynamic process of local attachment to extracellular matrix proteins, propagation of the cell over the fixed area, followed by dissociation of the rear end of the cell.

The biological significance of ADHERENCE and migration is clearly demonstrated by the clinical symptoms of patients with leukocyte adhesion deficiency type 1 (LAD-1). This condition is caused by a genetic defect in β2-INTEGRINS and is associated with serious recurrent bacterial infections, retarded wound healing, persistent leukocytosis, and a strong deficiency in the generation of inflammatory reactions. In view of the high incidence of death in LAD-1 patients, aggressive management of infections is required. The use of prophylactic treatment with trimethoprim-sulfamethoxazole antibiotics appears beneficial, and when a suitable donor is available, BONE MARROW transplantation may provide a permanent solution.

8.4 Recognition of Pathogens by Phagocytes

PHAGOCYTES are specialized in uptake and intracellular killing of a large variety of bacteria, yeasts, fungi, and mycoplasmata. Unlike cells of the adaptive IMMUNE SYSTEM, which recognize ANTIGENS via an immense REPERTOIRE of RECEPTORS generated by somatic RECOMBINATION from a set of encoded gene segments, cells of the INNATE IMMUNE SYSTEM depend on the use of products of a limited number of germ-line encoded RECEPTORS that have evolved to recognize relatively conserved microbial components, termed PATHOGEN-ASSOCIATED MOLECULAR PATTERNS (PAMPs). The RECEPTORS used by cells of the INNATE IMMUNE SYSTEM to recognize "nonself" are collectively termed PATTERN RECOGNITION RECEPTORS (PRRs) [2]. Examples of PAMPs are LIPOPOLYSACCHARIDE (LPS), β-glucan, and PEPTIDOGLYCAN (PGN). PRRs transmit signals that can lead to generation of INFLAMMATORY CYTOKINES and CHEMOKINES and to activation of microbicidal systems, such as the production of REACTIVE OXYGEN SPECIES (ROS) and the release of ANTIMICROBIAL PEPTIDES. PHAGOCYTES and sometimes also other cells express a number of PRRs (Table 8.1) that can influence their activation status, cytokine secretion, and life span.

Among the best characterized families of PRRs are the members of the TOLL-LIKE RECEPTORS (TLRs) (see also Chaps. 9 and 10). TLRs are transmembrane proteins consisting of an extracellular leucine-rich repeat (LRR) for the binding of PAMPs and a cytoplasmic tail that is responsible for the signal transduction after ligation of these RECEPTORS. TLRs are very conserved among animals, and the prototypic TLR molecule, Toll, was originally described in the fruit fly *Drosophila*. Eleven different TLRs have been identified within the human IMMUNE SYSTEM, and various microbial PAMP ligands have been identified for each of these (Table 8.1). TLRs are expressed on cells of the innate and the adaptive IMMUNE SYSTEMs but also on other cells. TLRs on the antigen-presenting DENDRITIC CELLS (see Chap. 5) act as sensors for PRR, in this context also known as "danger" signals, which trigger the expression of COSTIMULATORY MOLECULES. COSTIMULATORY MOLECULES on antigen-presenting cells, in addition to antigenic stimuli in the form of peptide in association with MHC molecules, provide a critical second signal for the activation of T cells. TLRs are expressed not only on the surface of PHAGOCYTES, enabling these cells to respond to PAMPs present in the extracellular milieu, but also on the phagosomal membrane that surrounds ingested microbes (Table 8.1), suggesting that PHAGOCYTES have the ability to "sense" the contents of the PHAGOSOME.

TLRs share part of their signal transduction pathway with the IL-1 RECEPTOR (IL-1R) family. Stimulation of both types of RECEPTOR ultimately leads to activation of the transcription factor NF-κB but also to activation of c-Jun N-terminal kinase (JNK) and p38 MITOGEN-activated protein kinase (MAPK) (Fig. 8.3). This cascade induces the expression of INFLAMMATORY CYTOKINES (see Chap. 6) and the differentiation of various immune cells into EFFECTOR CELLS. In addition to this general response to ligation of any TLR, responses exclusive to certain types of TLR also

Table 8.1 Pattern recognition receptors on phagocytes and other cells

	Localization	Ligand	Origin of the ligand
Membrane pattern recognition receptors			
Toll-like receptors (TLR)			
TLR1 (dimer with TLR2)	Plasma membrane	Triacyl lipoprotein	Bacteria
TLR2	Plasma membrane	Lipoprotein	Bacteria, viruses, parasites, host
TLR3	Endolysosome	dsRNA[a]	Virus
TLR4 (with MD2)	Plasma membrane	Lipopolysaccharide	Bacteria, viruses, host
TLR5	Plasma membrane	Flagellin	Bacteria
TLR6 (dimer with TLR2)	Plasma membrane	Diacyl lipoprotein	Bacteria, viruses
TLR7 (human TLR8)	Phagosomal membrane	ssRNA	Virus, bacteria, self
TLR9	Phagosomal membrane	CpG-DNA	Virus, bacteria, protozoa, host
TLR10	Phagosomal membrane	Unknown	Unknown
TLR11	Plasma membrane	Profilin-like molecule	Protozoa
NOD-like receptors (NLR)			
NOD1	Cytoplasm	iE-DAP	Bacteria
NOD2	Cytoplasm	Muramyl dipeptide	Bacteria
NLR1-14	Cytoplasm	Unknown	Microbes, host
RIG-like receptors (RLR)			
RIG-I	Cytoplasm	Short dsRNA, 5′ triphosphate dsRNA	RNA viruses, DNA virus RNA viruses (*Picornaviridae*)
MDA5	Cytoplasm	Long dsRNA	RNA viruses
LGP2	Cytoplasm	Unknown	
C-type lectin receptors (CLR)			
Dectin-1	Plasma membrane	β-glucan	Fungi
Dectin-2	Plasma membrane	β-glucan	Fungi
Mannose receptor	Plasma membrane	Mannose	Fungi, bacteria
DC-SIGN	Plasma membrane	Mannose	Bacteria, virus
Scavenger receptors			
SR-A	Plasma membrane	Polyanions	Bacteria, host
MARCO	Plasma membrane	Polyanions	Bacteria, host
CD163	Plasma membrane	Unknown	Bacteria, host
Secreted PRRs			
C-type lectins			
Mannose-binding lectin		Mannose	Fungi, bacteria
Ficolin			?
Pentraxins			
C-reactive protein		Phosphorylcholine	?
SAP		Phosphorylcholine	
PTX3			Microbes
LPS-binding protein (LBP)		Lipopolysaccharide	Gram-negative bacteria

[a]*ss/dsRNA* single-strand/double-strand RNA

exist, including, for instance, interferon-α/β (IFN-α/β) production after TLR3 and TLR4 ligation. In PHAGOCYTES, this pathway is involved in migration, DEGRANULATION, and NADPH OXIDASE activation (see next section of this chapter), three of the functions that are essential for proper finding and killing of microbes. These partially different effects of TLR signaling enable the IMMUNE SYSTEM to react in a specific way to various pathogens and to orchestrate the host response for efficient elimination of pathogen (see Chap. 10).

As opposed to the TLRs that are expressed on the plasma membrane, members of the NOD-LIKE RECEPTOR (NLR), such as NLR3, appear to act as intracellular sensors for infection. Together with other cytosolic components, including the protease caspase-1 and the adaptor

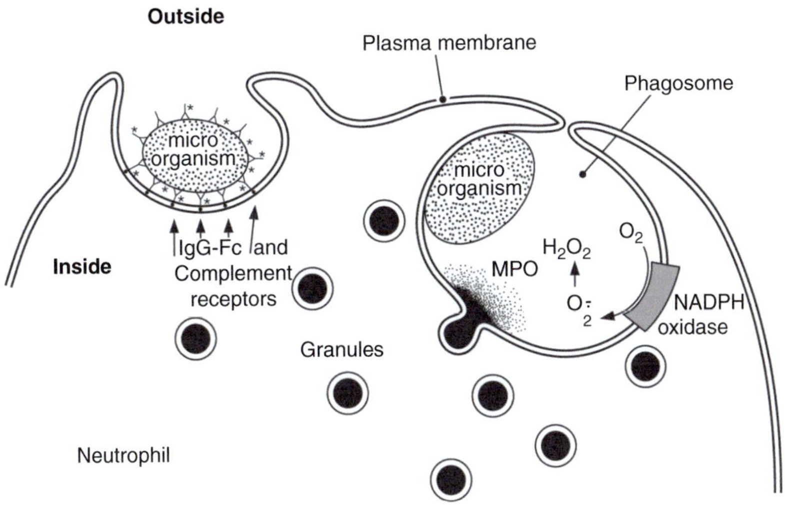

Fig. 8.3 Signal transduction pathways in TLR signaling. TLR2 forms heterodimers with either TLR1 or TLR6, the other TLRs act as monomers. The cytoplasmic tails of the TLRs contain a TIR domain also present in the adaptor protein MyD88. Upon ligand association with the TLRs, these TIR domains associate, thus coupling MyD88 to the activated TLRs. This leads to recruitment of the IRAK-1 protein kinase via the death domain in both MyD88 and IRAK-1. This recruitment is mediated by the Tollip protein. IRAK-1 is then phosphorylated, detaches from MyD88 and subsequently binds to, and activates a protein called TRAF-6. TRAF-6 triggers the activation of MKK6 and TAK1, which in turn activates JNK, p38 MAPK, and NF-κB, respectively. This general signal transduction route induces the expression of a set of genes that cause the induction of inflammatory cytokines and the differentiation of various cell types into effector cells. In addition, the MyD88 adapter protein, Mal-TIRAP, is involved in signaling through all TLRs except TLR3, TLR7, and TLR9. In addition, TRIF, another protein with a TIR domain, is involved in IFN production via TLR3 and TLR4. Figure reproduced with permission from [3]. *TIR* toll/IL-1 receptor, *IRAK* IL-1 receptor-associated kinase, *Tollip* toll/IL-1R-interacting protein, *TRAF* TNF receptor-associated factor, *MKK* MAPK kinase, *JNK* c-Jun N-terminal kinase, *Mal* MyD88-adapter-like, *TIRAP* TIR adapter protein, *TRIF* TIR domain-containing adapter inducing IFN-β

protein ASC, they form protein complexes known as INFLAMMASOMES, which upon LIGAND recognition by NLRs activate caspase activity and mediate cleavage and secretion of members of the IL-1 cytokine family that are synthesized as inactive precursors [4] (see Chaps. 6 and 9). Excessive activation of the NLR3 pathway occurs in gout where uric acid crystals are formed that provide a strong stimulus for INFLAMMASOME activation, for instance, in the joints (see Chap. 34). In addition, there are a number of rare genetic autoinflammatory disorders characterized by episodes of fever and INFLAMMATION that are caused by activating mutations in NRL3 [5]. In line with the important role of INFLAMMASOMES and IL-1β production, therapy with ANAKINRA, an IL-1 RECEPTOR antagonist, has shown to be effective in these situations.

8.5 Phagocytosis and Killing of Microorganisms

Most microorganisms can only be ingested efficiently after being covered with specific ANTIBODIES and/or complement fragments (a process called OPSONIZATION). ANTIBODIES bind with their variable regions to microbial ANTIGENS, and their exposed ANTIBODY Fc REGIONS promote activation of the classical complement pathway and the deposition of complement cleavage fragments such as C3b and iC3b onto the microbial surface (see below and Fig. 8.4). The ANTIBODY Fc REGIONS and the complement fragments can then bind to Fc RECEPTORS and COMPLEMENT RECEPTORS, respectively, which are expressed on the phagocyte surface. This binding of opsonized microorganisms to the PHAGOCYTES

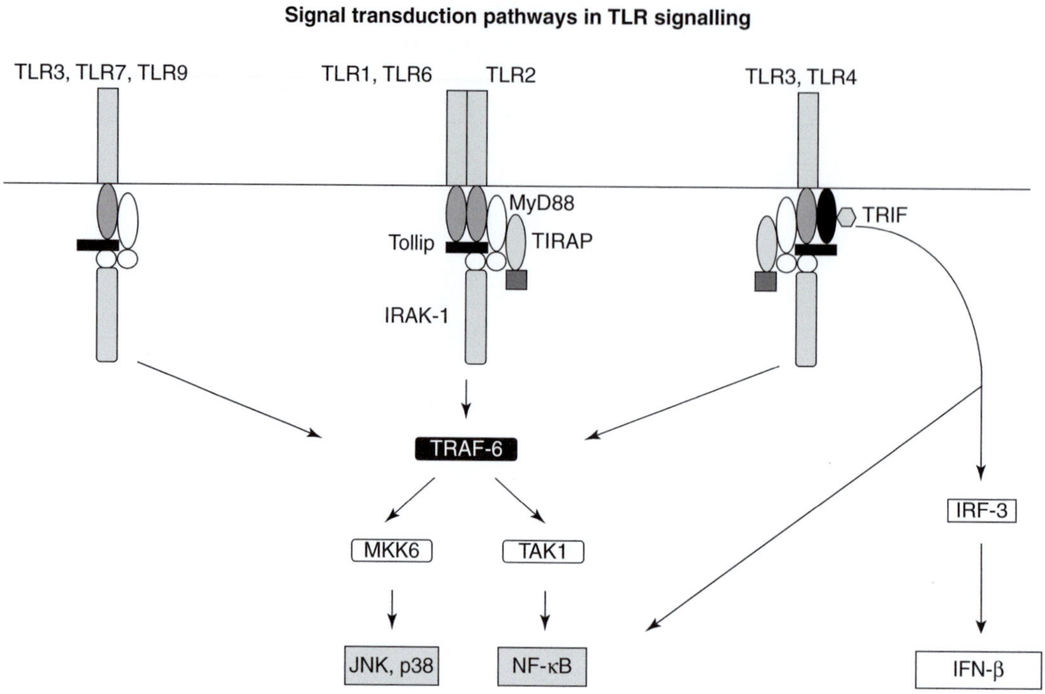

Fig. 8.4 Schematic representation of phagocytosis, degranulation and generation of reactive oxygen products. MPO, myeloperoxidase; *, complement fragments C3b or iC3b. Figure reproduced with permission from [6]

initiates three reactions in these cells: (1) rearrangement of cytoskeletal elements that results in folding of the plasma membrane around the microbes (i.e., PHAGOCYTOSIS); (2) fusion of intracellular granules with this PHAGOSOME (i.e., DEGRANULATION); and (3) generation of ROS within the PHAGOSOME (Fig. 8.4) by the phagocyte. The intracellular granules contain an array of microbicidal proteins, such as serine proteases, acid hydrolases, DEFENSINS, bactericidal permeability-increasing protein (BPI), and myeloperoxidase, as well as a number of microbiostatic proteins, such as metalloproteases, lactoferrin, and vitamin B12 binding protein [7]. In NEUTROPHILS, these proteins are divided among several distinct types of vesicles, i.e., azurophilic, specific and secretory granules. Simultaneously with the fusion of the granules with the phagosomal membrane, the NADPH OXIDASE enzyme in this membrane is activated. This enzyme complex, which is assembled from a number of membrane and cytosolic subunits upon appropriate phagocyte stimulation, pumps electrons donated by NADPH in the cytosol into the PHAGOSOME, which combine with molecular oxygen to form superoxide (O_2^-), an anion radical with high reactivity. This sudden increase in oxygen consumption is known as the RESPIRATORY BURST. To compensate for the negative charge delivered to the PHAGOSOME, protons and other cations are also pumped into the PHAGOSOME (Fig. 8.5). The influx of potassium ions is also instrumental in the release of CYTOTOXIC proteins from the proteoglycan matrix of the azurophil granules. The superoxide product of the NADPH OXIDASE enzyme is spontaneously converted into hydrogen peroxide, which then reacts with chloride anions in a myeloperoxidase-catalyzed reaction to form hypochlorous acid ($HOCl^-$). The latter compound is a highly reactive substance and very toxic to many bacteria and fungi. HOCl can also react with primary and secondary amines to form *N*-chloramines, which are as toxic as HOCl but much more stable. Thus, the NADPH OXIDASE enzyme is essential in the microbicidal action of PHAGOCYTES, both by liberating proteolytic enzymes and by generating reactive oxygen compounds.

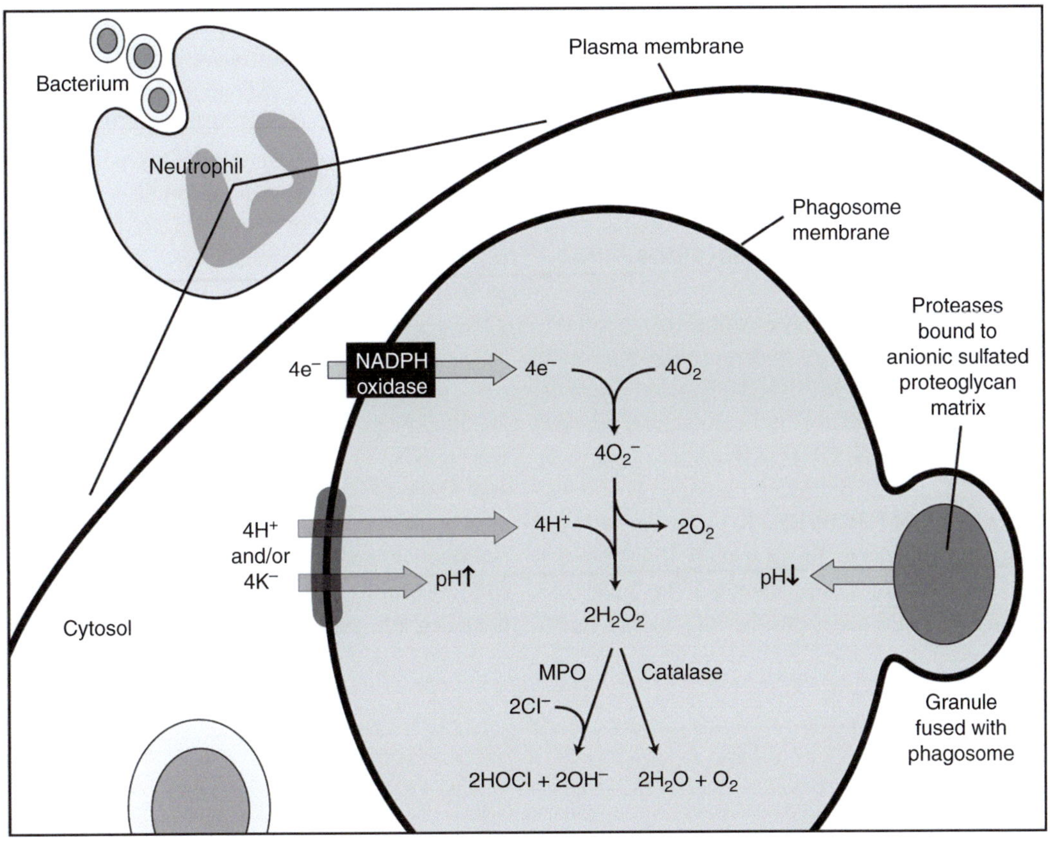

Fig. 8.5 Reactions in the phagosome. The active NADPH oxidase transports electrons across the phagosomal membrane into the phagosome, where the electrons combine with molecular oxygen to generate superoxide (O_2^-). The resulting charge separation is largely compensated for by protons (H^+), which are transported by a voltage-gated channel into the phagosome. However, this charge compensation by protons is not total, because the intraphagosomal pH rises during the first few minutes after phagosome formation. Potassium ions (K^+) also enter the phagosome, and these ions are instrumental in releasing proteases from their proteoglycan matrix in the azurophil granules that have fused with the phagosome. Superoxide combines with protons to form hydrogen peroxide (H_2O_2), which can combine either with chloride ions (Cl^-) in a myeloperoxidase (MPO)-catalyzed reaction to hypochlorous acid (HOCl), or with another molecule of H_2O_2 in a catalase-mediated reaction to water and molecular oxygen. Figure reproduced with permission from [28]

The biological significance of the microbicidal apparatus of PHAGOCYTES is again illustrated by the consequences of its failure. Patients with chronic granulomatous disease (CGD), whose PHAGOCYTE lack an active NADPH OXIDASE, suffer generally already at an early age from very serious infections caused by catalase-positive micro-organisms (catalase-negative organisms themselves secrete some hydrogen peroxide, which can be used by CGD PHAGOCYTES to kill these organisms). Patients with a deficiency of specific granules (a very rare disorder) suffer from recurrent infections with various microbes. Patients with the syndrome of Chédiak-Higashi are characterized by NEUTROPENIA and recurrent infections with purulent micro-organisms. The PHAGOCYTES (and many other cell types) of these patients contain aggregated granules, which decrease cell mobility and DEGRANULATION. Infections in CGD and Chédiak-Higashi patients are treated with intravenous antibiotics and surgical drainage or removal of resistant infections. Prophylactic treatment with trimethoprim-sulfamethoxazole is very successful. In addition, prophylaxis with high doses of vitamin C in Chédiak-Higashi patients and with IFN-γ in CGD patients may also be beneficial. BONE MARROW transplantation is at present the only curative therapy.

8.6 Inflammatory Reactions

PHAGOCYTES also play a critical role in inflammatory responses, e.g., by presenting microbial ANTIGENS to LYMPHOCYTES, by releasing inflammatory mediators (chemotactic peptides, LEUKOTRIENES, CYTOKINES; see Chaps. 1, 6 and 10), and by removing damaged host cells. Moreover, NEUTROPHILS in particular can also cause tissue damage; this is usually limited to the infectious period and intended to give the PHAGOCYTES access to the infectious agents. However, in chronically inflamed areas, such as those caused by autoimmune reactions, permanent inflammatory (M1-like)MACROPHAGE activation will occur, and NEUTROPHIL influx and activation will continue. This will lead to excessive release of toxic substances (i.e. ROS and proteases) from the NEUTROPHILS and pro-inflammatory mediators from the MACROPHAGES. Under normal conditions, these neutrophil proteases are quickly inactivated by serine protease inhibitors (serpins such as α1-antitrypsin) and α2-macroglobulin, which are abundantly present in plasma and tissue fluids. During NEUTROPHIL activation, however, ROS and elastase released from these cells will inactivate these protease inhibitors. Moreover, the reactive oxygen compounds will also activate METALLOPROTEASE precursors, which will then degrade tissue matrix proteins. Figure 8.6

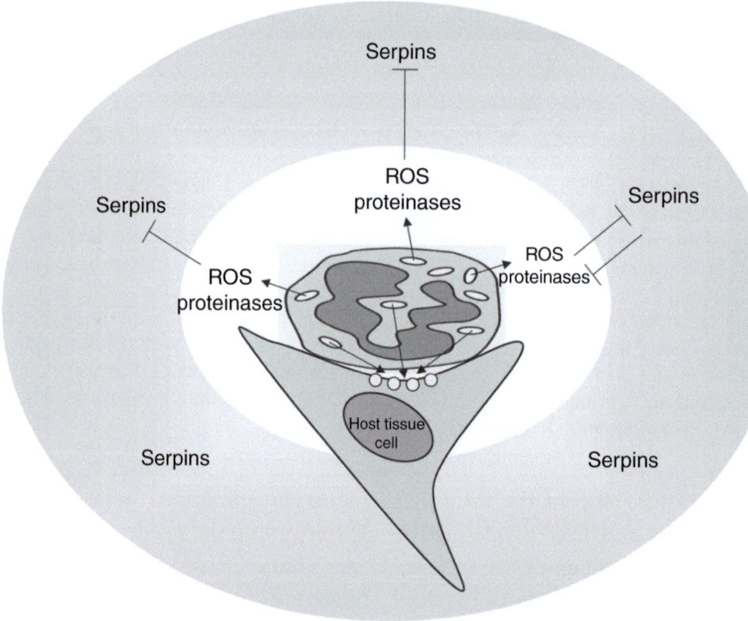

Fig. 8.6 Interplay between oxidative and proteolytic inflammatory reactions. During neutrophil activation, by pathogens, for instance, degranulation occurs, and proteinases are released. Among these proteinases, serine proteases (elastase, cathepsis G, proteinases 3, urokinase-type plasminogen activator) are released from the azurophilic granules. Normally, these proteinases are inactivated by serine protease inhibitors, collectively termed serpins [α1-proteinase, α2-macroglobulin, α1-chemotrypsin inhibitor, and secretory leukoproteinase inhibitor (SLPI)] produced in tissue fluids and plasma. However, elastase and reactive oxygen species (ROS), also produced during neutrophil activation, can inactivate these serpins permitting proteolytic activity towards pathogens, but also against normal tissue. ROS also induce the activation of metalloproteinases, such as collagenase and gelatinase released from specific granules, which in their activated form, can also directly contribute to tissue damage or indirectly via the inactivation of serpins. Thus, ROS and proteases collaborate to inflict tissue destruction during inflammation

provides an overview of these reactions. Clearly, when this process is not self-limiting, irreversible tissue damage may result. In addition, serpins involved in regulating the complement, coagulation, fibrinolytic, and the contact system cascades may also be inactivated, which will further add to the severity of the clinical symptoms. Well-known clinical conditions in which this may happen are septic shock, gout, RHEUMATOID, autoimmune vasculitis, some types of glomerulonephritis, adult respiratory distress syndrome, lung emphysema, acute myocardial infarction, burns, major trauma, and pancreatitis.

To limit the extent of these inflammatory reactions, phagocytes, especially NEUTROPHILS, have a very short survival time, because they are programmed to die within a few days after leaving the BONE MARROW by APOPTOSIS. This form of cell death prevents leakage of toxic compounds from the cells into the surroundings but instead leads to surface expression of molecules that induce binding, uptake, and degradation of the apoptotic cells by MACROPHAGES. Moreover, to prevent excessive phagocyte activation, nature has equipped these cells with a number of inhibitory mechanisms to dampen their activities. These include anti-inflammatory cytokines, such as IL10, for example (see Chap. 6), but also inhibitory IMMUNORECEPTORS expressed on the surface of phagocytes. The latter include, for instance, the inhibitory IMMUNORECEPTOR signal regulatory protein (SIRP)α. SIRPα association with the broadly expressed CD47 surface protein causes phosphorylation of its cytoplasmic IMMUNORECEPTOR TYROSINE-BASED INHIBITION MOTIFS (ITIM) and subsequent recruitment of the protein tyrosine phosphatases SHP-1 and SHP-2, which inhibit activatory signaling pathways dependent on tyrosine phosphorylation. Another example is the CD200R IMMUNORECEPTOR, also primarily expressed on PHAGOCYTES, which, upon binding to its LIGAND, the broadly expressed molecule CD200, provides inhibitory signals that suppress phagocyte function. Certain pathogens, such as pox or herpes viruses, encode CD200 homologues that, when expressed on virus-infected cells, act to suppress phagocyte functions via CD200 triggering and therefore to evade immunity.

MONOCYTES and MACROPHAGES also contain an inhibitory Fcγ RECEPTOR, FcγRIIb, which, in contrast to the activating Fcγ RECEPTORS, does not contain or associate with proteins with a cytoplasmic IMMUNORECEPTOR TYROSINE-BASED ACTIVATING MOTIF (ITAM) but instead contains again an ITIM. Probably, this FcγRIIb again serves to counterbalance phagocyte activities.

As discussed in chapters 1 and 7, MACROPHAGES are crucial to the limitation and resolution of inflammatory responses. In addition to phagocytosing apoptotic cells, including NEUTROPHILS (a process called efferocytosis), MACROPHAGES also take up and degrade various particles and tissue breakdown products. This cleaning up process is facilitated by the inherent plasticity (phenotype variability) of MACROPHAGES. During early stages of infection and inflammation, MACROPHAGES of the inflammatory (M1-like) phenotype are most frequent, which generate proinflammatory cytokines and enzymes and are able to process antigens for presentation to T cells. Under the influence of various hormonal (e.g. CORTICOSTEROID), protein (e,g, IL-10) and lipid mediators, such as SPECIALIZED PRO-RESOLVING MEDIATORS (SPMs), an anti-inflammatory (M2-like) MACROPHAGE phenotype is generated which is mainly phagocytic and promotes anti-inflammatory processes.

8.7 Natural Killer Cells

NK cells were described for the first time about 30 years ago. Operationally, these cells were defined by their ability to kill certain tumor cells *in vitro* without prior contact with these tumor cells. Development of NK cells does not require gene rearrangements as is the case for T LYMPHOCYTES, for example, but NK cells are nonetheless developmentally closely related to T LYMPHOCYTES (reviewed in [8]). Although clearly derived from HEMATOPOIETIC STEM CELLS (see Chap. 2), the anatomical site(s) of differentiation of NK cells is still unknown, though much of the NK cell development occurs in the bone marrow. CYTOKINES are critical for the development of NK cells both in humans and mice. For instance, NK cells are absent in mice with a deficiency of the gamma chain of the IL-2 RECEPTOR and in severe combined immunodeficiency patients who present with mutations in this gamma chain. The IL-2Rγ chain is shared by RECEPTORS for several CYTOKINES, including IL-15. This latter cytokine appears to be essential for optimal NK cell development [9].

NK cells are implicated in INNATE IMMUNITY against foreign tissue, tumor cells, and microbes such as parasites, intracellular bacteria, and viruses [10]. They appear to be important in early phases of the immune response, in which T LYMPHOCYTES are not yet involved. There is convincing evidence that, in man, NK cells are involved in defense against viral infections, in particular against herpes viruses [11]. The mechanisms by which NK cells mediate their effects in infections have not yet been fully elucidated, but it seems likely that CYTOKINES, like IFN-γ, produced by the NK cells are involved. In addition, NK cells can control virus infections by killing virus-infected cells. Excessive activation of NK cells may be deleterious. In animal models for lethal sepsis, it has been shown that elimination of NK cells prevents mortality and improves outcome. It should be noted that it has recently become clear that apart from NK cells, other groups of lymphoid cells exist that lack rearranging antigen receptors, and these are termed innate lymphoid cells (ILC). Although much less is known about these ILC than NK cells, it appears that they are primarily localized in the skin, lymphoid and mucosa-associated tissues and are important as a first-line INNATE defense against pathogens. Their activity depends largely on the production of different cytokines, which has led to their classification into three distinct ILC classes, including ILC1s which release Th1-like CYTOKINES, such as IFNγ facilitating defence against bacteria; ILC2s which generate Th2-like CYTOKINES, promoting defence against helminths; and ILC3s which generate mediators, such as IL17 to promote antibacterial immunity, and lymphoid tissue promoting CYTOKINES (see also Chap. 2), that mediate their effects via other immune cells.

NK cells can mediate ACUTE REJECTION of BONE MARROW grafts [12]. This is not the "raison d'être" of NK cells, of course, but this phenomenon has led to the concept that NK cells recognize cells in which one or more self-MHC class I ANTIGENS are lacking or modified, the "missing self hypothesis," which would explain why normal tissue is protected against NK cell-mediated lytic activity [12].

8.8 Cytokine Regulation of NK Cells and the Role of NK Cell Cytotoxicity in Immunity Against Infections

NK cells are intermingled in an intricate cytokine network; they respond to and produce CYTOKINES that play a role in immunity against infections [9, 11, 13]. NK cells respond to IL-15 produced by MONOCYTES and to IL-12 produced by infected MONOCYTES and DENDRITIC CELLS. These CYTOKINES induce growth of NK cells, and particularly IL-12 induces NK cells to produce IFN-γ rapidly after infection. IFN-γ not only has antiviral effects itself but is also a strong inducer of IL-12 production. Moreover, it has been shown convincingly that IFN-γ-activated MACROPHAGES are instrumental in the immune response against certain microorganisms such as *Listeria monocytogenes* [14]. Furthermore, IL-12 plays an essential role in induction of Th1 LYMPHOCYTES (producing IFN-γ but not IL-4; see Chap. A5). Thus, a complex interplay between DENDRITIC CELLS, MACROPHAGES, NK cells, and T LYMPHOCYTES ensures high levels of production of IFN-γ and IL-12, amplified through positive feedback loops. IL-10, a product of MACROPHAGES, LYMPHOCYTES, and other cell types, is a strong negative regulator of IL-12 production by PHAGOCYTES and of IFN-γ production by NK cells. IL-10 may be produced relatively late in an immune response, dampening the strong responses induced by IL-12 and IFN-γ.

Activation of NK cells by viral and microbial infections enhances CYTOTOXIC activity. This is mediated by IFN-α/β, produced by virus- or bacteria-stimulated NATURAL IFN-producing cells (IPC), also called plasmacytoid DENDRITIC CELLS (pDC) [15]. Bacteria may induce IFN-α/β production through unmethylated CpG motifs, which are prevalent in bacterial but not in vertebrate genomic DNA. Oligodeoxynucleotides (ODN) containing unmethylated CpG motifs activate host defense mechanisms, leading to innate and acquired immune responses. The recognition of CpG motifs requires TLR9 that is expressed on IPC/pDC [15]. Some microbial infections, however, activate NK CYTOTOXICITY directly without IFN-α/β induction; this appears to be dependent on IL-12 and IFN-γ.

8.9 Recognition by NK Cells

There are two mechanisms of cell-mediated cytolysis. One is mediated through perforin, a protein secreted by CYTOTOXIC LYMPHOCYTES that forms pores in the membranes of TARGET cells. TARGET cells can also be killed by an interaction of the Fas molecule on the TARGET cell with its LIGAND on the CYTOTOXIC cell. This interaction activates proteases in the TARGET cell, resulting in APOPTOSIS. Clearly NK cells mediate their CYTOTOXIC effects predominantly by a perforin-dependent mechanism, because little NK activity is present in perforin-deficient mice [16]. The remaining NK activity is probably mediated by Fas/FasL interaction [17–19].

The mechanisms of NK cell recognition and the RECEPTORS involved have remained elusive for a long time, but recent studies have shed light on the complex way in which NK cells recognize their TARGET cells. NK cells do not have one single NK RECEPTOR that accounts for all biological responses, such as cytokine production and CYTOTOXIC activity. Rather, it appears that NK cells utilize an array of RECEPTORS that induce their effector functions, which are often counterbalanced by inhibitory RECEPTORS specific for self-MHC class I ANTIGENS [10]. The positive and negative signaling pathways used by NK cells share many common features with IMMUNORECEPTORS expressed by LYMPHOCYTES and other immune cells. Signals are transmitted by small transmembrane adaptor proteins that possess the so-called ITAM in their cytoplasmic domains. One of these adaptors, DAP12, associates with numerous NK cell RECEPTORS, including Ly49, CD94/NKG2C, and CD94/NKG2E in mice, and in humans with several activating killer IMMUNOGLOBULIN-like RECEPTORS (KIR), with CD94/NKG2C and with NKp44. Other adapters, such as CD3ζ and the FcR common γ-chain, associate with the human NK cell RECEPTORS NKp30, NKp46, and CD16 [10]. The latter RECEPTORS appear to bind the a variety of ligands expressed on host cells, including tumor cells, as well as virus-infected cells and other pathogens [20]. The NKp44 and NKp46 RECEPTORS also recognize antigens expressed on tumor cells, but the nature of these ligands is unknown.

Another NK cell RECEPTOR, NKG2D, has spurred much interest recently, since it allows NK cells to recognize virus-infected and transformed cells [21]. NKG2D is associated with the adapter DAP10, expressed as a transmembrane-anchored disulfide homodimer. Ligands of NKG2D include cell surface ANTIGENS that are up-regulated on transformed or virus-infected cells, such as MICA and MICB, two MHC-like stress-dependent cell surface ANTIGENS. MONOCLONAL ANTIBODIES against many adhesion, activation, or COSTIMULATORY MOLECULES on NK cells are able to activate these cells in vitro. These molecules include CD2, CD27, CD28, CD44, CD69, LFA-1, and DNAM-1 [20, 22]. However, whether NK cells are activated through one or more of these RECEPTORS in the responses against infected cells or in graft rejection in vivo remains to be determined.

The strong CYTOTOXIC activities of activated NK cells raise the question as to how normal tissue is protected from attack by these cells. A solution to this conundrum came from studies on the phenomenon of hybrid resistance [12]. It was recognized in 1979 that NK cells mediate hybrid resistance to BONE MARROW or tumor grafts. This is a situation in which BONE MARROW or tumor grafts of parental origin (either A or B) are rejected by AxB F1 hosts (A and B designate the MHC genotype). This resistance cannot be mediated by T LYMPHOCYTES, because these cells are tolerant to the A and B MHC antigens of the parents. It is now clear that NK cells possess a sophisticated system of inhibitory RECEPTORS that account for their ability to reject BONE MARROW grafts that lack some MHC ANTIGENS present on the NK cells themselves. These inhibiting RECEPTORS interact with MHC ANTIGENS (Fig. 8.7). This feature of NK cells allows them to efficiently kill MHC class I-negative tumor cells and to remove infected cells with down-regulated self-MHC. What is more important, this provides for a mechanism by which normal tissue is protected against cytolytic activity by autologous NK cells. It is assumed that all NK cells express at least one RECEPTOR for self-MHC class I ANTIGENS. It is becoming clear that developing NK cells are subject to a continuous process of education that controls their responsiveness to target cells.

Two groups of these inhibitory MHC-binding RECEPTORS have now been identified (Table 8.2) [10]. One group consists of C-type

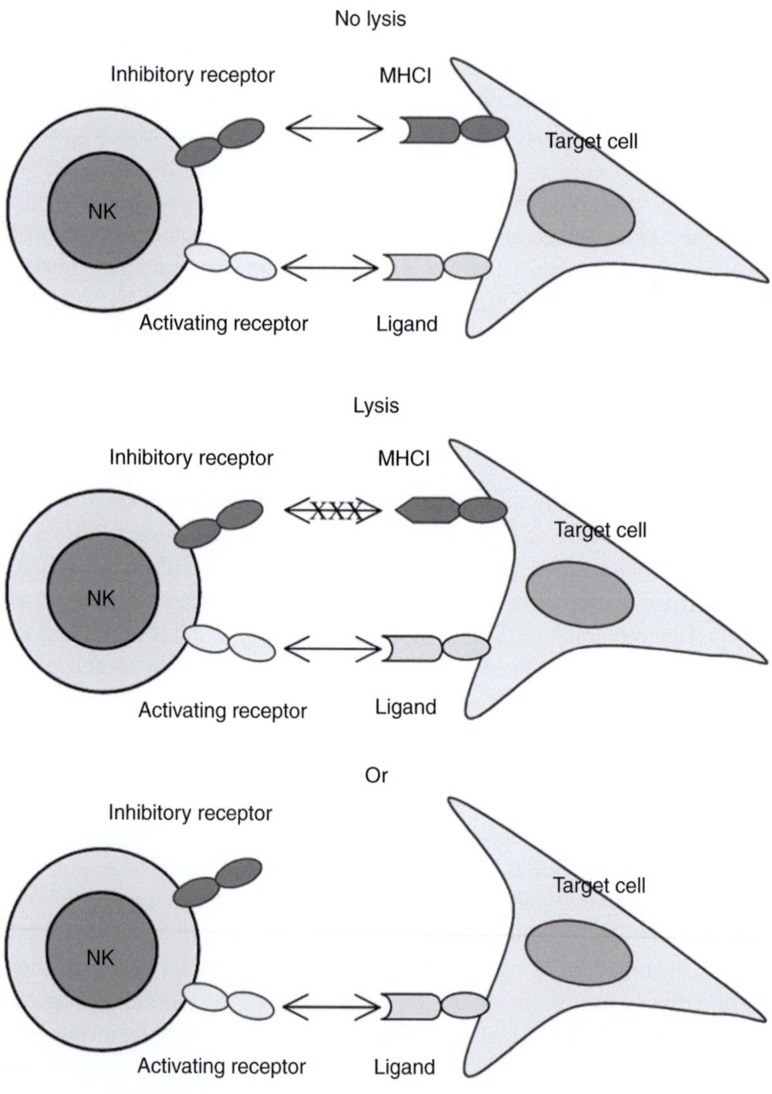

Fig. 8.7 Regulation of NK activity by MHC class I-binding receptors. Lysis by NK cells does not occur when an inhibitory receptor interacts with an MHC class I antigen on the target cell, despite the fact that a cytolysis-activating receptor also interacts with its ligand. NK cells lyse target cells when the interaction between the inhibitory receptor is not triggered, either because the receptor is not specific for the MHC antigen, when this interaction is blocked by antibodies, or when the target cell does not express MHC class I antigens at all

Table 8.2 Inhibitory NK cell receptors for MHC molecules

Species	Receptor	Ligand
Mouse	Ly49	H-2K, H-2D
	CD94/NK2A	Qa-1b
Human	KIR2DL2, KIR2DL3	HLA-C1, HLA-B*73, -B*46
	KIR2DL1	HLA-C2
	KIR2DL4	HLA-G
	KIR2DL5A+B	?
	KIR3DL2	HLA-A*3, -A*11
	KIR3DL1	HLA-Bw4
	KIR3DL3	?
	CD94/NKG2A	HLA-E

LECTIN molecules and is exemplified by the Ly49 gene family in the mouse. Ly49A is the best characterized gene and encodes a disulfide-bonded homodimer that binds to H-2D and H-2K molecules. As a consequence, TARGET cells that express these MHC ANTIGENS are not lysed by Ly49A-positive NK cells. The Ly49 family may comprise around ten members with different, though overlapping, MHC class I, H-2K, and H-2D, specificities (Table 8.2). A second murine inhibitory RECEPTOR is CD94/NKG2A, which recognizes the MHC class I-like molecule Qa-1b. Human homologues of Ly49 genes have not yet been identified. However, there is a human homologue of CD94/NKG2A that recognizes HLA-E, an antigen with limited POLYMORPHISM.

In humans, a second group of inhibitory RECEPTORS known as killer immunoglobulin-like receptors (KIR) appears to function on NK cells for the recognition of different HLA ANTIGENS (Table 8.2). Unlike the Ly49 RECEPTORS, KIR are type I glycoproteins and members of the immunoglobulin superfamily. In total there are 17 different KIR genes, of which some have inhibitory and others activating potential. Most of these are highly genetically diverse with a large variety of polymorphic variants present in the population. Although the exact roles of the various KIR family members are not well established, genetic association studies have indicated roles for human KIR receptors in the context of viral infection, tumor development, transplantation, and autoimmunity. KIR with Ig gene similarity that bind to MHC class I are not found in mice.

The cytoplasmic domains of all inhibitory NK RECEPTORS possess ITIM. These ITIM are phosphorylated upon engagement of the RECEPTORS and recruit phosphatases to counteract activating signals that induce phosphokinase activation. The src homology-containing phosphatases SHP-1 and SHP-2 are the predominant tyrosine phosphatases, but some Ly49 RECEPTORS can recruit the SH2 domain-containing inositol phosphatase SHIP.

8.10 The Complement System

The COMPLEMENT SYSTEM consists of more than 30 proteins. Most of these are synthesized in the liver and circulate in blood as inactive precursor proteins, also known as COMPLEMENT COMPONENTS. In addition, low levels of COMPLEMENT COMPONENTS are produced at extrahepatic sites such as immune cells and adipocytes. Some complement proteins are expressed as membrane proteins, which serve to dampen undesired activation on cell membranes. During activation, one factor activates the subsequent one by limited proteolysis and so on [23]. Because this activation process resembles a cascade, the COMPLEMENT SYSTEM is considered as one of the major plasma cascade systems, the others being the coagulation, the fibrinolytic, and the contact systems. The physiological role of the COMPLEMENT SYSTEM is to defend the body against invading microorganisms and to help remove immune complexes and tissue debris. Hence, deficiencies of complement may predispose to bacterial infections, and immune complex and AUTOIMMUNE DISEASES, depending on which COMPLEMENT COMPONENT is affected.

8.11 Activation

The COMPLEMENT SYSTEM can be activated by three pathways, which are initiated via separate mechanisms and eventually converge in a common TERMINAL PATHWAY. The CLASSICAL PATHWAY consists of the C1 complex (composed of the recognition molecule C1q and the associated proteases C1r and

C1s), C4 and C2. The CLASSICAL PATHWAY is activated by binding of C1q to IgG- or IgM-containing immune complexes or other repetitive structures such as the acute-phase protein CRP [24]. The LECTIN PATHWAY essentially uses the same molecules as the CLASSICAL PATHWAY, e.g., C4 and C2, except that its recognition molecules are mannan-binding LECTINS (MBL) or ficolins that bind to a wide array of carbohydrate structures on pathogenic surfaces and are associated with the serine proteases MASP-1 and MASP-2. The ALTERNATIVE PATHWAY, consisting of factors B, D, and properdin, can be initiated by spontaneous hydrolysis of complement component C3 into C3(H_2O) that subsequently binds factor B, which is then enzymatically cleaved by factor D, eventually resulting in a fluid phase C3 convertase C3(H_2O)Bb that is able to cleave multiple copies of C3 into the biologically active fragments C3b and C3a. AMPLIFICATION of the ALTERNATIVE PATHWAY only takes place on surfaces that are not protected by complement inhibitors. TARGET SPECIFICITY is mainly determined by the carbohydrate environment of bound C3b, which influences the outcome of the competition between factor B and factor H, the major ALTERNATIVE PATHWAY inhibitor. The ALTERNATIVE PATHWAY is not only an activation pathway by itself but also provides an AMPLIFICATION loop for the other two pathways. C3b, which is activated via the classical or LECTIN PATHWAY, may initiate the ALTERNATIVE PATHWAY, by forming a C3 convertase complex with factor B on the activating surface, resulting in rapid amplification of C3 activation. C3 convertases are macromolecular complexes formed during classical or ALTERNATIVE PATHWAY activation that cleave and activate C3.

The three pathways converge at the level of the third complement component, C3, to proceed as the common TERMINAL PATHWAY, which consists of C5, C6, C7, C8, and C9 (Fig. 8.8) and forms the MEMBRANE ATTACK COMPLEX. During activation some complement factors, in particular activated C3, covalently bind to the activator, thereby forming ligands for C3 RECEPTORS on phagocytic cells to facilitate PHAGOCYTOSIS. In case activation occurs at a cell membrane, the terminal pathway proteins (C5b to C9) form macromolecular complexes, consisting of polymerized C9 in addition to C5b, C6, C7, and C8, which insert into the membrane as pores. Insertion of complement pores into cell membranes will allow the exchange of ions and hence induce osmotic lysis of the cell. However,

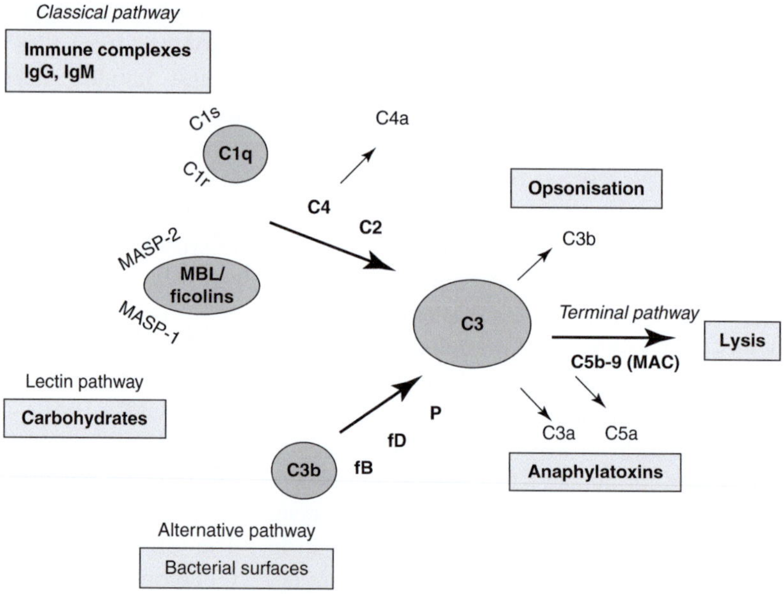

Fig. 8.8 Complement activation pathways. The complement system can be activated via three pathways, the classical, lectin, and alternative pathways, which eventually converge into the common terminal pathway. *MBL* mannan-binding lectin, *MASP* MBL-associated serine protease, *FB* factor B, *FD* factor D, *P* properdin, *MAC* membrane attack complex

under sublytic conditions, complement pores do not lyse cells but rather lead to signal transduction and hence to an altered activation state of the cell. In addition, complement pores may induce an exchange between phospholipids of the outer and inner leaflet of the cell membrane (a so-called flip-flop mechanism).

8.12 Biological Effects

Activation of complement not only fixes complement proteins onto the activator to promote PHAGOCYTOSIS, it also results in the generation of biologically active soluble peptides and macromolecular complexes [23]. Among these are C4a, C3a, and C5a, which are released from C4, C3, and C5, respectively, during activation and—because of their biological effects—are also known as the ANAPHYLATOXINS. For example, C5a, the most potent ANAPHYLATOXIN, is chemotactic for NEUTROPHILS and able to induce aggregation, activation, and DEGRANULATION of these cells. In addition, the ANAPHYLATOXINS may enhance vasopermeability, stimulate adhesion of NEUTROPHILS to endothelium, activate platelets and endothelial cells, and induce DEGRANULATION of MAST CELLS and the production of the vasoactive eicosanoid THROMBOXANE A2 and the peptidoleukotrienes LTC4, LTD4, and LTE4 by mononuclear cells (see Chap. 9). Moreover, they may stimulate or enhance the release of CYTOKINES such as TNF-α and IL-1β and IL-6 from mononuclear cells. The so-called TERMINAL COMPLEMENT COMPLEXES (TCC or C5b–C9), at sublytic concentrations, also induce cells to release mediators, such as CYTOKINES, proteinases, and EICOSANOIDS. Finally, complement activation products may induce the expression of tissue factor by cells and thereby initiate and enhance coagulation. Thus, complement activation products have a number of biological effects that may induce and enhance inflammatory reactions (Table 8.3).

Table 8.3 Biological effects of complement activation products

Complement product	Effect
C5a	Chemotaxis
C5a; C3a	Mast cell degranulation
C5a; C3a	Platelet degranulation
C5a	Phagocyte degranulation
C5a	Stimulation of O_2^- generation by phagocytes
C5a; C5b-9	Enhancement of cytokine release
C5b-9; C5a?	Expression of tissue factor
C5a; C5b-9	Induction of prostaglandin and leukotriene synthesis
C3b; iC3b; C4b	Opsonization of microorganisms
C5b-9	Cell lysis
C3d	Enhanced antibody response

8.13 Complement Regulation

The COMPLEMENT SYSTEM is tightly regulated to prevent excessive activation on a single TARGET, fluid phase activation, and activation on host cells. To this end, COMPLEMENT REGULATORY PROTEINS are present in plasma and on host cell membranes. The initiation step of both the classical and the LECTIN PATHWAY is inhibited by the soluble regulator, C1-inhibitor (C1-inh). Other fluid phase complement inhibitors are factor I, factor H, and C4bp.

To prevent lysis of innocent bystander cells, host cells are equipped with membrane proteins that inhibit complement activation at various levels. The two major mechanisms by which C3 convertases are inhibited are decay-accelerating activity (DAA) and cofactor activity (CA) for factor I. DAA is executed by the membrane regulatory proteins decay-accelerating factor (DAF; CD55), membrane cofactor protein (MCP; CD46, not to be confused with the chemokine MCP), and C3b RECEPTOR (CR1), as well as by the soluble inhibitor factor H. With the help of a cofactor, soluble regulatory protein factor I cleaves C3b into an inactive form iC3b and C4b into iC4b. Cofactors for factor I are the membrane regulatory proteins MCP and CR1, and soluble cofactors are factor H for the enzymatic inactivation of C3b and C4bp for the cleavage of C4b. Membrane inhibitor of reactive lysis (CD59) inhibits complement lysis by interfering with the formation of C5b–C9 complexes in the membrane.

8.14 Evaluation of the Complement System in Patients

The functional state of the COMPLEMENT SYSTEM in patients can be assessed in various ways. The overall activity of the system can be measured by so-called CH_{50} and AP_{50} HEMOLYTIC ASSAYS. In these assays, ANTIBODY-sensitized sheep ERYTHROCYTES (CH_{50}) or non-sensitized rabbit ERYTHROCYTES (AP_{50}) are incubated with dilutions of patient serum. ANTIBODY-sensitized sheep ERYTHROCYTES activate the CLASSICAL PATHWAY while the non-sensitized rabbit ERYTHROCYTES the ALTERNATIVE PATHWAY. The activity of the serum sample is then expressed in units representing the reciprocal of the dilution of serum that lyses 50% of the ERYTHROCYTES. The CH_{50} assay measures the overall activity of classical and common pathways, the AP_{50} assay that of the alternative and the common pathways. These HEMOLYTIC ASSAYS were the first to be applied in clinical studies. Decreased hemolytic activity of sera may occur during activation of complement in patients, because activated complement factors are cleared from the circulation more rapidly than non-activated (native) complement proteins. However, during an ongoing ACUTE-PHASE REACTION, a decrease in complement protein levels may be masked by increased synthesis. Immunochemical determination of individual complement proteins, for example, by nephelometry, has now largely replaced CH_{50} and AP_{50} determinations, the more so since the pattern of the relative decreases of complement proteins may provide important diagnostic and prognostic information. Nowadays, CH_{50} and AP_{50} determinations should only be used to screen for the presence of genetic deficiencies. Deficiencies of the CLASSICAL PATHWAY will yield decreased activity in the CH_{50} assay, while those of the ALTERNATIVE PATHWAY lead to decreased AP_{50} activity. Deficiencies of C3–C9 will yield decreased activity in either assay.

Next to HEMOLYTIC ASSAYS, novel ELISA-based assays have now become available to assess complement function of all three activation pathways. These assays make use of specific ligands fixed to a solid phase for each complement activation pathway (i.e., IgM for the CLASSICAL PATHWAY, mannan for the LECTIN PATHWAY, and LPS for the ALTERNATIVE PATHWAY). Completion of the activation pathways is detected with a monoclonal ANTIBODY against a NEO-EPITOPE on C9 [25].

Activation of complement in patients can best be assessed by measuring levels of specific complement activation products, such as levels of the ANAPHYLATOXINS (C3a and C5a), C3b, C4b, or circulating C5b–C9 complexes. The availability of MONOCLONAL ANTIBODIES specifically reacting with NEO-EPITOPES exposed on activation products and not cross-reacting with the native protein has greatly facilitated the development of specific, sensitive, and reproducible immunoassays for these activation products, which are now frequently used in clinical practice.

8.15 Clinical Relevance

The COMPLEMENT SYSTEM plays a pivotal role in human disease. On the one hand, complement activation has many functions in immunity, and deficiencies within the COMPLEMENT SYSTEM may lead to increased susceptibility to invasive bacterial infections or development of AUTOIMMUNE DISEASES. On the other hand, undesired or excessive complement activation is a major cause of tissue injury in many pathological conditions.

Deficiencies in the early components of the CLASSICAL PATHWAY (C1q, C4, and C2) are associated with an increased risk for the development of AUTOIMMUNE DISEASE, especially SYSTEMIC LUPUS ERYTHEMATOSUS (SLE). Recently, it has become clear that CLEARANCE of apoptotic cells is an important function of the CLASSICAL PATHWAY of complement. Deficiencies in this pathway may therefore lead to prolonged exposure of the specific IMMUNE SYSTEM to apoptotic cells and debris, leading to autoANTIBODY responses. Indeed most, if not all, autoANTIBODIES in patients with SLE are directed against EPITOPES exposed by apoptotic cells. This is known as the "waste disposal hypothesis" [26, 27].

Deficiencies of C3 are associated with recurrent infections by pyogenic microorganisms, because of lack of OPSONIZATION and inability to use the membrane attack pathway. Finally, deficiencies of C5–C8 may lead to an increased risk for *Neisseria* infections. Surprisingly, C9 deficiencies are not associated with an increased risk for infections. Hence, OPSONIZATION of microorganisms by C3 is apparently essential for defense against pyogenic bacteria, whereas the formation of complement pores contributes to defense against *Neisseria*. MBL levels may vary widely in the normal population, and 15–20% of the people in the Western world have strongly decreased levels of functional MBL. A number of studies have shown that individuals with low levels of MBL have an increased risk for infections, particularly when immunity is already compromised, e.g., in infants and in patients with cystic fibrosis and after chemotherapy or transplantation [28].

Activation of complement is considered to play an important role in the pathogenesis of a number of inflammatory disorders, including sepsis and septic SHOCK; toxicity induced by the in vivo administration of CYTOKINES or MONOCLONAL ANTIBODIES; immune complex diseases, such as RHEUMATOID ARTHRITIS, SLE, and vasculitis; multiple trauma; ischemia-reperfusion injuries; and myocardial infarction. The pathogenic role of complement activation in these conditions is probably related to the biological effects of its activation products (Table 8.3). Often, complement activation is not the primary cause of the disease, but excessive complement activation contributes to the severity of disease.

Additionally, deficiencies in complement regulatory proteins may cause disease, which is illustrated by the severe hematological condition, paroxysmal nocturnal hemoglobinuria (PNH). Several complement regulatory proteins, i.e., DAF and CD59, are anchored to the cell membrane via a glycan linkage to phosphatidylinositol. This link is defective in the blood cells of patients suffering from PNH. Hence, the red cells of these patients have strongly reduced levels of these complement-inhibiting membrane proteins and are therefore more susceptible to reactive complement lysis, which largely explains the clinical symptoms of PNH. Furthermore, reduction of membrane regulatory proteins has been found locally in the tissues exposed to complement activation, and this is assumed to contribute to complement-mediated tissue damage in INFLAMMATION. The mechanism of this reduced expression is not clear.

Inhibition of complement activation may be beneficial in conditions of excessive complement activation, which is substantiated by observations in animal models. The availability of clinically applicable complement inhibitors may help in the treatment of these diseases. A C5-blocking monoclonal ANTIBODY (eculizumab) has been approved for treatment of PNH, illustrating the critical role of C5 in this disease. Currently, clinical trials are being conducted to evaluate this novel complement-inhibiting drug in other diseases.

The first complement inhibitor that became available for clinical use was C1-inh, a major inhibitor of the CLASSICAL PATHWAY. It has been used largely due to the fact that a heterozygous deficiency state of this inhibitor is associated with the clinical picture of hereditary angioedema (HAE). This disease sometimes leads to the development of life-threatening edema of the glottis, which must be treated with intubation and the intravenous administration of plasma purified C1-inh. The pathogenesis of angioedema attacks associated with low C1-inh levels is not completely clear but probably involves the generation of BRADYKININ, resulting from the unopposed action of activated coagulation factor XII and kallikrein of the contact system (C1-inh is the main inhibitor of this system as well). Importantly, low levels of functional C1-inh may be caused by a genetic defect but may also be acquired. Acquired C1 inhibitor deficiency is often associated with the presence of autoANTIBODIES against C1-inh, which cause an accelerated consumption of C1-inh. These ANTIBODIES are usually produced by a malignant B-cell CLONE. HAE can be treated by attenuated androgens such as danazol, antifibrinolytic agents, or intravenously administered C1-inh.

8.16 Summary

As indicated in this chapter, the three branches of the INNATE IMMUNE SYSTEM cooperate to protect against PATHOGENIC MICROORGANISMS and to remove infected, dysregulated, damaged, or outdated cells. PHAGOCYTES act by migrating to infected areas and by ingesting and killing microorganisms. NK cells induce APOPTOSIS in virus-infected or tumor cells and other INNATE LYMPHOID cells facilitate the functions of specific effector Th cells. MACROPHAGES remove apoptotic and aged cells and promote the resolution of inflammation. Finally, the COMPLEMENT SYSTEM helps PHAGOCYTES to find and ingest microorganisms but also leads to direct lysis of microbes. Each of these systems is potentially dangerous to the host as well. Therefore, a very tight regulation of their activities exists to protect the host tissue against damage to innocent bystander cells and against excessive and long-lasting activation of these systems. Therapeutic intervention in the case of deficiencies or dysregulation is limited, but the possibilities may increase as we gain more insight into the basic principles.

Acknowledgments This chapter is a partial update of versions in previous editions of this book, originally created by D. Roos and C.E. Hack and subsequently modified by T.K. van den Berg and D. Wouters.

Selected Readings

Černý J, Stříž I. Adaptive innate immunity or innate adaptive immunity?Clin Sci (Lond). 2019;133:1549–1565.

Cerwenka A, Lanier LL. NKG2D ligands: unconventional MHC class I-like molecules exploited by viruses and cancer. Tissue Antigens. 2003;61:335–43.

Fearon DT, Locksley RM. The instructive role of innate immunity in the acquired immune response. Science. 1996;272:50–3.

Frank MM. Complement deficiencies. Pediatr Clin North Am. 2000;47:1339–54.

Janeway CA Jr, Medzhitov R. Innate immune recognition. Annu Rev Immunol. 2002;20:197–216.

Motwani MP, Gilroy DW. Macrophage development and polarization in chronic inflammation. Semin Immunol. 2015;27:257–66.

Worthylake RA, Burridge K. Leukocyte transendothelial migration: orchestrating the underlying molecular machinery. Curr Opin Cell Biol. 2001;13:569–77.

References

1. Springer TA. Traffic signals for lymphocyte recirculation and leukocyte emigration: the multistep paradigm. Cell. 1994;76:301–14.
2. Takeuchi O, Akira S. Pattern recognition receptors and inflammation. Cell. 2010;140:805–20.
3. Roos D, Dolman KM. Neutrophil involvement in inflammatory tissue damage. Neth J Med. 1990;36:89–94.
4. Schroder K, Tschopp J. The inflammasomes. Cell. 2010;140:821–32.
5. Hoffman HM. Therapy of autoinflammatory syndromes. J Allergy Clin Immunol. 2009;124:1129–38.
6. Roos D, Winterbourn CC. Immunology. Lethal weapons. Science. 2002;296:669–71.
7. Borregaard N, Sørensen OE, Theilgaard-Mönch K. Neutrophil granules: a library of innate immunity proteins. Trends Immunol. 2007;28:340–53.
8. Spits H, Lanier LL, Phillips JH. Development of human T and natural killer cells. Blood. 1995;85:2654–70.
9. Liu CC, Perussia B, Young JD. The emerging role of IL-15 in NK-cell development. Immunol Today. 2000;21:113–6.
10. Cerwenka A, Lanier LL. Natural killer cells, viruses and cancer. Nat Rev Immunol. 2001;1:41–9.
11. Biron CA, Nguyen KB, Pien GC. Innate immune responses to LCMV infections: natural killer cells and cytokines. Curr Top Microbiol Immunol. 2002;263:7–27.
12. Kärre K. NK cells, MHC class I antigens and missing self. Immunol Rev. 1997;155:5–10.
13. Trinchieri G. Cytokines acting on or secreted by macrophages during intracellular infection (IL-10, IL-12, IFN-gamma). Curr Opin Immunol. 1997;9:17–23.

14. Unanue E. Inter-relationship among macrophages, natural killer cells and neutrophils in early stages of Listeria resistance. Curr Opin Immunol. 1997;9:35–43.
15. Krieg AM. CpG motifs in bacterial DNA and their immune effects. Annu Rev Immunol. 2002;20:709–60.
16. Kagi D, Ledermann B, Burki K, Seiler P, Odermatt B, Olsen KJ, Podack ER, Zinkernagel RM, Hengartner H. Cytotoxicity mediated by T cells and natural killer cells is greatly impaired in perforin-deficient mice. Nature. 1994;369:31–7.
17. Oshimi Y, Oda S, Honda Y, Nagata S, Miyazaki S. Involvement of Fas ligand and Fas-mediated pathway in the cytotoxicity of human natural killer cells. J Immunol. 1996;157:2909–29116.
18. Kruse PH, Matta J, Ugolini S, Vivier E. Natural cytotoxicity receptors and their ligands. Immunol Cell Biol. 2014;92(3):221–9.
19. Cerwenka A, Lanier LL. NKG2D ligands: unconventional MHC class I-like molecules exploited by viruses and cancer. Tissue Antigens. 2003;61:335–43.
20. Lanier LL, Corliss B, Phillips JH. Arousal and inhibition of human NK cells. Immunol Rev. 1997;155:145–54.
21. Cooper NR. Biology of the complement system. In: Gallin JI, Snyderman R, editors. Inflammation: basic principles and clinical correlates. 3rd ed. New York, NY: Raven Press; 1999. p. 281–315.
22. Gewurz H, Ying SC, Jiang H, Lint TF. Nonimmune activation of the classical complement pathway. Behring Inst Mitt. 1993:138–47.
23. Seelen MA, Roos A, Wieslander J, Mollnes TE, Sjoholm AG, Wurzner R, Loos M, Tedesco F, Sim RB, Garred P, et al. Functional analysis of the classical, alternative, and MBL pathways of the complement system: standardization and validation of a simple ELISA. J Immunol Methods. 2005;296:187–98.
24. Hillmen P, Young NS, Schubert J, Brodsky RA, Socie G, Muus P, Roth A, Szer J, Elebute MO, Nakamura R, et al. The complement inhibitor eculizumab in paroxysmal nocturnal hemoglobinuria. N Engl J Med. 2006;355:1233–43.
25. Walport MJ. Complement. N Engl J Med. 2001;344:1058–66 and 1140–4.
26. Turner MW. The role of mannose-binding lectin in health and disease. Mol Immunol. 2003;40:423–9.
27. Roos D. The respiratory burst of phagocytic leukocytes. Drug Invest. 1991;3(Suppl 2):48–53.
28. van Bruggen R. Built for the kill: studies on the neutrophil NADPH oxidase. Academic Thesis, University of Amsterdam. 2004.

9 Inflammatory Mediators and Intracellular Signalling

Richard Korbut and Tomasz J. Guzik

9.1 Introduction

Inflammation is a protective response of the macroorganism to injury caused by trauma, noxious chemicals or microbiological toxins. This response is intended to inactivate or destroy invading organisms, remove irritants and set the stage for tissue repair. The inflammatory response consists of immunological and non-immunological reactions. The latter are triggered by the release from the injured tissues and migrating cells of lipid-derived autacoids, such as eicosanoids or "platelet-activating factor" (PAF); large peptides, such as interleukin-1; small peptides, such as bradykinin; and amines, such as HISTAMINE or 5-hydroxytryptamine. These constitute the chemical network of the inflammatory response and result in clinical and pathological manifestations of inflammation (Table 9.1). The concept of the inflammatory response was introduced over 2000 years ago with its description by Cornelius Celsus as "*rubor et tumor cum calore et dolore*" (redness and swelling with heat and pain). Centuries later, in the nineteenth century, this definition was extended by Rudolph Virchow to include loss of function ("*functio laesa*"). It was Virchow and his pupils, including J. Cohnheim, who explained the scientific basis for Celsus' description of inflammation. They found that the redness and heat reflected increased blood flow and that the swelling was related to the exudation of fluid and

Table 9.1 Symptoms of inflammation induced by inflammatory mediators

Symptom	Mediators
Vascular permeability	Vasoactive amines Bradykinin Leukotrienes C_4, D_4, E_4 PAF Complement (C3a and C5a) Substance P Nitric oxide
Vasodilatation	Nitric oxide PGI_2, PGE_1, PGE_2, PGD_2 Hydrogen peroxide
Vasoconstriction	Thromboxane A_2, Leukotrienes C_4, D_4, E_4 Superoxide
Chemotaxis and leukocyte adhesion	Chemokines LTB_4, HETE, lipoxins Complement (C5a) Bacterial antigens
Pain	Bradykinin Prostaglandins
Fever	IL-1, TNF, IL-6 Prostaglandins
Tissue and endothelial damage	Reactive oxygen species Nitric oxide Lysosomal enzymes

Final manuscript submitted on January 12, 2017.

R. Korbut (✉)
Jagiellonian University School of Medicine, Krakow, Poland
e-mail: mfkorbut@cyf-kr.edu.pl

T. J. Guzik
Department of Internal and Agricultural Medicine, Jagiellonian University School of Medicine, Krakow, Poland

M. J. Parnham et al. (eds.), *Nijkamp and Parnham's Principles of Immunopharmacology*,
https://doi.org/10.1007/978-3-030-10811-3_9

to the accumulation of cells, while pain follows [1]. The first understanding of the mechanism of inflammation was introduced by Elie Metchnikoff, who concluded in his book *Lectures on the Comparative Pathology of Inflammation* published in 1893 that "…inflammation is a local reaction, often beneficial, of living tissue against an irritant substance" [2]. This definition still stands today. For the first time, he observed that this reaction is mainly produced by phagocytic activity of the mesodermic cells and that it includes "the chemical action of the blood plasma and tissue fluids …", thus introducing the concept of mediators of inflammation [2]. Numerous further studies since then have identified the roles of individual mediators in inflammation, and we are beginning to understand the genetic and molecular aspects of the genesis of the inflammatory process. Inflammatory mediators include a plethora of cell-derived molecules (e.g. CHEMOKINES, CYTOKINES, antimicrobial peptides and reactive oxygen and nitrogen species) and of activated biochemical cascades originating in the vascular compartment (e.g. REACTIVE OXYGEN SPECIES, nitric oxide, COMPLEMENT, coagulation and fibrinolytic systems).

9.2 Eicosanoids

Arachidonic acid (AA) metabolites are formed rapidly from lipids of the cellular membrane, following activation of cells by numerous chemical and physical stimuli. They exert their effects locally (*autacoids*), affecting virtually every step of inflammation [3]. Eicosanoids encompass cyclic prostanoid structures, i.e. *prostaglandins* (PGs), *prostacyclin* (PGI_2) and *thromboxane* A_2 (TXA_2), and also straight-chain *leukotriene* structures (LTs), i.e. chemotactic LTB_4 and pro-inflammatory peptidolipids (LTC_4, LTD_4, LTE_4) [Fig. 9.1]. Subsequently, a new group of molecules was added to the family of eicosanoids,

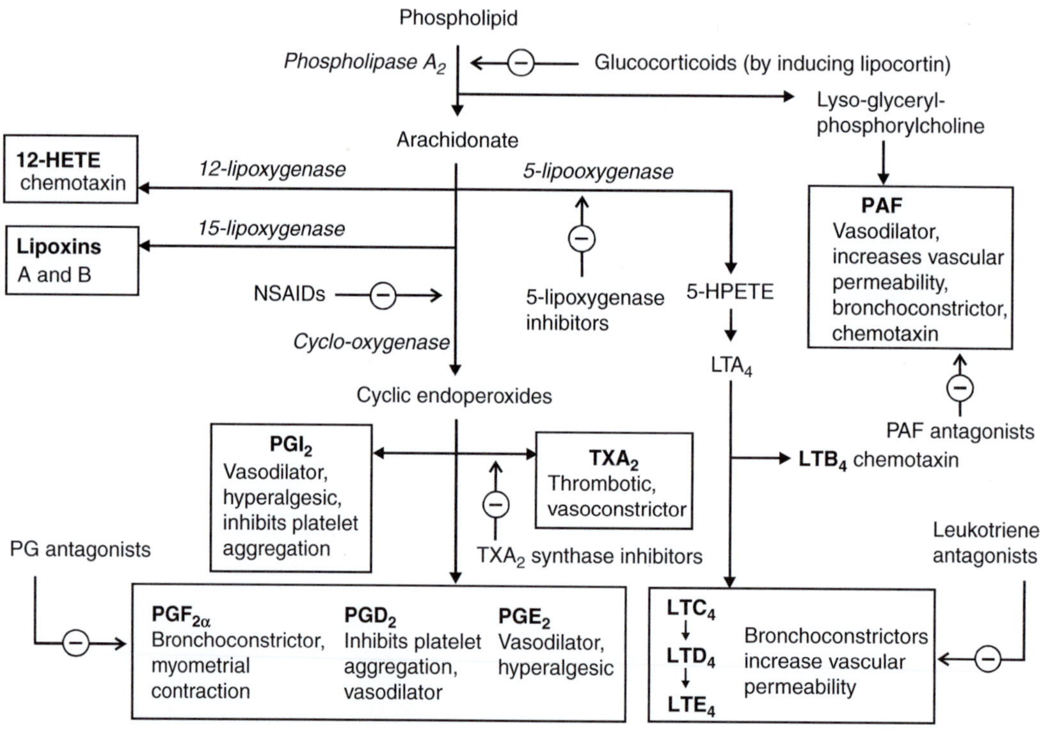

Fig. 9.1 Mediators derived from phospholipids and their actions, the sites of action of anti-inflammatory drugs

namely, *lipoxins* (LXA_4 and LXB_4), which are products of PLATELET 12-lipoxygenase metabolism of NEUTROPHIL LTA_4 (transcellular biosynthesis). Eicosanoids are synthesized by the cyclo-oxygenation (prostanoids) or lipoxygenation (leukotrienes) of a 20-carbon $\omega - 6$ polyunsaturated fatty acids (PUFAs)—5,8,11, 14-eicosatetraenoic acid (AA, arachidonic acid) [Fig. 9.1]. AA is an important structural constituent of cellular phospholipids and first must be liberated by acylhydrolases—directly by phospholipase A_2 (PLA_2) or indirectly by PLC before it becomes the substrate for the synthesis of eicosanoids.

9.2.1 Prostanoids

Prostanoids are produced by the *cyclooxygenase* pathway. Prostaglandin H synthase (PGHS) is a dimeric complex which contains cyclooxygenase (COX) and peroxidase (Px). The COX cyclizes the AA to an unstable cyclic 15-hydroperoxy prostaglandin endoperoxide (PGG_2), while the Px converts the 15-hydroperoxy to a 15-hydroxy group, in this way yielding PGH_2. Eventually, the end product of PGHS (the complex which contains either constitutive COX-1, inducible COX-2 or the COX-3) is an unstable cyclic prostaglandin endoperoxide (PGH_2), which in various types of cells is converted by corresponding isomerases or synthases to the stable prostanoids: PGD_2, PGE_2, $PGF_{2\alpha}$ and unstable prostanoids, i.e. PGI_2 or TXA_2. Special biological significance has been ascribed to PGI_2 synthase in vascular endothelial cells and TXA_2 synthase in blood platelets. The transcellular metabolism providing PGH_2 from activated platelets to endothelial cells is the main source of vascular PGI_2 [4]. The biological activity of stable prostanoids is terminated by catabolic enzymes, such as prostaglandin 15-hydroxy dehydrogenase (15-PGDH), D^{13}-reductase or α and ω oxidases which are present in high concentration in the lungs. These enzymes also break down inactive TXB_2 and 6-keto-$PGF_{1\alpha}$.

The role of individual cyclooxygenase enzymes in the development of inflammation remains unclear. The discovery of the inducible form, COX-2, led to the hypothesis that COX-1 is a constitutive enzyme responsible for the physiological activities of prostaglandins, while COX-2, which is expressed during inflammation, produces "bad" prostaglandins, which generate pain and fever. This hypothesis quickly turned out to be simplistic, and both enzymes show their activities under both physiological and pathological conditions [5]. Moreover COX-2 inhibitory drugs possess less analgesic properties than non-selective inhibitors. The picture became even more complicated in 2002 with the discovery of COX-3. This isoenzyme is not a separate genetic isoform (like COX-2), but a splice variant of COX-1. In fact, COX-1 mRNA gives rise to four different isoforms including classical COX-1, COX-3 (splice variant including intron 1) and two partially truncated, inactive PCOX-1a and 1b. COX-3 due to the presence of intron 1, which changes its conformational structure, shows significantly diminished activity (25%) [6]. It is expressed mainly in the human brain and the heart. It has been suggested that COX-3 is an isoform particularly involved in the mechanisms of pain and fever during inflammation. Some suggestions exist that this isoform is inhibited by paracetamol, which could explain its analgesic actions.

The biosynthesis of prostanoids is initiated by transductional mechanisms in an immediate response to the activation of various cell membrane receptors or to various physical and chemical stimuli. These lead to an increase in the cytoplasmic levels of calcium ions $[Ca^{2+}]i$, and in this way, they activate acyl hydrolases, which thereby release free AA for metabolism by the PGHS. Alternatively, these enzymes can be induced by delayed transcriptional mechanisms which are usually activated by CYTOKINES or bacterial toxins. The spectrum of prostanoids produced by individual tissues depends on the local expression of individual enzymes. For example, vascular endothelium possesses prostacyclin synthase and COX-2, but lacks thromboxane synthase, present in turn in the platelets. Accordingly, the major prostanoid released by endothelium is PGI_2, while platelets produce TXA_2.

Prostanoids regulate vascular tone and permeability in the development of inflammation. They (TX) also induce PLATELET aggregation and thrombus formation. Prostaglandins (in particular PGE_2) are also involved in the pathogenesis of the pain and fever which accompany inflammation.

Most actions of prostanoids appear to be brought about by activation of the cell surface receptors that are coupled by G proteins to either adenylate cyclase (changes in intracellular c-AMP levels) or PLC (changes in triphosphoinositol—IP_3 and diacylglycerol—DAG levels). The diversity of the effects of prostanoids is explained by the existence of a number of distinct receptors. The receptors have been divided into five main types, designated DP (PGD), FP (PGF), IP (PGI_2), TP (TXA_2) and EP (PGE). The EP receptors are subdivided further into EP_1 (smooth muscle contraction), EP_2 (smooth muscle relaxation), EP_3 and EP_4, on the basis of physiological and molecular cloning information. Subtype-selective receptor antagonists are under development. Only one gene for TP receptors has been identified, but multiple splice variants exist. PGI_2 binds to IP receptors and activates adenylate cyclase. PGD_2 interacts with a distinct DP receptor that also stimulates adenylate cyclase. PGE_1 acts through IP receptors, PGE_2 activates EP receptors, but it may also act on IP and DP receptors.

While most prostaglandins participate in the pathomechanism of inflammation, a more recently discovered member of this family—the 15-deoxy-Δ-12,14-prostaglandin J2 (15d-PGJ2)—is the dehydration end product of the PGD_2 and differs from other prostaglandins in several respects. 15d-PGJ2 has been shown to act via PGD_2 receptors (DP1 and DP2) and through interaction with intracellular targets. In particular, 15d-PGJ2 is recognized as the endogenous ligand for the intranuclear receptor PPARγ. This property is responsible for many of the anti-inflammatory functions of 15d-PGJ2.

9.2.2 Products of the Lipoxygenation of Arachidonic Acid

AA can be metabolized to straight-chain products by lipoxygenases (LOXs) which are a family of cytosolic enzymes that catalyse oxygenation of all polyenic fatty acids with two cis double bonds that are separated by the methylene group to corresponding lipid hydroperoxides [7] (Fig. 9.1). As in the case of AA, these hydroperoxides are called hydroperoxyeicosatetraenoic acids (HPETE's). Different LOX enzymes vary in their specificity for inserting the hydroperoxy group, and tissues differ in LOXs that they contain. Platelets contain only 12-LOX and synthesize 12-HPETE, whereas leukocytes contain both 5-LOX and 12-LOX producing both 5-HPETE and 12-HPETE. HPETE's are unstable intermediates, analogous to PGG_2 or PGH_2 and are further transformed by peroxidases or non-enzymatically to their corresponding hydroxy fatty acids (HETE's). 12-HPETE can also undergo catalysed molecular rearrangement to epoxy-hydroxyeicosatrienoic acids called *hepoxilins*. 15-HPETE may also be converted by lipoxygenation of LTA_4 to the trihydroxylated derivatives, the *lipoxins* (Fig. 9.1).

9.2.3 Leukotrienes

In activated leukocytes an increase in $[Ca^{2+}]$i binds 5-LOX to five-lipoxygenase-activating protein (FLAP), and this complex converts AA to 5-HPETE, which in turn is the substrate for LTA_4 synthase. In the course of transcellular metabolism between leukocytes and blood cells or endothelial cells, unstable LTA_4 is converted by corresponding enzymes to stable chemotactic LTB_4 or to cytotoxic cysteinyl-containing leukotrienes—C_4, D_4, E_4 and F_4 (also referred to as sulphidopeptide leukotrienes or peptidolipids) (Fig. 9.1). Note that the transcellular metabolism of AA can bring about either "protection", as is the case during the platelet/endothelium transfer of PGH_2 to make cytoprotective PGI_2 [1], or "damage", as in the case of the leukocyte/endothelium transfer of LTA_4 to make cytotoxic LTC_4 [6].

Consecutive splicing of amino acids from the glutathione moiety of LTC_4 occurs in the lungs, kidney and liver. LTE_4 is already substantially deprived of most of the biological activities of LTC_4 and LTD_4. Also LTC_4 may be inactivated by oxidation of its cysteinyl sulphur atom to a sulphoxide group. The principal route of inactivation of LTB_4 is by ω-oxidation. LTC_4 and LTD_4 comprise an important endogenous bronchoconstrictor, previously known as the "slow-reacting substance of anaphylaxis" (SRS-A) [8].

Three distinct receptors have been identified for LTs (LTB_4, LTC_4 and LTD_4/LTE_4). Stimulation of all of them appears to activate PLC. LTB_4, acting on specific receptors, causes adherence, chemotaxis and activation of polymorphonuclear leukocytes and monocytes, as well as promoting cytokine production in macrophages and lymphocytes. Its potency is comparable with that of various chemotactic peptides and PAF. In higher concentrations, LTB_4 stimulates the aggregation of PMN's and promotes DEGRANULATION and the generation of superoxide. It promotes adhesion of neutrophils to vascular endothelium and their transendothelial migration [9]. The cysteinyl-LTs are strongly cytotoxic and cause bronchoconstriction and vasodilatation in most vessels except the coronary vascular bed.

9.2.4 Lipoxins (Lipoxygenase Interaction Products)

Lipoxins are formed by a sequential transcellular metabolism of arachidonic acid by 15- and 5- or by 5- and 12-lipooxygenases [10]. The cellular context is critical for the synthesis of lipoxins (Fig. 9.2). Lipoxins have several anti-inflammatory properties as well as concomitant pro-inflammatory actions. Lipoxins inhibit the adhesion molecule expression on endothelium, cause vasodilatation and attenuate LTC_4-induced vasoconstriction by antagonism of $cysLT_1$ receptor. They also inhibit chemotaxis, adhesion and transmigration, IL-1β and superoxide production of polymorphonuclear leukocytes. On the other hand, lipoxins stimulate MONOCYTE adhesion and increase IL-4 formation [10, 11]. There is an inverse relationship between the amount of lipoxin and leukotriene production, which may indicate that lipoxins may be "endogenous regulators of leukotriene actions". High-affinity G-protein-coupled lipoxin receptors (ALXR) have been identified on numerous cells, including monocytes, PMNs fibroblasts and endothelial and epithelial cells. Receptor expression may be upregulated by interferon γ, IL-13 or even IL-1β. Activation of this receptor modulates phosphatidylinositol 3-kinase (PI3-kinase) activity. Lipoxins may also competitively bind and block the $cys\text{-}LT_1$ receptor. There are also suggestions that lipoxins may also bind within the cell, to ligand-activated transcription factors, therefore regulating gene expression in the nucleus.

A separate group of lipoxins was termed aspirin-triggered lipoxins (ATLs), as their synthesis is the result of acetylation of cyclooxygenase-2,

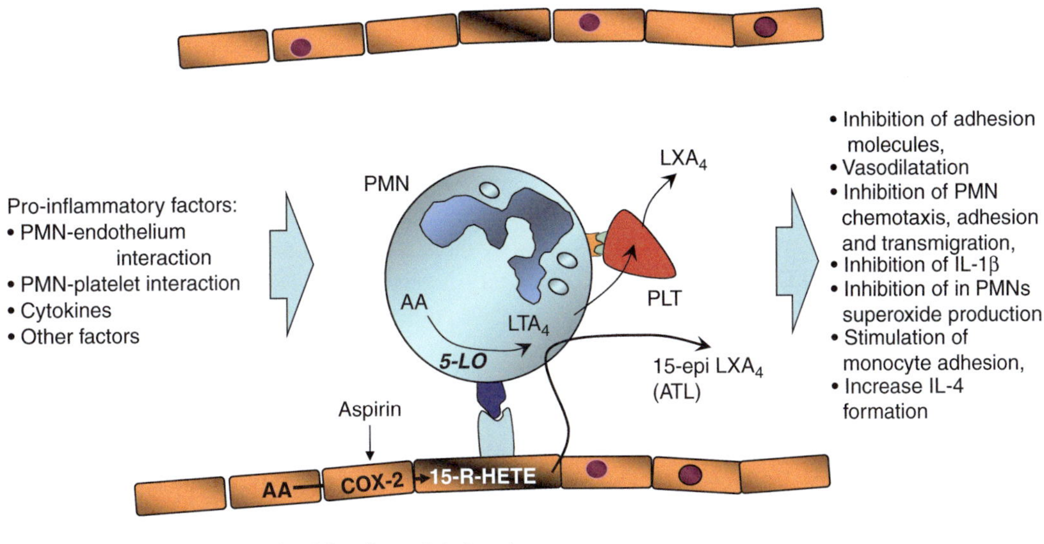

Fig. 9.2 Transcellular synthesis of lipoxins and their actions

which inhibits endothelial cell prostanoid formation and promotes synthesis of 15(R) HETE. These are then converted in PMNs to 15*R*-enantiomeres: 15-epi LXA_4 or 15-epi-LXB_4. ATLs share many actions of lipoxins, albeit with much greater potency [12]. Due to their anti-inflammatory properties, lipoxin analogues may find an important place in the treatment of inflammation [10, 11].

9.2.5 Other Pathways of Arachidonic Acid Metabolism

AA can be also metabolized by a NADPH-dependent cytochrome P-450-mediated monooxygenase pathway (MOX). The resulting 19-HETE, 20-HETE and a number of epoxyeicosatrienoic and dihydroxyeicosatrienoic acid isomers show vascular, endocrine, renal and ocular effects, the physiological importance of which remains to be elucidated [13].

Recently, a non-enzymatic, free radical-mediated oxidation of AA, while still embedded in phospholipids, has been discovered. Subsequently, acyl hydrolases give rise to a novel series of regioisomers of *isoprostanes*. Formed non-enzymatically, isoprostanes lack the stereospecificity of prostanoids. Highly toxic isoprostanes might contribute to the pathophysiology of inflammatory responses which are insensitive to currently available steroidal and non-steroidal anti-inflammatory drugs. The most thoroughly investigated regioisomer of isoprostanes is 8-epi-$PGF_{2\alpha}$. It has a potent vasoconstrictor action which is mediated by vascular TXA_2/PGH_2 receptors.

9.2.6 Actions and Clinical Uses of Eicosanoids

Eicosanoids produce a vast array of biological effects. TXA_2, $PGF_{2\alpha}$ and LTs represent cytotoxic, pro-inflammatory mediators. TXA_2 is strongly thrombogenic through aggregation of blood platelets. LTC_4 injures blood vessels and bronchi subsequent to activation of leukocytes. On a molecular level, their cytotoxicity is frequently mediated by stimulation of PLC or inactivation of adenylate cyclase. Cytoprotective, but not necessarily anti-inflammatory, actions are mediated by PGE_2 and PGI_2. They are both naturally occurring vasodilators. PGI_2 is the most comprehensive anti-platelet agent which is responsible for the thromboresistance of the vascular wall. PGE_2 through a similar adenylate cyclase-dependent mechanism inhibits the activation of leukocytes. PGE_2 is also responsible for protection of the gastric mucosa. PGE_2 and $PGF_{2\alpha}$ may play a physiological role in labour and are sometimes used clinically as abortifacients. Locally generated PGE_2 and PGI_2 modulate vascular tone, and the importance of their vascular actions is emphasized by the participation of PGE_2 and PGI_2 in the hypotension associated with septic shock. These prostaglandins also have been implicated in the maintenance of patency of the ductus arteriosus. Various prostaglandins and leukotrienes are prominent components released when sensitized lung tissue is challenged by the appropriate antigen. While both bronchodilator (PGE_2) and bronchoconstrictor ($PGF_{2\alpha}$, TXA_2, LTC_4) substances are released, responses to the peptidoleukotrienes probably dominate during allergic constriction of the airway. The relatively slow metabolism of the leukotrienes in lung tissue contributes to the long-lasting bronchoconstriction that follows challenge with antigen and may be a factor in the high bronchial tone that is observed in asthmatics in periods between attacks. Prostaglandins and leukotrienes contribute importantly to the genesis of the signs and symptoms of inflammation. The peptidoleukotrienes have effects on vascular permeability, while LTB_4 is a powerful chemoattractant for polymorphonuclear leukocytes and can promote exudation of plasma by mobilizing the source of additional inflammatory mediators. Prostaglandins do not appear to have direct effects on vascular permeability. However, PGE_2 and PGI_2 markedly enhance oedema formation and leukocyte infiltration by promoting blood flow in the inflamed region. PGEs inhibit the participation of lymphocytes in delayed hypersensitivity reactions. Bradykinin and the CYTOKINES (such as TNF-α, IL-1, IL-8) appear to liberate prostaglandins and probably other mediators that promote hyperalgesia (decreased pain threshold) and the pain of inflammation. Large doses of PGE_2 or $PGF_{2\alpha}$ given to women by intramuscular or subcutaneous injection to induce abortion cause intense local pain. Prostaglandins also can cause headache and vascular pain when infused intravenously. The capacity of prostaglandins to sensitize pain receptors to mechanical and chemical stimulation appears to result from a lowering of the threshold of the polymodal nociceptors of C fibres. Hyperalgesia also is produced by LTB_4. PGE_2 when infused into the cerebral ventricles or when injected into the hypothalamus produces fever. The mechanism of fever involves the enhanced formation of CYTOKINES that increase the synthesis of PGE_2 in circumventricular organs in and near to the preoptic hypothalamic area, and PGE_2, via increases in c-AMP, triggers the hypothalamus to elevate body temperature by promoting increases in heat generation and decreases in heat loss.

Synthetic PGE1, acting through IP and EP receptors, is given by infusion to maintain the patency of the ductus arteriosus in infants with transposition of large vessels until surgical correction can be undertaken. PGI_2 (epoprostenol) is occasionally used to prevent PLATELET aggregation in dialysis machines through inhibition of the thrombocytopenic action of heparin [14]. PGI_2 is also used for the treatment of primary and secondary pulmonary hypertension [15]. Stable analogues of PGI_2 (e.g. iloprost) as well as of PGE_1 are used in selected patients with peripheral vascular disease [14]. The PGE_1 analogue, misoprostol, is approved in the USA for the prevention of peptic ulcers, especially in patients who are required to take high doses of non-steroidal anti-inflammatory drugs (NSAID) for treatment of their arthritis.

9.2.7 Pharmacological Interference with Eicosanoid Synthesis and Actions

PLA_2 and COX are inhibited by drugs which are the mainstays in the treatment of inflammation. We discovered that GLUCOCORTICOSTEROIDS (hydrocortisone, dexamethasone) inhibit the generation of prostanoids in vivo through prevention of the release of AA from phospholipids [16]. This effect is mediated by intracellular steroid receptors which, when activated, increase expression of lipocortins which inhibit phospholipases. Many other actions of glucocorticosteroids on AA metabolism are known, one of them being inhibition of COX-2 transcription. These problems are further discussed in Chap. 32.

Aspirin selectively inhibits COX-1 explaining its inhibitory effect on the biosynthesis of TXA_2 in platelets (causing reduced thrombotic tending), of PGI_2 in endothelial cells and of PGE_2 in gastric mucosa (leading to gastric damage). This action of aspirin is more pronounced than that on the biosynthesis of prostanoids at the site of inflammation, where inducible COX-2 is most active. Consequently, aspirin at low doses seems to be a better antithrombotic than anti-inflammatory drug. Aspirin irreversibly acetylates the active centre of COX-1. Unlike endothelial cells, platelets lack the machinery required for de novo synthesis of COX-1, and, accordingly, aspirin-induced inhibition of TXA_2 synthesis in platelets is essentially permanent (until new platelets are formed), in contrast to the easily reversible inhibition of PGI_2 synthesis in vascular endothelium. The net effect of aspirin is, therefore, a long-lasting antithrombotic action. Unfortunately, most NSAIDs are more effective inhibitors of COX-1 than of COX-2. Meloxicam was the first clinically available drug which is claimed to be a selective COX-2 inhibitor—an anti-inflammatory drug with few side effects on the gastrointestinal tract, which causes no bleeding. However, population studies have verified that while protective for gastric mucosa, high doses of COX-2 selective inhibitors may induce cardiovascular (due to inhibition of endothelial COX-2) or renal side effects [17]. NSAID's usually are classified as mild analgesics, and they are particularly effective in settings in which inflammation has caused sensitization of pain receptors to normally painless mechanical or chemical stimuli. NSAID's do not inhibit fever caused by direct administration of prostaglandins, but they do inhibit fever caused by agents that enhance the synthesis of IL-1 and other CYTOKINES, which presumably cause fever at least in part by inducing the endogenous synthesis of prostaglandins.

9.3 Platelet-Activating Factor (PAF)

PAF (1-*O*-alkyl-2-acetyl-*sn*-glycero-3-phosphocholine) is a specialized phospholipid with an alkyl group (12–18C) attached by an ether bond at position 1 of glycerol and acetylated at position 2. PAF is not stored in cells, but it is synthesized from 1-*O*-alkyl-2-acyl-glycerophosphocholine as required (Fig. 9.3) [18]. Initially, PLA_2 converts the precursor to the inactive 1-*O*-alkyl-2-lysoglycerophosphocholine (lyso-PAF) with concomitant release of AA. Incidentally, in GRANULOCYTES, AA produced in this way represents a major source for the synthesis of PGs and LTA_4. In a second step, lyso-PAF is acetylated by acetyl coenzyme A in a reaction catalysed by lyso-PAF acetyltransferase. This is the rate-limiting step. The synthesis of PAF in different cells is stimulated during antigen-antibody reactions or by chemotactic peptides (e.g. f-MLP), CYTOKINES, thrombin, collagen and autacoids. PAF can also stimulate its own formation. Both PLA_2 and lyso-PAF acetyltransferase are calcium-dependent enzymes, and PAF synthesis is regulated by the availability of Ca^{2+}. The anti-inflammatory action of glucocorticosteroids is at least partially dependent on inhibition of the synthesis of PAF by virtue of the inhibitory effect of lipocortin on the activity of PLA_2.

Inactivation of PAF also occurs in two steps. Initially, the acetyl group of PAF is removed by PAF acetylhydrolase to form lyso-PAF; this

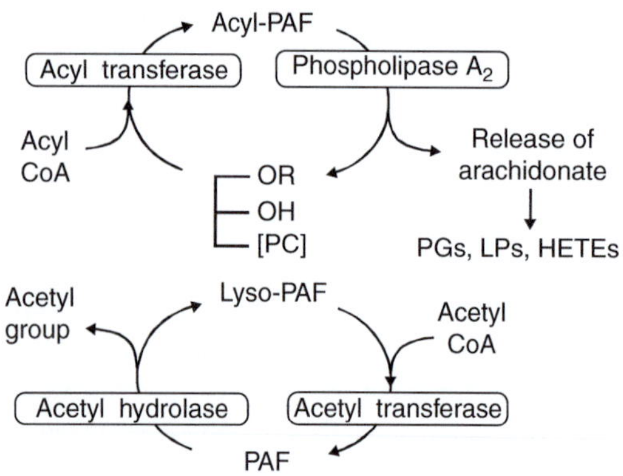

Fig. 9.3 The synthesis and metabolism of platelet-activating factor (PAF)

enzyme is present in both cells and plasma. Lyso-PAF is then converted to a 1-*O*-alkyl-2-acyl-glycerophosphocholine by an acyltransferase. This latter step is inhibited by Ca^{2+}.

PAF is synthesized by PLATELETS, NEUTROPHILS, MONOCYTES, BASOPHILS and MAST CELLS, EOSINOPHILS, renal mesangial cells, renal medullary cells and vascular endothelial cells. In most instances, stimulation of the synthesis of PAF results in the release of PAF and lyso-PAF from the cell. However, in some cells (e.g. endothelial cells), PAF is not released and appears to exert its effects intracellularly.

PAF exerts its actions by stimulating a single G protein-coupled, cell-surface receptor [19]. High-affinity binding sites have been detected in the plasma membranes of a number of cell types. Stimulation of these receptors triggers activation of phospholipases C, D, and A_2 and mobilization of $[Ca^{2+}]i$. Massive direct and indirect release of AA occurs with its subsequent conversion to PGs, TXA_2 or LTs. Eicosanoids seem to function as extracellular representatives of the PAF message. As its name suggests, PAF unmasks fibrinogen receptors on platelets, leading directly to platelet aggregation. In endothelial cells, the synthesis of PAF may be stimulated by a variety of factors, but here PAF is not released extracellularly. Accumulation of PAF intracellularly is associated with the adhesion of neutrophils to the surface of the endothelial cells and their diapedesis, apparently because it promotes the expression or exposure of surface proteins that recognize and bind neutrophils. Activated endothelial cells play a key role in "targeting" circulating cells to inflammatory sites. Expression of the various *adhesion molecules* varies among different cell types involved in the inflammatory response. For example, expression of E-selectin is restricted primarily to endothelial cells and is enhanced at sites of inflammation. P-selectin is expressed predominantly on platelets and on endothelial cells. L-selectin is expressed on leukocytes and is shed when these cells are activated. Cell adhesion appears to occur by recognition of cell surface glycoprotein and carbohydrates on circulating cells by the adhesion molecules whose expression has been enhanced on resident cells. Endothelial activation results in adhesion of leukocytes by their interaction with newly expressed L-selectin and P-selectin, whereas endothelial-expressed E-selectin interacts with glycoproteins on the leukocyte surface, and endothelial ICAM-1 interacts with leucocyte INTEGRINS.

PAF also very strongly increases vascular permeability. As with substances such as histamine and bradykinin, the increase in permeability is due to contraction of venular endothelial cells, but PAF is 1000–10,000 times more potent than histamine or bradykinin.

Intradermal injection of PAF duplicates many of the signs and symptoms of inflammation, including vasodilatation, increased vascular permeability, hyperalgesia, oedema, and infiltration of neutrophils. Inhaled PAF induces bronchoconstriction, promotes local oedema, accumulates EOSINOPHILS and stimulates secretion of mucus. In anaphylactic shock, the plasma concentration of PAF is high, and the administration of PAF reproduces many of the signs and symptoms of experimental anaphylactic shock. PAF receptor antagonists prevent the development of pulmonary hypertension in experimental septic shock. Despite the broad implications of these experimental observations, the clinical effects of PAF antagonists in the treatment of bronchial ASTHMA, septic shock and other inflammatory responses have been rather modest.

PAF receptor antagonists include PAF structural analogues, natural products (e.g. ginkgolides from *Ginkgo biloba*) and, interestingly, triazolobenzodiazepines (e.g. triazolam). The development of PAF receptor antagonists is currently at an early stage of clinical development, still leaving the hope that such antagonists may find future therapeutic application in inflammation and sepsis.

9.4 Innate Immune Signalling Receptors

Several families of innate immune signalling receptors are currently known. Their functionality and subcellular location vary. These receptors include the transmembrane Toll-like receptors (TLRs) and C-type lectin receptors, while other receptors are located in the cytosol, including the retinoic acid-inducible gene-I-like helicases (RLRs) and the nucleotide-binding domain leucine-rich repeat-containing receptors (NLRs). The common property and function of these receptors is to detect a broad variety of molecular entities including lipids, nucleic acids, proteins and their combinations. It is most likely that these innate immune receptors evolved to recognize specific molecules associated with microbial invasion and thus are designed to orchestrate antimicrobial defence—virtually all of these receptors can also detect molecular changes that occur during tissue damage during virtually every kind of inflammation. Moreover, their importance may be emphasized by the fact that they are located at the beginning of the inflammatory cascade.

9.4.1 Toll-Like Receptors

The Toll-like receptors (TLR1-10) are a part of the innate immune defence, recognizing conserved pathogen-associated molecular patterns (PAMPs) on microorganisms [20]. TLRs and their signalling pathways are present in mammals, fruit flies and plants. Ten members of the TLR family have been identified in humans, and several of them appear to recognize specific microbial products, including lipopolysaccharide, bacterial lipoproteins, peptidoglycan, bacterial DNA and viral RNA. Signals initiated by the interaction of TLRs with specific microbial patterns direct the subsequent inflammatory response including mononuclear phagocytic cell cytokine production. Thus, TLR signalling represents a key component of the innate immune response to microbial infection [20]. Interestingly, recent data indicates that TLRs play an important role not only in the modulation of innate immunity but also in the initiation of specific responses of adaptive immunity (Table 9.2). Moreover, T cells express certain types of TLRs during development and activation stages, and they participate in the direct regulation of adaptive immune

Table 9.2 Toll-like receptors in the regulation of adaptive immunity

Toll-like receptor	Recognized molecular pattern of microorganisms	Effect on T-cell function	Classical co-stimulatory effects
TLR1/2	Triacyl lipopeptides	Inhibition or reversal of regulatory T cells suppressive function	
TLR2	Peptidoglycan	Increase of regulatory T cell suppressive function	x
TLR3	ssRNA (viral), dsRNA, respiratory syncytial virus	Expressed; unclear function	x
TLR4	Lipopolysaccharide	Increase of regulatory T cell suppressive function	x
TLR5	Flagellin		x
TLR6	Lipoteichoic acid Zymosan, Diacyl lipopeptides		
TLR7	ssRNA (viral; inc influenza)	Inhibition or reversal of regulatory T cells suppressive function	
TLR8	ssRNA (viral)		
TLR9	Ds DNA (HSV); CpG dinucleotide motifs, haemozoin		x

response, possibly as co-stimulatory molecules. Co-stimulation of CD4+ effector cells with anti-CD3 mAb and TLR-5 ligand flagellin enhances T-cell proliferation and production of IL-2 levels equivalent to those achieved by co-stimulation with classical APC involving CD28. Moreover, CpG-containing oligodeoxynucleotides (CpG-ODN) can co-stimulate primary T cells in the absence of APCs. Finally, TLR activation on APCs may direct the development of immune responses into the regulatory T cells or Th17 pathway. These mechanisms are further discussed in Chaps. 2, 3 and 5.

9.4.1.1 Inflammasomes

Inflammasomes are cytoplasmic protein complexes critical in the regulation of inflammation. The term was first introduced in 2002 to describe a caspase-1-activating multimolecular complex consisting of caspase-1, caspase-5, apoptosis-associated speck-like protein containing a caspase recruitment domain (ASC) and NLRP1 (a NLR family member).

NLRs are a family of 20 intracellular immune receptors characterized by the presence of leucine-rich repeats (LRRs) near the C terminus and a central nucleotide-binding and oligomerization (NACHT) nucleotide-binding domain (NBD). The LRR domains of this family are thought to play a role in autoregulation, the recognition of PAMPs and/or protein-protein interactions. The NBDs can bind ribonucleotides, possibly regulating self-oligomerization. In spite of these similarities, different NLRs differ in their N-terminal domains. Most of these have an N-terminal pyrine domain (PYD) and are therefore called NLRP (NALP). Other NLRs have an N-terminal caspase recruitment domain (CARD) and include nucleotide-binding oligomerization domain-containing-1 (NOD1, also known as NLRC1), NOD2 (NLRC2). CARD domain-containing-4 NLRs are also vital (NLRC4, also known as CARD12 or IPAF). Other NLR family members have an acidic transactivation domain or a baculoviral inhibitory repeat-like domain, such as the member of NLR family, apoptosis inhibitory protein 5 (NAIP5).

The most important property of some NLR as discussed above is their ability to create inflammasomes. This involves particularly NLRP1, NLRP3 and NLRC4 which assemble multimolecular complexes in response to various activators, leading to the activation of inflammatory caspases. Through caspase-1 activation, the INFLAMMASOME controls the maturation of the cytokines of the IL-1 family.

For instance, an example is the NLRP3 inflammasome which can be activated by a myriad of microbial factors including Sendai virus, influenza A virus, adenoviruses, *Staphylococcus aureus*, *Listeria monocytogenes*, *E. coli*, *Mycobacterium marinum* and *Neisseria gonorrhoeae* as well as the fungal *Candida albicans*. NLRP3 is activated by products of the above microbes, including MDP, bacterial RNA, LPS, Pam2CysK4, poly(I:C) as well as bacterial toxins such as nigericin or listeriolysin O (from *L. monocytogenes*), or α-toxin and β- and γ-hemolysins (from *S. aureus*). In summary, activation of cytokine receptors or pattern recognition receptors such as TLRs leads to the induction of pro–IL-1β and NLRP3. In the next step, NLRP3 inflammasome assembly is triggered by low intracellular potassium levels (e.g. through formation of pores by bacterial toxins) which influences lysosomal stability and the binding of a putative ligand that is generated by proteolytic activity after lysosomal damage or by the action of ROS. The assembled NLRP3 inflammasome results in activation of caspase-1, which proteolytically activates IL-1β family cytokines. The produced pro-inflammatory IL-1β family cytokines can act on other cell types or act in a feed-forward loop. These mechanisms are excellently discussed in a recent review by Stutz et al. [21]

9.5 Cytokines

Cytokines are peptides produced by immune cells, which play key roles in regulating virtually all mechanisms of inflammation including innate immunity, antigen presentation, cellular differentiation, activation and recruitment as well as in repair processes (see Chap. 6). They are produced primarily by macrophages and lymphocytes, but also by other leukocytes, endothelial cells and fibroblasts. Substances considered to be cytokines include interleukins 1–35, interferons, tumour necrosis factors (TNF), platelet-derived growth factor (PDGF), transforming growth factor-β, CHEMOKINES (which will be discussed separately) and the COLONY-STIMULATING FACTORS. Major cytokine superfamilies are listed in Table 9.3, and the most important ones are discussed below. Further details are also given in Chap. A5. The cytokine production profile in response to immune insult determines the nature of the immune response (cell-mediated, humoral, cytotoxic or allergic) [22, 23].

Interleukin-1 is the term given to a family of four cytokines consisting of two active, *agonists*, IL-1α and IL-1β, an endogenous IL-1-receptor antagonist (IL-1ra), and the recently cloned cytokine IL-18, which is structurally related to IL-1. Both IL-1α and IL-1β as well as a related protein IL-18 are synthesized as a less active precursor. Their secretion in response to various stimuli (antigens, endotoxin, cytokines or microorganisms) depends on the cleavage of the pro-cytokines to their active forms by IL-1-converting enzyme (ICE or caspase 1). IL-1α remains cell-associated and is active mainly during cell-to-cell contact, while the soluble IL-1β is a form predominant in biological fluids. IL-1 is an important inflammatory mediator, and it is believed to be implicated in several acute (e.g. systemic inflammatory response syndrome—SIRS in sepsis) or chronic (e.g. rheumatoid arthritis) inflammatory diseases. IL-1 is also important in immune responses, facilitating interaction of both B and T cells with antigen.

One of the principal actions of IL-1 is activation of T lymphocytes and B cells by enhancing the production of IL-2 and expression of IL-2 receptors. In IL-1 knockout animals, diminished immune responses or state of tolerance is

Table 9.3 Main cytokine families

Cytokine family	Cytokines
IL-1 superfamily	IL-1, IL-18, IL-33
IL-6 like cytokines	IL-6, IL-11, IL-27, IL-30, IL-31 (oncostatins, cardiotrophin, etc.)
IL-10 family	IL-10, IL-19, IL-20, IL-22, IL-24, IL-26
Interferon type III (lambda)	IL-28, IL-29
Common gamma chain family	IL-2/15; IL-3; IL-4; IL-7; IL-9; IL-13; IL-21
IL-12 family	IL-12, IL-23, IL-27, IL-35
Other	IL-5, IL-8; IL14; IL-16; IL-17/25(A); IL-32; IL-34
IL-17 family	IL-17A, IL-17B, IL-17C, IL-17D, IL-17E (also called IL-25), and IL-17F.
TNF ligand superfamily	TNF-alpha, 4-1BB ligand, B-cell-activating factor, FAS ligand, TNF-beta (lymphotoxin), OX40L, RANKL, TRAIL
Interferons	I (alpha) II (gamma) III (IL-28; IL-29)

observed. In vascular endothelial cells, IL-1 increases the synthesis of leukocyte adhesion molecules (VCAM-1, ICAM-1 and E-selectin), stimulates NO production, releases "platelet-derived growth factor" (PDGF) and activates PLA_2, thus inducing the synthesis of prostanoids and PAF. It stimulates fibroblasts to proliferate, to synthesize collagen and to generate collagenase. It regulates the systemic inflammatory response by stimulating synthesis of acute phase proteins (C-REACTIVE PROTEIN, amyloid and COMPLEMENT), producing neutrophilia and causing fever by altering a set point of temperature in the hypothalamus. IL-1 also induces the generation of other cytokines such as the interferons, IL-3 and IL-6, and, in bone marrow, the COLONY-STIMULATING FACTORS. It synergises with tumour necrosis factor-α (TNF-α) in many of its actions, and its synthesis is stimulated by TNF-α. The therapeutic effects of glucocorticoids in rheumatoid arthritis and other chronic inflammatory and autoimmune diseases may well involve inhibition of both IL-1 production and IL-1 activity. Production of IL-1ra alleviates potentially deleterious effects of IL-1 in the natural course of the disease.

IL-18 although structurally close to IL-1 family exerts actions more related to IL-12. It was originally derived from liver but is produced by numerous cell types (including lung, kidney and smooth muscle cells) apart from lymphocytes. In contrast to other cytokines, IL-18 pro-cytokine is constitutively expressed, and therefore its activity is regulated primarily by caspase-1. It plays a critical role in cellular adhesion being the final common pathway leading to ICAM-1 expression in response to IL-1, TNF-α and other cytokines. It also synergises with IL-12 in stimulating IFN-gamma production. Soluble IL-18 receptor may be particularly interesting from an immunopharmacological point of view as it has lost its signalling domain and may therefore serve as a potent anti-inflammatory molecule.

Interleukin 17 family (IL-17A-F) includes cytokines that share a similar protein structure, with four highly conserved cysteine residues critical to their 3-dimensional shape which is unique for this cytokine. Production of IL-17A, which is characteristic for a specific subset of T helper CD4+ lymphocytes called Th17 cells, places this cytokine as one of the most important regulators of autoimmune processes. IL-17 is particularly important as it regulates expression and function of numerous other immune pro-inflammatory signalling molecules. The role of IL-17 is also commonly associated with allergic responses. IL-17A, the best characterized member of this family, induces the production of many other cytokines (such as IL-6, G-CSF, GM-CSF, IL-1β, TGF-β, TNF-α), CHEMOKINES (including IL-8, GRO-α and MCP-1) and prostaglandins (PGE_2) discussed in this chapter. IL-17 receptors, binding particularly IL-17A, are expressed on fibroblasts, endothelial cells, epithelial cells, keratinocytes and macrophages. As a result of these effects, the IL-17 family has been linked to many immune-/autoimmune-related diseases including rheumatoid ARTHRITIS, ASTHMA, SLE, allograft rejection and antitumour immunity.

9.5.1 Tumour Necrosis Factor-α and Tumour Necrosis Factor-β (TNF-α and TNF-β)

These cytokines are produced primarily in mononuclear phagocytes (TNF-α) and in lymphocytes (TNF-β) but also by numerous other cells. Activation of Toll-like receptors (TLR2 and TLR4) by LPS is the most commonly recognised intracellular pathway leading to production of TNF. TNF-α and TNF-β bind with similar affinity to the same cell surface receptors—TNFR 1 (p55) and TNFR 2 (p75). Therefore their activities are very similar. The generic name of these cytokines is based on tumour cytotoxic effects, but their pharmacological use in the treatment of tumours is limited by severe side effects. TNF is responsible for severe cachexia during chronic infections and cancer.

In endothelial cells these cytokines induce expression of adhesion molecules (ICAM-1 and VCAM-1) and synthesis of prostacyclin and of cytokines. TNFs act as chemoattractants, as well as potent activators for neutrophils and macrophages. TNF-α causes fever and releases acute phase proteins. TNF and IL-1 produce many of the same pro-inflammatory responses which include induction of cyclooxygenase and lipoxygenase enzymes as well as the activation of B cells and T cells. It is finally important to point out that TNF is the primary mediator of haemodynamic changes during septic shock through its negative inotropic effects as well as an increase in vascular permeability.

TGF-α (transforming growth factor-α) is a trophic regulator of cell proliferation and differentiation which is important in repair processes, it is involved in angiogenesis and in the organization of extracellular matrix, and it is chemotactic for monocytes.

PDGF (platelet-derived growth factors) cause proliferation of fibroblasts, vascular endothelial cells and smooth muscle. They are implicated in angiogenesis, atherosclerosis and possibly in chronic ASTHMA.

Interferons constitute a group of inducible cytokines which are synthesized in response to viral and other stimuli. There are three classes of interferons (IFN), termed type I (IFN γ), type II (IFNsα and β) and type III (IFN λ). IFN-α is not a single substance but a family of 15 proteins with similar activities. The interferons have antiviral activity, and interferon-γ has significant immunoregulatory function and only modest antiviral activity. Interferon-λ is also an antiviral cytokine, but it signals through a distinct receptor complex, composed of the IFN-λR1 and interleukin-10R2 (IL-10R2) receptor chains. Thus this interferon (type III) is functionally an interferon but structurally is related to the interleukin-10 family. The antiviral effects of interferons are achieved by inhibition of viral replication within infected cells as well as by stimulation of cytotoxic lymphocytes and NK cells. All interferons can be induced by other cytokines such as IL-1, IL-2, TNF and COLONY-STIMULATING FACTORS. IFN-α and IFN-β are produced in many cell types—macrophages, fibroblasts, endothelial cells, osteoblasts, etc., being strongly induced by viruses and less strongly by other microorganisms and bacterial products. Interferons induce the expression of the major histocompatibility molecules (MHC I and II) that are involved in antigen presentation to T cells. IFNs also stimulate the expression of Fc receptors on GRANULOCYTES, promote the differentiation of myeloid cells and modulate the synthesis of cytokines. Interferon γ is primarily made by T lymphocytes (T helper type 1) which may suggest that it is more of an interleukin than an interferon. Indeed, it functions as an inhibitor of IL-4-dependent expression of low affinity IgE receptors, therefore inhibiting IgE synthesis.

Colony stimulating factors. These include *IL-3* and *GM-CSF* (granulocyte macrophage-colony stimulating factor) and several other cytokines. They regulate haematopoiesis, are chemotactic for neutrophils, as well as activating neutrophils and macrophages.

Anti-inflammatory cytokines. It is important to point out that apart from pro-inflammatory actions, some cytokines may inhibit inflammatory processes. These include *IL-1ra*, mentioned above, as well as *TGF-β* or the *IL-10 family* (including *IL10*, *IL19*, *IL20*, *IL22* and *IL24*).

9.5.2 Intracellular Signalling by Cytokine Receptors

Binding of cytokines to their receptors leads to the activation of cytoplasmic tyrosine kinases. Janus kinases (JAKs), a recently described family of four related cytoplasmic protein tyrosine kinases, further transfer cytokine signalling. There are four JAKs, JAK1, JAK2, JAK3 and TYK2, which transduct signals from cytokine receptors to effector mechanisms. On binding of the cytokine, JAKs bind to the receptor and mediate tyrosine KINASE activity and phosphorylation of the receptor and of receptor-associated JAKs (Fig. 9.4). The next step in signal transduction involves tyrosine phosphorylation of *signal transducers and activators of transcription* (STATs) in the cytoplasm. Upon activation, STATs become phosphorylated, form homodimers and migrate to the nucleus, where they bind to regulatory sequences in the promoters of cytokine-responsive genes, e.g. ICAM-1 or other cytokine genes. In summary, cytokine

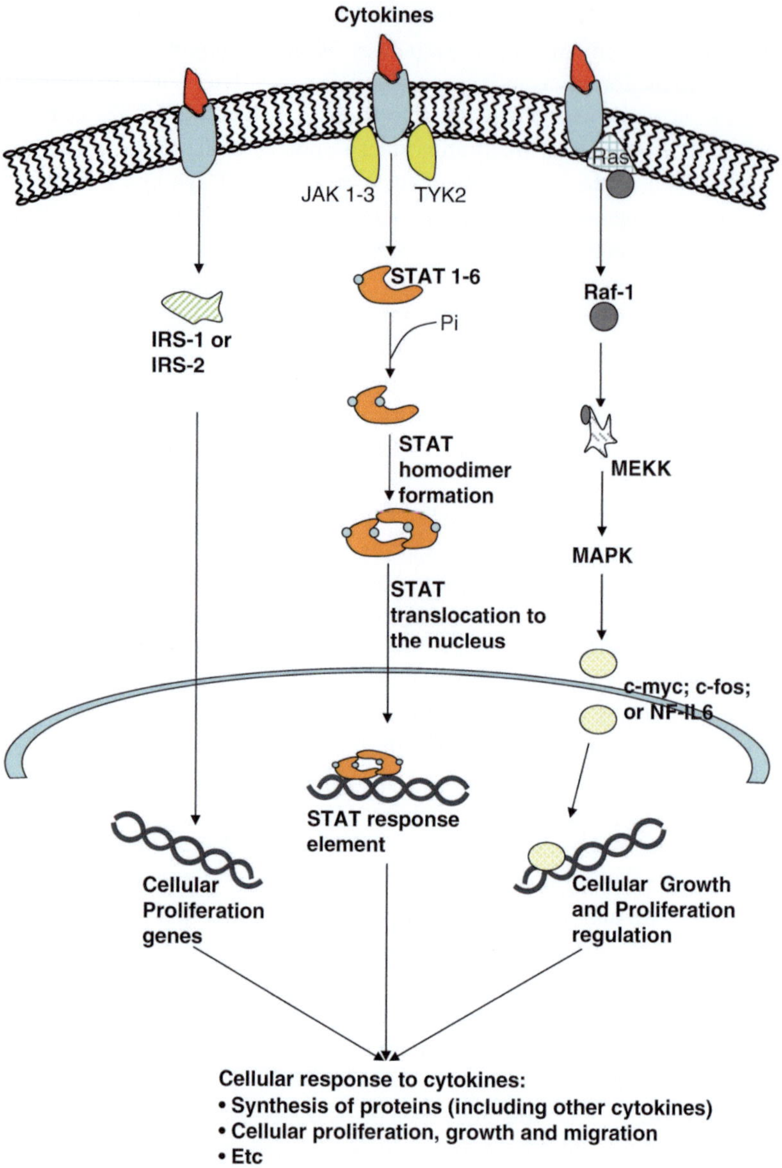

Fig. 9.4 Cytokine-induced intracellular signalling

signalling is based on a relatively small number of redundant tyrosine kinases. For instance, JAK-1 and JAK-3 transduct signals from cytokines such as IL-2 or IL-4, while JAK-2 is involved in IL-3, IL-6 and GM-CSF signalling. Similarly, the number of STATs is low when compared to the number of cytokines. Therefore, one can conclude that some additional mechanisms will guide different responses to various cytokines. An additional pathway used by many cytokine receptors includes *Ras*-dependent cascades. In this signal transduction cascade, Ras, Raf-1, Map/Erk kinase kinase (MEKK) and finally mitogen-activated protein kinases (MAPK) are sequentially activated and lead to regulation of cellular proliferation by growth factors and responses to IL-2 or IL-3. The activation of other signalling pathways, like insulin receptor substrates (IRS-1, IRS-2), can also mediate some other biological activities of cytokines, including proliferation and regulation of apoptosis. In conclusion, it becomes apparent that different combinations of the signalling mechanisms described above will lead to many distinct responses to different cytokines.

9.6 Chemokines and Their Intracellular Signalling

CHEMOKINES are a family of 8–12 kDa molecules, which induce chemotaxis of monocytes, lymphocytes, neutrophils, other GRANULOCYTES as well as vascular smooth muscle cells and variety of other cells. There are 47 chemokines, sharing 30–60% homology. Chemokines are characterised by the presence of 3–4 conserved cysteine residues. The new classification of chemokines is based on the positioning of the N-terminal cysteine residues (Table 9.4). Chemokines are usually secreted proteins, except for fractalkine (CX3CL1) which is the only membrane-bound chemokine, acting as an adhesion molecule. Most chemokines play roles in recruiting and activating immune cells to the site of inflammation, while others are important in maintaining homeostasis within the immune system (housekeeping chemokines: CCL5, CCL17-19, 21, 22, 25, 27, 28, CXCL13, CXCL14). Homeostatic chemokines are expressed in an organ-specific manner, while inflammatory chemokines can be produced by multiple cell types.

Their activities are achieved through interaction with chemokine receptors. There are 18 chemokine receptors currently known; therefore, some receptors may bind several ligands, which lead to overlapping functions of known chemokines. Moreover, a single cell may express several chemokine receptors. One of the key features of chemokine receptors, owing to their heptahelical transmembrane structure, is their ability to signal through different intracellular signalling pathways. Binding of chemokine to the receptor leads to activation of Gα protein and binding of GTP. Gα subunit activates Src kinases and subsequently mitogen-activated protein kinases (MAPKs) and protein kinase B (PKB). During activation of Gα protein, a Gβγ complex is liberated and may independently lead to activation of PKB and MAPKs (via PI_3), PKC activation via phospholipase C (PLC) and finally through Pyk-2 [24]. These pathways lead to up-regulation of membrane INTEGRINS and initiate rolling and adhesion of cells as well as their conformational changes. Some of these intracellular pathways (in particular PLC activation) may then lead to an increase in intracellular calcium and its consequences, including DEGRANULATION, NOS activation, etc. within the target cells.

It is difficult to accurately describe the relative importance of individual chemokines. The largest number of studies was conducted on the actions of *IL-8* as the most important chemoattractant for polymorphonuclear leukocytes, although it appears late during the inflammatory response. Other well-investigated members of this family include CCL3 (*MIP-1α*) or *RANTES* (CCL5).

Apart from effects on chemotaxis, chemokines have direct and indirect effects on T-cell differentiation into T helper 1 or 2 subclasses, therefore regulating the nature of immune responses [23, 24].

Due to the critical role of chemokines in inflammation, interest has focused on potential therapeutic effects of inhibiting their activity. Both peptide antagonists and gene transfer approaches have been successfully used to inhibit inflammation in various animal models (e.g. allergic inflammation models or ApoE-knockout atherosclerosis prone mice).

Table 9.4 Classes of chemokines

Subfamily	Chemokines	Characteristics
C-X-C	CXCL 1-16, includes IL-8 (CXCL8)	First two cysteines separated by a variable amino acid
C-C	CCL1-28 (include MIP-1 MCP and RANTES)	First two cysteines are adjacent to each other
C	XCL 1 (lymphotactin) and XCL 2	Lacks first and third cysteine residue
CX3CL1	CX3CL1 (Fractalkine)	Two N-terminal cysteine residues separated by three variable amino acids

9.7 Neuropeptides

Neuropeptides are released from sensory neurons, and in some tissues, they contribute to inflammatory reactions. For example, substance P and other tachykinins produce smooth muscle contraction and mucus secretion, cause vasodilation and increase vascular permeability. "Calcitonin gene-related peptide" (CGRP) is a potent vasodilator, acting on CGRP receptors leading to activation of adenylate cyclase. The overall pattern of effects of tachykinins is similar, though not identical, to the pattern seen with kinins.

9.7.1 Tachykinins

The mammalian tachykinins comprise three related peptides: substance P (SP), neurokinin A (NKA) also called substance K and neurokinin B (NKB). They occur mainly in the nervous system, particularly in nociceptive sensory neurons and in enteric neurons. They are released as neurotransmitters, often in combination with other mediators. SP and NKA are encoded by the same gene and they have a similar distribution. Three distinct types of tachykinin receptor are known: NK_1, NK_2 and NK_3. They are selective for three endogenous tachykinins with the following affinity: SP>NKA>. NKB for NK_1, NKA>NKB>SP for NK_2 and NKB>NKA>SP for NK_3 receptor. Receptor cloning has shown that tachykinin receptors belong to a family of G-protein-coupled receptors. Several potent antagonists of NK_1 and NK_2 and NK_3 receptors have been discovered [25], and novel therapeutic agents for various disease states (e.g. pain, ASTHMA, arthritis, headache) may be developed.

CGRP differs from other tachykinins. It is coded for by the calcitonin gene which also codes for calcitonin itself. Differential splicing allows cells to produce either procalcitonin (expressed in thyroid cells) or pro-CGRP (expressed in neurons) from the same gene. CGRP is found in non-myelinated sensory neurons, and it is a potent inducer of neurogenic inflammation.

9.8 Kinins

Kinins are polypeptides with vasodilator/hypotensive, thrombolytic, pro-inflammatory and algesic (painful) actions. The two best known kinins are bradykinin and kallidin, and they are referred to as plasma kinins. Since 1980, when Regoli and Barabe divided the kinin receptors into B_1 and B_2 classes, first- and second-generation kinin receptor antagonists have been developed, leading to a much better understanding of the actions of kinins.

Bradykinin is a nonapeptide; kallidin is a decapeptide and has an additional lysine residue at the amino-terminal position. These two peptides are formed from a class of α-2 globulins known as *kininogens* (Fig. 9.5). There are two *kininogens*: high molecular weight (*HMW*) and low molecular weight (*LMW*) kininogen which are products of a single gene that arises by alternative processing of mRNA. The highly specific proteases that release bradykinin and kallidin from the kininogens are termed *kallikreins*. Two distinct kallikreins, formed by different activation mechanisms from inactive *prekallikreins*, act on the kininogens. One of these is plasma kallikrein and the other is tissue kallikrein. LMW kininogen is a substrate only for the tissue kallikrein, and the product is kallidin, while HMW kininogen is cleaved by plasma and tissue kallikrein to yield bradykinin and kallidin, respectively.

Kallidin is similar in activity to bradykinin and need not be converted to the latter to exert its effects. However, some conversion of kallidin to bradykinin occurs in plasma due to the activity of plasma aminopeptidases.

The half-life of kinins in plasma is about 15 s, and concentrations of kinins found in the circulation are within the picomolar range. Bradykinin is inactivated by a group of enzymes known as *kininases*. The major catabolizing enzyme in the lung and in other vascular beds is *kininase II*, which is identical to peptidyl dipeptidase—known as angiotensin converting enzyme (ACE). *Kininase II* is inhibited by captopril, resulting in an increased concentration of circulating bradykinin, which contributes substantially to the

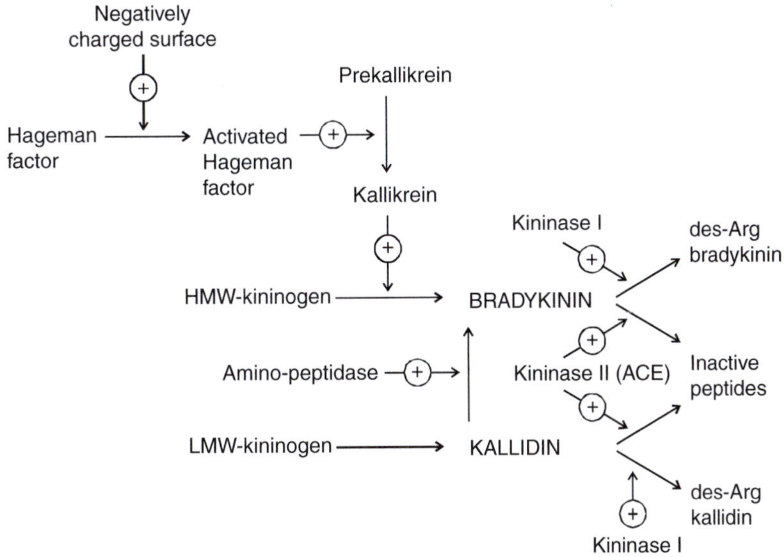

Fig. 9.5 The formation and metabolism of kinins

antihypertensive effect of captopril. On the other hand, *kininase I* is arginine carboxypeptidase, and it has a slower action than *kininase II*. It removes the carboxyl-terminal arginine residue producing des-Arg^9-bradykinin or des-Arg^{10}-kallidin, which are themselves potent B_1-kinin receptor agonists.

There are at least two distinct receptors for kinins, B_1 and B_2. The classical, constitutive bradykinin receptor, now designated the B_2 receptor, selectively binds bradykinin and kallidin and mediates a majority of the effects of bradykinin and kallidin in the absence of inflammation, such as the release of PGI_2 and NO from endothelial cells. On the other hand, inducible B_1 receptors are upregulated by inflammation. They bind des-Arg metabolites of bradykinin and kallidin. In contrast to B_1 receptors, the signalling mechanism of B_2 receptors has been well characterized. The B_2 receptor is coupled to G protein and activates both PLA_2 and PLC. While stimulation of the former liberates AA from phospholipids, with its subsequent oxidation to a variety of pro-inflammatory eicosanoids, the activation of PLC through IP_3 and DAG leads directly to pro-inflammatory effects.

During the last decade, the existence of other types of kinin receptors (B_3, B_4, B_5) has been suggested. However, recent studies indicate that some of them may actually represent functions of the B_2 receptor [15].

Kinins are among the most potent vasodilators known, acting on arteriolar beds of the heart, liver, skeletal muscle, kidney, intestines and ovaries. They are claimed to play a minor role in the regulation of blood pressure in health individuals, but they play a major vasodepressor regulatory role most likely mediated by arterial endothelium in hypertensive patients [26]. Indeed, kinins contract veins and non-vascular smooth muscle, such as gastrointestinal and bronchial muscle. Bradykinin and kallidin have similar contracting properties. At the level of the capillary circulation, kinins increase permeability and produce oedema. Stimulation of B_1 receptors on inflammatory cells such as macrophages can elicit the production of the inflammatory mediators such as IL-1 and TNF-α [27]. Kinins are also potent pain-inducing agents in both the viscera and skin. In acute pain, B_2 receptors mediate bradykinin-induced algesia. The pain of chronic inflammation appears to involve an increased expression of B_1 receptors.

As in the case of other autacoids, the therapeutic interest in kinins has focused particularly on attempts to modulate their formation or metabolism in vivo [28]. Blockade of kinin formation with a kallikrein inhibitor, aprotynin (Trasylol), has been used with some success to treat acute pancreatitis, carcinoid syndrome or Crohn disease. Experimentally, progress has been made in the development of selective antagonists of kinins. Currently, they are not available for clinical use. However, recent studies indicate that kinin receptor antagonists might be useful for the treatment of patients with septic shock, pancreatitis-induced hypotension bronchial ASTHMA and rhinovirus-induced symptoms and in fighting pain.

9.9 Nitric Oxide

In animal tissues, nitric oxide (NO) is generated enzymatically by NO synthases (NOS). The three NOS isoenzymes (neuronal, endothelial and inducible) are flavoproteins which contain tetrahydrobiopterin and haem, and they are homologous with cytochrome p 450 reductase [29]. Isoenzymes of NOS act as dioxygenases using molecular oxygen and NADPH to transform L-arginine to L-citrulline and NO (Fig. 9.6). NO formed by endothelial constitutive NOS (eNOS) is responsible for maintaining low vascular tone and preventing leukocytes and platelets from adhering to the vascular wall. eNOS is also found in renal mesangial cells. NO formed by neuronal constitutive NOS (nNOS) acts as a neuromodulator or neuromediator in some central neurons and in peripheral "non-adrenergic non-cholinergic" (NANC) nerve endings. NO formed by inducible NOS (iNOS) in macrophages and other cells plays a role in the inflammatory response.

NO was discovered by Furchgott and Zawadzki as "endothelium-derived relaxing factor" (EDRF) [30]. It soon became obvious that EDRF, like nitroglycerine, activates soluble guanylate cyclase in vascular smooth muscle by binding to its active haem centre. The rise in cyclic GMP achieved is responsible for vasodilatation and for other physiological regulatory functions of NO.

The activities of constitutive nNOS and eNOS are controlled by intracellular calcium/calmodulin levels. For instance, nNOS in central neurons is activated by glutamate binding to NMDA receptors with a subsequent rise in $[Ca^{2+}]i$ due to opening of voltage calcium channels, whereas eNOS is activated by blood shear stress or stimulation of endothelial muscarinic, purinergic, kinin, substance P or thrombin receptors. This triggers an increase in $[Ca^{2+}]i$ at the expense of the release of Ca^{2+} from endoplasmic reticulum.

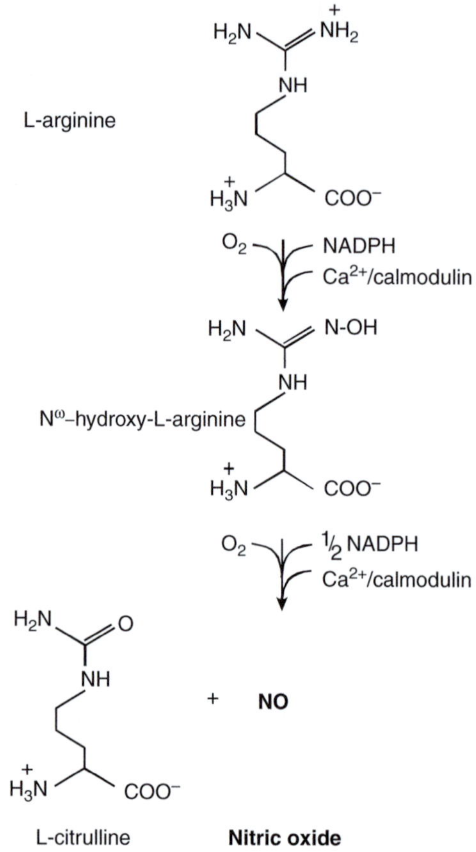

Fig. 9.6 The synthesis and metabolism of nitric oxide (NO)

Calcium ionophores (e.g. A23187) and polycations (e.g. poly-L-lysine) cause a rise in $[Ca^{2+}]i$ and activate eNOS, thereby bypassing the receptor mechanisms.

In contrast to the constitutive isoforms of NOS, iNOS does not require a rise in $[Ca^{2+}]i$ to initiate its activity. In macrophages, monocytes and other cells, the induction of iNOS and the presence of L-arginine are sufficient to initiate the generation of NO. Induction of iNOS can be initiated by IFN-γ, TNF-α or IL-1. However, the best recognized inducer is lipopolysaccharide (LPS) or endotoxin from *Escherichia coli* which is known to be responsible for the development of systemic inflammatory response syndrome (SIRS) in the course of sepsis due to gram-negative bacteria. Myeloid cells express a receptor for LPS on their cell membrane, m-CD14 protein. LPS, using an "LPS binding protein" (LBP), is anchored to m-CD14 and then triggers a chain of protein phosphorylation which eventually leads to the activation of the major transcription protein NF-κ-B. This is responsible for transcription of the message to make iNOS. In cells which lack m-CD14, the induction of iNOS is achieved by a complex of soluble s-CD14 with LBP and LPS itself. In a similar manner, LPS can also induce COX-2. Although NO fulfils more paracrine than autoendocrine functions, in the case of iNOS, large amounts of locally formed NO may inhibit iNOS itself as well as COX-2, in a negative feedback reaction. Glucocorticosteroids and some cytokines, such as TGF-β, IL-4 or IL-10, inhibit the induction of iNOS.

9.9.1 Nitric Oxide as an Effector of Inflammation

Kinetics of nitric oxide production by iNOS differ greatly from production by eNOS or nNOS (Fig. 9.7) [31]. Inducible NOS produces very large, toxic amounts of NO in a sustained manner, whereas constitutive NOS isoforms produce NO within seconds, and its activities are direct and short acting. There are multiple intracellular mechanisms through which nitric oxide may act as an inflammatory mediator [32]. Low levels of NO produced by constitutive synthases primarily interact directly with positively charged metal ions of guanylate cyclase, cytochrome p450 and NOS itself. Activation of guanylate cyclase leads to an increase in intracellular cyclic guanosine monophosphate (cGMP), which in turn activates cGMP-dependent protein kinases which mediate NO actions including vasorelaxation, increase of vascular permeability, as well as anti-proliferative, anti-platelet and antioxidant effects of nitric oxide. Recent data have also indicated that NO

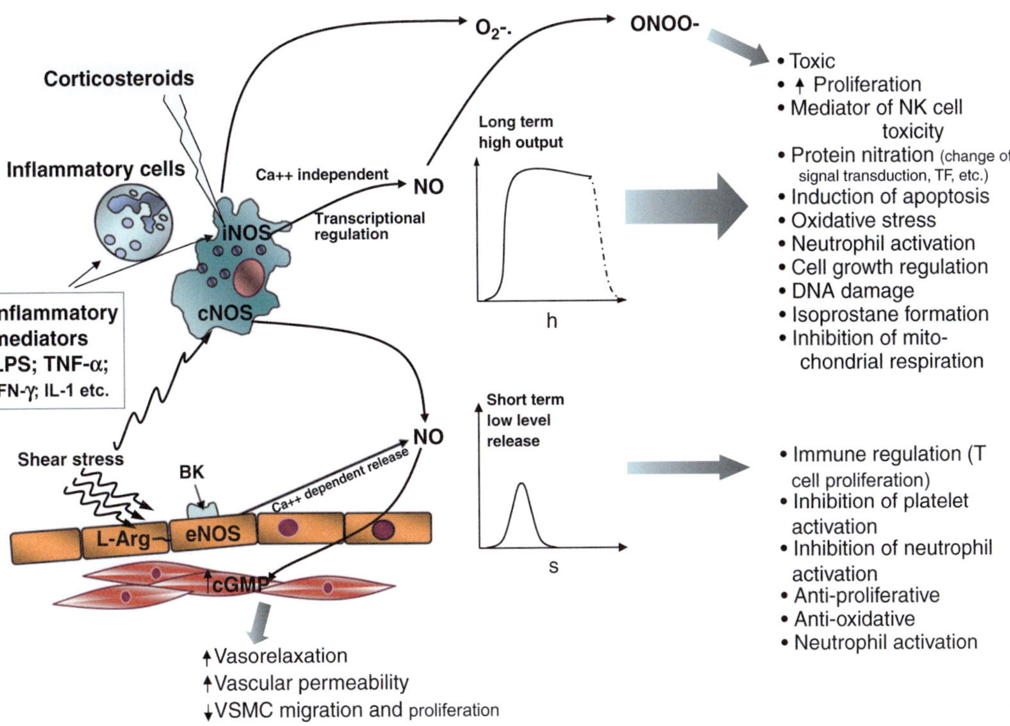

Fig. 9.7 Differences between kinetics of nitric oxide generation by eNOS and iNOS

produced by constitutive NOS enzymes may be involved in immune regulation of T helper cell proliferation and cytokine production.

During the course of an inflammatory response, the large amounts of NO formed by iNOS surpass the physiological amounts of NO which are usually made by nNOS or eNOS. The functions of iNOS-derived NO are also different. In immunologically or chemically activated macrophages, NO kills microorganisms and destroy macromolecules. NO, formed by constitutive isoforms of NOS, is stored as a nitrosothiol in albumin and acts physiologically as *N*-nitrosoglutathione and *N*-nitrosocysteine. Eventually, within a few seconds, NO is oxidized to nitrites or nitrates. Large amounts of "inflammatory NO" from myeloid cells are usually generated side by side with large amounts of superoxide anion (O_2^-). These two can form peroxynitrite (ONOO-) which mediates the cytotoxic effects of NO, such as DNA damage, LDL oxidation, isoprostane formation, tyrosine nitration, inhibition of aconitase and mitochondrial respiration. The discovery of this reaction opens new possibilities for the therapeutic use of superoxide dismutase (SOD). Indeed superoxide dismutase mimetics have been successfully used to limit the extent of inflammation. Interestingly, overstimulation of NMDA receptors by glutamate may activate nNOS to such an extent that NO itself exerts neurotoxic properties. NO formed by eNOS seems to be mostly cytoprotective, possibly due to its unusual redox properties.

Large amounts of NO and ONOO- may target numerous proteins and enzymes critical for cell survival and signalling. These include signalling molecules involved in cytokine signalling like JAK or STAT proteins, NKκB/IκB pathway as well as MAPK, some G proteins and transcription factors. Nitration of cysteines in these proteins may lead to their activation or inactivation.

NO is scavenged by haemoglobin, methylene blue and pyocyanin from *Pseudomonas coereleus*. These last two are also claimed to be inhibitors of guanylate cyclase. Glucocorticoids selectively inhibit the expression of iNOS. Arginine analogues, such as L-N^G-monomethyl arginine (L-NMMA) and L-N^G-nitroarginine methyl ester (L-NAME), inhibit inducible and constitutive NOS isoforms nonselectively. Selective iNOS inhibitors (e.g. alkylisothioureas or aminoguanidines) are being intensively investigated in the hope that selective inhibition of iNOS may prevent development of SIRS (systemic inflammatory response syndrome) or MODS (multiple organ dysfunction syndrome). Indeed, overproduction of NO by iNOS during septicaemia is claimed to be responsible for irreversible arterial hypotension, vasoplegia (loss of responses to noradrenaline), lactic acidosis, suffocation of tissues, their necrosis and apoptosis. However, it is important to remember that NO made by iNOS is of benefit to the host defence reaction by contributing to microbial killing.

Moreover, NO generated by eNOS is essential to maintain tissue perfusion with blood, to offer cytoprotection in the pulmonary and coronary circulation against toxic lipids which are released by LPS and to preserve red cell deformability which becomes reduced in septicaemia [33]. Preliminary clinical experience with L-NMMA has been reasonably encouraging, as long as a low dose of the NOS inhibitor is used. In animal models of endotoxic shock, nonselective NOS inhibitors were reported to decrease cardiac output, to increase pulmonary pressure, to decrease nutritional flow to organs, to damage gastric mucosa and to increase mortality rate. On the other hand, inhalation of NO gas (10 ppm) in septic patients has been found to prevent the mismatch of the ventilation/perfusion ratio in their lung. The exact role of NO in various stages of sepsis, SIRS and MODS still awaits further elucidation and evaluation.

9.9.2 Nitric Oxide in Immune Regulation

The exact role of NO in immune regulation is also unclear. Initial mouse studies suggested that ANTIGEN PRESENTING CELL-DERIVED NO may inhibit T-cell proliferation, particularly of the Th1 subset of T helper cells. Mouse Th1 cells were also shown to produce NO, suggesting that the above mechanism is a part of a negative feedback process. In this way, NO would inhibit Th1 and therefore promote Th2 type cytokine responses leading to humoral and allergic responses. Subsequent studies, however, indicate that both Th1 and Th2 produce similar amounts of NO, and both subsets respond similarly to nitric oxide. NO-induced changes in lymphocyte proliferation seem to be dependent more on the effects on the cell cycle proteins than due to changes in cytokine profile [31].

It is also important to recognize that cells which produce NO protect themselves against its toxic actions [34]. Recent studies show that GSH-GSSG anti-oxidative systems protect macrophages against large amounts of NO generated by iNOS. In addition, endothelial cells appear not to be primary responders to NO produced by eNOS because increases in intracellular calcium which mediate eNOS activation are also able to inhibit guanylate cyclase activity.

9.10 Reactive Oxygen Species

Reactive oxygen species (ROS) production plays an important role in modulation of inflammatory reactions. Major ROS produced within the cell are superoxide anion, hydrogen peroxide and hydroxyl radical [32]. Extracellular release of large amounts of superoxide anion produced by the respiratory burst in leukocytes is an important mechanism of pathogen killing and also leads to endothelial damage resulting in an increased vascular permeability as well as cellular death. However vast evidence has implicated intracellular ROS production as a key player in modulation of the release of other mediators of inflammation. This is related mainly to the constitutive expression of NAD(P)H oxidases (termed NOXs—non-phagocytic oxidases) in various tissues [31]. ROS produced by this family of enzymes can regulate adhesion molecule expression on endothelium and inflammatory cells, thus regulating cellular recruitment to the sites of inflammation. They also increase chemokine and cytokine expression. At least some of these effects result from the ability of ROS (in particular H_2O_2) to stimulate MAP kinase activities which lead to activation of several transcription factors. It is possible that intracellular ROS may act as second messengers in inflammatory signal transduction [31].

Inflammatory cytokines (like TNF-α) may in turn increase NAD(P)H oxidase activity and expression which closes the vicious circle of inflammation. While loss of NAD(P)H oxidase activity in cells leads to diminished inflammation in the vascular wall, several humoral factors may affect constitutive NAD(P)H oxidase expression in the vascular wall and therefore intracellular ROS production. These include angiotensin II, endothelins, high glucose or high cholesterol levels. Their effects on baseline ROS production may therefore mediate modulatory effects of these factors on inflammation which traditionally were not associated with inflammation. Interestingly, T and B LYMPHOCYTES at various stages of their development and activation express NADPH oxidases, mainly classical gp91phox containing NADPH oxidase, Nox2 (mature T cells), and a calcium-dependent Nox5 (during development).

Accordingly, attempts have been undertaken to inhibit intracellular ROS production in order to limit inflammatory responses. Apocynin, an NAD(P)H oxidase activation inhibitor, has been successfully used in limiting inflammation in animal models of rheumatoid arthritis, while decoy peptides preventing association of NAD(P)H oxidase subunits were shown to be effective in inflammation related to atherosclerosis.

9.11 Amines

Histamine, 2-(4-imidazolyl)-ethyl-amine, is an essential biological amine in inflammation and allergy. It is found mostly in the lung, skin and in the gastrointestinal tract. It is stored together with macroheparin in granules of mastocytes or basophils (0.01–0.2 mol per cell), from which it is released when COMPLEMENT components C3a and C5a interact with specific receptors or when antigen interacts with cell-fixed IgE.

These trigger a secretory process that is initiated by a rise in cytoplasmic Ca^{2+} from intracellular stores. Morphine and tubocurarine release histamine by a non-receptor action. Agents which increase cAMP formation inhibit histamine secretion, so it is postulated that, in these cells, c-AMP-dependent protein kinase is an intracellular restraining mechanism. Replenishment of the histamine content of MAST CELLS or basophils after secretion is a slow process, whereas turnover of histamine in the gastric histaminocyte is very rapid.

Histamine is synthesized from histidine by a specific decarboxylase and metabolized by histaminases and/or by imidazole N-methyltransferase. Histamine exerts its effects by acting on H_1-, H_2- or H_3 receptors on target cells [35]. It stimulates gastric secretion (H_2), contracts most of the smooth muscle other than that of blood vessels (H_1), causes vasodilatation (H_1) and increases vascular permeability by acting on the postcapillary venules [36]. Injected intradermally, histamine causes the triple response: local vasodilatation and wheal by a direct action on blood vessels and the surrounding flare which is due to vasodilatation resulting from an axon reflex in sensory nerves, thereby releasing a peptide mediator [36]. Of many functions of histamine, the stimulation of gastric acid secretion and mediation of type 1 hypersensitivity, such as urinary and hay fever, are among the most important. The full physiological significance of the H3 receptor has yet to be established [37]. Histamine may also be involved in T helper cell immune regulation (extensively reviewed in [38]).

5-Hydroxytryptamine (*5-HT*, *serotonin*) was originally isolated and characterized as a vasoconstrictor released from platelets in clotting blood. 5-HT occurs in chromaffin cells and enteric neurons of the gastrointestinal tract, in platelets and in the central nervous system. It is often stored together with various peptide hormones, such as somatostatin, substance P or "vasoactive intestinal polypeptide" (VIP). The biosynthesis and metabolism of 5-HT closely parallel that of catecholamines, except the precursor for decarboxylase of aromatic amino acids is 5-hydroxytryptophan instead of tyrosine (Fig. 9.8). 5-HT is inactivated mainly by the monoamine oxidases A or B (MAO A or B) to 5-hydroxyindoleacetic acid (5-HIAA) which is excreted in the urine. Some 5-HT is methylated to 5-methoxytryptamine, which is claimed to be involved in the pathogenesis of affective disorders.

The actions of 5-HT are numerous and complex, showing considerable variation between species [39]. For instance, in the inflammatory response, 5-HT seems to be more important in rats than in humans. 5-HT is known to increase gastrointestinal motility and to contract bronchi, uterus and arteries, although 5-HT may also act as a vasodilator through endothelial release of NO. In some species, 5-HT stimulates platelet aggregation, increases microvascular permeability and stimulates peripheral nociceptive nerve endings. A plethora of pathophysiological

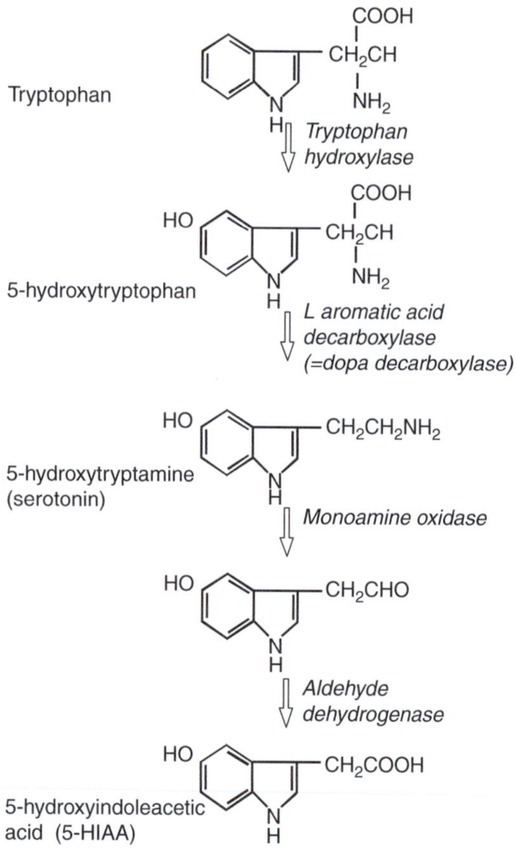

Fig. 9.8 The synthesis and breakdown of 5-HT

functions proposed for 5-HT includes control of peristalsis, vomiting, haemostasis, inflammation and sensitization of nociceptors by peripheral mechanisms or control of appetite, sleep, mood, stereotyped behaviour and pain perception by central mechanisms. Clinically, disturbances in the 5-HT regulation system have been proposed in migraine, carcinoid syndrome, mood disorders and anxiety [39].

These diverse actions of 5-HT are not mediated through one type of receptor. The amino acid sequence for many 5-HT receptor subtypes has been determined by cloning, and the transduction mechanisms to which these receptors are coupled have been explained. The four basic types of receptors are $5\text{-HT}_{1\text{-}4}$, 5-HT_1 and 5-HT_2 receptors which are further subdivided into A, B and C subtypes [40]. Types 1, 2 and 4 are G-protein-coupled receptors, while type 3 is a ligand-gated cation channel. 5-HT_1 receptors occur mainly in the CNS (all subtypes) and in blood vessels (5-HT_{1D} subtype). 5-HT_{1B} and 5-HT_{1D} receptors appear to be involved, at least in part, in the modulation of neurogenically induced (following electrical, chemical or mechanical depolarization of sensory nerves) vascular inflammation. 5-HT_2 receptors (5-HT_{2A} subtype being functionally the most important) are distributed more in the periphery than in the CNS, and they are linked to phospholipase C which catalyses phosphatidylinositol hydrolysis. The role of 5-HT_2 receptors in normal physiological processes is probably a minor one, but it becomes more prominent in pathological conditions, such as ASTHMA, inflammation or vascular thrombosis. 5-HT_3 receptors occur particularly on nociceptive sensory neurons and on autonomic and enteric neurons, on which 5-HT exerts an excitatory effect and evokes pain when injected locally.

Catecholamines. It has become increasingly recognized that the release of catecholamines at autonomic nerve endings and from the adrenal medulla may modulate the function of immunocompetent cells. The major lymphoid organs (spleen, lymph nodes, thymus and intestinal Peyer's patches) are extensively supplied by noradrenergic sympathetic nerve fibres. Sympathetic nervous system innervation of these lymphoid organs as well as the presence of adrenergic and dopamine receptors on immune cells provide the channels for noradrenergic signalling to lymphocytes and macrophages by sympathetic nerves [41]. Catecholamines have a wide range of direct effects on immune cells, particularly on macrophages and lymphocytes. Stimulation of β-adrenergic receptors on LPS-pretreated macrophages prevents the expression and release of pro-inflammatory TNF-α and IL-1, while the release of anti-inflammatory IL-10 is augmented. On the other hand, α-adrenergic stimulation augments phagocytic and tumouricidal activity of macrophages. Catecholamines acting through β-adrenergic and dopaminergic receptors, which are linked to adenylate cyclase through cyclic-AMP, modulate the function of immune cells. An increase in intracellular cyclic-AMP inhibits lymphocyte proliferation and production of pro-inflammatory cytokines. The demonstration of the presence of α_2-, β-adrenergic, D1 and D2 receptors on various immune cells has recently provided the basis for regulation of cytokine production, specifically interleukins and TNF, by these receptors in response to LPS [41]. Vasopressor and inotropic catecholamines seem to have potent immunomodulating properties which, as yet, have not been adequately explored and may contribute to the therapeutic effects of dobutamine or dopexamine in the treatment of septic shock and SIRS.

9.12 Summary

Inflammation is a protective response of the macroorganism to injury caused by trauma, noxious chemicals or microbiological toxins. This response is intended to inactivate or destroy invading organisms, remove irritants and set the stage for tissue repair. The inflammatory response consists of immunological and non-immunological reactions. The latter are triggered by the release from injured tissues and migrating cells of lipid-derived autacoids, such as eicosanoids or "platelet-activating factor" (PAF); large peptides, such as interleukin-1 and cytokines; small peptides, such as bradykinin; and amines, such as histamine 5-hydroxytryptamine. These constitute the chemical network of the inflammatory response and result in clinical and pathological manifestations of inflammation.

Prostanoids, as autacoids, are involved in virtually every stage of inflammation. They regulate vascular tone and permeability (PGs), induce platelet aggregation and thrombus formation (TX) and are involved in the pathogenesis of pain and fever (PGs) accompanying inflammation. The recently discovered lipoxins are important regulators of inflammatory reactions. PAF, cytokine and chemokine groups as well as kinins also play crucial pro-inflammatory roles. Recent studies have shed more light on our understanding of intracellular signalling mechanisms involved in the responses to pro-inflammatory cytokines such as IL-1, TNF, TGF and interferons. Toll-like receptors contribute to the mediation of effects of components of microorganisms on innate and adaptive immunity.

Nitric oxide and reactive oxygen species not only act as important effectors, causing damage to invading microorganisms (NO from iNOS or superoxide anion), but may also be very important in immunoregulation, in part by regulating redox-sensitive genes. Co-ordinated pharmacological interventions, which would modify different parallel pathways in the inflammatory cascade, are needed to treat inflammatory diseases.

References

1. Plytycz B, Seljelid R. From inflammation to sickness: historical perspective. Arch Immunol Ther Exp (Warsz). 2003;51(2):105–9.
2. Metchnikoff E. Secons sur la Pathologie comparee de l'inflammation. Paris: Masson; 1893.
3. Harris SG, et al. Prostaglandins as modulators of immunity. Trends Immunol. 2002;23(3):144–50.
4. Moncada S, et al. An enzyme isolated from arteries transforms prostaglandin endoperoxides to an unstable substance that inhibits platelet aggregation. Nature. 1976;263(5579):663–5.
5. Warner TD, Mitchell JA. Cyclooxygenase-3 (COX-3): filling in the gaps toward a COX continuum? Proc Natl Acad Sci U S A. 2002;99(21):13371–3.
6. Schwab JM, Schluesener HJ, Laufer S. COX-3: just another COX or the solitary elusive target of paracetamol? Lancet. 2003;361(9362):981–2.
7. Sigal E. The molecular biology of mammalian arachidonic acid metabolism. Am J Physiol. 1991;260 (2 Pt 1):L13–28.
8. Feldberg W, Kellaway C. Liberation of histamine and formation of lysocithin-like substances by cobra venom. J Physiol. 1938;94:187–226.
9. Yokomizo T, Izumi T, Shimizu T. Leukotriene B4: metabolism and signal transduction. Arch Biochem Biophys. 2001;385(2):231–41.
10. Serhan CN. Lipoxins and aspirin-triggered 15-epi-lipoxin biosynthesis: an update and role in anti-inflammation and pro-resolution. Prostaglandins Other Lipid Mediat. 2002;68-69:433–55.
11. McMahon B, et al. Lipoxins: revelations on resolution. Trends Pharmacol Sci. 2001;22(8):391–5.
12. Levy BD, et al. Lipoxins and aspirin-triggered lipoxins in airway responses. Adv Exp Med Biol. 2003;525:19–23.
13. Carroll MA, McGiff JC, Ferreri NR. Products of arachidonic acid metabolism. Methods Mol Med. 2003;86:385–97.
14. Kozek-Langenecker SA, et al. Effect of prostacyclin on platelets, polymorphonuclear cells, and heterotypic cell aggregation during hemofiltration. Crit Care Med. 2003;31(3):864–8.
15. Hache M, et al. Inhaled epoprostenol (prostacyclin) and pulmonary hypertension before cardiac surgery. J Thorac Cardiovasc Surg. 2003;125(3):642–9.
16. Sala A, et al. Morphological and functional changes of coronary vasculature caused by transcellular biosynthesis of sulfidopeptide leukotrienes in isolated heart of rabbit. Blood. 1996;87(5):1824–32.
17. Ray WA, et al. COX-2 selective non-steroidal anti-inflammatory drugs and risk of serious coronary heart disease. Lancet. 2002;360(9339):1071–3.
18. Chilton FH, et al. Metabolic fate of platelet-activating factor in neutrophils. J Biol Chem. 1983;258(10):6357–61.
19. Chao W, Olson MS. Platelet-activating factor: receptors and signal transduction. Biochem J. 1993;292. (Pt 3:617–29.
20. Takeda K, Kaisho T, Akira S. Toll-like receptors. Annu Rev Immunol. 2003;21:335–76.
21. Stutz A, Golenbock DT, Latz E. Inflammasomes: too big to miss. J Clin Invest. 2009;119(12):3502–11.
22. Liew FY. The role of innate cytokines in inflammatory response. Immunol Lett. 2003;85(2):131–4.
23. Borish LC, Steinke JW. 2. Cytokines and chemokines. J Allergy Clin Immunol. 2003;111(2 Suppl):S460–75.
24. Wong MM, Fish EN. Chemokines: attractive mediators of the immune response. Semin Immunol. 2003;15(1):5–14.
25. Patacchini R, Maggi CA. Peripheral tachykinin receptors as targets for new drugs. Eur J Pharmacol. 2001;429(1-3):13–21.
26. Pellacani A, Brunner HR, Nussberger J. Antagonizing and measurement: approaches to understanding of hemodynamic effects of kinins. J Cardiovasc Pharmacol. 1992;20(Suppl 9):S28–34.

27. Dray A, Perkins M. Bradykinin and inflammatory pain. Trends Neurosci. 1993;16(3):99–104.
28. Fein AM, et al. Treatment of severe systemic inflammatory response syndrome and sepsis with a novel bradykinin antagonist, deltibant (CP-0127). Results of a randomized, double-blind, placebo-controlled trial. CP-0127 SIRS and Sepsis Study Group. JAMA. 1997;277(6):482–7.
29. Cirino G, Fiorucci S, Sessa WC. Endothelial nitric oxide synthase: the Cinderella of inflammation? Trends Pharmacol Sci. 2003;24(2):91–5.
30. Furchgott RF, Zawadzki JV. The obligatory role of endothelial cells in the relaxation of arterial smooth muscle by acetylcholine. Nature. 1980;288:373–6.
31. Guzik TJ, Korbut R, Adamek-Guzik T. Nitric oxide and superoxide in inflammation and immune regulation. J Physiol Pharmacol. 2003;54(4):469–87.
32. Salvemini D, Ischiropoulos H, Cuzzocrea S. Roles of nitric oxide and superoxide in inflammation. Methods Mol Biol. 2003;225:291–303.
33. Korbut R, Trabka-Janik E, Gryglewski RJ. Cytoprotection of human polymorphonuclear leukocytes by stimulators of adenylate and guanylate cyclases. Eur J Pharmacol. 1989;165(1):171–2.
34. Coleman JW. Nitric oxide in immunity and inflammation. Int Immunopharmacol. 2001;1(8):1397–406.
35. Bakker RA, Timmerman H, Leurs R. Histamine receptors: specific ligands, receptor biochemistry, and signal transduction. Clin Allergy Immunol. 2002;17:27–64.
36. Repka-Ramirez MS, Baraniuk JN. Histamine in health and disease. Clin Allergy Immunol. 2002;17:1–25.
37. Dale M, Foreman J, Fan T. Textbook of immunopharmacology. Oxford: Blackwell Scientific Publications; 1994.
38. Akdis CA, Blaser K. Histamine in the immune regulation of allergic inflammation. J Allergy Clin Immunol. 2003;112(1):15–22.
39. Mossner R, Lesch KP. Role of serotonin in the immune system and in neuroimmune interactions. Brain Behav Immun. 1998;12(4):249–71.
40. Kroeze WK, Kristiansen K, Roth BL. Molecular biology of serotonin receptors structure and function at the molecular level. Curr Top Med Chem. 2002;2(6):507–28.
41. Madden KS, Sanders VM, Felten DL. Catecholamine influences and sympathetic neural modulation of immune responsiveness. Annu Rev Pharmacol Toxicol. 1995;35:417–48.

Immune Response in Human Pathology: Infections Caused by Bacteria, Viruses, Fungi, and Parasites

10

Jan Verhoef, Kok van Kessel, and Harm Snippe

10.1 Infections

In the middle of the nineteenth century, it became clear that micro-organisms could cause disease. Effective treatment, however, was not possible at that time; prevention and spread of infectious diseases depended solely on proper hygienic means. At the beginning of the twentieth century, passive and active vaccination procedures were developed against a number of these PATHOGENIC MICRO-ORGANISMS to prevent the diseases in question (rabies, diphtheria, tetanus, etc.). Thanks to the discovery of antimicrobial chemicals (by Paul Ehrlich) and antibiotics (by Sir Alexander Fleming), the threat of infectious diseases seemed to be minimised. Large-scale vaccination programmes against childhood diseases (diphtheria, whooping cough, and polio), started in the early 1950s, raised hopes of finally being able to eradicate these diseases from the planet. This approach was successful for smallpox (1980). However, new infectious diseases have emerged [e.g., *Legionella*, HUMAN IMMUNODEFICIENCY VIRUS (HIV), *Helicobacter*, SARS, etc.], and new vaccines and antibiotics are needed. Furthermore, due to intensive medical treatment with antibiotics and immunosuppressive drugs, hospital infections are a growing problem. Bacteria hitherto deemed harmless are causing OPPORTUNISTIC INFECTIONS in immunocompromised patients. The pathogens have developed resistance to many antibiotics, and sometimes no effective antibiotics are available to treat these patients.

To make the story even more serious, man is surrounded and populated by a large number of different non-PATHOGENIC MICRO-ORGANISMS. In the normal, healthy situation, there is a balance between the offensive capabilities of micro-organisms and the defences of the human body. The body's defences are based on vital non-specific and specific immunological defence mechanisms. An infection means that the micro-organism has succeeded in penetrating those lines of defence, signalling a partial or complete breakdown of the body's defence system.

10.2 Natural Resistance

The body's FIRST LINE OF DEFENCE comprises the intact cell layers of skin and mucous membrane, which form a physical barrier. The skin's low pH level and bactericidal fatty acids enhance the protection provided by this physical barrier. The defences in the respiratory tract and the gastrointestinal tract are mucus, the 'ciliary elevator' of the epithelium, and the motility of

Final manuscript submitted on March 10, 2017.

J. Verhoef · K. van Kessel (✉) · H. Snippe
Department of Medical Microbiology, University Medical Center Utrecht, Utrecht University, Utrecht, The Netherlands
e-mail: k.kessel@umcutrecht.nl

M. J. Parnham et al. (eds.), *Nijkamp and Parnham's Principles of Immunopharmacology*,
https://doi.org/10.1007/978-3-030-10811-3_10

the small intestine. The presence of normal microbial flora (colonisation resistance) in the intestine also plays a role in protection against colonisation by external bacteria.

The most important humoral NATURAL resistance factors are complement, ANTIMICROBIAL PEPTIDES, LYSOZYME, interferon, and a number of CYTOKINES (see Chaps. 6 and 8). ANTIMICROBIAL PEPTIDES are widely expressed as part of the professional phagocyte antimicrobial arsenal and rapidly induced at epithelial surfaces. They are found in mammals, invertebrates, and plants. In general, they are small, amphipathic molecules, contain positive charge, and can be structurally divided into several categories. The mode of action goes beyond their antimicrobial capacities, and they elicit a complex array of responses in different cell types. The most extensively studied mammalian gene families are the cathelicidins and DEFENSINS. In general the ANTIMICROBIAL PEPTIDES disrupt lipid membranes and thereby induce microbial killing. Micro-organisms have developed several countermeasures against ANTIMICROBIAL PEPTIDES, but the many structurally different peptide classes still provide protection against infection (Table 10.1).

LYSOZYME, which is found in almost all body fluids, degrades sections of the cell wall of Gram-positive and—in combination with complement—Gram-negative bacteria. This causes the otherwise sturdy cell wall to leak and the bacterium to burst.

Interferons are glycoproteins and may inhibit the replication of viruses. Within several hours after the onset of a virus infection, INTERFERONS are produced in the infected cell and help protect the neighbouring unaffected cells against infection. This protection is brief, but high concentrations of INTERFERONS are produced at a time when the primary immunological response is relatively ineffective.

CYTOKINES, such as INTERLEUKIN-2 (IL-2), granulocyte-macrophage colony-stimulating factor (GM-CSF), and tumour necrosis factor-α (TNF-α), stimulate non-specifically

Table 10.1 Some examples of important human pathogens

	Species	Disease/location	Treatment/prevention
Bacterium	*Streptococcus pneumoniae*	Pneumoniae/meningitis	Antibiotics/vaccination
	Mycobacterium tuberculosis	Lung tuberculosis	Antibiotics
	Vibrio cholerae	Severe diarrhoea	Antibiotics/liquid suppletion/sanitation
	Staphylococcus aureus MRSA	Wound infection/hospital infection	Antibiotics, MRSA not sensitive for standard antibiotics, difficult to treat
	Neisseria meningitidis	Meningitis	Antibiotics/vaccination
	Bacillus anthracis	Systemic infection (sepsis)	Antibiotics as early as possible
	Corynebacterium diphtheriae	Throat/heart	Antiserum, vaccination
	Campylobacter jejuni	Intestinal infections	Hygiene, especially food (chicken)
	Helicobacter pylori	Gastritis, ulcer	Antibiotics
DNA virus	*Poxviridae*	Smallpox	Vaccination, eradication
	Herpesviridae	Herpes genitalis	Antiviral agents
	Papovaviridae	Warts and cervical carcinoma	Surgery
	Hepadnaviridae	Hepatitis B	Vaccination
RNA virus	*Orthomyxoviridae*	Influenza	Vaccination
	Coronaviridae	SARS	Unknown
	Retroviridae	AIDS	Antiviral agents
	Caliciviridae	Gastrointestinal infection	Sanitation, hygiene
Parasites	*Plasmodium* species	Malaria	Prophylactic medication, antimalarial drugs
	Giardia species	Intestinal tract	Hygiene
	Trypanosoma cruzi	Sleeping sickness	Antiparasitic agents

the proliferation, maturation, and function of the cells involved in defence (see Chap. A6).

Innate immune cells recognise microbes by TOLL-LIKE RECEPTORS (TLR) (see section *Pathogenesis of* SHOCK), giving rise to the above production of CYTOKINES in the early phase of the response.

Micro-organisms that succeed in penetrating the FIRST LINE OF DEFENCE are ingested, killed, and degraded by phagocytic cells [POLYMORPHONUCLEAR LEUKOCYTES (PMN) or NEUTROPHILS, MONOCYTES, and MACROPHAGES], which are attracted to a microbial infection through CHEMOTAXIS. The ingestion by phagocytic cells of the micro-organism is enhanced by serum proteins (opsonins), such as ANTIBODIES and the C3b component of complement, which are recognised by specific RECEPTORS on the PHAGOCYTES. After ingestion, the particle is surrounded by the membrane of the phagocyte, forming a vacuole known as a PHAGOSOME. The PHAGOSOME then fuses with some of the countless granules in the phagocyte, thus allowing the lysosomal microbicidal agents and enzymes to do their work. The formation of toxic oxygen radicals greatly contributes to the killing and elimination of the ingested micro-organism (Fig. 10.1) (see Chap. 8).

A special role in cellular NATURAL resistance is reserved for the NATURAL KILLER CELLS (NK cells), which display considerable CYTOTOXIC activity against virus-infected cells. This NK activity is stimulated by INTERFERONS and, at a very early stage in the infection, serves to reinforce the non-specific defence mechanism.

10.3 Specific Resistance

In the specific immune response, elements of the NATURAL defence mechanism are directed against a specific enemy. Depending on the micro-organism, either the cellular defence mechanism (tuberculosis) or the humoral ANTIBODY-dependent defence mechanism (influenza) is of primary importance. In many cases, a joint cellular and humoral response is needed to provide an effective immune defence (typhus).

Both T LYMPHOCYTES and MACROPHAGES play a role in cellular defence. During the first contact with an antigen, MACROPHAGES process the antigen and present its protein fragments (T-cell EPITOPES) to T cells, which then proliferate and remain present for years in the body as memory cells. When a second

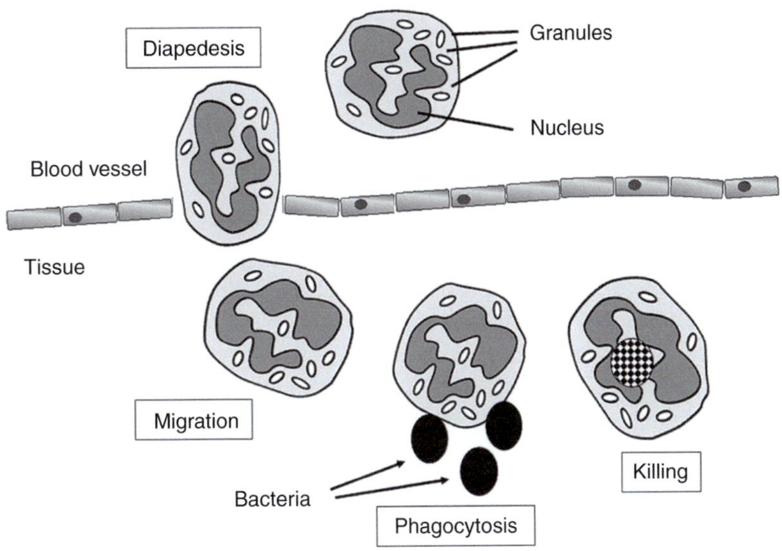

Fig. 10.1 Schematic representation of the progressive steps of phagocytic endocytosis

encounter occurs, T cells produce lymphokines, which activate the MACROPHAGES. These activated MACROPHAGES grow larger, produce more and better degrading enzymes, and are now able to eliminate micro-organisms, which otherwise would have survived intracellularly (tuberculosis, typhoid fever). MACROPHAGES from non-immune animals are not able to eliminate these micro-organisms.

Five different classes of ANTIBODIES can be distinguished in man, namely, IgG, IgA, IgM, IgD, and IgE. They differ from one another in size, charge, amino acid composition, and glycosylation (see Chap. 4). In principle, the structure of the ANTIBODIES is the same, i.e. two heavy and two light chains: it is the variable part of these chains that recognises the micro-organism. The biological function (see below) is determined by the constant part (Fc) of the heavy chain. With the exception of IgD, all these ANTIBODIES are important in antimicrobial activity.

- IgA, which is found in all external secretions, reacts with the surface of micro-organisms, preventing them from adhering to sensitive cells and mucous membranes.
- IgG neutralises microbial toxins.
- IgG, IgM, and C3b serve as opsonins, which promote phagocytosis.
- IgG, IgM, and to a lesser extent IgA activate the complement system after binding to the micro-organism. Activation products C3a and C5a ensure that the phagocytes are attracted to the inflammatory response.
- IgG and IgM, in combination with complement and lysozyme, have a lytic effect on Gram-negative bacteria and enveloped viruses.
- IgG and IgM inhibit the mobility of micro-organisms by attaching specifically to the flagellum. Thereby the chance of phagocytosis increases, and the chance of spreading of disease decreases.
- IgG, together with the killer or K cells, can eliminate infected host cells which carry viral or other foreign antigens on their surface.
- IgE is of importance in parasite infections. At the site of the infection, mast cells, bearing specific IgE, release large quantities of vasoactive amines, which cause the contraction of smooth muscle tissue and increase the permeability of the blood vessels. In the intestine, this results in worms being detached and eliminated.

10.4 Defence Against Bacteria, Viruses, Fungi, and Parasites

Several non-invasive bacteria, i.e. those that do not invade the body, cause disease through the production of EXOTOXINS (tetanus, diphtheria, cholera). The IMMUNE SYSTEM neutralises the toxin with the aid of ANTIBODIES (IgG, IgM). If the individual has not been inoculated, the toxin will act on certain cells in the body directly through a RECEPTOR. This bond is very strong (i.e. has a high AFFINITY) and is difficult to break by the administration of ANTIBODIES. In practice, if there are clinical symptoms of the disease, then large doses of antitoxins must be administered. If one is trying to prevent the development of the disease, then the presence of small quantities of specific ANTIBODIES (IgG) is sufficient.

The ADHERENCE of bacteria to cells is effectively blocked by IgA. Oral vaccination against cholera, for example, is aimed at obtaining sufficient specific IgA in the intestine, so that no colonisation of this bacterium can take place and the cholera toxin can no longer adhere to its RECEPTOR.

In general, defence against invasive bacteria is provided by ANTIBODIES (IgG, IgM) that are directed against bacterial surface ANTIGENS. In many cases, these bacteria have a CAPSULE, which interferes with effective PHAGOCYTOSIS. ANTIBODIES against the ANTIGENS of these capsules neutralise the interference, with subsequent elimination of the bacteria by PHAGOCYTES. ANTIBODIES (IgM, IgG, IgA) in combined action with complement kill bacteria by producing holes in the cell wall of the bacterium.

Although intracellular bacteria (tuberculosis, leprosy, listeriosis, brucellosis, legionellosis, and salmonellosis) are ingested by MACROPHAGES, they are able to survive and multiply. In these cases, cellular immunity alone provides the defence, since ANTIBODIES are not effective. Only activated MACROPHAGES are capable of killing and degrading these bacteria.

ANTIBODIES neutralise viruses directly and/or indirectly by destroying infected cells that carry the virus antigen on their surface. The mechanisms of this defence resemble those of humoral defence against bacterial surfaces. The ANTIBODY-dependent cellular CYTOTOXICITY reaction is specific for defence against viruses. Cells that carry an antigen encoded by the virus on their surface are attacked by CYTOTOXIC K cells, bearing ANTIBODIES that fit the antigen on the TARGET cell (K cells have Fc RECEPTORS for IgG) (Fig. 10.2).

Not only humoral but also cellular immunity plays an important role in virus infections. People with a genetic T-cell deficiency are highly susceptible to virus infections. In cellular defence, it is primarily the virus-infected cells that are attacked and eliminated. CYTOTOXIC T cells recognise MHC class I-presented T-cell EPITOPES on the surface of virus-infected cells and kill them.

The fungi responsible for human diseases can be divided into two major groups on the basis of their growth forms or on the type of infection they cause. Pathogens exist as branched filamentous forms or as yeasts, although some show both growth forms. The filamentous types (*Trichophyton*) form a 'mycelium'. In asexual reproduction, the fungus is dispersed by means of spores; the spores are a common cause of infection after inhalation. In yeast-like types (*Cryptococcus*), the characteristic form is the single cell, which reproduces by division or budding. Dimorphic types (*Histoplasma*) form a mycelium outside but occur as yeast cells inside the body. *Candida* shows the reverse condition and forms a mycelium within the body.

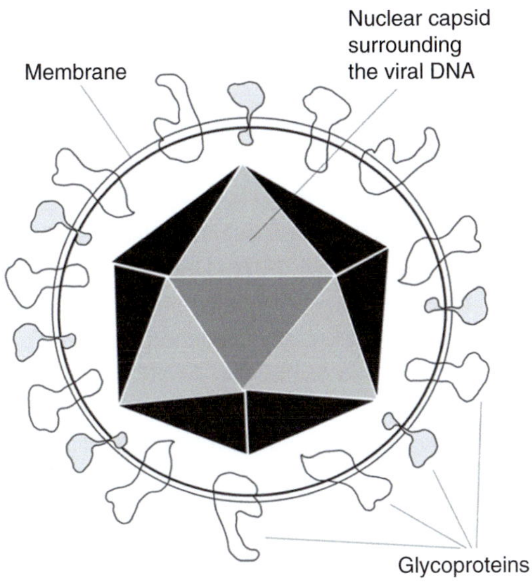

Fig. 10.2 Schematic illustration of an enveloped virus (Herpes simplex virus)

In superficial mycoses, the fungus grows on the body surface, for example, skin, hair, and nails (*Epidermophyton*, *Trichophyton*), the disease is mild, and the pathogen is spread by direct contact. In deep mycoses (*Aspergillus*, *Candida*, *Cryptococcus*, *Histoplasma*), internal organs are involved, and the disease can be life-threatening and is often the result of opportunistic growth in individuals with impaired immunocompetence.

Many of the fungi that cause disease are free-living organisms and are acquired by inhalation or by entry through wounds. Some exist as part of the normal body flora (*Candida*) and are innocuous unless the body's defences are compromised in some way. The filamentous forms grow extracellularly, while yeasts can survive and multiply within phagocytic cells. NEUTROPHILS kill yeasts by means of both intra- and extracellular factors. Some yeasts (*Cryptococcus neoformans*) form a thick polysaccharide CAPSULE to prevent phagocytic uptake. In addition, many cell-wall components of yeasts cause suppression of cell-mediated immune responses. The role of humoral and cellular immunity in controlling infections caused by fungi is not yet well defined, but cellular immunity is the cornerstone of host defence against (some) fungal infections. As a consequence, HIV INFECTION, which affects the cellular arm of the IMMUNE SYSTEM, results in previously uncommon infections such as those caused by *C. neoformans*.

The immunological defence systems against *parasites* are considerably more complex than those against bacteria and viruses. This is due to various factors. In the first place, each parasite has its own life cycle, consisting of various stages with specific antigen compositions. Moreover, parasites are able to avoid the host defence system (mimicry), to combat it (IMMUNOSUPPRESSION), or to mislead it (antigenic variation). Both humoral and cellular immunity are important for the defence against parasites growing intercellularly, as we have seen in the case of bacteria and viruses. ANTIBODY concentrations (IgM, IgG, IgE) are often elevated. IgE also plays a special role in the removal of parasites (especially worm infections) from the intestine (see above) (Table 10.2).

Table 10.2 Classification of micro-organisms

Classification of bacteria by:
Genotypical characteristics: chromosomal DNA fragment analysis, nucleic acid sequence analysis, probes **Phenotypical characteristics**: morphology, biotyping, serotyping, antibiotic resistance **Analytical characteristics**: cell-wall analysis, lipid and protein analysis, enzyme typing (catalase) **Gram staining positive or negative** **Aerobic, anaerobic**: Fermentation of different sugars
Naming and classification of viruses according to:
Structure: size, morphology (naked, enveloped), nucleic acid (RNA, DNA) **Molecular aspects**: mode of replication, assembly, and budding **Disease**: encephalitis, hepatitis **Means of transmission**: droplets, water, blood, insects **Host range**: animal, plant, bacteria
Classification of fungi according to:
Structure: macroscopic morphology of hyphae (mycelium); microscopic morphology of hyphae, conidiophores and conidia (spores); and shape and size **Cell features**: nucleus, cytosol, plasmalemma (cell membrane which contains cholesterol), physiology, staining properties **Sexual characteristics**: sexual and/or asexual reproduction, extended dikaryotic phase, basidium formation **Genotypical characteristics**: chromosomal DNA fragment analysis, nucleic acid sequence analysis, probes
Diagnosis of parasites by:
Macroscopical examination
Concentration of cysts and eggs by microscopic examination
Serological diagnosis: antibody response **Detection of parasite by serology and by nucleic acid hybridisation**: probes and amplification techniques

10.5 Pathogenesis of Shock

Sepsis is a systemic inflammatory response to presumed or known infection (see also Chap. 26). The resulting inflammatory response becomes over amplified, leading to multiple organ failure and death. PATHOGEN-ASSOCIATED MOLECULAR PATTERNS (PAMP) in bacteria, viruses, parasites, and fungi initiate the host response by triggering families of PATTERN RECOGNITION RECEPTORS (PRR). In Gram-negative (Fig. 10.3) bacterial infections, the interaction between bacterial ENDOTOXIN or LIPOPOLYSACCHARIDE (LPS; a major structural component of the cell wall) and various host-cell systems has been implicated in the pathogenesis of septic SHOCK. In particular, the release of TNF-α and INTERLEUKIN-1 (IL-1) after the activation of host cells by ENDOTOXIN induces haemodynamic SHOCK.

Biochemical and genetic evidence has identified TLR4 as the RECEPTOR that mediates cellular activation in response to LPS. This family of TLR proteins (Fig. 10.4), which resemble the antimicrobial Toll proteins of *Drosophila* (fruit fly), has been identified in humans and mice. TLR4 was identified as the missing link in LPS-induced cell signal transduction and responsiveness that is associated with MD-2 and CD14. It is known that C3H/HeJ mice are hyporesponsive to the biological effects of LPS. This proved to be the result of TLR4 deficiency. The TLR family members are coupled to a signalling adapter protein (MyD88) and form differential dimers that may explain the discrete responses to TLR ligands such as lipoproteins, heat SHOCK proteins, unmethylated CpG DNA, viral dsRNA, and bacterial flagellin. Intracellular signalling involves several kinases depending on the TLR involved and includes the MAP kinase and NF-κB pathways leading to a cellular response. PAMP for TLR2 include a variety of agonists derived from Gram-positive organisms such as PEPTIDOGLYCAN and lipoteichoic acid. Therefore, it seems that TLR2 and TLR4 are activated primarily by different PAMP to initiate the host response to Gram-positive and Gram-negative bacterial infection, respectively. This is

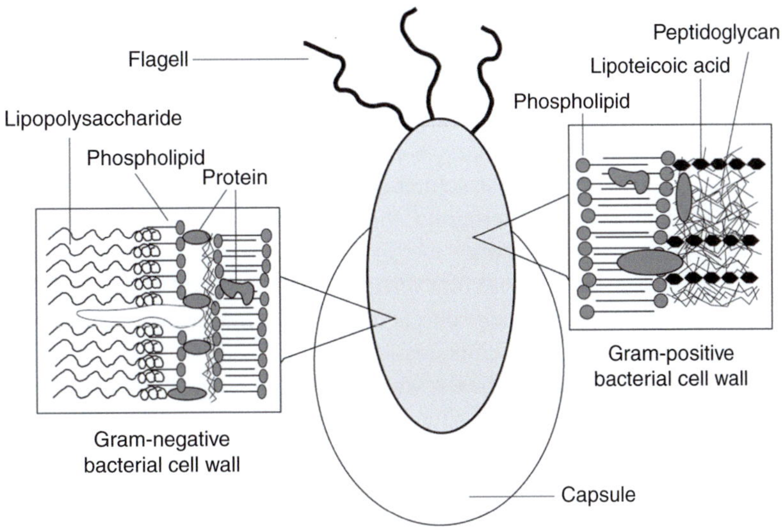

Fig. 10.3 Schematic illustration of the cell envelope of a Gram-negative and a Gram-positive bacterium

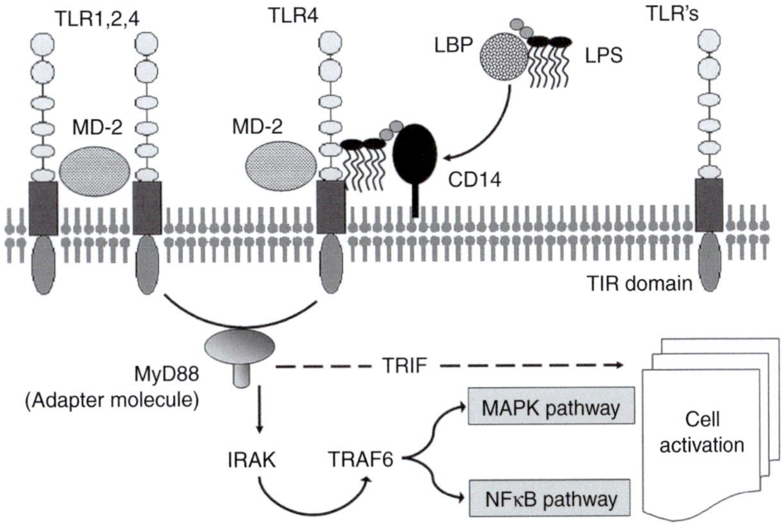

Fig. 10.4 Schematic illustration of cell activation through Toll-like receptors (TLRs)

further illustrated by the fact that TLR2 (but not TLR4) knockout mice are highly susceptible to GRAM-POSITIVE BACTERIA, like *Staphylococcus aureus*, whereas TLR4 (but not TLR2) knockout mice are highly susceptible to Gram-negative bacteria such as *Salmonella typhimurium*.

Several lines of evidence support the current hypothesis that the monocyte/macrophage is the principal cellular mediator of endotoxicity. LPS-hyporesponsive TLR4-deficient C3H/HeJ mice are made responsive by transfer of MACROPHAGES of a closely related LPS-sensitive strain. When the host is challenged with LPS, soluble factors are produced by MACROPHAGES that mediate fever and an acute-phase response. These factors include the pro-INFLAMMATORY CYTOKINES, IL-1, IL-6, IL-8, and TNF-α. Together, TNF-α and IL-1 stimulate endothelial cells to produce and express proteins on their membrane that have adhesive properties for LEUKOCYTES, promoting the margination and passage of neutrophils from blood vessels through the endothelial layer, leading to neutrophil influx into the tissue. ADHESION MOLECULEs that mediate the binding of neutrophils appear on the endothelium after an inflammatory stimulus, followed by molecules that are specific for adhesion of MONOCYTES or LYMPHOCYTES, which may be why NEUTROPHILS enter before mononuclear cells. Molecules that are currently known to be involved in leukocyte-endothelium interactions belong to three structural groups: the immunoglobulin gene superfamily, the integrin family, and the selectin family.

Concomitant with cytokine release, LPS induces the activation of neutrophils, MACROPHAGES, and many other cells, resulting in the release of toxic oxygen radicals, which lead to tissue damage. At the same time, membrane-associated phospholipases are activated, and products of the arachidonic acid cascade are released through the CYCLOOXYGENASE and/or LIPOXYGENASE pathways (see Chap. 9). PLATELET-ACTIVATING FACTOR (PAF) is also generated, partly in response to the same signals. All these products contribute to a generalised inflammatory state with influx of neutrophils, capillary-leak syndrome, disturbances in blood coagulation, and myocardial suppression.

ENDOTOXIN and TNF-α also trigger multiple abnormalities in coagulation and fibrinolysis, leading to microvascular clotting and diffuse intravascular coagulation. They also induce endothelial cells to produce plasminogen activator and IL-6, which is an important modulator of the production of acute-phase proteins by the liver. Interestingly, despite having important structural differences, TNF-α and IL-1 have multiple overlapping and few distinct biological activities, act synergistically, and mimic the whole spectrum of toxicity caused by LPS (see Chap. 6). IL-8 is an important chemoattractant and activator of NEUTROPHILS and is crucial in the early stages of INFLAMMATION.

Infusion of ENDOTOXIN in healthy humans leads to an early and transient increase in plasma levels of TNF-α (detectable after 30 min, peaking after 90–120 min, and undetectable after 4–6 h), which coincides with the development of clinical symptoms and pathophysiological responses encountered in Gram-negative septicaemia. TNF-α, IL-1, IL-6, and IL-8 levels are also increased in patients with sepsis syndrome, with high levels of these CYTOKINES being correlated with severity of disease.

All these observations support the concept that ENDOTOXIN largely acts by initiating an inflammatory response through the activation of MONOCYTES/MACROPHAGES and the subsequent release of CYTOKINES. It also activates the COMPLEMENT SYSTEM (leading to the generation of C5a, which induces aggregation of neutrophils and pulmonary vasoconstriction) and factor XII of the intrinsic coagulation pathway (Hageman factor). Finally, it induces the release of ENDORPHINS, which are also involved in the complex interactions of the inflammatory response in endotoxic septic SHOCK.

GRAM-POSITIVE BACTERIA are frequently and increasingly cultured from blood obtained from patients in SHOCK. Unlike the pathophysiology of SHOCK caused by Gram-negative bacteria, not much is known about the sequence of events that controls the signalling of MONOCYTES and MACROPHAGES that leads to the release of CYTOKINES. Cell-wall components, such as PEPTIDOGLYCAN and teichoic acid, are clearly important in the activation of these cells. EXOTOXINS, however, may also play a role in the pathogenesis of Gram-positive bacterial SHOCK.

Recently, another protein family was identified that also participates by sensoring microbial components derived from bacterial PEPTIDOGLYCAN. The NOD (NUCLEOTIDE-BINDING OLIGOMERISATION DOMAIN) proteins NOD1 and NOD2 have been implicated in intracellular recognition of the core structure, γ-D-glutamyl-meso-diaminopimelic acid, present in PEPTIDOGLYCAN.

A number of circulating inflammatory mediators have been investigated as marker tools to facilitate the early recognition of sepsis. These include IL-1, IL-6, TNF, pro-calcitonin, and triggering RECEPTOR on myeloid cells (TREM-1). TREM-1 is expressed on LEUKOCYTES, and TREM family members have been implicated in mounting the inflammatory response. Procalcitonin (PCT) and IL-6 have proved to be the most prominent biomarkers of early sepsis. PCT is the prohormone of the hormone calcitonin and can be produced by several cell types and many organs in response to pro-inflammatory stimuli, in particular by bacterial products. More recently HIGH MOBILITY GROUP BOX-1 (HMGB-1) has been implicated as a lethal mediator of systemic INFLAMMATION. HMGB-1 is a nuclear and cytosolic protein widely studied as a transcription and growth factor that is released into the extracellular environment. It has a weak pro-inflammatory activity by itself, and it may work in concert with other pro-INFLAMMATORY CYTOKINES. This molecule may also be useful as a biomarker in the stratification of sepsis.

Susceptibility to sepsis can be influenced by factors that include ethnicity, gender, age, genetic defects, and environmental factors. Single-nucleotide polymorphisms (involving single base-pair alterations) have been described in genes controlling the host response to infection such as alterations in TNF RECEPTORS, IL-1 RECEPTORS, coagulation factors, and TLR. It is now clear that sepsis is a complex, dynamic syndrome with great heterogeneity and not a distinct disease. Therefore, neutralisation of a single key mediator as a cure for all patients with sepsis is erroneous (see Chap. 26).

10.6 Human Immunodeficiency Virus Infection

The HIV is a retrovirus that infects cells bearing the CD4 antigen, such as T helper cells (Th), MACROPHAGES, and DENDRITIC CELLS. The CD4 molecule, together with other RECEPTOR molecules, like chemokine RECEPTOR CCR5, acts as a binding site for the gp120 envelope glycoprotein of the virus. In an attempt to respond to HIV ANTIGENS and concomitant secondary microbial infections, these cells are activated, thus inducing the replication of HIV in the infected CD4 T cells, which are finally destroyed. In contrast, HIV-1 infection of MACROPHAGES is self-sustained and results in an inexorable growth of chronic active inflammatory processes in many tissue compartments including the CENTRAL NERVOUS SYSTEM. Infected cells bear the fusion protein gp41 and may therefore fuse with other infected cells. This helps the virus to spread and accounts for the multinucleated cells seen in lymph nodes and brain. As a result of the decreased numbers of $CD4^+$ Th cells and defects in antigen presentation, depressed immune responses in these patients are observed. During the progression of the disease, OPPORTUNISTIC INFECTIONS by otherwise harmless micro-organisms can occur. These include *Candida albicans* oesophagitis, mucocutaneous herpes simplex, toxoplasma in the CENTRAL NERVOUS SYSTEM, and pneumonia caused by toxoplasma and *Pneumocystis carinii*; Kaposi's sarcoma also occurs frequently in these patients. This has been linked to the presence of a previously unknown type of herpes virus (HHV-8). This immune deficiency syndrome is called 'acquired immune deficiency syndrome' (AIDS). It has been suggested that infected MONOCYTES/MACROPHAGES carry the HIV virus into the brain where it replicates in microglia and infiltrating MACROPHAGES. As a consequence, many AIDS patients develop cognitive and motor brain impairments. However, the picture is complicated by the various persistent infections already present in these patients, which give rise to their own pathology in the brain. These include *Toxoplasma gondii*, *Cryptococcus neoformans*, and JC virus.

So far, no cure for HIV INFECTION has been achieved. The main effort in the prevention of HIV INFECTION lies in mass public education programmes. Treatment of infected individuals is possible but expensive. HIGHLY ACTIVE ANTIRETROVIRAL THERAPY (HAART) reduces morbidity and mortality among patients infected with HIV (see Fig. 10.5). Success is limited by the emergence of drug-resistance viruses that can be transmitted to newly infected individuals. Resistance, drug toxicity, and poor patient ADHERENCE lead to treatment failure and necessitate continuous development of alternative treatment strategies that intervene with the HIV replication cycle: i.e. on the level of virus entry, critical viral enzymes [reverse transcriptase (RT), integrase (IN), and proteases (PR)], or viral nucleocapsid (NC) protein. At this moment a triple therapy is being prescribed in Western countries (two RT inhibitors and one PR inhibitor,

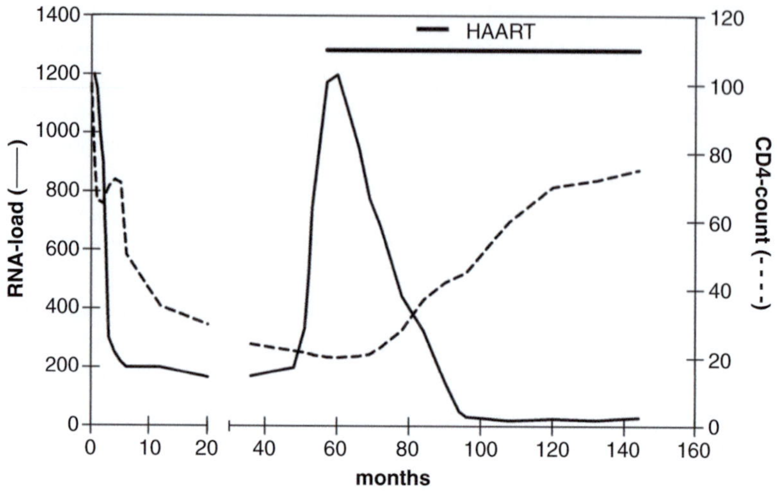

Fig. 10.5 The effect of single and triple therapy on viral load and CD4 cells over time in HIV-infected individuals

Fig. 10.4), each of which interfere with specific steps in the process of HIV replication. One major problem that has arisen is the increasing resistance to these drugs. Agents with novel MECHANISMS OF ACTION provide options for patients with DRUG-RESISTANT virus. Blocking of the chemokine RECEPTOR CCR5, a co-RECEPTOR on CD4 cells for HIV, is an alternative treatment for persons infected with the R5 HIV type. This notion is supported by a recent finding that a homozygous defect in this chemokine RECEPTOR accounts for resistance of multiple-exposed individuals to HIV-1 infection. Currently a commercially available drug is being used that specifically binds to CCR5 on the surface of the CD4 cell and selectively blocks HIV-1 binding. One remarkable HIV patient, known as the 'Berlin patient' and on an antiretroviral therapy for 10 years, was diagnosed with acute myeloid leukaemia and therefore received a haematopoietic stem cell transplant from a donor with the 'resistant' CCR5 receptor. He had to stop his antiretroviral medication, and after 3 months, the virus was undetectable, and his CD4 counts recovered. He is now considered cured from HIV.

10.7 Vaccines and Vaccination

Pasteur and Koch triggered the stormy development of vaccines (anthrax, rabies, cholera) at the end of the nineteenth century. While Pasteur remained faithful to the principle of attenuated micro-organisms in preparing his vaccines, Koch employed killed germs (cholera) as a vaccine. Since diphtheria and tetanus cause disease by means of toxins, the next logical step in the development of vaccines was the use of detoxified toxins to induce protection against these diseases [diphtheria (Von Behring) and tetanus (Kitasato)]. Von Behring and Kitasato were the first to demonstrate that the source of the protective activity induced by vaccines was present in blood serum. Von Behring was also the first to prove that protective immunity could be passed on via serum. The development of new vaccines had its ups (yellow fever) and downs (tuberculosis). With the arrival of antibiotics, all work on new bacterial vaccines was suspended or severely curtailed, although some researchers continued to work on viral vaccines, such as rubella, measles, polio, and mumps.

Since it has proved difficult to consistently develop new antibiotics to combat antibiotic-resistant bacteria, interest in vaccines has gradually increased over the last 15 years (see Chap. C1). Today, thanks to new insights into the IMMUNE SYSTEM and modern molecular biological and chemical techniques of analysis and synthesis, it is possible to produce well-defined vaccines. These contain only those determinants of the PATHOGENIC MICRO-ORGANISM that induce protection (EPITOPES). These EPITOPES are usually short peptide or oligosaccharide chains, which can be produced synthetically or by means of RECOMBINANT DNA techniques. The immunogenicity of these products can be enhanced by coupling them to a CARRIER (tetanus toxoid, liposomes) and/or by adding an adjuvant (a substance that strengthens the immune response non-specifically). The RECOMBINANT DNA technique can also be used to obtain attenuated strains of micro-organisms, which are fully immunogenic and thus provide protection but which are no longer virulent. One example of this is the development of a new cholera vaccine based on a bacterium that has all the characteristics of a virulent strain, except the toxin. The bacterium has retained all its ADHERENCE factors, which allow it to adhere to the intestinal mucosa; the length of time it spends in the intestine is sufficient to stimulate the local IMMUNE SYSTEM. The newest trend in vaccinology is immunisation by introducing plasmid DNA into the host. Success has been attained by this method for hepatitis B vaccination. An example that holds a combination of modern techniques has been developed to combat Malaria, a communicable disease caused by the *Plasmodium* parasite that is transmitted by female Anopheles mosquitoes. *P. falciparum* causes the highest rates of complications and mortality and is a serious public health concern in most countries in sub-Saharan Africa, especially young children. Besides many alternative strategies to combat the spread of malaria, vaccination is a long-going approach. A promising candidate vaccine combines the circumsporozoite coat protein fused to hepatitis B surface antigen exposed on the surface of self-assembled virus-like particles together with immunomodulatory adjuvants. It has been shown to induce an immunogenicity profile in controlled human malaria infection model and field studies in sub-Saharan Africa.

Not only are new vaccines being developed, but it is also possible to heighten NATURAL resistance for longer or shorter periods. Various INTERLEUKINS (IL-2, GM-CSF) and INTERFERONS are being studied in order to use them to combat infectious diseases (see Chap. C7). Passive ANTIBODY therapy with polyclonal or MONOCLONAL ANTIBODIES (mouse or human IgG with single SPECIFICITY) for infectious diseases is experiencing a renewed interest. Targeting soluble factors (like neutralisation of bacterial toxins and viruses) or common structures (like bacterial adhesins or viral entry factors) in high-risk patient groups may be beneficial alone or may enhance the therapeutic EFFICACY of other drugs. Targets for clinical development of MONOCLONAL ANTIBODIES include multiresistant staphylococci and enterococci, *Bacillus anthracis* toxin, HIV, HEPATITIS C VIRUS (HCV), and respiratory syncytial virus (RSV). Therapeutic application of human monoclonal antibodies is boosted by their proven success for various cancers. Molecular cloning and efficient in vitro production by cell lines enable customising antibody properties like IgG subclass, glycosylation pattern (to chance effector functions like complement activation, receptor triggering, or half-life), or coupling with toxins, antibiotics, and inhibitors. Renewed optimism for treatment of HIV is encouraged by uncovered naturally arising, broad, and potent HIV-1-neutralising antibodies. It is a challenge to select the best monoclonal antibody (or cocktail) that not only depends on its capacity to recognise HIV broadly, but may also relate to the kinetics of the antibody epitope expression on infected cells. Understanding the molecular complexity of the neutralising antibody repertoire towards the main antigenic determinant of the influenza virus, the HA surface glycoprotein, is important to design a future vaccine that can elicit long-term broadly protective antibodies to a variety of different strains. Combining new adjuvants with current H1N1 vaccine improved antibody affinity maturation and offers the promise of improved protection in vivo, particularly in naïve individuals.

The clustered regularly interspaced short palindromic repeats (CRISPR)/CRISPR-associated 9 (Cas9) system is a versatile RNA-guided DNA-targeting platform, which enables efficient and directed manipulation of genomes. It is a bacterial defence mechanism for degradation of foreign nucleic acids of bacteriophages and other genetic elements. This developing platform has been revolutionising our ability to modify and manipulate genomes, which profoundly changes biological research and therapeutics development. Targeting in vitro herpesviruses (HSV-1, HCMV, and EBV) with this new editing technique induced complete inhibition of viral replication and even eradication of viral genome from infected cells. Herpesviruses, hepatitis B virus, and HIV are known for their persistence or latency state where they are unrecognisable for the immune system. Therefore the DNA-targeting CRISPR/Cas system could be a powerful tool to tackle persistent viral infections. However, so far the technique proved is value for in vitro cell cultures, while application to the clinic will require a method of safe and efficient delivery.

10.8 Infections in the New Millennium

As outlined above for a number of bacteria and viruses, effective vaccines have been developed and applied worldwide. The eradication of smallpox (*Variola major*) virus in the 1970s was a milestone for the World Health Organisation. The next goal of the WHO is to eradicate poliovirus in the coming years. Major problems to be dealt with are the distribution of these vaccines, the costs involved, the registration and the compliance of the vaccines, and molecular techniques to trace the final bug. Meanwhile new unexpected microbiological threats become in focus. Hospital infections caused by multiresistant microorganism due to the abundance use of antibiotics and exchange of genetic material between microorganisms impose major problems on patients and healthcare workers. New antibiotics and/or vaccines should be developed and new strategies employed to contain these infections. Due to crowding and the high mobility of the world population, old and new pathogens, e.g. influenza and SARS, threaten our society. The recent influenza A H1N1 ('Mexican') flu outbreak in 2009 demonstrated how rapidly a new strain of flu can emerge and spread around the world. The sudden outbreak of this novel flu virus has tested the world's public health preparedness. In the Netherlands, Q fever emerged with large epidemic outbreaks. Q fever is a zoonotic disease that is passed from infected (farm) animals (cattle, sheep, and goats are the primary carriers) to humans. Q fever is a disease caused by the intracellular bacterium *Coxiella burnetii*. Organisms are excreted in milk, urine, and faeces of infected animals. Most importantly, during birthing the organisms are shed in high numbers within the amniotic fluids and the placenta. *C. burnetii* is resistant to heat, drying, and many common disinfectants, allowing it to survive for a long time in the environment. People can become infected by inhalation of the bacteria, but the risk of infection is low. Only about one-half of all people infected with *C. burnetii* show signs of clinical illness (e.g. flu-like illness, pneumonia, and hepatitis). Q fever can be treated with antibiotics, and most people will recover fully. As a control measurement, pregnant goats on dairy farms had to be killed, and a mandatory vaccination campaign was started. Since May 2015, Zika virus infection spread from Brazil to other countries in South and Middle America as well the Western Hemisphere. Although most infections are unnoticed or only follow mild symptoms like fever and rash, in pregnant women, the infection may cause microcephaly. This is a birth defect where a baby's head is smaller than expected and the brain is not developed properly. Zika virus was recently (1947) identified in a rhesus monkey and is spread by the bite of an infected yellow fever or dengue mosquito which are not endemic in the Netherlands. Currently no vaccine or specific medication is available, but new vaccine candidates elicit neutralising antibodies in mice and non-human primates providing protection against Zika virus infection. Since March 2014, the largest outbreak of Ebola virus in history has affected multiple countries in West Africa, especially Guinea and Sierra Leone. Ebola virus causes a grotesque haemorrhagic fever syndrome that is associated with high mortality in humans. The virus originates from an infected animal (like bats) and may jump to a person where it is easily transmitted by direct contact with infected blood or other body fluids. Infected people were placed in quarantine, but the lack of treatment capacity and isolation facilities hampered an efficient response. The infection caused major loss of life and socioeconomic disruption in the region. The international community responded with funds, specialists, equipment, and training. Two years later the outbreak has been brought under control, and only an occasional flare-up occurred. The frightening reputation of the disease has re-emerged the efforts to produce a vaccine. A recombinant, replication-competent vesicular stomatitis virus-based candidate vaccine, expressing a surface glycoprotein of Zaire ebolavirus, proved to be protective in animals including non-human primates. Although this vaccine candidate was not yet officially approved, an emergency stockpile was produced. A field 'ring vaccination' (after the outbreak) trial proved to be effective in localised infections. This encouraged new opportunities to develop robust and safe new vaccine strategies to combat Ebola virus outbreaks. High-throughput sequencing and epitope mapping, high-resolution proteomics, and single-cell antibody-cloning techniques combined with computational methods are applied to map both natural and elicited epitope targets of antibodies in a given individual. This guides both structural vaccinology approaches and development of protective immunisation with specific human monoclonal antibodies.

Besides developments in vaccinology, high-throughput sequencing (also known as next-generation sequencing) methods and computational analysis are becoming widely used to study complex bacterial (and viral) communities in various niches. It is estimated that 90% of cells in the human body are of microbial origin. This commensal relationship is well established, but the vast majority of the taxa present in the microbiome are unknown. Analysis of the various microbiome populations in the human body (gut, skin, oral cavity, saliva, breast milk, amniotic fluid, as well as from terrestrial and aquatic ecosystems) is just beginning and will shift to biological function in order to get insight in the bidirectional relationship between host and microbiota. The global initiative known as the Human Microbiome Project (started in 2007) aims to catalogue this balance. It is now realised that each person has a unique mix of bacteria and much of the diversity in healthy subjects remains unexplained. Nevertheless, communities of people have distinctive mixes of microbes and communities of microbes associated with a disease may be more or less diverse than controls. It is clear that Koch's postulates, which relate a given infectious disease to a single microbial species, are rarely applied for chronic diseases. Metagenome-wide association studies have demonstrated shifts that are associated with a variety of diseases like obesity, rheumatoid arthritis, and others. The recent advancements also reinforced research for identifying microbes responsible for causing different types of cancers.

On top of this, terrorists might intentionally use micro-organisms (smallpox, anthrax, plague, etc.) or bacterial toxins (botulism) to cause death and disease in humans or animals in a civilian setting. The recognition that an event was caused by a biological weapon presents a severe challenge to be prepared for such an attack, especially for medical care providers and public health officials. Strategies to combat bioterrorism have to be worked out, but with the experience of 100 years of combating micro-organisms with hygiene measures, vaccination, antibiotic, and antiviral treatment, there must be a way out.

Selected Readings

Arias CA, Murray BE. Antibiotic-resistant bugs in the 21st century – a clinical super-challenge. N Engl J Med. 2009;360:439–43.

Chappell KJ, Watterson D. Fighting ebola: a window for vaccine re-evaluation? PLoS Pathog. 2017;13(1):e1006037.

Franzosa EA, Hsu T, Sirota-Madi A, Shafquat A, Abu-Ali G, Morgan XC, Huttenhower C. Sequencing and beyond: integrating molecular 'omics' for microbial community profiling. Nat Rev Microbiol. 2015;13(6):360–72.

Henao-Restrepo AM, et al. Efficacy and effectiveness of an rVSV-vectored vaccine in preventing Ebola virus disease: final results from the Guinea ring vaccination, open-label, cluster-randomised trial (Ebola Ça Suffit!). Lancet. 2016;389(10068):505–18.

Hütter G, Nowak D, Mossner M, Ganepola S, Müßig A, Allers K, Schneider T, Hofmann J, Kücherer C, Blau O, Blau IW, Hofmann WK, Thiel E. Long-term control of HIV by CCR5 Delta32/Delta32 stem-cell transplantation. N Engl J Med. 2009;360:692–8.

Janeway CA, Travers P, Walport M, Shlomchik M. Immunobiology. the immune system in health and disease. New York, NY: Garland Science Publishing; 2001.

Jawetz E, Melnick JL, Adelberg EA, Brooks GF, Butel JS, Ornston LN. Medical microbiology. London: Prentice-Hall International; 2001.

Jiang W, Marraffini LA. CRISPR-cas: new tools for genetic manipulations from bacterial immunity systems. Annu Rev Microbiol. 2015;69:209–28.

Kazmin D, Nakaya HI, Lee EK, Johnson MJ, van der Most R, van den Berg RA, Ballou WR, Jongert E, Wille-Reece U, Ockenhouse C, Aderem A, Zak DE, Sadoff J, Hendriks J, Wrammert J, Ahmedan R, Pulendran B. Systems analysis of protective immune responses to RTS,S malaria vaccination in humans. Proc Natl Acad Sci U S A. 2017;114:2425–30.

Khurana S, Verma N, Yewdell JW, Hilbert AK, Castellino F, Lattanzi M, Del Giudice G, Rappuoli R, Golding H. MF59 adjuvant enhances diversity and affinity of antibody-mediated immune response to pandemic influenza vaccines. Sci Transl Med. 2011;3(85):85ra48.

Klein F, Mouquet H, Dosenovic P, Scheid JF, Scharf L, Nussenzweig MC. Antibodies in HIV-1 vaccine development and therapy. Science. 2013;341(6151):1199–204.

Mims CA, Playfair JHL, Roitt IM, Wakelin D, Williams R, Anderson RM. Medical microbiology. London: Mosby; 1998.

Montagnier L. 25 years after HIV discovery: prospects for cure and vaccine. Virology. 2010;397:248–54.

Prescott L, Harley J, Klein D. Microbiology. New York, NY: McGraw Hill; 2003.

Rittirsch D, Flierl MA, Ward PA. Harmful molecular mechanisms in sepsis. Nat Rev Immunol. 2008;8:776–8.

Roitt I, Brostoff J, Male D. Immunology. London: Mosby; 2001.

Saylora C, Dadachovaa E, Casadevall A. Monoclonal antibody-based therapies for microbial diseases. Vaccine. 2009;27S:G38–46.

Scheid JF, Horwitz JA, Bar-On Y, Kreider EF, Lu CL, Lorenzi JC, Feldmann A, Braunschweig M, Nogueira L, Oliveira T, Shimeliovich I, Patel R, Burke L, Cohen YZ, Hadrigan S, Settler A, Witmer-Pack M, West AP Jr, Juelg B, Keler T, Hawthorne T, Zingman B, Gulick RM, Pfeifer N, Learn GH, Seaman MS, Bjorkman PJ, Klein F, Schlesinger SJ, Walker BD, Hahn BH, Nussenzweig MC. HIV-1 antibody 3BNC117 suppresses viral rebound in humans during treatment interruption. Nature. 2016;535(7613):556–60.

Silverstein AM. A history of immunology. San Diego, CA: Academic Press; 1998.

van Belkum A, Melles DC, Nouwen J, van Leeuwen WB, van Wamel W, Vos MC, Wertheim HF, Verbrugh HA. Co-evolutionary aspects of human colonisation and infection by Staphylococcus aureus. Infect Genet Evol. 2009;9:32–47.

van der Poll T, Opal SM. Pathogenesis, treatment, and prevention of pneumococcal pneumonia. Lancet. 2009;374:1543–56.

Wu Y, Yoder A. Chemokine coreceptor signaling in HIV-1 infection and pathogenesis. PLoS Pathog. 2009;5:1–8.

Xiong X, Chen M, Lim WA, Zhao D, Qi LS. CRISPR/Cas9 for human genome engineering and disease research. Annu Rev Genomics Hum Genet. 2016;17:14.1–14.24.

Important Websites

Scientific Research

American Association of Immunology, http://www.aai.org.

American Society for Microbiologists, http://www.asm.org.

European Federation of Immunological Societies, http://www.efis.org.

Federation of European Microbiological Societies, http://www.fems-microbiology.org.

National Library of Medicine, http://www.ncbi.nlm.nih.gov.

Outbreaks of Infectious Diseases

Centers for Disease Control and Prevention, http://www.cdc.gov.

Daily update: http://www.promedmail.org.

International Society for Infectious Diseases, http://www.isid.org

The Marshall Protocol Knowledge Base, https://mpkb.org/home/pathogenesis/microbiota.

World Health Organisation, http://www.who.int.

Immune Response in Human Pathology: Hypersensitivity and Autoimmunity

11

Jacques Descotes

11.1 Introduction

The primary role of the immune system is to maintain the body's homeostasis by discriminating self from nonself through a complex network of specialized cells and molecules tuned to ensure coordinated immune responses under the control of mechanisms with either compensatory or opposing outcome.

When physical barriers, which are the very first line of the host's defense, are breached, protection against pathogenic microorganisms can be attained by a second line of defense consisting in non-antigen-specific mechanisms (it is of note that the term innate immunity encompasses both lines of defense). If protection is not adequately achieved, the sophisticated adaptive (or acquired) immunity aimed at mounting exquisitely specific and far more efficient immune responses against offending pathogen(s) or foreign antigen(s).

11.1.1 Nevertheless, Not All Immune Responses Are Beneficial

One example is chemical reactivity of molecular entities contributing to HYPERSENSITIVITY reactions. Moreover, immune responses against self-constituents of the host can result in AUTOIMMUNITY. A wealth of medicinal products and environmental or industrial chemicals facilitate and/or trigger hypersensitivity reactions, and autoimmune reactions to a lesser extent. The question was recently raised whether HYPERSENSITIVITY reactions are aimed at clearing a danger or instead merely accidental events with uncontrolled immunopathological consequences [1]. Although drug- or chemical-specific T-cell reactions could theoretically be the consequences of biological determinism to protect living organisms by eliminating hazardous chemicals, it is not at all clear why biological determinism would go awry when dealing with innocuous foreign substances, such as birch pollen, or beneficial medicinal products, such as penicillin.

11.2 Hypersensitivity

Nearly all chemicals within our health, household, labor, or environment milieu are and can be suspected to cause HYPERSENSITIVITY reactions. Drug-induced hypersensitivity reactions, the incidence, pathophysiology, clinical features, and severity of which widely vary, are a major cause for concern [2].

Noticeably, no comprehensive data on the epidemiology of clinical drug-induced HYPERSENSITIVITY are available [3]. Published data typically consist of small series of

---

Final manuscript submitted on June 21, 2018.

J. Descotes (✉)
Immunotoxicology Consulting Group, Cambridge, MA, USA

University of Lyon, Lyon, France
e-mail: jdescotes@itoxconsulting.com

M. J. Parnham et al. (eds.), *Nijkamp and Parnham's Principles of Immunopharmacology*,
https://doi.org/10.1007/978-3-030-10811-3_11

case reports, studies on focused groups of patients (e.g., regional or national groups, children, elderly people), selected clinical signs (such as anaphylaxis or toxidermia), or pharmacological classes (e.g., anesthetics or β-lactamins). HYPERSENSITIVITY might account for ≈10% of drug-induced adverse events even though this is only a rough estimate. Current estimates on the incidence of HYPERSENSITIVITY reactions linked to occupational and environmental exposures or "food allergies" are poorly documented as well.

11.2.1 Clinical Manifestations of Drug-Induced Hypersensitivity Reactions

The clinical features of HYPERSENSITIVITY reactions are extremely varied, which reflects the wide variety of underlying mechanisms. Although nearly every organ or tissue of the body can be affected, one organ or tissue is often a predominant target in one given reaction.

11.2.1.1 Anaphylactic Reactions

ANAPHYLAXIS (historically designated "immediate hypersensitivity") develops shortly—within minutes to a few hours depending on the route of administration—provided a previous and assumedly sensitizing contact with the causative antigen (e.g., a previous drug treatment) can be reasonably documented. The shortest delay is associated with the intravenous route. Patients' complaints usually include itching, urticaria, and/or angioedema often combined with nausea and vomiting and inconsistently with tachycardia and hypotension of variable intensity, sometimes progressing to cardiovascular collapse or shock with marked respiratory difficulties and cyanosis in the most severe cases [4, 5]. Anaphylactic shock is an absolute medical emergency; the death rate is estimated around 1%. Consequently, anaphylactic shock may account for approximately one death per one million people in the general population [6]. First-line supportive measures are primarily aimed at maintaining or restoring respiratory/circulatory functions. Subcutaneous, intramuscular, or intravenous injection of adrenaline (epinephrine) is the key treatment of anaphylactic shock [7] in sharp contrast with antihistamine drugs and/or corticosteroids.

11.2.1.2 Skin Reactions

Skin Reactions Are the Commonest Immune-Mediated Adverse Effects of Medicinal Products [8, 9]

Many of these reactions are thought to be T-cell-mediated. Indeed, either drug-specific IgE antibodies causing urticaria or angioedema or circulating immune complexes possibly associated with cutaneous vasculitis are comparatively uncommon [10–12]. The clinical presentation of drug-induced skin reactions is extremely varied including morbilliform or exanthematous rash and pustular (e.g., acute generalized exanthematous pustulosis) or bullous eruptions. Among the latter, the Stevens-Johnson syndrome (SJS) and toxic epidermal necrolysis (TEN) are the most severe—often life-threatening—cutaneous complications of drug treatments [13, 14]. It has been suggested that ≈80% of TEN are drug-induced with a lower incidence for SJS, so that 1–6 in one million persons per year could be globally affected. Early clinical manifestations, i.e., flu-like symptoms of variable severity and mucous membrane involvement, used to be observed within 7–21 days after the start of treatment. Skin lesions typically expand within 2–3 days to become purpuric macules and bullae leading to erosion of the epidermis. Painful erosions of mucous membranes account for dysphagia, conjunctivitis, keratitis, diarrhea, and/or respiratory distress. The prognosis and management are like those of severely burnt patients. The mortality rate is about 5% for SJS, and 30% for TEN.

DRESS—drug reaction with eosinophilia and systemic symptoms—is a recently categorized condition [15, 16]. It can be inferred that quite a few early case reports of the drug hypersensitivity syndrome (DHS) including "immunoallergic" hepatitis or nephritis according to the former terminology were actually undiagnosed DRESS [17]. Clinical manifestations of DRESS include fever and rash, often lymphadenopathy, arthritis or hepatitis (≈50% of patients), and less frequently kidney, heart, lung, thyroid, and/or brain involvement. High eosinophil count (eosinophilia) is a common but inconsistent laboratory finding in DRESS.

Allergic contact dermatitis with pruritic vesicles on an erythematous background is common in an occupational and environmental setting. It may also develop after topical application or, rarely, systemic administration of a medicinal product [10, 18]. Allergic contact dermatitis is genuinely a T LYMPHOCYTE DRESS-mediated reaction, which is in sharp contrast to irritant contact dermatitis involving non-antigen-specific mechanism(s).

11.2.1.3 Immunoallergic Cytopenias

The destruction of one or several hematological cell types can give rise to cytopenia. ANTIBODY-mediated destruction is one established causative mechanism. Even though drug treatments have formerly been suggested to account for 20–40% of cytopenias, the incidence is likely to be much lower [19].

Patients with agranulocytosis—i.e., a major destruction of white blood cells, primarily neutrophils—can be asymptomatic so that diagnosis is typically made after a routine blood examination. Alternatively, patients complain of clinical symptoms suggestive of infection, such as a sore throat. In this latter situation, their neutrophil count is often below $100/mm^3$. Drug-induced IMMUNOALLERGIC hemolytic anemias are rare [20, 21]. Clinical symptoms develop within hours after a subsequent drug intake (sensitizing contact) and often consist of abdominal and dorsal pain, headache, malaise, fever, nausea, and vomiting. Shock and acute renal failure have been noted in 30–50% of severe cases. Two immunological mechanisms can be involved: (1) non-specific binding of the causative drug, or a metabolite to ERYTHROCYTES, and then interaction with circulating ANTIBODIES triggering activation of the COMPLEMENT cascade, and (2) formation of IgM or IgG antigenic complexes provided the medicinal product is strongly bound to a plasma protein and antigenic complexes are passively fixed to ERYTHROCYTES. The reintroduction (challenge) of even a tiny amount of the drug can induce an ANTIGEN-ANTIBODY reaction with intravascular hemolysis.

With the notable exception of those linked to heparin, drug-induced IMMUNOALLERGIC thrombocytopenias are uncommon [21–23]. Thrombocytopenia is unlikely to cause bleeding as long as platelet counts are above $10–30{,}000/mm^3$. Most patients with heparin-induced immune thrombocytopenia have detectable ANTIBODIES against the platelet glycoproteins Ib/IX and IIb/IIIa [24].

11.2.1.4 Other Clinical Manifestations

Additionally, drug-induced HYPERSENSITIVITY reactions include interstitial hepatitis, nephritis, and pneumonitis. Hepatitis was claimed to be the leading cause of market drug withdrawal over the last decades [25]. Laboratory-proven liver injury associated with fever, rash, and eosinophilia within 1–8 weeks after the start of treatment is often held to be the proof of an IMMUNOALLERGIC reaction, even though this is still an unsolved cause for safety concern [26]. Underlying mechanisms leading to cytolytic, cholestatic, or mixed clinical and associated biological features are poorly elucidated [27, 28]. IDIOSYNCRATIC reaction is a term all but too often used to refer to drug-induced liver injury [29] as well as other ill-understood adverse effects. It is fair to say that such an oversold terminology has so far shed only limited light to apprehend causative mechanisms, except in very rare situations.

IMMUNOALLERGIC interstitial nephritis typically presenting as non-oliguric renal dysfunction associated with fever, rash, and/or eosinophilia is seldom reported [30, 31]. Furthermore, drugs can cause acute interstitial or eosinophilic pneumonia, and pneumonitis [32].

11.2.2 Mechanisms of Drug-Induced Hypersensitivity Reactions

11.2.2.1 Immunogenicity and Sensitization

An immune-mediated HYPERSENSITIVITY reaction, whatever the causative agent, is due to the exquisite capacity of the immune system to recognize nonself-constituents and retain immunological memory of this contact for a very long time. An unconditional prerequisite for any ANTIGEN-specific HYPERSENSITIVITY reaction to develop is that a first contact occurred prior to the eliciting contact. Substantiating the sensitizing nature of a prior contact is almost impossible. Thus, it can only be assumed that a previous treatment with the same medicinal product or, extremely rarely, a cross-reactive molecule including a nondrug chemical was actually sensitizing. A minimum of 5–7 days (≈1 week) is needed for sensitization to ensue without a prior contact. The accumulated clinical experience shows that most immune-mediated hypersensitivity reactions used to develop within the first month of treatment. For ill-elucidated reasons, a subsequent exposure (challenge) to the causative drug inconsistently leads to resurgence of the same adverse reaction. Therefore, in rare cases where drug challenge is ethically acceptable and medically feasible, only positive results can be claimed to be supportive [33].

Any compound is a potential immunogen provided if it is foreign ("nonself") and large enough [34]. Biopharmaceuticals ("biologics") are macromolecules that may induce the production of specific antibodies (ADAs, antidrug antibodies). Sensitization is rarely apparent in the early stage of treatment as ADAs develop more or less rapidly and then may impact the pharmacokinetics (either clearing or binding ADAs) and the therapeutic efficacy (neutralizing ADAs) and/or cause hypersensitivity reactions [35, 36]. A widely agreed threshold for direct immunogenicity is a molecular weight (MW) of about 5 kDa (e.g., tetracosapeptide MW, 4.5 kDa; insulin MW, 5.8 kDa). Moreover, the chemical structure plays a critical but ill-understood role. For instance, peptidic molecules made of a heterogeneous amino acid sequence used to be more immunogenic than much larger non-peptidic molecules, such as polysaccharides. The formation of aggregates [37], the glycosylation pattern [38], and the subcutaneous route [39] have also been shown to influence the immunogenicity of biopharmaceuticals.

Importantly, most medicinal products are far too small to be direct immunogens, but they can bring about specific sensitization when acting as HAPTENS [34]. Haptens are small molecules that strongly bind to carrier macromolecules and form complexes capable of triggering antigen-specific immune responses. In sharp contrast to many industrial chemicals, pharmaceutical molecules are not chemically reactive. This is why reactive metabolites are thought to be involved in most instances. As metabolites theoretically suspected to play a major role are short-lived, their bioanalytical identification is tricky, which is a major limitation to the development and qualification of in vitro tools for the diagnosis of "drug allergy" [40].

The hapten theory is unlikely to address all aspects of antigen-specific immune reactions against small molecules. In accordance with the danger theory [41], when an ANTIGEN is presented to T cells, a second signal (or danger signal), such as a viral infection or a damage caused by a reactive metabolite, is required for an antigen-specific adverse reaction to develop instead of tolerance. Studies on drug-specific human T-cell clones from patients with a history of HYPERSENSITIVITY evidenced a direct interaction of even poorly reactive native drugs with T-cell receptor resulting in T LYMPHOCYTE activation, cytokine release, and cytotoxicity. This led to the p-i concept, where hypersensitivity adverse reactions can occur upon first encounter with the offending drug [42].

11.2.2.2 Risk Factors

As immune-mediated HYPERSENSITIVITY reactions develop in a small percentage of treated/exposed patients, the role of risk factors linked to the patient or the drug is logically suspected.

Age, Gender, Atopy, and Genetic Predisposition Are Among Patient-Related Risk Factors [43]

Young adults seemingly tend to develop more reactions, even though a notable bias may be that less studies have been conducted in children and young adults than elderly patients. Women have been suggested to develop more drug-induced HYPERSENSITIVITY reactions than males, but this may merely reflect reporting biases. Atopy is defined as a condition characterized by the excessive production of IgE ANTIBODIES usually associated with clinical manifestations of asthma, hay fever, or constitutional dermatitis. There is a growing consensus that atopic patients do not have an increased risk for drug-induced HYPERSENSITIVITY reactions but develop more serious reactions. Genetic predispositions, particularly associations with well-identified HLA genotypes, have been evidenced in a few instances, notably carbamazepine-induced SJS/TEN and abacavir-induced hypersensitivity [44]. It remains to be convincingly documented to what extent one genetic predisposition can contribute to the development of one single drug-induced hypersensitivity reaction.

A number of *drug-related risk factors* are thought to play a role. Despite the likely involvement of specific motifs of the chemical structure, reliable structure-immunogenicity relationships are rarely identified. Every route of administration can be associated with sensitization, but the topical route has a greater potential. The oral route normally leads to TOLERANCE, but an ill-understood tolerance breakdown may ensue. In sensitized patients, the intravenous route is associated with more rapidly developing and more serious reactions. Finally, intermittent treatment regimens tend to facilitate sensitization as compared to daily administrations.

11.2.2.3 Pathophysiology of Drug-Induced Immune-Mediated Hypersensitivity Reactions

In 1963, Gell and Coombs proposed a classification of IMMUNOALLERGIC reactions. It is noteworthy that this antique classification was coined when, for instance, the mere existence of T lymphocytes and cytokines was totally ignored. For enigmatic reasons, this obsolete classification is still widely used. Anyway, the Gell and Coombs classification can be useful as a very basic introduction to the pathophysiology of drug-induced HYPERSENSITIVITY reactions provided its major inaccuracies and flaws are well understood and carefully weighed.

Conservatively, immunoallergic reactions are classified into four types (Fig. 11.1). Type I reactions (immediate HYPERSENSITIVITY) are caused by antigen-specific IgE (and to a lesser extent IgG_4 in man) [5]. IgE antibodies bind to high affinity receptors (FcεRI or CD64) on the cell membrane of mast cells and basophils. After a subsequent contact, the interaction between a divalent allergen and specific IgE may initiate the DEGRANULATION of mast cells and basophils resulting in the immediate release of preformed mediators stored within cytoplasmic granules, including HISTAMINE, neutral proteases (e.g., tryptase, cathepsin D, chymase), and heparin. DEGRANULATION triggers the de novo synthesis of mediators from membrane phospholipids including PROSTAGLANDINS and LEUKOTRIENES and their subsequent release. Clinical manifestations (ANAPHYLAXIS) are the direct consequences of the biological effects of the released mediators. Type II reactions are mediated by IgM or less often IgG ANTIBODIES. The encounter of a sensitizing drug bound to the surface of blood cells with circulating ANTIBODIES may lead to COMPLEMENT activation with the resulting destruction of blood cells (as seen in IMMUNOALLERGIC hemolytic anemias and thrombocytopenias) [21, 22]. Type III reactions are caused by circulating immune complexes, which may be formed when the ANTIGEN is in greater quantity in the serum than IgM or IgG ANTIBODIES. Large immune complexes can deposit in capillary vessels and activate the COMPLEMENT system, platelets, MACROPHAGES, and neutrophils, to release of a variety of mediators and free radicals damaging endothelial cells. If the ANTIGEN is present predominantly at one site, localized damage is seen as in the ARTHUS REACTION. When immune complexes are present in the circulation, they may

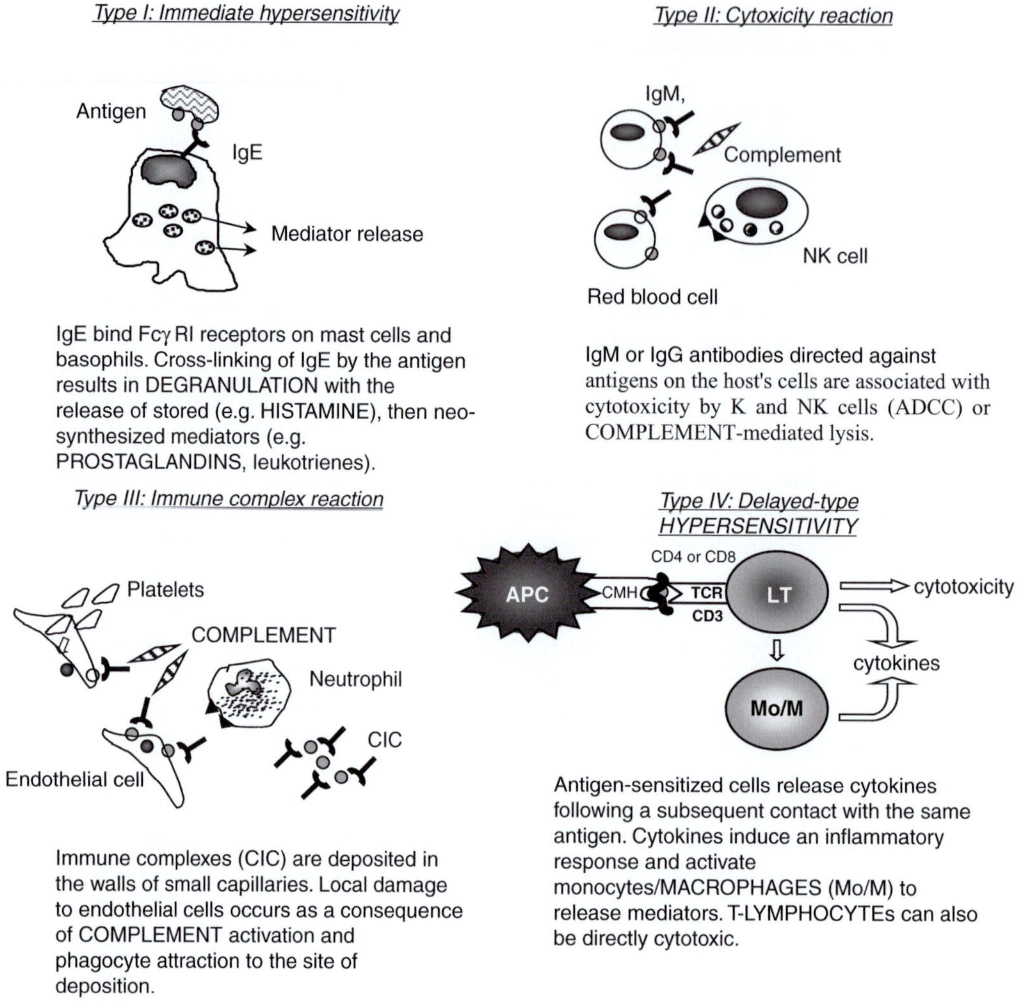

Fig. 11.1 Mechanisms of drug-induced hypersensitivity reactions according to the Gell and Coombs classification

cause serum sickness with fever, arthralgias, cutaneous eruption, gastrointestinal complaints, and proteinuria within 9–11 days after the injection of heterologous serum or ANTIBODIES [45]. Immune complex deposition is one possible mechanism of cutaneous vasculitis. It is noteworthy that no immune complexes either circulating in the blood or deposited in the glomeruli are seen after treatment with low-molecular-weight drugs, hence the preferred term "serum sickness-like disease." Similarly, Type III reactions are inappropriately claimed to be associated with biopharmaceuticals. Indeed, immune complexes have so far not been evidenced in published case reports [46].

Delayed HYPERSENSITIVITY (Type IV) reactions include allergic contact dermatitis and photoallergy. Contact dermatitis can be either nonimmune-mediated (irritant contact dermatitis) or immune-mediated (allergic contact dermatitis). Allergic contact dermatitis is characterized by the infiltration of T LYMPHOCYTES into the dermis and epidermis. After penetrating into the skin, drugs or their metabolites can play the role of haptens that bind to or complex with various cells, including Langerhans cells and keratinocytes. Langerhans cells process and present the drug antigen to T LYMPHOCYTES (see Chap. A2), which leads to the clonal proliferation of sensitized LYMPHOCYTES and to a clinically patent inflammatory reaction. Delayed HYPERSENSITIVITY reactions have been proposed to be subdivided into four categories [47]: type IV-a reactions involve a T_{H1} RESPONSE and closely correspond to the Gell and Coombs type IV reactions. Type IV-b reactions involve a T_{H2} RESPONSE in which the cytokine IL-5 is suspected to play a key role as in DRESS and drug-induced exanthemas. Type IV-c reactions are caused by cytotoxic T LYMPHOCYTEs and finally type IV-d reactions involving IL-8 lead to neutrophilic inflammation. This extended classification offers the advantage to more closely stick to the wide spectrum of drug-induced HYPERSENSITIVITY reactions, but it remains to be established whether it is fully applicable to non-cutaneous drug-induced reactions involving T LYMPHOCYTE-mediated mechanisms.

11.2.2.4 Pseudo-Allergic Reactions

An Immune-Mediated Mechanism Is Not Involved in Every Drug-Induced HYPERSENSITIVITY Reaction [42]

Clinical manifestations mimicking a genuine IgE-mediated reaction have indeed been reproducibly described in patients exposed for the first time to the same offending drug, hence the proposed term of PSEUDO-ALLERGIC or nonimmune-mediated HYPERSENSITIVITY reactions. Several mechanisms have been identified.

An IgE-independent (direct) mechanism involving a cytotoxic or osmotic effect can lead to HISTAMINE release by MAST CELLS and basophils (see Chap. 9). Clinical signs mimic a histaminic reaction with flush, redness of the skin, headache, cough, and abdominal pain. The red man syndrome induced by vancomycin is one typical example [48]. COMPLEMENT activation can be initiated by immunological as well as nonimmunological triggers by the pharmaceutical solvent Cremophor EL© or a few drug nanodelivery systems [49, 50]. The COMPLEMENT by-products C3a, C4a, and C5a (ANAPHYLATOXINS) are released during activation of this system and induce leukocyte chemotaxis, increased vascular permeability, contraction of bronchial smooth muscle, HISTAMINE release, LEUKOTRIENE generation, and IL-1 release. Direct (i.e., non-antigen-specific) COMPLEMENT activation is one mechanism of infusion reactions associated with therapeutic monoclonal ANTIBODIES, such as rituximab [51]. Aspirin and most NSAIDs can cause acute adverse reactions with asthma attack, possibly associated with rhinorrhea and conjunctival irritation, developing within 1 h after ingestion. Aspirin and the majority of NSAIDs inhibit the COX-1 isoform of the enzyme CYCLOOXYGENASE (COX) more potently than the COX-2 isoform (see Chaps. 9 and 33). Any NSAID with marked COX-1 inhibiting activity can precipitate asthma attacks in patients with a genetic predisposition resulting in excessive production of sulfido-leukotrienes (LTC4) [52]. Finally, angioedema associated with angiotensin-converting enzyme (ACE) inhibitors results from the decreased degradation of bradykinin, which increases vascular permeability, contracts smooth muscles, and elicits pain [53].

11.2.2.5 Autoimmunity

Autoimmunity Is Still a Mystery Even Though Autoimmune Diseases Are Collectively Rather Frequent

Few recent data are available. Estimates vary widely. Over one million new cases have been claimed to develop every 5 years in the USA with prevalence rates ranging from less than 5 per 100,000 to more than 500 per 100,000 depending on the autoimmune disease [54].

A wealth of factors including medicinal products and chemical exposures may be involved, but the underlying mechanisms are most often ill-understood. Formerly, one given medicinal product is used to be essentially associated with a single type of drug-induced autoimmune reaction, e.g., hydralazine and pseudolupus, α-methyldopa and autoimmune hemolytic anemia, or penicillamine and myasthenia [55]. Recently, the scene has changed dramatically with the rapid development of biologics. A wide range of autoimmune diseases that typically cannot be distinguished from spontaneous autoimmune diseases has been reported in patients treated with biologics including recombinant cytokines, such as rIL-2, interferons-α [56], therapeutic proteins (e.g., recombinant erythropoietin [57]), and monoclonal antibodies (most notably, immune checkpoint inhibitors [58]). The BIOGEAS registry recorded nearly 13,000 cases, which corresponds to an incidence of approximately 8/10,000 [59].

11.2.3 Clinical Manifestations of Autoimmunity

Autoimmune diseases are clinically very diverse so that the diagnosis is commonly based on the presence of several clinical signs and laboratory findings among a predefined list. The detection of AUTOANTIBODIES in the sera of treated patients is a prerequisite, although it is never sufficient to reach a firm medical diagnosis. Indeed, AUTOANTIBODIES have been detected in up to 25% of hydralazine-treated patients and 50% of procainamide-treated patients, but no clinical signs were associated in the majority of cases. Spontaneous autoimmune diseases as well as drug-induced autoimmune reactions are either systemic or organ-specific.

11.2.3.1 Systemic Autoimmune Diseases

Systemic lupus erythematosus (SLE) is estimated to affect 2–10/10,000 individuals. The causes are not known, but endocrine, genetic, and environmental factors are likely to be involved. The lupus syndrome or pseudolupus is a rare adverse event, although the most common drug-induced autoimmune reaction [60]. Medicinal products reported to induce lupus syndromes include hydralazine, procainamide, several anti-epileptic drugs, most β-blockers, chlorpromazine, isoniazid, and minocycline. In contrast to SLE, lupus syndromes are as frequent in men as in women. Arthritis, fever, weight loss, and muscular weakness with myalgias are the most typical clinical signs of pseudolupus. Cutaneous manifestations are often uncharacteristic. Renal involvement is inconsistent and usually mild. Neurological signs are usually lacking. One characteristic feature of drug-induced pseudolupus is the high incidence of pleural effusion and pericardial effusion. No biological abnormalities are characteristic. Antinuclear antibodies are always present. Anti-ds (double-stranded) or native DNA antibodies are found in 50–70% of patients with SLE but in less than 5% of those with the lupus syndrome. In contrast, antibodies to denatured DNA are relatively common in the lupus syndrome. No AUTOANTIBODIES have so far been qualified as reliable markers of drug-induced lupus syndromes. In contrast to SLE, drug-induced lupus syndromes used to have a favorable outcome after cessation of the causative drug treatment.

Scleroderma or systemic sclerosis is a relatively rare disease characterized by a more or less diffuse infiltration of the dermis and viscera by collagen with vascular abnormalities including vasospasm and microvascular occlusion. The pathogenesis of scleroderma is not elucidated. There is an overproduction of collagen by fibroblasts. T LYMPHOCYTES are suspected to play a pivotal role. Extremely few drugs have been reported to induce scleroderma-like diseases. The most serious example was the oculomucocutaneous syndrome induced by the β-blocker practolol withdrawn from the market in 1976.

11.2.3.2 Organ-Specific Autoimmune Diseases

In contrast to systemic drug-induced autoimmune reactions, organ-specific reactions are typically characterized by a homogeneous ANTIBODY response against a unique target and clinical symptoms closely mimicking those of the spontaneous autoimmune disease.

The *Guillain-Barré syndrome* presents with progressive lower extremity weakness potentially leading to autonomic dysfunction. A possible link with vaccination has long been suspected but remains to be substantiated [61]. The causative or contributing role of vaccines in *multiple sclerosis*, a multifocal demyelinating disease of the central nervous system, has been a matter of debate, but recent epidemiological findings did not support this claim [62]. *Myasthenia* is characterized by a loss of muscular strength due to impaired neuromuscular transmission. There is a predilection for certain cranial nerves, and virtually all patients complain of ocular symptoms.

Patients with a generalized form of the disease have IgG AUTOANTIBODIES against the nicotinic receptors of acetylcholine in the neuromuscular motor plates. Penicillamine is historically the most frequent cause of drug-induced myasthenia. Recently, myasthenia in patients treated with immune checkpoint inhibitors has been reported [58]. The underlying mechanism(s) is (are) not known [63].

Autoimmune thyroiditis typically presents as a slowly progressing atrophy of the thyroid gland due to a specific autoimmune response involving T cells and AUTOANTIBODIES. Rarely, biologics, such as rIL-2, the IFNs, and immune checkpoint inhibitors are suspected to be responsible.

11.2.4 Mechanisms of Drug-Induced Autoimmunity

Despite impressive progress in our current understanding of exquisite aspects of immunology and the immune system, the mechanisms of drug-induced autoimmunity are today restricted to hypotheses and assumptions.

Antithyroid AUTOANTIBODIES have been described in up to 30% of rIL-2-treated patients. One explanation is that thyroid cells under the influence of IL-2-induced production of IFN-γ may express MHC class II molecules and thus act as antigen-presenting cells with the production of antithyroid AUTOANTIBODIES [64]. Drug metabolites generated in the liver can bind to CYP450 isoforms, such as CYP1A2 (dihydralazine), CYP4E1 (halothane), or CYP2C9 (tienilic acid), and, then, result in (autoimmune or immunotoxic) hepatitis; extremely few drugs have so far been shown to be involved [65].

Molecular mimicry is another possible mechanism where part of a given protein closely resembles a part of another protein. Therefore, when a foreign protein enters the body, the immune system can mount an antigen-specific ANTIBODY response, and if the foreign protein closely resembles a self-protein of the body, AUTOANTIBODIES may be formed and then proved to be pathogenic. The involvement of molecular mimicry in drug-induced autoimmunity is largely assumptive. T LYMPHOCYTES are a major focus of research on autoimmunity. Recognition of closely related epitopes shared by either self or nonself molecules might trigger autoimmune responses through molecular mimicry (or more subtle and ill-established mechanisms). Alternatively, T-cell activation might be caused by drugs; the structure could mimic co-stimulatory molecules or MHC class II antigens essential to the immunological synapse.

11.3 Summary

Hypersensitivity and autoimmune reactions induced by medicinal products may be serious adverse events. HYPERSENSITIVITY reactions are more frequent. Even though the mechanisms involved are progressively better understood, a number of issues are still pending and could be actively addressed, including better diagnosis tools and procedure to ensure that drug-related hypersensitivity reactions are correctly identified, the role of T lymphocytes in these reactions, the influence of pharmacogenetics and immunogenetics traits, and finally, the development of new models and techniques to predict drug allergenicity at the preclinical stage. Unfortunately, very limited progress is foreseeable as far as drug-induced autoimmune reactions are concerned.

Selected Readings

Adkinson NF, Bochner BS, Burks W, Busse WW, Holgate ST, Lemanske RF, O'Behir RE. Middleton's allergy: principles and practice. 8th ed. Philadelphia, PA: Elsevier Saunders; 2014.

Baldo BA, Pham NH. Drug allergy: clinical aspects, diagnosis, mechanisms, structure-activity relationships. Berlin: Springer; 2013.

Shoenfeld Y, Meroni PL, Gershwin ME. Autoantibodies. 3rd ed. Amsterdam: Elsevier; 2014.

References

1. Pallardy M, Bechara R. Chemical or drug hypersensitivity: is the immune system clearing the danger? Toxicol Sci. 2017;158:14–22.
2. Baldo BA, Pham NH. Drug allergy: clinical aspects, diagnosis, mechanisms, structure-activity relationships. Berlin: Springer; 2013. 447 p.

3. Thong BY, Tan TC. Epidemiology and risk factors for drug allergy. Br J Clin Pharmacol. 2011;71:684–700.
4. Kuruvilla M, Khan DA. Anaphylaxis to drugs. Immunol Allergy Clin North Am. 2015;35:303–19.
5. Stone SF, Phillips EJ, Wiese MD, Heddle RJ, Brown SG. Immediate-type hypersensitivity drug reactions. Br J Clin Pharmacol. 2014;78:1–13.
6. Tejedor-Alonso MA, Moro-Moro M, Múgica-García MV. Epidemiology of anaphylaxis. Clin Exp Allergy. 2015;45:1027–39.
7. Song TT, Lieberman P. Epinephrine in anaphylaxis: doubt no more. Curr Opin Allergy Clin Immunol. 2015;15:323–8.
8. Marzano AV, Borghi A, Cugno M. Adverse drug reactions and organ damage: the skin. Eur J Intern Med. 2016;28:17–24.
9. Zalewska-Janowska A, Spiewak R, Kowalski ML. Cutaneous manifestation of drug allergy and hypersensitivity. Immunol Allergy Clin North Am. 2017;37:165–81.
10. Hausmann O, Schnyder B, Pichler WJ. Drug hypersensitivity reactions involving skin. Handb Exp Pharmacol. 2010;196:29–55.
11. Schrijvers R, Gilissen L, Chiriac AM, Demoly P. Pathogenesis and diagnosis of delayed-type drug hypersensitivity reactions, from bedside to bench and back. Clin Transl Allergy. 2015;5:31.
12. Pavlos R, Mallal S, Ostrov D, Buus S, Metushi I, Peters B, Phillips E. T cell-mediated hypersensitivity reactions to drugs. Annu Rev Med. 2015;66:439–54.
13. Harr T, French LE. Stevens-Johnson syndrome and toxic epidermal necrolysis. Chem Immunol Allergy. 2012;97:149–66.
14. Dodiuk-Gad RP, Chung WH, Valeyrie-Allanore L, Shear NH. Stevens-Johnson syndrome and toxic epidermal necrolysis: an update. Am J Clin Dermatol. 2015;16:475–93.
15. Spriet S, Banks TA. Drug reaction with eosinophilia and systemic symptoms syndrome. Allergy Asthma Proc. 2015;36:501–5.
16. Shiohara T, Kano Y. Drug reaction with eosinophilia and systemic symptoms (DRESS): incidence, pathogenesis and management. Expert Opin Drug Saf. 2017;16:139–47.
17. Avancini J, Maragno L, Santi CG, Criado PR. Drug reaction with eosinophilia and systemic symptoms/drug-induced hypersensitivity syndrome: clinical features of 27 patients. Clin Exp Dermatol. 2015;40:851–9.
18. Kostner L, Anzengruber F, Guillod C, Recher M, Schmid-Grendelmeier P, Navarini AA. Allergic contact dermatitis. Immunol Allergy Clin North Am. 2017;37:141–52.
19. Andrès E, Federici L, Weitten T, Vogel T, Alt M. Recognition and management of drug-induced blood cytopenias: the example of drug-induced acute neutropenia and agranulocytosis. Expert Opin Drug Saf. 2008;7:481–9.
20. Arndt PA. Drug-induced immune hemolytic anemia: the last 30 years of changes. Immunohematology. 2014;30:44–54.
21. Garratty G. Immune hemolytic anemia caused by drugs. Expert Opin Drug Saf. 2012;11:635–42.
22. Arnold DM, Nazi I, Warkentin TE, Smith JW, Toltl LJ, George JN, Kelton JG. Approach to the diagnosis and management of drug-induced immune thrombocytopenia. Transfus Med Rev. 2013;27:137–45.
23. Chong BH, Choi PY, Khachigian L, Perdomo J. Drug-induced immune thrombocytopenia. Hematol Oncol Clin North Am. 2013;27:521–40.
24. Dasararaju R, Singh N, Mehta A. Heparin induced thrombocytopenia: review. Expert Rev Hematol. 2013;6:419–28.
25. Onakpoya IJ, Heneghan CJ, Aronson JK. Post-marketing withdrawal of 462 medicinal products because of adverse drug reactions: a systematic review of the world literature. BMC Med. 2016;14:10.
26. Senior JR. Evolution of the Food and Drug Administration approach to liver safety assessment for new drugs: current status and challenges. Drug Saf. 2014;37(Suppl 1):S9–17.
27. Mosedale M, Watkins PB. Drug-induced liver injury: advances in mechanistic understanding that will inform risk management. Clin Pharmacol Ther. 2017;101:469–80.
28. Ju C, Reilly T. Role of immune reactions in drug-induced liver injury (DILI). Drug Metab Rev. 2012;44:107–15.
29. Roth AD, Lee MY. Idiosyncratic drug-induced liver injury (IDILI): potential mechanisms and predictive assays. Biomed Res Int. 2017;2017:9176937.
30. Krishnan N, Perazella MA. Drug-induced acute interstitial nephritis: pathology, pathogenesis, and treatment. Iran J Kidney Dis. 2015;9:3–13.
31. Nast CC. Medication-induced interstitial nephritis in the 21st century. Adv Chronic Kidney Dis. 2017;24:72–9.
32. Matsuno O. Drug-induced interstitial lung disease: mechanisms and best diagnostic approaches. Respir Res. 2012;13:39.
33. Brockow K, Przybilla B, Aberer W, Bircher AJ, et al. Guideline for the diagnosis of drug hypersensitivity reactions. Allergo J Int. 2015;24:94–105.
34. Schnyder B, Brockow K. Pathogenesis of drug allergy - current concepts and recent insights. Clin Exp Allergy. 2015;45:1376–83.
35. Chirmule N, Jawa V, Meibohm B. Immunogenicity to therapeutic proteins: impact on PK/PD and efficacy. AAPS J. 2012;14:296–302.
36. Vultaggio A, Nencini F, Pratesi S, Petroni G, Maggi E, Matucci A. Manifestations of antidrug antibodies response: hypersensitivity and infusion reactions. J Interferon Cytokine Res. 2014;34:946–52.
37. Ratanji KD, Derrick JP, Dearman RJ, Kimber I. Immunogenicity of therapeutic proteins: influence of aggregation. J Immunotoxicol. 2014;11:99–109.

38. Liu L. Antibody glycosylation and its impact on the pharmacokinetics and pharmacodynamics of monoclonal antibodies and Fc-fusion proteins. J Pharm Sci. 2015;104:1866–84.
39. Hamuro L, Kijanka G, Kinderman F, Kropshofer H, Bu DX, Zepeda M, Jawa V. Perspectives on subcutaneous route of administration as an immunogenicity risk factor for therapeutic proteins. J Pharm Sci. 2017;106:2946.
40. Thompson RA, Isin EM, Ogese MO, Mettetal JT, Williams DP. Reactive metabolites: current and emerging risk and hazard assessments. Chem Res Toxicol. 2016;29:505–33.
41. Pradeu T, Cooper EL. The danger theory: 20 years later. Front Immunol. 2012;3:287.
42. Pichler WJ, Hausmann O. Classification of drug hypersensitivity into allergic, p-i, and pseudo-allergic forms. Int Arch Allergy Immunol. 2016;171:166–79.
43. Warrington R, Silviu-Dan F. Drug allergy. Allergy Asthma Clin Immunol. 2011;7(Suppl 1):S10.
44. Usui T, Naisbitt DJ. Human leukocyte antigen and idiosyncratic adverse drug reactions. Drug Metab Pharmacokinet. 2017;32:21–30.
45. Bielory L, Gascon P, Lawley TJ, Young NS, Frank MM. Human serum sickness: a prospective analysis of 35 patients treated with equine anti-thymocyte globulin for bone marrow failure. Medicine. 1988;67:40–57.
46. Karmacharya P, Poudel DR, Pathak R, Donato AA, Ghimire S, Giri S, Aryal MR, Bingham CO III. Rituximab-induced serum sickness: a systematic review. Semin Arthritis Rheum. 2015;45:334–40.
47. Posadas SJ, Pichler WJ. Delayed drug hypersensitivity reactions - new concepts. Clin Exp Allergy. 2007;37:989–99.
48. Wallace MR, Mascola JR, Oldfield EC. Red man syndrome: incidence, etiology, and prophylaxis. J Infect Dis. 1991;64:1180–5.
49. Szebeni J. Complement activation-related pseudoallergy: a new class of drug-induced acute immune toxicity. Toxicology. 2005;216:106–21.
50. Szebeni J, Muggia F, Gabizon A, Barenholz Y. Activation of complement by therapeutic liposomes and other lipid excipient-based therapeutic products: prediction and prevention. Adv Drug Deliv Rev. 2011;63:1020–30.
51. van der Kolk LE, Grillo-López AJ, Baars JW, Hack CE, van Oers MH. Complement activation plays a key role in the side-effects of rituximab treatment. Br J Haematol. 2001;115:807–11.
52. de Weck AL, Gamboa PM, Esparza R, Sanz ML. Hypersensitivity to aspirin and other nonsteroidal anti-inflammatory drugs (NSAIDs). Curr Pharm Des. 2006;12:3347–58.
53. Byrd JB, Adam A, Brown NJ. Angiotensin-converting enzyme inhibitor-associated angioedema. Immunol Allergy Clin North Am. 2006;26:725–37.
54. Cooper GS, Stroehla BC. The epidemiology of autoimmune diseases. Autoimmun Rev. 2003;2:119–25.
55. Vial T, Nicolas B, Descotes J. Drug-induced autoimmunity: experience of the French pharmacovigilance system. Toxicology. 1997;119:23–7.
56. Vial T, Descotes J. Immune-mediated side-effects of cytokines in humans. Toxicology. 1995;105:31–57.
57. McKoy JM, Stonecash RE, Cournoyer D, Rossert J, Nissenson AR, Raisch DW, Casadevall N, Bennett CL. Epoetin-associated pure red cell aplasia: past, present, and future considerations. Transfusion. 2008;48:1754–62.
58. Day D, Hansen AR. Immune-related adverse events associated with immune checkpoint inhibitors. BioDrugs. 2016;30:571–84.
59. Pérez-De-Lis M, Retamozo S, Flores-Chávez A, Kostov B, Perez-Alvarez R, Brito-Zerón P, Ramos-Casals M. Autoimmune diseases induced by biological agents. A review of 12,731 cases (BIOGEAS Registry). Expert Opin Drug Saf. 2017;16:1255–71.
60. Rubin RL. Drug-induced lupus. Expert Opin Drug Saf. 2015;14:361–78.
61. Vellozzi C, Iqbal S, Broder K. Guillain-Barre syndrome, influenza, and influenza vaccination: the epidemiologic evidence. Clin Infect Dis. 2014;58:1149–55.
62. Langer-Gould A, Qian L, Tartof SY, Brara SM, Jacobsen SJ, Beaber BE, Sy LS, Chao C, Hechter R, Tseng HF. Vaccines and the risk of multiple sclerosis and other central nervous system demyelinating diseases. JAMA Neurol. 2014;71:1506–13.
63. Penn AS, Low BW, Jaffe IA, Luo L, Jacques JJ. Drug-induced autoimmune myasthenia gravis. Ann N Y Acad Sci. 1998;841:433–49.
64. Duntas LH. Environmental factors and autoimmune thyroiditis. Nat Clin Pract Endocrinol Metab. 2008;4:454–60.
65. Beaune PH, Lecoeur S. Immunotoxicology of the liver: adverse reactions to drugs. J Hepatol. 1997;26(Suppl 2):37–42.

12 Cancer Immunity

Javier Mora, Warner Alpízar-Alpízar, and Andreas Weigert

12.1 Introduction: The Immune System During Cancer Development

Tumors are highly complex tissues consisting not only of transformed cells but also of various stromal non-transformed cellular components. Tumor-associated cell subsets may include endothelial cells, fibroblasts, adipocytes, mesenchymal stem cells, cells of the peripheral nervous system, and immune cells. These cells, together with the tumor cells themselves, constitute the TUMOR MICROENVIRONMENT (TME), characterized by unique extracellular matrix (ECM), regions of divergent nutrient and oxygen availability, and the accumulation of catabolites that affect tumor biology. The resulting organ-specific TME plays a role in all aspects of tumorigenesis, from initiation to metastatic spread [1, 2].

A major proportion of research on TME's impact on tumorigenesis has been focused on the role of the immune system. The potential impact of immunity on tumor development was already proposed more than 100 years ago by scientists such as Rudolf Virchow and Paul Ehrlich and colleagues, who observed the presence of infiltrating mononuclear cells around or inside tumor lesions and noticed the pivotal contribution of the host to the growth of transplanted murine tumors [3]. Already in the late nineteenth century, William B. Coley, upon noticing a correlation between erysipelas and tumor remission in cancer patients, injected mixtures of live and inactivated bacteria directly into tumors of cancer patients and achieved significant clinical success in some patients suffering from different tumor types [4]. Despite these promising insights, clinicians chose to adopt surgery and radio- and chemotherapy approaches for treating cancer patients in the following century, mostly due to the vastly incomplete mechanistic understanding on how the immune system operates and the undoubtable risks associated with infecting patients with pathogenic bacteria.

This paradigm slowly started to change following the introduction of the theory of CANCER

Final manuscript submitted on May 27, 2018.

J. Mora
Faculty of Microbiology, Universidad de Costa Rica, San José, Costa Rica

Institut für Biochemie I: Pathobiochemie, Frankfurt am Main Universitätsklinikum, Johann Wolfgang Goethe-Universität, Frankfurt am Main, Germany
e-mail: mora@biochem.uni-frankfurt.de; javierfrancisco.mora@ucr.ac.cr

W. Alpízar-Alpízar
Department of Biochemistry, Centre for Research on Microscopic Structures (CIEMic for Its Acronym in Spanish), School of Medicine, Universidad de Costa Rica, San José, Costa Rica
e-mail: ciemic@ucr.ac.cr

A. Weigert (✉)
Institute of Biochemistry I, Faculty of Medicine, Goethe-University Frankfurt, Frankfurt, Germany
e-mail: weigert@biochem.uni-frankfurt.de

M. J. Parnham et al. (eds.), *Nijkamp and Parnham's Principles of Immunopharmacology*,
https://doi.org/10.1007/978-3-030-10811-3_12

IMMUNOTHERAPY by Thomas and Burnet in 1957 [5]. Further progress in unraveling the complexity of the immune system, and accumulating evidence, allowed Schreiber and colleagues, around the turn of the millennium, to reintroduce and reformulate the cancer immune surveillance theory, coining the term cancer immunoediting [5]. Today, epidemiological and experimental evidence, as well as the current clinical success of immune checkpoint blockade [6] (see section 12.5.2 and Chap. 25), have made it clear that the immune system is not a merely passive bystander but plays an active role in cancer by modulating tumor development on at least three levels.

First, chronic low-grade inflammation may favor tumorigenesis by promoting cellular transformation and altering the local microenvironment of the developing tumor to favor malignant outgrowth [7]. Besides chronic viral or bacterial infection, which is well established in stomach [8], cervical [9], and liver [10] cancer, chronic sterile inflammation resulting from lifestyle choices, including drug abuse, diet, obesity, exposure to radiation, etc., is associated with an increased risk of developing cancer [11, 12]. Cellular senescence may also contribute to inflammation-dependent tumor initiation and/or propagation in older individuals [13]. Chronic inflammation favors tumor growth in several ways, including the induction of somatic tumor-promoting mutations, as a consequence of the production of genotoxic agents (e.g., oxygen or nitrogen radicals) by activated immune cells [14], and promoting survival of early transformed cells by producing anti-apoptotic or proliferative signals [15].

Second, clinically detectable tumors contain various immune cell populations, lymphocytes, as well as bone marrow-derived cells that support tumor development through a variety of mechanisms, including the recruitment of blood supply and nutrients, remodeling of the ECM, and actively favoring tumor cell survival by producing growth factors. This response was described as being similar to the healing function of immune cells in wounds, leading to the comparison that tumors are “wounds that never heal” [16].

Third, immune cells can recognize cancer cells as nonself or altered-self and eliminate them. This discrimination between “self” and “nonself” by the immune system is the core of the immune surveillance hypothesis and has earned the 1960 Nobel Prize to Burnet and Medawar. It is now widely accepted that cytotoxic lymphocytes can recognize altered-self molecules in tumors, including TUMOR-ASSOCIATED ANTIGENS (TAA) and TUMOR-SPECIFIC ANTIGENS (TSA), which is followed by eradication of tumor cells [17]. This is probably the rule rather than the exception and the reason why we do not get more cancer [18]. Below we present the actual lines of evidence supporting an active role of the immune system in fighting cancer and, thus, arguments supporting the immune surveillance hypothesis:

- Human and murine tumors are immunogenic, and multiple tumor antigen-specific cytotoxic lymphocyte populations exist in cancer patients [19].
- Transplant recipients receiving immunosuppressive treatment and other immunodeficient individuals exhibit an increased risk of developing tumors [20, 21].
- The immune contexture in human tumors, i.e., the relative immune cell composition, is of prognostic and predictive value across multiple tumor entities [22].
- Autoimmunity in cancer patients can derive from a break of tolerance toward an endogenous antigen, triggered by its related TSA, which is indicative of active immunity against this particular tumor antigen [23].
- TAA and TSA can be used as vaccines, directly or via delivery through dendritic cells (DC), to induce tumor rejection in tumor-bearing animals and patients [24].
- The baseline cytotoxicity of lymphocytes correlates with the risk of developing cancer [25].
- Immunotherapeutic approaches including adoptive cellular immunotherapy and IMMUNE CHECKPOINT BLOCKADE can induce long-lasting remissions in cancer patients [6, 26] (see section 12.5.2 and Chap. 25).

In the following paragraphs, we summarize the importance of the TME and the interplay of immunity with transformed cells, focusing on the key elements involved in the generation of antitumor responses: the cellular and molecular components of the immune system. On this basis, we further review the current immunotherapeutic approaches to battle cancer and delineate future challenges.

12.2 The Tumor Microenvironment in Tumor Immunity

For a long time, cancer was regarded as a disease of mutationally corrupted cells that have lost the capacity to regulate their cell cycle and could, therefore, be understood—and cured—by examining their intrinsic genes and proteins. This reductionist model has radically changed throughout the years, since accumulating evidence has unequivocally demonstrated that malignant cell growth is severely compromised if these corrupted cells are not immersed in a "tumor nest," characterized by the presence of a heterogeneous set of nonneoplastic cells such as endothelial cells, fibroblasts, adipocytes, mesenchymal stem cells, cells of the peripheral nervous system, and immune cells [27]. These accessory cells secrete a vast array of mitogenic, growth, anti-apoptotic, and angiogenic factors, inflammatory mediators, metabolic intermediaries, ECM constituents, and ECM remodeling enzymes. All these cell subpopulations and secreted molecules constitute the TME, a very dynamic entity that influences the tumor growth in a number of ways, either promoting or inhibiting each of the hallmarks of cancer [15]. Reciprocally, the TME is constantly being reshaped as a result of the heterotypic signaling between the neoplastic and nonneoplastic cells that coexist in this milieu. Thus, malignant tumors are not just transformed cells but transformed tissues. It is not possible to understand the biology of cancer without considering the contribution of the TME and the immune components explained below to the malignant transformation and growth.

Although the cellular composition of TME varies according to the type of cancer and the degree of advance in the carcinogenic stepwise process, several populations of infiltrating immune cells (IICs) and mesenchymal and endothelial cells are typically present [15, 27]. As mentioned in the introduction of this chapter, malignant lesions are infiltrated by different kinds of immune cells such as macrophages, granulocytes, DC, natural killer (NK) lymphocytes, and different subsets of T lymphocytes and B lymphocytes. These are highly plastic cell populations, given their flexibility to respond with different functional programs to the ever-changing tumor microenvironmental conditions [28]. Macrophages residing in the tumors, for instance, adopt a very broad repertoire of phenotypes, ranging from the classically activated (M1) anti-tumorigenic macrophages to the alternatively activated (M2) pro-tumorigenic ones (the so-called tumor-associated macrophages (TAMs)), according to the specific TME conditions [29]. Similarly, the recruitment and functional polarization of T lymphocytes, and therefore their role as pro- or anti-tumorigenic effectors, are highly dependent on the tumor milieu [30]. Another cell type often found in the TME is cancer-associated fibroblasts (CAFs), a particular subpopulation of fibroblasts that when activated are characterized by expressing markers such as α-SMA and vimentin [31]. Endothelial cells that are present in the TME become activated in response to hypoxia or oncogenic signaling, which leads to the formation of new blood vessels that are usually disorganized and leaky [32], but crucial for providing cancer and accessory cells with nutrients and oxygen. These vessels also serve as passageways that can be used by neoplastic cells to disseminate from the primary tumor to distant organs [27].

As already mentioned, the TME is constantly modified throughout the progression of the disease, which is reflected in the composition of the ECM, as well as in the proportion and polarization state of the accessory cells residing there. Although not exclusively, the dynamicity of the TME is closely connected to the immune surveillance and cancer immunoediting phenomenon [33].

Thus, early in the malignant transformation, the TME is predominantly anti-tumorigenic, and, consequently, tumor growth is more or less contained. The strong pressure imposed by the immunogenic TME on the heterogeneous mass of neoplastic cells at this particular stage is the main evolutionary force by which some of the neoplastic cells harboring certain phenotypical characteristics are selected. As the cells that have survived the immune attack expand, they gradually modify the structural and cellular composition of the TME toward a less immunogenic milieu. At this point, equilibrium between tumor cell growth and elimination by the IICs is reached; however, in addition to the inflammatory response, new factors that limit malignant cell growth come into play, including scarcity of nutrients, oxygen, and growth and mitogenic signals [34]. All these interactions result in the selection of those cancer cells that are better adapted to survive under these hostile microenvironmental conditions. Eventually, most barriers imposed by the TME are overcome by the better-adapted neoplastic cells, and, as they proliferate and the heterotypic signaling between neoplastic and accessory cells is reinforced, the TME is reprogramed thus becoming primarily immunosuppressive and tumor-supportive, enabling the tumor mass to grow uncontrollably. This immunosuppressive character of the TME is a common feature of overt invasive and metastatic tumors [35].

IICs support cancer progression and dissemination in many different ways, which makes them almost mandatory components of the TME in nearly all solid tumors [27, 28]. Their roles in the TME, however, vary depending on their activation state, the type of tumor and its stage, their anatomical location within the neoplastic lesion, and a number of microenvironmental variables such as the availability of nutrients and oxygen and pH. Although, each subpopulation of IICs contributes differently, the heterotypic interaction of all coexisting cells is also a crucial factor to consider when it comes to unraveling the contribution of IICs to the malignant progression. Broadly speaking, IICs are an important source of growth, mitogenic and angiogenic factors, as well as ECM remodeling enzymes [15, 27]. However, they can also be crucial in the reprograming of metabolic pathways that enable cancer cells to adapt and grow better in their respective environment [34].

Besides IICs other cell populations such as CAFs play an important role on tumor progression. Via their active secretome, CAFs regulate cancer progression in a direct manner but also actively participate in the structural, immune, and metabolic reprograming of the TME [31]. CAFs secrete a number of growth and mitogenic factors that promote malignant cell survival and proliferation, both in primary and metastatic lesions. CAFs also release several molecules that mediate the recruitment of new capillaries and the accumulation of immune cells within the TME. The CAF secretory phenotype is in fact crucial in the creation of an immunomodulatory milieu, by releasing factors that induce immunosuppressive differentiation of the IICs embedded in the tumor stroma. CAFs exhibit increased deposition of collagens, fibrin, and other ECM constituents, creating ECM protein networks that restrict the access of immune cells to cancer cells, thus serving as a physical barrier to tumor infiltration by immune cells. The enhanced deposition of ECM proteins by CAF may also affect properties that have been related to cancer cell invasion and motility such as ECM composition, stiffness, and organization. That, however, is not the only contribution of CAFs to cancer cell invasion and motility, as these cells produce ECM-degrading proteolytic enzymes and molecules implicated in the activation of epithelial to mesenchymal transition (EMT) programs. Similar to what happens with some types of inflammatory cells, resting fibroblasts can differentiate into distinct subsets of CAFs, which could imply that these cells possess diverse activities depending on their differentiation state [31]. In summary, cancer biology and immunity are under heavy influence also by nonimmune stromal compartments. Pharmacological targeting of these components may therefore not only affect malignant growth directly but also improve antitumor immunity.

12.3 Key Players in the Immune Response Against Cancer

According to the concept of immune surveillance, IICs have the potential to attack tumor cells. However, as it was explained above, different factors in the TME modulate their effector capacity turning them into tumor-promoting cells. In this section, we will discuss the components of the immune system and its mechanistic capacity to target and eliminate tumor cells.

Both innate and adaptive components of the immune system interact to generate antigen-specific immune responses. As detailed in the previous chapters, the innate immune system constitutes the body's first line of defense against "foreign invaders." Innate immunity involves a large number of different cell populations such as epithelial cells, monocytes, macrophages, DC, polymorphonuclear leukocytes or granulocytes, and some lymphocyte subsets that are at the interface between innate and adaptive immunity [NK lymphocytes, cluster of differentiation (CD)5$^+$ B lymphocytes, T-cell receptor (TCR)-$\gamma\delta^+$ T lymphocytes, and natural killer T (NKT) lymphocytes]. The innate immune system also comprises a variety of humoral factors such as cytokines, chemokines, enzymes (e.g., lysozyme), metal-binding proteins, integral membrane ion transporters, complex carbohydrates, and complement. The phagocytic cells (monocytes, macrophages, DC, granulocytes) and the complement system constitute effector mechanisms by which the "invaders" can be destroyed. The production of cytokines and chemokines acts in concert with antigen presentation by DC and monocytes/macrophages to initiate adaptive immune responses. A major part of the innate immune response against cancer, besides antigen presentation and induction of adaptive immunity, depends on NK lymphocytes that are capable of recognizing and eliminating transformed cells directly. Figure 12.1 shows how the killing of target cells by NK lymphocytes is guided by the balance between activating and inhibitory signals.

Adaptive immunity makes use of a unique mechanism whereby genetic mutations occurring in two specialized cell populations, B and T

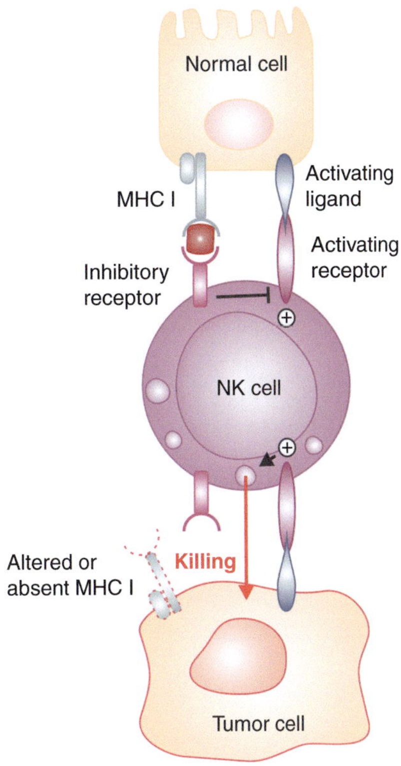

Fig. 12.1 Surface receptor interactions establishing the basis for tolerance and cytolytic responses by NK cells. NK cell activation is regulated by a balance of signals derived from activating and inhibitory surface receptors. Activating receptors recognize an array of surface molecules, some of which are expressed ubiquitously on surfaces of many cells of the body. The so-called killer-inhibitory receptors (KIR) recognize MHC class I molecules, which are expressed by most normal cells, while many tumor cells lack expression of these "self" molecules. Interaction between KIR and MHC class I triggers an inhibitory signal within the NK cell that dominantly suppresses activating receptor ligands, resulting in tolerance. Encounter of NK cells with tumor cells lacking MHC class I ("missing self") leads to rapid and direct release of cytolytic granules by the NK cell, resulting in specific cytolysis of the tumor cell.

lymphocytes, produce numerous molecular "shapes" that are expressed as antibodies (Ab) and TCR. Figure 12.2 provides a simplified overview of how the effector components of the adaptive immune system (T lymphocytes and Ab) are regulated and eliminate their targets. Antigen-specific immunity is generated when Ab and TCR are expressed and upregulated through the formation and release of cytokines and chemokines.

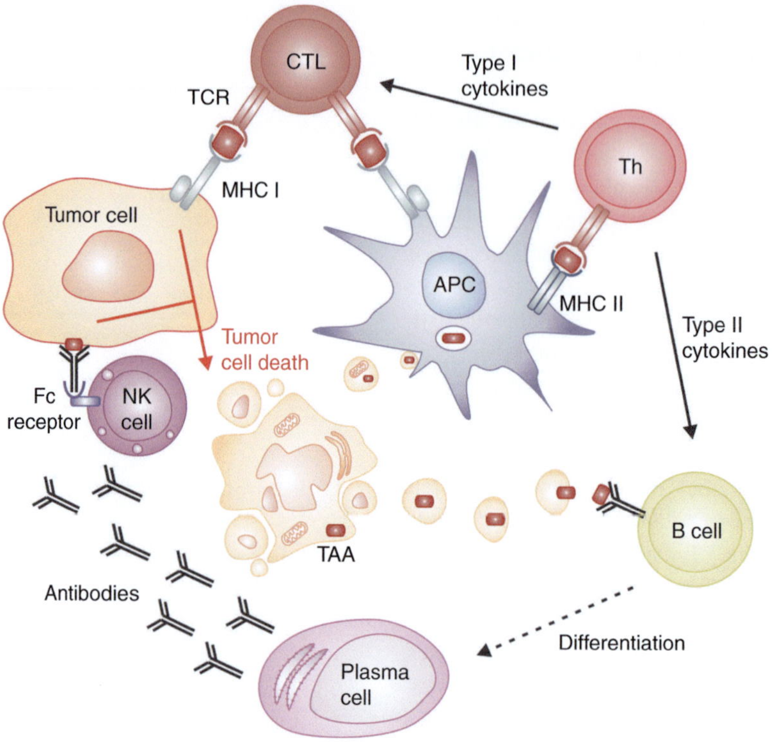

Fig. 12.2 Key players of the immune system in mounting an antitumor response. A "professional" antigen-presenting cell (APC) presents antigenic peptides (depicted as diamonds) to a helper T lymphocyte (HTL, Th) via its MHC class II molecules (MHC II) and to a cytotoxic T lymphocyte (CTL) via its MHC class I molecules (MHC I). Th lymphocytes may also recognize antigen on a tumor cell itself if the tumor cells express MHC class II molecules; similarly, a CTL may also recognize antigen on a tumor cell itself if the tumor cells express MHC class I molecules. The Th lymphocytes recognize antigen through their T-cell receptors for antigen (TCR), which are supported in this task by CD4 coreceptor molecules. By the same token, CTL recognize antigen by their TCR supported by CD8 coreceptor molecules. These coreceptors and other important accessory molecules and their receptors, as well as Th9, Th17, Tfh, and regulatory T cells, are not shown here (see text for details). The Th cells support CTL activation and proliferation by secreting the so-called Th1 cytokines, of which IL-2 and IFN-γ are the most important. Th2 cells stimulate B lymphocytes through specific cytokines (mainly IL-4 and IL-5). B cells recognize soluble antigen through their B-cell receptor for antigen; upon activation, B lymphocytes differentiate into plasma cells, which secrete antibodies specific for that antigen, and may in turn exert their effector function through complement-dependent cytotoxicity, in collaboration with NK lymphocytes through antibody-dependent cellular cytotoxicity (ADCC). Upon engagement of their TCR and accessory molecules, $CD8^+$ and also $CD4^+$ T cells can destroy tumor cells by secreting granzymes, perforins, and cytokines (e.g., TNF) or by upregulation of CD95 (Fas ligand) on tumor cells. The remnants of destroyed tumor cells can be taken up by APC, processed, and presented to T cells; in addition, they may be specifically recognized by antibodies and eliminated via Fc-receptor-expressing phagocytes

Thus, adaptive immunity involves a wide range of antigen receptors expressed on the surface of T and B lymphocytes to detect "foreign" molecules. B lymphocytes respond to antigens by secreting their own antigen receptors as Ab after having differentiated into plasma cells. Ab interact with effector mechanisms via binding of their constant part (Fc) to complement Fc-receptor-bearing phagocytes and Fc-receptor-bearing cytotoxic (NK and T) lymphocytes.

The major T lymphocyte subsets are characterized by expression of the differentiation markers CD4 or CD8. While Ab mostly react with intact proteins or carbohydrates, T cells mostly react with peptides expressed on the cellular surface via presentation by scaffolds, i.e., molecules of the major histocompatibility complex (MHC), on antigen-presenting cells (APCs). One of the most potent types of APC is the DC. DC pick up antigens in the peripheral or lymphoid tissues and migrate to the T lymphocyte zones of lymphoid organs where they stimulate naive CD4+ and CD8+ T lymphocytes.

Most CD4+ T lymphocytes are "T helper (Th) lymphocytes" and recognize antigens in the form of 15- to 25-mer peptides presented by MHC class II molecules (see Chap. 3). These molecules present peptides that are mainly derived from the extracellular compartment, as opposed to peptides presented by MHC class I molecules that are derived from endogenous proteins digested in the proteasome [36]. CD4+ T lymphocytes are important immunoregulatory cells. They recruit and activate other immune cells such as B lymphocytes, CD8+ T lymphocytes, macrophages, mast cells, neutrophils, eosinophils, and basophils. CD4+ Th lymphocytes are classified into a number of different subsets, including Th1, Th2, Th9, Th17, T follicular helper (Tfh), and regulatory T lymphocytes (Treg), based on their functions, their patterns of cytokine secretion, and their expression of specific transcription factors [37]. Importantly, subset specification is plastic, i.e., CD4+ Th lymphocytes can change identity and function in the context of a changing micro-milieu [37]. Th1 lymphocytes mediate immune responses against intracellular pathogens and play a particularly important role in resistance to mycobacterial infections. Th1 lymphocytes are important for cell-mediated immunity by phagocytes and secrete "inflammatory" type-1 cytokines such as interferon (IFN)-γ. Strong infiltration of T lymphocytes with a Th1 orientation is associated with a favorable patient prognosis in several tumor entities, including colorectal and mammary cancer [38]. Th2 lymphocytes are associated with neutralizing antibody responses, mediate host defense against extracellular parasites, and produce type-2 cytokines such as IL-4, IL-5, and IL-13, which additionally contribute to allergic responses. The more recently described Th9 lymphocytes are characterized by particularly high production of IL-9 and are involved in host defense against extracellular parasites. Th9 lymphocytes may promote autoimmune responses at mucosal sites and, given the association of tumor immunity with autoimmunity (see Sects. 12.1 and 12.5), play a part in antitumor immunity. Th17 lymphocytes are generated by cytokines such as IL-6 and TGF-β released by activated DC and propagate acute inflammation by activating the epithelium and promoting granulopoiesis through IL-17 secretion at sites of infection with extracellular pathogens (e.g., fungi). Propagating inflammation by Th17 bears the risk of developing chronic inflammation and autoimmunity but may be instrumental in antitumor immunity [39]. Tfh reside mainly in secondary lymphoid organs where they are involved in the formation of germinal centers and contribute to shaping B lymphocyte affinity maturation to provide "optimal antibodies" and B lymphocyte memory formation [40]. In cancer, Tfh lymphocytes may contribute to antitumor immunity through their role in tumor-associated tertiary lymphoid organ formation, which is often associated with good survival prognosis [38]. Finally, Treg can suppress both innate and adaptive immune responses, thus maintaining tolerance and limiting immunopathology. Treg are identified based on their high levels of surface CD25 (IL-2Rα) and expression of the transcriptional regulator forkhead box P3 (FoxP3). As observed in all other adaptive immune responses, Treg activation is antigen specific, which implies that suppressive activities of Treg are antigen dependent. Importantly, the presence of tumor-specific CD4+ Treg at tumor sites may inhibit T-cell responses against cancer [41].

CD8+ T lymphocytes differentiate into CTL that kill their target cells. Given their potentially destructive properties, they require costimulatory activity to become activated. This primarily comes from CD4+ T lymphocytes in addition to DC. CD8+ T lymphocytes react with eight- to ten-mer peptides presented on the surface of all nucleated cells by the scaffolds formed by MHC class I molecules. These peptides are usually of endogenous origin, derived from polypeptides that are routinely degraded in the proteasome. DC have the special capacity to load exogenous-derived peptides on their MHC-I molecules, through a process called cross presentation [36], which is essential to activate CD8+ T lymphocytes against tumor antigens. CD8+ T lymphocytes destroy their target cells, after attachment of (1) their TCR to the appropriate MHC-peptide complex and (2) their accessory and costimulatory molecules (such as CD8 and CD28) to the corresponding ligands, by perforating their membranes with enzymes (i.e., perforin, granzymes, and granulysin) or by triggering a process of self-destruction (i.e., apoptosis). In this way, CD8+ T lymphocytes can move from one tumor cell to another, as far as these express the same MHC-peptide complexes, kill them selectively, and thus can mount a very specific and robust antitumor response.

In the setting of ADOPTIVE IMMUNOTHERAPY of cancer (see below), the administration of autologous CD4+ Th lymphocytes, concurrently with CD8+ CTL, has shown to prevent exhaustion of the infused CD8+ CTL [42]. CD4+ T lymphocytes activate DC through cross-linking of CD40, which provides enhanced antigen presentation and costimulation and leads to priming of CD8+ CTL function [36]. Thus, CD4+ T lymphocyte help may result in de novo generation of tumor-specific CD8+ CTL and concomitant tumor destruction. The exact requirement for CD4+ T lymphocyte help during priming may depend on the nature of the stimulus, but the requirement for CD4+ T lymphocytes during memory responses, as outlined above for Tfh lymphocytes, is beyond dispute. Recent experimental studies, including vaccination approaches, suggest that CD4+ T lymphocytes play a critical role in antitumor immunity that goes beyond the mere induction and maintenance of tumor-specific CD8+ CTL. Specifically, CD4+ T lymphocytes exert direct antitumor immunity, recognize activated CD8+ CTL already present at the site of the tumor, enable other CD8+ CTL to migrate into the tumor, induce an antitumor Ab response, and contribute to tumor regression, e.g., through production of IFN-γ [43, 44].

Cytokines produced by CD4+ Th lymphocytes also regulate macrophages by stopping their migration after their engagement, allowing them to accumulate at a particular site. In this manner, more efficient PHAGOCYTOSIS may be stimulated to remove tumor cells. IFN-γ release by Th1 lymphocytes is capable of polarizing macrophages toward an M1 phenotype, as opposed to the M2 phenotype triggered by Th2 lymphocyte-derived IL-4. M1 macrophages are capable of direct and indirect antitumor activity by producing reactive oxygen and nitrogen radicals that kill tumor cells, as well as cytokines and chemokines to recruit and activate adaptive immune cells. In contrast, M2 macrophages often limit antitumor immunity, e.g., by recruiting immunosuppressive cells. Accordingly, the presence of M1 macrophages is associated with a favorable and the presence of M2 macrophages with a negative patient prognosis [38]. CD4+ T lymphocytes further amplify their own responses by secreting cytokines thus creating autocrine loops, of which IL-2 is a key example. This action enhances the Th lymphocyte response and thus the entire immune system's response to foreign antigens.

12.4 Tumor Antigens

To be able to mount an effective antitumor immune response, adaptive immune cells need to recognize and respond to antigens expressed by tumor cells. Such antigens can be classified in three broad categories: the abovementioned TSA (or neoantigens), TAA, and cancer/testis antigens (CTA) [24, 45]. The best characterized of those are probably TAA. TAA are derived from proteins that are expressed at low level in normal tissues, either at the site of tumor origin or in unrelated tissues, but whose expression is substantially upregulated in tumor cells as a result of activation of oncogenic signaling cascades. The promise of potentially targeting TAA is based on their expression in a large number of patients harboring the same type of tumor, and even across different tumor entities. CTA are proteins normally expressed in the testes, fetal ovaries, and/or trophoblasts, but largely absent in other healthy cells. Importantly, CTA are usually not presented to adaptive immunity at high levels since germline and trophoblastic cells do not display MHC class I molecules on their surface. The expression of CTA by tumors is the result of transcriptional reactivation of genes that are mostly silent in adult tissues. Expression of CTA by various tumor types therefore makes them attractive targets for cancer vaccination and ADOPTIVE IMMUNOTHERAPY approaches. Both TAA and CTA have already been targeted through immunotherapeutic means (see below, Adoptive Cell Transfer, CAR-T cells, etc.), with limited success. This can be attributed to their more or less abundant expression in normal tissues. In addition, there is a safety risk intrinsically associated with this strategy as immune cells may also harm normal tissues. On the other hand, both CTA and TAA are subjected to central and peripheral tolerance mechanisms, limiting immunogenicity [24, 45].

These problems are at much lower scale in the case of TSA, since they derive from proteins that are not found in normal tissues. Toxicity is therefore limited, and there is a chance of limited preexisting tolerance. TSA generally arise from diverse mechanisms. They can be generated as a consequence of oncogenic viral infections, for instance, the latent membrane protein (LMP)-1 of the Epstein-Barr virus (EBV), expressed by nasopharyngeal carcinoma and EBV^+ Hodgkin's lymphoma, or the human papilloma virus (HPV)-encoded E6 and E7 proteins expressed by HPV-16^+ cervical carcinoma. Also, TSA can be generated from the fusion of distant genes by chromosomal translocation occurring in tumor cells (e.g., bcr-abl, sarcoma translocation breakpoints, ERG, and ETV6-AML). Finally, TSA can be the result of point mutations in normal genes whose molecular changes may or may not be functionally associated with neoplastic transformation and tumor progression (e.g., EGFRvIII, ras, p53). The therapeutic utility of the latter ones in particular is limited because of their partly erratic nature, leading to patient-specific mutational landscapes. Moreover, different types of tumors show huge variety in the frequency and nature of somatic mutations [24]. The very robust whole-genome sequencing platforms currently available, in conjunction with the advances in bioinformatic tools to predict potent immunogenic MHC-I and MHC-II restricted neoantigens derived from mutated proteins, however, will allow to harvest the gold mine of TSA, in order to come out with personalized patient-specific vaccines, as demonstrated in two recent studies [46].

As new knowledge on the key players in antitumor immunity emerges, and new immunotherapeutic strategies arise, it is important to note that even the most sophisticated cancer vaccine or adoptive immunotherapy approach (see below) will face a major challenge upon arrival of the antigen-specific $CD4^+$ Th lymphocytes and $CD8^+$ CTL at the primary or secondary tumor site: the immunosuppressive TME. The TME can limit the activity of adaptive and innate antitumor immune cells in several ways, which include production of catabolites such as lactate, restricted access to essential amino acids or oxygen, and presence of "tumor-educated" tolerogenic immune cells [34, 47]. These educated tumor-infiltrating immune cells are the source of a number of factors that support growth, angiogenesis, and metastasis and actively suppress adaptive immune responses against the tumor. Depletion of IIC subsets such as Treg [48] or TAM [49] can partially reverse these tumor-supportive features and reduce tumor growth. However, since most immune cell subpopulations show a high degree of plasticity, they can all potentially be converted back into tumor-promoting cells. Taking innate immunity as an example, DC, macrophages, granulocytes, as well as immature myeloid cells, so-called myeloid-derived suppressor cells (MDSC), show a largely overlapping functional repertoire in tumors [47] and are all capable of suppressing $CD8^+$ CTL function. Immune suppression by tumor-infiltrating immune cells partly explains why invasive tumors actually arise, even in the presence of TSA, CTA, or TAA. Among the factors that educate immune cells in the TME is the presence of cells, mostly tumor cells, undergoing apoptotic cell death. Counterintuitively, apoptotic death of tumor cells is a frequently observed phenomenon, even under steady-state conditions without cytotoxic therapy [50]. The interaction between apoptotic cells and, predominantly, innate immune cell such as macrophages prevents inflammation and induces self-tolerance under physiological conditions, which is exploited by the tumor [51].

On top of these obstacles, autoregulatory circuits that normally serve to prevent autoimmunity limit the half-life of $CD4^+$ Th lymphocytes and $CD8^+$ CTL activity. Important components of this program are immune checkpoints. IMMUNE CHECKPOINTS consist of receptor-ligand interactions that determine the propagation versus termination balance of immune responses (see below). As an example, the immune checkpoint receptor programmed cell death 1 (PD-1) on lymphocytes interacts with the programmed death ligand-1 (PD-L1) expressed on myeloid or epithelial cells, to attenuate lymphocyte activity. Both molecules are usually not expressed under inflammatory conditions but are upregulated following lymphocyte activation and inflammatory reactions, thus acting as a negative feedback in order to terminate inflammation. The relevance of this process in tumors has been shown by the widespread expression of inhibitory IMMUNE CHECKPOINT receptors on tumor-infiltrating lymphocytes, the expression of their ligands by virtually all types of cells in the TME, and the clinical success of therapeutically breaking up immune checkpoint receptor/ligand interaction with neutralizing Ab [6].

In the following section, we describe the utilization of the described key components of tumor immunity for anticancer therapy in more detail.

12.5 Tumor Immunotherapy

During the last years, different strategies to (re-) activate immune defense mechanisms against tumors have been explored in preclinical and clinical studies. These approaches have demonstrated remarkably positive responses in different models, increasing the interest in immune modulation to treat cancer. Indeed, the long-term efficacy observed using antitumor immune activation strategies has dramatically shifted the strategies aiming on developing new cancer therapies [52]. In this section, we summarize the main immunotherapeutic approaches that are currently under investigation and/or already clinically exploited.

12.5.1 Adoptive Cell Transfer

Adoptive cell transfer involves the extraction of a certain immune cell population from one patient or an allogeneic donor, followed by ex vivo stimulation and expansion of antigen-specific lymphocytes, including NK lymphocytes and T LYMPHOCYTES, or APC which are subsequently infused back to the patient with the intention to treat disease.

Due to their unique mode of action to selectively identifying tumor cells from normal cells and their remarkable cytotoxic potential, NK lymphocytes have been considered suitable candidates for adoptive therapy. Major protocols include isolation of mature NK lymphocytes or hematopoietic stem cells that are differentiated to NK lymphocytes, which are then infused in cancer patients. The premise of this approach is that allogeneic NK lymphocytes do not recognize donor MHC class I molecules (due to their polymorphic receptors for MHC class I molecules) and, therefore, will lyse MHC class I-expressing tumor cells only, sparing normal cells that lack an activating signal (Fig. 12.1). Currently more than 200 clinical studies using NK lymphocytes have been conducted. So far, the outcomes of these trials were largely disappointing, which is partly due to a lack of consensus concerning selection, activation/expansion, and treatment protocols [53].

Adoptive transfer of APC, particularly DC, has also been investigated for several decades already. In this case, the premise is that patient DC loaded with TSA or TAA would deliver and present these antigens to T lymphocytes to mount an antitumor immune response. Similar to the situation with NK cells, no consensus has been reached so far on what specific DC subset, source, activation protocol, and treatment schedule should be used on each tumor entity. Currently more than 300 clinical studies are investigating DC vaccines, with mixed outcomes. To date, only one of the vaccines received approval by the FDA. Sipuleucel-T is approved to be used in prostate cancer and is based on patient-derived DC pulsed with a recombinant fusion protein of the TAA prostatic acid phosphatase and the immunostimulatory growth factor granulocyte-macrophage colony-stimulating factor (GM-CSF) [54].

Probably, the most promising cells for adoptive therapy so far are T lymphocytes isolated from a resected tumor or blood, which can subsequently be used in different settings. In the case of tumor-derived cells, T lymphocytes showing the highest affinity against cancer cell-derived TSA or TAA are re-injected into the patient in combination with activating factors such as IL-2. Another adoptive cell transfer approach is the development of CHIMERIC ANTIGEN RECEPTOR T (CAR-T) cells. The chimeric receptor is introduced into the patient's T lymphocytes by the fusion of the extracellular antigen-binding region of a monoclonal antibody with the intracellular domains of the T-cell receptor and/or costimulatory molecules. These CAR-T cells will react against particular tumor antigens with the antigen specificity being provided by the extracellular binding moiety and the cell activation being mediated through the signaling components included in the receptor [55, 56].

The fact that CAR-T cells recognize antigens in a MHC-independent manner represents an advantage of this particular strategy since some tumors downregulate the expression of MHC molecules as an immune escape mechanism. Furthermore, CAR-T cells specific for a certain antigen could be used in patients regardless of their specific human leukocyte antigen (HLA) type. Nevertheless, approximately 90% of potential tumor antigens are fragments of intracellular proteins, which need to be processed and presented in the correspondent MHC molecules. Therefore, CAR-T cells are limited to recognize molecules that are expressed on the surface of tumor cells only [55].

Distinct kinds of CAR-T cells have been developed, having different composition on their intracellular domain. The first-generation CAR-T cells contained only the tyrosine-based activating domain from the TCR/CD3 receptor complex, whereas the second generation includes an additional costimulatory molecule such as CD28, CD27, CD137, or OX40 (CD134). These molecules provide second-generation CAR-T cells with a higher activation and proliferation capacity, as well as superior in vivo persistence compared to the first generation. Additionally, third-generation CAR-T cells include two costimulatory molecules, further enhancing cell activation and proliferation. A fourth generation is currently under development; these cells additionally incorporate genetic information encoding for cytokines that are involved in lymphocyte activation/proliferation such as IL-15, IL-7, IL-12, and IL-21 [55, 57].

CAR-T cells have been explored for the treatment of acute lymphocytic leukemia (ALL), using the B-cell marker CD19 as specific antigen to attack malignant cells. CD19 CAR-T cells have shown high antitumor efficacy, with complete response rates of 70–90% in pediatric and adult patients with relapsed or refractory, chemotherapy-resistant acute B-cell leukemia [55]. Despite these good results, the use of this CD19 CAR-T cell therapy has been linked to side effects such as the CYTOKINE RELEASE SYNDROME and neurotoxicity. Moreover, a phase II clinical trial was recently put on hold due to fatal cerebral edema in five adult ALL patients [57]. Strategies to overcome the CYTOKINE RELEASE SYNDROME include neutralization of major inflammatory cytokines such as IL-6.

New efforts focus on increasing the safety of CAR-T cell therapy. Targeting more than one antigen is an interesting option that is currently under investigation. This approach will improve the specificity of the treatment and, additionally, limit relapse of tumor cell clones lacking the target antigen. Moreover, combinatorial antigen-targeting prevents off-tumor toxicity in the absence of true TSA. One possibility is to use one CAR that provides the activation signal, whereas a second receptor mediates the costimulatory signal, so that only the recognition of the two tumor antigens would lead to T-cell activation [57].

Application of CAR-T cell therapy in other oncological settings is limited by different factors. The identification of suitable target antigens that are expressed exclusively on the tumor cell surface represents one of the major challenges in

CAR design. In solid tumors, targeting antigens that are not expressed in all tumor cells would lead to selection of resistant clones and relapse. Additionally, solid tumors possess an immunosuppressive microenvironment, which can affect CAR-T cell antitumor activity. Moreover, the physical complexity of the TME in different tumor entities needs to be taken into consideration to guarantee T-cell infiltration and survival [57].

Despite the current limitations, CAR-T cell therapy represents a promising option for specific oncological applications. Further efforts to improve the specificity, safety, and efficiency of the treatment, as well as the extrapolation to other tumor types, are the main challenges of this immunotherapeutic approach.

12.5.2 Immune Checkpoint Inhibitors

T cells require more than one signal to acquire a fully activated status. First, TCR signaling is activated after T-cell receptor (TCR) recognition of the specific antigen presented in the corresponding MHC complex by APCs. Second, costimulatory signals such as the interactions of CD28 or OX40 with their respective ligands are necessary. Cytokines produced by the T cell and the APC provide an additional third signal to complete the T lymphocyte polarization and activation process. The activation of T lymphocytes is counter-regulated by the expression of IMMUNE CHECKPOINTS, which are feedback inhibitory pathways of the immune system that maintain homeostasis and prevent immune hyperactivation. Therefore, the fate of a T lymphocyte is defined by the degree of APC-antigen-mediated TCR activation, the number of costimulatory receptors, the presence of costimulatory ligands on APCs, the type and amount of the cytokines produced, and the presence or absence of IMMUNE CHECKPOINT effectors on the cell surface [58] (Fig. 12.3).

IMMUNE CHECKPOINTS consist of receptor and ligand proteins exposed on the surface of effector immune cells, which can be found in a variety of cell types. Individual IMMUNE CHECKPOINTS induce tolerogenic APC activation, priming of Treg, and T lymphocyte exhaustion/dysfunction through very specific mechanisms [58]. In cancer, several studies have identified the presence of exhausted T lymphocytes within the

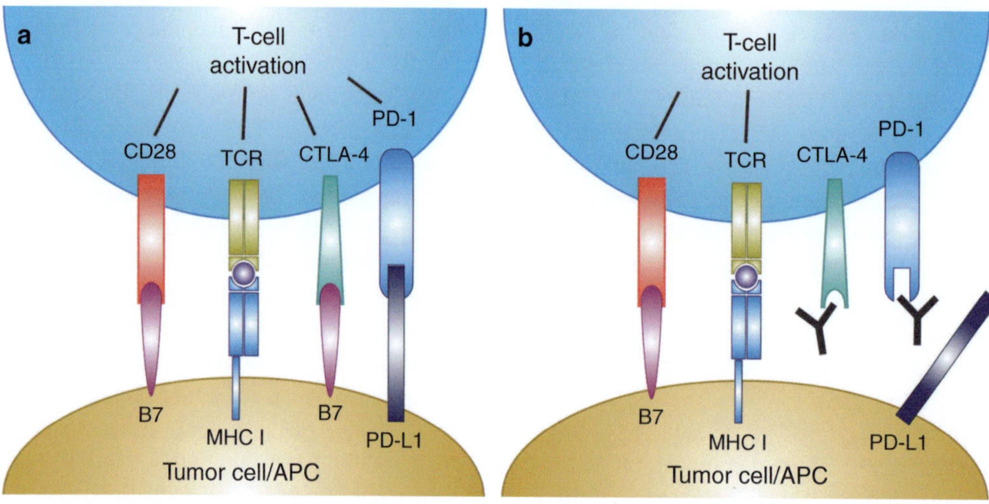

Fig. 12.3 Immunotherapy with IMMUNE CHECKPOINT INHIBITORS. (**a**) T lymphocyte activation is modulated by the balance between inducers and negative regulators. In tumors, high expression levels of IMMUNE CHECKPOINTS (CTLA-4, PD-1, PD-L1) lead to T-cell exhaustion. (**b**) By using antibodies as IMMUNE CHECKPOINT INHIBITORS, the inhibitory pathways are blocked, and antitumor immunity is induced. *APC* antigen-presenting cell, *TCR* T-cell receptor, *MHC I* major histocompatibility complex I, *CTLA-4* cytotoxic T lymphocyte antigen-4, *PD-1* programmed cell death protein-1, *PD-L1* programmed cell death ligand-1

TME. Therefore, the blockade of these immunosuppressive pathways using IMMUNE CHECKPOINT INHIBITORS reactivates T lymphocytes to induce antitumor immune activation.

Currently, six IMMUNE CHECKPOINT INHIBITORS are approved for treatment of melanoma, lung cancer, head and neck cancer, bladder cancer, Merkel cell cancer, as well as classic Hodgkin's lymphoma (Table 12.1). Additionally, several new IMMUNE CHECKPOINT INHIBITORS are in advanced stages of ongoing clinical testing.

Cytotoxic T lymphocyte antigen-4 (CTLA-4) is an IMMUNE CHECKPOINT protein mainly present on the surface of activated lymphocytes that interacts with ligands of the B7 family such as CD80 and CD86. These ligands usually interact with the costimulatory molecule CD28 during the T-cell activation process. Nevertheless, after binding to CTLA-4, inhibitory pathways are triggered limiting T lymphocyte activation, survival, and proliferation. CTLA-4 was the first IMMUNE CHECKPOINT that showed promising results as immunotherapeutic target. Currently a monoclonal antibody against CTLA-4, ipilimumab, is approved for treatment of unresectable stage III/IV metastatic melanoma and as adjuvant therapy for surgically treated "high-risk" melanoma patients [58].

Table 12.1 Approved indications for therapy targeting immune checkpoints

Target	Name	Approved indications
CTLA-4	Ipilimumab	Melanoma
PD-1	Nivolumab	Metastatic melanoma, metastatic non-small-cell lung cancer (NSCLC), renal cell carcinoma (RCC), classical Hodgkin's lymphoma, head and neck squamous cell carcinoma (HNSCC), metastatic urothelial carcinoma, hepatocellular carcinoma (HCC), colorectal cancer with MSI-H and MMR aberrations
PD-1	Pembrolizumab	Metastatic melanoma, metastatic NSCLC, classical Hodgkin's lymphoma, HNSCC, gastric cancer, solid tumors with MSI-H and MMR aberrations, metastatic urothelial carcinoma
PD-L1	Atezolizumab	Metastatic urothelial carcinoma, metastatic NSCLC
PD-L1	Avelumab	Merkel cell carcinoma, metastatic urothelial carcinoma
PD-L1	Durvalumab	Metastatic urothelial carcinoma

Modified from [58]

Programmed cell death protein-1 (PD-1) and its ligand programmed cell death ligand-1 (PD-L1) constitute another IMMUNE CHECKPOINT targeted with IMMUNE CHECKPOINT INHIBITORS to induce antitumor immune activation. PD-1 is expressed by different kinds of immune cells, including T lymphocytes, B lymphocytes, and other tumor-infiltrating lymphocytes, as well as APC such as monocytes, macrophages, and DC. PD-L1 is prominently expressed by tumor cells and APC. In the tumor context, PD-1 expression is elevated on activated T lymphocytes as a negative feedback. Moreover, cytokines produced by activated tumor-infiltrating lymphocytes induce PD-L1 expression on the surface of the tumor cells and other cells of the TME, facilitating the interaction of PD-1 with its ligand, leading to T-cell dysfunction and exhaustion, as well as production of the anti-inflammatory cytokine IL-10 and induction of immune tolerance. Indeed, the PD-1 expression level is considered as a marker for T-cell exhaustion [58, 59]. Therefore, both anti-PD-1 and anti-PD-L1 neutralizing antibodies have been introduced as therapeutic options to treat tumors with remarkably positive results (Table 12.1).

The efficacy of PD-1/PD-L1 IMMUNE CHECKPOINT INHIBITORS is influenced by intrinsic characteristics of the patients and the tumors. Factors such as patient gender, type of tumor, degree of lymphocyte infiltration, and mutation rate (i.e., the number of TSA) can affect the outcome of these treatments. PD-L1 expression levels have been proposed as a marker to define whether the inhibitor therapy will be effective.

Nonetheless, the PD-L1 expression pattern is heterogeneous and can change during the course of tumor development and progression. Therefore, other markers are needed, in addition to PD-L1, to predict a positive response to these treatments [59, 60]. Particularly, the relationship between tumor mutation rate and the response to IMMUNE CHECKPOINT INHIBITORS appears promising to refine prediction of patient response. Tumors with mismatch repair deficiency and microsatellite instability are more susceptible to immunotherapy. For instance, a subset of colorectal carcinomas with high microsatellite instability and high mutational rate have shown durable responses to PD-1 blockade [60]. Hypermutated tumors express abundant peptides that act as TSA, which in turn trigger immune mechanisms that ultimately lead to recruitment of infiltrating lymphocytes to the tumor. These attracted effector cells are often suppressed by regulatory factors such as IMMUNE CHECKPOINTS, which are highly expressed in these tumors. The use of IMMUNE CHECKPOINT INHIBITORS reverts this suppression, thus restoring an antitumor response. In fact, anti-PD-1 therapy was the first immunotherapeutic approach to receive approval for treating patients with solid tumors that are microsatellite instability-high and/or mismatch repair deficient [59, 61].

Immunotherapy using IMMUNE CHECKPOINT INHIBITORS has proven highly successful in different types of tumors. In certain types of cancer, the use of this approach translates into significantly longer relapse-free survival when compared with conventional therapy. As expected, due to suppression of immune-restricting mechanism, the main adverse effects of IMMUNE CHECKPOINT INHIBITORS are connected to an overactivated immune response, i.e., autoimmunity. Patients treated with combinations of more than one IMMUNE CHECKPOINT INHIBITOR show more severe complications such as diarrhea, colitis, myocarditis, and endocrine disorders [59]. Another caveat of IMMUNE CHECKPOINT inhibition in general, as outlined above specifically for the PD-1/PD-L1 IMMUNE CHECKPOINT, is that robust antitumor response is observed only in a fraction of patients. Hence, understanding the mechanisms that explain the unresponsiveness is an important focus of current research. Interestingly, it has been recently described that the gut microbiota plays a pivotal role determining the outcome of patients receiving IMMUNE CHECKPOINT INHIBITORS. For instance, a study showed that the composition of gut microbiota differs between melanoma patients responding and those not responding to IMMUNE CHECKPOINT INHIBITORS. More specifically, the bacterial species *Bifidobacterium longum*, *Collinsella aerofaciens*, and *Enterococcus faecium* were more abundant in responders. Interestingly, transferring the fecal material of these patients to germfree mice leads to improved tumor control [62]. Moreover, murine studies demonstrated that the presence of certain bacterial groups, such as *Bacteroides* spp. or *Bifidobacterium* spp., augmented immune-mediated tumor control and efficacy of anti-PD-L1 and anti-CTLA-4 therapy [63–65]. The fascinating interaction between gut microbiota and the response to immunotherapy is still far from being fully understood. Nevertheless, it emerges as one of the factors defining the response to IMMUNE CHECKPOINT INHIBITOR therapy in cancer.

Another of the current main challenges of IMMUNE CHECKPOINT INHIBITOR therapy is to develop combination therapies that improve the success rate of this treatment [66]. Exhausted T lymphocytes within the TME express multiple IMMUNE CHECKPOINTS; therefore combination therapy of more than one IMMUNE CHECKPOINT INHIBITOR is associated with better results, in terms of the percentage of patients responding to the therapy [58]. Combination of CTLA-4 and PD-1 blockade was approved for melanoma in 2016, thus improving the efficacy of the treatment. However, as outlined above, this approach resulted in more severe adverse effects as compared to individual therapies [66]. Furthermore, different combinations with other IMMUNE CHECKPOINT INHIBITORS or conventional therapy approaches have been proposed. The degree of T lymphocytes infiltrating the tumor is a major parameter of the response to IMMUNE CHECKPOINT INHIBITORS. Thus, combination of IMMUNE CHECKPOINT INHIBITORS with agents that increase tumor T lymphocyte infiltration may improve the clinical outcome [59]. Radiotherapy and some chemotherapeutic agents revert tumor immune suppression by inducing immunogenic cell death of tumor cells [66]. Immunogenic cell death pathways induce antigen release, maturation and activation of dendritic cells, and T-cell activation [67]. Moreover, immune activation induces the expression of PD-L1, which is one target of IMMUNE CHECKPOINT INHIBITOR therapy. Thus, an adequate combination of radio- or chemotherapy with blockade of the PD-1/PD-L1 pathway can lead to an increased response to immunotherapy, especially in less immunogenic tumors [66, 68].

12.5.3 Cancer Vaccines

Different attempts to induce antitumor immunity through vaccination have been developed in the last years. One approach explores the central role of DC in immune modulation (see Sect. 12.5.1 above). A second strategy is based on the identification of immunogenic TSA. Recent technological advances in genome-wide sequencing provide the platform necessary for the development of personalized tumor vaccines. Identification of TSA and analysis of RNA mutanome have yielded promising results for the treatment of melanoma alone or in combination of PD-1 blockade [66]. The rationale of this procedure lies on identifying the differences in the epitope repertoire between tumor and normal tissue. At the time of diagnosis, both tumor tissue and normal tissue are collected and subsequently subjected to genome sequencing to identify nonsynonymous somatic mutations in the tumor. These mutations are analyzed using MHC-I epitope prediction algorithms to identify potential targets that, according with the patient HLA, are efficiently presented to T lymphocytes. After in vitro target validation, selected neoantigens are integrated into RNA, DNA, or peptide-based vaccines that are delivered to patients. Currently, this approach is under investigation in phase I clinical trials [69–71]. Antitumor DNA vaccines may be a flexible, safe, and efficient option of the treatment of cancer. However, manufacturing requirements need to be developed in order to warrant an efficient workflow in such personalized approach. Interestingly, the flexibility of DNA vaccines enables genetic modifications of the selected TSA as well as multiple targeting of mutated peptides, which is known as polyepitope DNA vaccine. Furthermore, the possibility of integrating additional genes encoding cytokines or immune-activating factors can improve the results of antitumor vaccination [69].

12.6 Summary

It is now firmly established that the immune system is capable of recognizing and eliminating newly arising and established tumors, both spontaneously and therapy-induced, despite the fact that cancer cells are generally less immunogenic than microbial pathogens such as bacteria, fungi, and viruses. Nevertheless, immunotherapeutic modalities, such as cytokines and mAb, have long become components of several standard treatment regimens of human malignancies. Tumor vaccines, cellular and/or antigen based, have advanced through preliminary testing toward early clinical trials and have shown little toxicity but, in general, also have limited effects in patients with established tumors. Tumors, especially the TME, frequently interfere with the development and function of immune responses. The advent of IMMUNE CHECKPOINT INHIBITORS has provided an important means to overcome tumor-associated immunosuppression, capitalizing on basic knowledge concerning immune cell activation. In the near future, combination strategies including personalized vaccines together with IMMUNE CHECKPOINT inhibition will likely expand the number of cancer patients benefitting from cancer immunotherapy.

Selected Readings

Dunn GP, Bruce AT, Ikeda H, Old LJ, Schreiber RD. Cancer immunoediting: from immunosurveillance to tumor escape. Nat Immunol. 2002;3(11):991–8.

Palucka AK, Coussens LM. The basis of oncoimmunology. Cell. 2016;164(6):1233–47. https://doi.org/10.1016/j.cell.2016.01.049.

Pitt JM, Vétizou M, Daillère R, Roberti MP, Yamazaki T, Routy B, et al. Resistance mechanisms to immune-checkpoint blockade in cancer: tumor-intrinsic and -extrinsic factors. Immunity. 2016;44(6):1255–69. https://doi.org/10.1016/j.immuni.2016.06.001.

Yarchoan M, Johnson BA III, Lutz ER, Laheru DA, Jaffee EM. Targeting neoantigens to augment antitumour immunity. Nat Rev Cancer. 2017;17(9):569. https://doi.org/10.1038/nrc.2017.74.

References

1. Hanahan D, Weinberg RA. Hallmarks of cancer: the next generation. Cell. 2011;144(5):646–74.
2. Albini A, Sporn MB. The tumour microenvironment as a target for chemoprevention. Nat Rev Cancer. 2007;7(2):139–47. https://doi.org/10.1038/nrc2067.
3. Decker WK, da Silva RF, Sanabria MH, Angelo LS, Guimaraes F, Burt BM, et al. Cancer immunotherapy: historical perspective of a clinical revolution and emerging preclinical animal models. Front Immunol. 2017;8:829. https://doi.org/10.3389/fimmu.2017.00829.
4. Coley WB. The treatment of malignant tumors by repeated inoculations of erysipelas. With a report of ten original cases. 1893. Clin Orthop Relat Res. 1991;(262):3–11.
5. Dunn GP, Bruce AT, Ikeda H, Old LJ, Schreiber RD. Cancer immunoediting: from immunosurveillance to tumor escape. Nat Immunol. 2002;3(11):991–8.
6. Sharma P, Hu-Lieskovan S, Wargo JA, Ribas A. Primary, adaptive, and acquired resistance to cancer immunotherapy. Cell. 2017;168(4):707–23. https://doi.org/10.1016/j.cell.2017.01.017.
7. Trinchieri G. Cancer and inflammation: an old intuition with rapidly evolving new concepts. Annu Rev Immunol. 2012;30:677–706. https://doi.org/10.1146/annurev-immunol-020711-075008.
8. Uemura N, Okamoto S, Yamamoto S, Matsumura N, Yamaguchi S, Yamakido M, et al. Helicobacter pylori infection and the development of gastric cancer. N Engl J Med. 2001;345(11):784–9. https://doi.org/10.1056/NEJMoa001999.
9. De Strooper LM, Hesselink AT, Berkhof J, Meijer CJ, Snijders PJ, Steenbergen RD, et al. Combined CADM1/MAL methylation and cytology testing for colposcopy triage of high-risk HPV-positive women. Cancer Epidemiol Biomarkers Prev. 2014;23(9):1933–7. https://doi.org/10.1158/1055-9965.EPI-14-0347.
10. Horner SM, Gale M Jr. Regulation of hepatic innate immunity by hepatitis C virus. Nat Med. 2013;19(7):879–88. https://doi.org/10.1038/nm.3253.
11. Coussens LM, Zitvogel L, Palucka AK. Neutralizing tumor-promoting chronic inflammation: a magic bullet? Science. 2013;339(6117):286–91. https://doi.org/10.1126/science.1232227.
12. Kuraishy A, Karin M, Grivennikov SI. Tumor promotion via injury- and death-induced inflammation. Immunity. 2011;35(4):467–77.
13. Salama R, Sadaie M, Hoare M, Narita M. Cellular senescence and its effector programs. Genes Dev. 2014;28(2):99–114. https://doi.org/10.1101/gad.235184.113. 28/2/99 [pii].
14. Nathan C, Cunningham-Bussel A. Beyond oxidative stress: an immunologist's guide to reactive oxygen species. Nat Rev Immunol. 2013;13(5):349–61. https://doi.org/10.1038/nri3423.
15. Hanahan D, Coussens LM. Accessories to the crime: functions of cells recruited to the tumor microenvironment. Cancer Cell. 2012;21(3):309–22. https://doi.org/10.1016/j.ccr.2012.02.022. S1535-6108(12)00082-7 [pii].
16. Schafer M, Werner S. Cancer as an overhealing wound: an old hypothesis revisited. Nat Rev Mol Cell Biol. 2008;9(8):628–38.
17. Boon T, Cerottini JC, Van den Eynde B, van der Bruggen P, Van Pel A. Tumor antigens recognized by T lymphocytes. Annu Rev Immunol. 1994;12:337–65. https://doi.org/10.1146/annurev.iy.12.040194.002005.
18. Bissell MJ, Hines WC. Why don't we get more cancer? A proposed role of the microenvironment in restraining cancer progression. Nat Med. 2011;17(3):320–9. https://doi.org/10.1038/nm.2328.
19. Blankenstein T, Coulie PG, Gilboa E, Jaffee EM. The determinants of tumour immunogenicity. Nat Rev Cancer. 2012;12(4):307–13. https://doi.org/10.1038/nrc3246.
20. Roithmaier S, Haydon AM, Loi S, Esmore D, Griffiths A, Bergin P, et al. Incidence of malignancies in heart and/or lung transplant recipients: a single-institution experience. J Heart Lung Transplant. 2007;26(8):845–9. https://doi.org/10.1016/j.healun.2007.05.019.
21. Grulich AE, van Leeuwen MT, Falster MO, Vajdic CM. Incidence of cancers in people with HIV/AIDS compared with immunosuppressed transplant recipients: a meta-analysis. Lancet. 2007;370(9581):59–67. https://doi.org/10.1016/S0140-6736(07)61050-2.
22. Gentles AJ, Newman AM, Liu CL, Bratman SV, Feng W, Kim D, et al. The prognostic landscape of genes and infiltrating immune cells across human cancers. Nat Med. 2015;21(8):938–45. https://doi.org/10.1038/nm.3909.
23. Joseph CG, Darrah E, Shah AA, Skora AD, Casciola-Rosen LA, Wigley FM, et al. Association of the autoimmune disease scleroderma with an immunologic response to cancer. Science. 2014;343(6167):152–7. https://doi.org/10.1126/science.1246886.

24. Yarchoan M, Johnson BA III, Lutz ER, Laheru DA, Jaffee EM. Targeting neoantigens to augment antitumour immunity. Nat Rev Cancer. 2017;17(9):569. https://doi.org/10.1038/nrc.2017.74.
25. Imai K, Matsuyama S, Miyake S, Suga K, Nakachi K. Natural cytotoxic activity of peripheral-blood lymphocytes and cancer incidence: an 11-year follow-up study of a general population. Lancet. 2000;356(9244):1795–9. https://doi.org/10.1016/S0140-6736(00)03231-1.
26. Lim WA, June CH. The principles of engineering immune cells to treat cancer. Cell. 2017;168(4):724–40. https://doi.org/10.1016/j.cell.2017.01.016.
27. Quail DF, Joyce JA. Microenvironmental regulation of tumor progression and metastasis. Nat Med. 2013;19(11):1423–37. https://doi.org/10.1038/nm.3394.
28. Palucka AK, Coussens LM. The basis of oncoimmunology. Cell. 2016;164(6):1233–47. https://doi.org/10.1016/j.cell.2016.01.049.
29. Noy R, Pollard JW. Tumor-associated macrophages: from mechanisms to therapy. Immunity. 2014;41(1):49–61. https://doi.org/10.1016/j.immuni.2014.06.010.
30. van der Woude LL, Gorris MAJ, Halilovic A, Figdor CG, de Vries IJM. Migrating into the tumor: a roadmap for T cells. Trends Cancer. 2017;3(11):797–808. https://doi.org/10.1016/j.trecan.2017.09.006.
31. Kalluri R. The biology and function of fibroblasts in cancer. Nat Rev Cancer. 2016;16(9):582–98. https://doi.org/10.1038/nrc.2016.73.
32. De Palma M, Biziato D, Petrova TV. Microenvironmental regulation of tumour angiogenesis. Nat Rev Cancer. 2017;17(8):457–74. https://doi.org/10.1038/nrc.2017.51.
33. Schreiber RD, Old LJ, Smyth MJ. Cancer immunoediting: integrating immunity's roles in cancer suppression and promotion. Science. 2011;331(6024):1565–70.
34. Buck MD, Sowell RT, Kaech SM, Pearce EL. Metabolic Instruction of Immunity. Cell. 2017;169(4):570–86. https://doi.org/10.1016/j.cell.2017.04.004.
35. Joyce JA, Fearon DT. T cell exclusion, immune privilege, and the tumor microenvironment. Science. 2015;348(6230):74–80. https://doi.org/10.1126/science.aaa6204.
36. Heath WR, Carbone FR. Cross-presentation in viral immunity and self-tolerance. Nat Rev Immunol. 2001;1(2):126–34. https://doi.org/10.1038/35100512.
37. DuPage M, Bluestone JA. Harnessing the plasticity of CD4(+) T cells to treat immune-mediated disease. Nat Rev Immunol. 2016;16(3):149–63. https://doi.org/10.1038/nri.2015.18.
38. Fridman WH, Zitvogel L, Sautes-Fridman C, Kroemer G. The immune contexture in cancer prognosis and treatment. Nat Rev Clin Oncol. 2017;14(12):717–34. https://doi.org/10.1038/nrclinonc.2017.101.
39. Zou W, Restifo NP. T(H)17 cells in tumour immunity and immunotherapy. Nat Rev Immunol. 2010;10(4):248–56. https://doi.org/10.1038/nri2742.
40. Vinuesa CG, Linterman MA, Yu D, MacLennan IC. Follicular helper T cells. Annu Rev Immunol. 2016;34:335–68. https://doi.org/10.1146/annurev-immunol-041015-055605.
41. Wilke CM, Wu K, Zhao E, Wang G, Zou W. Prognostic significance of regulatory T cells in tumor. Int J Cancer. 2010;127(4):748–58. https://doi.org/10.1002/ijc.25464.
42. Hunziker L, Klenerman P, Zinkernagel RM, Ehl S. Exhaustion of cytotoxic T cells during adoptive immunotherapy of virus carrier mice can be prevented by B cells or CD4+ T cells. Eur J Immunol. 2002;32(2):374–82. https://doi.org/10.1002/1521-4141(200202)32:2<374::AID-IMMU374>3.0.CO;2-9.
43. Kreiter S, Vormehr M, van de Roemer N, Diken M, Lower M, Diekmann J, et al. Mutant MHC class II epitopes drive therapeutic immune responses to cancer. Nature. 2015;520(7549):692–6. https://doi.org/10.1038/nature14426.
44. Knocke S, Fleischmann-Mundt B, Saborowski M, Manns MP, Kuhnel F, Wirth TC, et al. Tailored tumor immunogenicity reveals regulation of CD4 and CD8 T cell responses against cancer. Cell Rep. 2016;17(9):2234–46. https://doi.org/10.1016/j.celrep.2016.10.086.
45. Zamora AE, Crawford JC, Thomas PG. Hitting the target: how T cells detect and eliminate tumors. J Immunol. 2018;200(2):392–9. https://doi.org/10.4049/jimmunol.1701413.
46. Melief CJM. Cancer: precision T-cell therapy targets tumours. Nature. 2017;547(7662):165–7. https://doi.org/10.1038/nature23093.
47. Gabrilovich DI, Ostrand-Rosenberg S, Bronte V. Coordinated regulation of myeloid cells by tumours. Nat Rev Immunol. 2012;12(4):253–68.
48. Facciabene A, Motz GT, Coukos G. T-regulatory cells: key players in tumor immune escape and angiogenesis. Cancer Res. 2012;72(9):2162–71. https://doi.org/10.1158/0008-5472.CAN-11-3687. 72/9/2162 [pii].
49. Weigert A, Sekar D, Brüne B. Tumor-associated macrophages as targets for tumor immunotherapy. Immunotherapy. 2009;1(1):83–95.
50. Willems JJ, Arnold BP, Gregory CD. Sinister self-sacrifice: the contribution of apoptosis to malignancy. Front Immunol. 2014;5:299. https://doi.org/10.3389/fimmu.2014.00299.
51. Weigert A, Mora J, Sekar D, Syed S, Brune B. Killing is not enough: how apoptosis hijacks tumor-associated macrophages to promote cancer progression. Adv Exp Med Biol. 2016;930:205–39. https://doi.org/10.1007/978-3-319-39406-0_9.
52. Chen DS, Mellman I. Elements of cancer immunity and the cancer-immune set point. Nature. 2017;541(7637):321–30. https://doi.org/10.1038/nature21349.
53. Vivier E, Ugolini S, Blaise D, Chabannon C, Brossay L. Targeting natural killer cells and natural killer T cells in cancer. Nat Rev Immunol. 2012;12(4):239–52. https://doi.org/10.1038/nri3174.
54. Kantoff PW, Higano CS, Shore ND, Berger ER, Small EJ, Penson DF, et al. Sipuleucel-T immunotherapy

for castration-resistant prostate cancer. N Engl J Med. 2010;363(5):411–22. https://doi.org/10.1056/NEJMoa1001294.

55. Gomes-Silva D, Ramos CA. Cancer immunotherapy using CAR-T cells: from the research bench to the assembly line. Biotechnol J. 2018;13(2) https://doi.org/10.1002/biot.201700097.
56. Fournier C, Martin F, Zitvogel L, Kroemer G, Galluzzi L, Apetoh L. Trial watch: adoptively transferred cells for anticancer immunotherapy. Oncoimmunology. 2017;6(11):e1363139. https://doi.org/10.1080/2162402x.2017.1363139.
57. Schubert ML, Hoffmann JM, Dreger P, Muller-Tidow C, Schmitt M. Chimeric antigen receptor transduced T cells: tuning up for the next generation. Int J Cancer. 2018;142:1738. https://doi.org/10.1002/ijc.31147.
58. Rotte A, Jin JY, Lemaire V. Mechanistic overview of immune checkpoints to support the rational design of their combinations in cancer immunotherapy. Ann Oncol. 2018;29:71. https://doi.org/10.1093/annonc/mdx686.
59. Alsaab HO, Sau S, Alzhrani R, Tatiparti K, Bhise K, Kashaw SK, et al. PD-1 and PD-L1 checkpoint signaling inhibition for cancer immunotherapy: mechanism, combinations, and clinical outcome. Front Pharmacol. 2017;8:561. https://doi.org/10.3389/fphar.2017.00561.
60. Chang L, Chang M, Chang HM, Chang F. Microsatellite instability: a predictive biomarker for cancer immunotherapy. Appl Immunohistochem Mol Morphol. 2018;26:e15. https://doi.org/10.1097/pai.0000000000000575.
61. Colle R, Cohen R, Cochereau D, Duval A, Lascols O, Lopez-Trabada D, et al. Immunotherapy and patients treated for cancer with microsatellite instability. Bull Cancer. 2017;104(1):42–51. https://doi.org/10.1016/j.bulcan.2016.11.006.
62. Matson V, Fessler J, Bao R, Chongsuwat T, Zha Y, Alegre ML, et al. The commensal microbiome is associated with anti-PD-1 efficacy in metastatic melanoma patients. Science. 2018;359(6371):104–8. https://doi.org/10.1126/science.aao3290.
63. Sivan A, Corrales L, Hubert N, Williams JB, Aquino-Michaels K, Earley ZM, et al. Commensal Bifidobacterium promotes antitumor immunity and facilitates anti-PD-L1 efficacy. Science. 2015;350(6264):1084–9. https://doi.org/10.1126/science.aac4255.
64. Vetizou M, Pitt JM, Daillere R, Lepage P, Waldschmitt N, Flament C, et al. Anticancer immunotherapy by CTLA-4 blockade relies on the gut microbiota. Science. 2015;350(6264):1079–84. https://doi.org/10.1126/science.aad1329.
65. Pitt JM, Vétizou M, Daillère R, Roberti MP, Yamazaki T, Routy B, et al. Resistance mechanisms to immune-checkpoint blockade in cancer: tumor-intrinsic and -extrinsic factors. Immunity. 2016;44(6):1255–69. https://doi.org/10.1016/j.immuni.2016.06.001.
66. Chowdhury PS, Chamoto K, Honjo T. Combination therapy strategies for improving PD-1 blockade efficacy: a new era in cancer immunotherapy. J Intern Med. 2018;283:110. https://doi.org/10.1111/joim.12708.
67. Kepp O, Tesniere A, Schlemmer F, Michaud M, Senovilla L, Zitvogel L, et al. Immunogenic cell death modalities and their impact on cancer treatment. Apoptosis. 2009;14(4):364–75. https://doi.org/10.1007/s10495-008-0303-9.
68. de Miguel-Luken MJ, Mansinho A, Boni V, Calvo E. Immunotherapy-based combinations: current status and perspectives. Curr Opin Oncol. 2017;29(5):382–94. https://doi.org/10.1097/cco.0000000000000391.
69. Zhang X, Sharma PK, Peter Goedegebuure S, Gillanders WE. Personalized cancer vaccines: targeting the cancer mutanome. Vaccine. 2017;35(7):1094–100. https://doi.org/10.1016/j.vaccine.2016.05.073.
70. Braunlein E, Krackhardt AM. Identification and characterization of neoantigens as well as respective immune responses in cancer patients. Front Immunol. 2017;8:1702. https://doi.org/10.3389/fimmu.2017.01702.
71. Capietto AH, Jhunjhunwala S, Delamarre L. Characterizing neoantigens for personalized cancer immunotherapy. Curr Opin Immunol. 2017;46:58–65. https://doi.org/10.1016/j.coi.2017.04.007.

Neuroimmunoendocrinology 13

Natalia P. Rocha, Moises E. Bauer,
and Antonio L. Teixeira

13.1 Introduction

The NERVOUS, the IMMUNE, and the ENDOCRINE SYSTEMS were described as independent components of the body for decades. In this regard, these three systems were characterized as having highly specialized signaling molecules: NEUROTRANSMITTERS, HORMONES, and CYTOKINES, for, respectively, neural, endocrine, and immune communication [1]. This highly didactical but artificial classification has been challenged since the advent of studies investigating the relationship among these components.

Immune-endocrine interactions have long been recognized, especially due to the immunosuppressive effects of GLUCOCORTICOIDS (GCs) (see also Chap. 32). A range of evidence demonstrated that immune cell products can profoundly affect endocrine function, and, then, it was accepted that immune-endocrine communications are bidirectional. This exciting idea was mostly put forth by J. Edwin Blalock during the 1980s [2]. The third component of the triad has been added more recently, arising the study of neuro-immune-endocrine interactions, focused mainly on the investigation of STRESS effects on these three physiological systems. Accordingly, the term "NEUROIMMUNOENDOCRINOLOGY" refers to the study of the interactions among behavioral, neural, neuroendocrine, and immunological processes of adaptation [3]. These systems can communicate closely with each other by several mechanisms that work in parallel.

The early investigations on the brain-IMMUNE SYSTEM interaction were performed more than 90 years ago by Metal'nikoff and Chorine and were based on classical conditioning of immune responses. This research derived from the Pavlovian perspective on the conditioning of behavioral and physiological responses, i.e., a conditioned stimulus (e.g., heat or scratching) was repeatedly paired with injections of antigens. After repeated stimuli, the exposure to the conditioned stimulus alone was capable of inducing ANTIBODY production in addition to a variety of nonspecific defensive responses. Other early studies performed by Szentivanyi and Filipp and Szentivanyi and Szekely 60 years ago demonstrated that hypothalamic lesions can prevent anaphylactic shock in animals. During the 1960s, another pioneer in the development

Final manuscript submitted on December 7, 2016.

N. P. Rocha · A. L. Teixeira (✉)
Neuropsychiatry Program, Department of Psychiatry and Behavioral Sciences, McGovern Medical School, University of Texas Health Science Center at Houston, Houston, TX, USA

Interdisciplinary Laboratory of Medical Investigation, Medical School, Federal University of Minas Gerais, Belo Horizonte, Brazil

M. E. Bauer
Laboratory of Immunosenescence, Institute of Biomedical Research, Pontifical Catholic University of Rio Grande do Sul, Porto Alegre, Brazil
e-mail: mebauer@pucrs.br

M. J. Parnham et al. (eds.), *Nijkamp and Parnham's Principles of Immunopharmacology*,
https://doi.org/10.1007/978-3-030-10811-3_13

of brain-to-immune communication research, George Solomon observed that psychological well-being presented protective effect in the face of the probable genetic predisposition to autoimmune disease. A great advance was achieved in the 1970s when Hugo Besedovsky investigated the effects of immune responses on neural and endocrine functioning, starting the description of a neuroendocrine-immune pathway. In the 1980s, Blalock and Smith suggested the existence of molecular pathways capable of explaining how behavior could influence the IMMUNE SYSTEM. This hypothesis would be true if the brain and IMMUNE SYSTEM used the same chemical language and thereby communicate with each other [4]. Indeed, they showed that IMMUNE CELLS (i.e., LYMPHOCYTES) could be a source of pituitary HORMONES. In addition, immune-derived CYTOKINES could function as HORMONES and hypothalamic-releasing factors [5]. Blalock's research was dedicated to show that the immune and the neuroendocrine systems share the same biochemical pathways and may influence each other. The concept of "NEUROIMMUNOENDOCRINOLOGY" arose at that time [6].

In the 1990s, studies advanced the knowledge on the mechanisms by which immune activation leads to behavioral changes, especially depressive-like behavior in rodents as a result of LIPOPOLYSACCHARIDE (LPS) treatment [7]. LPS is a product of the outer membrane of Gram-negative bacteria, well-known to activate immune function and to induce the release of several CYTOKINES both in the periphery and the brain. In the next decade, clinical studies showed that the patients treated with interferon (IFN)-α in different contexts (e.g., hepatitis C, melanoma) presented either mood/cognitive or neurovegetative symptoms [8]. Since then, several studies linking immune changes to PSYCHIATRIC DISORDERS, especially SCHIZOPHRENIA, MOOD DISORDERS, and STRESS-RELATED DISORDERS, have been conducted. Recently, the term "IMMUNOPSYCHIATRY" has emerged, and some authors proposed the concept that strategies targeting the IMMUNE SYSTEM should be considered in the development of new treatments of PSYCHIATRIC DISORDERS [7].

In this chapter, we shortly present basic and clinical evidence of the crosstalk between immune, endocrine, and nervous systems.

13.2 Regulation of the Immune System by Neuroendocrine Hormones

13.2.1 Pleiotropic Effects of Glucocorticoids on the Immune System

GCs are major steroid HORMONES synthetized and secreted by the adrenal cortex in response to STRESS. Upon exposure to STRESS, the hypothalamus is stimulated to release corticotrophin-releasing hormone (CRH), which then acts on the anterior pituitary gland to stimulate the synthesis of adrenocorticotropic hormone (ACTH). ACTH then acts on the adrenal cortex to induce the secretion of GCs—this is known as the HYPOTHALAMIC-PITUITARY-ADRENAL (HPA) AXIS (Fig. 13.1). GCs have pleiotropic effects on the immune cells including anti-inflammatory, immunosuppressive, regulatory, and immune-enhancing effects [9]. These contrasting effects are due to tissue-related differences, including differential expression of glucocorticoid receptors (GRs), GC signal transduction events, and metabolizing enzymes that limit final hormonal concentrations. Low GC concentrations are immune enhancing, whereas very high ones are known to be immunosuppressive. The timing of GC actions with respect to an immune response is thought to be critical: actions that coincide or follow are thought to be anti-inflammatory, while those that precede are likely to be pro-inflammatory [10]. After binding on membrane-bound or intracellular receptors, the complex HORMONE-GR may migrate to the nucleus where it is capable of altering the expression of several transcription factors and CYTOKINE

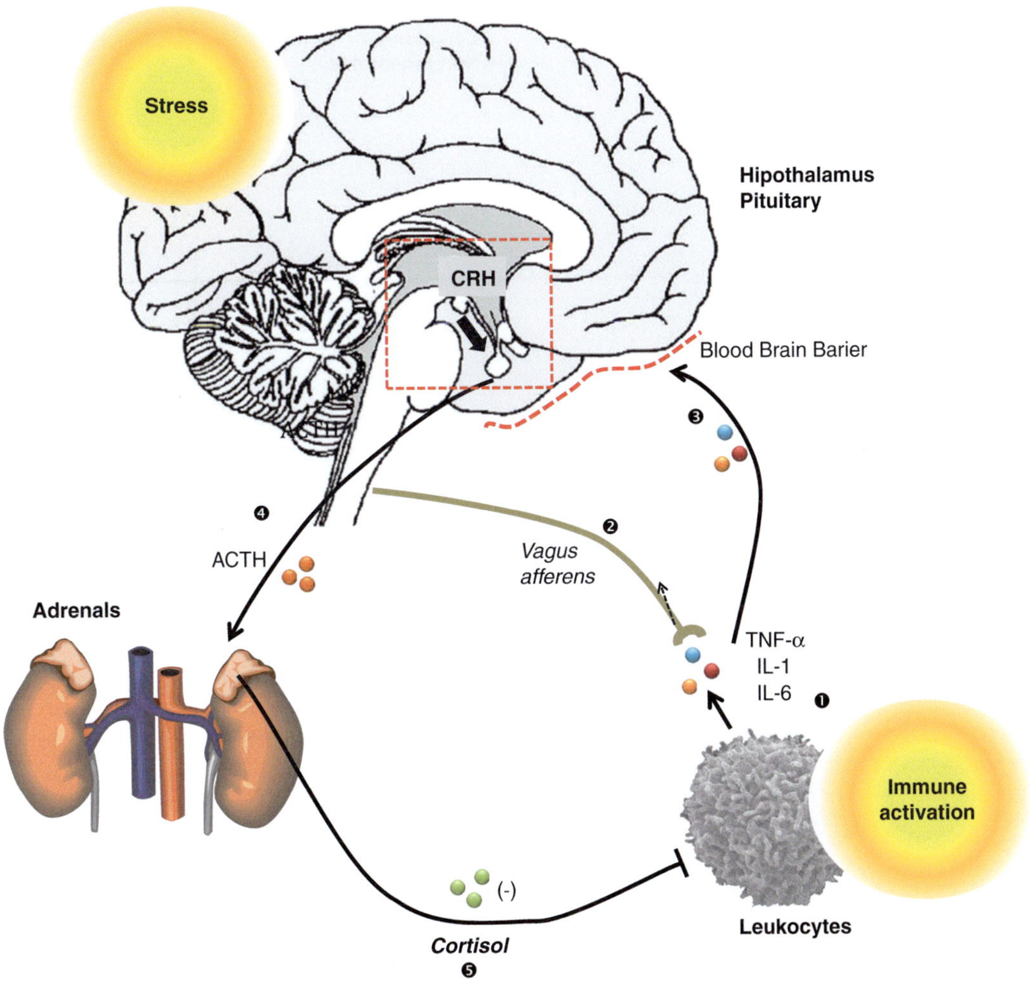

Fig. 13.1 The stress-related activation of the hypothalamic-pituitary-adrenal (HPA) axis. During stress, the hypothalamus is stimulated to release corticotrophin-releasing hormone (CRH), which then acts on the anterior pituitary gland to stimulate the synthesis of adrenocorticotropic hormone (ACTH). ACTH then acts on the adrenal cortex to induce the secretion of glucocorticoids. Glucocorticoids have pleiotropic effects on immune cells, including anti-inflammatory, immunosuppressive, regulatory, and immune-enhancing effects. These contrasting effects are due to tissue-related differences, including differential expression of glucocorticoid receptors (GRs), signal transduction events, and metabolizing enzymes that limit final hormonal concentrations

genes (Fig. 13.2). Among the transcription factors inhibited by glucocorticoids is the nuclear factor kappa-B (NF-κB), which is paramount in activating the expression of numerous pro-inflammatory CYTOKINES [e.g., tumor necrosis factor (TNF)-α, interleukin (IL)-1, and IL-6] (see Chaps. 6 and 9). However, the GC-related actions are not restricted to anti-inflammatory effects.

GCs are powerful modulators of CYTOKINES produced by CD4+ T cells, of keynote the Th1 and Th2 CYTOKINE profiles (Fig. 13.3). The Th1 CYTOKINE profile, characterized by IL-12, IFN-γ, and IL-2, may promote cellular immunity by activating MACROPHAGES, promoting LYMPHOCYTE proliferation as well as cytotoxic activity of NATURAL KILLER (NK) and CD8+ T CELLS. The Th2 CYTOKINE profile coordinates antibody responses and cellular differentiation of EOSINOPHILS and MAST CELLS and is characterized by the production of IL-4, IL-5, IL-10, and IL-13. GCs inhibit the

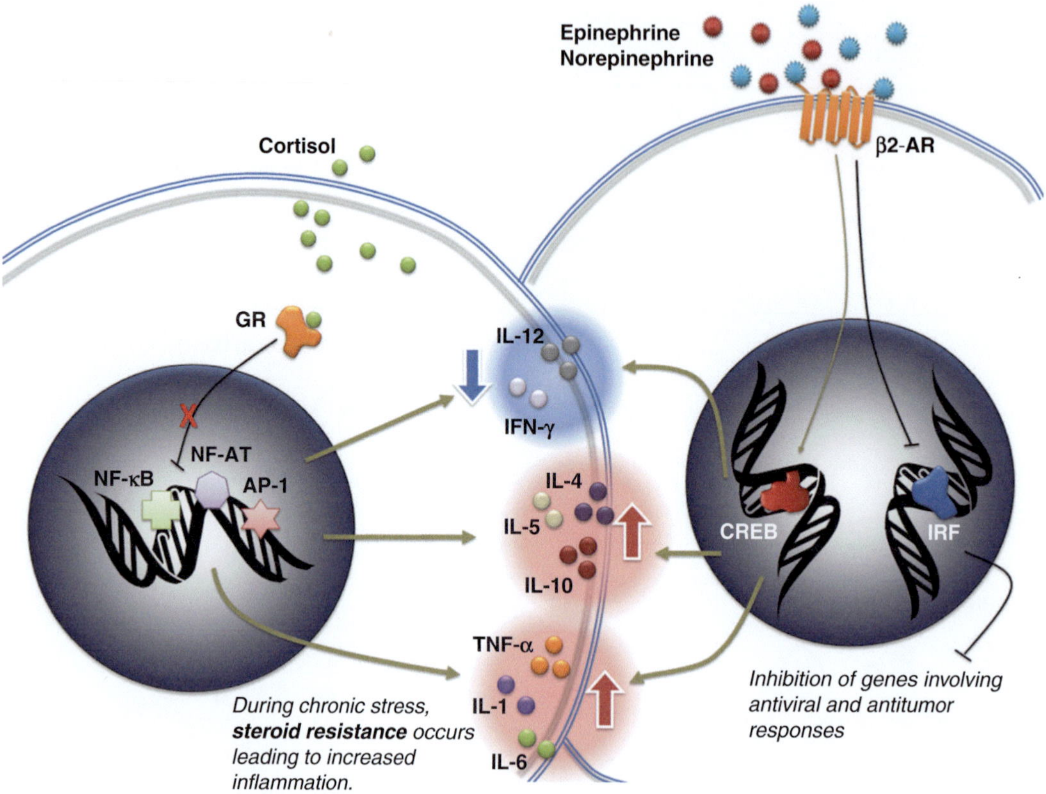

Fig. 13.2 The immunoregulatory actions of cortisol and catecholamines. After binding on membrane-bound or intracellular receptors, the complex hormone-glucocorticoid receptor (GR) migrates to the nucleus where it is capable of altering the expression of several transcription factors and cytokine genes. Among the transcription factors inhibited by glucocorticoids is the nuclear factor (NF)-κB, which is paramount in activating the expression of numerous pro-inflammatory cytokines (e.g., TNF-α, IL-1, and IL-6). However, during chronic stress, steroid resistance occurs, leading to increased peripheral inflammation. Glucocorticoids are also powerful modulators of cytokines produced by CD4+ T cells, inhibiting the production of Th1 cytokines (IL-12 and IFN-γ) and promoting the secretion of Th2 cytokines (IL-4, IL-5, and IL-10). Catecholamines (epinephrine and norepinephrine) mediate its immunosuppressive actions by stimulating the transcription of Th2-type cytokines while suppressing the Th1-type cytokine profile. β-adrenergic signaling may also impair innate immune responses, which involves suppression of type I IFN-mediated antiviral responses and upregulation of pro-inflammatory cytokine genes (*TNF*, *IL-1B*, and *IL-6*). These changes are mediated via activation of key transcription factors such as IRF (interferon regulatory factor) and CREB (cAMP-response element-binding protein), respectively

transcription of Th1-related CYTOKINES and upregulate the transcription of Th2 CYTOKINES [11]—this occurs by inhibiting IL-12 and stimulating IL-4 and IL-10 secretion, respectively [12, 13] (Fig. 13.3). The GC-related IL-12 inhibition occurs via signal transducer and activator of transcription (STAT)4, an essential transcription factor in Th1 cell differentiation, whereas the GC-induced secretion of IL-10 by B cells is mediated by STAT3 transcription factor [14]. GCs are capable of inhibiting the expression of T-bet, an essential factor for Th1 differentiation through a trans-suppression mechanism [15]. Glucocorticoids may also affect the survival of Th1/Th2 cells. For instance, the Th2 cells are resistant to apoptosis induced by dexamethasone, due to increased B-cell lymphoma 2 (Bcl-2) proto-oncogene [16].

GCs also modulate the differentiation of Th17 cells. The Th17 cells secrete CYTOKINES such as IL-17, IL-9, IL-21, and IL-23 and are importantly involved in promoting

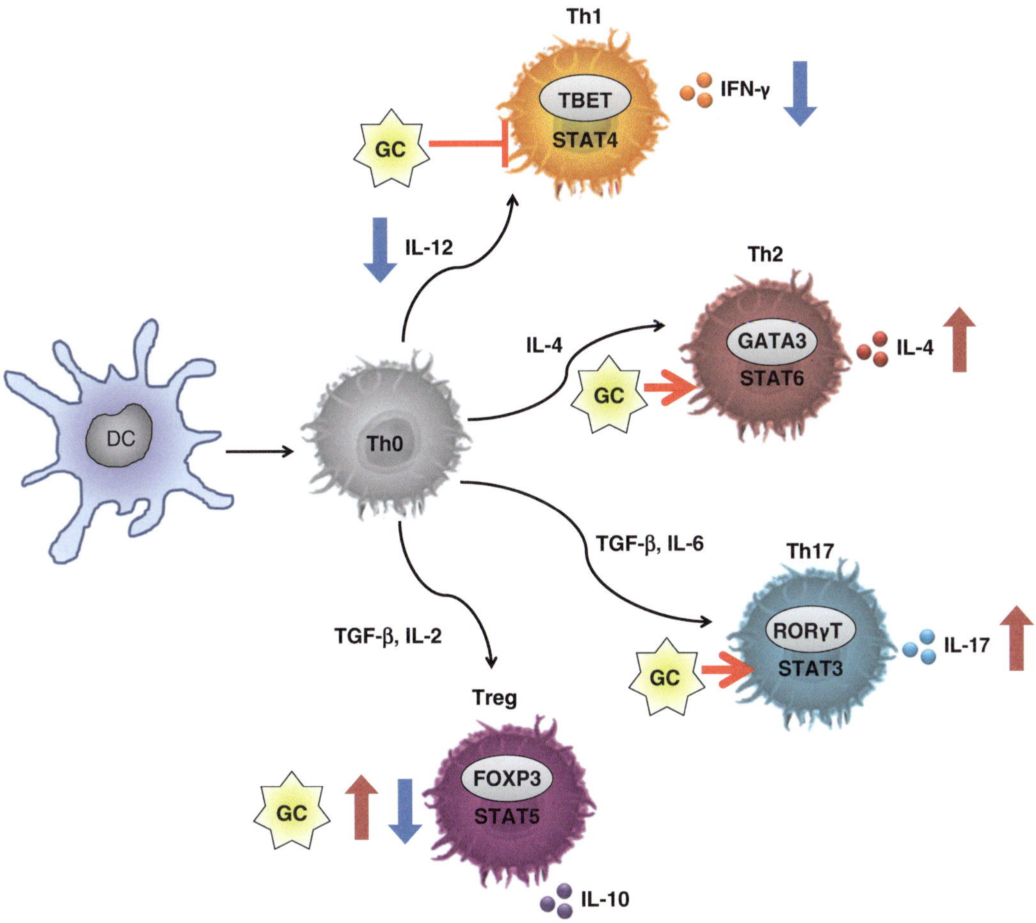

Fig. 13.3 Glucocorticoids (GC) are powerful regulators of T-cell differentiation programs. After antigen presentation provided by dendritic cells (DC) and elicited key cytokines, undifferentiated T cells (Th0) are capable of differentiating into canonical effector T cells, namely, Th1, Th2, Th17, and regulatory T cells (Treg). GCs impair the T helper 1 (Th1) differentiation and expansion via inhibition of T-bet and STAT4, whereas GC-induced Th2 polarization occurs via upregulation of GATA3 and STAT6 transcription factors. Under some pathological conditions, GCs may increase IL-17 secretion via STAT3 upregulation. Treg cells (CD4+CD25+FOXP3+) also exhibit altered differentiation in the presence of GCs. Increased levels of GCs, as observed during acute stress, may reduce the frequency of Treg cells in peripheral blood. However, this effect is reversed during chronic exposure to high levels of GCs, increasing the number of circulating Treg cells and promoting the immunosuppressive actions of GCs. In these conditions, GCs have been shown to increase FoxP3 expression—a key transcription factor involved in Treg differentiation

INFLAMMATION [17] (see Chap. 3). This LYMPHOCYTE subset is highly associated with development of autoimmune diseases. A previous study has shown that in patients with systemic lupus erythematosus, an autoimmune disease treated with synthetic GCs, the steroid treatment increased the levels of STAT3 and associated transcription of CYTOKINES that mediate the Th17 response. These results demonstrate that GCs act synergistically with certain transcription factors, modulating T-cell populations involved in inflammatory responses [18].

Regulatory T cells (CD4+CD25+FOXP3+; or Treg) also exhibit altered differentiation in the presence of GCs. Treg cells play pivotal roles in promoting peripheral immunological tolerance by engaging inhibitory receptors with effector (activated) T cells and by secreting

TGF-β and IL-10. Therefore, Treg cells are important for the control of inflammatory responses. Low number of regulatory T cells characterizes autoimmune diseases. Increased levels of natural GCs, as observed during acute STRESS, may also reduce the frequency of Treg cells in peripheral blood [19, 20]. This phenomenon may occur due to an adaptation of the body to respond more effectively to an acute STRESS mediated by a possible injury or tissue damage. However, this effect is reversed during chronic STRESS (high levels of GCs), increasing the number of circulating Treg cells in the blood and promoting the immunosuppressive actions of GCs. GCs may thus play their immunomodulatory action during STRESS by affecting the development of Treg cells.

Another well-known action of GCs is the powerful effect in redirecting LEUKOCYTES between tissues, known as cellular trafficking. Migration and distribution of LYMPHOCYTES in the body are extremely important to the development of effective immune responses. In general, GCs reduce LEUKOCYTE trafficking to INFLAMMATION. The GCs induce a rapid and significant increase in the number of neutrophils and NK CELLS in the circulation and significant reduction in LYMPHOCYTE counts (lymphopenia). These effects are due to the reduced expression of cellular adhesion molecules on LEUKOCYTES [e.g., intercellular adhesion molecule (ICAM)-1, vascular cell adhesion protein (VCAM)-1, endothelial-LEUKOCYTE adhesion molecule (ELAM)-1] and suppression of chemokines [regulated on activation, normal T cell expressed and secreted (RANTES), MONOCYTE chemoattractant protein (MCP)-1] necessary for cellular homing [21]. These changes are mediated by repression of NF-κB. GCs may also induce neutrophilia by inducing the release of bone marrow-derived neutrophils in the bloodstream [22]. The GC-related changes in cellular trafficking are temporary and occur with different intensities according to LYMPHOCYTE subsets, for instance, T more than B CELLS and CD4+ T cells more than CD8+ T cells and NK CELLS [10, 23].

13.2.1.1 Circadian Variation of Adrenal Hormones

HORMONES play an important influence on redistribution of LYMPHOCYTES in the body. CORTISOL levels are lower in the night (nadir at 23 h), whereas higher levels are observed in the early morning, preparing the body for energy demands during the period of greatest activity. B, T, and NK CELLS present circadian variations inversely correlated with plasma CORTISOL concentrations: lower blood counts in the morning when CORTISOL is high and higher counts at night when CORTISOL is decreased [24, 25]. CD4+ T cells or CD8+ naive (CD45RA+CD62L+), memory (CD45RA−CD62L+), and effector memory (CD45RA−CD62L−) T cells had enumerative variations inversely related to plasma concentrations of CORTISOL with higher counts during the night. These changes were associated with increased expression of CXCR4, a cell adhesion molecule responsible for cellular homing to the bone marrow. However, the fluctuation of CD8+ effector T cells (CD45RA+CD62L−) during the day was correlated with plasma epinephrine levels [26]. Adrenalectomized animals have no rhythmic variations in the number of peripheral LYMPHOCYTES [25].

Finally, GCs may also have pro-inflammatory actions [9]. Persistent elevations of GCs experienced during the lifespan, as seen during chronic STRESS and MOOD DISORDERS, may be prone to a systemic pro-inflammatory status. How does classic anti-inflammatory HORMONES promote this chronic low-grade INFLAMMATION? First, persistent increased GC levels would lead to increased abdominal fat (as seen during aging or GC treatment) and development of METABOLIC SYNDROME. Adipocytes and infiltrating MACROPHAGES secrete various adipokines (e.g., leptin, TNF-α, IL-6, IL-18) that reach the circulation and contribute to low-grade INFLAMMATION [27]. Secondly, chronic STRESS and persistent increased GC levels would promote immune cells more resistant to steroids, and, with less intracellular GC signaling, cells assume a pro-inflammatory phenotype [28]. Hypercortisolemia and GC resistance are common biological abnormalities described in patients with MAJOR DEPRESSION, and they have been associated with increased levels of pro-inflammatory markers [29–31]. The mechanisms underlying GC resistance include impaired GR expression and function, secondary to decreased GR translocation, decreased GR protein–protein interaction, or increased expression of the inert GR isoform, GRβ [32]. Lastly, GC may also increase the expression of innate-related genes [e.g., Toll-like receptors and nucleotide-binding oligomerization domain (NOD)-like receptors] and the purinergic receptor (P2Y2R) and potentiate TNF-α-regulated pro-inflammatory genes [9].

13.2.2 The Catecholaminergic Pathway

Another STRESS-related neuroimmunoendocrine regulation occurs through the release of CATECHOLAMINES [i.e., EPINEPHRINE, NOREPINEPHRINE (NE), and DOPAMINE (DA)] from the adrenal medulla or nerve endings of the autonomic nervous system. It has been shown that major lymphoid organs, such as the thymus, spleen, and lymph nodes, are richly innervated by autonomic nerve fibers [33] and all LYMPHOCYTES express β2-adrenergic receptors [34] (Fig. 13.4). It has been shown that these fibers make synapses with immune cells in secondary lymphoid organs [35]. During acute STRESS (psychological or physical nature), for example, there is a splenic contraction elicited by sympathetic innervated fibers which results in the

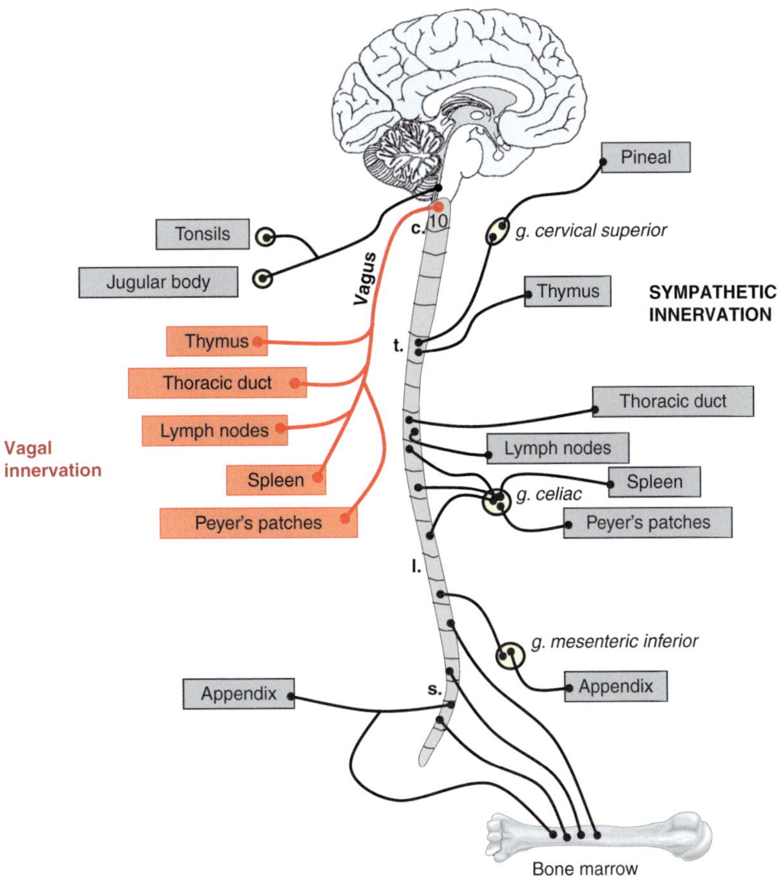

Fig. 13.4 The lymphoid organs are richly innervated by autonomic nerve fibers. Stress-related activation of sympathetic innervation leads to local release of norepinephrine (NE), with key regulatory actions including immunosuppression and pro-inflammatory responses. In secondary organs, these fibers make synapses with immune cells that express β2-adrenergic receptors. Most lymphoid organs are also innervated by the vagus nerve, providing a key efferent control of inflammation through secretion of acetylcholine (Ach)

expulsion of a large number of LYMPHOCYTES into the blood. During chronic STRESS, the production of NE in the bone marrow elicited the release of myeloid cells, including MONOCYTES and NEUTROPHILS [36, 37]. In this context, the phenotype of peripheral MONOCYTES seems to be less mature and more inflammatory [38]. Interestingly, the STRESS-induced trafficking of these "inflammatory MONOCYTES" to the brain may exacerbate both mental and physical conditions [37, 39–41].

Similarly to GC-mediated actions, the CATECHOLAMINES exert differential effects on LEUKOCYTES. For instance, the level of expression of the β2-adrenergic receptors varies between the LEUKOCYTE subsets: for example, B cells present more β2-adrenergic receptors than CD4+ T cells [34]. An activation of the SYMPATHETIC NERVOUS SYSTEM (SNS) via adrenaline infusion causes changes in LEUKOCYTE trafficking including significant increase in T-cell subsets (CD4+ and CD8+) and NK CELLS [42]. CATECHOLAMINES can regulate several immune functions, including cell proliferation, cytotoxic activity, CYTOKINE and antibody secretion, as well as chemotaxis [43]. It has been shown that NE may inhibit cytotoxic activity and increase LYMPHOCYTE proliferation [44]. NE mediates its immunosuppressive actions by stimulating the transcription of Th2-type CYTOKINES (IL-4, IL-5, and IL-10) while suppressing the Th1-type CYTOKINE profile (IFN-γ and IL-12) [45, 46]. The SNS may also participate of the homeostatic regulation of Treg activity and number. Removal of murine peripheral sympathetic innervation by 6-hydroxydopamine increased peripheral Tregs (CD4+FoxP3+) by a TGF-beta-dependent mechanism, as well as inhibited the induction of experimental autoimmune encephalomyelitis [47]. β-adrenergic signaling may also impair innate immune responses, which involves suppression of type I IFN-mediated antiviral responses [48] and upregulation of pro-inflammatory CYTOKINE genes (*TNF*, *IL-1B*, and *IL-6*) [49] (Fig. 13.2).

DA can inhibit the in vitro LYMPHOCYTE proliferation and cytotoxicity of human CD4+ and CD8+ T cells [50]. The DA-induced inhibition of NK CELL activity occurs via upregulation of D5 receptor [51]. In contrast, DA has been found to have anticancer activities in vivo, and this effect is mediated via inhibition of Gr-1+CD115+ myeloid-derived suppressor cells (MDSCs) through D1-like receptors [52]. DA and other D1 receptor agonists inhibited IFN-γ-induced NO production by MDSCs from tumor-bearing mice and cancer patients. It is noteworthy that the sympathetic regulation does not seem to be as relevant during chronic STRESS or depression, when only the GC levels are kept continuously elevated.

13.2.3 The Autonomic Nervous System: Vagus Nerve

INFLAMMATION can be controlled by two homeostatic mechanisms: immune-mediated mechanisms and brain-derived regulatory pathways (unconscious). Brain immunoregulatory actions can be mediated by the autonomic nervous system, through sympathetic and vagus nerve innervation. Most if not all lymphoid organs are innervated by vagus nerve (i.e., tenth cranial nerve, CN X). There are actually two ways of vagal bidirectional communication (the vagal reflex): an efferent pathway, mediated through the release of NEUROTRANSMITTERS by the vagus nerve on LEUKOCYTES, and afferent pathway modulated by the action of pro-inflammatory CYTOKINES on the vagus nerve. ACETYLCHOLINE (ACh) is the major efferent vagal NEUROTRANSMITTER that induces splenic nerve activation with increased production of NE in the spleen. The NE, in turn, regulates the secretion of ACh by T cells to refrain inflammatory responses [53] (Fig. 13.5). In nude mice, which lack T cells, vagal nerve stimulation fails to suppress TNF-α levels during septic shock. However, transfer of ACh-producing T cells, which repopulate the spleen in nude mice, restores the integrity of the neural circuit [35]. The ACh interacts with nicotinic ACh receptor (α7) expressed on various LEUKOCYTES, and intracellular cascades of such regulation have already been elucidated.

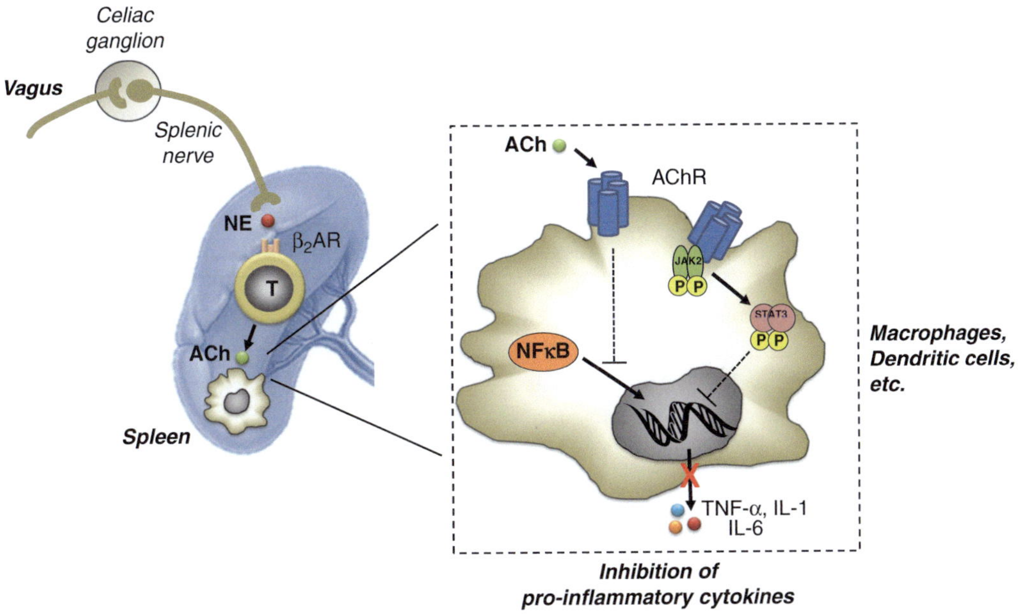

Fig. 13.5 Molecular mechanisms underlying the cholinergic control of inflammation. The efferent vagus stimulation results in direct acetylcholine (ACh) release from nerve endings in various organs (e.g., liver, intestine), yielding anti-inflammatory actions. In addition, the efferent vagus nerve may also stimulate noradrenergic-mediated activation of splenic T cells, which in turn secrete acetylcholine with important anti-inflammatory actions. The cholinergic control of inflammation initiates through binding of ACh in nicotinic acetylcholine receptors (α7nAChR) expressed by leukocytes. The downstream molecular pathways are known and include the inhibition of NF-κB nuclear translocation and activation of a JAK2–STAT3-mediated signaling cascade. Adapted from Pavlov & Tracey, 2012 [54]

The efferent vagal pathway acts, thus inhibiting inflammatory responses. In the animal model, it was also found that vagotomy increases the number of immature LYMPHOCYTES (CD4−CD8− or CD4+CD8+) in the spleen and lymph nodes [55]. The vagal afferent pathway of communication includes the binding of pro-inflammatory CYTOKINES on the vagus nerve, providing a representation (unconscious) in the CENTRAL NERVOUS SYSTEM (CNS) of a peripheral inflammatory signal [56]. In this sense, the IMMUNE SYSTEM would act as a "sensory organ." The two-way bidirectional property of the vagus nerve was defined as part of the inflammatory reflex: pro-inflammatory CYTOKINES signal INFLAMMATION in the CNS via vagal afferent, which induces the release of ACh, reducing peripheral inflammatory responses. Therefore, cholinergic neuron role in the inhibition of acute INFLAMMATION constitutes a "hardwire" neural mechanism of modulation of the immune response.

13.2.4 The Peptidergic Pathway: Neuropeptides

NEUROPEPTIDES play also a key role in regulating the IMMUNE SYSTEM. The activation of peripheral pain fibers leads to release of neuropeptides, namely, calcitonin gene-related peptide (CGRP), substance P (SP), adrenomedullin, neurokinins A and B, vasoactive intestinal peptide (VIP), neuropeptide Y (NPY), and gastrin-releasing peptide (GRP). They are released upon nociceptive stimulation by pain, mechanical, and chemical irritants to mediate skin responses to infection, injury, as well as wound healing. These NEUROPEPTIDES may act locally to recruit and regulate both innate and adaptive immune cells [43]. The immunological activity of these NEUROPEPTIDES is mediated through specific receptors expressed by immune cells. In addition, both activated innate cells and LYMPHOCYTES may also secrete NEUROPEPTIDES with known immunoregulatory actions.

Several NEUROPEPTIDES have anti-inflammatory actions. VIP, one of the best studied neuropeptides, affects both innate and adaptive immune responses and acts as major anti-inflammatory molecule [57]. VIP-ergic nerve fibers have been identified in both primary and secondary lymphoid organs. In activated MACROPHAGES, VIP inhibits the expression of pro-inflammatory CYTOKINES and chemokines [58, 59], sustaining the differentiation of CD4+ T cells in Th2 cells and promoting their proliferation and/or survival [57, 58]. VIP may also induce tolerogenic dendritic cells as well as CD4+ and CD8+ regulatory T cells with proved therapeutic effects on experimental autoimmune models, including the prevention or attenuation of arthritis, colitis, and experimental autoimmune encephalomyelitis [60–65]. VIP is also produced by trophoblast cells and modulates the maternal immune response toward a tolerogenic profile, via recruitment of induced regulatory T cells (iTregs), playing an active role in the decidualization process [66]. Several functions of the cellular IMMUNE SYSTEM are regulated by NPY, SP, and related-agouti protein (AgRP) [43]. NPY is a neuropeptide that increases food intake and storage of energy as fat, but it is also able to modulate LYMPHOCYTE proliferation, NK CELL activity, as well as IL-2 and TNF-α secretion [43]. In addition, AgRP is co-expressed with NPY and works by increasing appetite and decreasing metabolism and energy expenditure. Hypothalamic AgRP neurons are mandatory for feeding and survival [67] and are involved in the regulation of adaptive immune responses. Knockdown of Sirt1 in AgRP neurons induce a pro-inflammatory state characterized by a decrease in Treg functions in parallel to an increased effector T-cell activity, which determines an increased autoimmune disease susceptibility [68]. Sympathetic blockade in vivo was able to reverse the impaired suppressive function of Tregs, indicating that the SNS plays a pivotal role in mediating the effect of impaired AgRP neuronal activity on the IMMUNE SYSTEM. Substance P is involved in sensory and nociceptive pathways. In the periphery, substance P has been identified in C-type sensory nerve endings and autonomic afferents throughout the body. Substance P is present at sites of INFLAMMATION, and INFLAMMATION can enhance its expression [69]. During INFLAMMATION, SP stimulates LYMPHOCYTE migration, proliferation, and IgA secretion and promotes phagocytosis and chemotaxis in innate immune cells [70].

13.3 Regulation of Nervous System by Immune Factors

The IMMUNE SYSTEM was considered for many years as a system that protects us against pathogens. In this point of view, the IMMUNE SYSTEM was important only in pathogenic (or harmful) conditions. Recently this view has changed, and now the IMMUNE SYSTEM is regarded as an essential player in body homeostasis. For instance, a range of evidence pointed to the role played by the immune cellular and molecular processes in normal brain development. Nearly half of all the originally generated neurons are eliminated throughout CNS development. MICROGLIA, the CNS-resident immune cells, are key players in the process of eliminating excess neurons during development, as they are essential for programmed cell death and phagocytizing dead or dying neurons and associated debris. In contrast to their role in promoting cell death, MICROGLIA are also able to promote neuronal proliferation and survival [71].

The immune and nervous systems are intimately connected, and each can directly influence the other. Many lines of research have recognized the pathways by which the IMMUNE SYSTEM and the CNS communicate. As previously described, the CNS signals the IMMUNE SYSTEM using hormonal routes, including the HPA axis and STRESS-response HORMONES. In addition to hormonal routes, neural pathways can be used, mainly through the catecholaminergic pathway. In turn, CYTOKINES and immune cells are major effectors of the immune-to-brain communication, established by three independent pathways: the humoral, neural, and cellular routes (Fig. 13.6). The trafficking of LYMPHOID

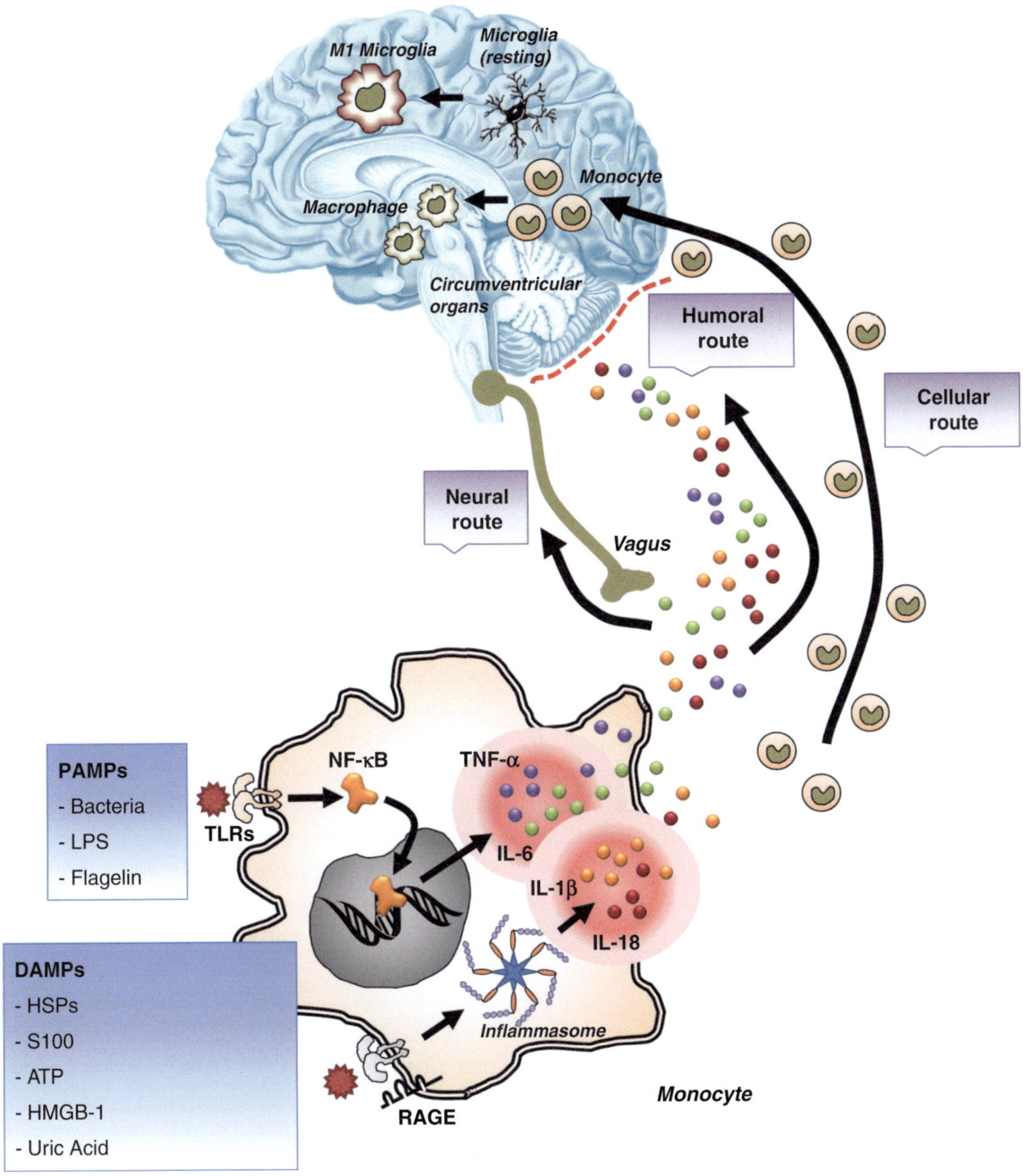

Fig. 13.6 Immune-to-brain communication. Three independent pathways establish the immune-to-brain communication: the humoral, neural, and cellular routes. Psychosocial stress may induce dysbiosis (leaky gut), providing key signaling to peripheral monocytes including pathogen-associated molecular patterns (PAMPs) or damage-associated molecular patterns (DAMPs). Both PAMPs and DAMPs (sterile injury) engage inflammatory signaling pathways such as nuclear factor-κB (NF-κB) and inflammasome. Pro-inflammatory cytokines (TNF-α, IL-1β, IL-6, and IL-18) are then promptly secreted and enter the circulation. Circulating cytokines can reach the brain (1) through active transport into the brain, crossing the brain-blood barrier (BBB) through leaky areas in the circumventricular organs (humoral route), or (2) through the activation of neural pathways such as the vagus nerve. Trafficking of peripheral leukocytes (i.e. monocytes, T cells, neutrophils) to brain constitutes the cellular route. Psychosocial stress can also lead to the activation of microglia to a M1 pro-inflammatory phenotype, which in turn attracts activated monocytes to the brain via a cellular route. Leukocytes are present in small numbers in brain circumventricular organs and choroid plexus. Under healthy conditions, these peripheral immune cells support neuronal function and scan the brain for pathogens or tissue damage. Adapted from Miller & Raison 2016 [72]

CELLS to the brain, including the meningeal space, has become an area of special interest with the recent description of a brain lymphatic system that heretofore had gone unrecognized [73].

Peripheral blood CYTOKINES can reach the brain through different pathways: (1) they can be actively transported into the brain, crossing the brain-blood barrier (BBB) through leaky areas in the circumventricular organs (humoral route) or (2) through the activation of neural pathways such as the vagal nerve [74].

CYTOKINES can affect brain functions, resulting in a constellation of physiological, behavioral, and hormonal changes. They can be produced locally by microglial cells and perivascular and meningeal MACROPHAGES [75].

13.3.1 Cytokines and Sickness Behavior

Following an immune challenge, the vertebrate organism will promptly start a set of changes in order to enhance host survival. This sickness response includes physiological (e.g., fever, alterations in blood composition), behavioral (e.g., decreased locomotion and food and water intake), and hormonal (e.g., release of HORMONE from the HPA axis) changes. This sickness response occurs due to immune-to-brain communication [75].

The immune-to-vagus-to-brain pathway seems to be involved in triggering rapid brain-mediated sickness responses. After being activated due to an infection or injury, immune cells produce and release pro-inflammatory CYTOKINES (especially TNF and IL-1). CYTOKINES are capable of either directly or indirectly exciting sensory nerves in the vagus. Accordingly, a rage of sickness responses are blocked by cutting the vagus nerve, including fever, decreased food-motivated behavior, increased sleep, decreased activity, decreased social interaction, changes in brain activity, and release of STRESS HORMONES [76].

The wording "SICKNESS BEHAVIOR" refers to a series of aforementioned CYTOKINE-induced symptoms. As explained, "SICKNESS BEHAVIOR" usually takes place in the course of an infection, in which patients present with lethargy, psychomotor slowness, depressive behavior, and appetite and sleep alterations. Nowadays, the term "SICKNESS BEHAVIOR" is also employed in the context of an existing INFLAMMATION-related profile observed in NEUROPSYCHIATRIC DISORDERS, mainly MAJOR DEPRESSION, but also ANXIETY and BIPOLAR DISORDERS, OBSESSIVE-COMPULSIVE DISORDER, and SCHIZOPHRENIA [77].

Some of the mechanisms that might be responsible for INFLAMMATION-mediated sickness and depressive behaviors have been elucidated. Peripheral CYTOKINES can communicate with the brain resulting in "SICKNESS BEHAVIOR" by several means that act in parallel: (1) CYTOKINES produced locally in infected tissues activate primary afferent nerves (vagal nerves during abdominal and visceral infections and the trigeminal nerves during oro-lingual infections); (2) MACROPHAGES residing in the circumventricular organs and the choroid plexus can produce pro-inflammatory CYTOKINES that enter the brain by volume diffusion; (3) overflowing circulating CYTOKINES can access the brain through saturable transporters at the BBB; and (4) IL-1 receptors located on perivascular MACROPHAGES and endothelial cells of brain venules, once activated by circulating CYTOKINES, result in the local production of prostaglandin E2 [78]. It has been demonstrated that "SICKNESS BEHAVIOR" can be reliably reproduced by administration of pro-inflammatory CYTOKINES or by treatment with CYTOKINE-production inducers, such as endotoxin or LPS.

IL-1β and TNF-α are the main CYTOKINES involved in SICKNESS BEHAVIOR. In rodents, both systemic and intracerebral administration of IL-1β or TNF-α resulted in signs of SICKNESS BEHAVIOR (i.e., decreased motor activity, social withdrawal, reduced food and water intake, and altered cognition) in a time- and dose-dependent manner. The same effects were not observed when IL-6 was injected either peripherally or inside the brain. However, IL-6 administration resulted in fever response. In addition, IL-6-deficient mice displayed less pronounced signs of LPS-induced SICKNESS BEHAVIOR in comparison with wild-type animals. Thus, brain IL-6 might contribute to the expression of other CYTOKINES (including IL-1β and TNF-α) after immune challenges [78]. The nuclear factor NF-κB, an intracellular signaling molecule, was described as an essential mediator of the periphery-to-brain communication. In rodents, central blockade of NF-κB inhibited IL-1β- and LPS-induced behavioral changes [79].

The balance between pro- and anti-inflammatory CYTOKINES is important for the control of SICKNESS BEHAVIOR responses. Intracerebroventricular injection of IL-10 or insulin-like growth factor I (IGF-I) blunted the signs of LPS-induced SICKNESS BEHAVIOR. Accordingly, IL-10-deficient mice presented exaggerated SICKNESS BEHAVIOR signs after peripheral LPS administration [78].

Clinical data corroborated the findings obtained from animal studies. In humans, intravenous injection of *Salmonella abortus equi* endotoxin resulted in a transient increase in the levels of ANXIETY and depressed mood and decrease in verbal and nonverbal memory functions. The administration of this endotoxin was also followed by an increase in circulating levels of TNF-α, soluble TNF receptors, IL-6, and CORTISOL [80]. Exposure to CYTOKINES was associated with SICKNESS BEHAVIOR in humans. For instance, up to 50% of patients receiving chronic IFN-α therapy for the treatment of infectious diseases or cancer developed clinically significant DEPRESSION. IFN-α-induced depressive syndrome is responsive to antidepressant drugs and clinically overlap idiopathic MAJOR DEPRESSION. Moreover, peripheral CYTOKINE administration results in the amplification of CNS inflammatory responses. Patients with hepatitis C receiving peripheral administration of IFN-α presented greater concentrations of IFN-α in the cerebrospinal fluid (CSF) which correlated with increased CSF levels of IL-6 and the chemokine MCP-1 [79].

Early-life trauma individuals, considered at high risk of developing DEPRESSION and other psychiatric disorders, showed increased inflammatory responses to laboratory-induced psychological STRESS compared with low-risk individuals [81]. It is worth emphasizing here that in this paradigm, the IMMUNE SYSTEM is capable of responding to a non-pathogen threat, i.e., a potential psychological harm. Subsequent experiments demonstrated that the failure to maintain a positive outlook under STRESS was associated with an increase in IL-1β reactivity, which in turn, prospectively predicted future increase in depressive symptoms [82].

Recently, the "pathogen host defense hypothesis of DEPRESSION" was postulated. According to this hypothesis, modern humans have inherited a genomic bias toward INFLAMMATION and the associated set of depressive symptoms as this response pattern enhanced host survival and increased the likelihood of reproduction in the highly dangerous and pathogenic environment where humans evolved [31].

13.3.2 Mechanisms Underlying Cytokine Effects on the Brain: The Influence on Neurotransmitter Synthesis and Reuptake

The depletion of serotonin has been proposed as an important mechanism by which CYTOKINES may influence behavior. Evidence from animal studies and clinical trials showed that the increase in synaptic availability of serotonin significantly reduced the development of CYTOKINE-induced depressive symptoms. Corroborating these findings, polymorphisms in the IL-6 and serotonin transporter genes predicted DEPRESSION during IFN-α treatment for hepatitis C [83].

Research investigating the effects of IFN-α on indolamine 2,3 dioxygenase (IDO) provided further evidence that the serotonin pathway is involved in IFN-α-induced DEPRESSION. IDO is an enzyme capable of breaking down tryptophan into kynurenine. Tryptophan is an essential amino acid and the primary precursor of serotonin. Usually, only a small portion of tryptophan is used for the synthesis of serotonin. The great majority of dietary tryptophan (>95%) is degraded in the liver through the kynurenine pathway by tryptophan dioxygenase (TDO). Tryptophan degradation into kynurenine can also occur extrahepatically by the enzyme IDO, but this pathway is not significant under physiological conditions. IDO is expressed in the brain, and is highly inducible by pro-inflammatory CYTOKINES, including IFN-γ and TNF-α. Under inflammatory conditions, tryptophan availability for serotonin synthesis decreases, while kynurenine levels increase due to an enhanced IDO activity [78]. In addition, kynurenine easily crosses the BBB and enters the brain, where it will be metabolized by two compartmentalized pathways: (1) MICROGLIA favorably generate 3-hydroxykynurenine (3-HK) and quinolinic acid (QA), and (2) astrocytes preferentially produce kynurenic acid (KA). 3-HK is an oxidative stressor, whereas QA is an *N*-methyl-D-aspartate (NMDA) receptor agonist. On the other hand, KA is an NMDA receptor antagonist that has been speculated to be neuroprotective [78]. In addition to pro-inflammatory CYTOKINES, IDO can be activated by multiple inflammatory signaling pathways, including STAT1a, interferon regulatory factor (IRF)-1, NF-κB, and p38 mitogen-activated protein kinase (MAPK) [79].

CYTOKINES can also interfere the synthesis of dopamine. Intramuscular injection of IFN-α in rats resulted in a significant decrease in the levels of tetrahydrobiopterin (BH_4) and dopamine in the amygdala and raphe areas. BH_4 is an important cofactor for tyrosine hydroxylase, the rate-limiting enzyme in dopamine synthesis [79].

Lastly, CYTOKINES and their associated signaling pathways might alter monoamine reuptake. MAPKs, including p38 and extracellular signal-regulated kinases (ERK) 1/2, were able to increase the activity of membrane transporters for serotonin and dopamine, as well as NE. IL-1 and TNF-α increased serotonin reuptake in rat brain synaptosomes through activation of p38 MAPK [79].

13.3.3 Infiltrating Monocytes: A Novel Axis of the Immune-to-Brain Communication

Recently, a growing body of evidence has pointed to a novel axis of immune-to-brain communication. This novel neuro-immune communication involves MICROGLIA activation and sympathetic outflow to the peripheral IMMUNE SYSTEM that further reinforces STRESS-related behaviors by facilitating the recruitment of inflammatory MONOCYTES to the brain (Fig. 13.6) [84]. In addition, preclinical studies have also identified other infiltrating LEUKOCYTES (e.g., granulocytes, dendritic cells, T cells) present in small numbers in brain vasculature, choroid plexus (CP), and meninges. Under healthy conditions, these peripheral immune cells support neuronal function and scan the brain for pathogens or tissue damage [85]. Evidence about this novel axis is based mainly on studies about STRESS and the associated depressive- and ANXIETY-like behaviors in rodents.

MICROGLIA are resident immune cells in the CNS. Most studies about MICROGLIA focused on their role as mediators of inflammatory response to infection and injury. After being challenged by a biological or physical insult, the resting MICROGLIA is transformed into activated, undergoing a series of changes, notably in shape, increased proliferation, and production of inflammatory mediators. The activated MICROGLIA are promptly recruited to injured sites where they will phagocyte debris and unwelcome dying cells. Once activated, MICROGLIA function similarly to peripheral MACROPHAGES [71].

Apart from the classical MICROGLIA function in brain infection and injury, recent studies have shown that MICROGLIA activation and the associated NEUROINFLAMMATION are involved in nonclassical inflammatory conditions, such as NEURODEGENERATIVE and PSYCHIATRIC diseases [86]. In rodents, prolonged STRESS exposure resulted in MICROGLIA activation in key STRESS-responsive brain areas, such as prefrontal cortex, hypothalamus, amygdala, and the CA3 and dentate gyrus of the hippocampus [84]. Noradrenergic signaling plays a pivotal role in STRESS-induced MICROGLIA activation. Pre-treatment with propranolol (a β-adrenergic receptor antagonist), known to prevent ANXIETY-like behavior, prevented STRESS-induced neuronal activation and MICROGLIA activation [87]. Conversely, the administration of isoproterenol, a β-adrenergic receptor agonist, was capable of inducing both peripheral and central (i.e., hippocampus) production of pro-inflammatory CYTOKINES [88]. Treatment with minocycline, an antibiotic known to limit MICROGLIA response, prevented or attenuated STRESS-induced MICROGLIA activation, pro-inflammatory CYTOKINES production inside the brain (especially IL-1β), and persistent neuronal activation. In addition, the treatment with minocycline diminished ANXIETY- and depressive-like behaviors and cognitive deficits following STRESS [84].

Following a STRESS-induced MICROGLIA activation, brain CYTOKINES in turn signal to increase neuroendocrine outflow, resulting in a cycle of STRESS-related responses and further MICROGLIA activation. IL-1β is thought to be the key CYTOKINE in inducing STRESS-related responses. The release of IL-1β in the hypothalamus activates the HPA axis by inducing the secretion of CRH from the paraventricular nucleus. CRH, in turn, induces the secretion of ACTH from the pituitary. Finally, ACTH stimulates the secretion of GCs from the adrenal cortex. In addition IL-1 is capable of directly inducing ACTH and GC secretion from the pituitary and adrenal, respectively. However, IL-1 type I receptors were found only in the pituitary, but not in the adrenal gland [89].

STRESS-induced brain-to-IMMUNE SYSTEM communication also occurs via SNS activation. Sympathetic nerve fibers innervate lymphoid tissues, and SNS activation results in NE release into immune organs, including the bone marrow. As peripheral immune cells express noradrenergic receptors, they undergo functional alterations after sympathetic activation. Under the circumstances of chronic STRESS, the prolonged or repeated sympathetic activation results in increased production and release of myeloid cells by the bone marrow. The enhanced cycling of myeloid cells culminates in a phenotype of less mature and more inflammatory circulating MONOCYTES, which will move throughout the tissues where they become effector cells. These inflammatory MONOCYTES traffic even to the brain, where they differentiate into brain MACROPHAGES, promoting INFLAMMATION [84]. The recruitment of blood MONOCYTES to the brain has been extensively described in classical neuroinflammatory diseases, such as trauma and infection. Recently, studies demonstrated that MONOCYTE trafficking into the brain is necessary for inducing ANXIETY-like behavior after neurophatic pain [90], SICKNESS BEHAVIOR due to INFLAMMATION [91], and cognitive decline following peripheral surgery [92]. Peripheral MONOCYTE migration to the brain was also demonstrated following psychological STRESS [93]. This MONOCYTE trafficking to the brain was associated with prolonged responses after STRESS. Using the repeated social defeat murine model of STRESS, a study showed that MONOCYTE infiltration into the brain is necessary for the development of STRESS-induced prolonged ANXIETY-like behavior [41, 93]. A postmortem study has also recently shown that more peripheral MONOCYTES are being recruited into the brains of depressed patients than in those of healthy controls [94].

13.3.4 The Role of T Cells During Immune-to-Brain Communication

Emerging data indicate that circulating T cells, especially those autoreactive to CNS antigens, may play an important role in neuroprotection and resilience against CNS pathologies, including neuropsychiatric disease. It has been estimated that human CSF contains as many as 5×10^5 T cells [95]. In healthy conditions, T cells do no penetrate the BBB and are rarely found in brain parenchyma. Previous studies indicated that peripheral T cells play a key role in the maintenance of brain plasticity, demonstrated by CNS-specific T cells required for supporting hippocampal-dependent cognitive abilities and neurogenesis [96]. Mice deficient in T cells or CNS-specific T cells had reduced spatial learning and memory functions, as well as reduced proliferation of neural progenitor cells and reduced neuronal differentiation leading to decreased neurogenesis in the adult brain. Furthermore, transgenic mice overexpressing a T-cell receptor to myelin basic protein (T_{MBP} mice) showed higher levels of adult neurogenesis relative to the wild type [96]. Moreover, CNS-specific pro-cognitive T cells are located at the borders of the CNS, at the meninges [95], and at the CP [97]. In the absence of T cells, meningeal myeloid cells acquire a pro-inflammatory phenotype (M1) with secretion of TNF, IL-1β, and IL-12, CYTOKINES that negatively affect brain function [78]. After learning and memory tasks in mice, the activated T cells found in their meningeal spaces express high levels of IL-4 [98] and maintain meningeal myeloid cells in a M2 anti-inflammatory state. In addition to maintaining the M2 phenotype of meningeal myeloid cells, IL-4 may also directly mediate an improvement in learning behavior via the upregulation of brain-derived NEUROTROPHIC FACTOR (BDNF) expression by neural cells [98]. BDNF is a key NEUROTROPHIC FACTOR supporting neuronal survival and brain plasticity and involved with memory and neurogenesis. Decreased levels of BDNF have been in found in psychiatric disorders, such as MAJOR DEPRESSION, and negatively correlated with memory performance [99].

Previous studies have also indicated the role of effector T cells in resilience to psychosocial STRESS. Following a stressful episode in mice, T cells were found to traffic to the CP [71]. T-cell-deficient mice (SCID) were more likely to develop PTSD than wild type. Reconstitution of SCID mice with T cells isolated from wild-type donors ameliorated the overactive STRESS response. In addition, when the T cell response was boosted in wild-type mice by vaccination with MBP, the long-term pathological response to STRESS was further diminished [100]. These studies indicated that T cells can attenuate response to STRESS [101] and inhibition of T cell function by STRESS and DEPRESSION may have profound consequences on key immunological elements of healthy brain activity. It has been speculated whether boosting memory T cells specific for brain antigens could be a novel treatment for DEPRESSION. More specifically, boosting effector T cells, for instance, by reducing Treg-mediated suppression of effector T cells, is an important mechanism for protection of CNS tissue in recovery from acute insult and also from mental STRESS [102, 103].

Autoreactive T cells play a key neuroprotective role in NEURODEGENERATIVE DISEASES. For instance, autoreactive T cells are beneficial and limit CNS damage in experimental autoimmune encephalomyelitis (EAE) in rodents [104]. Several studies have demonstrated the neuroprotective role of CNS autoreactive T cells (i.e., "protective autoimmunity") in NEURODEGENERATIVE DISEASES, including Alzheimer's disease (AD) and Parkinson's disease [101, 105]. In AD mouse model, transient depletion of Tregs was followed by amyloid-β plaque clearance, mitigation of the neuroinflammatory response, and reversal of cognitive decline [106]. It has been shown that transient Treg depletion affects the brain's CP, a selective gateway for immune cell trafficking to the CNS, and was associated with subsequent recruitment of immunoregulatory cells, including MONOCYTE-derived MACROPHAGES and Tregs, to cerebral sites of plaque pathology. These findings suggest targeting Treg-mediated systemic immunosuppression for treating AD. More recently, the same group demonstrated that immune checkpoint blockade against the programmed death-1 (PD-1) pathway evokes an interferon (IFN)-γ–dependent systemic immune response, followed by the recruitment of MONOCYTE-derived MACROPHAGES to the brain [107] (see also Chaps. 12 and 25). When induced in mice with established AD pathology, this immunological response leads to clearance of cerebral amyloid-β (Aβ) plaques and improved cognitive performance. Together these findings suggest that systemic immunity should be boosted, rather than suppressed, to drive an immune-dependent cascade needed for brain repair.

Recent evidence also supports the notion that a type I/II interferon balance is pivotal in the regulation of brain physiology and pathology [107]. Type I interferon consists of a family of pleiotropic CYTOKINES (e.g., IFN-α and IFN-β) produced by innate immune cells and is known to regulate several functions including the induction of antiviral innate signaling. Type II interferon (or IFN-γ) is produced mainly from Th1 CD4+ T cells and regulates several cell-mediated adaptive functions, including the activation of MACROPHAGES (see Chap. 6). The production of IFN-γ by CD4+ T cells at the CP is a key mechanism involved in LEUKOCYTE trafficking to the CSF. The absence of IFN-γ-signaling results in reduced LEUKOCYTE counts in the CSF and is correlated with premature cognitive decline in mice during adulthood [108]. In addition, it has been shown that a low IFN-γ content at the CP in neurodegenerative disorders, such as Parkinson's and AD, results in poorer immune support needed for repair [107]. During human aging, and in a mouse model of AD, insufficient IFN-γ signaling at the CP was found in parallel by local elevation of IFN-I produced by the CP epithelium [108]. Blockade of IFN-I signaling within the CNS in aged mice led to partial restoration of brain homeostasis, including attenuated age-related INFLAMMATION in the hippocampus [108]. IFN-I was shown to negatively affect CP epithelial expression of BDNF and IGF-1, NEUROTROPHIC FACTORS supporting neuronal differentiation, growth, survival, and synaptogenesis [108]. These findings indicate that blocking IFN-I signaling in the aged CNS/CP may have a positive effect on the brain aging and AD and that this effect is mediated, at least in part, by restoration of IFN-γ-dependent activity of the CP.

The meningeal immunity seems to be critical for social behavior. It was recently demonstrated that SCID mice or mice deficient in IFN-γ exhibit severe social deficits and hyper-connectivity of fronto-cortical brain regions [109]. It was shown that inhibitory neurons respond to IFN-γ and increase GABAergic (γ-aminobutyric-acid) currents in projection neurons, suggesting that IFN-γ is a molecular link between meningeal immunity and neural circuits recruited for social behavior. These data suggest that social deficits observed in several neurological and psychiatric disorders (e.g., autism, frontotemporal dementia, SCHIZOPHRENIA) may result from impaired circuitry homeostasis derived from impaired adaptive immunity.

In summary, these data indicate novel physiological roles for autoreactive T cells during health and disease. The autoimmune CNS-specific T cells that were previously considered as key inducers of pathology should now be considered as beneficial support system for the CNS (i.e., protective autoimmunity).

13.4 Summary

The classical view of the nervous, the immune, and the endocrine systems as independent components of the body has been challenged since the advent of studies investigating the relationship among them. The neuro-immune communication is bidirectional. The CNS signals the IMMUNE SYSTEM using hormonal routes, including the HPA axis and STRESS-response HORMONES. In addition to hormonal routes, neuronal pathways can be used, mainly through the catecholaminergic pathway. In turn, CYTOKINES and other immune mediators are the effectors of the immune-to-brain communication. Peripheral blood CYTOKINES can reach the brain through different pathways: (1) they can be actively transported into the brain, crossing the BBB through leaky areas in the circumventricular organs, or (2) through the activation of neural pathways such as the vagal nerve. CYTOKINES can affect brain functions, resulting in a constellation of physiological, behavioral, and hormonal changes. The wording "SICKNESS BEHAVIOR" refers to a series of CYTOKINE-induced symptoms, such as lethargy, psychomotor slowness, depressive behavior, and appetite and sleep alterations. Nowadays, the term "SICKNESS BEHAVIOR" is also employed in the context of an existing INFLAMMATION-related profile observed in neuropsychiatric disorders, mainly MAJOR DEPRESSION, but also ANXIETY, BIPOLAR DISORDER, OBSESSIVE-COMPULSIVE DISORDER, and SCHIZOPHRENIA. IL-1β and TNF-α are the main CYTOKINES involved in SICKNESS BEHAVIOR. Recently, a growing body of evidence has pointed to a novel axis of immune-to-brain communication. This novel neuro-immune communication involves MICROGLIA activation and sympathetic outflow to the peripheral IMMUNE SYSTEM that further reinforces STRESS-related behaviors by facilitating the recruitment of inflammatory MONOCYTES to the brain. The bidirectional communication between the nervous and IMMUNE SYSTEMs, through HORMONES, CYTOKINES, NEUROTRANSMITTERS, NEUROPEPTIDES, and immune cells, is essential for homeostasis maintenance and adequate response to infection and disease. Recently, neuro-immune interactions have been implicated in the pathophysiology of a series of PSYCHIATRIC DISORDERS, such as MAJOR DEPRESSION and ANXIETY disorders. Further studies are needed in order to investigate therapeutic approaches for psychiatric disorders targeting neuroimmunoendocrine communications.

Selected Readings

Dantzer R. Cytokine-induced sickness behavior: mechanisms and implications. Ann N Y Acad Sci. 2001;933:222–34.

Dantzer R, O'Connor JC, Freund GG, Johnson RW, Kelley KW. From inflammation to sickness and depression: when the immune system subjugates the brain. Nat Rev Neurosci. 2008;9(1):46–56.

Miller AH, Maletic V, Raison CL. Inflammation and its discontents: the role of cytokines in the pathophysiology of major depression. Biol Psychiatry. 2009;65(9): 732–41.

Miller AH, Raison CL. The role of inflammation in depression: from evolutionary imperative to modern treatment target. Nat Rev Immunol. 2016;16(1):22–34.

Wohleb ES, McKim DB, Sheridan JF, Godbout JP. Monocyte trafficking to the brain with stress and inflammation: a novel axis of immune-to-brain communication that influences mood and behavior. Front Neurosci. 2015;8:447.

References

1. Velázquez-Moctezuma J, Domínguez-Salazar E, Gómez-González B. Beyond the borders: the gates and fences of neuroimmune interaction. Front Integr Neurosci. 2014;8:26.
2. Azad N, Agrawal L, Emanuele MA, Kelley MR, Mohagheghpour N, Lawrence AM, Emanuele NV. Neuroimmunoendocrinology. Am J Reprod Immunol. 1991;26(4):160–72.
3. Weigent DA, Kraneveld AD, Blalock JE. Neuroimmunoendocrinology. In: Nijkamp FP, Parnham MJ, editors. Principles of immunopharmacology: 3rd revised and extended edition. Basel: Springer; 2011. p. 179–98.
4. Ader R. On the development of psychoneuroimmunology. Eur J Pharmacol. 2000;405(1-3):167–76.
5. Weigent DA, Blalock JE. Associations between the neuroendocrine and immune systems. J Leukoc Biol. 1995;58:137–50.
6. Blalock JE, Smith EM. The immune system: our mobile brain? Immunol Today. 1985;6(4):115–7.
7. Pariante CM. Psychoneuroimmunology or immunopsychiatry? Lancet Psychiatry. 2015;2(3):197–9.
8. Capuron L, Miller AH. Cytokines and psychopathology: lessons from interferon-alpha. Biol Psychiatry. 2004;56(11):819–24.
9. Cruz-Topete D, Cidlowski JA. One hormone, two actions: anti- and pro-inflammatory effects of glucocorticoids. Neuroimmunomodulation. 2015;22(1-2):20–32.
10. Sapolsky RM, Romero LM, Munck AU. How do glucocorticoids influence stress responses? Integrating permissive, suppressive, stimulatory, and preparative actions. Endocr Rev. 2000;21(1):55–89.
11. Elenkov IJ, Chrousos GP. Stress systemDOUBLEHYPHENorganization, physiology and immunoregulation. Neuroimmunomodulation. 2006;13(5-6):257–67.
12. Visser J, van Boxel-Dezaire A, Methorst D, Brunt T, de Kloet ER, Nagelkerken L. Differential regulation of interleukin-10 (IL-10) and IL-12 by glucocorticoids in vitro. Blood. 1998;91(11):4255–64.
13. Almawi WY, Melemedjian OK, Rieder MJ. An alternate mechanism of glucocorticoid anti-proliferative effect: promotion of a Th2 cytokine-secreting profile. Clin Transplant. 1999;13(5):365–74.
14. Unterberger C, Staples KJ, Smallie T, Williams L, Foxwell B, Schaefer A, Kempkes B, Hofer TP, Koeppel M, Lohrum M, Stunnenberg H, Frankenberger M, Ziegler-Heitbrock L. Role of STAT3 in glucocorticoid-induced expression of the human IL-10 gene. Mol Immunol. 2008;45(11):3230–7.
15. Liberman AC, Refojo D, Druker J, Toscano M, Rein T, Holsboer F, Arzt E. The activated glucocorticoid receptor inhibits the transcription factor T-bet by direct protein-protein interaction. FASEB J. 2007;21(4):1177–88.
16. Tamada K, Harada M, Abe K, Li T, Nomoto K. IL-4-producing NK1.1+ T cells are resistant to glucocor-

ticoid-induced apoptosis: implications for the Th1/Th2 balance. J Immunol. 1998;161(3):1239–47.
17. Elyaman W, Bradshaw EM, Uyttenhove C, Dardalhon V, Awasthi A, Imitola J, Bettelli E, Oukka M, van Snick J, Renauld JC, Kuchroo VK, Khoury SJ. IL-9 induces differentiation of TH17 cells and enhances function of FoxP3+ natural regulatory T cells. Proc Natl Acad Sci U S A. 2009;106(31):12885–90.
18. Prado C, de Paz B, Gómez J, López P, Rodríguez-Carrio J, Suárez A. Glucocorticoids enhance Th17/Th1 imbalance and signal transducer and activator of transcription 3 expression in systemic lupus erythematosus patients. Rheumatology (Oxford). 2011;50(10):1794–801.
19. Freier E, Weber CS, Nowottne U, Horn C, Bartels K, Meyer S, Hildebrandt Y, Luetkens T, Cao Y, Pabst C, Muzzulini J, Schnee B, Brunner-Weinzierl MC, Marangolo M, Bokemeyer C, Deter HC, Atanackovic D. Decrease of CD4(+)FOXP3(+) T regulatory cells in the peripheral blood of human subjects undergoing a mental stressor. Psychoneuroendocrinology. 2010;35(5):663–73.
20. Sbiera S, Dexneit T, Reichardt SD, Michel KD, van den Brandt J, Schmull S, Kraus L, Beyer M, Mlynski R, Wortmann S, Allolio B, Reichardt HM, Fassnacht M. Influence of short-term glucocorticoid therapy on regulatory T cells in vivo. PLoS One. 2011;6(9):e24345.
21. Cronstein BN, Kimmel SC, Levin RI, Martiniuk F, Weissmann G. A mechanism for the antiinflammatory effects of corticosteroids: the glucocorticoid receptor regulates leukocyte adhesion to endothelial cells and expression of endothelial-leukocyte adhesion molecule 1 and intercellular adhesion molecule 1. Proc Natl Acad Sci U S A. 1992;89(21):9991–5.
22. Baschant U, Tuckermann J. The role of the glucocorticoid receptor in inflammation and immunity. J Steroid Biochem Mol Biol. 2010;120(2-3):69–75.
23. Dhabhar FS, Miller AH, Stein M, McEwen BS, Spencer RL. Diurnal and acute stress-induced changes in distribution of peripheral blood leukocyte subpopulations. Brain Behav Immun. 1994;8(1):66–79.
24. Abo T, Kawate T, Itoh K, Kumagai K. Studies on the bioperiodicity of the immune response. I. Circadian rhythms of human T, B, and K cell traffic in the peripheral blood. J Immunol. 1981;126(4):1360–3.
25. Kawate T, Abo T, Hinuma S, Kumagai K. Studies of the bioperiodicity of the immune response. II. Co-variations of murine T and B cells and a role of corticosteroid. J Immunol. 1981;126(4):1364–7.
26. Dimitrov S, Benedict C, Heutling D, Westermann J, Born J, Lange T. Cortisol and epinephrine control opposing circadian rhythms in T cell subsets. Blood. 2009;113(21):5134–43.
27. Ouchi N, Parker JL, Lugus JJ, Walsh K. Adipokines in inflammation and metabolic disease. Nat Rev Immunol. 2011;11(2):85–97.
28. Cohen S, Janicki-Deverts D, Doyle WJ, Miller GE, Frank E, Rabin BS, Turner RB. Chronic stress, glucocorticoid receptor resistance, inflammation, and disease risk. Proc Natl Acad Sci U S A. 2012;109(16):5995–9.
29. Raison CL, Miller AH. When not enough is too much: the role of insufficient glucocorticoid signaling in the pathophysiology of stress-related disorders. Am J Psychiatry. 2003;160(9):1554–65.
30. Lopes RP, Grassi-Oliveira R, de Almeida LR, Stein LM, Luz C, Teixeira AL, Bauer ME. Neuroimmunoendocrine interactions in patients with recurrent major depression, increased early life stress and long-standing posttraumatic stress disorder symptoms. Neuroimmunomodulation. 2012;19(1):33–42.
31. Miller AH, Raison CL. The role of inflammation in depression: from evolutionary imperative to modern treatment target. Nat Rev Immunol. 2016;16(1):22–34.
32. Horowitz MA, Zunszain PA. Neuroimmune and neuroendocrine abnormalities in depression: two sides of the same coin. Ann N Y Acad Sci. 2015;1351:68–79.
33. Felten DL, Felten SY, Carlson SL, Olschowka JA, Livnat S. Noradrenergic and peptidergic innervation of lymphoid tissue. J Immunol. 1985;135(2 Suppl):755s–65s.
34. Sanders VM, Kavelaars A. Adrenergic regulation of immunity. In: Ader R, editor. Psychoneuroimmunology. 4th ed. New York, NY: Academic Press; 2007. p. 63–83.
35. Rosas-Ballina M, Olofsson PS, Ochani M, Valdés-Ferrer SI, Levine YA, Reardon C, Tusche MW, Pavlov VA, Andersson U, Chavan S, Mak TW, Tracey KJ. Acetylcholine-synthesizing T cells relay neural signals in a vagus nerve circuit. Science. 2011;334(6052):98–101.
36. Dhabhar FS, Malarkey WB, Neri E, McEwen BS. Stress-induced redistribution of immune cellsDOUBLEHYPHENfrom barracks to boulevards to battlefields: a tale of three hormones-DOUBLEHYPHENCurt Richter Award winner. Psychoneuroendocrinology. 2012;37(9):1345–68.
37. Heidt T, Sager HB, Courties G, Dutta P, Iwamoto Y, Zaltsman A, von Zur Muhlen C, Bode C, Fricchione GL, Denninger J, Lin CP, Vinegoni C, Libby P, Swirski FK, Weissleder R, Nahrendorf M. Chronic variable stress activates hematopoietic stem cells. Nat Med. 2014;20(7):754–8.
38. Engler H, Bailey MT, Engler A, Sheridan JF. Effects of repeated social stress on leukocyte distribution in bone marrow, peripheral blood and spleen. J Neuroimmunol. 2004;148(1-2):106–15.
39. Liezmann C, Stock D, Peters EM. Stress induced neuroendocrine-immune plasticity: a role for the spleen in peripheral inflammatory disease and inflammaging? Dermatoendocrinology. 2012;4(3):271–9.
40. Seifert HA, Hall AA, Chapman CB, Collier LA, Willing AE, Pennypacker KR. A transient decrease in spleen size following stroke corresponds to splenocyte release into systemic circulation. J Neuroimmune Pharmacol. 2012;7(4):1017–24.

41. Wohleb ES, Powell ND, Godbout JP, Sheridan JF. Stress-induced recruitment of bone marrow-derived monocytes to the brain promotes anxiety-like behavior. J Neurosci. 2013;33(34):13820–33.
42. Benschop RJ, Rodriguez-Feuerhahn M, Schedlowski M. Catecholamine-induced leukocytosis: early observations, current research, and future directions. Brain Behav Immun. 1996;10(2):77–91.
43. Procaccini C, Pucino V, De Rosa V, Marone G, Matarese G. Neuro-endocrine networks controlling immune system in health and disease. Front Immunol. 2014;5:143.
44. Madden KS. Catecholamines, sympathetic innervation, and immunity. Brain Behav Immun. 2003;17(Suppl 1):S5–10.
45. Panina-Bordignon P, Mazzeo D, Lucia PD, D'Ambrosio D, Lang R, Fabbri L, Self C, Sinigaglia F. Beta2-agonists prevent Th1 development by selective inhibition of interleukin 12. J Clin Invest. 1997;100(6):1513–9.
46. Cole SW, Korin YD, Fahey JL, Zack JA. Norepinephrine accelerates HIV replication via protein kinase A-dependent effects on cytokine production. J Immunol. 1998;161(2):610–6.
47. Bhowmick S, Singh A, Flavell RA, Clark RB, O'Rourke J, Cone RE. The sympathetic nervous system modulates CD4(+)FoxP3(+) regulatory T cells via a TGF-beta-dependent mechanism. J Leukoc Biol. 2009;86:1275–83.
48. Collado-Hidalgo A, Sung C, Cole S. Adrenergic inhibition of innate anti-viral response: PKA blockade of Type I interferon gene transcription mediates catecholamine support for HIV-1 replication. Brain Behav Immun. 2006;20(6):552–63.
49. Grebe KM, Takeda K, Hickman HD, Bailey AL, Embry AC, Bennink JR, Yewdell JW. Cutting edge: sympathetic nervous system increases proinflammatory cytokines and exacerbates influenza A virus pathogenesis. J Immunol. 2010;184(2):540–4.
50. Saha B, Mondal AC, Majumder J, Basu S, Dasgupta PS. Physiological concentrations of dopamine inhibit the proliferation and cytotoxicity of human CD4+ and CD8+ T cells in vitro: a receptor-mediated mechanism. Neuroimmunomodulation. 2001;9(1):23–33.
51. Mikulak J, Bozzo L, Roberto A, Pontarini E, Tentorio P, Hudspeth K, Lugli E, Mavilio D. Dopamine inhibits the effector functions of activated NK cells via the upregulation of the D5 receptor. J Immunol. 2014;193(6):2792–800.
52. Wu J, Zhang R, Tang N, Gong Z, Zhou J, Chen Y, Chen K, Cai W. Dopamine inhibits the function of Gr-1+CD115+ myeloid-derived suppressor cells through D1-like receptors and enhances anti-tumor immunity. J Leukoc Biol. 2015;97(1):191–200.
53. Tracey KJ. Reflex control of immunity. Nat Rev Immunol. 2009;9(6):418–28.
54. Pavlov VA, Tracey KJ. The vagus nerve and the inflammatory reflexDOUBLEHYPHENlinking immunity and metabolism. Nat Rev Endocrinol. 2012;8(12):743–54.
55. Antonica A, Ayroldi E, Magni F, Paolocci N. Lymphocyte traffic changes induced by monolateral vagal denervation in mouse thymus and peripheral lymphoid organs. J Neuroimmunol. 1996;64(2):115–22.
56. Goehler LE, Gaykema RP, Nguyen KT, Lee JE, Tilders FJ, Maier SF, Watkins LR. Interleukin-1beta in immune cells of the abdominal vagus nerve: a link between the immune and nervous systems? J Neurosci. 1999;19(7):2799–806.
57. Delgado M, Ganea D. Vasoactive intestinal peptide: a neuropeptide with pleiotropic immune functions. Amino Acids. 2013;45(1):25–39.
58. Delgado M, Gonzalez-Rey E, Ganea D. VIP/PACAP preferentially attract Th2 effectors through differential regulation of chemokine production by dendritic cells. FASEB J. 2004;18(12):1453–5.
59. Ran WZ, Dong L, Tang CY, Zhou Y, Sun GY, Liu T, Liu YP, Guan CX. Vasoactive intestinal peptide suppresses macrophage-mediated inflammation by downregulating interleukin-17A expression via PKA- and PKC-dependent pathways. Int J Exp Pathol. 2015;96(4):269–75.
60. Chorny A, Gonzalez-Rey E, Fernandez-Martin A, Pozo D, Ganea D, Delgado M. Vasoactive intestinal peptide induces regulatory dendritic cells with therapeutic effects on autoimmune disorders. Proc Natl Acad Sci U S A. 2005;102(38):13562–7.
61. Delgado M, Chorny A, Gonzalez-Rey E, Ganea D. Vasoactive intestinal peptide generates CD4+CD25+ regulatory T cells in vivo. J Leukoc Biol. 2005;78(6):1327–38.
62. Fernandez-Martin A, Gonzalez-Rey E, Chorny A, Ganea D, Delgado M. Vasoactive intestinal peptide induces regulatory T cells during experimental autoimmune encephalomyelitis. Eur J Immunol. 2006;36(2):318–26.
63. Gonzalez-Rey E, Chorny A, Fernandez-Martin A, Ganea D, Delgado M. Vasoactive intestinal peptide generates human tolerogenic dendritic cells that induce CD4 and CD8 regulatory T cells. Blood. 2006;107(9):3632–8.
64. Gonzalez-Rey E, Delgado M. Therapeutic treatment of experimental colitis with regulatory dendritic cells generated with vasoactive intestinal peptide. Gastroenterology. 2006;131(6):1799–811.
65. Gonzalez-Rey E, Fernandez-Martin A, Chorny A, Delgado M. Vasoactive intestinal peptide induces CD4+,CD25+ T regulatory cells with therapeutic effect in collagen-induced arthritis. Arthritis Rheum. 2006;54(3):864–76.
66. Grasso E, Paparini D, Agüero M, Mor G, Pérez Leirós C, Ramhorst R. VIP contribution to the decidualization program: regulatory T cell recruitment. J Endocrinol. 2014;221(1):121–31.

67. Gropp E, Shanabrough M, Borok E, Xu AW, Janoschek R, Buch T, et al. Agouti-related peptide-expressing neurons are mandatory for feeding. Nat Neurosci. 2005;8:1289–91.
68. Matarese G, Procaccini C, Menale C, Kim JG, Kim JD, Diano S, et al. Hunger-promoting hypothalamic neurons modulate effector and regulatory T-cell responses. Proc Natl Acad Sci U S A. 2013;110:6193–8.
69. Mashaghi A, Marmalidou A, Tehrani M, Grace PM, Pothoulakis C, Dana R. Neuropeptide substance P and the immune response. Cell Mol Life Sci. 2016;
70. Feistritzer C, Clausen J, Sturn DH, Djanani A, Gunsilius E, Wiedermann CJ, et al. Natural killer cell functions mediated by the neuropeptide substance P. Regul Pept. 2003;116:119–26.
71. Bilimoria PM, Stevens B. Microglia function during brain development: new insights from animal models. Brain Res. 2015;1617:7–17.
72. Miller AH, Raison CL. The role of inflammation in depression: from evolutionary imperative to modern treatment target. Nat Rev Immunol. 2016;16(1):22–34.
73. Louveau A, Smirnov I, Keyes TJ, Eccles JD, Rouhani SJ, Peske JD, et al. Structural and functional features of central nervous system lymphatic vessels. Nature. 2015;523:337–41.
74. Eskandari F, Webster JI, Sternberg EM. Neural immune pathways and their connection to inflammatory diseases. Arthritis Res Ther. 2003;5(6):251–65.
75. Dantzer R. Cytokine-induced sickness behavior: mechanisms and implications. Ann N Y Acad Sci. 2001;933:222–34.
76. Watkins LR, Maier SF. The pain of being sick: implications of immune-to-brain communication for understanding pain. Annu Rev Psychol. 2000;51:29–57.
77. Rocha NP, Reis HJ, Vanden Berghe P, Cirillo C. Depression and cognitive impairment in Parkinson's disease: a role for inflammation and immunomodulation? Neuroimmunomodulation. 2014;21(2-3):88–94.
78. Dantzer R, O'Connor JC, Freund GG, Johnson RW, Kelley KW. From inflammation to sickness and depression: when the immune system subjugates the brain. Nat Rev Neurosci. 2008;9(1):46–56.
79. Miller AH, Maletic V, Raison CL. Inflammation and its discontents: the role of cytokines in the pathophysiology of major depression. Biol Psychiatry. 2009;65(9):732–41.
80. Reichenberg A, Yirmiya R, Schuld A, Kraus T, Haack M, Morag A, Pollmächer T. Cytokine-associated emotional and cognitive disturbances in humans. Arch Gen Psychiatry. 2001;58(5):445–52.
81. Pace TW, Mletzko TC, Alagbe O, Musselman DL, Nemeroff CB, Miller AH, Heim CM. Increased stress-induced inflammatory responses in male patients with major depression and increased early life stress. Am J Psychiatry. 2006;163(9):1630–3.
82. Aschbacher K, Epel E, Wolkowitz OM, Prather AA, Puterman E, Dhabhar FS. Maintenance of a positive outlook during acute stress protects against pro-inflammatory reactivity and future depressive symptoms. Brain Behav Immun. 2012;26(2):346–52.
83. Miller AH, Lecture TWPNC. Mechanisms of cytokine-induced behavioral changes: psychoneuroimmunology at the translational interface. Brain Behav Immun. 2009;23(2):149–58.
84. Wohleb ES, McKim DB, Sheridan JF, Godbout JP. Monocyte trafficking to the brain with stress and inflammation: a novel axis of immune-to-brain communication that influences mood and behavior. Front Neurosci. 2015;8:447.
85. Schwartz M, Kipnis J, Rivest S, Prat A. How do immune cells support and shape the brain in health, disease, and aging? J Neurosci. 2013;33:17587–96.
86. Norden DM, Muccigrosso MM, Godbout JP. Microglial priming and enhanced reactivity to secondary insult in aging, and traumatic CNS injury, and neurodegenerative disease. Neuropharmacology. 2015;96(Pt A):29–41.
87. Wohleb ES, Hanke ML, Corona AW, Powell ND, Stiner LM, Bailey MT, Nelson RJ, Godbout JP, Sheridan JF. β-Adrenergic receptor antagonism prevents anxiety-like behavior and microglial reactivity induced by repeated social defeat. J Neurosci. 2011;31(17):6277–88.
88. Johnson JD, Campisi J, Sharkey CM, Kennedy SL, Nickerson M, Greenwood BN, Fleshner M. Catecholamines mediate stress-induced increases in peripheral and central inflammatory cytokines. Neuroscience. 2005;135(4):1295–307.
89. Goshen I, Yirmiya R. Interleukin-1 (IL-1): a central regulator of stress responses. Front Neuroendocrinol. 2009;30(1):30–45.
90. Sawada A, Niiyama Y, Ataka K, Nagaishi K, Yamakage M, Fujimiya M. Suppression of bone marrow-derived microglia in the amygdala improves anxiety-like behavior induced by chronic partial sciatic nerve ligation in mice. Pain. 2014;155(9):1762–72.
91. D'Mello C, Riazi K, Le T, Stevens KM, Wang A, McKay DM, Pittman QJ, Swain MG. P-selectin-mediated monocyte-cerebral endothelium adhesive interactions link peripheral organ inflammation to sickness behaviors. J Neurosci. 2013;33(37):14878–88.
92. Degos V, Vacas S, Han Z, van Rooijen N, Gressens P, Su H, Young WL, Maze M. Depletion of bone marrow-derived macrophages perturbs the innate immune response to surgery and reduces postoperative memory dysfunction. Anesthesiology. 2013;118(3):527–36.
93. Wohleb ES, McKim DB, Shea DT, Powell ND, Tarr AJ, Sheridan JF, Godbout JP. Re-establishment of anxiety in stress-sensitized mice is caused by monocyte trafficking from the spleen to the brain. Biol Psychiatry. 2014;75(12):970–81.

94. Torres-Platas SG, Cruceanu C, Chen GG, Turecki G, Mechawar N. Evidence for increased microglial priming and macrophage recruitment in the dorsal anterior cingulate white matter of depressed suicides. Brain Behav Immun. 2014;42:50–9.
95. Derecki NC, Cronk JC, Lu Z, Xu E, Abbott SB, Guyenet PG, et al. Wild-type microglia arrest pathology in a mouse model of Rett syndrome. Nature. 2012;484:105–9.
96. Ziv Y, Ron N, Butovsky O, Landa G, Sudai E, Greenberg N, et al. Immune cells contribute to the maintenance of neurogenesis and spatial learning abilities in adulthood. Nat Neurosci. 2006;9:268–75.
97. Baruch K, Ron-Harel N, Gal H, Deczkowska A, Shifrut E, Ndifon W, et al. CNS-specific immunity at the choroid plexus shifts toward destructive Th2 inflammation in brain aging. Proc Natl Acad Sci U S A. 2013;110:2264–9.
98. Derecki NC, Cardani AN, Yang CH, Quinnies KM, Crihfield A, Lynch KR, et al. Regulation of learning and memory by meningeal immunity: a key role for IL-4. J Exp Med. 2010;207:1067–80.
99. Grassi-Oliveira R, Stein LM, Lopes RP, Teixeira AL, Bauer ME. Low plasma brain-derived neurotrophic factor and childhood physical neglect are associated with verbal memory impairment in major depressionDOUBLEHYPHENa preliminary report. Biol Psychiatry. 2008;64:281–5.
100. Lewitus GM, Cohen H, Schwartz M. Reducing post-traumatic anxiety by immunization. Brain Behav Immun. 2008;22:1108–14.
101. Lewitus GM, Schwartz M. Behavioral immunization: immunity to self-antigens contributes to psychological stress resilience. Mol Psychiatry. 2009;14:532–6.
102. Cohen H, Ziv Y, Cardon M, Kaplan Z, Matar MA, Gidron Y, et al. Maladaptation to mental stress mitigated by the adaptive immune system via depletion of naturally occurring regulatory CD4+CD25+ cells. J Neurobiol. 2006;66:552–63.
103. Kipnis J, Mizrahi T, Hauben E, Shaked I, Shevach E, Schwartz M. Neuroprotective autoimmunity: naturally occurring CD4+CD25+ regulatory T cells suppress the ability to withstand injury to the central nervous system. Proc Natl Acad Sci U S A. 2002;99:15620–5.
104. Moalem G, Leibowitz-Amit R, Yoles E, Mor F, Cohen IR, Schwartz M. Autoimmune T cells protect neurons from secondary degeneration after central nervous system axotomy. Nat Med. 1999;5:49–55.
105. Avidan H, Kipnis J, Butovsky O, Caspi RR, Schwartz M. Vaccination with autoantigen protects against aggregated beta-amyloid and glutamate toxicity by controlling microglia: effect of CD4+CD25+ T cells. Eur J Immunol. 2004;34:3434–45.
106. Baruch K, Rosenzweig N, Kertser A, Deczkowska A, Sharif AM, Spinrad A, et al. Breaking immune tolerance by targeting Foxp3(+) regulatory T cells mitigates Alzheimer's disease pathology. Nat Commun. 2015;6:7967.
107. Deczkowska A, Baruch K, Schwartz M. Type I/II interferon balance in the regulation of brain physiology and pathology. Trends Immunol. 2016;37:181–92.
108. Baruch K, Deczkowska A, David E, Castellano JM, Miller O, Kertser A, et al. Aging. Aging-induced type I interferon response at the choroid plexus negatively affects brain function. Science. 2014;346:89–93.
109. Filiano AJ, Xu Y, Tustison NJ, Marsh RL, Baker W, Smirnov I, et al. Unexpected role of interferon-gamma in regulating neuronal connectivity and social behaviour. Nature. 2016;535:425–9.

Part II

Immunodiagnosis

Antibody Detection

14

Frank Redegeld, Klaus Hermann, Markus Ollert, and Johannes Ring

14.1 Introduction

The accurate and reliable determination of biologically active endogenous compounds in plasma, urine, and other body fluids for scientific or diagnostic purposes has been a challenge for many decades. In the past, the assessment of such compounds was difficult and tedious because specific and practical analytical tools were not available. In the early days, decisive and convincing conclusions about the significance of a biologically active molecule in disease or health could only be made after purification and isolation of the analyte and identification of its chemical structure. A definite improvement in the analysis of biologically active endogenous compounds was the introduction of bioassays using an intact animal model or *in vitro* tissue preparations. Although the bioassays possessed sufficient SENSITIVITY, there were problems with their lack of SPECIFICITY. Other analytical procedures such as liquid chromatography, electrophoresis, or photometric procedures have also been developed for *in vitro* diagnosis. However, these approaches are either tedious or time-consuming and require expensive equipment and specially trained personal. A landmark in diagnostics was the introduction of immunoassays, which are inexpensive and easy to perform with high reproducibility, SENSITIVITY, and SPECIFICITY.

Final manuscript submitted on July 24, 2017.

F. Redegeld
Division of Pharmacology, Department of Pharmaceutical Sciences, Faculty of Science, Utrecht University, Utrecht, Netherlands

Klinik und Poliklinik für Dermatologie und Allergologie am Biederstein, Technische Universität München, Munich, Germany
e-mail: F.A.M.Redegeld@uu.nl

K. Hermann
Klinik und Poliklinik für Dermatologie und Allergologie am Biederstein, Technische Universität München, Munich, Germany

M. Ollert (✉)
Luxembourg Institute of Health, Esch-sur-Alzette, Luxembourg

Department of Dermatology and Allergy, Am Biederstein, Technische Universität München, Munich, Germany
e-mail: ollert@lrz.tum.de; markus.ollert@lih.lu

J. Ring
Department of Dermatology and Allergy, Am Biederstein, Technische Universität München, Munich, Germany

Zentrum Allergie und Umwelt, Helmholtz Center Munich, Technische Universität München, Munich, Germany
e-mail: johannes.ring@tum.de

M. J. Parnham et al. (eds.), *Nijkamp and Parnham's Principles of Immunopharmacology*,
https://doi.org/10.1007/978-3-030-10811-3_14

14.2 Basic Principle of Immunoassays

Immunoassays are based on an antigen-ANTIBODY reaction utilizing the exceptional capability of the immune system to produce specific antibodies (Abs) which can recognize and discriminate between a practically infinite number of foreign compounds. The basic requirement for setting up an immunoassay is the production of a specific Ab to a given antigen or hapten by immunizing animals such as mice, rabbits, goats, horses, or others. High molecular weight compounds such as proteins are immunogenic and may serve as an antigen which can be injected into the host animal directly to produce Abs. In contrast, low molecular weight compounds such as drugs, amines, peptides, or steroids are haptens which do not induce immune responses. They can be rendered immunogenic after coupling to a carrier. The Abs obtained from the serum after several booster injections with the immunogen are heterogeneous polyclonal Abs (pAbs). A more sophisticated approach was the production of monoclonal Abs (mAbs) which was first introduced by Köhler and Milstein in 1975 [1]. This technique allows the production of homogeneous, single Abs in nearly infinite quantities with high SPECIFICITY for a certain antigen. The method involves the isolation of spleen cells from an immunized animal containing Ab-producing B-cells. The B-cells are then fused with myeloma cells. After selection and screening of the desired Ab-secreting cell line (hybridoma), the Ab can be harvested in the supernatant. These hybridomas can be grown in large volumes for the production of huge quantities of the mAb.

The essentials for the characterization of the Abs produced are high affinity, SPECIFICITY, and SENSITIVITY. The affinity is a measure of the strength of the binding interaction between the antigen and the Ab and can be experimentally determined by the dissociation constant (K_d) of the antigen to the Ab. The lower the K_d, the greater the affinity. The SPECIFICITY refers to the specific recognition of the ANALYTE by the Ab without crossreacting with closely related or structurally similar analytes. This in turn is closely related to the ability of the Ab to discriminate between negative and positive samples. The sensitivity describes the detection limit and is defined by the dose-response curve of the antigen to the Ab. The lower the detection limit, the higher the SENSITIVITY.

14.3 Antibody Structure

Abs, among other serum proteins such as albumins, belong to the gamma globulin or immunoglobulin (Ig) fraction according to their electrophoretic mobility. They are glycoproteins which are chemically very similar in structure and are constructed of two identical light chains with an approximate molecular weight (MW) of 50 kDa and two identical heavy chains with an approximate MW of 110 kDa. Each of the two light chains is attached to one heavy chain via disulfide bonds. Likewise, the heavy chains are bound to each other by disulfide bridging. Two locally distinct binding domains are prominent on Ig molecules. One is the antigen-binding site (Fab) for the binding and recognition of antigens, and the other is the receptor-binding site (Fc) for binding to specific receptors on various cells involved in immunological functions such as mononuclear phagocytes, NATURAL KILLER CELLS, mast cells, or basophil LEUKOCYTES (Fig. 14.1). Although Igs share an overall similarity, they can be divided into different classes and subclasses according to their size, charge, solubility, and their behavior as antigens. At present,

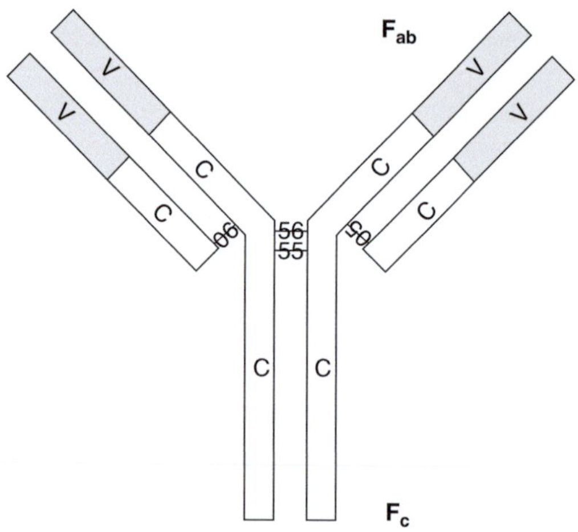

Fig. 14.1 Structure of immunoglobulins. For further details, see text

the classes of Ab molecules in humans can be divided into IgA, IgD, IgE, IgG, and IgM. IgA and IgG can be further subdivided into their subclasses IgA1 and IgA2 and IgG1, IgG2, IgG3, and IgG4 (see Chap. 4).

14.4 Clinical Relevance of Antibody Detection

Initially, the identification and characterization of specific Abs associated with pathophysiological conditions were generally constrained to scientific purposes. However, in many instances the measurement of specific Abs evolved into clinically relevant diagnostic markers in health and disease. Medical conditions in which the determination of specific Abs is used routinely are bacterial and viral infections, infestation with parasites, autoimmune diseases, and allergies.

14.4.1 Microbial Infections

The determination of Abs in infectious diseases has been known and used for a long time. Bacteria-specific Abs of the IgG and IgM class are routinely detected in the serum of patients infected with different bacteria such as *Borrelia burgdorferi* (lyme disease), *Chlamydia trachomatis* (sexually transmitted infection of the urogenital tract), *Legionella pneumophila* (legionnaire disease), *Staphylococcus* and *Streptococcus*, and *Treponema pallidum* (lues).

The assessment of IgG and IgM Abs is also a very valuable parameter in the diagnosis of viral infections such as hepatitis, measles, epidemic parotitis and rabies, and infections with the Epstein-Barr virus (mononucleosis), herpesvirus (herpes simplex and herpes zoster), and arbovirus (tick encephalitis).

Infestations with parasites leading to diseases such as leishmaniasis, amebiasis, malaria, toxoplasmosis, schistosomiasis, echinococcosis, trichinellosis, filariasis, or others can induce the formation of Ig molecules of different classes. In the diagnosis of parasitosis, it is recommended that the identity of the parasite be ascertained first. In addition, serological methods are available to identify circulating antigens, antigen-Ab complexes, or circulating Abs.

14.4.2 Autoimmune Diseases

Another area of pathophysiological abnormalities in which the measurement of Abs has predictive and diagnostic value is that of the numerous autoimmune diseases (ADs) in which the self-tolerance of the immune system against its own antigens is abrogated. Examples of an organ-specific AD are Hashimoto's thyroiditis with circulating Abs against thyroglobulin and myasthenia gravis with auto-Abs against the acetylcholine receptor. Examples of non-organ-specific ADs are Sjögren's syndrome, rheumatoid arthritis, scleroderma, and systemic lupus erythematosus affecting the skin, joints, and muscles with Abs against nuclear antigens such as DNA, RNA, or histones.

14.4.3 Allergy

The most important immunoglobulins for the *in vitro* diagnosis of allergic diseases, either immediate-type HYPERSENSITIVITY reactions such as rhinitis, conjunctivitis, allergic bronchial asthma, and ANAPHYLAXIS or late-phase reactions, e.g., allergic contact dermatitis, are Abs of the IgE class [2, 3]. Clinically relevant are the measurement of total IgE or allergen-specific IgE in the patients' serum for the determination of IgE-mediated sensitization. Total serum IgE levels of >100 (kilo units) kU/L in adults and children are a good indicator for atopy, a disease characterized by familial HYPERSENSITIVITY to exogenous environmental agents associated with high IgE Ab titers and altered reactivity against various pharmacological stimuli. However, high total IgE Abs can also be induced by parasitic worm infestations. Extremely high values, higher than 10,000 kU/L, are indicative of IgE-producing myelomas. The measurement of allergen-specific IgG Abs as IgE-blocking Abs for monitoring the success of immune therapy with insect venoms in patients with hymenoptera venom allergy has been used tentatively, but with discrepant results.

14.5 Antibody Detection Methods

The detection of Abs in the circulation or in tissue has become a useful analytical tool for the *in vitro* immunodiagnosis of numerous diseases. Several immunological techniques are available for the routine identification of IgA, IgD, IgE, IgG, or IgM Ab classes in the clinical chemistry laboratory. The most commonly used methods are discussed briefly.

14.5.1 Immunoprecipitation Assay

Immunoprecipitation is a very simple and easy to perform *in vitro* assay for the identification and semiquantitation of soluble Abs. The addition of the antigen to the Ab results in the formation of a three-dimensional, insoluble network of aggregates which precipitate and can be detected with a nephelometer. The assay is very similar to a volumetric acid/base titration. The bulk of precipitate, formed at equivalent concentrations of Ab and antigen, is a measure for the concentration of the Ab. The assay can also be used in reverse to measure the antigen concentration by adding Abs.

A variation of the immunoprecipitation assay is the hemagglutination test and the COMPLEMENT fixation test. The hemagglutination test allows the identification of Abs to red blood cell antigens or the detection of Abs to antigens which are covalently or noncovalently attached to the red cell surface. The COMPLEMENT fixation test is a three-step assay in which the Ab-containing serum is initially incubated with a fixed amount of antigen to form an immune complex. In the second step, complement is added which is firmly incorporated by the immune complexes. Finally, red blood cells are added as indicator cells. Red blood cells will only be lysed if immune complexes have been generated.

14.5.2 Immunocytochemistry

Immunocytochemistry is a technique for the detection of an Ab *in situ* in tissue slices. Frozen tissue or tissue embedded in various embedding media is cut into thin slices and then immobilized on a slide. After fixation of the tissue with formaldehyde, glutaraldehyde, alcohol, or acetone, the tissue is incubated with a specific primary Ab directed against the Ab to be detected. In the direct assay, the primary Ab is chemically coupled to a fluorescent dye (rhodamine, fluorescein), which allows the detection of the analyte by fluorescence microscopy. In the indirect assay, excess primary Ab is thoroughly washed off, and the tissue is incubated with a secondary Ab to form a sandwich. The secondary Ab can be fluoresceinated or coupled to an enzyme, e.g., alkaline phosphatase (ALP) or peroxidase, which allows the visualization of the analyte by fluorescence microscopy or by light microscopy, after addition of a colorless substrate which is enzymatically converted to a colored product. Only cells which contain the analyte will light up under the microscope.

14.5.3 Immunoblotting

The immunoblot or dot blot technique is similar to the immunoprecipitation assay. However, in immunoblotting the antigen-Ab reaction takes place in the solid phase, whereas in the immunoprecipitation assay, the Ab reacts with the antigen in solution. The assay utilizes the capability of nitrocellulose membranes to bind antigens. Antigens are applied in small dots, and the membranes are dried. The membranes are treated with ovalbumin, gelatin, or milk proteins to prevent nonspecific adsorption. After blocking, the membranes are incubated with the serum and dilutions of the serum which contain the Ab. The membranes are washed to remove abundant Ab. Next, the membranes are incubated with a secondary Ab raised against the Ab of interest which is conjugated with an enzyme. The formation of the antigen-Ab-secondary Ab complex can be visualized by adding a substrate which

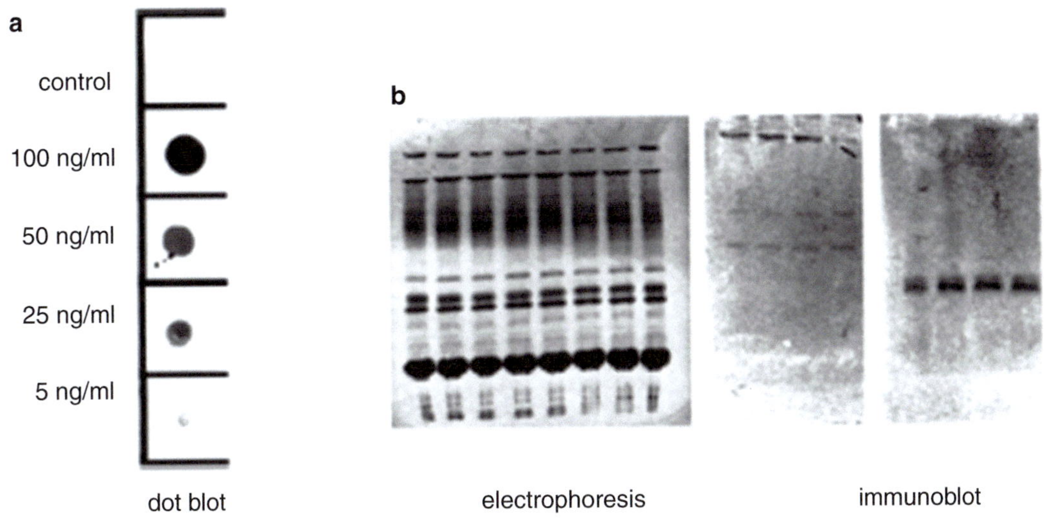

Fig. 14.2 (**a**, **b**) Dot blots and immunoblotting. Specific Abs to an antigen, e.g., an allergen extract, can be detected with the dot blot technique. Further characterization of the Ab or the antigens present in the allergen extract can be achieved by immunoblotting. For further details, see text; lane 1–8 are different allergen extracts

will be converted by the enzyme of the secondary Ab to yield a colored or light-emitting (chemiluminescent) spot. The intensity of the spots is proportional to the amount of Ab present in the serum samples (Fig. 14.2a). However, dot blot results alone do not reveal whether the antigen is made up of one or several antigenic components. Further characterization of the Ab can be achieved by separating the antigens electrophoretically. The separated components are transferred from the gel to a nitrocellulose membrane, a process which is called Western blotting. The membranes are then treated like dot blots as outlined above. This methodology combines the high resolving power of electrophoresis and the discriminating power of an immunological reaction. Components that are recognized by the Ab show up on the Western blot as colored bands (Fig. 14.2b).

14.5.4 Immunoadsorbent Assays

Immunoadsorbent assay (IAA) techniques are widely used for the measurement of serum IgE and IgG Abs. The concept of IAAs is basically very similar to immunoblotting. However, the major difference between immunoblotting and IAA is that the amount of Abs can be quantified. An antigen, e.g., an allergen extract from chicken meat, grass or tree pollen, or house dust mites, is attached to an inert matrix such as the wall of a reaction vial and microtiter plate wells or chemically coupled to a paper disk. The serum of an allergic patient is incubated in a first-step reaction with the allergen-carrying matrix. IgE molecules that recognize the allergen are bound. After removal of excess serum, a secondary Ab, in this case an antihuman IgE Ab raised in rabbits, goats, or horses, is added which forms an allergen-IgE-anti-IgE Ab complex (second-step reaction). Excess of the secondary Ab is also removed by washing. The formation of the allergen-IgE-anti-IgE Ab complex depends on the amount of specific IgE present in the serum sample. Since the secondary Ab or detecting Ab carries a covalently coupled label or tag, the formation of the allergen-IgE-anti-IgE Ab complex, a sandwich-like structure, can be monitored. Utilizing a standard curve with increasing concentrations of the allergens, the signal obtained with allergen-IgE-anti-IgE Ab complex in the serum sample can be compared with the signal of the standard curve, which permits the quantitation of the IgE Abs. The assay format of an IAA in general is summarized in Fig. 14.3.

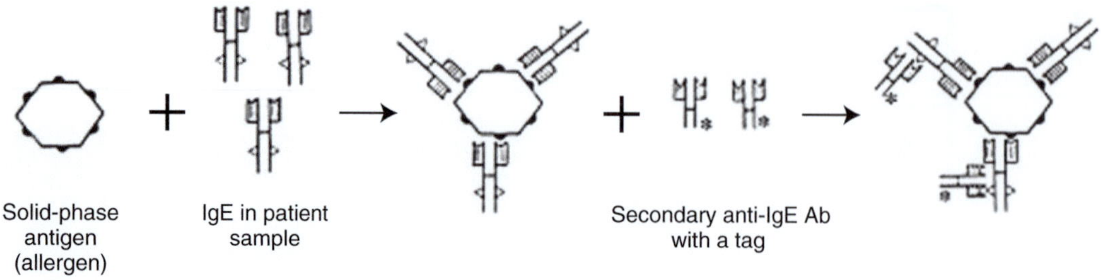

Fig. 14.3 Immuno-absorbent assay format. Reproduced with permission from DPC Biermann, Germany. For further details, see text

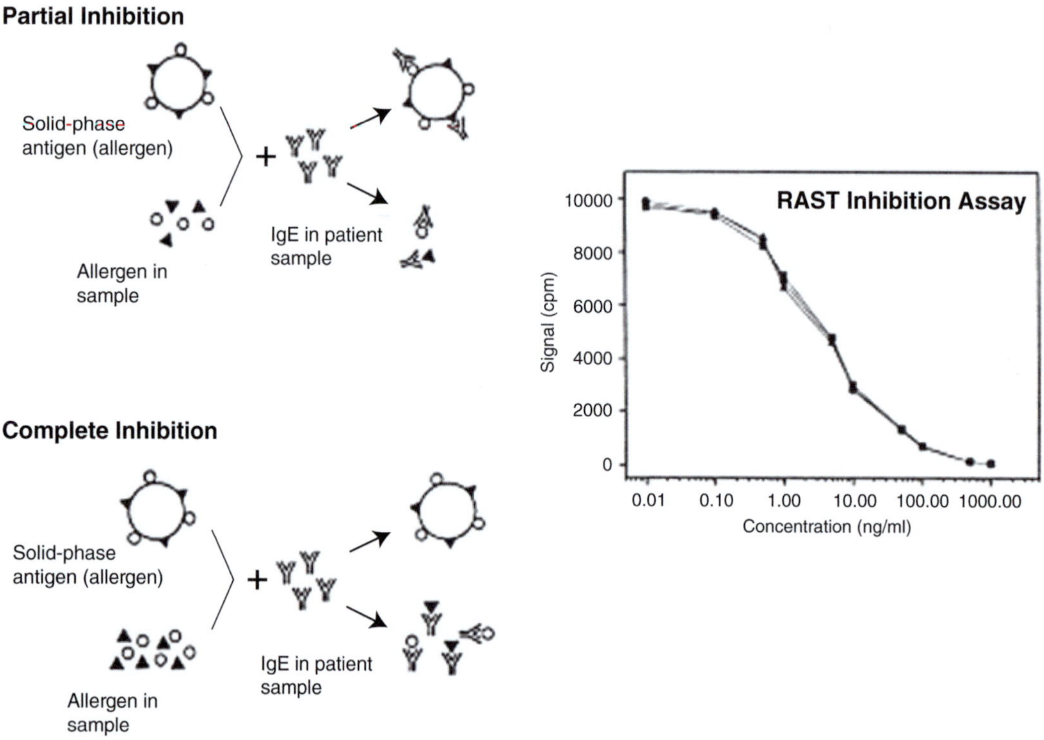

Fig. 14.4 RAST inhibition assay. Reproduced with permission from DPC Biermann, Germany. For further details, see text

The radioallergosorbent test (RAST) is a radioimmunoassay (RIA) [4] in which the allergens are chemically coupled to a paper disk and the secondary Ab is radioactively labeled with 125I. Similarly, the assay can be performed as an enzyme immunoassay (EIA) in which the secondary Ab is labeled with the enzyme β-galactosidase which can react with a colorless substrate to form a colored reaction product. In addition to the RAST or EIA, a RAST or EIA inhibition assay can be performed to confirm and validate the results [5]. The serum samples are first incubated *in vitro* with increasing concentrations of the allergens prior to the RAST or EIA. The binding inhibition of the Ab can be illustrated by a dose-response curve which inversely correlates with the concentration of allergens added; low allergen concentrations still give a high signal in the RAST or EIA, whereas the signal vanishes with high concentrations. The concentration of the allergen at 50% inhibition (IC50) can be calculated from the dose-response curve. A low IC50 is a good indicator for a high affinity and specificity of the Ab to the allergens (Fig. 14.4).

14.5.5 Automated Systems to Analyze Specific IgE Antibodies

Based on the RAST or EIA principle, different automated systems were developed to detect allergen-specific IgEs. Three-dimensional cellulose sponges are used in the ImmunoCAP system (Phadia Uppsala, Sweden). The allergens are covalently coupled to the sponges (in the form of a small cap), and detection of allergen-specific IgE in patient's serum takes place with an enzyme-coupled anti-IgE Ab. Other autoanalyzer platforms have been developed by Siemens (Immulite System) and Hycor-Agilent Technologies (Hytec-288 system). In the Immulite system, biotinylated allergens are used which are captured to an avidin solid phase, while the Hytec system uses a cellulose wafer to which the allergen is covalently coupled. All of these methods have enhanced reproducibility, precision, and sensitivity and are automated to prevent human error. However, although all three systems are calibrated to World Health Organization (WHO) serum standards, intramethod differences with respect to quantitative estimates of allergen-specific IgE concentrations make it difficult to compare analyses performed on different systems [6].

14.5.6 Microarray Systems

Several new microarray systems have emerged for the detection of specific antibodies in allergic or autoimmune diseases. Using a biochip, semiquantitative detection of IgE binding to more than 100 proteins derived from over 40 common allergens can be achieved in only 20 μL of serum or plasma (ImmunoCap Immuno solid-phase allergen chip (ISAC), Phadia, Sweden). Bound IgE is detected with a fluorescently labeled anti-IgE antibody [7]. Protein arrays to profile autoantibodies have been used in research on multiple sclerosis. In this approach, 37,000 different expression clones from a human fetal brain cDNA library were spotted on poly(vinylidene fluoride) (PVDF) membranes. Membranes are incubated with individual cerebrospinal fluid samples to detect binding of autoantibodies [8]. Other techniques have emerged to detect autoantibodies and autoantigens. Phage immunoprecipitation sequencing (PhIP-Seq) measures high-affinity interactions of immobilized autoantibodies with peptides expressed by T7 bacteriophages covering all known protein sequences. DNA of bound phages is subsequently sequenced to identify potentially novel autoantigens [9].

Microsphere-based flow cytometric analysis is a technique used in, e.g., clinical diagnosis [10]. The commercial available kits employing this technique allow simultaneous, quantitative detection of up to 100 different analytes (Luminex xMAP, VeraCode/BeadXpress, Cytometric Bead Array (CBA) BD Biosciences), while instruments to run 500-plex testing in single samples have now also entered the market (FLEXMAP 3D®, Luminex). In these platforms sets of antibody-coupled microspheres are used with different internal fluorescent labels. When passing the laser beam in the analysis cuvette, multiple microspheres can be differentiated from each other. In the xMAP technology, two different fluorescent dyes (one red and one far-red) are incorporated at different concentrations in the microspheres and are analyzed in a specialized analyzer (Luminex, USA). In the VeraCode technology, different antibody-labeled microbeads are encoded by a holographic inscription, and analysis is performed on Illumina reader. Analysis of CBA microbeads (BD Biosciences) can be performed on regular flow cytometers. Beads are differentiated based on their incorporated fluorescent dye intensity in FL3 (fluorescent channel) and forward scatter (FSC) and side scatter (SSC). Binding to bead-coupled antibodies is detected in channel FL2 (Fig. 14.5).

14.6 New Developments

Although classical ELISA-type assays with colorimetric or fluorescent detection are most commonly used in antibody-ligand interaction analysis, new ligand-binding assays with higher

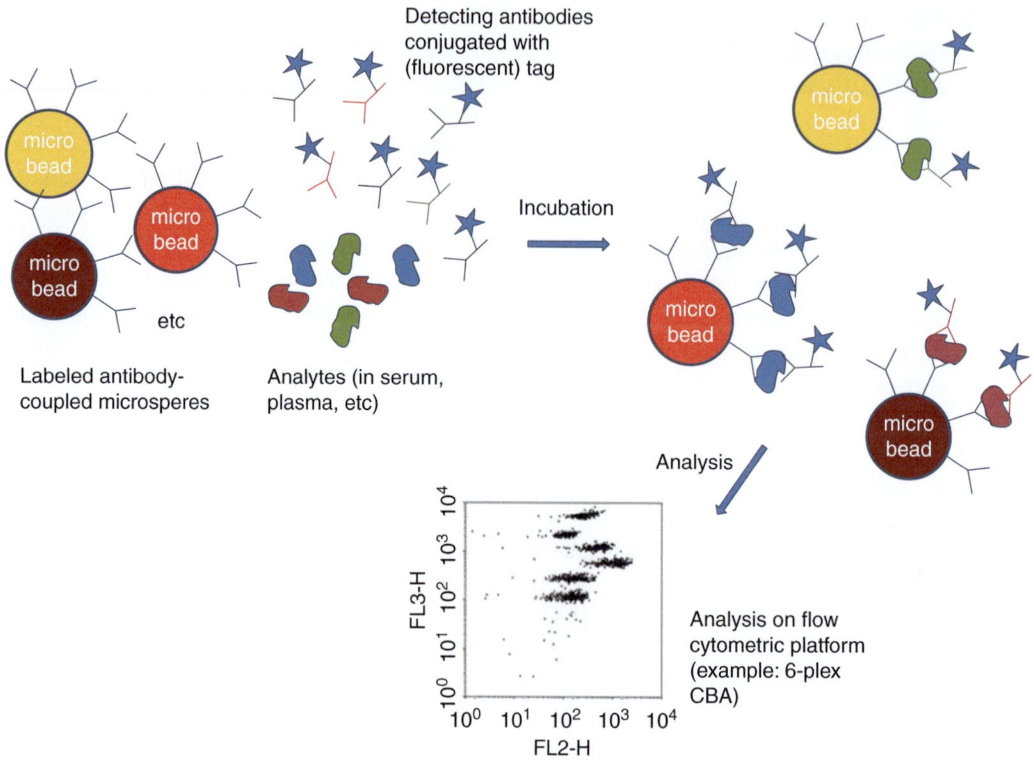

Fig. 14.5 Principle of microsphere-based flow cytometric analysis. Biological samples are incubated with a mixture of antibody-conjugated microbeads and tagged detecting antibodies. After incubation, samples are analyzed on a flow cytometric platform

sensitivity or label-free detection are surfacing [11, 12]. For example, sensitivity of ELISA assays can be increased from a µg/mL to ng/mL range using colorimetric/fluorimetric detection to a ng/mL to pg/mL range using (electro)chemiluminescence detection. Using immuno-PCR, where a DNA marker is conjugated to the detection antibody, the signal can be exponentially amplified by quantitative PCR and reach detection levels into the fg/mL range [11]. Label-free detection methods as surface plasmon resonance, oblique-incidence reflectivity difference, MALDI-TOF, and photonic crystal-based detection have the advantage that unmodified detection antibodies are used, real-time binding to ligands can be monitored, and also association and dissociation kinetics can be evaluated [13, 14]. These developments will be very helpful to design novel array platforms for high-content or high-throughput analyses in, e.g., biomarker research, drug development, and clinical practice.

Selected Readings

Abbas AK, Lichtman AH, Pillai S, editors. Cellular and molecular immunology. St. Louis, MO: Saunders/Elsevier; 2015.

Burtis C, Ashwood E, Bruns D, editors. Tietz textbook of clinical chemistry and molecular diagnostics. St. Louis, MO: Saunders/Elsevier; 2012.

Murphy K, Weaver C, editors. Janeway's immunobiology. New York, NY: Garland Science; 2016.

Uttamchandani M, Yao SQ, editors. Small molecule microarrays. New York, NY: Humana Press, Springer Nature; 2017.

References

1. Köhler G, Milstein C. Continuous cultures of cells fused secreting antibody of predefined specificity. Nature. 1975;256:495–7.
2. Ring J. Diagnostic methods in allergy. Behring Inst Mitt. 1981;68:141–52.
3. Ring J, editor. Angewandte allergologie. München: MMV Medizin Verlag; 1988.

4. Wide L, Bennich H, Johansson SGO. Diagnosis of allergy by an in vitro test for allergen antibodies. Lancet. 1967;2:1105.
5. Gleich GJ, Yunginger JW. Variations of the radioallergosorbent test for measurement of IgE antibody levels, allergens and blocking antibody activity. In: Ring J, Burg G, editors. New trends in allergy. Heidelberg: Springer; 1981. p. 98–107.
6. Hamilton RG. Clinical laboratory assessment of immediate-type hypersensitivity. J Allergy Clin Immunol. 2010;125:S284–96.
7. Hamilton RG. Proficiency survey-based evaluation of clinical total and allergen-specific IgE assay performance. Arch Pathol Lab Med. 2010;134:975–82.
8. Somers K, Govarts C, Stinissen P, Somers V. Multiplexing approaches for antibody profiling in multiple sclerosis. Autoimmun Rev. 2009;8:573–9.
9. Burbelo PD, O'Hanlon TP. New autoantibody detection technologies yield novel insights into autoimmune disease. Curr Opin Rheumatol. 2014;26:717–23.
10. Hsu H-Y, Joos TO, Koga H. Multiplex microsphere-based flow cytometric platforms for protein analysis and their application in clinical proteomics-from assays to results. Electrophoresis. 2009;30:4008–19.
11. Spengler M, Adler M, Niemeyer CM. Highly sensitive ligand-binding assays in pre-clinical and clinical applications: immuno-PCR and other emerging techniques. Analyst. 2015;140:6175–94.
12. Mahler M, Bentow C, Serra J, Fritzler MJ. Detection of autoantibodies using chemiluminescence technologies. Immunopharmacol Immunotoxicol. 2015;38:14–20.
13. Moore CD, Ajala OZ, Zhu H. Applications in high-content functional protein microarrays. Curr Opin Chem Biol. 2016;30:21–7.
14. Sun Y-S. Use of microarrays as a high-throughput platform for label-free biosensing. J Lab Autom. 2015;20:334–53.

Immunoassays

15

Martin Schwickart and Inna Vainshtein

15.1 Introduction

Immunoassays are biochemical tests that utilize antibodies as reagents to detect the presence and/or concentration of ANALYTES. This chapter focuses on immunoassays; however the principles apply also to other ligand binding assays that do not utilize antibodies but instead bind the ANALYTE with ligands, receptors, binding proteins, or other macromolecules. The basic principle of the immunoassay is antibody-antigen interaction, which typically is of high AFFINITY and specificity. This feature allows the detection and quantification of low-abundant ANALYTES. ANALYTES measured by immunoassays can be small molecule drugs, hormones, lipids, peptides, proteins, and even larger structures like viruses and bacteria. Immunoassays can measure ANALYTES not only in pure systems (e.g., buffer, cell medium) but also in complex MATRICES including cell lysate, serum, plasma, urine, synovial fluid, bronchoalveolar lavage, sputum, and amniotic fluid. The result can be available in a few minutes (e.g., pregnancy tests); however, it is typical that a few hours pass between sample preparation and result.

The first immunoassay was developed for quantification of insulin in 1959 [1]. Since then, immunoassays proved to be of such significance that Rosalyn Yalow was awarded the Nobel Prize in Medicine in 1977 for the "development of radioimmunoassays of peptide hormones." Immunoassay became widespread after the development of the hybridoma technique which enables the production of monoclonal antibodies [2], a key reagent in immunoassays. This discovery was a major advance in medicine, and Georges Köhler and César Milstein were awarded the Nobel Prize in Medicine in 1984 for "the principle for production of monoclonal antibodies." Since then, immunoassays were built to detect a plethora of ANALYTES.

In biological research and drug discovery, immunoassays are used for basic discovery research, identification of new therapeutic targets, and screening in search of new drug candidates. In pharmacology, immunoassays are integral to pharmacokinetic and toxicokinetic measurements as well as to pharmacodynamic biomarker and ANTIDRUG ANTIBODY analysis. In clinical practice, immunoassays are used for diagnostic tests and THERAPEUTIC DRUG MONITORING. Immunoassays still find new applications. As part of drug development, biomarkers are routinely evaluated to identify predictors of drug response. Although no FDA-approved immunoassay companion diagnostic yet exists,

Final manuscript submitted on September 07, 2017.

M. Schwickart (✉)
Translational Development and Diagnostics, Celgene, San Francsico, CA, USA
e-mail: mschwickart@celgene.com

I. Vainshtein
BioPharmaceuticals R&D, AstraZeneca, South San Francisco, CA, USA
e-mail: vainshteini@medimmune.com

M. J. Parnham et al. (eds.), *Nijkamp and Parnham's Principles of Immunopharmacology*,
https://doi.org/10.1007/978-3-030-10811-3_15

striking relationships between biomarkers and response have been reported (e.g., periostin [3]).

This chapter aims to give a practical introduction to immunoassays. The "Immunoassays Handbook" [4] provides more comprehensive and detailed information.

15.2 Principles of Immunoassays

An immunoassay detects an ANALYTE either by direct binding to an antibody (noncompetitive or sandwich immunoassay) or through competition between the ANALYTE and a labeled ANALYTE reagent for binding to an antibody (competitive immunoassay). Both sandwich and competitive immunoassays can be performed in a heterogeneous (with wash steps) or homogeneous (no wash) format.

15.2.1 Sandwich Immunoassay

Sandwich immunoassays measure an ANALYTE in an immunocomplex of two antibodies. In this frequently used format, a capture antibody immobilized on a SOLID PHASE is incubated with a sample containing the ANALYTE. After allowing time to bind, the wells are washed to remove all unbound materials, and the antibody-bound ANALYTE is detected with a detection antibody (Fig. 15.1). The detection antibody is labeled with a signal-producing molecule (e.g., radioactive isotope, enzyme, fluorescent or chemiluminescent tag, etc.). After washing, the amount of label remaining on the SOLID PHASE is directly related to the amount of the ANALYTE in the sample. By comparing the signal to a standard curve (Fig. 15.1), a concentration of the ANALYTE can be determined. In sandwich immunoassay, both the capture and the detection antibody are in excess of the ANALYTE. Therefore, this type of assay is also called reagent-excess immunoassay. For a sandwich immunoassay to work, the ANALYTE must bind to two antibodies simultaneously. Therefore this type of assay tends to be very specific as it is unlikely that *two* distinct antibodies would bind a MATRIX component by chance. On the other hand, smaller molecules, such as organic compounds or short peptides that are too small to contain two nonoverlapping EPITOPES, cannot be detected with this assay format. Sandwich assays are highly sensitive and measure ANALYTE concentration over a wide concentration range.

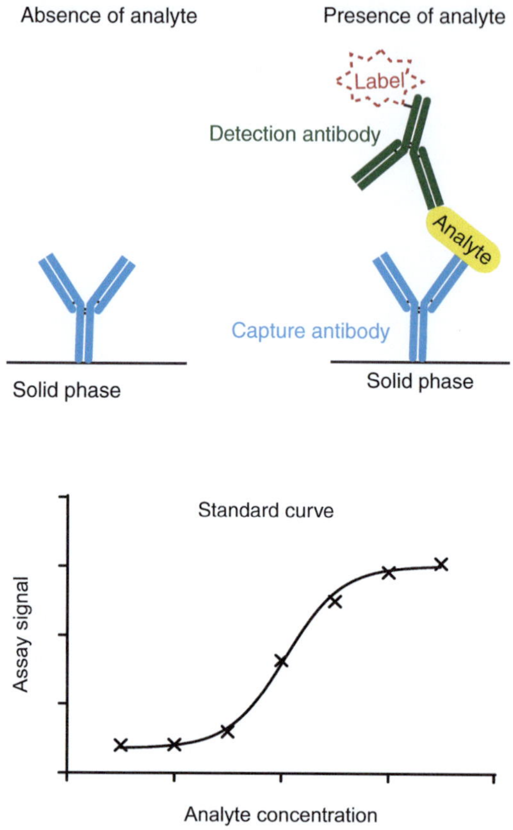

Fig. 15.1 Heterogeneous noncompetitive assay. The capture antibody is immobilized on a solid phase. In the presence of analyte, a "sandwich" is formed between the capture antibody, the analyte, and the detection antibody, which persists after washing away sample and surplus reagents. In absence of analyte, the detection antibody cannot bind and is washed away. The label on the detection antibody (e.g., enzyme or fluorescent molecule) produces a detectable signal which is directly related to the analyte concentration. The standard curve consists of a series of known concentrations of standard material, which is identical or similar to the analyte, to etablish a signal-to-concentration relationship

15.2.2 Competitive Assays

In the competitive assay format, a capture antibody immobilized on a solid surface is incubated with a sample and a labeled reagent analyte.

The labeled reagent analyte and the ANALYTE from the sample compete for binding to the antibody. The more ANALYTE is present in the sample, the less labeled reagent analyte will bind to the antibody. After allowing time to bind, the plates are washed to remove all unbound material, and the remaining antibody-bound labeled reagent analyte is detected by its label (Fig. 15.2). Because of the competition, the assay signal is inversely related to the amount of the ANALYTE in the sample. Unlike reagent-excess sandwich assays, competition assays are reagent-limited. The concentration of the labeled reagent ANALYTE and the ANALYTE has to be in a similar range for effective competition, and the concentration of the antibody has to be limited. If the antibody concentration is too high, the ANALYTE and the labeled reagent analyte bind to the antibody, and the signal is no longer dependent on the ANALYTE concentration in the sample.

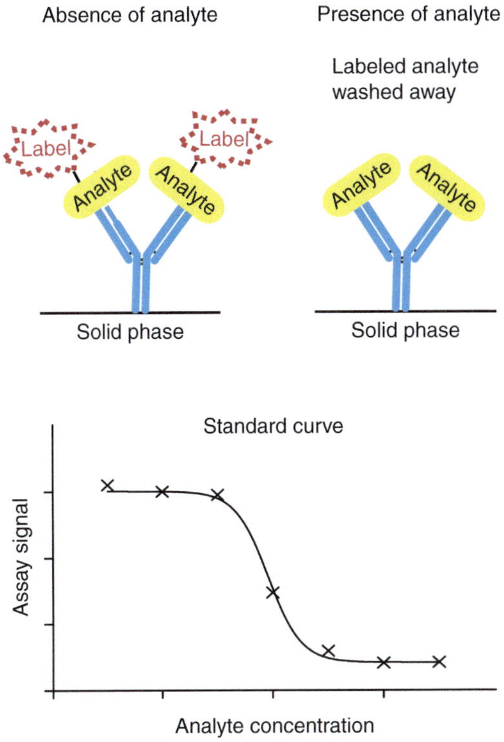

Fig. 15.2 Heterogeneous competitive assay. The analyte in the sample and labeled reagent analyte compete for binding to the capture antibody. In absence of analyte, all available labeled reagent analyte is bound, and the assay signal is maximal. In presence of analyte in the sample, binding sites are occupied with analyte. In this scenario, after washing, labeled analyte is not retained and no signal is detected. The assay signal is inversely related to the analyte concentration in the sample

Competitive assays use only one antibody, and this is an advantage for ANALYTEs where an antibody pair is not available/possible. Small ANALYTEs, which do not have two independent EPITOPES and therefore do not accommodate simultaneous binding of two antibodies, are often measured with competitive assays. Competitive assays tend to have a narrow quantitative range and are less sensitive than sandwich immunoassays.

15.2.3 Homogeneous and Heterogeneous Immunoassays

In order to illustrate the principle of immunoassays, we described both the competitive and sandwich approaches as heterogeneous immunoassay. However, both formats are also possible as homogeneous immunoassay. In contrast to heterogeneous immunoassays where reagents are added and washed away or separated, homogeneous immunoassays do not require washing or separation. In homogeneous immunoassay, reagent and sample are mixed together, and the signal is generated as a result of antibody binding to the ANALYTE. Figures 15.3 and 15.4 show examples for competitive and noncompetitive homogeneous immunoassays (see also Sect. 15.3.3). Although microtiter plates (e.g., 96-well plates) have facilitated wash and reagent-adding steps tremendously, washing and reagent addition is still compara tively labor intensive. Homogeneous immunoassays are therefore easier to automate. Homogeneous immunoassays are fast and fairly simple procedures and deliver results in a much shorter period of time. Numerous clinical chemistry and toxicology laboratories have automated clinical analyzers that are based on homogeneous detection of ANALYTES. Another advantage of the homogeneous format is detection of low-AFFINITY interactions that would be washed away in a heterogeneous format. One of the drawbacks

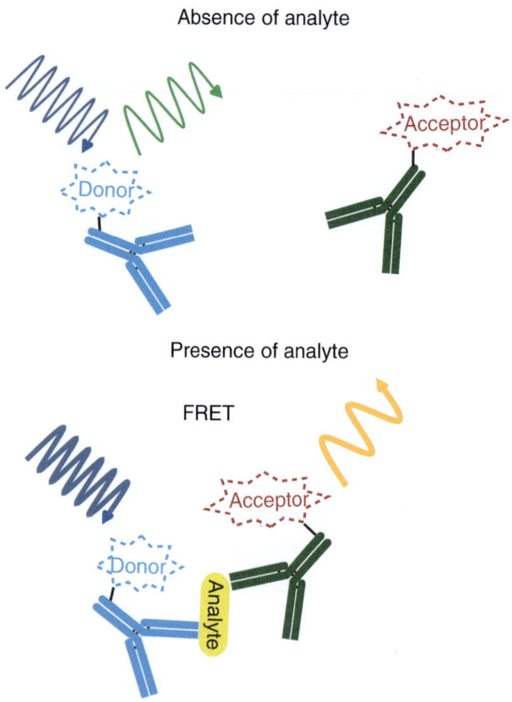

Fig. 15.3 Homogenous noncompetitive assay. The example depicted here is a fluorescence resonance energy transfer (FRET) assay. The donor fluorophore initially absorbs the energy, and the acceptor is the fluorophore to which the energy is subsequently transferred. The transfer of energy leads to a reduction in the donor's fluorescence intensity and an increase in the acceptor's emission intensity. In presence of analyte, through binding to the analyte, donor and acceptor are in close enough proximity for FRET. In absence of analyte, donor and acceptor are too far apart for FRET

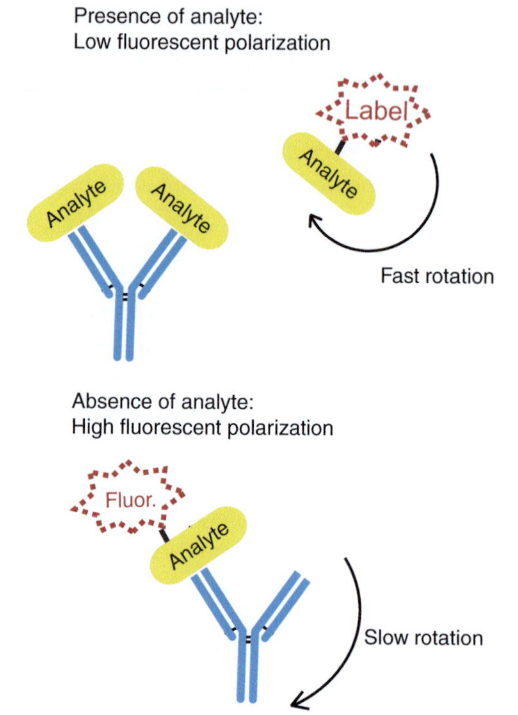

Fig. 15.4 Homogeneous competitive assay. The example depicted here is a fluorescence polarization assay. The fluorophore is excited with polarized light, and the degree of polarization of the emitted light is detected. The fluorophore on the labeled analyte rotates quickly, driven by Brownian motion. This fast rotation leads to reduced polarization of the emitted light. The slow-rotating larger analyte-antibody complex emits polarized light

of homogeneous format is that it is more prone to non-specific effects of constituents in the sample MATRIX [5, 6], which are washed away in homogeneous immunoassays.

15.3 Components of Immunoassays

15.3.1 Detection Systems

In most immunoassays, the antibodies or reagent ANALYTE is chemically linked with some kind of detectable label. A vast diversity of labels is used in different immunoassays. Labels are detectable because they produce an observable signal such as radiation, a color change in solution or fluorescence. A plethora of labels has been developed in the last decades; for more detailed information, see [7, 8].

15.3.1.1 Radioactive Labels

Early immunoassays utilized radioactive labels such as iodine-125. Ionizing radiation from iodine-125 and other isotopes is measured in scintillation counters. In the counter, a photomultiplier tube detects flashes of light when radioactivity is absorbed by a scintillator, which can be a transparent crystal or organic liquid. Radioactive assays confer high sensitivity. However, radioactivity is inherently dangerous, and therefore labs conducting radioactive work need to follow appropriate safety protocols. Also, since radioactivity decays over time, the shelf life of radioactive reagents is limited. Radioactive labels became increasingly unpopular when sensitive nonradioactive alternatives were developed.

15.3.1.2 Enzymes

Many immunoassays use enzymes as labels. The enzymes convert substrates into chromogenic, fluorescent, or chemiluminescent products. Absorbance, fluorescence, or luminescence is measured with simple instrumentation such as spectrophotometers or fluorescence/luminescence readers. One very common type of enzymatic immunoassay is the enzyme-linked immunosorbent assay (ELISA). The combination of the enzyme HORSERADISH PEROXIDASE (HRP) with the chromogenic substrate 3,3′,5,5′-tetramethylbenzidine (TMB) is a mainstay ELISA technique in many labs, and a multitude of kits and reagents using this detection system are commercially available. ELISA is so well-established that newer systems often benchmark their assay performance against ELISA. An example of an ELISA assay standard curve is shown in Fig. 15.5.

15.3.1.3 Fluorescent Labels

Fluorescent molecules are either produced by an enzyme (see above) or conjugated directly to a reagent ANALYTE or a detection antibody. Fluorescence is measured with instruments called fluorometers (also called fluorimeters). A fluorometer generates light of a particular wavelength range to excite fluorescent molecules and detects emitted light of a different wavelength with a photomultiplier tube or photodiode. Biological samples often have some degree of autofluorescence, and washing steps are useful to remove fluorescent sample constituents. Fluorophores with high quantum yield (bright fluorophores) like phycoerythrin (PE) or allophycocyanin (APC) are currently used in commercially available platforms (e.g., Luminex, Gyros, and Singulex). The sensitivity is frequently similar or better than comparable ELISA assays.

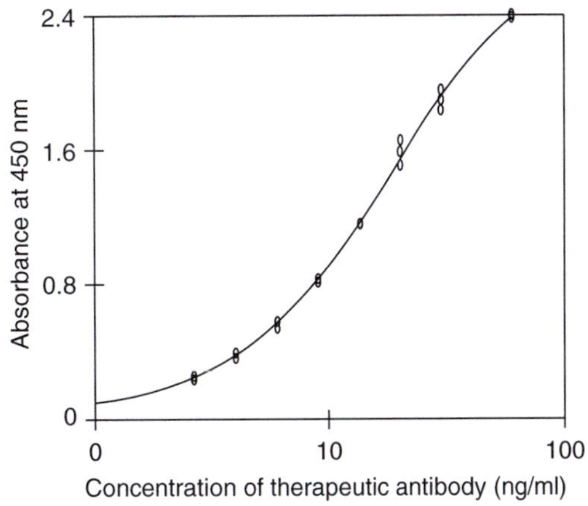

Fig. 15.5 Standard curve of a universal assay to detect therapeutic human monoclonal antibody. In preclinical pharmacology and toxicology studies, therapeutic human monoclonal antibodies in serum are measured often by a "universal assay." In this assay, two different anti-human IgG capture and detection antibodies form a sandwich with the therapeutic antibody analyte. The universal assay is only feasible in nonhuman specimens. In human specimens, this assay would detect the sum of endogenous and therapeutic IgG, and since the endogenous IgG concentration is very high compared to therapeutic antibody, the assay signal would be mostly defined by the endogenous IgG concentration. In animals, endogenous IgG does not bind to the reagent antibodies and therefore only measures the therapeutic antibody. In this example, serum samples from monkeys dosed with a human monoclonal antibody were tested. The sample matrix is 10% cynomolgus monkey serum and 90% buffer. In case that the result was above the quantitative range, the sample was retested at a higher dilution in 10% pooled cynomolgus monkey serum. In brief, standards, controls, and test samples were incubated with sheep antihuman IgG immobilized on a microtiter plate. After sample incubation, unbound materials were washed away, and therapeutic antibody was detected using goat antihuman IgG-HRP conjugate. After washing, TMB substrate solution was added. The optical density at 450 nm was measured after the stop of the reaction with 2 N phosphoric acid

15.3.1.4 Chemiluminescence

Analogous to fluorescence, luminescence can be produced by enzymes, or by chemiluminescent molecules, which are directly attached to the detection antibody. HORSERADISH PEROXIDASE (HRP) and alkaline phosphatase (AP) catalyze the conversion of substrates into luminescent products which then spontaneously decompose to produce light. Electricity can also be used to induce a luminescent reaction; an immunoassay platform is commercially available from Meso Scale Discovery. See [9] for further information.

15.3.1.5 Other Labels

In principle, immunoassays can make use of any label that generates a measurable signal. In lateral flow assays, larger particles (e.g., colored beads or gold particles) migrate through a test strip and provide a color signal that can be detected by the eye or with readers [10].

DNA has successfully been used as label. The amount of DNA conjugated to the detection antibody is quantified with a real-time PCR instrument. Since only minute amounts of DNA are required to generate a signal, DNA as a label can improve assay sensitivity [11]. Reagents are commercially available from Chimera Biotec.

It is also possible to detect the binding of antibodies just by their mass. Change in mass can be detected by surface plasmon resonance. In this technique, the sensor is a thin metal strip (e.g., gold) on a glass support. The metal surface is decorated with the antibody. A microfluidics system brings the ANALYTE in contact with the metal surface. Polarized light is reflected on the backside of the metal strip. At the resonant angle, surface plasmon resonance causes photons to be transformed to surface plasmons. At this angle, the intensity of the reflected light is at a minimum. When the ANALYTE binds to the surface, the resonant angle changes, and this is the assay signal. Surface plasmon resonance has been mainly used for the characterization of binding constants (i.e., on-rate and off-rate). Surface plasmon resonance immunoassays have also been used for analysis of clinical samples; however this assay platform not commonly used for routine testing and assay is not (yet) commercially available [12]. Mass detection is also achieved with vibrating surfaces. The frequency of the vibration changes when antibody binds to it. The commercially available Proterix Vibe platform makes use of this principle.

More recently, capture with antibodies has been combined with mass spectrometry to allow mass spectrometric detection of low-abundant proteins. Here, the ANALYTE is first captured by an immobilized antibody, eluted, often enzymatically digested, and then analyzed in a mass spectrometer. In comparison with immunoassays, mass spectrometry readily detects modifications of proteins. The combination of immunocapture and mass spectrometric detection is especially useful for quantification of antibody-drug conjugates and might expand to more applications in the future.

Often, the signal-generating label is not directly conjugated to the antibody but linked to STREPTAVIDIN, a protein of approximately 52 kDa. STREPTAVIDIN binds with exceptional AFFINITY to biotin, a small molecule of 244 Da. The detection antibody is labeled typically with many biotin molecules leading to an amplification of signal with this system. Many biotinylated antibodies and STREPTAVIDIN-conjugated labels are already commercially available. This speeds up assay development. Biotinylation tends to not adversely affect the properties of the antibody. In contrast, direct labeling of antibodies with enzymes, fluorophores, etc. is often prone to affect the function of antibodies and requires often longer optimization.

15.3.2 Separation Systems

Heterogeneous immunoassays rely on separation of bound antibody-ANALYTE complexes from unbound ANALYTE and excess reagents. In radioimmunoassays, precipitation and centrifugation are often used to separate bound complexes. In ELISA assays, the capture antibody is immobilized on a SOLID PHASE (i.e., 96-well polystyrene plate), which allows for easy washing of 96 samples at the same time. Plates pre-coated with the capture antibody are often part of commercially available biomarker assay kits. Some plate-based platforms allow the determination of several ANALYTEs in the same run by multiplexing. Capture antibodies to each ANALYTE are coated in discrete spots in one well. Each of these antibody spots will bind its respective ANALYTE in the sample. A mixture of detection antibodies against all ANALYTEs allows detection of ANALYTE-associated signals in the individual spots. Aushon, a CHEMILUMINESCENCE platform, and the ELECTROCHEMILUMINESCENCE platform Meso Scale Discovery offer plate-based multiplexing.

A similar approach is to immobilize antibody reagents on polystyrene or magnetic beads. Magnetic particles are easily separated just by applying a magnetic field, while polystyrene beads require centrifugation or filtering. The proprietary platform from Luminex, Singulex, and Quanterix uses magnetic beads. Singulex and Quanterix have optimized the technology for high sensitivity. Luminex has the capability to multiplex ANALYTES; an example of a Luminex assay is shown in Fig. 15.6. The Luminex platform uses fluorescent beads with different emission spectra. Each bead type is coated with an antibody to a particular ANALYTE. By mixing different fluorescent beads (with different ANALYTE specificities), several ANALYTES can be captured at the same time. After washing, a mixture of PE-labeled detection antibodies completes the sandwich. The signals are read in a flow cytometer, and the median fluorescent intensity in the PE channel is directly related to the concentration of each ANALYTE. Due to different fluorescent coloring of the beads, the signal for each ANALYTE can be distinguished.

Small AFFINITY COLUMN can be used to capture the ANALYTE. The commercial Gyros system combines this principle with microfluidics to wash the column and deliver assay reagents. This system has the advantage that the sample incubation time is very short.

Separation can also be achieved by membranes. Western blot is an important analytical technique to characterize antibodies or identify proteins. Proteins in a complex sample (e.g., cell lysates) are separated by gel electrophoresis, blotted onto a membrane, and then detected with one specific antibody. Western blot is not quantitative and not amenable to automation.

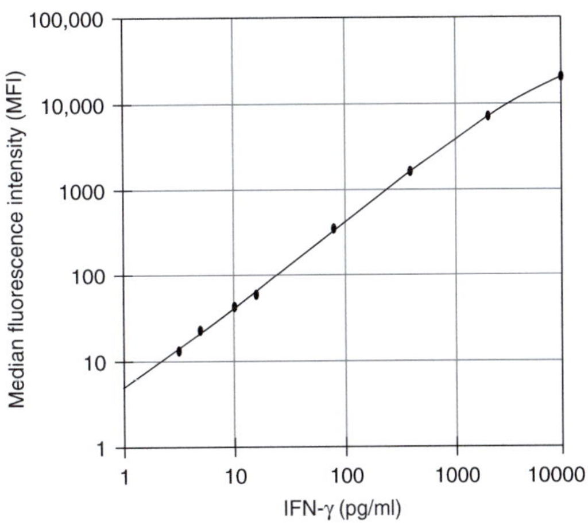

Fig. 15.6 IFN-γ standard curve from a 4-Plex Luminex assay. Cytokines are produced mostly by cells of the immune system and play a key role in the differentiation and activation of immune cells (see Chap. 6). Cytokines are measured in many in vitro assays. In vivo, cytokines are markers of immune activation and also of a potentially life-threatening condition—cytokine release syndrome (or cytokine-associated toxicity). Cytokine release syndrome can arise after injection of immunomodulatory therapeutic drugs, which activate the immune system. Standards, controls, and twofold diluted test samples were incubated with the fluorescent Luminex 4-Plex beads mixture containing immobilized antibodies to TNF-α, IFN-γ, IL-2, and IL-6. During sample incubation each analyte binds to its corresponding beads. The beads were washed, and a mixture of biotinylated detection antibodies to the four analytes was added to the beads. After incubation, beads were washed, and streptavidin-labeled phycoerythrin was added to the beads. After incubation, beads were washed again before analysis in the Luminex instrument. After a minimum of 100 beads for each analyte was collected, mean fluorescent intensity for each analyte was determined. Only the standard curve for IFN-γ is depicted. Standard curves for TNF-α, IL-2, and IL-6 are not shown

15.3.3 Homogeneous Assays Systems

Homogeneous assays are truly mix-and-read systems. Several approaches have been implemented that distinguish bound from unbound ANALYTE without separation and washes [8]. One of the first homogeneous assays was the agglutination immunoassay. In this assay, large particles (e.g., latex beads) are coated with antibodies. The presence of ANALYTE leads to cross-linking of the particles or beads and is observed as agglutination. Agglutination can be detected by the naked eye or measured by TURBIDIMETRY, as the reduction of light passing through the sample. Alternatively, NEPHELOMETRY measures the scattered light. This simple technique, however, is affected by particulate matter in the sample.

ANALYTE-induced proximity assays such as fluorescence resonance energy transfer (FRET) use the property of particular combinations of fluorescent molecules to generate a signal only when they are in proximity to each other. The fluorophore pair is selected in such way that the emission spectrum of the one fluorophore (donor) overlaps with the excitation wavelength of the other fluorophore (acceptor). These fluorescent molecules are conjugated to capture and detect antibody. When an immunocomplex is formed with the ANALYTE, the two fluorophores are brought close together, and energy can be transferred from the donor to the acceptor. Upon energy transfer the acceptor emits fluorescence, and the fluorescence intensity is directly related to the concentration of the ANALYTE (Fig. 15.3). Time-resolved fluorescence resonance transfer (TR-FRET) is a refinement of FRET where the time delay between excitation and emission is utilized to distinguish background fluorescence from FRET fluorescence. This allows short-lived fluorescence (background) to decay and do not interfere with the acceptor signal detection. A CHEMILUMINESCENCE bead-based homogeneous system is AlphaScreen. This assay utilizes another type of donor-acceptor bead proximity measurements. The donor bead produces singlet oxygen, which produces luminescence of the acceptor bead when reacts with the bead surface. The system capitalizes on the property of singlet oxygen to travel only short distance (less 200 nm) before decay. Chemical conjugation of binding partners to donor and acceptor beads allows to run immunoassays in either sandwich or competition formats without washing or separation.

Fluorescence polarization is another elegant principle used in homogeneous assays. Fluorescent molecules in solution emit mostly non-polarized light upon excitation with polarized light due to their fast rotation, driven by BROWNIAN MOTION. When these fluorescent ANALYTES are bound to large molecules like antibodies, they rotate slower and mostly emit polarized light (Fig. 15.4). The degree of emitted polarized light is related to the amount of antibody-fluorescent ANALYTE complex. The assay is typically run in the competitive format.

Reconstitution assays are another type of homogeneous systems intensely used to detect drug abuse substances and also for high-throughput screening (HTS). One of the examples of this group is the cloned enzyme donor immunoassay (CEDIA). The assay is based on the spontaneous assembly of two fragments of BETA GALACTOSIDASE. Together, the two fragments complement and form an active beta-galactosidase complex, which catalyzes hydrolysis of chromogenic, fluorescent, or chemiluminescent substrates. One fragment is conjugated to an ANALYTE of interest without impacting enzyme complementation. Antibody in the assay binds to the fragment-conjugated ANALYTE and inhibits the enzyme complementation. When ANALYTE is present in the sample, it will bind to the antibody and prevent binding of antibody to the conjugated fragment, the enzyme can again assemble, and the signal is restored. Similarly, in the enzyme multiplied immunoassay technique (EMIT), an antibody binding to an ANALYTE-conjugated enzyme blocks the enzymatic activity. Both the CEDIA and EMIT are competitive homogeneous assays.

15.3.4 Binding Reagents

Antibodies are the predominant binding reagent in immunoassays. Antibodies are either polyclonal or monoclonal. A monoclonal antibody only binds to one defined EPITOPE, whereas a POLYCLONAL ANTIBODY binds typically to several EPITOPES on any one antigen. A POLYCLONAL ANTIBODY is a mixture of antibodies with different affinities, whereas a monoclonal antibody has one defined AFFINITY.

Polyclonal antibodies are produced in batches from blood of animals that were immunized with an antigen. Polyclonal antibodies in immunoassays are mostly produced in rabbits; however, for larger batches, goats are the animal of choice. Monoclonal antibodies are mainly produced by hybridoma technique and mostly originate from mice and rats. After immunization, the spleen is harvested and fused with a myeloma cell line, and single antibody-producing hybridoma cells are expanded and produce antibodies only from one clone—the monoclonal antibody.

Because of the differences in antibody generation, a POLYCLONAL ANTIBODY has a limited supply. A new immunization or bleed is not identical to a previous batch of POLYCLONAL ANTIBODY. In contrast, an established monoclonal cell line can produce the same antibody theoretically forever, leading to an unlimited supply of one antibody. Monoclonal antibody production is more elaborate, a usable amount of monoclonal antibody can be expected approximately 6 months after immunization; a POLYCLONAL ANTIBODY can be produced in a time frame of approximately 2 months. Compared to monoclonal antibodies, polyclonal antibodies are often more stable and less sensitive to labeling. Polyclonal antibodies are often used as detection antibodies in sandwich immunoassays. The ability to bind to several EPITOPES on the same ANALYTE molecule leads to an amplification of signal and therefore higher sensitivity. The uniformity of monoclonal antibodies is beneficial when specificity to one particular EPITOPE is of relevance (e.g., PROTEIN ISOFORMS).

15.3.5 Standard Material

The standard curve or calibration curve translates the assay signal into a concentration. Quantitative immunoassays rely on a standard curve, whereas qualitative (e.g., pregnancy test) or semiquantitative assays do not require a standard curve. A strictly quantitative immunoassay requires the standard curve material to behave identical to the endogenous ANALYTE in the sample. Especially for proteins, this is rarely the case. Protein calibrators are mostly derived from recombinant purified material and may differ from the endogenous ANALYTE (e.g., by isoform, fragment, posttranslational modifications, protein tags, purity, etc.). Even if the standard curve material and endogenous ANALYTE do not behave identically, relative quantification is still possible (see ACCURACY).

15.3.6 Buffers

Buffers are used to dilute the standard curve and samples. A multitude of buffers are used in immunoassays. Most often, buffers contain a pH buffer system based on phosphate or Tris(hydroxymethyl) aminomethane (Tris). Additionally, buffers often contain components that reduce non-specific interaction of the ANALYTE or assay reagents. This serves to minimize the background signal or prevent spurious high signals. Bovine serum albumin and casein in buffers are used to "block" non-specific interactions with the assay well. DETERGENTS also help to reduce non-specific binding. In some assays, MATRIX effects cannot be eliminated by dilution. In this case, it is desirable to create a buffer that behaves similar to the sample MATRIX. This is achieved with animal proteins or animal sera.

15.4 Evaluation of Assay Performance

During assay development, assay performance criteria are incorporated into the design of the assay. After the assay development is finalized, assay performance is formally evaluated in validation. Validation consists of a set of experiments that demonstrates key performance parameters. Not all assay parameters need to be tested in each validation; however the validation of an assay should always be as extensive as necessary to ensure reliable and interpretable results [13].

15.4.1 Precision

Precision is the ability of the assay to deliver consistent results. Precision is most commonly expressed as the coefficient of variation (CV):

$$\%\mathrm{CV} = \frac{\text{Standard deviation of measurements}}{\text{Mean of mesurements}} \times 100$$

Precision of the assay is evaluated through intra-assay precision and inter-assay precision. Samples, controls, or calibrators are evaluated for precision. For inter-assay precision, replicates from different assays are evaluated. For immunoassays a sample with a CV of 25% is usually considered acceptable. During testing, samples are typically tested in duplicate, and samples with CVs higher than 25% are tested again. For assays that are used for a long time and across multiple reagent lots, it is advisable to prepare control samples that can be tested to detect DRIFT or sudden changes of the assay results.

15.4.2 Accuracy

ACCURACY is the ability of the assay to measure true "true" ANALYTE concentrations. True ACCURACY can only be determined if an accepted reference method exists or if the ANALYTE is manufactured and as such the standard curve material is identical to the analyte (e.g., drug product). For most ANALYTES in immunoassays, accepted reference methods are not available. However, indirect evaluation of ACCURACY is still useful and an integral part of method evaluation/validation. To that end, a known amount of ANALYTE (i.e., standard curve material) is added (=spiked) to a few samples to determine recovery. The sample is tested with (spiked) and without addition (unspiked) of the ANALYTE. In the optimal case, 100% recovery, the difference between these two measurements should be identical to the spiked concentration. Considering some degree of imprecision, the spiked concentration should typically be a few fold (three- to fivefold) greater than the endogenous ANALYTE concentration. The degree of recovery is calculated by:

$$\%\,\text{Recovery} = \frac{(\text{Measured conc. spiked} - \text{measured conc.unspiked})}{\text{Nominal conc. spiked}} \times 100$$

Additionally, linearity of dilution can complement recovery data. A serial dilution of a sample is prepared and tested, and dilution-adjusted results are compared. Complex samples like serum or plasma at low dilutions tend to have MATRIX effects. At higher dilutions, when MATRIX effects are negligible, dilution should have no influence on the dilution-adjusted results. Should that not be the case, the standard curve material and the endogenous ANALYTE are different, and the assay is not strictly quantitative.

15.4.3 Sensitivity/Quantitative Range

The sensitivity of an assay can be determined in different ways. The limit of detection (LOD) represents the lowest concentration of ANALYTE that can be reliably detected. A common approach is to test negative or blank samples repeatedly and define the limit of detection equal to their mean plus two or three standard deviations. For a quantitative assay, the lower limit of quantification (LLOQ) is of more practical relevance. The LLOQ is the lowest concentration for which a certain precision criteria (e.g., a CV of 25%) and ACCURACY (e.g., recovery between 75% and 125% of the nominal concentration) can be achieved. Similarly, the upper limit of quantification (ULOQ) is the highest concentration at which the precision and ACCURACY criteria can be achieved. The sensitivity is often an assay development focus for low-level ANALYTES, whereas the quantitative range of the assay is important for ANALYTES that are present over a broad range in the samples.

15.4.4 Stability

The stability assessment should mimic the conditions that a sample will be exposed to. The results of the stability assessment will influence how samples will be stored and tested. Samples left at room temperature are tested to evaluate whether the time of sample preparation has an impact. Storage at −80 °C, −20 °C, or 4 °C is evaluated to determine whether samples are stable in storage. Evaluation of ANALYTE stability after freeze/thaw cycles determines how often a sample can be retested from frozen storage.

15.5 Summary

Immunoassays are a core analytical technique in biological sciences. In the drug development process, from target discovery to clinical testing, immunoassays are a crucial bioanalytical technique. As long as specific antibodies can be generated, an immunoassay is possible to virtually any ANALYTE.

Immunoassays can be easily categorized by their configuration. In homogeneous assays, all assay components are present at the same time, whereas in heterogeneous immunoassays, the ANALYTE-antibody complex needs to be separated from other assay reagents. In competitive assays, the ANALYTE in the sample is measured by its competition with labeled reagent ANALYTE. In the noncompetitive sandwich format, the ANALYTE is measured by simultaneous binding of two different antibodies.

Immunoassays can be automated, and even without automation, a hundred samples can be typically tested by a single analyst in a day. With multiplex techniques, many ANALYTEs can be tested simultaneously, further enhancing the information generated from a single test.

References

1. Yalow RS, Berson SA. Assay of plasma insulin in human subjects by immunological methods. Nature. 1959;184. (Suppl 21:1648–9.
2. Kohler G, Milstein C. Continuous cultures of fused cells secreting antibody of predefined specificity. Nature. 1975;256(5517):495–7.
3. Corren J, Lemanske RF, Hanania NA, Korenblat PE, Parsey MV, Arron JR, et al. Lebrikizumab treatment in adults with asthma. N Engl J Med. 2011;365(12):1088–98.
4. Wild D, editor. The immunoassay handbook. 4th ed. Amsterdam: Elsevier Science; 2013.
5. Dinis-Oliveira RJ. Heterogeneous and homogeneous immunoassays for drug analysis. Bioanalysis. 2014;6(21):2877–96.
6. Schwickart M, Vainshtein I, Lee R, Schneider A, Liang M. Interference in immunoassays to support therapeutic antibody development in preclinical and clinical studies. Bioanalysis. 2014;6(14):1939–51.
7. Ian Weeks LK, Wild D. Signal generation and detection systems (excluding homogenous assays). In: The immunoassay handbook. Amsterdam: Elsevier Science; 2013.
8. Ullman EF. Homogenous immunoassays, Chapter 2.3. In: The immunoassay handbook. 4th ed. Amsterdam: Elsevier Science; 2013.
9. Scott RAW. Enhanced chemiluminescent immunoassays. In: The protein protocols handbook. Basel: Springer Nature Switzerland AG; 2009.
10. Huang X, Aguilar ZP, Xu H, Lai W, Xiong Y. Membrane-based lateral flow immunochromatographic strip with nanoparticles as reporters for detection: a review. Biosens Bioelectron. 2016;75:166–80.
11. Chang L, Li J, Wang L. Immuno-PCR: an ultrasensitive immunoassay for biomolecular detection. Anal Chim Acta. 2016;910:12–24.
12. Riedel T, Majek P, Rodriguez-Emmenegger C, Brynda E. Surface plasmon resonance: advances of label-free approaches in the analysis of biological samples. Bioanalysis. 2014;6(24):3325–36.
13. Lee JW, Devanarayan V, Barrett YC, Weiner R, Allinson J, Fountain S, et al. Fit-for-purpose method development and validation for successful biomarker measurement. Pharm Res. 2006;23(2):312–28.

Flow Cytometry

16

John F. Dunne and Holden T. Maecker

16.1 Introduction

Various technologies exist for analyzing cells in suspension using optical, electrical, or other techniques. This chapter will focus on flow cytometers, tools that can sequentially interrogate single cells for expression of proteins or other target molecules with high speed and precision. In conventional flow cytometers, the target molecules on these cells are detected with probes containing fluorescent tags and excited by lasers or other light sources. Alternately, in mass cytometry, the probes are labeled with heavy metal ions, and the readout is by mass spectrometry. Other devices, including scanning cytometers, optical microscopes, and counting devices based on electrical impedance measurements, will be outside the scope of this discussion.

While the fundamental fluidic constraints of flow cytometers have not changed, the available equipment has been enhanced dramatically with developments in lasers, digital electronics, fluorescence chemistries, and computer power. Analysis rates have migrated from hundreds to tens of thousands of cells per second, but more importantly, the number of colors that can be simultaneously measured has gone from 1 or 2 in the 1970s to 12 or more in recent years [1]. These polychromatic experiments create a new level of information about blood cells, as well as a new set of technical challenges. The optimization of instrument and experimental protocols and the efficient analysis of this kind of data become critical to successful implementation of this technology.

Final manuscript submitted on September 03, 2015.

J. F. Dunne
Miltenyi Biotec, San Diego, CA, USA

H. T. Maecker (✉)
Stanford University, Stanford, CA, USA
e-mail: maecker@stanford.edu

16.2 Mechanistic Principles

Fundamentally, flow cytometers measure fluorescent or scattered light emitted by a cell during its illumination in bright light, typically a highly focused laser beam [2]. In most cases, the cell stream is positioned using hydrodynamic focusing, in which a thin core of cell suspension is limited to the center of a larger flowing sheath fluid. The cells thus arrive sequentially into the laser beam at thousands of cells per second.

16.2.1 Light Scatter Measurements

Light is scattered by the cells intersecting the laser, and this scattered light is usually measured at narrow angles just above and below the laser beam (commonly called "FORWARD SCATTER") and by a separate detector at wider angles orthogonal to the beam (commonly called "SIDE SCATTER"). Forward scatter is descriptive of the size of the cell, while side scatter is proportional to size and granularity. Thus in commonly analyzed populations of blood cells, platelets, lymphocytes, monocytes, and granulocytes can be distinguished reasonably well simply on the basis of their intrinsic light scattering properties (Fig. 16.1).

M. J. Parnham et al. (eds.), *Nijkamp and Parnham's Principles of Immunopharmacology*,
https://doi.org/10.1007/978-3-030-10811-3_16

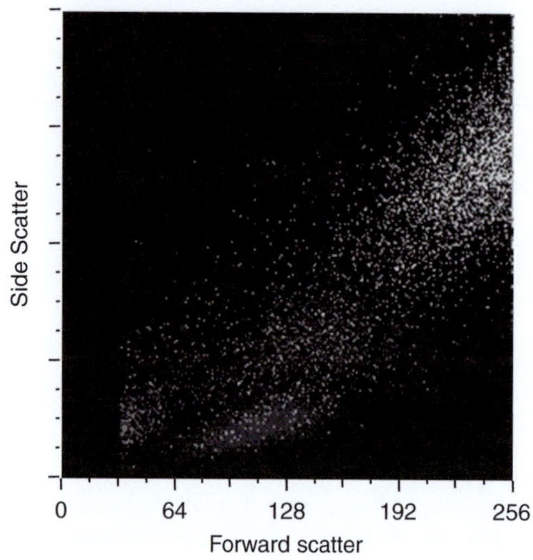

Fig. 16.1 Discrimination of leukocyte subsets by light scatter. The labeled subsets of leukocytes were fluorescently stained with specific monoclonal antibodies (not shown), and the specific antibody-stained populations were colored as indicated. These populations were then displayed in a dot plot of forward versus side scatter. A threshold was set on forward scatter, excluding much of the platelet population, in order to prevent the collection of small debris that would otherwise interfere with the analysis. Note that the colored populations are reasonably well-resolved from each other. However, the relative positions of these populations will depend upon the treatment of the sample (in this example, formaldehyde-fixed and detergent-permeabilized cells were used)

Box 16.1: Why Use Flow Cytometry as an Analytic Method?

Many different methods exist for quantifying the proteins expressed by a population of cells. In addition to flow cytometry, Western blotting, immunoprecipitation, ELISA, RIA, enzyme-linked immunospot (ELISPOT), fluorescence microscopy, and immunohistochemical staining are all methods used to quantify cellular proteins. Of these, only flow cytometry, ELISPOT, fluorescence microscopy, and immunohistochemical staining provide information on a single-cell basis. The others are bulk assays that do not quantify the number or phenotype of cells that are expressing the protein of interest (though they may provide other information, like the size of the targeted protein). Of the single-cell assays, flow cytometry is unique in the number of fluorescence parameters that have been combined in a single assay. Thus, it provides the richest information, on a per-cell basis, of the current generation of assays for cellular protein analysis.

16.2.2 Fluorescence Measurements

More valuably, cells can be stained with fluorescent reporter molecules, and the binding of these reporters can describe extremely subtle phenotypes. The most common class of reporters is fluorescently conjugated monoclonal antibodies. Thousands of these products are commercially available. The antibody binds very specifically to particular epitopes in proteins present on the cell surface, or if the cell is permeabilized, the antibody can enter the cell and bind to epitopes within. Since the antibody is covalently linked to a fluorescent molecule, the antibody binding is directly correlated to the fluorescent signal emanating from the cell during laser excitation. In a well-developed assay, the brightness associated with antibody binding can be a direct measure of the abundance of the relevant antigenic protein (see Box 16.1). By using several monoclonal antibodies, each conjugated to fluorescent dyes with characteristic colors, various properties of multiple populations of cells can be identified in a single sample, using optical detectors that filter light to detect only specific wavelengths.

16.2.3 Fluorescence Spillover and Compensation

Fluorescent reporter molecules, whether small organic dyes or larger protein molecules, have characteristic excitation and emission curves within the optical spectrum. Because these curves are relatively broad, overlap is inevitable and leads to detection of more than one fluorochrome per optical detector. This is referred to as "spillover" (Fig. 16.2). To properly deconvolute the spillover, fluorescence "COMPENSATION" is applied to the signals in multicolor experiments. Most commonly, a spillover matrix is calculated based on the observed fluorescence in all detectors of samples each stained with a single fluorochrome. Such

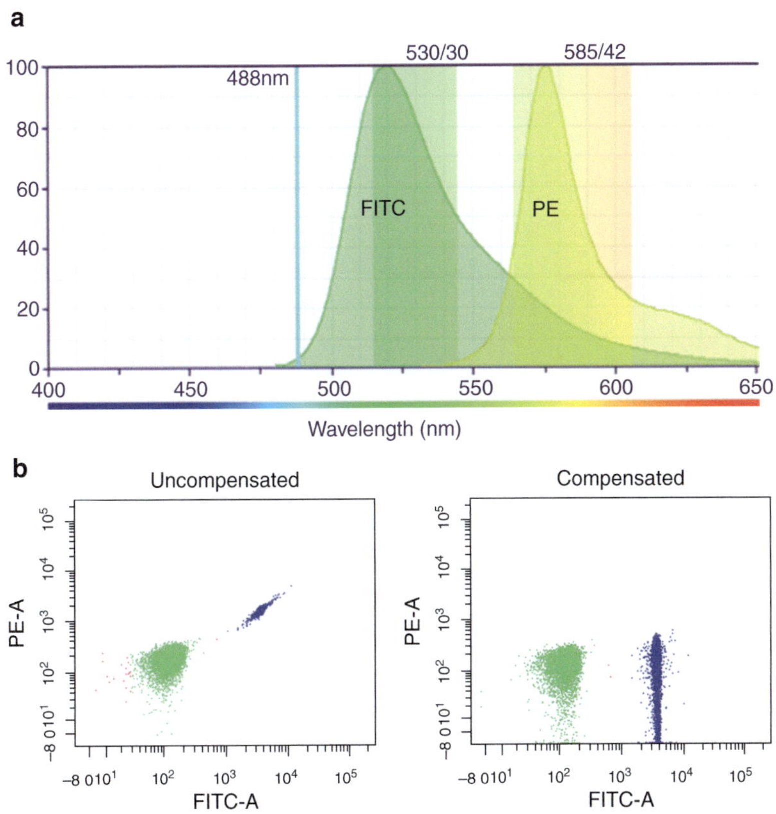

	FITC mean fluorescence		PE mean fluorescence	
	Negative	Positive	Negative	Positive
Uncompensated	125	3540	185	1650
Compensated	125	3560	135	135

$$\% \text{ Spillover} = \frac{1650 - 185}{3540 - 125} \times 100$$

Fig. 16.2 Spectral overlap and optical spillover. (**a**) Detection of FITC in the PE detector is a consequence of the overlapping emission spectra of FITC and PE. (**b**) By calculating the signal in both the FITC and PE detectors from a sample stained only with FITC, the relative spillover can be calculated and a compensation factor applied, so that single-stained populations in fact appear so in the data display

samples can be single-stained cells, antibody capture beads, or dye-embedded beads. The inverse of the spillover matrix thus calculated becomes the compensation matrix, which is applied to all experimental sample measurements. In newer digital flow cytometers, these calculations are automated and are done using a software algorithm (referred to as "software compensation"). This has the advantage that one can view and change the applied compensation at any later time during the analysis of the data. Earlier analog systems calculated compensation during the actual fluorescence measurement and stored only the compensated result (referred to as "hardware compensation").

16.2.4 Mass Cytometry

An alternative platform for flow cytometry, termed mass cytometry or CyTOF™, uses heavy metal ion labels and mass spectrometry in lieu of fluorescent labels and optical detection. This platform has the dual advantages of allowing many more simultaneous labels, with little or no spillover between detector channels. It has been used to simultaneously measure over 40 parameters on immune cells, including functional readouts like phosphoproteins and cytokines [3]. Its main drawbacks, other than cost, include a much slower acquisition speed (around 500 events per second)

Table 16.1 Commonly used fluorochromes for antibody-coupled flow cytometry

Fluorochrome	Type of molecule	Typical excitation laser (nm)	Approximate emission peak (nm)
Fluorescein isothiocyanate (FITC)	Small organic	488	518
Alexa Fluor 488	Small organic	488	518
BB515	Organic polymer	488	515
Phycoerythrin (PE)	Protein	488 or 532	574
PE-Texas Red	Protein tandem	488 or 532	615
PE-Cy5	Protein tandem	488 or 532	665
Peridinin chlorophyll protein (PerCP)	Protein	488 or 532	676
PerCP-Cy5.5	Protein tandem	488 or 532	695
PE-Cy7	Protein tandem	488 or 532	776
Allophycocyanin (APC)	Protein	633	659
Alexa Fluor 647	Small organic	633	667
Alexa Fluor 700	Small organic	633	718
APC-Cy7 or APC-H7	Protein tandem	633	784
Brilliant Violet (BV) 421	Organic polymer	405	421
BV510	Organic polymer	405	510
BV605	Organic polymer	405	605
Quantum dot (Qdot) 605	Molecular nanocrystal	405	605
BV650	Organic polymer	405	650
Qdot 655	Molecular nanocrystal	405	655
Qdot 705	Molecular nanocrystal	405	705
BV711	Organic polymer	405	711
BV786	Organic polymer	405	786
BUV395	Organic polymer	355	395
BUV496	Organic polymer	355	496
BUV661	Organic polymer	355	661
BUV737	Organic polymer	355	737
BUV805	Organic polymer	355	805

and loss of cells prior to ionization, which results in acquisition of data on only about half of labeled cells. It also cannot match the sensitivity of the very brightest FLUOROCHROMES used in fluorescence FLOW CYTOMETRY. Additionally, there are no light scatter parameters collected, but also no need to apply compensation, due to the relative lack of spillover (Table 16.1).

16.3 Classification/Types of Assay

16.3.1 Immunophenotyping

In the 1980s, with the early propagation of flow cytometry and monoclonal antibodies, there was some expectation that the protein surfaces of hematopoietic cells could be mapped with ever-increasing complexity to reveal changes in the types of cells present, or their relative abundance, that might correlate with clinical disease onset or progression, especially under the influence of therapeutic intervention. In the intervening 30 years, with thousands of fluorescent antibodies commercially available, the management of only two patient classes is substantially driven by these kinds of flow assays: HIV disease and leukemia/lymphoma.

16.3.2 Flow Immunophenotyping in HIV Disease

One of the most proximal and dramatic effects of HIV infection is an eventually lethal loss of

CD4+ T cells [4]. This was one of the first easily recognized manifestations of infection, indeed the first clinical test that helped identify AIDS [5], and has been one of the most useful analytical tools available to characterize mechanisms of HIV pathology as well as therapeutic benefit of anti-HIV treatment. Often augmented by assays that quantitate viral load, CD4 counting remains the standard method of monitoring disease progression [5].

The CD4 counting assay is relatively simple among flow cytometry assays. Most commonly, fluorescently conjugated CD4 monoclonal antibodies are added to whole blood, generally as part of a cocktail of two to four antibodies which together uniquely identify those T cells that bear the CD4 protein. With the addition of a known number of fluorescent calibrating beads per volume of blood, or by sampling a known absolute volume of cell suspension, the cytometer can report the number of CD4 cells present per microliter of blood. Normal ranges center around 1000 CD4 cells per microliter.

With the increasing availability of antiretroviral therapies in developing countries, the need for low-cost CD4 monitoring that can be robustly carried out in remote settings has become a new challenge. Smaller dedicated flow cytometers with preformatted reagents for CD4 counting are now available, but other technologies are also being tested, including light microscopy and related imaging techniques requiring minimal technology investment and training.

16.3.3 Flow Immunophenotyping in Autoimmune Disease

Using similar technical strategies, the frequency of circulating B cells is of use in monitoring the long-term efficacy of rituximab (anti-CD20) therapy in rheumatoid arthritis, multiple sclerosis, and other inflammatory diseases [6]. Various aspects of response to treatment including duration of B cell depletion, details of B cell subset frequencies, resolution of individual patient resistance, and linkages to clinical relapse have been described.

16.3.4 Flow Immunophenotyping in Leukemia/Lymphoma

The observation that the major IMMUNOPHENOTYPES found in peripheral blood are relatively stable may have been disappointing in some regards, but it was helpful in allowing the recognition of unusual phenotypes characteristic of cancer cells. Especially in the case of transformed cells of hematopoietic lineages, these unusual phenotypes have now been mapped into broadly recognized categories of malignant disease, and flow cytometric assays are thus useful to help diagnose and monitor leukemias and lymphomas.

Several clinical research consortia have published consensus documents describing the general utility, practical aspects, and interpretation of combinations of markers and their distribution on cancer cells [7–9]. Still, no commercially available or otherwise standard kit has been approved for diagnostic use, and common practice is for clinical centers to use their own discretion to implement and validate such assays. Nevertheless, the standard of care of patients suffering from these malignancies has been improved, sometimes dramatically, based on practitioners' ability to detect and characterize leukemias and lymphomas by flow cytometry.

16.3.5 Detection of Circulating Tumor Cells

Flow cytometry is a powerful platform for detecting rare cells in blood samples. In leukemias and lymphomas, for example, flow has been used to detect minimal residual disease [10, 11] and in the case of solid tumors for the detection of cells with an epithelial (or endothelial) phenotype normally not present in blood [12]. In the latter case, immunomagnetic and microscopic techniques have been developed as an IVD platform to help manage solid tumor treatment. Despite many attempts, research in the flow sorting of circulating epithelial cells presumably shed from primary or metastatic loci for expression profiling has not yet led to a clear strategy for customized therapies.

16.3.6 Quantitation of Specific Antigen Expression by Normal Blood Cells

As the various blood cell types and their frequencies have become better characterized, so too has the magnitude of expression of various antigens on their surfaces. With calibration against a bead standard, or by using ratiometric methods to report relative changes in the signal strength associated with labeled antigens present on cell subtypes, flow cytometry can report the dynamics of protein expression in response to treatment or disease. The magnitude of CD64 expression on monocytes and neutrophils has been useful to monitor various aspects of systemic inflammation including sepsis and has been developed as an IVD product [13].

16.3.7 Immunophenotyping in Immunotoxicology

While most common disease states do not change the frequency of major blood cell components, the toxicity associated with novel drugs or industrial pollutants can have profound effects. Recent work establishing consensus protocols for the evaluation of immunotoxicity of such chemical entities documents a growing utility for this class of assays [14]. The foci of these assays are typically rodent, dog, or nonhuman primate models, and monoclonal antibody reagents are less commonly available or less well-characterized. The normal ranges of various lymphoid cell types, other environmental factors that can influence these ranges, and the magnitude of changes in these cell types that should be used as sentinels for concern or prohibit the use of such chemicals in various exposure scenarios all represent ongoing work.

16.3.8 Functional Assays

An important additional class of flow assays consists of those assays that experimentally induce a functional response in cells of interest and then use flow cytometry to measure the response. One of the most common assays of this type is the measurement of CYTOKINE production in lymphocytes in response to antigenic stimulation. The production of a cytokine can be easily measured using well-established reagents and protocols and is a hallmark of lymphocyte function. With multicolor flow cytometry, the frequency of responsive cells, the amount and kinetics of cytokine production, and the combination of various cytokines can very richly describe an immunological response qualitatively and quantitatively. These assays will be discussed in detail below, and by way of comparison, other functional assays are also overviewed (see Box 16.2).

Box 16.2: Different Assays for Measuring T Cell Function

In this chapter, measurement of intracellular cytokines by flow cytometry is described in some detail as a method for determining T cell functional responses after short-term stimulation. However, alternative methods are also in common use. Some are flow cytometry assays, such as MHC multimer staining [15], or cytokine capture assays [16]; others use different analytical platforms, such as ELISPOT [17]. The main advantage of flow cytometry as an analytical platform is its multiparameter capability (see Box 16.1). Of the flow cytometry methods available for measuring T cells in a short-term assay, MHC multimer staining is unique in that it measures T cell specificity rather than function and thus requires no activation at all. Multimeric forms of MHC molecules with bound antigenic peptides are produced and labeled with a fluorochrome such as PE. These can then be used to stain T cells (via their T cell receptor) in much the same way as a fluorochrome-conjugated antibody. However, their use requires a knowledge of the particular peptide and MHC restriction pattern of a T cell immune response; as such, they are not useful for quantitating the overall T cell response to a pathogen, especially in an

MHC-diverse population. Cytokine capture assays, like intracellular cytokine staining, measure responses without regard to MHC restriction. The cytokine capture assay is especially useful if one wishes to maintain the cells in a viable state for sorting and further analysis (since it does not require fixation and permeabilization). Still newer assays, such as measures of granule exocytosis [18], can also be used on viable cells and add to the armamentarium of tools available for dissecting T cell responses.

16.3.9 Proliferation

One hallmark of lymphocyte function is cellular proliferation. Several flow assays have been well-established in the literature that have been correlated to traditional 3H-thymidine incorporation associated with DNA synthesis. Perhaps the most directly correlated of these flow assays is the use of bromodeoxyuridine (BrdU), an ortholog of thymidine used by DNA synthase and incorporated in the nuclei of proliferating cells. BrdU is detected by fluorochrome-conjugated specific monoclonal antibodies after nuclear permeabilization [19].

Another proliferation assay utilizes a cytosolic dye, commonly carboxyfluorescein diacetate succinimidyl ester (CFSE), which is used to label cells prior to in vitro stimulation. The dye then becomes diluted with each sequential cell division, resulting in successive generations of cells with half the fluorescence of their parent [20]. By calculating the frequency of cells in each generation, an estimate of the original progenitor frequency can be made. This calculation is flawed by not accounting for cell death. Still, one can differentiate between a large progenitor population that has only divided once or twice and a small progenitor population that has divided many times. By contrast, this difference would be difficult to see using 3H-thymidine or BrdU incorporation.

One of the earliest flow assays for proliferation, which is still in common use, is the direct measurement of DNA content using a quantitative DNA stain such as propidium iodide or the vital dye Hoechst 33258 [21, 22]. This is especially useful for peripheral blood lymphocytes, which are natively arrested with a 2c DNA content in the G0 phase of the cell cycle. Upon stimulation, they enter S phase and progress to the 4c DNA content typical of G2/M. This doubling of DNA is easily recognized flow cytometrically, and the percentage of cells with more than 2c DNA can be roughly related to traditional metrics like mitotic index.

Another nucleic acid stain, acridine orange, has the unusual property of shifting its emission spectrum depending on whether it is bound to single- or double-stranded nucleic acid (roughly speaking, RNA or DNA) and has been used to simultaneously report DNA doubling and the increase in mRNA characteristic of lymphoid stimulation [23].

16.3.10 PhosphoEPITOPE Flow Cytometry

One intriguing class of functional assays is the quantitation by flow cytometry of signaling molecule phosphorylation states. This has been made possible by the development of monoclonal antibodies specific for particular phosphoepitopes of common signaling intermediates. By using optimal permeabilization strategies, these phosphoepitopes can be exposed for intracellular staining by the relevant antibodies, in combination with staining for phenotypic markers or cell subsets [24]. While not yet clinically approved tests, this technique (often termed "phospho-flow") has been shown to be useful in predicting disease outcome for certain types of cancers [25]. It can also be used to test the potency of kinase inhibitors in vitro [26] or in clinical trial monitoring.

16.3.11 Bead Matrix Immunoassays

Recently bead-based IMMUNOASSAYS have been developed for flow cytometric assessment of various soluble analytes. Commonly, each member of a set of beads is identified by specific fluorescence levels and/or size and acts as an analyte capture platform using covalently bound specific monoclonal antibodies. The amount of captured analyte is measured using a second antibody specific for an alternate site of the same analyte, this second antibody being conjugated to a fluorescent reporter. Since the various beads can be recognized independently, and since each captured analyte can be quantitated independently, this assay format is well suited to multiplexed analysis. Thus small volumes of biological fluids (e.g., as little as 10 μL of tears [27]) can be inspected for quantitation of soluble proteins or other ligands for which specific antibody pairs are available. Common implementations of this strategy include the simultaneous quantitation of 5–50 different immunomodulatory proteins including cytokines, chemokines, growth factors, and hormones. Commercially available kits range from completely integrated systems to basic platforms that can be developed for custom analyte sets and are used broadly in basic research and drug discovery proteomics projects. A clinical application of this technology is the tracking of antibodies to allo-HLA, used in the setting of solid organ transplantation [28].

Other flow cytometric assays that may be of interest in the field of immunopharmacology but which have not yet achieved broad clinical utility include reporters of mitochondrial function, plasma membrane potential, oxidative stress, dendritic cell activation, and the abundant microparticle environment recently being characterized in the subcellular serum.

16.4 Components/Construction of Assays

Traditionally, flow cytometry sample preparation and processing have been constrained by the fact that flow cytometers were designed to accept tubes, and those tubes were loaded onto the instrument manually, one at a time. A second constraint came from instrument setup and data analysis, which were typically time-consuming and required a certain knowledge base. Both of these constraints are now changing with new instrumentation and software. Current cytometers can often handle racks of tubes that are automatically run in a walkaway mode. Plate loaders are also available that can feed samples from 96-well plates directly to the cytometer, again in a walkaway mode. These developments have been complemented by software that can either (1) perform data analysis simultaneously with acquisition or (2) perform a batch analysis routine that analyzes all of the data from an experiment at once. The usefulness of such analysis routines is further augmented by flexible analysis templates that can accommodate changing data without repetitive adjustment by the operator [29].

The impact of these changes in flow cytometry hardware and software has opened up the use of flow cytometry to larger preclinical and clinical studies that might involve hundreds of specimens. However, the steps involved in sample preparation can still be complex, as detailed below. Automation of these steps will further allow the use of flow cytometry in high-volume settings where it was previously considered too cumbersome.

16.4.1 Validation of Flow Cytometric Assays Is of Great Interest for Evaluation of Immunomodulatory Therapies

The use of flow cytometry in HIV and leukemia/lymphoma diagnoses has established an expectation regarding assay validation and best practices. Consensus publications have addressed issues and recommendations for the translation of basic research to well-characterized and validated assays that withstand regulatory review. Sample handling, multicenter data acquisition standardization, and the use of expert-intensive data analysis are all factors to be controlled [30].

16.4.2 Antibodies for Cell Staining

Monoclonal antibodies have now virtually replaced polyclonal antisera for most applications, including flow cytometry. In addition, these antibodies are now commonly available pre-conjugated to a wide variety of fluorescent labels, negating the need for indirect staining with secondary reagents that carry the fluorescent tag. Most strikingly, the breadth and brightness of available fluorochromes have increased dramatically in the last several years, due largely to new chemistries such as conductive organic polymers that have been used to create a whole family of bright dyes with tunable excitation and emission spectra [31].

Not all fluorochromes are compatible with all cytometers, or even with all other fluorochromes, as can be predicted from their excitation and emission spectra (see, e.g., www.bdbiosciences.com/spectra). Minimizing spillover between detectors is a major consideration when choosing fluorochromes for multicolor experiments. Relative brightness and compatibility with certain experimental parameters (e.g., permeabilization protocols) will also steer the choice of fluorochromes. For a more complete discussion of the issues of designing multiparameter experiments, see references [32, 33].

16.4.3 Cell Types

FLOW CYTOMETRY can be performed on any cell type that can be rendered into a single-cell suspension. Since blood cells already exist in this state, they have been most widely studied by the technique. However, adherent cells or tissues can also be analyzed if they are dissociated from each other and/or their substrate using either enzymatic (e.g., trypsin, Accutase) or nonenzymatic (e.g., EDTA) treatments. Where possible, nonenzymatic dissociation protocols are preferable, because they do not cleave cell surface proteins that might be targets of the flow cytometric analysis.

16.4.4 Erythrocyte Lysis

Whole blood consists of relatively homogeneous erythrocytes and a much more complex collection of leukocytes. These can be stained with fluorochrome-conjugated antibodies in the context of whole blood and resolved based on fluorescence. But analysis of whole blood is difficult because of the light scattering properties of the erythrocytes, which are so numerous as to obscure the illumination of the leukocytes. Fortunately, erythrocytes are differentially sensitive to hypotonic lysis and can be removed by incubation of blood with ammonium chloride in water. Leukocytes are more resistant to osmotic damage than erythrocytes, which cannot exclude this salt, and subsequently take up water and are lysed. Typically, whole blood is treated for 10 min with a large volume of ammonium chloride solution either before or immediately after staining with fluorochrome-conjugated antibodies.

An alternative to ammonium chloride lysis involves hypotonic lysis in the presence of a fixative (e.g., formaldehyde), whereby the erythrocytes are preferentially lysed, while leukocytes are fixed in a single incubation of 10 min or so. Solutions for this procedure are commercially available. After lysis/fixation, the cells may be directly analyzed by flow cytometry ("no-wash" assays); or they may be subjected to washing to remove the residual red cell debris and unbound fluorochrome-conjugated antibodies before analysis ("washed" assays). Washed assays typically allow better signal-to-noise discrimination of fluorescently stained populations and better visualization of lymphocyte scatter properties. However, no-wash assays have become popular in clinical laboratories because of their simplicity; and by detecting cells based on threshold staining for a common leukocyte antigen, such as CD45, these assays can adequately resolve subpopulations of lymphocytes.

16.4.5 Density Gradient Separation

As an alternative to erythrocyte lysis of whole blood, mononuclear leukocytes (lymphocytes and monocytes) can be directly isolated from whole blood prior to staining. Solutions of high molecular weight carbohydrates, such as FICOLL, are used for this separation, which is accomplished on the basis of density. By underlaying whole blood (usually diluted 1:1 with buffer or tissue culture media) with a Ficoll solution, a density gradient is formed. This biphasic solution is then subjected to centrifugation. Erythrocytes and granulocytes, which have the greatest buoyant density, are pelleted to the bottom of the Ficoll layer. Lymphocytes and monocytes, which are less dense, collect at the interface of the plasma and Ficoll, where they are collected by pipetting. Platelets, the least dense leukocytes, will remain in the plasma layer. Successive centrifugation of the mononuclear cells collected from the interface is then carried out to remove residual Ficoll and further deplete the mononuclear cells of platelets.

Density gradient separation techniques are a standard, albeit time-consuming method for the isolation of mononuclear cells from small to large volumes of blood. However, simpler alternatives are also available that consist of a gel matrix pre-dispensed into a blood collection tube or centrifuge tube. By adding whole blood and centrifuging at a prescribed speed, the mononuclear cell layer can be collected from the top of the gel interface, while erythrocytes and granulocytes are forced through the gel plug. Such systems are available from commercial vendors and provide generally equivalent results with greater convenience than the Ficoll separation technique [34]. Mononuclear cells isolated by either method are often cryopreserved in liquid nitrogen, allowing them to be banked for later studies, including flow cytometry.

16.4.6 Activation

As described briefly above, functional assays are those in which the cell types of interest are stimulated in vitro and allowed to manifest some kind of response, the quantity and quality of which are measured in the flow cytometer. Lymphocyte activation is a particularly valuable class of functional assays and typically requires specific antigen, polyclonal MITOGEN, or a cytokine as a stimulus. Short-term simulation (5–30 min) can be optimal for detecting phosphorylation changes in signaling proteins such as Erk, Mek, Akt, the Stat proteins, and a host of other phosphoproteins [24]. Longer stimulation (4–24 h) can be used to detect cytokine production, such as IL-2, IFNγ, TNF-α, etc. [35]. Detection of cytokine production is aided by addition of a secretion inhibitor, such as brefeldin A [36] or monensin [37], which allows for intracellular accumulation of the cytokines and thus brighter staining. Finally, stimulation for periods of 2–7 days is typical for detecting cell proliferation, whether by CFSE dye dilution, BrdU incorporation, or other assays.

16.4.7 Fixation/Permeabilization

For staining of intracellular epitopes, cells are first fixed (usually with formaldehyde) and then permeabilized (usually with detergent or in some cases methanol). Commercial reagents containing fixatives and detergents are readily available and allow for reproducible fixation and permeabilization protocols.

16.4.8 Cell Staining

For assays designed to measure cytokines or signaling proteins, intracellular staining is required; but cell surface staining is usually also performed in the same sample to allow phenotyping of the responding cell population. Depending upon the antibodies and epitopes targeted, this surface staining can sometimes be done together with the intracellular cytokine staining, if the targeted surface epitopes are resistant to the conditions of fixation and permeabilization [38]. If they are not, a surface staining step is carried out after activation but before fixation and permeabilization. In certain assays where proteins are transiently expressed on the cell surface (e.g., CD107, CD154), the antibodies to these markers can be added during the stimulation phase of the assay.

For most epitopes, surface and/or intracellular staining can be done by incubation with a cocktail of fluorochrome-conjugated antibodies for 30–60 min at room temperature. Titration of antibodies is required for optimal staining (although many manufacturers offer pre-titered antibodies and cocktails of antibodies). The optimal titer for surface staining (unfixed cells) is often different than that for intracellular staining (fixed and permeabilized cells), even for the same cell surface target. This is due to increased non-specific binding of fixed and permeabilized cells.

As noted above, simple surface staining of whole blood can be done using a no-wash assay format. But more complex assays, such as intracellular cytokine staining, require washing, both after fixation and permeabilization steps and after antibody staining.

16.4.9 Increasing Automation and Throughput

The relative complexity of sample preparation, especially for the more complex functional assays, has inhibited their application to studies of large numbers of animals or large clinical trials. However, robotic sample preparation devices are available, and more are in development that would significantly aid in this process. For example, robotic workstations are commercially available that can aliquot whole blood into staining tubes, add antibodies, incubate, and add erythrocyte lysis buffer. There are also workstations that can perform cell washing and thus can automate most of the steps of a washed assay. Integration of these robotic workstations can not only increase the number of samples handled in these applications but can also lend considerable standardization to processes that otherwise are highly operator-dependent.

Truly high-throughput sample processing, however, is best accomplished with multiwell plates. Both phenotypic and functional assays can be performed in 96-well plates; in fact, protocols for activation, processing, and analysis of samples for intracellular cytokine staining in a single 96-well plate have been published [29]. With the availability of plate-based loaders for flow cytometers, much larger numbers of samples can be processed in a single run, with minimal incremental technician time. Further automation of plate-based sample processing can also be accomplished using preformatted lyophilized reagent plates [39]. These have the advantage of long shelf life, allowing a single reagent batch to be used for a longitudinal study. They also tend to prevent errors of reagent addition that are otherwise difficult to prevent with complex multicolor experiments.

16.4.10 Data Acquisition and Analysis

Although sample preparation can be made parallel using multiwell plates and automation, sample acquisition is still a serial process. Fortunately, tube and plate loaders are available that can automate acquisition, and software can be used to annotate data files before they are run.

Data analysis involves the setting of "gates" or regions in one- or two-dimensional data space, followed by further analysis of the fluorescence properties of cells within those gates. Typically, viable cells of interest are first identified by their light scatter properties, using forward and side scatter parameters. This may be followed by gating on subsets of lymphocytes, for example, CD3+ (T cells), CD19+ (B cells), etc. Subsets of T cells (CD4+, CD8+) may be identified through further gating. In the case of functional assays, a responsive population, e.g., cytokine-positive, is quantitated from the subset of interest, e.g., CD8+ T cells.

The precise placement of fluorescence and light scatter gates is historically done by the eye, based upon recognition of typical patterns of staining (negative, dim, bright, etc.). In the simplest case, serial samples with identical staining properties can be analyzed using rigid data processing templates to export various statistical summaries for cross-sample comparisons. More commonly, small variations arising from real sample differences or sample preparation inconsistencies need to be handled by experts who occasionally make small adjustments to the analysis template for each sample. This becomes more challenging as the complexity of the flow measurements increases and has been a primary source of variance in large multicenter studies

[39]. Automated analysis tools have been developed, both open source and commercial, that use various data-driven adaptive techniques to process complex data. Applications aimed at specific assays (e.g., clinical tests) tend to be the most robust and easiest to develop (see, e.g., [40]). Visual inspection of the gates as applied to each data file is, however, still highly recommended.

A final component of data analysis that is often overlooked is the incorporation of flow cytometric data into a database that may also link it with other types of measurements (patient clinical data, etc.). Fundamentally, this requires the ability to extract measurements of interest from the flow cytometry data files into a spreadsheet along with any annotation associated with the data files. This can be done with modern flow cytometry software packages, such that manual entry of data into a spreadsheet is not necessary. The most recent data standard adopted by the flow community [FCS3.1 [41]] enables current database technologies to manage both the experimental annotation, which may include read/write interfaces, and experimental measurements, which are read-only. Proprietary and open-source tools are developing, so that enterprise-level flow experiments, covering years, and many clinical sites can be managed more successfully. Such tools include data normalization, sample quality assurance metrics and filters, pattern recognition algorithms, and linkages with annotation information to enable data visualization and trend analyses [see, e.g., [42]].

16.5 Examples and Their Application

16.5.1 Immunophenotyping Assays

An example of an IMMUNOPHENOTYPING assay to quantitate T cell subsets is shown in Fig. 16.3. Note that this was done as a "no-wash" assay in four colors, along with fluorescent beads to allow absolute counting. Fifty microliters of whole blood were stained with antibodies to CD45, CD3, CD4, and CD8. The blood was subjected to fixation/erythrocyte lysis and then run on a four-color flow cytometer. An acquisition threshold was set on CD45 fluorescence, in order to identify leukocytes. All cells above this threshold were collected, with acquisition halted at 20,000 CD45+ events.

Analysis of this assay was done as follows. An initial gate was set on CD3+ cells, in order to identify all T cells. CD4+ and CD8+ T cells were then quantitated from a plot gated on CD3+ cells. Note that this sample also contained counting beads which are identified by very high fluorescence in all colors. Because a known number of these beads were added to a known volume of the blood sample, a simple calculation can convert the percentages of each cell subpopulation to an absolute count of cells per microliter of blood. Since percentages are relative to other populations, absolute counts have become a standard readout for reporting these types of results. Commercial software packages can perform these calculations in an automated fashion for this particular application.

16.5.2 Functional Assays

An example of a functional assay (identifying intracellular cytokine production) is shown in Fig. 16.3 [43]. This assay was done by stimulation of whole blood with peptides derived from human cytomegalovirus (HCMV), a common herpesvirus that causes chronic, latent infection of various tissues. HCMV causes non-pathological infection in immunocompetent hosts but is an opportunistic pathogen in immunocompromised hosts [44].

The peptides used for stimulation in this assay were a mixture of 138 15-mers, overlapping by 11 amino acid residues each, and spanning the sequence of the pp65 glycoprotein of HCMV. Pp65 is an immunodominant protein which contains epitopes that stimulate CD4 and CD8 T cell responses in a variety of HLA haplotypes [45]. By using such overlapping peptides, both CD4 and CD8 T cell responses can be stimulated with relative efficiency [46]. The blood was stimulated with this peptide mixture for 6 h

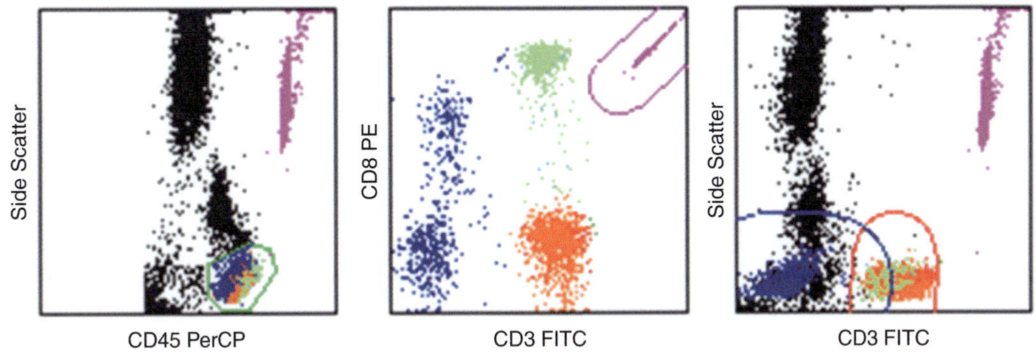

Lymphocyte Events	3002
Bead Events	2135
CD3+% of Lymphocytes	59
CD3+Absolute Count	823
CD3+CD8+% of Lymphocytes	24
CD3+CD8+ Absolute Count	336
CD3+CD4+% of Lymphocytes	32
CD3+CD4+ Absolute Count	457
CD3+CD4+CD8+ % of Lymphocytes	0
CD3+CD4+CD8+ Absolute Count	6
CD45+ Absolute Count	1406
CD4/CD8 Ratio	1.36

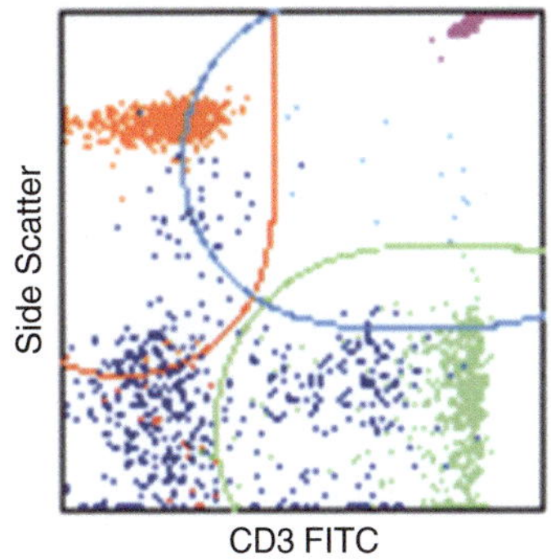

Fig. 16.3 CD4 T cell enumeration assay from whole blood. Cells and counting beads were successively gated as described in the text, and the percentages and absolute counts of various lymphocyte subsets were automatically reported by the analysis software

in the presence of brefeldin A, followed by EDTA treatment, surface staining, fixation/erythrocyte lysis, permeabilization, and intracellular staining. The antibodies used included anti-IFNγ, anti-IL-2, CD3, CD4, CD8, CD27, CD28, and CD45RA. The latter three markers identify memory and effector subsets of CD4+ and CD8+ T cells. A threshold was set on CD3 to allow collection of only CD3+ cells, reducing the file size. Acquisition was stopped at 100,000 CD3+ cells.

Analysis of this assay is outlined in the figure and includes "snap-to" gates on major cell subsets [29]. These gates are able to move according to small changes in the staining pattern from one sample to the next. Cytokine-positive gates are rectangles that are "tethered" to the negative population, so that they shift in response to changes in the background staining. In this way, even very rare populations of cytokine-positive cells can be quantitated, whether they form a recognizable cluster or not (Fig. 16.4).

If desired, one could combine the information from the above two assays, in order to express the HCMV pp65-responsive cell populations as absolute numbers of cells per microliter (or milliliter). This might provide more standardized results in populations that have varying numbers of CD4 and CD8 T cells (such as HIV patients).

What is the usefulness of identifying and quantitating intracellular cytokine responses? One major application is the development of new vaccines that are designed to elicit cellular immunity [47]. These include both prophylactic and therapeutic vaccines for HIV, cancer, and certain other viral and intracellular bacterial pathogens. Establishing biomarkers of immunogenicity of vaccines (and eventually surrogate markers of protection) would greatly facilitate the comparison of different vaccine constructs and strategies and allow more rapid improvement of vaccines for these diseases.

Similarly, phospho-flow is another powerful way to measure functional changes in particular

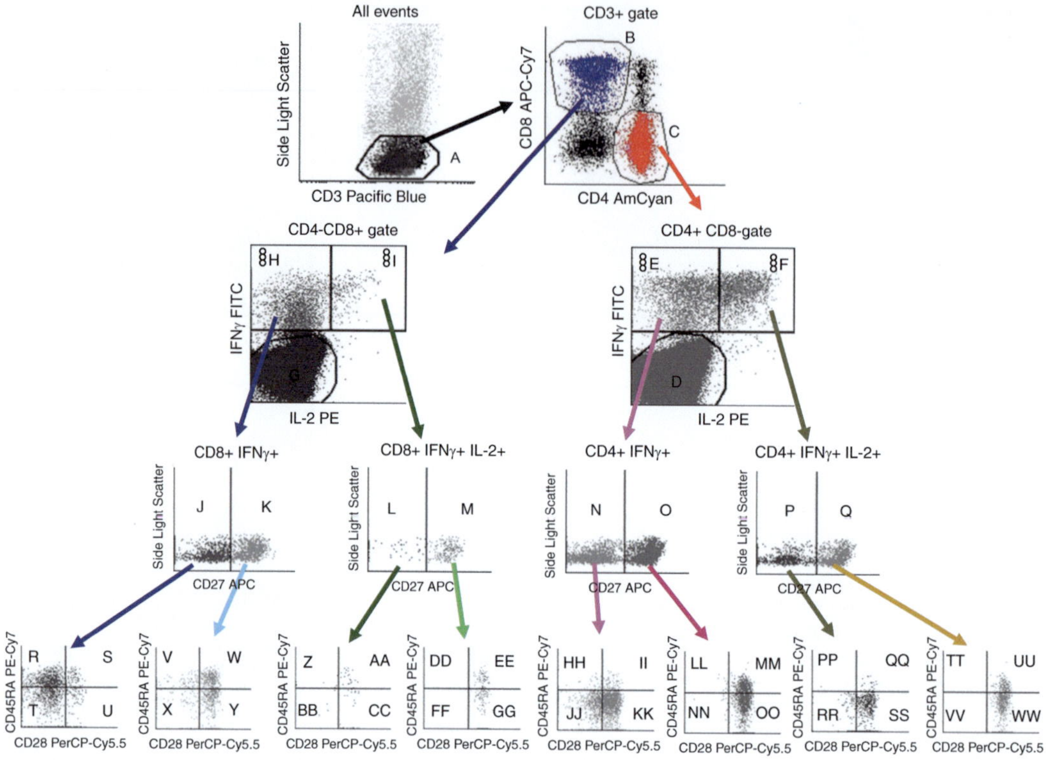

Fig. 16.4 Eight-color intracellular cytokine staining [43]. Peripheral blood mononuclear cells were isolated from whole blood, stimulated with a peptide pool from human cytomegalovirus, and stained with antibodies to eight markers that include two cytokines (IFNγ and IL-2), T cell identification and subsetting markers (CD3, CD4, and CD8), and differentiation markers (CD27, CD28, CD45RA) that identify various forms of memory and effector T cells. The scheme for gating and identification of populations of interest is shown and described further in the text

types of immune cells. While less relevant for antigen-specific T cell populations, phosphoflow can be used in any disease setting where immune signaling patterns may be altered. Clinical relevance of signaling in B cell lymphomas has been demonstrated [48], and investigation in other diseases is ongoing.

Quantification of functional immune cell responses, whether to specific antigens or to mitogens, cytokines, or other activators, could also provide valuable information in the area of immunotoxicology [49]. While immunophenotyping can potentially uncover gross changes in cellular subsets in response to a drug, functional assays can uncover much more subtle changes. Suppression (or augmentation) of antigen or mitogen responses could give important clues to immunological side effects of drugs. This could be done as part of the early screening of drug candidates, before expensive animal studies are undertaken. Functional flow cytometry assays could also be used to monitor the dosing of immunomodulatory drugs in settings such as transplantation and autoimmunity [50, 51]. By standardizing, automating, and increasing the ease and throughput of these assays, they can become ever more potent tools for immunological research, immunotoxicology, and clinical medicine.

16.6 Summary

FLOW CYTOMETRY is a powerful technique for analyzing cells in suspension and has been extensively applied to the analysis of lymphocytes and their subsets. Flow cytometric assays can be divided into immunophenotyping assays and functional assays, the latter requiring in vitro stimulation of cells in order to read out a response, such as phosphorylation or cytokine production. Such assays are being applied to the monitoring of clinical disease states and responses to vaccination or immunomodulation, and they have great potential to be used in measuring immunotoxicology as well.

Suggested Readings and Websites

Givan AL. Flow cytometry: first principles. New York, NY: J. Wiley; 2001.

Purdue cytometry bulletin board. http://www.cyto.purdue.edu/hmarchiv/cytomail.htm

Robinson JP, editor. Current protocols in cytometry. New York, NY: J. Wiley.; 2005. (Current protocols series)

References

1. Perfetto SP, Chattopadhyay PK, Roederer M. Seventeen-colour flow cytometry: unravelling the immune system. Nat Rev Immunol. 2004;4:648–55.
2. Givan AL. Flow cytometry: first principles. New York, NY: J. Wiley; 2001.
3. Atkuri KR, Stevens JC, Neubert H. Mass cytometry: a highly multiplexed single-cell technology for advancing drug development. Drug Metab Dispos. 2015;43:227–33.
4. Phillips AN, Elford J, Sabin C, Janossy G, Lee CA. Pattern of CD4+ T cell loss in HIV infection. J Acquir Immune Defic Syndr. 1992;5:950–1.
5. Mandy F, Nicholson J, Autran B, Janossy G. T-cell subset counting and the fight against AIDS: reflections over a 20-year struggle. Cytometry A. 2002;50:39–45.
6. Trouvin AP, Jacquot S, Grigioni S, Curis E, Dedreux I, Roucheux A, et al. Usefulness of monitoring of B cell depletion in rituximab-treated rheumatoid arthritis patients in order to predict clinical relapse: a prospective observational study. Clin Exp Immunol. 2015;180:11–8.
7. Davis BH, Foucar K, Szczarkowski W, Ball E, Witzig T, Foon KA, et al. U.S.-Canadian Consensus recommendations on the immunophenotypic analysis of hematologic neoplasia by flow cytometry: medical indications. Cytometry A. 1997;30:249–63.
8. Stewart CC, Behm FG, Carey JL, Cornbleet J, Duque RE, Hudnall SD, et al. U.S.-Canadian Consensus recommendations on the immunophenotypic analysis of hematologic neoplasia by flow cytometry: selection of antibody combinations. Cytometry A. 1997;30:231–5.
9. Rothe G, Schmitz G. Consensus protocol for the flow cytometric immunophenotyping of hematopoietic malignancies. Working Group on Flow Cytometry and Image Analysis. Leukemia. 1996;10:877–95.
10. Wood BL. Principles of minimal residual disease detection for hematopoietic neoplasms by flow cytometry. Cytometry B Clin Cytom. 2016;90:47.
11. Van Dongen JJ, van der Velden VH, Brüggemann M, Orfao A. Minimal residual disease (MRD) diagnostics in acute lymphoblastic leukemia (ALL): need for sensitive, fast and standardized technologies. Blood. 2015;125:3996.
12. Caceres G, Puskas JA, Magliocco AM. Circulating tumor cells: a window into tumor development and therapeutic effectiveness. Cancer Control. 2015;22:167–76.
13. Hoffmann JJ. Neutrophil CD64: a diagnostic marker for infection and sepsis. Clin Chem Lab Med. 2009;47:903–16.
14. Burchiel SW, Kerkvliet NL, Gerberick GF, Lawrence DA, Ladics GS. Assessment of immunotoxicity by multiparameter flow cytometry. Fundam Appl Toxicol. 1997;38:38–54.
15. Altman JD, Moss PA, Goulder PJ, Barouch DH, McHeyzer-Williams MG, Bell JI, et al. Phenotypic analysis of antigen-specific T lymphocytes. Science. 1996;274:94–6.
16. Brosterhus H, Brings S, Leyendeckers H, Manz RA, Miltenyi S, Radbruch A, et al. Enrichment and detection of live antigen-specific CD4(+) and CD8(+) T cells based on cytokine secretion. Eur J Immunol. 1999;29:4053–9.
17. Hutchings PR, Cambridge G, Tite JP, Meager T, Cooke A. The detection and enumeration of cytokine-secreting cells in mice and man and the clinical application of these assays. J Immunol Methods. 1989;120:1–8.
18. Betts M, Brenchley J, Price D, De Rosa S, Douek D, Roederer M, et al. Sensitive and viable identification of antigen-specific CD8+ T cells by a flow cytometric assay for degranulation. J Immunol Methods. 2003;281:65–78.
19. Mehta BA, Maino VC. Simultaneous detection of DNA synthesis and cytokine production in staphylococcal enterotoxin B activated CD4+ T lymphocytes by flow cytometry. J Immunol Methods. 1997;208:49–59.
20. Lyons AB, Parish CR. Determination of lymphocyte division by flow cytometry. J Immunol Methods. 1994;171:131–7.
21. Taylor I, Milthorpe B. An evaluation of DNA fluorochromes, staining techniques, and analysis for flow cytometry. I. Unperturbed cell populations. J Histochem Cytochem. 1980;28:1224–32.
22. Braylan RC, Diamond LW, Powell ML, Harty-Golder B. Percentage of cells in the S phase of the cell cycle in human lymphoma determined by flow cytometry. Cytometry. 1980;1:171–4.
23. Darzynkiewicz Z. Simultaneous analysis of cellular RNA and DNA content. Methods Cell Biol. 1994;41:401–20.
24. Perez OD, Mitchell D, Campos R, Gao GJ, Li L, Nolan GP. Multiparameter analysis of intracellular phosphoepitopes in immunophenotyped cell

populations by flow cytometry. Curr Protoc Cytom. 2005;Chapter 6:Unit 6.20.
25. Irish JM, Hovland R, Krutzik PO, Perez OD, Bruserud Ø, Gjertsen BT, et al. Single cell profiling of potentiated phospho-protein networks in cancer cells. Cell. 2004;118:217–28.
26. Krutzik PO, Crane JM, Clutter MR, Nolan GP. High-content single-cell drug screening with phosphospecific flow cytometry. Nat Chem Biol. 2008;4:132–42.
27. Cook EB, Stahl JL, Lowe L, Chen R, Morgan E, Wilson J, et al. Simultaneous measurement of six cytokines in a single sample of human tears using microparticle-based flow cytometry: allergics vs. non-allergics. J Immunol Methods. 2001;254:109–18.
28. Billen EV, Voorter CE, Christiaans MH, van den Berg-Loonen EM. Luminex donor-specific crossmatches. Tissue Antigens. 2008;71:507–13.
29. Suni MA, Dunn HS, Orr PL, de Laat R, Sinclair E, Ghanekar SA, et al. Performance of plate-based cytokine flow cytometry with automated data analysis. BMC Immunol. 2003;4:9.
30. Davis BH, Wood B, Oldaker T, Barnett D. Validation of cell-based fluorescence assays: practice guidelines from the ICSH and ICCS - part I - rationale and aims. Cytometry B Clin Cytom. 2013;84:282–5.
31. Chattopadhyay PK, Gaylord B, Palmer A, Jiang N, Raven MA, Lewis G, et al. Brilliant violet fluorophores: a new class of ultrabright fluorescent compounds for immunofluorescence experiments. Cytometry A. 2012;81:456–66.
32. Maecker HT, Frey T, Nomura LE, Trotter J. Selecting fluorochrome conjugates for maximum sensitivity. Cytometry A. 2004;62:169–73.
33. McLaughlin BE, Baumgarth N, Bigos M, Roederer M, De Rosa SC, Altman JD, et al. Nine-color flow cytometry for accurate measurement of T cell subsets and cytokine responses. Part I: Panel design by an empiric approach. Cytometry A. 2008;73:400–10.
34. Ruitenberg JJ, Mulder CB, Maino VC, Landay AL, Ghanekar SA. VACUTAINER CPT and Ficoll density gradient separation perform equivalently in maintaining the quality and function of PBMC from HIV seropositive blood samples. BMC Immunol. 2006;7:11.
35. Nomura LE, Walker JM, Maecker HT. Optimization of whole blood antigen-specific cytokine assays for CD4(+) T cells. Cytometry. 2000;40:60–8.
36. Waldrop SL, Pitcher CJ, Peterson DM, Maino VC, Picker LJ. Determination of antigen-specific memory/effector CD4+ T cell frequencies by flow cytometry: evidence for a novel, antigen-specific homeostatic mechanism in HIV-associated immunodeficiency. J Clin Invest. 1997;99:1739–50.
37. Prussin C, Metcalfe DD. Detection of intracytoplasmic cytokine using flow cytometry and directly conjugated anti-cytokine antibodies. J Immunol Methods. 1995;188:117–28.
38. Babcock GF, Dawes SM. Immunophenotyping using fixed cells. Methods Cell Biol. 1994;41:81–93.
39. Maecker HT, Rinfret A, D'Souza P, Darden J, Roig E, Landry C, et al. Standardization of cytokine flow cytometry assays. BMC Immunol. 2005;6:13.
40. Wong L, Hill BL, Hunsberger BC, Bagwell CB, Curtis AD, Davis BH. Automated analysis of flow cytometric data for measuring neutrophil cd64 expression using a multi-instrument compatible probability state model. Cytometry B Clin Cytom. 2015;88:227.
41. Spidlen J, Moore W, Parks D, Goldberg M, Bray C, Bierre P, et al. Data file standard for flow cytometry, version FCS 3.1. Cytometry A. 2010;77:97–100.
42. Klinke DJ, Brundage KM. Scalable analysis of flow cytometry data using R/Bioconductor. Cytometry A. 2009;75:699–706.
43. Nomura LE, Emu B, Hoh R, Haaland P, Deeks SG, Martin JN, et al. IL-2 production correlates with effector cell differentiation in HIV-specific CD8+ T cells. AIDS Res Ther. 2006;3:18.
44. Drago F, Aragone MG, Lugani C, Rebora A. Cytomegalovirus infection in normal and immunocompromised humans. A review. Dermatology (Basel). 2000;200:189–95.
45. Kern F, Bunde T, Faulhaber N, Kiecker F, Khatamzas E, Rudawski IM, et al. Cytomegalovirus (CMV) phosphoprotein 65 makes a large contribution to shaping the T cell repertoire in CMV-exposed individuals. J Infect Dis. 2002;185:1709–16.
46. Maecker HT, Dunn HS, Suni MA, Khatamzas E, Pitcher CJ, Bunde T, et al. Use of overlapping peptide mixtures as antigens for cytokine flow cytometry. J Immunol Methods. 2001;255:27–40.
47. Macchia I, Urbani F, Proietti E. Immune monitoring in cancer vaccine clinical trials: critical issues of functional flow cytometry-based assays. Biomed Res Int. 2013;2013:726239.
48. Irish JM, Myklebust JH, Alizadeh AA, Houot R, Sharman JP, Czerwinski DK, et al. B-cell signaling networks reveal a negative prognostic human lymphoma cell subset that emerges during tumor progression. Proc Natl Acad Sci U S A. 2010;107:12747–54.
49. Maecker HT, Maino VC, Picker LJ. Immunofluorescence analysis of T-cell responses in health and disease. J Clin Immunol. 2000;20:391–9.
50. Sindhi R, Allaert J, Gladding D, Koppelman B, Dunne JF. Cytokines and cell surface receptors as target end points of immunosuppression with cyclosporine A. J Interferon Cytokine Res. 2001;21:507–14.
51. Sindhi R, Allaert J, Gladding D, Haaland P, Koppelman B, Dunne J, et al. Modeling individual variation in biomarker response to combination immunosuppression with stimulated lymphocyte responses-potential clinical implications. J Immunol Methods. 2003;272:257–72.

Gene Expression

17

Julia Tree, Angela Essex-Lopresti, Stefan Wild, Ute Bissels, Barbara Schaffrath, Andreas Bosio, and Michael Elmore

17.1 Introduction

The phenotype of a cell is determined by the amount, the proportion and the condition of proteins present in this cell. Although every cell in an organism possesses the same genetic information, only certain genes are transcribed into MESSENGER RIBONUCLEIC ACID (mRNA) according to the function and demands of the cell. Based on the information provided by the mRNA, the information is translated into the corresponding protein, contributing to a distinctive set of proteins for every cell and every status of the cell, defining its phenotype. The mapping of the whole human genome was completed in 2004 [1]. Researchers are focusing now on the illumination of functions and interactions of genes and gene products by measuring, for example, the number of activated genes. An established method is DNA MICROARRAY technology, which, as well as other established DNA and RNA detection methods, utilises the characteristic of RNA strands to form helices due to complementary sequences. This process of combining two RNA strands to form a double helix is called HYBRIDISATION. Since Southern introduced the blotting technique [2] for DNA, the HYBRIDISATION process has been used in a wide range of techniques for the recognition and quantification of DNA or RNA. Such "classical" HYBRIDISATION techniques measure one DNA or RNA sequence per HYBRIDISATION using a specific probe. In contrast, DNA microarrays consist of several thousands of specific probes arrayed in a two-dimensional pattern allowing the parallel investigation of thousands of genes. A more recent development in measuring the expression levels of genes, using next-generation sequencing technology, is RNA-Seq. In this method, the entire transcriptome (mRNA content) of the sample is sequenced. The read depth, or number of sequence reads, corresponding to each gene is used as a proxy of the expression level of that particular gene. RNA-Seq analysis is still in its infancy but has distinct advantages over traditional microarray.

Final manuscript submitted on October 06, 2016.

J. Tree (✉) · M. Elmore
Public Health England (PHE), Salisbury, UK
e-mail: julia.tree@phe.gov.uk;
mike.elmore@phe.gov.uk

A. Essex-Lopresti
Defence Science and Technology Laboratory, Salisbury, UK
e-mail: aeelopresti@dstl.gov.uk

S. Wild · U. Bissels · A. Bosio
Miltenyi Biotec GmbH, Bergisch Gladbach, Germany
e-mail: macs@miltenyibiotec.de;
ute.bissels@miltenyibiotec.de;
andreasbo@miltenyibiotec.de

B. Schaffrath
Eppendorf Nordic A/S, Hørsholm, Denmark
e-mail: schaffrath.b@eppendorf.dk

M. J. Parnham et al. (eds.), *Nijkamp and Parnham's Principles of Immunopharmacology*,
https://doi.org/10.1007/978-3-030-10811-3_17

17.2 Principle of Microarray Technology

Microarrays are tiny devices made for the analysis of targets of interest with a high degree of parallelism. Initially, the technology evolved around the analysis of mRNA levels in cells in different states, taking "classical" HYBRIDISATION-based technologies to a new level. For "classical" HYBRIDISATION-based analysis, genomic DNA (Southern) or RNA (Northern), extracted from the tissue of interest, is immobilised on a membrane. A single specific nucleotide sequence (the probe) that is complementary to the sequence of interest is labelled and applied to the membrane to subsequently detect the corresponding gene or gene transcript (Fig. 17.1). For array analysis, this principle is reversed and applied to thousands of sequences of interest by immobilising DNA fragments (probes) with distinct sequences on a SUBSTRATE (a membrane, glass, silicon or plastic slides) at defined positions (see Box 17.1). Nucleic acids from the cells of interest are labelled and applied to the SUBSTRATE for HYBRIDISATION, and the hybridised nucleic acids are identified by their position on the array.

Box 17.1: Production of Microarrays

A variety of different array substrates (membranes, plastics, glass), in combination with a range of different coatings, are used as the solid phase for microarray production. Coatings permit the functionalisation of substrates with reactive groups, like aldehyde, epoxy or isothiocyanate moieties, to bind DNA probes on the substrates.

The DNA probes can be directly synthesised on the microarray substrate (in situ synthesis) or the complete DNA probes are spotted on the substrate. The in situ synthesis, by photomediated synthesis or inkjet technology, allows a parallel production of

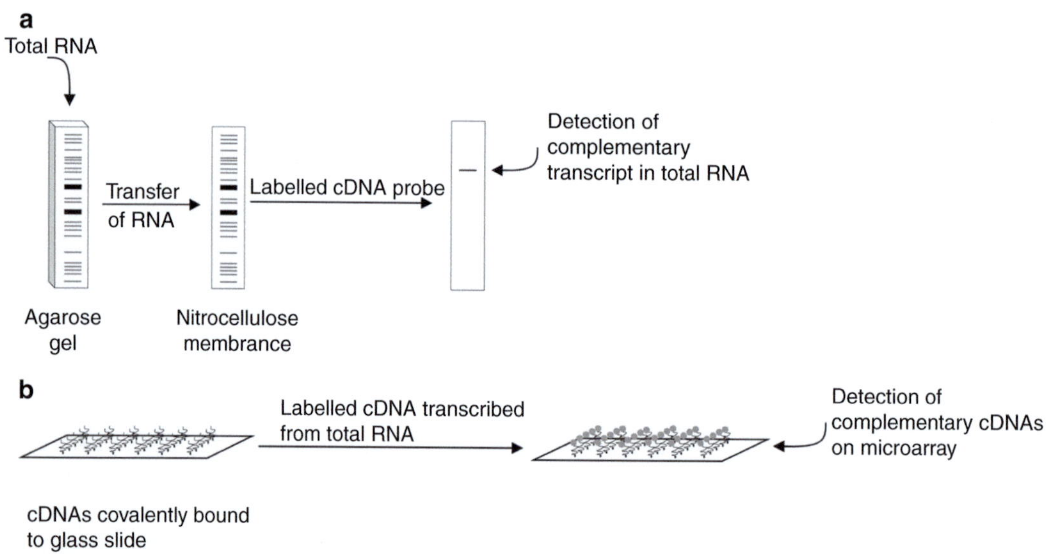

Fig. 17.1 Comparison of traditional Northern-blot and DNA microarray. (**a**) Total RNA of the tissue of interest is separated by gel electrophoresis and is blotted to a membrane. A labelled cDNA probe complementary to the transcript of interest is labelled and hybridised to the membrane. If the transcript is present in the total RNA, a signal can be detected due to hybridisation of probe and transcript. One experiment—one gene using a single labelled probe. (**b**) Several cDNAs (hundreds to thousands) complementary to mRNA transcripts of selected genes are covalently bound to a glass slide at defined positions (spots). Total RNA from the tissue of interest is transcribed into cDNA and labelled by reverse transcription. The labelled cDNA is hybridised to the bound cDNAs. Signals can be detected after hybridisation of two complementary cDNAs

OLIGONUCLEOTIDE ARRAYS, comprising oligonucleotides of 20–60 nucleotides in length [3, 4]. The use of short oligonucleotides (20–30 base pairs) is suitable to differentiate between perfectly matched duplexes and single-base or two-base mismatches [5–7]. When working with short oligonucleotide probes, the use of several different oligonucleotides corresponding to a single gene is typically required to enhance the reliability of the hybridisation signals [8].

Alternatively, cDNA fragments or presynthesised oligonucleotides with a length of up to 70 base pairs are spotted on the functionalised substrate in two manners: contact printing and non-contact printing.

CONTACT PRINTING typically involves rigid pins dipping into the spotting buffer containing the DNA probes. The drop at the tip of the pin is brought close to the surface at a given position, and a tiny drop remains on the surface. Non-contact printing methods are based on inkjet technology. The spotting buffer containing the DNA probes is dispensed as tiny droplets from the print head. Independent of the spotting mode, binding of the DNA probes occurs at the position of the drop. After the actual spotting process is completed, unbound DNA is removed, and the reactive substrate is blocked to avoid non-specific (independent of the provided sequence) binding of nucleic acids during hybridisation. The microarrays are now ready for processing.

The workflow of this process is illustrated by means of a DNA MICROARRAY experiment: In a typical scenario, GENE EXPRESSION of tumour cells, for instance, is compared to that in normal cells. RNA from tumour and normal cells is extracted from the respective tissue (Fig. 17.2). The RNA is transcribed into its reverse complimentary copy, the so-called cDNA. The cDNA derived from tumour cells and normal cells is labelled and applied to the DNA array. During the HYBRIDISATION step, the labelled nucleic acids bind to the complementary sequences of the respective probes. After washing away all unbound labelled nucleic acids, the signal intensities for each probe position are determined. After signal intensities have been generated for all probes on the array, signals derived from normal cells and tumour cells are compared, and differences in GENE EXPRESSION are identified. The altered expression of certain genes in the tumour, such as oncogenes, can help to typify the tumour. Combining the expression profile with clinical data may then be used to decide on the prognosis and the best therapy for the patient.

In addition to the described GENE EXPRESSION PROFILING, microarrays are also used to investigate other nucleic acids like genomic DNA [9] or non-coding RNAs [10] including MICRORNAS (miRNAs) [11–15]. In addition, the array principle has also been adapted to other ANALYTES such as proteins [16] or carbohydrates [17].

Due to the parallel measurement of up to thousands of ANALYTES, microarrays offer the opportunity to observe complex biological systems while using minimal amounts of sample material. Although in the following sections specifications and workflow procedures are mainly related to DNA microarrays for GENE EXPRESSION PROFILING, the general aspects hold true for other MICROARRAY-based technologies as well.

17.3 Application of Microarrays

17.3.1 Preparation and Quality of RNA

The first crucial step to achieve reliable GENE EXPRESSION results is RNA isolation. RNA is susceptible to chemical hydrolysis and to RNases, widespread enzymes that digest RNA molecules into small pieces. If the RNA is slightly degraded or contaminated by residual genomic DNA, for instance, the results may be biased and irreproducible (see also Box 17.2). Commonly, RNA is

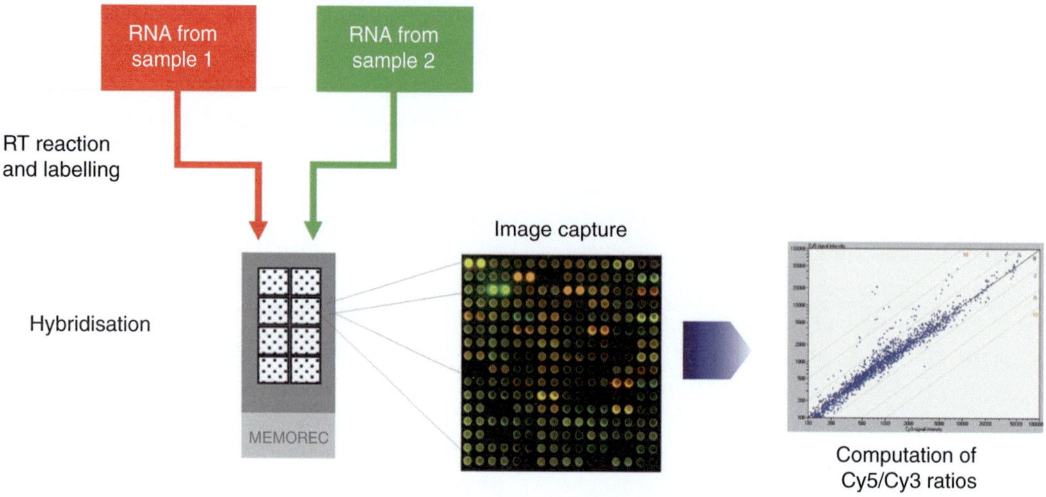

Fig. 17.2 Workflow diagram of microarray analysis

extracted from cells or tissues using organic solvents or silica filter-based methods. Since RNA extraction protocols may influence the outcome of the expression analysis, the same extraction procedure should be used for all samples analysed in one set of experiments.

Box 17.2: Quality of Total RNA

Integrity and purity are the most critical factors for the quality of RNA.

- Ratio of 28S rRNA and 18S rRNA should be 2, reflecting the higher molar mass of 28S rRNA compared to 18S rRNA. A more precise quality measure is given by the RNA integrity number (RIN) calculated by the Agilent Bioanalyzer.
- Ratio of the extinction 260 nm/280 nm should be between 1.8 and 2.0.
- The sample can be treated with RNase-free DNase to avoid contamination of genomic DNA.
- Protocols for RNA extraction have to be adapted according to the analysed tissue (e.g. high fat content or fibrous tissue).
- The choice of the preparation protocol may have an influence on the range of transcript lengths present in the extracted RNA (e.g. silica filters usually have a cut-off size of about 50–100 bases). Therefore, preparations derived in this way do not contain the whole range of fragment lengths. This might have an impact on the subsequent steps (labelling, amplification or hybridisation).

17.3.2 Amplification of RNA

The SENSITIVITY of MICROARRAY experiments strongly depends on the amount of material used for HYBRIDISATION. As the amount of RNA is usually limited, different AMPLIFICATION methods are available. The most common method utilises T7 DNA-dependent RNA polymerase to amplify RNA. The mRNA is first reverse transcribed to cDNA. The primer used for the reverse transcription additionally comprises the sequence of the T7 promotor. After the second strand synthesis, the T7 promotor is used by the T7 DNA-dependent RNA polymerase for in vitro transcription. The T7

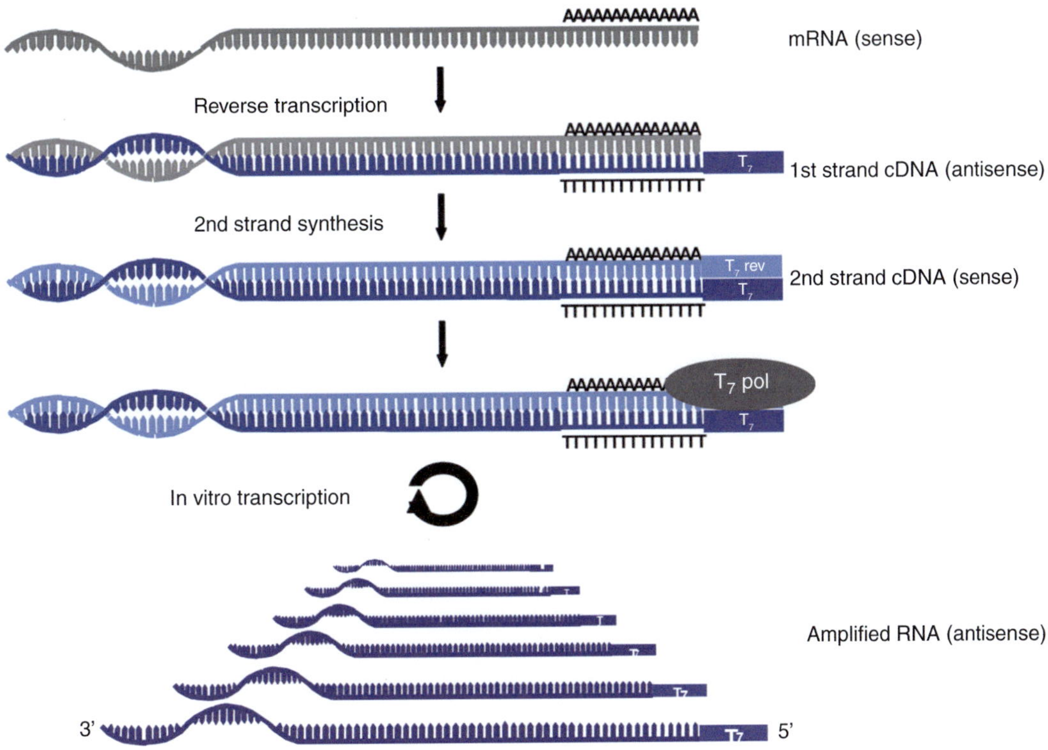

Fig. 17.3 Schematic diagram of T7 polymerase-based mRNA amplification

DNA-dependent RNA polymerase repeatedly transcribes the same cDNA thereby amplifying the original RNA (Fig. 17.3) [18]. In case even higher SENSITIVITY is needed, the amplified RNA can again be used as SUBSTRATE for cDNA synthesis and a second round of T7-based AMPLIFICATION. Alternatively, a variety of other AMPLIFICATION methods like PCR-based AMPLIFICATION methods have been developed (Fig. 17.4) [19]. Due to the slightly different properties of the different RNAs, such as length, sequence or GC content, the AMPLIFICATION efficiency can vary for different RNAs, again depending on the AMPLIFICATION method. Therefore, to allow comparison of different RNA samples, it is advisable to use the same AMPLIFICATION method for all samples. The most sensitive AMPLIFICATION methods allow MICROARRAY experiments from as little as a single cell (see also Sect. 5.1 and Fig. 17.4).

17.3.3 Dyes, Labelling and Hybridisation Methods

Most commonly, fluorescent dyes are used to detect the hybridised samples on microarrays, but alternative labelling methods using radioactivity or silver particles, for example, can also be applied.

In DIRECT LABELLING protocols, the labelled nucleotides are incorporated during the cDNA synthesis or the T7 DNA-dependent RNA polymerase-based AMPLIFICATION. Since the incorporation rate of labelled nucleotides is compromised by the partly bulky fluorescent dye, two-step labelling protocols (INDIRECT LABELLING) have also been established. During a two-step labelling procedure, nucleotides, labelled with a small molecule like biotin or an aliphatic amine, are incorporated by the polymerase. In a second step, the fluorescent dyes are linked to the modified nucleotides via

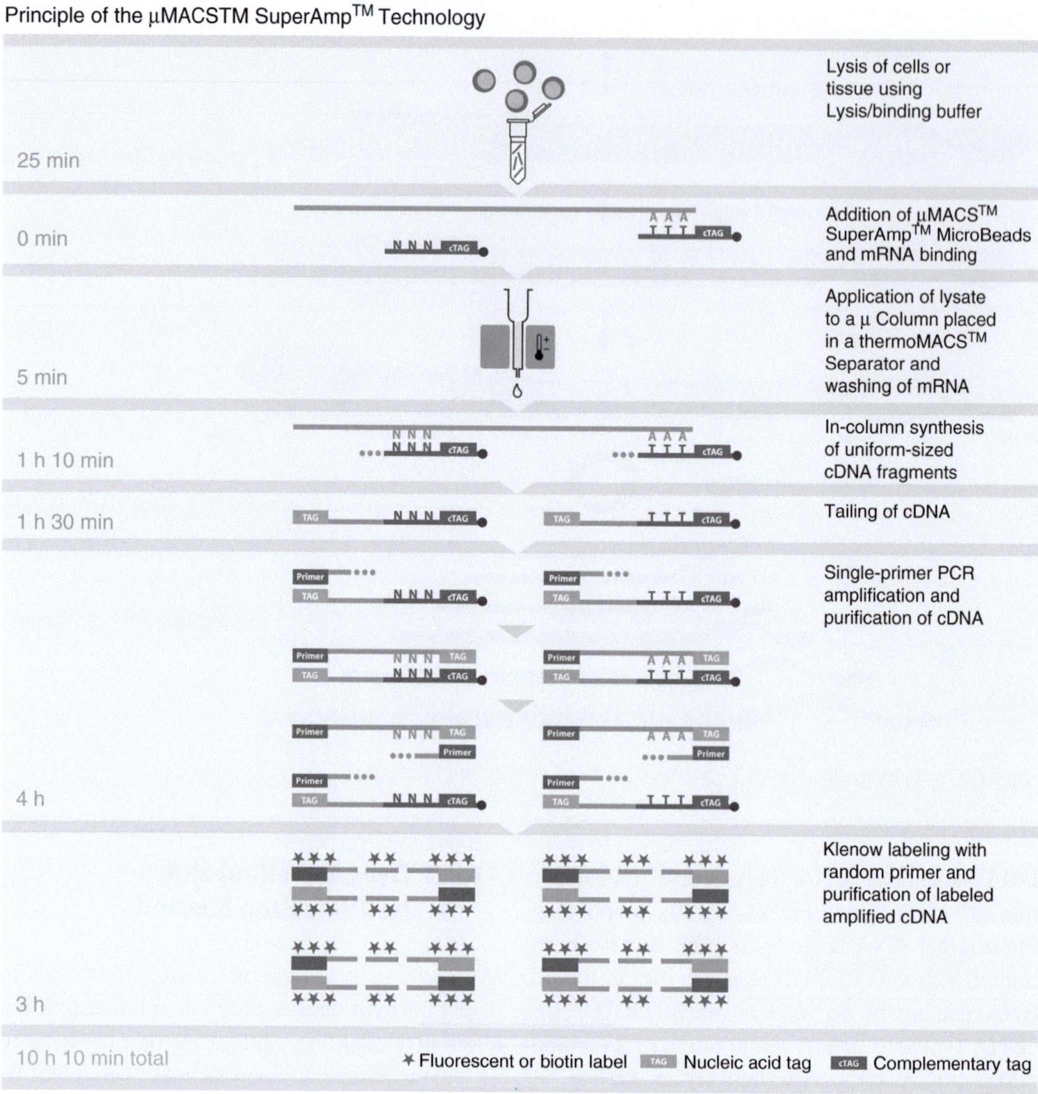

Fig. 17.4 Schematic diagram of a global PCR-based mRNA amplification

streptavidin or amine reactive groups like NHS esters. Depending on the system, the second step of the labelling protocol can also be performed after the HYBRIDISATION step (on-chip labelling).

After the labelling, the samples are hybridised on the MICROARRAY. The HYBRIDISATION can either be achieved by simple diffusion of the TARGET DNA molecules to the corresponding probes, or probe TARGET interaction can be assisted by moving the HYBRIDISATION mixture on top of the array. After the HYBRIDISATION step has been completed, unbound labelled TARGET molecules are removed by washing the array. Finally, the array is dried.

To minimise experimental variance caused by some of the processing steps, like the labelling or HYBRIDISATION, it is advisable to perform replicate MICROARRAY experiments using the same sample.

The HYBRIDISATION is usually performed as a one- or two-colour experiment. For one-colour experiments, each sample is hybridised on one array, and the signal intensities derived from different arrays are compared. When using two colours, the two samples to be compared are labelled with different dyes and hybridised together on the same array. The direct comparison of the two samples on one array has the advantage that any experimental bias related to the array or the HYBRIDISATION step will affect both samples, therefore reducing detection of artefacts. When working with fluorescent dyes, however, the integrated dyes not only differ in their emission wavelength but also in the fluorescence intensity gained per dye, due to wavelength-dependent scanner properties, diverse quantum yields of the dyes or different stabilities of the dyes. Therefore, the raw data gained by two-colour MICROARRAY experiments has to be corrected for such dye effects. The methods used to centre or normalise the signal intensities for both wavelengths are based on the assumption that some of the genes, like housekeeping genes, are not regulated (see Box 17.3). The differences found for these genes can therefore be used to calculate a factor reflecting the different dye properties. As the reproducibility of array production and MICROARRAY HYBRIDISATIONS has dramatically improved, there is a trend in favour of single-colour HYBRIDISATIONS.

Box 17.3: Normalisation of Microarray Data

Integrity and purity are the most critical factors for the quality of RNA.

The main idea of NORMALISATION for dual-labelled samples is to adjust differences in the intensity of the two labels. Such differences result from the efficiency of dye integration, differences in amount of sample and label used and settings of laser power and photomultiplier. NORMALISATION of one channel arrays mainly corrects spatial heterogeneity. Although NORMALISATION alone cannot control all systematic variations, NORMALISATION plays an important role in the earlier stage of microarray data analysis because expression data can vary significantly due to different NORMALISATION procedures. A number of NORMALISATION methods have been proposed, but there is no general rule which method performs best. The NORMALISATION method strongly depends on several factors like the number of detectable genes, the number of regulated genes, signal intensities, quality of the hybridisation, etc.

For a rough classification, global NORMALISATION can be distinguished from local (signal intensity-dependent) NORMALISATION and NORMALISATION via transcripts known to be nonregulated or spike-in controls.

If global NORMALISATION is used, a single NORMALISATION factor is applied to all detectable genes, leading to a linear shift of all signal intensities. The underlying assumption is that constant systematic variations occur, including a lower integration rate of one dye in respect to the second dye. However, global NORMALISATION based on the median of all detected genes can only be used if a sufficient number of genes are nonregulated. If it is expected that most of the genes are regulated (which is of special interest regarding miRNA arrays), a set of "housekeeping genes" or spike-in controls should be included in the array configuration. Because housekeeping genes (by definition) are not regulated, the signal intensities of those genes should be the same on dual-labelled arrays. Using local NORMALISATION, a different NORMALISATION factor is calculated for every gene. Local NORMALISATION offers the opportunity of a signal intensity-dependent NORMALISATION. Some variations (e.g. laser settings) have different impacts on detected genes depending on their signal intensity. Thus, a non-linear shift of the signal intensities can be achieved based on the signal intensity of each single spot.

In the field of miRNA microarray research, NORMALISATION via spike-in controls is preferably used, as global NORMALISATION methods may fail due to (a) missing housekeeping miRNA, (b) limited number of expressed miRNAs and (c) a general up- or downregulation of many miRNAs. The used spike-ins represent a set of synthetic RNAs, which have no similarity to any known miRNA. The spike-ins are added to all experimental and control samples, and all signal intensities of the investigated samples are normalised using the median of the spike-ins.

17.3.4 Control Samples

The measurement of GENE EXPRESSION in a given sample is usually referred to the GENE EXPRESSION in other samples, here referred to as "control". Obviously, it is very important to choose the right control in order to gain valuable data. The best controls in most experiments are untreated cells or unaffected tissue of the same origin as treated cells or affected tissue, respectively. However, for practical or ethical reasons, it is not always possible to receive untreated cells or healthy tissue of the same origin, which is especially true for material derived from patients. If it is impossible to get matched control samples, a "related" control can be established, for instance, by pooling RNA from different individuals to reduce the effects of particular properties of single individuals in the control. In some cases, cell lines might also be a sufficient control. Alternatively, a pool of all samples used in an experimental series can work as the control (see Box 17.4). However, a sample pool carries the risk of missing genes that are consistently expressed differentially in all samples. In general, controls should either be case-matched to the samples of interest or consist of pooled material to compensate for individual differences.

Box 17.4: The Reference Strategy for Two-Colour Hybridisations

In microarray experiments, the direct comparison of absolute signal intensities of different microarrays can be critical due to different hybridisation efficiencies. To avoid this obstacle, two-colour microarray hybridisations can be performed. In two-colour microarray hybridisations, the sample, labelled with Cy5, for instance, and the control, labelled with Cy3, are hybridised on the same microarray. As the labelled molecules compete for the same probes on the microarray, the hybridisation efficiency is also the same and allows a direct comparison of sample versus control. Therefore, the ratio of the signal intensities of the two dyes represents the proportion of the analyte in the sample compared to the control. The principle of two-colour hybridisation can be extended to compare more than two samples by applying a reference scheme. For a microarray reference experiment, each of several samples and controls is hybridised versus the reference. The reference can then be used to compensate differences of the hybridisation efficiency for each microarray and allows standardisation and cross-referencing of microarray experiments. For the analysis of mRNA expression profiles, references consisting of total RNA mixtures are used [20]. For miRNA analysis, universal references consisting of known amounts of synthetic miRNAs are available [21]. Besides the cross-referencing of array experiments, such a reference allows the absolute quantification of miRNAs. The universal reference, consisting of an equimolar pool of about 1000 miRNAs, is labelled and hybridised versus each sample in a two-colour microarray approach. In this way, each single miRNA is quantified in comparison to an identical standard, compensating the bias related to sequence, labelling, hybridisation or signal detection.

17.4 Array Data: Acquisition, Analysis and Mining

17.4.1 Data Acquisition

Data acquisition of MICROARRAY experiments consists of two parts: the read-out of the MICROARRAY, meaning the detection of the signals, and the following image analysis. Whereas films have been used to detect radioactive signals, nowadays predominantly MICROARRAY scanners are used to excite commonly used dyes and measure the emitted fluorescence signals. The picture derived from the read-out of the MICROARRAY is saved as greyscale TIFF images for further analysis. During the next step, the signal intensity of each spot is determined and assigned to the gene represented by the given spot using appropriate image analysis software. In addition, the background signal, usually gained from the surrounding area of each spot, is subtracted from the signal to receive the net signal intensity. Spots of poor quality (empty or negative spots, irregular shape, spots showing background smears) can be excluded from further analyses. The set of data that results from the data acquisition step is referred to as primary data.

17.4.2 Data Analysis and Mining

For the analysis of the primary data, weak signals are excluded as non-reliable. The minimum reliable signal intensity of a spot can be determined by setting a minimum threshold for signal intensities, which is either dependent on the background or on negative controls. For some microarrays, *p*-values giving an estimate of the likelihood of the signal differing from background signals are used to indicate the reliability of the detected genes. To compare different samples, ratios of the signal intensities gained, such as for sample versus control, are computed for every detected gene. To correct for different labelling and HYBRIDISATION efficiencies, as well as for potential dye bias in two-colour MICROARRAY hybridisations, the signal intensities are centred or normalised prior to calculating the ratios (see Box 17.3).

Because of the multiparametric nature of MICROARRAY experiments, data mining and bioinformatics analysis are essential for interpretation of the numerical data produced by (series of) MICROARRAY experiments. Starting from relatively simple demands for appropriate visualisation of the data, bioinformatics tools are necessary to focus on candidate genes and reveal subtle changes in expression patterns.

A reliable identification of candidate genes by statistical methods is only possible if a sufficient number of replicate experiments have been performed. Technical replicates using the same starting material are usually performed to define the overall reproducibility of MICROARRAY experiments. Biological replicates are important to discriminate individual differences (e.g. patient specific) from general changes of GENE EXPRESSION (e.g. disease specific).

Additional bioinformatics methods can be used to identify groups of genes showing a comparable regulation. One method commonly used is the HIERARCHICAL CLUSTER ANALYSIS where genes and arrays are ordered by similarity in expression [22]. Due to the overwhelming amount of data, it is often difficult to understand MICROARRAY results in the light of certain biological questions. To assist researchers in interpreting the results, MICROARRAY data can be combined with knowledge stored in diverse databases like pathway information, genomic localisation or protein family classification.

Different data analysis tools can be applied to identify genes that may be related to a disease or treatment of interest. Linking the data to biological knowledge can also elucidate possible functions of the genes interest. Succeeding experiments using mostly molecular biology techniques like RT-PCR, in situ HYBRIDISATION, RNAi, knockout experiments, etc. are commonly performed to validate and corroborate the biological function concluded from the MICROARRAY data.

17.5 An Example of a Microarray Experiment

17.5.1 Global RNA Amplification and Microarray Analysis of T Cell Subpopulations

Naïve T cells differentiate in response to pathogens into multiple CD4⁺ and CD8⁺ subsets (see Chap. 3). To improve the understanding of this differentiation process as well as the nature of the different subsets, GENE EXPRESSION PROFILING has been used. As an example, MICROARRAY experiments were performed from ten different subpopulations covering the major stages of post-thymic CD4⁺ and CD8⁺ T cell differentiation (Fig. 17.5) [23]. The CD4⁺ and CD8⁺ subsets were isolated by immunomagnetic and flow cytometric cell sorting (see Chap. 16) based on the expression of CD4/CD8, CD27, CD28, CD45RA and CCR7. These markers characterise the major steps of T cell differentiation from naïve to highly differentiated cells in humans [24, 25]. The GENE EXPRESSION profiles were generated from multiple T cell subsets independently gained from two blood samples. As only limited cell numbers can be isolated from 20 mL of blood, a global PCR AMPLIFICATION method was applied allowing MICROARRAY experiments from 1000 cells per T cell population.

For the AMPLIFICATION of RNA from small cell numbers, loss of material is critical, and the pipetting of samples from one tube to another should be avoided as much as possible. For the global AMPLIFICATION, the cells were collected in a small volume of buffer and lysed (Fig. 17.4). Then, superparamagnetic oligo dT microbeads were directly added to the cell lysate binding the poly(A) residues of the mRNA. The labelled cell lysate was applied to a column that was placed in the magnetic field of a heatable permanent magnet. The magnetically labelled mRNA was retained in the strong magnetic field, while effective washing steps removed all other cell components. In-column cDNA synthesis and purification was performed in the same column used for mRNA isolation to avoid loss of material. Oligo dT and randomer oligonucleotides coupled to microbeads were used as primers for the cDNA synthesis. Thereby, cDNA fragments of uniform size were generated, and each transcript was represented by several cDNA

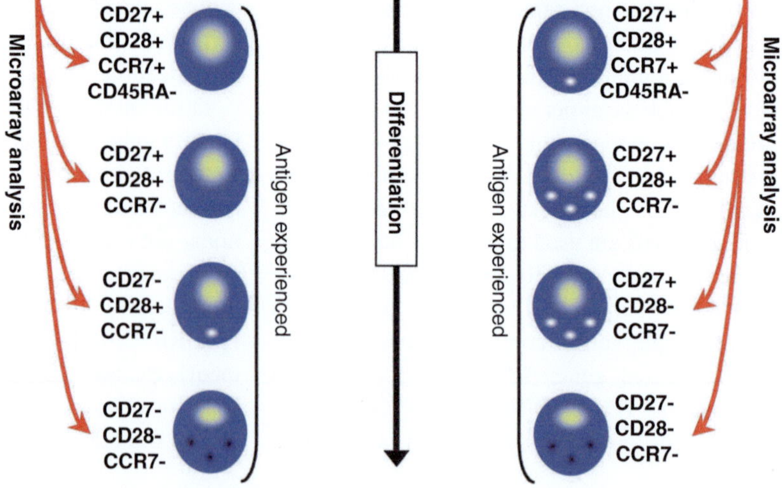

Fig. 17.5 Microarray analysis of CD8⁺ and CD4⁺ T cell subpopulations defining distinct stages of differentiation

fragments enabling uniform AMPLIFICATION during PCR. After eluting the cDNA fragments from the column, a tag was added to the 3′ end of each cDNA fragment by utilising a terminal deoxynucleotidyl transferase. A global PCR amplified the uniform-sized cDNA fragments 10^6-fold, resulting in sufficient TARGET material for MICROARRAY HYBRIDISATION. The PCR performed with a single primer enabled unbiased AMPLIFICATION due to the uniform annealing temperature. The primer binding site at the 3′ end was added during cDNA tailing. The complementary sequence of the tag was inserted at the 5′ end of the cDNA fragments during cDNA synthesis. After purification of the PCR products, a Klenow fragment labelling procedure with random primers in the presence of labelled nucleotides, in this case, Cy3-dCTP or Cy5-dCTP, yielded labelled DNA fragments that were used for MICROARRAY HYBRIDISATION.

All differentiated T cell subsets were hybridised against the corresponding naïve T cells as control in two-colour MICROARRAY experiments. Therefore, the genes found differentially expressed on the microarrays represented potential genes related to the differentiation from naïve to antigen-experienced T cells.

For the first differentiation stage ($CD27^+$/$CD28^+$/$CCR7^+$/$CD45RA^+$), about 15% of the detected genes were found to be differentially expressed, and this proportion increased for stages 2–5 to about 50%, which is consistent with the differentiation process.

A detailed analysis of the differentially detected genes revealed the acquisition of a cytolytic program by the highly differentiated T cells represented by the expression of genes encoding for the lytic granule membrane protein LAMP-3 and the CYTOTOXIC factors granzyme B and perforin. The up-regulation of these genes giving rise to lytic and CYTOTOXIC proteins supported the idea of CYTOTOXIC T cells as late differentiation state.

Another interesting set of genes was found downregulated in highly differentiated T cells. These genes encode for proteins involved in cell cycle entry and/or cell proliferation, as well as anti-apoptotic factors, suggesting a quiescence state and limited survival potential for the highly differentiated T cells under stress or upon activation.

Overall, during the differentiation process, the changes in GENE EXPRESSION for the differentiated T cells compared to the naïve T cells became increasingly similar between $CD4^+$ and $CD8^+$ T cells. So despite the clear differences between naïve $CD4^+$ and $CD8^+$ T cells, the differentiation process might be orchestrated by analogous changes in the GENE EXPRESSION profile.

In summary, the GENE EXPRESSION analysis using global RNA AMPLIFICATION for MICROARRAY experiments suggested functional changes especially during the late differentiation state pointing to CYTOTOXIC potential and limited lifespan. In addition, common changes in the GENE EXPRESSION pattern pointed to a similar differentiation process for $CD4^+$ and $CD8^+$ T cells.

17.6 RNA Sequencing

17.6.1 Introduction

Microarrays are currently the most popular choice for studying changes in the transcriptome, and significant advances in medical research have been made possible in the last 20 years by applying this technique [26]. Despite this, however, microarray technology does have limitations. There can be difficulties with probe design/performance, for instance, some probes cross hybridise with other genes, while some non-specifically hybridise [27]. The dynamic range of a probe can be restricting [28]; when an mRNA is abundantly expressed, a DNA microarray shows saturation, while at the low end of abundance, it suffers a loss of signal. Another disadvantage of microarray technology is that it is generally limited only to those genomes that have been previously sequenced [26].

In the last 5–10 years, a new technology for studying changes in the host transcriptome has emerged; this is known as RNA sequencing (RNA-Seq). Instead of using molecular hybridization to "capture" transcript molecules of interest, RNA-Seq samples transcripts in the starting material by direct sequencing using next-generation sequencing (NGS) technologies (Box 17.5). Once detected, transcript sequences are then mapped back to a reference. Reads that map back to the reference are then counted to assess the level of gene expression, the number of mapped reads being the measure of expression level for that gene or genomic region [26].

A comparison of RNA-Seq versus microarray technology is presented in Table 17.1. RNA-Seq has many advantages compared to microarray analysis, for example, it can detect novel transcripts, allele-specific expression and splice junctions, and it can also be applied to any species even if the reference genome is unknown [27]. RNA-Seq has a larger dynamic range than microarrays and can detect more accurately transcripts in low abundance in the presence of highly abundant transcripts [30]. Another significant advantage offered by RNA-Seq is the need for a lower input of RNA starting material [29]. Currently, the challenges associated with RNA-Seq technology are the complexity of the data analyses and storage of large amounts of data, which should not be underestimated [26]. Despite this, RNA-Seq technology is revolutionising transcriptomic analysis and provides a powerful tool to decipher global gene expression patterns far beyond the limitations of microarrays.

Box 17.5: Next-Generation Sequencing (NGS)

Next-generation sequencing is a term used to describe a number of different modern sequencing technologies which have in general replaced the traditional Sanger-based platform. NGS is also known as "high-throughput" or "deep" sequencing which reflects the vast increase in the number of sequenced bases per run (typically 10^3- to 10^6-fold greater), with a corresponding reduction in cost per sequenced base. There are a number of competing platforms each with different characteristics (e.g. read length, sequencing capacity, error rate, cost per base), with perhaps Illumina being the most widely used at present (a typical run generating 500 Gb of data), in preference to Ion Torrent, SOLiD and 454 platforms. These are characterised by relatively short reads (35–1000 bases) and also require significant sample preparation and amplification. More recently, single-molecule real-time (SMRT) methods have been introduced, which require less sample preparation time, and yield much longer read lengths ('000's of bases per run), although the error rate is relatively high and they are at present more expensive, currently limiting the widespread adoption of this technology. Examples of this include PacBio and Oxford Nanopore systems.

Table 17.1 Comparison of microarray and RNA-Seq technologies

	Microarray	RNA-Seq
Amount of RNA required	High	Low
Resolution	Several to 100 bp	Single base
Distinguish splice forms?	Limited	Yes
Discover new genes?	No	Yes
Strandedness?	No	Yes
Dynamic range	Few hundredfold	>8000-fold
Reproducibility	Yes	Yes
Cost	Medium	High (due to computation)

Adapted from Bauer et al., BMC Bioinformatics 2014, 15(Suppl 11):S3 [29]

17.6.2 RNA-Seq Experimental Workflow

A typical RNA-Seq workflow, including important factors for consideration, is shown in Fig. 17.6. Experimental planning is one of the most important factors to consider, including determining if RNA-Seq is the most appropriate technique to use. RNA-Seq generates a huge, potentially bewildering, amount of data, and it is

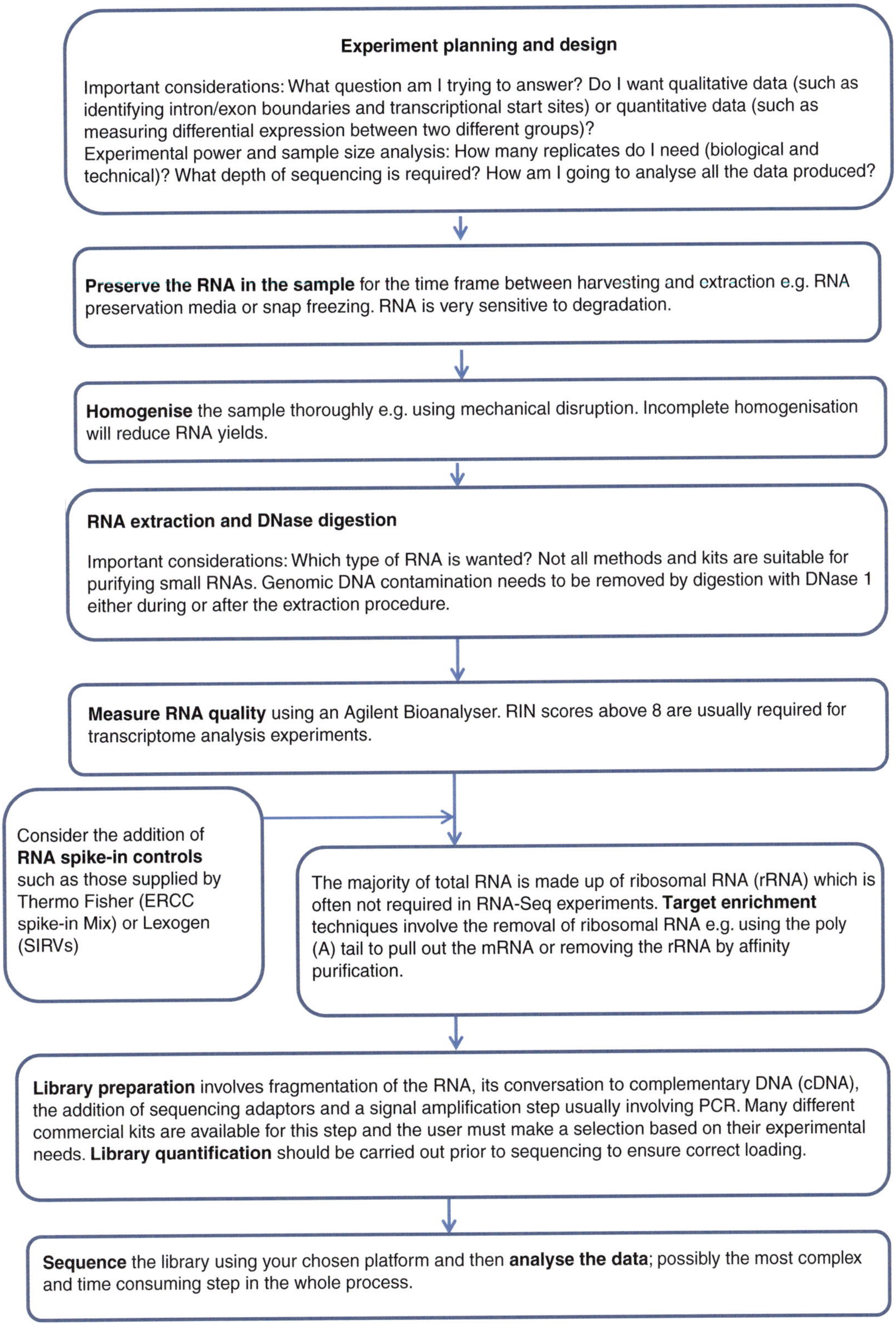

Fig. 17.6 Flow diagram for RNA-Seq experiment

vitally important to determine whether a more targeted approach such as reverse-transcription PCR would be preferable. The experimental design must ensure that it has the capability to answer the research question and is sufficiently powered including an appropriate number of biological replicates [31]. The success of RNA-Seq experiments is highly dependent upon recovering pure and intact RNA from samples which is free from DNA contamination. RNA is more labile than DNA, and RNases are very stable enzymes, so extra care should be taken when purifying and working with RNA as differential degradation of samples will adversely affect the experimental outcome. Several commercial kits exist for RNA extraction from various sample types ranging from bacteria and viruses to human and animal blood and tissue samples. Depending on the source of the RNA, it is vital to ensure the sample is thoroughly homogenised to ensure maximal isolation of RNA. Once the RNA has been isolated, the quality is usually assessed using an Agilent Bioanalyzer (see Box 17.2). In order to understand the biological variation in RNA-Seq results, it is important to understand and control for the technical variation that can be introduced at every step in the procedure from the operator carrying out the work, the method of homogenisation used to the methods of data analysis. To do this, spike-in external controls can be added to RNA-Seq experiments such as those designed by the External RNA Controls Consortium (ERCC; Thermo Fisher) or Spike-In RNA Variants (SIRVs, Lexogen).

The RNA isolated is usually the total RNA though most differential transcriptome analysis experimenters only want to look at the messenger RNA (mRNA). The majority (>95%) of the total RNA is made up of ribosomal RNA (rRNA). Before library preparation, many researchers choose to perform target enrichment to maximise the amount of their target RNA fraction in the final sample. Several commercial kits exist for either removing the rRNA from the sample or pulling out the mRNA using the poly-A tail.

The exact method of library preparation will depend on the sequencing platform being used and the experimental question being answered. Companies involved in sequencing sell a variety of library preparation kits. It is possible using RNA-Seq, if the appropriate library preparation method has been used, to obtain information from both the sense and antisense strands of the RNA template. This information is important for analysis of transcript orientation and the detection of overlapping transcripts. It is essential to accurately quantify the sequencing library before loading on the sequencing platform to ensure optimal performance and success of the sequencing run. Methods of sequencing library quantification include quantitative real-time PCR and spectrophotometry.

17.6.3 Choice of Sequencing Platform

A number of different NGS sequencing platforms (Box 17.5) are currently available for RNA-Seq experiments (see NGS review [32, 33]). Each platform has different characteristics (e.g. read length, number of reads per run, base-calling accuracy), and the choice will depend on the experimental question being asked. Typically, however, the Illumina platform (which yields relatively short reads (~150 bases), but has a low error rate and a high read depth) is the most commonly used. For more specialised applications where longer reads are required (e.g. novel transcriptome assembly), the PacBio RS instrument is more appropriate.

17.6.4 Data Analysis Pipeline

Once raw reads from the RNA-Seq experiment have been generated, they are processed and analysed through a series of software analysis steps. A schematic of a typical data analysis pipeline is shown in Fig. 17.7 and described as follows:

1. *Preprocessing and quality control of raw reads*: For samples that have been multiplexed (combined samples on one sequencing run), they must first be demultiplexed. Then adapter sequences are removed, generic quality

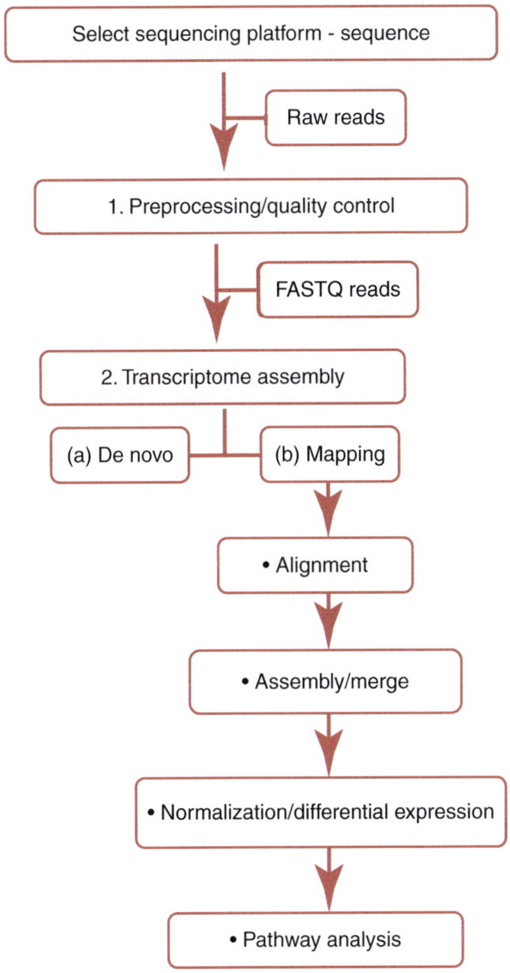

Fig. 17.7 Schematic data analysis pipeline

control steps may be performed (e.g. FastQC software), and reads are trimmed and filtered by base quality score to remove low-quality or contaminating reads (using "trimmomatic" for Illumina reads [34]). Sequence reads (base identity along with a quality score) are provided in a format known as FASTQ.

2. *Transcriptome assembly*: Assembly of the transcriptome falls into two methods depending on whether a reference genome or transcriptome of the organism under study exists. If a reference genome does not exist, then the de novo approach is taken (a); if a reference genome exists, a mapping approach is taken (b).

   (a) *De novo approach*—This approach is more computer resource intensive than the mapping approach as each read has to be compared with every other read in order to generate a set of contigs, instead of just to one reference genome. Typically de novo transcriptome assembly software use a graph-based approach—examples are Velvet/Oases [35, 36], Trans-ABySS [37] and Trinity [38].

   (b) *Mapping approach*—Assuming that a reference genome, transcriptome or annotation is available, many analyses and those leading to differential expression will use the mapping approach. As an example, typical programs from the "Tuxedo" protocol are used [39]: a representative pipeline could be:

   - *Alignment*—Using TopHat software—Aligns an RNA-Seq read to the reference using the Bowtie short read aligner and then analyses the mapping results to identify splice junctions between exons.
   - *Assembly/merge assemblies*—Using Cufflinks (assembles transcripts, estimates abundance and tests for differential expression and regulation in RNA-Seq samples) and Cuffmerge (merges together assemblies).
   - *Normalization and differential expression*—The Cuffdiff software contains methods for normalization and calculation of statistically significantly changes in transcript expression. Numerous other methods exist, and those written for the free statistical software environment "R" are commonly used (e.g. edgeR [40], DESeq [41]).
   - *Differentially expressed (DE) gene prediction*, using Cuffdiff, searches for significant changes in transcript expression. These genes may then be entered into a pathway analysis software package to identify key biological pathways.

An example of an RNA-Seq experiment which has used the described workflow is included in Box 17.6.

Box 17.6: Differential Transcriptome Analysis in an Animal Model of Bacterial Disease

The response a host makes to an infection is very complex. A full understanding of this response can lead to the development of new ideas for treatment [42]. RNA-Seq can be used to gain an understanding of this response by studying RNA isolated from an animal model over a time course of infection. BALB/c mice were exposed to an inhalational challenge of the Gram-negative pathogen *Burkholderia pseudomallei*, with a mean retained dose of 30 colony-forming units per mouse [43]. Animals were culled at predetermined time points, and tissues were harvested and stored immediately in RNA*later*. An identical number of control mice received a phosphate buffered saline exposure and handled in exactly the same manner as the infected groups. The mRNA was isolated from the lungs and sequenced using the workflow described above. The transcriptome obtained from the control group acted as the baseline comparator for the infected group. Once the transcriptome had been generated, Ingenuity Pathways Analysis (IPA; Qiagen, http://www.ingenuity.com) was used to interpret the data and apply biological meaning (Essex-Lopresti et al., personal communication).

When a host is exposed to a dose of bacteria, an important part of the response process is the body's ability to recognise the presence of the bacteria. The primary way of doing this is by the activation of pattern recognition receptors through the innate immune system [44]. Figure 17.8 shows the pattern recognition receptor pathway which has been overlaid with the transcriptome data from day 3, postexposure, of the *B. pseudomallei* aerosol infection. This transcriptome data shows that the mouse host has upregulated expression of several toll-like receptors (TLRs) which in turn leads ultimately to the release of cytokines via transcription factors such as NFκB. The cytokines released as a result of this pathway then circulate around the host influencing other response pathways and cell populations. Detailed knowledge of this and other response pathways helps researchers, trying to design novel drugs, to identify places where medical interventions might assist the host in its fight against the pathogen. © Crown copyright (2016), Dstl. This material is licensed under the terms of the Open Government Licence except where otherwise stated. To view this licence, visit http://www.nationalarchives.gov.uk/doc/open-government-licence/version/3 or write to the Information Policy Team, The National Archives, Kew, London TW9 4DU, or email: psi@nationalarchives.gsi.gov.uk.

17.7 The Future of RNA-Seq

Developing technologies are likely to help shape the future of RNA-Seq for transcriptome profiling. For instance, advances in long-read sequencing technologies will impact upon RNA-Seq analysis. Currently, most RNA-Seq experiments are undertaken on platforms (e.g. Illumina) which yield relatively short-read lengths. Long-read sequencing technology (e.g. Pacific Biosciences (PacBio) single-molecule real-time (SMRT) sequencing approach) can produce reads matching longer transcripts. The advantage of longer reads is that a lot of mapping errors that occur after sequencing will be eliminated, and as a consequence the accuracy of sequencing will greatly improve.

Hand-held sequencing technologies are likely to impact on transcriptome profiling too.

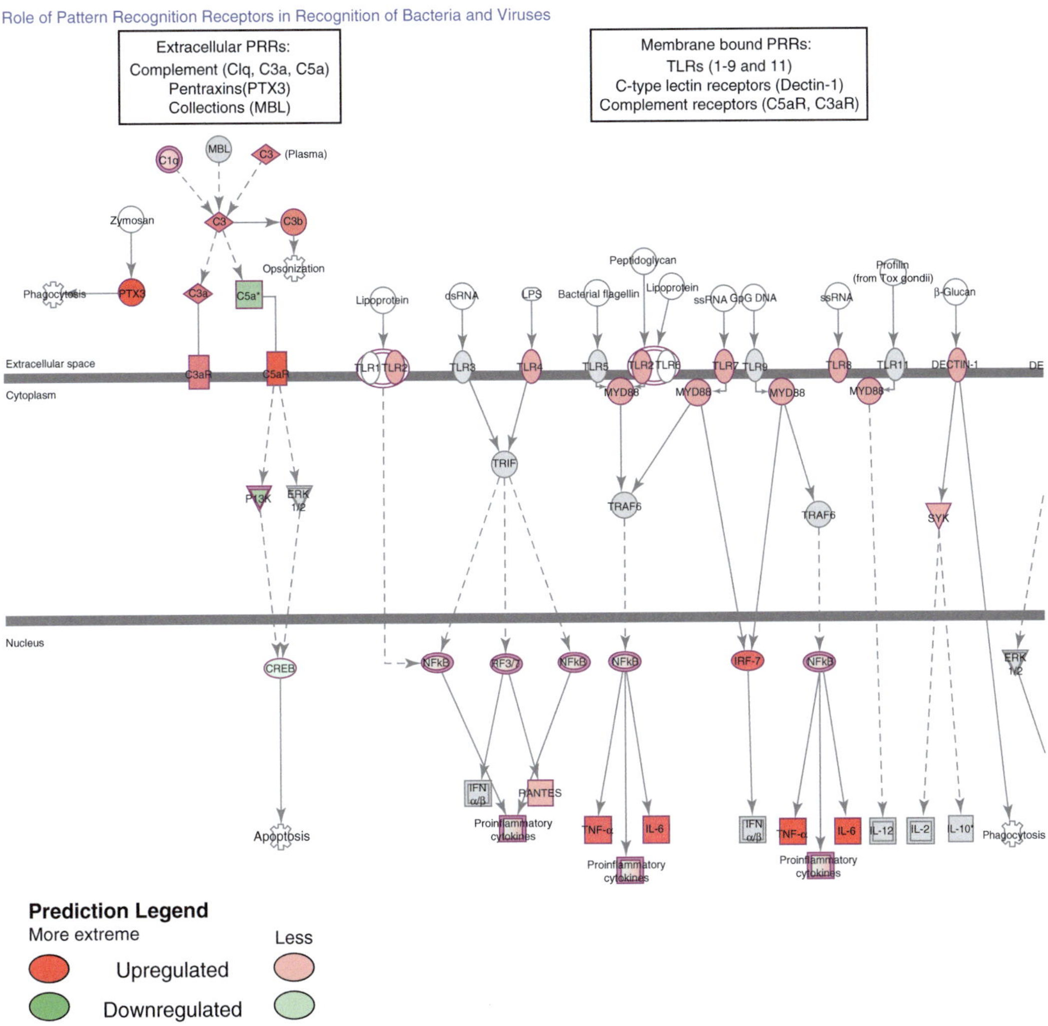

Fig. 17.8 Canonical pathway analysis. Canonical pathway, found within IPA, showing the role of pattern recognition receptors in recognition of bacteria and viruses overlaid with the transcriptome data from day 3 of the *B. pseudomallei* mouse aerosol infection model. Fold changes of up- and downregulated genes are indicated by red and green shading according to the prediction legend.

Currently MinIONs, palm-sized NGS sequencing devices from Oxford Nanopore (https://www.nanoporetech.com/), offer read lengths of tens of kilobases, limited only by the length of DNA molecules presented to it [45]. These devices were used to undertake the genomic surveillance of Ebola virus, in the field, in 2015 in West Africa [46]. Results were generated in less than 24 h after receiving an Ebola-positive sample where the sequencing process took as little as 15–60 min. Pathogen genomes are usually not very large and thus are suitable for use with this "miniaturised" technology. In the future however, RNA-Seq analysis might well become portable, and this might be aided by the use of direct RNA sequencing (DRS). DRS is another new pioneering technology, with which it is possible to carry out direct single RNA sequencing without prior conversion of RNA to cDNA. The advantage of DRS over RNA-Seq is that it removes the technical artefacts introduced by having to create a cDNA library or by having amplification steps [47, 48].

RNA-Seq technology is currently evolving; some researchers have moved the science forward already and developed methods for analysing the transcriptome of individual cells, otherwise known as single-cell RNA sequencing (scRNA-Seq) [49]. This allows the complex analyses of heterogeneous samples and profiling of cell-to-cell variables on a genomic scale [50]. There are significant challenges with this technology (RNA losses, differences in strand-specificity and difficulty in distinguishing between noise and variability for low abundance transcripts), but it is hoped that advances in sequencing technology (such as nanopores, mentioned above) will overcome these barriers [51]. Given the high anticipated value of single-cell transcriptomics, explosive growth of scRNA-Seq data is expected in the next 5–10 years [49].

17.8 Overview

The ability to monitor changes in the mRNA expression of multiple genes by using microarray technology is firmly established. This technology is routinely used, reasonably cost-effective, reliable and highly reproducible. Many scientific advances in medical research and diagnosis have been made possible using this technique. RNA-Seq technology, however, is closely following in the footsteps of microarray technology, and once sequencing costs become lower, data analyses become more streamlined, and data storage issues are resolved, RNA-Seq is likely to significantly impact upon transcriptomic research in the future.

Selected Readings

Bosio A, Gerstmayer B, editors. Microarrays in inflammation (Progress in inflammation research). Basel: Birkhäuser Verlag; 2008.

Goodwin S, McPherson JD, McCombie WR. Coming of age: ten years of next-generation sequencing technologies. Nat Rev Genet. 2016;17(6):333–51.

Müller UR, Nicolau DV, editors. Microarray technology and its applications. Berlin: Springer-Verlag; 2005.

Stekel D. Microarray bioinformatics. Cambridge, UK: Cambridge University Press; 2003.

Wang Z, Gerstein M, Snyder M. RNA-Seq: a revolutionary tool for transcriptomics. Nat Rev Genet. 2009;10(1):57–63.

Recommended Websites

Functional Genomics Data Society – FGED Society: Minimum information about a microarray experiment – MIAME. http://fged.org/projects/miame/. Accessed Aug 2016.

Genetic Information Research Institute: Repbase. http://www.girinst.org. Accessed Aug 2016.

Illumina website – http://www.illumina.com/. Accessed Aug 2016.

miRBase: microRNA database. http://www.mirbase.org/. Accessed Aug 2016.

National Center for Biotechnology Information: Gene Expression Omnibus. http://www.ncbi.nlm.nih.gov/geo. Accessed Aug 2016.

RNA-seqlopedia – http://rnaseq.uoregon.edu/. Accessed Aug 2016.

Swiss-Prot: Curated protein sequence database. http://www.expasy.org/sprot/. Accessed Aug 2016.

Unigene: http://www.ncbi.nlm.nih.gov/unigene. Accessed Aug 2016.

References

1. International Human Genome Sequencing Consortium. Finishing the euchromatic sequence of the human genome. Nature. 2004;431(7011):931–45.
2. Southern EM. Detection of specific sequences among DNA fragments separated by gel electrophoresis. J Mol Biol. 1975;98(3):503–17.
3. McGall GH, Fidanza JA. Photolithographic synthesis of high-density oligonucleotide arrays. Methods Mol Biol. 2001;170:71–101.
4. Pease AC, Solas D, Sullivan EJ, Cronin MT, Holmes CP, Fodor SP. Light-generated oligonucleotide arrays for rapid DNA sequence analysis. Proc Natl Acad Sci U S A. 1994;91(11):5022–6.
5. Hacia JG, Collins FS. Mutational analysis using oligonucleotide microarrays. J Med Genet. 1999;36(10):730–6.
6. Hacia JG, Fan JB, Ryder O, Jin L, Edgemon K, Ghandour G, Mayer RA, Sun B, Hsie L, Robbins CM, et al. Determination of ancestral alleles for human single-nucleotide polymorphisms using high-density oligonucleotide arrays. Nat Genet. 1999;22(2):164–7.
7. Okamoto T, Suzuki T, Yamamoto N. Microarray fabrication with covalent attachment of DNA using bubble jet technology. Nat Biotechnol. 2000;18(4):438–41.
8. Lockhart DJ, Dong H, Byrne MC, Follettie MT, Gallo MV, Chee MS, Mittmann M, Wang C, Kobayashi

M, Horton H, et al. Expression monitoring by hybridization to high-density oligonucleotide arrays. Nat Biotechnol. 1996;14(13):1675–80.

9. Hester SD, Reid L, Nowak N, Jones WD, Parker JS, Knudtson K, Ward W, Tiesman J, Denslow ND. Comparison of comparative genomic hybridization technologies across microarray platforms. J Biomol Tech. 2009;20(2):135–51.
10. He H, Cai L, Skogerbo G, Deng W, Liu T, Zhu X, Wang Y, Jia D, Zhang Z, Tao Y, et al. Profiling *Caenorhabditis elegans* non-coding RNA expression with a combined microarray. Nucleic Acids Res. 2006;34(10):2976–83.
11. Barad O, Meiri E, Avniel A, Aharonov R, Barzilai A, Bentwich I, Einav U, Gilad S, Hurban P, Karov Y, et al. MicroRNA expression detected by oligonucleotide microarrays: system establishment and expression profiling in human tissues. Genome Res. 2004;14(12):2486–94.
12. Krichevsky AM, King KS, Donahue CP, Khrapko K, Kosik KS. A microRNA array reveals extensive regulation of microRNAs during brain development. RNA. 2003;9(10):1274–81.
13. Landgraf P, Rusu M, Sheridan R, Sewer A, Iovino N, Aravin A, Pfeffer S, Rice A, Kamphorst AO, Landthaler M, et al. A mammalian microRNA expression atlas based on small RNA library sequencing. Cell. 2007;129(7):1401–14.
14. Liu CG, Calin GA, Meloon B, Gamliel N, Sevignani C, Ferracin M, Dumitru CD, Shimizu M, Zupo S, Dono M, et al. An oligonucleotide microchip for genome-wide microRNA profiling in human and mouse tissues. Proc Natl Acad Sci U S A. 2004;101(26):9740–4.
15. Thomson JM, Parker JS, Hammond SM. Microarray analysis of miRNA gene expression. Methods Enzymol. 2007;427:107–22.
16. Wulfkuhle J, Espina V, Liotta L, Petricoin E. Genomic and proteomic technologies for individualisation and improvement of cancer treatment. Eur J Cancer. 2004;40(17):2623–32.
17. Wang Z, Gao J. Microarray-based study of carbohydrate-protein binding. Methods Mol Biol. 2010;600:145–53.
18. Eberwine J. Amplification of mRNA populations using aRNA generated from immobilized oligo(dT)-T7 primed cDNA. Biotechniques. 1996;20(4):584–91.
19. Singh R, Maganti RJ, Jabba SV, Wang M, Deng G, Heath JD, Kurn N, Wangemann P. Microarray-based comparison of three amplification methods for nanogram amounts of total RNA. Am J Physiol Cell Physiol. 2005;288(5):C1179–89.
20. Novoradovskaya N, Whitfield ML, Basehore LS, Novoradovsky A, Pesich R, Usary J, Karaca M, Wong WK, Aprelikova O, Fero M, et al. Universal reference RNA as a standard for microarray experiments. BMC Genomics. 2004;5(1):20.
21. Bissels U, Wild S, Tomiuk S, Holste A, Hafner M, Tuschl T, Bosio A. Absolute quantification of microRNAs by using a universal reference. RNA. 2009;15(12):2375–84.
22. Eisen MB, Spellman PT, Brown PO, Botstein D. Cluster analysis and display of genome-wide expression patterns. Proc Natl Acad Sci U S A. 1998;95(25):14863–8.
23. Appay V, Bosio A, Lokan S, Wiencek Y, Biervert C, Kusters D, Devevre E, Speiser D, Romero P, Rufer N, et al. Sensitive gene expression profiling of human T cell subsets reveals parallel post-thymic differentiation for CD4+ and CD8+ lineages. J Immunol. 2007;179(11):7406–14.
24. van Lier RA, ten Berge IJ, Gamadia LE. Human CD8(+) T-cell differentiation in response to viruses. Nat Rev Immunol. 2003;3(12):931–9.
25. Appay V, Rowland-Jones SL. Lessons from the study of T-cell differentiation in persistent human virus infection. Semin Immunol. 2004;16(3):205–12.
26. Malone JH, Oliver B. Microarrays, deep sequencing and the true measure of the transcriptome. BMC Biol. 2011;9:34.
27. Zhao S, Fung-Leung WP, Bittner A, Ngo K, Liu X. Comparison of RNA-Seq and microarray in transcriptome profiling of activated T cells. PLoS One. 2014;9(1):e78644.
28. Wang Z, Gerstein M, Snyder M. RNA-Seq: a revolutionary tool for transcriptomics. Nat Rev Genet. 2009;10(1):57–63.
29. Bauer MA, Chavan SS, Peterson EA, Heuck CJ, Johann DJ. Leveraging the new with the old: providing a framework for the integration of historic microarray studies with next generation sequencing. BMC Bioinformatics. 2014;15(Suppl 11):S3.
30. Weber AP. Discovering new biology through sequencing of RNA. Plant Physiol. 2015;169(3):1524–31.
31. Hart SN, Therneau TM, Zhang YJ, Poland GA, Kocher JP. Calculating sample size estimates for RNA sequencing data. J Comput Biol. 2013;20(12):970–8.
32. Mardis ER. Next-generation sequencing platforms. Annu Rev Anal Chem. 2013;6:287–303.
33. Goodwin S, McPherson JD, McCombie WR. Coming of age: ten years of next-generation sequencing technologies. Nat Rev Genet. 2016;17(6):333–51.
34. Bolger AM, Lohse M, Usadel B. Trimmomatic: a flexible trimmer for Illumina sequence data. Bioinformatics. 2014;30(15):2114–20.
35. Zerbino DR, Birney E. Velvet: algorithms for de novo short read assembly using de Bruijn graphs. Genome Res. 2008;18(5):821–9.
36. Schulz MH, Zerbino DR, Vingron M, Birney E. Oases: robust de novo RNA-seq assembly across the dynamic range of expression levels. Bioinformatics. 2012;28(8):1086–92.
37. Robertson G, Schein J, Chiu R, Corbett R, Field M, Jackman SD, Mungall K, Lee S, Okada HM, Qian JQ, et al. De novo assembly and analysis of RNA-seq data. Nat Methods. 2010;7(11):909–12.
38. Grabherr MG, Haas BJ, Yassour M, Levin JZ, Thompson DA, Amit I, Adiconis X, Fan L, Raychowdhury R, Zeng Q, et al. Full-length transcriptome assembly from RNA-Seq data without a reference genome. Nat Biotechnol. 2011;29(7):644–52.

39. Trapnell C, Roberts A, Goff L, Pertea G, Kim D, Kelley DR, Pimentel H, Salzberg SL, Rinn JL, Pachter L. Differential gene and transcript expression analysis of RNA-seq experiments with TopHat and Cufflinks. Nat Protoc. 2012;7(3):562–78.
40. Robinson MD, McCarthy DJ, Smyth GK. edgeR: a Bioconductor package for differential expression analysis of digital gene expression data. Bioinformatics. 2010;26(1):139–40.
41. Anders S, Huber W. Differential expression analysis for sequence count data. Genome Biol. 2010;11(10):R106.
42. Hancock RE, Nijnik A, Philpott DJ. Modulating immunity as a therapy for bacterial infections. Nat Rev Microbiol. 2012;10(4):243–54.
43. Lever MS, Nelson M, Stagg AJ, Beedham RJ, Simpson AJ. Experimental acute respiratory Burkholderia pseudomallei infection in BALB/c mice. Int J Exp Pathol. 2009;90(1):16–25.
44. Akira S, Uematsu S, Takeuchi O. Pathogen recognition and innate immunity. Cell. 2006;124(4):783–801.
45. Laver T, Harrison J, O'Neill PA, Moore K, Farbos A, Paszkiewicz K, Studholme DJ. Assessing the performance of the Oxford Nanopore Technologies MinION. Biomol Detect Quantif. 2015;3:1–8.
46. Quick J, Loman NJ, Duraffour S, Simpson JT, Severi E, Cowley L, Bore JA, Koundouno R, Dudas G, Mikhail A, et al. Real-time, portable genome sequencing for Ebola surveillance. Nature. 2016;530(7589):228–32.
47. Ozsolak F, Platt AR, Jones DR, Reifenberger JG, Sass LE, McInerney P, Thompson JF, Bowers J, Jarosz M, Milos PM. Direct RNA sequencing. Nature. 2009;461(7265):814–8.
48. Ozsolak F. Attomole-level genomics with single-molecule direct DNA, cDNA and RNA sequencing technologies. Curr Issues Mol Biol. 2016;18:43–8.
49. Yu P, Lin W. Single-cell transcriptome study as big data. Genomics Proteomics Bioinformatics. 2016;14(1):21–30.
50. Dey SS, Kester L, Spanjaard B, Bienko M, van Oudenaarden A. Integrated genome and transcriptome sequencing of the same cell. Nat Biotechnol. 2015;33(3):285–9.
51. Saliba AE, Westermann AJ, Gorski SA, Vogel J. Single-cell RNA-seq: advances and future challenges. Nucleic Acids Res. 2014;42(14):8845–60.

Imaging Inflammation

18

Thomas H. Craven, Philippe M. D. Potey, David A. Dorward, and Adriano G. Rossi

18.1 Introduction

The cardinal features of inflammation were identified many centuries ago, and the subsequent demonstration of the cellular basis of these features first relied upon light microscopy and the development of histological processes. As the complexity of inflammation becomes more understood, tools are required to access ever smaller and deeper compartments of tissue, in vitro and in vivo, in animals and in man, at the same time maintaining specificity for the particular process under investigation. Whilst imaging modalities can rely solely upon the endogenous features of the tissue under investigation, commonly some form of contrast agent is added, and the qualities of the tissue, the imaging technique and the contrast agent can be exploited together to answer specific questions about inflammation.

Molecular imaging refers to "the visualization, characterization, and measurement of biological processes at the molecular and cellular levels in humans and other living systems" [1]. The imaging of individual molecules is beyond the resolution of the vast majority of imaging technologies, but the application of specific "contrast" agents can amplify molecular processes, so they can be detected by systems otherwise not equipped to resolve at the molecular level.

Classically, in clinical practice, diseases have been observed as the late physiological manifestations of much earlier biological and molecular changes, and imaging has consisted primarily of detecting structural change [2]. Molecular imaging has the potential for earlier detection, translating laboratory understanding of disease processes into the clinical arena where strategies may be used for diagnosis and surveillance, as well as treatment planning and monitoring.

The vast majority of imaging modalities employ electromagnetic radiation, which is transmitted by the elementary particle: the photon. Like all elementary particles, the photon exhibits wave-particle duality, that is, it possesses the characteristics of both a wave and a particle. Travelling as a wave, it has a wavelength, a frequency and an intensity or energy. As wavelength elongates, frequency and intensity drop. The electromagnetic spectrum covers all possible types of electromagnetic radiation and ranges from short wavelength, high-frequency gamma rays, through the visible portion of the spectrum, through microwaves to radio waves, which are low frequency and long wavelength. The majority of imaging technologies described utilize energy from the electromagnetic spectrum and have characteristics that rely upon the features of the wave that is used to obtain the image (Fig. 18.1).

Final manuscript submitted on April 18, 2018.

T. H. Craven · P. M. D. Potey · D. A. Dorward
The University of Edinburgh, Edinburgh, UK

A. G. Rossi (✉)
Centre for Inflammation Research, The Queen's Medical Research Institute, The University of Edinburgh, Edinburgh, UK
e-mail: Adriano.Rossi@ed.ac.uk; a.g.rossi@ed.ac.uk

M. J. Parnham et al. (eds.), *Nijkamp and Parnham's Principles of Immunopharmacology*,
https://doi.org/10.1007/978-3-030-10811-3_18

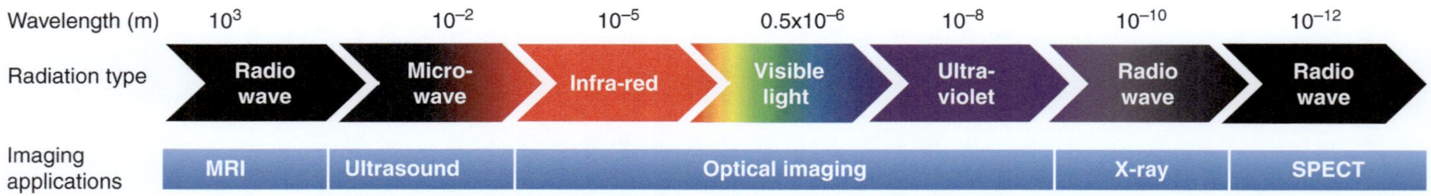

Fig. 18.1 The electromagnetic spectrum. A variety of imaging tools have been designed which span across the electromagnetic spectrum. γ = gamma

18.2 Considerations of Choosing an Imaging Modality

Each imaging modality has its own unique characteristics that make it more or less suitable for the application at hand. A few characteristics are outlined below.

18.2.1 Resolution

RESOLUTION refers to the ability of an imaging system to separate two distinct points when illustrating an image to the user and can be classified into spatial (or angular) resolution, temporal resolution and contrast resolution. *Spatial resolution* refers to the ability of an imaging system to resolve two points in space and is divided into two components, lateral and axial resolution; lateral resolution referring to two points side-by-side in a plane perpendicular to the wave used to obtain the image, and axial resolution referring to two points side-by-side along the axis of wave. *Temporal resolution* requires two points separated only in time to be resolved from one another and is especially important in real-time imaging and where the source material is mobile relative to the detection device. *Contrast resolution* refers to the ability of an imaging system to detect different intensities of signal from a sample and then portrays them accurately to the investigator. An example of contrast resolution is the dynamic range of a system, which refers to the ratio of the smallest and largest usable signal that is transmitted from object to user. Whilst extremes of resolution are fundamentally limited by the physics of the wave used to obtain an image, they can often be adjusted over a large working range to meet the requirements of a particular experiment.

18.2.2 Depth of Penetration

Photons travelling in a vacuum will travel unimpeded with no change in direction or energy. If a particle is placed in the path of the beam of photons, some will be absorbed and some will be scattered. The beam of photons continuing on from the particle will have been attenuated. The rate at which a beam is attenuated depends upon the energy of the beam, the number of barrier particles and the capability of those particles to absorb or scatter that particular beam. For a beam or wave to be absorbed, there must be subatomic particles within the medium with energy levels that correspond to the photon energy contained within the wave. Low-frequency waves such as radio waves pass straight through the human body because the body contains no material with the appropriate energy level to absorb them. Visible light is readily absorbed or scattered, and so waves are attenuated very rapidly, and penetration depth is limited to just a few millimetres. As the wavelength drops even shorter towards X-rays and gamma rays, there is less and less material to absorb or scatter the high-energy photons, and penetration depth again becomes very high.

18.2.3 Safety

Higher frequency waves in the electromagnetic spectrum such as ultraviolet, X- and gamma rays have sufficient energy to ionize molecules within the sample, which can lead to biological toxicity including DNA mutation. Some lasers used in optical imaging could potentially cause tissue destruction through the generation of heat in an extremely small area. In general, the power of lasers used to investigate biological tissue is not in the order of magnitude required to produce tissue damage. Whilst there are specific safety concerns with some of the imaging modalities detailed below such as eye safety when using magnets and lasers, they predominantly concern the investigator rather than the biological sample under investigation.

18.2.4 Image Processing, Storage and Analysis

Apart from the simplest light microscopes, most device detectors convert analogue continuous signal into a digital signal, which is used for storage and analysis. Because our eyes and brains cannot interpret digital images directly, they are converted back to analogue for visual interpretation. Each stage in the processing may lead to the introduction of error or noise, which may lead to misinterpretation. Analysis is inherently subjective, and methodologies to entirely remove sources of bias are fraught with technical difficulty, and procedures of image acquisition, processing and analysis should be subject to critical appraisal. Imaging devices are also capable of producing an enormous quantity of data in a short space of time. Consideration should be given to file format used for storage where no international standard is established, as data loss can occur with some lossy compression algorithms (which reduce file size by eliminating unnecessary information) such as JPEG and MPEG. Even if data are economically stored, analysts can easily become overwhelmed by the acquired data, and even the simplest interpretation can become time-consuming.

18.3 Contrast Agents for Molecular Imaging

Whilst many insights into inflammation and other biological processes can be gained during label-free imaging, the development of contrast agents has vastly increased the capability of molecular imaging. Each molecular imaging agent comprises a targeting segment that confers localization and a component that facilitates imaging by the chosen imaging modality [3]. The primary function is to identify a particular biological process occurring at a molecular or cellular level and amplify it to produce a signal which is detectable by modalities which otherwise do not possess an adequate resolution. The scope of molecular imaging agents is vast, from progressing the understanding of disease right across the experimental spectrum of in vitro to in vivo work, to high throughput drug discovery and toxicity assays, to clinical applications in the search of person-specific disease detection, stratification and treatment. Careful design of the contrast agent is imperative and is a multidisciplinary undertaking. There are four broad considerations in the selection and design of a contrast agent for a particular experiment (Box 18.1).

Box 18.1: Considerations in the Selection and Design of a Contrast Agent

What is the method of localization to the target of interest?

There are three main types of agent. Ones that:

- Bind directly to the target molecule.
- Accumulate by the molecular or cellular activity of the target.
- Become detectable only upon activation by the target.

What is the nature of the biological sample?

- Will the agent be able to access the target?
- Is the sample fixed?
- For in-vivo work, will pharmacokinetic issues permit sufficient exposure of the target to the agent?

What are the limitations of the imaging system in question?

- Does the resolution of the system deliver adequate information to answer the question at hand?
- Is there adequate depth of penetration to image the molecular agent?

Will the agent cause any adverse affects?

- Will the agent cause artifact that leads to false interpretation?
- Are there are tissue pharmacodynamic or toxicological issues?

18.4 Imaging Modalities

There are numerous molecular imaging modalities available, each with advantages and disadvantages. They can be classified into optical, magnetic resonance imaging, nuclear medicine and ultrasound techniques [4]. An overview of some of the more common imaging modalities is provided in Table 18.1.

18.4.1 Optical Imaging

OPTICAL IMAGING relies upon light from the ultraviolet, visible and infrared portions of the electromagnetic spectrum. Ultraviolet light has a short wavelength in the region of about 100–400 nm, whilst infrared light has a wavelength of up to approximately 1 mm. The spatial resolution of an optical system depends upon the wavelength of the light in use and the characteristics of the lens used to focus the image on the detector.

18.4.1.1 An Overview of Light Microscopy

As light from an infinitely sharp point on a sample is focused through an objective to be perfectly focused on a detector, the propagating waves diffract and will blur to a finitely sized disk on the detector. The pattern of blurring as light propagates from the objective to the detector is known as the point spread function (PSF), and it forms a series of concentric light and dark rings as it strikes the detector. This two-dimensional representation of the infinitely sharp point is known as the Airy disk. As two distinct points on the sample get closer together, their Airy disks on the detector will begin to overlap and there comes a point (known as the Rayleigh criterion) where they cannot be resolved from one another. For an

Table 18.1 Overview of commonly used imaging modalities in the molecular imaging of inflammation

Imaging modality	Main application	Spatial resolution	Advantages	Disadvantages	Clinical application	Cost
Bioluminescent	In vivo	Millimetres	High signal to noise ratio Multiplex imaging Good temporal resolution	Requires exogenous substrate and genetic manipulation of target cell or organism	Limited, likely to remain a pre-clinical tool	+
Fluorescent	In vivo, in vitro	Micrometres	Multi-channel imaging NIRF allows greater depth of penetration Wide range of molecular probes Significant area of current probe development	Scatter Autofluorescence Photobleaching Surface weighting of images	Mainly pre-clinical use at present Clinical applications in surgery, endoscopy and surface tissue imaging	+
Confocal microscopy	In vivo, in vitro	Micrometres	In vivo imaging with microendoscopic techniques Intravital imaging	Small fields of view	Early stages of clinical use—especially bronchoscopy and endoscopy	++
PET	In vivo	3–4 mm (microPET and high-resolution PET 1–2 mm)	Highly sensitive in hybrid imaging system (e.g. CT)	Short radionuclide half-life requires on-site cyclotron	In clinical use	+++
SPECT	In vivo	8–15 mm (small animal SPECT <1 mm)	Ability to perform multiplex imaging Longer half-life of commonly used radionuclides	Limited number of target-specific radionuclides available	In clinical use	+++

(continued)

Table 18.1 (continued)

Imaging modality	Main application	Spatial resolution	Advantages	Disadvantages	Clinical application	Cost
MRI	In vivo	100 μm	Non-ionising radiation Excellent spatial resolution	Relatively few specific molecular probes Slow acquisition time Low sensitivity for MI High concentration of contrast agents required	In clinical use—mainly structural imaging	++
Ultrasound	In vivo	Millimetres (dependent on frequency used)	Inexpensive Highly portable Useful in vascular imaging	Variable depth of penetration Operator-dependent quality of image acquisition Few molecular applications due to lack of contrast agents	In clinical use—mainly structural imaging	+
CT	In vivo	100 μm	Simultaneous acquisition of detailed structural information	Few contrast agents available for molecular imaging High concentrations of contrast agents required	In clinical use—mainly structural imaging	+++
Hybrid systems	In vivo	Variable (application dependent)	Improves anatomical accuracy of information	Increases expense	PET/CT in clinical use	++

optical system with no aberration, resolution is therefore said to be diffraction limited, and Abbe's diffraction limit states:

$$d = \frac{\lambda}{2n\sin\theta}$$

where d is the maximum resolvable feature size, λ is the wavelength, n is the refractive index of the medium in which images are obtained and θ is the angular aperture of the objective lens. $n \sin \theta$ is known as the numerical aperture (NA) and is determined by the technical specification of the microscope objective. Numerical apertures rarely fall below approximately 1.4 [5]. Considering a system with NA = 1, then $d = \lambda/2$, and for green light with a wavelength of 500 nm, the maximum resolvable feature size is approximately 250 nm. Axial resolution is typically worse than lateral resolution for a given wavelength because light is usually only collected from one side of the sample, and so a large amount of information is lost. This limitation can be overcome by illuminating and collecting light from additional locations [5].

The main advantages conferred by using the visible spectrum to obtain microscopic images are the low cost and simplicity of the system, the ability to multiplex many different wavelengths into one image and the inherent safety, although ultraviolet wavelengths do cause tissue damage. Another restriction placed upon optical imaging by the properties of the light wave is the lack of penetration depth, which is just a few millimetres for visible light. Infrared microscopes extend the penetration depth to several centimetres at the expense of a reduction in resolution. Penetration depth may also be increased by measuring ballistic photons, which are not scattered or absorbed on their passage through a turbid medium such as biological tissue, but this requires detectors that are extremely sensitive and time resolved especially at tissue depths beyond a few millimetres.

18.4.1.2 An Overview of Fluorescence Microscopy

FLUORESCENCE occurs when an orbital electron of a substance (known as a FLUOROPHORE) becomes excited by the absorbance of a photon (or photons) and subsequently relaxes to its resting state through the emission of a photon. For absorbance to occur, the photon must have energy approximately equal to the difference between the excited and ground state of the electron. Once an orbital electron is excited, there are several potential fates. By far the most common is a return to the ground state through radiation-free dissipation of energy in the form of heat. Occasionally, however, the electron will return to its ground state by emitting a photon. The emitted photon will have a lower energy and longer wavelength than the absorbed photon due to the liberation of some energy through non-radiative relaxation by the excited electron before photon emission, a phenomenon known as Stokes shift. Each fluorophore has characteristic excitation and emission spectra, which describe the efficiency of photon absorption and emission as the wavelength of stimulating light varies. The time taken for a photon to be emitted after the fluorophore has absorbed energy is known as the fluorescence lifetime and is independent of concentration of fluorophore or the intensity of the stimulating light, unlike the quantity of liberated photons (Fig. 18.2).

Lasers are often used to stimulate fluorophores as they produce monochromatic light, and the appropriate part of the emission spectrum of a fluorophore can be targeted without spilling over into other wavelengths. Lasers also produce collimated light, which means each point on a biological sample can be precisely stimulated without affecting adjacent tissue. Light-emitting diodes (LEDs) provide a cheaper and more reliable alternative to lasers at the expense of producing non-collimated (i.e. naturally diverging) wave of light, less power and an emission of light over a slightly broader wavelength.

The field of fluorescence, the precision of the behaviour of the fluorophores, the technical accuracy and simplicity of fluorescence excitation and emission detection systems all combine to make this an extremely potent field for the multiplexing of detectable information from within the same biological sample.

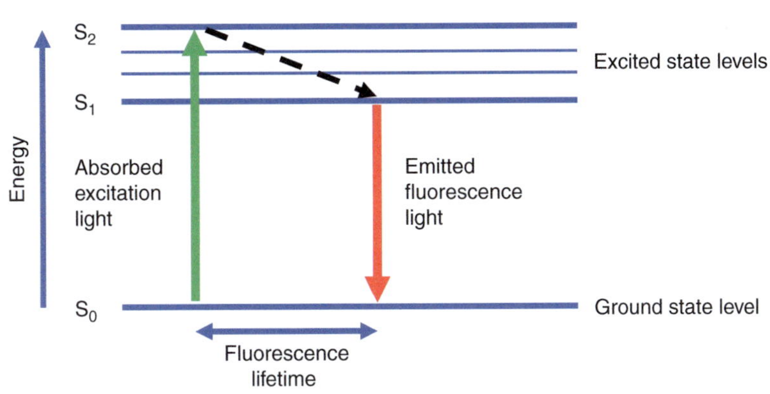

Fig. 18.2 Jablonski's diagram of energy states of a molecule. For an electron to leave the ground state and enter an excited state, absorbance has to occur. This event happens when the energy of the photon has approximately the equal energy to the difference between the ground and the excited state. The excited electron will return to the ground state the majority of times, emitting a photon at a lower energy and longer wavelength. The time it takes for the emitted photon to be released from the moment the electron was excited is known as the fluorescence lifetime

18.4.1.3 Some Strategies and Examples for Molecular Imaging Using Fluorescent Agents

Fluorophores conjugated to a molecule that binds a particular target presents by far the commonest and simplest method for molecular targeting. There are thousands of biological ligand and fluorophore combinations available commercially, and typically samples must be washed before imaging in order to remove unbound fluorophore. Examples include cell surface marker and nucleic acid stains used in cellular imaging.

Fluorescent imaging agents can be designed specifically to accumulate in the tissue of interest. Examples include fluorescent agents conjugated to large molecular dendrimeric constructs. Such agents have been used to provide spatiotemporal imaging of the process of phagocytosis [6].

Fluorescent imaging agents can be designed to remain silent until reaching the target of interest. Strategies include relying upon environmental changes [7] or employing a quenching mechanism. An example of quenching would be a Förster resonance energy transfer (FRET) quenched system. FRET refers to the radiationless transfer of photon energy from one fluorophore to another through dipole-dipole interactions [8]. A FRET quenched system employs photon donor and acceptor moieties in close proximity. The donor moiety absorbs a photon as normal, but the energy is transferred to the acceptor, which then either emits a photon with wavelength outside of the range of interest or returns to a ground state through non-radiative means [9]. The efficiency of a FRET system is proportional to $1/R^6$ where R = the donor-acceptor distance, so the effect disappears rapidly as the two moieties become separated. FRET systems have been employed in many biological applications. One example is the investigation of spatiotemporal function in enzyme biology. A molecular agent for a specific protease may comprise of a FRET quenched system that is separated by a specific peptide and does not yield fluorescence until the peptide sequence is digested by a specific protease and the components of the FRET system become separated in space [6, 10].

In addition to the general considerations of selecting a molecular agent for a particular application, the following additional concerns apply to fluorescent imaging agents:

- What is the process of fluorophore excitation and emission and how can this be managed amongst other competing exogenous fluorophores?
- What are the autofluorescent characteristics of endogenous biological molecules and how can these be de-convolved from the molecular imaging agents?

Whilst exploiting the ability of optical platforms to multiplex numerous different fluorophores into one investigation, there is a requirement to maintain specificity. One problem is broadness of the spectra of many fluorophores that confers a degree of overlap from the excitation of one fluorophore to detection of the other, thereby introducing a potential source of error. There are several potential strategies that may be employed to overcome this such as the use of tandem dyes, quantum dots or the temporal separation of the interrogation of each fluorophore. Tandem dyes use a FRET system to transfer exciting photons from one wavelength from a donor to an acceptor that emits a photon at a different wavelength than the donor would otherwise be capable of doing. The development of quantum dots has led to brighter fluorophores with much narrower spectra, which allows multiple dots to be used without spilling over into the emission of other fluorophores, although these dots are known to exhibit cellular toxicity [11].

18.4.1.4 Confocal Microscopy

One of the main limitations in normal (or wide-field) light microscopy is the collection of out-of-focus light onto the detector, which reduces the axial resolution of the collected image as this light is overlaid on the focused image. CONFOCAL MICROSCOPY overcomes this limitation by placing a pinhole in the path of collected light, therefore eliminating any out-of-focus light from proceeding to the detector. Modern confocal microscopy combines widefield light microscopy with laser scanning confocal microscopy, whereby lasers of a particular wavelength are sequentially scanned across the entire field of view so the fluorescence of each point, at each wavelength, can be built up to yield a multi-faceted image relying on the endogenous fluorescence of the sample and on the properties of exogenously applied fluorescent compounds. 3D images can be developed by accurately moving the objective up and down, the so-called *z*-stacking.

18.4.1.5 Multiphoton Microscopy

MULTIPHOTON MICROSCOPY or multiphoton excitation microscopy is a non-linear optical imaging technique. Two-photon microscopy is an example of multiphoton microscopy and refers to the excitation of a fluorophore by the simultaneous arrival of two relatively low-energy photons. Both photons stimulate the same electron to a higher energy level than a single photon alone. When the electron relaxes, the fluorophore emits a photon that has a shorter wavelength than the exciting photons. Such non-linear excitation requires enormous photon flux, and therefore stimulating photons from an infrared laser are focused to a point in the sample, which also minimizes photo damage. It is only at this point that there are a sufficient number of photons to produce appropriate stimulation of the fluorophore. In this way off-target signal is rejected by restricting fluorescence excitation volume rather than by restricting fluorescence collection volume, which is the case for confocal microscopy [12]. Two-photon microscopy is well established for in vitro molecular imaging and has been used during intravital imaging, typically through a body surface window, such as the investigation of neutrophil infiltration during lung injury [13] and imaging of the gastrointestinal tract [14] (Fig. 18.3).

18.4.1.6 Fluorescence Molecular Tomography

There remains an experimental need to study thick or whole biological tissue samples. Even if fluorescent light can be detected at the surface of a sample, it cannot be assigned to a precise location deep to the surface because of the aforementioned effects of scatter and absorption, which cannot be predicted or modelled in

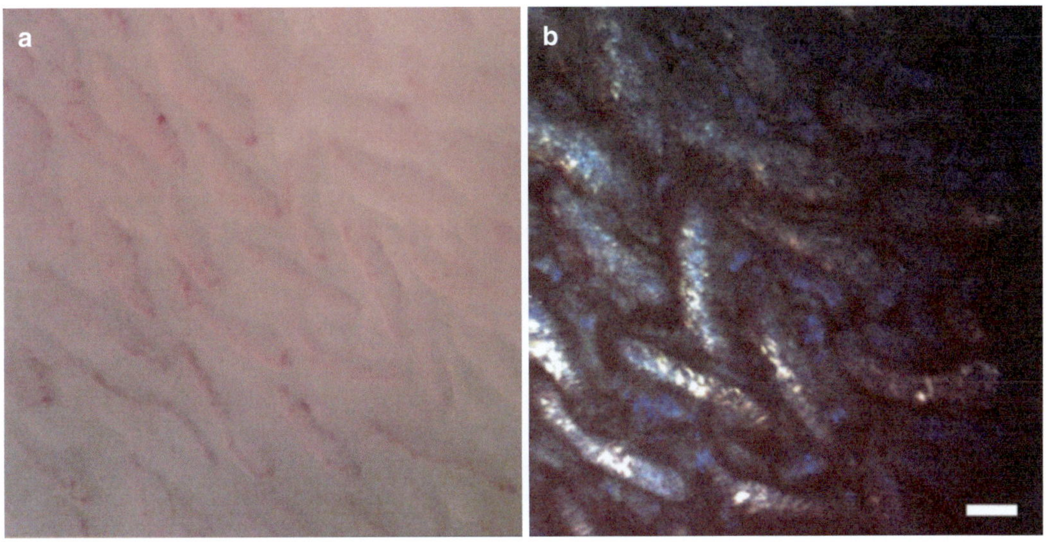

Fig. 18.3 Two-photon microscopy. Images taken from normal small intestine with white light imaging (**a**) and two-photon microscopy (**b**). Two-photon microscopy generates a more detailed image, providing an insight to the cellular morphology of the small intestine as opposed to the tissue outline of whilst light imaging. Adapted from Grosberg et al. [14]

anything but the most uniform of transmission media. FLUORESCENCE MOLECULAR TOMOGRAPHY comprises a detector and narrow excitation beam, which is scanned over the sample in order to collect photons which have propagated through the tissue [15]. Through the collection of multiple data points from multiple orientations, a tomographic three-dimensional image can be developed, albeit with relatively low resolution. The images are still limited by the depth of photon penetration, and so infrared light and infrared-stimulated fluorophores are often used which can be detected at a depth of several centimetres [16]. Examples of the use of this technology have been the detection of specific protease activity [17, 18] and the tracking of inflammatory cells in inflammatory disorders [19]. The technique is often limited to small animal work because of depth penetration restrictions but has found use in some clinical applications especially in breast cancer because of the low attenuation of infrared light by breast tissue [20].

18.4.1.7 Raman Spectroscopy

In contrast to fluorescence microscopy where molecules absorb photons to become electronically excited, Raman scattering occurs when a molecule interacts with a photon and changes its vibrational excitation. Typically when this occurs, the molecule will scatter a photon at the same wavelength of the absorbed photon (elastic or Rayleigh scattering). Much less frequently, the molecule will scatter a photon with a different wavelength (fewer than 1 in 1^9 of all incident photons), either longer (Stokes shifted) or shorter (anti-Stokes shifted). The precise shift is unique to each molecule under investigation and can be used to build a unique "fingerprint" spectrum for each. The information derived from the Raman spectrum from a particular point in the sample can be overlaid on traditional light microscopy images to give additional label-free information about the nature of the biological sample [21]. RAMAN SPECTROSCOPY has been employed to provide label-free live imaging of inflammatory cells [22] and stochastic surveillance of apoptosis [23] (Fig. 18.4).

18.4.1.8 Fibre-Based Imaging

One method of overcoming the limited penetration depth of optical imaging modalities is simply to access the tissue under investigation using an optical fibre. White light widefield endoscopes have been used for many years to provide macroscopic imaging of any endoscopically accessible biological tissue. A fibre optic cable is a bundle of many thousand individual optical fibres, with each fibre employing an inner core and outer cladding, each with a different refractive index. The interface of the outer cladding and the inner core provides the opportunity for total internal reflection to propagate light from one end of the fibre to the other. It is possible to add a gradient index (GRIN) lens to add focusing ability to the distal tip of the fibre bundle, but this adds bulk and reduces tissue accessibility. Fibre bundles without lenses therefore have a fixed and very short working distance. The lateral resolution of the transmitted image in FIBRE-BASED IMAGING is restricted by the diameter of the fibre cladding, as the inner cores that conduct light move further apart. Thinner cladding leads to optical cross talk as light leaks from one core to an adjacent core although this problem can be overcome by randomly aligning the cores or illuminating each core in random sequence at the proximal end of the fibre. Each fibre can be calibrated to adjust for intrinsic autofluorescence, and each individual fibre is mapped before use to improve image accuracy. Fibre bundles may include a channel for the local delivery of contrast agents and molecular agents. Optical endomicroscopy has been used in human patients for label-free imaging of gastrointestinal tract [25] or lung architecture [26] but also for the detection of neutrophilic inflammation in the diseased human lung [27]. Fibres have long lent themselves to fluorescent imaging [28], but two-photon microscopy [29, 30], fluorescent lifetime and Raman spectroscopy [31] as well as hybrid devices [32] have also been demonstrated.

18.4.1.9 Super-Resolution Microscopy

Abbe's diffraction limit remains the theoretical limit for spatial resolution, but several technologies have been developed to penetrate the

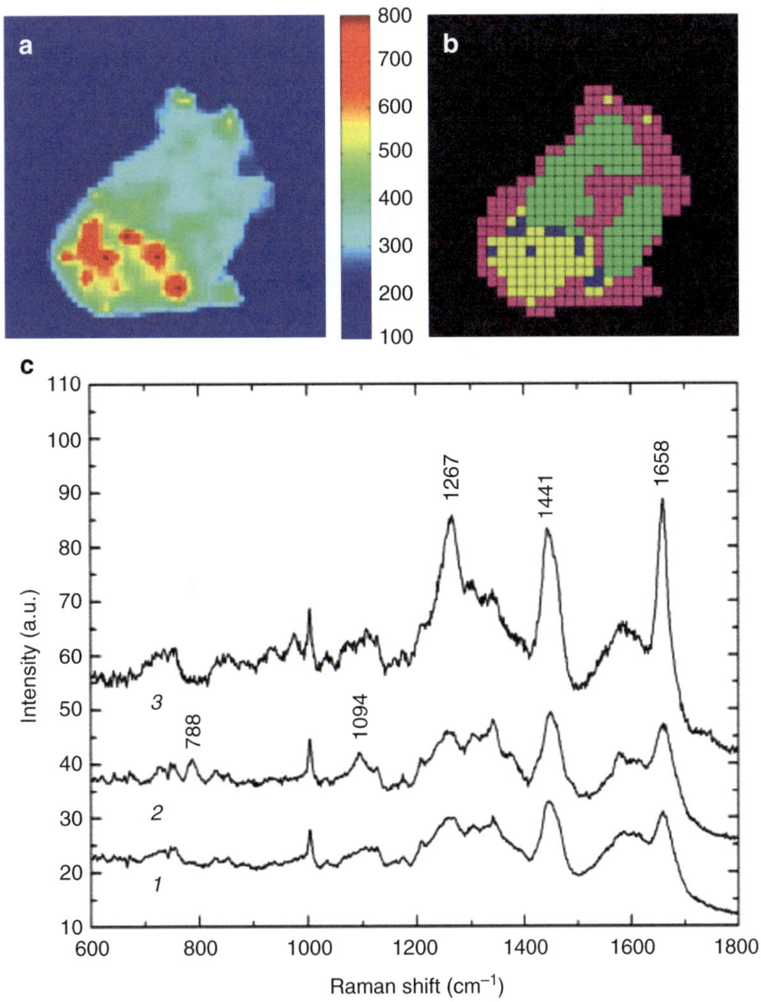

Fig. 18.4 Confocal Raman microscopy (CRM). CRM is a cellular imaging technique which provides an insight to the molecular composition of a cell. Figure (**a**) is a CRM image of a neutrophil with (**b**) a hierarchical cluster that visualizes regions with high Raman spectral similarities. Sections with high lipid content in (**a**) and (**b**) are seen as red and blue regions, respectively. Lipids have been shown to be powerful mediators of inflammation in leukocytes. Figure (**c**) is an average Raman spectra from the various colours seen in (**b**). The strong bands (1267, 1441, 1658) are Raman shift peaks associated with lipids. Adapted by van Manen et al. [24]

diffraction limit for SUPER-RESOLUTION MICROSCOPY and provide novel insights into molecular biology.

18.4.1.9.1 Near-Field Scanning Optical Microscopy

Because diffraction only occurs when light has propagated into the far field (i.e. beyond a distance equal to one wavelength of the illuminating light), near-field imaging has been used to avoid the diffraction limit entirely [33, 34]. The sample is placed in the near field, adjacent to the device aperture, and light is collected before it begins to diffract. Lateral resolutions in the order of 20–50 nm have been achieved [35], far in excess of a typical diffraction-limited light microscope. However, because imaging is limited to the diffraction-free near field, the depth of penetration into the sample is incredibly small, and because the aperture used to illuminate the sample and collect light is tiny, image construction is time-consuming.

18.4.1.9.2 Super Resolution by Spatially Patterned Excitation

Spatially patterned excitation super-resolution microscopy covers several different techniques including STIMULATED EXCITATION DEPLETION (STED) MICROSCOPY. Although the stimulation laser excites a large number of fluorophores, a second laser is employed to drive all off-target fluorophores (i.e. those away from the centre of the stimulation area) into their dark state. The remaining fluorescence emanates only from those fluorophores in the very centre of the excitation field [36]. Lateral resolutions of approximately 30 nm have been demonstrated, but image construction is again relatively time-consuming.

Typically imaging takes place in fixed, prepared tissue although it is possible to perform imaging on live tissue, for example, combination with a two-photon technique facilitates a better depth of penetration [37].

18.4.1.9.3 Super Resolution by Single-Molecule Imaging

By using fluorescent agents that are able to switch rapidly and reversibly from fluorescent to dark states, it is possible to resolve two points that would otherwise be too close, by distinguishing them in time and collating images to generate a resolution which beats the diffraction limit [5]. Examples of this approach include stochastic optical reconstruction microscopy [38] (STORM) and photoactivated localization microscopy [39] (PALM). In order to accept the assumption that each collected source of light does not overlap with another, there must only be a few activated fluorophores per imaging frame. Several thousand frames are therefore required to build up a resolved image of even the smallest field of view, which increases the time taken for image acquisition. Brighter fluorophores (less time required to produce a detectable number of photons), rapidly switchable lasers and high-density data analysis will lead to shorter acquisition times and facilitate the imaging of live cellular process which occur over short time scales (Fig. 18.5) [41].

18.4.2 Non-optical Imaging

18.4.2.1 Magnetic Resonance Imaging

Magnetic resonance imaging (MRI) uses powerful magnetic fields to align magnetically susceptible spinning protons in the specimen, before deflecting them from their axis with a radio wave. As the protons return to their alignment, they emit a radio signal that can be detected by multiple receiver coils placed throughout the scanner. The signal from a relaxing proton has two dimensions, the time taken to relax to their original vector (T1 relaxation) and the time taken for axial spin to return to normal (T2 relaxation). Different tissues relax at different time constants, and the received information is used to build up a greyscale image composed of multiple tissue slices [42]. Whilst there are no known biological hazards associated with the use of MRI, the equipment remains bulky and expensive, and there are safety dangers associated with the inadvertent juxtaposition of loose

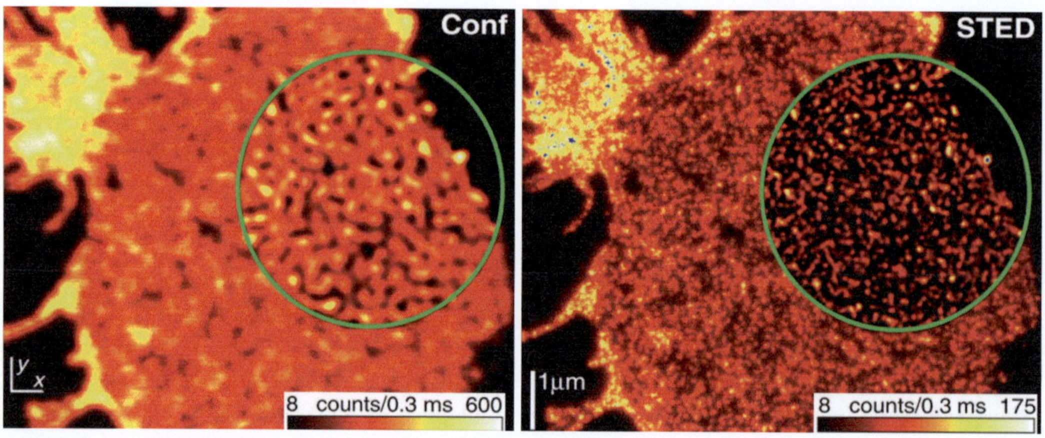

Fig. 18.5 Stimulated emission depletion (STED) microscopy. The images show the tagging and detection of SNAP-25, a plasma membrane protein, on a neuronal cell with confocal (conf) and stimulated emission depletion (STED) microscopy. STED provides super resolution by spatially patterned excitation and unlimited diffraction, resulting in a clearer image of SNAP-25 in comparison to confocal microscopy. Adapted from Willig et al. [40]

ferrous metal within the scanning zone. MRI remains invaluable for the imaging of tissue structure and diseases associated with structural deviation, but relative to optical technologies, the resolution is poor, and there are few contrast agents to target molecular and cellular processes. Contrast agents that do exist need to be paramagnetic in order to be susceptible to the magnetic field of the scanner and most work by shortening the T1 relaxation time of nearby protons [43]. Gadolinium is commonly used in clinical investigations to provide additional structural contrast. Super-paramagnetic particles of iron oxide (SPIO) or ultrasmall super-paramagnetic particles of iron oxide (USPIO) have been used as contrast agents [44]. They are taken up by macrophages when injected in order to track cells to sites of inflammation including atherosclerotic plaque formation [45] and rupture [46] and can be conjugated in a variety of ways to study biological processes [47].

18.4.2.2 Nuclear Medicine

18.4.2.2.1 Single-Photon Emission Computed Tomography (SPECT)

SPECT functions in similar way to simple planar gamma camera imaging (scintigraphy), in that it relies upon the emission of high energy photons (or gamma rays) from a decaying radionuclide which has been conjugated to a biologically relevant ligand. Unlike scintigraphy, which has a fixed camera, the rotational gantry used in SPECT provides additional photon vector information, which allows the production of 3D images and images with a better resolution. SPECT scanners can use radionuclides with long half-lives and which have been widely available since the advent of nuclear imaging, such as Technetium-99 and Gallium-67, which have been extensively used to study inflammation when paired with a biologically functional conjugate [48]. SPECT and CT scanners can be incorporated onto the same gantry to provide accurate structural information about the location of the biological process of interest (Fig. 18.6).

18.4.2.2.2 Positron Emission Tomography (PET)

Like SPECT, PET relies upon the production of gamma rays from a decaying radionuclide that has been conjugated to a biologically active molecule. The radionuclide is transmitted to its target of interest by the biologically active conjugate and emits two positrons as it undergoes radioactive decay. Within a small distance of emission, the positrons annihilate with electrons resulting in the emission of two gamma rays (high-energy photons) in opposite directions, which are detected by a rotating gamma camera [50, 51]. Because two photons are produced within close proximity to the location of positron emission and because they are exactly coincident, the

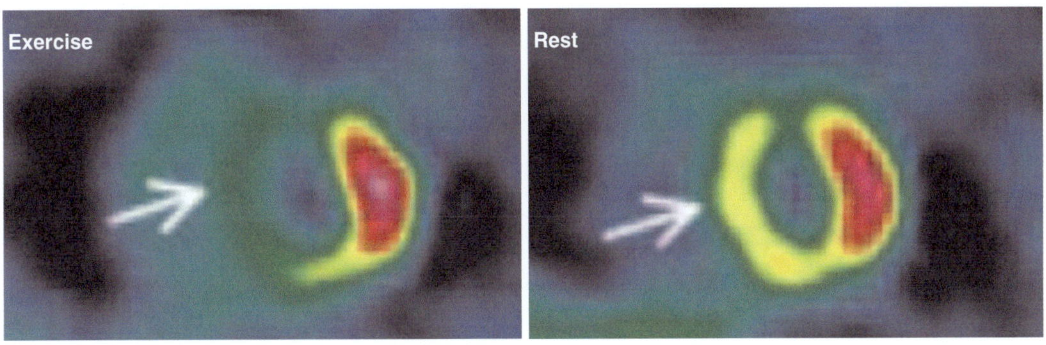

Fig. 18.6 Single-photon emission computed tomography (SPECT). SPECT can be used to evaluate myocardial perfusion and viability to assess inducible ischemia due to flow-limiting coronary stenosis. This is achieved by intravenously administering minute levels of radioactive trace and recording images after exercise. Stenosis is observed in the first image of a man with angina after exercise, which is reversed after a period of rest (arrows). Adapted from He et al. [49]

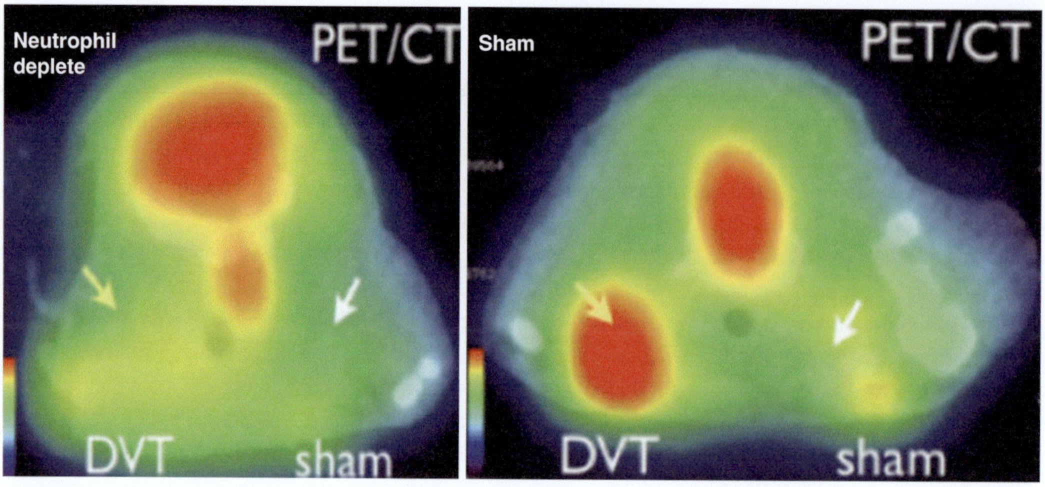

Fig. 18.7 Positron emission tomography and computed tomography (PET/CT). The images show the incorporation of computed tomography (CT) and positron emission tomography (PET) imaging of a deep vein thrombosis (DVT) mouse model. Neutrophil-depleted mice (left) showed a decreased signal (yellow arrow) in comparison to the control (right), suggesting the importance of neutrophils in thrombus-associated inflammation. White arrow = sham-operated jugular vein. Adapted from Hara et al. [53]

location of emission can be detected with a higher degree of accuracy compared to SPECT imaging. The distance travelled by the positrons before annihilation occurs is the predominant limiting factor of spatial resolution. Radionuclides suitable for PET typically have a very short half-life, so they need to be manufactured in an on-site cyclotron, increasing expense and restricting use to research or specialist centres. Several suitable radionuclides exist, but by far the most commonly used is fluorine-18, which is usually incorporated into a biologically inactive analogue of glucose (18-fluorodeoxyglucose, ^{18}FDG) and so accumulates in rapidly metabolizing tissues such as cancer or inflammatory cells and has a wide repertoire of clinical uses [52]. Like SPECT, PET scanners can be incorporated with CT scanners to provide excellent co-localized structural information (Fig. 18.7) [54].

18.4.3 Non-electromagnetic Imaging

18.4.3.1 Ultrasound

Sound waves, unlike radiation on the electromagnetic spectrum, are physical waves and require a medium through which to be propagated. ULTRASOUND refers to any sound frequency above the limit of human hearing, about 20 kHz, but medical ultrasound machines typically use sound with frequencies in the range of 1–20 MHz. In the head of the ultrasound transducer, the application of a precise electric current across a piezoelectric crystal transduces electrical energy to mechanical energy in the form of sound waves. The sound waves are propagated through the tissue of interest and scattered at the interfaces of tissues with different acoustic impendence. Some scattered sound reflects back to the transducer that converts the returning mechanical energy back to an electric current to be processed into a visible image. The speed and intensity with which sound is reflected back will depend upon the type and depth of tissue in the path of the beam. Higher-frequency, shorter wavelength sound waves provide better spatial resolution than lower-frequency waves but lack penetration depth because of acoustic attenuation (Fig. 18.8). Ultrasound is low cost, low complexity, low size and readily transportable and has no biological safety concerns, but there is a paucity of biological contrast agents. Microbubbles have long

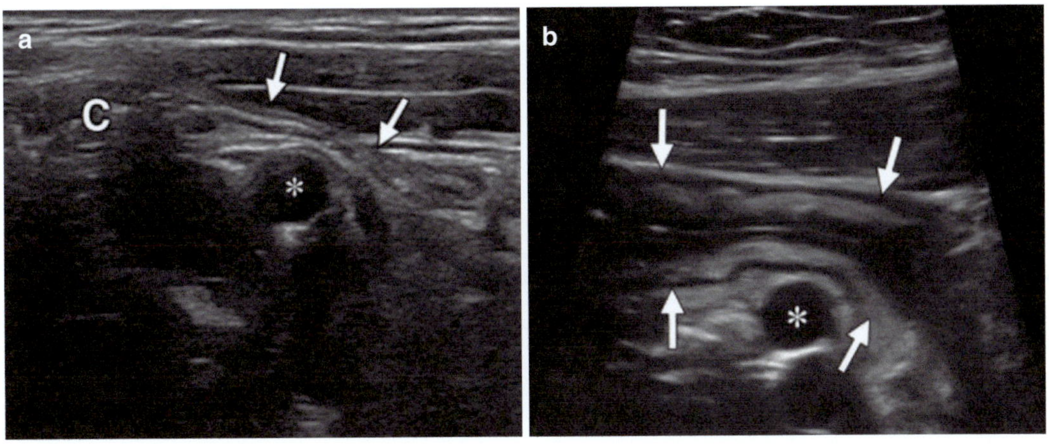

Fig. 18.8 Ultrasound. The images show ultrasound imaging of the ileum of a healthy individual (**a**) and a patient with Crohn's disease (**b**). In a healthy individual the ileum has a thin wall (arrows) as opposed to the thick wall seen in a diseased ileum. *C* caecum, * iliac vessel. Adapted from Dillman et al. [55]

been used as molecular contrast agents in ultrasound, and recent advances in manufacturing technology have enabled the conjugation of microbubbles to a biological ligand, increasing the specificity to particular biological targets [56].

18.4.3.2 Hybrid Systems

Virtually any of the modalities described can be hybridized to produce techniques that confer additional advantages whilst overcoming some of the disadvantages if each technique were used alone. Images from two or more different technologies need to be co-registered in order to provide accurate and precise information that it is both reliably quantified and localized. As the speed of acquisition increases, the size of the devices falls and processing power increases, these devices become more and more useful. Molecular agents can be designed to offer detection through more than one technique, reducing the number of the agents that need to be applied [57]. Examples include PET-Optical [58, 59] and MRI-Optical [60, 61]. If structural information is not already available, then any single or hybrid technique can be coupled with a structural imaging technique such as CT or MRI. One hybrid technique that does not result from the co-registration of images is photoacoustic imaging.

18.4.3.3 Photoacoustic

Unlike hybrid systems which co-register two images from two independent techniques, PHOTOACOUSTIC IMAGING generates a single image by combining optical excitation and acoustic detection [62]. Tissue is excited by a pulsed laser, which induces a small temperature change. Thermoelastic expansion occurs and creates a pressure wave, which can be detected by one or more ultrasonic transducers outside the specimen that process the information to build up a tomographic image. By relying on ultrasound for detection, only the absorption spectrum of the target tissue is important, not the emission spectrum, and smaller numbers of photons are required to be absorbed to produce an acoustic effect. Photoacoustic imaging can be implemented in a variety of ways, including via endoscopy [63], with spatial resolution being approximately equal to 1/200th of the imaging depth, which can reach 7 cm [62]. Contrast agents that absorb infrared light avidly can be used, and their structure can be manipulated to alter their physical and biological effects [64]. Commercially available fluorescent dyes, active in the infrared region, have also been used for their photoacoustic properties, for example, for the detection of matrix metalloproteinase activity in atherosclerotic plaque formation (Fig. 18.9) [65].

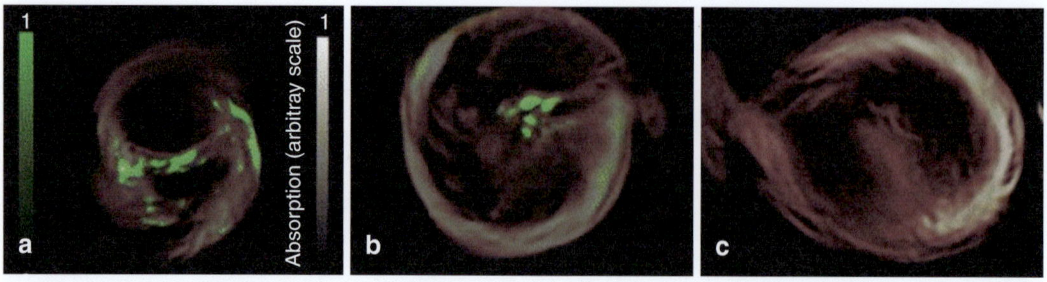

Fig. 18.9 Multispectral photoacoustic tomography. The images show human carotid plaques obtained by multi-spectral photoacoustic tomography technology. A matrix metalloproteinase (MMP)-sensitive, activatable fluorescent probe was applied to samples (**a**) and (**b**) in order to detect MMP levels as elevated expression is associated with plaque instability compared to control (**c**). Adapted from Razansky et al. [65]

18.5 Conclusion

Imaging has progressed from the simplest light microscope to complex and multiplexed hybrid systems with capability across the entire experimental spectrum from the bench to the clinic. Remarkable insights into inflammatory processes have been achieved, and the discipline continues to develop rapidly as funders recognize the ability of such techniques to improve understanding of disease and clinical care. One emerging field of molecular imaging is the possible merger of therapeutic and diagnostic imaging agents, so investigators might visualize the delivery of a novel therapeutic and determine its disease-modulating ability [56].

References

1. Mankoff D. A definition of molecular imaging. J Nucl Med. 2007;48(18N):21N.
2. Weissleder R, Mahmood U. Molecular imaging. Radiology. 2001;219:316–33.
3. Miller JC, Thrall JH. Clinical molecular imaging. J Am Coll Radiol. 2004;1:4–23.
4. Pysz MA, Gambhir SS, Willmann JK. Molecular imaging: current status and emerging strategies. Clin Radiol. 2010;65:500–16.
5. Huang B, Bates M, Zhuang X. Super-resolution fluorescence microscopy. Annu Rev Biochem. 2009;78:993–1016.
6. Avlonitis N, Debunne M, Aslam T, et al. Highly specific, multi-branched fluorescent reporters for analysis of human neutrophil elastase. Org Biomol Chem. 2013;11:4414–8.
7. Akram AR, Avlonitis N, Lilienkampf A, et al. A labelled-ubiquicidin antimicrobial peptide for immediate in situ optical detection of live bacteria in human alveolar lung tissue. Chem Sci. 2015;6:6971–9.
8. Ding S, Cargill A, Das S, et al. Biosensing with Förster resonance energy transfer coupling between fluorophores and nanocarbon allotropes. Sensors. 2015;15:14766–87.
9. Le Reste L, Hohlbein J, Gryte K, Kapanidis AN. Characterization of dark quencher chromophores as nonfluorescent acceptors for single-molecule FRET. Biophys J. 2012;102:2658–68.
10. Gehrig S, Mall MA, Schultz C. Spatially resolved monitoring of neutrophil elastase activity with ratiometric fluorescent reporters. Angew Chem Int Ed Engl. 2012;51:6258–61.
11. Hoshino A, Hanada S, Yamamoto K. Toxicity of nanocrystal quantum dots: the relevance of surface modifications. Arch Toxicol. 2011;85:707–20.
12. Oheim M, Michael DJ, Geisbauer M, et al. Principles of two-photon excitation fluorescence microscopy and other nonlinear imaging approaches. Adv Drug Deliv Rev. 2006;58:788–808.
13. Kreisel D, Nava RG, Li W, et al. In vivo two-photon imaging reveals monocyte-dependent neutrophil extravasation during pulmonary inflammation. Proc Natl Acad Sci U S A. 2010;107:18073–8.
14. Grosberg LE, Radosevich AJ, Asfaha S, et al. Spectral characterization and unmixing of intrinsic contrast in intact normal and diseased gastric tissues using hyperspectral two-photon microscopy. PLoS One. 2011;6:e19925.
15. Stuker F, Ripoll J, Rudin M. Fluorescence molecular tomography: principles and potential for pharmaceutical research. Pharmaceutics. 2011;3:229–74.
16. Ntziachristos V. Fluorescence molecular imaging. Annu Rev Biomed Eng. 2006;8:1–33.
17. Kossodo S, Zhang J, Groves K, et al. Noninvasive in vivo quantification of neutrophil elastase activity in acute experimental mouse lung injury. Int J Mol Imaging. 2011;2011:581406.

18. Ntziachristos V, Tung C-H, Bremer C, Weissleder R. Fluorescence molecular tomography resolves protease activity in vivo. Nat Med. 2002;8:757–61.
19. Larmann J, Frenzel T, Hahnenkamp A, et al. In vivo fluorescence-mediated tomography for quantification of urokinase receptor-dependent leukocyte trafficking in inflammation. Anesthesiology. 2010;113:1.
20. Ntziachristos V, Bremer C, Weissleder R. Fluorescence imaging with near-infrared light: new technological advances that enable in vivo molecular imaging. Eur Radiol. 2003;13:195–208.
21. Movasaghi Z, Rehman S, Rehman IU. Raman spectroscopy of biological tissues. Appl Spectrosc Rev. 2007;42:493–541.
22. Bauer M, Popp J. Toward a spectroscopic hemogram: Raman spectroscopic differentiation of the two most abundant leukocytes from peripheral blood. Anal Chem. 2012;84:5335–42.
23. Zoladek A, Pascut FC, Patel P, Notingher I. Non-invasive time-course imaging of apoptotic cells by confocal Raman micro-spectroscopy. J Raman Spectrosc. 2011;42:251–8.
24. van Manen H-J, Kraan YM, Roos D, Otto C. Single-cell Raman and fluorescence microscopy reveal the association of lipid bodies with phagosomes in leukocytes. Proc Natl Acad Sci U S A. 2005;102:10159–64.
25. Napoléon B, Lemaistre A-I, Pujol B, et al. A novel approach to the diagnosis of pancreatic serous cystadenoma: needle-based confocal laser endomicroscopy. Endoscopy. 2014;47:26–32.
26. Thiberville L, Salaün M, Lachkar S, et al. Confocal fluorescence endomicroscopy of the human airways. Proc Am Thorac Soc. 2012;6:444–9.
27. Craven T, Walton T, Akram A, et al. In-situ imaging of neutrophil activation in the human alveolar space with neutrophil activation probe and pulmonary optical endomicroscopy. Lancet. 2016;387:S31.
28. Flusberg BA, Cocker ED, Piyawattanametha W, et al. Fiber-optic fluorescence imaging. Nat Methods. 2005;2:941–50.
29. Gu M, Kang H, Li X. Breaking the diffraction-limited resolution barrier in fiber-optical two-photon fluorescence endoscopy by an azimuthally-polarized beam. Sci Rep. 2014;4:3627.
30. Helmchen F, Denk W, Kerr JND. Miniaturization of two-photon microscopy for imaging in freely moving animals. Cold Spring Harb Protoc. 2013;2013:904–13.
31. Santos LF, Wolthuis R, Koljenović S, et al. Fiber-optic probes for in vivo Raman spectroscopy in the high-wavenumber region. Anal Chem. 2005;77:6747–52.
32. Dochow S, Ma D, Latka I, et al. Combined fiber probe for fluorescence lifetime and Raman spectroscopy. Anal Bioanal Chem. 2015;407:8291–301.
33. Dürig U, Pohl DW, Rohner F. Near-field optical-scanning microscopy. J Appl Phys. 1986;59:3318.
34. de Lange F, Cambi A, Huijbens R, et al. Cell biology beyond the diffraction limit: near-field scanning optical microscopy. J Cell Sci. 2001;114:4153–60.
35. Huckabay HA, Armendariz KP, Newhart WH, et al. Near-field scanning optical microscopy for high-resolution membrane studies. Methods Mol Biol. 2013;950:373–94.
36. Blom H, Widengren J. STED microscopy—towards broadened use and scope of applications. Curr Opin Chem Biol. 2014;20:127–33.
37. Takasaki KT, Ding JB, Sabatini BL. Live-cell super-resolution imaging by pulsed STED two-photon excitation microscopy. Biophys J. 2013;104:770–7.
38. Rust MJ, Bates M, Zhuang X. Sub-diffraction-limit imaging by stochastic optical reconstruction microscopy (STORM). Nat Methods. 2006;3:793–5.
39. Betzig E, Patterson GH, Sougrat R, et al. Imaging intracellular fluorescent proteins at nanometer resolution. Science. 2006;313:1642–5.
40. Willig KI, Keller J, Bossi M, Hell SW. STED microscopy resolves nanoparticle assemblies. New J Phys. 2006;8:106.
41. Cox S. Super-resolution imaging in live cells. Dev Biol. 2015;401:175–81.
42. Berger A. How does it work?: magnetic resonance imaging. BMJ. 2002;324:35.
43. Shokrollahi H. Contrast agents for MRI. Mater Sci Eng C Mater Biol Appl. 2013;33:4485–97.
44. Jin R, Lin B, Li D, Ai H. Superparamagnetic iron oxide nanoparticles for MR imaging and therapy: design considerations and clinical applications. Curr Opin Pharmacol. 2014;18:18–27.
45. Metz S, Beer AJ, Settles M, et al. Characterization of carotid artery plaques with USPIO-enhanced MRI: assessment of inflammation and vascularity as in vivo imaging biomarkers for plaque vulnerability. Int J Cardiovasc Imaging. 2011;27:901–12.
46. Qi C, Deng L, Li D, et al. Identifying vulnerable atherosclerotic plaque in rabbits using DMSA-USPIO enhanced magnetic resonance imaging to investigate the effect of atorvastatin. PLoS One. 2015;10:e0125677.
47. Bull E, Madani SY, Sheth R, et al. Stem cell tracking using iron oxide nanoparticles. Int J Nanomedicine. 2014;9:1641–53.
48. Basu S, Zhuang H, Torigian DA, et al. Functional imaging of inflammatory diseases using nuclear medicine techniques. Semin Nucl Med. 2009;39:124–45.
49. He Z-X, Shi R-F, Wu Y-J, et al. Direct imaging of exercise-induced myocardial ischemia with fluorine-18-labeled deoxyglucose and Tc-99m-sestamibi in coronary artery disease. Circulation. 2003;108:1208–13.
50. Ter-Pogossian MM, Phelps ME, Hoffman EJ, Mullani NA. A positron-emission transaxial tomograph for nuclear imaging (PETT). Radiology. 1975;114:89–98.
51. Nolting DD, Nickels ML, Guo N, Pham W. Molecular imaging probe development: a chemistry perspective. Am J Nucl Med Mol Imaging. 2012;2:273–306.
52. Vaidyanathan S, Patel CN, Scarsbrook AF, Chowdhury FU. FDG PET/CT in infection and inflammation-current and emerging clinical applications. Clin Radiol. 2015;70:787–800.

53. Hara T, Truelove J, Tawakol A, et al. 18F-fluorodeoxyglucose positron emission tomography/computed tomography enables the detection of recurrent same-site deep vein thrombosis by illuminating recently formed, neutrophil-rich thrombus. Circulation. 2014;130:1044–52.
54. Townsend DW, Cherry SR. Combining anatomy and function: the path to true image fusion. Eur Radiol. 2001;11:1968–74.
55. Dillman JR, Smith EA, Sanchez RJ, et al. Pediatric small bowel Crohn disease: correlation of US and MR enterography. Radiographics. 2015;35:835–48.
56. Gessner R, Dayton PA. Advances in molecular imaging with ultrasound. Mol Imaging. 2010;9:117–27.
57. Heidt T, Nahrendorf M. Multimodal iron oxide nanoparticles for hybrid biomedical imaging. NMR Biomed. 2013;26:756–65.
58. Nahrendorf M, Keliher E, Marinelli B, et al. Hybrid PET-optical imaging using targeted probes. Proc Natl Acad Sci U S A. 2010;107:7910–5.
59. Phillips E, Penate-Medina O, Zanzonico PB, et al. Clinical translation of an ultrasmall inorganic optical-PET imaging nanoparticle probe. Sci Transl Med. 2014;6:260ra149.
60. Lee WW, Marinelli B, van der Laan AM, et al. PET/MRI of inflammation in myocardial infarction. J Am Coll Cardiol. 2012;59:153–63.
61. Kim HS, Cho HR, Choi SH, et al. In vivo imaging of tumor transduced with bimodal lentiviral vector encoding human ferritin and green fluorescent protein on a 1.5T clinical magnetic resonance scanner. Cancer Res. 2010;70:7315–24.
62. Wang LV, Hu S. Photoacoustic tomography: in vivo imaging from organelles to organs. Science. 2012;335:1458–62.
63. Yang J-M, Favazza C, Chen R, et al. Simultaneous functional photoacoustic and ultrasonic endoscopy of internal organs in vivo. Nat Med. 2012;18:1297–302.
64. Zackrisson S, van de Ven SMWY, Gambhir SS. Light in and sound out: emerging translational strategies for photoacoustic imaging. Cancer Res. 2014;74:979–1004.
65. Razansky D, Harlaar NJ, Hillebrands JL, et al. Multispectral optoacoustic tomography of matrix metalloproteinase activity in vulnerable human carotid plaques. Mol Imaging Biol. 2012;14:277–85.

Part III

Immunotherapeutics

Vaccines 19

Gideon F. A. Kersten and Wim Jiskoot

19.1 Introduction

Vaccines are the most commonly administered immunotherapeutics. Supported by great improvements in sanitation facilities such as safe drinking water, vaccination has been the most effective measure to control a diversity of life-threatening infectious diseases in the twentieth century. The most impressive success of vaccination has been the global eradication of smallpox in the 1970s. Moreover, the incidence of many other infectious diseases, such as diphtheria, tetanus, pertussis, poliomyelitis, measles, mumps and rubella, has been drastically reduced, thanks to extensive vaccination programmes.

Upon a natural infection with a pathogen, an unprotected person usually falls ill before the immunological defence system is able to respond adequately. Vaccination aims to stimulate the specific immune response against a pathogen by the administration of attenuated or inactivated organisms, or fractions thereof. If vaccination is successful and the host comes into contact with the pathogen afterwards, the specific immune response will be immediate and sufficiently strong to kill the invading organism before it would get the opportunity to multiply and cause disease. Thus, in a strict sense vaccines are immunoprophylactics rather than therapeutics.

In most cases, repeated doses of the vaccine are given to boost the immune response. Apart from the number of doses, several other factors determine vaccine efficacy, as summarised in Table 19.1. If a high enough proportion of a population is immunised, vaccination not only protects the immunised individuals but also may help to protect the community as it decreases the chance that non-immunised persons encounter the pathogen. This is referred to as herd immunity [1].

A brief history of vaccination will be given below. Next, current vaccine categories for human use will be addressed. Finally, new developments in vaccinology will be outlined.

Final manuscript submitted on August 30, 2015.

G. F. A. Kersten (✉)
Institute for Translational Vaccinology, Bilthoven, The Netherlands

Division of BioTherapeutics, Leiden Academic Centre for Drug Research (LACDR), Leiden University, Leiden, The Netherlands
e-mail: gideon.kersten@intravacc.nl

W. Jiskoot
Division of BioTherapeutics, Leiden Academic Centre for Drug Research (LACDR), Leiden University, Leiden, The Netherlands

19.2 Historical Background

Vaccination has a long history [1, 2]. The most prominent milestones of vaccinology are listed in Table 19.2. The first attempts to become immune probably date back to as early as the seventh century, when Indian Buddhists drank snake venom and may thus have become immune against this toxin. Written reports bear witness to the practice of variolation, i.e. the administration

M. J. Parnham et al. (eds.), *Nijkamp and Parnham's Principles of Immunopharmacology*,
https://doi.org/10.1007/978-3-030-10811-3_19

Table 19.1 Factors determining vaccine efficacy

Pathogen dependent	Host dependent	Vaccine dependent	Vaccination schedule dependent
Port of entry	Species	Nature of antigenic component(s)	Route of administration
Localisation in host	Age	Antigen content	Number of doses
Antigenic variation	Genetic factors	Delivery systems	Immunisation intervals
Mutation frequency	Physical state	Adjuvants	Simultaneous administration of other vaccines (administered separately)
	Immune status	Combination with other vaccine components (in one vial or syringe)	

Table 19.2 Milestones in vaccine history[a]

Year	Event[b]
ca. 1000	Intranasal administration of preparations of scabs from smallpox patients in China
Sixteenth to seventeenth century	Parenteral variolation in India by Hindus
Seventeenth century	Oral administration of white cow flea pills for smallpox prevention in China
1796	Immunisation of 8-year-old boy with cowpox virus and challenged with smallpox virus (Edward Jenner)
1798	Initiation of general cowpox immunisation with Jenner's variola vaccine
1870s	Discovery of attenuation of fowl cholera bacteria (Louis Pasteur)
1884	Attenuated *Vibrio cholerae*: the first bacterial vaccine used in humans (Robert Koch)
1885	First administration to humans of attenuated rabies vaccine (Louis Pasteur)
1896–1897	Introduction of the first heat-inactivated vaccines against typhoid, cholera and plague
1923	Introduction of the first subunit vaccine: formaldehyde-treated diphtheria toxin
1927	Introduction of BCG, attenuated tuberculosis vaccine
1955	Introduction of inactivated poliovirus vaccine (Salk): the first vaccine developed with tissue culture technique
1961	Attenuated polio vaccine (Sabin) as the first licensed oral vaccine
1980	Declaration of the eradication of smallpox by the WHO
1986	Licensing of the first rDNA vaccine: recombinant HBsAg
1987	Licensing of the first conjugate vaccine against Hib: PRP-T
2006	Licensing of HPV vaccines, preventing cervical cancer
2008	Oncophage, first therapeutic (kidney) cancer vaccine approved in Russia
2012	MenAfriVac, first vaccine that can be used in a 'controlled temperature chain' instead of a cold chain (40 °C for up to 4 days)
2013	Licensing of a meningococcal group B vaccine developed by reverse vaccinology

[a]Sources: [1, 2]

[b]Abbreviations: *BCG* bacille Calmette-Guérin, *HBsAg* hepatitis B surface antigen, *Hib Haemophilus influenzae* type b, *HPV* human papilloma virus, *PRP-T* polyribosylribitol phosphate-tetanus toxoid conjugate vaccine, *WHO* World Health Organisation

of scabs or pustule preparations obtained from patients recovered from smallpox, ever since about 1000 A.D. in various parts of the world, among others in China, India, North Africa and England.

Variolation was widely applied until Edward Jenner introduced cowpox vaccination at the end of the eighteenth century. His practice was based on the recognition that milkmaids were frequently subjected to mild pox infection acquired from the cows they milked but were spared from disease during smallpox epidemics. The first demonstration that the principle of immunisation works was Jenner's anecdotal experiment with an 8-year-old boy who remained healthy when challenged with smallpox material from a patient after he had been immunised with cowpox virus. It was Jenner who introduced the terms 'vaccine' for cowpox preparations (derived from the Latin vacca = cow) and 'vaccination' for the administration thereof. Later, in honour of Jenner, Louis Pasteur generalised the meaning of vaccination to immunisation with agents other than cowpox. During the nineteenth century, vaccination with live cowpox virus became common practice. In the twentieth century, vaccinia virus, which is closely related to cowpox virus [1], became widely used as a live vaccine until smallpox was eradicated.

Pasteur gave a new impetus to vaccinology in the last quarter of the nineteenth century. He showed that the virulence (i.e. infectivity) of pathogens could be reduced by successive passage in culture. Vaccination with attenuated strains thus obtained could confer protection without causing disease. The efforts of Louis Pasteur and others led to the development of live ATTENUATED VACCINES against cholera, anthrax and rabies. Along with the introduction of ATTENUATED VACCINES, it became apparent that infection with live material was not essential to induce immunity. The procedure of killing bacteria by heat and subsequent stabilisation with phenol was developed, resulting in the introduction of heat-inactivated whole-cell vaccines against cholera, typhoid and plague at the end of the nineteenth century.

In the first half of the twentieth century, the development and introduction of new live (tuberculosis (TB), yellow fever) and inactivated vaccines (pertussis, influenza, rickettsia) followed. Moreover, it was being recognised that some components of a microorganism were more relevant for protection than others, and the concept of SUBUNIT VACCINES was born. This and the discovery of chemical inactivation of bacterial toxins with formaldehyde led to the introduction of SUBUNIT VACCINES against diphtheria (1923) and tetanus (1927).

In the early 1950s, tissue culture techniques for virus propagation were developed. This resulted in the licensing of Salk's inactivated polio vaccine (IPV) in 1955. In the same period, Sabin developed an oral polio vaccine (OPV) consisting of live attenuated viruses, which became available in the USA in 1961. Several other viral vaccines derived from tissue cultures followed. Furthermore, several bacterial SUBUNIT VACCINES based on purified proteins or polysaccharides were introduced since the 1970s. The first vaccine based on rDNA technology, hepatitis B vaccine, was marketed in 1986.

19.3 Current Vaccine Categories

19.3.1 Classification

The currently available prophylactic vaccines for human use are of either bacterial or viral origin and can be divided into several categories (see Table 19.3). These categories will be discussed below. For a more detailed description of individual vaccines currently in practice, the reader is referred to the textbook of Plotkin et al. [1].

19.3.2 Live Attenuated Vaccines

The first and most successful vaccine was a live vaccine. Cowpox virus was used to eradicate smallpox. It was naturally attenuated because humans are not easily infected by cowpox virus and do not get ill. When human pathogens are used, attenuation through serial passage and selection of less virulent and less toxic variants has been applied to obtain safe vaccine strains. Genetic approaches to attenuate are the use of reassortment techniques and reverse genetics (Box 19.1). Once a suitable strain has been obtained, master and working seedlots are prepared. The seedlot system provides the basis for the reproducible production of live (and other) vaccines. The dose of live vaccines is determined on the basis of the number of viable organisms.

Live vaccines have a number of advantages over nonliving vaccines. Although attenuation generally means reduced infectivity, attenuated strains will replicate to some extent in the recipient. This furnishes a sustained antigen dose, inducing strong immune responses even after a single dose. In general, live vaccines generate higher cell-mediated immune responses than

Table 19.3 Classification and examples of current vaccines

Category	Example	Vaccine characteristics
Live attenuated organisms		
Viral	Poliovirus (Sabin)	Attenuated viruses, serotypes 1–3; oral vaccine
	Measles virus	Attenuated virus
	Mumps virus	Attenuated virus
	Rubella virus	Attenuated virus
	Yellow fever virus	Attenuated virus
	Influenza	Attenuated virus, intranasal
Bacterial	Bacille Calmette-Guérin	Attenuated *Mycobacterium bovis*
	Salmonella typhi	Attenuated bacteria, oral vaccine
Inactivated whole organisms		
Viral	Influenza virus	β-Propiolactone-inactivated virus
	Poliovirus (Salk)	Formaldehyde-inactivated viruses, serotypes 1–3
	Rabies virus	β-Propiolactone-inactivated virus
	Hepatitis A virus	Formaldehyde-inactivated virus
	Japanese B encephalitis virus	Formaldehyde-inactivated virus
Bacterial	*Bordetella pertussis*	Heat-inactivated bacteria
	Vibrio cholerae	Phenol-inactivated bacteria
	Salmonella typhi	Heat-inactivated bacteria
Subunit vaccines		
Viral	Influenza virus	Influenza surface antigens
	Hepatitis B virus	Recombinant hepatitis B surface antigen
	Human papilloma virus	Virus-like particles
Bacterial	*Corynebacterium diphtheriae*	Formaldehyde-treated toxin
	Clostridium tetani	Formaldehyde-treated toxin
	Bordetella pertussis	Mixture of purified proteins
	Neisseria meningitidis (A,C,W,Y)	Purified capsular polysaccharides
	Neisseria meningitidis (B)	Recombinant proteins, outer membrane vesicles
	Streptococcus pneumoniae	Purified capsular polysaccharides
	Haemophilus influenzae type b	Polysaccharide-protein conjugates

inactivated vaccines. Immunisation with a live vaccine can provide lifelong immunity.

The major drawback of live vaccines is the risk of reversion to pathogenicity. For instance, the occurrence of vaccine-associated paralytic poliomyelitis after the introduction of OPV has been reported [1, 3]. Furthermore, live vaccines sometimes cause mild symptoms resembling the disease caused by the pathogen. Live vaccines should never be given to immunosuppressed persons, because they lack the ability to respond even to infections by attenuated organisms.

19.3.2.1 Attenuated Viral Vaccines

Examples of live viral vaccines are polio, measles, mumps, rubella and rotavirus vaccines. Attenuated polio vaccine is administered orally.

Box 19.1: Reassortment and Reverse Genetics

Genetic reassortment is a method to obtain hybrid virus strains by coinfection of cells. The coinfection will lead to progeny virus containing many gene combinations. The virus strain with the desired properties (high immunogenicity, low pathogenicity) is selected for vaccine production. The method is a standard technique to obtain influenza vaccine strains (also for inactivated vaccines). In that case reassortants are made by injecting an egg with two virus strains: the pathogenic strain and a strain which is harmless for humans but grows well in eggs.

In reverse genetics the genes responsible for pathology or infectivity are changed or deleted. Host cells are transfected with plasmids containing viral genes. In this way progeny virus with the desired properties can be produced. This is opposed to forward genetics, which is the classical way of obtaining mutants by selection pressure, resulting in phenotypic mutants, in this case attenuation of pathogenicity.

OPV contains the three existing serotypes, which differ from each other in a number of distinct epitopes relevant for protection. OPV plays an important role in the eradication of poliomyelitis, pursued by the WHO, because it is very effective and relatively cheap. Of the three serotypes, wild type 2 virus has been eliminated. However, the vaccine has a tendency to revert to neurovirulence causing rare cases of vaccine-associated poliomyelitis. Therefore OPV is being replaced by IPV (see Sect. 3.3).

Attenuated mumps, measles and rubella viruses are often formulated in a COMBINATION VACCINE (MMR vaccine, sometimes combined with *Varicella zoster* virus as fourth component). These attenuated RNA viruses vary in size and number of structural proteins. Measles, mumps and rubella vaccines, whether separate or combined, are lyophilised preparations. The three vaccine components have in common that one single s.c. administration is probably sufficient for decades, perhaps lifelong protection. Nevertheless, in some countries the first dose given at 12–15 months of age is followed by a second vaccination at the age of 4–6 or 11–12 years. Although antibody formation is a correlate of protection, also cell-mediated immunity probably contributes. Vaccine efficacy is estimated to be at least 90%, and combining the components does not seem to influence their effectiveness. Side effects are generally mild and usually occur 7–12 days after vaccination. MMR vaccines are not indicated for infants below the age of 1 year, because circulating maternal antibodies impair vaccine efficacy in this age group.

The only licensed vaccine that is applied intranasally is an influenza vaccine containing four live attenuated virus strains. The strains are cold adapted, attenuated strains obtained by classical reassortment (see Box 19.1).

The above-mentioned vaccines are attenuated by classical methods, i.e. repeated passage in cell culture or reassortment of strains. This is also the case for one of the two live attenuated rotavirus vaccines on the market. Rotarix is based on a single rotavirus strain from human origin attenuated by multiple passages in cell culture. Its competitor, RotaTeq, contains five attenuated strains obtained by genetic reassortment (Box 19.1) of human and bovine strains. The first rotavirus vaccine, based on a human-resus reassortant and licensed in 1998, was withdrawn due to a rare but serious side effect: intussusception of the gut. In order to rule out this possibility, the new-generation products were tested in very large clinical trials.

19.3.2.2 Attenuated Bacterial Vaccines

The best-known attenuated bacterial vaccine is TB vaccine, which has been incorporated in many immunisation programmes as of the 1930s. The vaccine is based on *Mycobacterium bovis* bacteria, which primarily infect cattle but can also infect humans. The vaccine consists of lyophilised attenuated *M. bovis*, known as bacille Calmette-Guérin (BCG), and is administered i.d. to infants and older children. The vaccine is not effective in preventing pulmonary TB, but it can protect against disseminated TB in young children. Current vaccine strains vary in the extent of attenuation, and the dosage varies among vaccine suppliers. The immunisation schedule varies significantly among nations. The nature of the immune response is not known in detail, but cell-mediated immune mechanisms are probably involved in protection, whereas antibodies do not seem to play a substantial role. There is a great need for effective TB vaccines because strains emerge that are resistant to all known antibiotics. The development of genetically modified live attenuated strains is one of them [4].

Oral attenuated *Salmonella typhi* vaccines are indicated for high-risk groups, such as children in endemic areas and travellers, to prevent typhoid fever. The only licensed strain is Ty21a, whose attenuation has been stimulated by using nitrosoguanidine, a chemical mutagenic agent. Strain Ty21a lacks the ability to synthesise capsular polysaccharides, which are essential for virulence. In order to protect the bacteria against peptic digestion, the vaccine is formulated as lyophilised bacteria in enteric-coated capsules. Protection is achieved through three to four doses administered every other day. The vaccine provides significant protection by inducing relatively strong intestinal IgA and cell-mediated responses, and a weak systemic antibody response. Protective antibodies are directed against flagelli and lipopolysaccharide (LPS). The duration of the protection is relatively short (3–5 years).

19.3.3 Inactivated Whole Organisms

Inactivated bacterial and viral vaccines are obtained from virulent strains by heat treatment or by chemical inactivation, usually with formaldehyde or beta-propiolactone. Since inactivated pathogens are not able to propagate after administration, these vaccines usually are less immunogenic than live vaccines, and higher antigen doses are needed. An advantage over the latter is the inability to revert to virulence. On the other hand, deficient inactivation has caused vaccine-related accidents. For instance, immunisation with insufficiently inactivated polio vaccine in 1955 resulted in cases of paralytic disease [1, 2]. Adaptations in the inactivation procedure, like transfer to a new container to guarantee that all liquid will be in contact with the formaldehyde and a filtration step to remove possible viral aggregates, guarantee complete inactivation. Examples of this category include inactivated polio vaccine (IPV) and whole-cell pertussis vaccine, which are discussed below.

19.3.3.1 Inactivated Polio Vaccine

IPV is currently used in an increasing number of countries, at the expense of OPV. This is a necessary step in the endgame of polio eradication [5]. The vaccine consists of formaldehyde-inactivated poliovirus and includes the three serotypes. Advantages of IPV over OPV are a better temperature stability, the absence of vaccine-related disease and the possibility of combination with diphtheria, tetanus and pertussis components in one formulation (DTP-IPV vaccine). In contrast to OPV, IPV does not elicit secretory IgA antibodies, but its effect relies on the induction of virus-neutralising serum IgG. Because IPV does not replicate in vivo, the dose needs to be about ten times higher as compared to OPV, increasing cost and decreasing production capacity. For that reason the number of IPV-producing vaccine manufacturers is expected to increase.

19.3.3.2 Whole-Cell Pertussis Vaccine

Whole-cell pertussis (whooping cough) vaccine consists of heat-inactivated *B. pertussis* cells. The dose is determined on the basis of the opacity of the inactivated cell suspension. The vaccine potency is tested by protection assays in mice. This is a crude and animal unfriendly test involving intracerebral challenge and death as end point. Alternatives, like serological models, are under development. The protective efficacy of whole-cell pertussis vaccines is probably based on antibodies against several pertussis antigens, such as pertussis toxin, filamentous hemagglutinin and LPS. Whole-cell pertussis vaccines are notorious for their frequent side reactions, mostly fever and irritability. Therefore, the vaccine has been replaced in many western countries by safer acellular pertussis vaccines (see next section). Whole-cell pertussis vaccines are still used in developing countries, mainly for economical reasons. Whole-cell vaccines can be produced relatively easily by local manufacturers and are free of intellectual property rights. Another advantage of whole-cell pertussis vaccine is that it may induce better protection [6]. Because it contains many antigens, it also protects against *B. bronchiseptica* infections and probably is less prone to vaccine-induced selection of circulating *B. pertussis* strains. The adverse effects of whole-cell pertussis vaccines are largely due to the LPS present in *B. pertussis*' outer membrane. The adverse effects are stronger in older children and adults, so that whole-cell pertussis vaccines are not indicated for these age groups. Limited duration of protection is a problem. Although infected adolescents and adults in general have relatively mild symptoms, they can spread infections to very young non-immune infants. For this reason vaccination of adults or pregnant women is considered in order to reduce whooping cough in infants. The immunised mother passively protects her child by maternal antibodies. The vaccine contains colloidal aluminium salt as ADJUVANT and is usually combined with diphtheria and tetanus vaccine components (DTP vaccine). The vaccine is given in 4–5 i.m. doses.

19.3.4 Subunit Vaccines

SUBUNIT VACCINES contain one or more selected antigens (subunits) significant for protection against the pathogen they are derived from. SUBUNIT VACCINES have better-defined physicochemical characteristics and show less side effects than vaccines consisting of attenuated or inactivated organisms. Antigens used for current SUBUNIT VACCINES include viral and bacterial proteins as well as bacterial capsular polysaccharides.

19.3.4.1 Proteins

Protection against *Corynebacterium diphtheriae* or *Clostridium tetani* is based on the presence of antibodies directed against the respective toxins. These toxins are water-soluble proteins and form the basis of diphtheria and tetanus vaccines. In order to eliminate the toxicity of diphtheria and tetanus toxin, they are incubated with formaldehyde. This process is called toxoidation, and the resulting products are referred to as toxoids. Formaldehyde forms covalent bonds with the toxin, which is initiated by a reversible reaction of formaldehyde with primary amino groups, followed by an irreversible reaction with other amino acid residues [2]:

$$R-NH_2 + CH_2O \leftrightarrow R-NH-CH_2OH \leftrightarrow R-N=CH_2 + H_2O \quad (1)$$

$$R-N=CH_2 + R'-H \rightarrow R-NH-CH_2-R' \quad (2)$$

where R is the toxin and R′ can be a reactive amino acid residue (e.g. lysine, arginine, tryptophan, tyrosine, histidine) in the toxin molecule (or possibly a neighbouring toxin molecule) or a free amino acid present in the matrix. Thus, stable cross-links are formed, yielding a heterogeneous product with respect to number and sites of formaldehyde adducts and molecular weight. The degree of toxoidation is highly dependent on the reaction conditions, including formaldehyde concentration, pH, temperature and the presence of other components. The toxoidation process must be a compromise between sufficient detoxification and preservation of relevant epitopes. To enhance the relatively poor immunogenicity of toxoids, they are adsorbed to colloidal aluminium salt.

Acellular pertussis vaccines were introduced in Japan in the early 1980s as alternatives to the whole-cell vaccines for the immunisation of older children. Although pertussis toxoid alone may be sufficient for protection, most acellular pertussis vaccines contain at least two proteins important for the virulence of *B. pertussis*, including (inactivated) pertussis toxin, filamentous haemagglutinin, fimbriae and pertactin. Clinical trials indicate that the efficacy of these vaccines may be lower than that of whole-cell vaccines, but the acellular vaccines induce virtually no adverse effects [6]. The latter makes them suitable for immunisation of older children and adults. Acellular pertussis vaccines have now been introduced in many national immunisation programmes, mostly combined with diphtheria and tetanus components (DTaP vaccine).

19.3.4.2 Virus-Like Particles, Virosomes and Vesicles

To induce an antibody response, an antigen has to be accessible to B-cells. Therefore, antigens eliciting protective antibody responses are often membrane proteins. In the presence of lipids, these antigens can spontaneously form virus-like particles (VLPs). VLPs are usually very immunogenic because of the multimeric presentation form.

Hepatitis B vaccine was the first marketed recombinant vaccine, which has replaced hepatitis B vaccines obtained from plasma of infected humans. Recombinant hepatitis B vaccines are composed of hepatitis B surface antigen (HBsAg) derived from yeast or mammalian cells. Purified HBsAg and lipids self-assemble to 22-nm particles identical to those excreted by cells infected with the native virus. The ease of production has made the vaccine available worldwide, and in many countries, the vaccine is given to all infants.

Virosomes are lipid vesicles extracted from membrane viruses by detergent treatment. They contain membrane-associated viral antigens and often induce potent immune responses.

Currently two human papillomavirus (HPV) vaccines are on the market: Gardasil from Merck and Cervarix from GSK. Both are based on recombinant VLPs. Gardasil is a quadrivalent vaccine, and Cervarix contains VLPs from two virus types. The vaccines protect against cervical cancer. It has been shown in phase III clinical trials that the bivalent Cervarix vaccine not only protects against the two virus types present in the vaccine but also cross protects against three other important cancer-inducing strains [7]. Despite this, the introduction of the vaccine was difficult, and vaccine coverage remains below expectations. Risk perception (chance of getting cervical cancer vs potential adverse effects due to the vaccine), moral objections (providing vaccines to children against sexually transmitted disease) and anti-vaccination sentiments on social media probably contribute to this.

19.3.4.3 Capsular Polysaccharides

Many bacteria have a capsule consisting of high-molecular-weight polysaccharides, which act as virulence factors. Capsule-forming species include both Gram-positive (e.g. pneumococci) and Gram-negative bacteria (e.g. meningococci). The polysaccharides of the different species are composed of linear repeat oligosaccharide units that vary in sugar composition and chain length. The host defence against encapsulated bacteria relies on anti-polysaccharide antibodies interacting with complement to opsonise the organisms and prepare them for phagocytosis and clearance. Licensed capsular polysaccharide vaccines include meningococcal (serogroups A, C, W-135, Y), pneumococcal (up to 23 serotypes) and *Haemophilus influenzae* type b (Hib) vaccines.

The main disadvantage of capsular polysaccharide vaccines is their T-cell independency, which implies that they do not elicit immunological memory. Moreover, infants up to 2 years of age show very weak, non-protective immune responses, whereas they belong to the highest risk groups for infections with the encapsulated bacteria mentioned above.

19.3.4.4 Polysaccharide-Protein Conjugate Vaccines

The poor immunogenicity of plain polysaccharides can be overcome by covalent coupling to carrier proteins containing T-cell epitopes. These helper epitopes make them T-cell dependent and enables the induction of strong immune responses and immunological memory in all age groups, including infants. Conjugate vaccines licensed so far are Hib, meningococcal group C and pneumococcal vaccines. The Hib polysaccharide consists of repeat units of ribosyl(1-1)ribitol phosphate. An effective *Haemophilus influenzae* vaccine is relatively easy to produce, because—in contrast to the diversity of pathogenic meningococcal and pneumococcal strains—Hib is responsible for about 95% of infections with *Haemophilus* species, so only one polysaccharide type has to be included in the vaccine. Table 19.4 shows that the four licensed Hib conjugate vaccines vary in composition, owing to differences in polysaccharide length, carrier protein, coupling procedure and polysaccharide-to-protein ratio. As a result, these vaccines differ with respect to immunogenicity and efficacy [8]. The vaccines are incorporated in many childhood immunisation programmes and are normally administered i.m. in a multi-dose schedule with DTP, either separate or as a combined DTP-Hib vaccine. In some countries Hib is combined in a pentavalent (DTP-IPV-Hib) or hexavalent (DTP-IPV-Hib-HepB) combination vaccine. Another example of a

Table 19.4 Characteristics of licensed *Haemophilus influenzae* type b conjugate vaccines[a]

	Vaccine[b]			
Property	PRP-D	HbOC	PRP-OMP	PRP-T
Polysaccharide size	Medium	Small	Medium	Large
Polysaccharide content (μg)	25	10	15	10
Carrier protein	Diphtheria toxoid	Diphtheria toxin mutant	Meningococcal group B outer membrane proteins	Tetanus toxoid
Protein content (μg)	18	20	250	20
Linkage	Via spacer	Direct	Via spacer	Via spacer
Formulation	Aqueous solution	Aqueous solution	Lyophilised, reconstituted with alum salt suspension	Lyophilised, reconstituted with aqueous buffer

[a]Abbreviations: *PRP* polyribosylribitol phosphate; *PRP-D* PRP-diphtheria toxoid conjugate vaccine; *HbOC Haemophilus* type b oligosaccharide conjugate vaccine, *PRP-OMP* PRP-outer membrane protein conjugate vaccine, *PRP-T* PRP-tetanus toxoid conjugate vaccine
[b]Source: [1]

licensed conjugate vaccine is a mixture of polysaccharides from 13 types of pneumococci conjugated to diphtheria toxin. Finally, several manufacturers produce meningococcus group C conjugate vaccine.

19.4 Pharmacological Effects of Vaccination

The efficacy of a vaccine is difficult to estimate, because the relationship between immune response and degree of protection is not straightforward. Seroconversion, i.e. the increase in the level of specific circulating Ab, is commonly determined as a measure for the immunogenicity. Moreover, the protective quality of these antibodies can be measured with assays for bactericidal activity, i.e. their ability to kill bacteria in the presence of complement (e.g. meningococcal vaccines), virus-neutralising activity (e.g. polio vaccines) or toxin-neutralising activity (e.g. diphtheria vaccines). However, it is hard to correlate the level and persistence of circulating antibodies to protective efficacy. For some vaccines, like tetanus and diphtheria toxoid, Hib and meningococcal type C vaccine, serological correlates of protection have been established [9]. However, the extent of cell-mediated immunity may in some instances be a better measure for protection, e.g. against TB, some viral diseases and therapeutic tumour vaccines. Advances in the measurement of cell-mediated immune responses enable vaccine developers to measure cellular immunity in small blood samples. T-cells can be detected with sensitive methods like flow cytometry after cytokine staining or staining with fluorescent HLA tetramers. Antigen-specific lymphocytes can be detected by the Elispot technique, in which cytokine-producing cells are detected in plates coated with anti-cytokine antibodies. The effectiveness of vaccination is most clearly demonstrated by the reduction of disease after introduction of a vaccine in national immunisation programmes. Recent examples are the drastic reduction in incidence of Hib infections observed in those areas where routine vaccination in infants was introduced and a similar effect after the introduction of meningococcus group C vaccination in the UK. There is much indirect evidence of vaccine efficacy. For instance, in The Netherlands, where the use of IPV has effectively protected most of the population, two significant outbreaks of poliomyelitis in 1978 and 1992 were restricted to communities which refuse vaccination on religious grounds. Outbreaks of measles in the USA in 2015 were linked to non-immunised individuals: the majority of measles cases (80%) occurred in non-immunised people [10].

Since the target groups of vaccines in many cases include healthy infants and young children, vaccine safety is of particular importance. The occurrence of side effects may be due to the antigenic components (e.g. LPS in whole-cell pertussis vaccine), impurities derived from the production process (e.g. chick protein from the cell substrate used for measles vaccine production) or additives used in a vaccine formulation (e.g. neomycin or gelatin in MMR vaccines, aluminium salts in SUBUNIT VACCINES). Before a new vaccine candidate is licensed, its safety is investigated in animals and in humans in phase I, II and III clinical trials. Phase I trials include a small number of healthy adults and serve to collect preliminary safety data. In phase II studies, safety and immunogenicity are determined in a larger number of volunteers, usually in the target population. Vaccine dosage is also assessed. Phase III trials are meant to evaluate safety, including rare adverse effects, and efficacy in large target populations.

19.5 New Developments

19.5.1 Introduction

Notwithstanding the success of vaccination, several infectious diseases remain against which an effective vaccine is not yet available. New vaccines against bacterial (e.g. group *Shigella*), viral (e.g. respiratory syncytial virus) and parasitic (there is only one licensed parasitic vaccine, a malaria vaccine with modest efficacy) infections are under development. Ideally, these vaccines should provide lifelong protection in any individual of any age, be absolutely safe, easy to produce in unlimited quantities, stable under varying conditions, easy to administer and cheap. As yet the design of a vaccine with all these ideal characteristics combined remains an important challenge for developers of new and better vaccines.

Apart from new prophylactic vaccines, current research is also focused on the development of therapeutic vaccines, especially for the treatment of chronic diseases such as AIDS and cancer. The rationale of administering vaccines to patients already suffering from disease is to specifically boost the immune system weakened by the disease or to prime the immune system with naturally low immunogenic or low abundant antigens.

The number of vaccines routinely applied is expected to increase, which demands efforts to reduce the number of injections. An obvious way to achieve this is combining separate vaccine components into one vial or syringe. Examples of such COMBINATION VACCINES have been given before. Simply mixing vaccine components, however, may not only pose pharmaceutical problems (e.g. incompatibility of vaccine components and/or excipients) but also bears the risk of immunological interference. For instance, hepatitis B vaccine was reported to be less immunogenic when mixed with DTaP-IPV-HiB vaccine [11].

19.5.2 Modern Technologies

Whereas traditional vaccine development has largely been dependent on empirical methods, a better insight into immune mechanisms and immunogenic structures of infectious organisms has led to a better understanding of what would be the optimal vaccine composition as related to the desired immunological effect. Recent advances in genomics, proteomics and bioinformatics are now being applied to identify putative antigen and to unravel immune mechanisms (Box 19.2) [12]. Moreover, the advent of (bio) technological advances has enabled scientists to translate the improved immunological knowledge into the rational design of new vaccines against a variety of life-threatening and chronic diseases. Several classical and modern approaches to the development of a variety of new vaccines are currently being explored, some of which are schematically shown in Fig. 19.1. Most of these approaches offer the following common advantages over classical vaccines: (1) relevant epitopes of pathogenic organisms or

Box 19.2: Vaccinomics

To understand how vaccines work, vaccinologists need to know details about the composition of vaccines and the interaction of vaccine components with the immune system. Omics technologies are being used to do this.

GENOMICS One of the difficult tasks in the development of subunit vaccines is the selection of the relevant antigen(s). Apart from being able to induce protecting immmune responses, they need to be conserved in different strains of a pathogen in order to provide broad protection. Moreover the antigens must be stable under immune pressure in order to prevent vaccination-induced selection.

A genomics approach to screen for leads was coined reversed vaccinology. Once the genome of a pathogen, for instance, a bacterium, is sequenced, possible vaccine candidates can be selected by in silico searching for sequences coding for transmembrane sequences or sequences coding signal peptides for extracellular transport. This will reveal proteins that would be exposed to the immune system, containing B-cell epitopes. These genes are cloned in plasmids, and recombinant proteins are produced. Immunisation studies will reveal whether the proteins are immunogenic and induce functional antibodies. The number of candidates is further reduced by looking for sequence variability within clinical isolates of the pathogen, selecting for conserved antigens. Bexsero, a vaccine against *Neisseria meningitides*, was developed in this way.

TRANSCRIPTOME AND PROTEOME ANALYSIS Gene expression profiles are being used to unravel immune responses during infection or after vaccination [12]. Methods to measure mRNA are well established, using gene arrays or PCR primers. The goals are to correlate gene signatures to protective responses, to identify and understand adverse effects or to judge how an animal model compares with human responses. Although mRNA analysis is informative, the final product of a gene is protein. High mRNA expression may not correlate with the amount of protein being produced. Also, post-translational modifications may be important for a functional gene product. Therefore proteome analysis is being performed. Proteome analysis is less straightforward, but state-of-the-art mass spectrometers can identify and quantify thousands of proteins in one run.

Although the technology is complex, the most difficult part of vaccinomics approaches is to extract meaningful data. To achieve this pathway analysis, bioinformatics and biostatistics are of key importance.

cancer cells are obtained by safer means, and (2) in greater quantities, (3) the products are better defined and (4) epitopes of a single or multiple pathogenic agents can be combined easily in one vaccine. Approaches not yet addressed before are briefly discussed below. For more detailed information about modern vaccine technology, the reader is referred to specialised textbooks [1, 2].

19.5.2.1 Adjuvants and Innate Immunity

Active immunisation is based on the presence of unique antigens recognised by the adaptive immune system, i.e. B-cells and T-cells. However, the presence of only an antigen will usually not lead to an adaptive immune response. It is essential to have other components than antigens in the vaccine. These components must stimulate the innate immune system. This is the non-specific part of the immune system, like phagocytes and the complement system. These components are called adjuvants. Pathogens and vaccines containing whole viruses or bacteria already contain adjuvants by themselves. These adjuvants are evolutionary conserved molecular patterns unique

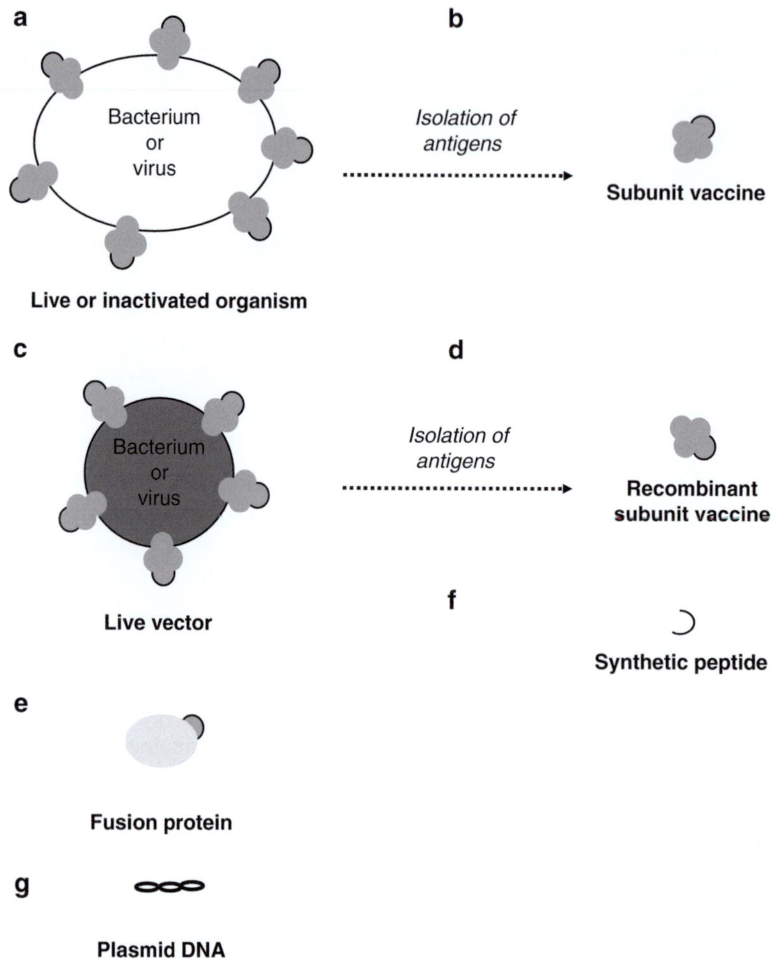

Fig. 19.1 Schematic representation of (**a**, **b**) classical vaccine components and (**c**–**g**) new-generation vaccine components: (**a**) whole bacterium or virus, live attenuated or inactivated, with protein antigens (grey objects) containing protective epitope (black semi-circle); (**b**) subunit vaccine, antigen isolated from pathogenic organism; (**c**) live vector, antigenic proteins derived from pathogenic organism expressed by live, non-pathogenic bacterium or virus; (**d**) recombinant subunit vaccine, antigenic protein isolated from heterologous expression system; (**e**) fusion protein, non-toxic protein containing epitope of protein from pathogen isolated from non-pathogenic organism; (**f**) synthetic peptide with amino acid sequence mimicking epitope of antigen from pathogen; (**g**) nucleic acid vaccine, plasmid DNA containing the gene encoding antigenic protein or epitope from pathogen

for pathogens. These are called pathogen-associated molecular patterns (PAMPs), and examples are bacterial LPS and methylated C- and G-rich DNA sequences (CpG). PAMPs interact with danger sensing receptors on cells of the innate immune system. These are called pattern recognition receptors, and examples are toll-like receptors (TLRs). TLR activation may result in the induction of cytokines which in turn initiate an adaptive response if an antigen is also present.

The second type of adjuvants is delivery systems (see Box 19.3). These are microparticles or nanoparticles designed to deliver associated antigen to antigen-presenting cells (APCs).

When developing subunit vaccines, it became clear that vaccines cannot exist without adjuvants. On the other hand, adjuvants can significantly contribute to adverse effects. These may range from the induction of fever and erythema (caused by a strong pro-inflammatory

Box 19.3: Presentation Forms

ADJUVANTS comprise a large number of substances of variable chemistry and origin. Examples are colloidal salts, lipid matrices, surface-active compounds and emulsions of mineral, bacterial, vegetable or synthetic nature. Adjuvants have in common that they are not immunogenic per se but enhance the immunogenicity of co-administered Ag.

VACCINE DELIVERY SYSTEMS are colloidal carriers with a size which can vary from ~50 nm to the micrometer range, allowing multimeric Ag presentation at their surface. Thereby they mimic the natural presentation of Ag on viral or bacterial surfaces. In general, multimeric presentation of Ag strongly improves their immunogenicity. Colloidal carriers can function as a depot at the administration site, resulting in sustained delivery and a reduction of the number of doses required. Moreover, they can enhance humoral and/or cellular immune reactions, because colloidal particles are taken up more efficiently by APCs (in particular dendritic cells) as compared to free Ag. Uptake of colloidal particles by dendritic cells can be further promoted by coupling targeting moieties, specifically recognising dendritic cell receptors, to their surface. As M cells present in mucosal membranes are specialised in the uptake of particulate material and subsequent presentation to immune cells, colloidal particles are also suitable Ag carriers when mucosal immunisation is pursued. In addition, they may protect the Ag from proteolytic attack (e.g. in oral vaccine formulations). Besides Ag, adjuvants are sometimes incorporated into these carrier systems, and many carrier systems have intrinsic adjuvant activity.

response) to the induction of autoimmunity (because of stimulation of responses against self-antigens).

Because adjuvants operate in a small window of mimicking danger and causing damage, it is difficult to license them for mass vaccination purposes. In the last decades, many adjuvants have been tested in clinical trials, but few have made it to licensure [13]. Aluminium phosphate and aluminium hydroxide have been and still are the first choice for human vaccines, mainly because they have a convincing safety record. Their adjuvant effect was initially explained by assuming a depot of antigen adsorbed to the insoluble alum salt particles at the site of injection, resulting in slow release of antigen as well as some local tissue damage resulting in the attraction of APC. Today, the mechanism of action has been unravelled in much more detail [14]. More recent adjuvant introductions are monophosphoryl lipid A (MPL), which is a TLR4 ligand, and particulate adjuvants (see Box 19.3) like squalene-in-water emulsions and virosomes, which are reconstituted viral membranes including the membrane proteins.

Mosquirix, the licensed malaria vaccine, contains MPL, liposomes and plant triterpene glycoside QS21 as adjuvants, indicating a trend to the use of several adjuvants in one vaccine formulation.

19.5.2.2 Recombinant Live Vaccines

Non-pathogenic or attenuated organisms can be used as carriers for heterologous protein Ag. Such live carriers are called vectors (Fig. 19.1c). They are obtained by cloning the desired gene and introducing it in an appropriate carrier organism. Both viral (e.g. vaccinia virus, adenoviruses) and bacterial vectors (e.g. *Salmonella* species, BCG) are being explored as carrier to express a variety of Ag. The properties of recombinant live vaccines are comparable to those of classical attenuated vaccines.

Some of the Ebola vaccine concepts that were clinically evaluated as a response to the Ebola epidemic in 2014 are vector based. The use of vectors prevents the necessity to handle high containment pathogens but still uses advantages of live pathogens, such as 'natural' antigen processing (resulting in a balanced immune response) and the relatively low dose needed.

19.5.2.3 Fusion Proteins

Fusion proteins are non-toxic proteins containing inserted epitopes, larger protein fragments or even entire proteins derived from pathogenic species (Fig. 19.1e). They are obtained by the insertion of DNA sequences encoding epitopes in the gene of the carrier protein, such as HBsAg or a fusion partner with the capability to target the fused antigen to APCs [15]. The recombinant gene is expressed in a suitable organism such as yeast and the fusion protein is then purified. A drawback of this genetic fusion technology is potential misfolding of the epitope when incorporated in the carrier protein, which would lead to irrelevant immune responses.

19.5.2.4 Synthetic Peptide Vaccines

Chemically synthesised peptides belong to the best-defined vaccine components presently under investigation. The synthetic peptide technology allows for the design of vaccines consisting of selected epitopes free from irrelevant or unwanted structures. Large amounts of linear peptides resembling T-cell or B-cell epitopes (Fig. 19.1g) can be prepared by automated methods. The immunogenicity of synthetic peptide antigen is weak but can be enhanced by conjugation to carrier proteins (analogous to polysaccharide-protein conjugates) or to lipids or by the construction of multiple different B- and T-cell epitopes. These options offer the possibility to render synthetic B-cell epitopes T-cell dependent. Furthermore, the conformational freedom of small linear peptides can be restricted by cyclisation, which aims to force them into a conformation reflecting the native structure, which is especially important for peptide analogues of B-cell epitopes. However, the most promising results obtained with peptide vaccines have been achieved with linear peptides representing T-cell epitopes. By using long peptides of about 30 amino acids, still protein-like processing by APC occurs, resulting in efficient T-cell responses, although powerful adjuvants must be included [16].

19.5.2.5 Nucleic Acid Vaccines

A development of potential clinical use is genetic immunisation, i.e. direct administration of DNA encoding one or more antigens of interest (Fig. 19.1g). Upon i.m. immunisation with non-replicating plasmid DNA, the protein encoded is produced and expressed by the host cell. RNA may also be used but is less suitable because it is rapidly degraded in vivo and more expensive to produce. Nucleic acid vaccines are capable of eliciting both humoral and cellular immunity. Induction of cellular immunity is particularly important for immunisation against retroviruses, such as HIV and hepatitis C virus.

To induce strong immune responses, it may be needed to develop heterogeneous prime-boost strategies, combining the use of DNA for priming and protein antigen for boosting.

19.5.3 Route of Administration

The efficacy of vaccination programmes would be enhanced greatly through the availability of needle-free immunisation methods. Injections with needle and syringe are not appreciated by most people, and about 10% of the population avoids injections. Furthermore, needlestick injuries and reuse of needles still cause many infections of hepatitis B virus and HIV. Several alternative routes of administration have therefore been explored, including mucosal (oral, intranasal, pulmonary) or needle-free dermal immunisation.

Mucosal immunisation routes are attractive alternatives to parenteral routes, not only because of the ease of administration but also because both systemic and mucosal (secretory IgA) responses are induced. The latter is advantageous, because mucosal surfaces are the common port of entrance of many organisms and a strong local immune response may hamper entry into the host by preventing adherence to and colonisation in mucosal surfaces. However, although mucosal immunisation is among the oldest means of vaccination (see Table 19.2), the number of vaccines suitable for mucosal immunisation is limited to a few oral vaccines and a live attenuated influenza vaccine called FluMist. Poliovirus can be given orally because the virus is relatively resistant to low pH, whereas oral typhoid vaccine is protected from gastric breakdown by formulation in enteric-coated capsules. Other mucosal administration routes, such as nasal or pulmonary delivery, have the advantage that the harsh conditions in the gastrointestinal tract are circumvented. Approaches to augment the immunogenicity of future mucosal vaccines include the use of vectors (e.g. *Salmonella*) or VACCINE DELIVERY SYSTEMS and co-administration of ADJUVANTS.

Topical administration of antigens is an appealing delivery route because the skin is a very immune active organ containing a large number of specialised APC, the Langerhans cells and dermal dendritic cells [17]. The skin was for a long time considered almost impermeable for large antigens, and the only way to deliver to the skin was by shallow injection. Recently a syringe with a 1.5 mm needle for intradermal delivery of 0.1 ml vaccine became available. The small needle length allows injection perpendicular to the skin, making true dermal delivery easier and more reliable. Many other techniques, avoiding classical needles and syringes, are under development. Chemical penetration enhancers (e.g. surfactants and liposomes) as well as physical approaches (e.g. iontophoresis) to enhance the permeability of the skin for macromolecules have shown some success.

Methods to bypass the stratum corneum are the use of microneedles [18]. Arrays of needles of 0.5 mm or less can be used to puncture the skin without causing pain. Microneedles can be coated with vaccine, are hollow and thus capable of injecting a small volume or are dissolvable with the antigen dispersed in the needle matrix.

Finally, jet injectors are a way of delivering vaccine. These devices do not contain needles, and the vaccine is delivered by forcing it under high pressure through a nozzle. The fluid or powder jet will penetrate the skin and, depending on the jet speed and nozzle design, will be deposited intradermally, subcutaneously or intramuscularly.

19.6 Summary

Vaccines have been very successful in the prevention of infectious diseases. Traditional vaccines consist of whole (live or inactivated) bacteria or viruses or components thereof and are among the oldest biotechnological pharmaceuticals. Several modern approaches, most of which are based on rDNA technologies, are emerging with the aim to generate more effective and safer vaccines. As a result of the tendency to design smaller, better-defined antigenic components, proper antigen presentation forms and the use of adjuvants are becoming increasingly important. New adjuvants are now routinely used in some vaccines, and more are under development. Moreover, to limit the increasing number of injections, needle-free immunisation routes are being explored. It is expected that vaccines to be marketed in the near future will be based on some of the modern vaccine technologies discussed in this chapter and—like the conventional vaccines—will have a significant contribution to the improvement of public health.

Selected Readings

Levine MM, Dougan G, Good MF, Liu MA, Nable GJ, Nataro JP, Rappuoli R, editors. New generation vaccines. 4th ed. New York: Informa Healthcare; 2009.

Plotkin SA, Orenstein WA, Offit PA, Edwards KM, editors. Plotkin's vaccines. 7th ed. Saint Louis: Elsevier; 2017.

Recommended Websites

College of Physicians of Philadelphia: http://www.historyofvaccines.org/

FDA: http://www.fda.gov/BiologicsBloodVaccines/Vaccines/

Vaccine page: http://vaccines.org/

WHO: http://www.who.int/immunization/

References

1. Plotkin SA, Orenstein WA, Offit PA, Edwards KM, editors. Plotkin's vaccines. 7th ed. Saint Louis: Elsevier; 2017.
2. Levine MM, Dougan G, Good MF, Liu MA, Nable GJ, Nataro JP, et al., editors. New generation vaccines. 4th ed. New York: Informa Healthcare; 2009.
3. Minor P. Vaccine-derived poliovirus (VDPV): impact on poliomyelitis eradication. Vaccine. 2009;27:2649–52.
4. da Costa C, Walker B, Bonavia A. Tuberculosis vaccines – state of the art, and novel approaches to vaccine development. Int J Infect Dis. 2015;32:5–12.
5. World Health Organization. Polio Global Eradication Initiative: polio eradication and endgame strategic plan 2013–2018, report 2013. Accessible via www.polioeradication.org.

6. Herzog C. Changing from whole-cell to acellular pertussis vaccines would trade superior tolerability for inferior protection. Expert Rev Vaccines. 2015;14:1065–72.
7. Paavonen J, Naud P, Salmerón J, Wheeler CM, Chow SN, Apter D, et al. Efficacy of the HPV-16/18 AS04-adjuvanted vaccine against cervical infection and pre-cancer caused by oncogenic HPV types: final event-driven analysis in young women (the PATRICIA trial). Lancet. 2009;374:301–14.
8. Madore DC. Impact of immunization on *Haemophilus influenzae* type b disease. Infect Agents Dis. 1996;5:8–20.
9. Pichichero ME. Booster vaccinations: can immunological memory outpace disease pathogenesis? Pediatrics. 2009;124:1633–41.
10. Clemmons NS, Gastanaduy PA, Fiebelkorn AP, Redd SB, Wallace GS. Measles—United States, January 4–April 2, 2015. MMWR. 2015;64:373–6.
11. Mallet E, Belohradsky BH, Lagos R, Gothefors L, Camier P, Carrière JP, et al. A liquid hexavalent combined vaccine against diphtheria, tetanus, pertussis, poliomyelitis, *Haemophilus influenzae* type B and hepatitis B: review of immunogenicity and safety. Vaccine. 2004;22:1343–57.
12. Zak DE, Aderem A. Systems integration of innate and adaptive immunity. Vaccine. 2015; https://doi.org/10.1016/j.vaccine.2015.05.098.
13. Amorij J-P, Kersten GFA, Saluja V, Tonnis WF, Hinrichs WLJ, Slütter B, Bal SM, Bouwstra JA, Huckriede A, Jiskoot W. Towards tailored vaccine delivery: needs, challenges and perspectives. J Control Release. 2012;161:363–76.
14. Maughana CN, Preston SG, Williams GR. Particulate inorganic adjuvants: recent developments and future outlook. J Pharm Pharmacol. 2015;67:426–49.
15. Léonetti M, Thai R, Leroy JC, Drevet P, Ducancel F, Boulain JC, et al. Increasing immunogenicity of antigens fused to Ig-binding proteins by cell surface targeting. J Immunol. 1998;160:3820–7.
16. Melief CJ, van Hall T, Arens R, Ossendorp F, van der Burg SH. Therapeutic cancer vaccines. J Clin Invest. 2015; https://doi.org/10.1172/JCI80009.
17. Theunissen MBM, Haniffa M, Collin MP. Insight into the immunobiology of human skin and functional specialization of skin dendritic cell subsets to innovate intradermal vaccination design. Curr Top Microbiol Immunol. 2012;351:25–76.
18. van der Maaden K, Jiskoot W, Bouwstra JA. Microneedle technologies for (trans)dermal drug and vaccine delivery. J Control Release. 2012;161:645–55.

Plasma-Derived Immunoglobulins

20

Adrian W. Zuercher, Mel Berger, Reinhard Bolli, Cédric Vonarburg, Martin Spycher, Amgad Shebl, Rolf Spirig, Christoph Kempf, Fabian Käsermann, and Sylvia Miescher

20.1 Humoral Immunity

When the human body is invaded by foreign organisms, it defends itself either through the propagation of a cellular response (cell-mediated immunity) or by the production of soluble proteins. This production of soluble proteins is known as humoral immunity, so called because it refers to substances found within the humours, or body fluids. While there are a number of components that have been identified as important in the humoral immune response, including complement proteins [1], the key molecules necessary for mediating specific humoral immunity are immunoglobulins (Ig), also known as antibodies (Ab) (see also Chap. 4).

Ig molecules are complex proteins whose basic structure consists of two identical heavy chains and two identical light chains, linked by disulphide bonds and arranged in a "Y" shape (Fig. 20.1a) [2]. They are expressed on the surface of B cells or secreted, recognising and binding to ANTIGENS (Ag), i.e. "foreign" or "dangerous" structures, such as bacterial or viral surface proteins and surface polysaccharides as well as secreted bacterial toxins. Variable domains in the N-terminal regions of the heavy and light chains recognise and bind Ag, while the constant regions of the heavy chain C-terminal domains mediate effector functions and define the class or isotype of Ig. An Ag is recognised by shape and charge. While this may be a continuous peptide or surface polysaccharide, more often it is the three-dimensional shape made when the Ag is folded into its native structure. The specific part of the Ag that is recognised by the Ig is called an EPITOPE.

Each B cell can only produce one specific Ab, which defines its IDIOTYPE. The total number of Ag structures a human body can recognise is vast; it is estimated that the body can produce up to 10^{15} different Ab [3]. The number of IDIOTYPES a body can produce is known as the immune repertoire. Due to the huge variety of Ab structures present in the human body, it is not

Final manuscript submitted on June 25, 2019.

A. W. Zuercher · R. Bolli · C. Vonarburg · M. Spycher · R. Spirig · F. Käsermann · S. Miescher (✉)
CSL Behring AG, Bern, Switzerland
e-mail: sylvia.miescher@hispeed.ch

M. Berger
CSL Behring AG, Bern, Switzerland

CSL Behring LLC, King of Prussia, PA, USA

A. Shebl
Global Clinical Safety and Pharmacovigilance, CSL Behring GmbH, Marburg, Germany

C. Kempf
CSL Behring AG, Bern, Switzerland

Departement für Chemie und Biochemie, Universität Bern, Bern, Switzerland
e-mail: christoph.kempf@dcb.unibe.ch

M. J. Parnham et al. (eds.), *Nijkamp and Parnham's Principles of Immunopharmacology*,
https://doi.org/10.1007/978-3-030-10811-3_20

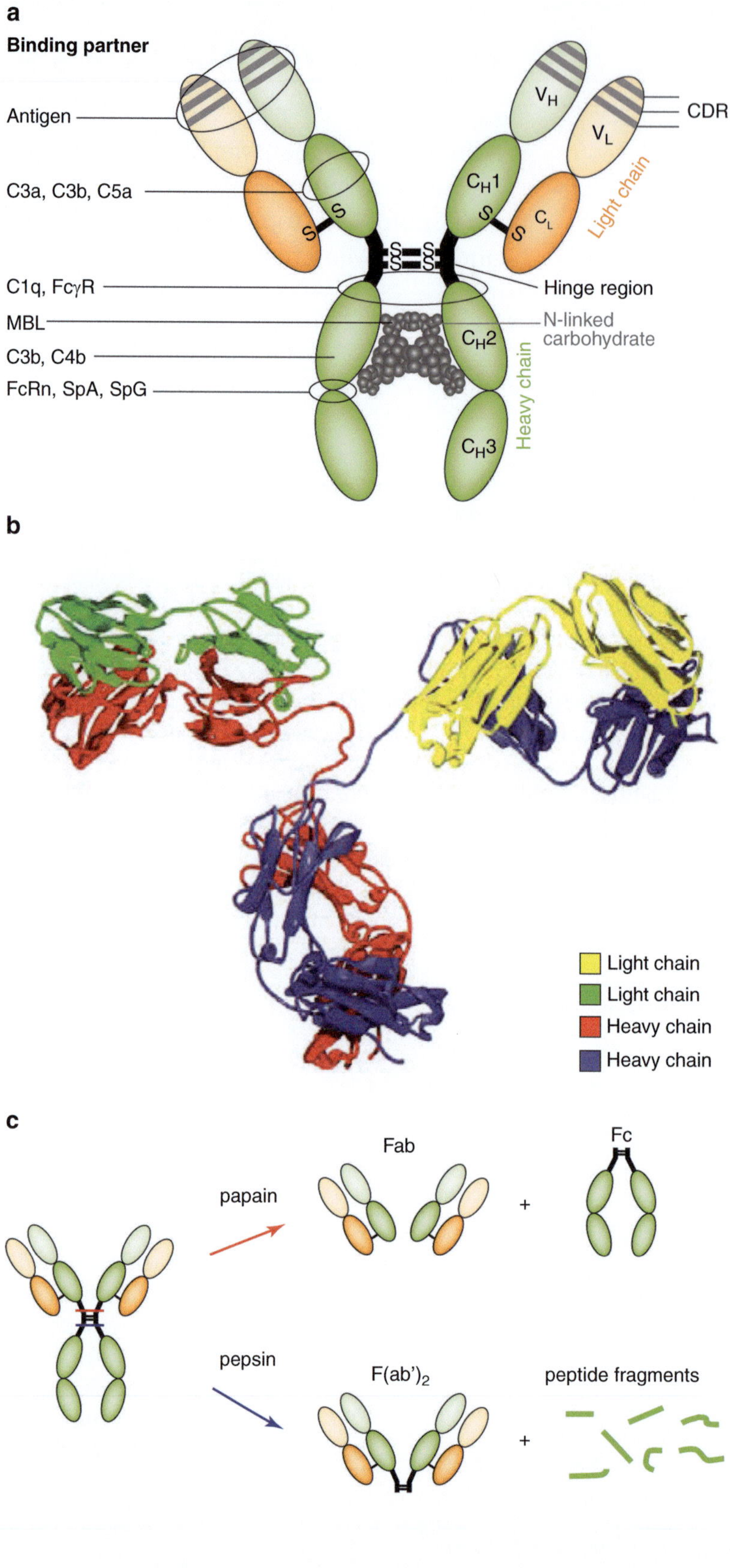

Fig. 20.1 Structure of IgG. (**a**) Detailed structure of an IgG1 molecule. CDR, complementarity-determining region; C_H, constant, heavy; C_L, constant, light; V_H, variable, heavy; V_L, variable, light; S-S, disulphide bonds; C3a, C3b, C4b, C5a, C1q, complement proteins; FcγR, Fcγ receptor; MBL, mannan-binding lectin; FcRn, neonatal Fc receptor, SpA, *Staphylococcus* protein A; SpG, *Streptococcus* protein G. (**b**) Ribbon structure of an IgG_2 antibody. This image has been released worldwide into the public domain by its author, Tim Vickers, at the Wikipedia project. Carbohydrates are not shown in this figure. (**c**) IgG fragments

possible for large numbers of each type of B cell to be present under normal conditions. Therefore, a process of clonal selection takes place once the closest matching Ab molecules are found for a specific Ag. The corresponding B cells then undergo rapid proliferation and differentiation into PLASMA CELLS, which within days start to secrete large amounts of the specific Ab.

The therapeutic use of Ig has been known for many years. The founder of serum therapy and passive immunotherapy was Emil von Behring who was awarded the Nobel Prize in 1901 [4]. HYPERIMMUNE SERA collected from immunised animals, or now more often from immunised or naturally exposed donors, have long been used as a PASSIVE IMMUNOTHERAPY treatment to provide specific Ab to an individual immediately, for example, after snake envenomation, *C. tetani* infection or rabies infection (Table 20.1) [5, 6]. Similarly, anti-rhesus D (anti-RhD) Ig has been used successfully in the prevention of haemolytic disease in neonates due to rhesus incompatibility between mother and child or after mistransfusion of blood components containing RhD-positive erythrocytes to RhD-negative recipients. More details on hyperimmune Ig are provided later in this chapter.

For over half a century, Ig has also been used therapeutically in patients with immune deficiencies [7], primarily ANTIBODY deficiencies. Initially administered subcutaneously (s.c.) by Bruton [8] for a case of agammaglobulinaemia, thereafter intramuscularly (i.m.), Ig were later given intravenously (i.v.), referred to as IVIG, with benefits in efficacy and safety. More recent developments in product formulation and concentrations have been explored enabling s.c. administration for a wide range of applications [9]. These are discussed in more detail later in the chapter.

IVIG (intravenous immunoglobulin) is prepared using IgG isolated from pooled plasma samples usually obtained from 10,000 to 40,000 normal healthy donors [10]. Patients with primary immunodeficiency (PID) are unable to produce sufficient amounts of Ig and are therefore at increased risk of bacterial, fungal or viral infections. Thus, IVIG as a replacement therapy is nowadays the standard of care in the treatment of a number of primary immunodeficiencies such as severe combined immunodeficiency (SCID), X-linked agammaglobulinaemia and

Table 20.1 High-titre immunoglobulin preparations against selected antigens

	Origin	Administration	Manufacturer (selection)
Anti-virus			
Cytomegalovirus	Human	i.m., i.v.	CSL Behring, Biotest
Hepatitis A	Human	i.m.	CSL Behring
Hepatitis B	Human	i.m., i.v.	CSL Behring, Biotest, Emergent, Grifols, Kedrion
Rabies	Human	i.m.	CSL Behring, Haffkine
Vaccinia	Human	i.v.	Emergent
Varicella zoster	Human	i.m., i.v.	CSL Behring, Biotest, Emergent, Cangene
Antibacterial toxins			
Botulism	Human	i.v.	Baxter, Cangene[a]
Diphtheria	Horse	i.m., i.v.	Haffkine
Tetanus	Human	i.m., i.v.	CSL Behring, Grifols, Kedrion
Antivenoms			
Scorpion	Horse	i.v.	Haffkine, SAIMR
Snakes	Horse, sheep	i.v.	bioCSL, Haffkine, SAIMR, Savage Laboratories
Spiders	Horse	i.m./i.v.	bioCSL, SAIMR
Jellyfish	Sheep	i.v.	bioCSL
Stonefish	Horse	i.m. (i.v.)	bioCSL
Other Ag			
Rhesus D	Human	i.m., i.v.	Baxter, CSL Behring, Emergent, Grifols, Kedrion
Digitalis	Sheep	i.v.	Roche, Savage Laboratories
Human thymocytes	Rabbit	i.v.	Genzyme

[a]Manufactured for the California Department of Public Health

Wiskott-Aldrich syndrome as well as secondary/acquired immunodeficiencies primarily associated with haematological malignancies [11].

IVIG is also indicated for the treatment of some autoimmune diseases including immune thrombocytopaenia, Kawasaki disease and Guillain-Barré syndrome (see Sect. 20.7 for more details) [12–14]. The MECHANISMS OF ACTION of IVIG in the treatment of autoimmune diseases have been the subject of many investigations but are still not fully characterised. Interference with complement components and the cytokine network, fragment crystallisable (Fc) receptor blockade, presence of anti-idiotypic Ab and modulation of B-cell, T-cell and dendritic cell functions are all thought to play a role in IVIG efficacy, and these will be discussed further in a later section of this chapter.

In recent years increased efforts have been made to develop recombinant alternatives of plasma-derived IVIG, especially of recombinant Fc fragments, and these will be briefly reviewed later in the chapter.

20.2 Antibody Structure and Mechanisms of Action

As mentioned above, all Ig molecules have a common symmetrical core structure made up of two covalently linked heavy chains as well as two light chains, each linked to a heavy chain. Within each chain there are two to five sequences of approximately 110 amino acids, which are independently folded and make up the Ig domains, termed VH, CH1, CH2, CH3, CH4, VL and CL (Fig. 20.1a, b). These Ig domains contain conserved regions and intrachain disulphide bonds. The Fab (fragment ANTIGEN-binding; Fig. 20.1c) portion of the Ig (the top of the "Y" shape) is the region responsible for Ag recognition. The Ag is recognised by three hypervariable regions within the N-terminal, or the top, of the heavy and light chains. These hypervariable regions consist of approximately ten amino acids and are collectively known as the complementarity-determining region (CDR; Fig. 20.1a). It is the large variability within the CDR that primarily accounts for the wide range of Ags recognised by Ig molecules. Once the Ag is bound to the CDR, the C-terminal regions at the base of the heavy chain, which have a uniform or constant structure within each isotype (see below), can mediate the effector functions of the Ig molecule. The effector portion of the Ig, known as the Fc, can mediate a number of different reactions (Table 20.2; Fig. 20.1c). Digestion by papain or pepsin yields either two Fab fragments and an Fc fragment or a single bivalent Ag-binding fragment [F(ab′)2], respectively (Fig. 20.1c).

Ig molecules are classified into five isotypes based on the structure of their heavy chain C-terminal regions: IgM, IgD, IgG, IgA and IgE (Fig. 20.2). Each of the Ig isotypes has distinct characteristics and functions within the overall humoral immune response. Triggering of the B cell via its surface Ig-receptor complex leads to activation and production of Ab. The Ab produced has the same variable domains as the Ab produced by the immature B cell and expressed on its surface, but the constant domains of the heavy chains vary depending on a number of factors. This process by which different Ig isotypes can be produced by daughter cells from the same activated B cell is called Ab isotype class switch.

20.2.1 Immunoglobulins of Class G (IgG)

IgG is the most abundant Ig in human plasma, where it accounts for approximately 75% of all Ig molecules (Table 20.2). The class switch from IgM to IgG often occurs in response to cell surface molecules, e.g. CD40 ligand (CD40L) and soluble factors expressed on and derived from Th1-polarised T cells, which are stimulated by viral and bacterial structures. IgG is found in both intra- and extravascular pools in monomeric form. IgG molecules are involved predominantly in the secondary Ab response, namely, the response observed after repeated exposure to an Ag. IgG molecules bind to toxins and many types of pathogens including viruses, bacteria and fungi. Once bound, they mediate the neutralisation of toxins

Table 20.2 Biological functions of human Ig classes (isotypes) and subclasses

Immunoglobulin	IgG_1	IgG_2	IgG_3	IgG_4	IgM	IgA_1	IgA_2	IgD	IgE
Adult serum level range, g/L (mean g/L)	4.9–11.4 (6.98)	1.5–6.4 (3.8)	0.20–1.10 (0.51)	0.08–1.40 (0.56)	0.7–1.7	1.5–2.6[a]		0.04	0.0003
Proportion of total IgG (%)	43–75	16–48	1.7–7.5	0.8–11.7	na	na		na	na
Half-life (days)	21	21	7.5–9	21	5	6[a]		3	2.5
Proportion intravascular (%)	45–52 (all IgG)				76	40–42[a]		75	na
Placental transfer	++	+	++	++	–	–	–	–	–
Antibody response to:									
Proteins	++	±	++	±	++	+	+	+	+
Polysaccharides	+	++	(–)	(–)	++	+	+	+	+
Allergens	+	(–)	(–)	++	+	+	+	nd	+++
Complement activation:									
Classical pathway activation (C1q binding)	++	+	+++	–	+++	–	–	nd	nd
Alternative pathway activation	–	–	–	–	–	+	+	nd	nd
Binding to Fcγ receptors:									
FcγRI (CD64): monocytes, macrophages, neutrophils, eosinophils, dendritic cells	+++	–	+++	+	–	–	–	–	–
FcγRIIa (CD32): monocytes, macrophages, neutrophils, eosinophils, platelets, dendritic cells, endothelial cells	++	[b]	++	–	–	–	–	–	–
FcγRIIa (CD32)-H131	++	+++	++	–	–	–	–	–	–
FcγRIIa (CD32)-R131	++	–	++	–	–	–	–	–	–
FcγRIIb (CD32): B cells, basophils, dendritic cells	++	–	++	(+)	–	–	–	–	–
FcγRIIIa (CD16): NK cells, eosinophils, macrophages, subsets of T cells, mast cells[c]	++	–	+++	–	–	–	–	–	–
FcγRIIIb (CD16): neutrophils									
FcγRIIIb (CD16)-NA1	+++	–	+++	–	–	–	–	–	–
FcγRIIIb (CD16)-NA2	++	–	++	–	–	–	–	–	–

(continued)

Table 20.2 (continued)

Immunoglobulin	IgG_1	IgG_2	IgG_3	IgG_4	IgM	IgA_1	IgA_2	IgD	IgE
FcRn (at acidic pH): monocytes, macrophages, subsets of dendritic cell, endothelial cells, epithelial cells in various tissues	++	++	±	+	–	–	–	–	–
FcεRI: mast cells, eosinophils, basophils	–	–	–	–	–	–	–	–	+++
FcεRII (CD23): B cells, T cells, follicular dendritic cells, eosinophils, platelets, Langerhans' cells	–	–	–	–	–	–	–	–	++
FcαR (CD89): monocytes, neutrophils, eosinophils	–	–	–	–	–	++	++	–	–
Fcα/μR: follicular dendritic cells in tonsil, plasma cells, macrophages, Paneth cells in lamina propria and intestinal germinal centres	–	–	–	–	+[d]	+[d]	–	–	–
Transferrin R (CD71): ubiquitous cell expression	–	–	–	–	–	+[e]	–	–	–
FcμR: T cells, macrophages	–	–	–	–	+	–	–	–	–
FcδR: T cells, B cells	–	–	–	–	–	–	–	+	–
Poly Ig receptor, mucosal transport	–	–	–	–	+	+	+	–	–
Binding to:									
Staph. protein A	++	++	–	+	–	–	–	–	–
Strept. protein G	++	++	++	++	–	–	–	–	–
Virus/toxin neutralisation	+	+	+	+	+	+	+	–	–
Functions	Opsonisation, complement activation, ADCC, neonatal immunity, feedback inhibition of B cells				Naïve B cell Ag receptor, complement activation	Mucosal immunity in plasma: anti-inflammatory		Naïve B cell Ag receptor	Defence against helminth parasites; immediate hypersensitivity

ADCC antibody-dependent cellular cytotoxicity, *Ag* antigen, *NA* neutrophil antigen, *NK* natural killer, *Staph. Staphylococcus*, *Strept. Streptococcus*, *nd* not determined, *na* not applicable

Receptors for IgG: CD64, high affinity; CD32 and CD16 are both of low affinity. From Abbas [15], Gillis et al. [16], Male et al. [17], Murphy et al. [18]

[a]Refers to plasma IgA; IgA_1 is the predominant form, making up 90% of total IgA

[b]FcγRII allotype-dependent

[c]Polymorphisms in FcRγIIIa gene Phe158Val affect IgG binding and functional activities

[d]Monomeric IgM and IgA_1

[e]Polymeric IgA

Fig. 20.2 Structure of all Ig isotypes and subclasses. N-linked carbohydrates are shown as red pins; clusters of O-linked carbohydrates are shown as blue hexagonals. Please note that the number of O-linked carbohydrates in a cluster varies. Hinge length correlates to the number of amino acids. Disulphide bridges are indicated by black bars between the heavy chains; there is no disulphide bridge between the heavy and light chains in IgA2. IgD is mainly expressed as a membrane-bound monomer. Please also note that the IgA2m(2) allotype is shown; the IgA2m(1) allotype (not shown) only has one Fab glycosylation and no disulphide bridge between the heavy and light chains

and inactivation or removal of these pathogens via mechanisms such as opsonisation leading to complement activation and phagocytosis.

IgG molecules are also involved in Ab-dependent cellular cytotoxicity (ADCC) by binding to natural killer cells and other leukocytes via FcγRIIIa. Other effector cells include monocytes, macrophages and granulocytes acting via their Fcγ receptors (FcγR): FcγRI (CD64), FcγRIIa (CD32), FcγRIIIa (CD16) and FcγRIIIb (Table 20.2). IgG can also bind to FcγR on B cells and dendritic cells, especially FcγRIIb, which is the only FcγR that mediates inhibitory signalling and thus fulfils important functions in the downregulation of ongoing immune responses.

The Fc portion of IgG can also bind to the neonatal Fc receptor (FcRn), which is involved in the transfer of IgG across the placenta to the foetus, providing humoral immunity before birth and during the first 6 months of life; IgG is the only Ig to pass through the placenta (Table 20.2).

IgG molecules are further divided into four subclasses based on differences in amino acid sequence of the heavy chains, especially in the hinge region (Fig. 20.2). Numbered based on their relative abundance in human serum, IgG1 accounts for 60–70% of the total IgG, IgG2 14–20%, IgG3 4–8% and IgG4 2–6%. The variations in the hinge region of the IgG subclasses result in differences in their flexibility; the flexibility decreases in the order IgG3, IgG1, IgG4 and IgG2. These subclasses also differ in their ability to bind effector molecules. For example, IgG3 binds C1q with high affinity, IgG1 with intermediate affinity and IgG2 with low affinity, while IgG4 is unable to activate complement (Table 20.2). Similarly, variation in the affinity of binding to FcγRI, FcγRIIa, FcγRIIb, FcγRIIIa or FcγRIIIb contributes to the differences in FcγR-mediated effector functions of IgG subclasses.

20.2.2 Immunoglobulin of Class A (IgA)

Serum IgA molecules are predominantly present as monomers and are produced by bone marrow PLASMA CELLS. They are the second most common Ig after IgG, making up 15–20% of the total serum Ig (Table 20.2). Typically, IgA class switching is induced by transforming growth factor-β or by interleukin 5. IgA is also found as a dimer, synthesised by mucosal PLASMA CELLS of the intestinal lamina propria, the upper and lower respiratory tract or the urogenital tract. Polymerisation involves the association of two monomers with a joining chain (J chain) in PLASMA CELLS. Importantly, dimeric IgA is mainly released on mucosal surfaces in a secretory form (SIgA), which is detailed later in this chapter.

There are two subclasses of IgA: IgA_1 and IgA_2 (Fig. 20.2). These two subclasses mainly differ in their hinge region. While IgA_1 has a 26 amino acid hinge region, IgA_2 has only 13 amino acids. In serum, IgA_1 is the predominant form, making up 90% of the total IgA, whereas SIgA consists mainly of IgA_2 (which comprises up to 62% of SIgA in the colon).

It has been proposed that binding of monomeric serum IgA to the FcαRI present on MONOCYTES, MACROPHAGES, NEUTROPHILS and EOSINOPHILS induces inhibitory signalling pathways [19, 20]; thus serum IgA may serve an anti-inflammatory function in the systemic circulation. In aggregated form, serum IgA can activate the lectin pathway of the complement system as a result of binding to the carbohydrate recognition domain of mannan-binding lectin (MBL) [21] and induces activating signals in phagocytes via CD89 [19].

20.2.3 Immunoglobulins of Class M (IgM)

IgM molecules make up 10% of the total serum Ig content (Table 20.2). They are confined predominantly to the intravascular pool and are part of the primary, antigen-specific, humoral immune responses. Phylogenetically and ontogenetically, they are the earliest Ab molecules and may be secreted before exposure to Ag, leading to the term "natural antibodies". IgM has an extra constant domain in the heavy chain (C_H4) (Fig. 20.2), and when secreted, it exists predominantly as a pentamer of identical five-chain Ig protomers joined by a "J" chain and arranged into a planar structure (Fig. 20.3). Electron microscope studies showed a starlike structure, which, when bound to Ag, assumes a crablike conformation. Occasionally, IgM can also be found in a hexameric form (Fig. 20.3) [22].

IgM is a very potent activator of the classical complement pathway by its efficient binding of

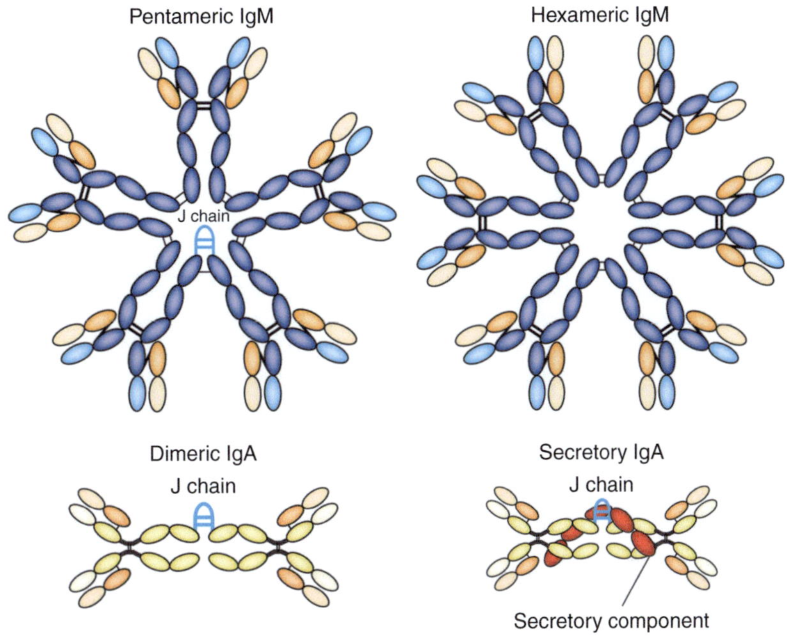

Fig. 20.3 Structures of polymeric IgM, IgA and secretory IgA (SIgA). The type of association of the secretory component with IgM is still unknown, as no crystal structure of SIgM is currently available

C1q through the C_H3 domain. IgM can also enhance primary and memory Ab responses as well as affinity maturation. However, it requires a functional complement system to achieve this [23].

Naturally occurring antibodies, termed natural antibodies (NAb), utilise germline-encoded genes in the variable region, are generated in the absence of external Ag stimulation and are the predominant form of IgM in the serum of uninfected newborns (note that newborns with in utero infections may make specific IgM against the pathogens, which are useful in diagnosis of prenatal infection). The natural antibodies are often broadly specific since they have not undergone affinity maturation, and may react with self and altered self components [24]. NAb are often but not exclusively of IgM isotype and throughout life fulfil important functions in tissue homeostasis (clearance of oxidatively damaged structures, apoptotic cells, intracellular proteins released from necrotic cells) and tumour surveillance (recognition of newly emerging carbohydrate, glycolipid or glycoprotein pattern on malignant cells). Other actions of IgM NAb molecules include proteolysis and modulation of B- and T-cell immune responses, as well as an accidental promotion of diseases arising from a sudden insult, such as in infarction or systemic inflammatory response syndrome which result in exposure/release of normally hidden self-antigens [25].

20.2.4 IgA and IgM Multimerisation

Multimerisation of immunoglobulins in vivo is generally mediated by association with the J chain. This peptide is expressed in almost all PLASMA CELLS. However, in mammals, only IgA and IgM are able to multimerise. Heavy chain tail pieces from other immunoglobulin isotypes do not allow for J chain association [26]. Pentameric IgM requires the J chain for secretion but not formation, while hexameric IgM has no requirement for the J chain. In IgM pentamers, the exact details of how the J chain associates with the IgM monomers are still not known, as the crystal structure is currently lacking (Fig. 20.3).

Regulation of J chain expression in B cells is not yet understood. But tissue-specific factors seem to contribute to its expression as dimeric IgA-producing PLASMA CELLS are only found at mucosal sites [26].

20.2.5 Secretory Properties of IgA and IgM

Polymeric IgA and IgM, as well as IgG, are found on mucosal surfaces, where dimeric IgA and pentameric IgM are associated with the so-called secretory component (SC), a 70 kDa polypeptide (Fig. 20.3).

To form SIgA, dimeric IgA released from mucosal PLASMA CELLS binds to the polymeric Ig receptors (pIgR) expressed on the basolateral side of mucosal epithelial cells. The pIgR-IgA complex is transported across mucosal epithelial cells, and, when released into the secretions at the apical side of the epithelial cells, the pIgR is cleaved, leaving the SC fragment covalently bound with the IgA dimer. Association with the SC protects IgA from proteolytic degradation. SIgA is the predominant Ig in seromucous secretions such as saliva, nasal and tracheobronchial secretions, colostrum, milk, tear fluid, intestinal secretions and genito-urinary secretions. It is the most prominent Ig produced at mucosal linings (and thus in the human body); approximately 3–5 g of IgA is secreted daily into the intestinal lumen and subsequently lost from the body.

SIgA is present as high- and low-affinity Ab. High-affinity SIgA helps defend mucous membranes of the intestine and nose against viral or bacterial infections and takes part in the neutralisation of bacterial toxins. Low-affinity SIgA has a more "homeostatic" function in shaping and controlling the commensal microbiota in the intestinal lumen by "immune exclusion". SIgA can neither bind nor activate complement, but it has been shown recently that by interacting with a dendritic cell receptor termed DC-SIGN, SIgA can induce tolerogenic mucosal immune responses [27].

Complementary to the immune function of IgA and IgM, SC has been reported to act as a microbial scavenger in a glycan-dependent manner. SC alone can prevent the binding and infectivity of *Escherichia coli* on epithelial cells as well as neutralisation of *Clostridium difficile* toxin A in vitro [28].

20.2.6 Immunoglobulin of Class D (IgD)

IgD is present at very low concentrations in serum; it accounts for less than 1% of the total serum Ig (Table 20.2). It is mostly found as a membrane-bound monomer, co-expressed with IgM, on naïve B cells. Upon activation of naïve B cells with specific Ag, class-switch recombination occurs, and the expression of IgD is lost. IgD class-switched B cells (IgM− IgD+) found in the upper respiratory mucosa secrete IgD that binds pathogenic respiratory bacteria and viruses. Mucosal immune responses are enhanced by cross-linking IgD bound to basophils or mast cells in the respiratory mucosa [29, 30].

20.2.7 Immunoglobulin of Class E (IgE)

IgE is found only in trace amounts in human serum (Table 20.2). Class switch to IgE production is classically induced by IL-4 derived from Th2-polarised T cells, but alternative pathways have also been described. Similar to IgM, IgE has an extra constant domain in the heavy chain, C_H4 (Fig. 20.2). IgE is found predominantly bound to the high-affinity Fcε receptor (FcεRI) on basophils and mast cells even prior to the interaction with its cognate Ag. When Ag/allergen binds to the IgE on mast cells and basophils, it causes the aggregation of the Fcε receptors and subsequent degranulation of the cell, with the release of vasoactive and chemotactic mediators. This results in allergic reactions such as hay fever, extrinsic asthma and the Prausnitz-Küstner skin reaction; in severe cases anaphylaxis can even be life-threatening. As well as its role in atopic allergy, IgE is important in protection against helminths and parasites. Indeed, elevation of IgE levels can be used as a diagnostic tool for parasitic infections. Circulating IgE cannot activate complement via the classic pathway (Table 20.2) [31].

IgE can bind to two Fc receptors, namely, FcεRI and FcεRII. IgE is mostly bound to the FcεRI; the FcεRI binding site on IgE is located at the interface between C_H2 and C_H3. The IgE binding affinity for the FcεRII (CD23), which is an Fc receptor found on monocytes, B cells and platelets, is lower than for FcεRI.

20.3 Glycosylation of Ig

Glycosylation of Ig has a number of important roles including the maintenance of structure and stability, receptor binding, Fc effector function, intracellular transport, secretion and clearance and cross-talk between the innate and adaptive immune pathways. The five classes of Ig are highly diverse in terms of the location and number of conserved N-linked glycosylation sites situated on the Fc and Fab portions (Figs. 20.1a and 20.2) [32]. Abnormalities in the glycosylation profiles of Ig molecules have been linked to certain diseases. For example, increases in agalactosyl glycoforms of IgG have been isolated from patients with rheumatoid arthritis, and the pathogenesis of IgA nephropathy is influenced by abnormal IgA_1 O-glycosylation and hence reduced IgA clearance. Furthermore, the glycosylation of all glycoproteins is also affected in congenital disorders of glycosylation; abnormal IgG glycoforms are used in the diagnosis of these disorders [32, 33].

20.3.1 IgG Glycosylation

All IgG molecules have a single conserved N-linked glycosylation site, Asn297 (Kabat numbering), in the Fc region, which is important in maintaining Fc effector function (Figs. 20.1a and 20.2). The oligosaccharides present in this region are of the complex bi-antennary type. Comprised of a core heptasaccharide, these oligosaccharides have the variable addition of fucose, galactose, bisecting *N*-acetylglucosamine and sialic acid. IgG_3 has an additional conserved glycosylation site in the C_H3 domain at Asn471 and additional O-glycosylations in the elongated hinge region [34]. Sialylation tends to be present in less than 10% of IgG. Glycosylation is critical for effector functions mediated via the FcγR, the C1q component of complement and MBL. For example, the absence of core fucose on the Fc glycans leads to increased binding to FcγRIIIa and enhanced ADCC [35]. Furthermore, although some studies suggest terminal sialylation of the Fc glycans is responsible for the anti-inflammatory properties of human IgG in mice [36, 37], this issue is still controversial. In contrast, Fc glycosylation does not appear to play a role in FcRn interactions [35]. In addition to Fc glycosylation, approximately 20–30% of normal plasma IgG is glycosylated in the Fab region. In contrast to the Fc oligosaccharides, Fab oligosaccharides are bisected, extensively galactosylated and substantially sialylated. This difference may be due to a lack of accessibility of the Fc portion to specific transferases [38].

Various factors influence the glycosylation of IgG, including age and gender. Age dependency is particularly clear; galactosylation and sialylation increases up to the age of 25 years and then decreases throughout life. Moreover, during pregnancy there is a temporary increase in IgG galactosylation and sialylation [39].

20.3.2 Glycosylation of Other Ig Molecules

IgA molecules possess two conserved N-glycosylation sites in the heavy chains, one in the C_H2 domain and one in the tailpiece of the Fc region (Fig. 20.2). In addition, there are further two to three N-glycosylation sites in IgA_2, up to five O-glycosylation sites in the hinge region of IgA_1, seven N-glycosylation sites in the secretory component and a further one in the J chain. In contrast to IgG, over 90% of IgA molecules are sialylated [40, 41]. The O-linked glycans in IgA_1 protect the extended hinge region from proteases. IgA_1 has a T-shaped structure that exposes the O-glycans for efficient interaction with bacterial adhesins.

The μ-chain of IgM molecules has five N-linked glycosylation sites; three are of the complex type and two of the oligomannose type. These contain predominantly monosialylated, bi-antennary structures. In addition, there is one N-linked glycosylation site in the J chain containing mainly sialylated bi-antennary glycans. Circulating predominantly as pentamers, the glycans clustered on the antigen-binding face are thought to allow IgM to agglutinate microorganisms in the serum [32].

IgE is the most heavily glycosylated Ig. It contains seven N-linked glycosylation sites in the ε-chain. These N-linked glycans result in a reduction in the flexibility of the IgE molecules, which is thought to control Ag binding and prevent unwanted and potentially fatal immune responses through inappropriate Ag binding [32].

IgD has three N-linked glycosylation sites in the δ-chain, and site-directed mutagenesis has shown that the presence of oligomannose glycans at Asn354 is of vital importance for the production of IgD. Absence of this site results in incomplete assembly and lack of secretion. The O-glycans in the hinge region of IgD expand the range of hinge conformation by increasing the angle between the Fab regions [32].

20.4 Pharmacokinetics of Ig

Catabolism provides the main route of elimination of circulating Igs [42]. Due to its large size, very little Ig is excreted in the urine, and only small amounts have been identified in the bile; it is assumed that approximately one-third of Ig is broken down in the liver, with another third in the intestines. The cellular compartment in close proximity to the blood and lymph, such as endothelial cells, constitutes other sites important in Ig catabolism. There are differences in sites of catabolism between Ig classes, with IgA, particularly SIgA, being different from other classes as it is present in high amounts in body excretions such as intestinal secretions and saliva, with a resulting high turnover.

20.4.1 Pharmacokinetics of IgG

In pharmacokinetic studies with IVIG administration, a two-compartment model has been proposed to describe the time course of IgG serum concentrations [43, 44]. After the initial rapid rise in serum IgG concentration with an IVIG infusion, a comparably rapid decrease follows (Fig. 20.4). This early, or α, phase which in humans occurs roughly during the first 5 days after infusion is primarily the result of distribution between the intravascular and extravascular spaces, although some catabolism also takes place during this time interval. The following β phase shows a slower decrease in IgG serum concentration, which is caused by the catabolism of IgG and thus determines the serum half-life of IgG. This general model has been confirmed using various IVIG preparations [44, 45]. A high inter-individual variability of pharmacokinetic parameters was found, which is not unexpected considering the complex mechanisms of catabolism and distribution of IgG [45, 46].

Studies using radioiodinated IgG have shown that IgG, except for IgG_3, can survive longer in the blood than any other serum protein, including other Ig classes, with a mean half-life of about 3 weeks (Table 20.2) for the subclasses IgG_1,

Fig. 20.4 Two-compartment model of IgG pharmacokinetics. (**a**) IgG is in equilibrium between the vascular and extravascular space. Synthesised in the bone marrow, IgG_{ENDO} diffuses into the lymph and then into the blood. Intravenously infused IgG (i.v. dose) enters the intravascular space directly, whereas subcutaneously administered IgG (s.c. dose) is absorbed from the subcutaneous tissue. The catabolism of IgG occurs in the vascular endothelium, liver and other areas. IgG may also be lost from the vascular space by other mechanisms such as protein loss in the intestines or urinary tract. Please note that IgG_{ENDO} refers to endogenous IgG. Reprinted from Postgraduate Medicine, 125, Landersdorfer CB, Bexon M, Edelman J, et al. Pharmacokinetic modelling and simulation of biweekly subcutaneous immunoglobulin dosing in primary immunodeficiency, page 55, Copyright 2013, with permission from Taylor & Francis Ltd. (www.tandfonline.com). (**b**) Serum IgG concentration over time after IVIG administration. The initial early α phase is due to the passage of IgG from the vasculature to the lymph and extracellular fluid compartments in addition to catabolism. The late β phase is due to IgG catabolism. From Bonilla [43], with modifications. (**c**) Serum IgG concentration over time after SCIG administration. Fluctuations in IgG levels are essentially absent with SCIG administration, and mean trough levels are higher compared with IVIG treatments

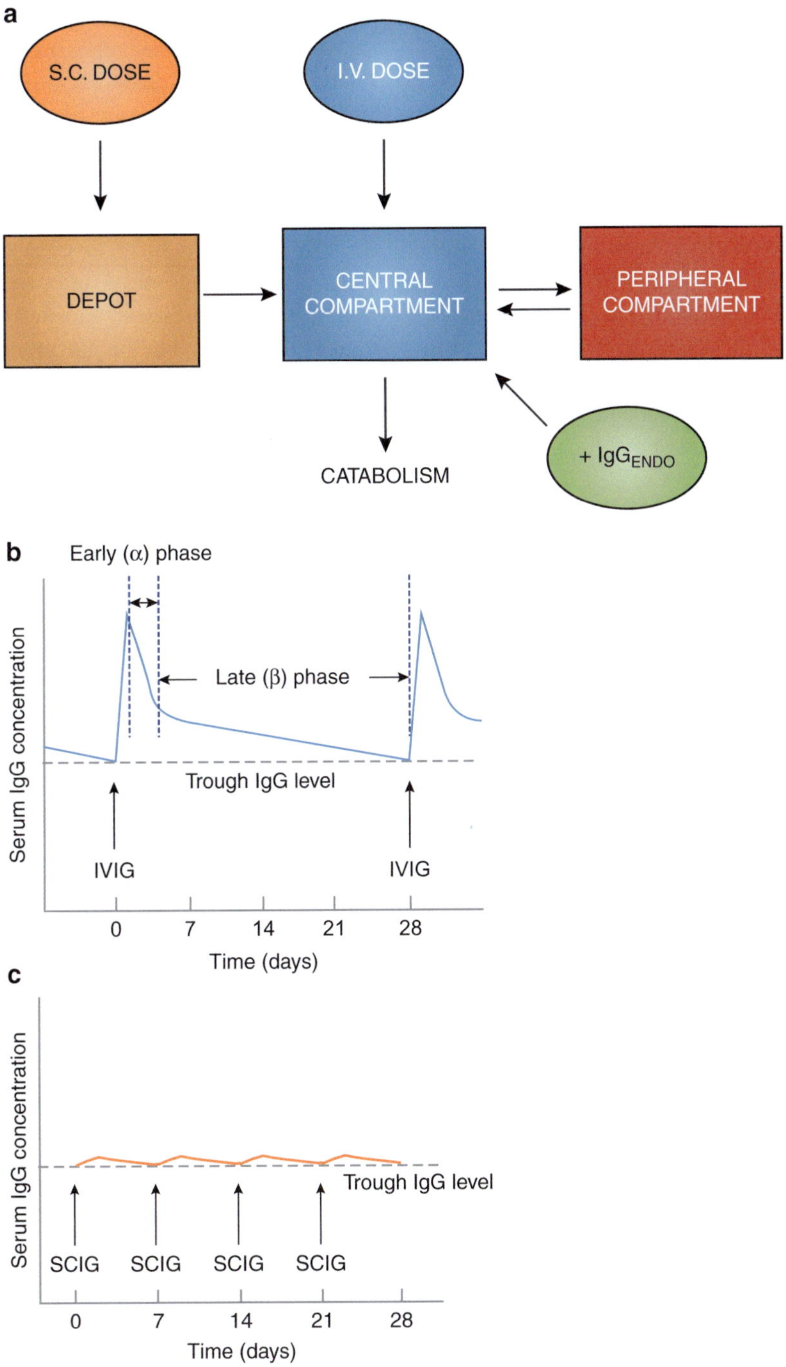
a
S.C. DOSE
I.V. DOSE
DEPOT
CENTRAL COMPARTMENT
PERIPHERAL COMPARTMENT
CATABOLISM
+ IgG$_{ENDO}$
b
Early (α) phase
Late (β) phase
Serum IgG concentration
Trough IgG level
IVIG
IVIG
0
7
14
21
28
Time (days)
c
Serum IgG concentration
Trough IgG level
SCIG
SCIG
SCIG
SCIG
0
7
14
21
28
Time (days)

IgG_2 and IgG_4 and a shorter half-life for IgG_3 (7.5–9 days) [42, 43].

The long half-life of IgG_1, IgG_2 and IgG_4 is due to their binding to the FcRn. Initially identified for its role in the transfer of IgG from mother to foetus across the placenta, the FcRn has also been found to prolong the half-life of IgG (and albumin) by recycling these proteins back into the circulation after endocytosis. Thus, as it is expressed on a multitude of cells and tissues such as leukocytes, endothelium, epithelium, interstitium, kidneys, lungs and blood-brain barrier cells [47, 48], it helps to maintain high IgG concentrations in the circulation at a comparably low synthesis rate [47–49].

The proposed cycle of IgG protection through FcRn binding is summarised in Fig. 20.5a. In brief, various cells, as mentioned above, take up IgG (monomeric or antigen-bound as immune complexes) by macropinocytosis or by a saturable FcRn-dependent mechanism. Unbound IgG is directed to lysosomes where degradation occurs; IgG bound to FcRn enters recycling endosomes which are kept away from lysosomes, preventing degradation. In these endosomes, the IgG remains intact and is subsequently returned to the plasma or interstitial fluid and is finally released from the FcRn [51]. This pathway may explain why IgG half-life decreases as IgG serum concentrations rise: increasing concentrations of IgG result in FcRn saturation, and hence more IgG molecules enter lysosomes where they are catabolised [47]. Humans may have polymorphisms in the promoter region which lead to different levels of expression of FcRn, affecting catabolism of IgG; and when mutations occur which prevent expression of FcRn, a state of hypercatabolic hypoproteinaemia (with low levels of albumin and IgG) occurs [52–54]. The importance of FcRn saturation level on the length of IgG half-lives has also been demonstrated in a mouse model of autoimmune disease [55]. Monoclonal mouse IgG was applied in constant amounts as a model autoantibody. Treatment of mice with high-dose IVIG resulting in high IgG serum concentration reduced the half-life, and thus the serum concentration, of the monoclonal IgG. This provides evidence for one of several mechanisms of action of high-dose IVIG treatment, i.e. the reduction of autoantibody serum concentrations in autoimmune disease. On the other hand, in diseases such as PID, treatment with human IgG generally results in lower than normal IgG serum concentrations, with resulting serum half-lives longer than normal [46].

IgG synthesis can be inhibited in multiple ways; abnormal increases in IgG concentrations may result in reduced activation, lifespan and proliferation of B cells and downmodulation of their differentiation to plasma cells [56]. In contrast, a reduction in IgG concentration, for instance, as a result of plasmapheresis, may increase Ig synthesis.

20.4.2 Ig Distribution

Close to half of the total body pool of IgG can be found in the intravascular space (Table 20.2), and the amount of IgG transferred from the intravascular pool to the extravascular space has been reported to be between 27% and 39% per day [42]. As IgG does not cross the blood-brain barrier under normal conditions, IgG concentrations in the cerebrospinal fluid are low, in the range of 25 mg/L (compared with about 10 g/L in serum) [57]. However, in certain neurological and inflammatory diseases such as bacterial meningitis, IgG can readily enter the cerebrospinal fluid [48, 49] due to a transient opening of the blood-brain barrier. In these situations, FcRn is thought to play a role in removal of IgG from the central nervous system. Although movement of IgG from intravascular space to immunologically privileged compartments, such as the brain, is comparably low, FcRn may play a role in the transport of IgG to such compartments [48, 49]. Polarised epithelial cells expressing FcRn can be exploited, e.g. to supply Fc-containing chimeric molecules to the bloodstream by inhalation and into the CNS in inflammatory conditions. In addition, bidirectional transport has been shown in a human FcRn-expressing mouse model. Antigens present in the intestinal lumen were bound by specific antibodies previously transcytosed by FcRn from the lamina propria to the lumen, and then back, as antigen-antibody complexes to the lamina propria, for delivery to antigen-presenting cells (Fig. 20.5b).

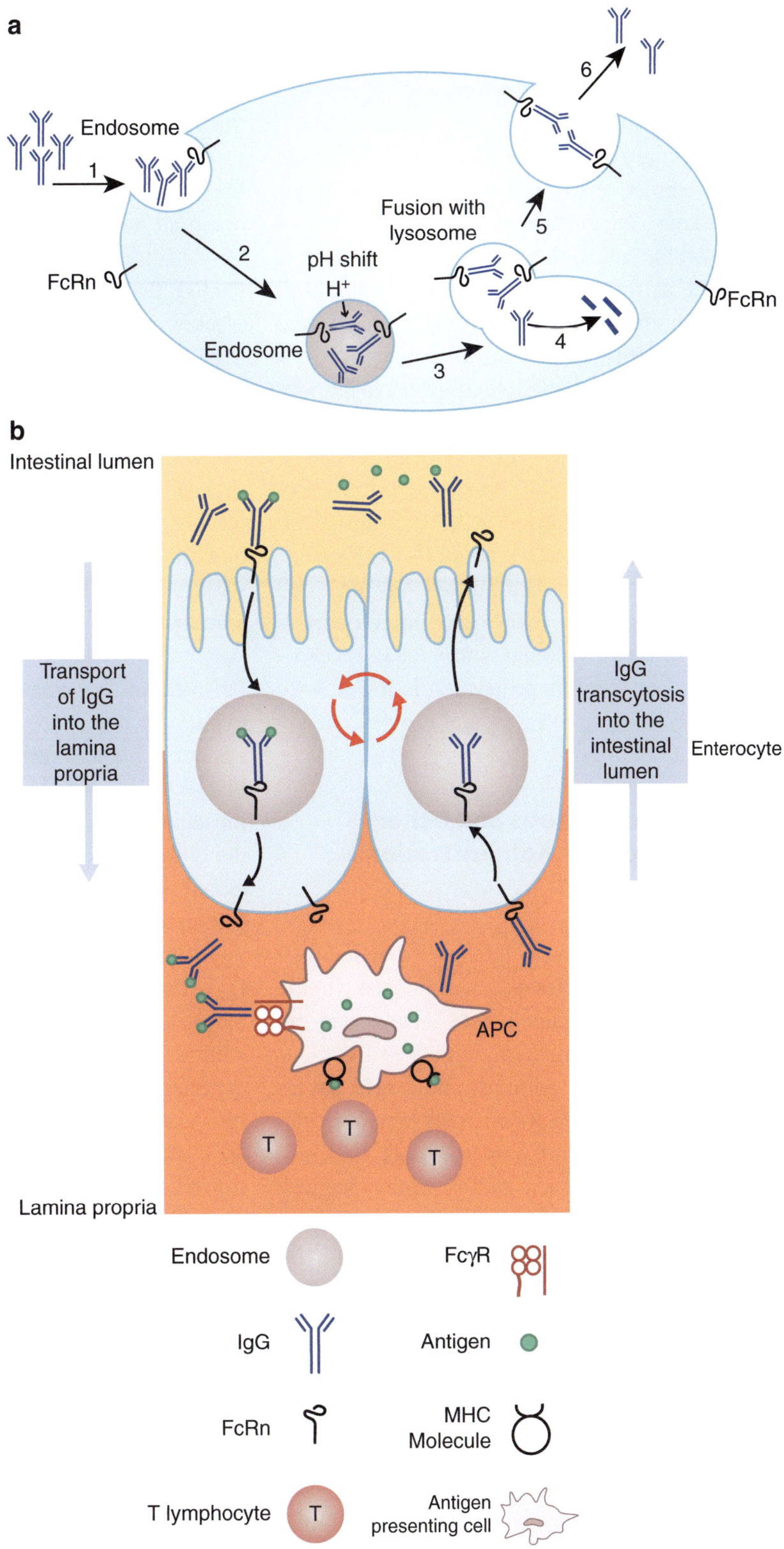

Fig. 20.5 Recycling of IgG. (**a**) Proposed mechanism for IgG recycling by FcRn. (1) IgG molecules enter cells (e.g. vascular endothelial cells, muscle cells, etc.) via pinocytosis (e.g. fluid-phase endocytosis); (2) H^+ enters the endosome, reducing pH, and IgG binds with FcRn; (3) the endosome fuses with the lysosome, but FcRn-bound IgG is not released to the lysosome; (4) unbound IgG is released into the lysosome and catabolised by proteases; (5) the endosome fuses with the plasma membrane, and the FcRn-IgG complex is exposed to neutral pH; and (6) IgG is released into plasma (or interstitial fluid). From Lobo et al. [50], with modifications. (**b**) Gut transcytosis of IgG. IgG molecules are trancytosed into the gastrointestinal lumen by FcRn located in enterocytes. In the lumen, IgG molecules bind to antigen molecules to form antigen-IgG immune complexes. FcRn then trancytoses the immune complexes back to the lamina propria where they are delivered to antigen-presenting cells (APC). APCs loaded with antigen migrate to draining lymph nodes to prime a T-cell response

This kind of transport, although small with respect to the amount of IgG transported, has been shown to be physiologically relevant [48].

IgG constitutes one of the immunoglobulin classes excreted into milk and colostrum in humans and rodents, with some species differences. Thus, an FcRn-dependent absorption of IgG from the gut to the blood of the newborn is observed in rodents and cows, but only minimally in humans. In humans, IgG is transferred prenatally solely across the placenta. FcRn is responsible for placental IgG transport in humans [48, 49]. At term, three- to fivefold higher serum IgG levels are found in the neonate than in the mother. In the neonate, serum concentrations of IgG_1 and IgG_4 are higher, IgG_3 equal and IgG_2 lower than in the serum of the mother [58, 59].

The intravascular distributions of IgM and IgA vary from IgG. More than three-quarters (76%) of IgM can be found within the intravascular space. For IgA this proportion is lower, with only 40–42% being intravascular (Table 20.2).

20.4.3 Subcutaneous and Other Routes for Administration of IgG

The s.c. administration of IgG has expanded with the prospect of increased convenience (home therapy), more sustained IgG levels and an improved safety profile. In contrast to i.v. treatment, following s.c. administration, IgG passes first into the lymphatic system and from there into the bloodstream [60]. Comparisons of IgG bioavailability using the i.v. and s.c. routes suggest that the bioavailability by the s.c. route is about 2/3 of that by the IV route [61]. Pharmacometric modelling was performed to simulate Ig administration through different routes using different doses and frequencies of application [62] and has been performed based on pharmacokinetic results from multiple clinical studies with i.v. or s.c. administration of IgG to PID patients [63]. The model confirmed clinical results showing that fluctuations in IgG levels are essentially absent with weekly or more frequent subcutaneous immunoglobulin (SCIG) administration, and mean trough levels are higher compared with IVIG treatments at the same monthly doses (Fig. 20.4b, c) [43]. This is due to the higher frequency of SCIG administration (once per week vs. once per 3–4 weeks) and the slow release from the s.c. depot into the intravascular space, with maximal plasma concentrations being reached 36–72 h after infusion [64]. This model demonstrated that administering the equivalent of a weekly maintenance SCIG dose in 1, 2, 3, 5 or 7 instalments per week resulted in overlapping steady-state concentration-time profiles [62]. With simulation of biweekly administration of twice the weekly dose, there was only a minimal reduction in the trough serum IgG compared with weekly administration, and the FDA allowed labelling for biweekly administration of 2× the weekly dose. This enables flexibility of SCIG administration frequency/dose to patients. Another modelling approach took into account FcRn-dependent transcytosis from the interstitial to intravascular space, in addition to the lymphatic system in the absorption of monoclonal IgG antibodies administered by the subcutaneous or intramuscular route; it was shown that absorption via the lymphatic system is the major route of monoclonal antibody delivery, with FcRn-mediated transcytosis playing a negligible role [65].

Little is known about oral absorption of IgG. The acid conditions of the stomach provide a hostile environment for any protein. Upon oral administration of human IgG to PID patients experiencing rotavirus infections, about 50% of IgG was found in stool samples, approximately 40% of which was in a higher molecular form, suggesting immune complexes with antigen in the lumen [66]. At the same time, free rotavirus protein decreased in the stools. Similar results were obtained using bovine IgG containing anti-*Clostridium difficile* antibodies [67, 68]. Adsorption from the gut into the blood is minimal in humans, as compared to mice, either by FcRn mediated as discussed earlier or by diffusion, due to the large molecular size and high polarity of IgG.

Aerosol delivery of Ig directly to the upper and lower respiratory tracts might be another relevant route to target local infections.

20.4.4 Pharmacokinetics of Other Isotypes

Approximately half of all IgA molecules retained in the body are found in the intravascular space. They have an average half-life of 6 days, which is shorter than that of IgG molecules (see Table 20.2) [42, 43, 69]. One of the main reasons for this shorter half-life is that IgA molecules do not bind to FcRn and are not protected from degradation [70]. Considering this shorter half-life of IgA, and the lower serum concentration of IgA compared to IgG, the synthesis rates of plasma IgA and plasma IgG are similar. However, intravascular IgA represents only about one-third of the total IgA, with two-thirds produced at mucosal sites and lost from the body in secretions. Thus, in humans, more IgA is synthesised per day than IgG and IgM combined [71]. IgA at mucosal sites is present primarily as IgA dimers with an associated secretory component (see Sect. 20.2). This secretory component protects IgA from the proteolytic degradation which would be found with other immunoglobulin classes such as plasma IgA or IgG in these environments [72].

Very little is known about the pharmacokinetics of the other Ig classes. In terms of distribution, unlike IgG, both IgM and IgD are found predominantly in the intravascular space (Table 20.2); for IgM this is due to its large molecular weight [42]. IgE has a very short half-life because it is rapidly sequestered by the high-affinity FcεR.

20.5 Ig Preparations for Medical Use

20.5.1 The History of Ig Preparations

Preparations containing Ab for "serum therapy" were introduced by Emil von Behring and Shibasaburo Kitasato more than 100 years ago [73]. The first "antitoxins" were heterologous serum preparations from hyperimmunised animals with Ab specificity to diphtheria and tetanus toxins. Unfortunately, the initial success, particularly with the diphtheria antisera, was marred by the increasing incidence of side effects from the heterologous animal serum proteins, known as serum sickness. As a result, specific human convalescent and hyperimmune preparations were developed.

The initial Ig preparations derived from human plasma were for intramuscular (i.m.) injection only and were not suitable for i.v. administration due to adverse reactions, which were often severe and even fatal. Such adverse reactions were attributed to the presence of aggregates in the Ig preparations [7]. Ig aggregates can activate the classical complement pathway and cause generation of anaphylatoxins (C3a, C5a). In addition, the presence of endogenous prekallikrein activator (PKA), known to induce hypotension, further contributed to the unfavourable adverse event profile, making these intramuscular IgG (IMIG) preparations unsuitable for i.v. administration.

Initial IVIG preparations developed in the 1970s underwent plasma fractionation by cold ethanol precipitation followed by chemical modification or protease digestion, including trypsin, plasmin and pepsin to improve tolerability following i.v. administration. Unfortunately, these chemical treatments led to modifications of the Ig molecule and the loss of fully active Fab and Fc functions. The efficacy of these early preparations was reduced due to shortened half-lives and a marked reduction in inherent complement binding [74]. Although effective in the treatment of some diseases, they were not as functional as native, intact Ig. As a result, they became obsolete for all markets. Protease-treated Ig preparations contained predominantly fragmented Ig, including bivalent $F(ab')_2$ (Fig. 20.1c), and therefore were mainly devoid of Fc-mediated activity, which is essential for the full spectrum of IVIG effector functions. Moreover, although better tolerated, these preparations had a very short half-life of 24 h or less, making them generally unsuitable for Ab replacement therapy. Subsequent Ig preparations, which were obtained by treatment with only trace amounts of pepsin, resulted in decreased aggregates and more than 90% intact, fully functional immunoglobulins. Such preparations are well tolerated when i.v. administered and display the full spectrum of IVIG activities. These Ig preparations are not stable in solution but when lyophilised have an extended shelf life of more than 3 years, even at temperatures up to 30 °C. Such IVIG products have been successfully used for more than three decades to treat a variety of immune system diseases and continue to be used today in some markets.

20.5.2 Purification Process and Ig Formulations Today

More recent developments in the production of IVIG rely on more gentle isolation and purification procedures [75]. In current protocols, fewer steps of cold ethanol fractionation are used, followed by a series of individual purification procedures that may also be used in combination. Polymers such as polyethylene glycol, which support selective precipitation and stabilisation, may also be used. A highly selective purification of Ig is obtained by fractionation using octanoic acid (caprylate). The procedure efficiently removes/inactivates serine proteases. This is particularly important as it depletes coagulation factors [76] which are responsible for potential thromboembolic side effects (procoagulant activity) of IVIG (see Sect. 20.9). State-of-the-art production processes use cation and/or anion exchange chromatography as a final IgG purification step, particularly to remove non-IgG immunoglobulins (i.e. IgA, IgM and IgE) and protein aggregates. Recently developed immunoaffinity chromatography (IAC) techniques are also used to specifically bind and remove antibodies specific for blood group A and B antigens (A/B isoagglutinins) which are considered at least partially responsible for rarely observed haemolytic side effects after high-dose treatment with IVIG (see Sect. 20.9.4) [77–79]. All these techniques yield highly purified IVIG preparations with nearly 100% unaltered and functionally intact IgG, an IgG subclass distribution characteristic of human serum and a normal half-life of IgG.

Virus inactivation and removal steps are integrated into the IVIG production process. More detail on this topic is given later in the chapter (Sect. 20.10).

Lyophilised IVIG products offer excellent stability and have a shelf life of up to several years at room temperature. A disadvantage of lyophilised products is that they are less convenient because they require reconstitution prior to administration. In addition, once reconstituted, these preparations are stable only for a matter of hours.

With this in mind, more recently developed IVIG products have been formulated as ready-to-use liquids. While the first liquid preparations of IVIG required refrigeration and had a limited shelf life, more recent formulations are stable at room temperature for extended time periods. Studies have shown that by reducing the pH of the formulation to moderately acidic and adding appropriate stabilisers, especially amino acids such as glycine or proline, to the formulation, the stability of the liquid IVIG can be maintained for up to 3 years [80, 81].

Liquid IVIG preparations should not only have appropriate long-term stability but also need to be clinically well tolerated during their entire shelf life. Idiotype/anti-idiotype Ab dimers are formed in liquid IVIG isolated from pooled plasma of numerous donors [82, 83]. These dimers are formed when the Ag-binding region of one IgG molecule binds to the variable region of another. High levels of dimeric IgG in IVIG have been described to be associated with adverse reactions such as headache, fever and flushing observed during i.v. infusion [84, 85]. Therefore, in order to maintain good clinical tolerability of liquid IVIG preparations, the formation of IgG dimers has to be controlled by choosing appropriate formulations and storage conditions.

Tables 20.3 and 20.4 provides characteristics of a selection of currently available IVIG preparations.

For certain patients, such as children, patients requiring frequent dosing or with adverse effects of IVIG infusions or individuals with poor venous access, s.c. administration of Ig is more convenient [64]. Moreover, a number of validated health-related quality of life questionnaires have found that many patients prefer SCIG to IVIG because of its convenience [64, 86]. Although some of the current IVIG preparations have also been licensed for s.c. administration, specific

Table 20.3 Various i.v. immunoglobulin products

Product	Manufacturing process	Viral safety	Formulation/stabiliser	Storage	Manufacturer
Bivigam	Cold ethanol fractionation, IEX	Precipitation/filtration, S/D, virus filtration	Liquid 10% protein/glycine, polysorbate 80	Refrigerated (2–8 °C), 2 years	Biotest
Clairyg	Cold ethanol fractionation, octanoic acid fractionation, IEX, IAC	Precipitation/filtration, S/D, virus filtration	Liquid 5% protein/mannitol, glycine, polysorbate 80	Room temperature, 2 years	LFB
Flebogamma DIF	Cold ethanol fractionation, PEG precipitation, IEX	Precipitation/filtration, S/D, virus filtration, low pH, pasteurisation	Liquid 5% and 10% protein/sorbitol	Room temperature, 2 years	Grifols
Gammagard Liquid (Kiovig)	Cold ethanol fractionation, IEX	Precipitation/filtration, S/D, virus filtration, low pH	Liquid 10% protein/glycine	Room temperature, 2 years; refrigerated (2–8 °C), 3 years	Shire/Takeda
Gammagard SD	Cold ethanol fractionation, AIEX	Precipitation/filtration, S/D	Lyophilised 5 g and 10 g/glucose, albumin, glycine, PEG, polysorbate 80	Room temperature, 2 years	Shire/Takeda
Gammaked	Cold ethanol fractionation, octanoic acid fractionation, AIEX	Precipitation/filtration, AIEX	Liquid 10% protein/glycine	Refrigerated (2–8 °C), 3 years; room temperature (<25 °C), 6 months	Kedrion BioPharma
Gammaplex	Cold ethanol fractionation, IEX	Precipitation/filtration S/D, low pH, virus filtration	Liquid 5% protein/glycine, sucrose, sodium acetate, sodium n-octanoate	Refrigerated (2–8 °C), 4 months; room temperature (<25 °C), 3 months	Bio Products Laboratory
Gamunex-C	Cold ethanol fractionation, octanoic acid fractionation, AIEX	Precipitation/filtration, low pH, AIEX	Liquid 10% protein at low pH/glycine	Refrigerated (2–8 °C), 3 years; room temperature (<25 °C), 6 months	Grifols
Intragam P	IEX	Pasteurisation, low pH	Liquid 6% protein at low pH/maltose	Refrigerated (2–8 °C), 2 years	CSL Behring
Intragam 10 NF	IEX	Pasteurisation, low pH, virus filtration	Liquid 10% protein at low pH/glycine	Refrigerated (2–8 °C), 2 years	CSL Behring

(continued)

Table 20.3 (continued)

Product	Manufacturing process	Viral safety	Formulation/stabiliser	Storage	Manufacturer
Intratect	Cold ethanol fractionation, octanoic acid fractionation, IEX	Precipitation/filtration, S/D, virus filtration	Liquid 5% and 10% protein/glycine	Room temperature (<25 °C), 2 years	Biotest
Octagam	Cold ethanol fractionation, octanoic acid fractionation, IEX	Precipitation/filtration, S/D AIEX chromatography	Liquid 5% and 10% protein/maltose	Room temperature (<25 °C), 2 years	Octapharma
Privigen	Cold ethanol fractionation, octanoic acid fractionation, AIEX, IAC	Precipitation/filtration, low pH, virus filtration	Liquid 10% protein/L-proline	Room temperature (<25 °C), 3 years	CSL Behring
Sandoglobulin, Carimune NF	Cold ethanol fractionation, low pH/pepsin	Precipitation/filtration, low pH, virus filtration	Lyophilised 3, 6, 12 protein/sucrose	Room temperature (<30 °C), 3 years	CSL Behring
Tégéline	Cold ethanol fractionation, low pH/pepsin	Precipitation/filtration, low pH, virus filtration	Lyophilised 0.5, 2.5, 5, 10 g protein/sucrose	Room temperature, 3 years	LFB Biomanufacturing

(A)IEX (an)ion-exchange chromatography, *IAC* immunoaffinity chromatography, *i.v.* intravenous, *PEG* polyethylene glycol, *S/D* solvent/detergent treatment

Table 20.4 Various s.c. immunoglobulin preparations

Product	Manufacturing process	Viral safety	Formulation/stabiliser	Storage	Manufacturer
Gammagard liquid	Cold ethanol fractionation, IEX	Precipitation/filtration, S/D, virus filtration, low pH	Liquid 10% protein/glycine	Room temperature, 2 years; refrigerated (2–8 °C), 3 years	Shire/Takeda
Gammanorm	Cold ethanol fractionation	Precipitation/filtration, S/D	Liquid 16.5% protein/glycine, polysorbate 80	Refrigerated (2–8 °C), 3 years	Octapharma
Gamunex-C	Cold ethanol fractionation, octanoic acid fractionation, AIEX	Precipitation/filtration, AIEX, low pH	Liquid 10% protein at low pH/glycine	Refrigerated (2–8 °C), 3 years; room temperature (<25 °C), 6 months	Grifols
Hizentra	Cold ethanol fractionation, octanoic acid fractionation, AIEX, IAC	Precipitation/filtration, low pH incubation, virus filtration	Liquid 20% protein/L-proline, polysorbate 80	Room temperature (<25 °C), 30 months	CSL Behring
HyQvia	Cold ethanol fractionation, IEX	Precipitation/filtration, S/D, virus filtration, low pH	IgG: Liquid 10% protein/glycine Recombinant human hyaluronidase: Liquid 160 U/ml	Refrigerated (2–8 °C), 2 years	Shire/Takeda
Subcuvia	Cold ethanol fractionation	Precipitation/filtration, S/D	Liquid 16% protein/glycine	Refrigerated (2–8 °C), 30 months	Shire/Takeda
Subgam	Cold ethanol fractionation	N/A	Liquid 16% protein/glycine, polysorbate 80	Refrigerated (2–8 °C), 2 years	Bio Products Laboratory

(A)IEX (an)ion-exchange chromatography, *IAC* immunoaffinity chromatography, *s.c.* subcutaneous, *S/D* solvent/detergent treatment

formulations of Ig for s.c. administration have been developed and approved by the US Food and Drug Administration (FDA) and other authorities worldwide. To compensate for the limited volume that can be administered by the s.c. route, these SCIG preparations contain higher Ig concentrations of up to 20% (Tables 20.3 and 20.4) [87].

The starting material for the production of polyvalent human Ig (SCIG or IVIG) is a pool of plasma from thousands of healthy human donors. The plasma used for producing therapeutic Ig is obtained either by plasmapheresis ("source" plasma) or from whole blood donations ("recovered" plasma) [87]. Despite the use of a large number of donors, Ig preparations still demonstrate some variability in epitope specificity and antibody concentration to certain antigens. The antibodies contained in each Ig preparation reflect the cumulative Ag experience of the donors. In addition, the specificities and antibody titres of a preparation depend on the donor population in the region where and when the plasma was collected. For example, vaccination programmes and decreases in antibody responses to natural infections have significantly influenced some antibody titres in pooled Ig preparations. Thus, infection with the measles virus triggers a higher Ab titre than the vaccine. This has resulted in a decline of mean measles titres since the introduction of a widely accepted vaccination programme. In comparison, mean Ab titres to hepatitis B virus have increased significantly since the introduction of a vaccine in the mid-1990s [88].

High-titre antibodies to defined antigens may be isolated from the plasma of specifically immunised donors, from the plasma of convalescents or from screened plasma containing high titres of a desired specificity. Such hyperimmune plasma from volunteers immunised with approved vaccines against diseases such as hepatitis B, varicella zoster, tetanus toxoid or rabies provides a therapeutic Ig enriched for higher titres against certain pathogens or toxins [89]. The use of such products for passive prophylaxis against certain infectious diseases is of particular interest. While the use of animal hyperimmune sera and antitoxins often result in unwanted side effects due to their non-human origin [5], hyperimmune Ig from human sources are similar to regular IVIG regarding tolerability but contain high levels of Ab against specific Ag. In certain conditions, such hyperimmune Ig preparations can be more beneficial than normal Ig [89]. A good example is anti-RhD IgG prepared from the plasma of RhD-negative human donors who agree to be immunised with RhD-positive red blood cells. These specific anti-RhD Ig preparations have been successfully used for decades to prevent RhD incompatibility between mother and child in the course of pregnancy or during delivery or after mistransfusion of RhD-positive blood to RhD-negative recipients. For one product a dedicated production process has been developed that selectively concentrates anti-D IgG and provides higher yields and purity of this hyperimmune Ig than conventional manufacturing processes for total plasma-derived IgG [90].

Table 20.1 provides details of a selection of currently available hyperimmune Ig preparations.

20.6 Potential Alternatives to Plasma-Derived Immunoglobulins

Alternative approaches to IVIG/SCIG are under active evaluation, with the aim of achieving similar benefits in immunodeficiency or autoimmune inflammatory diseases. These include the testing of a new generation of gene therapy vectors to correct genetic mutations causing immunodeficiency [91, 92] and alternative compounds able to modulate an immune system in a chronic inflammatory state, such as helminthic probiotics [93] or subunits/specific proteins thereof.

Early stage research is underway to capture and express the complete immunoglobulin repertoire of human donors, in order to create recombinant IVIG. With recent advances in high-throughput single cell analysis and mass sequencing, it is now possible to clone complete immunoglobulin repertoires [94]. However, complete and reproducible expression of large quantities of recombinant, polyclonal immunoglobulins has not yet been achieved. Many technical and regulatory hurdles will need to be overcome before the first testing of recombinant IVIG in humans can be carried out.

While the ANTIGEN-binding Fab portion of IgG is essential in replacement therapy, many anti-inflammatory effector functions of IgG are mediated by the Fc portion [14] (see also Sect. 20.8). Thus, various strategies are applied to recreate one or multiple Fc-mediated effector functions with molecules based on IgG-Fc fragments. For example, De Groot et al. [95] have described conserved sequences in the Fc (and the constant region of the Fab) portion of human IgG that activate natural regulatory T cells. These peptides, termed "Tregitopes", can be expressed synthetically or as recombinant fusion proteins and downmodulate inflammatory responses in vitro and in some in vivo animal models [96]. It is hypothesised that these are some of the active components of IVIG that mediate anti-inflammation. Consequently, they may represent a potential synthetic pharmaceutical product.

An alternative approach is fusion of the human IgG_2 hinge region to human IgG_1 Fc or mouse IgG_{2a} Fc [97], leading to expression of multimerised Fc fragments (called "stradomers") that bind Fc receptors with high avidity. These have demonstrated efficacy in animal models of arthritis and immune thrombocytopaenia [97], in a model of inflammatory neuropathy (experimental autoimmune neuritis) [98] and in *Myasthenia gravis* [99]. Interestingly, efficacy was achieved with low doses of approximately 50 mg protein/kg body weight, compared to the standard dose of 1–2 g/kg body weight for IVIG in acute anti-inflammatory indications. If translated to humans, this significant reduction in dosing could represent an advantage of multimerised Fc over traditional IVIG therapy in selected indications. Additional efforts to generate multimerised recombinant Fc fragments are underway, e.g. as described in Mekhaiel et al. [100].

Other recombinant molecules include full-length IgG or IgG-Fc fragments engineered to bind FcRn with increased affinity, testing the hypothesis that blocking FcRn will increase the elimination of pathogenic autoantibodies [101].

After showing preclinical efficacy [102], a soluble Fc-gamma receptor IIb (sFcγRIIb; termed SM101), able to compete with activating Fcγ receptors for pathogenic immune complexes, has been tested in Phase II clinical trials in patients suffering from systemic lupus erythematosus and immune thrombocytopaenia. It is important to note that, apart from the last example, all other recombinant IVIG alternatives are in preclinical evaluation. In this highly dynamic field, it is expected that additional molecules will be evaluated in human trials over the next few years.

20.7 Indications for Ig

IVIG was first introduced as a replacement therapy for immunodeficiency diseases [8, 103]. Following the observation that IVIG replacement therapy raised the platelet count in patients with antibody deficiency and concomitant immune thrombocytopaenia, and the success of IVIG treatment in resolving acute systemic symptoms and late inflammatory damage in Kawasaki syndrome [104], IgG has also become the standard treatment for a wide variety of autoimmune and inflammatory diseases. IgG is administered i.v. or s.c. at relatively low doses when used as antibody replacement therapy (usually 400–800 mg/kg/month) [105–108] and at high doses when used as an "immunomodulatory/anti-inflammatory" therapy (up to 1–2 g/kg/dose) [109, 110].

20.7.1 Ig Replacement Therapy in Primary Immunodeficiency Diseases (PIDDs)

IgG, administered i.v. or s.c., is indicated as replacement therapy in patients with PID characterised by reduced or absent antibody production and recurrent infections [111]. More than 140 distinct, inherited PID disorders have been identified [111–113]. IgA deficiency is the most common antibody deficiency (for which IgG therapy is not indicated) followed by common variable immunodeficiency (CVID) [111].

20.7.1.1 X-Linked Agammaglobulinaemia (XLA)

XLA, first described by Bruton in 1952 [8], is caused by mutations in an intracellular tyrosine kinase important in early B-cell differentiation and development, resulting in B-cell deficiency and severely impaired antibody production. The hallmark of XLA is repeated bacterial and viral infections, particularly of the respiratory tract [114]. Chronic sinusitis occurs in nearly all patients, and many develop bronchiectasis, even if acute infections are successfully prevented and/or treated. Intestinal infections caused by bacteria such as *Campylobacter jejuni* and *Giardia lamblia* are also common, as is chronic arthritis [114]. IgG treatment is of clear benefit, reducing both acute and chronic infections. Meta-analyses of large datasets have found that in PID, in general, the number and severity of acute infections are inversely proportional to the dose of IgG [115, 116]. However, individual patients vary in terms of the IgG levels and doses required to remain free from infection [117, 118]. With current therapy, most XLA patients are now living into adulthood. However, delays in diagnosis and initiation of IgG replacement therapy and lower than optimal IgG doses can result in bronchiectasis and slowly progressive chronic obstructive lung disease [119]. Higher doses of IgG replacement are generally used in patients with chronic lung and/or sinus disease, and careful monitoring with pulmonary function tests and high-resolution chest computed tomography scans should be performed to be sure that chronic, as well as acute, infections are adequately controlled by the IgG replacement regimen [106, 108, 119].

20.7.1.2 Hypogammaglobulinaemia

Hypogammaglobulinaemia may occur due to a large number of defects and most commonly presents as the heterogeneous group of disorders termed "common variable immunodeficiency (CVID)" [120–122]. CVID patients generally have readily detectable B cells, but they do not function properly. CVID patients present with recurrent and often severe infections such as sinusitis, otitis media, pneumonia and meningitis. In addition, CVID patients are distinguished from XLA patients by the frequent occurrence of non-infectious complications such as autoimmune cytopenia, other types of autoimmunity, granulomatous disease, hepatosplenomegaly/lymphadenopathy and an increased incidence of malignancies [120, 122]. The benefits of IgG replacement in reducing the incidence and severity of infections are clear. In recent years there has been growing recognition that the optimal dose and serum level of IgG differ between patients, and thus therapeutic regimens should be individualised [117, 118, 123]. Although high-dose IgG treatment has anti-inflammatory effects and is used in many autoimmune diseases, IVIG is not effective in treating granulomatous/lymphocytic interstitial lung disease in CVID patients [124].

Another family of PID, referred to as hyper-IgM syndromes, occur when class switching of immunoglobulin heavy chain gene segments is impaired by mutations in lymphocyte surface signalling molecules that induce class switching or in the genes for DNA processing enzymes necessary for class-switch recombination [125, 126]. Significant IgG and specific antibody deficiency may occur in the presence of normal or elevated IgM in patients with hyper-IgM syndromes, and those with mutations in CD40L expression on T cells may have also a broader spectrum of "opportunistic infections". Antibody replacement is the standard of care for most of these patients, although HSCT may result in a "cure" in some cases. Wiskott-Aldrich syndrome and ataxia telangiectasia are caused by mutations in multifunctional intracellular proteins which result in defects in antibody production, in addition to other immunologic and non-immunologic abnormalities [127, 128]. IgG replacement is also a mainstay of treatment in these diseases, in which increased incidence of infection is also a major problem.

Full reconstitution of B-cell production and function is often delayed or may not occur at all in patients undergoing HSCT for SCID or other serious immune deficiencies. Such patients are routinely given IgG replacement for at least a year after HSCT, but many require it for the rest of their lives [129].

20.7.1.3 Other PID

In recent years there has been an explosion in the understanding of molecular defects underlying complex syndromes previously described by a single aspect of their phenotype, such as "hyper IgE syndrome" or "chronic mucocutaneous candidiasis". Even when categorised as molecular entities, these disorders may have variable penetrance and/or modifier genes [130]. It is, therefore, important to carefully consider the pattern of infections and laboratory findings for each individual patient. The decision to initiate IgG replacement therapy should include a careful evaluation of the clinical situation and should not be based solely on laboratory values or identification of a mutation per se. In each situation, a plan for monitoring the efficacy and outcomes of replacement therapy should be included at the time of initiation.

20.7.1.4 Specific (Polysaccharide) Antibody Deficiency

Some patients, most often presenting with recurrent/chronic sinusitis and other respiratory infections, seem unable to mount normal antibody responses to bacterial antigens, as evidenced by failure to respond or to maintain memory to pneumococcal and other polysaccharide vaccines [131–133]. Such patients may have normal numbers and subtypes of B cells and total IgG levels within the normal range, although frequently towards the lower limit of normal. Some may have relatively low levels of one or more IgG subclasses. These patients may require full-dose IgG replacement therapy to supply the amounts of the specific antibodies they require.

20.7.1.5 Ig Replacement Therapy in Secondary Immunodeficiencies

With the increasing use of monoclonal antibodies and other anti-B-cell therapies for autoimmune/rheumatologic diseases in combination with chemotherapy regimens, the prevalence of iatrogenic or "secondary" hypogammaglobulinaemia and symptomatic antibody deficiency is increasing [134]. Such therapies include rituximab (a recombinant monoclonal anti-CD20 antibody), chemotherapy regimens in patients with non-Hodgkin's lymphoma and prolonged rituximab maintenance regimens for follicular lymphoma. Antibody replacement therapy for hypogammaglobulinaemia secondary to haematologic malignancies is the fastest-growing category of IgG use recorded by the Australian National Blood Authority in its 2013 report. This accounts for more IgG use than for any other indication in Australia [135].

Increasing use of rituximab for non-malignant autoimmune conditions, with or without antimetabolites such as azathioprine or cyclophosphamide and/or corticosteroids, is also recognised as a cause of persistent hypogammaglobulinaemia [136]. Secondary hypogammaglobulinaemia is also increasingly recognised in solid organ transplant recipients, particularly in those receiving regimens including mycophenolate mofetil (MMF) [137].

In all these situations, principles governing who should receive IgG replacement therapy, dosing and choice of route and regimen should be similar to those discussed above for primary antibody deficiency. It is recommended that patients have quantitative baseline immunoglobulin and IgG subclass measurements before beginning any therapeutic regimen containing an anti-B-cell monoclonal antibody or MMF. This should be followed by careful review of infection history and antibiotic use at regular intervals and by periodic measurements of peripheral blood B cells and quantitative immunoglobulins. In any patient with subnormal or falling IgG levels, subclass and specific antibody titres should also be obtained. The response to vaccines can be assessed to assist in determining the need for IgG therapy [132]. It is important to recognise that in secondary immunodeficiency, as well as in PID, there may be subclinical chronic lung and/or sinus infection even in the absence of acute episodes of pneumonia or other serious infections. Patients at risk should be carefully monitored for chronic low-grade infection and considered for IgG replacement therapy even if acute severe infections have not occurred. The development of bronchiectasis in 2.4% of MMF-treated renal transplant recipients, as opposed to 0% in non-MMF-treated transplant patients, provides just one example of the importance of such monitoring [137].

20.7.2 Ig and Immunomodulation in Autoimmune Diseases

20.7.2.1 Immune Thrombocytopaenia

Immune thrombocytopaenia is an autoimmune disease characterised by low platelet counts ($<30 \times 10^9$/L) due to antibody-mediated destruction in the reticuloendothelial system [138–140]. Many patients have no symptoms, while others present with petechiae, purpura or frank bleeding. The hallmark presentation of immune thrombocytopaenia is mucocutaneous bleeding that manifests as purpura (petechiae, ecchymosis), epistaxis, menorrhagia or oral mucosal, gastrointestinal or intracranial haemorrhage. The rate of fatal bleeding is estimated at 0.02–0.04 cases per patient year, with the highest rate (0.13 cases per patient year) among patients older than 60 years. Paediatric immune thrombocytopaenia is most commonly acute and, in most cases, a postinfectious condition in an otherwise normal child. In most cases, IVIG is not necessarily required. Chronic immune thrombocytopaenia is seen primarily in adults but may also occur in children with CVID.

Immune thrombocytopaenia was the first reported instance in which IVIG was used to treat an autoimmune disease. Practice guidelines suggest initiating treatment with oral prednisone or prednisolone at a dose of 1 mg/kg per day [139, 140]. IVIG is suggested as a second-line treatment which may be preferred if the patient does not tolerate steroids and/or if a rapid rise in platelet count is necessary due to bleeding or surgery. A total of up to 2 g/kg of IVIG is generally given over 2–5 days. In most studies, this has resulted in a rise of platelet count to $>50 \times 10^9$/L within 7 days in more than 60–80% of subjects, with prompt resolution of bleeding or clinical signs of thrombocytopaenia in most subjects [141, 142]. Multiple mechanisms likely contribute to the dramatic effects of IVIG seen in patients with immune thrombocytopaenia. Inhibition of binding and/or phagocytosis of IgG-sensitised platelets by Fc receptors in the reticuloendothelial system is likely the most important mechanism. This explains the ability to achieve the same effect by administering small amounts of anti-RhD globulin in patients whose erythrocytes are D+ [138]. Effects of IVIG in immune thrombocytopaenia are usually transient, wearing off after the 30 day half-life of IgG. Alternative approaches to treatment of immune thrombocytopaenia include splenectomy, rituximab treatment to remove the B cells producing the anti-platelet antibodies and the use of thrombopoietin receptor agonists.

20.7.2.2 Kawasaki Disease

Kawasaki disease is an acute systemic syndrome of high fever, vasculitis and lymphadenopathy which primary affects children under 5 years of age. The vasculitis causes coronary artery aneurysms, in up to 25% of untreated children, which may result in fatalities [143, 144]. High-dose IVIG, 2 g/kg over 8–12 h, in combination with acetylsalicylic acid (aspirin), is a well-established standard of care in the initial treatment of children [145]. A meta-analysis of studies in which different doses of IVIG were co-administered with aspirin showed a correlation between the dose of IVIG and its efficacy in preventing coronary artery aneurysms, which were present at the convalescent phase in only 3.8% of patients who received the highest dose studied, 2 g/kg [144]. Furthermore, a randomised trial in 549 subjects and the meta-analysis cited above both reported that administering IVIG as a single infusion over 8–12 h led to more rapid resolution of symptoms and was more effective than giving five consecutive daily doses of 400 mg each. Thus, the commonly used first-line treatment for Kawasaki syndrome is 2 g/kg of IVIG given as a single large bolus over 8–12 h [145]. The aetiology of Kawasaki disease is unknown. The predominant mechanisms of action of IVIG in treating it are also unknown, but the general anti-inflammatory effects are dramatic and occur rapidly, with many infants defervescing within hours.

20.7.2.3 Guillain-Barré Syndrome (GBS)

GBS refers to a group of acute autoimmune polyneuropathies presenting with flaccid paralysis, frequently following resolution of an infectious disease, most notably gastroenteritis due to *Campylobacter jejuni* [146]. The leading hypothesis for the pathogenesis of GBS is that antibodies produced in response to the lipooligosaccharide of the infectious agent cross-react with ganglioside antigens on myelin, activating complement and causing dysfunction and demyelination of peripheral nerves [146–149]. As such, GBS is frequently cited as a leading example of the *molecular mimicry* theory of autoimmune disease. Although GBS is generally considered an acute, monophasic disease, as many as 5% of patients succumb during the acute stage, and up to 20% may have slow recovery and/or prolonged disability [146, 147, 150, 151]. First-line treatments for GBS include high-dose IVIG and/or plasma exchange [147, 150, 151]. Laboratory evidence supports the hypothesis that IVIG acts predominantly by blocking the binding of autoantibodies to nerves, presumably due to the presence of anti-idiotypes, and/or by blocking complement deposition [148, 149]. Analogously, plasmapheresis probably acts primarily by removing autoantibodies. Autoantibodies against a variety of antigens/ epitopes induced by different infectious agents may cause different clinical variants in individual patients. Neither measurements of autoantibodies nor in vitro evaluation of the ability of IVIG to neutralise them is used to select or adjust the dose or schedule of IVIG treatment. Practice parameters of the American Academy of Neurology consider plasma exchange and IVIG as equally effective first-line treatments [151]. A recent Cochrane review reached the same conclusion but noted that the full course of IVIG treatments was more likely to be completed than the series of plasma exchanges and that there was little additional benefit of combining the two modalities [150]. The most common dose of IVIG is 2 g/kg, usually given over 2–5 days. A more recent study showed that serum IgG levels achieved by patients receiving 0.4 g/kg/day for 5 days were variable and that patients with highest increments in IgG level from baseline to day 14 showed the best response in terms of the proportion able to walk unaided 6 months after the acute episode and vice versa [152].

20.7.2.4 Chronic Inflammatory Demyelinating Polyneuropathy (CIDP)

CIDP is the most common autoimmune peripheral neuropathy. In many surveys, CIDP accounts for more IVIG use than any other single diagnostic category [153, 154]. In many ways CIDP resembles a chronic form of GBS. However, it generally has a more indolent onset, and few patients recall preceding infections or other triggering events [154]. CIDP is considered by many to be an autoantibody-mediated disease, and plasma exchange is very effective. However, no single major target antigen(s) has been identified in CIDP, and it is likely that different clinical variants represent attack on different targets. Based upon results of the ICE study, a randomised, placebo-controlled trial of IVIG in 117 CIDP patients, the FDA approved IVIG for CIDP using a loading dose of 2 g/kg followed by maintenance dosing of 1 g/kg every 3 weeks [153]. In many patients, the effects of each dose of IVIG in CIDP are transient, suggesting a mechanism of action involving competition with autoantibodies rather than inhibition of their production, and "wear-off" effects are common [155, 156]. In this context, it has been proposed that functional effects of autoantibodies on conduction through the nodes of Ranvier, rather than, or in addition to, demyelination per se, may be the main target for IVIG therapy [155, 157]. Prescribing regimens other than those used in the ICE trial is prevalent, and guidelines call for individualised therapy [156, 158]. A recent prospective study found that 60% of CIDP patients required IVIG more often than once every 2 weeks for stable maintenance of optimal strength [159]. Anecdotal reports and small case series suggest that maintaining high serum IgG levels with SCIG may result in more stable maintenance of strength [160]. A large, randomised, controlled multicentre study of SCIG therapy is now underway to test this hypothesis (chronic idiopathic polyneuropathy and treatment with subcutaneous immunoglobulin; NCT 01545076). CIDP also responds to corticosteroid therapy, particularly high-dose intravenous methylprednisolone. Some studies suggest that although IVIG is better tolerated than high-dose steroids, there is higher rate of relapse after IVIG therapy than methylprednisolone [161].

20.7.2.5 Multifocal Motor Neuropathy (MMN)

MMN is a rare disease of the peripheral nervous system which presents with progressive asymmetrical muscle weakness in one or more limbs [162, 163]. Although the pathophysiology of MMN is not understood, autoantibodies against the ganglioside GM1 are identified in nearly 50% of patients [163]. A paradox in understanding MMN is that, unlike GBS, CIDP and other autoimmune neuropathies, MMN does not respond well to steroids or plasma exchange [164, 165], leaving IgG as the mainstay of treatment [158, 166, 167]. IVIG doses of 1–2 g/kg are generally given every 3–4 weeks. A successful randomised, double-blinded trial of IVIG for MMN has resulted in licensing of IVIG for this disease in the USA and EU [166]. Interestingly, even early reports noted improvement in strength and conduction within a few days following IVIG, but this lasted only 1–2 months at best. As with CIDP, many patients complain that their strength deteriorates in the third or fourth week after an IVIG dose. Small case series suggest that SCIG may help to ameliorate these "wear-off effects" and promote more constant activity, but long-term follow-up studies of SCIG are needed to determine if this will prevent the slow deterioration which now characterises most MMN patients.

20.7.2.6 Myasthenia Gravis (MG)

MG refers to a family of diseases in which antibodies against components of the neuromuscular junction interfere with stimulation of the muscles, resulting in weakness [168]. The most common forms of MG are associated with antibodies against acetylcholine receptors. However, distinct forms with antibodies against muscle-specific kinases (MuSK) and proteins such as low-density lipoprotein receptor-related protein 4 (LRP4) which complex with MuSK in the post-synaptic membrane have been identified as well [169, 170]. There are two major patterns of presentation of MG: In "early-onset" disease, females predominate and thymoma is uncommon. In "late-onset" disease, there is no gender predominance and thymoma is more common. Although the classical presentation begins with weakness of the extraocular muscles, the extent of generalised muscle involvement and severity varies [168, 171, 172]. Cholinesterase inhibitors and corticosteroids are first-line treatments, but most patients eventually require other immunosuppressives. Plasma exchange and IVIG have been found to be equally effective but are generally considered "rapid immunomodulators" which are used to control exacerbations (myasthenic crisis) and/or to prepare the patient for surgery [168, 171–173]. In the instance of thymoma, surgical removal is usually beneficial. IVIG may be preferred as a maintenance therapy in patients who do not tolerate steroids or other immunosuppressive regimens. Recent studies suggest that IVIG acts by inhibiting the effects of the autoantibodies, rather than modulating their production or the underlying autoimmune process, and thus it should not be considered a "disease-modifying" therapy [173].

Lambert-Eaton myasthenic syndrome (LEMS) shares features with MG but is a rare paraneoplastic condition associated with autoantibodies against presynaptic voltage-gated calcium channels, most commonly occurring in patients with small cell lung cancer. IVIG is effective in controlling weakness and other symptoms of LEMS, but removal of the tumour generally results in resolution of the disease.

20.7.2.7 Autoimmune Mucocutaneous Bullous (Blistering) Diseases

Autoimmune mucocutaneous bullous (blistering) diseases are a family of conditions, including multiple subtypes of pemphigus and pemphigoid, in which intraepidermal or subepidermal layers of skin separate due to antibodies against intercellular adhesion molecules such as desmoglein

[174–176]. Most of these disorders have relapsing-remitting courses, with subtypes characterised by differences in the location and depth of the blisters, possibly related to differences in the specific targets of the autoantibodies. Corticosteroids are generally first-line therapy, and most patients are also given cyclophosphamide, azathioprine and/or other cytotoxic drugs. "High-dose" IVIG is often added as an alternative or addition to steroids and is regarded as a "steroid-sparing" therapy [174–176]. IVIG may also be preferentially used in patients with poor tolerance of cytotoxic agents. IVIG's major effect seems to be reduction of autoantibody titre, although this may take months. In cases of severe mucosal and/or ocular involvement, high-dose IVIG may be used to induce rapid remission of acute attacks/exacerbations.

20.7.2.8 IVIG in Solid Organ Transplantation

IVIG has several uses in solid organ transplantation. These include regimens that involve plasma exchange and rituximab to reduce levels of preformed donor-specific anti-HLA antibodies (also called panel reactive antibodies) in patients who have been pre-sensitised by previous transplants and/or exposure to foetal cells during pregnancy [177]. IVIG is also widely used for prevention and treatment of antibody-mediated rejection [178] and as replacement therapy in patients who develop hypogammaglobulinaemia due to their immunosuppressive regimen [179, 180].

20.8 Mechanisms of Action of IVIG

Historically, the primary use of IVIG was as a reconstitution therapy for people lacking the capacity to produce antibodies in sufficient amounts to protect them from microbial infection. In these patients, protection is mediated by the neutralisation and opsono-phagocytic effects of the four different IgG subtypes described above. However, when IVIG is used in autoimmune or chronic inflammatory diseases, beneficial effects are thought to be due to the interaction of specific Ab binding to, and thus modulating, humoral and cellular constituents of the dysregulated immune system in the patient and to the interactions with FcRn and other FcRs.

Effects on soluble factors in the patient's circulation include:

- The direct binding and neutralisation of pathological autoantibodies by their anti-idiotypes in IVIG, leading to a reduction in autoantibody-mediated phagocytosis and damage to self-structures [181, 182].
- The direct binding and neutralisation of immune mediators, such as cytokines, chemokines, soluble cytokine receptors or superantigens, metalloproteinases and adhesion molecules [183].
- The binding of potentially detrimental complement components like C3b, C4b, as well as anaphylatoxins C3a and C5a is a mechanism termed complement scavenging. This binding is believed to attenuate complement deposition and the ensuing damage inflicted by complement-mediated cytotoxicity and concurrent inflammation [184] (see Chap. 8).

Mechanisms directed towards immune cells include:

- Blockade of access of sensitised platelets or other antibody-coated cells to phagocytic FcγRs on tissue macrophages and neutrophils, NK cells and monocytes. Shifts in expression levels of activating and inhibitory FcγR trigger different signalling pathways and result in changes in thresholds of activation and inhibition for different immune effector cells, e.g. reticuloendothelial system phagocytes, dendritic cells and monocytes. Also, restoration of FcγRIIb expression has been observed on B cells and monocytes after IVIG therapy in CIDP patients [185, 186].
- Saturation of FcRn leading to reduced half-life and increased catabolism of pathological autoantibodies [55, 187].
- Interaction of NAb in IVIG with immunoregulatory molecules, such as B-/T-cell receptors, cytokine receptors, human leukocyte antigen

(HLA) molecules and Siglecs [188], thereby reducing proinflammatory cytokine production, the expansion of autoreactive B cells and the number of potentially harmful effector T cells.

- The modification of dendritic cell function in favour of an anti-inflammatory state, including the production of IL-10, downregulation of co-stimulatory molecules and the prevention of dendritic cell-mediated self-reactive T-cell activation and general T-cell priming [189] (see Chap. 5).
- The induction/expansion of regulatory T cells, possibly via Tregitopes (see Sect. 20.6). Other mechanisms include interaction of IVIG with DC-SIGN on dendritic cells and induction of COX-2-dependent PGE-2 secretion in human dendritic cells resulting in expansion of regulatory T cells [190].
- Reduction of oligoclonal expansions of both CD4 and CD8 populations in certain autoimmune disease conditions [191].

Indeed, many more mechanisms of action, observed in vitro and in animal models, are described in the scientific literature and have recently been reviewed [14, 37, 110, 155, 192–195]. The extrapolation of in vitro experiments and observations in animal models to delineate mechanisms of action that function in human patients should nevertheless be interpreted with caution.

20.9 Adverse Reactions to IgG Therapy

Infusions of IVIG and SCIG are considered safe and usually well tolerated. Occasionally, however, they can be associated with side effects (adverse drug reactions, ADRs) which are most often transient and mild in nature. ADRs occur more frequently at the initiation of therapy and are often related to the infusion rate [196, 197] and the total dose applied. Systemic ADRs after SCIG administration are less common than after IVIG administration, whereas local adverse reactions such as bruising, swelling and erythema at the infusion site are more common after SCIG administration.

20.9.1 Intravenous Infusion-Related Adverse Reactions

Systemic adverse reactions after IVIG can be immediate, occurring within 6 h of an infusion, or delayed, occurring 6 h to 1 week after an infusion. Late reactions, occurring weeks and months after an infusion, are very rare. Immediate systemic reactions such as head and body aches, chills and fever are usually mild and readily treatable. True immediate anaphylactic reactions are extremely rare, but "anaphylactoid" reactions are infusion rate-related and may be more common, especially in IVIG-naïve patients. The most common delayed systemic reaction is headache; less common but more serious ADRs include aseptic meningitis, renal failure, thromboembolism and haemolytic reactions [196]. Adverse reactions may be due to the presence of large molecular weight protein aggregates and/or formation of complexes between administered antibodies and circulating microbial antigens or self-antigens. Likewise, the presence of small amounts of other bioactive proteins (such as factor XIa) in the IVIG product may cause ADR.

ADRs to IVIG vary in severity. Lowering the infusion rate and temporarily interrupting the infusion are important therapeutic measures which often reduce their severity. Premedication (e.g. paracetamol, corticosteroids or non-steroidal anti-inflammatory drugs or anti-histamine) is often used to prevent systemic adverse reactions. In addition, experienced clinicians point out that the tolerated infusion rate differs widely between individuals and may depend on the medical history of the patient. In a clinical study, which followed "experienced" immunodeficiency patients who had received at least six IVIG infusions over 2 years, an overall ADR rate of only 0.8% of the infusions was reported [198]. These findings indicate that systemic ADRs mainly occur in naïve patients during the first few infusions and thereafter any ADRs are less severe. The same publication also confirmed that the occurrence of systemic reactions in immunodeficient patients is frequently associated with intercurrent infections [198].

20.9.2 Hypersensitivity and Anaphylactoid Reactions

True allergic/anaphylactic reactions to human IgG preparations are very rare and may be due to IgE or IgG antibodies against IgG. A very similar type of ADR, termed "anaphylactoid" reactions, occurs more commonly. Although both types of reactions include flushing and tachycardia, the anaphylactoid reactions are often accompanied by *hyper*- rather than *hypo*tension, and adrenaline (epinephrine) is usually not needed. Anaphylactoid reactions may be due to mediators released by therapeutic IgG reacting with foreign antigens and/or endogenous molecules in the recipient. These reactions often respond to temporarily slowing or pausing the infusion and are further distinguished from true anaphylaxis because they tend to become milder with subsequent exposures, rather than more severe as would be expected with true IgE-mediated anaphylaxis.

IgA deficiency has been reported to be a very rare cause for true anaphylactic reactions in patients with IgE or IgG antibodies against IgA [199, 200]. The true frequency of anti-IgA mediated reactions remains uncertain and a source of continued controversy [200, 201]. There are reports that replacing i.v. treatment with s.c. administration of IgG may reduce the risk of such reactions [197, 199, 202].

20.9.3 Aseptic Meningitis

Headache is the most common adverse reaction to IVIG treatment. Severe headache, in combination with nuchal rigidity, vomiting and photophobia, may occur as an infrequent complication of IVIG. This phenomenon, which is called acute aseptic meningitis, usually occurs with a latency of several hours to 2 days following IVIG treatment. Analysis of the cerebrospinal fluid is negative for pathogens, although invasion of white blood cells, predominantly granulocytes, is frequently observed in patients suffering from aseptic meningitis. The aetiology of this adverse reaction is unclear. However, patients with a history of migraine seem to have a higher risk for aseptic meningitis [203].

20.9.4 Haemolytic Reactions

IVIG products can contain anti-blood group antibodies, which may act as haemolysins and induce in vivo coating of red blood cells (RBC) with IgG. Haemolytic anaemia can develop subsequent to high-dose IVIG therapy in patients with blood groups A, B and AB due to enhanced RBC sequestration or intravascular lysis in rare cases. An underlying inflammatory status constitutes an additional risk factor, predisposing some patients to this complication. The majority of reports on haemolytic reactions are associated with high cumulative total doses of IVIG treatment [204, 205].

Isoagglutinins (anti-A and anti-B antibodies) from donor plasma contained in IgG products are a potential causative factor of haemolysis [206]. Therefore, IgG manufacturers have started to explore and implement approaches to reduce anti-A/B isoagglutinin in IgG products. Measures such as screening and exclusion of plasma donors with high anti-A antibody titres and the introduction of a specific immunoaffinity chromatography step for isoagglutinin reduction into the manufacturing process have been implemented by some manufacturers [77, 207, 208]. These may reduce the risk of haemolysis in patients receiving IgG therapy (see Sect. 20.5).

20.9.5 Thromboembolic Events

There is evidence of an association between IgG administration and rare thromboembolic events (TEE) such as myocardial infarction, stroke, pulmonary embolism and deep vein thrombosis in at-risk patients. A common hypothesis relates to a relative increase in blood viscosity through the high influx of immunoglobulin. However, the recent correlation of TEE with the administration of specific lots of certain IVIG products points to other mechanisms: the occurrence of TEE might have been caused by unwanted impurities, e.g. residual coagulation factor (mainly FXIa)-associated procoagulant activity present in selected lots of IgG preparations produced mainly by a previous generation production process [209]. Sensitive assays to detect procoagulant activity in Ig preparations have been developed [210] and methods implemented to remove this activity from Ig products [76], ensuring that Ig treatment-related TEE are very rare and limited.

20.9.6 Skin Reactions

There are different types of skin reactions, which might be subclassified into immediate and delayed reactions. Immediate skin reactions (reported as rash or urticaria) may be part of acute hypersensitivity reactions and are often their only symptom (see Chap. 11). A particular type of excematoid rash on the palms has been reported in some neurology patients receiving high-dose IVIG.

For SCIG, although many patients may have some local discomfort associated with swelling and redness at the site of the infusions, the swelling and local symptoms usually subside within 24–48 h and do not usually deter patients from continuing with their SCIG regimen. Systemic adverse events with SCIG are rare and premedication is seldom necessary, nor is close monitoring required during the infusion. SCIG has thus emerged as an ideal route for home use in many patients [209].

20.9.7 Renal Failure

Osmotic nephrosis leading to renal failure may develop after IVIG therapy due to sugar excipients in some IVIG products, particularly older lyophilised products [211]. Products containing sucrose account for a disproportionate share of the total number of such cases, but products containing maltose or glucose have also been associated with renal failure [212].

The pathogenesis of most cases is believed to involve osmotic damage to tubular cells, which take up sucrose, then water, leading to swelling that can obstruct the tubules. Sucrose-containing products should be avoided in patients with pre-existing renal disease or who are at risk, and all patients should be well hydrated before an IVIG infusion is started.

However, new state-of-the-art Ig products do not contain sugar excipients and thus have a very low risk of renal failure [213, 214]; nevertheless, acute severe haemolysis, a separate complication (see Sect. 20.9.4), can also lead to haemoglobinuria and renal failure [215].

20.10 Viral Safety of Ig

The importance of viral safety has long been realised with all products originating from human plasma. Pathogen safety is of particular importance with IVIG preparations; patients receiving these products have impaired immune defences and often receive these preparations for many years and sometimes at high or very high doses. The increase in reported cases of acute hepatitis C in patients receiving IVIG therapy in the early 1980s led to the development of more rigorous donor screening and to manufacturing processes for IVIG production which specifically inactivate or remove potential blood-borne pathogens [216].

Viral safety is generally based on three measures, namely, donor selection, donation testing and manufacturing. Of critical importance is the manufacturing step. In addition to regulatory requirements, voluntary industry standards play a key role in maintaining a strong safety record. These are the International Quality Plasma Program (IQPP) for collectors of plasma and the Quality Standards of Excellence, Assurance and Leadership (QSEAL) for manufacturers.

20.10.1 Donor Selection

Prospective donors are very carefully screened prior to donation. Whole blood donors whose plasma can be used for further manufacturing must meet the requirements of donation for transfusion. Source plasma donors who donate plasma only for further manufacture undergo annual medical examination. In particular, concerns regarding the spread of variant Creutzfeldt-Jakob disease or spongiform encephalopathies has meant that the questionnaires include sections aimed at eliminating those donors that may have been exposed to either of these diseases. In addition, the IQPP viral marker standard assures that centres are located appropriately to assure donors are low risk for infectious disease. Other standards further enhance this safety.

20.10.2 Donation Testing

Donor sera are tested by serology for hepatitis B antigen, hepatitis C antibody and antibody to HIV. Nucleic acid amplification testing (NAT), such as polymerase chain reaction tests with enhanced sensitivity, reduces the window period between infection and test positivity, thereby reducing the residual risk of a virus in a manufacturing pool.

20.10.3 Ig Manufacture

Manufacturers of Ig are required to show their manufacturing processes inactivate and remove known blood-borne pathogens. Furthermore, the combination of different pathogen elimination mechanisms also provides confidence that the process could cope with newly emerging viruses. By developing scaled-down versions of the production processes, manufacturers can spike samples with test viruses and demonstrate their inactivation and/or removal. Virus removal or inactivation is expressed in the form of logarithmic reduction factors, calculated as the decimal logarithm of the ratio of virus titre spiked to the virus titre recovered after the manufacturing step. Validation of virus partitioning relies on the ability to establish an even balance in virus titre or to substantiate the mechanism of virus reduction. For example, in a fractionation step, the sum of virus titres recovered from the supernatant and the precipitate should equal the titre of the spiked virus to assure validity of the assays. For virus inactivation processes, the kinetics of decline in infectivity and robustness of the conditions must be shown. Virus reduction steps may include virus elimination, virus inactivation or even a combination of both. Either is acceptable as long as multiple, complimentary steps are validated and used.

The three mechanisms applicable within a manufacturing process are partitioning, inactivation and virus filtration (see Table 20.5).

Table 20.5 Virus reduction methods used in large-scale manufacturing of plasma products

Mechanism	Principle (examples)	Methods (examples)	Effective against
Partitioning (separation into various physical phases)	Solubility Adsorption Ionic interactions	Cryoprecipitation Cold ethanol fractionation Octanoic acid fractionation Depth filtration Chromatography	Depends on virus characteristics such as charge, hydrophobicity, etc.
Inactivation (alteration of structures that are essential for the virus infectivity)	Disintegration of virus (membrane) integrity Alteration of viral structural proteins (DNA, RNA strand breaks)	Solvent/detergent treatment Pasteurisation Dry heat treatment pH 4 incubation Octanoic acid incubation UV- or γ-irradiation	Enveloped viruses Pasteurisation (heat) UV- and γ-irradiation are effective against a variety of non-enveloped viruses. Low pH incubation also inactivates some non-enveloped viruses such as B19V and encephalomyocarditis virus
Elimination based on size	Virus (nano-) filtration	Tangential or dead-end filtrations	All viruses; depending on virus size and selected membrane pore size

DNA deoxyribonucleic acid, *pH* the decimal logarithm of the reciprocal of the hydrogen ion activity in a solution, *RNA* ribonucleic acid, *UV* ultraviolet

20.10.3.1 Partitioning

Protein precipitation steps followed by filtration or centrifugation are widely used during the purification of plasma proteins and can be very successful at removing the viruses to the waste fractions. This includes the removal of small, non-enveloped viruses that may be resistant to pasteurisation or solvent/detergent treatment [217]. It is also one of the key methods of reducing theoretical contamination of IVIG by misfolded prion proteins (PrP^{Sc}) [218]. One commonly used partitioning process is depth filtration in the presence of filter aids (adsorbents), which is usually applied after precipitation of plasma proteins. The use of column chromatography also relies on partitioning as the principal mechanism. When using chromatography, a careful validation of the columns, taking into account the resin's expected lifespan in production, is vital to avoid any loss of partitioning efficacy over time.

20.10.3.2 Inactivation

The most commonly employed methods of virus inactivation are low pH, pasteurisation, octanoic acid incubation or solvent/detergent treatment [219]. While octanoic acid incubation and solvent/detergent treatment are effective against enveloped viruses only, low pH has been demonstrated to inactivate B19V and picornavirus encephalomyocarditis virus too. Pasteurisation even inactivates a broader range of non-enveloped viruses. Validation of virus inactivation must include data on the inactivation kinetics.

20.10.3.3 Virus Filtration

Also known as nanofiltration, this involves the removal of viruses based on size; it is effective at removing viruses regardless of any other physicochemical properties. While filters of 35 nm meet current production standards, additional studies have shown that reducing the size of the filters to approximately 20 nm enhances the capacity to remove smaller viruses and even prions [220, 221]. The latter is particularly important, as size exclusion, along with partitioning, forms the principal methods of removing PrP^{Sc} from plasma products. This is also an important consideration when looking to the future and the removal of yet unknown viruses.

20.11 Summary

Humoral immunity is a vital part of the human immune response. Ig molecules are fundamental to this humoral immune response. When the humoral immune response is compromised, homologous, polyclonal Ig preparations may be applied using various therapeutic regimens depending on the disease. IVIG, which is comprised mainly of IgG, has been shown to possess anti-inflammatory and immunomodulatory properties which are used for a growing number of autoimmune diseases, especially neuropathies. Considerable advances in the last century have enabled significant and important improvements in the therapeutic Ig formulations available today. Continual improvements in the manufacturing processes of therapeutic Ig have led to the development of preparations devoid of aggregates, with well-maintained Fab and Fc effector functions. These developments have allowed changes in the route of administration from s.c., which was originally used, to i.m. and then to i.v. administration and more recently improved Ig preparations for s.c. administration, offering significant advantages for patients. Further research led to the development of ready-to-use Ig solutions with Ig concentrations ranging from 10% to 20%, which are stable at room temperature and offer increased convenience for the patient and healthcare provider alike. Over the coming years it is envisaged that additional products will be made available.

Acknowledgements The authors thank Toby Simon, CSL Behring LLC, King of Prussia, PA, USA, for his advice and assistance in the preparation of this book chapter. Editorial assistance was provided by Fishawack Communications GmbH, a member of the Fishawack Group of Companies.

References

1. Baerenwaldt A, Biburger M, Nimmerjahn F. Mechanisms of action of intravenous immunoglobulins. Expert Rev Clin Immunol. 2010;6:425–34.
2. Edelman GM. Antibody structure and molecular immunology. Science. 1973;180:830–40.
3. Delves PJ, Roitt IM. The immune system. First of two parts. N Engl J Med. 2000;343:37–49.

4. von Behring E A. Speech at the Nobel Banquet in Stockholm, December 10, 1901. 2015. http://www.nobelprize.org/nobel_prizes/medicine/laureates/1901/behring-facts html.
5. Heard K, O'Malley GF, Dart RC. Antivenom therapy in the Americas. Drugs. 1999;58:5–15.
6. Newcombe C, Newcombe AR. Antibody production: polyclonal-derived biotherapeutics. J Chromatogr B Analyt Technol Biomed Life Sci. 2007;848:2–7.
7. Eibl MM. History of immunoglobulin replacement. Immunol Allergy Clin North Am. 2008;28:737–64.
8. Bruton OC. Agammaglobulinemia. Pediatrics. 1952;9:722–8.
9. Chapel H, Gardulf A. Subcutaneous immunoglobulin replacement therapy: the European experience. Curr Opin Allergy Clin Immunol. 2013;13:623–9.
10. Cai K, Gierman TM, Hotta J, et al. Ensuring the biologic safety of plasma-derived therapeutic proteins: detection, inactivation, and removal of pathogens. Biodrugs. 2005;19:79–96.
11. Jolles S, Sewell WA, Misbah SA. Clinical uses of intravenous immunoglobulin. Clin Exp Immunol. 2005;142:1–11.
12. Arnson Y, Shoenfeld Y, Amita H. Intravenous immunoglobulin therapy for autoimmune diseases. Autoimmunity. 2009;42:553–60.
13. Danieli MG, Gambini S, Pettinari L, et al. Impact of treatment on survival in polymyositis and dermatomyositis. A single-centre long-term follow-up study. Autoimmun Rev. 2014;13:1048–54.
14. Schwab I, Nimmerjahn F. Intravenous immunoglobulin therapy: how does IgG modulate the immune system? Nat Rev Immunol. 2013;13:176–89.
15. Abbas AK. Cellular and molecular immunology. In: Abbas AK, Lichtman AH, Pillai S, editors. Cellular and molecular immunology. 6th ed. Philadelphia, PA: Saunders; 2009. p. 326–8.
16. Gillis C, Gouel-Cheron A, Jonsson F, et al. Contribution of human FcgammaRs to disease with evidence from human polymorphisms and transgenic animal studies. Front Immunol. 2014;5:254.
17. Male D, Brostoff J, Roth D, et al. Immunology. 7th ed. Maryland Heights, MO: Mosby Elsevier; 2006.
18. Murphy KM, Travers P, Walport M. Janeway's immunobiology. 7th ed. New York: Garland Publishing; 2008. p. 410.
19. Monteiro RC. Role of IgA and IgA fc receptors in inflammation. J Clin Immunol. 2010;30:1–9.
20. Rossato E, Ben MS, Kanamaru Y, et al. Reversal of arthritis by human monomeric IgA through receptor mediated SHP-1 inhibitory pathway. Arthritis Rheumatol. 2015;67:1766–77.
21. Roos A, Bouwman LH, van Gijlswijk-Janssen DJ, et al. Human IgA activates the complement system via the mannan-binding lectin pathway. J Immunol. 2001;167:2861–8.
22. Feinstein A, Munn EA. Conformation of the free and antigen-bound IgM antibody molecules. Nature. 1969;224:1307–9.
23. Hjelm F, Carlsson F, Getahun A, et al. Antibody-mediated regulation of the immune response. Scand J Immunol. 2006;64:177–84.
24. Schwartz-Albiez R, Monteiro RC, Rodriguez M, et al. Natural antibodies, intravenous immunoglobulin and their role in autoimmunity, cancer and inflammation. Clin Exp Immunol. 2009;158(Suppl 1):43–50.
25. Lutz HU, Binder CJ, Kaveri S. Naturally occurring auto-antibodies in homeostasis and disease. Trends Immunol. 2008;30:43–51.
26. Castro CD, Flajnik MF. Putting J chain back on the map: how might its expression define plasma cell development? J Immunol. 2014;193:3248–55.
27. Corthesy B. Role of secretory IgA in infection and maintenance of homeostasis. Autoimmun Rev. 2013;12:661–5.
28. Perrier C, Sprenger N, Corthesy B. Glycans on secretory component participate in innate protection against mucosal pathogens. J Biol Chem. 2006;281:14280–7.
29. Chen K, Cerutti A. The function and regulation of immunoglobulin D. Curr Opin Immunol. 2011;23:345–52.
30. Edholm ES, Bengten E, Wilson M. Insights into the function of IgD. Dev Comp Immunol. 2011;35:1309–16.
31. Bell RG. IgE, allergies and helminth parasites: a new perspective on an old conundrum. Immunol Cell Biol. 1996;74:337–45.
32. Arnold JN, Wormald MR, Sim RB, et al. The impact of glycosylation on the biological function and structure of human immunoglobulins. Annu Rev Immunol. 2007;25:21–50.
33. Maverakis E, Kim K, Shimoda M, et al. Glycans in the immune system and The Altered Glycan Theory of Autoimmunity: a critical review. J Autoimmun. 2015;57:1–13.
34. Plomp R, Dekkers G, Rombouts Y, et al. Hinge-region O-glycosylation of human immunoglobulin G3 (IgG3). Mol Cell Proteomics. 2015;14:1373–84.
35. Jefferis R. Recombinant antibody therapeutics: the impact of glycosylation on mechanisms of action. Trends Pharmacol Sci. 2009b;30:356–62.
36. Kaneko Y, Nimmerjahn F, Ravetch JV. Anti-inflammatory activity of immunoglobulin G resulting from Fc sialylation. Science. 2006;313:670–3.
37. Nagelkerke SQ, Kuijpers TW. Immunomodulation by IVIg and the role of Fc-gamma receptors: classic mechanisms of action after all? Front Immunol. 2014;5:674.
38. Jefferis R. Glycosylation of antibody therapeutics: optimisation for purpose. Methods Mol Biol. 2009a;483:223–38.
39. van de Geijn FE, Wuhrer M, Selman MH, et al. Immunoglobulin G galactosylation and sialylation are associated with pregnancy-induced improvement of rheumatoid arthritis and the postpartum flare: results from a large prospective cohort study. Arthritis Res Ther. 2009;11:R193.

40. Royle L, Roos A, Harvey DJ, et al. Secretory IgA N- and O-glycans provide a link between the innate and adaptive immune systems. J Biol Chem. 2003;278:20140–53.
41. Woof JM, Russell MW. Structure and function relationships in IgA. Mucosal Immunol. 2011;4:590–7.
42. Waldmann TA, Strober W. Metabolism of immunoglobulins. Prog Allergy. 1969;13:1–110.
43. Bonilla FA. Pharmacokinetics of immunoglobulin administered via intravenous or subcutaneous routes. Immunol Allergy Clin North Am. 2008b;28:803–19.
44. Morell A. Pharmacokinetics of intravenous immunoglobulin preparations. In: Lee ML, Strand V, editors. Intravenous immunoglobulins in clinical practice. New York: Marcel Dekker Inc.; 1997. p. 1–18.
45. Koleba T, Ensom MHH. Pharmacokinetics of intravenous immunoglobulin: a systematic review. Pharmacotherapy. 2006;26:813–27.
46. Wasserman RL, Church JA, Peter HH, et al. Pharmacokinetics of a new 10% intravenous immunoglobulin in patients receiving replacement therapy for primary immunodeficiency. Eur J Pharm Sci. 2009;37:272–8.
47. Kontermann RE. Strategies to extend plasma half-lives of recombinant antibodies. Biodrugs. 2009;23:93–109.
48. Roopenian DC, Akilesh S. FcRn: the neonatal Fc receptor comes of age. Nat Rev Immunol. 2007;7:715–25.
49. Rath T, Kuo TT, Baker K, et al. The immunologic functions of the neonatal Fc receptor for IgG. J Clin Immunol. 2013;33(Suppl 1):S9–17.
50. Lobo ED, Hansen RJ, Balthasar JP. Antibody pharmacokinetics and pharmacodynamics. J Pharm Sci. 2004;93:2645–68.
51. Weflen AW, Baier N, Tang QJ, et al. Multivalent immune complexes divert FcRn to lysosomes by exclusion from recycling sorting tubules. Mol Biol Cell. 2013;24:2398–405.
52. Passot C, Azzopardi N, Renault S, et al. Influence of FCGRT gene polymorphisms on pharmacokinetics of therapeutic antibodies. MAbs. 2013;5:614–9.
53. Sachs UJ, Socher I, Braeunlich CG, et al. A variable number of tandem repeats polymorphism influences the transcriptional activity of the neonatal Fc receptor alpha-chain promoter. Immunology. 2006;119:83–9.
54. Wani MA, Haynes LD, Kim J, et al. Familial hypercatabolic hypoproteinemia caused by deficiency of the neonatal Fc receptor, FcRn, due to a mutant beta2-microglobulin gene. Proc Natl Acad Sci U S A. 2006;103:5084–9.
55. Bleeker WK, Teeling JL, Hack E. Accelerated autoantibody clearance by intravenous immunoglobulin therapy: studies in experimental models to determine the magnitude and time course of effect. Blood. 2001;98:3136–42.
56. Seite JF, Hillion S, Harbonnier T, et al. Review: intravenous immunoglobulin and B cells: when the product regulates the producer. Arthritis Rheumatol. 2015;67:595–603.
57. Wurster U, Haas J. Passage of intravenous immunoglobulin and interaction with the CNS. J Neurol Neurosurg Psychiatry. 2009;57:21–5.
58. Gasparoni A, Avanzini A, Ravagni PF, et al. IgG subclasses compared in maternal and cord serum and breast milk. Arch Dis Child. 1992;67:41–3.
59. Kane SV, Acquah LA. Placental transport of immunoglobulins: a clinical review for gastroenterologists who prescribe monoclonal antibodies to women during conception and pregnancy. Am J Gastroenterol. 2009;104:228–33.
60. Berger M. Subcutaneous administration of IgG. Immunol Allergy Clin North Am. 2008b;28:779–802.
61. Berger M, Jolles S, Orange JS, et al. Bioavailability of IgG administered by the subcutaneous route. J Clin Immunol. 2013a;33:984–90.
62. Sidhu J, Rojavin M, Pfister M, et al. Enhancing patient flexibility of subcutaneous immunoglobulin G dosing: pharmacokinetic outcomes of various maintenance and loading regimens in the treatment of primary immunodeficiency. Biol Ther. 2014;4:41–55.
63. Landersdorfer CB, Bexon M, Edelman J, et al. Pharmacokinetic modeling and simulation of biweekly subcutaneous immunoglobulin dosing in primary immunodeficiency. Postgrad Med. 2013;125:53–61.
64. Moore ML, Quinn JM. Subcutaneous immunoglobulin replacement therapy for primary antibody deficiency: advancements into the 21st century. Ann Allergy Asthma Immunol. 2008;101:114–21.
65. Zhao L, Ji P, Li Z, et al. The antibody drug absorption following subcutaneous or intramuscular administration and its mathematical description by coupling physiologically based absorption process with the conventional compartment pharmacokinetic model. J Clin Pharmacol. 2013;53:314–25.
66. Losonsky GA, Johnson JP, Winklestein JA, et al. Oral administration of human serum immunoglobulin in immunodeficient patients with viral gastroenteritis. A pharmacokinetic and functional analysis. J Clin Invest. 1985;76:2362–7.
67. Kelly CP, Chetham S, Keates S, et al. Survival of anti-*Clostridium difficile* bovine immunoglobulin concentrate in the human gastrointestinal tract. Antimicrob Agents Chemother. 1997;41:236–41.
68. Warny M, Fatimi A, Bostwick EF, et al. Bovine immunoglobulin concentrate-*Clostridium difficile* retains *C difficile* toxin neutralising activity after passage through the human stomach and small intestine. Gut. 1999;44:212–7.
69. Morell A, Skvaril F, Noseda G, et al. Metabolic properties of human IgA subclasses. Clin Exp Immunol. 1973;13:521–8.
70. Dechant M, Valerius T. IgA antibodies for cancer therapy. Crit Rev Oncol Hematol. 2001;39:69–77.
71. Conley ME, Delacroix DL. Intravascular and mucosal immunoglobulin A: two separate but related systems of immune defense? Ann Intern Med. 1987;106:892–9.

72. Longet S, Miled S, Lotscher M, et al. Human plasma-derived polymeric IgA and IgM antibodies associate with secretory component to yield biologically active secretory-like antibodies. J Biol Chem. 2013;288:4085–94.
73. Behring EA, Kitasato S. Über das Zustandekommen der Diphtherie-Immunität und der Tetanus-Immunität bei Tieren. Dtsch Med Wochenschr. 1890;49:1113–4.
74. Stephan W. Beseitigung der Komplementfixierung von Gamma-Globulin durch chemische Modifizierung mit Beta-Propiolacton. Z Klin Chem Klein Biochem. 1969;7:282–6.
75. Radosevich M, Burnouf T. Intravenous immunoglobulin G: trends in production methods, quality control and quality assurance. Vox Sang. 2010;98:12–28.
76. Komenda M, Stadler D, Malinas T, et al. Assessment of the ability of the Privigen(R) purification process to deplete thrombogenic factor XIa from plasma. Vox Sang. 2014;107:26–36.
77. Hoefferer L, Glauser I, Gaida A, et al. Isoagglutinin reduction by a dedicated immunoaffinity chromatography step in the manufacturing process of human immunoglobulin products. Transfusion. 2015;55(Suppl 2):S117–21.
78. Romberg V, Hoefferer L, El Menyawi I. Effects of the manufacturing process on the anti-A isoagglutinin titers in intravenous immunoglobulin products. Transfusion. 2015;55(Suppl 2):S105–9.
79. Siani B, Willimann K, Wymann S, et al. Donor screening reduces the isoagglutinin titer in immunoglobulin products. Transfusion. 2015;55(Suppl 2):S95–7.
80. Bolli R, Woodtli K, Bartschi M, et al. L-Proline reduces IgG dimer content and enhances the stability of intravenous immunoglobulin (IVIG) solutions. Biologicals. 2010;38:150–7.
81. Cramer M, Frei R, Sebald A, et al. Stability over 36 months of a new liquid 10% polyclonal immunoglobulin product (IgPro10, Privigen©) stabilized with L-proline. Vox Sang. 2009;96:219–25.
82. Tankersley DL. Dimer formation in immunoglobulin preparations and speculations on the mechanism of action of intravenous immune globulin in autoimmune disease. Immunol Rev. 1994;139:159–72.
83. Tankersley DL, Preston MS, Finlayson JS. Immunoglobulin G dimer: an idiotype/anti-idiotype complex. Mol Immunol. 1988;25:41–8.
84. Kroez M, Kanzy EJ, Gronski P, et al. Hypotension with intravenous immunoglobulin therapy: importance of pH and dimer formation. Biologicals. 2003;31:277–86.
85. Spycher MO, Bolli R, Hodler G, et al. Well-tolerated liquid intravenous immunoglobulin G preparations (IVIG) have a low immunoglobulin G dimer (IgG-dimer) content. J Autoimmun. 1996;96(Suppl 1):96.
86. Misbah S, Sturzenegger MH, Borte M, et al. Subcutaneous immunoglobulin: opportunities and outlook. Clin Exp Immunol. 2009;158(Suppl 1):51–9.
87. Berger M. Principles of and advances in immunoglobulin replacement therapy for primary immunodeficiency. Immunol Allergy Clin North Am. 2008a;28:413–37.
88. Fernandez-Cruz E, Kaveri SV, Peter HH, et al. 6th International Immunoglobulin Symposium: poster presentations. Clin Exp Immunol. 2009;158(Suppl 1):60–7.
89. Stiehm ER, Keller MA, Vyas GN. Preparation and use of therapeutic antibodies primarily of human origin. Biologicals. 2008;36:363–74.
90. Stucki M, Moudry R, Kempf C, et al. Characterisation of a chromatographically produced anti-D immunoglobulin product. J Chromatogr B Biomed Sci Appl. 1997;700:241–8.
91. Gaspar HB, Cooray S, Gilmour KC, et al. Hematopoietic stem cell gene therapy for adenosine deaminase-deficient severe combined immunodeficiency leads to long-term immunological recovery and metabolic correction. Sci Transl Med. 2011;3:97ra80.
92. Hacein-Bey-Abina S, Pai SY, Gaspar HB, et al. A modified gamma-retrovirus vector for X-linked severe combined immunodeficiency. N Engl J Med. 2014;371:1407–17.
93. Elliott DE, Weinstock JV. Helminthic therapy: using worms to treat immune-mediated disease. Adv Exp Med Biol. 2009;666:157–66.
94. DeKosky BJ, Kojima T, Rodin A, et al. In-depth determination and analysis of the human paired heavy- and light-chain antibody repertoire. Nat Med. 2015;21:86–91.
95. De Groot AS, Moise L, McMurry JA, et al. Activation of natural regulatory T cells by IgG Fc-derived peptide "Tregitopes". Blood. 2008;112:3303–11.
96. Cousens LP, Su Y, McClaine E, et al. Application of IgG-derived natural Treg epitopes (IgG Tregitopes) to antigen-specific tolerance induction in a murine model of type 1 diabetes. J Diabetes Res. 2013;2013:621693.
97. Jain A, Olsen HS, Vyzasatya R, et al. Fully recombinant IgG2a Fc multimers (stradomers) effectively treat collagen-induced arthritis and prevent idiopathic thrombocytopenic purpura in mice. Arthritis Res Ther. 2012;14:R192.
98. Niknami M, Wang MX, Nguyen T, et al. Beneficial effect of a multimerized immunoglobulin Fc in an animal model of inflammatory neuropathy (experimental autoimmune neuritis). J Peripher Nerv Syst. 2013;18:141–52.
99. Thiruppathi M, Sheng JR, Li L, et al. Recombinant IgG2a Fc (M045) multimers effectively suppress experimental autoimmune myasthenia gravis. J Autoimmun. 2014;52:64–73.
100. Mekhaiel DN, Czajkowsky DM, Andersen JT, et al. Polymeric human Fc-fusion proteins with modified effector functions. Sci Rep. 2011;1:124.
101. Patel DA, Puig-Canto A, Challa DK, et al. Neonatal Fc receptor blockade by Fc engineering

ameliorates arthritis in a murine model. J Immunol. 2011;187:1015–22.
102. Werwitzke S, Trick D, Sondermann P, et al. Treatment of lupus-prone NZB/NZW F1 mice with recombinant soluble Fc gamma receptor II (CD32). Ann Rheum Dis. 2008;67:154–61.
103. Berger M. A history of immune globulin therapy, from the Harvard crash program to monoclonal antibodies. Curr Allergy Asthma Rep. 2002;2: 368–78.
104. Imbach P. 30 years of immunomodulation by intravenous immunoglobulin. Immunotherapy. 2012;4:651–4.
105. Berger M. Choices in IgG replacement therapy for primary immune deficiency diseases: subcutaneous IgG vs. intravenous IgG and selecting an optimal dose. Curr Opin Allergy Clin Immunol. 2011;11:532–8.
106. Eijkhout HW, van Der Meer JW, Kallenberg CG, et al. The effect of two different dosages of intravenous immunoglobulin on the incidence of recurrent infections in patients with primary hypogammaglobulinemia. A randomized, double-blind, multicenter crossover trial. Ann Intern Med. 2001;135:165–74.
107. Jolles S, Orange JS, Gardulf A, et al. Current treatment options with immunoglobulin G for the individualization of care in patients with primary immunodeficiency disease. Clin Exp Immunol. 2015;179:146–60.
108. Roifman CM, Levison H, Gelfand EW. High-dose versus low-dose intravenous immunoglobulin in hypogammaglobulinaemia and chronic lung disease. Lancet. 1987;1:1075–7.
109. Ballow M. The IgG molecule as a biological immune response modifier: mechanisms of action of intravenous immune serum globulin in autoimmune and inflammatory disorders. J Allergy Clin Immunol. 2011;127:315–23.
110. Gelfand EW. Intravenous immune globulin in autoimmune and inflammatory diseases. N Engl J Med. 2012;367:2015–25.
111. Buckley RH, editor. Immune Deficiency Foundation diagnostic and clinical care guidelines for primary immunodeficiency diseases. 3rd ed. Towson, MD: Immune Deficiency Foundation; 2015. p. 4–7.
112. Hernandez-Trujillo V. New genetic discoveries and primary immune deficiencies. Clin Rev Allergy Immunol. 2014;46:145–53.
113. Parvaneh N, Casanova JL, Notarangelo LD, et al. Primary immunodeficiencies: a rapidly evolving story. J Allergy Clin Immunol. 2013;131:314–23.
114. Winkelstein JA, Marino MC, Lederman HM, et al. X-linked agammaglobulinemia: report on a United States registry of 201 patients. Medicine (Baltimore). 2006;85:193–202.
115. Orange JS, Belohradsky BH, Berger M, et al. Evaluation of correlation between dose and clinical outcomes in subcutaneous immunoglobulin replacement therapy. Clin Exp Immunol. 2012b;169:172–81.
116. Orange JS, Grossman WJ, Navickis RJ, et al. Impact of trough IgG on pneumonia incidence in primary immunodeficiency: a meta-analysis of clinical studies. Clin Immunol. 2010;137:21–30.
117. Bonagura VR, Marchlewski R, Cox A, et al. Biologic IgG level in primary immunodeficiency disease: the IgG level that protects against recurrent infection. J Allergy Clin Immunol. 2008;122:210–2.
118. Lucas M, Lee M, Lortan J, et al. Infection outcomes in patients with common variable immunodeficiency disorders: relationship to immunoglobulin therapy over 22 years. J Allergy Clin Immunol. 2010;125:1354–60.
119. Chen Y, Stirling RG, Paul E, et al. Longitudinal decline in lung function in patients with primary immunoglobulin deficiencies. J Allergy Clin Immunol. 2011;127:1414–7.
120. Jolles S. The variable in common variable immunodeficiency: a disease of complex phenotypes. J Allergy Clin Immunol Pract. 2013;1:545–56.
121. Orange JS, Glessner JT, Resnick E, et al. Genome-wide association identifies diverse causes of common variable immunodeficiency. J Allergy Clin Immunol. 2011;127:1360–7.
122. Resnick ES, Moshier EL, Godbold JH, et al. Morbidity and mortality in common variable immune deficiency over 4 decades. Blood. 2012;119:1650–7.
123. Gathmann B, Mahlaoui N, Gerard L, et al. Clinical picture and treatment of 2212 patients with common variable immunodeficiency. J Allergy Clin Immunol. 2014;134:116–26.
124. Chase NM, Verbsky JW, Hintermeyer MK, et al. Use of combination chemotherapy for treatment of granulomatous and lymphocytic interstitial lung disease (GLILD) in patients with common variable immunodeficiency (CVID). J Clin Immunol. 2013;33:30–9.
125. Conley ME, Dobbs AK, Farmer DM, et al. Primary B cell immunodeficiencies: comparisons and contrasts. Annu Rev Immunol. 2009;27:199–227.
126. Qamar N, Fuleihan RL. The hyper IgM syndromes. Clin Rev Allergy Immunol. 2014;46:120–30.
127. Ambrose M, Gatti RA. Pathogenesis of ataxia-telangiectasia: the next generation of ATM functions. Blood. 2013;121:4036–45.
128. Buchbinder D, Nugent DJ, Fillipovich AH. Wiskott-Aldrich syndrome: diagnosis, current management, and emerging treatments. Appl Clin Genet. 2014;7:55–66.
129. Dvorak CC, Cowan MJ, Logan BR, et al. The natural history of children with severe combined immunodeficiency: baseline features of the first fifty patients of the primary immune deficiency treatment consortium prospective study 6901. J Clin Immunol. 2013;33:1156–64.
130. Boisson B, Quartier P, Casanova JL. Immunological loss-of-function due to genetic gain-of-function in humans: autosomal dominance of the third kind. Curr Opin Immunol. 2015;32:90–105.
131. Kashani S, Carr TF, Grammer LC, et al. Clinical characteristics of adults with chronic rhinosinusitis and

specific antibody deficiency. J Allergy Clin Immunol Pract. 2015;3:236–42.
132. Orange JS, Ballow M, Stiehm ER, et al. Use and interpretation of diagnostic vaccination in primary immunodeficiency: a working group report of the Basic and Clinical Immunology Interest Section of the American Academy of Allergy, Asthma & Immunology. J Allergy Clin Immunol. 2012a;130:S1–24.
133. Stevens WW, Peters AT. Immunodeficiency in chronic sinusitis: recognition and treatment. Am J Rhinol Allergy. 2015;29:115–8.
134. Makatsori M, Kiani-Alikhan S, Manson AL, et al. Hypogammaglobulinaemia after rituximab treatment-incidence and outcomes. QJM. 2014;107:821–8.
135. National Blood Authority Australia. Annual report 2013–14. 2014. Available at: http://www.blood.gov.au/nba-annual-report-2013-14. Accessed 31 July 2015.
136. Roberts DM, Jones RB, Smith RM, et al. Immunoglobulin G replacement for the treatment of infective complications of rituximab-associated hypogammaglobulinemia in autoimmune disease: a case series. J Autoimmun. 2015;57:24–9.
137. Boddana P, Webb LH, Unsworth J, et al. Hypogammaglobulinemia and bronchiectasis in mycophenolate mofetil-treated renal transplant recipients: an emerging clinical phenomenon? Clin Transpl. 2011;25:417–9.
138. Bussel JB, Graziano JN, Kimberly RP, et al. Intravenous anti-D treatment of immune thrombocytopenic purpura: analysis of efficacy, toxicity, and mechanism of effect. Blood. 1991;77:1884–93.
139. Neunert C, Lim W, Crowther M, et al. The American Society of Hematology 2011 evidence-based practice guideline for immune thrombocytopenia. Blood. 2011;117:4190–207.
140. Provan D, Stasi R, Newland AC, et al. International consensus report on the investigation and management of primary immune thrombocytopenia. Blood. 2010;115:168–86.
141. Bussel JB, Hilgartner MW. The use and mechanism of action of intravenous immunoglobulin in the treatment of immune haematologic disease. Br J Haematol. 1984;56:1–7.
142. Fehr J, Hofmann V, Kappeler U. Transient reversal of thrombocytopenia in idiopathic thrombocytopenic purpura by high-dose intravenous gamma globulin. N Engl J Med. 1982;306:1254–8.
143. Rowley AH, Shulman ST. Pathogenesis and management of Kawasaki disease. Expert Rev Anti Infect Ther. 2010;8:197–203.
144. Terai M, Shulman ST. Prevalence of coronary artery abnormalities in Kawasaki disease is highly dependent on gamma globulin dose but independent of salicylate dose. J Pediatr. 1997;131:888–93.
145. Newburger JW, Takahashi M, Beiser AS, et al. A single intravenous infusion of gamma globulin as compared with four infusions in the treatment of acute Kawasaki syndrome. N Engl J Med. 1991;324:1633–9.
146. Yuki N, Hartung HP. Guillain-Barre syndrome. N Engl J Med. 2012;366:2294–304.
147. Hughes RA, Swan AV, Raphael JC, et al. Immunotherapy for Guillain-Barre syndrome: a systematic review. Brain. 2007;130:2245–57.
148. van Doorn PA, Brand A, Vermeulen M. Anti-neuroblastoma cell line antibodies in inflammatory demyelinating polyneuropathy: inhibition in vitro and in vivo by IV immunoglobulin. Neurology. 1988;38:1592–5.
149. Zhang G, Lopez PH, Li CY, et al. Anti-ganglioside antibody-mediated neuronal cytotoxicity and its protection by intravenous immunoglobulin: implications for immune neuropathies. Brain. 2004;127:1085–100.
150. Hughes RA, Swan AV, van Doorn PA. Intravenous immunoglobulin for Guillain-Barre syndrome. Cochrane Database Syst Rev. 2014;(9):CD002063.
151. Hughes RA, Wijdicks EF, Barohn R, et al. Practice parameter: immunotherapy for Guillain-Barre syndrome: report of the Quality Standards Subcommittee of the American Academy of Neurology. Neurology. 2003;61:736–40.
152. Kuitwaard K, de GJ, Tio-Gillen AP, et al. Pharmacokinetics of intravenous immunoglobulin and outcome in Guillain-Barre syndrome. Ann Neurol. 2009;66:597–603.
153. Hughes RA, Donofrio P, Bril V, et al. Intravenous immune globulin (10% caprylate-chromatography purified) for the treatment of chronic inflammatory demyelinating polyradiculoneuropathy (ICE study): a randomised placebo-controlled trial. Lancet Neurol. 2008;7:136–44.
154. Peltier AC, Donofrio PD. Chronic inflammatory demyelinating polyradiculoneuropathy: from bench to bedside. Semin Neurol. 2012;32:187–95.
155. Berger M, McCallus DE, Lin CS. Rapid and reversible responses to IVIG in autoimmune neuromuscular diseases suggest mechanisms of action involving competition with functionally important autoantibodies. J Peripher Nerv Syst. 2013b;18:275–96.
156. Broyles R, Rodden L, Riley P, et al. Variability in intravenous immunoglobulin G regimens for autoimmune neuromuscular disorders. Postgrad Med. 2013;125:65–72.
157. Lin CS, Krishnan AV, Park SB, et al. Modulatory effects on axonal function after intravenous immunoglobulin therapy in chronic inflammatory demyelinating polyneuropathy. Arch Neurol. 2011;68:862–9.
158. Joint Task Force of the EFNS and the PNS. European Federation of Neurological Societies/Peripheral Nerve Society guideline on management of multifocal motor neuropathy. Report of a joint task force of the European Federation of Neurological Societies and the Peripheral Nerve Society—first revision. J Peripher Nerv Syst. 2010;15:295–301.

159. Kuitwaard K, van Doorn PA, Vermeulen M, et al. Serum IgG levels in IV immunoglobulin treated chronic inflammatory demyelinating polyneuropathy. J Neurol Neurosurg Psychiatry. 2013; 84:859–61.
160. Berger M, Allen JA. Optimizing IgG therapy in chronic autoimmune neuropathies: a hypothesis driven approach. Muscle Nerve. 2015;51:315–26.
161. Nobile-Orazio E, Cocito D, Jann S, et al. Intravenous immunoglobulin versus intravenous methylprednisolone for chronic inflammatory demyelinating polyradiculoneuropathy: a randomised controlled trial. Lancet Neurol. 2012;11:493–502.
162. Muley SA, Parry GJ. Multifocal motor neuropathy. J Clin Neurosci. 2012;19:1201–9.
163. Vlam L, van der Pol WL, Cats EA, et al. Multifocal motor neuropathy: diagnosis, pathogenesis and treatment strategies. Nat Rev Neurol. 2012;8: 48–58.
164. Carpo M, Cappellari A, Mora G, et al. Deterioration of multifocal motor neuropathy after plasma exchange. Neurology. 1998;50:1480–2.
165. Donaghy M, Mills KR, Boniface SJ, et al. Pure motor demyelinating neuropathy: deterioration after steroid treatment and improvement with intravenous immunoglobulin. J Neurol Neurosurg Psychiatry. 1994;57:778–83.
166. Hahn AF, Beydoun SR, Lawson V, et al. A controlled trial of intravenous immunoglobulin in multifocal motor neuropathy. J Peripher Nerv Syst. 2013;18:321–30.
167. Nobile-Orazio E, Cappellari A, Meucci N, et al. Multifocal motor neuropathy: clinical and immunological features and response to IVIg in relation to the presence and degree of motor conduction block. J Neurol Neurosurg Psychiatry. 2002;72:761–6.
168. Alabdali M, Barnett C, Katzberg H, et al. Intravenous immunoglobulin as treatment for myasthenia gravis: current evidence and outcomes. Expert Rev Clin Immunol. 2014;10:1659–65.
169. Higuchi O, Hamuro J, Motomura M, et al. Autoantibodies to low-density lipoprotein receptor-related protein 4 in myasthenia gravis. Ann Neurol. 2011;69:418–22.
170. Zhang B, Tzartos JS, Belimezi M, et al. Autoantibodies to lipoprotein-related protein 4 in patients with double-seronegative myasthenia gravis. Arch Neurol. 2012;69:445–51.
171. Eienbroker C, Seitz F, Spengler A, et al. Intravenous immunoglobulin maintenance treatment in myasthenia gravis: a randomized, controlled trial sample size simulation. Muscle Nerve. 2014;50:999–1004.
172. Liew WK, Powell CA, Sloan SR, et al. Comparison of plasmapheresis and intravenous immunoglobulin as maintenance therapies for juvenile myasthenia gravis. JAMA Neurol. 2014;71:575–80.
173. Hellmann MA, Mosberg-Galili R, Lotan I, et al. Maintenance IVIg therapy in myasthenia gravis does not affect disease activity. J Neurol Sci. 2014;338:39–42.
174. Chee SN, Murrell DF. The use of intravenous immunoglobulin in autoimmune bullous diseases. Immunol Allergy Clin North Am. 2012;32:323–30, viii.
175. Gurcan HM, Jeph S, Ahmed AR. Intravenous immunoglobulin therapy in autoimmune mucocutaneous blistering diseases: a review of the evidence for its efficacy and safety. Am J Clin Dermatol. 2010;11:315–26.
176. Sami N, Bhol KC, Ahmed RA. Influence of intravenous immunoglobulin therapy on autoantibody titers to desmoglein 3 and desmoglein 1 in pemphigus vulgaris. Eur J Dermatol. 2003;13:377–81.
177. Vo AA, Lukovsky M, Toyoda M, et al. Rituximab and intravenous immune globulin for desensitization during renal transplantation. N Engl J Med. 2008;359:242–51.
178. Burton SA, Amir N, Asbury A, et al. Treatment of antibody-mediated rejection in renal transplant patients: a clinical practice survey. Clin Transpl. 2015;29:118–23.
179. Farmer DG, Kattan OM, Wozniak LJ, et al. Incidence, timing, and significance of early hypogammaglobulinemia after intestinal transplantation. Transplantation. 2013;95:1154–9.
180. Mawhorter S, Yamani MH. Hypogammaglobulinemia and infection risk in solid organ transplant recipients. Curr Opin Organ Transplant. 2008;13:581–5.
181. Cavill D, Waterman SA, Gordon TP. Antiidiotypic antibodies neutralize autoantibodies that inhibit cholinergic neurotransmission. Arthritis Rheum. 2003;48:3597–602.
182. Sultan Y, Rossi F, Kazatchkine MD. Recovery from anti-VIII:C (antihemophilic factor) autoimmune disease is dependent on generation of antiidiotypes against anti-VIII:C autoantibodies. Proc Natl Acad Sci U S A. 1987;84:828–31.
183. Abe Y, Horiuchi A, Miyake M, et al. Anti-cytokine nature of natural human immunoglobulin: one possible mechanism of the clinical effect of intravenous immunoglobulin therapy. Immunol Rev. 1994;139:5–19.
184. Basta M, Van Goor F, Luccioli S, et al. F(ab)′2-mediated neutralization of C3a and C5a anaphylatoxins: a novel effector function of immunoglobulins. Nat Med. 2003;9:431–8.
185. Tackenberg B, Jelcic I, Baerenwaldt A, et al. Impaired inhibitory Fcgamma receptor IIB expression on B cells in chronic inflammatory demyelinating polyneuropathy. Proc Natl Acad Sci U S A. 2009;106:4788–92.
186. Tjon AS, van Gent R, Jaadar H, et al. Intravenous immunoglobulin treatment in humans suppresses dendritic cell function via stimulation of IL-4 and IL-13 production. J Immunol. 2014;192: 5625–34.
187. Yu Z, Lennon VA. Mechanism of intravenous immune globulin therapy in antibody-mediated autoimmune diseases. N Engl J Med. 1999;340: 227–8.

188. Crocker PR, Paulson JC, Varki A. Siglecs and their roles in the immune system. Nat Rev Immunol. 2007;7:255–66.
189. Bayry J, Lacroix-Desmazes S, Carbonneil C, et al. Inhibition of maturation and function of dendritic cells by intravenous immunoglobulin. Blood. 2003;101:758–65.
190. Trinath J, Hegde P, Sharma M, et al. Intravenous immunoglobulin expands regulatory T cells via induction of cyclooxygenase-2-dependent prostaglandin E2 in human dendritic cells. Blood. 2013;122:1419–27.
191. Mausberg AK, Dorok M, Stettner M, et al. Recovery of the T-cell repertoire in CIDP by IV immunoglobulins. Neurology. 2013;80:296–303.
192. Lunemann JD, Nimmerjahn F, Dalakas MC. Intravenous immunoglobulin in neurology—mode of action and clinical efficacy. Nat Rev Neurol. 2015;11:80–9.
193. Nimmerjahn F, Gordan S, Lux A. FcgammaR dependent mechanisms of cytotoxic, agonistic, and neutralizing antibody activities. Trends Immunol. 2015;36:325–36.
194. Nimmerjahn F, Ravetch JV. Anti-inflammatory actions of intravenous immunoglobulin. Annu Rev Immunol. 2008;26:513–33.
195. Tjon AS, Tha-In T, Metselaar HJ, et al. Patients treated with high-dose intravenous immunoglobulin show selective activation of regulatory T cells. Clin Exp Immunol. 2013;173:259–67.
196. Ballow M. Safety of IGIV therapy and infusion-related adverse events. Immunol Res. 2007;38:122–32.
197. Bonilla FA. Intravenous immunoglobulin: adverse reactions and management. J Allergy Clin Immunol. 2008a;122:1238–9.
198. Brennan VM, Salome-Bentley NJ, Chapel HM. Prospective audit of adverse reactions occurring in 459 primary antibody-deficient patients receiving intravenous immunoglobulin. Clin Exp Immunol. 2003;133:247–51.
199. Horn J, Thon V, Bartonkova D, et al. Anti-IgA antibodies in common variable immunodeficiencies (CVID): diagnostic workup and therapeutic strategy. Clin Immunol. 2007;122:156–62.
200. Rachid R, Bonilla FA. The role of anti-IgA antibodies in causing adverse reactions to gamma globulin infusion in immunodeficient patients: a comprehensive review of the literature. J Allergy Clin Immunol. 2012;129:628–34.
201. Sandler SG, Eder AF, Goldman M, et al. The entity of immunoglobulin A-related anaphylactic transfusion reactions is not evidence based. Transfusion. 2015;55:199–204.
202. Sundin U, Nava S, Hammarstrom L. Induction of unresponsiveness against IgA in IgA-deficient patients on subcutaneous immunoglobulin infusion therapy. Clin Exp Immunol. 1998;112:341–6.
203. Sekul EA, Cupler EJ, Dalakas MC. Aseptic meningitis associated with high-dose intravenous immunoglobulin therapy: frequency and risk factors. Ann Intern Med. 1994;121:259–62.
204. Daw Z, Padmore R, Neurath D, et al. Hemolytic transfusion reactions after administration of intravenous immune (gamma) globulin: a case series analysis. Transfusion. 2008;48:1598–601.
205. Kahwaji J, Barker E, Pepkowitz S, et al. Acute hemolysis after high-dose intravenous immunoglobulin therapy in highly HLA sensitized patients. Clin J Am Soc Nephrol. 2009;4:1993–7.
206. Desborough MJ, Miller J, Thorpe SJ, et al. Intravenous immunoglobulin-induced haemolysis: a case report and review of the literature. Transfus Med. 2014;24:219–26.
207. Dhainaut F, Guillaumat PO, Dib H, et al. In vitro and in vivo properties differ among liquid intravenous immunoglobulin preparations. Vox Sang. 2013;104:115–26.
208. Siani B, Willimann K, Wymann S, et al. Isoagglutinin reduction in human immunoglobulin products by donor screening. Biol Ther. 2014;4:15–26.
209. Berger M. Adverse effects of IgG therapy. J Allergy Clin Immunol Pract. 2013;1:558–66.
210. Funk MB, Gross N, Gross S, et al. Thromboembolic events associated with immunoglobulin treatment. Vox Sang. 2013;105:54–64.
211. Dickenmann M, Oettl T, Mihatsch MJ. Osmotic nephrosis: acute kidney injury with accumulation of proximal tubular lysosomes due to administration of exogenous solutes. Am J Kidney Dis. 2008;51:491–503.
212. Gaines R, Varricchio F, Kapit R, et al. Renal insufficiency and failure associated with immune globulin intravenous therapy – United States, 1985–1998. MMWR Morb Mortal Wkly Rep. 1999;48:518–21.
213. Chapman SA, Gilkerson KL, Davin TD, et al. Acute renal failure and intravenous immune globulin: occurs with sucrose-stabilized, but not with D-sorbitol-stabilized formulation. Ann Pharmacother. 2004;38:2059–67.
214. Winward DB, Brophy MT. Acute renal failure after administration of intravenous immunoglobulin: review of the literature and case report. Pharmacotherapy. 1995;15:765–72.
215. Welles CC, Tambra S, Lafayette RA. Hemoglobinuria and acute kidney injury requiring hemodialysis following intravenous immunoglobulin infusion. Am J Kidney Dis. 2010;55:148–51.
216. Yap PL. The viral safety of intravenous immune globulin. Clin Exp Immunol. 1996;104(Suppl 1):35–42.
217. Dichtelmuller HO, Biesert L, Fabbrizzi F, et al. Contribution to safety of immunoglobulin and albumin from virus partitioning and inactivation by cold ethanol fractionation: a data collection from Plasma Protein Therapeutics Association member companies. Transfusion. 2011;51:1412–30.
218. Cai K, Groner A, Dichtelmuller HO, et al. Prion removal capacity of plasma protein manufacturing processes: a data collection from PPTA member companies. Transfusion. 2013;53:1894–905.

219. Dichtelmuller HO, Biesert L, Fabbrizzi F, et al. Robustness of solvent/detergent treatment of plasma derivatives: a data collection from Plasma Protein Therapeutics Association member companies. Transfusion. 2009;49:1931–43.
220. Soluk L, Price H, Sinclair C, et al. Pathogen safety of intravenous Rh immunoglobulin liquid and other immune globulin products: enhanced nanofiltration and manufacturing process overview. Am J Ther. 2008;15:435–43.
221. Stucki M, Boschetti N, Schaefer W, et al. Investigations of prion and virus safety of a new liquid IVIG product. Biologicals. 2008;36:239–47.

Monoclonal Antibodies for Immune System-Mediated Diseases

21

Frank R. Brennan, Emma Smith, and Sherri Dudal

21.1 Introduction

Over the last two decades, large molecule biologics such as MONOCLONAL ANTIBODIES (mAbs) have emerged as alternatives to small molecule chemical drugs for treating a wide range of human diseases, including AUTOIMMUNITY and cancer. There are currently over 50 mAbs and 3 Fc-fusion proteins (consisting of an endogenous RECEPTOR molecule linked to the Fc fragment of human IgG) currently approved for human use. Among the 31 of these target immunological diseases are RHEUMATOID ARTHRITIS (RA), PSORIASIS, MULTIPLE SCLEROSIS (MS), SYSTEMIC LUPUS ERYTHEMATOSUS (SLE), inflammatory bowel disease (IBD), ASTHMA and organ transplant rejection [1]. The success of mAb-based therapeutics is due to their exquisite specificity, which, in combination with their multi-functional properties, high potency, long HALF-LIFE (permitting intermittent dosing and prolonged pharmacological effects) and general lack of off-target toxicity, makes them ideal therapeutics. Over time, knowledge has been acquired to understand what may limit the EFFICACY of mAbs such as immunogenicity and the challenges of certain routes of administration and to have a better understanding of the mechanism of action. This chapter aims to give a general introduction to the immune diseases and inflammatory pathways being targeted by mAbs, focusing on those that are approved or in late-stage clinical development.

Final manuscript submitted on January 25, 2017.

F. R. Brennan (✉) · S. Dudal
UCB Pharma, New Medicines, Non-Clinical Development, Slough, UK
e-mail: frank.brennan@ucb.com; sherri.dudal@roche.com

E. Smith
UCB Pharma, New Medicines, Translational Medicine, Slough, UK
e-mail: emma.smith@ucb.com

21.2 General Overview of the Immune System and Key Pathways Driving Inflammatory Diseases

The IMMUNE SYSTEM is composed of a diverse array of immunocompetent cells such as T- and B-LYMPHOCYTES (T and B cells), MONOCYTES/MACROPHAGES, granulocytes, NATURAL KILLER (NK) CELLS, MAST CELLS and DENDRITIC CELLS (DCs). They secrete pro-inflammatory and anti-inflammatory mediators (CYTOKINES, CHEMOKINES, GROWTH FACTORS, COMPLEMENT components) that co-exist in regulated networks. These components interact through cascades and utilize positive and negative feedback loops to maintain normal INFLAMMATION and immunosurveil-

M. J. Parnham et al. (eds.), *Nijkamp and Parnham's Principles of Immunopharmacology*,
https://doi.org/10.1007/978-3-030-10811-3_21

lance against invading pathogens. However during immune diseases, a foreign or self (auto)-ANTIGEN or allergen might trigger the dysregulation of these networks leading to altered immunity, persistent INFLAMMATION and ultimately pathologic sequelae [2]. These diseases can be broadly categorized into those inflammatory/AUTOIMMUNE DISEASES which exhibit an immunological profile characterized by a T HELPER(TH)1/TH17 cell response (e.g. RA, PSORIASIS, MS, IBD) versus allergic diseases (e.g. ASTHMA, atopic dermatitis) with a predominantly TH2 cell profile. However, for certain diseases and patient subsets, this classification is too simplistic.

The initiation of an adaptive cell-mediated immune response begins with the production of pro-INFLAMMATORY CYTOKINES from innate immune cells such as DCs, MONOCYTES/MACROPHAGES and granulocytes and from tissue cells such as fibroblasts, smooth muscle cells and epithelial and endothelial cells in response to a stimulus (Fig. 21.1). This stimulus can be an infection, tissue injury or exposure to ANTIGEN. This then drives the activation of naive CD4+ T CELLS and their differentiation into specific T HELPER CELL subsets with distinct effector functions. Autocrine production of INTERFERON-γ (IFNγ) (driven by IL-12 and IL-18), IL-4 and IL-21 promotes the differentiation of TH1, TH2 and T follicular helper cells, respectively, whilst IL-6, IL-23, IL-1 and TRANSFORMING GROWTH FACTOR-β (TGF-β) produced by DCs and other non-T CELLS create an environment that favours TH17 cell development. IFNγ and TUMOUR NECROSIS FACTOR-α (TNF-α) secreted by TH1 cells activate MACROPHAGES and DCs to generate REACTIVE OXYGEN SPECIES and reactive nitrogen species and recruit NEUTROPHILS, all of which are pivotal in mediating host defence against microbial species but also can induce collateral tissue damage. In addition, TH1 memory cells are key for survival and function of memory CD8+ T CELLS. IL-21 secreted by TFH cells promotes germinal centre formation and B CELL development, activation and IMMUNOGLOBULIN production. CYTOKINES such as IL-4 and IFN-γ promote IMMUNOGLOBULIN class switching. IL-17A, IL-17F and IL-22 produced by TH17 cells and γδ T CELLS mediate an antibacterial activity at epithelial barriers by promoting recruitment

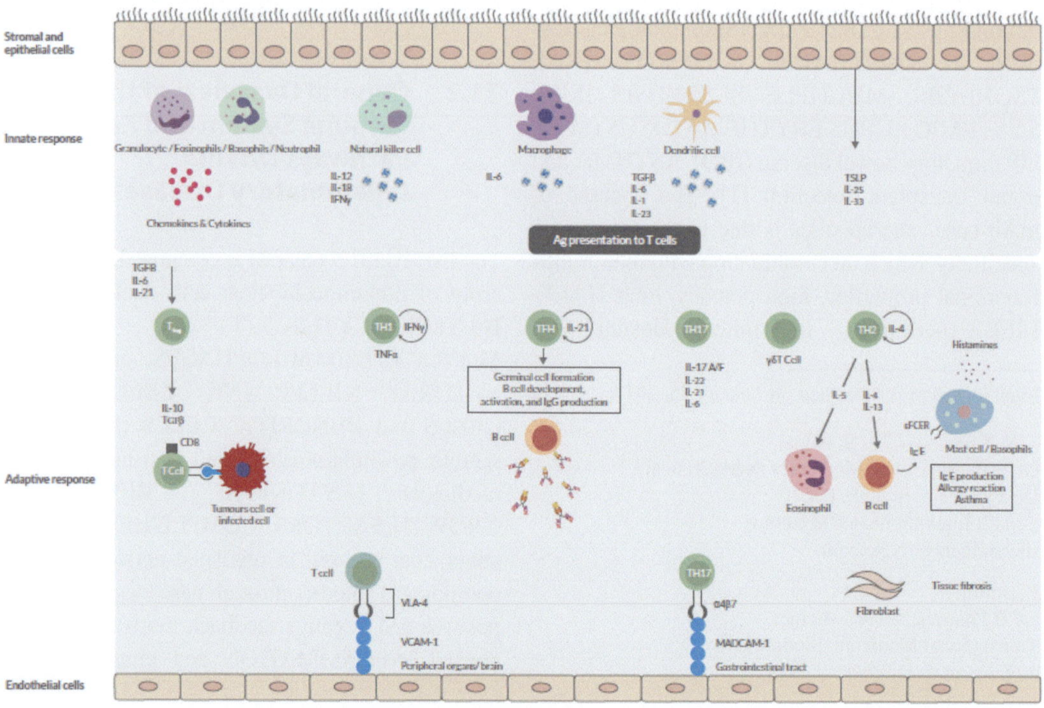

Fig. 21.1 Overview of innate and adaptive immune responses

and activation of NEUTROPHILS and MACROPHAGES (type 1 effector) and epithelial cells.

However, dysregulated TH17 cells can also drive organ-specific AUTOIMMUNITY such as RA and PSORIASIS via secretion of IL-17A/F, IL-1, IL-6, IL-12, IL-18 and COLONY-STIMULATING FACTOR-2 (CSF2) from MACROPHAGES, DCs and other cells that promote these autoinflammatory TH1 and TH17 responses. TH2 cells are responsible for ALLERGY, ASTHMA and tissue fibrosis by the secretion of IL-4, IL-5 and IL-13. CYTOKINES such as thymic stromal lymphopoietin (TSLP), IL-25 and IL-33 produced by stromal cells and other T CELLS induce the production of these CYTOKINES and serve to induce and amplify this TH2 allergic environment. IL-5 is key for eosinophilic INFLAMMATION, whereas IL-4 and IL-13 promote B CELL IgE production (that binds Fc RECEPTORS for IgE (FcεRI/II)) and drives MAST CELL activation following cross-linking by allergen, goblet cell mucus secretion, fibrosis and alternative activation of MACROPHAGES (type 2 MACROPHAGES). REGULATORY T CELLS (Tregs) interacting with specific tolerogenic DC subsets can efficiently control EFFECTOR T CELL response to self and non-self-antigens through the expression of anti-inflammatory molecules such as IL-10, TGF-β and/or inhibitory RECEPTORS such as CTLA-4 [2]. These Tregs and DCs are frequently dysregulated during inflammatory diseases.

In view of their pivotal role in the development, maintenance and chronicity of autoimmune and allergic disease, T and B CELLS along with their soluble mediators and cell surface RECEPTORS represent the major targets of mAbs for immunological disease therapy (Fig. 21.2; Tables 21.1 and 21.2). However, since these same responses are involved in combatting infections and in tumour immunosurveillance, suppressing or modulating these responses has to be weighed in terms of EFFICACY/benefit since this could increase the risk of infection and cancer development.

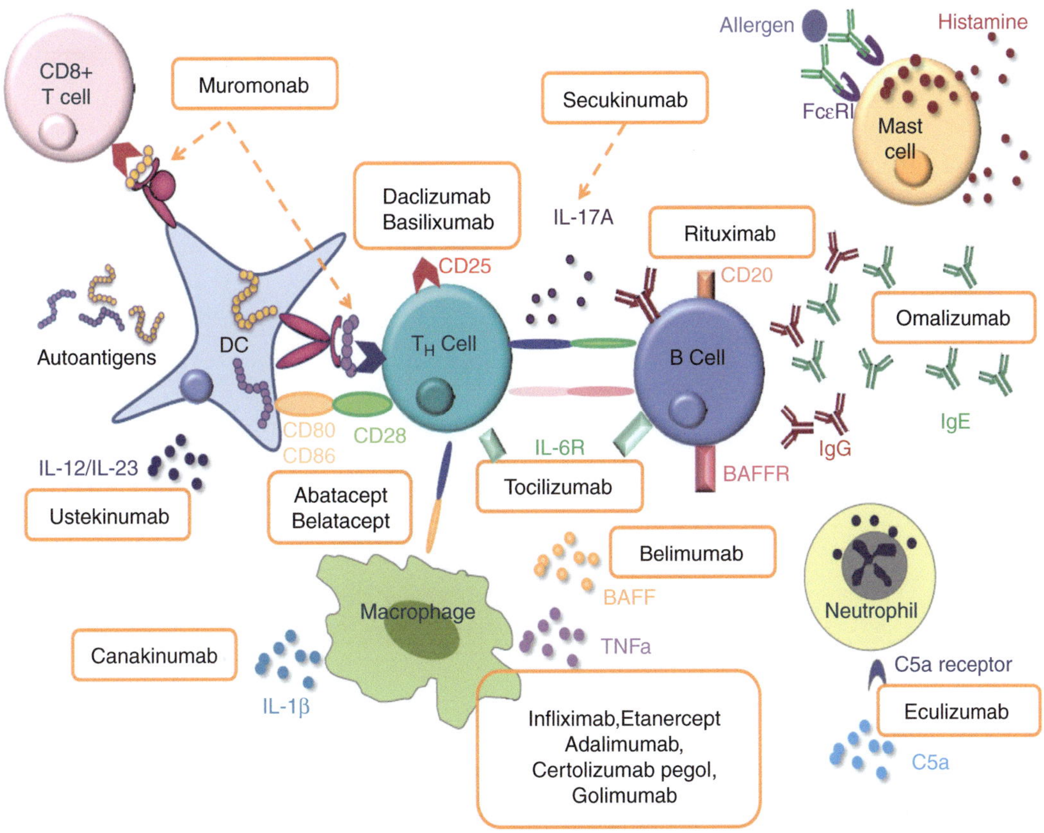

Fig. 21.2 Overview of approved mAbs and Fc-fusion proteins and their immune targets

Table 21.1 mAbs/Fc-fusion proteins approved or under review[a] by FDA and/or EMA for immune diseases

INN name	Trade name	Company	Species/ isotype	Target	Inflammatory indication(s)	Dosing regimen/RoA	Pharmacology/MoA
Muromonab	Orthoclone-OKT3®	Janssen-Cilag	Mouse IgG2a	CD3	Organ rejection (renal, heart, liver)	iv bolus 5 mg per day for 10–14 days	Inhibits T cell function
Daclizumab[b]	Zenapax®	Roche	Humanized IgG1	IL-2R	Organ rejection (renal)	1 mg/kg iv infusion; −1 day before organ transplant and 4 days post-transplant every 14 days afterwards	Inhibits T cell function
Rituximab[a]	Rituxan™, MabThera™	Biogen-IDEC, Roche	Chimeric IgG1	CD20	RA[a]	2× 1000 mg 2 weeks apart, again 24 weeks later; 100 mg methylprednisolone 30 min before infusion	B cell depletion by ADCC, CDC and apoptosis
Basiliximab	Simulect®	Novartis	Chimeric IgG1	IL-2R	Organ rejection (renal)	Bolus or iv infusion 2× 20 mg at −1 day before organ transplant and 4 days post-transplant	Inhibits T cell function
Infliximab	Remicade®	Johnson & Johnson	Chimeric IgG1	TNF-α	RA, Crohn's disease (CrD), ulcerative colitis (UC), ankylosing spondyitis (AS), psoriasis, psoriatic arthritis (PsA)	iv infusion **CrD** 5 mg/kg at 0, 2 and 6 weeks, then every 8 weeks. Max 10 mg/kg **UC** 5 mg/kg at 0, 2 and 6 weeks, and then every 8 weeks **RA** In conjunction with methotrexate, 3 mg/kg at 0, 2 and 6 weeks, and then every 8 weeks. Max 10 mg/kg and every 4 weeks **AS** 5 mg/kg at 0, 2 and 6 weeks, and then every 6 weeks **Psoriasis and PsA** 5 mg/kg at 0, 2 and 6 weeks, and then every 8 weeks	Blocks TNF-α-mediated leukocyte migration, inhibition of apoptosis, macrophage, osteoclast and EC activation
Etanercept	Enbrel®	Amgen, Pfizer	Human TNFR-FcIgG1	TNF-α	RA, psoriasis, psoriatic arthritis (PsA), ankylosing spondylitis (AS); juvenile idiopathic arthritis (JIA); Crohn's disease (CrD)	sc injection **RA, PsA**: 50 mg once weekly with or without methotrexate (MTX) **AS**: 50 mg once weekly **Psoriasis**: 50 mg twice weekly for 3 months, followed by 50 mg once weekly **JIA**: 0.8 mg/kg weekly, with a maximum of 50 mg per week	Blocks TNF-α-activity

Alemtuzumab	Campath® Lemtrada® (for MS)	Genzyme, Bayer	Humanized IgG1	CD52	RA, MS	iv infusion 12 mgl day 5× per week year 1; 3× per week year 2	Depletes leukocytes (T and B cells monocytes, macrophages) by ADCC
Adalimumab	Humira™/ Trudexa	Abbott	Human IgG1	TNF-α	RA, Crohn's disease (CrD), ankylosing spondylitis (AS), psoriasis, psoriatic arthritis (PsA), juvenile idiopathic arthritis (JIA)	sc injection **RA**, **PsA**, **psoriasis**, **AS** 40 mg every other week. Some patients with RA not receiving methotrexate may benefit from increasing the frequency to 40 mg every week. **JIA** 15–30 kg 20 mg every other week. Over 30 kg 40 mg every other week. **CrD** Initial dose (day 1) is 160 mg (four 40 mg injections in 1 day or two 40 mg injections per day for 2 consecutive days), followed by 80 mg 2 weeks later (day 15). Two weeks later (day 29) begin a maintenance dose of 40 mg every other week. **Ps** 80 mg initial dose, followed by 40 mg every other week starting 1 week after initial dose	Blocks TNF-α-activity
Omalizumab	Xolair®	Genentech, Roche, Novartis	Humanized IgG1	IgE	Allergic Asthma	150–375 mg is administered SC every 2 or 4 weeks	Blocks IgE binding to, and cross-linking of FcεRI, mast cells
Efalizumab[b]	Raptiva®	Genentech, Merck Serono	Humanized IgG1	CD11a (LFA-1)	Psoriasis	A single 0.7 mg/kg SC conditioning dose followed by weekly SC doses of 1 mg/kg (maximum single dose not to exceed a total of 200 mg)	Blocks inflammatory T cell migration
Alefacept	Amevive®	Astellas	Human LFA-3-FcIgG1	CD2	Psoriasis	7.5 mg given once weekly as an IV bolus or 15 mg given once weekly as an IM injection for 12 weeks Retreatment for 12 weeks if CD4+ T lymphocyte counts are within the normal and a 12 weeks interval since previous treatment	Inhibition of T cell function (binds CD2 on T cells blocking interaction with LFA-3 on APC)

(continued)

Table 21.1 (continued)

INN name	Trade name	Company	Species/ isotype	Target	Inflammatory indication(s)	Dosing regimen/RoA	Pharmacology/MoA
Natalizumab	Tysabri®	Biogen IDEC, Elan	Humanized IgG4	CD49d (VLA-4)	MS, Crohn's disease	300 mg iv infusion every 4 weeks	Blocks inflammatory T cell migration
Abatacept	Orencia®	Bristol-Myers Squibb	Human CTLA4-FcIgG1	CD80, CD86	RA, juvenile idiopathic arthritis	iv infusion at 2 weeks and 4 weeks; and then every 4 weeks; 500–1000 mg	Inhibition of T cell function (binds to CD80 and CD86 on APCs leading to blocking of CD28 interaction and T cell activation)
Eculizumab	Soliris®	Alexion	Humanized IgG2/4	C5	Paroxysmal nocturnal haemoglobinuria (PNH), atypical haemolytic uremic syndrome (aHUS)	iv infusion; **PNH** 600 mg weekly for 4 weeks; 900 mg for 1 week; then 900 mg every 2 weeks; **aHUS** 900 mg weekly for 4 weeks; 1200 mg for 1 week; then 1200 mg every 2 weeks	Blocks complement C5, inhibiting its cleavage to C5a and C5b and generation of membrane attack complex C5b–C9
Certolizumab pegol	Cimzia®	UCB	Humanized Fab-PEG	TNF-α	Crohn's disease (CrD), RA, psoriatic arthritis (PsA), ankylosing spondylitis (AS)	sc injection **CrD**, **AS** 400 mg at 2 weeks and 4 weeks, then 400 mg every 4 weeks if response. **RA**, **PsA** and **AS** 400 mg at 2 and 4 weeks, and then 200 mg every other week; for maintenance dosing, 400 mg every 4 weeks can be considered	Blocks TNF-α activity
Golimumab	Simponi™	Johnson & Johnson	Human IgG1	TNF-α	RA, ankylosing spondylitis, psoriatic arthritis	50 mg sc injection once a month	Blocks TNF-α activity
Canakinumab	Ilaris®	Novartis	Human IgG1	IL-1β	Cryopyrin-associated periodic syndrome (CAPS)	sc injection; 150 mg for CAPS patients with body weight greater than 40 kg and 2 mg/kg for CAPS patients with body weight greater than or equal to 15 kg and less than or equal to 40 kg	Blocks IL-1β-mediated leukocyte migration, macrophage and osteoclast activation, DC activation, Th17 differentiation
Tocilizumab	Actemra®, RoActemra®	Roche, Chugai	Humanized IgG1	IL-6R	RA, juvenile idiopathic arthritis (JIA)	iv infusion **RA** 4 mg/kg for 4 weeks than 8 mg/kg for 4 weeks; sc 162 mg every week or every other week. **JIA** 8–12 mg/kg every 4 weeks (PJIA) or 2 weeks (SJIA)	Blocks IL-6-mediated B/T cell and osteoclast activation, inhibition of T cell apoptosis, Th17 differentiation

Ustekinumab	Stelara™	Johnson & Johnson	Human IgG1	IL-12/23 (p19)	Psoriasis, psoriatic arthritis (PsA)	sc injection 45 mg 1 and 4 weeks and 45 mg every 12 weeks; later dose is doubled for over 100 kg	Blocks IL-12-mediated TH1/NK cell activation and TNF-α/IFN-γ release; blocks IL-23-mediated TH17 cell expansion and IL-17/IL-22 release
Belimumab	Benlysta® Lymphostat B	Human Genome Sciences	Human IgG1	BLyS (BAFF)	SLE	iv infusion 10 mg/kg at 2-week intervals for the first three doses and at 4-week intervals thereafter	B cell depletion by apoptosis (block Blys binding to BCMA, BAFFR and TACI required for B cell maturation/survival)
Belatacept	Nulojix®	Bristol-Myers Squibb	Human CTLA-4-IgG Fc-IgG4	CD80, CD86	Organ rejection (renal)	iv infusion −1 day and at 4 days 10 mg/kg; then 2, 4, 8 and 12 weeks, 10 mg/kg; then 16 weeks and every 4 weeks after at 5 mg/kg	Inhibition of T cell function (binds to CD80 and CD86 on APCs leading to blocking of CD28 interactions and T cell activation)
Siltuximab	Sylvant®	Janssen	Chimeric IgG1	IL-6	Castleman disease	iv 11 mg/kg every 3 weeks	Inhibits IL-6 activity
Vedolizumab[a]	Entyvio®	Takeda-Millennium	Humanized IgG1	α4β7	Ulcerative colitis, Crohn's disease	iv infusion of 300 mg at 0, 2, 6 weeks and every 8 weeks thereafter	Inhibits lymphocyte migration to gut
Secukinumab	Cosentyx®	Novartis	Human IgG1	IL-17A	Psoriasis, psoriatic arthritis, ankylosing spondylitis	sc injection 300 mg at weeks 0, 1, 2, 3 and 4 followed by 300 mg every 4 weeks	Blocks IL-17A-mediated pro-inflammatory cytokine/chemokine production, granulocyte and monocyte mobilization, TNF-α/IL-1β production, ICAM-1 expression, DC maturation, EC and osteoclast activation
Mepolizumab	Nucala®	GlaxoSmithKline	Humanized IgG1	IL-5	Asthma (eosinophilic)	sc 100 mg every 4 weeks for 12 months Also, iv 75–750 mg every 4 weeks	Inhibits IL-5-mediated eosinophil maturation, activation and migration (binds IL-5 thereby blocking binding to IL-5R on eosinophils)

(continued)

Table 21.1 (continued)

INN name	Trade name	Company	Species/ isotype	Target	Inflammatory indication(s)	Dosing regimen/RoA	Pharmacology/MoA
Reslizumab	Cinqair®	Teva (Ception Therapeutics)	Humanized IgG4	IL-5	Asthma (eosinophilic)	iv 0.3 or 3 mg/kg every 4 weeks	Inhibits IL-5 activity
Ixekizumab	Talz®	Eli Lilly	Humanized IgG4	IL-17A	Psoriasis	sc 2× 80 mg and then 80 mg thereafter	Inhibits IL-17A activity
Sarilumab	Kevzara®	Sanofi (from Regeneron)	Human IgG1	IL-6R	RA	sc 150 or 200 mg every 2 weeks	Inhibits IL-6 activity
Brodalumab	Siliq™/ Kyntheum®	Valeant, AstraZeneca (from Amgen)	Human IgG2	IL-17RA	Psoriasis	sc 210 mg or 140 mg every 2 weeks	Inhibits IL-17A receptor (bocks activity of IL-17A, C, E and F)
Ocrelizumab	OCREVUS®	Roche	Humanized IgG1	CD20	MS	iv infusion of 600 mg every 24 weeks	Induce B cell depletion by ADCC, CDC and apoptosis induction
Dupilumab	Dupixent®	Sanofi	Human IgG4	IL-4Ra	Atopic dermatitis	sc 150 or 300 mg every week	Blocks IL-4-mediated Th2 cell generation; blocks IL-13-mediated B cell proliferation/IgE production, mast cell and eosinophil recruitment, M2 macrophage activation, eotaxin production, collagen synthesis, mucus production and bronchoconstriction
Benralizumab		AZ-Medimmune	Humanized IgG1 (afucosylated)	IL-5Rα	Asthma (eosinophilic)		Inhibits IL-5 activity
Guselkumab		J&J-Janssen	Human IgG1	IL-23 (p19)	Psoriasis, psoriatic arthritis		Blocks IL-23-mediated activation of IL-17-producing cells
Tildrakizumab		Merck, Sun Pharmaceuticals	Humanized IgG1	IL-23 (p19)	Psoriasis		Blocks IL-23-mediated activation of IL-17-producing cells

ADCC antibody-dependent cellular cytotoxicity, *aHUS* atypical haemolytic uremic syndrome, *AS* ankylosing spondylitis, *APC* antigen-presenting cell, *BAFF* B cell-activating factor, *BCMA* B cell maturation protein, *CDC* complement-dependent cytotoxicity, *CrD* Crohn's disease, *CTLA-4* cytotoxic T lymphocyte antigen-4, *DC* dendritic cell, *EC* endothelial cell, *ICAM-1* intercellular adhesion molecule-1, *IFN* interferon, *IL* interleukin, *INN* international non-proprietary name, *iv* intravenous, *JIA* juvenile idiopathic arthritis, *LFA-1/3* leukocyte function antigen-1/3, *MS* multiple sclerosis, *PNH* paroxysmal nocturnal haemoglobinuria, *PsA* psoriatic arthritis, *RA* rheumatoid arthritis, *sc* subcutaneous, *SLE* systemic lupus erythematosus, *TACI* transmembrane activator and calcium modulator and cyclophilin ligand interactor, *TNF-α* tumour necrosis factor-α, *UC* ulcerative colitis, *VLA-4* very late antigen-4

[a]Also approved in non-Hodgkin's lymphoma

[b]Products now withdrawn

Table 21.2 Selected mAbs in Phase 3 clinical trials for immune diseases

INN name	Company	Species	Target	Indication(s)	Pharmacology/MoA
Gevokizumab	Xoma/Servier	Humanized IgG2	IL-1β	Pyodermas gangrenosum	Blocks IL-1β activity
SA237	Chugai/Roche	Humanized IgG2	IL-6Rα	Neuromyelitis optica	Inhibits IL-6 activity
Lebrikizumab	Roche	Humanized IgG4	IL-13	Asthma	Blocks IL-13-mediated activity
Tralokinumab	AZ-Medimmune	Human IgG4	IL-13	Asthma	Blocks IL-13-mediated activity
Anifrolumab	AZ-Medimmune	Human IgG1	IFN-α/βR	Systemic lupus erythematosus (SLE)	Blocks type 1 IFN receptor
Etrolizumab	Roche-Genentech	Humanized IgG1	β7 integrin	Ulcerative colitis, Crohn's disease	Inhibits lymphocyte migration to gut
Begedina	Adienne	Murine IgG2b	CD26	Graft-versus-host disease, transplant rejection	Reduces the CD26 on CD4+ T lymphocytes and inhibits the immune response
Lampalizumab	Roche-Genentech	Humanized Fab	Complement Factor D	Dry age-related macular degeneration	Inhibits complement activation

21.3 The Structure of IgG and Mechanisms of Action (MoA)

The vast majority of all mAb therapeutics are based on IgG due to its long HALF-LIFE and Fc effector function. The structure of IgG, highlighting the main structural and functional components, is shown in Fig. 21.3 (see also Chaps. 4 and 14). The ANTIBODY-binding domains bind the intended target, and the Fc REGION binds both the neonatal Fc receptor (FcRn), giving it a long HALF-LIFE, and IgG RECEPTORS (FcγR) on immune cells. The multiple actions of mAbs are shown in Fig. 21.4. The primary MoA of most mAbs is to directly or indirectly modulate the IMMUNE SYSTEM (HUMORAL, CELL-MEDIATED and INNATE IMMUNITY), with the potential to induce immunosuppression or immune modulation [3]. mAbs can also bind directly to RECEPTORS on immune cells to deplete these cells and thereby inhibit the antigen-presenting and autoantibody-producing function of B CELLS (e.g. **rituximab**) and autoimmune T cell activity (**alemtuzumab**) [4]. mAbs of the IgG_1 ISOTYPE have high FcγR binding, which promotes killing of B and T cells by triggering Fc-mediated immune effector mechanisms such as ANTIBODY-DEPENDENT CYTOTOXICITY (ADCC) by NK cells and COMPLEMENT-DEPENDENT CYTOTOXICITY (CDC). Other mAbs are designed to induce cell killing indirectly by ANTIBODY-DEPENDENT CELLULAR PHAGOCYTOSIS (ADCP). Furthermore some mAbs bind to cell surface RECEPTORS, not to deplete the cells but to block and/or downregulate activation RECEPTORS (thereby reducing LIGAND binding) to suppress or modify their function (e.g. **tocilizumab**) or prevent their homing to LYMPHOID ORGANS and inflammatory sites (e.g. **vedolizumab**). Other mAbs are designed to neutralize pleiotropic mediators such as CYTOKINES, CHEMOKINES, COMPLEMENT proteins, enzymes and

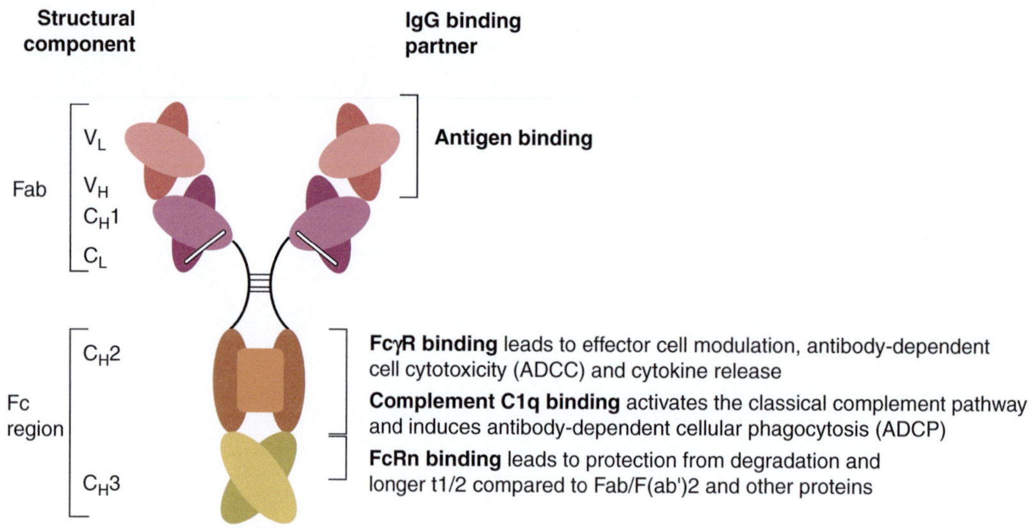

Fig. 21.3 The structure of IgG

IMMUNOGLOBULIN molecules or GROWTH FACTORS that have multiple contributions to inflammatory disease pathogenesis (e.g. **adalimumab**, **secukinumab**, **ustekinumab**, **belimumab**, **omalizumab**). The vast majority of these mAbs are target antagonists designed to cause direct immunosuppression. However, others such as teplizumab have the same overall immunosuppressive goal, but they achieve this through direct agonism (activation) of a target receptor resulting in IMMUNOMODULATION through TCR complex downregulation, T cell anergy, induction of IL-10-producing CD4+ T cells and Treg induction [5].

21.3.1 Optimizing mAbs for Efficacy and Safety

The majority of new mAbs in clinical development have been rationally designed to increase potency, optimize desirable pharmacology, enhance PHARMACOKINETICS (PK) and minimize potential immunogenicity whilst avoiding undesirable effects that could increase safety concerns [6]. Immunogenicity occurs when the body detects the mAb as a foreign protein and produces antibodies (antidrug antibody, ADA) against it to eliminate it. This occurred more frequently with chimeric antibodies that contained mouse antibody-binding domains engrafted on to a human backbone and with mouse antibodies. Nowadays, most mAbs are fully human or humanized (95% human) proteins. However, even fully human proteins can be immunogenic, eliciting predominantly anti-idiotypic antibodies that block the target binding site. It is accepted that the nature of the target, glycosylation and aggregation profile of the ANTIBODY, route of administration, treatment regimen, co-administration of other immunosuppressive agents and individual patient factors can all impact the immunogenicity of a mAb.

21.3.2 Functional Optimization

IgGs can be functionally optimized in both their antigen-binding domains and Fc domains to improve EFFICACY and minimize safety concerns [7, 8] (see also Chap. 14). Different approaches can be applied such as increasing the affinity of IgGs to give low picomolar affinity for their specific targets or by using the modular nature of ANTIBODY to produce new highly potent and sometimes multimeric mAb constructs. A wide range of structures has been explored since the field of BIOTHERAPEUTICS

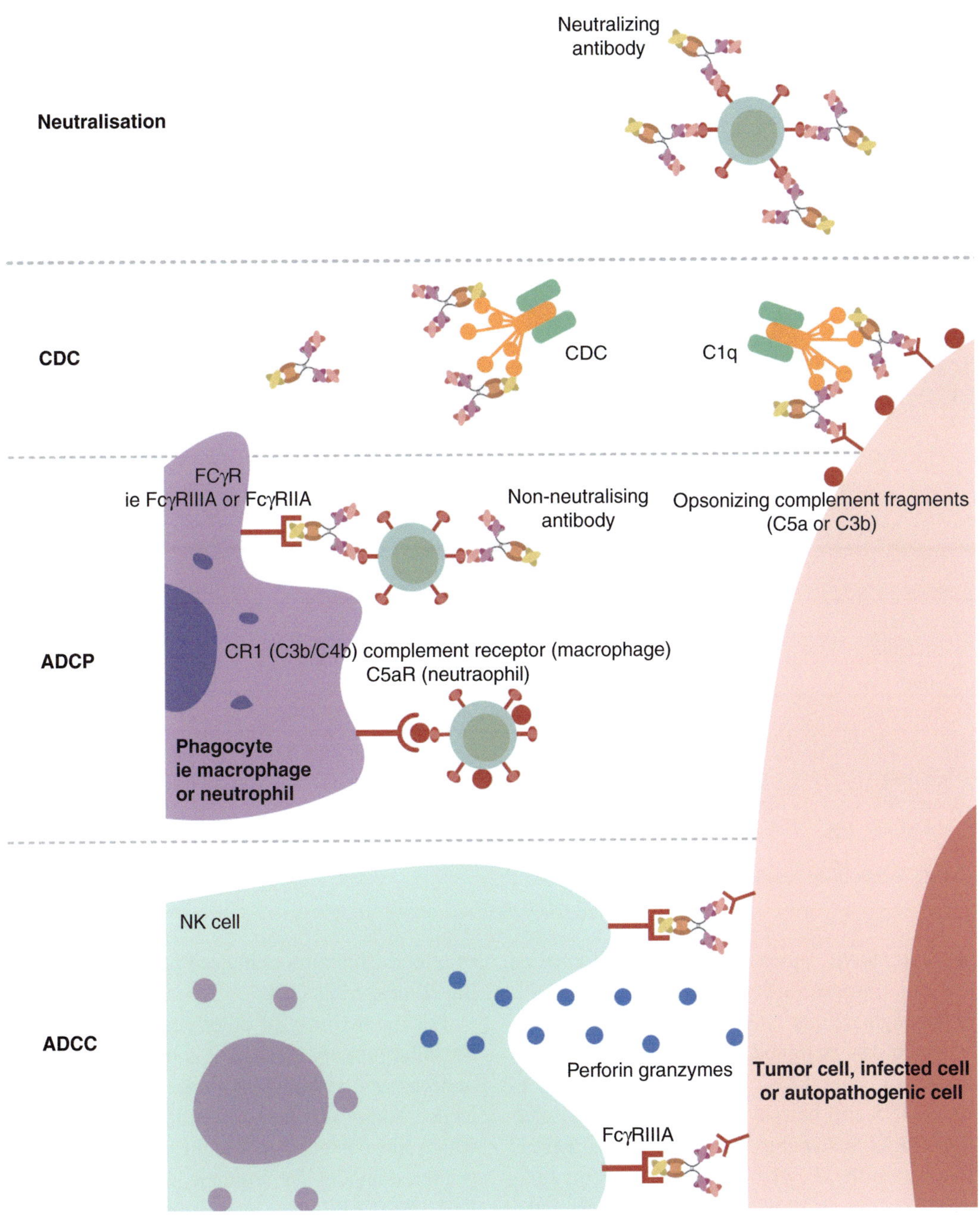

Fig. 21.4 Mechanisms of action of an IgG1 mAb: target neutralization, CDC, ADCP and ADCC

began, comprising classic whole IgGs and related fragment such as $F(ab)_2$, Fabs, scFvs and single domains (Fig. 21.5, top). Antibody domains can also be combined into multivalent or multi-specific molecules (with increased avidity and the ability to bind two (bi-specific) or more different targets, to bind two EPITOPES on the same target (bi-paratopic) or to co-target other functional domains) (Fig. 21.5, bottom). Antibodies derived from other species, e.g. camel (nanobodies) or shark (IgNAR), with favourable physico-chemical properties, as well

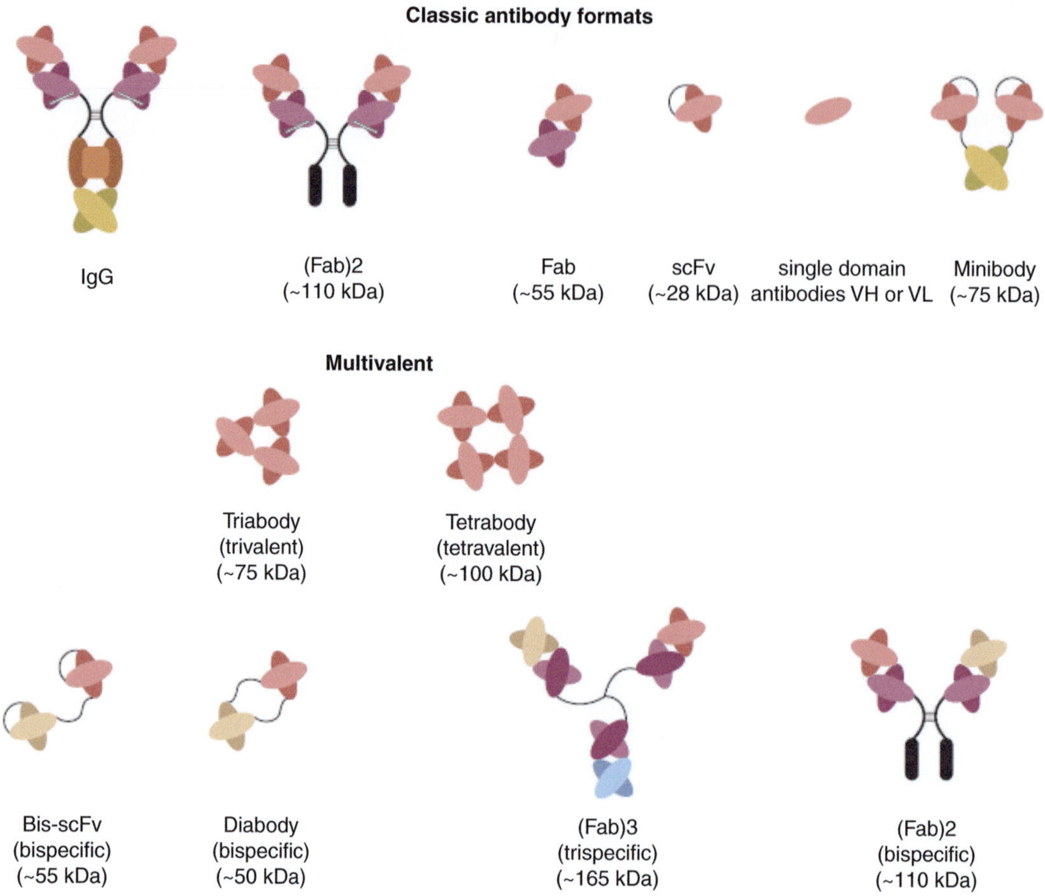

Fig. 21.5 Antibody-based formats

as alternative protein scaffolds (lipocalins, ankyrin, fibronectin, etc.), are also being used to create high-affinity binding protein therapeutics.

For mAbs designed to deplete immune cells by ADCC and CDC (see Fig. 21.4), the Fc REGIONS can be optimized through the selection of IgG_1 with naturally high COMPLEMENT fixation ability, FcγR binding and effector function. This can also be achieved through the introduction of mutations in the Fc domain that increases binding to FcγRIIIA and IIA as well as the use of mAbs with altered glycosylation (e.g. defucosylation) [9]. The increased ADCC, CDC and cell killing leads to the release of pro-INFLAMMATORY CYTOKINES such as TNF-α, IFN-γ, IL-6 and COMPLEMENT activation that frequently leads to infusion reactions that are often controlled with co-medication. A number of effector function-enhanced IgGs are undergoing clinical trials [10].

For the majority of other nondepleting mAbs for inflammatory disease, it is desirable to avoid unnecessary activation of immune effector cells. Hence, many of these anti-inflammatory mAbs, which are designed to inhibit immune function (rather than deplete immune cells), are of the IgG_1 ISOTYPE that have been preselected or engineered for low/no FcγR binding and effector function or are based on the inert IgG_2 or IgG_4 ISOTYPES with low/no COMPLEMENT activation and FcγR binding capacity [11]. Loss of effector function can be achieved through structural changes including mutations in the CH2 domain to reduce FcγR interaction [12] and mAb aglycosylation [13], which both completely remove effector function. However, immunogenicity of any non-natural mutation or structure needs to be considered, as increasing the number of point mutations can increase the risk of immunogenicity. Avoidance of effector function can also be achieved through non-Fc-bearing antibody fragments (Fig. 21.5). Monovalent antibody fragments might also be used to block RECEPTORS whilst avoiding receptor cross-linking (Fig. 21.5).

21.3.3 Optimizing Pharmacokinetics (PK)

The PK and HALF-LIFE of mAbs can also be tailored to a specific indication, depending on the exposure required to achieve EFFICACY (see also Chap. 14). For mAbs intended for chronic dosing, the HALF-LIFE of an IgG can be extended by incorporating mutations that increase affinity for natural IMMUNOGLOBULIN RECEPTORS through binding of FcRn [14]. FcRn binding confers mAbs with a longer HALF-LIFE compared to the majority of proteins. Generally, antibodies are taken into cells via pinocytosis, and the Fc portion of the mAb binds to FcRn RECEPTORS in vesicles with high affinity at low pH. This protects the ANTIBODY from enzymatic degradation. Once the vesicle recycles to the membrane surface, the ANTIBODY is then released externally as the affinity for FcRn is lowered. Specific mutations on the Fc region of a mAb can increase the binding to the FcRn RECEPTORS at physiological pH as well as low pH. Consequently, a fine balance is needed to allow an increase in binding at the low pH found in the lysosome to allow recycling of the mAb to the surface and its dissociation but to avoid binding to FcRn on the cell surface where normally mAbs do not bind with a high affinity. A very high-affinity mAb would alter the distribution and eventually interfere with its binding to the target [15]. Other methods of extending HALF-LIFE such as the use of albumin fusion proteins (since albumin binds FcRn) and polyethylene glycol (PEG), which slows protein distribution and CLEARANCE, have also been used [16, 17].

Alternatively, short-acting Fab fragments might be used for acute indications where only short-term suppression of the target is required, and hence prolonged exposure is undesirable (since, for immunosuppressant mAbs, it could increase infection risk).

Alternative routes of administration are also being used to improve EFFICACY and mitigate against safety risks. Local delivery of IgG and antibody fragments, e.g. by intraocular, intra-articular (joint) and inhalation administration, should maximize mAb delivery to the intended site of action in the inflamed tissue whilst minimizing systemic pharmacology (at non-intended sites), safety concerns and immunogenicity. The use of fragments in these cases is preferred to IgG as volume is often limiting, and the use of smaller fragments allows significantly higher molar quantities per mg of protein, thus increasing the number of binding sites of material to be delivered to the target site. Strategies have also been adopted to increase the targeting and retention of mAbs and other proteins at local sites, e.g. through the incorporation of binding domains specific for molecules such as transferrin receptor for brain targeting [18] and the use of novel drug formulation and delivery systems for inhaled and oral delivery [19].

One important aspect of biologics is their dependence on their target for drug disposition. Binding of mAbs to targets can affect their PK profile since they will be cleared together with the target, increasing their rate of elimination compared to the mAb on its own. This is typically referred to as target-mediated drug disposition [20]. The simplest example is when a mAb binds a receptor, it is internalized and degraded with its target. Thus, less mAb is measured in the blood. As this has a higher impact on the mAb concentrations when the ratio of mAb to target is less than equivalent, i.e. 0.1:1, the rapid CLEARANCE has a strong effect on the mAb concentrations, and a sharp drop in the PK curve (exposure) is observed. At levels a log or more above the target levels that saturate the target, this effect is not visible due to the high amounts of mAb present. For soluble targets, CLEARANCE becomes more complex, but it is important to understand for optimal dosing and dosing frequency. A mAb may bind its target and form a target-mAb complex that is then cleared more quickly than a mAb but slower than the target itself. A target that turns over quickly will require high, frequent dosing and may prove to be untargetable, as has been seen with some CHEMOKINES such as MCP-1 and CCL21 [21, 22]. In contrast, COMPLEMENT fragment C5 is a known soluble target that has a slow turnover rate, and even though it is present in high levels in the blood, it can be suppressed, which contributes to the success of biologics such as eculizumab [23]. In most cases, PK/PD (PHARMACOKINETIC-PHARMACODYNAMIC) modelling can help in determining whether a target can be suppressed (and thereby inform the clinical dosing regimen) by taking into account the turnover rate of the target, its basal levels and levels in the disease setting, the access to the target, PHARMACOKINETIC characteristics of the biologics and any other information related to the MoA. An additional benefit of PK/PD modelling is to tailor treatment to patients on an individual basis [24].

21.4 Overview of mAbs Approved or in Phase 3 Clinical Studies for Immune-Mediated Diseases

The following sections describe the biology of the major immunological targets and immune diseases targeted by mAbs. mAbs approved or in Phase 3 clinical trials are listed in Tables 21.1 and 21.2, respectively. mAbs listed in these tables are highlighted in bold throughout the text.

21.4.1 MAbs Targeting Soluble Mediators

Soluble mediators initiate, promote and sustain immune responses, including both desirable immune responses to pathogenic challenge and also undesirable inflammatory and autoimmune responses. The success of mAbs inhibiting upstream CYTOKINES, such as TNF-α in RA, CROHN'S DISEASE and PSORIASIS, has prompted an explosion in the development of novel mAbs that either (1) seek to exceed the EFFICACY and improve on the safety profiles of TNF-α inhibitors in the aforementioned diseases and other diseases in which the TNF-α pathway may play a role, (2) aim to block other soluble pro-inflammatory mediators believed to be active in TNF nonresponders (e.g. IL-6, IL-17, IL-23, CSF2) with these same diseases or (3) target other novel mediators linked to the pathogenesis of other immune diseases.

21.4.1.1 Cytokine Inhibitors

The early release of CYTOKINES shapes the nature of inflammatory responses, and at the top of the inflammatory cytokine cascade are molecules such as TNF-α, IL-1, IL-6, IL-12, IL-17 and IL-23 (see Figs. 21.1 and 21.2). They are secreted mainly by myeloid cells (but also by endothelial and epithelial cells), and they all have fundamental effects on multiple components of the IMMUNE SYSTEM. Together with the local environment, these CYTOKINES drive the differentiation of cell types, which will produce combinations of CYTOKINES to aid the CLEARANCE of invading pathogens or, in some cases, to induce inflammatory disorders. Thus, these early CYTOKINES are at key rate-limiting steps of disease development and have been a big focus of therapeutic mAb development.

Biological treatments such as **TNF-α inhibitors** have revolutionized RA treatment, improving the signs and symptoms of disease, functional status and quality of life and slowing radiographic progression in patients with established RA who have suboptimal responses to traditional disease-modifying antirheumatic drugs such as METHOTREXATE [25] (see also Chap. 34). TNF-α-blocking drugs are typically administered with METHOTREXATE, which improves their EFFICACY and probably also impairs ADA development. If inadequate responses are achieved to one or two TNF inhibitors, other biologics with a different mechanism of action such as **abatacept** (CTLA-4-Ig; blocks CD28-mediated T cell activation), **rituximab** (anti-CD20; depletes B CELLS involved in ANTIGEN presentation and autoantibody production), **tocilizumab** (anti-IL6R) or anakinra (IL-1 receptor antagonist) are used [26]. Despite the success of the aforementioned biologicals, new treatments are needed for the significant proportion of subjects who fail to meet the minimum response criteria (e.g. ACR20) or experience significant toxicities (e.g. serious and opportunistic infection with TNF-α inhibitors [24, 27]), tachyphylaxis or development of resistance through the production of ADA [28] and to increase the likelihood of achieving disease remission. Blocking multiple upstream CYTOKINES may prove beneficial in these cases, though the greater immunosuppression achieved (and increased risk of infection) may prove to be no better than using current broad-spectrum immunosuppressive drugs.

More focus is now being placed on neutralizing downstream pro-INFLAMMATORY CYTOKINES to allow disease modulation with greater specificity and with fewer side effects. Rather than blocking early pleiotropic CYTOKINES, future success may lie in the combined neutralization of effector CYTOKINES with narrower ranges of defined activity to increase the risk-benefit ratio to patients and to overcome both disease heterogeneity between individuals and the numerous levels of redundancy that have evolved in cytokine networks. Our improved understanding of cytokine networks, together with the development of both biomarkers of disease and dual-/multi-specific mAb as well as other biologic technologies allowing neutralization of multiple CYTOKINES, should allow the production of highly target-specific mAb therapeutics. These can be aimed at pathogenic disease pathways operating in individuals or in certain patient cohorts within a specific inflammatory disease. This could be achieved by developing more potent, long-lasting, less-immunogenic and locally delivered mAbs against the same targets. Alternatively, mAbs against novel targets could be developed that would COMPLEMENT or replace these existing therapies (e.g. anti-IL-17 mAbs described below), as well as developing mAbs with multiple specificities.

The timing of administration of different therapies is also important as this can be key in designing IMMUNOTHERAPY schedules. Timing needs to go hand in hand with a clear understanding of the disease sequence of events in patients to be able to effectively target the key mediator of their disease at the right time. Thus, administration of mAbs with multiple specificities has an additional challenge in terms of dosing two targets at the same time with one drug. But if one considers that a multi-specific mAb remains in the body for weeks, this should allow efficient targeting during the immunological events occurring during disease progression. The affinity, potency and target compartment of each mAb needs to be taken into account to tailor multi-specific antibodies for a specific MoA, thus adding further complexity to PK/PD analysis.

21.4.1.1.1 Anti-TNF-α

TNF-α has fundamental roles in lymphoid organogenesis, INFLAMMATION, antitumour activity and host defence against intracellular pathogens (see also Chap. 6). However, dysregulated TNF-α production can promote INFLAMMATION associated with sepsis and several other inflammatory and autoimmune diseases, including IBD and several inflammatory arthritides (e.g. RA, PSORIATIC ARTHRITIS) [29]. It is produced by MACROPHAGES, MONOCYTES, DCs, NEUTROPHILS, T cells and NK cells and is primarily produced as a biologically active membrane-bound pro-form 51 kDa homotrimer, which is cleaved by TNF-α-converting enzyme (TACE) to release the 17 kDa soluble TNF-α. Its trimeric RECEPTORS, TNF receptor 1 (TNFR1; p60) and TNFR2 (p75), are widely expressed and upregulated in inflamed tissue. TNF-α can also exist as a 26 kDa membrane form (mTNF). The TNF-α biological pathway and the actions of the approved TNF blockers are shown in Fig. 21.6. TNF-α binding mediates signal transduction involving recruitment of TNF receptor-associated factors (TRAFs) to their cytoplasmic domain that activate the transcription factors NF-kB and AP-1. The multiple actions of TNF-a are shown in Box 21.1.

Box 21.1 The Pleiotropic Activities of TNF-α

- Regulates cell survival, death, activation and proliferation of various cell types (inhibits apoptosis of inflammatory cells)
- Promotes accumulation of immunocompetent cells at sites of inflammation by activation of vascular endothelium and upregulation of adhesion molecules
- Upregulates MHC expression important in T cell activation
- Stimulates synthesis of other pro-inflammatory cytokines (e.g. IL-1, IL-6), chemokines (e.g. IL-8) and other mediators that drive neutrophil and macrophage recruitment and activation
- Mediates bone and cartilage destruction through activation of osteoclasts (e.g. via receptor activator for nuclear factor kB ligand (RANKL) and colony-stimulating factor-1 (CSF1)), macrophages and fibroblast-like synoviocytes (FLS) to release destructive mediators (e.g. MMPs, collagenase, and prostaglandins) that destroy joint structure [30]

Transgenic mice expressing high concentrations of TNF-α spontaneously developed arthritis and colitis, and anti-TNF-α antibodies ameliorate disease in rodent models of these diseases. There are five approved TNF-α-blocking BIOLOGICALS (Fig. 21.6), all of which neutralize both soluble and mTNF. The IgG1 mAbs **infliximab**, **adalimumab** and **golimumab** also induce CDC, ADCC and APOPTOSIS of mTNFα-expressing cells [29, 31]. The TNFR2-Fc-fusion protein **etanercept** has a lower capacity to induce CYTOTOXICITY of mTNFα-expressing cells, and **certolizumab pegol** (a Fab-PEG conjugate) does not mediate CYTOTOXICITY of mTNFα-expressing cells [31]. There is evidence that mTNFα can mediate reverse signalling back into the cell, and this is a potential mechanism by which infliximab, adalimumab and etanercept can further increase the proportion of apoptotic cells. A potential anti-inflammatory role for reverse signalling has been proposed in the subsequent downregulation of cytokine production in response to LPS stimulation by MONOCYTES. Interestingly, both infliximab and certolizumab pegol (but not etanercept) inhibit LPS-induced cytokine production mediated through mTNF, and this mechanism has been proposed to be important in Crohn's disease in which etanercept is not efficacious [31].

These biologics are most commonly used for the treatment of RA, CROHN'S DISEASE, PSORIASIS, PSORIATIC ARTHRITIS and ankylosing spondylitis (see Chaps. 34 and 35). However TNF-α may also play a pathogenic role in osteoarthritis, uveitis and acute lung injury. There are significant numbers of nonresponders to TNF-α inhibitors, which may be due in part to an increase in IL-17, prompting the development of dual TNF-α-IL-17 blockers. TNF-α is involved in immune responses to both tumours and infectious pathogens; hence TNF-α blockade raises concerns for increased risk of cancer and infection. Despite conflicting observational clinical study results on the impact of anti-TNF-α treatment on tumour development [24, 32], and the fact that some diseases such as RA are associated with an increased cancer risk, the FDA has issued a "black box" warning to the label of these products, in particular the risk of developing lymphoma in children/adolescents. Nevertheless, many tumours produce TNF-α themselves, and there is strong evidence that TNF-α can promote tumour growth directly or indirectly by induction of other pro-INFLAMMATORY CYTOKINES and angiogenic factors involved in cancer development, as well as promoting the influx of myeloid-derived suppressor cells to the tumour microenvironment. Hence TNF-α blockers may actually reduce cancer risk depending on the type of tumour. Treatment with TNF-α blockers has been associated with reactivation of *Mycobacterium tuberculosis* and HEPATITIS B VIRUS infections [33].

Another strategy is to target TNFR1 rather than TNF-α. Inflammatory responses are mainly mediated by TNFR1, whereas TNFR2 has beneficial effects including immune regulation (and protection from infection), anti-apoptotic and cell proliferative effects, tissue homeostasis and regeneration and neuroprotection [34]. Hence selective inhibition of TNFR1 (as opposed to targeting TNF-α LIGAND which neutralizes the activities of both RECEPTORS) might neutralize the pro-inflammatory activity of TNFR1 whilst maintaining the advantageous responses mediated by TNFR2. This might improve therapeutic EFFICACY by minimizing side effects, such as recurrence of tuberculosis and other infections and risk of malignancies or increased relapse rates in MS, probably due to lack of TNFR2-mediated myelin regeneration. This concept was demonstrated using a TNF mutant (R1antTNF), which is capable of forming inactive complexes with sTNF and selectively neutralizing TNFR1, administered either as unmodified or PEGYLATED protein (PEG-R1antTNF), which showed therapeutic EFFICACY in the mouse experimental allergic encephalomyelitis (EAE) model of MS and the collagen-induced arthritis (CIA) model of RA [35].

21.4.1.1.2 Anti-IL-1

IL-1 is produced in response to inflammatory stimuli and plays an important role in inflammatory responses and host defence, often by acting in concert with IL-6 and TNF-α (see Chap. 6). IL-1 is synthesized primarily by MACROPHAGES, MONOCYTES, fibroblasts and DCs but also by B CELLS, NK cells and epithelial cells [30]. Both IL-1α and IL-1β bind to the IL-1R type I expressed by a wide range of cells, inducing the formation of a high-affinity complex with the IL-1R accessory protein (IL-1RAcP) and the recruitment of the intracellular adaptor protein myeloid differentiation factor 88 (MYD88) and IL-1R-associated kinase-1 (IRAK-1), which drive IL-1 signalling. IL-1R antagonist (IL-1Ra) is an endogenously secreted inhibitor that competes with IL-1α and IL-1β for binding to IL-1RI without transducing a signal. The activities of IL-1 are summarized in Box 21.2.

Box 21.2 Major Activities of IL-1

- Upregulates adhesion molecule expression on endothelial cells to promote leukocyte migration
- Promotes DC maturation
- Promotes induction and expansion of TH17 cells (suggesting that IL-1 has a major role at the onset rather than at the effector stage of inflammatory disease [36])
- Impacts thermoregulatory centres in the hypothalamus, inducing fever hyperalgesia, vasodilation and hypertension
- Activates osteoclasts and macrophages that in RA and osteoarthritis patients contribute to joint degradation through the release of mediators that stimulate bone resorption and cartilage degradation as a result of loss of proteoglycans

Although there is an increase in IL-1Ra in RA, the levels are apparently insufficient to counteract the effects of IL-1. Mice lacking IL-1RI or both IL-1α and IL-1β are protected from the development of collagen-induced arthritis possibly due to decreased IL-17 and IFN-γ production from T cells [37].

Both IL-1α and IL-1β exist as an intracellular inactive pro-form that is activated by cleavage. Pro-IL-1α is cleaved by calpain into mature IL-1α, which is thought to act mainly bound to the cell surface. Pro-IL-1β (and Pro-IL-18) is cleaved into its active form by caspase 1 following interaction with the activated NLRP3 (NOD-like receptor family, pyrin domain containing 3) INFLAMMASOME [38] (see also Chap. 9). INFLAMMASOME over-activation leading to increased IL-1 and inflammatory sequelae is observed in patients with gain-of-function mutations in NLRP3 (cryopyrin) who have cryopyrin-associated periodic syndromes (CAPS), rare disorders of varying severity characterized by excess IL-1β levels [39]. INFLAMMASOME activation with increased IL-1/IL-18 levels is also observed in diseases where there is an increase in NLRP3-activating damage-associated molecular patterns (DAMPS), including gout (urate crystals), atherosclerosis, type 2 diabetes (lipids) and age-related macular degeneration (drusen). Hence these diseases may benefit from IL-1 inhibition. IL-1 is also involved in lipid metabolism where it adversely affects β cells in the pancreas, perturbing insulin production in models of type 2 diabetes. IL-1 has also been implicated in the pathogenesis of chronic obstructive pulmonary disease (COPD) through its ability to promote pulmonary INFLAMMATION, emphysema and small airways remodelling (see Chap. 23).

Different IL-1 blockers have been tested in RA, CAPS and type 2 diabetes among others [40]. Anakinra (Kineret; Amgen), a RECOMBINANT non-glycosylated version of IL-1Ra (IL-1 Ra), was only modestly efficacious in RA and was inferior to TNF-α blockers, perhaps because it has less of a role in established disease. Anakinra did however improve vascular and left ventricular function in patients with RA by reducing APOPTOSIS, consistent with a role of IL-1 in arterial INFLAMMATION and cardiac hypertrophy. In contrast to RA, anakinra potently inhibited the clinical symptoms of systemic onset juvenile idiopathic arthritis (SOJIA). Anakinra is approved in RA, SOJIA and CAPS. Anakinra treatment of type 2 diabetes patients improved glycaemic indices and insulin secretion. Rilonacept (Arcalyst, IL-1 Trap; Regeneron), also approved in CAPS, is a dimeric fusion protein comprising the ligand-binding domains of IL-1RI and IL-1RAcP fused to human IgG1 that neutralizes both IL-1α and IL-β [41]. Both anakinra and rilonacept are also being tested in chronic gout and AOSD. **Canakinumab** (Ilaris, Novartis) is a human IL-1β-specific mAb approved for CAPS and gout and in late-stage clinical trials for SOJIA, type 2 diabetes, COPD and osteoarthritis (see Table 21.1) [42] . **Gevokizumab** (Xoma/Servier) is a humanized IgG_2 anti-IL-1β mAb in Phase 3 for pyoderma gangrenosum and uveitis [43]. Other anti-IL-1R and dual IL-1α and IL-1β blockers are in early clinical trials for COPD and osteoarthritis [44].

21.4.1.1.3 Anti-IL-6/IL-6R

IL-6 is produced mainly by T cells, MACROPHAGES and endothelial cells in response to infection and tissue damage (see also Chap. 6); however, dysregulation is associated with RA and other inflammatory diseases [45]. IL-6 binds the IL-6R which occurs both in a membrane-bound and a soluble form and requires the accessory molecule gp130 (IL-6Rβ) for signal transduction. Expression of membrane-bound IL-6Rα is restricted mainly to cells of the IMMUNE SYSTEM and to hepatocytes, whereas gp130 is ubiquitously expressed. Besides classical signalling involving membrane-bound IL-6Rα and gp130, IL-6 bound to soluble IL-6Rα can induce trans-signalling by associating with gp130 in a multitude of cell types that do not express membrane-bound IL-6Rα. The major activities of IL-6 are summarized in Box 21.3.

Box 21.3 Major Activities of IL-6

- Induces liver acute-phase protein production
- Promotes optimal B cell and T cell effector responses to pathogens [46]
  - IL-6 KO mice highly susceptible to infection with bacterial and viral pathogens [47]
- Regulates fever through the hypothalamic-pituitary-adrenal axis
- Accelerates mononuclear cell accumulation at sites of inflammation (via MCP-1 production)
- Promotes B and T cell differentiation and maturation
- Anti-apoptotic for T cells (prolongs T cell survival)
- Induces proliferation of vascular endothelial cells (via VEGF) promoting angiogenesis
- Key driver of osteoclast differentiation/activation
- Promotes TH17 cell and Treg cell differentiation (with TGF-β)

Elevated levels of IL-6 or soluble IL-6R are seen in several inflammatory diseases including RA, PSORIASIS and colitis. IL-6 promotes animal models of inflammatory disease such as arthritis, SLE and colitis through pathogenic TH17 induction. Arthritis development in IL-6 KO mice is completely inhibited, whereas up to 20% of TNF-α KO mice still develop the disease, suggesting possible nonredundant functions of TNF-α and IL-6 in RA and therapeutic potential of IL-6 inhibition in TNF-α nonresponders. The role of IL-6 in the differentiation of B CELLS into plasma cells and T cells into effector cells has made it a target in SLE.

Tocilizumab (Actemra) is a humanized mAb that binds to both soluble and membrane-bound IL-6Rα and blocks IL-6 binding [48]. It is approved in the USA for RA in which it was shown to rapidly improve the signs and symptoms of disease, including anti-TNF-α nonresponders, especially when combined with METHOTREXATE (see also Chap. 34). Tocilizumab is associated with a slightly higher rate of bacterial and viral infections, mainly of the respiratory and gastrointestinal tracts. Transient decreases in neutrophil counts and increases in serum lipids and liver enzymes have been observed. **Siltuximab** is a chimeric anti-IL-6 IgG1 approved for the Castleman disease. **Sarilumab** (Sanofi/Regeneron) is a human IgG_1 anti-IL-6Rα [49], currently under FDA and EMA review for RA. Sarilumab has shown better EFFICACY in RA in a head-to-head study with adalimumab. **SA237** (Chugai/Roche) is a humanized IgG2 anti-IL-6Rα in Phase 3 for neuromyelitis optica. A number of other anti-IL-6 and anti-IL-6R mAbs are in earlier clinical trials [44].

21.4.1.1.4 Anti-IL-17

IL-17A (formerly IL-17) is a pro-inflammatory cytokine that belongs to a family of six homologous CYTOKINES termed IL-17A, IL-17B, IL-17C, IL-17D, IL-17E (IL-25) and IL-17F. IL-17A is produced by a distinct population of T HELPER CELLS termed TH17 cells which can be either $CD4^+$ T cells, $CD8^+$ T cells, gamma-delta $(\gamma\delta)^+$ T cells or NKT cells, some of which are activated effector memory T cells of the $CD45RO^+$ subset [50] (see also Chap. 6). Whereas TGF-β, IL-6 and IL-21 cooperate to achieve commitment of naive T cells to the TH17 phenotype, IL-23 is the key inducer of IL-17A production by committed TH17 memory T cells. Although T cells are the major source of IL-17A, NK cells, NEUTROPHILS and MACROPHAGES can also produce IL-17. Two isoforms of IL-17, IL-17A and IL-17F, are expressed by TH17 cells both as homodimeric proteins and as an IL-17A/17F heterodimer. These three CYTOKINES are similar in function and signal through the same IL-17RA/IL-17RC receptor complex (Fig. 21.7). TH17 cells also secrete IL-21, IL-22, TNF-α and IL-6, which have overlapping and distinct functions in INFLAMMATION and host defence. IL-17A (and IL-17-F) can induce the expression of diverse pro-INFLAMMATORY CYTOKINES and CHEMOKINES from epithelial and endothelial cells, fibroblasts, osteoblasts and MONOCYTES/MACROPHAGES. The major function of IL-17 is shown on Box 21.4. There is considerable overlap in the function of IL-17A and IL-17F but likely also nonredundant features.

Box 21.4 Major Activities of IL-17

- Induces and/or enhances expression of colony-stimulating factors (e.g. CSF2 and CSF1), CXC chemokines (e.g. CXCL-8 (IL-8), CXCL1 (Gro-α) and CXCL10) and MCP-1
  - Promotes the mobilization, production and expansion of granulocytes and monocytes
  - Critical role in mucosal-epidermal immunity and defence from pathogens
- Induces TNF-α and IL-1β production from macrophages and keratinocytes and synergizes with TNF-α and IL-1β in some of their activities
- Contributes to T cell responses through ICAM-1 and HLA-DR expression and DC maturation

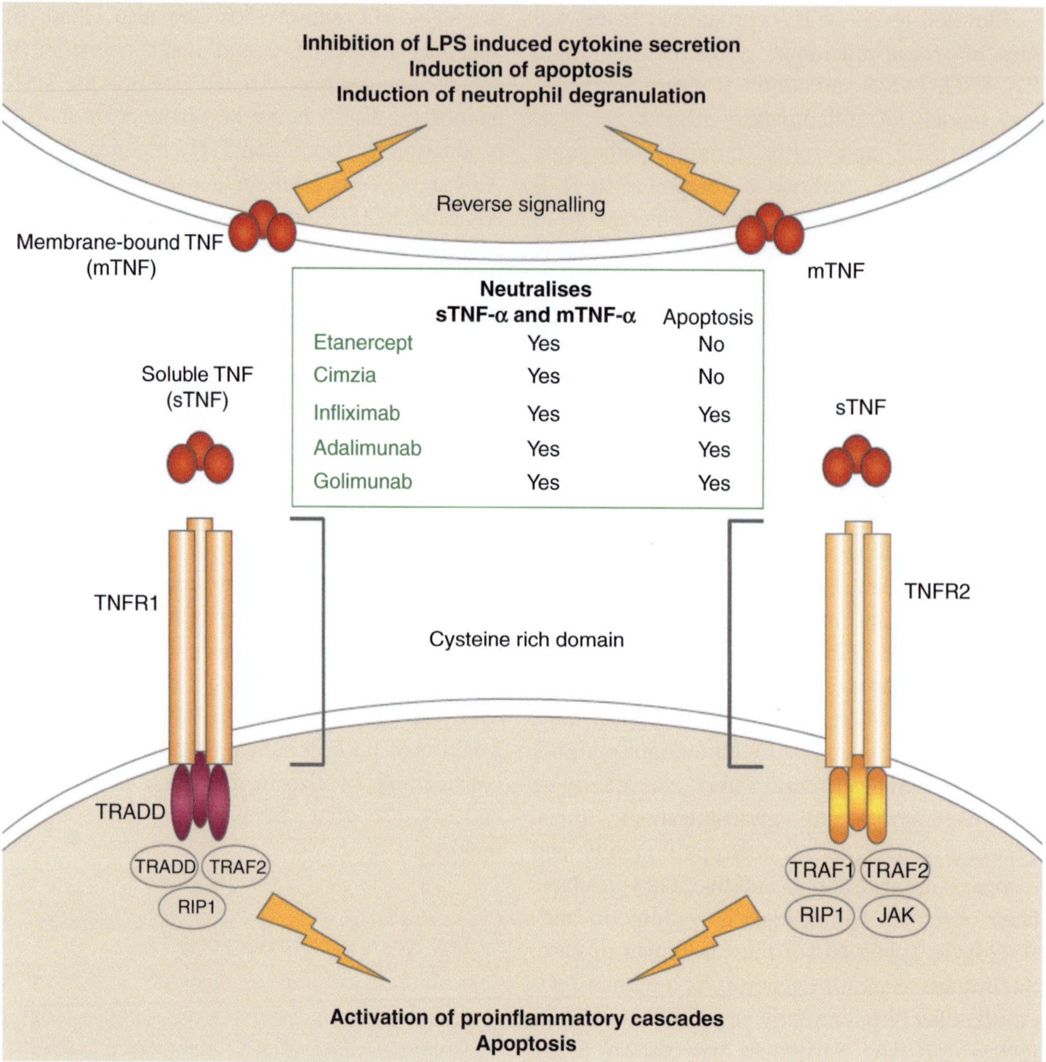

Fig. 21.6 TNF-α biology and the activities of the approved anti-TNF-α biologics

Although IL-17A probably developed to aid host defence against extracellular bacteria and other pathogens at mucosal epithelial barriers (IL-17 KO mice are susceptible to many pathogens [51]), IL-17 dysregulation has been linked to PSORIASIS, RA, MS and IBD, and inhibition of IL-17 ameliorates these diseases in animal models. IL-17A (and IL-17E) may also have role in neutrophilic airways INFLAMMATION and in airway remodelling.

Two humanized mAbs neutralizing IL-17A (**secukinumab** [Cosentyx], Novartis; **ixekizumab**, Eli Lilly) and an anti-IL-17R mAb (**brodalumab**, Amgen) which inhibits IL-17A, IL-17E and IL-17F have shown impressive results in Phase 3 clinical trials for PSORIASIS and PSORIATIC ARTHRITIS (Fig. 21.7) (see also Chap. 34). Both secukinumab and ixekizumab are approved in the USA and EU for PSORIASIS, and the approval of brodalumab is currently under review. PSORIASIS might also be an indication that could benefit from combined IL-17A and IL-17F blockade, since IL-17F rather than IL-17A has been shown to be a selective neutrophil chemoattractant and inducer of IL-6 in PSORIASIS [52].

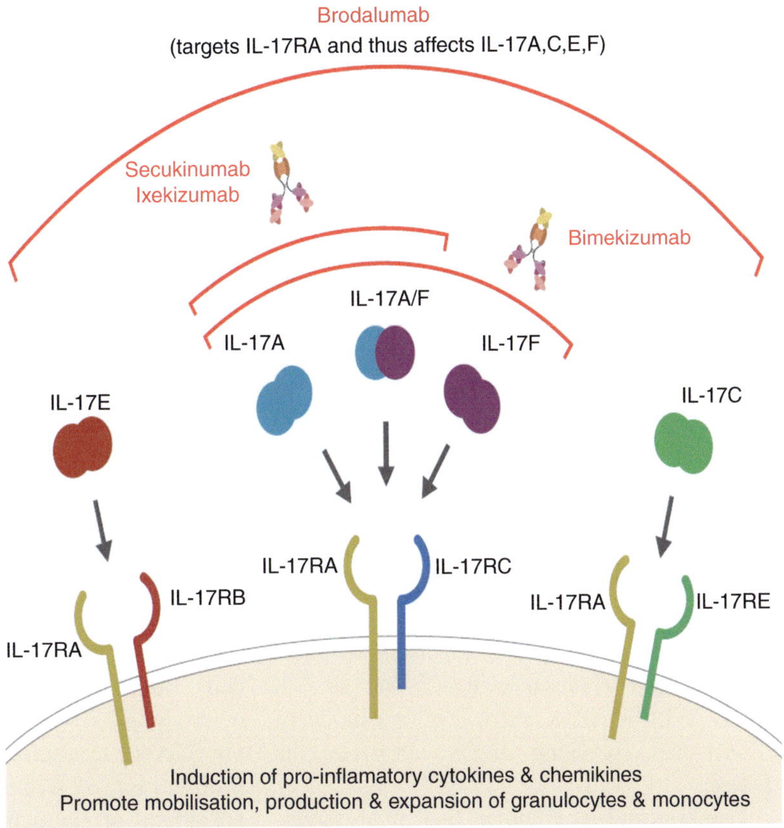

Fig. 21.7 Neutralization of different IL-17 cytokine subtypes by anti-IL-17 mAbs

Bimekizumab (UCB) is a dual IL-17A/F-specific humanized mAb in Phase 2 for PSORIASIS and PSORIATIC ARTHRITIS. In RA, the response rates for the anti-IL-17A mAbs were lower compared to anti-TNF-α or anti-IL-6 and only secukinumab is continuing in RA. In RA, TNF-α appears to dominate over IL-17, whilst in PSORIASIS, IL-17 dominates over TNF-α. Treatment with anti-TNF-α in human RA results in increases in TH17 cells and higher serum IL-17 levels, especially in RA patients least responsive to anti-TNF-α therapy [53]. Similarly, anti-TNF-α-treated arthritic mice show strong correlation between IL-17 production and clinical score. Thus, transgenic overexpression of both IL-17 and TNF-α results in a more erosive arthritis than expression of each cytokine alone, and combined blockade of TNF-α and IL-17 is more effective than monotherapy [54]. Hence the full effect of TNF inhibitors in RA may be compromised by an increase in pathogenic TH17 cells. Dual TNF-α/IL-17 bi-specific mAbs are in early development [44].

21.4.1.1.5 Anti-IL-23/IL-12

IL-23 is a heterodimeric cytokine composed of the p19 and the p40 subunits that is produced primarily by MACROPHAGES and DCs (see Chap. 5). The p40 subunit can also associate with the IL-12p35 subunit to form biologically active IL-12. It exerts its biological activity through binding to a heterodimeric receptor comprising the IL-12Rβ1 and IL-23R subunits. The primary role of IL-23 is to maintain populations of potentially pathogenic TH17 cells [55]. IL-23 does not initiate commitment of naive CD4+ T cells to TH17 cells but does induce TH17 cell expansion and the subsequent IL-17 and IL-22 secretion by these cells drives pro-inflammatory cytokine induction, leukocyte migration (particularly polymorphonuclear LEUKOCYTES), and osteoclast and epithelial cell activation. IL-23 KO mice show decreased HUMORAL and CELL-MEDIATED IMMUNITY, with reduced levels of both IL-17A- and IL-17F-producing cells.

The immune phenotype of IL-23 KO mice is more pronounced compared with mice lacking the TH17-secreted CYTOKINES IL-17A, IL-17F, IL-22 or IL-21, suggesting that the role of IL-23 goes beyond TH17 differentiation or the existence of a yet unidentified pathogenic TH17 cell effector cytokine. IL-23R polymorphisms are associated with PSORIASIS, IBD and other disease. IL-23-deficient mice are resistant to the development of several autoimmune inflammatory disease, and this protection was associated with a decreased number of TH17 cells, whereas there was no difference in TH1 cells or impact of IL-12 deficiency. In psoriatic lesions, DCs and keratinocytes produce IL-23 and encourage TH17 cell development and subsequent production of IL-17, IL-22 and IL-23 itself. In mice, IL-23 causes epidermal hyperplasia and a PSORIASIS-like state, and in an animal model of PSORIASIS, blocking IL-23 alleviates the disease.

Ustekinumab is a human IgG_1 mAb targeting the p40 subunit, thereby neutralizing both IL-23 and IL-12 (induces TH1 cell differentiation and NK and $CD8^+$ cytotoxic T cell activation resulting in TNF-α and IFN-γ production) and inhibiting both TH17 and TH1 responses. It is approved for the treatment of PSORIASIS, is at preregistration for PSORIATIC ARTHRITIS and is in Phase 2 for RA, CROHN'S DISEASE and sarcoidosis. Interestingly, the duration of clinical benefit after a few injections is prolonged and appears to far exceed the PK profile of the mAb. Many patients who failed to respond to anti-TNFs subsequently respond to ustekinumab [56]. Ustekinumab has also shown benefits in moderate-to-severe CROHN'S DISEASE, especially in anti-TNF-α nonresponders [36]. Cardiovascular events have also been observed with ustekinumab, and hence it remains to be seen if this is a class effect of IL-12/23 suppression [38]. Given the pivotal role of the IL-12-TH1-mediated pathway in the defence against facultative intracellular bacterial (e.g. mycobacteria), fungal, viral and protozoan parasites, the targeting of IL-23 alone by blocking only the p19 subunit and leaving IL12 function intact is probably the safer strategy. However ustekinumab-treated patients have not shown an increased risk of infection and cancer after several years of treatment. **Guselkumab** (Janssen), a human IgG1, and **Tildrakizumab** (Merck, Sun Pharma), a humanized IgG1, are anti-IL-23 p19 α subunit mAbs in Phase 3 for PSORIASIS [43], and other anti-IL-23 p19 mAbs are in earlier clinical trials for psoriasis, RA and CROHN'S DISEASE [44].

21.4.1.1.6 Anti-IL-4 and Anti-IL-13

IL-4 and IL-13 play key roles in the allergic cascade [57, 58] (see Chaps. 11, 22–24). Both are produced primarily by CD4+ TH2 cells as well as by MACROPHAGES, EOSINOPHILS, BASOPHILS, MAST CELLS and NK and NKT cells. Both bind to a common (type II) receptor heterodimer, IL-4Rα/IL-13Rα1, which signals via STAT6, to elicit IgE class production and proliferation of B CELLS, MAST CELL and eosinophil recruitment, alternative M2 macrophage activation, adhesion molecule expression and CC chemokine production (e.g. exotoxin) from structural cells of the lung. In addition, IL-4 but not IL-13 induces TH2 cell differentiation and MAST CELL proliferation through binding to IL-4Rα heterodimerized with the common cytokine gamma chain (γc), whilst IL-13 but not IL-4 binds to a second high-affinity receptor IL-13Rα2 on fibroblasts, epithelial cells and airways smooth muscle cells to induce collagen synthesis, mucus production and bronchoconstriction, respectively. These activities make them key players in the development and maintenance of ASTHMA. Overexpression of either IL-4 or IL-13 induces ASTHMA in mice. Elevated IL-13/IL-13R levels have been detected in the airways, bronchoalveolar lavage fluid and sputum of asthma patients, and are induced by allergen challenge, and mast cell-derived IL-13 and IL-4 have been observed in the airway smooth muscle cells of these patients.

Blocking either IL-4 or IL-13 has shown EFFICACY in animal asthma models (see Chaps. 23 and 24). Current therapies target both CYTOKINES or IL-13 alone to try and ablate TH2-mediated INFLAMMATION, reduce lung remodelling and prevent progression to the more chronic severe form of the disease. There is also interest in targeting IL-13 for fibrotic diseases such as idiopathic pulmonary fibrosis (IPF) as well as in chronic obstructive pulmonary disease (COPD) and in ulcerative colitis where IL-13 may play a role in goblet cell hyperplasia and excess mucus production as well as villus shortening and epithelial cell hyperproliferation (see Chaps. 23 and 35). **Lebrikizumab** (Roche-Genentech) is a humanized IgG4 anti-IL-13 mAb, currently in Phase 3 for ASTHMA [59], which has demonstrated an improvement in forced expiratory volume in 1 min (FEV_1) measurements and asthma exacerbations in patients with moderate-to-severe ASTHMA, including those on long-acting β-agonists. The greatest response was in patients with high serum periostin levels (periostin is highly expressed by airway epithelial cells in response to IL-13 and therefore has been proposed as a serum TH2 biomarker). **Tralokinumab** (AZ-Medimmune) is a human IgG_4 mAb specific for IL-13 in Phase 3 for ASTHMA [60]. **Dupilumab** (Regeneron) is a human anti-IL-4Rα IgG4 mAb under FDA and EMA review for atopic dermatitis, in Phase 3 for ASTHMA and in Phase 2 for eosinophilic esophagitis [61, 62]. It neutralizes both IL-4 and IL-13 signalling. A number of other anti-IL4 and/or IL-13-specific mAbs are in earlier clinical trials in ASTHMA, idiopathic pulmonary fibrosis and eosinophilic esophagitis [44].

21.4.1.1.7 Anti-IL-5R/IL-5

IL-5 is produced primarily by $CD4^+$ TH2 cells and has multiple effects on EOSINOPHILS through binding to the IL-5R which comprises a heterodimer of the IL-5Rα chain and the common beta chain shared with the IL-3Rα and CSF2Rα [63] (see also Chap. 6). All three CYTOKINES contribute to eosinophil development, endothelial adhesion, activation and survival, acting both on progenitor cells within the bone marrow as well as on mature cells at distal sites. However IL-5 is the most selective and potent for differentiation, proliferation and maturation of EOSINOPHILS within the bone marrow [64]. Only IL-5 and the eotaxins selectively regulate eosinophil trafficking. Increased EOSINOPHILS are observed in both bone marrow, peripheral blood and lung tissue of atopic asthmatics, and upon allergen challenge in asthmatics and in murine models of airway INFLAMMATION. Following activation by IL-5, EOSINOPHILS degranulate and release lipid mediators, CYTOKINES, cytotoxins, LEUKOTRIENES and PLATELET-ACTIVATING FACTOR (PAF) and pro-fibrogenic factors such as TGF-α, TGF-β, PDGF and MMP-9 that induce mucus hypersecretion, remodelling and AHR.

Therefore, therapeutic strategies that target IL-5 may be effective in treating eosinophilic asthma phenotypes that become refractory to CORTICOSTEROIDS and omalizumab (see Chaps. 23 and 32). IL-5 KO mice and treatment with an anti-IL-5 mAb abolished eosinophilic responses in the lungs of allergen-sensitized mice following allergen challenge. Two anti-IL-5 mAbs, **mepolizumab** (Nucala; GSK) and **reslizumab** (TEVA) showed EFFICACY in Phase 3 clinical trials for allergic (eosinophilic) ASTHMA and eosinophilic esophagitis [65, 66]. Both have recently been approved in the USA and EU for eosinophilic ASTHMA. Early trials of these mAbs in mild-to-moderate asthmatic subjects showed substantial reductions in sputum and blood eosinophilia but failed to demonstrate significant clinical or functional improvement in outcomes such as FEV1, quality of life scores, use of β2 agonists and airways HYPERSENSITIVITY response (AHR). However, not all asthmatic INFLAMMATION is eosinophilic. Eosinophilic ASTHMA tends to be adult-onset and more symptomatic, with frequent exacerbations and persistent airflow limitation. Further studies in patients with persistent eosinophilic ASTHMA showed that both mepolizumab and reslizumab could reduce blood and lung eosinophilia, had steroid-sparing effects and reduced exacerbation rates. Hence anti-IL-5 mAbs may only be active in small subgroups of severe asthma patients. **Benralizumab** (AZ-Medimmune) is a humanized afucosylated anti-IL-5Rα IgG_1 mAb in Phase 3 for ASTHMA and COPD [67]. It not only induces APOPTOSIS in EOSINOPHILS and BASOPHILS but also efficiently depletes tissue EOSINOPHILS (by ADCC) which may be incompletely blocked by anti-IL-5 mAbs as a result of the sustained actions of IL-3 and CSF2. Depleting activated EOSINOPHILS should also overcome any eotaxin-mediated activity not dependent on the IL-5 pathway.

21.4.1.1.8 Anti-IFN-α

Type I INTERFERONS comprise IFN-α, IFN-β, IFN-ε and IFN-κ [68] (see Chap. 6). There are 13 different IFNa protein with 77–98% sequence identity, whereas there is only 1 IFN-β, IFN-ε and IFN-κ. All of the type I IFNs bind to a single receptor (IFN-α/βR) expressed by many cell types. IFN-α is primarily secreted by plasmacytoid DCs and MACROPHAGES following activation of viral pattern recognition RECEPTORS (PRRs) such as the endosomal Toll-like RECEPTORS (TLRs)s and cytosolic nucleic acid sensors in response to double-stranded viral RNA. It plays a key role in protecting cells from virus infection and boosts antiviral and antitumour immune responses. IFN-α signalling results in a wide range of effects upon the INNATE and ADAPTIVE IMMUNE SYSTEMS, including upregulation of major histocompatibility complex (MHC) molecules; activation of APCs, DC maturation, T cell, and NK cell; macrophage activation and cytokine/chemokine release; and B cell antibody production [68]. Increased expression of IFN-α (but not other type I INTERFERONS) correlates with several autoimmune diseases [69] including SLE, type 1 diabetes and dermatomyositis. IFN-α and/βR inhibition ameliorates SLE in animals. Anti-interferon mAb intervention in SLE would require the neutralization of most, if not all, of the IFN-αs, but not the neutralization of the other type I INTERFERONS, thereby retaining significant antiviral activity. The anti-IFN-α/βR **anifrolumab**, a human IgG_1 mAb, is in Phase 3 for SLE and in early trials for systemic sclerosis [46]. It should inhibit the activity of all type I and type II INTERFERONS. The same company is also developing **sifalimumab**, a human IgG_1 mAb specific for IFN-α in clinical trials for SLE and muscle INFLAMMATION (myositis). It neutralizes a majority of the subtypes of human IFN-α. Promising results of a Phase 2b SLE study have been published [47].

21.4.1.2 Other Cytokines, Chemokines and Growth Factor Targets

mAbs targeting a range of other CYTOKINES (lymphotoxin-alpha (LTα), TWEAK (TNF-like weak inducer of APOPTOSIS), IL-18, IL-20, IL-21, IFNγ, TSLP and IL-31), CHEMOKINES (eotaxins, CXCR5, CCR4, RANTES, MIF) and GROWTH FACTORS (CSF2R, CSF1, CSF1R) are in earlier Phase 1 and 2 clinical trials [44].

21.4.2 T Cell Inhibitors

Activated T cells drive the initiation and progression of a number of inflammatory diseases following their recognition of autoantigen or allergen. These activated T cells, including TH1, TH2 and TH17 cells, support B cell activation and autoantibody production and produce a wide array of cytokine, CHEMOKINES and GROWTH FACTORS which promote further T cell activation (see Chap. 3). This includes cytotoxic T cells that cause direct tissue damage and attract and activate other LEUKOCYTES and non-haematopoietic cells, serving to amplify the inflammatory response. Not surprisingly, mAbs have been used to directly modulate the activity of these cells. There is a considerable unmet medical need for a more specific approach, utilizing the increasing knowledge of the plasticity of lymphocyte subsets, to prevent acute organ rejection, without unnecessarily exposing the patient to non-specific or open-ended immune suppression.

The T CELL RECEPTOR (TCR) for ANTIGEN on MHC-restricted $CD4^+$ T HELPER CELLS and $CD8^+$ CTLs is a heterodimer consisting of two covalently linked transmembrane polypeptide chains, designated TCR α and β. Ligation of the TCR by MHC-peptide LIGANDS results in the clustering of co-RECEPTORS (CD3 and ζ chains) with the antigen receptor and phosphorylation of ITAM tyrosine residues [70]. CD4 is a T cell co-receptor expressed on the surface of peripheral T cells and thymocytes that binds to non-polymorphic regions of MHC class II molecules and facilitates TCR signalling. The cytoplasmic tail of CD4 binds the src family kinase lck. The ability of CD4 to bind MHC II helps the lck to be brought close to the adjacent TCR that contacts the same MHC-peptide complex on APCs. Lck is drawn close to the ITAMs in CD3 and zeta chain thus facilitating subsequent recruitment and activation of the kinase ZAP-70. Assembly of multiple ZAP-70s on CD3 ζ results in the phosphorylation of other adaptor molecules required for initiation of T cell activation.

Productive CD4 T cell responses require not only binding antigen-specific TCR activation but also a second signal from costimulatory molecules. If T cells do not receive the second signal, then TOLERANCE or ignorance of the ANTIGEN ensues, and a productive immune response is not generated. Activation of T cells following simultaneous engagement of the TCR and CD28 (following binding to its LIGANDS CD80 and CD86 on APCs) results in T cell expansion and survival through the production of IL-2 and the generation of survival signals. In addition to the costimulatory properties of CD28, other molecules of the Ig and TNFR superfamilies, as well as other cytokine RECEPTORS, play instrumental roles in positively and negatively regulating the activation and survival of T cells following antigenic stimulation [71]. CD28, ICOS and CD2 typify costimulatory molecules of the Ig superfamily and play important roles in many types of T cell responses. Signals through a number of TNFR family members have also been shown to augment T cell responses in various settings, and these include CD40L, OX40 (CD134), 4-1BB (CD137), CD27, CD30, herpes virus entry mediator (HVEM) and glucocorticoid-induced TNFR family-related gene (GITR). Cytokine RECEPTORS that can control T cell growth or survival in some situations are also numerous and include IL-2R, IL-7R, IL-15R, IL-1R and IL-6R. CD6 is another nonclassical costimulatory molecule expressed on the majority of T cells (and some B CELLS) and is thought to be involved in antigen presentation by B CELLS and subsequent T cell proliferation [72]. Antibody-mediated CD6 cross-linking potentiates proliferative T cell responses, and soluble, monomeric CD6 inhibits antigen-specific T cell responses. CD6 has multiple LIGANDS, including activated leukocyte CELL ADHESION MOLECULE (ALCAM), its functional LIGAND in AUTOIMMUNE DISEASE [73] (ALCAM overexpression enhance lymphocyte recruitment to inflammatory sites such as the RA synovium). There is considerable overlap in the activities of costimulatory molecules activities; however, it is the timing, context and intensity of these costimulatory signals that might determine the functional consequence of their activity (and effects of mAb-induced modulation in disease). T cells also express a range of co-inhibitory RECEPTORS, termed immune checkpoint inhibitors (e.g. CTLA-4, PD-1), that suppress T cell responses. These are being primarily targeted by mAbs that block their function to increase tumour-specific T cell responses in cancer patients [74]. However they could also be targeted by activating (agonist) mAbs to increase their inhibitory effect and suppress pro-inflammatory T cell responses in inflammatory and autoimmune diseases.

Anti-CD3 mAbs were the first T cell-targeting mAbs to be studied, initially in transplant rejection to inhibit effector T cell-signalling, activation and cytokine release, primarily for acute organ rejection following transplantation where broad-spectrum immunosuppression is required [75] (see Chap. 31). However, due to the first-dose cytokine storms, observed with mAbs such as the murine **muromonab** (OKT3) (Table 21.1), they have been more recently studied in AUTOIMMUNITY, primarily to induce T cell TOLERANCE via Treg induction. Tregs play an important role in maintaining peripheral TOLERANCE by suppressing responses to self-antigen, as evidenced by AUTOIMMUNE

DISEASE development in mice and humans lacking FoxP3 [76]. In patients with type 1 diabetes, MS and RA, the persistence of INFLAMMATION suggests that Tregs are unable to suppress ongoing disease, perhaps due to an inhibition of their functions by pro-INFLAMMATORY CYTOKINES such as IL-6. Alternatively, the increased number of activated EFFECTOR T CELLS may be responsible or autoimmune T cells may be resistant to Treg-mediated control. Such regulation is dependent on the local concentrations of Treg-associated inhibitory CYTOKINES, TGF-β, IL-10 and IL-35, in facilitating Treg suppressive activity and promoting Treg generation [77].

Humanized anti-CD3 mAbs with Fc mutations to eliminate FcγR binding (and agonism) are teplizumab [75] and otelixizumab [78]. Teplizumab (based on OKT3) was tested in acute kidney rejection and PSORIATIC ARTHRITIS, demonstrating EFFICACY and an acceptable safety. Both teplizumab and otelixizumab showed promising results in Phase 2 studies in early onset Type 1 diabetes, halting disease progression for up to 4 years (teplizumab, preserving β cell function and driving patients towards insulin independency that lasted for 1.5–3 years after a single course of treatment [75, 78]). Transient decreases in circulating T cells with evidence of Treg induction were observed. Side effects were mild-to-moderate including "flu-like" symptoms that waned at subsequent doses, with no EBV-related symptoms. No higher incidence of infections and no lymphoproliferative or other types of cancer were observed. However both teplizumab and otelixizumab failed to meet their primary endpoint in Phase 3 studies in type 1 diabetes. Otelixizumab is currently in clinical trials for RA.

T cell proliferation induced by autocrine IL-2 was the target of the anti-IL-2R (CD25) mAbs **daclizumab** (a humanized IgG1) and **basiliximab** (a chimeric IgG1), two of the earliest mAbs to be approved. T cell costimulatory pathways are targeted by the IgG1 fusion proteins abatacept, belatacept and alefacept. **Abatacept**, approved for RA, is a soluble protein consisting of the extracellular domain of CTLA-4 linked to the Fc portion of IgG_1 that binds CD80 and CD86 on APCs thereby blocking CD28 binding and T cell activation [79]. Abatacept was well-tolerated in clinical trials, with a similar EFFICACY to anti-TNFs in some studies and a slight increase in the incidence of infections, especially among those with underlying COPD [79]. Abatacept is used in RA, for patients who have had inadequate responses to one or two TNF inhibitors. The highly related **belatacept** against the same target is approved for kidney rejection. **Alefacept**, approved for PSORIASIS, is a fusion protein of a soluble form of the extracellular domain of lymphocyte function-associated antigen-3 (LFA-3) attached to the Fc portion of IgG_1. It binds $CD2^+$ T cells, blocking T cell activation, inflammatory cytokine production and migration to plaques, and induces memory T cell APOPTOSIS [80]. Itolizumab is a humanized IgG1 nondepleting anti-CD6 antibody (Biocon) approved in India only for PSORIASIS. It inhibits CD6-ALCAM-induced proliferation of peripheral blood mononuclear cell and reduces pro-INFLAMMATORY CYTOKINES [81].

A number of other T cell-targeting mAbs and Fc-fusion proteins are in early clinical trials in organ transplantation and associated GRAFT-VERSUS-HOST disease (GvHD), RA, PSORIASIS, MS, SLE, IBD and ASTHMA. These are designed to directly deplete or modulate/suppress the function of EFFECTOR T CELLS or to induce the generation of Tregs by targeting either the T CELL RECEPTOR (TCR) complex, including the TCR itself or CD3, the T cell activation-associated accessory molecule CD4, or a range of costimulatory molecules. These include CD28, CD40/CD40L, ICOS LIGAND (ICOSL), LIGHT (lymphotoxin-like, exhibits inducible expression, and competes with HSV glycoprotein D for HVEM), CD26 and NKG2A (natural killer cell group 2A) [82]. TGN-1412 is a superagonist of CD28 (stimulating T cells independently of T cell receptor [83] that promotes the preferential expansion and activation of naturally occurring $CD4^+CD25^+CTLA\text{-}4^+Foxp3+$ Tregs cells over conventional T cells). Catastrophic adverse effects were induced by TGN1412 during initial Phase 1 trials [84]. This was a result of the unexpected TGN1412-mediated activation (proliferation and cytokine release) of predominantly tissue resident CD28-expressing $CD4^+CD45RO^+$ effector memory T cells [85], highlighting the importance of understanding the translational biology when modulating these T cell activation pathways. Begelomab (Begedina; Adienne) is a murine IgG2b anti-CD26 mAb which was under EMA review for GvHD but recently withdrawn (insufficient data to support EFFICACY claim).

Agonists of T CELL RECEPTORS require understanding of the PK/PD relationship compared to antagonistic biologics. Due to their stimulatory component, it is important to understand the type of agonistic effect. There are three main types of effects: a Cmax effect, where a single peak concentration can stimulate a response; a Cmin effect, where a threshold concentration needs to be reached in order to maintain an effect; or an AUC effect which is observed when exposure over a period of time is required to provide EFFICACY. Dose fractionation studies are often performed to better understand the PK/PD relationship and allow improved predictions of preclinical to clinical dosing. Another aspect is the temporal relationship of the immune cells in relation to the biologic drug. The temporal and PK/PD relationship of otelixizumab was studied in an in vitro system, demonstrating the relationship for otelixizumab, between binding to human CD3 antigen and expression of CD3/TCR complexes on LYMPHOCYTES [86].

21.4.3 B Cell Inhibitors

B CELLS contribute to the pathogenicity of a number of inflammatory diseases through the production of AUTOANTIBODIES, presentation of AUTOANTIGENS to T cells, generation of ectopic lymphoid FOLLICLES and the production of CYTOKINES and CHEMOKINES. The success of the B cell-depleting mAbs **rituximab** (anti-CD20) and **belimumab** (anti-Blys) for the treatment of RA and SLE, respectively, has validated the important role of B CELLS in the pathogenesis of these diseases and promoted the search for mAbs with improved EFFICACY and comparable safety for the treatment of these and other diseases where B CELLS are believed to play a role (see also Chap. 34).

The first B cell-targeting mAb was the anti-CD20 chimeric IgG_1 mAb **rituximab**, initially approved for B cell non-Hodgkin's lymphoma (NHL) and later for anti-TNF refractory RA and antineutrophil cytoplasmic antibody (ANCA)-associated vasculitis. CD20 is expressed on the surface of pre-B CELLS through activated mature B CELLS and memory B CELLS. It is absent on stem cells, early pre-B cells and terminally differentiated antibody-producing plasma cells [87]. It has no known LIGAND nor function. On mAb binding, CD20 is neither shed nor internalized, making it an ideal target for B cell-directed therapy. Rituximab is thought to induce lysis of $CD20^+$ B CELLS through several mechanisms, including CDC, ADCC and APOPTOSIS induction. Depletion of B CELLS can last up to 9 months or longer after a single course of therapy. Rituximab has been shown to improve the signs and symptoms of disease, functional status, quality of life and slow radiographic progression of disease in patients with RA [88], being seen most frequently in combination with disease-modifying antirheumatic drugs and in rheumatoid factor-positive patients. Rituximab has shown promising results in other autoimmune diseases, such as SLE, Sjogren's syndrome, immune thrombocytopenia purpura (ITP), chronic inflammatory demyelinating polyneuropathy and pemphigus vulgaris (PV).

Rituximab has also been well-tolerated, with infusion reactions (more common in NHL than RA due to higher B cell burden) being the most common adverse event. Despite the depletion of mature B CELLS, no significant increases in serious or opportunistic infections were reported in treated RA or NHL patients except a few occurrences of progressive multifocal leukoencephalopathy (PML). Serum IMMUNOGLOBULIN levels generally remain stable during treatment, probably because plasma cells are not depleted; however chronic treatment might become more of a safety concern because plasma cells are not replenished by memory B CELLS. Frequent infusion reactions, potentially neutralizing anti-chimeric antibody development, as well as variable EFFICACY in diseases such as SLE (possibly related to defects associated with apoptotic machinery, COMPLEMENT insufficiency, Fc receptor expression variations and genetic polymorphisms) have prompted development of several humanized and fully human anti-CD20 mAbs, some with improved ADCC activity.

Ocrelizumab (Roche) is a humanized version of rituximab that is currently under review by FDA and EMA for MS [89] (see Table 21.1). Ofatumumab is a human IgG_1 mAb against CD20 but binds to a different EPITOPE than rituximab that confers a longer binding time and stronger CDC activity compared with rituximab, without inducing APOPTOSIS. It was in Phase 2 for RA and in Phase 2 for MS. It is currently in Phase 1/2 for graft-versus-host disease. Both ocrelizumab and ofatumumab have shown promising results in RA patients who are poor responders to TNF-α inhibitors, with less immunogenicity and fewer infusion reactions than rituximab [90, 91]. However ocrelizumab has been discontinued in both RA and SLE due to an increased rate of serious infections.

The anti-CD22 mAb epratuzumab (UCB) recently failed in a Phase 3 trial in SLE. Veltuzumab (Takeda) is an anti-CD20 "biobetter" (improved version) of rituximab, with antiproliferative, apoptotic and ADCC effects on B CELLS and B cell lymphoma in vitro similar to those of rituximab but with significantly slower off-rates and increased CDC activity [92]. It has shown favourable results in early studies for B cell lymphoproliferative malignancies with lower immunogenicity and fewer infusion reactions than rituximab. It can be given either intravenously or subcutaneously (an advantage over rituximab).

B cell-activating factor (BAFF) or B lymphocyte stimulator (BLyS) and a proliferation-inducing LIGAND (APRIL) are members of the TNF cytokine family and are important survival and GROWTH FACTORS for both normal and pathogenic B CELLS. **Belimumab**, a human anti-BAFF IgG_1 mAb preventing the binding of BAFF to BAFFR on B CELLS, is approved for SLE (Table 21.1). Belimumab plus standard of care therapies were well-tolerated, induced a significant reduction of B CELLS in peripheral blood and achieved mild-to-moderate improvement in disease remission and time to flare. Its impact in RA, however, was modest [93]. It is currently in Phase 3 for ANCA-associated vasculitis and LYMPH NODES. Tabalumab (LY2127399) is a fully human IgG_4 mAb specific for BAFF recently discontinued in both SLE and RA [94]. Blisibimod (AMG 623), comprising four BAFFR domains fused to IgG Fc and which binds to BAFF inhibiting interaction with BAFF RECEPTORS, is in Phase 3 for SLE and ITP [95]. Atacicept is a fusion protein composed of the extracellular domain of TACI fused with the Fc fragment of human IgG_1 that blocks both BAFF and APRIL and, hence, unlike BAFF-specific therapies, might also target plasma cells responding to APRIL. Several early studies in patients with RA and SLE have shown that atacicept can reduce peripheral blood B CELLS and serum IMMUNOGLOBULIN without associated toxicity. It lacked EFFICACY in RA despite depleting B CELLS, RF, and anticyclic citrullinated AUTOANTIBODIES [96]. It is currently in Phase 2 for SLE. VAY736 (Novartis) is a fully human afucosylated IgG_1 mAb specific for BAFFR in Phase 2 for pemphigus vulgaris, RA and Sjogren's syndrome [97]. It has potent ADCC-mediated B cell depletion activity coupled with its ability to block BAFF-mediated B cell growth and survival.

Anti-CD20 therapy together with BAFF neutralization/inhibition may have synergistic effects and could be an option in humans to optimize EFFICACY. Another attractive target on B CELLS is CD19 which shares a similar cell expression pattern as CD20 but is also expressed on some plasma cell populations not depleted by current B cell therapies. Inebilizumab (AZ-Medimmune) is a humanized IgG1 anti-CD19 mAb in Phase 2b for neuromyelitis optica, and other anti-CD19 mAbs are in early clinical trials in SLE, RA and MS [97].

21.4.4 Inhibitors of Leukocyte Adhesion and Migration

LEUKOCYTES circulate in the blood vessels; however at specific locations, such as in normal high endothelial venules (HEV) in LYMPHOID ORGANS, or in inflamed venules in almost any organ, they are able to leave the blood and extravasate into the tissue. This process is of major importance in the maintenance of normal innate and acquired immunity; however, inappropriate migration of activated LEUKOCYTES such as T cells to target organs is a key pathogenic factor in a wide number of inflammatory diseases [98, 99] (see Chaps. 1 and 8). The extravasation cascade starts by tethering and rolling of the leukocyte on the endothelial cells. If followed by suitable activation stimuli, the leukocyte can then firmly adhere to the endothelial cell and finally transmigrate through the inter-endothelial junctions, or through the endothelial cell, into the tissue to exert its effector functions. The distinct steps of the extravasation cascade are typically mediated by different receptor-ligand pairs on the two opposing cell types. Thus, selectins and their oligosaccharide-based LIGANDS presented on sialomucin-like proteins are key mediators of the transient first contacts, whereas CHEMOKINES and their RECEPTORS are heavily involved at the activation step. Thereafter, INTEGRINS on the LEUKOCYTES and their LIGANDS belonging to the IMMUNOGLOBULIN superfamily on the endothelium, together with certain homotypically interacting molecules such as CD31 and CD99, play essential roles in the firm adhesion and transmigration. Adhesion molecules targeted by mAbs are shown in Fig. 21.8.

Particularly important for T cell migration to LYMPHOID ORGANS and inflammatory sites are the INTEGRINS, very late antigen-4 (VLA-4; $\alpha4\beta1$) and its counter-receptor, vascular CELL ADHESION MOLECULE 1 (VCAM-1), as well as leukocyte function antigen-1 (LFA-1) and its counter-RECEPTORS, intercellular adhesion molecule (ICAM)-1 and ICAM-2. The INTEGRINS $\alpha4\beta7$ and $\alpha E\beta7$ integrin are the key integrins required for trafficking and retention in LYMPHOCYTES in the GI tract [100]. MAdCAM-1, expressed on intestinal postcapillary venules, supports the specific rolling, adhesion and diapedesis of $\alpha4\beta7^+$ T cells to the specialized endothelium of the GI tract. $\alpha E\beta7$ is expressed mainly on gut intraepithelial LYMPHOCYTES and subsets of peripheral tissue DCs and binds E-cadherin on intestinal epithelial cells. Activated effector memory CD45RO$^+$ CD4$^+$ T cells expressing $\alpha4\beta7$ and $\alpha E\beta7$ are increased in the GI tract (and decreased in the

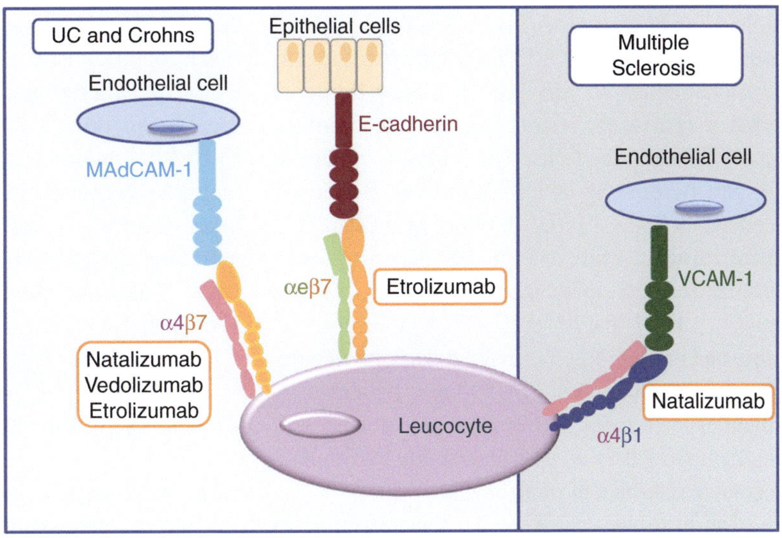

Fig. 21.8 Overview of mAbs targeting leukocyte adhesion molecules

periphery) of ulcerative colitis and CROHN'S DISEASE patients [100]. In several preclinical colitis models in rodents and NHPs, blockade of the individual α4 and β7 chains, the α4β7 heterodimer or MAdCAM-1 by genetic deletion or using mAbs reduced T cell recruitment, mucosal destruction and clinical signs, suggesting that targeting this axis might have utility in the treatment of IBD. αEβ7 is expressed mainly on gut intraepithelial LYMPHOCYTES and subsets of peripheral tissue DCs and binds E-cadherin on intestinal epithelial cells. Blocking of αEβ7 with a mAb is also effective in blocking T cell migration to/retention in the GI tract in colitis models.

Natalizumab is a humanized IgG_4 mAb specific, for the α4 subunit of VLA-4/α4β1 [101]. It also binds to and inhibits the function of the a4β7 INTEGRINS which bind MAdCAM-1. α4β1 is highly expressed on activated $CD45RO^+$ $CD4^+$ T effector memory T cells and directs migration of LYMPHOCYTES into the brain and intestinal parenchyma. It also induces T cell APOPTOSIS and anergy and prevents T cell binding to osteopontin and fibronectin, thereby attenuating T cell-mediated INFLAMMATION. Natalizumab, either alone or in combination with IFN-β, was associated with significantly lower relapse rates and disability and fewer new brain lesions as determined by magnetic resonance imaging (MRI) [101]. However it is associated with the development of PML caused by reactivation of John Cunningham virus, limiting its use (approved only in the USA for CROHN'S DISEASE and as a monotherapy in MS [102]). Susceptibility to PML may arise as a result of disruption of immune surveillance in the CNS, which may be mediated in part by homing of α4β1-positive cytotoxic T cells via VCAM-1 [102, 103] (see also Chap. 35). **Efalizumab**, approved in PSORIASIS, is a humanized IgG_1 mAb specific for CD11a, a subunit of LFA-1 that blocks T cell homing into the skin and inhibits T cell activation. Despite impacting symptoms of skin PSORIASIS, the development of PML among several patients treated with efalizumab led to its withdrawal due to an unacceptable risk-benefit in psoriasis.

The observed EFFICACY of natalizumab in IBD, most likely by blocking α4β7 interaction, is driving the developing of new mAb therapies to inhibit T cell migration to the intestine [104]. By specifically blocking α4β7 and not α4β1, lymphocyte homing to the GI tract should be inhibited without affecting lymphocyte trafficking to non-mucosal tissues such as the brain, thereby reducing the risk of PML. Several therapeutic agents have been developed that specifically target the α4β7/MAdCAM-1 axis. **Vedolizumab** is a humanized mAb specific for α4β7 (Table 21.1). It demonstrated promising EFFICACY in Phase 3 trials in both ulcerative colitis and CROHN'S DISEASE and was subsequently approved in the USA and EU [105]. **Etrolizumab**, also in Phase 3 for both these IBD indications, is specific for β7 subunit of the INTEGRINS α4β7 and αEβ7, blocking interaction with their LIGANDS MadCAM-1 and E-cadherin, respectively (see Table 21.2). Inhibition of binding of αEβ7 to E-cadherin might provide superior EFFICACY to the inhibition of binding of α4β7 to MAdCAM-1 alone [106]. Other mAbs in earlier clinical trials for IBD include PF-00547659 specific for MadCAM-1 [107] and AMG 181, another anti-α4β7 [108]. These hold promise for a more acceptable risk-benefit profile compared to natalizumab (Tysabri) but also to TNF blocking mAbs. Although the latter reduce exacerbations and corticosteroid use in CROHN'S DISEASE patients [109], there are a substantial number of poor/nonresponders and an enduring concern over the long-term safety and elevated risk of opportunistic infections with these mAbs. mAbs in earlier clinical trials are specific. mAbs specific for other INTEGRINS such as VLA-1 and VLA-2 which interact with the extracellular matrix and not endothelium (in an effort to block inflammatory T cell migration without an increased risk of PML) and selectins, as well as novel leukocyte adhesion RECEPTORS such VAP-1, are in early clinical trials [110].

21.4.5 IgE Inhibitors

ALLERGY begins in the early stages of life, and the inflammatory response increases as an individual is repeatedly exposed to allergens (see Chaps. 11 and 22). Early response to aerosolized antigen leads to a transient rise in allergen-specific IgE, followed by an increase in the IgE level as exposure continues [111]. MAST CELLS and BASOPHILS are central to the immediate HYPERSENSITIVITY reaction following IgE binding to their high-affinity receptor, FcεRI. Multivalent allergens cross-link receptor-bound IgE, and trigger the DEGRANULATION of these cells, resulting in the release of HISTAMINE, TH2 CYTOKINES, LEUKOTRIENES and TNF-α mediators, intimately associated with pathophysiological responses such as bronchoconstriction, airway hyperresponsiveness and airway INFLAMMATION. B CELLS, DCs and other APCs can also bind IgE through their low-affinity receptor, FcεRII, which can result in enhanced presentation of allergens to T cells and enhanced allergen-specific antibody production. There is a consensus that the antigen-specific IgE/MAST CELL axis is crucial for the development of the acute manifestations of these allergic disorders; however, there is less agreement about the role of IgE and MAST CELLS in the long-term tissue remodelling following chronic allergen exposure that account for much of the morbidity of allergic disease.

The most direct approach to inhibit the MAST CELL/IgE axis is to inhibit the function and/or production of IgE. **Omalizumab** is a humanized IgG_1 ANTI-IgE mAb that is approved for the treatment of moderate-to-severe allergic ASTHMA (with raised serum IgE levels) where it reduces disease exacerbations, improves symptom control and reduces overall use of inhaled CORTICOSTEROIDS (ICS) and β-agonists [112] (see Chap. 23). It binds to the Fcε3 domain of IgG forming tri- and hexameric IgE-containing immune complexes that are unable to bind to the FcεRI on MAST CELLS or FcεRII on APCs. It is non-anaphylactogenic (does not cause ANAPHYLACTIC SHOCK), being unable to bind IgE prebound to FceRI, and hence is unable to mediate IgE receptor cross-linking and HISTAMINE release. This loss in free circulating IgE (i.e. that can bind to FεRI) leads to progressive FcεR1 internalization in MAST CELLS, BASOPHILS and DCs thereby enhancing its therapeutic EFFICACY and inducing disease modification. Decreases in circulating IgE and reduced FcεR1 expression markedly attenuate MAST CELL and basophil release of preformed and newly generated mediators. However, omalizumab is not the optimal therapeutic. It has a relatively poor affinity for IgE, and this, coupled with the very high affinity of IgE for FcεRI, means that omalizumab has to be used in significant molar excess of IgE to be effective. This high-dose requirement excludes patients with high levels of IgE or large body mass from treatment with omalizumab due to the high subcutaneous dosing volume required. Also two-thirds of treated patients have an inadequate or no improvement in disease. This may be due to a relative lack of correlation between free IgE levels and FcεR1 expression, differences in intrinsic cellular sensitivity and the importance of the ratio of antigen-specific IgE to total IgE. Higher affinity anti-IgE antibodies such as ligelizumab (Novartis), a humanized IgG1 currently in Phase 2 in ASTHMA [113] and chronic urticaria, are likely to improve the cost-benefit ratio for ANTI-IGE therapy and improve compliance. None of these antibodies will modulate the production of IgE as there is no diminishment of IgE synthesis, only blockage of its binding. mAbs designed to inhibit the production of IgE by B CELLS are in early clinical trials.

21.4.6 Complement Inhibitors

The COMPLEMENT SYSTEM comprises more than 30 different plasma proteins, produced mainly by the liver, that serves as a potent innate immune defence cascade, protecting the host from microorganisms such as bacteria and other foreign and abnormal cells [114, 115] (see also Chap. 8). Various COMPLEMENT components interact with each other to form three canonical pathways of activation which are the classical, alternative and lectin-binding pathways. In the absence of specific triggers, COMPLEMENT proteins circulate in an inactive form. However in the presence of triggers such as pathogens, antibody bound to pathogens and traumatic or surgical tissue damage, they become activated, serving to opsonize bacteria for killing by antibody and to generate chemotactic effector proteins (C3a, C5a) that attract innate inflammatory cells such as NEUTROPHILS and MACROPHAGES. COMPLEMENT has many other important physiological functions, including sensing and CLEARANCE of apoptotic cells and immune complexes and other roles related to tissue homeostasis. However, excessive and/or inappropriate COMPLEMENT activation has been shown to play a significant role in the pathogenesis of a number of inflammatory diseases, including RA, PSORIASIS, SLE, ASTHMA, COPD, sepsis, ischemia/reperfusion (I/R) injury, cardiovascular diseases, age-related macular degeneration (AMD), idiopathic thrombocytopenia purpura (ITP), paroxysmal nocturnal haemoglobinuria (PNH) and many others [115, 116]. COMPLEMENT activation is capable of inducing all classical signs of INFLAMMATION, with the occurrence of pain, swelling, reddening, hyperthermia and impaired function. Hence, there is considerable interest in this system as a potential therapeutic target [116], and research has focused on how the complement-mediated tissue damage in these diseases can be contained.

Many COMPLEMENT therapeutics are mAbs, since the critical steps in the COMPLEMENT cascade are fundamentally based on large protein-protein interactions, which are challenging to influence with small molecules. mAbs in clinical development that target COMPLEMENT components are focusing on AMD, RA and COPD as primary indications. Strategies involve blocking COMPLEMENT early at the stage of C3 or later by blocking C5. **Lampalizumab** is an anti-complement factor D (CFD) humanized IgG1 Fab in Phase 3 for geographic atrophy (dry AMD), which aims to block assembly of the C3 convertase that generates C3a and C3b and downstream pathway activation [117].

C5 is produced in hepatocytes, MACROPHAGES, pulmonary epithelium and astrocytes and is present at high concentration in plasma [118]. COMPLEMENT activation leads to C5 convertase-mediated cleavage of C5 in to C5a (anaphylatoxin) and C5b, a component of the MEMBRANE ATTACK COMPLEX (MAC). The C5b-9 MAC plays a key role in killing of pathogens. However, although C5a is important in driving CR1-mediated PHAGOCYTOSIS of C3b-opsonized bacteria, many of the pathological effects seen in inflammatory diseases are attributable to excessive C5a levels. C5a is thought to contribute to disease initiation and progression by increasing vascular permeability and recruitment of NEUTROPHILS and other cells to the site of damage and then stimulating neutrophil activation and release of pro-inflammatory mediators and CYTOKINES e.g. IL-1, IL-6, IL-8 and TNF-α, which ultimately leads to tissue damage. The C5a/C5R pathway is upregulated in inflammatory tissues, including the RA synovium and in AMD, and disease-modifying activity has been achieved in animal models of INFLAMMATION (RA, AMD, I/R injury, sepsis, IBD, etc.) using antibodies and small molecule inhibitors of C5a and C5aR or genetic ablation [114, 115]. This has led to the development of C5 inhibitors for clinical use [116–119], primarily in indications where complement-mediated lysis and/or NEUTROPHILS are driving the pathology.

Eculizumab (Soliris) is a humanized IgG_1 anti-complement C5 neutralizing mAb that blocks the cleavage of C5 into C5a and C5b. It is approved in PNH and haemolytic uremic syndrome and is in Phase 2 for dry AMD, ASTHMA, myasthenia gravis and transplant rejection [119]. A newer version of eculizumab, ALXN1210, with enhanced FcRn recycling is currently being developed and is in Phase 1 [120]. A number of anti-C5, C5a and C5aR mAbs are in early clinical trials [121]. Considering safety, an advantage of blocking C5 is that it will not block C3b-mediated OPSONIZATION of pathogens (the case with blocking the upstream C3 or CFD); however blocking C5 will also not inhibit the generation of C3b nor C3a anaphylatoxin which may have a role in disease pathogenesis. An anti-C5 mAb will interfere with C5b-9 formation and subsequent assembly of the lytic MAC, which is important in host defence against encapsulated bacteria. Increased *Neisseria meningitidis* infection is observed in C5-deficient humans and in PNH patients treated with eculizumab, although the rates are low due to prior meningococcal vaccination [119]. Neutralizing mAbs specific for C5a or C5aR is more preferable for systemic administration in chronic diseases leaving the C5b-C9 MAC intact and reducing meningococcal infection risk concerns.

21.4.7 Other Targets of mAbs for Immune Diseases

mAbs in early Phase 1 and 2 clinical trials against other immune targets [122] include those targeting toll-like RECEPTORS (TLRs, e.g. TLR2, TLR3, TLR4), APOPTOSIS (anti-CD95/Fas), angiogenesis (anti-VEGF; vascular endothelial growth factor; anti-S1P; sphingosine 1-phosphate) and tissue injury (anti-MMP-9; matrix METALLOPROTEINASE-9). Others are designed to promote neuronal repair (anti-LINGO-1; leucine-rich repeat and Ig domain containing NOGO receptor-interacting protein-1) and inhibit virus-induced injury (anti-MRSV; MS-associated retrovirus) during brain/neuronal INFLAMMATION, whilst an anti-fibronectin mAb fused to IL-10 aims to target this immunoregulatory cytokine to inflamed joints.

21.5 Summary

BIOTHERAPEUTICS will continue to evolve as we understand the existing targets better, introduce new targets, and as we apply new strategies for delivering them effectively to the site of action. This continuous improvement is paralleled in bioengineering with new construct designs, a better understanding of what characteristics may contribute to immunogenicity and stability, and an increase in delivery options. PHARMACOKINETIC advancements also play a role by designing dosing strategies in the clinic that are linked to the effect observed using PK/PD and data from the preclinical setting. With more knowledge on the immunopathogenesis of the disease, the link of PK to biomarkers and how to stratify patients better, trial designs can be optimized, and medicine can become more personalized. In all cases, obtaining better EFFICACY needs to be balanced against safety considerations. BIOTHERAPEUTICS have potential for optimization to be easier to use, longer-lasting, less immunogenic, more potent and less costly in order to make them more attractive for treatment of chronic diseases.

References

1. Reichert JM. Marketed therapeutic antibodies compendium. MAbs. 2012;4(3):413–5.
2. Kopf M, Bachmann MF, Marsland BJ. Averting inflammation by targeting the cytokine environment. Nat Rev Drug Discov. 2010;9(9):703–18.
3. Chan AC, Carter PJ. Therapeutic antibodies for autoimmunity and inflammation. Nat Rev Immunol. 2010;10(5):301–16.
4. Bielekova B, Becker BL. Monoclonal antibodies in MS: mechanisms of action. Neurology. 2010;74(Suppl 1):S31–40.
5. Bisikirska B, Colgan J, Luban J, Bluestone JA, Herold KC. TCR stimulation with modified anti-CD3 mAb expands CD8+ T cell population and induces CD8+ CD25+ Tregs. J Clin Invest. 2005;115(10):2904–13.
6. Brennan FR, Morton LD, Spindeldreher S, Kiessling A, Allenspach R, Hey A, et al. Safety and immunotoxicity assessment of immunomodulatory monoclonal antibodies. MAbs. 2010;2(3):233–55.
7. Campbell J, Lowe D, Sleeman MA. Developing the next generation of monoclonal antibodies for the treatment of rheumatoid arthritis. Br J Pharmacol. 2011;162(7):1470–84.
8. Nimmerjahn F, Ravetch JV. Translating basic mechanisms of IgG effector activity into next generation cancer therapies. Cancer Immun. 2012;12(1):13.
9. Kubota T, Niwa R, Satoh M, Akinaga S, Shitara K, Hanai N. Engineered therapeutic antibodies with improved effector functions. Cancer Sci. 2009;100(9):1566–72.
10. Kolbeck R, Kozhich A, Koike M, Peng L, Andersson CK, Damschroder MM, et al. MEDI-563, a humanized anti–IL-5 receptor α mAb with enhanced antibody-dependent cell-mediated cytotoxicity function. J Allergy Clin Immunol. 2010;125(6):1344–53.e2.
11. Bruhns P, Iannascoli B, England P, Mancardi DA, Fernandez N, Jorieux S, et al. Specificity and affinity of human Fcγ receptors and their polymorphic variants for human IgG subclasses. Blood. 2009;113(16):3716–25.
12. Labrijn AF, Aalberse RC, Schuurman J. When binding is enough: nonactivating antibody formats. Curr Opin Immunol. 2008;20(4):479–85.

13. Jefferis R. Antibody therapeutics: isotype and glycoform selection. Expert Opin Biol Ther. 2007;7(9):1401–13.
14. Kelley RF, Meng YG. Methods to engineer and identify IgG 1 variants with improved FcRn binding or effector function. Methods Mol Biol. 2012;901:277–93.
15. Igawa T, Maeda A, Haraya K, Tachibana T, Iwayanagi Y, Mimoto F, et al. Engineered monoclonal antibody with novel antigen-sweeping activity in vivo. PLoS One. 2013;8(5):e63236.
16. Elsadek B, Kratz F. Impact of albumin on drug delivery—new applications on the horizon. J Control Release. 2012;157(1):4–28.
17. McDonnell T, Ioannou Y, Rahman A. PEGylated drugs in rheumatology—why develop them and do they work? Rheumatology (Oxford). 2013;53(3):391–6.
18. Yu YJ, Zhang Y, Kenrick M, Hoyte K, Luk W, Lu Y, et al. Boosting brain uptake of a therapeutic antibody by reducing its affinity for a transcytosis target. Sci Transl Med. 2011;3(84):84ra44.
19. Schüle S, Schulz-Fademrecht T, Garidel P, Bechtold-Peters K, Frieß W. Stabilization of IgG1 in spray-dried powders for inhalation. Eur J Pharm Biopharm. 2008;69(3):793–807.
20. Luu KT, Bergqvist S, Chen E, Hu-Lowe D, Kraynov E. A model-based approach to predicting the human pharmacokinetics of a monoclonal antibody exhibiting target-mediated drug disposition. J Pharmacol Exp Ther. 2012;341(3):702–8.
21. Fetterly GJ, Aras U, Meholick PD, Takimoto C, Seetharam S, McIntosh T, et al. Utilizing pharmacokinetics/pharmacodynamics modeling to simultaneously examine free CCL2, total CCL2 and carlumab (CNTO 888) concentration time data. J Clin Pharmacol. 2013;53(10):1020–7.
22. Dudal S, Subramanian K, Flandre T, Law W, Lowe P, Skerjanec A, et al. Integrated pharmacokinetic, pharmacodynamic and immunogenicity profiling of an anti-CCL21 monoclonal antibody in cynomolgus monkeys. mAbs. 2015;7(5):829–37.
23. Gatault P, Brachet G, Ternant D, Degenne D, Récipon G, Barbet C, et al. Therapeutic drug monitoring of eculizumab: rationale for an individualized dosing schedule. mAbs. 2015;7(6):1205–11.
24. Bongartz T, Sutton AJ, Sweeting MJ, Buchan I, Matteson EL, Montori V. Anti-TNF antibody therapy in rheumatoid arthritis and the risk of serious infections and malignancies: systematic review and meta-analysis of rare harmful effects in randomized controlled trials. JAMA. 2006;295(19):2275–85.
25. Smolen JS, Landewé R, Breedveld FC, Dougados M, Emery P, Gaujoux-Viala C, et al. EULAR recommendations for the management of rheumatoid arthritis with synthetic and biological disease-modifying antirheumatic drugs. Ann Rheum Dis. 2010;69(6):964–75.
26. Balanescu A, Wiland P. Maximizing early treatment with biologics in patients with rheumatoid arthritis: the ultimate breakthrough in joints preservation. Rheumatol Int. 2013;33(6):1379–86.
27. Buch MH, Rubbert-Roth A, Ferraccioli G. To switch or not to switch after a poor response to a TNFα blocker? It is not only a matter of ACR20 OR ACR50. Autoimmun Rev. 2012;11(8):558–62.
28. Rosenblum H, Amital H. Anti-TNF therapy: safety aspects of taking the risk. Autoimmun Rev. 2011;10(9):563–8.
29. Tracey D, Klareskog L, Sasso EH, Salfeld JG, Tak PP. Tumor necrosis factor antagonist mechanisms of action: a comprehensive review. Pharmacol Ther. 2008;117(2):244–79.
30. Dinarello CA. Immunological and inflammatory functions of the interleukin-1 family. Annu Rev Immunol. 2009;27:519–50.
31. Nesbitt A, Fossati G, Bergin M, Stephens P, Stephens S, Foulkes R, et al. Mechanism of action of certolizumab pegol (CDP870): in vitro comparison with other anti-tumor necrosis factor alpha agents. Inflamm Bowel Dis. 2007;13(11):1323–32.
32. Zidi I, Bouaziz A, Mnif W, Bartegi A, Amor NB. Golimumab and malignancies: true or false association? Med Oncol. 2011;28(2):641–8.
33. Wallis RS. Tumour necrosis factor antagonists: structure, function, and tuberculosis risks. Lancet Infect Dis. 2008;8(10):601–11.
34. Van Hauwermeiren F, Vandenbroucke RE, Libert C. Treatment of TNF mediated diseases by selective inhibition of soluble TNF or TNFR1. Cytokine Growth Factor Rev. 2011;22(5):311–9.
35. Abe Y, Nomura T, Yoshioka Y, Kamada H, Tsunoda S-i, Tsutsumi Y. Anti-inflammatory effects of a novel TNFR1-selective antagonistic TNF mutant on established murine collagen-induced arthritis. Adv Exp Med Biol. 2011;691:493–500.
36. Tillack C, Ehmann LM, Friedrich M, Laubender RP, Papay P, Vogelsang H, et al. Anti-TNF antibody-induced psoriasiform skin lesions in patients with inflammatory bowel disease are characterised by interferon-gamma-expressing Th1 cells and IL-17A/IL-22-expressing Th17 cells and respond to anti-IL-12/IL-23 antibody treatment. Gut. 2014;63(4):567–77. https://doi.org/10.1136/gutjnl-2012-302853.
37. Heymann MC, Rösen-Wolff A. Contribution of the inflammasomes to autoinflammatory diseases and recent mouse models as research tools. Clin Immunol. 2013;147(3):175–84.
38. Tzellos T, Kyrgidis A, Trigoni A, Zouboulis CC. Association of ustekinumab and briakinumab with major adverse cardiovascular events: an appraisal of meta-analyses and industry sponsored pooled analyses to date. Dermatoendocrinol. 2012;4(3):320–3. https://doi.org/10.4161/derm.23100.
39. Goldbach-Mansky R. Immunology in clinic review series; focus on autoinflammatory diseases: update on monogenic autoinflammatory diseases: the role of interleukin (IL)-1 and an emerging role

for cytokines beyond IL-1. Clin Exp Immunol. 2012;167(3):391–404.
40. Zhao R, Zhou H, Su SB. A critical role for interleukin-1β in the progression of autoimmune diseases. Int Immunopharmacol. 2013;17(3):658–69.
41. Mitha E, Schumacher HR, Fouche L, Luo S-F, Weinstein SP, Yancopoulos GD, et al. Rilonacept for gout flare prevention during initiation of uric acid-lowering therapy: results from the PRESURGE-2 international, phase 3, randomized, placebo-controlled trial. Rheumatology. 2013;52(7):1285–92.
42. Wulffraat NM, Woo P. Canakinumab in pediatric rheumatic diseases. Expert Opin Biol Ther. 2013;13(4):615–22.
43. Reichert JM. Antibodies to watch in 2016. MAbs. 2016;8(2):197–204. https://doi.org/10.1080/19420862.2015.1125583.
44. Brennan FR. Chapter 31: Mabs targeting soluble mediators in phase 1 and 2 clinical studies for immunological disorders. In: Dubel S, Reichert J, editors. Handbook of therapeutic antibodies. New York: John Wiley and Sons Inc.; 2014. p. 969–1078.
45. Garbers C, Hermanns HM, Schaper F, Müller-Newen G, Grötzinger J, Rose-John S, et al. Plasticity and cross-talk of interleukin 6-type cytokines. Cytokine Growth Factor Rev. 2012;23(3):85–97.
46. Guo X, Higgs BW, Bay-Jensen AC, Karsdal MA, Yao Y, Roskos LK, et al. Suppression of T cell activation and collagen accumulation by an anti-IFNAR1 mAb, anifrolumab, in adult patients with systemic sclerosis. J Investig Dermatol. 2015;135(10):2402–9.
47. Khamashta M, Merrill JT, Werth VP, Furie R, Kalunian K, Illei GG, et al. Sifalimumab, an anti-interferon-α monoclonal antibody, in moderate to severe systemic lupus erythematosus: a randomised, double-blind, placebo-controlled study. Ann Rheum Dis. 2016;75(11):1909–16.
48. Tanaka T, Kishimoto T. Targeting interleukin-6: all the way to treat autoimmune and inflammatory diseases. Int J Biol Sci. 2012;8(9):1227–36.
49. Genovese MC, Fleischmann R, Kivitz AJ, Rell-Bakalarska M, Martincova R, Fiore S, Rohane P, van Hoogstraten H, Garg A, Fan C, van Adelsberg J, Weinstein SP, Graham NM, Stahl N, Yancopoulos GD, Huizinga TW, van der Heijde D. Sarilumab plus methotrexate in patients with active rheumatoid arthritis and inadequate response to methotrexate: results of a phase III study. Arthritis Rheumatol. 2015;67:1424–37.
50. Basu R, Hatton RD, Weaver CT. The Th17 family: flexibility follows function. Immunol Rev. 2013;252(1):89–103.
51. Bowman EP, Chackerian AA, Cua DJ. Rationale and safety of anti-interleukin-23 and anti-interleukin-17A therapy. Curr Opin Infect Dis. 2006;19(3):245–52.
52. Watanabe H, Kawaguchi M, Fujishima S, Ogura M, Matsukura S, Takeuchi H, et al. Functional characterization of IL-17F as a selective neutrophil attractant in psoriasis. J Investig Dermatol. 2009;129(3):650–6.
53. Alzabin S, Abraham SM, Taher TE, Palfreeman A, Hull D, McNamee K, et al. Incomplete response of inflammatory arthritis to TNFα blockade is associated with the Th17 pathway. Ann Rheum Dis. 2012;71(10):1741–8.
54. Koenders MI, Marijnissen RJ, Devesa I, Lubberts E, Joosten LA, Roth J, et al. Tumor necrosis factor–interleukin-17 interplay induces S100A8, interleukin-1β, and matrix metalloproteinases, and drives irreversible cartilage destruction in murine arthritis: rationale for combination treatment during arthritis. Arthritis Rheum. 2011;63(8):2329–39.
55. Toussirot E. The IL23/Th17 pathway as a therapeutic target in chronic inflammatory diseases. Inflamm Allergy Drug Targets. 2012;11(2):159–68.
56. Viguier M, Livideanu C, Beylot-Barry M, Richard MA, Paul C, Bachelez H, et al. Observational case series on a group of psoriasis patients who failed to respond to any TNF blockers. J Dermatolog Treat. 2014;25(1):75–7. https://doi.org/10.3109/09546634.2013.806766.
57. Oh CK, Geba GP, Molfino N. Investigational therapeutics targeting the IL-4/IL-13/STAT-6 pathway for the treatment of asthma. Eur Respir Rev. 2010;19(115):46–54. https://doi.org/10.1183/09059180.00007609.
58. Wills-Karp M. Interleukin-13 in asthma pathogenesis. Immunol Rev. 2004;202:175–90. https://doi.org/10.1111/j.0105-2896.2004.00215.x.
59. Hanania NA, Noonan M, Corren J, Korenblat P, Zheng Y, Fischer SK, et al. Lebrikizumab in moderate-to-severe asthma: pooled data from two randomised placebo-controlled studies. Thorax. 2015;70(8):748–56. https://doi.org/10.1136/thoraxjnl-2014-206719.
60. Brightling CE, Chanez P, Leigh R, O'Byrne PM, Korn S, She D, et al. Efficacy and safety of tralokinumab in patients with severe uncontrolled asthma: a randomised, double-blind, placebo-controlled, phase 2b trial. Lancet Respir Med. 2015;3(9):692–701. https://doi.org/10.1016/s2213-2600(15)00197-6.
61. Wenzel S, Ford L, Pearlman D, Spector S, Sher L, Skobieranda F, et al. Dupilumab in persistent asthma with elevated eosinophil levels. N Engl J Med. 2013;368(26):2455–66. https://doi.org/10.1056/NEJMoa1304048.
62. Thaçi D, Simpson EL, Beck LA, Bieber T, Blauvelt A, Papp K, et al. Efficacy and safety of dupilumab in adults with moderate-to-severe atopic dermatitis inadequately controlled by topical treatments: a randomised, placebo-controlled, dose-ranging phase 2b trial. Lancet. 2016;387(10013):40–52.
63. Garcia G, Taillé C, Laveneziana P, Bourdin A, Chanez P, Humbert M. Anti-interleukin-5 therapy in severe asthma. Eur Respir Rev. 2013;22(129):251–7.
64. Uhm TG, Kim BS, Chung IY. Eosinophil development, regulation of eosinophil-specific genes, and role of eosinophils in the pathogenesis of asthma. Allergy, Asthma Immunol Res. 2012;4(2):68–79.

65. Nair P, Pizzichini MM, Kjarsgaard M, Inman MD, Efthimiadis A, Pizzichini E, et al. Mepolizumab for prednisone-dependent asthma with sputum eosinophilia. N Engl J Med. 2009;360(10):985–93.
66. Bjermer L, Lemiere C, Maspero J, Weiss S, Zangrilli J, Germinaro M. Reslizumab for inadequately controlled asthma with elevated blood eosinophil levels: a randomized phase 3 study. Chest. 2016;150(4):789–98.
67. Ghazi A, Trikha A, Calhoun WJ. Benralizumab—a humanized mAb to IL-5Rα with enhanced antibody-dependent cell-mediated cytotoxicity—a novel approach for the treatment of asthma. Expert Opin Biol Ther. 2012;12(1):113–8.
68. Pestka S, Krause CD, Walter MR. Interferons, interferon-like cytokines, and their receptors. Immunol Rev. 2004;202(1):8–32.
69. Meyer O. Interferon-alpha as a treatment target in systemic lupus erythematosus. Joint Bone Spine. 2012;79(2):113–6.
70. Brownlie RJ, Zamoyska R. T cell receptor signalling networks: branched, diversified and bounded. Nat Rev Immunol. 2013;13(4):257–69.
71. Kinnear G, Jones ND, Wood KJ. Costimulation blockade: current perspectives and implications for therapy. Transplantation. 2013;95(4):527.
72. Gimferrer I, Calvo M, Mittelbrunn M, Farnós M, Sarrias MR, Enrich C, et al. Relevance of CD6-mediated interactions in T cell activation and proliferation. J Immunol. 2004;173(4):2262–70.
73. Bowen MA, Patel DD, Li X, Modrell B, Malacko AR, Wang W-C, et al. Cloning, mapping, and characterization of activated leukocyte-cell adhesion molecule (ALCAM), a CD6 ligand. J Exp Med. 1995;181(6):2213–20.
74. Topalian SL, Drake CG, Pardoll DM. Immune checkpoint blockade: a common denominator approach to cancer therapy. Cancer Cell. 2015;27(4):450–6.
75. Herold KC, Burton JB, Francois F, Poumian-Ruiz E, Glandt M, Bluestone JA. Activation of human T cells by FcR nonbinding anti-CD3 mAb, hOKT3γ1 (Ala-Ala). J Clin Invest. 2003;111(3):409–18.
76. Shih FF, Mandik-Nayak L, Wipke BT, Allen PM. Massive thymic deletion results in systemic autoimmunity through elimination of CD4+ CD25+ T regulatory cells. J Exp Med. 2004;199(3):323–35.
77. Bettini M, Vignali DA. Regulatory T cells and inhibitory cytokines in autoimmunity. Curr Opin Immunol. 2009;21(6):612–8.
78. Keymeulen B, Walter M, Mathieu C, Kaufman L, Gorus F, Hilbrands R, et al. Four-year metabolic outcome of a randomised controlled CD3-antibody trial in recent-onset type 1 diabetic patients depends on their age and baseline residual beta cell mass. Diabetologia. 2010;53(4):614–23.
79. Cutolo M, Sulli A, Paolino S, Pizzorni C. CTLA-4 blockade in the treatment of rheumatoid arthritis: an update. Expert Rev Clin Immunol. 2016;12(4):417–25. https://doi.org/10.1586/1744666x.2016.1133295.
80. Elyoussfi S, Thomas BJ, Ciurtin C. Tailored treatment options for patients with psoriatic arthritis and psoriasis: review of established and new biologic and small molecule therapies. Rheumatol Int. 2016;36(5):603–12.
81. Nair P, Ghorai A, Adhikary L, Melarkode R, Montero E. The inhibition of T cell proliferation in a mixed lymphocyte reaction by Itolizumab (T1h) is associated with reduction in pro inflammatory cytokines and CD6 internalization. J Immunol. 2011;186(Meeting Abstracts 1):52.27.
82. Brennan FR. Chapter 32: T cell inhibitors in phase 1 and 2 clinical studies for immunological disorders. In: Dubel S, Reichert J, editors. Handbook of therapeutic antibodies. New York: John Wiley and Sons Inc.; 2014. p. 1079–113.
83. Tacke M, Hanke G, Hanke T, Hünig T. CD28-mediated induction of proliferation in resting T cells in vitro and in vivo without engagement of the T cell receptor: evidence for functionally distinct forms of CD28. Eur J Immunol. 1997;27(1):239–47.
84. Suntharalingam G, Perry MR, Ward S, Brett SJ, Castello-Cortes A, Brunner MD, et al. Cytokine storm in a phase 1 trial of the anti-CD28 monoclonal antibody TGN1412. N Engl J Med. 2006;355(10):1018–28.
85. Eastwood D, Findlay L, Poole S, Bird C, Wadhwa M, Moore M, et al. Monoclonal antibody TGN1412 trial failure explained by species differences in CD28 expression on CD4+ effector memory T-cells. Br J Pharmacol. 2010;161(3):512–26.
86. Page KR, Mezzalana E, MacDonald AJ, Zamuner S, De Nicolao G, van Maurik A. Temporal pharmacokinetic/pharmacodynamic interaction between human CD3ε antigen-targeted monoclonal antibody otelixizumab and CD3ε binding and expression in human peripheral blood mononuclear cell static culture. J Pharmacol Exp Ther. 2015;355(2):199–205.
87. Leandro MJ. B-cell subpopulations in humans and their differential susceptibility to depletion with anti-CD20 monoclonal antibodies. Arthritis Res Ther. 2013;15(Suppl 1):S3.
88. Buch MH, Smolen JS, Betteridge N, Breedveld FC, Burmester G, Dörner T, et al. Updated consensus statement on the use of rituximab in patients with rheumatoid arthritis. Ann Rheum Dis. 2011;70(6):909–20.
89. Sorensen PS, Blinkenberg M. The potential role for ocrelizumab in the treatment of multiple scerosis: current evidence and future prospects. Ther Adv Neurol Disord. 2016;9:44–52.
90. Taylor PC, Quattrocchi E, Mallett S, Kurrasch R, Petersen J, Chang DJ. Ofatumumab, a fully human anti-CD20 monoclonal antibody, in biological-naive, rheumatoid arthritis patients with an inadequate response to methotrexate: a randomised, double-blind, placebo-controlled clinical trial. Ann Rheum Dis. 2011;70(12):2119–25.
91. Stohl W, Gomez-Reino J, Olech E, Dudler J, Fleischmann RM, Zerbini CA, et al. Safety and

efficacy of ocrelizumab in combination with methotrexate in MTX-naive subjects with rheumatoid arthritis: the phase III FILM trial. Ann Rheum Dis. 2012;71(8):1289–96.

92. Goldenberg DM, Morschhauser F, Wegener WA. Veltuzumab (humanized anti-CD20 monoclonal antibody): characterization, current clinical results, and future prospects. Leuk Lymphoma. 2010;51(5):747–55.
93. Stohl W, Merrill JT, McKay JD, Lisse JR, Zhong ZJ, Freimuth WW, et al. Efficacy and safety of belimumab in patients with rheumatoid arthritis: a phase II, randomized, double-blind, placebo-controlled, dose-ranging study. J Rheumatol. 2013;40(5):579–89.
94. Genovese M, Bojin S, Biagini I, Mociran E, Cristei D, Mirea G, et al. Tabalumab in rheumatoid arthritis patients with an inadequate response to methotrexate and naive to biologic therapy: a phase II, randomized, placebo-controlled trial. Arthritis Rheum. 2013;65(4):880–9.
95. Stohl W, Merrill JT, Looney RJ, Buyon J, Wallace DJ, Weisman MH, et al. Treatment of systemic lupus erythematosus patients with the BAFF antagonist "peptibody" blisibimod (AMG 623/A-623): results from randomized, double-blind phase 1a and phase 1b trials. Arthritis Res Ther. 2015;17(1):1–14.
96. Genovese M, Kinnman N, de La Bourdonnaye G, Pena Rossi C, Tak P. Atacicept in patients with rheumatoid arthritis and an inadequate response to tumor necrosis factor antagonist therapy: results of a phase II, randomized, placebo-controlled, dose-finding trial. Arthritis Rheum. 2011;63(7):1793–803.
97. Brennan FR. Chapter 33: B cell inhibitors in phase 1 and 2 clinical studies for immunological disorders. In: Dubel S, Reichert J, editors. Handbook of therapeutic antibodies. New York: John Wiley and Sons Inc.; 2014. p. 1115–25.
98. Marelli-Berg F, Cannella L, Dazzi F, Mirenda V. The highway code of T cell trafficking. J Pathol. 2008;214(2):179–89.
99. Sigmundsdottir H, Butcher EC. Environmental cues, dendritic cells and the programming of tissue-selective lymphocyte trafficking. Nat Immunol. 2008;9(9):981–7.
100. Thomas S, Baumgart DC. Targeting leukocyte migration and adhesion in Crohn's disease and ulcerative colitis. Inflammopharmacology. 2012;20(1):1–18. https://doi.org/10.1007/s10787-011-0104-6.
101. Stuve O. The effects of natalizumab on the innate and adaptive IMMUNE SYSTEM in the central nervous system. J Neurol Sci. 2008;274(1–2):39–41. https://doi.org/10.1016/j.jns.2008.03.022.
102. Stuve O, Marra CM, Jerome KR, Cook L, Cravens PD, Cepok S, et al. Immune surveillance in multiple sclerosis patients treated with natalizumab. Ann Neurol. 2006;59(5):743–7. https://doi.org/10.1002/ana.20858.
103. Aksamit AJ. Review of progressive multifocal leukoencephalopathy and natalizumab. Neurologist. 2006;12(6):293–8.
104. McLean LP, Cross RK. Integrin antagonists as potential therapeutic options for the treatment of Crohn's disease. Expert Opin Investig Drugs. 2016;25(3):263–73.
105. Singh H, Grewal N, Arora E, Kumar H, Kakkar AK. Vedolizumab: a novel anti-integrin drug for treatment of IBD. J Nat Sci Biol Med. 2016; 7(1):4.
106. Makker J, Hommes DW. Etrolizumab for ulcerative colitis: the new kid on the block? Expert Opin Biol Ther. 2016;16(4):567–72.
107. Vermeire S, Ghosh S, Panes J, Dahlerup JF, Luegering A, Sirotiakova J, et al. The mucosal addressin cell adhesion molecule antibody PF-00547659 in ulcerative colitis: a randomised study. Gut. 2011;60(8):1068–75.
108. Pan W, Hsu H, Rees W, Lear S, Lee F, Foltz I, et al. Pharmacology of AMG 181, a human anti-$\alpha 4\beta 7$ antibody that specifically alters trafficking of gut-homing T cells. Br J Pharmacol. 2013;169(1): 51–68.
109. Behm BW, Bickston SJ. Tumor necrosis factor-alpha antibody for maintenance of remission in Crohn's disease. Cochrane Database Syst Rev. 2008;(1):CD006893.
110. Brennan FR. Chapter 34: Inhibitors of leukocyte adhesion and migration in phase 1 and 2 clinical studies for immunological disorders. In: Dubel S, Reichert J, editors. Handbook of therapeutic antibodies. New York: John Wiley and Sons Inc.; 2014. p. 1127–44.
111. MacGlashan DW. IgE-dependent signaling as a therapeutic target for allergies. Trends Pharmacol Sci. 2012;33(9):502–9.
112. McKeage K. Omalizumab: a review of its use in patients with severe persistent allergic asthma. Drugs. 2013;73(11):1197–212.
113. Arm J, Bottoli I, Skerjanec A, Floch D, Groenewegen A, Maahs S, et al. Pharmacokinetics, pharmacodynamics and safety of QGE031 (ligelizumab), a novel high-affinity anti-IgE antibody, in atopic subjects. Clin Exp Allergy. 2014;44(11):1371–85.
114. Ehrnthaller C, Ignatius A, Gebhard F, Huber-Lang M. New insights of an old defense system: structure, function, and clinical relevance of the complement system. Mol Med. 2011;17(3):317.
115. Mastellos D, Germenis A, Lambris J. Complement: an inflammatory pathway fulfilling multiple roles at the interface of innate immunity and development. Curr Drug Targets Inflamm Allergy. 2005;4(1):125–7.
116. Ricklin D, Lambris JD. Complement in immune and inflammatory disorders: therapeutic interventions. J Immunol. 2013;190(8):3839–47.
117. Jack LS, Sadiq MA, Do DV, Nguyen QD. Emixustat and lampalizumab: potential therapeutic options for geographic atrophy. Dev Ophthalmol. 2016;55:302–9.
118. Laursen NS, Magnani F, Gottfredsen RH, Petersen SV, Andersen GR. Structure, function and control of complement C5 and its proteolytic fragments. Curr Mol Med. 2012;12(8):1083–97.

119. Keating GM, Lyseng-Williamson KA, McKeage K. Eculizumab: a guide to its use in paroxysmal nocturnal hemoglobinuria. BioDrugs. 2012;26(2):125–30.
120. Sahelijo L, Mujeebuddin A, Mitchell D, Larouche R, Yu Z-X, Zhang Y, et al. First in human single-ascending dose study: safety, biomarker, pharmacokinetics and exposure-response relationships of ALXN1210, a humanized monoclonal antibody to C5, with marked half-life extension and potential for significantly longer dosing intervals. Blood. 2015;126(23):4777.
121. Brennan FR. Chapter 37: Complement inhibitors in phase 1 and 2 clinical studies for immunological disorders. In: Dubel S, Reichert J, editors. Handbook of therapeutic antibodies. New York: John Wiley and Sons Inc.; 2014. p. 1165–73.
122. Brennan FR. Chapter 30: Monoclonal antibodies in phase 1 and 2 clinical studies for immunological disorders. In: Dubel S, Reichert J, editors. Handbook of therapeutic antibodies. New York: John Wiley and Sons Inc.; 2014. p. 927–67.

22 Anti-allergic Drugs

Eva M. Valesky, Sue McKay, Antoon J. M. van Oosterhout, and Michael J. Parnham

22.1 Introduction

ALLERGY is defined as a disease following a response by the IMMUNE SYSTEM to an otherwise innocuous antigen. Allergic diseases include allergic rhinitis, atopic dermatitis, systemic ANAPHYLAXIS, food ALLERGY, allergic ASTHMA and acute urticaria and are mediated by unwanted type-I HYPERSENSITIVITY reactions to extrinsic ALLERGENS like pollen, house dust, animal dander, drugs and insect venom. These diseases are characterised by the production of IgE ANTIBODIES to the allergen that binds to the high-AFFINITY IgE RECEPTOR, FcεRI, on mast cells and BASOPHILS. Binding of allergen to IgE cross-links these RECEPTORS and causes the release of chemical mediators from MAST CELLS, leading to the development of a type-I HYPERSENSITIVITY reaction (Fig. 22.1). This acute response is often followed by a late and more sustained inflammatory response characterised by the recruitment of other EFFECTOR CELLS such as EOSINOPHILS and T helper type-2 (Th2) LYMPHOCYTES. Among the mainstays in the drug treatment of allergic INFLAMMATION, glucocorticosteroids remain the most potent inhibitors, and the reader is referred to Chap. 32 for detailed discussion of these drugs. The use of SPECIFIC IMMUNOTHERAPY in severe allergies is considered in Chap. 24. This present chapter focuses on anti-allergic drugs that specifically TARGET the activation of the mast cell or block the effects of its chemical mediators, in particular HISTAMINE.

22.2 Disodium Cromoglycate and Nedocromil Sodium (Cromones)

In the 1960s, Dr. Roger E. C. Altounyan (1922–1987), a British physician and pharmacologist at the pharmaceutical company Fisons, first discovered, in an unusual manner, that disodium cromoglycate (also called cromolyn sodium) possessed

Final manuscript submitted on February 01, 2017.

E. M. Valesky (✉)
Department of Dermatology, Venerology and Allergology, University Hospital Frankfurt, Frankfurt am Main, Germany
e-mail: Eva.Valesky@kgu.de

S. McKay
Unilever Research & Development, Vlaardingen, The Netherlands
e-mail: sue.mckay@unilever.com

A. J. M. van Oosterhout
Department of Pathology and Medical Biology, University Medical Center Groningen (UMCG), Groningen, The Netherlands
e-mail: a.j.m.van.oosterhout@umcg.nl

M. J. Parnham
Translational Drug Validation, Fraunhofer IME-TMP and Goethe University Frankfurt, Frankfurt am Main, Germany
e-mail: parnham@med.uni-frankfurt.de; michael.parnham@ime.fraunhofer.de

M. J. Parnham et al. (eds.), *Nijkamp and Parnham's Principles of Immunopharmacology*, https://doi.org/10.1007/978-3-030-10811-3_22

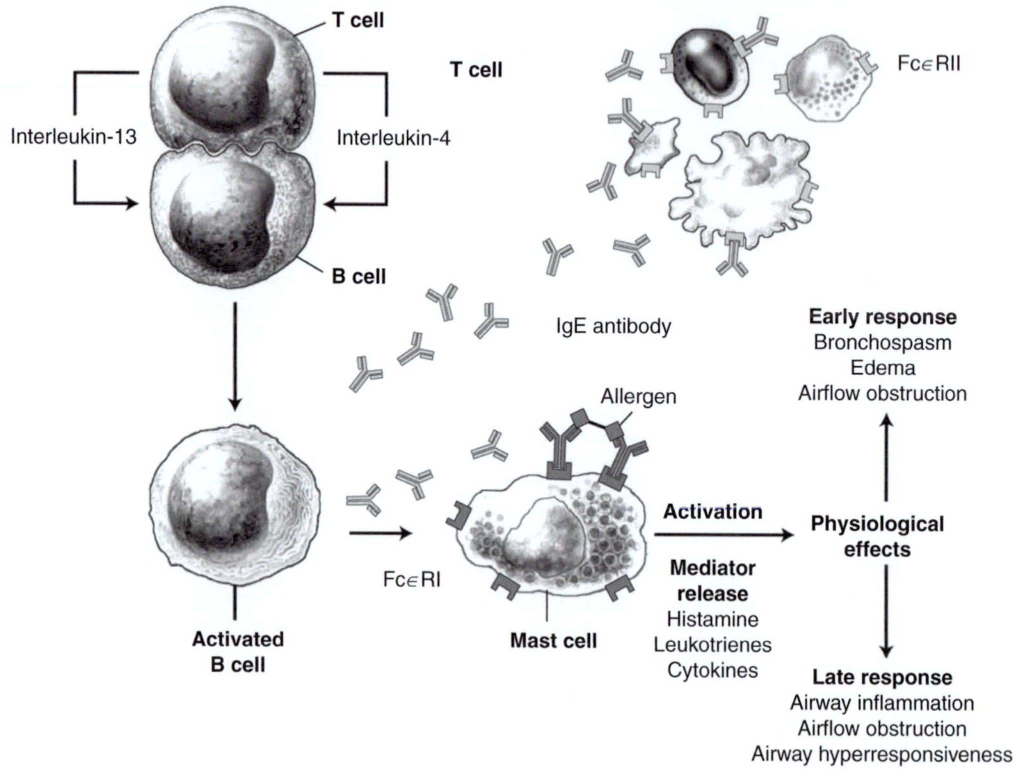

Fig. 22.1 Schematic representation of the induction of IgE synthesis by B lymphocytes. Once formed, IgE antibody circulates in the blood, eventually binding to high-affinity IgE receptors (FcεRI) on mast cells and low-affinity IgE receptors (FcεRII or CD23) on eosinophils and macrophages. After subsequent encounters with allergen, cross-linking of the high-affinity IgE receptors causes the release of preformed and newly generated mediators. Once present in tissues, mediators may produce various physiological effects, depending on the target organ [1]

an antiASTHMA action. He induced an ASTHMA attack in himself by inhaling animal dander ANTIGENS and showed that cromoglycate afforded protection against this bronchial provocation. Disodium cromoglycate was introduced as an anti-allergic drug in 1968. Many companies tried to find improved versions of this compound, using the chemical structure as a starting point, but most of these attempts failed. Nedocromil was discovered and introduced 20 years later by Eady.

To this day, the precise mechanism of action of disodium cromoglycate and the related drug, nedocromil sodium, remains unclear, but their clinical activity probably represents a combination of effects [2]. It was originally suggested that these non-steroidal anti-inflammatory drugs act as mast cell stabilisers [3]. However, although these drugs can prevent HISTAMINE release from MAST CELLS, it has been demonstrated that this effect is not the basis of their action in allergic ASTHMA. Sodium cromoglycate and nedocromil sodium also partly inhibit the IgE-mediated release of other mediators from MAST CELLS, such as prostacyclins and LEUKOTRIENES [4]. In addition, they have been described to exhibit suppressive effects on inflammatory cells, such as MACROPHAGES, MONOCYTES, NEUTROPHILS and EOSINOPHILS, but do not have any direct effects on smooth muscle and they do not inhibit the actions of smooth muscle contractile agonists [5, 6]. Sodium cromoglycate and nedocromil sodium inhibit the influx of inflammatory cells and the release of inflammatory mediators following provocation with non-specific agents, such as cold air and air pollutants

Table 22.1 Comparison of effects of sodium cromoglycate and nedocromil sodium on inflammatory cells

Effect	Nedocromil sodium	Sodium cromoglycate
Mast cells from BAL, lung, conjunctiva, nasal mucosa, gastric mucosa and basophils		
Mediator release following Ascaris Ag or α-IgE Ab inhibited (histamine, PGD_2, LTC_4)	↓	
Release of cytokines (TNF-α)	↓	↓
Release of histamine	↓	↓
Numbers	↓	↓
Macrophages/monocytes		
Release of cytokines (IL-6)	↓	
Release of lysosomal enzymes and oxygen radicals	↓	
Numbers	↓	
Eosinophils		
Numbers in BAL	↓	↓
Number of activated eo's in submucosa	↓	
Release of mediators (preformed and newly generated)	↓	↓
Chemotactic response to PAF and LTB_4	↓	
Chemotactic response to zymosan-activated serum		↓
Activation	↓	
Survival time in presence of IL-5	↓	
Neutrophils		
Activation	↓	↓
Chemotactic response	↓	↓
Release of mediators (TNF-α, IL-6)	↓	↓
Numbers		↓
Platelets		
Release of cytotoxic mediators	↓	
IgE-mediated activation	↓	
Generation of thromboxane B_2 and IP_3	↓	
Epithelial cells		
Release of 15-HETE	↓	
Release of cytokines (TNF-α, IL-8, GM-CSF) and ICAM-1	↓	
Expression of ICAM-1, VCAM-1, E-selectin	↓	↓
B cells		
IgE Ab formation	↓	↓
T cells		
Numbers	↓	↓
Proliferation (allergen or mitogen induced)	±	±
Endothelial cells		
Expression of ICAM-1, VCAM-1, E-selectin	↓	↓
Sensory nerve (C fibres) activation		
Release of neuropeptides	↓	↓

↓ reduction, ± no change

[7–9]. Furthermore, they have been reported to depress the exaggerated neuronal reflexes that are triggered by the stimulation of "irritant" RECEPTORS by decreasing neuropeptide release from C fibres and via antagonism of tachykinin RECEPTORS [10–13].

Furthermore these drugs act by inducing the release of the anti-inflammatory protein Annexin-A1 from mast cells [14]. Moon et al. have shown that mast cells even in the same tissue respond differently to different stimuli due to their heterogeneity [15]. For example, the existence of difference in mast cell types may explain the unresponsiveness of skin mast cells to sodium cromoglycate and nedocromil sodium [16]. A comparison of the activities of sodium cromoglycate and nedocromil sodium on a variety of inflammatory cell types is shown in Table 22.1. The chemical structures of sodium cromoglycate and nedocromil sodium are depicted in Fig. 22.2.

22.2.1 Mechanisms of Action

Many studies have been performed to determine the mechanisms by which cromones inhibit the activation of cells. Since these compounds affect a wide variety of cells, it has been assumed that a common mechanism must exist. Experiments have been performed to investigate whether this mechanism is regulated through a specific

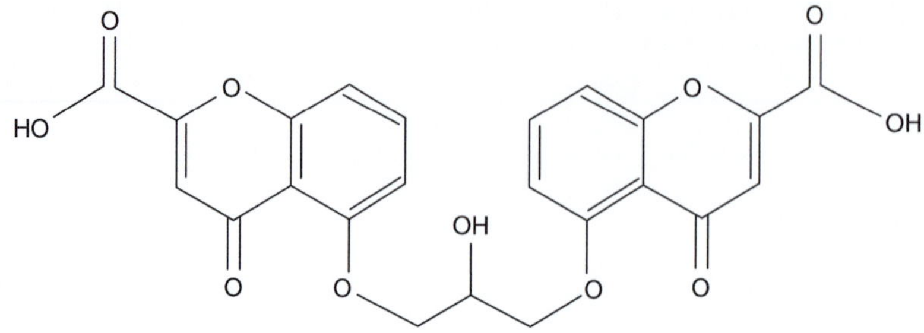

Sodium cromoglycate

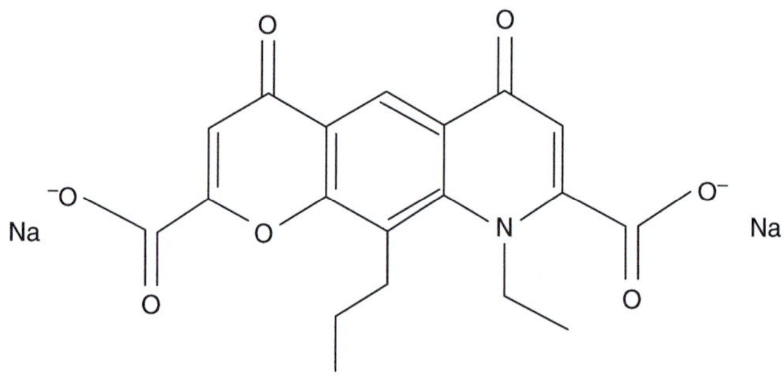

Nedocromil sodium

Fig. 22.2 Chemical structures of sodium cromoglycate and nedocromil sodium

RECEPTOR or if it is due to the modulation of a second messenger signal.

Initially Ca^{2+} ions were implicated as the TARGET for sodium cromoglycate [17], but this was discounted after it was shown that sodium cromoglycate could also inhibit mast cell activation in the presence of Ca^{2+}-chelating agents [18]. Sodium cromoglycate and nedocromil sodium have been shown to reduce Ca^{2+} influx into cells, although it is believed that they do not interfere directly with Ca^{2+} channels [17]. These compounds have also been shown to phosphorylate intracellular proteins preceding mediator release from rat peritoneal MAST CELLS; however, this does not hold true for MAST CELLS isolated from the macaque [3]. Additionally, activation of protein kinase C (PKC) has been suggested as the molecular TARGET for sodium cromoglycate [17, 19, 20], although other researchers report an inhibition of PKC activity [21]. The ability of sodium cromoglycate and nedocromil sodium to affect intracellular targets directly is unlikely if we consider the physical properties of these compounds. Both compounds are extremely polar and hydrophilic at pH 7.4 and therefore unlikely to penetrate the cell membrane. Consequently, it was assumed that they must function via a cell membrane component, possibly a RECEPTOR. Mazurek and colleagues [22–24] have described a "sodium cromoglycate-binding protein" in rat basophil leukaemia (RBL) cells that may be involved in Ca^{2+} mobilisation. Eady and coworkers [3, 4] also identified proteins on rat peritoneal MAST CELLS and Chinese hamster ovary cells that may act as RECEPTORS. Until now, the identification of a specific RECEPTOR has not yet been established. However both cromones were reported to be potent agonists at the G protein-coupled RECEPTOR 35 (GPR35), which is up-regulated on MAST CELLS, EOSINOPHILS and BASOPHILS upon challenge with IgE [25]. GPR35 is a newly discovered RECEPTOR on neurons, which inhibits calcium channels and regulates neuronal excitability and synaptic neurotransmitter release, including nociception (pain perception) in DORSAL ROOT GANGLIA in the spinal cord. Agonists at GPR35 include kynurenic acid, a metabolite of tryptophan, and zaprinast, a phosphodiesterase 5 (PDE5) inhibitor. The cromones are both more potent agonists than zaprinast.

There is also an increasing amount of evidence suggesting that modulation of chloride channel activity is possibly a common mechanism to explain the effects of sodium cromoglycate and nedocromil sodium. An accumulation of intracellular Ca^{2+} (Ca^{2+}_i) often precedes cell activation and mediator secretion in many cells. This accumulation of Ca^{2+}_i can result from a Ca^{2+} influx due to a negative membrane potential, which in turn is the result of an inward flow of Cl^- ions through chloride channels. DEGRANULATION is dependent on a sustained elevation of intracellular Ca^{2+}, due to release of Ca^{2+} from intracellular stores and influx of Ca^{2+} ions. A small-conductance chloride channel (0.5–1 pS), identified in rat peritoneal MAST CELLS, can realise this by providing the negative membrane potential necessary for maintaining Ca^{2+} influx and its sustained elevation. The Ca^{2+} current activated by this mechanism is described as I_{CRAC} (Ca^{2+} release-activated Ca^{2+} current). By replacing extracellular Cl^- ions with non-permeant isethionate or gluconate anions, Friis and colleagues [26] were able to inhibit antigen-stimulated HISTAMINE secretion from rat peritoneal MAST CELLS, although some HISTAMINE secretion still occurred. Sodium cromoglycate can also block intermediate-conductance chloride channels on RBL cell membranes [17].

Studies on epithelial cells provide more evidence that sodium cromoglycate and nedocromil sodium affect chloride transport. Alton and Norris [27] showed that these compounds are able to block the activity of a chloride channel present on the mucosal surface of airway epithelial cells. Moreover, epithelial cells are sensitive to the concentration of solutes in their environment, and chloride currents are believed to be involved in the regulation of cell volume. Nedocromil and cromoglycate can inhibit the chloride current induced in epithelial cells in response to osmotic changes, thereby inhibiting cell swelling [28]. Furthermore, Paulmilchl and colleagues [29] showed that sodium cromoglycate and nedocromil sodium inhibit hypotonic saline-induced activation of a chloride channel in mouse 3T3 fibroblasts. It is now becoming apparent that this chloride channel inhibition by cromones may well be related to their activity as GPR35 agonists, since the chloride channel blocker 5-nitro-2-(3-phenylpropylamino) benzoic acid is also an agonist of GPR35 in kidney cells [30].

Further, evidence from in vitro studies suggests that these compounds affect neuronal chloride transport—chloride efflux from sensory nerves leads to depolarisation and the generation of action potentials. Nedocromil sodium prevents the contraction of guinea pig bronchus that is induced by electric field stimulation in the presence of atropine [12]. Bronchoconstriction is probably mediated by the release of NEUROPEPTIDES from C fibre terminals. This is supported by other studies that show nedocromil inhibition of substance P-induced potentiation of the cholinergic neural responses in rabbit trachea [31] as well as inhibition of tachykinin release [32]. Once again, the presence of GPR35 on neuronal membranes suggests a unifying RECEPTOR-mediated mechanism for cromone action.

Chloride channel activation is a mechanism that occurs when cells are activated. By preventing chloride channel activation, sodium cromoglycate and nedocromil sodium would be expected to maintain cells in a normal resting physiological state, and this is associated with the relative lack of toxicity of these compounds.

22.2.2 Biochemical and Pharmacological Effects

The anti-inflammatory effects of sodium cromoglycate and nedocromil can result in a number of biological effects:

- Inhibition of mediator release from human MAST CELLS isolated from bronchoalveolar lavage (BAL) fluid and from MAST CELLS derived from the lung, conjunctiva and nasal mucosa. Human skin-derived MAST CELLS, however, do not respond to cromoglycate or nedocromil. HISTAMINE secretion by enterochromaffin-like cells from the gastric mucosa can also be inhibited by sodium cromoglycate. Mediator release from EOSINOPHILS, NEUTROPHILS and platelets is also inhibited.
- The numbers of inflammatory cells such as EOSINOPHILS, NEUTROPHILS,

BASOPHILS (though not unequivocally confirmed for nedocromil sodium), MAST CELLS, MACROPHAGES and T LYMPHOCYTES are reduced, in both tissues and blood.
- Inflammatory cell infiltration depends on the activating effects of chemotactic factors that are often released by infiltrating inflammatory cells. Cromoglycate and nedocromil can completely suppress the activating effects of chemoattractant peptides on human EOSINOPHILS, NEUTROPHILS and MONOCYTES.
- Inflammatory cell infiltration also depends on the expression of ADHESION MOLECULEs. These compounds can inhibit the expression of various ADHESION MOLECULEs, such as ICAM-1, VCAM-1 and E-SELECTIN, which are crucial for the passage of inflammatory cells from the blood to peripheral tissues.
- Cell activation and cytokine release from inflammatory cells such as T LYMPHOCYTES, MACROPHAGES and MAST CELLS are inhibited.
- Inhibition of IL-4-induced IgE isotype switching and suppression of IgG4 production, without further effects on B cells that have already undergone switching.
- Sensory nerve (C fibre) activation is inhibited resulting in reduced release of NEUROPEPTIDES such as substance P and tachykinins.
- Survival of platelets can be increased and these compounds inhibit IgE activation of platelets.
- MICROVASCULAR LEAKAGE is reduced, presumably through functional antagonism of tachykinin RECEPTORS.

22.2.3 Pharmacokinetics

Disodium cromoglycate and nedocromil are poorly absorbed from the gastrointestinal tract and are therefore given locally per inhalation, as either an aerosol (a nebulised solution or in powder form) or as eye drops. Nedocromil and disodium cromoglycate are not metabolised and are excreted unchanged. Their plasma half-life is approximately 90 min.

22.2.4 Clinical Indications

Therapeutic studies have revealed and confirmed the clinical EFFICACY. Considering disodium cromoglycate is not therapeutically effective in all patients, it has mainly an adjunctive role in asthma and allergic rhino-conjunctivitis. Its excellent safety profile is predisposing for the use in children and during pregnancy.

The diverse clinical effects, including the anti-inflammatory character, of these drugs have been described in detail.

Sodium cromoglycate and nedocromil sodium have been reported to demonstrate protective effects on the immediate ASTHMATIC RESPONSE (IAR) as well as the late ASTHMATIC RESPONSE (LAR) induced by bronchial challenge with allergen. The delayed ASTHMATIC RESPONSE, however, was altered by nedocromil but not by cromoglycate. These compounds not only reduce the numbers of inflammatory cells in the BAL fluid, they also decrease the activation and/or stimulation state of these cells. They also reduce the number of circulating LEUKOCYTES (EOSINOPHILS, NEUTROPHILS and BASOPHILS) during the IAR and LAR following allergen challenge as well as decreasing the activation of circulating T LYMPHOCYTES. Cromoglycate and nedocromil can also inhibit the IAR that occurs during exercise-induced ASTHMA and reduce bronchoconstriction due to non-specific hyperreactivity mechanisms. However, not all ASTHMA subjects respond to these drugs, and children respond more often than adults [3, 33, 34].

Similar to the treatment of allergic ASTHMA, sodium cromoglycate and nedocromil sodium have demonstrated protective effects on the immediate nasal response (INR) as well as the late nasal response (LNR) induced by nasal challenge with allergen. The delayed nasal response, however, was not altered by cromoglycate and only partially prevented by nedocromil. Prophylactic treatment with these compounds significantly reduces inflammatory cell infiltration and epithelial cell numbers as determined by cytological analysis of nasal secretions following allergen challenge in patients with allergic rhinitis [35]. DEGRANULATION of MAST CELLS,

BASOPHILS and EOSINOPHILS is inhibited, and the expression of ICAM-1 on epithelial cells is downregulated [36, 37].

The immediate and late responses to allergen challenge are prevented in the eye. Sodium cromoglycate and nedocromil sodium inhibit the emergence of conjunctival oedema and erythema and reduce mast cell DEGRANULATION as well as vascular leakage [38–40].

Adverse reactions to food, including ALLERGY, can lead to unwanted organ responses. The various organ responses to food ingestion challenge can be inhibited by sodium cromoglycate treatment. This compound significantly inhibits immediate and late types of asthmatic, nasal, paranasal sinus, middle ear, conjunctival, migraine, atopic eczema, urticarial and Quincke's oedema responses to food ingestion ALLERGY [3, 35, 41, 42].

It can be concluded that sodium cromoglycate and nedocromil sodium are effective drugs in the prophylaxis of allergic bronchial ASTHMA, allergic rhinitis, allergic conjunctivitis and related allergic disorders, with little difference between the two agents [43]. However, some authors suggest that it is not justified to recommend sodium cromoglycate as a first-line prophylactic agent in childhood ASTHMA, apart from the treatment of mild symptoms [44, 45].

22.2.5 Unwanted Effects

Unwanted effects are infrequent and consist predominantly of the effects of irritation in the upper airway. HYPERSENSITIVITY reactions have been reported and include urticaria, bronchospasm, angio-oedema and ANAPHYLAXIS, but these are uncommon.

22.3 Histamine Receptor Antagonists

HISTAMINE was first identified in 1910 by Sir Henry Hallett Dale, and the first HISTAMINE RECEPTOR antagonists were synthesised over 20 years later. Early ANTI-HISTAMINE studies were qualitative, for example, by demonstrating their effectiveness in protecting against HISTAMINE-induced bronchospasm. Nevertheless, these studies introduced compounds, such as mepyramine, that remain major ligands to define HISTAMINE RECEPTORS. The first-generation ANTI-HISTAMINEs were introduced to clinical use in the 1940s. It became apparent in the 1950s, however, that there were multiple HISTAMINE RECEPTORS, and research still continues to identify novel HISTAMINE RECEPTORS. The early ANTI-HISTAMINEs were found to be non-selective antagonists of H1 RECEPTORS, and all exerted CENTRAL NERVOUS SYSTEM (CNS) side effects, including drowsiness and psychomotor impairment. The development of second- and third-generation H1 RECEPTOR antagonists resulted in anti-allergic drugs with increased RECEPTOR SPECIFICITY and potency and reduced CNS penetration. In addition, these newer agents exert broader anti-allergic effects.

22.3.1 Histamine

HISTAMINE [2-(4-imidazolyl)ethylamine or 5-amino-ethylimidazole] plays a significant role in the regulation of physiological processes and is an important mediator during allergic reactions (see also Chaps. 6 and 11). It is synthesised from L-histidine by histidine decarboxylation and stored in various cells, including MAST CELLS, BASOPHILS, neurones and enterochromaffin-like cells. Other cells, predominantly from the haematopoietic lineage, can also synthesise and secrete HISTAMINE, although these cells lack specific storage granules. HISTAMINE can closely mimic the anaphylactic response that usually results from an antigen-ANTIBODY reaction in sensitised tissue. Once released, HISTAMINE can be metabolised by diamine oxidase (DAO) and HISTAMINE *N*-methyltransferase (HMT). The effect of HISTAMINE is produced by its action on specific RECEPTORS, which are subdivided into several groups (H_1, H_2, H_3 and H_4 RECEPTORS). All subtypes are members of the seven membrane-spanning G protein-coupled RECEPTOR (GPCR) family.

22.3.2 Characterisation of Histamine Receptors

The RECEPTOR subtype determines the biological effects of HISTAMINE. H_1 RECEPTORS are expressed on most smooth muscle cells, endothelial cells, adrenal medulla, the heart and CNS. They have also been reported to be expressed on bronchial epithelial cells, fibroblasts, T cells, MONOCYTES, MACROPHAGES, DENDRITIC CELLS and B cells. Their stimulation leads to smooth muscle contraction, stimulation of NO formation, endothelial cell contraction, stimulation of hormone release, negative inotropism, depolarisation, increased neuronal firing and increased vascular permeability. Stimulation of H_1 RECEPTORS can also lead to pro-inflammatory reactions, such as induction of the expression of ADHESION MOLECULEs on endothelial cells, the production of CYTOKINES by these cells and the induction of COSTIMULATORY MOLECULES on DENDRITIC CELLS. H_1 RECEPTOR expression can be modified during inflammatory reactions. H_2 RECEPTORS are expressed on parietal cells in the gut, vascular smooth muscle cells, heart, suppressor T cells, NEUTROPHILS, CNS and BASOPHILS. Their stimulation triggers gastric acid secretion, vascular smooth muscle relaxation, positive chronotropic and ionotropic effects on cardiac muscle, inhibition of lymphocyte function, basophil CHEMOTAXIS and other immune responses. H_3 RECEPTORS are found mainly on cells in the CNS and PERIPHERAL NERVOUS SYSTEM as presynaptic RECEPTORS. They have also been identified on endothelium and enterochromaffin cells. These RECEPTORS control release of HISTAMINE and other neurotransmitters, such as ACETYLCHOLINE and dopamine, from neurones. H_4 RECEPTORS, sharing 37% amino acid homology with H_3 RECEPTORS, have been described more recently and are primarily expressed on cells of the haemopoietic lineage and on immunocompetent cells such as MAST CELLS, BASOPHILS, T cells, DENDRITIC CELLS, NEUTROPHILS and EOSINOPHILS, as well as in the spleen and thymus. Stimulation of H_4 RECEPTORS mediates CHEMOTAXIS of MAST CELLS, NEUTROPHILS, EOSINOPHILS and DENDRITIC CELLS and regulates cytokine release from $CD8^+$ T cells [46–53].

H_1 RECEPTOR antagonists are clinically effective when used to treat inflammatory and allergic reactions. The main clinical effect of H_2 RECEPTOR antagonists is on gastric secretion. The clinical relevance of H_3 RECEPTOR antagonists is still being explored, although they appear to be effective in the treatment of CNS disorders. Neither H_2 RECEPTOR antagonists nor H_3 RECEPTOR antagonists are considered to be clinically effective in anti-allergic therapies. H_4 RECEPTOR antagonists, such as JNJ7777120, are being developed and are showing considerable potential for the treatment of allergic diseases such as allergic rhinitis, ASTHMA and atopic dermatitis as well as for AUTOIMMUNE DISEASES like RHEUMATOID ARTHRITIS and chronic pain in the future [49, 52, 54].

22.3.3 Mechanisms of Action of H_1 Receptor Antagonists

The term ANTI-HISTAMINE conventionally refers to H_1 RECEPTOR antagonists, and these drugs are discussed in this section. The EFFICACY of these drugs is attributed principally to downregulation of H_1 RECEPTOR activity. In Table 22.2, a number of first-, second- and third-generation H1 RECEPTOR antagonists are shown.

Signal transduction by H_1 RECEPTORS (and also for H_4 RECEPTORS) occurs through the hydrolysis of phosphatidylinositols. HISTAMINE binds to the RECEPTOR, which in turn activates the $G_{\alpha q}$ protein ($G_{i/o}$ protein of H_4 RECEPTORS). Activation of these G proteins precedes activation of phospholipase C (PLC), which cleaves phosphatidylinositol bisphosphate (PIP_2) to form inositol triphosphate (IP_3) and diacylglycerol (DAG). IP_3 activates IP_3 RECEPTORS on the endoplasmic reticulum, causing the release of $Ca^{2+}{}_i$, as depicted in Fig. 22.3. The various biological effects follow the rise in Ca^{2+}.

Table 22.2 First-, second- and third-generation H_1 receptor antagonists

Antagonist	Disorder
First generation	
Clemastine	Acute anaphylactic reactions to insect bites or food allergy
Dexchlorpheniramine	Allergic conditions
Dimethindene	Allergic conditions
Diphenhydramine	Rhinitis/urticaria
Emedastine	Conjunctivitis/rhinitis
Hydroxyzine	Pruritus/chronic urticaria
Mebhydroline	Allergic conditions
Oxatomide	Allergic conditions
Promethazine	Allergic conditions/anaphylactic shock
Second generation	
Acrivastine	Allergic rhinitis/hay fever
Astemizole	Urticaria
Cetirizine	Allergic rhinitis/conjunctivitis/urticaria
Fexofenadine	Allergic rhinitis/chronic urticaria
Ketotifen	Allergic rhinitis/allergic skin conditions/prophylactic for asthma
Levocabastine	Allergic rhinitis/conjunctivitis
Levocetirizine	Allergic rhinitis/chronic urticaria
Terfenadine	Allergic rhinitis/conjunctivitis/allergic skin disorders
Third generation	
Azelastine	Allergic rhinitis/conjunctivitis
Bilastine	Allergic rhinitis/conjunctivitis/urticaria
Desloratadine	Allergic rhinitis/urticaria
Ebastine	Allergic rhinitis/conjunctivitis/urticarial/pruritus
Loratadine	Allergic rhinitis/conjunctivitis/chronic urticaria/pruritus
Mizolastine	Allergic rhinitis/conjunctivitis/urticaria
Rupatadine	Allergic rhinitis/conjunctivitis/urticaria

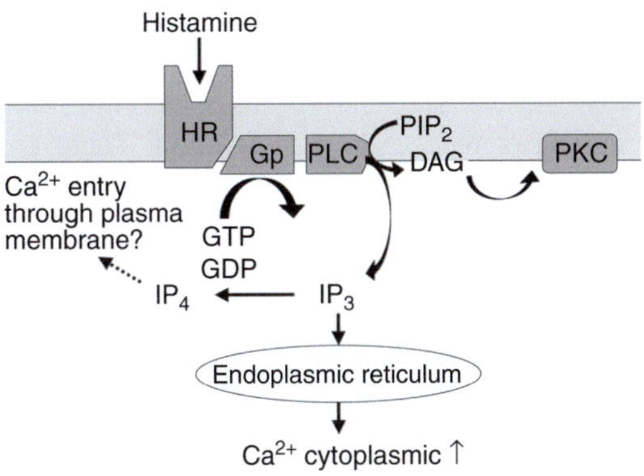

Fig. 22.3 Signalling pathway of H_1 receptor (and possibly H_4 receptor). Histamine binds to the receptor, which in turn activates the $G_{\alpha q}$ protein ($G_{i/o}$ protein of H_4 receptors). Activation of these G proteins precedes activation of PLC, which hydrolyses IP_3. IP_3 activates IP_3 receptors on the endoplasmic reticulum, causing the release of intracellular Ca^{2+}. The rise in Ca^{2+} is followed by the biological effect. *HR* histamine receptor, *Gp* G protein, *PLC* phospholipase C, *PIP_2* phosphatidylinositol bisphosphate, *DAG* diacylglycerol, *PKC* protein kinase C, *GTP* guanosine triphosphate, *GDP* guanosine diphosphate, *IP_3* inositol triphosphate, *Ca^{2+}* calcium ions

22.3.4 Pharmacological Effects of H_1 Receptor Antagonists

H_1 RECEPTORS modulate inflammatory and allergic responses by controlling NO formation and smooth muscle and endothelial cell contraction, which subsequently result in increased vascular permeability. HISTAMINE can also stimulate sensory nerve endings, thereby causing itching of the mucosa and skin through stimulation of C fibres. Regulation of the transcription factor NF-κB and subsequent generation of ADHESION MOLECULEs (ICAM-1 and P-selectin) and CYTOKINES (IL-6, IL-8, GM-CSF, RANTES) are also H_1 RECEPTOR dependent. The pharmacological actions of H_1 RECEPTOR antagonists are therefore useful for the inhibition of contraction of the smooth muscle and the inhibition of HISTAMINE-induced vascular permeability. Additionally, H_1 RECEPTOR antagonists inhibit the constitutive activation of NF-κB, which results in an inhibition of cytokine production. H_1 RECEPTOR antagonists inhibit HISTAMINE-induced bronchospasm in the guinea pig, but they are not effective in decreasing allergen-induced bronchospasm in human airways. Furthermore, first-generation H_1 RECEPTOR antagonists can exhibit anti-serotonergic, anti-emetic and/or anticholinergic characteristics, depending on the particular H_1 RECEPTOR antagonist used. Second-generation H_1 RECEPTOR antagonists, however, are more selective, have minimal sedative effects and have little AFFINITY for muscarinic cholinergic, α-adrenergic or serotoninergic RECEPTORS, although they still have dose-related adverse effects at high doses. Third-generation H_1 RECEPTOR antagonists are either active metabolites or enantiomers of second-generation compounds and show reduced adverse effects.

Suppression of mediator release from MAST CELLS and BASOPHILS by second- and third-generation H_1 RECEPTOR antagonists is thought to occur independently of the H_1 RECEPTOR. Neither MAST CELLS nor BASOPHILS express H_1 RECEPTORS on their surface. However, MAST CELLS, BASOPHILS and EOSINOPHILS all express the H_4 RECEPTOR on their surface. Recently, a variety of known H_1 RECEPTOR antagonists were shown to be weaker partial agonists at H_4 RECEPTORS, which would help to explain their broader spectrum of activity [55].

Rupatadine is a novel non-sedating H1 receptor antagonist with additional antagonist effects on platelet-activating factor (PAF) [56]. Rupatadine inhibits PAF- and LTB4-induced human neutrophil chemotaxis. In vitro rupatadine was more effective than other anti-histamines such as cetirizine, fexofenadine, loratadine and mizolastine [57, 58].

22.3.5 Pharmacokinetics

Most H_1 RECEPTOR antagonists are given orally, some are given as nose sprays or eye drops, and they are well absorbed and reach their peak effect in 1–2 h. The duration of activity depends largely on whether first-, second- or third-generation drugs are administered and can vary between 2 h and a few days. Most of these drugs are widely distributed throughout the body, but the third-generation drugs do not pass the blood-brain barrier. They are largely metabolised in the liver and excreted in the urine.

22.3.6 Clinical Indications

H_1 RECEPTOR antagonists can effectively control allergic disorders with mild symptoms, especially of the upper airways and skin. Allergic disorders with more severe symptoms and a complicated clinical picture, such as severe ASTHMA, require other therapies, but H_1 RECEPTOR antagonists may be supplemental.

H_1 RECEPTOR antagonists, administered either orally or topically to mucosal surfaces, are the most frequently used first-line medication for intermittent (seasonal) and persistent (perennial) allergic rhinitis. Non-sedating second-generation H_1 RECEPTOR antagonists such as cetirizine, fexofenadine and loratadine have been proven effective in short- and long-term studies. Also desloratadine, levocetirizine and tecastemizole have been found to be effective in allergic rhinitis. H_1 RECEPTOR antagonists reduce sneezing and rhinorrhoea as well as itchy, watery, red eyes. They reduce itchy nose, palate, or throat and sometimes reduce nasal congestion. There are very few clinical differences between the H_1 RECEPTOR antagonists. Some have reported that H_1 RECEPTOR antagonists reduce the influx of inflammatory cells into nasal secretion, whereas other investigators report a lack of inhibitory effects.

Cetirizine, desloratadine and loratadine have been reported to improve mild "seasonal" ASTHMA symptoms, reduce the amount of β_2-agonist usage and improve pulmonary function. These compounds prevent and relieve allergic INFLAMMATION in both the upper and lower airways. H_1 RECEPTOR antagonists have not been reported to be clinically effective in the treatment of bronchial ASTHMA and are therefore not a first choice for the treatment of this disorder. They do not affect HISTAMINE- or methacholine-induced bronchospasm, but they may be useful as additional or supplemental therapy.

This disorder usually occurs as part of an allergic syndrome, i.e. together with allergic rhinitis. If acute symptoms arise, the eyes can be treated locally with azelastine, Emadine, ketotifen, levocabastine or olopatadine. The main symptoms are red, itchy, watery eyes.

This disorder can be treated with oral, H_1 RECEPTOR antagonists. These compounds reduce itching as well as the number, size and duration of urticarial lesions. Erythema may not be completely inhibited because the vascular effects of HISTAMINE are also mediated via H_2 RECEPTORS as well as by other vasoactive substances such as proteases, EICOSANOIDS (LEUKOTRIENES, PROSTAGLANDIN E_1) and NEUROPEPTIDES (substance P). First- and second-generation H_1 RECEPTOR antagonists have been shown to be equally effective. H1 receptor antagonists (second generation) are licensed as first-line treatment of chronic urticaria [59], although many patients receive these drugs at up to four times higher than the approved dose to an adequate disease control [60].

The initial treatment of choice is adrenaline (epinephrine), and treatment of anaphylactic SHOCK can be accomplished with intramuscular or subcutaneous injections of adrenaline. The H_1 RECEPTOR antagonist clemastine may be given intravenously (2 mg) as an adjuvant. H_1 RECEPTOR antagonists may also be useful in the ancillary treatment of pruritus, urticaria and angio-oedema.

This is often treated with oral H_1 RECEPTOR antagonists that also exhibit a sedative action, in conjunction with TOPICAL GLUCOCORTICOIDS to relieve itching. Second-generation H_1 RECEPTOR antagonists are generally less effective than first-generation drugs.

The next-generation antihistaminic agents processing H_1 and H_4 receptor antagonistic actions may be more effective in the treatment of chronic allergic dermatitis and pruritus. But unfortunately further development is necessary [54].

22.3.7 Unwanted Effects

Most H_1 RECEPTOR antagonists have few unwanted effects when used at the recommended doses, although adverse CNS effects have been observed. The first-generation H_1 RECEPTOR antagonists have marked sedative effects due to the fact that they can cross the blood-brain barrier. The second-generation H_1 RECEPTOR antagonists are more specific for the H_1 RECEPTOR than first-generation drugs and therefore have little AFFINITY for muscarinic cholinergic, α-adrenergic or serotoninergic RECEPTORS; these compounds have less sedative effects. Third-generation H_1 RECEPTOR antagonists have been developed more recently to further reduce adverse effects. Dry mouth, urinary dysfunction, constipation, tachycardia and other unwanted effects are consequently not often observed. Allergic dermatitis following TOPICAL application has been reported.

22.4 H_4 Receptor Antagonists

Increased numbers of MAST CELLS are found in the airways of allergic ASTHMA and allergic rhinitis patients. Hofstra and colleagues [48] showed that stimulation of H_4 RECEPTORS with HISTAMINE mediates cell signalling and mast cell CHEMOTAXIS in a dose-dependent manner. Redistribution of MAST CELLS during allergic episodes may be mediated by this mechanism, indicating that specific H_4 RECEPTOR antagonists may prove useful in the treatment of allergic diseases in the future. Since EOSINOPHILS have also been demonstrated to express H_4 RECEPTORS, it is likely that specific H_4 RECEPTOR antagonists will also inhibit their migration and infiltration of tissues during allergic reactions. Furthermore, H_4 RECEPTORS appear to play a role in the control of cytokine release from $CD8^+$ T cells, monocyte-derived DENDRITIC CELLS and possibly other cells that express the H_4 RECEPTOR, during inflammatory disorders such as ASTHMA [46, 52].

Some of the currently available H_3 RECEPTOR agonists and antagonists are also recognised by the H_4 RECEPTOR, although they are much less potent for the H_4 RECEPTORS. Using cells transfected with the H_4 RECEPTOR, it has been shown that specific H_1 and H_2 RECEPTOR agonists and antagonists do not bind to the H_4 RECEPTOR. Specific antagonists for the H_4 RECEPTOR are already under development. The non-imidazole, neutral antagonist, JNJ7777120, has a 1000-fold selectivity for H_4 RECEPTORS in comparison to other HISTAMINE RECEPTOR subtypes and inhibits experimental allergic INFLAMMATION in animals. However, it has a short half-life ($t_{1/2} = 0.8$ h) after oral administration [49, 52]. Several other novel compounds are in early clinical trials for allergic respiratory and skin INFLAMMATION [54].

22.5 Anti-leukotrienes

In addition to HISTAMINE, IgE-stimulated MAST CELLS release a large number of different inflammatory mediators, including proteases, CHEMOKINES, CYTOKINES and EICOSANOIDS (Fig. 22.1). Currently, apart from the ANTI-HISTAMINEs, the anti-LEUKOTRIENES are the only other class of mast cell mediator inhibitors that are in clinical use. As presented in Chap. 6, the LEUKOTRIENES (LTs) exert a number of effects that are relevant for allergic respiratory INFLAMMATION, including bronchoconstriction and vasodilation by the cysteinyl-LTs and CHEMOTAXIS and activation of POLYMORPHONUCLEAR LEUKOCYTES by LTB_4. Inhibition of these effects by anti-LT agents, including the 5-LIPOXYGENASE (5-LO) inhibitor, zileuton, and the cysteinyl-LEUKOTRIENE (cys-LT_1) RECEPTOR antagonists, montelukast, zafirlukast and pranlukast, exerts beneficial effects on allergic INFLAMMATION, including ASTHMA, as discussed in greater detail in Chap. 23.

In addition to its use in ASTHMA, oral montelukast exhibits comparable EFFICACY and add-on effects to H_1 RECEPTOR antagonists in daytime nasal congestion due to allergic rhinitis but is less effective than intranasal glucocorticosteroids. Cys-LT_1 RECEPTOR antagonists are ineffective for treatment of allergic inflammatory symptoms outside the airways (conjunctivitis, oral symptoms, eczema and/or urticaria). Montelukast is well tolerated, although rare cases of neuropsychiatric events (e.g. agitation, aggression, anxiousness, dream abnormalities) have been reported [61].

22.6 Anti-IgE

Ever since the discovery of the function of IgE more than three decades ago [62], research has been focussed on the selective inhibition of either the activity or the production of IgE (see also Chap. 21). Omalizumab, marketed as Xolair, is a monoclonal ANTIBODY that targets IgE. Omalizumab is a RECOMBINANT DNA-derived humanised IgG1κ monoclonal ANTIBODY that selectively binds to human IgE. It is a humanised mouse ANTIBODY that contains only 5% amino acid sequence derived from the mouse. The ANTIBODY has a molecular mass of ~149 kDa and is produced by Chinese hamster ovary cells. Omalizumab has been approved by the US Food and Drug Administration (FDA) in 2003 and the European Medicine Agency (EMA) in 2005 for the treatment of adults and adolescents (12 years of age and older) with moderate to severe persistent ASTHMA who have a positive skin test or in vitro reactivity to a perennial aeroallergen and whose symptoms are inadequately controlled with inhaled corticosteroids. Omalizumab given by subcutaneous injection at a dose of 75–600 mg every 2–4 weeks based upon pretreatment serum IgE levels and total body weight.

In 2014 omalizumab was approved for the treatment of H1-anti-histamine-refractory chronic idiopathic urticaria. Treatment regimen is independent of serum IgE levels and total body weight. Patients receive 300 mg omalizumab subcutaneously every 4 weeks.

22.6.1 Biochemical and Pharmacological Effects

Omalizumab binds to free IgE at the FcεRI-binding site on the Cε3 domain of the IgE ANTIBODY, thereby inhibiting the binding to FcεRI on the surface of mast cells, BASOPHILS and other $FcεRI^+$ cells (Fig. 22.4). The reduction of surface-bound IgE on FcεRI-bearing cells prevents or limits the release of chemical mediators of the allergic response. Levels of serum-free IgE decrease by >90% from pretreatment values within 24 h after subcutaneous administration of omalizumab. In addition, omalizumab reduces the expression of FcεRI on BASOPHILS AND DENDRITIC CELLS [64].

22.6.2 Clinical Trials

Several clinical studies using RECOMBINANT humanised monoclonal ANTIBODY to human IgE (E25, Xolair, omalizumab) have been conducted and published (reviewed in [63]).

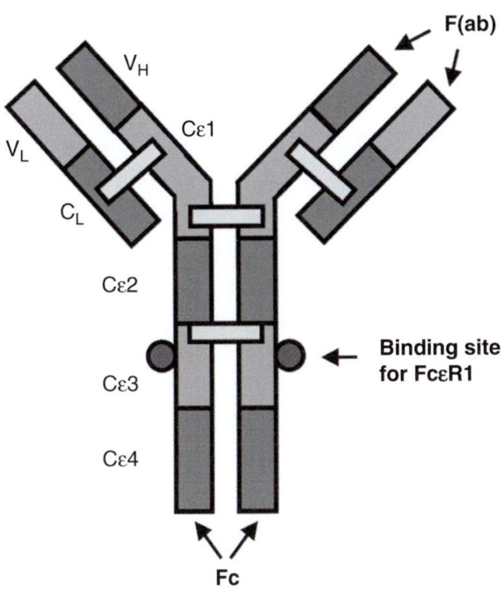

Fig. 22.4 Structure of IgE and position of binding site for FcεRI [63]

The approval of omalizumab was based upon EFFICACY data from several multicentre placebo-controlled phase III clinical trials with symptomatic patients with moderate to severe persistent ASTHMA who had a positive skin test reaction to a perennial aeroallergen [1, 65–68]. All patients were also being treated with inhaled corticosteroids and short-acting β-agonists. EFFICACY in these trials was based upon the number of ASTHMA exacerbations per patient, defined as a worsening of ASTHMA that required treatment with systemic corticosteroids or a doubling of the baseline dose of inhaled corticosteroid. Study results showed that the number of exacerbations per patient was reduced in patients receiving omalizumab compared with placebo. Reduction of ASTHMA exacerbations was not observed in omalizumab-treated patients who had baseline $FEV_1 > 80\%$ or in patients who required oral steroids as maintenance therapy. Omalizumab is not indicated for the treatment of bronchospasm or status asthmaticus.

The following effects of omalizumab have been observed in clinical trials with allergic ASTHMA patients:

- Reduction of severe asthma exacerbation [67].
- Reduction of asthma symptom scores, increase of quality-of-life scores and improvement of pulmonary function testing, decreased use of β-agonists after high dose of omalizumab, and reduction of oral or inhaled steroid use [1, 67].
- Reduction of asthma exacerbations during stable- and steroid-reduction phase and steroid-sparing effect [1, 66, 69].
- Efficacy of omalizumab is independent of age, gender, initial IgE serum levels and pretreatment with inhaled steroid use.
- Reduction of severe asthma exacerbation in children with inadequately controlled allergic (IgE-mediated) asthma [70, 71].

The following effects of omalizumab have been observed in clinical trials with H1-antihistamine-refractory chronic idiopathic URTICARIA [72–74]:

- Significant reduction of ISS (Itch Severity Score) and wheels after 12 weeks of treatment with 300 mg omalizumab every 4 weeks [73].
- 44.3% of treated patients had neither hives nor itching [73].

Off-label uses of omalizumab in which IgE may certainly play a major role are under investigation. Indications are, e.g. allergic rhinitis, allergic bronchopulmonary aspergillosis, anaphylaxis, keratoconjunctivitis, food allergy, drug allergy, urticarial angio-oedema, non-atopic asthma, atopic dermatitis, nasal polyps, Churg-Strauss syndrome, eosinophilic otitis media, chronic rhinosinusitis, bullous pemphigoid, contact dermatitis and pre-co-administration with specific immunotherapy [75].

22.6.3 Unwanted Effects

Single- and multiple-dose trials in adults with and without allergic diseases have demonstrated that omalizumab is well tolerated. Immune complex formation of omalizumab with IgE leads to relatively small complexes (<1000 kDa) that are not complement fixing and do not accumulate within any organ system [76]. The immune complexes have a serum half-life of approximately 20 days and are cleared via Fcγ RECEPTORS of the reticuloendothelial system [76]. The most commonly reported adverse effect is an urticarial rash, with an incidence of 0.5–0.7%. The rash develops within 1 h of receipt of the first dose and responds to ANTI-HISTAMINE therapy. Other adverse effects associated with omalizumab are pruritus, headache, syncope, paraesthesia and anaphylaxis. Generally the drug is well tolerated. A higher risk of suffering from thrombotic events caused by the therapy with omalizumab was reported [77]. But further evaluation in epidemiological studies is necessary to confirm this hypothesis.

Ligelizumab, MEDI-4212 and quilizumab are new potential drugs in the pipeline for asthma treatment and IgE as drug target [78].

22.6.4 Conclusions

The IgE-mast cell pathway plays a central role in the pathogenesis of allergic diseases. It is therefore not surprising that many anti-allergic drugs are targeted at this pathway. These drugs either prevent or reverse allergen-induced mast cell activation by blocking IgE (omalizumab) or stabilising mast cells (cromones) or block the biological effects of the HISTAMINE released upon allergen-induced mast cell activation (ANTI-HISTAMINEs). ANTI-HISTAMINEs have proven to be safe and successful, although they may not block all effects of mast cells. Cromones can be used as prophylactic drugs for allergic diseases, but their clinical EFFICACY is not undisputed. Omalizumab is very effective in reducing serum IgE levels and the consequent mast cell activation. However, this protein drug needs to be administered parenterally, and the cost of treatment is very high compared to other anti-allergic treatments. Thus, omalizumab is not the first line of prophylactic treatment for allergic diseases. Anti-LTs have limited therapeutic application beyond their use in ASTHMA.

22.7 Summary

ALLERGY is defined as a disease that develops following a response by the IMMUNE SYSTEM to an otherwise innocuous agent. This chapter describes the actions of anti-allergic drugs and therapeutics on diseases such as allergic rhinitis, atopic dermatitis, allergic conjunctivitis, systemic ANAPHYLAXIS, food ALLERGY, allergic ASTHMA and acute urticaria. The putative MECHANISMS OF ACTION of disodium cromoglycate and nedocromil sodium (cromones) are discussed in detail as well as the biochemical and pharmacological effects, clinical applications and unwanted effects of these drugs. The characterisation of HISTAMINE RECEPTORS is described, and the MECHANISMS OF ACTION as well as the observed biological effects of H_1 and the relatively new H_4 RECEPTORS are explained. Further, the biochemical and pharmacological effects of anti-IgE therapy are described, and recent clinical trials with these drugs are briefly reviewed.

Selected Readings

Carr WW, Nelson MR, Hadley JA. Managing rhinitis: strategies for improved patient outcomes. Allergy Asthma Proc. 2008;29:349–57.

Hill SJ, Ganellin CR, Timmerman H, Schwartz JC, Shankley NP, Young JM, Schunack W, Levi R, Haas HL. International Union of Pharmacology. XIII. Classification of histamine receptors. Pharmacol Rev. 1997;49:253–78. (see also: http://pharmrev.aspetjournals.org/cgi/reprint/49/3/253. Accessed Sept 2010).

Holgate S, Buhl R, Bousquet J, Smith N, Panahloo Z, Jimenez P. The use of omalizumab in the treatment of severe allergic asthma: a clinical experience update. Respir Med. 2009;103:1098–113.

Lieberman P. Intranasal antihistamines for allergic rhinitis: mechanism of action. Allergy Asthma Proc. 2009;30:345–8.

References

1. Busse W, Corren J, Lanier BQ, McAlary M, Fowler-Taylor A, Cioppa GD, van As A, Gupta N. Omalizumab, anti-IgE recombinant humanized monoclonal antibody, for the treatment of severe allergic asthma. J Allergy Clin Immunol. 2001;108:184–90.
2. Zhang T, Finn DF, Barlow JW, Walsh JJ. Mast cell stabilisers. Eur J Pharmacol. 2016;778:158–68.
3. Eady RP, Norris AA. Nedocromil sodium and sodium cromoglycate: pharmacology and putative modes of action. In: Kay AB, editor. Allergy and allergic diseases. Oxford: Blackwell; 1997. p. 584–95.
4. Eady RP. The pharmacology of nedocromil sodium. Eur J Respir Dis Suppl. 1986;147:112–9.
5. Janssen LJ, Wattie J, Betti PA. Effects of cromolyn and nedocromil on ion currents in canine tracheal smooth muscle. Eur Respir J. 1998;12:50–6.
6. Rang HP, Dale MM, Ritter JM. The respiratory system. In: Hunter L, editor. Pharmacology. London: Churchill Livingston; 1999. p. 347.
7. Dixon CM, Fuller RW, Barnes PJ. Effect of nedocromil sodium on sulphur dioxide induced bronchoconstriction. Thorax. 1987;42:462–5.
8. Dixon CMS, Ind PW. Inhaled sodium metabisulphite induced bronchoconstriction: inhibition by nedocromil sodium and sodium cromoglycate. Br J Clin Pharmacol. 1990;30:371–6.
9. Bigby B, Boushey H. Effects of nedocromil sodium on the bronchomotor response to sulfur dioxide in asthmatic patients. J Allergy Clin Immunol. 1993;92:195–7.
10. Yamawaki I, Tamaoki J, Takeda Y, Nagai A. Inhaled cromoglycate reduces airway neurogenic inflammation *via* tachykinin antagonism. Res Commun Mol Pathol Pharmacol. 1997;98:265–72.
11. Dixon M, Jackson DM, Richards IM. The action of sodium cromoglycate on 'C' fibre endings in the dog lung. Br J Pharmacol. 1980;70:11–3.
12. Verleden GM, Belvisi MG, Stretton CD, Barnes PJ. Nedocromil sodium modulates nonadrenergic, noncholinergic bronchoconstrictor nerves in guinea pig airways *in vitro*. Am Rev Respir Dis. 1991;143:114–8.
13. Edwards AM, Norris AA. Cromoglycate and asthma. Lancet. 1994;343:426.
14. Yazid S, Sinniah A, Solito E, Calder V, Flower RJ. Anti-allergic cromones inhibit histamine and eicosanoid release from activated human and murine mast cells by releasing Annexin A1. PLoS One. 2013;8:e58963.
15. Moon TC, St Laurent CD, Morris KE, Marcet C, Yoshimura T, Sekar Y, Befus AD. Advances in mast cell biology: new understanding of heterogeneity and function. Mucosal Immunol. 2010;3:111–28.
16. Okayama Y, Benyon RC, Rees PH, Lowman MA, Hillier K, Church MK. Inhibition profiles of sodium cromoglycate and nedocromil sodium on mediator release from mast cells of human skin, lung, tonsil, adenoid and intestine. Clin Exp Allergy. 1992;22:401–9.
17. Foreman JC, Hallett MB, Mongar JL. Site of action of the antiallergic drugs cromoglycate and doxantrazole. Br J Pharmacol. 1977;59:473P–4P.
18. Ennis M, Atkinson S, Pearce FL. Inhibition of histamine release induced by Compound F48/80 and peptide 401 in the presence and absence of calcium: implications for the mode of action of anti-allergic compounds. Agents Actions. 1980;10:222–8.
19. Sagi-Eisenberg R. The role of protein kinase C in histamine secretion: implications for the mode of action of the antiasthmatic drug cromoglycate. Curr Top Pulm Pharmacol Toxicol. 1987;2:24–42.
20. Sagi-Eisenberg R, Mazurek N, Pecht I. Calcium fluxes and protein phosphorylation in stimulus-secretion coupling of basophils. Mol Immunol. 1984;21:175–81.
21. Lucas AM, Shuster S. Cromolyn inhibition of protein kinase C activity. Biochem Pharmacol. 1987;36:562–5.
22. Mazurek N, Bashkin P, Loyter A, Pecht I. Restoration of Ca^{2+} influx and degranulation capacity of variant RBL-2H3 cells upon implantation of isolated cromolyn binding protein. Proc Natl Acad Sci U S A. 1983;80:6014–8.
23. Mazurek N, Geller-Bernstein C, Pecht I. Affinity of calcium ions to the anti-allergic drug, dicromoglycate. FEBS Lett. 1980;111:194–6.
24. Mazurek N, Schindler H, Schurholz T, Pecht I. The cromolyn binding protein constitutes the Ca^{2+} channel

of basophils opening upon immunological stimulus. Proc Natl Acad Sci U S A. 1984;81:6841–5.
25. Yang Y, Lu JY, Wu X, Summer S, Whoriskey J, Saris C, Reagan JD. G-protein-coupled receptor 35 is a target of the asthma drugs cromolyn disodium and nedocromil sodium. Pharmacology. 2010;86:1–5.
26. Friis UG, Johansen T, Hayes NA, Foreman JC. IgE-receptor activated chloride uptake in relation to histamine secretion from rat mast cells. Br J Pharmacol. 1994;111:1179–83.
27. Alton EW, Norris AA. Chloride transport and the actions of nedocromil sodium and cromolyn sodium in asthma. J Allergy Clin Immunol. 1996;98:S102–6.
28. Anderson SD, Rodwell LT, Daviskas E, Spring JF, du Toit J. The protective effect of nedocromil sodium and other drugs on airway narrowing provoked by hyperosmolar stimuli: a role for the airway epithelium? J Allergy Clin Immunol. 1996;98:S124–34.
29. Paulmilchl M, Norris AA, Rainey DK. Role of chloride channel modulation in the mechanism of action of nedocromil sodium. Int Arch Allergy Appl Immunol. 1995;107:416.
30. Taniguchi Y, Tonai-Kachi H, Shinjo K. 5-Nitro-2-(3-phenylpropylamino)benzoic acid is a GPR35 agonist. Pharmacology. 2008;82:245–9.
31. Armour CL, Johnson PRA, Black JL. Nedocromil sodium inhibits substance P-induced potentiation of the cholinergic neural responses in the isolated innervated rabbit trachea. J Auton Pharmacol. 1991;11:167–72.
32. Javdan P, Figini M, Emanueli C, Geppetti P. Nedocromil sodium reduces allergen-induced plasma extravasation in the guinea-pig nasal mucosa by inhibition of tachykinin release. Allergy. 1995;50:825–9.
33. Pelikan Z, Pelikan-Filipek M, Remeijer L. Effects of disodium cromoglycate and beclomethasone dipropionate on the asthmatic response to allergen challenge II. Late response (LAR). Ann Allergy. 1988;60:217–25.
34. Pelikan Z, Pelikan-Filipek M, Schoemaker MC, Berger MP. Effects of disodium cromoglycate and beclomethasone dipropionate on the asthmatic response to allergen challenge I. Immediate response (IAR). Ann Allergy. 1988;60:211–6.
35. Pelikan Z. Late type of the nasal allergic response review. Scr Med (Brno). 2001;74:303–44.
36. Larsson K, Larsson BM, Sandstrom T, Sundblad BM, Palmberg L. Sodium cromoglycate attenuates pulmonary inflammation without influencing bronchial responsiveness in healthy subjects exposed to organic dust. Clin Exp Allergy. 2001;31:1356–68.
37. Hoshino M, Nakamura Y. The effect of inhaled sodium cromoglycate on cellular infiltration into the bronchial mucosa and the expression of adhesion molecules in asthmatics. Eur Respir J. 1997;10:858–65.
38. James IG, Campbell LM, Harrison JM, Fell PJ, Ellers-Lenz B, Petzold U. Comparison of the efficacy and tolerability of topically administered azelastine, sodium cromoglycate and placebo in the treatment of seasonal allergic conjunctivitis and rhino-conjunctivitis. Curr Med Res Opin. 2003;19:313–20.
39. Katelaris CH, Ciprandi G, Missotten L, Turner FD, Bertin D, Berdeaux G. A comparison of the efficacy and toler-ability of olopatadine hydrochloride 0.1% ophthalmic solution and cromolyn sodium 2% ophthalmic solution in seasonal allergic conjunctivitis. Clin Ther. 2002;24:1561–75.
40. Tauber J. Nedocromil sodium ophthalmic solution 2% twice daily in patients with allergic conjunctivitis. Adv Ther. 2002;19:73–84.
41. Pelikan Z, Pelikan-Filipek M. Effects of oral cromolyn on the nasal response due to foods. Arch Otolaryngol Head Neck Surg. 1989;115:1238–43.
42. Yokooji T, Matsuo H. Sodium cromoglycate prevents exacerbation of IgE-mediated food-allergic reaction induced by aspirin in a rat model of egg allergy. Int Arch Allergy Immunol. 2015;167:193–202.
43. Parnham MJ. Sodium cromoglycate and nedocromil sodium in the therapy of asthma, a critical comparison. Pulm Pharmacol. 1996;9:95–105.
44. Tasche MJ, Uijen JH, Bernsen RM, de Jongste JC, van der Wouden JC. Inhaled disodium cromoglycate (DSCG) as maintenance therapy in children with asthma: a systematic review. Thorax. 2000;55:913–20.
45. Kaditis AG, Gourgoulianis K, Winnie G. Anti-inflammatory treatment for recurrent wheezing in the first five years of life. Pediatr Pulmonol. 2003;35:241–52.
46. Gantner F, Sakai K, Tusche MW, Cruikshank WW, Center DM, Bacon KB. Histamine h(4) and h(2) receptors control histamine-induced interleukin-16 release from human CD8(+) T cells. J Pharmacol Exp Ther. 2002;303:300–7.
47. Takeshita K, Sakai K, Bacon KB, Gantner F. Critical role of histamine H4 receptor in LTB4 production and mast cell-dependent neutrophil recruitment induced by zymosan *in vivo*. J Pharmacol Exp Ther. 2003;307:1072–8.
48. Hofstra CL, Desai PJ, Thurmond RL, Fung-Leung WP. Histamine H4 receptor mediates chemotaxis and calcium mobilization of mast cells. J Pharmacol Exp Ther. 2003;305:1212–21.
49. Smits RA, Leurs R, de Esch IJ. Major advances in the development of histamine H4 receptor ligands. Drug Discov Today. 2009;14:745–53.
50. Hill SJ, Ganellin CR, Timmerman H, Schwartz JC, Shankley NP, Young JM, Schunack W, Levi R, Haas HL. International Union of Pharmacology. XIII. Classification of histamine receptors. Pharmacol Rev. 1997;49:253–78.
51. Schneider E, Rolli-Derkinderen M, Arock M, Dy M. Trends in histamine research: new functions during immune responses and hematopoiesis. Trends Immunol. 2002;23:255–63.
52. Leurs R, Chazot PL, Shenton FC, Lim HD, de Esch IJ. Molecular and biochemical pharmacology of the histamine H4 receptor. Br J Pharmacol. 2009;157:14–23.
53. Thurmond RL, Gelfand EW, Dunford PJ. The role of histamine H1 and H4 receptors in allergic inflammation: the search for new antihistamines. Nat Rev Drug Discov. 2008;7:41–53.

54. Ohsawa Y, Hirasawa N. The role of histamine H1 and H4 receptors in atopic dermatitis: from basic research to clinical study. Allergol Int. 2014;63:533–42.
55. Deml KF, Beermann S, Neumann D, Strasser A, Seifert R. Interactions of histamine H1-receptor agonists and antagonists with the human histamine H4-receptor. Mol Pharmacol. 2009;76:1019–30.
56. Shamizadeh S, Brockow K, Ring J. Rupatadine: efficacy and safety of a non-sedating antihistamine with PAF-antagonist effects. Allergo J Int. 2014;23: 87–95.
57. Ramis I, Giral M, Ferrando R, Merlos M. Inhibition of PAF- and LTB4-induced human neutrophil chemotaxis by rupatadine using a new fluorescent chemotaxis assay. Allergy. 2000;55:94–5.
58. Barrón S, Roman J, Michelena P, Ramis I, Merlos M. Rupatadine inhibits cytokine production and NF-κB activity by histamine H1 receptor-dependent mechanism. Rev Rinol. 2006;6:18.
59. Zuberbier T, Asero R, Bindslev-Jensen C, Walter Canonica G, Church MK, Giménez-Arnau AM, Grattan CE, Kapp A, Maurer M, Merk HF, Rogala B, Saini S, Sánchez-Borges M, Schmid-Grendelmeier P, Schünemann H, Staubach P, Vena GA, Wedi B, Dermatology Section of the European Academy of Allergology and Clinical Immunology; Global Allergy and Asthma European Network; European Dermatology Forum; World Allergy Organization. EAACI/GA(2)LEN/EDF/WAO guideline: management of urticaria. Allergy. 2009;64:1427–43.
60. Staevska M, Popov TA, Kralimarkova T, Lazarova C, Kraeva S, Popova D, Church DS, Dimitrov V, Church MK. The effectiveness of levocetirizine and desloratadine in up to 4 times conventional doses in difficult-to-treat urticaria. J Allergy Clin Immunol. 2010;125:676–82.
61. Meltzer EO, Caballero F, Fromer LM, Krouse JH, Scadding G. Treatment of congestion in upper respiratory diseases. Int J Gen Med. 2010;3:69–91.
62. Ishizaka K, Ishizaka T. Biological function of gamma E antibodies and mechanisms of reaginic hypersensitivity. Clin Exp Immunol. 1970;6:25–42.
63. Hamelmann E, Rolinck-Werninghaus C, Wahn U. From IgE to anti-IgE: where do we stand? Allergy. 2002;57:983–94.
64. Chanez P, Contin-Bordes C, Garcia G, Verkindre C, Didier A, De Blay F, de Lara MT, Blanco P, Moreau JF, Robinson P, Bourdeix I, Trunet P, Le Gros V, Humbert M, Molimard M. Omalizumab-induced decrease of FcξRI expression in patients with severe allergic asthma. Respir Med. 2010;104:1608–17.
65. Finn A, Gross G, van Bavel J, Lee T, Windom H, Everhard F, Fowler-Taylor A, Liu J, Gupta N. Omalizumab improves asthma-related quality of life in patients with severe allergic asthma. J Allergy Clin Immunol. 2003;111:278–84.
66. Soler M, Matz J, Townley R, Buhl R, O'Brien J, Fox H, Thirlwell J, Gupta N, Della Cioppa G. The anti-IgE antibody omalizumab reduces exacerbations and steroid requirement in allergic asthmatics. Eur Respir J. 2001;18:254–61.
67. Humbert M, Beasley R, Ayres J, Slavin R, Hébert J, Bousquet J, Beeh KM, Ramos S, Canonica GW, Hedgecock S, Fox H, Blogg M, Surrey K. Benefits of omalizumab as add-on therapy in patients with severe persistent asthma who are inadequately controlled despite best available therapy (GINA 2002 step 4 treatment): INNOVATE. Allergy. 2005;60:309–16.
68. Busse WW, Massanari M, Kianifard F, Geba GP. Effect of omalizumab on the need for rescue systemic corticosteroid treatment in patients with moderate-to-severe persistent IgE-mediated allergic asthma: a pooled analysis. Curr Med Res Opin. 2007;23:2379–86.
69. Bousquet J, Rabe K, Humbert M, Chung KF, Berger W, Fox H, Ayre G, Chen H, Thomas K, Blogg M, Holgate S. Predicting and evaluating response to omalizumab in patients with severe allergic asthma. Respir Med. 2007;101:1483–92.
70. Kulus M, Hébert J, Garcia E, Fowler Taylor A, Fernandez Vidaurre C, Blogg M. Omalizumab in children with inadequately controlled severe allergic (IgE-mediated) asthma. Curr Med Res Opin. 2010;26:1285–93.
71. Lanier B, Bridges T, Kulus M, Taylor AF, Berhane I, Vidaurre CF. Omalizumab for the treatment of exacerbations in children with inadequately controlled allergic (IgE-mediated) asthma. J Allergy Clin Immunol. 2009;124:1210–6.
72. Kaplan A, Ledford D, Ashby M, Canvin J, Zazzali JL, Conner E, Veith J, Kamath N, Staubach P, Jakob T, Stirling RG, Kuna P, Berger W, Maurer M, Rosén K. Omalizumab in patients with symptomatic chronic idiopathic/spontaneous urticaria despite standard combination therapy. J Allergy Clin Immunol. 2013;132:101–9.
73. Maurer M, Rosén K, Hsieh HJ, Saini S, Grattan C, Gimenéz-Arnau A, Agarwal S, Doyle R, Canvin J, Kaplan A, Casale T. Omalizumab for the treatment of chronic idiopathic or spontaneous urticaria. N Engl J Med. 2013;368:924–35.
74. Saini SS, Bindslev-Jensen C, Maurer M, Grob JJ, Bülbül Baskan E, Bradley MS, Canvin J, Rahmaoui A, Georgiou P, Alpan O, Spector S, Rosén K. Efficacy and safety of omalizumab in patients with chronic idiopathic/spontaneous urticaria who remain symptomatic on H1 antihistamines: a randomized, placebo-controlled study. J Invest Dermatol. 2015;135:925.
75. El-Qutob D. Off-label uses of omalizumab. Clin Rev Allergy Immunol. 2016;50:84–96.
76. Fox JA, Hotaling TE, Struble C, Ruppel J, Bates DJ, Schoenhoff MB. Tissue distribution and complex formation with IgE of an anti-IgE antibody after intravenous administration in cynomolgus monkeys. J Pharmacol Exp Ther. 1996;279:1000–8.
77. Ali AK, Hartzema AG. Assessing the association between omalizumab and arteriothrombotic events through spontaneous adverse event reporting. J Asthma Allergy. 2012;5:1–9.
78. Heck S, Nguyen J, Le DD, Bals R, Dinh QT. Pharmacological therapy of bronchial asthma: the role of biologicals. Int Arch Allergy Immunol. 2015;168:241–52.

23 Drugs for the Treatment of Airway Disease

Gaetano Caramori, Ian M. Adcock, Peter J. Barnes, and Kian Fan Chung

23.1 Introduction

Both ASTHMA and CHRONIC OBSTRUCTIVE PULMONARY DISEASE (COPD) are characterized by airflow obstruction and chronic inflammation of the lower airways, but there are important differences in inflammatory mechanisms and response to therapy between these diseases [1, 2]. This chapter discusses the pharmacology of the drugs used in the treatment of these obstructive airway diseases. These drugs include bronchodilators, which act mainly by reversing AIRWAY SMOOTH MUSCLE (ASM) contraction, and anti-inflammatory drugs, which in asthma suppress the inflammatory response in the airways. The most effective anti-inflammatory treatment for asthma is corticosteroids, which are covered in a separate chapter (Chap. 32). In COPD, no effective anti-inflammatory drugs are available, but several new classes of drug are now in development. We will not discuss the drug treatment of airway disease in the paediatric population for which we refer the interested readers to specific excellent textbooks (please see below the selected readings).

Final manuscript submitted on February 12, 2018.

G. Caramori
Unità Operativa Complessa di Pneumologia, Dipartimento di Scienze Biomediche, Odontoiatriche e delle Immagini Morfologiche e Funzionali (BIOMORF), University of Messina, Messina, Italy
e-mail: gcaramori@unime.it

I. M. Adcock · P. J. Barnes · K. F. Chung (✉)
Airways Disease Section, National Heart and Lung Institute, Imperial College London, London, UK
e-mail: k.f.chung@imperial.ac.uk

23.2 Bronchodilators

Bronchodilator drugs have an "anti-bronchoconstrictor" effect, which may be demonstrated directly *in vitro* by a relaxant effect on precontracted airways. Bronchodilators cause immediate reversal of airflow obstruction in asthma *in vivo* which is believed to be due to a direct effect on ASM. However, additional pharmacological effects on other airway cells (such as reduced microvascular leakage and reduced release of bronchoconstrictor mediators from inflammatory cells) may contribute to the reduction in airway narrowing.

Three main classes of bronchodilator are in current clinical use:

- B_2-ADRENERGIC AGONISTS (sympathomimetics)
- THEOPHYLLINE (dimethylxanthine)
- INHALED ANTIMUSCARINICS (MUSCARINIC RECEPTOR ANTAGONISTS OR ANTICHOLINERGICS)

For both β_2-adrenergic agonists and antimuscarinics, the inhaled route of administration is preferable to the oral route because side effects are less and it may be more effective.

M. J. Parnham et al. (eds.), *Nijkamp and Parnham's Principles of Immunopharmacology*,
https://doi.org/10.1007/978-3-030-10811-3_23

When inhaled from a PRESSURIZED METERED DOSE INHALER (pMDI), DRY POWDER INHALERS (DPIs), SOFT MIST INHALERS or NEBULIZERS, they are convenient, easy to use and without significant side effects. The trend is to develop bronchodilators with increasing duration and rapid onset of action. Thus, we have progressed from short-acting, slow onset drugs to fast-acting, long duration and even ultra-long acting, fast-acting drugs.

Drugs, such as sodium cromoglycate, which prevent bronchoconstriction, have no direct bronchodilator action and are ineffective once bronchoconstriction has occurred. Anti-leukotrienes have a small bronchodilator effect in some patients and appear to act more to prevent bronchoconstriction.

23.3 β_2-Adrenergic Agonists

23.3.1 Chemistry

The development of β_2-agonists was a logical development of substitutions in the catecholamine structure of noradrenaline and adrenaline (epinephrine). The catechol ring consists of hydroxyl groups in the three and four positions of the benzene ring (Fig. 23.1). Noradrenaline differs from adrenaline only in the terminal amine group, which therefore indicates that modification at this site confers β-receptor selectivity. Further substitution of the terminal amine resulted in β_2-receptor selectivity, as in SALBUTAMOL (albuterol) and TERBUTALINE. Catecholamines are rapidly metabolized by the enzymes catechol-*o*-methyltransferases (COMTs), which methylate at the 3-hydroxyl position and account for the short duration of action of catecholamines. Modification of the catechol ring, as in salbutamol and terbutaline, prevents this degradation and therefore prolongs their effect. Catecholamines are also broken down in sympathetic nerve terminals and in the gastrointestinal tract by monoamine oxidase (MAO) enzymes which cleave the side chain. Isoprenaline, which is a substrate for MAO, is therefore metabolized in the gut, making absorption variable. Substitution in the amine group confers resistance to MAO and ensures reliable absorption.

Many other β_2-selective agonists have now been introduced (Fig. 23.1), and, while there may be differences in potency, there are no clinically significant differences in selectivity. Inhaled β_2-selective drugs in current clinical use (apart from rimiterol which is broken down by COMTs) have a similar duration of action of 3–6 h. The inhaled long-acting β_2-agonists (LABAs) SALMETEROL and FORMOTEROL have a much longer duration of effect, providing bronchodilation and bronchoprotection for over 12 h [3]. Formoterol has a bulky substitution in the aliphatic chain and has a moderate lipophilicity, which keeps the drug within the membrane close to the receptor, so it behaves as a slow-release drug. Salmeterol has a long aliphatic chain, and its long duration may be due to binding within the receptor binding cleft that anchors the drug in the binding cleft. Long-acting oral β_2-agonists include slow-release formulations of salbutamol or terbutaline and the once-daily β_2-agonist bambuterol (a prodrug that is slowly converted to terbutaline in the body). Their use has been discontinued in most countries due to their systemic side effects.

In a few countries, including Japan, a transdermal preparation of the β_2-agonist TULOBUTEROL (CAS 41570-61-0) is commercially available that can maintain effective blood tulobuterol levels for 24 h when applied once daily. Once-daily inhaled β_2-agonists, such as INDACATEROL, VILANTEROL and OLODATEROL, with a duration of action >24 h, have been introduced into clinical practice [4]. The onset and duration of action of various inhaled β_2-agonists are summarized in Table 23.1.

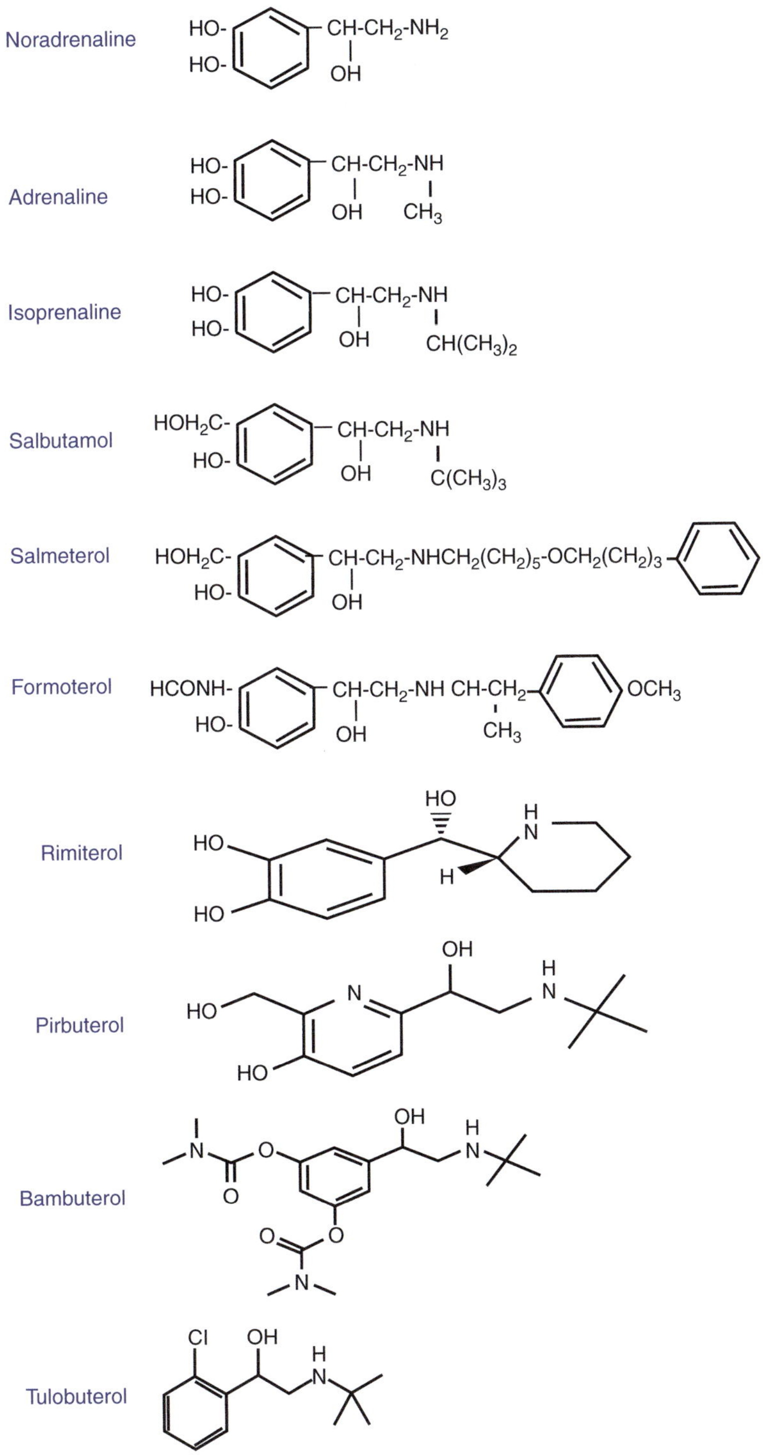

Fig. 23.1 Chemical structure of some inhaled β_2-agonists showing development from catecholamines

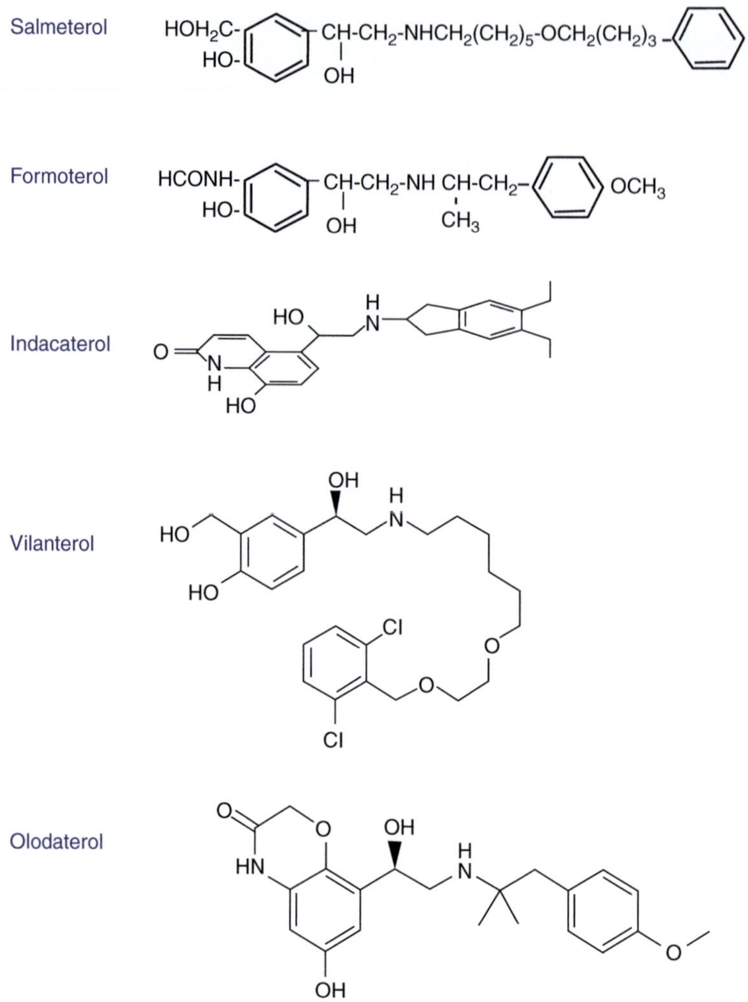

Fig. 23.1 (continued)

Table 23.1 Inhaled bronchodilators for the treatment of asthma and COPD

Short-acting (3–8 h duration of the effect):	
β_2 agonists (SABAs)	Salbutamol (albuterol), levalbuterol, terbutaline, fenoterol, pirbuterol, rimiterol (3–6 h duration, all fast onset of action)
Muscarinic antagonists (SAMAs)	Ipratropium (4–6 h duration, fast onset of action)
	Oxitropium (6–8 h duration, fast onset of action)
Long-acting (12 h duration of the effect):	
β_2 agonists (LABAs)	Salmeterol (slow onset of action)
	Formoterol and arformoterol (both fast onset of action)
Muscarinic antagonists (LAMAs)	Aclidinium (fast onset of action)
Ultra-long-acting (24 h duration of the effect):	
β_2 agonists (ULABAs)	Indacaterol, vilanterol and olodaterol (all fast onset of action)
Muscarinic antagonists (ULAMAs)	Tiotropium (slow onset of action)
	Glycopyrronium, umeclidinium (both fast onset of action)

23.3.2 Mode of Action

Inhaled β_2-agonists target B$_2$-ADRENERGIC RECEPTORS (β_2-ARs) which are members of the G PROTEIN-COUPLED RECEPTOR (GPCR) superfamily that are expressed in high numbers (3–4 × 10^4 per cell) on the ASM cells of the large and small airways (their number increases with increasing airway generation) and also on other important target cells, including lower airway epithelium (with the highest levels expressed on type 1 and 2 alveolar cells) and submucosal glands, vascular smooth muscle, macrophages, neutrophils, lymphocytes, eosinophils, mast cells and the endothelium of the postcapillary venules [4].

Ligand binding to the β_2-AR results in activation of receptor-associated G STIMULATORY (G_S) PROTEIN and enhanced coupling with the membrane-bound enzyme adenylyl cyclase. Gs protein is a heterotrimeric protein with G alpha (α), G beta (β) and G gamma (γ) subunits. Among these, only Gα is in direct contact with β_2-AR. The inactive Gs protein is bound to guanosine diphosphate (GDP). The activated β_2-AR of the complex with Gs protein induces a conformational change in the Gs protein to release GDP and bind guanosine triphosphate (GTP). The GTP-bound Gα subunit dissociates from Gβγ subunits, and the separate Gα and Gβγ subunits interact with effectors, such as adenylate cyclase and calcium channel, to propagate the GPCR signalling [5]. The coupling of activated G_s and adenylyl cyclase leads to increased production of cyclic ADENOSINE 3′,5′-MONOPHOSPHATE (cAMP) and subsequent activation of cAMP-DEPENDENT PROTEIN KINASE A (PKA), which phosphorylates several target proteins within the cell, leading to relaxation (Fig. 23.2). In fact, cAMP binds the regulatory subunits of the inactive PKA to release the active PKA catalytic subunits.

β-Agonists produce bronchodilation by directly stimulating β_2-AR in ASM and *in vitro* relax human bronchi and lung strips (indicating an effect on peripheral airways), and *in vivo* there

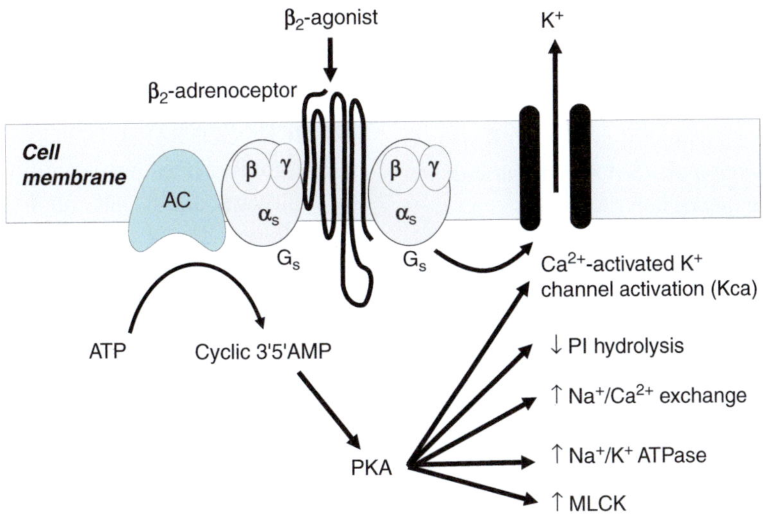

Fig. 23.2 Molecular mechanism of action of inhaled β_2-agonists on airway smooth muscle cells. Ligand binding to the β_2-receptors (β_2AR) results in activation of receptor-associated G-stimulatory (G_s) proteins and enhanced coupling with the enzyme adenylyl cyclase (AC). The coupling of activated G_s and adenylyl cyclase leads to increased production of cyclic 3′5′adenosine monophosphate (cAMP) and subsequent activation of cAMP-dependent protein kinase A (PKA), which then phosphorylates several target proteins which result in opening of calcium-activated potassium channels (K_{Ca}) or maxi-K channels, decreased phosphoinositide (PI) hydrolysis, increase sodium/calcium ion (Na^+/Ca^{2+}) exchange, increase Na^+/K^+ ATPase and decrease myosin light chain kinase (MLCK) activity. In addition, β_2-receptors may be coupled directly via G_s to K_{Ca}. *ATP* adenosine triphosphate. For a more detailed explanation of the molecular mechanisms of action of β_2-agonists, please refer to the text

is a rapid decrease in lower airways resistance. β_2-ARs have been localized to ASM of all lower airways.

The molecular mechanisms by which β_2-agonists induce relaxation of ASM include:

- Lowering of intracellular calcium ion (Ca^{2+}) concentration by active removal of Ca^{2+} from the cell and into intracellular stores. Activation of Ca^{2+}-Mg^{2+}-ATPases in the endoplasmic reticulum and plasma membrane decreases Ca^{2+} levels, thereby reducing Ca^{2+}-dependent actin-myosin interactions and leading to relaxation of ASM.
- An inhibitory effect on PHOSPHOINOSITIDE (PI) hydrolysis, which leads to intracellular Ca^{2+} release.
- Inhibition of MYOSIN LIGHT CHAIN KINASE (MLCK) that is phosphorylated and thus inactivated, preventing myosin phosphorylation.
- Concomitant activation of myosin light chain phosphatase.
- Opening of a LARGE CONDUCTANCE CALCIUM-ACTIVATED POTASSIUM CHANNEL (K_{CA}) which repolarizes the smooth muscle cell and may stimulate the sequestration of Ca^{2+} into intracellular stores.
- β_2-ARs are also *directly* coupled to K_{Ca} via G_s so that relaxation of ASM may occur independently of an increase in cAMP.

Several actions of β_2-agonists are not mediated via PKA, and these effects are transduced through other cAMP-regulated proteins, such as EXCHANGE PROTEIN ACTIVATED BY CAMP (EPAC), and EPAC-selective activators may be sufficient to cause ASM relaxation [6]. However, direct inhibition of PKA in human ASM cells or murine airways *ex vivo* inhibits the vast majority of the relaxant effect of the β_2-agonists, demonstrating that PKA is the principal effector of β_2-agonist-mediated ASM relaxation [7].

β_2-Agonists may also influence gene transcription through elevation of cAMP and activation of PKA. cAMP mediates the hormonal stimulation of many genes through a conserved CAMP RESPONSE ELEMENT (CRE). cAMP causes translocation of the catalytic subunit of PKA to the nucleus, where it phosphorylates and activates the transcription factor CAMP RESPONSE ELEMENT BINDING PROTEIN (CREB), enhancing its DNA-binding and transactivating activity. Activated CREB may persist for prolonged periods within the nucleus, and therefore, even a brief exposure to β_2-agonist may lead to a prolonged effect on mRNA transcription. Activation of the β_2AR can also lead to stimulation of the EXTRACELLULAR SIGNAL-REGULATED KINASES (ERKs)/ mitogen-activated protein kinases (MAPKs) pathway. Activation of the β_2AR can also induce p38 MAPK activation in a manner similar to that seen with ERKs/MAPKs activation. These effects may contribute to the inhibition of ASM cell proliferation and inflammatory mediator release [4].

β_2-Agonists act as *functional antagonists* and reverse bronchoconstriction irrespective of the contractile agent. This is an important property for the treatment of asthma, as many BRONCHOCONSTRICTOR mechanisms (through the effect of neurotransmitters and mediators) are likely to be contributory. In COPD, the major mechanism of action is likely to be a reduction of cholinergic reflex BRONCHOCONSTRICTION (see below).

β_2-Agonists may have additional effects on lower airways, and β_2-ARs are localized on several different lower airway cells. β_2-Agonists may therefore cause BRONCHODILATION *in vivo* not only via a direct action on lower ASM but also indirectly by inhibiting the release of bronchoconstrictor mediators from inflammatory cells and of bronchoconstrictor neurotransmitters from nerves in the lower airways (Table 23.2). These mechanisms include:

- Prevention of mediator release from isolated human lung mast cells (via β_2-AR).
- Attenuation of neutrophil granulocytes recruitment and activation.
- Prevention of microvascular leakage and thus the development of bronchial mucosal oedema after exposure to pro-inflammatory mediators, such as histamine and leukotriene D_4.

Table 23.2 Main effects of inhaled β-adrenergic agonists on lower airways

• Relaxation of airway smooth muscle (central and peripheral lower airways)
• Inhibition of mast cell and neutrophils activation
• Inhibition of plasma exudation and airway oedema
• Increased mucociliary clearance
• Increased mucus secretion
• Decreased cholinergic neurotransmission
• Decreased cough
• No effect on *chronic* inflammation

- Weak increase in *mucus secretion* from submucosal glands and *ion transport* across airway epithelium; these effects may enhance mucociliary clearance and therefore reverse the defective clearance found in asthma and COPD.
- Reduction of the bronchial epithelial damage caused by bacterial toxins, such as lipopolysaccharides (LPS).
- Reduction in *neurotransmission* in human airway CHOLINERGIC NERVES by an action at pre-junctional β_2-receptors to inhibit ACETYLCHOLINE release. This may contribute to their bronchodilator effect by reducing CHOLINERGIC reflex bronchoconstriction.

Although these additional effects of β_2-agonists may be relevant to the prophylactic use of these drugs against various challenges, their rapid bronchodilator action is likely attributable to a direct effect on bronchial/bronchiolar smooth muscle cells.

In the lower airways of patients with asthma and COPD, β_2-agonists may potentiate the molecular mechanism of GLUCOCORTICOID actions, with increased nuclear localization of GLUCOCORTICOID RECEPTORS (GRS) and additive or sometimes synergistic suppression of inflammatory mediator release [4]. Glucocorticoids are able to activate the expression of a number of anti-inflammatory genes due to GR binding to specific GREs in the promoter regions of responsive genes. This occurs in concert with other transcription factors, for example, in ASM cells, activation of CCAAT/ENHANCER-BINDING PROTEIN (C/EBPA), which in turn leads to an optimal antiproliferative action via $p21^{WAF1/Cip1}$, has been proposed to account, at least in part, for the enhanced effect of this combination treatment in asthmatic and COPD patients [8].

Inhaled glucocorticoids on the other hand increase the cellular expression of β_2-AR by increasing gene transcription acting through glucocorticoid response elements and, importantly, may regulate β_2-AR function by restoring G protein-β_2-AR coupling and inhibiting β_2-AR downregulation, thereby preventing its desensitization [4, 8].

As a regulatory protective mechanism, the activation of the β_2-AR concomitantly activates rapid and slow feedback mechanisms that lead to the inhibition of β_2-AR signalling, with the rapid mechanisms involving receptor phosphorylation by kinases that leads to uncoupling of the β_2-AR from the G_s protein. For example, agonist-occupied β_2-AR is rapidly phosphorylated by GPCR kinases (GRKs)2/3 which not only diminish the ability of the β_2-AR to couple to G_s but also promote the association of BETA-ARRESTIN-2 to the β_2-AR that sterically inhibit β_2-AR-G_s association [9]. Mice with β-arrestin-2 knocked down show a selective increase (40–60%) in the β2-AR signalling and function in the lower airways, whereas control mice show no effect on signalling and function of β2-AR but a selective inhibition (~50%) of M3 MUSCARINIC ACETYLCHOLINE RECEPTOR signalling and function [10]. Other kinases, including PKA and PKC, can also phosphorylate the β_2-AR to promote uncoupling, without the β_2-AR necessarily being occupied by agonist.

More extended mechanisms of β_2-AR desensitization include receptor internalization (mediated by arrestins for the agonist-occupied receptor) which sequesters the receptor from G proteins and receptor downregulation (loss of total cellular β_2-AR protein) which can occur slowly over the course of hours when internalized receptors traffic to lysosomes for degradation [9].

On the other hand, there are also data, mainly in animals however, showing that β_2-ARs can also stimulate a G protein-independent, β-arrestin-dependent signalling pro-inflammatory pathway. Airway epithelial expression of membrane-bound β_2-AR and cytosolic β-arrestin-2 is significantly elevated relative to that seen in ASM cells or T lymphocytes, and β-arrestins expressed within the structural and epithelial cells of the lungs, as well as infiltrating leukocytes, may contribute to the inflammatory responses during asthma [11].

Another important rapid feedback mechanism istherapidactivationofPHOSPHODIESTERASES (PDEs) that occurs as a result of intracellular PKA activation. PDEs break down cAMP, thus removing the stimuli for PKA activation [9]. In addition, PKA can phosphorylate GPCRs, Gα subunits as well as other effectors (PLC, IP3R, MLCK, HSP20, K^+ Ca channels) to exert negative feedback (of G_s-coupled receptor signalling), inhibitory crosstalk (of G_q-coupled receptors) and ASM relaxation [9].

23.3.3 β_2-AR Gene Polymorphisms and Response to β_2-Agonists

The β_2-AR is coded by the ADRβ2 gene, located on the long arm of chromosome 5, which has many single nucleotide polymorphism sites, which may affect the activity of β_2-AR. The mutated β_2-AR shows reduced adenylate cyclase activity suggesting that the mutation decreases the efficacy of signal transduction [5]. The most common variants are $Gly^{16}Arg$ and $Gln^{27}Glu$ which have *in vitro* effects on receptor desensitization. However, clinical studies have shown inconsistent effects on the bronchodilator responses to inhaled β_2-agonists [12, 13]. Some studies have shown that asthmatic patients with the common homozygous $Arg^{16}Arg$ variant have more frequent adverse effects and a poorer response to SABAs than heterozygotes or $Gly^{16}Gly$ homozygotes. Overall these differences are small, and there appears to be no clinical value in measuring the ADRβ2 genotype/haplotype. No differences have been found in terms of the airway responses to LABAs in the patients with asthma between these different genotypes/haplotypes [12, 13].

Similarly, the role of genetic variants in determining bronchodilator responsiveness to short-acting β2-agonists in COPD patients is still uncertain [13].

23.3.4 Anti-inflammatory Effects in the Lower Airways of Patients with Asthma or COPD

Whether inhaled β_2-agonists have anti-inflammatory effects in asthma is controversial. The inhibitory effects of β_2-agonists on mast cell mediator release and microvascular leakage are clearly anti-inflammatory, suggesting that β_2-agonists may modify *acute* inflammation. However inhaled β_2-agonists do not appear to have a significant inhibitory effect on the *chronic* inflammation of asthmatic airways, which is suppressed by corticosteroids. This has now been confirmed by several biopsy and bronchoalveolar lavage studies in patients with asthma who are taking regular inhaled β_2-agonists (including LABAs), which demonstrate no significant reduction in the number or activation in inflammatory cells in the airways, in contrast to resolution of the inflammation which occurs with inhaled corticosteroids. This is likely to be related to the fact that inhaled β_2-agonists effects on macrophages, eosinophils and lymphocytes are rapidly desensitized due to the low density of β_2-AR on these cells.

Similarly, the few airway biopsy studies performed in patients with stable COPD have demonstrated that the long-acting inhaled β_2-agonists have little significant *in vivo* anti-inflammatory activity despite demonstrating their known bronchodilator properties [1, 4, 14].

23.3.5 Clinical Use

23.3.5.1 Short-Acting β_2-Agonists

Several short-acting β_2-agonists are available. With the exception of RIMITEROL (which retains the catechol ring structure and is therefore susceptible to rapid enzymatic degradation), they have a longer duration of action because they are resistant to uptake and enzymatic degradation by COMT and MAO. There is little to choose between the various short-acting β_2-agonists currently available; all are used by the inhaled or oral route, have a similar duration of action (usually 3–4 h) and have similar side effects.

Differences in β_2-selectivity have been claimed, but these are not clinically important. Drugs in clinical use include *salbutamol*, LEVALBUTEROL (*see below*), *terbutaline*, FENOTEROL, PIRBUTEROL and *rimiterol* (Table 23.1).

Inhaled β_2-agonists are the most widely used and effective bronchodilators in the treatment of asthma due to their functional antagonistic effects. In addition to an acute bronchodilator effect, they are effective in protecting against various challenges, such as exercise, cold air and allergen. SABAs should be used "as required" by symptoms and not on a regular basis in the treatment of mild asthma, as increased usage serves as an indicator for the need for more anti-inflammatory therapy. It has proposed their as needed use in the treatment of asthmatics in combination with inhaled glucocorticoids only. In combination with rapid-acting inhaled antimuscarinics, SABAs are usually the bronchodilators of choice in treating asthmatic and COPD exacerbations, when the inhaled route of administration is as effective as and easier and safer than intravenous administration.

23.3.5.2 Long-Acting and Ultra-Long-Acting Inhaled β_2-Agonists

The long-acting inhaled β_2-agonists (LABAs) *salmeterol* and *formoterol* have proved to be a major advance in asthma and COPD therapy. Both drugs have a bronchodilator action of >12 h and also protect against bronchoconstriction for a similar period [3]. Both improve asthma control (when given twice daily) compared with regular treatment with SABAs four times daily and are well tolerated. While both drugs have a similar duration of effect in clinical studies, there are differences. Formoterol has a more rapid onset of action and is an almost full agonist, whereas salmeterol is a partial agonist with a slower onset of action. The rapid onset of action of formoterol means that it can also be used as a reliever, whereas salmeterol cannot [3]. However, no significant clinical differences between salmeterol and formoterol have been found in the chronic treatment of patients with persistent asthma [ginasthma.org].

Indacaterol, olodaterol and vilanterol are newly introduced LABAs characterized by a rapid onset and a 24-h duration of action (Table 23.1). They have been termed ULTRA-LABAS (ULABAs). Indacaterol is already approved for clinical use in stable COPD patients alone or in combination with inhaled antimuscarinics, and it is in advanced clinical development in combination with mometasone for the regular treatment of patients with persistent asthma. Vilanterol is already approved for clinical use in stable COPD patients in combination with inhaled antimuscarinics or the once-daily inhaled corticosteroid fluticasone furoate [4], and this latter combination is also approved for the regular treatment of patients with persistent asthma, whereas olodaterol is already approved for its clinical use in stable COPD patients alone or in combination with inhaled antimuscarinics.

In asthmatic patients, LABAs/ULABAs are an effective add-on therapy to INHALED CORTICOSTEROIDS (ICS) and are more effective than increasing the dose of ICS when asthma is not controlled at low doses. In asthmatic patients, LABAs/ULABAs should never be used alone as they do not treat the underlying chronic inflammation so should always be used in combination with ICS (preferably in a fixed combination inhaler) ([15]; see also Chap. 32 and ginasthma.org).

In stable COPD patients, LABAs/ULABAs are effective bronchodilators which may be used alone or in combination with antimuscarinics and/or ICS [goldcopd.org]. LABAs/ULABAs improve symptoms and exercise tolerance by reducing air trapping and also reduce COPD exacerbations although they are slightly less effective compared with the inhaled antimuscarinics alone [16].

23.3.5.3 Stereoselective Inhaled β_2-Agonists

Salbutamol (albuterol) is a racemic mixture of active *R* and inactive *S* isomers. Animal studies have suggested that the *S* isomer may increase airway responsiveness, providing a rationale for the development of *R*-albuterol (levalbuterol). Although the *R* isomer is more potent than racemic *RS*-albuterol in some studies, dose-response analysis shows no advantage in terms of efficacy and no evidence that *S*-albuterol is detrimental in asthmatic patients [17]. As levalbuterol is more expensive than normally used racemic albuterol, this therapy has no clear clinical advantage [18]. Stereoselective formoterol (*R, R*-formoterol, ARFORMOTEROL) has now been developed as a nebulized solution of arformoterol tartrate and is approved in the USA for twice-daily administration in patients with COPD but again offers no clinical advantage over racemic formoterol [19].

23.3.5.4 Combination Inhalers

Combination inhalers that contain a LABA/ULABA and an ICS (FLUTICASONE PROPIONATE/SALMETEROL, BUDESONIDE/FORMOTEROL, BECLOMETHASONE DIPROPIONATE/FORMOTEROL, FLUTICASONE PROPIO-NATE/FORMOTEROL, MOMETASONE FUROATE/FORMOTEROL, FLUTICASONE FUROATE/VILANTEROL) are now widely used, twice or once daily, in the treatment of persistent asthma and stable COPD [4, 15]. In asthma there is a strong scientific rationale for combining a LABA/ULABA with an ICS as these treatments have complementary actions and may also interact positively with the corticosteroids enhancing the effect of the LABA/ULABA and the LABA/ULABA potentiating the effect of the corticosteroid ([20]; see also Chap. C13). The combination inhaler is more convenient for patients, simplifies therapy and improves compliance with ICSs as the patients perceive clinical benefit, but there may be an additional advantage as delivering the two drugs in the same inhaler ensures that they are delivered simultaneously to the same cells in the airways, allowing the beneficial molecular interactions between LABAs/ULABAs and ICSs to occur. It is likely that these different combination inhalers will become the preferred therapy for all patients with persistent asthma. These combination inhalers are also more effective in stable COPD patients than LABAs/ULABAs or ICS alone, but the mechanisms accounting for this beneficial interaction are less well understood than in patients with asthma. In addition, new combinations of inhaled LABA/ULABAs, an ICS and an inhaled ANTIMUSCARINIC in a single inhaler have already been approved (beclomethasone dipropionate/formoterol/glycopyrronium twice daily and fluticasone furoate/vilanterol/umeclidinium once daily) by the European Medicines Agency's (EMA) Committee for Medicinal Products for Human Use (CHMP) and/or the Food and Drug Administration (FDA) for the regular treatment of COPD or under advanced clinical development (budesonide/formoterol/glycopyrronium twice daily and mometasone/indacaterol/glycopyrronium) for both asthma and COPD with already some evidence that long-term regular treatment of stable COPD with this triple therapy may further decrease the risk of COPD exacerbations, including hospital admissions [4, 21]. However, it is important to assess the risk/benefit ratio of triple therapy on an individual basis and to identify the patients most likely to benefit. The role of elevated blood eosinophils as a biomarker for the identification of candidates for ICS treatment is currently debated, but further prospective evidence is required [22].

Combination inhalers that contain budesonide/formoterol or beclomethasone dipropionate/formoterol or fluticasone propionate/formoterol have been shown to be more effective as a reliever for the treatment of asthmatic symptoms than SABAs or formoterol alone, suggesting that the ICS may also be contributing to the benefit [15]. This makes it now possible to control asthma with a single inhaler both for maintenance and relief of symptoms (see below).

23.3.6 Side Effects

Unwanted effects are dose-related and are usually due to stimulation of extrapulmonary β-receptors (Table 23.3). Side effects are not common with inhaled therapy, but more common with oral or parenteral (especially intravenous) administration.

- *Muscle tremor and/or restlessness* is due to stimulation of β_2-AR in skeletal muscles and is the commonest side effect. It may be more troublesome with elderly patients so is a greater problem in COPD patients. It is often prevented if the patients are instructed to rinse and clean their mouth after inhalation of the drug.

Table 23.3 Side effects of inhaled β_2-agonists

• Muscle tremor and/or restlessness
• Tachycardia and palpitations
• Hypokalaemia
• Hypoxemia
• Paradoxical bronchoconstriction

- *Tachycardia* and *palpitations* are due to reflex cardiac stimulation secondary to peripheral vasodilatation via endothelial β_2-receptors, from direct stimulation of atrial β_2-AR (human heart has a relatively high proportion of β_2-AR) and possibly also from stimulation of myocardial β_1-AR as the doses of β_2-agonist are increased. These side effects tend to disappear with continued use of the drug, reflecting the development of tolerance. There is a dose-related prolongation of the corrected QT interval (QTc) (https://crediblemeds.org/).
- *Metabolic effects* (increase in blood concentration of free fatty acid, insulin, glucose, pyruvate and lactate) are usually seen only after large systemic doses.
- *Hypokalaemia* is a potentially more serious side effect. This is due to β_2-AR stimulation of potassium uptake into skeletal muscle, which may be secondary to a rise in insulin secretion. Hypokalaemia might be serious in the presence of severe hypoxemia, as in asthmatic or COPD exacerbations, when there may be a predisposition to cardiac dysrhythmias. In practice, however significant cardiac arrhythmias after administration of high doses of nebulized β_2-agonist have been reported only exceptionally in asthmatic or COPD exacerbations.
- *Ventilation-perfusion (V/Q) mismatching* can be caused by pulmonary arterial vasodilatation in blood vessels previously constricted by hypoxemia, resulting in the shunting of blood to poorly ventilated areas and a fall in arterial oxygen tension. Although, in practice, the effect of β_2-agonists on P_{AO_2} is usually very small (<5 mm Hg fall), occasionally in severe COPD it is large, although it may be prevented by giving additional inspired oxygen.
- PARADOXICAL BRONCHOCONSTRICTION in response to inhaled β_2 agonists is a rare event, and it is more common in African-Americans [23].

23.3.7 Tolerance

Continuous treatment with an agonist often leads to tolerance (desensitization, subsensitivity), which may be due to downregulation of the β_2-adrenergic receptor signalling pathway. For this reason, there have been many studies of bronchial β_2-AR function after prolonged therapy with β_2-agonists. Tolerance of non-airway β_2-AR-mediated responses, such as tremor and cardiovascular and metabolic responses, is readily induced in normal and asthmatic subjects. Tolerance of human ASM to β_2-agonists *in vitro* has been demonstrated, although the concentration of agonist necessary is high and the degree of desensitization is variable. Animal studies suggest that ASM β_2-AR may be more resistant to desensitization than β_2-AR elsewhere due to a high receptor reserve. In normal subjects, bronchodilator tolerance has been demonstrated in some, but not all, studies after high dose inhaled salbutamol.

Tolerance to the bronchodilator effects of β_2-agonists is not usually reported in asthmatic patients. However, tolerance develops to the bronchoprotective effects of β_2-agonists, and this is more marked with indirect bronchoconstrictors, such as adenosine, allergen and exercise (that activate mast cells), than with direct bronchoconstrictors, such as histamine and methacholine. The reason for the relative resistance of ASM β_2-AR to desensitization remains uncertain, but may reflect the fact that there is a large receptor reserve, so that >90% of β_2-receptors may be lost without any reduction in the relaxation response. The higher level of β_2-AR expression in central ASM cells compared to the peripheral lung may also contribute to the resistance to tolerance since there is likely to be a high rate of β_2-AR synthesis. Another potential mechanism is the low expression of GRK2, which phosphorylates and inactivates occupied β_2-receptors, in ASM [24]. By contrast, there is no receptor reserve in inflammatory cells, and GRK2 expression is high, so that indirect effects of β_2-agonists are more readily lost and tolerance to β_2-agonists rapidly develops.

Tolerance to the bronchodilator effect of LABAs/ULABAs and the bronchoprotective effects of LABAs/ULABAs has been demonstrated, but this is a small loss of protection that does not appear to be progressive and is of doubtful clinical significance. Experimental studies have shown that corticosteroids prevent the development of tolerance in ASM and prevent and reverse the fall in pulmonary β_2-receptor density [25].

23.3.8 Long-Term Safety of Inhaled β_2-Agonists

Because of a possible relationship between adrenergic drug therapy and the rise in asthma deaths in several countries during the early 1960s, doubts were cast on the long-term safety of β_2-agonists. A causal relationship between β_2-agonist use and mortality has never been firmly established, although in retrospective studies, this would not be possible. A particular β_2-agonist, fenoterol, was linked to the rise in asthma deaths in New Zealand in the early 1990s since significantly more of the fatal cases were prescribed fenoterol than the case-matched control patients. This association was strengthened by two subsequent studies, and once fenoterol ceased to be available, the asthma mortality fell dramatically. An epidemiological study examined the links between drugs prescribed for asthma and death or near-death from asthmatic exacerbations, based on computerized records of prescriptions. There was a marked increase in the risk of death with high doses of all inhaled β_2-agonists [26]. The risk was greater with fenoterol, but when the dose is adjusted to the equivalent dose for salbutamol, there is no significant difference in the risk for these two drugs. The link between high β_2-agonist usage and increased asthma mortality does not prove a causal association, since patients with more severe and poorly controlled asthma, and who are therefore more likely to have an increased risk of fatal asthmatic exacerbations, are more likely to be using higher doses of β_2-agonist inhalers and less likely to be using effective anti-inflammatory treatment. Indeed, in the patients who used regular inhaled corticosteroids, there was a significant reduction in risk of death [27]. Regular use of inhaled β_2-agonists has also been linked to increased asthma morbidity. Regular use of fenoterol was associated with worse asthma control and a small increase in AIRWAY HYPERRESPONSIVENESS compared to patients using fenoterol "on demand" for symptom control over a 6-month period [28]. However, this was not found in a study with regular salbutamol [29]. There is some evidence that regular inhaled β_2-agonists may increase allergen-induced asthma and sputum eosinophilia [30]. A possible mechanism is that β_2-agonists activate phospholipase C via coupling through Gq, resulting in augmentation of the bronchoconstrictor responses to cholinergic agonists and other inflammatory mediators [31]. For all these reasons, SABAs should only be used "on demand" for symptom control, and if they are required frequently (more than two times weekly), then an ICS is needed. It has even proposed their as needed use in the treatment of asthmatics in combination with inhaled glucocorticoids only.

The safety of LABAs/ULABAs in asthma has now been more firmly established. A large study of the safety of salmeterol showed an excess of respiratory deaths and near-deaths in patients prescribed salmeterol, but these deaths occurred mainly in African-Americans living in inner cities who were not taking ICSs [32]. Similar data have also raised concerns about formoterol. This may be predictable since LABAs/ULABAs do not treat the underlying chronic inflammation of asthma. However, in large clinical trials, the concomitant treatment with ICSs appears to obviate any risk [33, 34], so *it is recommended that in patients with asthma, LABAs/ULABAs should only be used when ICSs are also prescribed (preferably in the form of a combination inhaler so that the LABAs/ULABAs can never be taken without the ICSs).* There are no safety concerns with LABA/ULABAs use in stable COPD, and no adverse effects were reported in a large study over 3 years in COPD patients and in several other long-term studies, even in patients at increased cardiovascular risk [35, 36].

23.3.9 Future Developments

Inhaled β_2-agonists will continue to be the bronchodilators of choice for asthma for the foreseeable future, as they are effective in all patients and have few or no side effects when used in low doses. It would be difficult to find a bronchodilator that improves on the efficacy and safety of inhaled β_2-agonists. Although some concerns have been expressed about the long-term effects of inhaled β_2-agonists, when used as required for symptom control, SABAs are safe. In patients who are using large doses, their asthma must be assessed and appropriate controller medication used. LABAs/ULABAs are very useful for long-term control in asthma and COPD. In asthmatic patients, LABAs/ULABAs should only be used if the patient is receiving concomitant ICS. There is little advantage to be gained by improving β_2-receptor selectivity, since most of the side effects of β-agonists are due to β_2-receptor stimulation (muscle tremor, tachycardia, hypokalaemia). New ULABAs under clinical development, like ABEDITEROL, may have reduced potential for class-related cardiac side effects [37].

23.4 Inhaled Antimuscarinics

Datura stramonium plants, which contain MUSCARINIC ANTAGONISTS, mainly L-hyoscyamine (also known as daturine or L-atropine), were smoked for asthma relief two centuries ago. Atropine (D-L-hyoscyamine), a related naturally occurring tertiary ammonium alkaloid obtained from the plant *Atropa belladonna*, was also introduced for treating asthma, but these compounds gave side effects, particularly drying of secretions, so less soluble quaternary ammonium compounds, such as atropine methyl nitrate and ipratropium bromide, were developed. These compounds are topically active when given by inhalation, but are highly hydrophilic and are not significantly absorbed from the respiratory or gastrointestinal tracts and do not cross the blood-brain barrier.

The main clinically relevant differences between the various inhaled antimuscarinics in current clinical use are summarized in Fig. 23.3 and Tables 23.1

Datura stramonium

Atropa belladonna

Hyoscyamine

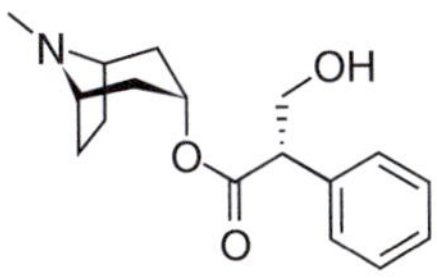

Atropine

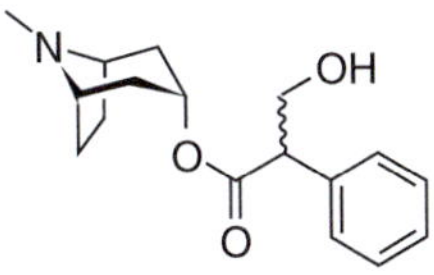

Ipratropium bromide

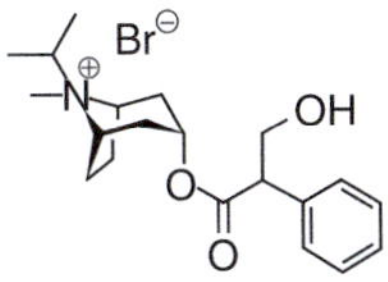

Oxitropium bromide

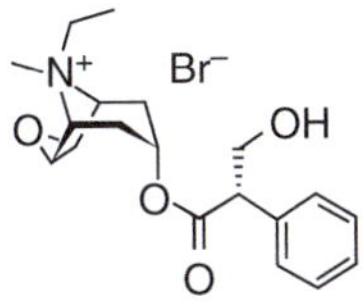

Fig. 23.3 Chemical structure of some inhaled antimuscarinics showing development from plant extract compounds. The colour image of *Datura stramonium* is from the atlas Piccoli F, Pellizzari M, Alessandrini A. Flora del Ferrarese. IBC Emilia-Romagna. 2014, Longo Editore, Ravenna, Italy, p. 314, by kind permission of the authors. The colour image of *Atropa belladonna* is a courtesy of Dr. Alessandro Alessandrini (Regione Emilia-Romagna, Italy)

Fig. 23.3 (continued)

Tiotropium bromide

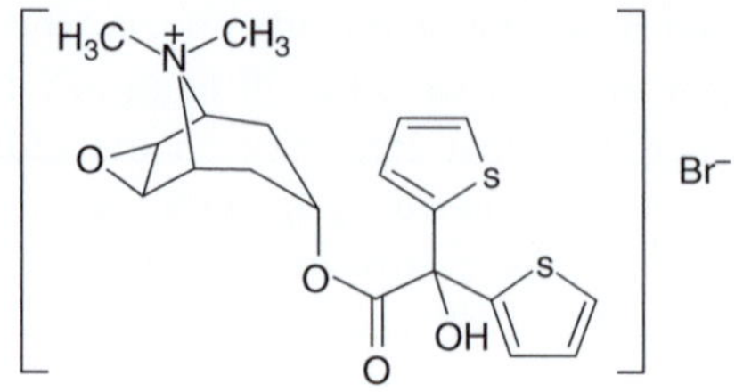

Aclidinium bromide

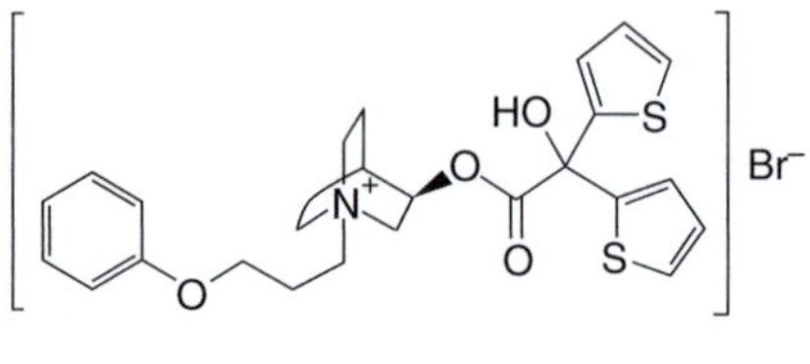

Glycopyrronium bromide

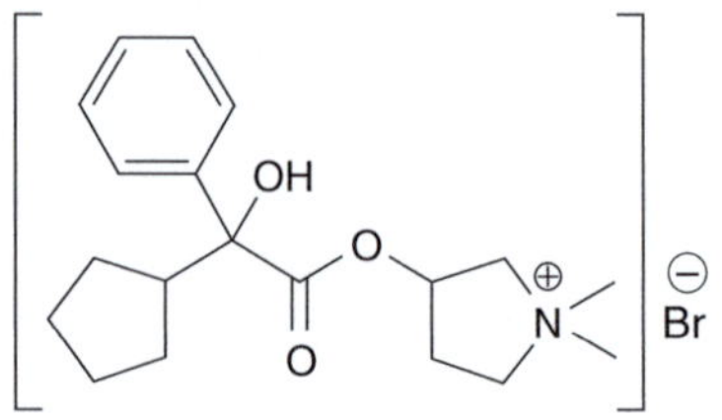

Umeclidinium bromide

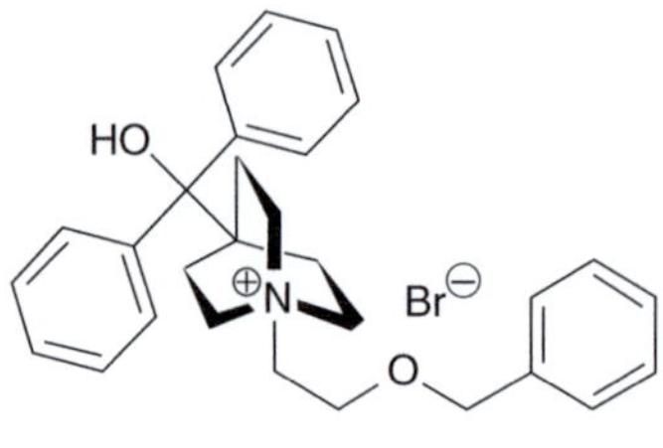

and 23.4. They are all quaternary ammonium compounds, and differences between them relate to their duration of action with SHORT-ACTING ANTIMUSCARINICS (SAMAs) lasting until 6 h and LONG-ACTING ANTIMUSCARINICS (LAMAs) lasting in excess of 12 h. There is DEBATE as to whether some of the current ultra-long acting antimuscarinics (ULAMAs) (such as GLYCOPYRRONIUM) should really be considered to act for 24 h duration.

Ipratropium hydrobromide is the *N*-isopropyl analogue of atropine that has non-selective affinity for all five muscarinic (M) receptors. It is available as a pMDI and nebulized preparation. The onset of bronchodilation is relatively slow and is usually maximal 30–60 min after inhalation but may persist for 4–8 h. It is usually given by MDI three to four times daily on a regular basis, rather than intermittently for symptom relief, in view of its slow onset of action. *Combination inhalers* of ipratropium and salbutamol are popular, particularly in patients with COPD. Several studies have demonstrated additive effects of these two drugs, thus providing an advantage over increasing the dose of SABA in patients who have side effects. *OXITROPIUM BROMIDE* is similar to ipratropium bromide in terms of muscarinic receptors blockade and onset and duration of action. It is also available as a pMDI and nebulized preparation.

TIOTROPIUM BROMIDE is the dithienyl derivative of *N*-methyl scopolamine, a quaternary

Table 23.4 Clinical comparison of inhaled antimuscarinics for the therapy of COPD

	Ipratropium	Oxitropium	Aclidinium	Tiotropium	Umeclidinium	Glycopyrronium
Onset of action (vs placebo)	~30 min	30–60 min	~5 min	~30 min	~5 min	~5 min
Peak bronchodilation	1.2 h	1–2 h	30 min	2–2.5 h	1–2 h	1–2 h
Duration of action	4–6 h	6–8 h	12 h	24 h	24 h	8–12 h
Effect on COPD-related hospitalizations (vs placebo)	None	No	Decrease	Decrease	Decrease	No
Effect on COPD mortality (vs placebo)	No	No studies	No studies	No	No studies	No studies
Anti-inflammatory effect in COPD lower airways (vs placebo)	No	No studies	No studies	No	No studies	No studies

analogue of scopolamine. It is a ULAMA that is suitable for once-daily dosing as a dry powder inhaler or via a soft mist mini-nebulizer inhaler [38]. It binds with equal affinity to all muscarinic receptor subtypes but dissociates very slowly from M_3 and M_1 receptors, giving it a degree of kinetic receptor selectivity for these receptors compared to M_2 receptors, from which it dissociates more rapidly.

ACLIDINIUM BROMIDE has non-selective affinity for all five muscarinic receptors, but with kinetic selectivity for M_3 receptors over M_2 with a rapid onset of action and a shorter duration of action compared with tiotropium bromide, delivered by a breath-actuated multiple-dose DPI has been approved for the regular treatment twice daily of stable COPD patients in many countries [38].

Glycopyrronium bromide or GLYCOPYRROLATE[3-[2-cyclopentyl(hydroxy) phenylacetoxy]-1,1-dimethylpyrrolidinium bromide, previously termed NVA237], has non-selective affinity for all five muscarinic receptors, but the dissociation time at the M_3 receptor is shorter than that of aclidinium bromide, but it still has kinetic selectivity for M_3 over M_2. It has a rapid onset of action; however, it has a shorter duration of action than either aclidinium bromide or tiotropium bromide in vitro. Delivery by a single-dose DPI has been approved for the regular once-daily treatment of stable COPD patients in many countries [38]. It has also been approved in combination with formoterol as a pMDI for the regular treatment twice daily of stable COPD patients in many countries.

UMECLIDINIUM BROMIDE [diphenyl-[1-(2-phenylmethoxyethyl)-1-azoniabicyclo[2.2.2] octan-4-yl]methanol bromide] has non-selective affinity for all five muscarinic receptors but has kinetic selectivity for M_3 over M_2. It has a rapid onset of action and *in vitro* has a duration of action similar to those of tiotropium bromide [39].

Combination of SABAs/LABAs/ULABAs with SAMAs/LAMAs/ULAMAs: Many fixed dose combinations (as MDI and/or DPI and/or nebulized formulations) of SABAs/SAMAs (IPRATROPIUM/ALBUTEROL for both the regular and as needed use), LABAs/LAMAs (ACLIDINIUM/FORMOTEROL, GLYCOPYRRONIUM/FORMOTEROL, both for regular use only), ULABAs/ULAMAs (INDACATEROL/GLYCOPYRRONIUM, UMECLIDINIUM/VILANTEROL, TIOTROPIUM/OLODATEROL, all for regular use only) have been approved for the treatment of COPD patients. Their clinical advantage compared with a single SAMA or LAMA/ULAMA alone is still unclear, and their regular use is suggested only in patients with stable moderate COPD with exercise dyspnoea despite regular treatment with one SAMA or one LAMA/ULAMA alone (goldcopd.org). SABAs/SAMAs combinations are also approved for the prn treatment of asthma [38].

23.4.1 Mode of Action

Inhaled antimuscarinics are antagonists at MUSCARINIC RECEPTORS and in therapeutic use have no other significant pharmacological effects. The predominant innervation of the human lower airways is via the CHOLINERGIC VAGAL PARASYMPATHETIC SYSTEM, but this diminishes peripherally with no motor innervation of small airways and lung parenchyma [40]. Acetylcholine (ACh) is the endogenous neurotransmitter released from cholinergic vagal nerve terminals and has strong bronchoconstrictor activity mediated by the muscarinic M_3 receptors (Fig. 23.4). Since muscarinic receptors are expressed in ASM of small airways which do not appear to be innervated by cholinergic nerves, this might be an important mechanism of cholinergic narrowing in peripheral airways in COPD. In normal human subjects, there is a small degree of resting BRONCHOMOTOR tone which is probably due to tonic VAGAL NERVE impulses which release acetylcholine in the vicinity of ASM since it is blocked by anticholinergics [41] (Fig. 23.5). Cholinergic pathways may play an important role in regulating acute bronchomotor responses in animals, and there are a wide variety of mechanical, chemical and immunological stimuli which elicit reflex bronchoconstriction via vagal pathways. This suggested that cholinergic mechanisms might underlie airway hyperresponsiveness and acute bronchoconstrictor responses in asthma, with the implication that anticholinergic drugs would be effective bronchodilators. While these drugs may afford protection against acute challenge by sulphur dioxide, inert dusts, cold air and emotional factors, they are less effective against antigen challenge, exercise and fog. This is not surprising, as anticholinergic drugs will only

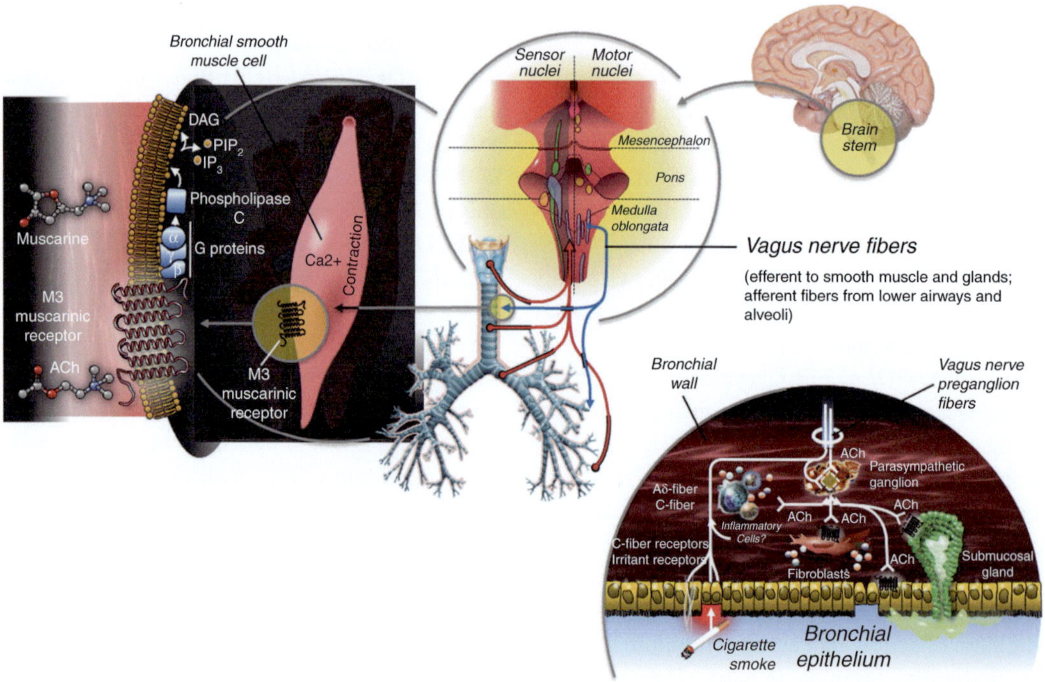

Fig. 23.4 Motor efferent vagal nerve fibres originate in the vagal nuclei in the brainstem and innervate mainly the central lower airways. Their activation causes bronchoconstriction by releasing the neurotransmitter acetylcholine (ACh) that stimulates the M_3 muscarinic (so termed because are also activated by the mushroom toxin muscarine) receptors located on the cell membrane of the bronchial smooth muscle cells. M_3 stimulation by their ligand causes smooth muscle cell contraction by activating the G protein-phospholipase C (PLC) pathway with hydrolysis of polyphosphoinositides [with release of inositol-1,4,5-triphosphate (IP_3)], release of diacylglycerol (DAG), activation of the protein kinase C (PKC) and mobilization of intracellular free calcium (Ca^{2+}). ACh also stimulates M_1 and M_2 muscarinic receptors that are involved in mucus hypersecretion, lower airway remodelling and modulation of the inflammatory response. Sensitive afferent vagal nerve fibres located inside the bronchial walls can be activated by cigarette smoking and by inflammatory mediators released by structural and inflammatory cells and may further amplify the release of ACh from both the vagal nerves and the airway extra-neuronal cholinergic system

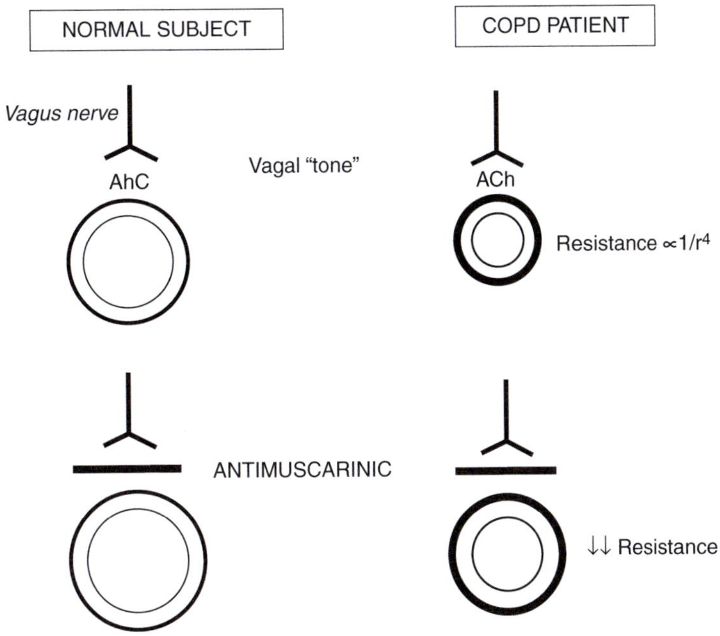

Fig. 23.5 Inhaled antimuscarinics inhibit vagally mediated lower airway bronchomotor tone leading to bronchodilation. This effect is small in the lower airways of normal subjects but is greater in the lower airways of patients with chronic obstructive pulmonary disease (COPD) which are structurally narrowed as airway resistance (*R*) is inversely related to the fourth power of the radius (*r*)

inhibit reflex cholinergic bronchoconstriction and have no blocking effect on the *direct* effects of inflammatory mediators, such as histamine and leukotrienes, on bronchial smooth muscle. Furthermore, cholinergic antagonists probably have little or no effect on mast cells, microvascular leak or the chronic inflammatory response [38, 41].

It is unclear whether the potential anti-inflammatory effects of inhaled antimuscarinics have any clinical relevance. Although mucus secretion is mediated by M_1 and M_3 receptors [41] and, theoretically, inhaled antimuscarinics may reduce airway mucus secretion and reduce lung mucociliary clearance, this is not apparent in clinical studies [38]. Only a few small studies have investigated the effects of inhaled antimuscarinics on cough and cough reflex sensitivity. Methodological differences make interpretation of these studies difficult [38].

ACh is also produced by non-neuronal cells, including bronchial epithelial cells, endothelial cells, smooth muscle, lymphocytes and macrophages in the lower airways. This defines an airway-specific EXTRA-NEURONAL CHOLINERGIC SYSTEM intrinsic that uses ACh as an autocrine/paracrine mediator. This may be activated by cigarette smoking and inflammatory mediators [43, 44]. *In vitro*, the synthesis of ACh in human bronchial epithelial cells is also increased by rhinovirus infection which increases the expression of choline acetyltransferase (CAT), and this could therefore contribute to cholinergic effects during exacerbations of asthma and COPD. This area requires controlled clinical trials.

23.4.2 Clinical Use

Nebulized SAMAs are effective in the treatment of asthmatic exacerbations, although when used in monotherapy, they are less effective than SABAs alone in this situation. Because of its relatively slow onset of action, inhaled ipratropium is not recommended as monotherapy but can be added to SABAs for a greater and longer-lasting bronchodilator effect [45]. In particular, combination inhaled therapy with SABAs/SAMAs is more effective in preventing hospitalization in adults with severe asthma exacerbations who are at increased risk of hospitalization, compared to those with mild-moderate exacerbations, who are at a lower risk to be hospitalized. A single dose of combination therapy and multiple doses both show reductions in the risk of hospitalization among adults with acute asthma. However, adults receiving combination therapy are more likely to experience adverse events, such as tremor, agitation and palpitations, compared to patients receiving SABA alone [45].

Nebulized SAMAs are often used in combination with SABAs for the treatment of COPD exacerbations, but there is no evidence that the degree of bronchodilation achieved with ipratropium bromide is greater than that using a SABAs and also the combination of a SABAs and ipratropium did not appear to increase the effect on FORCED EXPIRATORY VOLUME IN ONE SECOND (FEV_1) more than either used alone [46].

Inhaled antimuscarinics are less effective as bronchodilators than β_2-agonists for the regular treatment of asthmatic patients, and they also offer less efficient protection against various bronchial challenges. However, tiotropium as an add-on therapy slightly improves the control of some asthmatic patients not well controlled on ICS/LABAs(ULABAs) combination therapy, and it is approved for clinical use with this indication [42, 47, 48]. The tiotropium/ICS combination is also useful in African-Americans adults with asthma that do not respond well to ICS/LABAs (ULABAs) combination therapy [49].

In stable COPD, regular treatment with tiotropium is slightly more effective in reducing the number of COPD exacerbations compared with salmeterol [16]. The relatively greater effect of inhaled antimuscarinics in stable COPD patients than in asthma may be explained by an inhibitory effect on vagal tone which is increased in stable COPD patients and in addition is exaggerated by geometric factors in the narrowed airways of COPD patients (Fig. 23.5). By contrast, β_2-agonists acting as functional antagonists reverse all bronchoconstrictor mechanisms, including the bronchoconstrictor effect of multiple inflammatory mediators in asthma.

Tiotropium is now the bronchodilator in monotherapy of choice for the regular treatment of stable COPD patients. It is an effective bronchodilator in patients with stable COPD and is more effective than ipratropium four times daily in reducing the number of hospital admissions for COPD exacerbations without any loss of efficacy over years of treatment period. Tiotropium reduces air trapping and improves exercise tolerance [38]. Over a 4-year period, tiotropium improves lung function and health status and reduces exacerbations and all-cause mortality, although there is no effect on COPD progression [50].

23.4.3 Side Effects

Atropine has systemic side effects that are dose-related and are due to cholinergic antagonism in other systems, which may lead to dryness of the mouth, blurred vision and urinary retention, whereas systemic side effects with all the inhaled antimuscarinics are very uncommon because there is little systemic absorption.

Inhaled antimuscarinics are usually well tolerated, and there is no evidence for any decline in responsiveness with continued use. On stopping inhaled antimuscarinics, a small rebound increase in responsiveness has been described, but the clinical relevance of this is uncertain. The most common local effect to all the inhaled antimuscarinics is dryness of the mouth in 10–15% of patients, but this usually disappears, particularly if the patients clean their mouth with water immediately after the inhalation of the drug, and it is rarely a reason for discontinuing therapy. This dryness is due to an effect on the minor submucosal salivary glands in the mouth. Cholinergic PARASYMPATHETIC AUTONOMIC NERVES stimulate salivary secretion, signalling through muscarinic M_3 receptors on salivary acinar cells, leading to secretion of fluid and salivary proteins [51]. This local side effect of the inhaled antimuscarinics is used therapeutically to decrease the disabling sialorrhoea in advanced Parkinson disease, for example, using oral glycopyrrolate [52].

Inhaled antimuscarinics may also rarely precipitate anisocoria and/or narrow-angle *glaucoma* particularly in elderly patients and when these drugs are nebulized directly on the eye [53]. This may be prevented by nebulization with a mouthpiece rather than a face mask. The unpleasant *bitter taste* of inhaled ipratropium may contribute to poor compliance with long-term treatment with this drug [54].

Reports of *paradoxical bronchoconstriction* with ipratropium bromide as well as with other bronchodilators including salbutamol, particularly when given by nebulizer, were largely explained by the hypotonicity of the nebulizer solution and by antibacterial additives, such as benzalkonium chloride and ethylenediaminetetraacetic acid (EDTA). Nebulizer solutions free of these problems are less likely to cause bronchoconstriction. Occasionally, however, bronchoconstriction may also occur with these drugs given by MDIs or DPI inhalers. It is possible that this is due to blockade of pre-junctional M_2 receptors on airway cholinergic nerves which normally inhibit acetylcholine release [55].

Because cholinergic agonists stimulate *mucus secretion*, there have been several studies of mucus secretion following inhaled antimuscarinics as there has been a concern that they may result in a more viscous mucus. Atropine reduces lung mucociliary clearance in normal subjects and in patients with asthma and COPD, but inhaled antimuscarinics, even in high doses, have no detectable effect in either normal subjects or in patients with asthma or COPD [42].

ACh released from postganglionic efferent cholinergic (parasympathetic) nerves is the main contractile transmitter in the human bladder detrusor muscle where acting on muscarinic receptors (~70% M_1 and ~20% M_3) produces the contraction that empties the bladder [56]. Acute urinary retention is occasionally seen, therefore, after administration of inhaled antimuscarinics. Older patients with benign prostatic hyperplasia seem to be at the highest risk of this adverse effect which tends to occur soon after treatment initiation [57]. Paralytic ileus has been reported on very rare occasions after swallowing inhaled antimuscarinics; the highest risk is likely in bedridden patients with concomitant severe neurological disorders with impairment of the autonomic nerves function and/or severe hypoxemia and/or in the post-operative period.

Tiotropium is excreted predominantly in an unchanged form by the kidneys, and tiotropium plasma concentrations increase as renal impairment worsens [58]; for this reason, *we do not recommend its use in patients with moderate-to-severe renal failure (creatinine clearance <50 ml/min). In these patients aclidinium that has a rapid plasma hydrolysis to inactive metabolites and a very low urinary excretion of unchanged aclidinium may represent a safer alternative.*

23.4.4 Future Developments

Inhaled antimuscarinics are the bronchodilators of choice in COPD. A nebulized formulation of glycopyrrolate (EP-101) is in advanced clinical development for the regular treatment of stable patients with COPD [59]. Few new LAMAs/ULAMAs compounds are now in clinical development, including AZD8683 [38]. Revefenacin, is a once-daily, long-acting muscarinic antagonist that has been recently approved for nebulized therapy of stable COPD patients. Dual-action drugs that are both MUSCARINIC ANTAGONISTS and B_2-AGONISTS (MABAs) are also in early clinical development particularly for the regular treatment of stable COPD patients; the most advanced compound in this class is represented by the BATEFENTEROL [60].

23.5 Theophylline

METHYLXANTHINES, such as theophylline, which are related to caffeine, have been used in the treatment of asthma since 1930, and theophylline is still widely used in developing countries because it is inexpensive. Theophylline became more useful with the availability of rapid plasma assays and the introduction of reliable slow-release preparations. The frequency of side effects and the relative low efficacy of theophylline have recently led to reduced usage in many countries, since inhaled β_2-agonists are far more effective as bronchodilators and ICS have a greater anti-inflammatory effect. However, in patients with severe asthma and advanced stable COPD, it still remains a very useful drug. There is increasing evidence that theophylline has anti-inflammatory or immunomodulatory effects and may enhance the anti-inflammatory effects of CORTICOSTEROIDS ([61], see also Chap. 32).

23.5.1 Chemistry

Theophylline (1,3-dimethyl-xanthine) is a methylxanthine similar in structure to the common dietary xanthines caffeine and theobromine. Several substituted derivatives have been synthesized, but none has any advantage over theophylline, apart from the 3-propyl derivative, ENPROFYLLINE, which is more potent as a bronchodilator and may have fewer toxic effects as it does not antagonize adenosine receptors. Many salts of theophylline have also been marketed, the most common being AMINOPHYLLINE, which is the ethylenediamine salt, used to increase its solubility at neutral pH. Other salts, such as choline theophyllinate, do not have any advantage, and others, such as acepifylline, are virtually inactive, so that theophylline remains the major methylxanthine in clinical use. DOXOFYLLINE (7-[1,3-dioxalan-2-ylmethyl] theophylline) is a methylxanthine with a dioxolane group at position 7 which is available in some countries. It has a similar inhibitory effect on phosphodiesterases to theophylline but is less active as an adenosine antagonist so has a better side effect profile [62]. BAMIFYLLINE (a methylxanthine derivative with two side chains in positions 7 and 8, distinguishing it from theophylline, with non-selective phosphodiesterase inhibitory activity and selective adenosine A1 receptor antagonism) and ACEBROPHYLLINE (ambroxol + theophylline-7-acetate) are other synthetic xanthine derivatives, available in some countries [62].

23.5.2 Pharmacodynamics

The mechanism of action of theophylline is still not completely clarified. In addition to its bronchodilator action, theophylline has many non-bronchodilator cellular effects that may be relevant to its beneficial clinical effects in asthma and COPD patients. Many of these effects are seen only at high concentrations that exceed the therapeutic range, however.

23.5.2.1 Bronchodilator Effects

Theophylline was primarily used as a bronchodilator, and it relaxes large and small human airways in vitro, acting as a functional antagonist by increasing intracellular cAMP concentrations. However, it is a relatively weak bronchodilator at therapeutic concentrations, with little bronchodilator effect at plasma concentrations of less than 10 mg/L. *In vivo* intravenous aminophylline has an acute bronchodilator effect in patients with asthma, which is most likely to be due to a relaxant effect on ASM and has a small protective effect of theophylline on histamine-, methacholine- or exercise-induced bronchoconstriction [61]. Oral theophylline reduces air trapping in patients with stable COPD, indicating an effect on peripheral airways [63].

23.5.2.2 Non-bronchodilator Effects

Theophylline has several anti-inflammatory effects in asthma and COPD, and these may be seen at lower plasma concentrations than are required for its bronchodilator actions [61]. *In vitro* theophylline inhibits mediator release from mast cells and reactive oxygen species from neutrophils, although this is significant only at relatively high concentrations. At high concentrations, theophylline inhibits proliferation of $CD4^+$ and $CD8^+$ lymphocytes and the chemotactic response of T lymphocytes, effects that are mediated through PDE inhibition [64]. Chronic oral treatment with theophylline inhibits the late response to inhaled allergen and a reduced infiltration of eosinophils and $CD4^+$ lymphocytes into the airways after allergen challenge [65]. In patients with mild asthma, low doses of theophylline (mean plasma concentration ~5 mg/L) reduce the numbers of eosinophils in bronchial biopsies, bronchoalveolar lavage (BAL) and induced sputum [66], whereas in severe asthma, withdrawal of theophylline results in increased numbers of activated $CD4^+$ cells and eosinophils in bronchial biopsies [67]. In patients with severe asthma, low-dose theophylline treatment results in an increase in activated blood $CD4^+$ and $CD8^+$ T cells but a decrease in these cells in the lower airways, suggesting that it may reduce the trafficking of activated T cells into the airways [67]. It also reduces BAL neutrophil influx in patients with nocturnal asthma [68]. In patients with stable COPD, theophylline reduces the total number and proportion of neutrophils in induced sputum and the concentration of CXCL8 and neutrophil chemotactic responses, suggesting an anti-inflammatory effect unlike corticosteroids [69].

23.5.2.3 Extrapulmonary Effects

Aminophylline increases diaphragmatic contractility and reverses diaphragm fatigue [70], but this effect has not been observed by all investigators. However, there are doubts about the relevance of these observations to the clinical benefit provided by theophylline [71]. Whether theophylline has any effects on systemic effects or comorbidities in patients with COPD has not yet been established.

23.5.3 Molecular Mechanism of Action

Several molecular mechanisms of action have been proposed for theophylline (Table 23.5).

- *Inhibition of phosphodiesterases.* Phosphodiesterases (PDEs) break down cyclic nucleotides in the cell, thereby leading to a decrease in intracellular cAMP and CYCLIC GUANOSINE 3′5 MONOPHOSPHATE (cGMP) concentrations. Theophylline is a weak non-selective inhibitor of all PDE isoenzymes (Fig. 23.6), but the degree of inhibition is relatively minimal at concentrations of theophylline which are within the "therapeutic range". Theophylline relaxes ASM by inhibition mainly of PDE3 activity, but relatively high concentrations are needed for maximal relaxation, and its inhibitory effect on mediator release from alveolar macrophages is mediated by inhibition of PDE4 activity [61]. Inhibition of PDE should lead to synergistic interaction with β_2-agonists through an increase in cAMP, but this has not been convincingly demonstrated in clinical studies. Inhibition of PDEs accounts for the most frequent side effects of theophylline [61].
- *ADENOSINE RECEPTOR antagonism.* Theophylline antagonizes adenosine A_1 and A_2 receptors at therapeutic concentrations but is less potent at A_3 receptors, suggesting that this could be the basis for its bronchodilator effects. Although adenosine has little effect on normal human ASM *in vitro*, it constricts airways of patients with asthma via the release of histamine and leukotrienes, suggesting that adenosine releases mediators from mast cells of patients with asthma via A_{2B} receptors [61, 72]. Inhaled adenosine causes bronchoconstriction in subjects with asthma via release of histamine from airway mast cells, and this is prevented by therapeutic concentrations of theophylline, although this does not signify that this is important for its antiasthma effect. However, adenosine antagonism is likely to account for the serious side effects of theophylline, such as seizures and cardiac arrhythmias, via blockade of A_1 receptors [72].
- *INTERLEUKIN-10 release.* Interleukin (IL)-10 has a broad spectrum of anti-inflammatory effects, and there is evidence that its secretion is reduced in asthma. IL-10 release is increased by theophylline, and this effect may be mediated via PDE inhibition, although this has not been seen at the low doses that are effective in asthma [73].

Table 23.5 Mechanisms of action of theophylline

Mechanisms
• Phosphodiesterase inhibition (non-selective)
• Adenosine receptor antagonism (A_1, A_{2A}, A_{2B}-receptors)
• Increased interleukin-10 release
• Stimulation of catecholamine (adrenaline) release
• Mediator inhibition (prostaglandins, tumour necrosis factor-α)
• Inhibition of intracellular calcium release
• Inhibition of nuclear factor-κB (↓ nuclear translocation)
• Increased apoptosis
• ↑ Histone deacetylase 2 activity (↑ efficacy of corticosteroids)
• ↓ Excitability of the airway sensory (activation of calcium-activated K^+ channels)

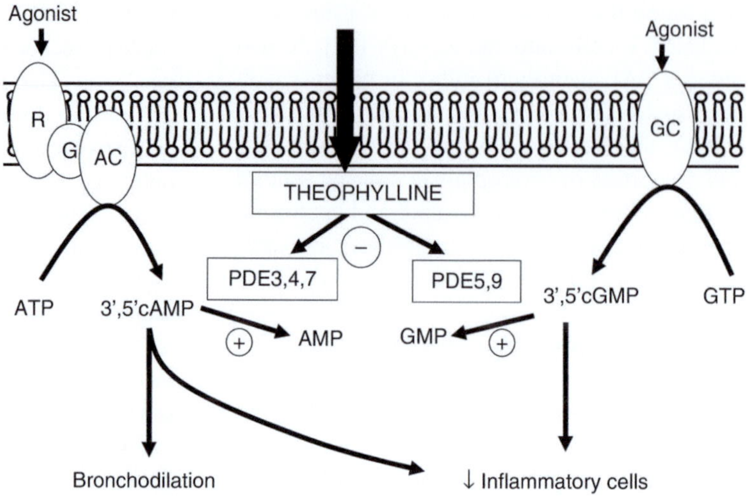

Fig. 23.6 The inhibitory effect of theophylline on phosphodiesterases (PDE) may result in bronchodilation and inhibition of the functional activity many different inflammatory cells. *ATP* adenosine triphosphate, *AMP* adenosine monophosphate, *PKA* protein kinase A, *GTP* guanosine triphosphate, *GMP* guanosine monophosphate, *PKG* protein kinase G

- *Effects on gene transcription*. Theophylline prevents the translocation of the pro-inflammatory TRANSCRIPTION FACTOR NUCLEAR FACTOR-KB (NF-KB) into the nucleus, through preventing the degradation of the inhibitory protein I-κBα, thus potentially reducing the expression of inflammatory genes in asthma and COPD [74]. However, these effects are seen at high concentrations and may be mediated by inhibition of PDE.
- *Effects on cell survival*. Prolonged survival of granulocytes due to a reduction in apoptosis may be important in perpetuating chronic inflammation in asthma (eosinophils) and COPD (neutrophils). Theophylline promotes apoptosis in eosinophils and neutrophils in vitro. This is associated with a reduction in the anti-apoptotic protein *B-cell lymphoma 2* (Bcl-2) [75]. This effect is not mediated via PDE inhibition but in neutrophils may be mediated by antagonism of adenosine A_{2A}-receptors [76]. Theophylline also induces apoptosis of T lymphocytes, thus reducing their survival, and this effect appears to be mediated via PDE inhibition [77]. Theophylline also inhibits, in the airway epithelium, the enzyme POLY(ADP-RIBOSE) POLYMERASE-1 (PARP-1), which is activated by oxidative stress and leads to a reduction in nicotinamide adenine dinucleotide (NAD) levels, resulting in an energy crisis that leads to cell death [78].
- *Histone deacetylase activation*. Recruitment of HISTONE DEACETYLASE 2 (HDAC2) by GR switches off inflammatory genes (see below). Theophylline, in low therapeutic concentrations (~5 mg/L) that do not inhibit phosphodiesterase (PDE)4 activity or antagonize adenosine receptors, enhances HDAC2 activity, especially when its activity is reduced by oxidative stress, and functionally this enhances glucocorticoid effects ([61, 79]; see also Chap. 32). In macrophages obtained from the lower airways of COPD patients, in which HDAC2 activity and expression are markedly reduced, theophylline (10^{-6} M) restores HDAC2 activity to normal and thus reverses the corticosteroid resistance in these cells, an effect that is blocked by a pan-inhibitor of HDAC activity trichostatin A [80]. This effect is due to selective inhibition of PHOSPHOINOSITIDE-3-KINASE (PI3K)δ that is activated by oxidative stress and involved in the inhibition of HDAC2 activity via phosphorylation and is upregulated in peripheral lung tissue of patients with COPD where it is activated by oxidative stress [81–83]. Increased reactive oxygen species and nitric oxide from increased expression of inducible nitric oxide synthase result in the formation of peroxynitrite radicals, which nitrate tyrosine residues in HDAC2, resulting in its inactivation and degradation [84]. Theophylline reduces the formation of

peroxynitrite, thus providing a further mechanism for increasing HDAC2 function in asthma and COPD [85]. In a small clinical study of 30 patients with stable COPD, treatment with low-dose slow-release theophylline plus low-dose inhaled fluticasone significantly reduced sputum neutrophils and improved lung function, and this was associated with an increase in HDAC2 activity in circulating peripheral blood monocytes [86]. Low doses of theophylline increase HDAC activity in bronchial biopsies of stable asthmatic patients and correlate with the reduction in eosinophil numbers in the bronchial mucosa [87].

- *Other effects*. Several other effects of theophylline have been described, including an increase in circulating catecholamines, inhibition of calcium influx into inflammatory cells, inhibition of prostaglandin effects and antagonism of tumour necrosis factor-α (TNF)-α [61]. These effects are generally seen only at high concentrations of theophylline that are above the therapeutic range in asthma and are therefore unlikely to contribute to the anti-inflammatory actions of theophylline.

Theophylline alone does not regulate sputum levels of the gasotransmitter, hydrogen sulphide (H_2S), but the combination of low-dose theophylline plus ICS maintains sputum H_2S levels in patients with stable GOLD II or III grades [88]. Theophylline inhibits capsaicin-induced cough by decreasing the excitability of the airway sensory nerves through activation of small- and intermediate-conductance calcium-activated potassium channels [89].

23.5.4 Pharmacokinetics

There is a close relationship between the acute improvement in airway function and serum theophylline concentrations. Below 10 mg/L bronchodilator, effects are small, and above 25 mg/L, additional benefits are outweighed by side effects, so that the therapeutic range was usually taken as 10–20 mg/L (10–20 μg/μl, 55–110 μM). However, non-bronchodilator effects of theophylline may be seen at plasma concentrations of less than 10 mg/L, so it is preferable to redefine the therapeutic range as 5–15 mg/L [61]. The dose of theophylline required to give these therapeutic concentrations varies between subjects, largely because of differences in clearance of the drug. Theophylline is rapidly and completely absorbed, but there are large interindividual variations in clearance, due to differences in its hepatic metabolism.

Theophylline is metabolized in the liver by the cytochrome P450 microsomal enzymes system, and a large number of factors may influence hepatic metabolism. Theophylline is predominantly metabolized by CYP1A2, whereas at higher plasma concentrations, CYP2E1 is also involved [61]. *Increased* theophylline clearance is seen in cigarette and marijuana smokers due to induction of CYP1A2. Concurrent administration of phenytoin, phenobarbital (phenobarbitone) or rifampicin increases activity of P450 isoenzymes (especially CYP1A2), resulting in increased metabolic breakdown, so that higher doses of theophylline may be required [61].

Reduced theophylline clearance is found in liver disease, pneumonia and heart failure, and doses need to be reduced to half and plasma levels monitored carefully. Decreased clearance is also seen with several drugs, including erythromycin, certain quinolone antibiotics (ciprofloxacin, but not ofloxacin), allopurinol, cimetidine (but not ranitidine), serotonin uptake inhibitors (fluvoxamine), the 5′-LIPOXYGENASE inhibitor ZILEUTON and the LEUKOTRIENE RECEPTOR ANTAGONIST ZAFIRLUKAST, all of which interfere with CYP1A2 function. Thus, if a patient on maintenance theophylline requires a course of erythromycin, the dose of theophylline should be halved. However, there is no such drug interaction with clarithromycin or azithromycin [90].

Because of these potentially wide variations in clearance, individualization of theophylline dosage is required, and plasma concentrations should be measured 4 h after the last dose with slow-release preparations when steady state has been achieved, usually after 2–3 days of regular treatment.

23.5.5 Routes of Administration and Dosing

Intravenous aminophylline has been used for many years in the treatment of severe exacerbations of asthma and COPD but is used much less now as it is less effective than rapid-acting inhaled β_2-agonists alone or in combination with SAMAs. The recommended dose is 6 mg/kg given intravenously slowly over 20–30 min, followed by a maintenance dose of 0.5 mg/kg/h. If the patient is already taking theophylline, or there are any factors which decrease clearance, these doses should be halved and the plasma level checked more frequently.

Oral plain theophylline tablets or elixir is rapidly absorbed but gives wide fluctuations in plasma levels and is not recommended. Several sustained-release preparations of theophylline and aminophylline are absorbed at a constant rate and provide steady plasma concentrations over a 12–24 h period. The recommended doses for bronchodilation are 200–400 mg twice daily, but one-half of these doses may be effective as an anti-inflammatory treatment. Although there are differences between preparations, these are relatively minor. Once optimal doses have been established, plasma concentrations usually remain stable, providing no factors that alter clearance are introduced [61].

Aminophylline may also be given as a suppository, but rectal absorption is unreliable, and proctitis may occur so should be avoided. Inhalation of theophylline is irritant and ineffective [91]. Intramuscular injections of theophylline are very painful and should never be given.

23.5.6 Clinical Use

23.5.6.1 Exacerbations of Asthma and COPD

Intravenous aminophylline was previously widely used in the management of exacerbations of asthma and COPD. However, in patients with asthmatic exacerbations, a systematic review has shown no evidence of benefit when added to nebulized β_2-agonists for any outcome measure, whereas there was an increased risk of side effects [92]. Aminophylline similarly has no place in the management of COPD exacerbations [93].

23.5.6.2 Regular Treatment of Patients with Persistent Asthma or Stable COPD

Currently, theophylline is recommended as an additional bronchodilator if asthma remains difficult to control despite high doses of ICS plus LABAs combinations [94], but it is still uncertain whether it is more effective than an oral anti-leukotriene for this indication [95]. Although LABAs/ULABAs are more effective as an add-on therapy at steps 3 and 4 of the Global Initiative for Asthma (GINA) guidelines, theophylline is considerably less expensive and may be the only affordable add-on treatment when the costs of medication are limiting [ginasthma.org]. Several studies have demonstrated that adding low-dose theophylline to ICSs in patients whose asthma is not controlled gives slightly better asthma control than doubling the dose of ICS [96]. Interestingly, there is a greater degree of improvement in forced vital capacity (FVC) than in FEV_1, suggesting an effect on air trapping and peripheral airways. The improvement in lung function is relatively slow, suggesting an anti-inflammatory rather than a bronchodilator effect of theophylline [61]. Low-dose theophylline is also effective in smoking patients with asthma, who have a poor response to ICSs, and this may be through increasing HDAC2 activity, which is reduced in the lower airways of smoking asthmatics ([97–99]; see also Chap. 32).

Theophylline is still used as a second-line oral bronchodilator in stable COPD, but inhaled anticholinergics and β_2-agonists are preferred in most of the patients [goldcopd.org]. A theoretical advantage of theophylline is that its systemic administration may have effects on small airways, resulting in reduction of hyperinflation and thus a reduction in dyspnoea. Theophylline increases exercise tolerance in patients with stable COPD [100] and reduces air trapping [63] although its effects are small. Theophylline withdrawal in severe stable COPD patients results in worsening of disease in ~50% of the patients [101]. Low-dose theophylline reduces exacerbations in

Table 23.6 Side effects of theophylline

	Mechanism
• Nausea and vomiting	PDE4 inhibition
• Headaches	PDE4 inhibition
• Gastric discomfort	PDE4 inhibition
• Diuresis	A_1-repector antagonism
• Cardiac arrhythmias	A_1-repector antagonism
	PDE3 inhibition
• Epileptic seizures	A_1-repector antagonism

patients with moderate-to-severe stable COPD by ~50% compared to placebo when used as single therapy over 1 year [102]. Low-dose theophylline increases the recovery from acute exacerbations of COPD, and this is associated with reduced inflammation and increased HDAC activity [103].

In patients with moderate stable COPD, low-dose theophylline has a greater anti-inflammatory effect and improvement in FEV_1 when added to an ICS than either drug alone [86]. This suggests that theophylline may be useful in reversing corticosteroid resistance in patients with COPD, and long-term clinical trials are currently underway in patients with COPD to investigate this. Low-dose theophylline added to ICS/LABA combination reduces dyspnoea in stable COPD [104]. The Theophylline With Inhaled Corticosteroids (TWICS) is a large controlled clinical trial that has demonstrated that the addition of "low-dose" theophylline to ICS reduces the number of severe COPD exacerbations causing hospital admission during 1 year of treatment [105]. A small, underpowered, pilot trial has not shown any significant clinical efficacy when low-dose theophylline has been added to the regular ICS/LABA treatment of patients with stable COPD during 1 year of treatment [106].

23.5.7 Side Effects

The main limitation to the use of theophylline in conventional doses has been the relatively high frequency of adverse effects. Unwanted effects of theophylline are usually related to plasma concentration and tend to occur when plasma levels exceed 20 mg/L, although patients develop side effects at low plasma concentrations [61]. Use of low doses of theophylline which give plasma concentrations of 5–10 mg/L largely avoids side effects and drug interactions and makes it unnecessary to monitor plasma concentrations (unless checking for compliance) or if there are concerns about compliance. The plasma theophylline concentrations should also be checked if there are any adverse effects; most frequent side effects are headache, nausea and vomiting, increased acid secretion and gastroesophageal reflux, which may be explained by PDE inhibition. Diuresis may be due to adenosine receptor antagonism. At high concentrations, epileptic seizures and cardiac arrhythmias may occur and may be due to adenosine A_{1A}-receptor antagonism (Table 23.6). Doxofylline appears to have less effect on adenosine receptors so may be safer [62, 107]. Theophylline should never be given with ROFLUMILAST (or other PDE4 inhibitors; see below), as both inhibit PDE4.

23.5.8 Future Developments

Theophylline use has been declining, partly because of the problems with side effects, but mainly because more effective anti-inflammatory therapy (ICSs) has been introduced. Now that the molecular mechanisms for the anti-inflammatory effects of theophylline are better understood, there is a strong scientific rationale for combining low-dose theophylline with ICSs, particularly in patients with more severe asthma and in stable COPD. Oral theophylline is still a very useful treatment in some patients with severe asthma and appears to have anti-inflammatory effects beyond those provided by ICSs. Rapid-release theophylline preparations are cheap and are the only affordable antiasthma medication in some developing countries. In the treatment of stable COPD patients, low-dose theophylline is the first drug to demonstrate clear anti-inflammatory effects, and thus it may even have a role in preventing progression of the disease. At present, there are no fixed combination therapies available. In the future, a combination of low-dose theophylline and an oral corticosteroid might be useful in COPD, particularly in less affluent countries. The results of the ongoing TASC (theophylline and steroids in COPD) 5-year study are awaited with interest. Doxofylline may be more effective and safer than theophylline, but we need large controlled clinical trials of comparison in both asthma and COPD patients [62, 107, 108].

23.6 Novel Classes of Bronchodilators

Several new classes of bronchodilator are under development, but it is difficult to envisage a more effective bronchodilator than inhaled β_2 agonists for asthma and inhaled antimuscarinics for COPD. It has been difficult to find new classes of bronchodilators, and several new potential drugs have often had problems with vasodilator side effects since they relax vascular smooth muscle to a greater extent than ASM.

23.6.1 Magnesium Sulphate

There is increasing evidence that MAGNESIUM SULPHATE ($MGSO_4$) is useful as an additional bronchodilator in patients with severe asthmatic exacerbations. In the emergency department, a single infusion of 1.2 or 2 g IV $MgSO_4$ over 15–30 min slightly reduces hospital admissions and improves lung function in adults with asthmatic exacerbations who have not responded sufficiently to oxygen, nebulized SABA and IV corticosteroids, whereas inhaled magnesium sulphate is not effective [109, 110]. The treatment is cheap and well tolerated. Side effects include flushing and nausea but are usually minor. It appears to act as a bronchodilator and may reduce cytosolic Ca^{2+} concentrations in ASM cells. The concentration of magnesium is lower in serum and erythrocytes in asthmatic patients compared to normal controls and correlates with airway hyperresponsiveness [111], although the improvement in severe asthmatic exacerbations after magnesium does not correlate with plasma concentrations. More studies are needed in patients not responding to standard maximal treatment, including inhaled β_2-agonists and ipratropium bromide and systemic corticosteroids, before intravenous magnesium sulphate is routinely recommended for the management of severe asthmatic exacerbations. There are too few well-designed controlled studies investigating the role of magnesium in COPD exacerbations to make any firm recommendation [112].

23.6.2 K^+ Channel Modulators

Potassium (K^+) channels are involved in recovery of excitable cells after depolarization and therefore are important in stabilization of cells. In ASM cells large-conductance Ca^{2+}-activated K^+ (KCa) channels are activated by β2-agonists, via Gs, and suppressed by antimuscarinics via Gi. ATP-dependent K^+ (KATP) channel openers, big-conductance K^+ (BKCa or Maxi-K) channel openers and intermediate-conductance K^+ (IKCa) channel blockers may be the most effective agents for treating asthma and COPD [113]. Modulation of potassium channels by these agents may produce beneficial effects such as bronchodilation, a reduction in airway hyperresponsiveness, a reduction in cough and mucus production and an inhibition in airway inflammation and remodelling [113]. However, clinical studies in both asthma and COPD have been disappointing. The cardiovascular side effects of these drugs (postural systemic arterial hypotension, skin flushing) limit the oral dose. Inhaled potassium channel modulators also have side effects.

23.6.3 Natriuretic Peptides

ATRIAL NATRIURETIC PEPTIDE (ANP) and URODILATIN, the renal form of natriuretic peptide type A, activate the enzyme GUANYLYL CYCLASE and increase cyclic guanosine 3′5 monophosphate (cGMP), leading to bronchodilation. ANP and urodilatin are bronchodilators in stable asthma and give comparable effects to SABAs [114, 115]. Since they work via a different mechanism from β_2-agonists, they may give additional bronchodilation that may be useful in severe asthmatic exacerbations. BRAIN

NATRIURETIC PEPTIDE (BNP) also activates the NATRIURETIC PEPTIDE RECEPTOR A (NPR-A) and is a bronchorelaxant in human airways in vitro through an autocrine loop elicited by the activation of NPR-A, localized on bronchial epithelium with release of ACh that stimulates via M2 receptors on the same cells the production of nitric oxide that has a relaxant effect on the surrounding ASM cells, which does not express NPR-A [116].

PL-3994(Hept-cyclo(Cys-His-Phe-d-Ala-Gly-Arg-d-Nle-Asp-Arg-Ile-Ser-Cys)-Tyr-[Arg mimetic]-NH(2)) is a novel natriuretic peptide receptor-A (NPR-A) agonist, resistant to degradation by human NEUTRAL ENDOPEPTIDASE (92% remaining after 2 h), whereas the natural ligands, ANP and C-TYPE NATRIURETIC PEPTIDE (CNP), are rapidly metabolized ($\leq$1% remaining after 2 h). PL-3994 has longer clinical bronchodilator activity than observed previously with ANP and in the future may be useful in the treatment of asthmatic exacerbations [117].

23.6.4 VIP Analogues

VASOACTIVE INTESTINAL POLYPEPTIDE (VIP) is a potent bronchodilator of human airways *in vitro*, but is not effective in the treatment of asthmatic patients as it is rapidly metabolized and also causes vasodilator side effects. More stable analogues of VIP, such as Ro 25-1533, which selectively stimulates VIP receptors (VPAC2) in ASM, have been synthesized. Inhaled Ro 25-1533 has a rapid bronchodilator effect in asthmatic patients, but this is not as prolonged as with formoterol [118]. Novel more stable long-acting inhaled VIP analogues, in combination with appropriate drug delivery systems, are under preclinical development.

23.6.5 Bitter Taste Receptor Agonists

Human ASM cells express six BITTER TASTE RECEPTOR (TAS2R) subtypes, previously thought only to exist in taste buds of the tongue. Agonists acting at TAS2Rs evoke profound bronchodilation via a Ca^{2+}-dependent mechanism. TAS2Rs function is not altered in animal models of asthma models. TAS2R agonists relax bronchi even under extreme desensitization of bronchorelaxation by β-agonists, with minimal tachyphylaxis upon repetitive dosing [119]. This suggests that TAS2Rs agonists may be developed in the future for the treatment of asthma and COPD.

23.6.6 Airway Smooth Muscle Contraction Inhibitors

Drugs that inhibit the contractile machinery in ASM, such as inhibitors of the myosin light chain kinase and direct smooth muscle myosin inhibitors, are also in development. As these agents also cause vasodilatation, it will be necessary to administer them by inhalation. RHO-ASSOCIATED KINASE (ROCK)1 and ROCK2 are key regulators of actin cytoskeleton dynamics downstream of Rho GTPases that participate in the control of important physiologic functions including cell contraction and inflammation. In animal models, both ROCK1 and ROCK2 are independently required for allergen-induced airway hyperresponsiveness, likely through their effects in ASM cell contraction and mast cell degranulation [120]. Many ROCK inhibitors, such as FASUDIL, are under development [121]. In a small clinical trial, fasudil slightly reduced the pulmonary arterial hypertension associated with COPD [122].

23.7 Cromones (See Also Chap. C3)

Sodium CROMOGLYCATE is a derivative of khellin, an Egyptian herbal remedy, and was found to protect against allergen challenge without any bronchodilator effect. A structurally related drug, NEDOCROMIL SODIUM, which has a similar pharmacological profile to cromoglycate, was subsequently developed. Although cromoglycate was popular in the past because of its good safety profile, its use has sharply declined with the more widespread use of the more effective ICS.

23.7.1 Mode of Action

Initial investigations indicated that cromoglycate inhibited the release of mast cells mediators by allergen following stabilization of the mast cell membrane and thus cromoglycate was classified as a mast cell stabilizer. However, cromoglycate has a rather low potency in stabilizing human lung mast cells, and other drugs which are more potent in this respect have little or no effect in clinical asthma. This has raised doubts about mast cell stabilization as the mode of action of cromoglycate.

Cromoglycate and nedocromil potently inhibit bronchoconstriction induced by sulphur dioxide, metabisulphite and bradykinin, which are believed to act through activation of SENSORY NERVES in the airways. In dogs, cromones suppress firing of unmyelinated C-FIBRE NERVE endings, reinforcing the view that they might be acting to suppress sensory nerve activation and thus NEUROGENIC INFLAMMATION. Cromones have variable inhibitory actions on other inflammatory cells which may participate in allergic inflammation, including macrophages and eosinophils. *In vivo* cromoglycate is capable of blocking the early response to allergen (which may be mediated by mast cells) but also the late response and airway hyperresponsiveness, which are more likely to be mediated by other inflammatory cells. There is no convincing evidence that cromones reduce inflammation in asthmatic airways. The molecular mechanism of action of cromones is not understood, but some evidence suggests that they may block a particular type of chloride channel that may be expressed in sensory nerves, mast cells and other inflammatory cells [123]. Their effects are mimicked in clinical challenge studies by furosemide, which also has an effect on chloride channel function (in addition to its diuretic action).

23.7.2 Current Clinical Use

Cromoglycate is a prophylactic treatment and needs to be given regularly. Cromoglycate protects against various indirect bronchoconstrictor stimuli, such as exercise. It is only effective in mild persistent asthma, but is not effective in all patients, and there is no way to predict which patients will respond. Drug delivery aspects, such as hygroscopicity and the poor performance of delivery systems, may also have been important determinants of therapeutic failures [124]. In adults with asthma, ICS are preferable to cromones, as they are effective in all patients. Cromoglycate has a very short duration of action and has to be given four times daily to provide good protection, which makes it much less useful than ICS which may be given once or twice daily. It may also be taken prior to exercise in exercise-induced asthma that is not blocked by an inhaled β_2-agonist [125]. In clinical practice, nedocromil has a similar efficacy to cromoglycate, but has the disadvantage of an unpleasant taste. The introduction of oral anti-leukotrienes has further eroded the market for cromones as these drugs are of comparable or greater clinical efficacy and are taken orally.

23.7.3 Side Effects

Cromones are among the safest drugs available, and side effects are extremely rare. The dry powder inhaler may cause throat irritation, coughing and, occasionally, wheezing, but this is usually prevented by prior administration of a SABA. Very rarely a transient skin rash and urticaria are seen, and a few cases of pulmonary eosinophilia have been reported, all of which are due to hypersensitivity reactions.

23.8 Phosphodiesterase Inhibitors

Phosphodiesterases (PDEs) break down cyclic nucleotides that inhibit cell activation, and at least 11 families of enzymes have been characterized. Theophylline, long used as an asthma treatment, is a weak but non-selective PDEs inhibitor, as well as doxofylline and bamifylline. PDE4 degrades cAMP and is the predominant family of PDEs in inflammatory cells, including mast cells, eosinophils, neutrophils, T lymphocytes, macrophages and structural cells such as sensory nerves and epithelial cells of the lower airways [126]. This has suggested that PDE4 inhibitors would be useful as an anti-inflammatory treatment in both asthma and COPD (Fig. 23.7). In animal models of asthma, PDE4 inhibitors reduce eosinophil infiltration and AHR responses to allergen, and they are also effective in reducing cigarette smoke-induced inflammation and emphysema in animal models of COPD. Several PDE4 inhibitors, such as CILOMILAST, have been tested clinically, but with disappointing results, as the dose has been limited by side effects, particularly nausea, vomiting, headache and diarrhoea. In patients with severe stable COPD, the addition to an ICS/LABA of the oral PDE4 inhibitor roflumilast has some effect in reducing exacerbations and improving lung function, and this drug has been approved for the market, but its use is limited by the common side effects. For every 100 COPD patients treated with PDE_4 inhibitors, 7 more suffered from diarrhoea during the study period compared with placebo [number needed to treat to harm (NNTH) = 15]. Roflumilast is also associated with weight loss and an increase in insomnia and depressive mood symptoms. Longer-term trials are needed to determine whether or not PDE_4 inhibitors modify FEV_1 decline, hospitalization or mortality in COPD [127–129].

The potential role of roflumilast as another add-on therapy in the treatment of severe asthma is still under investigation [129, 130]. There are four subfamilies of PDE4, and it now seems that PDE4D is the major enzyme mediating vomiting, whereas PDE4B is important for anti-inflammatory effects. This suggests that selective PDE4B inhibitors might have a greater therapeutic ratio. Another approach has been to develop inhaled PDE4 inhibitors, but this has so far proved to be ineffective [126]. PDE3 degrades both cAMP and cGMP. It is expressed on ASM cells and acts as a bronchoconstrictor. Combined PDE3/PDE4 inhibition is often additive and synergistic. Dual PDE3/PDE4 inhibitors, such as RPL554, are under clinical development for the treatment of asthma and COPD [131]. PDE5 promotes pulmonary arterial vasoconstriction, and vessel wall hypertrophy and PDE5 inhibitors (such as SILDENAFIL and TADALAFIL) are vasodilators that are used in the treatment of idiopathic pulmonary arterial hypertension (IPAH) but have not been demonstrated to be particularly effective in the treatment of patients with COPD-associated PAH [132, 133].

23.9 Mediator Antagonists

Many inflammatory mediators have been implicated in asthma and COPD (see also Chaps. 6 and 11), suggesting that inhibition of synthesis or receptors of these mediators may be beneficial

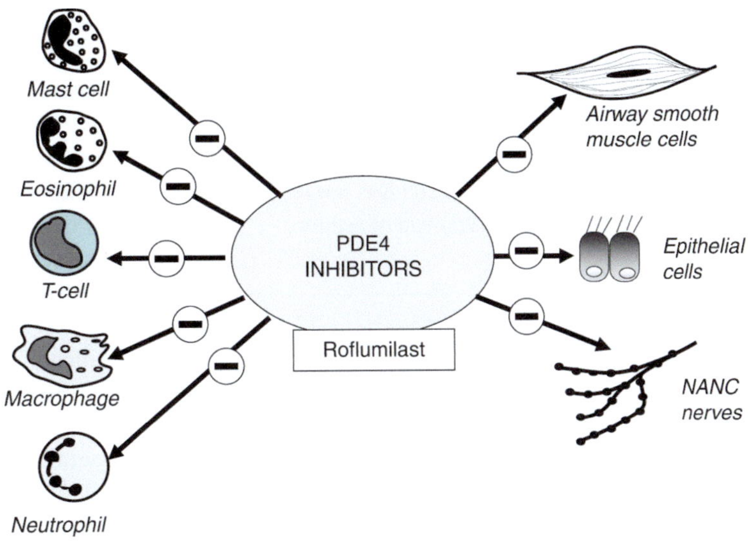

Fig. 23.7 Inhibitors of phosphodiesterase (PDE)-4 may be useful anti-inflammatory treatments in COPD and asthma as they inhibit the functional activity of several inflammatory and structural cells of the lower airways involved in the pathogenesis of these diseases

[1, 2, 134, 135]. However, because many different mediators have often similar effects, specific inhibitors have usually been disappointing in both asthma and COPD treatment.

23.10 Antihistamines

HISTAMINE mimics many of the features of asthma and is released from mast cells in acute asthmatic responses, suggesting that antihistamines may be useful in asthma therapy (see also Chap. 9). Many trials of HISTAMINE H_1-RECEPTOR ANTAGONISTS have been conducted, but there is little evidence of any useful clinical benefit, as demonstrated by a meta-analysis [136]. Newer antihistamines, including cetirizine, loratadine and azelastine, have some beneficial effects, but this may be unrelated to their H_1-receptor antagonism. In addition, histamine H_1-receptor antagonists are effective in controlling rhinitis, and this may help to improve overall asthma control [137] (see also Chap. 22). Antihistamines are not recommended in the routine management of asthma.

More recently, the use of selective antagonists in animal models of asthma has shown a potential role for the histamine H4 receptor in mediating lung function and inflammation [138]. TOREFORANT (JNJ 39758979) has entered clinical trials for other chronic inflammatory diseases; however, its further development was halted during phase II because of the observation of drug-induced agranulocytosis in a few subjects. KETOTIFEN is described as a prophylactic antiasthma compound. Its predominant effect is H_1-receptor antagonism, and it is this antihistaminic effect that accounts for its sedative effect. Ketotifen has little or no effect in placebo-controlled trials in clinical asthma [139], and it is not recommended in the routine management of asthma.

23.11 Anti-leukotrienes

CYSTEINYL-LEUKOTRIENES (Cys-LTs) are produced in asthma and that they have potent effects on airway function, inducing bronchoconstriction, airway hyperresponsiveness, plasma exudation and mucus secretion, and possibly on eosinophilic inflammation (Fig. 23.8) [140]. This suggested that blocking the leukotriene pathways with LEUKOTRIENE MODIFIERS may be useful in the treatment of asthma, leading to the development of 5′-lipoxygenase [5-LO (ALOX5)] enzyme inhibitors (of which zileuton is the only drug marketed) and several antagonists of the cys-LT_1-receptor, including MONTELUKAST, ZAFIRLUKAST and PRANLUKAST (see also Chap. 22).

23.11.1 Clinical Studies

LEUKOTRIENE RECEPTOR ANTAGONISTS inhibit the bronchoconstrictor effect of inhaled LTD_4 in normal and asthmatic volunteers. They also inhibit bronchoconstriction induced by a variety of challenges, including allergen, exercise and cold air, with approximately 50% inhibition. With aspirin challenge, in aspirin-sensitive asthmatic patients, there is almost complete inhibition of the response [141]. Similar results have been obtained with the 5-LO inhibitor zileuton. This suggests that there may be no additional advantage in blocking LTB_4 in addition to cysteinyl-LTs. These drugs are active by oral administration, and this may confer an important advantage in chronic treatment, particularly in relation to compliance.

Leukotriene receptor antagonists (LTRAs) have been intensively investigated in clinical studies, whereas there are less studies with 5-LO inhibitors. In patients with mild-to-moderate persistent asthma, there is a significant improvement in lung function and asthma symptoms, with a reduction in the use of rescue inhaled β_2-agonists. In several studies there is evidence for a bronchodilator effect, with an improvement in baseline lung function, suggesting that leukotrienes are contributing to the baseline bronchoconstriction in asthma, although this varies between patients [142]. However, anti-leukotrienes are considerably less effective than ICS in the treatment of mild persistent asthma and are not the treatment of first choice as monotherapy but are an alternative when patients are not compliant with inhaled drugs [143]. Anti-leukotrienes

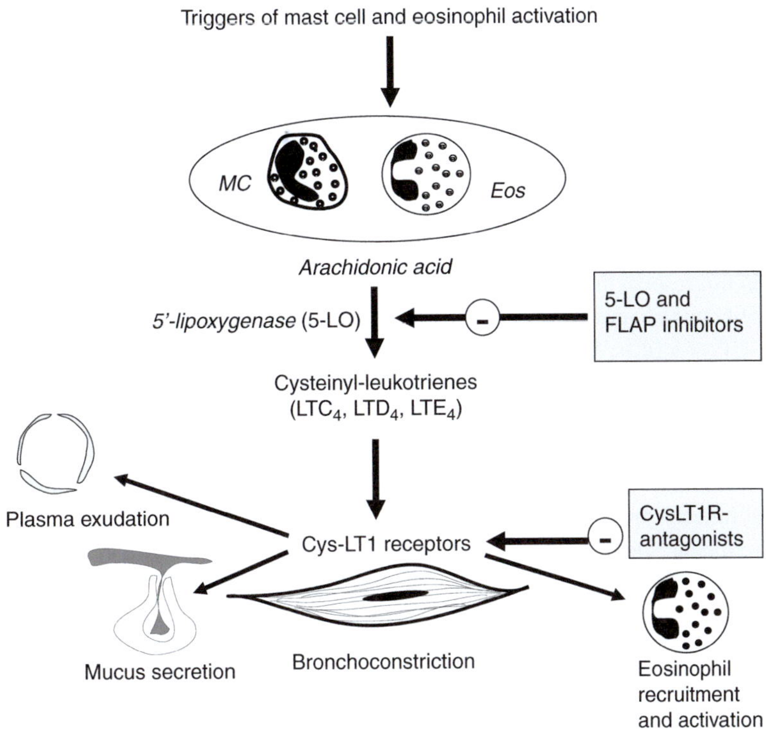

Fig. 23.8 Effects of cysteinyl-leukotrienes on the lower airways and their inhibition by anti-leukotrienes. *PAF* platelet-activating factor, *LT* leukotriene; *5-LO* 5′-lipoxygenase; *FLAP* 5′-lipoxygenase activating protein

are also indicated more as an add-on therapy in patients not well controlled on ICS. In adults with persistent asthma that is inadequately controlled by low-dose ICS with significant bronchodilator reversibility, the addition of LABA to ICS is modestly superior to the addition of LTRA in improving lung function and reducing oral corticosteroid-treated exacerbations and, to a lesser extent, in rescue medication use, symptoms and quality of life [143]. Anti-leukotrienes are also indicated as an add-on therapy (as an alternative or in adjunct to tiotropium) in patients not well controlled on ICS/LABAs (or ULABAs) combinations [142].

ASPIRIN-EXACERBATED RESPIRATORY DISEASE (AERD) is characterized by asthma, often associated with chronic rhinosinusitis with nasal polyposis, and pathognomonic respiratory reactions to aspirin (SAMTER'S TRIAD OR FERNAND WIDAL-LERMOYEZ SYNDROME). It has been estimated that this syndrome affects 7% of adults with asthma and 14% of those who have severe asthma [144]. A spectrum of abnormalities in inducible cyclooxygenase (COX)-2-dependent prostaglandin (PG)E_2 production and E prostanoid 2 receptor expression, potentially due to subtle genetic or epigenetic functional perturbations, could compromise homeostasis to permit exaggerated cysteinyl-leukotriene production, with activation of platelets and mast cells in patients with aspirin sensitivity [144]. Theoretically anti-leukotrienes should be of particular value in these patients as they block the airway response to aspirin challenge, but although ANTI-LEUKOTRIENES have some benefit in these patients, it is not any greater than in other types of asthma [141]. Anti-leukotrienes are also effective in preventing exercise-induced bronchoconstriction and are similar in efficacy to the LABAs/ULABAs in this respect [145]. Anti-leukotrienes also have a weak effect in allergic persistent rhinitis that may be additive to the effects of histamine 1 receptor antagonists.

Montelukast has some efficacy in the treatment of asthmatic exacerbations, but it is inferior to SABAs, and there is no benefit from the addition of oral montelukast over conventional treatment in the management of asthmatic exacerbations [146].

Studies have demonstrated weak anti-inflammatory effects of anti-leukotrienes in reducing eosinophils in sputum or in biopsies [14], but this is much less marked than with an ICS, and there is no additional anti-inflammatory effect when added to an ICS [147]. Anti-leukotrienes, therefore, appear to act mainly as anti-bronchoconstrictor drugs, although they are clearly less effective in this respect than β_2-agonists, as they antagonize only one of several bronchoconstrictor mediators.

Cys-LTs have no role in the pathogenesis of COPD, and cys-LT1 receptor antagonists have no role in therapy. By contrast, LTB_4, a potent neutrophil chemoattractant, is elevated in COPD indicating that 5-LO inhibitors that inhibit LTB4 synthesis may have some potential benefit by reducing neutrophil inflammation. However, small clinical trials did not indicate any clear benefit of 5-LO inhibitors in COPD patients either during the stable phase or during exacerbations of the disease [148, 149].

23.11.2 Side Effects

Anti-leukotrienes are usually well tolerated. Zileuton can cause an increased level of liver enzymes, so that monitoring of liver enzymes is necessary with this drug and high doses of zafirlukast may be associated with abnormal liver function. Montelukast is usually well tolerated, with no significant adverse effects. The lack of side effects implies that leukotrienes do not appear to be essential for any physiological functions.

Several cases of eosinophilic granulomatosis with polyangiitis (EGPA or, as it was traditionally termed, Churg-Strauss syndrome) have been associated with the use of leukotriene receptor antagonists. EGPA is a rare systemic vasculitis that may also affect the heart, peripheral nerves and kidney and is associated with blood eosinophilia, asthma, chronic rhinosinusitis and increased serum total levels of immunoglobulin E and in a fraction of patients with autoantibodies anti-myeloperoxidase [a perinuclear anti-neutrophil cytoplasmic antibody (ANCAs)]. It is uncertain whether the cases so far reported are due to a reduction in oral or ICS dose unmasking an underlying undiagnosed EGPA, rather than as a direct effect of the drug, although cases of EGPA have been described in patients on anti-leukotrienes who were not on concomitant corticosteroids therapy, suggesting that there is a causal link [150–152]. Few case reports have also been published of montelukast-associated agitation, anxiety, depression, sleep disturbance, hallucinations, suicidal thinking and suicidality, tremor, dizziness, drowsiness, neuropathies and seizures [152].

23.11.3 Future Developments

One of the major advantages of anti-leukotrienes is that they are effective by oral administration. This may increase compliance with chronic therapy. Montelukast is effective as a once-daily preparation (10 mg in adults) and is therefore easy for patients to use. In addition, oral administration may treat concomitant allergic rhinitis. However, the currently available clinical studies indicate a relatively modest effect on lung function and symptom control, which is less for every clinical parameter measured than with ICS. This is not surprising, as there are many more mediators than cys-LTs involved in the pathophysiology of asthma, and it is unlikely that antagonism of a single mediator could ever be as effective as corticosteroids, which inhibit all aspects of the inflammatory process in asthma. Similarly, if anti-leukotrienes are functioning in asthma as bronchodilators and anti-constrictors, it is unlikely that they will be as effective as a β_2-agonist, which will counteract bronchoconstriction, irrespective of the bronchoconstrictor stimulus. It is likely that anti-leukotrienes will be used less in the future as once-daily combination inhalers are now available as the mainstay of asthma therapy.

An interesting feature of the clinical studies with anti-leukotrienes is that some asthmatic patients appear to show better responses than others, suggesting that leukotrienes may play a more important role in some patients. The variability in response to anti-leukotrienes may reflect differences in production of or responses to leukotrienes in different patients, and this in turn may be related to genetic polymorphisms of 5-LO, LTC_4 synthase or cys-LT_1-receptors that are involved in the synthesis of leukotrienes [146, 153].

It is unlikely that further advances can be made in cys-LT_1-receptor antagonists as montelukast is a once-daily drug that probably gives maximal receptor blockade. Cys-LT_2-receptors may be important in vascular and ASM proliferative effects of cys-LT and are not inhibited by current cys-LT_1-receptor antagonists [154]. The role of cys-LT_2-receptors in asthma is unknown so the value of also blocking cys-LT_2 is uncertain, and no specific antagonist is available so far. 5-LO inhibitors may have some potential in severe neutrophilic persistent asthma as LTB_4 may contribute to neutrophil chemoattraction in the lungs. GSK2190915, an ALOX5-ACTIVATING PROTEIN [ALOX5AP (FLAP)] inhibitor, inhibits the production of cysteinyl-leukotrienes and LEUKOTRIENE B4 and 5-oxo-6,8,11,14-eicosatetraenoic acid and it is in early clinical development for asthma [155].

23.12 Anti-IgE

Increased specific immunoglobulin (Ig)E is a fundamental feature of allergic asthma (see also Chaps. 4, 11 and 14). OMALIZUMAB is a humanized monoclonal antibody (MAb) that blocks the binding of IgE to high-affinity IgE receptors (FcεR1) on mast cells and thus prevents their activation by allergens (Fig. 23.9) [156]. It also blocks binding of IgE to low-affinity IgE receptors (FcεRII, CD23) on other inflammatory cells, including T and B lymphocytes, macrophages and possibly eosinophils to inhibit chronic inflammation. It also results in a reduction of serum IgE levels. The antibody has a high affinity and blocks IgE receptors by over 99% which is necessary because of the amplification of these receptors. Interestingly, FcεR1 expression on blood plasmacytoid dendritic cells correlates with COPD grading of severity [157].

23.12.1 Clinical Use

Omalizumab is used for the treatment of patients with severe asthma. The antibody is administered by subcutaneous injection every 2–4 weeks, and the dose is determined by the titre of venous blood IgE. Omalizumab reduces the requirement for ICS and reduces asthma exacerbations and hospitalizations in adults. It is also beneficial in treating a concomitant severe allergic rhinitis. Because of its very high cost, this treatment should probably be used only in patients with very severe asthma who are poorly controlled by ICS/LABAs/ULABAs in combination and inhaled antimuscarinics and montelukast and low-dose theophylline. However, it remains to be tested prospectively whether the addition of omalizumab has a systemic glucocorticoid-sparing effect. It is also not clear whether there is a threshold level of baseline serum IgE for optimum efficacy of omalizumab. Given the high cost of the drug, identification of biomarkers predictive of response is of major importance for future research [158]. A pilot clinical trial of omalizumab in patients with stable COPD and elevated serum IgE levels

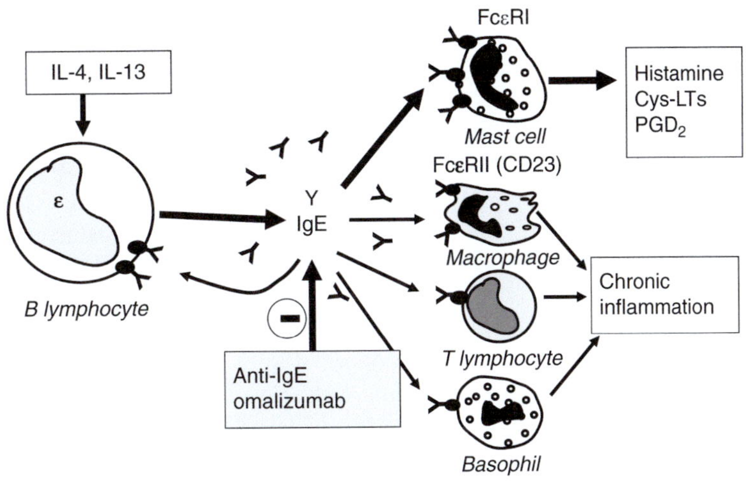

Fig. 23.9 IgE plays a central role in the pathogenesis of allergic diseases, and blocking IgE using an antibody, such as omalizumab, is a logical approach. IgE may activate high-affinity receptors (FcεRI) on mast cells as well as low-affinity receptors (FcεRII, CD23) on other inflammatory cells. *IL* interleukin, *cys-LT* cysteinyl-leukotriene, *PG* prostaglandin

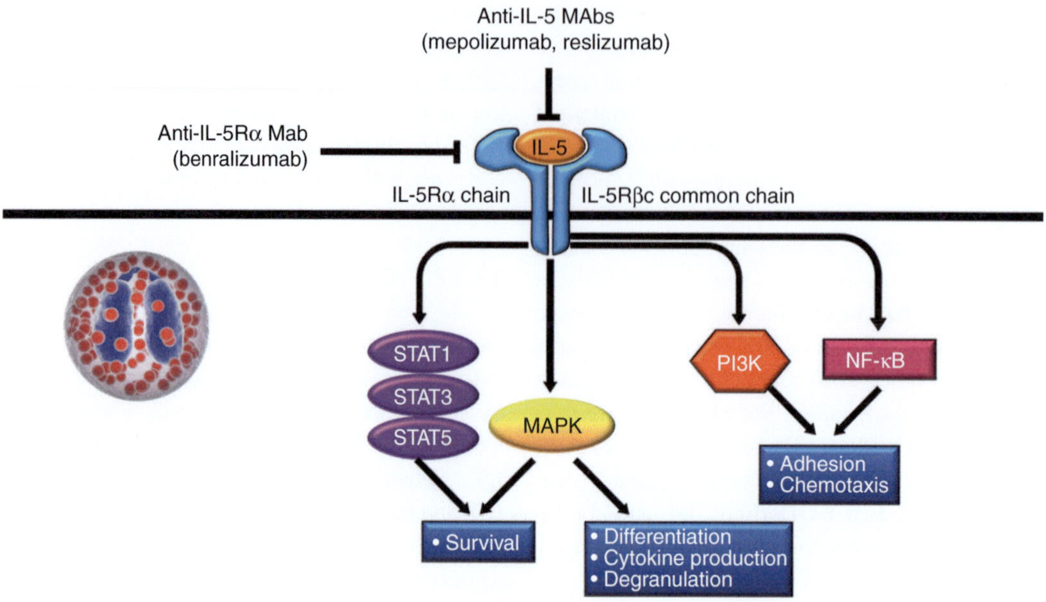

Fig. 23.10 Monoclonal antibodies for clinical use inhibiting the IL-5-mediated signalling in the eosinophils and other cells

was withdrawn due to lack of patients meeting inclusion criteria [NCT00851370].

23.12.2 Side Effects

Omalizumab is usually well tolerated, sometimes causing injection site local reactions, whereas systemic anaphylactic responses are very uncommon (<0.1%). The majority of anaphylactic reactions to omalizumab are observed during the first two (69%) or first three (72%) doses. The reaction occurred within 60 min of omalizumab administration in the majority of cases (64%); median time to anaphylactic reaction is 60 min. Omalizumab was prescribed for asthma in most (80%) of the patients with anaphylactic reactions [158, 159].

23.12.3 Future Developments

LIGELIZUMAB (QGE031) is another anti-IgE antibody, which binds IgE with higher affinity than omalizumab, under early clinical development [160].

23.13 IL-5-Targeted Therapy in Asthma (See Also Chap. A7)

The T2 cells cytokine, IL-5, plays an important role in eosinophil granulocyte maturation, differentiation, recruitment and survival. IL-5 knockout mice appeared to confirm a role in asthma models where eosinophilia and AHR are markedly suppressed. The humanized anti-IL-5 monoclonal antibodies (MAbs) MEPOLIZUMAB (formerly termed SCH55700) and RESLIZUMAB (formerly Res-5-0010) have already both been approved for clinical use for difficult to control severe persistent asthma with peripheral blood eosinophilia of different levels. Another fully humanized Mab, BENRALIZUMAB (previously named MEDI-563), instead targets eosinophils by binding IL-5 receptor α inhibiting IL-5 binding and inducing eosinophil apoptosis through antibody-dependent cell-mediated cytotoxicity and has also been approved for the treatment of severe persistent asthma with peripheral blood eosinophilia. Mepolizumab, reslizumab and benralizumab all markedly reduce blood eosinophils, but benralizumab results in almost complete depletion, whereas a small number remains with mepolizumab and reslizumab. The implications of this for their efficacy and/or risk of adverse events are unclear [134, 161–163].

These anti-IL-5 pathway mAbs (Fig. 23.10) mepolizumab, reslizumab (both anti-IL-5) and benralizumab (anti-IL5R alpha chain) are all currently used as an adjunct to standard of care in people with severe eosinophilic asthma and poor control. These treatments roughly halve the rate of asthma exacerbations in this population. There is limited evidence for improved quality of life scores and lung function, which may not meet clinically detectable levels. There are no safety concerns regarding mepolizumab or reslizumab and no excess serious adverse events with benralizumab, although there remains a question over adverse events significant enough to prompt discontinuation.

Further research is needed on biomarkers for assessing treatment response, optimal duration and long-term effects of treatment, risk of relapse on withdrawal and non-eosinophilic patients and comparing anti-IL-5 treatments to each other and, in people eligible for both, to anti-immunoglobulin E. For benralizumab, future studies should closely monitor rates of adverse events prompting discontinuation [134, 161, 162]. No genetic associations that may predict response to mepolizumab treatment in patients with severe asthma have been identified so far [163].

The results of the first clinical trials with mepolizumab or benralizumab in the treatment of stable COPD have been disappointing [134].

23.14 Anti-IL-4/IL-13-Targeted Therapies

As part of the T2 hypothesis of asthma, IL-13 has been shown to play a critical role in various aspects of airway inflammation and epithelial remodelling, including mucus cell metaplasia and epithelial-mesenchymal signalling (see Chaps. 6, 11 and 14). This leads to increased subepithelial fibrosis or airway smooth muscle hyperplasia. Blocking IL-13, but not IL-4, in animal models of asthma prevents the development of AHR after allergen challenge, despite a strong eosinophilic response [134]. In addition, soluble IL-13Rα2 is effective in blocking the actions of IL-13, including IgE production, pulmonary eosinophilia and AHR in animal models of asthma [134].

LEBRIKIZUMAB, a humanized anti-IL-13 monoclonal antibody, improves lung function in adults with asthma inadequately controlled despite ICS therapy. Overall, the patients on lebrikizumab have a 5.5% increase in FEV_1 compared to the placebo group. When patients are divided according to high or low serum periostin (considered a biomarker of T2-like inflammation) levels, those with high periostin have greater improvement in lung function with lebrikizumab (8.2%) than those with low periostin levels (1.6%). Patient selection, however, is the key to success with this drug since lebrikizumab fails to increase FEV_1 in adult asthmatics not receiving ICS, irrespective of periostin levels [134]. The clinical development of lebrikizumab for the treatment of asthma has apparently suspended.

There is an ongoing clinical trial of lebrikizumab in patients with stable COPD (NCT02546700).

Similarly, subcutaneous TRALOKINUMAB (formerly CAT-354), an IL-13-neutralizing IgG4 MAb, improves lung function, but not global asthma control scores, in adults with moderate-to-severe uncontrolled asthma despite controller therapies [134].

IL-4 analogues that act as antagonists have been developed which fail to induce signal transduction and block IL-4 effects in vitro. These IL-4 antagonists prevent the development of asthma *in vivo* in animal models [134].

DUPILUMAB (SAR231893/REGN668) is a fully humanized anti-IL-4Rα (which is involved in both IL-4 and IL-13 signalling) MAb which improves, when administered by subcutaneous route, the control of asthma in adults with uncontrolled persistent asthma despite use of medium-to-high dose ICS/LABAs and irrespective of the presence or absence of peripheral blood eosinophilia and is another promising MAb for the treatment of severe uncontrolled asthma and nasal polyposis [164]. It has already been approved by the FDA for the treatment of these diseases.

23.15 Immunomodulatory Therapies

Immunosuppressive therapy with immunomodulatory drugs different from the corticosteroids may be considered in GINA step 5 level of asthma treatment, when other treatments have been unsuccessful or to reduce the dose of oral corticosteroids required, but mostly they have a greater propensity to side effects than oral corticosteroids and, therefore, cannot be routinely recommended. The role of immunosuppressive drugs different from the corticosteroids in COPD has not been yet sufficiently evaluated.

23.15.1 Methotrexate

Low-dose methotrexate (15 mg weekly) has a corticosteroid-sparing effect in asthma and may be indicated when oral corticosteroids are contraindicated because of unacceptable side effects (e.g. in postmenopausal women when osteoporosis is a problem) and omalizumab and other MAbs are ineffective in controlling severe asthma [165]. Some patients show better responses than others, but whether a patient will exhibit a useful corticosteroid-sparing effect is unpredictable. Overall, methotrexate has a small steroid-sparing effect that is insufficient to significantly reduce side effects of systemic corticosteroids, and this needs to be offset against the relatively high risk of side effects [166]. Side effects of methotrexate are relatively common and include nausea (reduced if methotrexate is given as a weekly *injection*), blood dyscrasias and hepatic damage. Careful monitoring of such patients (monthly blood counts and liver enzymes) is essential. Pulmonary infections and/or diffuse pulmonary fibrosis may rarely occur and even death (see also Chaps. 30 and 34).

23.15.2 Gold

GOLD has long been used in the treatment of rheumatoid arthritis. A controlled trial of an oral gold preparation (AURANOFIN) demonstrated some corticosteroid-sparing effect in asthmatic patients maintained on oral corticosteroids. Side effects such as skin rashes and nephropathy are a limiting factor. Overall gold provides little benefit in view of its small therapeutic ratio and is not recommended [167].

23.15.3 Cyclosporin A

CYCLOSPORIN A is active against $CD4^+$ lymphocytes and, therefore, has been tested in asthma, in which these cells are implicated in the pathogenesis (see also Chap. 31). A trial of low-dose oral cyclosporin A in patients with systemic corticosteroid-dependent asthma indicates that it can improve control of symptoms in patients with severe asthma on oral corticosteroids, but other trials have been unimpressive, and overall its poor efficacy is outweighed by its side effects [168]. Side effects, such as nephrotoxicity and systemic arterial hypertension, are common, and there are concerns about the risk of infections and cancers with long-term immunosuppression, so the treatment is not recommended. A randomized, double-blinded, placebo-controlled trial of oral cyclosporin A safety in a small number of patients with advanced stage COPD has been recently completed, but the results have not been promising [NCT00974142].

23.15.4 Intravenous Immunoglobulin

Intravenous immunoglobulin (IVIG) has been reported to have corticosteroid-sparing effects in severe corticosteroid-dependent asthma when high doses are used each month. IVIG reduces the production of IgE from B lymphocytes, and this may be the rationale for its use in severe asthma. Due to the poor effectiveness and very high cost of immunoglobulin replacement therapy, it is not routinely recommended in asthmatic patients without primary antibody immunodeficiency ([169]; see Chaps. 4 and 20).

23.15.5 Other Immunosuppressant Drugs

Many other immunosuppressant therapies, such as AZATHIOPRINE, COLCHICINE, KELIXIMAB and TACROLIMUS, have been tested in asthma, but currently their use is not recommended [ginasthma.org].

23.15.6 Specific Allergen Immunotherapy

Theoretically, specific IMMUNOTHERAPY with common allergens should be effective in preventing asthma. While this treatment is effective in allergic rhinitis due to single allergens (see Chaps. 23 and 24), there is little evidence that desensitizing subcutaneous injections to common allergens are very effective in controlling persistent asthma. Double-blind placebo-controlled studies have demonstrated poor effect in chronic asthma in adults. Overall, the benefits of allergen-specific immunotherapy by the subcutaneous route are small in asthmatics. Because there is a risk of anaphylactic and local reactions and because the course of treatment is time consuming, this allergen-specific immunotherapy cannot be routinely recommended for asthma [ginasthma.org].

SUBLINGUAL ALLERGEN IMMUNOTHERAPY (SLIT) is safer, but its role in asthma therapy has not yet been defined, despite having some clinical efficacy in house dust mite sensitized mild-to-moderate asthmatic patients [170].

The cellular mechanisms of allergen-specific immunotherapy are of interest as this might lead to safer and more effective approaches in the future. Allergen-specific immunotherapy induces the production of blocking IgG/IgG_4 antibodies that can inhibit IgE-dependent activation mediated through both FcεRI on mast cells and basophils and FcεRII on B cells. Suppression of T2 immunity can occur as a consequence of either deletion or anergy of antigen-specific T cells, induction of antigen-specific regulatory T cells (Treg) or immune deviation in favour of T1 responses. Anti-inflammatory cytokine IL-10-producing regulatory B cells and "protective" antibodies likely contribute to long-term tolerance [171]. In the future, the efficacy and safety of allergen-specific immunotherapies may be improved altering the method of delivery of allergen immunotherapy including epicutaneous, intralymphatic, intranasal and oral mucosal immunotherapy; activation of the innate immune system through Toll-like receptor agonists to stimulate T1 immunity and suppress T2 immunity; and the use of chemically altered allergens, allergoids, recombinant allergens and relevant T-cell epitope peptides [172].

23.16 New Drugs for Asthma and COPD

Several new classes of drug are now in development for asthma and COPD that are directed at the underlying chronic inflammatory process (Table 23.7). The inflammation responses in asthma and COPD are different, so that different approaches are needed, but there are some common inflammatory mechanisms [1, 2]. Indeed, patients with severe asthma have an inflammatory process that becomes more similar to that in COPD, suggesting that drugs that are effective in COPD may also be useful in patients with severe asthma that is not well controlled with corticosteroids [2]. In asthma, many new therapies have targeted eosinophilic inflammation. In COPD, a better understanding of the inflammatory process has highlighted several new therapeutic targets [1]. In all cases, it is likely that the novel therapies will not treat all subjects and some degree of subphenotyping will be required to determine which patients will respond optimally to each treatment regime. This will require the development of good clinical or readily accessible biomarkers to distinguish these subjects.

Table 23.7 New anti-inflammatory drugs for asthma and COPD

• New glucocorticoids (see Chap. C13)
• New immunomodulators
• Phosphodiesterases inhibitors
• Kinase inhibitors
• NF-κB inhibitors
• Adhesion molecule blockers
• Pro-inflammatory cytokine inhibitors
• Anti-inflammatory cytokines
• Chemokine receptor antagonists
• New allergen-specific immunotherapies

23.16.1 The Need for New Treatments

23.16.1.1 Asthma

Current asthma therapy is highly effective, and the majority of patients can be well controlled with ICS alone or associated to LABAs/ULABAs in combination inhalers. These treatments are not only effective but safe and relatively inexpensive. This poses a challenge to the development of new treatments, since they will need to be safer and at least as effective as existing treatments or offer some other advantage in long-term disease management. However, there are problems with existing therapies:

- Existing therapies have side effects, as they are not specific for asthma. Inhaled β_2-agonists may have side effects, and there is some evidence for the development of tolerance, especially to their bronchoprotective effects. ICSs also may have local and systemic side effect at high doses, and there is still a fear of using long-term corticosteroid treatment in many patients. Other treatments, such as low-dose theophylline, inhaled antimuscarinics and anti-leukotrienes, are less effective and are largely used as add-on therapies. Omalizumab and the new monoclonal antibodies targeting the IL-5 pathway are very expensive drugs, and again their use is limited as add-on therapies.
- There is still a major problem with poor compliance (adherence) in the long-term management of asthma, particularly as symptoms come under control with effective therapies [173], despite the recent introduction on the market of once-daily inhaled combinations which may improve the long-term adherence of the patients and the control of asthma in real life practice [174]. It is likely that a once-daily tablet or even an infrequent subcutaneous injection (see the above section on MAbs) may give improved compliance. However, oral therapy is associated with a much greater risk of systemic side effects and, therefore, needs to be specific for the abnormality in asthma. The development of smart inhalers linked to real-time remote electronic assessment of inhaler use and technique will enable better objective determination of compliance.
- Patients with severe asthma (approximately 5–10% of total, to be distinguished from uncontrolled asthma and difficult-to-control asthma) are often not controlled on maximal doses of inhaled therapies associated with low-dose theophylline and anti-leukotrienes or may have serious side effects from therapy, especially oral corticosteroids. These patients are relatively resistant to the anti-inflammatory actions of corticosteroids and require some other class of therapy to control the asthmatic process. The treatment of these patients with omalizumab and/or the new monoclonal antibodies targeting the IL-5 or the IL-4/IL-13 pathways is very expensive, and cheaper therapies should be developed [175].
- None of the existing treatments for asthma is disease-modifying, which means that the disease recurs as soon as treatment is discontinued.
- None of the existing treatments is curative, although it is possible that therapies which prevent the immune aberration of allergy may carry the prospects for a cure in the future [175].

23.16.1.2 COPD

In sharp contrast to asthma, there are few effective therapies for COPD, despite the fact that it is a common disease that is increasing worldwide [1, 2, 134].

The neglect of COPD is probably the result of several factors:

- COPD is regarded as largely irreversible and is treated as poorly responsive asthma.
- COPD is self-inflicted and, therefore, does not deserve investment.
- There are few satisfactory animal models that closely mimic the human disease.
- Relatively little is understood about the cellular and molecular pathogenesis of this disease.

None of the treatments currently available prevent the progression of the disease, and yet the disease is associated with an active inflammatory process that results in progressive obstruction of

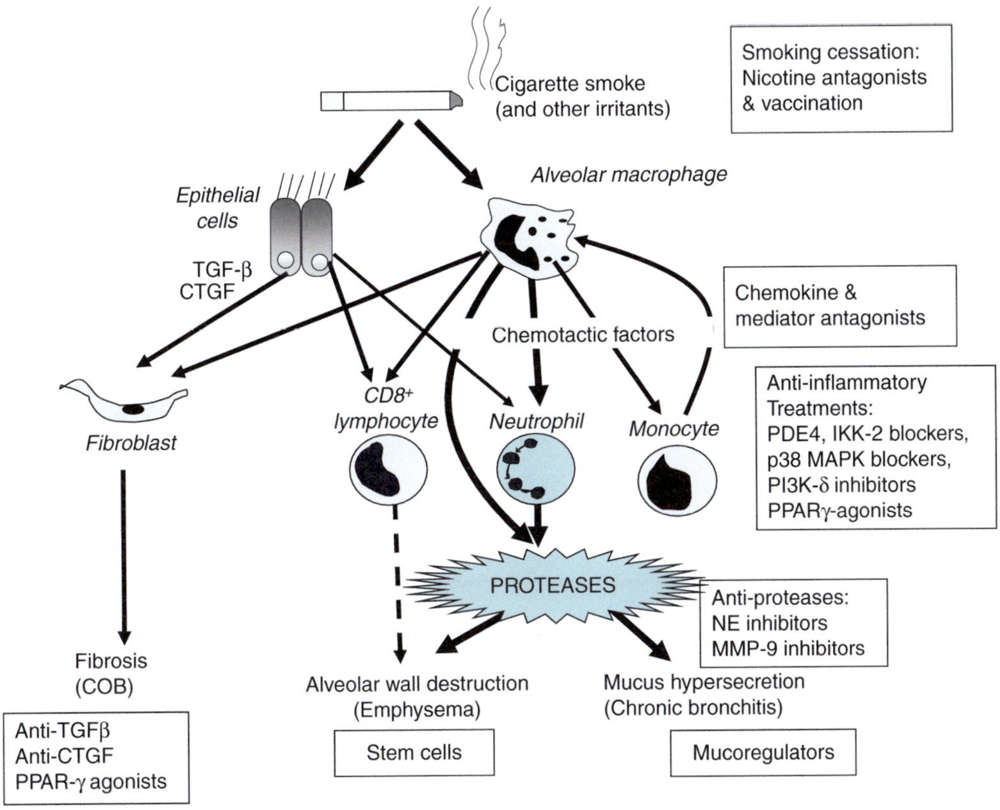

Fig. 23.11 Cigarette smoke (and other irritants) activates macrophages in the respiratory tract that releases multiple chemotactic factors that attract neutrophils, monocytes and T lymphocytes (particularly $CD8^+$ cells). Several cells also release proteases, such as neutrophil elastase (NE) and matrix metalloproteinase-9 (MMP-9), which break down connective tissue in the lung parenchyma (pulmonary emphysema) and also stimulate mucus hypersecretion (chronic bronchitis). $CD8^+$ cells may also be involved in alveolar wall destruction. Transforming growth factor (TGF)-βs and connective tissue growth factor (CTGF) released from inflammatory cells may modulate small airway fibrosis/autoimmune responses. The inflammatory process may be inhibited at several stages (shown in the boxes). Other abbreviations: *PDE* phosphodiesterase, *IKK* inhibitor of nuclear factor-κB kinase, *MAPK* mitogen-activated protein kinase, *PI3K* phosphoinositide-3-kinase, *PPAR* peroxisome proliferator activated receptor, *COB* chronic obstructive bronchitis

small airways and destruction of lung parenchyma. Increased understanding of COPD pathogenesis will identify novel targets for future therapy [1] (Fig. 23.11).

23.16.2 Development of New Therapies

Several strategies have been adopted in the search for new therapies:

- *Improvement of existing classes of drug*. This is well exemplified by the increased duration of β_2-agonists with the development of the ULABAs and of the inhaled antimuscarinics with the development of the ULAMAs, with the improved pharmacokinetic profile of new ICS with increased first-pass metabolism and, therefore, with reduced systemic absorption and the development of drugs with once-daily duration of action.
- *Development of novel therapies through better understanding of the disease process.* Examples of this are reported below.
- *Serendipitous observations, often made in other therapeutic areas.* Examples are the clinical trials with TNF-α blockers or statins for the treatment of asthma and/or COPD, derived from observations in other chronic inflammatory

diseases. So far, this approach has not led to the development of any useful therapies.

- *Identification of novel targets through "omics"*. This approach will be increasingly used to identify the abnormal expression of genes (molecular genomics), RNAs (transcriptomics), proteins (proteomics) and metabolites (metabolomics) from diseased fluids/cells/tissues that contribute to the disease process (diseasome). So far, this approach has not led to the development of any useful therapies.

23.16.3 Novel Inflammatory Mediator Modulators

Blocking the receptors or synthesis of inflammatory mediators is a logical approach to the development of new treatments for asthma and COPD. However, in both diseases many different mediators are involved, and therefore, blocking a single mediator is unlikely to be very effective, unless it plays a key role in the disease process [1, 2]. Several specific mediator antagonists have been found to be ineffective in asthma and COPD, including antagonists/inhibitors of thromboxane, platelet-activating factor, bradykinin, tachykinins and adhesion molecules (see Chap. 6).

23.16.3.1 CRTh2 Antagonists

Chemotactic factor for Th2 cells has been identified as prostaglandin D_2, which acts on a DP_2-receptor termed chemokine receptor homologous molecule expressed on Th2 lymphocytes (CRTh2). Several oral CRTH2 ANTAGONISTS, such as FEVIPIPRANT and TIMAPIPRANT, are now in clinical development for persistent asthma with promising initial results [176].

23.16.3.2 Other Cytokine Modifiers

Other cytokines, besides IL-4, IL-5 and IL-13 (see above), may play a critical role in the pathogenesis of asthma and COPD, suggesting that anti-cytokines may be beneficial as therapies [2, 177]. Numerous drugs targeting specific pro-inflammatory cytokines are under investigation for asthma and/or COPD, and these include monoclonal antibodies and/or other blockers directed against IL-6, IL-17 (or IL-17R), IL-18, IL-22, IL-23, IL-25, IL-33 (or against IL-33R, such as CNTO7160), thymic stromal lymphopoietin (TSLP, such as tezepelumab) and granulocyte-macrophage colony-stimulating factor (GM-CSF) [177, 178]. Although most attention has been focused on inhibition of cytokines, some cytokines are anti-inflammatory and might have therapeutic potential. However, the administration of the anti-inflammatory cytokine IL-10 has not been effective for the treatment of asthma. The administration of the immunomodulatory interferon beta for persistent asthma has also been investigated with disappointing results [NCT02491684].

23.16.3.3 Chemokine Receptor Antagonists

Many chemokines are involved in the pathogenesis of asthma and COPD and play a key role in recruitment of inflammatory cells, such as eosinophils, neutrophils, macrophages and lymphocytes into the lungs. CHEMOKINE RECEPTORS are attractive targets for new drugs, as they are GPCRs, like β_2-ARs, and small molecule inhibitors are now in development for both asthma and COPD [179, 180]. In asthma, CCR3 antagonists that should block eosinophil recruitment into the airways have failed because of toxicology problems. Likewise, both in neutrophilic severe persistent asthma and stable COPD patients, CXCR2 antagonists, which prevent neutrophil and monocyte chemotaxis due to CXC chemokines such as CXCL1 and CXCL8, were ineffective [181, 182]. There is a strong scientific rationale to test CCR5 inhibitors (already used for the treatment of HIV infection) for the treatment of stable COPD [179].

23.16.3.4 Endothelin Antagonists

Endothelins have been implicated in some of the structural changes that occur in the COPD lungs. ENDOTHELIN RECEPTOR ANTAGONISTS are approved for the treatment of idiopathic arterial pulmonary hypertension and might be useful in treating the structural changes that occur in COPD, but so far have been proven ineffective [183].

23.16.3.5 Antioxidants

Oxidative stress is important in severe asthma and COPD and may contribute to corticosteroid resistance. Existing ANTIOXIDANTS have weak effects, but more potent antioxidants are in development [184, 185].

23.16.3.6 Inducible NO Synthase Inhibitors

Nitric oxide (NO) production is increased in asthma and COPD as a result of increased inducible NO synthase (iNOS or NOS2) expression in the lower airways. NO and oxidative stress generate peroxynitrite which may nitrate proteins, leading to altered cell function. Several selective NOS2 inhibitors have been developed, but found to be ineffective in asthma, however [186].

23.16.3.7 Protease Inhibitors

Several proteolytic enzymes (proteases) are involved in the pathogenesis of chronic inflammation and remodelling observed in the lower airways of asthma and COPD patients. Mast cell tryptase has several effects on the lower airways, including increasing responsiveness of ASM to constrictors, increasing plasma exudation, potentiating eosinophil recruitment and stimulating fibroblast proliferation. Some of these effects are mediated by activation of the proteinase-activated receptor PAR2. TRYPTASE INHIBITORS have so far proved to be disappointing in clinical studies in asthmatic patients [187].

Several other proteases are involved in the degradation of the extracellular matrix in COPD, particularly enzymes that break down elastin fibres, such as neutrophil elastase and matrix metalloproteinases (MMPs). NEUTROPHIL ELASTASE INHIBITORS are ineffective in COPD patients, but more selective inhibitors have been developed. MMP9 and MMP12 appear to be the predominant elastolytic enzymes in the pathogenesis of COPD and pulmonary emphysema, but selective dual inhibitors are ineffective in the treatment of stable COPD [1, 2].

23.16.3.8 Transcription Factor Modulation

Therapeutic targeting of GATA3, an important transcription factor of the Th2 pathway, may be beneficial in the treatment of asthma. SB010, a novel DNA enzyme (DNAzyme) that is able to cleave and inactivate GATA3 messenger RNA, has some efficacy in the prevention of allergen-induced asthmatic responses [188].

23.16.4 New Anti-inflammatory Drugs

ICSs are by far the most effective therapy for asthma yet are much less effective in COPD and less effective in severe asthma. This has led to the search for alternative broad-spectrum anti-inflammatory treatments for severe asthma and COPD (Table 23.7).

23.16.4.1 NF-κB Inhibitors

NF-κB plays an important role in the orchestration of chronic inflammation, and many of the inflammatory genes that are expressed in asthma and COPD are regulated by this transcription factor [189]. This has prompted a search for specific blockers of these transcription factors. NF-κB is naturally inhibited by the inhibitory protein IκB, which is degraded after activation by specific kinases. Small molecule inhibitors of the IκB KINASE IKK2 (or IKKβ) are in early clinical development [190]. However, there are concerns that inhibition of NF-κB may cause side effects such as increased susceptibility to infections, which has been observed in gene disruption studies when components of NF-κB are inhibited. These drugs may be of particular value during COPD exacerbations in which corticosteroids are largely ineffective and during rhinovirus-induced asthma and COPD exacerbations in which NF-κB contributes to the generation of a relative corticosteroid resistance [191].

23.16.4.2 MAP Kinase Inhibitors

There are three major mitogen-activated protein (MAP) kinase pathways, and there is increasing recognition that these pathways are involved in chronic inflammation of the lower airways [190].

There has been particular interest in the p38MAP kinase pathway inhibitors that may reverse corticosteroid resistance in COPD. However, clinical studies with oral antagonists of the p38MAPK, such as losmapimod, in the treatment of stable COPD have been unsuccessful [192], so that inhaled delivery is being explored. These drugs may prove more useful in the treatment of severe persistent asthma and COPD exacerbations [193].

Theophylline, at concentrations that do not inhibit phosphodiesterase (PDE)4 activity, enhances HDAC2 activity, and functionally this enhances glucocorticoid effects. This effect may be via phosphoinositide-3-kinase (PI3K) δ-induced hyperphosphorylation of HDAC2, particularly since PI3Kδ is upregulated in peripheral lung tissue of patients with COPD. The results of an ongoing long-term phase IIa controlled study [NCT03345407] on the clinical efficacy of the highly selective inhaled PI3Kδ inhibitor nemiralisib (GSK2269557) in patients with COPD are awaited with interest. Due to the efficacy of PI3Kδ and γ inhibitors in animal models and against primary human cells, it is possible that these and also JAK-STAT inhibitors will be effective in patients where these pathways are activated [190].

23.17 Summary

- Pharmacological therapies are the mainstay of management of asthma and COPD. Current bronchodilators relax ASM and include inhaled β_2-adrenergic agonists, inhaled antimuscarinics and theophylline. Long- and ultra-long-acting inhaled β_2-agonists are an important advance, and these are very effective add-on therapy to controller drugs in asthma. Long- and ultra-long-acting inhaled antimuscarinics are an important long-term treatment for reducing symptoms and preventing COPD exacerbations. Although new classes of bronchodilators have been explored, they are unlikely to be as effective in asthmatics as inhaled β_2-agonists which act as functional antagonists to counteract all bronchoconstrictor mechanisms. Controller drugs act on the underlying disease by suppressing the inflammatory process and the most effective drugs in this class are represented by the corticosteroids. Cromones are much less effective in asthma compared with ICS and are now little used. Oral low-dose theophylline may also have anti-inflammatory effects, and it is still a useful add-on therapy in more severe asthma and advanced COPD. Many inflammatory mediators are involved in asthma and COPD, so blocking a single mediator is unlikely to have a major beneficial effect. Oral anti-leukotrienes have relatively weak effects in asthma compared to ICS and are not effective during asthmatic exacerbations or in COPD patients. Omalizumab and the new monoclonal antibodies targeting the IL-4, and IL-5 pathways are very expensive drugs, and their use will be limited as add-on parenteral therapies for severe asthma.

  Oral roflumilast as add-on therapy for severe stable COPD has some benefits but with many side effects. There are several new promising classes of controller drugs now in development for asthma and COPD, including phosphodiesterase inhibitors, cytokine and chemokine receptor antagonists and inhaled kinase blockers.

Selected Readings

Barnes PJ, Drazen JM, Rennard SI, Thomson NC. Asthma and COPD. 2nd ed. Amsterdam: Elsevier; 2009. p. 1–878.

Chung KF, Barnes PJ, editors. Pharmacology and therapeutics of airways diseases, Lung biology in health and disease, vol. 234. 2nd ed. New York: Informa Healthcare; 2009. p. 1–431.

Seyberth HW, Rane A, Schwab M. Pediatric clinical pharmacology. In: Handbook of experimental pharmacology, vol. 205. Berlin: Springer-Verlag; 2011. p. 1–384.

Selected Websites

ginasthma.org
goldcopd.org
http://www.cochranelibrary.com/
https://clinicaltrials.gov/
https://www.ncbi.nlm.nih.gov/pubmed

References

1. Caramori G, Casolari P, Barczyk A, Durham AL, Di Stefano A, Adcock I. COPD immunopathology. Semin Immunopathol. 2016;38:497–515.
2. Barnes PJ. Cellular and molecular mechanisms of asthma and COPD. Clin Sci (Lond). 2017;131:1541–58.
3. Cazzola M, Page CP, Calzetta L, Matera MG. Pharmacology and therapeutics of bronchodilators. Pharmacol Rev. 2012;64:450–504.
4. Caramori G, Ruggeri P, Casolari P, Chung KF, Girbino G, Adcock IM. Fluticasone furoate and vilanterol for the treatment of chronic obstructive pulmonary disease. Expert Rev Respir Med. 2017;11:955–67.
5. Bang I, Choi HJ. Structural features of β2 adrenergic receptor: crystal structures and beyond. Mol Cells. 2015;38:105–11.
6. Zieba BJ, Artamonov MV, Jin L, Momotani K, Ho R, Franke AS, Somlyo AV. The cAMP-responsive Rap1 guanine nucleotide exchange factor, Epac, induces smooth muscle relaxation by down-regulation of RhoA activity. J Biol Chem. 2011;286:16681–92.
7. Morgan SJ, Deshpande DA, Tiegs BC, Misior AM, Yan H, Hershfeld AV, Rich TC, Panettieri RA, An SS, Penn RB. β-Agonist-mediated relaxation of airway smooth muscle is protein kinase A-dependent. J Biol Chem. 2014;289:23065–74.
8. Chung KF, Caramori G, Adcock IM. Inhaled corticosteroids as combination therapy with β-adrenergic agonists in airways disease: present and future. Eur J Clin Pharmacol. 2009;65:853–71.
9. Pera T, Penn RB. Bronchoprotection and bronchorelaxation in asthma: new targets, and new ways to target the old ones. Pharmacol Ther. 2016;164:82–96.
10. Pera T, Hegde A, Deshpande DA, Morgan SJ, Tiegs BC, Theriot BS, Choi YH, Walker JK, Penn RB. Specificity of arrestin subtypes in regulating airway smooth muscle G protein-coupled receptor signaling and function. FASEB J. 2015;29:4227–35.
11. Walker JK, DeFea KA. Role for β-arrestin in mediating paradoxical β2AR and PAR2 signaling in asthma. Curr Opin Pharmacol. 2014;16:142–7.
12. Chung LP, Waterer G, Thompson PJ. Pharmacogenetics of β2 adrenergic receptor gene polymorphisms, long-acting β-agonists and asthma. Clin Exp Allergy. 2011;41:312–26.
13. Hardin M, Cho MH, McDonald ML, Wan E, Lomas DA, Coxson HO, MacNee W, Vestbo J, Yates JC, Agusti A, Calverley PM, Celli B, Crim C, Rennard S, Wouters E, Bakke P, Bhatt SP, Kim V, Ramsdell J, Regan EA, Make BJ, Hokanson JE, Crapo JD, Beaty TH, Hersh CP, ECLIPSE and COPDGene Investigators; COPDGene Investigators—Clinical Centers. A genome-wide analysis of the response to inhaled β2-agonists in chronic obstructive pulmonary disease. Pharmacogenomics J. 2016;16:326–35.
14. Caramori G, Adcock IM. Pharmacology of airway inflammation in asthma and COPD. Pulm Pharmacol Ther. 2003;16:247–77.
15. Chantaphakul H, Ruxrungtham K. Fixed-dose combination of the inhaled corticosteroid and long-acting beta2-agonist therapy in adults with persistent asthma. Expert Opin Pharmacother. 2016;17:631–42.
16. Vogelmeier C, Hederer B, Glaab T, Schmidt H, Rutten-van Mölken MP, Beeh KM, Rabe KF, Fabbri LM, POET-COPD Investigators. Tiotropium versus salmeterol for the prevention of exacerbations of COPD. N Engl J Med. 2011;364:1093–103.
17. Lotvall J, Palmqvist M, Arvidsson P, Maloney A, Ventresca GP, Ward J. The therapeutic ratio of R-albuterol is comparable with that of RS-albuterol in asthmatic patients. J Allergy Clin Immunol. 2001;108:726–31.
18. Jat KR, Khairwa A. Levalbuterol versus albuterol for acute asthma: a systematic review and meta-analysis. Pulm Pharmacol Ther. 2013;26:239–348.
19. Loh CH, Donohue JF, Ohar JA. Review of drug safety and efficacy of arformoterol in chronic obstructive pulmonary disease. Expert Opin Drug Saf. 2015;14:463–72.
20. Barnes PJ. Scientific rationale for combination inhalers with a long-acting β2-agonists and corticosteroids. Eur Respir J. 2002;19:182–91.
21. Single inhaler triple therapy for COPD. Drug Ther Bull. 2017;55:138–41.
22. Calverley PMA, Magnussen H, Miravitlles M, Wedzicha JA. Triple therapy in COPD: what we know and what we don't. COPD. 2017;14:648–62.
23. Bhatt SP, Wells JM, Kim V, Criner GJ, Hersh CP, Hardin M, Bailey WC, Nath H, Kim YI, Foreman MG, Stinson DS, Wilson CG, Rennard SI, Silverman EK, Make BJ, Dransfield MT, COPDGene Investigators. Radiological correlates and clinical implications of the paradoxical lung function response to β_2 agonists: an observational study. Lancet Respir Med. 2014;2:911–8.
24. McGraw DW, Liggett SB. Heterogeneity in beta-adrenergic receptor kinase expression in the lung accounts for cell-specific desensitization of the beta2-adrenergic receptor. J Biol Chem. 1997;272:7338–44.
25. Mak JCW, Nishikawa M, Shirasaki H, Miyayasu K, Barnes PJ. Protective effects of a glucocorticoid on down-regulation of pulmonary β_2-adrenergic receptors *in vivo*. J Clin Invest. 1995;96:99–106.
26. Spitzer WO, Suissa S, Ernst P, Horwitz RI, Habbcek B, Cockroft D, Boivin J-F, McNutt M, Buist AS, Rebuck AS. The use of β-agonists and the rate of death and near-death from asthma. N Engl J Med. 1992;326:503–6.
27. Suissa S, Ernst P, Benayoun S, Baltzan M, Cai B. Low-dose inhaled corticosteroids and the prevention of death from asthma. N Engl J Med. 2000;343:332–6.
28. Sears MR, Taylor DR, Print CG, Lake DG, Li Q, Flannery EM, Yates DM, Lucas MK, Herbison

GP. Regular inhaled beta-agonist treatment in bronchial asthma. Lancet. 1990;336:1391–6.
29. Dennis SM, Sharp SJ, Vickers MR, Frost CD, Crompton GK, Barnes PJ, Lee TH. Regular inhaled salbutamol and asthma control: the TRUST randomised trial. Lancet. 2000;355:1675–9.
30. Mcivor RA, Pizzichini E, Turner MO, Hussack P, Hargreave FE, Sears MR. Potential masking effects of salmeterol on airway inflammation in asthma. Am J Respir Crit Care Med. 1998;158:924–30.
31. McGraw DW, Almoosa KF, Paul RJ, Kobilka BK, Liggett SB. Antithetic regulation by β-adrenergic receptors of Gq receptor signaling via phospholipase C underlies the airway β-agonist paradox. J Clin Invest. 2003;112:619–26.
32. Nelson HS, Weiss ST, Bleecker ER, Yancey SW, Dorinsky PM. The Salmeterol Multicenter Asthma Research Trial: a comparison of usual pharmacotherapy for asthma or usual pharmacotherapy plus salmeterol. Chest. 2006;129:15–26.
33. Stempel DA, Raphiou IH, Kral KM, Yeakey AM, Emmett AH, Prazma CM, Buaron KS, Pascoe SJ, AUSTRI Investigators. Serious asthma events with fluticasone plus salmeterol versus fluticasone alone. N Engl J Med. 2016;374:1822–30.
34. Peters SP, Bleecker ER, Canonica GW, Park YB, Ramirez R, Hollis S, Fjallbrant H, Jorup C, Martin UJ. Serious asthma events with budesonide plus formoterol vs. budesonide alone. N Engl J Med. 2016;375:850–60.
35. Calverley PM, Anderson JA, Celli B, Ferguson GT, Jenkins C, Jones PW, Yates JC, Vestbo J. Salmeterol and fluticasone propionate and survival in chronic obstructive pulmonary disease. N Engl J Med. 2007;356:775–89.
36. Vestbo J, Anderson JA, Brook RD, Calverley PM, Celli BR, Crim C, Martinez F, Yates J, Newby DE, SUMMIT Investigators. Fluticasone furoate and vilanterol and survival in chronic obstructive pulmonary disease with heightened cardiovascular risk (SUMMIT): a double-blind randomised controlled trial. Lancet. 2016;387:1817–26.
37. Aparici M, Gómez-Angelats M, Vilella D, Otal R, Carcasona C, Viñals M, Ramos I, Gavaldà A, De Alba J, Gras J, Cortijo J, Morcillo E, Puig C, Ryder H, Beleta J, Miralpeix M. Pharmacological characterization of abediterol, a novel inhaled β(2)-adrenoceptor agonist with long duration of action and a favorable safety profile in preclinical models. J Pharmacol Exp Ther. 2012;342:497–509.
38. Matera MG, Cazzola M. Muscarinic receptor antagonists. Handb Exp Pharmacol. 2017;237:41–62.
39. Wang L, Zhai CJ, Liu Y, Liu Y, Jiang SJ. Umeclidinium plus vilanterol versus placebo, umeclidinium, or vilanterol monotherapies for chronic obstructive pulmonary disease: a meta-analysis of randomized controlled trials. Clin Drug Investig. 2016;36: 865–75.
40. Barnes PJ. Neurogenic inflammation in the airways. Respir Physiol. 2001;125:145–54.
41. Barnes PJ. Distribution of receptor targets in the lung. Proc Am Thorac Soc. 2004;1:345–51.
42. Cazzola M, Ora J, Rogliani P, Matera MG. Role of muscarinic antagonists in asthma therapy. Expert Rev Respir Med. 2017;11:239–53.
43. Beckmann J, Lips KS. The non-neuronal cholinergic system in health and disease. Pharmacology. 2013;92:286–302.
44. Pieper MP. The non-neuronal cholinergic system as novel drug target in the airways. Life Sci. 2012;91:1113–8.
45. Kirkland SW, Vandenberghe C, Voaklander B, Nikel T, Campbell S, Rowe BH. Combined inhaled beta-agonist and anticholinergic agents for emergency management in adults with asthma. Cochrane Database Syst Rev. 2017;1:CD001284.
46. McCrory DC, Brown CD. Anti-cholinergic bronchodilators versus beta2-sympathomimetic agents for acute exacerbations of chronic obstructive pulmonary disease. Cochrane Database Syst Rev. 2002;4:CD003900.
47. Kew KM, Dahri K. Long-acting muscarinic antagonists (LAMA) added to combination long-acting beta2-agonists and inhaled corticosteroids (LABA/ICS) versus LABA/ICS for adults with asthma. Cochrane Database Syst Rev. 2016;1:CD011721.
48. Bulkhi A, Tabatabaian F, Casale TB. Long-acting muscarinic antagonists for difficult-to-treat asthma: emerging evidence and future directions. Drugs. 2016;76:999–1013.
49. Wechsler ME, Yawn BP, Fuhlbrigge AL, Pace WD, Pencina MJ, Doros G, Kazani S, Raby BA, Lanzillotti J, Madison S, Israel E, BELT Investigators. Anticholinergic vs long-acting β-agonist in combination with inhaled corticosteroids in black adults with asthma: the BELT randomized clinical trial. JAMA. 2015;314:1720–30.
50. Tashkin DP, Celli B, Senn S, Burkhart D, Kesten S, Menjoge S, Decramer M. A 4-year trial of tiotropium in chronic obstructive pulmonary disease. N Engl J Med. 2008;359:1543–54.
51. Proctor GB. The physiology of salivary secretion. Periodontol 2000. 2016;70:11–25.
52. Arbouw ME, Movig KL, Koopmann M, Poels PJ, Guchelaar HJ, Egberts TC, Neef C, van Vugt JP. Glycopyrrolate for sialorrhea in Parkinson disease: a randomized, double-blind, crossover trial. Neurology. 2010;74:1203–7.
53. Geller DE. Clinical side effects during aerosol therapy: cutaneous and ocular effects. J Aerosol Med. 2007;20(Suppl 1):S100–8; discussion S109.
54. Cuvelier A, Muir JF, Benhamou D, Weitzenblum E, Zuck P, Delacenserie R, Taytard A, Iacono P. Dry powder ipratropium bromide is as safe and effective as metered-dose inhaler formulation: a cumulative dose-response study in chronic obstructive pulmonary disease patients. Respir Care. 2002;47:159–66.
55. Facchini G, Antonicelli L, Cinti B, Bonifazi F, Massei V. Paradoxical bronchospasm and cutaneous rash after metered-dose inhaled bronchodilators. Monaldi Arch Chest Dis. 1996;51:201–3.

56. Giglio D, Tobin G. Muscarinic receptor subtypes in the lower urinary tract. Pharmacology. 2009;83:259–69.
57. Loke YK, Singh S. Risk of acute urinary retention associated with inhaled anticholinergics in patients with chronic obstructive lung disease: systematic review. Ther Adv Drug Saf. 2013;4:19–26.
58. Türck D, Weber W, Sigmund R, Budde K, Neumayer HH, Fritsche L, Rominger KL, Feifel U, Slowinski T. Pharmacokinetics of intravenous, single-dose tiotropium in subjects with different degrees of renal impairment. J Clin Pharmacol. 2004;44:163–72.
59. Santus P, Radovanovic D, Cristiano A, Valenti V, Rizzi M. Role of nebulized glycopyrrolate in the treatment of chronic obstructive pulmonary disease. Drug Des Devel Ther. 2017;11:3257–327.
60. Ambery CL, Wielders P, Ludwig-Sengpiel A, Chan R, Riley JH. Population pharmacokinetics and pharmacodynamics of GSK961081 (batefenterol), a muscarinic antagonist and β2-agonist, in moderate-to-severe COPD patients: substudy of a randomized trial. Drugs R D. 2015;15:281–91.
61. Barnes PJ. Theophylline. Am J Respir Crit Care Med. 2013;188:901–6.
62. Spina D, Page CP. Xanthines and phosphodiesterase inhibitors. Handb Exp Pharmacol. 2017;237:63–91.
63. Chrystyn H, Mulley BA, Peake MD. Dose response relation to oral theophylline in severe chronic obstructive airways disease. BMJ. 1988;297:1506–10.
64. Hidi R, Timmermans S, Liu E, Schudt C, Dent G, Holgate ST, Djukanović R. Phosphodiesterase and cyclic adenosine monophosphate-dependent inhibition of T-lymphocyte chemotaxis. Eur Respir J. 2000;15:342–9.
65. Sullivan P, Bekir S, Jaffar Z, Page C, Jeffery P, Costello J. Anti-inflammatory effects of low-dose oral theophylline in atopic asthma. Lancet. 1994;343:1006–8.
66. Lim S, Tomita K, Carramori G, Jatakanon A, Oliver B, Keller A, Adcock I, Chung KF, Barnes PJ. Low-dose theophylline reduces eosinophilic inflammation but not exhaled nitric oxide in mild asthma. Am J Respir Crit Care Med. 2001;164:273–6. Erratum in: Am J Respir Crit Care Med. 2004;170:345. Carramori G [corrected to Caramori G].
67. Kidney J, Dominguez M, Taylor PM, Rose M, Chung KF, Barnes PJ. Immunomodulation by theophylline in asthma: demonstration by withdrawal of therapy. Am J Respir Crit Care Med. 1995;151:1907–14.
68. Kraft M, Torvik JA, Trudeau JB, Wenzel SE, Martin RJ. Theophylline: potential anti-inflammatory effects in nocturnal asthma. J Allergy Clin Immunol. 1996;97:1242–6.
69. Culpitt SV, de Matos C, Russell RE, Donnelly LE, Rogers DF, Barnes PJ. Effect of theophylline on induced sputum inflammatory indices and neutrophil chemotaxis in COPD. Am J Respir Crit Care Med. 2002;165:1371–6.
70. Aubier M, De Troyer A, Sampson M, Macklem PT, Roussos C. Aminophylline improves diaphragmatic contractility. N Engl J Med. 1981;305:249–52.
71. Moxham J. Aminophylline and the respiratory muscles: an alternative view. Clin Chest Med. 1988;9:325–36.
72. Caruso M, Alamo A, Crisafulli E, Raciti C, Fisichella A, Polosa R. Adenosine signaling pathways as potential therapeutic targets in respiratory disease. Expert Opin Ther Targets. 2013;17:761–72.
73. Oliver B, Tomita K, Keller A, Caramori G, Adcock I, Chung KF, Barnes PJ, Lim S. Low-dose theophylline does not exert its anti-inflammatory effects in mild asthma through upregulation of interleukin-10 in alveolar macrophages. Allergy. 2001;56:1087–90.
74. Ichiyama T, Hasegawa S, Matsubara T, Hayashi T, Furukawa S. Theophylline inhibits NF-kB activation and IkBa degradation in human pulmonary epithelial cells. Naunyn Schmiedebergs Arch Pharmacol. 2001;364:558–61.
75. Chung IY, Nam-Kung EK, Lee NM, Chang HS, Kim DJ, Kim YH, Park CS. The downregulation of bcl-2 expression is necessary for theophylline-induced apoptosis of eosinophil. Cell Immunol. 2000;203:95–102.
76. Yasui K, Agematsu K, Shinozaki K, Hokibara S, Nagumo H, Nakazawa T, Komiyama A. Theophylline induces neutrophil apoptosis through adenosine A_{2A} receptor antagonism. J Leukoc Biol. 2000;67:529–35.
77. Ohta K, Yamashita N. Apoptosis of eosinophils and lymphocytes in allergic inflammation. J Allergy Clin Immunol. 1999;104:14–21.
78. Moonen HJ, Geraets L, Vaarhorst A, Bast A, Wouters EF, Hageman GJ. Theophylline prevents NAD+ depletion via PARP-1 inhibition in human pulmonary epithelial cells. Biochem Biophys Res Commun. 2005;338:1805–10.
79. Ito K, Lim S, Caramori G, Cosio B, Chung KF, Adcock IM, Barnes PJ. A molecular mechanism of action of theophylline: induction of histone deacetylase activity to decrease inflammatory gene expression. Proc Natl Acad Sci U S A. 2002;99:8921–6.
80. Cosio BG, Tsaprouni L, Ito K, Jazrawi E, Adcock IM, Barnes PJ. Theophylline restores histone deacetylase activity and steroid responses in COPD macrophages. J Exp Med. 2004;200:689–95.
81. Marwick JA, Caramori G, Stevenson CC, Casolari P, Jazrawi E, Barnes PJ, Ito K, Adcock IM, Kirkham PA, Papi A. Inhibition of PI3Kd restores glucocorticoid function in smoking-induced airway inflammation in mice. Am J Respir Crit Care Med. 2009;179:542–8.
82. To M, Ito K, Kizawa Y, Failla M, Ito M, Kusama T, Elliot M, Hogg JC, Adcock IM, Barnes PJ. Targeting phosphoinositide-3-kinase-δ with theophylline reverses corticosteroid insensitivity in COPD. Am J Respir Crit Care Med. 2010;182:897–904.

83. Marwick JA, Caramori G, Casolari P, Mazzoni F, Kirkham PA, Adcock IM, Chung KF, Papi A. A role for phosphoinositol 3-kinase delta in the impairment of glucocorticoid responsiveness in patients with chronic obstructive pulmonary disease. J Allergy Clin Immunol. 2010;125:1146–53.
84. Osoata GO, Yamamura S, Ito M, Vuppusetty C, Adcock IM, Barnes PJ, Ito K. Nitration of distinct tyrosine residues causes inactivation of histone deacetylase 2. Biochem Biophys Res Commun. 2009;384:366–71.
85. Hirano T, Yamagata T, Gohda M, Yamagata Y, Ichikawa T, Yanagisawa S, Ueshima K, Akamatsu K, Nakanishi M, Matsunaga K, Minakata Y, Ichinose M. Inhibition of reactive nitrogen species production in COPD airways: comparison of inhaled corticosteroid and oral theophylline. Thorax. 2006;61:761–6. Erratum in: Thorax 2008;63:1029.
86. Ford PA, Durham AL, Russell RE, Gordon F, Adcock IM, Barnes PJ. Treatment effects of low-dose theophylline combined with an inhaled corticosteroid in COPD. Chest. 2010;137:1338–44.
87. Ito K, Caramori G, Lim S, Oates T, Chung KF, Barnes PJ, Adcock IM. Expression and activity of histone deacetylases in human asthmatic airways. Am J Respir Crit Care. 2002;166:392–6.
88. Kirkham PA, Whiteman M, Winyard PG, Caramori G, Gordon F, Ford PA, Barnes PJ, Adcock IM, Chung KF. Impact of theophylline/corticosteroid combination therapy on sputum hydrogen sulfide levels in patients with COPD. Eur Respir J. 2014;43: 1504–6.
89. Dubuis E, Wortley MA, Grace MS, Maher SA, Adcock JJ, Birrell MA, Belvisi MG. Theophylline inhibits the cough reflex through a novel mechanism of action. J Allergy Clin Immunol. 2014;133:1588–98.
90. Bolhuis MS, Panday PN, Pranger AD, Kosterink JG, Alffenaar JW. Pharmacokinetic drug interactions of antimicrobial drugs: a systematic review on oxazolidinones, rifamycins, macrolides, fluoroquinolones, and beta-lactams. Pharmaceutics. 2011;3:865–913.
91. Cushley MJ, Holgate ST. Bronchodilator actions of xanthine derivatives administered by inhalation in asthma. Thorax. 1985;40:176–9.
92. Nair P, Milan SJ, Rowe BH. Addition of intravenous aminophylline to inhaled beta(2)-agonists in adults with acute asthma. Cochrane Database Syst Rev. 2012;12:CD002742.
93. Duffy N, Walker P, Diamantea F, Calverley PM, Davies L. Intravenous aminophylline in patients admitted to hospital with non-acidotic exacerbations of chronic obstructive pulmonary disease: a prospective randomised controlled trial. Thorax. 2005;60:713–7.
94. Tee AK, Koh MS, Gibson PG, Lasserson TJ, Wilson AJ, Irving LB. Long-acting beta2-agonists versus theophylline for maintenance treatment of asthma. Cochrane Database Syst Rev. 2007;(3):CD001281.
95. Chen X, Kang YB, Wang LQ, Li Y, Luo YW, Zhu Z, Chen R. Addition to inhaled corticosteroids of leukotriene receptor antagonists versus theophylline for symptomatic asthma: a meta-analysis. J Thorac Dis. 2015;7:644–552.
96. Celis P, Rada G. High-dose inhaled corticosteroids or addition of theophylline in patients with poorly controlled asthma? Medwave. 2015;15(Suppl 2):e6224.
97. Spears M, Donnelly I, Jolly L, Brannigan M, Ito K, McSharry C, Lafferty J, Chaudhuri R, Braganza G, Adcock IM, et al. Effect of low-dose theophylline plus beclometasone on lung function in smokers with asthma: a pilot study. Eur Respir J. 2009;33: 1010–7.
98. Ito K, Ito M, Elliott WM, Cosio B, Caramori G, Kon OM, Barczyk A, Hayashi S, Adcock IM, Hogg JC, Barnes PJ. Decreased histone deacetylase activity in chronic obstructive pulmonary disease. N Engl J Med. 2005;352:1967–76.
99. Marwick JA, Wallis G, Meja K, Kuster B, Bouwmeester T, Chakravarty P, Fletcher D, Whittaker PA, Barnes PJ, Ito K, et al. Oxidative stress modulates theophylline effects on steroid responsiveness. Biochem Biophys Res Commun. 2008;377:797–802.
100. Murciano D, Auclair MH, Pariente R, Aubier M. A randomized, controlled trial of theophylline in patients with severe chronic obstructive pulmonary disease. N Engl J Med. 1989;320:1521–5.
101. Kirsten DK, Wegner RE, Jörres RA, Magnussen H. Effects of theophylline withdrawal in severe chronic obstructive pulmonary disease. Chest. 1993;104:1101–7. Erratum in: Chest. 1994; 106:328.
102. Zhou Y, Wang X, Zeng X, Qiu R, Xie J, Liu S, Zheng J, Zhong N, Ran P. Positive benefits of theophylline in a randomized, double-blind, parallel-group, placebo-controlled study of low-dose, slow-release theophylline in the treatment of COPD for 1 year. Respirology. 2006;11:603–10.
103. Cosio BG, Iglesias A, Rios A, Noguera A, Sala E, Ito K, Barnes PJ, Agusti A. Low-dose theophylline enhances the anti-inflammatory effects of steroids during exacerbations of chronic obstructive pulmonary disease. Thorax. 2009;64:424–9.
104. Subramanian S, Ragulan, Jindal A, Viswambhar V, Arun Babu V. The study of efficacy, tolerability and safety of theophylline given along with formoterol plus budesonide in COPD. J Clin Diagn Res. 2015;9:OC10–3.
105. Devereux G, Cotton S, Barnes P, Briggs A, Burns G, Chaudhuri R, Chrystyn H, Davies L, De Soyza A, Fielding S, Gompertz S, Haughney J, Lee AJ, McCormack K, McPherson G, Morice A, Norrie J, Sullivan A, Wilson A, Price D. Use of low-dose oral theophylline as an adjunct to inhaled corticosteroids in preventing exacerbations of chronic obstructive pulmonary disease: study protocol for a randomised controlled trial. Trials. 2015;16:267.

106. Cosío BG, Shafiek H, Iglesias A, Yanez A, Córdova R, Palou A, Rodriguez-Roisin R, Peces-Barba G, Pascual S, Gea J, Sibila O, Barnes PJ, Agusti A. Oral low-dose theophylline on top of inhaled fluticasone-salmeterol does not reduce exacerbations in patients with severe COPD: a pilot clinical trial. Chest. 2016;150:123–30.
107. Shukla D, Chakraborty S, Singh S, Mishra B. Doxofylline: a promising methylxanthine derivative for the treatment of asthma and chronic obstructive pulmonary disease. Expert Opin Pharmacother. 2009;10:2343–56.
108. Lal D, Manocha S, Ray A, Vijayan VK, Kumar R. Comparative study of the efficacy and safety of theophylline and doxofylline in patients with bronchial asthma and chronic obstructive pulmonary disease. J Basic Clin Physiol Pharmacol. 2015;26:443–51.
109. Kew KM, Kirtchuk L, Michell CI. Intravenous magnesium sulfate for treating adults with acute asthma in the emergency department. Cochrane Database Syst Rev. 2014;5:CD010909.
110. Rowe BH. Intravenous and inhaled $MgSO_4$ for acute asthma. Lancet Respir Med. 2013;1:276–7.
111. Emelyanov A, Fedoseev G, Barnes PJ. Reduced intracellular magnesium concentrations in asthmatic subjects. Eur Respir J. 1999;13:38–40.
112. Shivanthan MC, Rajapakse S. Magnesium for acute exacerbation of chronic obstructive pulmonary disease: a systematic review of randomised trials. Ann Thorac Med. 2014;9:77–80.
113. Malerba M, Radaeli A, Mancuso S, Polosa R. The potential therapeutic role of potassium channel modulators in asthma and chronic obstructive pulmonary disease. J Biol Regul Homeost Agents. 2010;24:123–30.
114. Angus RM, Mecallaum MJA, Hulks G, Thomson NC. Bronchodilator, cardiovascular and cyclic guanylyl monophosphate response to high dose infused atrial natriuretic peptide in asthma. Am Rev Respir Dis. 1993;147:1122–5.
115. Fluge T, Forssmann WG, Kunkel G, Schneider B, Mentz P, Forssmann K, Barnes PJ, Meyer M. Bronchodilation using combined urodilatin-albuterol administration in asthma: a randomized, double-blind, placebo-controlled trial. Eur J Med Res. 1999;4:411–5.
116. Calzetta L, Orlandi A, Page C, Rogliani P, Rinaldi B, Rosano G, Cazzola M, Matera MG. Brain natriuretic peptide: much more than a biomarker. Int J Cardiol. 2016;221:1031–8.
117. Edelson JD, Makhlina M, Silvester KR, Vengurlekar SS, Chen X, Zhang J, Koziol-White CJ, Cooper PR, Hallam TJ, Hay DW, Panettieri RA Jr. In vitro and in vivo pharmacological profile of PL-3994, a novel cyclic peptide (Hept-cyclo(Cys-His-Phe-d-Ala-Gly-Arg-d-Nle-Asp-Arg-Ile-Ser-Cys)-Tyr-[Arg mimetic]-NH(2)) natriuretic peptide receptor-A agonist that is resistant to neutral endopeptidase and acts as a bronchodilator. Pulm Pharmacol Ther. 2013;26:229–38.
118. Linden A, Hansson L, Andersson A, Palmqvist M, Arvidsson P, Lofdahl CG, Larsson P, Lotvall J. Bronchodilation by an inhaled VPAC(2) receptor agonist in patients with stable asthma. Thorax. 2003;58:217–21.
119. Liggett SB. Bitter taste receptors in the wrong place: novel airway smooth muscle targets for treating asthma. Trans Am Clin Climatol Assoc. 2014;125:64–74; discussion 74–5.
120. Loirand G. Rho kinases in health and disease: from basic science to translational research. Pharmacol Rev. 2015;67:1074–95.
121. Feng Y, LoGrasso PV, Defert O, Li R. Rho kinase (ROCK) inhibitors and their therapeutic potential. J Med Chem. 2016;59:2269–300.
122. Liu P, Zhang HM, Tang YJ, Sheng CF, Liu JX, Zeng YJ. Influence of Rho kinase inhibitor fasudil on late endothelial progenitor cells in peripheral blood of COPD patients with pulmonary artery hypertension. Bratisl Lek Listy. 2015;116:150–3.
123. Norris AA. Pharmacology of sodium cromoglycate. Clin Exp Allergy. 1996;26(Suppl 4):5–7.
124. Keller M, Schierholz J. Have inadequate delivery systems hampered the clinical success of inhaled disodium cromoglycate? Time for reconsideration. Expert Opin Drug Deliv. 2011;8:1–17.
125. Dryden DM, Spooner CH, Stickland MK, Vandermeer B, Tjosvold L, Bialy L, Wong K, Rowe BH. Exercise-induced bronchoconstriction and asthma. Evid Rep Technol Assess (Full Rep). 2010;189:1–154.
126. Parikh N, Chakraborti AK. Phosphodiesterase 4 (PDE4) inhibitors in the treatment of COPD: promising drug candidates and future directions. Curr Med Chem. 2016;23:129–41.
127. Chong J, Leung B, Poole P. Phosphodiesterase 4 inhibitors for chronic obstructive pulmonary disease. Cochrane Database Syst Rev. 2017;9:CD002309.
128. Yan JH, Gu WJ, Pan L. Efficacy and safety of roflumilast in patients with stable chronic obstructive pulmonary disease: a meta-analysis. Pulm Pharmacol Ther. 2014;27:83–9.
129. Kawamatawong T. Roles of roflumilast, a selective phosphodiesterase 4 inhibitor, in airway diseases. J Thorac Dis. 2017;9:1144–54.
130. Bateman ED, Goehring UM, Richard F, Watz H. Roflumilast combined with montelukast versus montelukast alone as add-on treatment in patients with moderate-to-severe asthma. J Allergy Clin Immunol. 2016;138:142–9.
131. Franciosi LG, Diamant Z, Banner KH, Zuiker R, Morelli N, Kamerling IM, de Kam ML, Burggraaf J, Cohen AF, Cazzola M, Calzetta L, Singh D, Spina D, Walker MJ, Page CP. Efficacy and safety of RPL554, a dual PDE3 and PDE4 inhibitor, in healthy volunteers and in patients with asthma or chronic obstructive pulmonary disease: findings from four clinical trials. Lancet Respir Med. 2013;1:714–27.
132. Goudie AR, Lipworth BJ, Hopkinson PJ, Wei L, Struthers AD. Tadalafil in patients with chronic

obstructive pulmonary disease: a randomised, double-blind, parallel-group, placebo-controlled trial. Lancet Respir Med. 2014;2:293–300.
133. Vitulo P, Stanziola A, Confalonieri M, Libertucci D, Oggionni T, Rottoli P, Paciocco G, Tuzzolino F, Martino L, Beretta M, Callari A, Amaducci A, Badagliacca R, Poscia R, Meloni F, Refini RM, Geri P, Baldi S, Ghio S, D'Alto M, Argiento P, Sofia M, Guardamagna M, Pezzuto B, Vizza CD. Sildenafil in severe pulmonary hypertension associated with chronic obstructive pulmonary disease: a randomized controlled multicenter clinical trial. J Heart Lung Transplant. 2017;36:166–74.
134. Durham AL, Caramori G, Chung KF, Adcock IM. Targeted anti-inflammatory therapeutics in asthma and chronic obstructive lung disease. Transl Res. 2016;167:192–203.
135. Barnes PJ. Inflammatory mechanisms in patients with chronic obstructive pulmonary disease. J Allergy Clin Immunol. 2016;138:16–27.
136. van Ganse E, Kaufman L, Derde MP, Yernault JC, Delaunois L. Effects of antihistamines in adult asthma: a meta-analysis of clinical trials. Eur Respir J. 1997;10:2216–24.
137. Mener DJ, Lin SY. Improvement and prevention of asthma with concomitant treatment of allergic rhinitis and allergen-specific therapy. Int Forum Allergy Rhinol. 2015;5(Suppl 1):S45–50.
138. Thurmond RL, Venable J, Savall B, La D, Snook S, Dunford PJ, Edwards JP. Clinical development of Histamine H4 receptor antagonists. Handb Exp Pharmacol. 2017;241:301–20.
139. Grant SM, Goa KL, Fitton A, Sorkin EM. Ketotifen. A review of its pharmacodynamic and pharmacokinetic properties, and therapeutic use in asthma and allergic disorders. Drugs. 1990;40:412–48.
140. Peters-Golden M, Henderson WR Jr. Leukotrienes. N Engl J Med. 2007;357:1841–54.
141. Simon RA, Dazy KM, Waldram JD. Aspirin-exacerbated respiratory disease: characteristics and management strategies. Expert Rev Clin Immunol. 2015;11:805–17.
142. Tamada T, Ichinose M. Leukotriene receptor antagonists and antiallergy drugs. Handb Exp Pharmacol. 2017;237:153–69.
143. Chauhan BF, Ducharme FM. Addition to inhaled corticosteroids of long-acting beta2-agonists versus anti-leukotrienes for chronic asthma. Cochrane Database Syst Rev. 2014;1:CD003137.
144. Laidlaw TM, Boyce JA. Aspirin-exacerbated respiratory disease—new prime suspects. N Engl J Med. 2016;374:484–8.
145. Boulet LP, O'Byrne PM. Asthma and exercise-induced bronchoconstriction in athletes. N Engl J Med. 2015;372:641–8.
146. Zhang HP, Jia CE, Lv Y, Gibson PG, Wang G. Montelukast for prevention and treatment of asthma exacerbations in adults: systematic review and meta-analysis. Allergy Asthma Proc. 2014;35:278–87.
147. O'Sullivan S, Akveld M, Burke CM, Poulter LW. Effect of the addition of montelukast to inhaled fluticasone propionate on airway inflammation. Am J Respir Crit Care Med. 2003;167:745–50.
148. Bernstein JA, Liu N, Knorr BA, Smugar SS, Hanley WD, Reiss TF, Greenberg S. MK-0633, a potent 5-lipoxygenase inhibitor, in chronic obstructive pulmonary disease. Respir Med. 2011;105: 392–401.
149. Woodruff PG, Albert RK, Bailey WC, Casaburi R, Connett JE, Cooper JA Jr, Criner GJ, Curtis JL, Dransfield MT, Han MK, Harnden SM, Kim V, Marchetti N, Martinez FJ, McEvoy CE, Niewoehner DE, Reilly JJ, Rice K, Scanlon PD, Scharf SM, Sciurba FC, Washko GR, Lazarus SC, COPD Clinical Research Network. Randomized trial of zileuton for treatment of COPD exacerbations requiring hospitalization. COPD. 2011;8:21–9.
150. Nathani N, Little MA, Kunst H, Wilson D, Thickett DR. Churg-Strauss syndrome and leukotriene antagonist use: a respiratory perspective. Thorax. 2008;63:883–8.
151. Bibby S, Healy B, Steele R, Kumareswaran K, Nelson H, Beasley R. Association between leukotriene receptor antagonist therapy and Churg-Strauss syndrome: an analysis of the FDA AERS database. Thorax. 2010;65:132–8.
152. Calapai G, Casciaro M, Miroddi M, Calapai F, Navarra M, Gangemi S. Montelukast-induced adverse drug reactions: a review of case reports in the literature. Pharmacology. 2014;94:60–70.
153. Vijverberg SJH, Farzan N, Slob EMA, Neerincx AH, Maitland-van der Zee AH. Treatment response heterogeneity in asthma: the role of genetic variation. Expert Rev Respir Med. 2018;12:55–65.
154. Back M. Functional characteristics of cysteinyl-leukotriene receptor subtypes. Life Sci. 2002;71:611–22.
155. Kent SE, Bentley JH, Miller D, Sterling R, Menendez R, Tarpay M, Pearlman DS, Norris V. The effect of GSK2190915, a 5-lipoxygenase-activating protein inhibitor, on exercise-induced bronchoconstriction. Allergy Asthma Proc. 2014;35:126–33.
156. Avila PC. Does anti-IgE therapy help in asthma? Efficacy and controversies. Annu Rev Med. 2007;58:185–203.
157. Stoll P, Bähker A, Ulrich M, Bratke K, Garbe K, Christian Virchow J, Lommatzsch M. The dendritic cell high-affinity IgE receptor is overexpressed in both asthma and severe COPD. Clin Exp Allergy. 2016;46:575–83.
158. Normansell R, Walker S, Milan SJ, Walters EH, Nair P. Omalizumab for asthma in adults and children. Cochrane Database Syst Rev. 2014;1:CD003559.
159. Lieberman PL, Jones I, Rajwanshi R, Rosén K, Umetsu DT. Anaphylaxis associated with omalizumab administration: risk factors and patient characteristics. J Allergy Clin Immunol. 2017;140: 1734–1736.e4.

160. Gauvreau GM, Arm JP, Boulet LP, Leigh R, Cockcroft DW, Davis BE, Mayers I, FitzGerald JM, Dahlen B, Killian KJ, Laviolette M, Carlsten C, Lazarinis N, Watson RM, Milot J, Swystun V, Bowen M, Hui L, Lantz AS, Meiser K, Maahs S, Lowe PJ, Skerjanec A, Drollmann A, O'Byrne PM. Efficacy and safety of multiple doses of QGE031 (ligelizumab) versus omalizumab and placebo in inhibiting allergen-induced early asthmatic responses. J Allergy Clin Immunol. 2016;138:1051–9.
161. Farne HA, Wilson A, Powell C, Bax L, Milan SJ. Anti-IL-5 therapies for asthma. Cochrane Database Syst Rev. 2017;9:CD010834.
162. Tian BP, Zhang GS, Lou J, Zhou HB, Cui W. Efficacy and safety of benralizumab for eosinophilic asthma: a systematic review and meta-analysis of randomized controlled trials. J Asthma. 2017;6:1–10. https://doi.org/10.1080/02770903.2017.1379534.. [Epub ahead of print].
163. Condreay L, Chiano M, Ortega H, Buchan N, Harris E, Bleecker ER, Thompson PJ, Humbert M, Gibson P, Yancey S, Ghosh S. No genetic association detected with mepolizumab efficacy in severe asthma. Respir Med. 2017;132:178–80.
164. Pelaia C, Vatrella A, Gallelli L, Terracciano R, Navalesi P, Maselli R, Pelaia G. Dupilumab for the treatment of asthma. Expert Opin Biol Ther. 2017;17:1565–72.
165. Domingo C, Moreno A, Mirapeix R. Rationale for the use of immunomodulatory therapies in the global initiative for asthma (GINA) step V asthma other than oral glucocorticosteroids. Intern Med J. 2011;41:525–36.
166. Davies H, Olson L, Gibson P. Methotrexate as a steroid sparing agent for asthma in adults. Cochrane Database Syst Rev. 2000;2:CD000391.
167. Evans DJ, Cullinan P, Geddes DM. Gold as an oral corticosteroid sparing agent in stable asthma. Cochrane Database Syst Rev. 2001;2:CD002985.
168. Evans DJ, Cullinan P, Geddes DM. Cyclosporin as an oral corticosteroid sparing agent in stable asthma (Cochrane Review). Cochrane Database Syst Rev. 2001;2:CD002993.
169. Krivan G, Jolles S, Granados EL, Paolantonacci P, Ouaja R, Cissé OA, Bernatowska E. New insights in the use of immunoglobulins for the management of immune deficiency (PID) patients. Am J Clin Exp Immunol. 2017;6:76–83.
170. Virchow JC, Backer V, Kuna P, Prieto L, Nolte H, Villesen HH, Ljørring C, Riis B, de Blay F. Efficacy of a house dust mite sublingual allergen immunotherapy tablet in adults with allergic asthma: a randomized clinical trial. JAMA. 2016;315: 1715–25.
171. Shamji MH, Durham SR. Mechanisms of allergen immunotherapy for inhaled allergens and predictive biomarkers. J Allergy Clin Immunol. 2017;140:1485–98.
172. Ricketti PA, Alandijani S, Lin CH, Casale TB. Investigational new drugs for allergic rhinitis. Expert Opin Investig Drugs. 2017;26:279–92.
173. Normansell R, Kew KM, Stovold E. Interventions to improve adherence to inhaled steroids for asthma. Cochrane Database Syst Rev. 2017;4:CD012226.
174. Woodcock A, Vestbo J, Bakerly ND, New J, Gibson JM, McCorkindale S, Jones R, Collier S, Lay-Flurrie J, Frith L, Jacques L, Fletcher JL, Harvey C, Svedsater H, Leather D, Salford Lung Study Investigators. Effectiveness of fluticasone furoate plus vilanterol on asthma control in clinical practice: an open-label, parallel group, randomised controlled trial. Lancet. 2017;390:2247–55. Erratum in: Lancet. 2017;390:e40.
175. Adcock IM, Caramori G, Chung KF. New targets for drug development in asthma. Lancet. 2008;372:1073–87.
176. Bel EH, Ten Brinke A. New anti-eosinophil drugs for asthma and COPD: targeting the trait! Chest. 2017;152:1276–82.
177. Caramori G, Chung KF, Adcock IM. Cytokine inhibition in the treatment of COPD. Int J Chron Obstruct Pulmon Dis. 2014;9:397–412. eCollection 2014.
178. Chung KF. Targeting the interleukin pathway in the treatment of asthma. Lancet. 2015;386:1086–96.
179. Caramori G, Di Stefano A, Casolari P, Kirkham PA, Padovani A, Chung KF, Papi A, Adcock IM. Chemokines and chemokine receptors blockers as new drugs for the treatment of chronic obstructive pulmonary disease. Curr Med Chem. 2013;20:4317–49.
180. Pease JE. Targeting chemokine receptors in allergic disease. Biochem J. 2011;434:11–24.
181. O'Byrne PM, Metev H, Puu M, Richter K, Keen C, Uddin M, Larsson B, Cullberg M, Nair P. Efficacy and safety of a CXCR2 antagonist, AZD5069, in patients with uncontrolled persistent asthma: a randomised, double-blind, placebo-controlled trial. Lancet Respir Med. 2016;4:797–806.
182. Kirsten AM, Förster K, Radeczky E, Linnhoff A, Balint B, Watz H, Wray H, Salkeld L, Cullberg M, Larsson B. The safety and tolerability of oral AZD5069, a selective CXCR2 antagonist, in patients with moderate-to-severe COPD. Pulm Pharmacol Ther. 2015;31:36–41.
183. Chen X, Tang S, Liu K, Li Q, Kong H, Zeng X, Xie W, Wang H. Therapy in stable chronic obstructive pulmonary disease patients with pulmonary hypertension: a systematic review and meta-analysis. J Thorac Dis. 2015;7:309–19.
184. Barnes PJ. Identifying molecular targets for new drug development for chronic obstructive pulmonary disease: what does the future hold? Semin Respir Crit Care Med. 2015;36:508–22.
185. Moreno-Macias H, Romieu I. Effects of antioxidant supplements and nutrients on patients with asthma and allergies. J Allergy Clin Immunol. 2014;133:1237–44.

186. Hesslinger C, Strub A, Boer R, Ulrich WR, Lehner MD, Braun C. Inhibition of inducible nitric oxide synthase in respiratory diseases. Biochem Soc Trans. 2009;37:886–91.
187. Caughey GH. Mast cell proteases as pharmacological targets. Eur J Pharmacol. 2016;778:44–55.
188. Krug N, Hohlfeld JM, Kirsten AM, Kornmann O, Beeh KM, Kappeler D, Korn S, Ignatenko S, Timmer W, Rogon C, Zeitvogel J, Zhang N, Bille J, Homburg U, Turowska A, Bachert C, Werfel T, Buhl R, Renz J, Garn H, Renz H. Allergen-induced asthmatic responses modified by a GATA3-specific DNAzyme. N Engl J Med. 2015;372:1987–95.
189. Caramori G, Casolari P, Adcock I. Role of transcription factors in the pathogenesis of asthma and COPD. Cell Commun Adhes. 2013;20:21–40.
190. Barnes PJ. Kinases as novel therapeutic targets in asthma and chronic obstructive pulmonary disease. Pharmacol Rev. 2016;68:788–815.
191. Papi A, Contoli M, Adcock IM, Bellettato C, Padovani A, Casolari P, Stanciu LA, Barnes PJ, Johnston SL, Ito K, Caramori G. Rhinovirus infection causes steroid resistance in airway epithelium through nuclear factor κB and c-Jun N-terminal kinase activation. J Allergy Clin Immunol. 2013;132:1075–85.
192. Watz H, Barnacle H, Hartley BF, Chan R. Efficacy and safety of the p38 MAPK inhibitor losmapimod for patients with chronic obstructive pulmonary disease: a randomised, double-blind, placebo-controlled trial. Lancet Respir Med. 2014;2:63–72.
193. Vallese D, Ricciardolo FL, Gnemmi I, Casolari P, Brun P, Sorbello V, Capelli A, Cappello F, Cavallesco GN, Papi A, Chung KF, Balbi B, Adcock IM, Caramori G, Di Stefano A. Phospho-p38 MAPK expression in COPD patients and asthmatics and in challenged bronchial epithelium. Respiration. 2015;89:329–42.

24 Immunotherapy in the Management of Asthma and Other Allergic Conditions

Ahmad Hamad and Timothy P. Moran

24.1 Introduction

Over the past few decades, a number of pharmacological agents have been developed for the treatment of ASTHMA and ALLERGY, including antihistamines, leukotriene receptor antagonists, and GLUCOCORTICOIDS. While these medications are effective for controlling symptoms of allergic disease in most patients, they provide only temporary relief and do not change the underlying course of disease. Moreover, some patients experience intolerable side effects from ASTHMA and ALLERGY drugs, thus limiting their efficacy. For these patients, ALLERGY immunotherapy offers a viable treatment option. ALLERGY immunotherapy can be as effective as pharmacological agents for the treatment of allergic rhinitis and can significantly improve symptoms in patients with allergen-driven ASTHMA. In contrast to pharmacological treatments, ALLERGY immunotherapy has the potential to provide long-lasting clinical benefits through the induction of TOLERANCE. The advancement of research in allergic disease pathogenesis has also resulted in the development of biological agents that target key cellular and molecular mediators of allergic inflammation. While many of these biological agents remain investigational, a few have been recently approved for the treatment of patients with certain subtypes of ASTHMA. Greater understanding of which patients are most likely to benefit from a specific immunotherapy will further improve the feasibility of personalized medicine for the treatment of allergic disease.

24.2 General Aspects of Immunotherapy in Allergy and Asthma

In clinical medicine, the term "immunotherapy" describes the manipulation of the immune system for the treatment of disease. Immunotherapies can be designed to either enhance or suppress immune responses. Immunotherapies that stimulate the immune system are typically employed for cancer treatment, where the goal is to promote antitumor immunity in an effort to destroy malignant cells. In contrast, immunotherapies used for the treatment of ASTHMA and allergic disorders seek to suppress or modify maladaptive immune responses that contribute to allergic disease pathogenesis.

To understand the rationale for what pathways are targeted by immunotherapies for ASTHMA and atopic diseases, it is important to understand how allergic responses are initiated and propagated (Fig. 24.1). In individuals genetically

Final manuscript submitted on May 15, 2017.

A. Hamad · T. P. Moran (✉)
Department of Pediatrics, University of North Carolina School of Medicine, Chapel Hill, NC, USA

UNC Hospitals Children's Specialty Clinic, N.C. Children's Hospital, Chapel Hill, NC, USA
e-mail: tmoran@email.unc.edu

M. J. Parnham et al. (eds.), *Nijkamp and Parnham's Principles of Immunopharmacology*,
https://doi.org/10.1007/978-3-030-10811-3_24

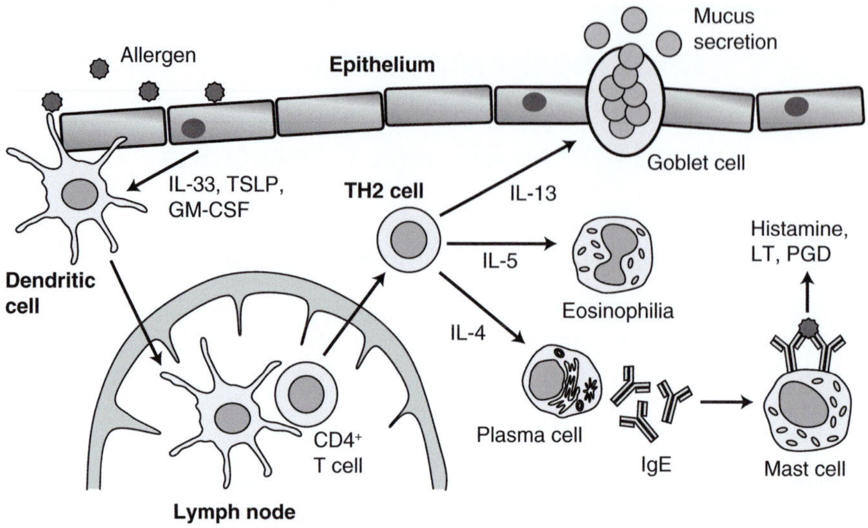

Fig. 24.1 Initiation and progression of allergic immune responses. Allergen exposure initially occurs at barrier epithelial surfaces such as the lung, gut, or skin. Allergen penetrating the epithelium is taken up by dendritic cells (DCs), which subsequently undergo maturation in response to epithelial-derived cytokines such as interleukin-33 (IL-33), thymic stromal lymphopoietin (TSLP), and granulocyte-macrophage colony-stimulating factor (GM-CSF). Maturate DCs migrate to tissue-draining lymph nodes, where they present allergen-derived peptides to naïve $CD4^+$ T cells. Upon antigen recognition, T cells differentiate into T helper 2 (TH2) effector cells, which produce the proallergic cytokines IL-4, IL-5, and IL-13. These cytokines have multiple effects, including induction of mucus secretion by goblet cells (IL-13), activation and recruitment of eosinophils (IL-5), and the development of IgE-producing plasma cells (IL-4). Allergen-specific IgE produced by plasma cells enters circulation and binds to receptors (FcεR) on tissue mast cells and other effector cells. Binding of allergen to membrane-associated IgE results in receptor cross-linking, which triggers cell degranulation and release of inflammatory mediators including histamine, leukotrienes (LT), and prostaglandins (PGD)

predisposed to atopy, exposure to environmental allergens results in allergic sensitization, which is characterized by the development of allergen-specific IMMUNOGLOBULIN E (IgE). Allergic sensitization can occur in the respiratory tract, gut, or skin, where allergens are taken up by antigen-presenting cells known as dendritic cells (DCs). DCs become activated directly by allergens or indirectly by epithelial-derived CYTOKINES, such as interleukin 33 (IL-33), thymic stromal lymphopoietin (TSLP), and granulocyte-macrophage colony-stimulating factor (GM-CSF). Once activated, DCs migrate to tissue-draining lymph nodes, where they present allergen-derived peptides on major histocompatibility (MHC) class II molecules to antigen-naïve $CD4^+$ T cells. Upon recognition of allergen-derived peptides, $CD4^+$ T cells proliferate and differentiate into T helper 2 (TH2) effector cells, which produce the CYTOKINES IL-4, IL-5, and IL-13. IL-4 instructs allergen-specific B cells to undergo isotype class switching and become IgE-producing plasma cells. IL-5 and IL-13 promote tissue eosinophilia and mucous gland hyperplasia, respectively, which are classic hallmarks of allergic inflammation. Allergen-specific IgE circulates through the body and binds to high-affinity IgE receptors (FcεRI) on MAST CELLS, BASOPHILS, and other effectors cells in peripheral tissues. Subsequent encounter with ALLERGEN leads to cross-linking of membrane-bound IgE, resulting in cell degranulation and the release of inflammatory mediators such as HISTAMINE, LEUKOTRIENES, PROSTAGLANDINS, and proteolytic enzymes. These factors are responsible for the immediate HYPERSENSITIVITY reaction to allergens, which is characterized by increased vascular permeability, vasodilation, and bronchial and visceral smooth muscle contraction. The immediate responses are then followed by a late-phase reaction where eosinophils, BASOPHILS, and allergen-specific TH2 cells are recruited to the site of ALLERGEN exposure. These cells release a number of inflammatory CYTOKINES that promote mucus hypersecretion, goblet cell hyperplasia, smooth muscle proliferation, and tissue remodeling [1].

Immunotherapy for allergic disease can be broadly divided into allergen-specific and allergen-nonspecific treatments. As the name implies, allergen-specific immunotherapy (AIT) aims to modulate immune responses against a particular ALLERGEN or allergens that are causing disease symptoms. AIT has the potential to permanently alter the maladaptive immune response contributing to allergic disease and thus provides a curative option for patients. However, AIT requires knowledge of what allergen(s) are the main drivers of disease in a particular patient and may take several months before therapeutic benefit is attained. Allergen-nonspecific immunotherapy involves strategies that inhibit key mediators of allergic inflammation, such as proallergic CYTOKINES and IgE. While these nonspecific immunotherapies may be beneficial for a variety of allergic diseases, they are generally expensive, require continuous treatment, and have the potential to inhibit cells and molecular pathways that are important for normal immune function.

24.3 Allergen-Specific Immunotherapy

AIT is a potentially disease-altering treatment for allergic diseases. Initially described over 100 years ago, AIT involves the repeated administration of ALLERGEN to a sensitized individual with the goal of reducing symptoms upon natural ALLERGEN exposure. Subcutaneous immunotherapy (SCIT), or "ALLERGY shots," has traditionally been the predominant form of AIT used in clinical practice. Over the past 30 years, sublingual immunotherapy (SLIT) has grown in popularity, particularly in Europe. Other ALLERGEN delivery routes under investigation include oral, epicutaneous and intralymphatic injection. Several controlled trials have demonstrated AIT to be efficacious for patients with allergic rhinoconjunctivitis, allergic ASTHMA, and stinging insect ALLERGY [2, 3]. AIT may also be effective for atopic dermatitis in individuals sensitized to aeroallergens and is being investigated as a potential treatment for food allergies [4]. In addition to reducing allergic disease symptoms, AIT can also prevent new sensitization to allergens and the development of ASTHMA in patients with allergic rhinitis [2]. AIT is only indicated for patients who have allergic disease symptoms caused by ALLERGEN exposure and the presence of specific IgE to that ALLERGEN. While generally safe, AIT can cause life-threatening reactions and therefore should only be prescribed by appropriately trained healthcare providers.

24.3.1 Mechanisms of AIT

AIT results in several immunological changes that likely contribute to its clinical efficacy (Fig. 24.2) [2, 5]. Shortly after initiation of AIT, there is an early desensitization phase characterized by decreased mast cell and basophil activity. This is followed by the development of allergen-specific T cell TOLERANCE, which is thought to be necessary for the long-term clinical benefits of AIT. Allergen-specific IgE levels initially rise during AIT but then progressively decline as specific IgG4 levels increase. After months to years of treatment, there is a gradual reduction in tissue MAST CELLS and eosinophils, which is mirrored by a decrease in skin test reactivity to allergens. The effects of AIT on innate and adaptive immune responses are summarized below.

24.3.1.1 Mast Cell, Basophil, and Eosinophil Responses

AIT rapidly leads to a decrease in IgE-mediated degranulation of MAST CELLS and BASOPHILS. Desensitization of MAST CELLS and BASOPHILS occurs despite the presence of high levels of allergen-specific IgE in patients. The mechanisms for the early desensitization of MAST CELLS and BASOPHILS during AIT are not fully known but may involve upregulation of HISTAMINE receptor 2, which can suppress IgE-

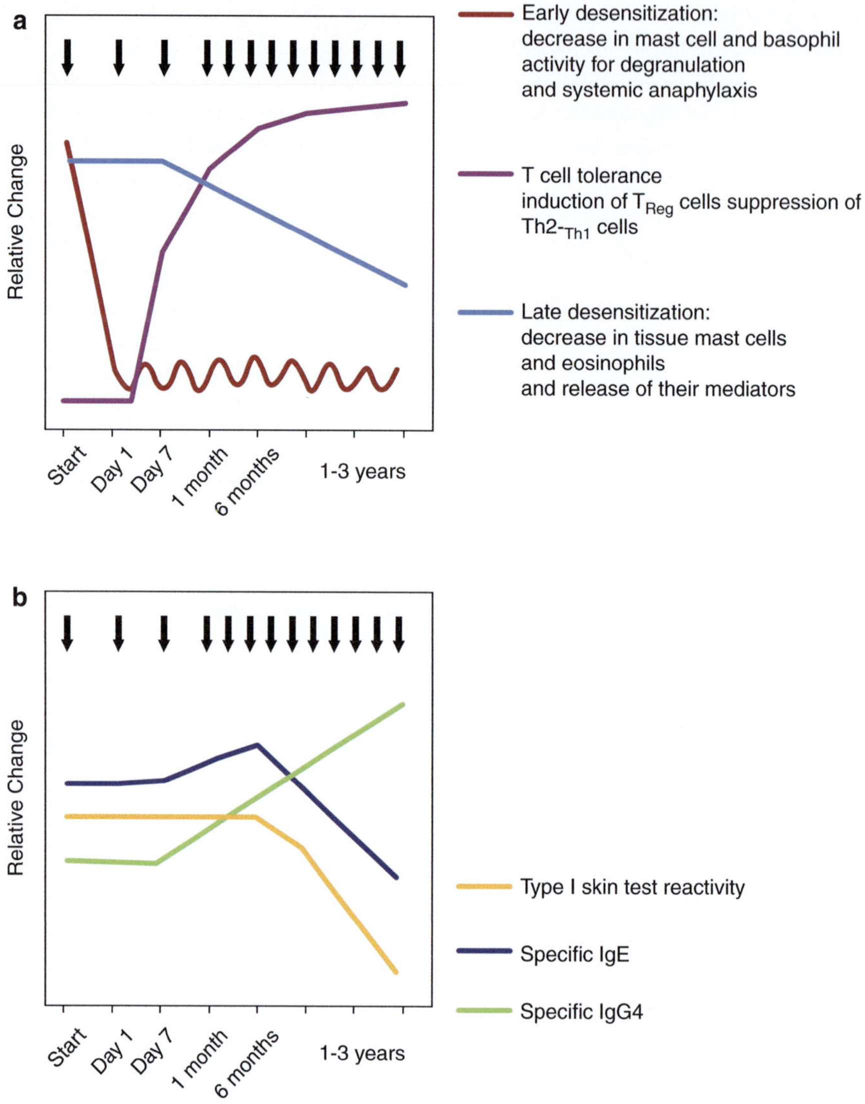

Fig. 24.2 Immunological changes during AIT. (**a**) Immediately after initiation of AIT, there is an early desensitization phase characterized by decreased mast cell and basophil reactivity to allergen. This is followed by the induction of regulatory T cells (Treg), which suppress allergen-specific TH2 responses and promote the development of tolerance. After several months, there is a late desensitization phase where tissue mast cell and eosinophil numbers are gradually reduced. (**b**) Humoral immune changes during AIT are characterized by an initial increase then decrease in allergen-specific IgE. This is mirrored by increase in specific IgG4, which continues to rise during the course of AIT. A significant decrease in type I skin reactivity to allergens is also observed relatively late during AIT. Reprinted from Journal of Allergy and Clinical Immunology, A. W. Burks et al., Update on allergy immunotherapy: American Academy of Allergy, Asthma & Immunology/European Academy of Allergy and Clinical Immunology/PRACTALL consensus report, pages 1288–1296, 2013©, with permission from Elsevier

mediated activation and degranulation of BASOPHILS [6]. Tissue eosinophils and MAST CELLS are decreased after several months of AIT, which is associated with inhibition of immediate and late responses to ALLERGEN skin testing [7].

24.3.1.2 Allergen-Specific Antibody Responses

AIT initially causes a transient increase in allergen-specific IgE levels, which is followed by a gradual decrease over several months of therapy [8, 9].

AIT can also blunt the rise in allergen-specific IgE that occurs during natural exposure to seasonal allergens [9]. AIT is associated with increases in allergen-specific IgG1 and IgG4, which occurs after the initial rise in IgE and can persist for several years following cessation of therapy [10]. IgG4 may compete for ALLERGEN with IgE, thereby preventing IgE-mediated mast cell degranulation following ALLERGEN exposure [11]. IgG4 may also inhibit binding of IgE to the low-affinity IgE receptor, resulting in decreased ALLERGEN uptake and presentation by antigen-presenting cells [12]. Functional rather than quantitative measurements of allergen-specific IgG4 in patients undergoing AIT could be a useful marker for clinical efficacy [13].

24.3.1.3 Cellular Immune Responses

AIT modifies allergen-specific T cell responses, which is thought to be critical for the development of long-term clinical TOLERANCE (see Chap. 3). During AIT, there is a gradual decrease in allergen-specific TH2 cells producing IL-4 [14]. This is associated with the development of allergen-specific regulatory T cells, which are important for suppressing allergic inflammation and maintaining TOLERANCE to innocuous antigens [5]. AIT results in increased numbers of IL-10-secreting Treg, which inhibit TH2 responses and induce allergen-specific IgG4 production by B cells [15, 16]. $CD4^+CD25^+$ Treg expressing the transcription factor forkhead box protein 3 (FoxP3) are also increased in patients undergoing AIT and are associated with decreased allergic inflammation at mucosal sites [17]. The importance of $FoxP3^+$ T cells in maintaining TOLERANCE to allergens is further underscored by the observation that patients with mutations in FoxP3 suffer from severe allergic disease in addition to autoimmunity [18].

24.3.2 Selection and Administration of Allergens for AIT

24.3.2.1 Allergen Extracts

ALLERGEN extracts are complex mixtures of allergenic and nonallergenic compounds including proteins, polysaccharides, lipids, nucleic acids, and metabolites. Most clinically relevant allergens are proteins or glycoproteins. ALLERGEN extracts are typically aqueous extractions of biological materials, such as pollens, arthropods (cockroaches or dust mites), insect venoms, animals (dander, pelt, or feathers), fungi (mycelia or spores), and foods. Glycerin (up to 50% volume by volume) is frequently added to ALLERGEN extracts to increase stability and extend the shelf life but can result in increased pain at the injection site during SCIT.

The majority of commercially available ALLERGEN extracts are labeled using a weight by volume (w/v) or protein nitrogen unit (PNU) designation. These measurements provide little information on the biological potency of the extract, as the actual ALLERGEN content of extracts can vary significantly between manufacturers. To address the problem of ALLERGEN extract variability, methods for standardizing extracts to a reference extract of known potency and composition have been developed [19]. In the USA, standardized ALLERGEN extracts are available for dust mite (*Dermatophagoides pteronyssinus* and *Dermatophagoides farinae*), cat hair or pelt, grasses (Bermuda, timothy, meadow fescue, sweet vernal, orchard, redtop, perennial ryegrass, and Kentucky bluegrass), short ragweed, and Hymenoptera venoms (wasp, honey bee, yellow jacket, yellow hornet, and white-faced hornet). Standardized extracts in the USA are most commonly labeled in bioequivalent ALLERGY units (BAU), and each extract is compared to a standard of potency that is established by the Center for Biological Evaluation and Research (CBER). In Europe, each ALLERGEN manufacturer uses an in-house reference extract for standardization, which ensures batch-to-batch uniformity but does not allow direct comparison of potency between different manufacturers.

Selection of allergens for AIT is based upon the clinical history and the results of skin tests or in vitro assays for allergen-specific IgE. AIT preparations should only include allergens to which the patient is sensitized and that are likely contributing to the patient's symptoms. Other important considerations include the patterns of cross-reactivity between different allergens and the compatibility of extracts when mixed. Some ALLERGEN extracts, such as cockroach and molds, contain proteases that can degrade other allergens and therefore should not be mixed with pollens, dust mites, or animal dander [20]. It should be noted that while the majority of patients with allergic rhinitis or ASTHMA are sensitized to multiple allergens, nearly all controlled studies of AIT efficacy have been performed with single-allergen extracts.

24.3.2.2 Route of Administration

For the past century, subcutaneous injection of ALLERGEN extracts has been the most common administration route for AIT. While SCIT has proven efficacious for the treatment of allergic respiratory diseases and venom HYPERSENSITIVITY, it requires frequent injections that are given in a clinician's office due to the risk of ANAPHYLAXIS. Because of these limitations, sublingual administration of liquid ALLERGEN extracts or dissolvable ALLERGEN tablets has been growing in popularity. SLIT has been used for the treatment of allergic respiratory diseases in Europe for several years but was only recently approved for the treatment of grass pollen and ragweed ALLERGY in the USA [21]. Systematic reviews have found SLIT to be safe and effective for the treatment of allergic rhinitis [22], whereas its efficacy for allergic ASTHMA requires further study [23]. The advantages of SLIT include self-administration at home and reduced risk of ANAPHYLAXIS relative to SCIT [24]. Few studies have directly compared the efficacy of SLIT and SCIT, although a meta-analysis suggested that SCIT may be more effective for the treatment of grass pollen ALLERGY [25].

Other routes of ALLERGEN administration for AIT are under investigation [26]. Oral immunotherapy (OIT), which involves ingesting increasing amounts of ALLERGEN, is being evaluated for the treatment of food ALLERGY. Several studies have shown that OIT can result in desensitization to food allergens, but the long-term efficacy is unclear [27]. Administration of ALLERGEN through an intact skin barrier (epicutaneous immunotherapy) may allow sustained ALLERGEN delivery at low concentrations, thereby improving the efficacy and safety of AIT [28]. Epicutaneous immunotherapy is currently being investigated for the treatment of food and respiratory allergies. Because ALLERGEN presentation to T and B cells primarily occurs in lymphatic tissues, there has been interest in directly administering modified allergens to lymph nodes in an effort to enhance AIT. Intralymphatic injection of allergens has been shown to increase the efficiency of AIT, and therefore as few as three ALLERGEN injections can provide prolonged symptom relief in patients with allergic rhinoconjunctivitis [29]. While these alternative delivery routes appear promising, more research on their efficacy and safety is required before they can be employed for routine clinical use.

24.3.2.3 Dosage and Administration Schedules

The doses of ALLERGEN extracts used for AIT depends upon whether the extracts are standardized or nonstandardized, the treatment schedule, whether single or multiple allergens are used, and the route of delivery. Most European practices prescribe single-allergen AIT, whereas multi-allergen AIT is commonly used in the USA For SCIT, the effective maintenance doses for several major allergens have been characterized in controlled trials (Table 24.1) [30, 31]. However, it is important to note that these studies were not performed with commercial standardized extracts in the USA, which are standardized by their bioactivity and not their major ALLERGEN content [31]. Nevertheless, probable effective doses for standardized ALLERGEN extracts available in the USA have been proposed [30]. With few exceptions, the effective maintenance doses of nonstandardized extracts for SCIT are unknown. For nonstandardized pollen extracts, 0.5 ml of a 1:100 w/v or 1:200 w/v solution is likely efficacious, whereas cockroach and fungal extracts may require higher dosing. Fewer studies are available regarding the effective ALLERGEN doses for SLIT, but generally the cumulative doses of ALLERGEN used for SLIT are 20–30 times greater than the dose used in SCIT [2].

Table 24.1 Probable effective maintenance doses for SCIT

Allergen extract	Major allergen	Dose range (μg)
Dermatophagoides (dust mite)	Der p 1, Der f 1	7–10
Short ragweed	Amb a 1	6–24
Birch	Bet v 1	3–12
Timothy grass	Phl p 5	15–20
Cat hair or pelt	Fel d 1	11–17
Dog dander	Can f 1	15
Hymenoptera venom	Not applicable (based on total protein)	100

The conventional schedule for SCIT involves an initial buildup phase, during which patients receive injections of increasing ALLERGEN doses once or twice weekly for several weeks. This is then followed by a maintenance phase, where patients receive the maintenance dose of ALLERGEN every 2–4 weeks over several years. For the buildup phase, serial tenfold dilutions of the maintenance dose are prepared. Most buildup phases begin with a 1:1000 volume to volume (v/v) dilution of ALLERGEN extract, but lower concentrations may be required for highly sensitive patients. A typical administration schedule for SCIT is shown in Box 24.1. Protocols that accelerate the buildup phase, such as cluster or rush schedules, have been described but are associated with increased risk for systemic reactions [3, 30]. In SLIT, a buildup phase is generally not required or is significantly shortened.

Box 24.1: Example of a Conventional Schedule for Aeroallergen SCIT

Vial #4 1:1000 v/v	Vial #3 1:100 v/v	Vial #2 1:10 v/v	Vial #1 Maintenance concentration
0.05 ml	0.05 ml	0.05 ml	0.05 ml
↓	↓	↓	↓
0.1 ml	0.1 ml	0.1 ml	0.07 ml
↓	↓	↓	↓
0.2 ml	0.2 ml	0.15 ml	0.1 ml
↓	↓	↓	↓
0.4 ml	0.3 ml	0.2 ml	0.15 ml
	↓	↓	↓
	0.4 ml	0.25 ml	0.2 ml
	↓	↓	↓
	0.5 ml	0.3 ml	0.25 ml
		↓	↓
		0.35 ml	0.3 ml
		↓	↓
		0.4 ml	0.35 ml
		↓	↓
		0.45 ml	0.4 ml
		↓	↓
		0.5 ml	0.45 ml
			↓
			0.5 ml (maintenance dose)

Injections are administered once weekly beginning with Vial #4 and progressing to Vial #1. Once the maintenance dose is reached, it is typically given once weekly for 4–6 weeks and then every 2–4 weeks for minimum of 3 years. Note that the indicated dilutions are relative to the maintenance concentration (Vial #1)

While the optimal duration of AIT is unknown, there is general consensus that 3–5 years of therapy is most likely to be efficacious. A prospective study of asthmatic patients undergoing SCIT for dust mite ALLERGY suggests that at least 3 years of therapy is required for prolonged remission of symptoms after cessation of therapy [32]. Similarly, a prospective study of SLIT for house dust mite ALLERGY indicated that 3–4 years of treatment can provide long-term symptom relief [33].

24.3.3 Indications and Efficacy

AIT is indicated for the treatment of allergic rhinitis, allergic conjunctivitis, and allergic ASTHMA (Table 24.2). AIT is also indicated for patients with a history of systemic reaction to Hymenoptera stings. Atopic dermatitis associated with aeroallergen sensitivity is a potential indication for AIT, although this recommendation is based on a limited number of studies. AIT is not currently indicated for the treatment of food allergies. The efficacy of AIT for various allergic diseases is described in greater detail below.

24.3.3.1 Allergic Asthma

Several controlled studies have shown AIT (specifically SCIT) to be beneficial for carefully selected patients with allergic ASTHMA. A systematic review of 88 clinical trials concluded that AIT reduced ASTHMA symptoms, decreased ASTHMA medication usage, and improved bronchial hyperreactivity [34]. Additionally, AIT can reduce the dose of corticosteroids required for maintaining ASTHMA control [35, 36]. Importantly, SCIT has been reported to prevent the development of ASTHMA in children with

Table 24.2 Current indications for AIT

Allergic rhinitis (seasonal or perennial)
Allergic conjunctivitis (seasonal or perennial)
Allergic asthma (seasonal or perennial)
Systemic allergic reactions to Hymenoptera stings (e.g., bees, wasps, hornets, yellow jackets)
Potential indication
Atopic dermatitis associated with aeroallergen sensitivity

allergic rhinitis [37]. According to the National ASTHMA Education and Prevention Program guidelines for ASTHMA management in 2007, SCIT is recommended as an adjunct to standard pharmacotherapy for patients with allergic ASTHMA [38]. However, the Global Initiative for ASTHMA report concluded that evidence for the efficacy of AIT in ASTHMA is limited and the potential benefits must be carefully weighed against the risks of adverse effects and cost [38]. Overall, AIT is most likely to benefit sensitized patients for whom ALLERGEN exposure is an important trigger of ASTHMA symptoms.

24.3.3.2 Allergic Rhinitis and Conjunctivitis

AIT is indicated for the treatment of seasonal or perennial allergic rhinitis and conjunctivitis, particularly in patients who cannot tolerate pharmacotherapy [30]. Systematic reviews have found that both SCIT and SLIT can significantly reduce symptom scores and medication use in patients with allergic rhinitis [2, 39]. AIT has similar efficacy to pharmacological medications in controlling allergic rhinitis symptoms [40] and may be more cost-effective [41]. In contrast to pharmacotherapy, AIT can result in long-lasting clinical benefits after discontinuation [37, 42]. Furthermore, AIT may prevent the development of new allergies in monosensitized patients [43].

24.3.3.3 Hymenoptera Venom Allergy

Venom immunotherapy is indicated for adults and children with a history of systemic reactions to stinging Hymenoptera, as these patients have a 30–60% risk of experiencing another systemic reaction with each future sting [3]. Several trials have found that venom immunotherapy is highly effective in reducing the risk of future systemic reactions in venom-allergic individuals [44]. Venom immunotherapy appears to provide long-lasting protection in patients who have completed at least 3 years of therapy [45], although some patients with significant risk factors for recurrent systemic reactions may benefit from indefinite therapy. Only SCIT is approved for the treatment of stinging insect ALLERGY.

24.3.3.4 Atopic Dermatitis

AIT may be beneficial for some patients with atopic dermatitis who are sensitized to inhalant allergens, although this remains an area of debate. Both SCIT and SLIT with house dust mite ALLERGEN have been reported to improve atopic dermatitis in sensitized children [46, 47]. However, a recent systematic review found limited evidence to support the efficacy of AIT for atopic dermatitis [48]. While AIT appears to be safe in patients with atopic dermatitis, further studies are needed to determine which patients are most likely to respond to this therapy.

24.3.3.5 Food Allergy

There is growing interest in using AIT for treating patients with food allergies. Early studies investigating the efficacy of SCIT for food ALLERGY were abandoned due to the high rates of adverse reactions [49]. Recently, studies of OIT or SLIT to peanut, milk, and egg have yielded more encouraging results [27]. There is good evidence that OIT or SLIT to these food allergens can result in desensitization. However only a minority of patients experience long-term TOLERANCE after therapy is stopped [50, 51]. OIT appears to be more efficacious than SLIT for the treatment of food allergies but is associated with a higher risk of serious adverse events [52]. While studies of AIT for the treatment of food allergies have been promising, larger randomized clinical trials are needed before it can be routinely offered in the clinical setting.

24.3.4 Safety and Contraindications

AIT is associated with both local and systemic reactions that can range in severity from mild generalized itching to ANAPHYLAXIS. The risk of systemic reactions depends on the ALLERGEN used, the route of administration, dosing protocols, and patient characteristics. SCIT has a higher risk of systemic reactions compared to SLIT and thus should be administered in the clinical setting by trained personnel who are capable of managing ANAPHYLAXIS [30]. Patients receiving SCIT are generally observed for 30 min after each injection, as many but not all adverse reactions will occur during this period. While local symptoms such as oral pruritus are common with SLIT, systemic reactions are exceptionally rare, and therefore SLIT is typically administered by patients at home. Fatal reactions to SCIT are estimated to be 1 in every 2.5 million injections [53], whereas no SLIT-associated fatalities have been reported to date. Factors associated with fatal ANAPHYLAXIS include uncontrolled ASTHMA, prior history of systemic reactions during AIT, and a failure to administer epinephrine [53]. AIT for aeroallergens is contraindicated in patients with poorly controlled ASTHMA or significant cardiovascular disease that may reduce the patient's ability to survive a systemic reaction. In Europe, active autoimmune disorders, malignancy, and acquired immunodeficiency syndrome are also considered contraindications for AIT [54]. There are no absolute contraindications for venom immunotherapy, although certain preexisting medical conditions, such as severe cardiac disease, should be considered relative contraindications. AIT should be administered with caution to patients receiving beta-adrenergic blockers or angiotensin-converting enzyme inhibitors, as these medications have been associated with more severe systemic reactions [4]. While initiation of AIT during pregnancy should be avoided, ongoing therapy may be continued with caution. There are no defined age limits for SCIT or SLIT, although most guidelines recommend that AIT should be not be used in children under the age of 5 years [30, 54].

24.4 Nonspecific Immunotherapy

As the name implies, nonspecific immunotherapies aim to modulate allergic responses in an allergen-independent manner. Many nonspecific immunotherapies are biological agents or "biologics," which are synthesized in living organisms using methods that typically involve recombinant DNA technology. These therapies, which include monoclonal antibodies, target specific mediators of allergic inflammation such as IgE- and TH2-associated CYTOKINES. Other therapies stimulate innate immune signaling pathways that antagonize TH2 responses. A significant number of biologics have been developed over the past several years [55]. Most of these agents remain investigational, but a few have been approved for patients with certain types of ASTHMA, thus allowing more "personalized" treatment of allergic disease.

24.4.1 Specific Mediator Antagonists

A variety of biological agents have been developed to inhibit specific mediators involved with allergic inflammation. Most of the agents are being investigated for the treatment of moderate-to-severe ASTHMA, although it is likely that some may prove beneficial for patients with other allergic diseases such as atopic dermatitis and allergic rhinitis.

24.4.1.1 Anti-IgE Therapy

It is well established that IgE antibodies play a central role in the pathogenesis of allergic diseases, including ASTHMA (see Chaps. 4 and 21). The development of humanized monoclonal antibodies that inhibit IgE activity has provided a promising new approach for the treatment of allergic ASTHMA and other IgE-mediated diseases.

Omalizumab (Xolair®) is currently the only anti-IgE MONOCLONAL ANTIBODY available for clinical use. Omalizumab is a recombinant humanized IgG1 MONOCLONAL ANTIBODY that binds to a conserved domain in the Fc portion of the IgE molecule, thereby preventing binding of IgE to FceRI on effector cells [56]. Omalizumab only binds to free IgE molecules and therefore does not cause cross-linking of membrane-bound IgE and subsequent degranulation of effector cells. Serum concentrations of free IgE are rapidly decreased by 99% following administration of omalizumab [57]. The dramatic reduction in serum IgE results in a gradual decrease in FcεRI expression on effector cells, which in turn suppresses their responsiveness to ALLERGEN [57, 58]. Omalizumab treatment also reduces markers of allergic inflammation, including exhaled nitrogen oxide, blood and sputum eosinophilia, and serum levels of IL-13 [59].

Omalizumab is approved for the treatment of allergic ASTHMA and chronic idiopathic urticaria in both the USA and Europe. In the USA, omalizumab is indicated as add-on therapy for patients 12 years or older with moderate-to-severe allergic ASTHMA that is uncontrolled on inhaled corticosteroids [60]. Omalizumab therapy significantly decreases the rate of ASTHMA exacerbations, reduces the risk of hospitalization for ASTHMA, and allows for a reduction in inhaled glucocorticoid use [61]. Dosing for patients with ASTHMA is based on the patient's weight and pre-treatment total IgE level. The recommended dose is 0.016 mg/kg body weight per international unit of IgE every 4 weeks, which is administered subcutaneously every 2 or 4 weeks. Peak serum concentrations occur about 7 days after administration, and the serum half-life is 26 days [60]. Patients who respond to therapy typically do so within 12–16 weeks. The duration of treatment is indefinite, and there is limited data on whether patients experience persistent benefit after therapy is stopped.

Omalizumab has a favorable safety profile, although serious adverse effects have been described. Minor local cutaneous reactions at the injection site are the most common side effects. Interestingly, ANAPHYLAXIS has been reported in about 0.1% of patients receiving omalizumab, prompting the US Food and Drug Administration to recommend adding a boxed warning to the package insert. ANAPHYLAXIS may occur after any dose, although most episodes happen after the first few injections. Based on these findings, it has been recommended that all patients receiving omalizumab be prescribed an epinephrine autoinjector and should be observed for 2 h after each of the first three doses and then for 30 min after each subsequent injection [62]. Early trials suggested a possible association between omalizumab treatment and cancer; however, subsequent studies have not found an increased risk of malignancy in patients receiving omalizumab [63]. There is limited data on the safety of omalizumab during pregnancy, but a recent observational study found no increased risk of adverse neonatal outcomes in women receiving omalizumab [64]. While omalizumab should not be initiated during pregnancy, it can be continued if the therapeutic benefits are deemed to outweigh potential harms.

24.4.1.2 Anti-IL-5 Therapy

IL-5 is a homodimeric cytokine that is critical for the development, recruitment, and activation of eosinophils. Given the importance of eosinophils in allergic airway inflammation, agents targeting IL-5 have been developed for the treatment of ASTHMA, including monoclonal anti-IL-5 antibodies (mepolizumab and reslizumab) and anti-IL-5 receptor antibodies (benralizumab). Mepolizumab was recently approved for clinical use, while reslizumab and benralizumab are still undergoing clinical trials [65].

Mepolizumab is a fully humanized IgG1 MONOCLONAL ANTIBODY that binds IL-5 and blocks its interaction with the alpha chain of the IL-5 receptor expressed on eosinophils [66]. A single dose of mepolizumab dramatically reduces blood eosinophils for several weeks [67]. It was therefore surprising that early studies of mepolizumab failed to demonstrate significant clinical benefit in patients with ASTHMA [67, 68]. However, more recent trials that specifically focused on patients with markers of eosinophilic ASTHMA, such as blood or sputum eosinophilia, found that mepolizumab reduced the risk of ASTHMA exacerbations and had a significant steroid-sparing effect [69, 70]. These studies underscore the importance of phenotyping patients with ASTHMA to help identify those most likely to benefit from specific therapies.

Mepolizumab (Nucala®) has been approved in the USA and Europe as add-on therapy for patients with severe ASTHMA who are 12 years of age or older and have an eosinophilic phenotype [66]. What exactly constitutes an eosinophilic phenotype has not been clearly specified, but clinical trials suggest that patients should have a blood eosinophil count ≥150 cells per microliter prior to therapy [71]. Mepolizumab is administered subcutaneously at a dose of 100 mg every 4 weeks. Pharmacodynamic studies indicate that blood eosinophils are reduced by 84% within 4 weeks of treatment and remain so while on therapy. The mean terminal half-life is 16–22 days following subcutaneous administration. Common side effects include injection site reactions and headache. HYPERSENSITIVITY reactions have also been reported and may occur hours or days after administration. Herpes zoster infections have occurred in patients receiving mepolizumab, and therefore varicella vaccination should be considered prior to starting therapy [66]. Although no adverse events were observed in animal reproduction studies, there is insufficient data regarding the safety of mepolizumab during human pregnancy. Like all IgG monoclonal antibodies, mepolizumab is expected to cross the placenta, but the potential effects on the fetus are unknown.

24.4.1.3 Anti-IL-4 and -IL-13 Therapy

IL-4 and IL-13 are prototypic TH2 CYTOKINES that play a central role in allergic ASTHMA pathogenesis (see Chaps. 6 and 11). Both CYTOKINES bind to the alpha subunit of the IL-4 receptor, which forms heterodimeric complexes with either the common gamma chain (IL-4 type I receptor) or the IL-13 receptor alpha 1 chain (IL-4 type 2 receptor). IL-4 can bind to both of these receptor complexes, whereas IL-13 binds exclusively to the IL-4 type 2 receptor. Ligation of these receptors activates signaling pathways that induce transcription of genes involved with TH2 differentiation, airway hyperreactivity, and mucus production [72].

Several agents that block IL-4 and IL-13 activity are being investigated in clinical trials, but none have yet to be approved for use in patients [55]. Lebrikizumab is a humanized IgG4 MONOCLONAL ANTIBODY that specifically binds to and inhibits the activity of IL-13. In a randomized control trial of patients with moderate-to-severe ASTHMA, lebrikizumab was found to significantly improve airflow obstruction but did not decrease ASTHMA exacerbations in treated patients [73]. Interestingly, improvement in airflow was greatest in patients with elevated levels of periostin, which may be a potential biomarker for a TH2 ASTHMA phenotype. However, a subsequent study found no significant improvement in airflow obstruction or bronchodilator usage in patients receiving lebrikizumab, regardless of periostin levels [74]. Tralokinumab is another humanized anti-IL-13 MONOCLONAL ANTIBODY that has been evaluated in patients with moderate-to-severe ASTHMA [75]. While tralokinumab was well tolerated, it did not improve ASTHMA control as determined by a validated patient questionnaire, which was the primary end point of the study. Dupilumab is a MONOCLONAL ANTIBODY that binds to the IL-4 receptor alpha chain and thus inhibits both IL-4 and IL-13 activity. In patients with moderate-to-severe ASTHMA and elevated blood eosinophils, dupilumab treatment was associated with improved lung function and fewer ASTHMA exacerbations when ASTHMA controller medications were gradually weaned [76]. Pitrakinra is a recombinant human IL-4 protein that has been genetically modified to act as an antagonist of the IL-4 receptor alpha chain and therefore inhibits both IL-4 and IL-13 signaling. Inhalational treatment with pitrakinra was shown to decrease airflow obstruction after ALLERGEN challenge in asthmatic patients [77]. However, a subsequent trial found no effect of inhaled pitrakinra on the primary clinical end point of ASTHMA exacerbations [78]. While some therapies targeting IL-4 and IL-13 appear promising, more research is needed to evaluate their efficacy in ASTHMA.

24.4.1.4 Antitumor Necrosis Factor (TNF) Agents

TNF is a proinflammatory cytokine that has been implicated in the pathogenesis of several chronic inflammatory disorders, including ASTHMA (see Chap. 6). TNF is increased in the airways of severe asthmatics and is associated with airway remodeling, suggesting that antagonists of TNF signaling may have a role in ASTHMA treatment [79]. Several trials have evaluated the use of ETANERCEPT—a TNF receptor fusion protein that binds to and inhibits TNF—for the treatment of ASTHMA. While initial small studies suggested some efficacy of ETANERCEPT for moderate-to-severe ASTHMA, these findings were not replicated in a larger randomized trial [80]. In a large trial of patients with severe ASTHMA, treatment with the anti-TNF MONOCLONAL ANTIBODY golimumab did not improve clinical outcomes and was associated with a number of severe adverse events [81]. Current evidence does not support a role for TNF antagonists in the treatment of ASTHMA, although it remains to be determined if anti-TNF therapy may be beneficial for certain ASTHMA phenotypes.

24.4.1.5 Other Biological Therapies for Allergic Disease

Biologic agents targeting several other CYTOKINES involved with allergic inflammation, including IL-1, IL-9, IL-12, IL-17A, IL-22, IL-23, IL-31, and TSLP, are in various stages of development for the treatment of ASTHMA, allergic rhinitis, and atopic dermatitis [55]. Strategies aimed at inhibiting chemokine receptors involved with immune cell migration are also under investigation. As observed in other therapeutic trials, it is likely that these biological agents will be most beneficial for certain subgroups of patients with a particular disease phenotype. More studies are needed to identify patient characteristics and disease biomarkers that help predict responses to specific biological therapies.

24.4.2 Immunological Adjuvants

The development of adaptive immune responses is greatly influenced by signals from innate immune cells. Innate immune cells express a variety of pattern recognition receptors, including Toll-like receptors (TLR), C-type lectin receptors, and scavenger receptors that recognize microbes and allergens. Stimulation of TLR by immunological ADJUVANTS can promote the development of TH1 or Treg and inhibit TH2 responses that drive allergic inflammation. Immunological ADJUVANTS have been investigated for use in both allergen-specific and allergen-nonspecific immunotherapies [82].

Mixing immunological ADJUVANTS with allergens is a potential strategy for enhancing allergen-specific immunotherapy. Combination of the TLR4 agonist monophosphoryl lipid A (MPL) with grass modified ALLERGEN tyrosine adsorbate (MATA) reduced allergic symptoms, increased allergen-specific IgG levels, and blunted the seasonal rise in allergen-specific IgE levels after only four subcutaneous injections [83]. An ultrashort course of ragweed MATA and MPL also significantly reduced symptoms in patients with ragweed ALLERGY [84]. Modified ALLERGEN vaccines containing MPL are available in Europe but not in the USA. Immunotherapy with a conjugate of ragweed ALLERGEN and a TLR9 agonist was reported to decrease nasal symptoms and suppress ragweed-specific IgE levels in a small group of patients [85]. However, a subsequent larger trial was inconclusive due to a lack of significant allergic symptoms during the ragweed pollen season [82].

TLR agonists have also been tested as standalone therapy for allergic diseases. VTX-1463 is a small molecule TLR8 agonist that has been evaluated for the treatment of allergic rhinitis due to grass pollen. Four weekly intranasal administrations of VTX-1463 decreased nasal symptoms in allergic patients challenged to grass pollen in an exposure chamber [86]. QbG10 is a novel TLR9 agonist with TH1-adjuvant activity. In a randomized controlled trial involving patients with mild-to-moderate persistent ASTHMA, injections of QbG10 resulted in improved clinical outcomes after withdrawal of inhaled corticosteroids [87]. The treatment was well tolerated with injection site reactions being the most commonly reported adverse event. While more studies are needed, the ability to use immunological ADJUVANTS as stand-alone therapy for allergic disease is a highly appealing attribute.

24.5 Conclusions

AIT is an effective treatment for patients with allergic ASTHMA, allergic rhinoconjunctivitis, and stinging insect ALLERGY. AIT has several effects on the immune system that are likely important for its clinical efficacy, including reduced mast cell and basophil activity, an increase in allergen-specific IgG4 antibody levels, and the induction of Treg. AIT can result in prolonged clinical TOLERANCE to allergens and therefore has the ability to change the clinical course of allergic disease. While AIT may ultimately prove to be an effective treatment for atopic dermatitis and food allergies, more research is needed before it can be routinely offered to patients with these conditions. When used in appropriately selected patients, AIT is generally safe, but serious adverse effects can occur, and therefore AIT should only be prescribed by appropriately trained healthcare providers. For patients with moderate-to-severe ASTHMA who are not suitable candidates for AIT, allergen-independent therapies that antagonize allergic immune responses may be beneficial. While most biological therapies are still under investigation, some of these agents, such as omalizumab and mepolizumab, have been recently approved for use in patients with specific ASTHMA phenotypes. While there is general excitement regarding the use of biologics in allergic disease, questions about their cost-effectiveness and long-term efficacy and safety remain to be fully addressed. Identifying disease biomarkers that predict responsiveness to specific immunotherapies will further improve the benefit-to-risk ratio for these interventions.

Selected Readings

Boyman O, Kaegi C, Akdis M, Bavbek S, Bossios A, Chatzipetrou A, et al. EAACI IG Biologicals task force paper on the use of biologic agents in allergic disorders. Allergy. 2015;70(7):727–54.

Burks AW, Calderon MA, Casale T, Cox L, Demoly P, Jutel M, et al. Update on allergy immunotherapy: American Academy of Allergy, Asthma & Immunology/European Academy of Allergy and Clinical Immunology/PRACTALL consensus report. J Allergy Clin Immunol. 2013;131(5):1288–96.

Casale TB, Stokes JR. Immunotherapy: what lies beyond. J Allergy Clin Immunol. 2014;133(3):612–9.

References

1. Paul WE, Zhu J. How are T(H)2-type immune responses initiated and amplified? Nat Rev Immunol. 2010;10(4):225–35. https://doi.org/10.1038/nri2735.
2. Burks AW, Calderon MA, Casale T, Cox L, Demoly P, Jutel M, et al. Update on allergy immunotherapy: American Academy of Allergy, Asthma & Immunology/European Academy of Allergy and Clinical Immunology/PRACTALL consensus report. J Allergy Clin Immunol. 2013;131(5):1288–96.e3. https://doi.org/10.1016/j.jaci.2013.01.049.
3. Golden DB, Moffitt J, Nicklas RA, Freeman T, Graft DF, Reisman RE, et al. Stinging insect hypersensitivity: a practice parameter update 2011. J Allergy Clin Immunol. 2011;127(4):852–4.e1–23. https://doi.org/10.1016/j.jaci.2011.01.025.
4. Jutel M, Agache I, Bonini S, Burks AW, Calderon M, Canonica W, et al. International consensus on allergy immunotherapy. J Allergy Clin Immunol. 2015;136(3):556–68. https://doi.org/10.1016/j.jaci.2015.04.047.
5. Akdis M, Akdis CA. Mechanisms of allergen-specific immunotherapy: multiple suppressor factors at work in immune tolerance to allergens. J Allergy Clin Immunol. 2014;133(3):621–31. https://doi.org/10.1016/j.jaci.2013.12.1088.
6. Novak N, Mete N, Bussmann C, Maintz L, Bieber T, Akdis M, et al. Early suppression of basophil activation during allergen-specific immunotherapy by histamine receptor 2. J Allergy Clin Immunol. 2012;130(5):1153–8.e2. https://doi.org/10.1016/j.jaci.2012.04.039.
7. Iliopoulos O, Proud D, Adkinson NF Jr, Creticos PS, Norman PS, Kagey-Sobotka A, et al. Effects of immunotherapy on the early, late, and rechallenge nasal reaction to provocation with allergen: changes in inflammatory mediators and cells. J Allergy Clin Immunol. 1991;87(4):855–66.
8. Gleich GJ, Zimmermann EM, Henderson LL, Yunginger JW. Effect of immunotherapy on immunoglobulin E and immunoglobulin G antibodies to ragweed antigens: a six-year prospective study. J Allergy Clin Immunol. 1982;70(4):261–71.
9. Lichtenstein LM, Ishizaka K, Norman PS, Sobotka AK, Hill BM. IgE antibody measurements in ragweed hay fever. Relationship to clinical severity and the results of immunotherapy. J Clin Invest. 1973;52(2):472–82. https://doi.org/10.1172/jci107204.
10. Peng ZK, Naclerio RM, Norman PS, Adkinson NF Jr. Quantitative IgE- and IgG-subclass responses during and after long-term ragweed immunotherapy. J Allergy Clin Immunol. 1992;89(2):519–29.
11. James LK, Shamji MH, Walker SM, Wilson DR, Wachholz PA, Francis JN, et al. Long-term tolerance after allergen immunotherapy is accompanied by selective persistence of blocking antibodies. J Allergy Clin Immunol. 2011;127(2):509–16.e1–5. https://doi.org/10.1016/j.jaci.2010.12.1080.
12. Wachholz PA, Soni NK, Till SJ, Durham SR. Inhibition of allergen-IgE binding to B cells by IgG antibodies after grass pollen immunotherapy. J Allergy Clin Immunol. 2003;112(5):915–22. https://doi.org/10.1016/s0091.
13. Shamji MH, Ljorring C, Francis JN, Calderon MA, Larche M, Kimber I, et al. Functional rather than immunoreactive levels of IgG4 correlate closely with clinical response to grass pollen immunotherapy. Allergy. 2012;67(2):217–26. https://doi.org/10.1111/j.1398-9995.2011.02745.x.
14. Majori M, Caminati A, Corradi M, Brianti E, Scarpa S, Pesci A. T-cell cytokine pattern at three time points during specific immunotherapy for mite-sensitive asthma. Clin Exp Allergy. 2000;30(3):341–7.
15. Akdis M, Verhagen J, Taylor A, Karamloo F, Karagiannidis C, Crameri R, et al. Immune responses in healthy and allergic individuals are characterized by

a fine balance between allergen-specific T regulatory 1 and T helper 2 cells. J Exp Med. 2004;199(11):1567–75. https://doi.org/10.1084/jem.20032058.
16. Mobs C, Ipsen H, Mayer L, Slotosch C, Petersen A, Wurtzen PA, et al. Birch pollen immunotherapy results in long-term loss of Bet v 1-specific TH2 responses, transient TR1 activation, and synthesis of IgE-blocking antibodies. J Allergy Clin Immunol. 2012;130(5):1108–16.e6. https://doi.org/10.1016/j.jaci.2012.07.056.
17. Radulovic S, Jacobson MR, Durham SR, Nouri-Aria KT. Grass pollen immunotherapy induces Foxp3-expressing CD4+ CD25+ cells in the nasal mucosa. J Allergy Clin Immunol. 2008;121(6):1467–72, 72.e1. https://doi.org/10.1016/j.jaci.2008.03.013.
18. Torgerson TR, Ochs HD. Immune dysregulation, polyendocrinopathy, enteropathy, X-linked: forkhead box protein 3 mutations and lack of regulatory T cells. J Allergy Clin Immunol. 2007;120(4):744–50; quiz 51–2. https://doi.org/10.1016/j.jaci.2007.08.044.
19. Esch RE, Plunkett GA. Immunotherapy preparation guidelines, rules, and regulation. Curr Allergy Asthma Rep. 2013;13(4):406–13. https://doi.org/10.1007/s11882-013-0358-8.
20. Esch RE. Allergen immunotherapy: what can and cannot be mixed? J Allergy Clin Immunol. 2008;122(3):659–60. https://doi.org/10.1016/j.jaci.2008.07.018.
21. Cox L. Sublingual immunotherapy for aeroallergens: status in the United States. Allergy Asthma Proc. 2014;35(1):34–42. https://doi.org/10.2500/aap.2014.35.3708.
22. Radulovic S, Wilson D, Calderon M, Durham S. Systematic reviews of sublingual immunotherapy (SLIT). Allergy. 2011;66(6):740–52. https://doi.org/10.1111/j.1398-9995.2011.02583.x.
23. Normansell R, Kew KM, Bridgman AL. Sublingual immunotherapy for asthma. Cochrane Database Syst Rev. 2015;(8):Cd011293. https://doi.org/10.1002/14651858.CD011293.pub2.
24. Tabatabaian F, Casale TB. Selection of patients for sublingual immunotherapy (SLIT) versus subcutaneous immunotherapy (SCIT). Allergy Asthma Proc. 2015;36(2):100–4. https://doi.org/10.2500/aap.2015.36.3830.
25. Di Bona D, Plaia A, Leto-Barone MS, La Piana S, Di Lorenzo G. Efficacy of subcutaneous and sublingual immunotherapy with grass allergens for seasonal allergic rhinitis: a meta-analysis-based comparison. J Allergy Clin Immunol. 2012;130(5):1097–107.e2. https://doi.org/10.1016/j.jaci.2012.08.012.
26. Moingeon P, Mascarell L. Novel routes for allergen immunotherapy: safety, efficacy and mode of action. Immunotherapy. 2012;4(2):201–12. https://doi.org/10.2217/imt.11.171.
27. Moran TP, Vickery BP, Burks AW. Oral and sublingual immunotherapy for food allergy: current progress and future directions. Curr Opin Immunol. 2013;25(6):781–7. https://doi.org/10.1016/j.coi.2013.07.011.
28. Senti G, von Moos S, Tay F, Graf N, Johansen P, Kundig TM. Determinants of efficacy and safety in epicutaneous allergen immunotherapy: summary of three clinical trials. Allergy. 2015;70(6):707–10. https://doi.org/10.1111/all.12600.
29. Senti G, Prinz Vavricka BM, Erdmann I, Diaz MI, Markus R, McCormack SJ, et al. Intralymphatic allergen administration renders specific immunotherapy faster and safer: a randomized controlled trial. Proc Natl Acad Sci U S A. 2008;105(46):17908–12. https://doi.org/10.1073/pnas.0803725105.
30. Cox L, Nelson H, Lockey R, Calabria C, Chacko T, Finegold I, et al. Allergen immunotherapy: a practice parameter third update. J Allergy Clin Immunol. 2011;127(1 Suppl):S1–55. https://doi.org/10.1016/j.jaci.2010.09.034.
31. Nelson HS. Subcutaneous injection immunotherapy for optimal effectiveness. Immunol Allergy Clin N Am. 2011;31(2):211–26, viii. https://doi.org/10.1016/j.iac.2011.02.010.
32. Des Roches A, Paradis L, Knani J, Hejjaoui A, Dhivert H, Chanez P, et al. Immunotherapy with a standardized Dermatophagoides pteronyssinus extract. V. Duration of the efficacy of immunotherapy after its cessation. Allergy. 1996;51(6):430–3.
33. Marogna M, Spadolini I, Massolo A, Canonica GW, Passalacqua G. Long-lasting effects of sublingual immunotherapy according to its duration: a 15-year prospective study. J Allergy Clin Immunol. 2010;126(5):969–75. https://doi.org/10.1016/j.jaci.2010.08.030.
34. Abramson MJ, Puy RM, Weiner JM. Injection allergen immunotherapy for asthma. Cochrane Database Syst Rev. 2010;(8):Cd001186. https://doi.org/10.1002/14651858.CD001186.pub2.
35. Zielen S, Kardos P, Madonini E. Steroid-sparing effects with allergen-specific immunotherapy in children with asthma: a randomized controlled trial. J Allergy Clin Immunol. 2010;126(5):942–9. https://doi.org/10.1016/j.jaci.2010.06.002.
36. Mosbech H, Deckelmann R, de Blay F, Pastorello EA, Trebas-Pietras E, Andres LP, et al. Standardized quality (SQ) house dust mite sublingual immunotherapy tablet (ALK) reduces inhaled corticosteroid use while maintaining asthma control: a randomized, double-blind, placebo-controlled trial. J Allergy Clin Immunol. 2014;134(3):568–75.e7. https://doi.org/10.1016/j.jaci.2014.03.019.
37. Jacobsen L, Niggemann B, Dreborg S, Ferdousi HA, Halken S, Host A, et al. Specific immunotherapy has long-term preventive effect of seasonal and perennial asthma: 10-year follow-up on the PAT study. Allergy. 2007;62(8):943–8. https://doi.org/10.1111/j.1398-9995.2007.01451.x.
38. Global Initiative for Asthma. Global Strategy for Asthma Management and Prevention. 2015. www.ginasthma.org.
39. Calderon MA, Alves B, Jacobson M, Hurwitz B, Sheikh A, Durham S. Allergen injection immunotherapy for seasonal allergic rhinitis. Cochrane Database Syst Rev. 2007;(1):Cd001936. https://doi.org/10.1002/14651858.CD001936.pub2.
40. Matricardi PM, Kuna P, Panetta V, Wahn U, Narkus A. Subcutaneous immunotherapy and pharmacotherapy

in seasonal allergic rhinitis: a comparison based on meta-analyses. J Allergy Clin Immunol. 2011;128(4):791–9.e6. https://doi.org/10.1016/j.jaci.2011.03.049.
41. Hankin CS, Cox L, Bronstone A, Wang Z. Allergy immunotherapy: reduced health care costs in adults and children with allergic rhinitis. J Allergy Clin Immunol. 2013;131(4):1084–91. https://doi.org/10.1016/j.jaci.2012.12.662.
42. Durham SR, Walker SM, Varga EM, Jacobson MR, O'Brien F, Noble W, et al. Long-term clinical efficacy of grass-pollen immunotherapy. N Engl J Med. 1999;341(7):468–75.
43. Des Roches A, Paradis L, Menardo JL, Bouges S, Daures JP, Bousquet J. Immunotherapy with a standardized Dermatophagoides pteronyssinus extract. VI. Specific immunotherapy prevents the onset of new sensitizations in children. J Allergy Clin Immunol. 1997;99(4):450–3.
44. Boyle RJ, Elremeli M, Hockenhull J, Cherry MG, Bulsara MK, Daniels M, et al. Venom immunotherapy for preventing allergic reactions to insect stings. Cochrane Database Syst Rev. 2012;(10):CD008838. https://doi.org/10.1002/14651858.CD008838.pub2.
45. Lerch E, Muller UR. Long-term protection after stopping venom immunotherapy: results of re-stings in 200 patients. J Allergy Clin Immunol. 1998;101(5):606–12. https://doi.org/10.1016/s0091-6749(98)70167-8.
46. Pajno GB, Caminiti L, Vita D, Barberio G, Salzano G, Lombardo F, et al. Sublingual immunotherapy in mite-sensitized children with atopic dermatitis: a randomized, double-blind, placebo-controlled study. J Allergy Clin Immunol. 2007;120(1):164–70. https://doi.org/10.1016/j.jaci.2007.04.008.
47. Novak N, Bieber T, Hoffmann M, Folster-Holst R, Homey B, Werfel T, et al. Efficacy and safety of subcutaneous allergen-specific immunotherapy with depigmented polymerized mite extract in atopic dermatitis. J Allergy Clin Immunol. 2012;130(4):925–31.e4. https://doi.org/10.1016/j.jaci.2012.08.004.
48. Tam H, Calderon MA, Manikam L, Nankervis H, Garcia Nunez I, Williams HC, et al. Specific allergen immunotherapy for the treatment of atopic eczema. Cochrane Database Syst Rev. 2016;(2):Cd008774. https://doi.org/10.1002/14651858.CD008774.pub2.
49. Nelson HS, Lahr J, Rule R, Bock A, Leung D. Treatment of anaphylactic sensitivity to peanuts by immunotherapy with injections of aqueous peanut extract. J Allergy Clin Immunol. 1997;99(6 Pt 1):744–51.
50. Burks AW, Jones SM, Wood RA, Fleischer DM, Sicherer SH, Lindblad RW, et al. Oral immunotherapy for treatment of egg allergy in children. N Engl J Med. 2012;367(3):233–43. https://doi.org/10.1056/NEJMoa1200435.
51. Vickery BP, Scurlock AM, Kulis M, Steele PH, Kamilaris J, Berglund JP, et al. Sustained unresponsiveness to peanut in subjects who have completed peanut oral immunotherapy. J Allergy Clin Immunol. 2014;133(2):468–75. https://doi.org/10.1016/j.jaci.2013.11.007.
52. Narisety SD, Frischmeyer-Guerrerio PA, Keet CA, Gorelik M, Schroeder J, Hamilton RG, et al. A randomized, double-blind, placebo-controlled pilot study of sublingual versus oral immunotherapy for the treatment of peanut allergy. J Allergy Clin Immunol. 2015;135(5):1275–82.e1–6. https://doi.org/10.1016/j.jaci.2014.11.005.
53. Bernstein DI, Wanner M, Borish L, Liss GM. Twelve-year survey of fatal reactions to allergen injections and skin testing: 1990–2001. J Allergy Clin Immunol. 2004;113(6):1129–36. https://doi.org/10.1016/j.jaci.2004.02.006.
54. Pitsios C, Demoly P, Bilo MB, Gerth van Wijk R, Pfaar O, Sturm GJ, et al. Clinical contraindications to allergen immunotherapy: an EAACI position paper. Allergy. 2015;70(8):897–909. https://doi.org/10.1111/all.12638.
55. Boyman O, Kaegi C, Akdis M, Bavbek S, Bossios A, Chatzipetrou A, et al. EAACI IG Biologicals task force paper on the use of biologic agents in allergic disorders. Allergy. 2015;70(7):727–54. https://doi.org/10.1111/all.12616.
56. Easthope S, Jarvis B. Omalizumab. Drugs. 2001;61(2):253–60; discussion 61.
57. MacGlashan DW Jr, Bochner BS, Adelman DC, Jardieu PM, Togias A, McKenzie-White J, et al. Down-regulation of Fc(epsilon)RI expression on human basophils during in vivo treatment of atopic patients with anti-IgE antibody. J Immunol. 1997;158(3):1438–45.
58. Lin H, Boesel KM, Griffith DT, Prussin C, Foster B, Romero FA, et al. Omalizumab rapidly decreases nasal allergic response and FcepsilonRI on basophils. J Allergy Clin Immunol. 2004;113(2):297–302. https://doi.org/10.1016/j.jaci.2003.11.044.
59. Segal M, Stokes JR, Casale TB. Anti-immunoglobulin e therapy. World Allergy Organ J. 2008;1(10):174–83. https://doi.org/10.1097/WOX.0b013e318187a310.
60. Xolair (omalizumab) [package insert]. South San Francisco, CA: Genentech Inc.; 2015.
61. Normansell R, Walker S, Milan SJ, Walters EH, Nair P. Omalizumab for asthma in adults and children. Cochrane Database Syst Rev. 2014;(1):Cd003559. https://doi.org/10.1002/14651858.CD003559.pub4.
62. Cox L, Platts-Mills TA, Finegold I, Schwartz LB, Simons FE, Wallace DV. American Academy of Allergy, Asthma & Immunology/American College of Allergy, Asthma and Immunology Joint Task Force Report on omalizumab-associated anaphylaxis. J Allergy Clin Immunol. 2007;120(6):1373–7. https://doi.org/10.1016/j.jaci.2007.09.032.
63. Busse W, Buhl R, Fernandez Vidaurre C, Blogg M, Zhu J, Eisner MD, et al. Omalizumab and the risk of malignancy: results from a pooled analysis. J Allergy Clin Immunol. 2012;129(4):983–9.e6. https://doi.org/10.1016/j.jaci.2012.01.033.
64. Namazy J, Cabana MD, Scheuerle AE, Thorp JM Jr, Chen H, Carrigan G, et al. The Xolair Pregnancy Registry (EXPECT): the safety of omalizumab

use during pregnancy. J Allergy Clin Immunol. 2015;135(2):407–12. https://doi.org/10.1016/j.jaci.2014.08.025.
65. Varricchi G, Bagnasco D, Borriello F, Heffler E, Canonica GW. Interleukin-5 pathway inhibition in the treatment of eosinophilic respiratory disorders: evidence and unmet needs. Curr Opin Allergy Clin Immunol. 2016;16(2):186–200. https://doi.org/10.1097/aci.0000000000000251.
66. Nucala (mepolizumab) [package insert]. Research Triangle Park, NC: GlaxoSmithKline; November 2015.
67. Leckie MJ, ten Brinke A, Khan J, Diamant Z, O'Connor BJ, Walls CM, et al. Effects of an interleukin-5 blocking monoclonal antibody on eosinophils, airway hyper-responsiveness, and the late asthmatic response. Lancet. 2000;356(9248):2144–8.
68. Flood-Page P, Swenson C, Faiferman I, Matthews J, Williams M, Brannick L, et al. A study to evaluate safety and efficacy of mepolizumab in patients with moderate persistent asthma. Am J Respir Crit Care Med. 2007;176(11):1062–71. https://doi.org/10.1164/rccm.200701-085OC.
69. Bel EH, Wenzel SE, Thompson PJ, Prazma CM, Keene ON, Yancey SW, et al. Oral glucocorticoid-sparing effect of mepolizumab in eosinophilic asthma. N Engl J Med. 2014;371(13):1189–97. https://doi.org/10.1056/NEJMoa1403291.
70. Pavord ID, Korn S, Howarth P, Bleecker ER, Buhl R, Keene ON, et al. Mepolizumab for severe eosinophilic asthma (DREAM): a multicentre, double-blind, placebo-controlled trial. Lancet. 2012;380(9842):651–9. https://doi.org/10.1016/s0140-6736(12)60988-x.
71. Ortega HG, Liu MC, Pavord ID, Brusselle GG, FitzGerald JM, Chetta A, et al. Mepolizumab treatment in patients with severe eosinophilic asthma. N Engl J Med. 2014;371(13):1198–207. https://doi.org/10.1056/NEJMoa1403290.
72. Chung KF. Targeting the interleukin pathway in the treatment of asthma. Lancet. 2015;386(9998):1086–96. https://doi.org/10.1016/s0140-6736(15)00157-9.
73. Corren J, Lemanske RF, Hanania NA, Korenblat PE, Parsey MV, Arron JR, et al. Lebrikizumab treatment in adults with asthma. N Engl J Med. 2011;365(12):1088–98. https://doi.org/10.1056/NEJMoa1106469.
74. Noonan M, Korenblat P, Mosesova S, Scheerens H, Arron JR, Zheng Y, et al. Dose-ranging study of lebrikizumab in asthmatic patients not receiving inhaled steroids. J Allergy Clin Immunol. 2013;132(3):567–74.e12. https://doi.org/10.1016/j.jaci.2013.03.051.
75. Piper E, Brightling C, Niven R, Oh C, Faggioni R, Poon K, et al. A phase II placebo-controlled study of tralokinumab in moderate-to-severe asthma. Eur Respir J. 2013;41(2):330–8. https://doi.org/10.1183/09031936.00223411.
76. Wenzel S, Ford L, Pearlman D, Spector S, Sher L, Skobieranda F, et al. Dupilumab in persistent asthma with elevated eosinophil levels. N Engl J Med. 2013;368(26):2455–66. https://doi.org/10.1056/NEJMoa1304048.
77. Wenzel S, Wilbraham D, Fuller R, Getz EB, Longphre M. Effect of an interleukin-4 variant on late phase asthmatic response to allergen challenge in asthmatic patients: results of two phase 2a studies. Lancet. 2007;370(9596):1422–31. https://doi.org/10.1016/s0140-6736(07)61600-6.
78. Slager RE, Otulana BA, Hawkins GA, Yen YP, Peters SP, Wenzel SE, et al. IL-4 receptor polymorphisms predict reduction in asthma exacerbations during response to an anti-IL-4 receptor alpha antagonist. J Allergy Clin Immunol. 2012;130(2):516–22.e4. https://doi.org/10.1016/j.jaci.2012.03.030.
79. Brightling C, Berry M, Amrani Y. Targeting TNF-alpha: a novel therapeutic approach for asthma. J Allergy Clin Immunol. 2008;121(1):5–10; quiz 1–2. https://doi.org/10.1016/j.jaci.2007.10.028.
80. Holgate ST, Noonan M, Chanez P, Busse W, Dupont L, Pavord I, et al. Efficacy and safety of etanercept in moderate-to-severe asthma: a randomised, controlled trial. Eur Respir J. 2011;37(6):1352–9. https://doi.org/10.1183/09031936.00063510.
81. Wenzel SE, Barnes PJ, Bleecker ER, Bousquet J, Busse W, Dahlen SE, et al. A randomized, double-blind, placebo-controlled study of tumor necrosis factor-alpha blockade in severe persistent asthma. Am J Respir Crit Care Med. 2009;179(7):549–58. https://doi.org/10.1164/rccm.200809-1512OC.
82. Casale TB, Stokes JR. Immunotherapy: what lies beyond. J Allergy Clin Immunol. 2014;133(3):612–9; quiz 20. https://doi.org/10.1016/j.jaci.2014.01.007.
83. Drachenberg KJ, Wheeler AW, Stuebner P, Horak F. A well-tolerated grass pollen-specific allergy vaccine containing a novel adjuvant, monophosphoryl lipid A, reduces allergic symptoms after only four preseasonal injections. Allergy. 2001;56(6):498–505.
84. Patel P, Holdich T, Fischer von Weikersthal-Drachenberg KJ, Huber B. Efficacy of a short course of specific immunotherapy in patients with allergic rhinoconjunctivitis to ragweed pollen. J Allergy Clin Immunol. 2014;133(1):121–9.e1–2. https://doi.org/10.1016/j.jaci.2013.05.032.
85. Creticos PS, Schroeder JT, Hamilton RG, Balcer-Whaley SL, Khattignavong AP, Lindblad R, et al. Immunotherapy with a ragweed-toll-like receptor 9 agonist vaccine for allergic rhinitis. N Engl J Med. 2006;355(14):1445–55. https://doi.org/10.1056/NEJMoa052916.
86. Horak F. VTX-1463, a novel TLR8 agonist for the treatment of allergic rhinitis. Expert Opin Investig Drugs. 2011;20(7):981–6. https://doi.org/10.1517/13543784.2011.583237.
87. Beeh KM, Kanniess F, Wagner F, Schilder C, Naudts I, Hammann-Haenni A, et al. The novel TLR-9 agonist QbG10 shows clinical efficacy in persistent allergic asthma. J Allergy Clin Immunol. 2013;131(3):866–74. https://doi.org/10.1016/j.jaci.2012.12.1561.

25 Immunostimulants in Cancer Therapy

James E. Talmadge

25.1 Introduction

During the past few decades, immunotherapy has become a clinical reality, and an ever-increasing number of cancer patients are receiving immunologic intervention(s). The therapeutic focus has been on the development of interventions that initiate or boost an existing immune response against a patient's tumor. Indeed, in 2013, the clinical success of immunotherapy was recognized by the editors of *Science Magazine* with the designation of "Breakthrough of the Year" [1]. Nonetheless, we have only just begun to understand and develop the potential of immunotherapy. Ongoing clinical studies are testing the safety and efficacy of biologic, molecular, and cellular therapeutic regimens either as stand-alone interventions or in combination with standard of care therapy. Immunoaugmenting drugs have been used to treat disease since early in the twentieth century when William B. Coley treated cancer patients with mixed bacterial toxins [2]. These studies stimulated the clinical use of microbial substances, such as Bacillus Calmette-Guérin (BCG) (bladder cancer, United States), Krestin, Picibanil and lentinan (gastric and other cancers, Japan), and Biostim and Broncho-Vaxom (recurrent infections, Europe). While these unsophisticated drugs can induce various immunopharmacological activities, their use is associated with regulatory obstacles due to impurity, lot-to-lot variability, unreliability, and adverse side effects. Similarly, traditional herbal medicines (Asia) can act as a source of active substances for immunotherapy but require purification, characterization, and synthetic production of the active moieties. Purified entities derived from natural products that are in "routine" clinical use includes Bestatin®, Taxol®, FK-506, rapamycin, deoxyspergualin, and CYCLOSPORINE, all of which are derived from natural products. The current immunotherapy focus is on the use of MONOCLONAL ANTIBODIES such as checkpoint inhibitors and RECOMBINANT proteins (cytokines), although the utility of these drugs can be limited due to immunotoxicity and pharmacological deficiencies. However, there remains a potential utility for BIOLOGICAL RESPONSE MODIFIERS (BRM) due to their ability to induce multiple CYTOKINES for immune augmentation and hematopoietic restoration.

In 2002, the 20th anniversary of the first approved biopharmaceutical, RECOMBINANT insulin (Humulin: Genentech, South San Francisco, CA, USA) was observed. Today, biotechnological drugs incorporate not only immunoregulatory proteins, enzymes, and biologicals derived from natural sources but also engineered (manipulated) MONOCLONAL ANTIBODIES

Final manuscript submitted on October 06, 2016.

J. E. Talmadge (✉)
986495 Nebraska Medical Center, University of Nebraska Medical Center, Omaha, NE, USA
e-mail: jtalmadg@immpharm.com; jtalmadg@unmc.edu

M. J. Parnham et al. (eds.), *Nijkamp and Parnham's Principles of Immunopharmacology*,
https://doi.org/10.1007/978-3-030-10811-3_25

(mAbs) and CYTOKINES, in addition to gene therapy and cell and cell and tissue engineering strategies. In 2014, there were 246 biopharmaceuticals approved in the United States and/or European Union (EU) for use in humans with 174 approved between January 1995 and June 2007 [3]. In addition, some 400 biologicals are currently undergoing clinical evaluation [4]. Perhaps the most telling statement regarding the maturation of the field is that the EU, as of March 2005, had approved six biosimilars and one in the United States [3]. These are biologicals that are comparable in quality, safety, and efficacy to a reference product [5].

The overall approach in this chapter is to limit the discussion of biotherapeutics to the RECOMBINANT, natural, and synthetic drugs that are currently approved for clinical use against cancer (Box 25.1), although in association with their recent focus, we have also discussed the checkpoint inhibits that are making a significant impact on cancer therapy. We have focused on RECOMBINANT proteins (except antithrombotic, vaccines, and mAbs) and will not discuss nucleic-acid-based or cell-tissue-engineered gene therapeutic products. A brief discussion follows regarding combination therapy and cellular therapy as future prospects. Recently, the immune-augmenting effects of conventional chemotherapeutics have also provided a new therapeutic strategy. Indeed, several chemotherapeutics have shown off-target effects, in particular when administered as low dose, metronomic schedules [6].

Box 25.1: Types of Cancer Amenable to Immunotherapy

- Melanoma and renal carcinoma. These tumors are highly antigenic and frequently have a significant histiocytic and lymphocytic cellular infiltration. As such, they have demonstrated significant responsiveness to intervention with biotherapeutics.
- Bladder cell carcinoma. Superficial transitional cell carcinoma has proven highly responsive to therapy with Bacillus Calmette-Guérin (BCG), as well as cytokines, such as interferon-alpha (IFN-α).
- Head and neck cancer. Head and neck cancer is responsive to cytokine therapy, due in part to its availability for direct therapeutic intervention, and has been shown to be highly responsive to low-dose paralymphatic administration of interleukin-2 (IL-2).
- Solid tumors. A number of solid tumors have shown a response to immune-augmenting agents, but in general, phase III trials have not been undertaken to demonstrate responsiveness.
- Leukemias and lymphomas. In general, these "liquid" tumors have shown responsiveness to immune-augmenting agents, including intervention with IFN.

25.2 Recombinant Proteins

RECOMBINANT proteins have emerged as an important class of drugs for the treatment of cancer, immunosuppression, myeloid dysplasia, and infectious diseases. However, our limited understanding of their pharmacology and mechanism of action (MOA) has hindered their development (Table 25.1), although numerous RECOMBINANT proteins have been approved by the US Food and Drug Administration (FDA). To facilitate advancement, information is needed on their pharmacology, MOA, and toxicology [7, 8]. One approach to the development of biotherapeutics is to identify a clinical hypothesis based upon therapeutic surrogate(s) identified during preclinical pharmacological studies [9]. A surrogate for clinical efficacy may be a phenotypical, biochemical, enzymatic, functional (immunological, molecular, or hematological), or quality-of-life measurement that is believed to be associated with therapeutic activity. Phase I clinical trials can then be designed to identify the optimal immunomodulatory dose (OID) and treatment schedule that maximizes the augmentation of surrogate end point(s). Subsequent phase II/III trials can be established to determine if the changes in the surrogate levels correlate with therapeutic activity. Table 25.1 lists the

Table 25.1 Recombinant drugs with multiple targets that are FDA approved

Drug	Indications	Approval date
AMD-3100 antagonist of CSCR4 (Mozobil®)	Mobilization of non-Hodgkin's lymphoma (NHL) and malignant melanoma (MM) patients in combination with granulocyte colony-stimulating factor (G-CSF)	December 2008
Bone morphogenetic protein-2 (BMP-2)	Treatment of spinal degenerative disc disease	July 2002
Denosumab	Osteoporosis, monoclonal antibody (mAb) to receptor activator of nuclear factor kappa-B ligand (RANKL)	February 2009
Interferon alpha N 1 (IFN-αN1) Wellferon®	Treatment of hepatitis C	March 1999
IFN-αN3 (Alferon-N®)	Genital warts	July 1997
IFN-α2a (Roferon A®)	Chronic hepatitis C, Ph-positive chronic myelogenous leukemia (CML), hairy cell leukemia	November 1999
IFN-α2b (Intron A®)	Hairy cell leukemia, malignant melanoma (MM), *condyloma acuminatum*, acquired immune deficiency syndrome (AIDS)-related Kaposi's sarcoma (KS), chronic hepatitis C, chronic hepatitis B	March 1997
IFN-β1a (Avonex®)	MS	May 1996
IFN-βlb (Rebif)	Relapsing multiple sclerosis (MS)	August 2009
Interferon-gamma (IFN-γ)	Management of chronic granulomatous disease (CGD) and osteoporosis	December 1990
Interleukin-2 (IL-2)	Cutaneous T-cell lymphoma (CTCL), metastatic melanoma, renal cell carcinoma (RCC)	May 1992
Leukocyte function-associated antigen-1 (LFA)-1/immunoglobulin GI (IgG1)	Moderate to severe chronic plaque psoriasis	January 2003
Nplate® (TPO-R)	Idiopathic thrombocytopenia purpura	August 2008
Promacta® thrombopoietin (TPO) R agonist	Idiopathic thrombocytopenia purpura, thrombocytopenia secondary to chronic hepatitis C, severe aplastic anemia	November 2008
Recombinant erythropoietin (rEPO)	Anemia caused by chemotherapy, anemia, chronic renal failure, anemia in Retrovir®-treated HIV-infected, dialysis, surgical blood loss	June 1989
Recombinant granulocyte colony-stimulating factor (rG-CSF)	Acute myelogenous leukemia (AML), autologous or allogeneic bone marrow transplantation (BMT), chemotherapy-induced neutropenia, chronic severe neutropenia, peripheral blood (PB) progenitor cell transplantation	February 1991
Recombinant human interleukin-11 (rHu IL-11)	Chemotherapy-induced thrombocytopenia	November 1997
Recombinant human platelet-derived growth factor BB (rHPDGF-BB) Regranex®	Diabetic neuropathy, foot ulcers	December 1997
Recombinant human granulocyte-macrophage colony-stimulating factor (rHuGM-CSF)	Allogeneic and autologous BMT, neutropenia resulting from chemotherapy, PB progenitor cell mobilization	March 1991
Stem cell factor (SCF) Carticel®	Mobilization of hematopoietic stem cells	August 1997
Tumor necrosis factor receptor/Fc fusion protein (TNFR/Fc)	Moderate to severe active rheumatoid arthritis (RA) and juvenile RA, active ankylosing spondylitis	November 1998

immunologically and hematological active CYTOKINES that are approved for use in the United States. This information is expanded to include RECOMBINANT proteins under study as shown in Appendix 2.

In contrast to strategies based on the identification of surrogates for therapeutic efficacy, protocols for RECOMBINANT proteins are often identified based on practices developed for conventional drugs. But these may not be advantageous to identify therapeutic efficiency in response to CYTOKINES. Further, due to pharmacologic deficiencies, novel formulations and administration approaches are also used (Box 25.2).

Box 25.2: Pharmacological and Dose Relationship Considerations with Cytokines

Cytokines have, in several instances, shown increased bioactivity following delivery by slow release [10–12]. The covalent attachment of polyethylene glycol (PEG) to cytokines (pegylation) (Table 25.2), including interferon alpha (IFN-α) and granulocyte colony-stimulating factor (G-CSF), has significant enhanced biological activity due, in part, to their improved pharmacokinetic profile [10, 12]. Thus, strategies to limit pharmacological deficiencies are critical to the development of recombinant biotherapeutics. The pharmacological attributes of recombinant biotherapeutics are improved with targeted delivery, which prolongs their short half-life [13]. In addition, there can be unexpected relationships between the dose administered and the biological effect of recombinant biotherapeutics, including a nonlinear dose-response relationship, described as "bell-shaped" [14, 15]. This lack of a linear dose-response relationship may be due to nonlinear dispersal throughout the body, poor ability to enter into a saturable receptor-mediated transport process, chemical instability, sequence of administration with other agents, an incorrect time of administration, inappropriate location, and/or response of the large T cells. A "bell-shaped" dose-response curve may be associated with receptor tachyphylaxis expression or a signal transduction mechanism, whereby the cells become refractory to subsequent receptor-mediated augmentation.

Table 25.2 Slow release variant drugs

Drug	Indication	Approval date
Aranesp®, darbepoetin alfa	Anemia associated with chronic kidney disease and chemotherapy-induced anemia	June 2001 to July 2002
Interferon alpha-2a (IFN-α-2a), Pegasys®, polyethylene glycol (PEG)	First-line treatment of chronic hepatitis C	October 2, 2014
Interferon beta-1a (IFN-β-1a) (Plegridy™)	Relapsing forms of multiple sclerosis	August 5, 2014
Peginterferon alpha-2b (PEG-Intron™)	Treatment of chronic hepatitis C in patients not previously treated with IFN-α	January 1, 2014
Pegfilgrastim, Neulasta®, granulocyte colony-stimulating factor polyethylene glycol (GM-CSF-PEG)	Febrile neutropenia in patients receiving chemotherapy	January 2, 2014

25.3 Interferon-Alpha (IFN-α)

25.3.1 Clinical Activity of IFN-α Against Cancer

The initial, non-randomized, clinical studies with IFN-α suggested that it had therapeutic activity for malignant melanoma (MM), osteosarcoma, and various lymphomas [16]. Subsequent randomized trials, however, demonstrated significant therapeutic activity against less common tumor histiotypes, including hairy cell and chronic myelogenous leukemia (CML) [16, 17] and a few types of lymphoma, including low-grade non-Hodgkin's lymphoma [18] and cutaneous T-cell lymphoma [19]. Currently, the list of responding indications has expanded to include MM [20], acquired immune deficiency syndrome (AIDS), and Kaposi's sarcoma [21], genital warts, and hepatitis B and C.

25.3.2 Pharmacological Actions: Dose Response

It has taken almost three decades to translate the concept of IFN-α as an antiviral to its routine utility in clinical oncology and infectious diseases. Despite extensive study, the development of IFN-α is still ongoing, and basic parameters, including optimal dose and therapeutic schedule, remain to be clarified [22, 23]. The MOA is also controversial since IFN-α has been shown to have dose-dependent antitumor activities in vitro, yet be active at low doses for hairy cell leukemia [17, 22]. Immunomodulation, as the mechanism of therapeutic activity of IFN-α, is perhaps best supported by its action against this disease. Treatment with IFN-α is associated with a 90–95% response rate; however, this is not fully achieved until the patients have been on the protocol for a year, and it appears that low doses of IFN-α are as active as higher doses [24]. It should be noted that the clinical use of IFN-α has been supplanted by other more effective drugs, but its approval for hairy cell leukemia precipitated expanded studies for other diseases.

Initial dose-finding studies determined that a dose of 12×10^6 U/M^2 of IFN-α was not tolerable in patients with hairy cell leukemia [17]. Subsequently, it was reported that natural IFN-α (2×10^6 U/M^2) was both well tolerated and effective when administered on a schedule of three times per week for 28 days [25]. However, natural IFN-α retained some toxicity, including myelosuppression, as well as neurotoxicity and cardiotoxicity. In these studies, a lower dose (2×10^5 U/M^2) was also administered and discovered to be better tolerated while inducing equivalent improvements in NEUTROPHIL and PLATELET counts. Further, substantial clinical improvement was reported within the first 4–8 weeks of treatment, as well as an improved quality of life including a decrease in cardiac and neurologic toxicity, myelosuppression, flu-like syndrome, PLATELET transfusions, and bacterial infections. However, IFN-α also has a dose-response effect, whereby higher doses of IFN-α will induce a quantitatively greater antileukemic response than that observed with low doses of IFN-α. Thus, therapy with IFN-α may be initiated at 2×10^5 U/M^2, allowing patients to become tolerant to the acute toxicity of IFN-α, with a subsequent dose increase to 3×10^6 U/M^2 to obtain a greater antileukemic effect [26]. It is noted that cladribine (2-chlorodeoxyadenosine, 2-CdA) and pentostatin provide a higher complete response (CR) rate and have largely replace the use of IFN-α as an induction regimen. Further, a randomized comparison of pentostatin to IFN-α demonstrated significantly higher and more durable responses to pentostatin [27]. In CML, sustained therapeutic responses with IFN-α are found in more than 75% of patients [28]. In addition to reducing leukemic-cell mass, there is also a gradual reduction in the frequency of cells bearing a 9–22 chromosomal translocation [29].

The unique cellular and molecular activities of IFN-α can potentially complement the MOA of other therapies [30]. At present, therapeutic applications for IFN-α are focused on synergistic or additive effects with IFN-gamma (-γ), granulocyte macrophage-colony stimulating factor (GM-CSF), and INTERLEUKIN (IL)-2. A recent meta-analysis of 12 clinical studies for high-risk melanoma showed a significant recurrence-free survival (RFS) following treatment with IFN-α. However, the benefit of IFN-α therapy on overall survival (OS) is less clear [31]. There is a significant trend toward a correlation between increasing dose and RFS, but not OS. Similarly, studies on immune augmentation in melanoma patients receiving high- or low-dose IFN-α revealed no association between immune response and baseline phenotypical and functional immunity [32]. However, numerous immune surrogates are augmented by IFN-α treatment and are associated with dosage. Administration of IFN-α has been shown to significantly upregulate class II MAJOR HISTOCOMPATIBILITY COMPLEX (MHC) and INTERCELLULAR ADHESION MOLECULE (ICAM)-1 expression on tumor cells in a dose-dependent manner. NATURAL KILLER (NK) cell and T-cell functions are also augmented in a dose-dependent manner by IFN-α as are changes in T-cell phenotypes. Further, high-dose IFN-α regulates immune parameters more rapidly than low-dose IFN-α.

25.4 Interferon-Gamma (IFN-γ)

25.4.1 Effect in Animal Models and MOA

IFN-γ has multiple, potential mechanisms that may be involved in tumor protection and therapy. These include, but are not limited to, (1) antiproliferative activity for tumor growth/survival, (2) induction of angiostasis, and (3) augmentation of both innate and adoptive immunities (Fig. 25.1). However, it is unclear which, if any, of these potential mechanisms are critical for the activity of this pleiotropic cytokine as elaborated below:

1. Antiproliferative effects on tumor growth/survival: IFN-γ has direct antiproliferative and antimetabolic effects on tumor cells. It can also induce the apoptosis of tumor cells via conventional signaling mechanisms, resulting in the induction of genes that promote cellular apoptosis, including caspase-1 (IL-1α-converting enzyme or ICE) [33] and Fas ligand (FasL) [34].
2. Induction of angiostasis: The antitumor activity of IFN-γ can also be mediated by the inhibition of neo-angiogenesis. The growth of solid tumors requires new blood vessel formation, which is supported by tumor-induced angiogenesis [35]. Pro-angiogenic molecules are secreted by tumors, including vascular endothelial cell growth factor (VEGF) and basic fibroblast growth factor (FGF). However, IFN-γ induces CHEMOKINES with potent angiostatic actions, including inducible protein (IP)-10 [36], which belongs to a family of CXC non-ELR CHEMOKINES that all have angiostatic activity.
3. Augmentation of both innate and adoptive immune responses: IFN-γ is a potent macrophage-activating cytokine capable of augmenting tumoricidal activity in vitro and

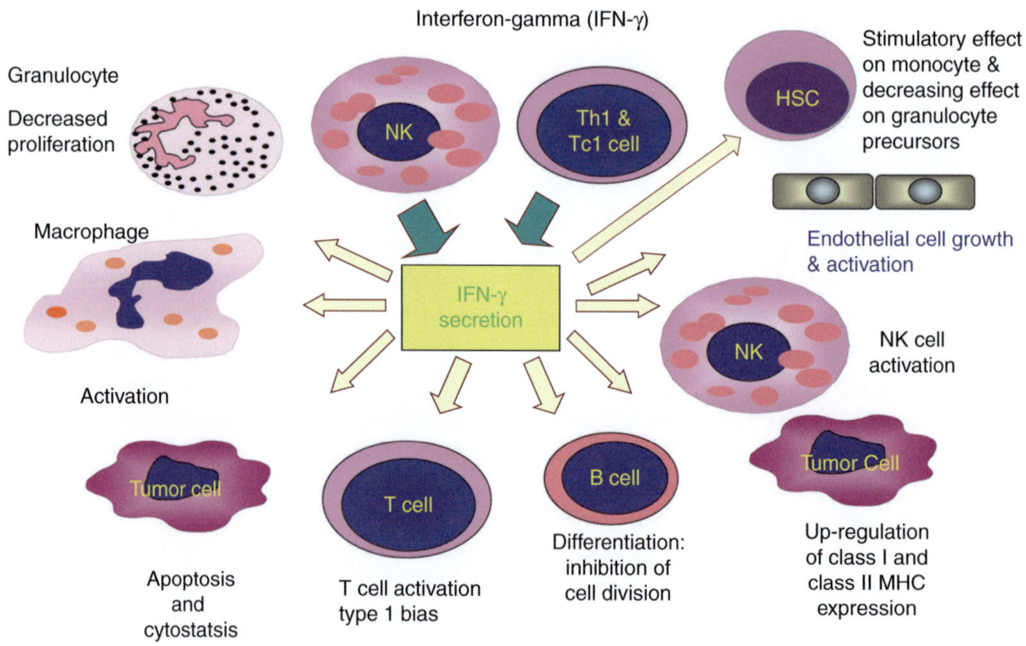

Fig. 25.1 Regulation of immune responses by interferon-gamma (IFN-γ). IFN-γ is produced predominately by Th1, Tc1, and natural killer (NK) cells, resulting in the activation of T cells, NK cells, macrophages, and granulocytes. In addition, it upregulates class I and class II major histocompatibility complex (MHC) expression on tumor cells, as well as the expression of a wide variety of receptors on both tumor cells and epithelial cells. *IFN-γ* interferon-gamma, *NK* natural killer cell, *Th1* T-helper cell type 1, *Tc* T cytotoxic cell type 1, *MHC* major histocompatibility complex

in vivo [37, 38]. IFN-γ-activated MACROPHAGES express multiple tumoricidal mechanisms, including the production of reactive oxygen and/or nitrogen intermediates and upregulated expression of cytotoxic ligands including tumor necrosis factor (TNF)-α, FasL [39], and TNF-related apoptosis-inducing ligand (TRAIL) [40]. In addition, IFN-γ significantly enhances IL-12 secretion by MACROPHAGES and dendritic cells (DCs) [41]. It can also activate NK cells that also have a potential role in promoting antitumor responses, via at least two mechanisms. NK cells are important sources of IFN-γ and also exert direct cytocidal activity against tumors through mechanisms involving perforin [42] and TRAIL [43]. In addition, IFN-γ can markedly enhance adaptive T-cell responses, as well as an important role in regulating the Th1/Th2 balance during the host response to a tumor [43].

25.4.2 Dose-Response Relationship

Preclinical studies have suggested that IFN-γ has significant therapeutic activity with a bell-shaped dose-response curve [44]. Studies of immune response in normal animals have revealed the same bell-shaped dose-response curve to augment macrophage tumoricidal activity [25, 44]. Thus, optimal therapeutic activity has been observed with the same dose and protocol of IFN-γ but has significantly less therapeutic activity at lower and higher doses. A significant correlation between macrophage augmentation and therapeutic efficacy has been reported [25], suggesting that immunological augmentation provides one mechanism for the therapeutic activity of IFN-γ and supports the hypothesis that treatment with the maximum tolerated dose (MTD) of IFN-γ may not be optimal in an adjuvant setting. The preclinical hypothesis of a bell-shaped dose-response curve for IFN-γ has been confirmed in clinical studies examining the immunoregulatory effects of IFN-γ, which defined an OID [15, 45, 46]. In general, the OID for IFN-γ has been found to be between 0.1 and 0.3 mg/M^2 [47]. In contrast, the MTD for IFN-γ may range from 3 to 10 mg/M^2, depending upon the source of IFN-γ and/or the clinical center. The identification of an OID for IFN-γ in patients with minimal tumor burden resulted in clinical trials to test the hypothesis that the immunological enhancement induced by IFN-γ will result in prolongation of the disease-free period and OS of patients in an adjuvant setting [45].

25.4.3 Clinical Therapeutic Activity: MOA

IFN-γ was found on an empirical basis to have therapeutic activity in chronic granulomatous disease (CGD) [48], and it was for this indication that the FDA first approved IFN-γ. In addition to its approval for CGD, IFN-γ has also been approved for the treatment of RHEUMATOID ARTHRITIS (RA) in Germany and in the United States; thereafter it was approved for delaying time to disease progression in patients with severe malignant osteoporosis in association with enhanced superoxide production by phagocytes. In a randomized phase III trial, IFN-γ was also reported to have activity in women receiving first-line platinum-based CHEMOTHERAPY against ovarian cancer [49]. In this study, there was a significantly higher response rate and longer progression free survival (PFS) in women receiving IFN-γ plus CHEMOTHERAPY as compared to CHEMOTHERAPY alone. However, there was no statistically significant improvement in OS. The IFN-γ MOA is unknown for this study, although the authors speculated that it may be associated with the inhibition of HER-2/neu expression (Box 25.3).

Box 25.3: IFN-γ in CGD

Studies in patients with chronic granulomatous disease (CGD) suggest that the mechanism of therapeutic activity by interferon-gamma (IFN-γ) is associated with enhanced phagocytic oxidase and superoxide activity in polymorphonuclear

(PMN) cells. However, more recent data suggests that the majority of CGD patients obtain clinical benefit by prolonging IFN-γ therapy and that the MOA may not be due to enhanced PMN oxidase activity but rather the correction of a respiratory-burst deficiency in monocytes [50]. Further, IFN-γ administration could induce nitric oxide (NO) synthetase activity by PMN in patients with CGD [51]. Following 2 days of IFN-γ administration, a significant increase in PMN-produced NO is observed in association with an increase in the bactericidal capacity of PMN [51]. As PMN in patients with CGD lack the capacity to produce superoxide anions, it is possible that the increased NO release and in vitro bactericidal activity are instrumental in augmenting host defenses and reducing the morbidity of CGD [51]. Similarly, IFN-γ increases PMN expression of Fc gamma receptor I (FcγRI) and improves FcγR-mediated phagocytosis, at least in normal subjects [52]. Therefore, the MOA of IFN-γ critical in reducing the frequency of infections in patients with CGD may not be increased oxidase and superoxide production, but rather effects on other mechanisms of granulocyte activity, including NO production.

25.5 Interleukin-2 (IL-2)

25.5.1 Pharmacological Actions Relevant to Cancer

IL-2, a T-cell growth factor, has a significant role in regulating immunity to infectious and neoplastic diseases (Fig. 25.2). IL-2 is produced primarily by activated DC, NK, and T cells and induces pleiotropic biological responses after binding to one of three receptors (see also Chap. 6). IL-2 stimulates the growth of naïve T cells following ANTIGEN (Ag) activation and later induces activation-induced cell death (AICD) [53, 54]. IL-2 also has effects on several other immune cells, including NK cells [55], B cells [56], MONOCYTE/MACROPHAGES [57], and POLYMORPHONUCLEAR CELLS (PMN) [58]. The ability of IL-2 to stimulate NK and T-cell lysis of tumor cells stimulated clinical interest in IL-2 [59]. NK cells are part of the innate immune system and comprise 10–15% of peripheral blood LYMPHOCYTES (PBL). Functionally, NK cells are characterized by NK and lymphokine-activated killing (LAK), antibody (Ab)-dependent cellular cytotoxicity (ADCC), and immunoregulatory cytokine production (see Chap. 6). NK cells also express NK receptors (NKR) that recognize MHC class I ligands and regulate, inhibit, and activate a response to large T cells [60].

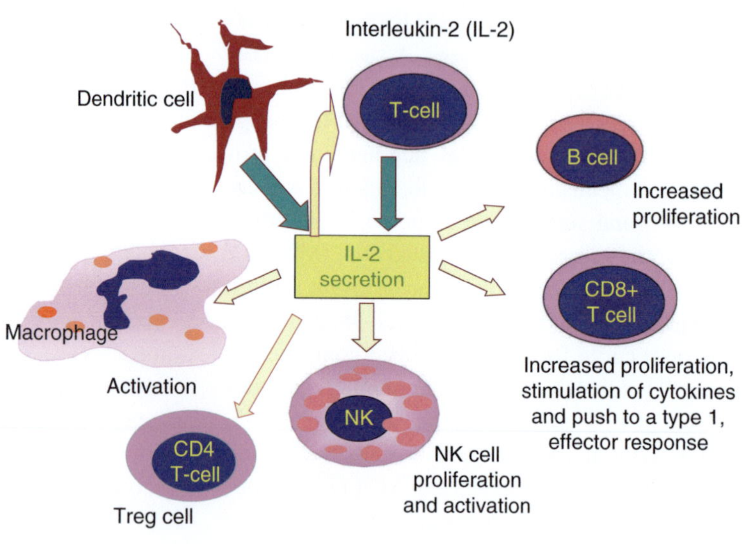

Fig. 25.2 Regulation of immune responses by interleukin-2 (IL-2). IL-2 production by T cells and dendritic cells (DCs) supports the proliferation of T cells, B cells, and natural killer (NK) cells, in addition to establishing a bias toward a type 1 T-cell response. IL-2 can also activate monocytes and NK cells, resulting in increased cytotoxicity. *DC* dendritic cell, *IL-2* interleukin-2, *NK* natural killer cell

One important in vivo role of IL-2 for T-cell responses is the promotion of thymic development and peripheral expansion of $CD4^+CD25^+$ T cells known as REGULATORY T CELLS (Treg) cells (see Chaps. 2 and 3). Loss of Treg activity in IL-2- or IL-2R-deficient mice results in lymphadenopathy and autoimmune disease. It appears that IL-2-dependent Treg cells regulate homeostatic and Ag-induced T-cell proliferative responses, resulting in pathological autoreactivity [61]. IL-2 and its high-affinity receptor are rapidly upregulated following binding with a cognate Ag, and the absence of either IL-2 or IL-2R can result in a loss of T-cell reactivity [62]. Further, the induction of $CD8^+$ T-cell responses depends upon $CD4^+$ T-cell help and their secretion of IL-2 [63]. Recently, IL-2 production by DCs has also been shown to be essential for the initiation of both $CD4^+$ and $CD8^+$ T-cell responses [64].

A functionally similar cytokine to IL-2, IL-15, was identified based upon its ability to stimulate proliferation of IL-2-dependent T cells in the presence of neutralizing IL-2 Abs [65]. IL-15 shares two of the three IL-2Rs including IL-2/15R beta (β) and the γc receptor subunits [65]. Similar to IL-2, resting T cells do not respond to IL-15, and T-cell Ag receptor (TCR) ligation induces IL-15Rα expression, allowing a response to IL-15. Both IL-2 and IL-15 can induce the proliferation of activated T cells and the differentiation of CYTOTOXIC T LYMPHOCYTES (CTL) and LAK cells expressing IL-2R or IL-15R in vitro. However, despite their shared receptor usage, distinct roles for IL-2 and IL-15 are observed in the proliferation and survival of $CD4^+$ and $CD8^+$ T cells [54]. Both IL-2 and IL-15 promote the proliferation of $CD4^+$ T cells, but continued stimulation with IL-2 promotes AICD. IL-15 has the opposite effect and can even inhibit IL-2-induced AICD of $CD4^+$ T cells [66]. IL-15 can also selectively stimulate the proliferation of memory $CD8^+$ T cells, in contrast to IL-2, which inhibits $CD8^+$ memory T-cell proliferation [67]. Thus, IL-15 and IL-2 have similar biological activity in vitro, while their critical and nonredundant functions in vivo are distinct [68]. Further, IL-15 is better than IL-2 in mediating NK-cell differentiation and promoting memory $CD8^+$ T-cell survival. Given IL-15's role in the maintenance and proliferation of T cells, including memory $CD8^+$ T cells, clinical investigations are ongoing with a high priority [69].

25.5.2 Clinical Effects Against Cancer

IL-2 has been approved for use as a single agent in the treatment of renal cell carcinoma (RCC), metastatic melanoma, and hepatitis C. It has also been administered in conjunction with LAK or T-cell-infiltrating LYMPHOCYTES (TIL) in adoptive cellular therapy protocols. TIL are T cells obtained from a tumor and, when expanded in vitro with low levels of IL-2 and in the presence of tumor Ag, result in a population of tumor-specific cytotoxic T cells. However, it has been questioned whether the adoptive transfer of LAK or TIL cells is necessary or adds to the clinical efficacy of IL-2. Indeed, there has been little indication of an improved therapeutic effect of IL-2 plus LAK cells versus IL-2 alone [70]. When the clinical trials with IL-2 are rigorously examined, neither strategy has impressive (as opposed to significant) therapeutic activity [70, 71]. The overall response rate with IL-2 is 7–14% and is associated with considerable toxicity [72]. However, it should be remarked that most of these responses are durable. In one of the first clinical studies [73], partial responses were observed in 4 out of 31 patients. Interestingly, these partial responders did not correspond to patients with increased LAK or NK-cell activity. The antitumor effect of both TIL and LAK cells could be due to either a direct effect or secondary to the induction of other cytokine mediators. This is suggested by the observation that IL-2-stimulated LYMPHOCYTES produce IFN-γ and TNF, as well as other CYTOKINES, and that the therapeutic activity of IL-2 may be synergistic with these CYTOKINES [73]. IL-2 has also been examined as an adjuvant to augment the tumor host response to human immunodeficiency virus (HIV) [74] and anticancer vaccines [75].

Many of the IL-2 clinical trials in metastatic RCC have used the MTD of IL-2. A study by

Fefer et al. [76] compared maintenance IL-2 therapy at the MTD or a dose 60% lower. They found that it was possible to maintain patients on therapy for a median of 4 days at the IL-2 MTD but in the presence of severe hypertension and capillary leak syndrome. In the lower-dose protocol, none of the patients experienced severe hypertension or capillary leak syndrome, and the median duration of maintenance IL-2 therapy was 9 days. Further, there was a total response rate of 41% in the lower-dose protocol compared to a 22% response rate for the higher-dose protocol and a shorter duration of administration. These investigators suggest that there may be an improved therapeutic activity associated with a longer IL-2 maintenance protocol that can be achieved at lower doses.

The treatments for the two primary immunotherapy targets, melanoma and RCC (due to the durable responses noted historically with immunotherapy and the tumor cell antigenicity), have changed dramatically in the past 5 years with the US FDA approval of checkpoint inhibitors (ipilimumab, pembrolizumab, and nivolumab). While CHEMOTHERAPY is still used to treat melanoma, the low overall survival (OS) has stimulated efforts to identify new therapeutic strategies. This included molecular targeted therapeutics for melanoma (vemurafenib, dabrafenib, and trametinib) that are useful only for patients whose disease has a BRAF mutation. While these drugs provide a comparatively high objective response rate (ORR) of approximately 50%, resistance and disease recurrence remain common. In contrast to CHEMOTHERAPY and targeted molecular therapies, the recently approved checkpoint inhibitors have been associated with durable long-term responses, although response rates, particularly with ipilimumab, are relatively low (11–40%) [77]. As discussed in the later section of this review on checkpoint inhibitors the combination of pembrolizumab and ipilimumab, checkpoint inhibitors with differing MOAs have shown durable response rates that were significantly greater than their monotherapy, CHEMOTHERAPY, molecular therapy, and IL-2 therapy dramatically changing the landscape for melanoma and RCC therapy.

25.5.3 Dose-Response, Toxicity, and Pharmacological Studies

Several IL-2 dose-response studies have examined the effect of IL-2 administration on cytokine messenger ribonucleic acid (mRNA) levels in the PBL of cancer patients. The results from one IL-2 dose-response study suggested that (1) doses of IL-2 as low as 3×10^4 U/day administration by continuous infusion could augment T-cell function; and (2) doses of IL-2 $\geq 1 \times 10^5$ U/day increases not only T-cell but also macrophage function [78]. The latter was measured as an upregulation of TNF mRNA, which was observed at the higher dose of IL-2. The increased TNF mRNA levels combined with the increased levels of IFN-α observed at the lower dose of IL-2 may be responsible for IL-2 toxicity [79]. The effect of low-dose, subcutaneous (SC) IL-2 administration was studied in healthy males at 1000 or 10,000 international units (IU)/kg [80]. No consistent changes were observed with 1000 IU/kg; however, phenotypic and immunoregulatory changes were observed following administration of 10,000 IU/kg of IL-2. This dose significantly depressed the number of circulating LYMPHOCYTES, including $CD4^+$, $CD8^+$, and activated T, B, and NK cells. In contrast, the number of neutrophils and monocytes was increased. There was also a significant increase in IL-4 serum levels, while the levels of IFN-γ and IL-2 receptor were significantly depressed. Kinetic studies revealed that these effects varied with time but occurred at IL-2 serum levels sufficient to saturate the high-affinity receptor and by 3 h, following injection of 10,000 IU/kg of IL-2, remained elevated for approximately 12 h. Together, these studies suggest that IL-2-mediated immune augmentation can occur with low doses of IL-2 administered SC or by continuous infusion.

The potential for IL-2 to be active at low doses was also shown in patients with squamous cell carcinoma (SCC) of the oral cavity and oropharynx [81]. In this study, 202 patients were randomly assigned to receive either surgery and radiotherapy or surgery, radiotherapy, and daily IL-2. The IL-2 was injected perilymphatically at 5000 units of IL-2, daily for 10 days prior to and following surgery. IL-2 was also given post-surgery for 12 monthly cycles, each consisting of five daily injections. This study revealed a significant increase in disease-free survival (DFS) and OS in the IL-2 treated patients.

In a study of RCC patients, cohorts were randomized to receive either a high-dose intravenous (IV) IL-2 regimen or one using one-tenth the dose (72,000 IU/kg/8 h) administered by the same schedule [82]. A third arm using the low dose of IL-2, given daily by SC administration, was added later. In the latest report [71], 156 patients were assigned to the high-dose arm, 150 patients to the low-dose IV arm, and 94 patients to the low-dose SC arm. Toxicities were less frequent when IL-2 was given at the low doses, especially hypotension; however, there were no IL-2 associated mortality in any arm. A higher response rate was observed in the patients given the high dose of IL-2 (21%) versus low-dose IV IL-2 (13%), but no OS differences were observed. The response rate of SC IL-2 (10%) was similar to the low-dose IV IL-2 but was significantly less than the high-dose IL-2 therapy ($P = 0.033$). The response duration and survival in the complete responders were significantly better in patients receiving high-dose IL-2, as compared to low-dose IL-2 IV therapy ($P = 0.04$). Thus, tumor regressions, as well as complete responses, were seen with all IL-2 regimes. Thus, IL-2 appears to be more clinically active at higher dose given IV, although this did not provide an OS benefit, and only a small percentage of patients achieved durable clinical response. Currently, it is suggested that one reason the low IL-2 doses have lower therapeutic activity is the preferential effect of low-dose IL-2 on Treg relative to cytotoxic T cells and the associated immunosuppression [83].

25.6 Granulocyte Macrophage-Colony Stimulating Factor (GM-CSF)

25.6.1 Pharmacological Activity Related to Cancer Therapy

GM-CSF was initially defined by its ability to support the growth of both granulocyte and macrophage colonies from hematopoietic precursor cells [84]. However, GM-CSF-deficient mice have no obvious deficiency in myeloid cells [85], suggesting the presence of redundant growth factor(s). GM-CSF can also potentiate the functions of mature granulocytes and MACROPHAGES [86], in addition to its role as a hematopoietic regulator (Fig. 25.3) [86].

Similar to other pro-inflammatory CYTOKINES, the production and activity of GM-CSF occur at the site of inflammation, and increased levels of GM-CSF mRNA are observed in skin biopsies from allergic patients with cutaneous reactions [87] and in arthritic joints [88]. At present, GM-CSF is considered an important regulator (proliferation, maturation, and activation) of granulocyte and macrophage lineage populations at all stages of maturation (see Chap. 5). GM-CSF in combination with TNF-α or IL-4 can differentiate monocytes into DC in vitro [89], which are then used as vaccines. Treatment of mice with pegylated GM-CSF has also been shown to expand DC in the spleen [90], and clinically, the number of circulating DC is expanded in the peripheral blood (PB) by GM-CSF and IL-4 administration [91].

25.6.2 Clinical Use in Cancer Patients

GM-CSF was approved in 1991 by the FDA to support transplant-associated neutropenia and to mobilize stem cells. In Europe, it is also approved for prophylactic treatment following dose-intensive CHEMOTHERAPY. However, the rate of absolute NEUTROPHIL count (ANC) recovery in response to treatment with GM-CSF in patients receiving myelosuppressive CHEMOTHERAPY, or in the mobilization of stem cell into the peripheral blood (PB) with GM-CSF, is 1 day slower than that observed with granulocyte colony-stimulating factor (G-CSF). It also has a toxicity profile, including low-grade fever, myalgias, and bone pain that may be slightly greater than that observed with G-CSF. Because of the real and perceived problems, GM-CSF has been used to a lesser extent than G-CSF. However, patients receiving chronic GM-CSF therapy post marrow graft failure have been shown to have significantly improved survival in comparison to historically matched controls [92]. Similarly, significantly reduced hospital and

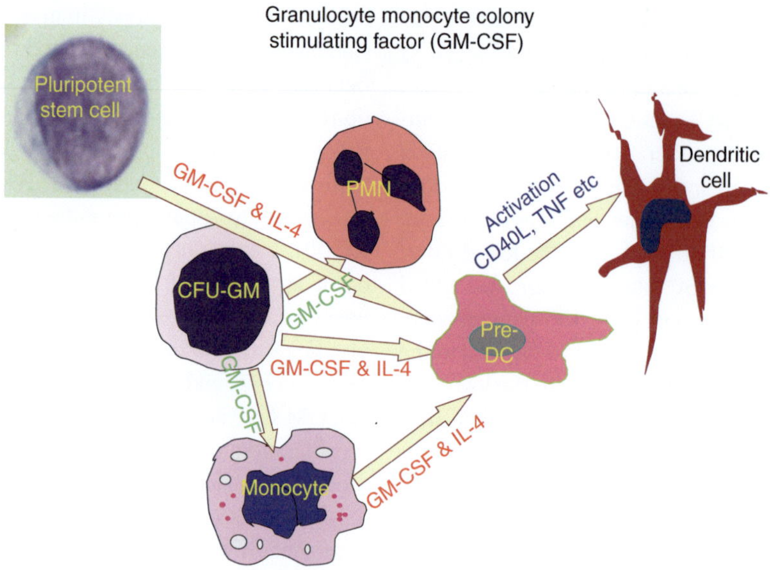

Fig. 25.3 Maturation of precursors into dendritic cells (DC) by granulocyte macrophage-colony stimulating factor (GM-CSF). GM-CSF and interleukin-4 (IL-4) can drive pluripotent stem cells, polymorphonuclear neutrophil (PMN), monocytes, and colony-forming unit-granulocyte-monocyte (CFU-GM) genetic precursors into pre-DC, which following activation can become mature activated DC. *DC* dendritic cell, *GM-SCF* granulocyte macrophage-colony stimulating factor, *IL-4* interleukin-4, *PMN* polymorphonuclear neutrophil, *CFU-GM* colony-forming unit-granulocyte monocyte, *CD40L* CD40 ligand; *TNF* tumor necrosis factor

antibiotic therapy duration have been reported for several phase II and III clinical trials [93]. Despite more than several decades experience using growth factors following stem cell transplantation (SCT), a consensus on their impact on clinical outcome, other than myeloid recovery, remains controversial [94]. More recent studies have suggested that G-CSF may provide immunoregulatory activity, including skewing T-cell cytokine secretion to a T-helper type 2 response, to promote regulatory T-cell and tolerogenic DC differentiation [95], as well as myeloid-derived suppressor cell (MDSC) expansion [96]. Thus, patients undergoing autologous bone marrow transplantation (BMT) or SCT and receiving prophylactic GM-CSF administration have demonstrated significant improvements in ANC recovery, fewer cases of infection, and fewer days spent in the hospital [97].

25.6.3 GM-CSF Effects on Histiocytes

In addition to an effect on HEMATOPOIESIS, administration of GM-CSF can also affect MACROPHAGES and DC. MACROPHAGES are functionally activated following stimulation with GM-CSF, and preclinical and clinical data suggests that GM-CSF can act as an adjuvant for vaccines [98]. However, due to the ability of GM-CSF to expand MDSC, the vaccine adjuvant activity is dose dependent [99]. DC are Ag-presenting cells that play a major role in the induction of primary and secondary T-cell immune responses against cancer (see Chap. 4). Because GM-CSF is a mediator of proliferation, maturation, and migration of DC [100], it has been used to enhance the induction of Ag-specific cytotoxic T cells. Due to the effect of GM-CSF alone or in combination with IL-4 on DC, it has also been used both ex vivo and in vivo to expand DC for use as vaccines [91, 101] following transfection or priming with peptide Ag [91] (see Chap. 19). Further, preclinical studies have suggested that GM-CSF-transfected tumor cells can provide prophylactic and, in some instances, therapeutic anticancer activity [47, 101]. Thus, GM-CSF has a potential not only to address neutropenia but also as a direct therapeutic agent within an adjuvant protocol or as a vaccine adjuvant.

Cancer patients have been shown to have defective macrophage and DC function [102]. Preclinical studies have shown that monocytes and tumor-associated histiocytes can be stimulated by GM-CSF to become cytotoxic against tumor cells [103]. Because of the functional attributes of GM-CSF, a few clinical trials have studied GM-CSF either alone or in combination with IL-2 for the potential to improve anticancer immunity. In one study, 48 cancer patients with surgically resected stage III or stage IV melanomas received GM-CSF SC for 14 days followed by treatment monthly for 1 year or until disease recurrence and the outcomes compared to matched historical controls. A median survival time of 37.5 months was observed with GM-CSF therapy versus 12.2 months for historical controls [104]. In a retrospective study by Grotz et al. [105], 317 surgically resected stage III melanoma patients managed with observation or adjuvant GM-CSF and a median follow-up of 34 months supported therapeutic activity by GM-CSF. In this study adjuvant GM-CSF was associated with improved melanoma-specific survival, but did not reach statistical significance ($P = 0.08$). However, patients with stage IIIC melanoma derived significant benefit, with a 52% risk reduction in melanoma-specific death (hazard ratio 0.48; 95% confidence interval, 0.27–0.87; $P = 0.02$).

In addition to the FDA-approved CYTOKINES and indications (Appendix 1), a number of additional RECOMBINANT CYTOKINES are in clinical development (Appendix 2). The Translational Research Working Group, created as a national initiative by the National Cancer Institute (NCI), developed a ranked list of 30 immune response modifiers for development [106]. These drugs were listed as part of a proceeding of the NCI Agent Workshop and in a published report [107, 108]. It should be noted that several of these drugs have subsequently been FDA approved for the treatment of neoplasia, although not necessary within the rank order identified. Other biologics that are under development include RECOMBINANT IL-21, which has a role in NK- and T-cell activation and which has demonstrated antitumor activity for metastatic melanoma [109]. RECOMBINANT IL-7, a T-cell "growth factor" with a wide range of activities, including a role in T-cell development, has been shown to accelerate lymphoid recovery following cytoreductive therapy [110, 111].

25.7 Engineered Recombinant Proteins

Several RECOMBINANT biopharmaceuticals have been engineered to improve their pharmacological properties. The primary strategy has been to improve the pharmacological half-life by pegylation. This improves the bioactivity of a cytokine by decreasing systemic clearance, resulting in a reduced dosing frequency, improved patient compliance, and lower costs. Many growth factors require daily administration, often requiring a visit to the clinic, and a weekly or bimonthly administration of a pegylated protein is preferred. Indeed, there are a few reports suggesting improved outcomes for neutropenia, with pegylated G-CSF, as compared to the parent protein [112], although there are concerns associated with splenomegalia and hyperleukocytosis [113]. Similarly, genetic manipulation to increase the glycosylation of erythropoietin (EPO) has been shown to prolong the pharmacological half-life. Several pegylated IFN have been studied, including native IFN that have a relatively short half-life, typically 1–4 h. In contrast, pegylation increases this to 24 h or longer. As an example, IntronA (IFN-α2b) is typically administered three times a week, whereas pegylated IFN is administered as a single weekly injection. Similarly, thrice weekly injections of Roferon (IFN-α2a) can be replaced by a single weekly injection of Pegasys, resulting in an increased efficacy for the treatment of hepatitis C. Most recently, pegylated G-CSF has allowed a single injection of pegylated G-CSF, replacing the requirement for daily or twice daily injections of G-CSF for 5 or more days (Box 25.4).

Box 25.4: Pegylation

Colony-stimulating factors have significant bioactivity in the treatment or prevention of neutropenia. These cytokines were used originally as recombinant proteins, although recently pegylated granulocyte colony-stimulating factor (G-CSF) has shown improved pharmacologic properties. In addition to significant clinical efficacy, these recombinant and pegylated growth factors have an economic impact within dose-dense, standard-of-care, and myeloablative chemotherapy and stem cell transplantation (SCT). In general, growth factor administration is associated with a shorter duration of severe neutropenia, antibiotic treatment, and, in myeloablative therapy, a reduced incidence of transfusions and days in the intensive care unit.

- Prolongs half-life and bioavailability
- Protects against binding to monocyte and polymorphonuclear cells (PMN)
- Protects against enzyme degradation and antibody (Ab) induction
- Reduces number of injections required
- Improved patient compliance
- May result in increased bioactivity or a new profile of activity

Similarly, the sequence of EPO was mutated to alter oligosaccharide sequences, resulting in improved pharmacokinetics. The sialic acid content of the carbohydrate component of glycoproteins has a significant effect on a protein's half-life, which was exploited to create this novel EPO sequence (Aranesp®). This resulted in increased sialic acid content from the addition of two extra glycosylation sites in the EPO backbone and a longer half-life. Additional strategies are being used to prolong the half-life of proteins, including nanoparticles, liposomes, and poloxamer matrixes, which allow the slow release of a protein. These formulations provide not only a slow release but also the potential for specific targeting to organs or tumors via modification of the formulation.

25.8 Natural and Chemically Defined BRM

25.8.1 Natural BRM

The use of BRM to treat human disease has its origins in the use of bacterial toxins to treat cancer as suggested by Coley [2]. This form of therapy can act by multiple mechanisms as shown in Box 25.5. These early studies resulted in the clinical use of microbially derived substances, such as BCG vaccine or Picibanil; carbohydrates from plants or fungi, such as Krestin and lentinan; and other products such as Biostim and Broncho-Vaxom, as well as thymic extracts.

Box 25.5: Biological Response Modifiers

- Stop, control, or suppress processes that permit cancer growth, inflammatory, or autoimmune process, including overcoming immunosuppressive processes
- Induce processes that result in cancer cells being more recognizable and susceptible to destruction by the immune system
- Boost the cytotoxic activity of effector cells, such as T cells, natural killer (NK) cells, and macrophages
- Alter the growth patterns of cancer cells to promote a cellular behavior like that of healthy cells
- Block or reverse the process that changes a normal cell or a precancerous cell into a cancerous cell
- Enhance the body's ability to restore normal cells damaged or destroyed by treatment, such as chemotherapy or radiation
- Prevent cancer cells from spreading (metastasizing) to other parts of the body

However, there is considerable lot-to-lot variation in the purity and bioactivity of these compounds. In addition, due to the particulate nature of some of the BRM, IV injections can result in

Table 25.3 Natural BRM

Agent	Chemical nature	Action	Clinical use
Bacillus Calmette-Guérin (BCG) (USA and EU)	Live mycobacteria	Macrophage activator	Bladder cancer
Biostim® (EU)	Extract *Klebsiella penum*	Macrophage activator	Chronic or recurrent infections
Krestin®, polysaccharide-K (PSK) (Japan)	Fungal polysaccharide	Macrophage activator	Gastric/other cancers
Lentinan (Japan)	Fungal polysaccharide	Macrophage activator	Gastric/other cancers
Picibanil, OK432 (Japan)	Extract *Strep. pyogenes*	Macrophage activator	Gastric/other cancers
T-activin (Russia)	Thymic peptide extract	T-cell stimulant	Cancer and infection
Thym-Uvocal® (FRG)	Thymic peptide extract	T-cell stimulant	Cancer and infection
Thymostimulin (EU)	Thymic peptide extract	T-cell stimulant	Cancer and infection

pulmonary thrombosis and respiratory distress, as well as in the development of granulomas following dermal administration and scarification (Table 25.3).

25.8.1.1 BCG

The most commonly used BRM for cancer therapy is BCG. It has been used systemically for the treatment of metastatic disease or adjuvant therapy, intralesionally, especially for cutaneous MM, topically for superficial bladder cancer, and in combination with other immune modulators, tumor vaccines, and CHEMOTHERAPY. Its use is FDA approved in the United States for intravesical administration in the treatment of superficial bladder cancer, residual disease, and adjuvant activity [114]. Several randomized studies have shown prolonged disease-free interval and time to progression in patients treated with intradermal and intravesical BCG as compared to controls [115]. Intravesical therapy with BCG or a chemotherapeutic agent, including thiotepa, mitomycin, or valrubicin, is used in bladder cancer patients with multiple tumors and recurrent tumors or as a prophylactic approach in high-risk patients after transurethral resection. Treatment with BCG delays progression to muscle-invasive and/or metastatic disease, improves bladder preservation, and decreases the risk of death from bladder cancer [116]. In one randomized study of patients with superficial bladder cancer, tumor recurrence was reduced with intravesical and percutaneous BCG administration as compared with controls [117]. Although intravesical BCG may not prolong OS for carcinoma in situ, it results in complete response rates of about 70% decreasing the need for salvage cystectomy [118] and delays tumor recurrence and tumor progression [119]. Results from one prospectively randomized trial suggested that maintenance BCG, when given to patients who are disease-free after a 6-week induction course, may also have improved survival [120]. In summary, meta-analyses have shown that BCG reduces recurrence and progression rates [121] and, in comparison to IFN-α within prospectively randomized trials, demonstrates similar efficacy [122]. It has been suggested that BCG plus IFN-α is a viable alternative in patients who fail intravesical BCG [123].

The mechanism by which BCG mediates its antitumor response is not known, but BCG treatment induces granulomatous inflammation [124] and elevates IL-2 levels in the urine of treated patients [125], suggesting that an augmented local immune response may be important. Clinical studies have shown that intravesical installation of BCG in patients with superficial bladder cancer results in a significant increase in IL-1β, IL-2, IL-6, TNF-α, IFN-γ, and macrophage colony-stimulating factor (M-CSF) with a concomitant and significant increase in serum levels of IL-2 and IFN-α [126]. There appears to be a relationship between cytokine production and therapeutic efficacy, since a multivariant logistic analysis demonstrated that IL-2 induction was a discriminating parameter for remission in patients receiving BCG treatment for superficial bladder carcinoma [127]. This association of increased IL-2 levels was recently confirmed in a study of 20 bladder cancer in situ patients [128].

25.8.2 Chemically Defined BRM

The use of nonspecific immunostimulants has also been extensively studied (Table 25.4). The microbial-derived agents have in common widespread effects on the immune system and side effects akin to the inflammatory response to infection (e.g., fever, malaise, myalgia, etc.). These agents can enhance nonspecific resistance to microbial or neoplastic challenge when administered prior to challenge (immunoprophylactic) but rarely when administered following challenge (immunotherapeutic). This is an important distinction in that the primary objective for the oncologist is the treatment of preexistent metastatic disease.

25.8.2.1 Levamisole

Following a long history of experimental use in many different cancers and diseases, LEVAMISOLE, a chemically defined, orally active immunostimulant, demonstrated significant therapeutic activity (meta-analysis) [129] and was approved for the treatment of Duke's C colon cancer in combination with 5-fluorouracil (5-FU). LEVAMISOLE promotes T-lymphocyte, macrophage, and NEUTROPHIL function, suggesting multiple MOAs. LEVAMISOLE stimulates T-cell function in vivo, particularly in immunodeficient individuals, presumably through the action of its sulfur moiety. One study with LEVAMISOLE demonstrated a significant increase in the frequency of peripheral blood mononuclear cells (PBMC) expressing the NK-cell Ag CD16 at all dose levels, although lower toxicity was observed at the lower doses of LEVAMISOLE [130]. The authors suggested that short-term LEVAMISOLE administration was only minimally immunomodulatory and that chronic administration at low doses may be better tolerated and provide similar levels of immune modulation as that observed with higher doses [130]. It is relatively nontoxic (flu-like symptoms, gastrointestinal upset, metallic taste, skin rash, and Antabuse reaction) but can produce an agranulocytosis, particularly in human leukocyte Ag (HLA) B-27^{+} patients with RA where its use has been discontinued. The adjuvant therapeutic activity of LEVAMISOLE has been questioned in recent years. In one phase III trial, comparing 5-FU with leucovorin to 5-FU with LEVAMISOLE, it was found that the 5-FU and LEVAMISOLE significantly prolonged DFS and OS in patients with type III colon cancer who had undergone curative resection relative to adjuvant therapy with LEVAMISOLE [131]. In a recent report from the Norwegian Gastrointestinal Cancer Group [132], it was found in a randomized phase III study of adjuvant CHEMOTHERAPY with 5-FU and LEVAMISOLE for the treatment of stage II and III colon and rectum cancer that there was no significant survival difference between treatments with or without LEVAMISOLE. However, there was a subgroup of colon cancer patients with stage III disease who had a statistically significant difference in their DFS and cancer-specific survival in favor of adjuvant CHEMOTHERAPY. It was concluded that colon cancer patients with lymph node metastases benefit from adjuvant CHEMOTHERAPY with 5-FU in combination with LEVAMISOLE.

25.8.2.2 Amino-Bisphosphonates

Bone is a frequent site of metastasis by breast and prostate tumors, often resulting in loss of bone mineral density (BMD). In addition, treatment with chemotherapeutics and aromatase inhibitors can result in reduced BMD. Bisphosphonates, synthetic analogues of naturally occurring pyrophosphate compounds, are the standard treatment for BMD, in addition to their potential to reduce bone pain and bone metastases. Randomized trials have shown that bisphosphonates can reduce the risk of bone fracture, cord compression, hypercalcemia, and the need for palliative radiation [133, 134]. Amino-bisphosphonates also directly affect tumor cells, in addition to their inhibition of osteoclast bone resorption [135]. They have been suggested to have anti-angiogenic activity [136], in part due to the inhibition of osteoblastic cell secretion of VEGF [137]. Santini reported that the oral bisphosphonate, pamidronate, reduced VEGF levels in cancer patients with bone metastases, and pamidronate and zelondratic acid (ZA) reduced beta-fibroblast growth factor (β-FGF) and VEGF-induced proliferation of vascular tissue in mice [137]. In addition, ZA has been reported to suppress

Table 25.4 Chemically defined BRMs

Agent	Chemical nature	Action	Clinical use
Adenosine	Purine nucleotide	Innate immunostimulant	Asthma, sepsis, cancer
All-trans retinoic acid (ATRA), tretinoin		Differentiation of myeloid-derived suppressor cells (MDSC) and other myeloid cells	Differentiation of acute promyelocytic leukemia (APL)
Aranesp®	Glycosylated erythropoietin (EPO)	Erythropoiesis -stimulating agent	Anemia
Apremilast, Otezla®	$C_{22}H_{24}N_2O_7S$	Phosphodiesterase 4 (PDE4) inhibitor	Active psoriatic arthritis
Azathioprine, Imuran®	Purine antimetabolite	Immunosuppressant	Graft-versus-host disease (GVHD), allograft
Bestatin® (Japan)	Dipeptide	Macrophage and T-cell stimulant	Acute myeloid leukemia (AML)
Celebrex®	Celecoxib	Cyclooxygenase 2 (COX-2) inhibitor	Anti-inflammatory and familial adenomatous polyposis
Cialis®	Tadalafil	PDE5 inhibitor	MDSC inhibition
Cyclosporine, Sandimmune®	Cyclic undecapeptide, a metabolite of soil fungus	Inhibits nuclear factor of activated T-cell (NFAT)-mediated IL-2 production	GVHD, allografts, rheumatoid arthritis (RA), psoriasis
Cytosine-phosphate-guanine (CpG) 7909 (astuprotimut)	Nucleotide, Toll-like receptor 9 (TLR9) agonist	Binding to TLR9 and dendritic cell (DC) activity	Vaccine adjuvant, solid tumors, primarily melanoma
Dimethyl fumarate (DMF), Tecfidera®	Methyl ester of fumaric acid	Immunomodulatory properties in the absence of immunosuppression	Reduced relapse rate and increased time to progression in multiple sclerosis (MS)
Deoxyspergualin	Peptide fermentation product of *Bacillus laterosporus*	Immunosuppressant	Acute renal rejection
Epanova®	Ω-3 carboxylic acids (EPA and DHA)	Reduced lipid and inflammatory biomarkers	Reduce triglyceride (TG) levels in patients with hypertriglyceridemia
Everolimus, Certican®	Target of rapamycin (TOR) kinase inhibitor	Immunosuppressant	Cardiac transplant
Fingolimod, Gilenya®	Sphingosine-1-phosphate (S1P) receptor modulator	Immunosuppressant that sequesters lymphs in lymph nodes	Relapsing-remitting MS
Glatiramer acetate (Copaxone®)	Polymer containing four amino acids found in the myelin basic protein	Induces $CD8^-$ Tregs that suppress proliferation of pathogenic $CD4^+CD25^-$ T-cells	Relapsing-remitting MS
Imiquimod, Aldara®	Toll-like receptor 7 (TLR7) agonist	Immune augmenting including stimulation of interferon alpha (IFN-α), interleukin 6 (IL-6) and tumor necrosis factor (TNF) secretion	Actinic keratosis, *Condylomata acuminat*, breast cancer and melanoma and superficial basal cell carcinoma (BCC) FDA approved
Isoprinosine (EU)	Inosine: salt complex	T-cell stimulant	Infection

(continued)

Table 25.4 (continued)

Agent	Chemical nature	Action	Clinical use
Leflunomide, Arabloc®, Arava®, Lunava®, Repso®	Pyrimidine synthesis inhibitor	Inhibits the mitochondrial enzyme, dihydroorotate dehydrogenase	RA and psoriatic arthritis
Lenalidomide	Thalidomide analogue	T-cell stimulant	Myelodysplastic syndrome with 5q deletion; multiple myelomas (MM)
Levamisole (USA)	Phenylimidothiazole	T-cell stimulant	Cancer
Lovaza®	Ω-3 fatty acid, ethyl esters of EPA and docosahexaenoic acid (DHA)	Reduced lipid and inflammatory biomarkers	Reduce TG levels in patients with hypertriglyceridemia
Metformin®	1,1-Dimethylbiguanide hydrochloride	Increases the peripheral uptake of glucose and decreases hepatic glucose production	Anti-inflammation, decreases TNF, transcription factors (TF), IL-6, IL-8, monocyte chemotactic protein-1 (MCP-1), C reactive protein (cRP)
Mifamurtide	Muramyl tripeptide phosphatidylethanolamine	Macrophage stimulating	Osteosarcoma
Monophosphoryl lipid A	TLR4 agonist	Vaccine adjuvant	Prevention of human papillomavirus (HPV)-associated cervical carcinoma
Mycophenolate mofetil (MMF), Celleept®	2-Morpholine ethylester of mycophenolic acid	Immunosuppressant	
Poly-ICLC (Hiltonol)	TLR3 agonist	NK/monocyte/macrophage/DC activation	Solid and hematologic malignancies, viral infections
Pomalidomide	Transcription factor Ikaros (IKZF) degradation and immunomodulation	Suppresses tumor necrosis factor (TNF) and adhesion molecule expression	MM
Rapamycin, Rapamune®, sirolimus	Mammalian target of rapamycin (mTOR)	Immunosuppressant	Solid organ transplantation, GVHD
Resiquimod	TLR7/8 agonist	Monocyte/macrophage/DC activation	Solid and hematologic malignancies
Revlimid®, lenalidomide	Thalidomide analogue	Suppresses TNF and adhesion molecule expression	Low- and intermediate-risk myelodysplastic syndrome (MDS)
Rintatolimod (Ampligen®)	TLR3 agonist	Natural killer (NK)/monocyte/macrophage/dendritic cell (DC) activation, vaccine adjuvant	Ovarian, brain, and fallopian tube cancers
Roflumilast	Daliresp®	PDE4 inhibitor	Chronic obstructive pulmonary disease (COPD)
Romurtide (Japan)	Muramyl dipeptide (MDP)-Lys18	Macrophage stimulant	Bone marrow (BM) recovery
Statins	Various	Targets 3-hydroxy-3-methyl-glutaryl-CoA (HMG-CoA) reductase stimulates bone morphogenetic protein 2 (BMP-2)	Anti-inflammatory, decreased nitric oxide synthase 2 (NOS-2), growth factors (GF), chemokines, interleukins (IL)
Sunitinib, Sutent®	Tyrosine kinase inhibitor (TKI)	Inhibits MDSC	Renal cell carcinoma (RCC)

Tacrolimus, FK506, Prograf®	Macrolide lactone, a fermentation product of *Streptomyces*	Inhibits NFAT-mediated IL-2 production	Eczema, solid organ transplantation, GVHD
Tecfidera®	Dimethyl fumarate (DMF)	Promotes the transcription of genes downstream of nuclear transcription factor related to nuclear factor erythroid 2 (NF-E2) NF-E2-related factor 2 (NRF2)	Relapsing-remitting MS
Teriflunomide	Active metabolite of leflunomide	Inhibits the mitochondrial enzyme, dihydroorotate dehydrogenase	Relapsing-remitting MS
Teriflunomide, Aubagio®	(Z)-2-cyano-3-hydroxy-but-2-enoic acid-(4-tri-fluoromethylphenyl)-amide	Reversibly inhibits the mitochondrial enzyme, dihydroorotate (DHODH), required for de novo pyrimidine synthesis in rapidly dividing lymphocytes	Relapsing-remitting MS
Thalidomide (USA)	α-(*N*-phthalimido)glutarimide	Suppresses TNF and adhesion molecule expression	*Erythema nodosum leprosum*
Thymopentin TP-5 (Italy and FRG)	Pentapeptide	T-cell stimulant	RA infection and cancer
Tofacitinib	Janus kinase 3 (JAK3) inhibitor	Inhibits cytokine activity via common gamma chain receptor	RA
Trabectedin	Tetrahydroisoquinoline alkaloid	Selectively depletes monocytes and tumor-associated macrophages (TAM)	Soft tissue sarcoma and ovarian carcinoma
Tretinoin, ATRA	Carboxylic acid form of vitamin A	Cellular differentiation	Acne, keratosis pilaris, photoaging, APL
Vascepa® (AMR101)	ω-3 fatty acid, eicosapentaenoic acid (EPA) ethyl ester	Reduced lipid and inflammatory biomarkers	Treatment of patients with high TG levels
Vitamin D3	Various analogues	Myeloid differentiation most notably MDSC, inhibition of IL-1, IL-6, TNF, and TLR	Head and neck squamous cell carcinoma (HNSCC), psoriasis, and AML
Zelondratic acid (ZA)	Bisphosphonate	Inhibits vascular endothelial growth factor (VEGF), matrix metallopeptidase 9 (MMP9), and MDSC	Bone mineral density, bone metastases, and Paget's disease

myelopoiesis, including MDSC, helping to overturn the tumor-induced immune suppression and stimulate antitumor immune responses [138]. The growth of tumors and the administration of chemotherapeutic drugs have been shown to increase the frequency of myeloid cells, including those which can suppress T-cell number and function [139–141]. These cells, identified in mice as MDSC (Fig. 25.4), can suppress T-cell function [148] and are also found in cancer patients [149, 150]. In humans, the phenotype of MDSC has been reported to be DR^- and Lin^- and $CD11b^+$ and $CD33^+$ [151–154]. Several strategies have been used in rodent models clinically to reduce their function and decrease their accumulation resulting in an improved T-cell responses and apparent therapeutic activity [155–157].

25.8.2.3 Sunitinib

SU11248 or Sutent® is another drug that has been identified with the potential to regulate MDSCs. It was developed as an oral, multi-targeted receptor tyrosine kinase (RTK) inhibitor with antitumor and anti-angiogenic activity. It targets platelet-derived growth factor receptor (PDGFR), vascular endothelial growth factor receptor (VEGFR), KIT, and Fms-like tyrosine kinase 3 (Flt3). Thus, sunitinib has antitumor activity due to its inhibition of tumor cell proliferation, survival, and vascularization via an effect on endothelial cells [158]. In addition, it has been suggested to have therapeutic activity via the inhibition of MDSC expansion [159–162]. Preclinical studies using autochthonous rodent tumor models and human tumor xenografts have demonstrated inhibition of target RTK and concurrent tumor growth effects with sunitinib [159, 163, 164]. Sunitinib monotherapy has also been shown to induce the regression of mammary tumors in MMTV-v-Ha-ras transgenic mice (82% regression) and 7,12-dimethylbenz(a)anthracene-induced mammary tumors in rats [164]. Sunitinib is currently approved for the treatment of RCC patients who have a high expression level of VEGF and PDGF [164]. It is also FDA approved to treat gastrointestinal stromal tumors (GIST) in patients intolerant of, or unresponsive to, imatinib [165]. Sunitinib can also reduce Flt3-driven phosphorylation, induce apoptosis in vitro, and inhibit Flt3-induced VEGF production [166]. This is an important attribute of the activity of sunitinib as ligand-mediated activation of the Flt3 receptor is important for normal proliferation of primitive hematopoietic cells. Indeed, RTK inhibitors with Flt3 affinity have been shown to suppress PLASMACYTOID DCs (pDC) and DC development in Flt3L-supplemented mouse BM-cell cultures (Box 25.6).

Box 25.6: Sunitinib

Sunitinib is a multi-targeted receptor tyrosine kinase inhibitor that can selectively inhibit vascular endothelial growth factor receptor 1 (VEGFR1), 2 (VEGFR2), and 3 (VEGFR3), platelet-derived growth factor receptor α (PDGFRα) and β (PDGFβ), Fms-like tyrosine kinase 3 ligand (Flt3L), and stem cell factor (SCF) [167, 168]. Clinical studies have explored the anti-angiogenic activity of sunitinib [167, 168]. In vivo, sunitinib treatment can reduce tumor microvessel density, inhibit neovascularization, and prevent metastases [169]. It has been approved for use as an anti-angiogenic drug in patients with gastrointestinal stromal tumors (GIST) and renal cell carcinoma (RCC). In the latter disease, it has become frontline therapy [170, 171]. Although sunitinib has been explored in the clinic mainly for its anti-angiogenesis effects, it has also been suggested to induce marked tumor cell necrosis, not associated with an effect on tumor vasculature [172]. Recent preclinical [173] and clinical studies have also shown that sunitinib treatment results in reduced numbers of myeloid-derived suppressor cells (MDSC) [161, 174, 175], improved Th1 response, and diminished Treg cells in RCC patients. These studies suggest that mechanisms mediating sunitinib's therapeutic activity may extend beyond its anti-angiogenic activity.

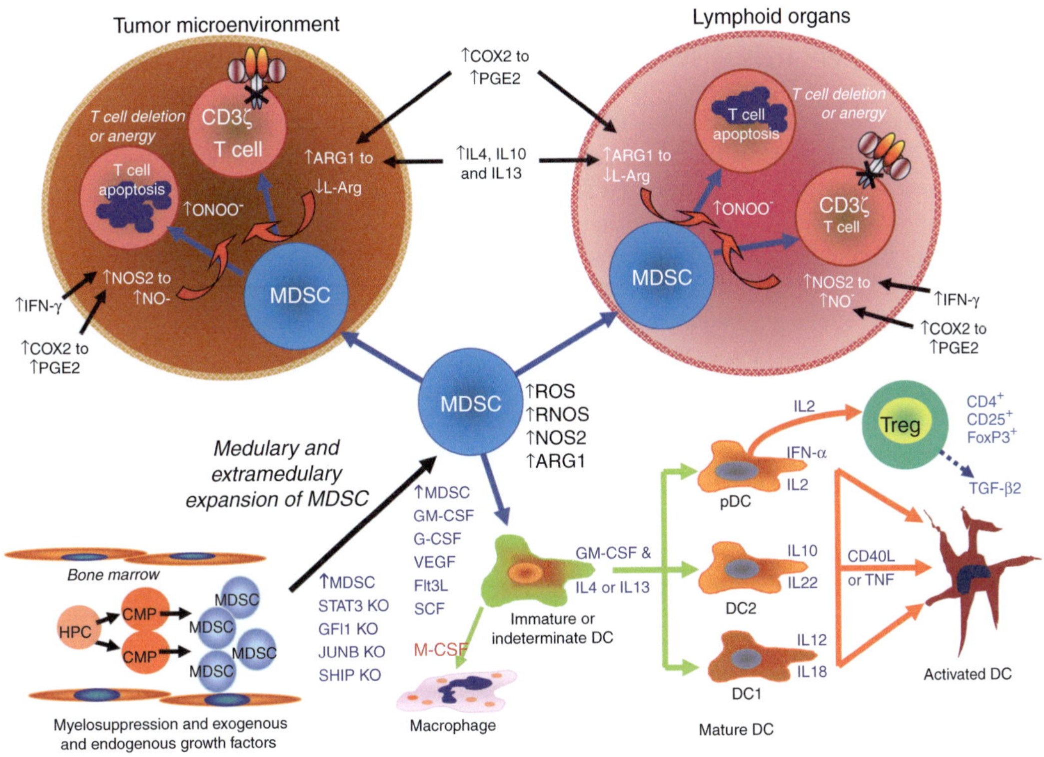

Fig. 25.4 Expansion and regulation of myeloid-derived stem cells (MDSC). Hematopoietic progenitor cells (HPC) proliferate, differentiate, and commit to various hematopoietic lineages, including committed myeloid progenitors (CMP). Under conditions of myelosuppression or exogenous administration of growth factors (GF), significant increases in MDSCs occur in the peripheral blood (PB) and spleen and infiltrate tumors. MDSC, during expansion, are also mobilized into the circulation, lymphoid organs, and tumor microenvironments. The expansion of MDSC is associated with increased levels of GF or vascular endothelial growth factor (VEGF). In addition to proliferation, MDSC can also be activated, secreting high levels of arginase 1 (ARG1) and nitric oxide synthase 2 (NOS2). ARG1 decreases L-arginine levels, resulting in a translational blockade of the CD3ζ chain. This is T-cell suppressive, affecting multiple pathways, including the kinases, GCN2, and mTOR. In addition, high levels of NOS2 and NO are induced, which nitrosylate cysteine residues in target proteins and affect the production of cyclic guanosine monophosphate (GMP). This affects IL-2 receptor signaling by blocking the phosphorylation of signal-inducing molecules coupled to IL-2R and altering the stability of IL-2 messenger ribonucleic acid (mRNA). Upregulation of both enzymes, in addition to affecting the two pathways described above, can increase the production of other reactive oxygen species (ROS) and reactive nitrogen oxide species (RNOS), including O_2^-, NOO^-, and H_2O_2. This occurs either by nitrating tyrosine residues or by controlling BCL-2 and CD95L expression, resulting in T-cell apoptosis. Signaling elements that have been shown to regulate this process include signal transducer and activator of transcription 3 (STAT3) [142–144], GFL-1 [145], JunB [146], and SHIP [147]. Furthermore, MDSC can differentiate into DCs, which can be matured by M-CSF into macrophages or into DCs by granulocyte-macrophage colony-stimulating factor (GM-CSF) and IL-4 or IL-13. This includes myeloid DCs, which are immunosuppressive via the production of IL-10 and IL-21. These may differentiate into lymphoid or myeloid DC with the production of immune-augmenting levels of IL-12 and IL-18. Alternatively plasmacytoid DCs (pDC) may be formed, which have important roles in the response to viral infections via the production of IL-2 and interferon (IFN), the high levels of IL-2 increasing the frequency of Treg cells. All of these DC types can be activated by a variety of cytokines, including CD40L and TNF-α, to express increased levels of co-stimulatory molecules. *MDSC* myeloid-derived stem cell, *HPC* hematopoietic progenitor cell, *CMP* committed myeloid progenitor, *GF* growth factor, *VEGF* vascular endothelial growth factor, *ARG1* arginase 1, *NOS2* nitric oxide synthase 2, *cyclic GMP* cyclic guanosine monophosphate, *mRNA* messenger ribonucleic acid, *ROS* reactive oxygen species, *STAT3* signal transducer and activator of transcription 3, *GM-CSF* granulocyte-macrophage colony-stimulating factor, *IFN* interferon

25.8.2.4 Muramyl Dipeptides (MDP): Synthetic Natural BRM

One of the largest and best studied classes of synthetic agents is based on MURAMYL DIPEPTIDES (MDP) discovered as the minimally active substitute for intact BCG in Freund's adjuvant [176–178]. Unfortunately, as with many of the polypeptides with low molecular weights, MDP has a short serum half-life and requires frequent administration at high doses to be active. In addition, agents such as MDP are strongly pyrogenic, presumably in association with their ability to induce IL-1. MDP analogues have also been incorporated into multilamellar vesicles (MLV) for higher stability and to facilitate monocytic phagocytosis.

The first MDP approved for clinical use was romurtide (Japan), which induces BM recovery following cancer CHEMOTHERAPY [179] via growth factor induction. Its MOA is the activation of MACROPHAGES to secrete colony-stimulating factors (CSF), IL-1, and TNF, resulting in the expansion of marrow precursors and subsequent commitment and differentiation into mature granulocytes and monocytes. Therefore, the period of granulocytopenia and the risk of secondary infections are reduced, allowing more frequent and/or intense CHEMOTHERAPY. Murabutide (Table 25.4), an orally active form of MDP that does not induce fevers, is currently in clinical trials in patients with cancer and infectious disease (France). In order to further stabilize the incorporation of MDP into MLVs, lipophilic analogues of MDP, such as muramyl tripeptide phosphatidylethanolamine [MTP-PE (mifamurtide)], have been developed. MTP-PE has shown significant therapeutic activity in pediatric patients with osteosarcoma [176, 177, 180] and was recently approved in Europe to treat nonmetastatic osteosarcoma following surgical resection. Preclinically, MTP-PE has also shown protection of the mucosal epithelium from cytoreduction therapy [176, 177, 181]. The MDP are also potent adjuvants, either alone or in oil emulsion, and are under consideration for use as adjuvants with HIV peptide vaccines.

25.8.2.5 Bestatin®: Engineered Synthetic Natural BRM

Bestatin® (ubenimex) is a potent inhibitor of aminopeptidase N and aminopeptidase B [182]. Bestatin® was isolated from a culture filtrate of *Streptomyces olivoreticuli* during the search for specific inhibitors of enzymes present on the membrane of eukaryotic cells [183]. Inhibitors of aminopeptidase activity are associated with macrophage activation and differentiation, and Bestatin® has shown significant therapeutic effects in several clinical trials [184]. In one multi-institutional study, patients with acute nonlymphocytic leukemia (ANLL) were randomized to receive either Bestatin® or control [185] orally after completion of induction and consolidation therapy and concomitant with maintenance CHEMOTHERAPY. Remission duration was prolonged in the Bestatin® group, although this difference did not reach statistical significance. However, OS was prolonged in the Bestatin® group. Recently, a confirmatory phase III trial in ANLL was reported which extended the observation to a significant prolongation of remission [186]. Bestatin® has also shown adjuvant activity when administered to acute leukemia and CML patients who did not develop graft-versus-host disease (GVHD) within 30 days following BMT [187]. Bestatin®-treated acute leukemia patients had an increased incidence of chronic low-grade GVHD as compared to the control arm and a lower relapse rate. Recently, a phase III study of resected stage 1 squamous cell lung cancer patients treated with either Bestatin® or placebo daily per OS for 2 years revealed that 5-year cancer-free survival was significantly greater in the Bestatin® group as compared to the placebo group. In this study, the 5-year, cancer-free survival was 71% for the Bestatin® cohort and 62% for the placebo group. OS was also significantly improved as was cancer-free survival [188]. Recent studies in patients with non-small cell lung cancer (NSCLC) suggest that it also has anti-angiogenic activity [189].

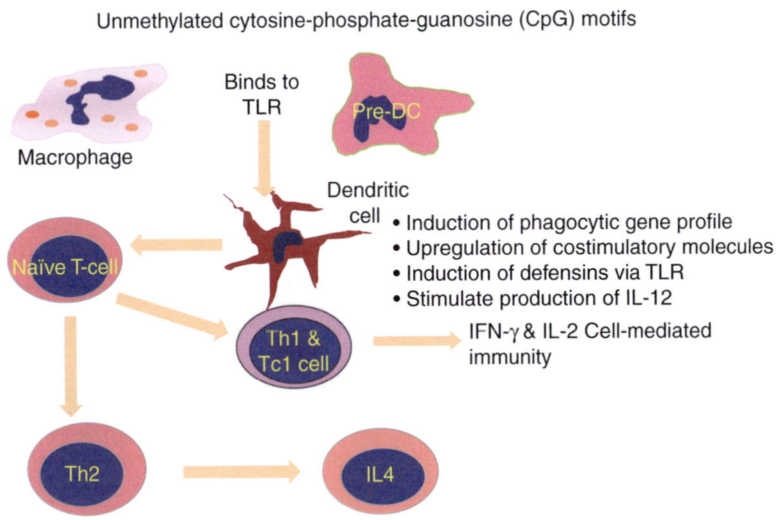

Fig. 25.5 Regulation of innate and adaptive immunity by cytosine-phosphate-guanine (CpG) motifs. The presence or injection of unmethylated CpG motifs results in their binding to Toll-like receptor (TLR9), leading to the activation and maturation of dendritic cells (DC), which can interact with Th1 or Tc1 cells and upregulate cell-mediated immunity. In addition, CpG can induce phagocytosis, upregulation of co-stimulatory molecules, defensins, and interleukin-12 (IL-12). *CpG* cytosine-phosphate-guanine, *TLR9* Toll-like receptor 9, *DC* dendritic cell, *Th1* T-helper cell type 1, *Tc1* T cytotoxic cell type 1, *IFN-γ* interferon-gamma, *IL-2* interleukin-2, *IL-12* interleukin-12

25.8.2.6 Oligonucleotides (ODN) as Natural/Synthetic BRM

Bacterial extracts can activate both innate and adaptive immunities. The molecular receptors that regulate how the innate immune system detects infectious agents and distinguish different classes of pathogens have recently been elucidated. The immune system uses pattern recognition receptors that are expressed on certain innate immune cells, to trigger cellular activation when they recognize conserved microbial-specific molecules [190]. These molecules, originally thought of as nonspecific immune activators as discussed above, are now known to be recognized by receptors that are expressed in a cell-specific and compartmentalized manner (see Chap. 6). The best-characterized family of pattern recognition receptors is the TOLL-LIKE RECEPTOR (TLR) family. One of these, TLR9, is expressed in the endosomal compartment of PLASMACYTOID DC and B cells [191] and is essential for the recognition of viral and intracellular bacterial DNA [192]. Based on the identification of these specific ligands, immunotherapy has begun to grow beyond the nonspecific effects of whole bacterial extracts, using synthetic TLR ligands (TLRL). One example of such synthetic immunomodulatory molecules are the short ODN that mimic the innate immune response to microbial DNA and which contain one or more cytosine-phosphate-guanine (CpG) dinucleotide-containing motifs with unmethylated cytosine residues that are recognized by TLR9. The immune effects of CpG ODN occur in two stages: an early stage of innate immune activation and a later stage of enhanced adaptive immunity (Fig. 25.5). Within minutes of the exposure of B cells or PLASMACYTOID DC (pDC) to CpG, the expression of co-stimulatory molecules, resistance to apoptosis, upregulation of the chemokine receptor type 7 (CCR7) (associated with trafficking to the T-cell zone of the lymph nodes), and secretion of T-helper 1 (Th1)-promoting CHEMOKINES and CYTOKINES including macrophage inflammatory protein-1 (MIP-1) and other IFN-inducible genes [193] are observed. pDC secrete IFN-α and mature into highly effective Ag-presenting cells (APC) [194]. The CpG-induced secretion of IFN-α, TNF-α, and other CYTOKINES and CHEMOKINES induces, within hours, secondary effects, including NK-cell activation and enhanced expression of Fc receptors, resulting in increased ADCC. This innate immune activation and pDC maturation into myeloid DC is followed by the induction of adaptive immune responses. B cells are strongly co-stimulated if they bind specific Ag at the same time as CpG, which selectively enhances the development of Ag-specific Ab [195]. CpG binding also activates B cells to proliferate, secrete IL–6, and differentiate into plasma cells [196]. The CpG-enhanced APC function occurs via the upregulated expression of co-stimulatory molecules, including CD40, CD80, and CD86 [197]. In mixed cell populations, a cascade of secondary responses, including activation of MACROPHAGES and NK cells and the induction of IFN-γ by Th1 cells, also occurs [197, 198]. The efficient activation of APC and induction of IL-12, IL-18, IFN-α, and IFN-γ are directly associated with the ability of ODN to induce Th1 polarization, adjuvant activity [199], and inhibition of Th2 responses [200] and stimulate CD8 T-cell responses [201].

CpG ODNs appear also to be effective as vaccine adjuvants to enhance adaptive Th1 cellular immune responses [202]. In mice, CpG ODNs can trigger strong Th1 responses [202, 203], enhancing the number and function of tumor-specific CTLs and IFN-γ-secreting T cells [204]. CpG ODN also enhance therapeutic responses to vaccines, including DC vaccines [205], proteins, irradiated cells transduced with GM-CSF [206], and long peptide vaccines [204]. This has resulted in therapeutic vaccines in mouse tumor models where no other approach has shown comparable efficacy, even with 1-cm established tumors [205, 207]. Even without a vaccine, CpG ODN can induce $CD8^+$ T-cell-mediated regression of established tumors with durable memory responses [208]. By the stimulation of innate immunity, TLR activation by CPG analogues has been shown to enhance Ag-specific humoral and cellular immune responses such that CPG ODN have seen increasing utility as vaccine adjuvants [209, 210]. In one study, melanoma patients, which were randomized to preoperative administration of either PF-3512676 or saline, developed IFN-γ^+-specific $CD8^+$ T-cell responses as determined by Enzyme-Linked ImmunoSpot (ELISpot) analysis from sentinel lymph nodes (SLN) and PB against melanoma-associated antigens (MAA). In this study, $CD8^+$ T-cell responses to MAA were 0 of 11 for the saline group, versus 5 of 10 in the patients who received PF-3512676 which was significantly higher. The $CD8^+$ T-cell response was also found to correlate with the activation of pDCs in the SLN [211]. In another clinical study, CpG ODN PF-3512676 in combination with taxane/platinum CHEMOTHERAPY administration to chemotherapy-naïve patients with stage IIIb/IV NSCLC [212] resulted in a significant, higher overall response rate (38%) versus patients randomized to standard CHEMOTHERAPY (19%). The median OS was 6.8 months in the CHEMOTHERAPY arm, versus 12.3 months in the combination arm. In contrast, a phase III study combining PF-3512676 with paclitaxel/carboplatin or gemcitabine/cisplatin versus CHEMOTHERAPY alone as first-line treatment for patients with advanced NSCLC [213] resulted in no improvement in OS or PFS when PF-3512676 was added to standard CHEMOTHERAPY.

25.9 Checkpoint Inhibitors

As mentioned above, in recent years, immunotherapy has provided new and frequently less toxic therapeutic approaches for cancer, which have included four FDA-approved, checkpoint inhibitors (Table 25.5) as of June 2016, each of which has multiple clinical targets (see also Chap. 12). IMMUNE CHECKPOINT INHIBITORS provide a new approach to cancer treatment that contrasts with chemotherapies and targeted agents that interfere with key tumor signaling, cell growth, or cell division drugs. IMMUNE CHECKPOINT INHIBITORS are designed to restore a patient's antitumor immune responses that were suppressed by tumor growth. However, critical questions remain regarding these novel therapeutic drugs, including the identity of the patient subset that will be responsive including tumor type, stage, and optimal duration and combination for therapy. Additional questions include the identification of potentially responsive patients, based not only on tumor checkpoint expression but also systemic

Table 25.5 FDA-approved checkpoint inhibitors

Target	Drug name	Other name	FDA-approved indication	Date FDA approval
CTLA4	Ipilimumab	Yervoy	Unresectable or metastatic melanoma	3/25/2011
PD-1	Pembrolizumab	Keytruda	Advanced or unresectable melanoma	9/4/2014
PD-1	Nivolumab	Opdivo	Unresectable or met melanoma that has progression following ipilimumab and a BRAF inhibitor	12/22/2014
PD-1	Nivolumab	Opdivo	Metastatic squamous NSCLC with progression on platinum	3/4/2015
CTLA4 and PD-1	Ipilimumab and nivolumab	Yervoy and Opdivo	BRAF V600 wild-type melanoma	9/30/2015
PD-1	Pembrolizumab	Keytruda	Metastatic NSCLC	10/2/2015
PD-1	Nivolumab	Opdivo	Metastatic NSCLC progressed on platinum	10/9/2015
CTLA4	Ipilimumab	Yervoy	Adjuvant therapy with stage III melanoma	10/28/2015
PD-1	Nivolumab	Opdivo	Metastatic RCC with prior anti-angiogenic therapy	11/23/2015
PD-1	Pembrolizumab	Keytruda	First-line therapy for unresectable or metastatic melanoma	12/18/2015
PD-1	Nivolumab	Opdivo	Hodgkin's disease	5/17/2016
PD-L1	Atezolizumab	Tecentriq	Bladder Ca if the patient fails platinum	5/18/2016
CTLA4 and PD-1	Ipilimumab and nivolumab	Yervoy and Opdivo	BRAF V600 wild-type melanoma and BRAF V600 mutation+ or unresectable/metastatic melanoma	12/23/2016

and infiltrating leukocytes, biologic surrogates such as the association of checkpoint expression with clinical responses and how to limit the associated toxicities.

The immune system can recognize and defend against foreign attack, including microbial pathogens and tumors, while also avoiding self-reactivity [214]. The PD-1/PD-L1 pathway is one of many peripheral tolerance mechanisms that has evolved to prevent autoimmunity by regulating T-cell activation and associated tissue damage [215]. Further, checkpoint interactions have an important role in regulating persistent antigenic stimulation, induced by chronic viral infections and tumors growth, and may contribute to T-cell exhaustion [216] and have been implicated as a strategy by which microbes and tumors evade immune responses.

Tumors take advantage of immune checkpoints to evade immune recognition and limit cytotoxic T-cell functions [217]. The activation of T cells requires two signals controlling T-cell survival, proliferation, and responsiveness to antigens. The first signal is initiated by the T-cell receptor (TCR) via ANTIGEN recognition, while the second signal is mediated by an interaction between co-stimulatory and co-inhibitory signals, also known as immune checkpoints [218]. Under physiologic conditions, there is a balance between co-inhibitory and co-stimulatory signals, which is crucial for the maintenance of self-tolerance and immune homeostasis [219]. Tumors may express checkpoint inhibitory molecules that attenuate T-cell responses by integrating both adaptive and innate effector mechanisms, such that agonists of co-stimulatory receptors or antagonists of inhibitory receptors can amplify antigen-specific T-cell responses [220]. Indeed, the blockade of immune checkpoints using respective monoclonal mAbs can induce antitumor responses not only in classical "immunogenic" tumor types, such as melanoma and renal cell carcinoma [221, 222], but also other less antigenic solid tumors.

Monoclonal Abs to cytotoxic T-lymphocyte-associated ANTIGEN 4 (CTLA-4), programmed death receptor-1 (PD-1), and programmed death ligand 1 (PD-L1) have been approved by the US FDA (Table 25.5), although numerous other molecular targets are in preclinical and clinical development. These neutralizing/blocking monoclonal Abs are immunomodulatory and can limit immune escape by tumor cells using "normal" immune suppressive mechanisms to prevent autoimmunity and tissue damage in response to acute infection in otherwise healthy individuals and inhibit tumor progression in cancer patients [223].

Ipilimumab, a fully humanized, immunoglobulin (Ig) G1 monoclonal Ab targeting CTLA-4, was the first immune checkpoint inhibitor approved by the FDA. This occurred in 2011 for the treatment of unresectable or metastatic melanoma based on improved overall survival in two phase III clinical trials [222]. Nivolumab and pembrolizumab, both mAbs against PD-1, received accelerated FDA approval in 2014 for the treatment of unresectable or metastatic melanoma patients and disease progression following treatment with ipilimumab and, if BRAF V600 mutation positive, a BRAF inhibitor [224].

In 2015, *nivolumab* [225] and *pembrolizumab* [226] received FDA approval for the treatment of metastatic squamous and non-squamous non-small cell lung cancer (NSCLC), with progression on or after platinum-based therapy. In the case of patients with EGFR or ALK genomic tumor mutations, they were required to have disease progression while receiving FDA-approved therapy prior to nivolumab therapy. In contrast to nivolumab, which was approved for PD-L1-expressing and PD-L1-non-expressing tumors, pembrolizumab was only approved for patients whose tumors expressed PD-L1. Thus, pembrolizumab was approved concurrent with a companion diagnostic test for PD-L1 immunohistochemistry (22C3 pharmDx) for patient selection. Nivolumab is also approved as a monotherapy for BRAF V600 wild-type advanced melanoma patients and second-line treatment of advanced renal cell carcinoma patients.

PD-L1, the cognate ligand for PD-1, is expressed on the tumor cell surface or on neighboring host immune cells in the tumor microenvironment and engages the PD-1 receptor on activated cytotoxic T cells downregulating the tumor-directed host immune response. In a recent report [227] of an international, multicenter, single-arm, phase II trial, a therapeutic role for atezolizumab (anti-PD-L1) in patients with locally advanced or metastatic urothelial carcinoma was reported. Overall, atezolizumab was found to be safe and well tolerated with grade 3–4 treatment-related AEs occurred in 16% of the patients, with fatigue being the most common, and grade 3–4 immune-mediated AEs occurred in 5% of patients. Based on this and other clinical studies, atezolizumab was approved by the US FDA in 2016.

Thus, both ipilimumab and nivolumab have demonstrated clinical efficacy, and we now have almost a decade of experience using these drugs in clinical trials and as FDA-approved drugs. However, there remain questions regarding their optimal administration for both efficacy and to limit toxicity [228]. Ipilimumab as a single agent demonstrated significant therapeutic efficacy, but with an undesirable toxicity profile [229], and with the FDA approval of nivolumab and pembrolizumab, its use as a single agent has diminished as nivolumab and pembrolizumab monotherapies have shown improved efficacy and a better AE profile. However, there is a rationale for combination therapy with PD1 (nivolumab) and CTLA-4 inhibition (ipilimumab) as these drugs are active on differing and nonredundant regulatory pathways. Further, CTLA-4 acts on the lymphoid system, while PD-1 inhibition is active downstream in the tumor microenvironment including antigen-presenting cells and MDSCs [230]. Thus, the mechanisms of resistance to PD-1 and CTLA-4 inhibitor monotherapy may be circumvented by combination therapy, such that the CTLA-4 blockade mitigates the CTLA-4 upregulation associated with resistance to PD-1 blockade. However, these mechanisms of resistance are not yet well understood, and combination therapy may not completely resolve therapeutic resistance, suggesting additional mechanisms of actions (MOAs) to overcome resistance may be needed.

The first phase I combination trial of nivolumab and ipilimumab [231] examined co-administration versus a sequence of ipilimumab followed by nivolumab [231]. At the maximum tolerated doses of the combination (nivolumab and ipilimumab), a response rate of 53% and grade 3/4 adverse event (AE) rate of 53% were observed. The sequential regimen cohort had a response rate of 18% when nivolumab was given alone followed by ipilimumab and an AE incidence of 20%. Based on this observation, additional combination studies were initiated.

Building on the maximum tolerated dose, a double-blind, placebo-controlled trial assessing the combination of nivolumab and ipilimumab versus ipilimumab alone in untreated melanoma patients was initiated [232]. Patients were randomized to receive ipilimumab in combination with nivolumab or placebo once every 3 weeks for four doses followed by nivolumab or placebo every 2 weeks until disease progression or unacceptable toxicity occurred. Objective response rates for the combination were reported as 61% for BRAF wild-type patients and 52% for those with BRAF V600 mutation as compared to 11% and 10% for the ipilimumab alone arm. Drug-related AEs were the most common cause of protocol discontinuation in the combination arm occurring in about 45% of patients. In the combination group, 59% and 57% of patients received the four doses of ipilimumab and nivolumab, but did not go on to receive the single agent nivolumab. In this study 54% of the patients experienced grade 3 and 4 treatment-related AEs in the combination group versus 24% in the ipilimumab alone cohort.

In a similar study, Larkin et al. evaluated the combination of nivolumab and ipilimumab versus each as monotherapy in untreated melanoma patients [233]. In these trials previously untreated patients with unresectable stage 3 or 4 melanoma patients were randomized to the combination versus ipilimumab alone or nivolumab alone. Overall and complete response rates were 57.6% and 11.5% in the combination, 19% and 2.2% for ipilimumab monotherapy, and 43.7% and 8.9% for nivolumab monotherapy. However, the time to objective response was similar across all cohorts. Thus, combination therapy with nivolumab and ipilimumab is more effective than as monotherapies providing a meaningful improvement in PFS and/or overall survival (OS) vs. standard of care. However, this increased efficacy is associated with an increase in AEs [233], which is not unexpected given the potential for overlapping toxicities. Treatment-related AEs of any grade were high with 95.5% in the nivolumab-plus-ipilimumab cohort, 86.2% in the ipilimumab cohort, and 82.1% in the nivolumab cohort. The most common AEs in the combination group were diarrhea, fatigue, and pruritus. Grade 3 and 4 treatment-related AEs were 55% in the combination cohort, 27.3% with ipilimumab, and 16.3% in the nivolumab alone cohort. Similar to the Wolchok experience, the combination cohort had, as the most common reason for drug discontinuation, treatment-related AEs, which occurred in 36.4% of the patients. However, the benefit/risk assessment resulted in the accelerated FDA approval of this combination.

Studies to determine if there was a subset of patients that benefited most from the combination versus single drug checkpoint inhibition, a subgroup analysis stratifying for PD-LI expression, were undertaken. It had been previously demonstrated that PD-LI expression correlated with PD-I Ab response [234]. This observation was supported in the combination study, although significance was not achieved. The ORR stratified for PD-LI expression was 57.5% vs. 41.3% for nivolumab alone and 72.1% vs. 54.8% for nivolumab plus ipilimumab [233]. Thus, there is a debate over the use of PD-LI as a biomarker. This is consistent with the observation that although tumors that are PD-LI positive tend to respond better to PD-I Ab therapy, those who are negative can also respond. Furthermore, the location of staining (tumor vs. immune cell), marker (PD-I vs. PD-LI), tumor location (primary vs. metastasis), or timing (before vs. after treatment) provides additional parameters that require consensus [235]. At present, the general concept of using PD-LI expression as an integral biomarker for determining who may benefit from what types of therapy, and the way in which we test it, is yet to be clarified.

Despite success in the metastatic setting, approximately 40% of patients have no response to checkpoint inhibitor therapy, providing a challenge for future development. Although patient selection through predictive biomarkers is a reasonable approach, challenges such as a lack of a standardized method and tissue availability remain. Combination therapies, as discussed with nivolumab and ipilimumab, are a current clinical focus with additional tumor targets and checkpoint targets (mAbs to PDL2, lymphocyte-activation gene (LAG)-3 and Tim-3) as examples are being developed [236]. Further, combination therapy of checkpoint inhibitors with targeted therapy, radiotherapy, anti-VEGF, or other immunotherapy, such as adoptive cellular and vaccine therapy, is also appropriate providing strategies to overcome failed responses to monotherapy with checkpoint inhibitors. However, the discovery of effective combinations is challenging [237]. This is associated with the multitude of immunotherapy agents approved or under development and the number of vaccines under development. Experimentally testing every possible combination of these drugs is unfeasible, and a mechanistic basis for the selection of agents with complementary MOAs that target multiple mechanisms of resistance and immune escape will be required, as well as the identification of bio-surrogate markers.

25.10 Vaccine Adjuvants

Immune adjuvants are drugs that enhance the host's humoral and/or cellular immune response when co-administered with an Ag. Recent progress in our understanding of innate immunity and its potential in shaping adaptive responses has driven a search for a new generation of vaccine adjuvants, largely based on studies using RECOMBINANT CYTOKINES and synthetic/molecular immunomodulatory agents. Currently, the adjuvants that are used for human vaccines

Table 25.6 Adjuvants used for human vaccines

Alum (1924)	Mineral salts	Aluminum phosphate or aluminum hydroxide	Various
Adjuvants licensed for use in human vaccines			
AS03 (MF59) (GlaxoSmithKline; 2009)	Oil-in-water emulsion	Squalene, Tween 80, α-tocopherol	Pandremix (pandemic influenza), Prepandrix (pre-pandemic influenza)
AS04 (GlaxoSmithKline; 2005)	Alum-absorbed TLR4 agonist (MPL) and VLPs	Aluminum hydroxide, MPL	Fendrix (hepatitis B), Cervarix (human papilloma virus)
MF59 (Novartis; 1997)	Oil-in-water emulsion	Squalene, polysorbate 80 (Tween 80; ICI Americas), sorbitan trioleate (Span 85; Croda International)	Fluad (seasonal influenza), Focetria (pandemic influenza), Aflunov (pre-pandemic influenza)
Virosomes (Berna Biotech; 2000)	Liposomes	Lipids, hemagglutinin	Inflexal (seasonal influenza), Epaxal (hepatitis A)
Vaccine adjuvants tested in humans but not licensed for use			
AF03	Oil-in-water emulsion	Squalene, Montane 80, Eumulgin B1 PH	–
AS01	Combination	Liposome, MPL, saponin (QS21)	–
AS02	Combination	Oil-in-water emulsion, MPL, saponin (QS21)	–
CAF01	Combination	Liposome, DDA, TDB	–
CpG 7909, CpG 1018	TLR9 agonist	CpG oligonucleotides alone or combined with alum/emulsions	–
Flagellin	TLR5 agonist	Bacterial protein linked to antigen	
IC31	Combination	Oligonucleotide, cationic peptides	–
Imidazoquinolines	TLR7 and TLR8 agonists	Small molecules	–
Iscomatrix	Combination	Saponin, cholesterol, dipalmitoylphosphatidylcholine	–
Pam3Cys	TLR2 agonist	Lipopeptide	–
PolyILC	TLR3 agonist	Double-stranded RNA analogues	–

(Table 25.6) are few, each with a number of deficiencies. For example, aluminum-based compounds enhance Th2 humoral immune responses but fail to stimulate cellular Th1 responses [238]. Oil-in-water emulsions induce a more balanced immune response but remain weak inducers of Th1 immunity. In addition, all of the adjuvants that are part of licensed products fail to induce CD8 T-cell responses in humans. Consequently, there is a critical need for new, safe, and nontoxic adjuvants effective in inducing long-lasting protective responses incorporating CD8 cellular immunity. Thus, extensive preclinical and clinical studies are ongoing to identify adjuvants resulting in a focus on novel immune modulators as part of vaccine formulations.

Vaccine adjuvants exert their effects through different mechanisms. Some adjuvants, such as alum and emulsions (MF59®), function in part by generating depots that trap Ag at the injection site, providing slow release and chronic stimulation of immunity. These adjuvants also increase recruitment and activation of APC. The first approved and most commonly used vaccine adjuvant, alum, has the capability to form multi-molecular Ag aggregates which increase uptake and presentation by APC [239]. It was developed by Glenny et al. [240] in 1926, with diphtheria toxoid absorbed to alum, and has been in use for over eight decades. Alum is used as the adjuvant for almost all vaccines and provokes a strong Th2 response [241] but is rather ineffective against pathogens that require Th1-cell-mediated immunity. Alum induces the immune response by a depot effect and activation of APC. Other adjuvants used either alone or combined with emulsions are essentially ligands for pattern recognition receptors (PRR) and act by inducing innate immunity, predominantly by targeting the APC. One of these, monophospholipid (MPL) A, is a chemically modified derivative of lipopolysaccharide (LPS) that has greatly reduced toxicity while maintaining most of the immunostimulatory activity of LPS and expresses potent stimulatory activity for T-cell and Ab responses. MPL is also the first and only TLR ligand approved for clinical use in the United States in the form of AS04.

25.11 Off-Label Use of Drugs as Immune Regulatory Agents

A number of drugs have been used off label as part of immune-augmenting strategies. This has included, as discussed above, metronomic cyclophosphamide, several TKI most notably Sutent®, and bisphosphonates, such as ZA, all of which have been shown to decrease MDSCs [242]. As shown in Table 25.4, numerous additional drugs are used off label as immune-augmenting agents. This includes the three FDA-approved omega 3 (ω-3) polyunsaturated fatty acids (PUFAs)-based capsules, which were originally approved for patients with hypertriglyceridemia and to reduce myocardial infarctions [243]. The ω-3 PUFAs bind to G-protein-coupled receptor 120 (GPR120) [244] and peroxisome proliferator-activated receptor gamma (PPAR-γ) [244], which results in anti-inflammatory activity [245] with apparent cardioprotective and chemopreventive activity for neoplasia [243]. Additional drugs with activity for coronary artery disease, including Metformin® [246] and the statins [247], have also been suggested to have anti-inflammatory activity with immunotherapeutic potential in neoplasia [246]. PDE5 inhibitors, including Cialis® (Tadalafil) and Viagra®, in addition to their cyclic adenosine monophosphate (AMP) regulatory activity, have been shown to have anti-inflammatory activity predominantly via the downregulation of MDSC [248]. This decrease in MDSC by PDE5 inhibitors has been associated with therapeutic activity for multiple myeloma and head and neck SCC cancer patients [249]. Additional studies using all-trans retinoic acid (ATRA) and vitamin D3 have documented MDSC reducing activity [250, 251] that has been associated with antineoplastic activity.

25.11.1 Combination Immunotherapy and Cellular SCT

25.11.1.1 Demonstration of T-Cell Therapeutic Activity

Because CYTOKINES have unique mechanisms of action, they are ideal candidates for combination therapy with chemotherapeutic agents. However, increased knowledge and consideration of the potential interactions are necessary for successful clinical use. The use of high-dose CHEMOTHERAPY (HDT), which is myeloablative (destructive to the BM) and requires stem cell rescue, provides the ultimate in cytoreductive therapy. Further, SCT provides one of the few statistically supported demonstrations of clinical therapeutic efficacy by T cells based on the survival of patients receiving an allogeneic versus an autologous transplant [252]. Thus, strategies to upregulate T-cell function post-autologous SCT provide one focus for cytokine therapy. This is important as the return of immunological function in transplant patients is slow and accompanied by depressed numbers of $CD4^+$ T cells, a low CD4/CD8 T-cell ratio, and suppressed T-cell responses [176, 253]. The role of T cells in controlling neoplastic disease is described as a graft-versus-tumor (GVT) response. This role of T cells in treating neoplasia is supported by the significantly higher risk of relapse in patients receiving an allogeneic SCT, which has been T-cell-depleted (TCD) or who receive CYCLOSPORINE A (CSA) to prevent GVHD [254]. However, GVHD also has unfavorable effects on transplant-related morbidity and mortality. In the first remission, the decreased relapse rate with acute and/or chronic GVHD is more than offset by the increased risk of death from other causes. Consequently, patients with GVHD have a lower risk of treatment failure but an increased risk of morbidity due to GVHD (Box 25.7).

Box 25.7: Combination Chemo-Immunotherapy (Biochemotherapy)

- Immunotherapy is most active against minimal residual disease.
- Immunotherapy uses mechanisms of action (MOA) that differ from chemotherapy and, as such, has a different resistance profile.
- Chemotherapy generally results in a reduction in the extent of neoplastic disease, but frequently does not remove all residual tumor cells.
- Tumor resistance to chemotherapy frequently occurs such that high doses are often needed to assure maximal effect.
- In part, because immunotherapy uses different mechanisms of action, additional efficacy can be expected.
- Chemotherapy and other cytoreductive therapies are generally toxic, while immunotherapy is, generally, minimally toxic.
- Chemotherapy can reduce the number of host effector cells capable of responding to a tumor and limit the extent of immunotherapy.
- In contrast, chemotherapy, often in a dose-dependent manner, can be immune augmenting by reducing immune suppressor cells.
- Based on the above considerations, significant insight into the pharmacology and toxicology of the therapeutic agents is needed in order to successfully combine chemotherapy and immunotherapy.

TCD markedly reduces the incidence of severe GVHD. However, as discussed above, TCD is also associated with an increased rate of severe, and often, fatal infections, a higher incidence of graft rejection, and an increased risk of disease recurrence. The increased risk of infectious complications is associated with the slow recovery of $CD4^+$ and $CD8^+$ T cells that occurs following SCT, as the initial T-cell recovery that occurs with an unmanipulated stem cell product is associated with the T cells transplanted with the stem cells [252]. Similarly, the increased graft failure observed following transplantation with a TCD product likely reflects the contribution that infused T cells make toward the eradication of residual host T cells following the transplant preparative regimen. Due to the increased incidence of infections and relapse, donor leukocyte infusions (DLI) may be used to reduce the incidence of graft loss, disease relapse, and secondary infections [255]. However, DLI is also associated with an increased risk of GVHD, and thus, alternatives to TCD and DLI, such as strategies that can induce Ag-specific tolerance shortly after allogeneic SCT, are appealing as they might prevent GVHD without resulting in a requirement for post-graft immunosuppression.

25.11.1.2 SCT and Immunotherapy

Adjuvant studies in patients receiving HDT and SCT include a focus on immunotherapy. The dose-intense preparative regimens, commonly referred to as HDT, are administered before transplantation, and a number of cytokine- and/or vaccine-associated protocols are given following transplant with a TCD product or intact product in an attempt to improve immunological function, particularly those directed against tumor cells. One therapeutic strategy is the use of vaccines capable of inducing Ag-specific effector T cells. In addition, T cells from the donor may be stimulated ex vivo, expanded, and then reinfused. Strategies have also focused on the initiation of CTL response to viruses, which can reduce the incidence of treatment-related Epstein-Barr virus (EBV)-associated lymphomas or infections, such as cytomegalovirus (CMV) [256].

25.11.1.3 SCT Combination Therapy with IL-2

One approach to improving survival of cancer patients has been the use of IL-2 immunotherapy following HDT and SCT to induce an autologous GVT response. Based on this strategy, studies using IL-2 alone following SCT have shown an increase in NK-cell phenotype and function [176, 257, 258]. In one study [259] with 18 evaluable patients, three responses were observed. In another study, IL-2 was infused for a median of 85 days following both autologous and allogeneic SCT [258]. Toxicity was minimal and the treatment was undertaken in the outpatient setting via a Hickman catheter. In this study, no patient developed any signs of GVHD, hypertension, or pulmonary capillary leak syndrome. Despite the administration of low-dose IL-2, significant immunological changes were noted with a 5- to 40-fold increase in NK-cell number. In addition, there was a significant augmentation of ex vivo cytotoxicity against K-562 and colon tumor targets. In a similar study, it was shown that following continuous infusion of IL-2 in patients receiving autologous SCT, the $CD3^+$ and $CD16^+$ cells secreted increased levels of IFN-γ and TNF following in vitro culture, and there was a significant increase in serum levels of IFN-γ, but not TNF [258]. Recently, posttransplantation IL-2 administration has been extended to include the use of IL-2 or IL-2 and G-CSF for the mobilization of stem cells [260, 261]. The objective of using IL-2 in this context is to mobilize T cells or change the population of T cells to those that may improve antitumor activity, as well as the potential to reduce secondary infections. It is noted that low-dose IL-2 administration following allogeneic stem cell transplant has been shown to increase the frequency and number of Treg, reducing GVHD with retention of GVT activity [262].

25.11.1.4 SCT Combination Therapy with IFN-α

Similar posttransplantation strategies with IFN-α have been undertaken with the suggestion of a reduced risk of relapse and an increase in myelosuppression [263]. In an early study of the prophylactic use of IFN-α following allogeneic BMT, the Seattle group [263] found that adjuvant treatment with IFN-α had no effect on the probability or severity of CMV infections or GVHD in acute lymphocytic leukemia (ALL) patients who were in remission at the time of transplantation. In this large study, there was a significant reduction in the probability of relapse in the IFN-α recipients ($P = 0.004$) as compared to transplant patients who did not receive IFN-α, although survival rates did not differ between the two groups. It was suggested that the administration of IFN-α following transplantation reduced the risks of relapse, but did not affect CMV infection, perhaps because IFN-α was not initiated until a median of 18 days following transplantation and was not administered chronically.

25.11.1.5 SCT and Vaccine or Adoptive Cellular Therapy

Immunization and adoptive cellular therapy strategies target the activation and expansion of tumor reactive T-cell populations in hosts with an intact immune system. However, immunity, within cancer patients, is dysregulated in association with tumor-induced suppression and iatrogenic manipulation. As a consequence, immunosuppressive CYTOKINES and cells are increased, potentially limiting the effectiveness of vaccine-induced tumor-specific T cells and/or expansion of adoptively transferred T cells [264]. One approach to address this challenge has been to induce lymphopenia or reduce/oblate cellular suppressors in patients, allowing residual host or adoptively transferred, naive, or Ag-specific T cells to undergo homeostasis-driven proliferation and activation to restore memory and effector T-cell compartments. This approach has several potential advantages including the elimination of inhibitory immune cells, such as Tregs, allowing lymphoid reconstitution to overcome inherent defects in T-cell signaling and to improve Ag presentation by APC via an upregulation of co-stimulatory factors [265]. Prior studies have reported that immunomodulatory doses of cyclophosphamide can enhance vaccine-induced antitumor immune responses by inhibiting suppressor T-cell activity [266, 267]. The administration of fludarabine, which is lymphodepleting, but minimally myelodepleting, provides another approach either as a single agent or in combination with cyclophosphamide; it has been used as a non-myeloablative preoperative regime with adoptive transfer of tumor reactive T cells in patients with metastatic melanoma [268]. Lymphopenia also results in heightened secretion and availability of immunostimulatory CYTOKINES including IL-7 and IL-15 that can enhance T-cell function [269]. Some studies have shown enhanced T-cell trafficking into tumors during and following the induction of lymphopenia [270]. In association with the expansion of T-cell trafficking into tumors, there is an enhanced intratumoral proliferation of effector cells following vaccination after non-myeloablative therapy and vaccination [271]. Thus, it is hypothesized that vaccination during homeostasis-driven T-cell proliferation, secondary to lymphodepletion, may facilitate education of the T-cell repertoire, resulting in enhanced T-cell memory induction and maturation or differentiation to effector cells to tumor-associated Ag [272, 273]. In clinical studies, the transfer of activated tumor-infiltrating LYMPHOCYTES, ex vivo expanded T cells, or vaccination after myeloablative or non-myeloablative lymphopenia has been shown to induce significant clinical responses in patients with melanoma and non-small cell leukemia [268, 274, 275].

25.12 Summary

The goal of regulating the host's immune responses, as a therapeutic strategy for neoplastic, infectious, autoimmune, and inflammatory diseases, has been achieved for some indications. Optimism for this approach has fluctuated, but at present, numerous immunoregulatory drugs have been approved, and, currently, immunotherapeutics represent a quarter of all drug approvals in the United States. During the last two decades, we have observed an explosion in the cloning of immunoregulatory genes and their receptors, as well as the development of novel therapeutic approaches. These critical advances represent the culmination of efforts with crude and fractionated natural products, supernatants, and cell products.

In the last 20 years, nonspecific immunostimulation has progressed from initial trials with crude microbial mixtures and extracts to sophisticated use of a large collection of targeted immune pharmacologically active compounds (only a few of which are discussed here) having diverse actions on immunity. Further, a body of immunopharmacological knowledge has evolved with these BRM, which has shown substantial divergence from conventional pharmacology, particularly in terms of the relationship of the dosing schedule to immunopharmacodynamics. This knowledge is important in evaluating agents and predicting appropriate use and efficacy. While much remains to be learned and new compounds to be extracted and/or cloned, the future of immunotherapy seems bright. A number of CYTOKINES have been approved, as well as under study, in numerous supplemental indications [276] in the United States, Europe, and Asia. However, it is apparent that CYTOKINES and BRM will have optimal activity when used as adjuvants with more traditional therapeutic modalities. Nevertheless, combination immunotherapy must address several challenges including (a) tumor burden, (b) tumor-induced immunosuppressive of local and systemic environments, and (c) pharmacokinetic and toxicological challenges. We posit, an optimal immuno-chemotherapy approach would combine (1) CHEMOTHERAPY or surgery to decrease the tumor burden and reduce the expansion of immunosupppressive cellular mediators, (2) blockade of both T-cell and myeloid cellular inhibitors, and (3) induction and expansion of multiple epitope, tumor-specific cytotoxic T cells (please see Chap. 6 for further details on CYTOKINES).

Selected Websites

The National Cancer Institute. http://www.cancer.gov/cancertopics/factsheet/Therapy/biological

Cytokines Online Pathfinder Encyclopedia. http://www.copewithCYTOKINEs.de/cope.cgi

References

1. Couzin-Frankel J. Breakthrough of the year 2013. Cancer immunotherapy. Science. 2013;342(6165):1432–3.
2. Coley WB. The treatment of malignant tumors by repeated inoculations of erysipelas. With a report of ten original cases. 1893. Clin Orthop Relat Res. 1991;(262):3–11.
3. Walsh G. Biopharmaceutical benchmarks 2014. Nat Biotechnol. 2014;32(10):992–1000.
4. Battelle Memorial Institute. Technology, talent and capital: state bioscience initiatives 2008. Columbus, OH: Battelle Technology Partnership Practice; 2008.
5. Schellekens H, Moors E. Clinical comparability and European biosimilar regulations. Nat Biotechnol. 2010;28(1):28–31.
6. Chen CS, Doloff JC, Waxman DJ. Intermittent metronomic drug schedule is essential for activating antitumor innate immunity and tumor xenograft regression. Neoplasia. 2014;16(1):84–96.
7. Mihich E. Future perspectives for biological response modifiers: a viewpoint. Semin Oncol. 1986;13(2):234–54.
8. Talmadge JE, Herberman RB. The preclinical screening laboratory: evaluation of immunomodulatory and therapeutic properties of biological response modifiers. Cancer Treat Rep. 1986;70(1):171–82.
9. Holden C. FDA okays surrogate markers. Science. 1993;259:32.
10. Jen JF, et al. Population pharmacokinetic analysis of pegylated interferon alfa-2b and interferon alfa-2b in patients with chronic hepatitis C. Clin Pharmacol Ther. 2001;69(6):407–21.
11. Talmadge JE, et al. Systematic preclinical study on the therapeutic properties of recombinant human interleukin 2 for the treatment of metastatic disease. Cancer Res. 1987a;47(21):5725–32.
12. van Der Auwera P, et al. Pharmacodynamics and pharmacokinetics of single doses of subcutaneous pegylated human G-CSF mutant (Ro 25-8315) in healthy volunteers: comparison with single and multiple daily doses of filgrastim. Am J Hematol. 2001;66(4):245–51.
13. Tomlinson E. Site-specific proteins. In: Hider RC, Barlow D, editors. Polypeptide and protein drugs: production, characterization and formulation. Chichester: Ellis Horwood Ltd.; 1991. p. 251–364.
14. Ehrke MJ, et al. The effect of PS-K, a protein bound polysaccharide, on immune responses against allogeneic antigens. Int J Immunopharmacol. 1983;5(1):35–42.
15. Kopp WC, et al. Immunomodulatory effects of interferon-gamma in patients with metastatic malignant melanoma. J Immunother Emphasis Tumor Immunol. 1993;13(3):181–90.
16. Misset JL, et al. Treatment of leukemias and lymphomas by interferons: II. Phase II of the trial treatment of chronic lymphoid leukemia by human interferon alpha+. Biomed Pharmacother. 1982;36(2):112–6.
17. Quesada JR, et al. Alpha interferon for induction of remission in hairy-cell leukemia. N Engl J Med. 1984;310(1):15–8.
18. O'Connell MJ, et al. Clinical trial of recombinant leukocyte A interferon as initial therapy for favorable histology non-Hodgkin's lymphomas and chronic lymphocytic leukemia. An Eastern Cooperative Oncology Group pilot study. J Clin Oncol. 1986;4(2):128–36.
19. Bunn PA Jr, et al. Recombinant leukocyte A interferon: an active agent in advanced cutaneous T-cell lymphomas. Ann Intern Med. 1984;101(4):484–7.
20. Kirkwood JM, et al. Interferon alfa-2b adjuvant therapy of high-risk resected cutaneous melanoma: the Eastern Cooperative Oncology Group Trial EST 1684. J Clin Oncol. 1996;14(1):7–17.
21. Lane HC, et al. Anti-retroviral effects of interferon-alpha in AIDS-associated Kaposi's sarcoma. Lancet. 1988;2(8622):1218–22.
22. Golomb HM, et al. Report of a multi-institutional study of 193 patients with hairy cell leukemia treated with interferon-alfa2b. Semin Oncol. 1988;15(5 Suppl 5):7–9.

23. Pfeffer LM, et al. Biological properties of recombinant alpha-interferons: 40th anniversary of the discovery of interferons. Cancer Res. 1998;58(12):2489–99.
24. Teichmann JV, et al. Modulation of immune functions by long-term treatment with recombinant interferon-alpha 2 in a patient with hairy-cell leukemia. J Interf Res. 1988;8(1):15–24.
25. Black PL, et al. Antitumor response to recombinant murine interferon gamma correlates with enhanced immune function of organ-associated, but not recirculating cytolytic T lymphocytes and macrophages. Cancer Immunol Immunother. 1993;37(5):299–306.
26. Capnist G, et al. Long term results of interferon treatment in hairy cell leukemia. Italian Cooperative Group of Hairy Cell Leukemia (ICGHCL). Leuk Lymphoma. 1994;14(5–6):457–64.
27. Grever M, et al. Randomized comparison of pentostatin versus interferon alfa-2a in previously untreated patients with hairy cell leukemia: an intergroup study. J Clin Oncol. 1995;13(4):974–82.
28. Alimena G, et al. Interferon alpha-2b as therapy for Ph′-positive chronic myelogenous leukemia: a study of 82 patients treated with intermittent or daily administration. Blood. 1988;72(2):642–7.
29. Guilhot F, et al. Interferon alfa-2b combined with cytarabine versus interferon alone in chronic myelogenous leukemia. French Chronic Myeloid Leukemia Study Group. NEnglJ Med. 1997;337(4):223–9.
30. Wadler S, Schwartz EL. Antineoplastic activity of the combination of interferon and cytotoxic agents against experimental and human malignancies: a review. Cancer Res. 1990;50:3473–86.
31. Wheatley K, et al. Does adjuvant interferon-alpha for high-risk melanoma provide a worthwhile benefit? A meta-analysis of the randomised trials. Cancer Treat Rev. 2003;29(4):241–52.
32. Kirkwood JM, et al. Immunomodulatory effects of high-dose and low-dose interferon α2b in patients with high-risk resected melanoma. Cancer. 2002;95(5):1101–12.
33. Detjen KM, et al. Interferon gamma inhibits growth of human pancreatic carcinoma cells via caspase-1 dependent induction of apoptosis. Gut. 2001;49(2):251–62.
34. Xu X, et al. IFN-gamma induces cell growth inhibition by Fas-mediated apoptosis: requirement of STAT1 protein for up-regulation of Fas and FasL expression. Cancer Res. 1998;58(13):2832–7.
35. Folkman J. Angiogenic zip code. Nat Biotechnol. 1999;17(8):749.
36. Luster AD, Ravetch JV. Biochemical characterization of a gamma interferon-inducible cytokine (IP-10). J Exp Med. 1987;166(4):1084–97.
37. Fogler WE, Talmadge JE, Fidler IJ. The activation of tumoricidal properties in macrophages of endotoxin responder and nonresponder mice by liposome-encapsulated immunomodulators. J Reticuloendothel Soc. 1983;33(3):165–74.
38. Key ME, et al. Isolation of tumoricidal macrophages from lung melanoma metastases of mice treated systemically with liposomes containing a lipophilic derivative of muramyl dipeptide. J Natl Cancer Inst. 1982;69(5):1198.
39. Singh RK, et al. Fas-FasL-mediated CD4+ T-cell apoptosis following stem cell transplantation. Cancer Res. 1999;59(13):3107–11.
40. Griffith TS, et al. Monocyte-mediated tumoricidal activity via the tumor necrosis factor-related cytokine, TRAIL. J Exp Med. 1999;189(8):1343–54.
41. Jackson JD, et al. Interleukin-12 enhances peripheral hematopoiesis in vivo. Blood. 1995;85(9):2371–6.
42. Smyth MJ, et al. Perforin-mediated cytotoxicity is critical for surveillance of spontaneous lymphoma. J Exp Med. 2000;192(5):755–60.
43. Takeda K, et al. Involvement of tumor necrosis factor-related apoptosis-inducing ligand in surveillance of tumor metastasis by liver natural killer cells. Nat Med. 2001;7(1):94–100.
44. Talmadge JE, et al. Immunomodulatory and immunotherapeutic properties of recombinant gamma-interferon and recombinant tumor necrosis factor in mice. Cancer Res. 1987b;47(10):2563–70.
45. Jaffe HS, Herberman RB. Rationale for recombinant human interferon-gamma adjuvant immunotherapy for cancer. J Natl Cancer Inst. 1988;80(9):616–8.
46. Maluish AE, et al. The determination of an immunologically active dose of interferon-gamma in patients with melanoma. J Clin Oncol. 1988;6(3):434–45.
47. Colombo MP, et al. Granulocyte colony-stimulating factor gene transfer suppresses tumorigenicity of a murine adenocarcinoma in vivo. J Exp Med. 1991;173(4):889–97.
48. The International Chronic Granulomatous Disease Cooperative Study Group. A controlled trial of interferon gamma to prevent infection in chronic granulomatous disease. N Engl J Med. 1991;324:509–16.
49. Windbichler GH, et al. Interferon-gamma in the first-line therapy of ovarian cancer: a randomized phase III trial. Br J Cancer. 2000;82(6):1138–44.
50. Woodman RC, et al. Prolonged recombinant interferon-gamma therapy in chronic granulomatous disease: evidence against enhanced neutrophil oxidase activity. Blood. 1992;79(6):1558–62.
51. Ahlin A, et al. Gamma interferon treatment of patients with chronic granulomatous disease is associated with augmented production of nitric oxide by polymorphonuclear neutrophils. Clin Diagn Lab Immunol. 1999;6(3):420–4.
52. Schiff DE, et al. Increased phagocyte Fc gammaRI expression and improved Fc gamma-receptor-mediated phagocytosis after in vivo recombinant human interferon-gamma treatment of normal human subjects. Blood. 1997;90(8):3187–94.
53. Smith KA. Interleukin-2: inception, impact, and implications. Science. 1988;240(4856):1169–76.
54. Waldmann TA, Dubois S, Tagaya Y. Contrasting roles of IL-2 and IL-15 in the life and death of lymphocytes: implications for immunotherapy. Immunity. 2001;14(2):105–10.

55. Robertson MJ, Ritz J. Biology and clinical relevance of human natural killer cells. Blood. 1990;76(12):2421–38.
56. Mingari MC, et al. Human interleukin-2 promotes proliferation of activated B cells via surface receptors similar to those of activated T cells. Nature. 1984;312(5995):641–3.
57. Espinoza-Delgado I, et al. Interleukin-2 and human monocyte activation. J Leukoc Biol. 1995;57(1):13–9.
58. Ferrante A. Activation of neutrophils by interleukins-1 and -2 and tumor necrosis factors. Immunol Ser. 1992;57:417–36.
59. Rosenberg SA. Progress in human tumour immunology and immunotherapy. Nature. 2001;411(6835):380–4.
60. Moretta A, et al. Activating receptors and coreceptors involved in human natural killer cell-mediated cytolysis. Annu Rev Immunol. 2001;19:197–223.
61. Nelson BH. IL-2, regulatory T cells, and tolerance. J Immunol. 2004;172(7):3983–8.
62. Gillis S, Smith KA. Long term culture of tumour-specific cytotoxic T cells. Nature. 1977;268(5616):154–6.
63. Keene JA, Forman J. Helper activity is required for the in vivo generation of cytotoxic T lymphocytes. J Exp Med. 1982;155(3):768–82.
64. Andrews DM, et al. Infection of dendritic cells by murine cytomegalovirus induces functional paralysis. Nat Immunol. 2001;2(11):1077–84.
65. Grabstein KH, et al. Cloning of a T cell growth factor that interacts with the beta chain of the interleukin-2 receptor. Science. 1994;264(5161):965–8.
66. Marks-Konczalik J, et al. IL-2-induced activation-induced cell death is inhibited in IL-15 transgenic mice. Proc Natl Acad Sci U S A. 2000;97(21):11445–50.
67. Zhang X, et al. Potent and selective stimulation of memory-phenotype CD8+ T cells in vivo by IL-15. Immunity. 1998;8(5):591–9.
68. Lodolce JP, et al. IL-15 receptor maintains lymphoid homeostasis by supporting lymphocyte homing and proliferation. Immunity. 1998;9(5):669–76.
69. Fehniger TA, Cooper MA, Caligiuri MA. Interleukin-2 and interleukin-15: immunotherapy for cancer. Cytokine Growth Factor Rev. 2002;13(2):169–83.
70. West WH, et al. Constant-infusion recombinant interleukin-2 in adoptive immunotherapy of advanced cancer. N Engl J Med. 1987;316(15):898–905.
71. Yang JC, et al. Randomized study of high-dose and low-dose interleukin-2 in patients with metastatic renal cancer. J Clin Oncol. 2003;21(16):3127–32.
72. Lotze MT, et al. High-dose recombinant interleukin 2 in the treatment of patients with disseminated cancer. Responses, treatment-related morbidity, and histologic findings. JAMA. 1986;256(22):3117–24.
73. Heslop HE, et al. In vivo induction of gamma interferon and tumor necrosis factor by interleukin-2 infusion following intensive chemotherapy or autologous marrow transplantation. Blood. 1989;74(4):1374–80.
74. Barouch DH. Augmentation of immune responses to HIV-1 and simian immunodeficiency virus DNA vaccines by IL-2/Ig plasmid administration in rhesus monkeys. Proc Natl Acad Sci U S A. 2000;97(8):4192–7.
75. Oosterwijk-Wakka JC, et al. Vaccination of patients with metastatic renal cell carcinoma with autologous dendritic cells pulsed with autologous tumor antigens in combination with interleukin-2: a phase 1 study. J Immunother. 2002;25(6):500–8.
76. Thompson JA, et al. Prolonged continuous intravenous infusion interleukin-2 and lymphokine-activated killer-cell therapy for metastatic renal cell carcinoma. J Clin Oncol. 1992;10(6):960–8.
77. Kaufman HL, et al. The Society for Immunotherapy of Cancer consensus statement on tumour immunotherapy for the treatment of cutaneous melanoma. Nat Rev Clin Oncol. 2013;10(10):588–98.
78. Hladik F, et al. Biologic activity of low dosage IL-2 treatment in vivo. Molecular assessment of cytokine network interaction. J Immunol. 1994;153(4):1449–54.
79. Mier JW, et al. Induction of circulating tumor necrosis factor (TNF-alpha) as the mechanism for the febrile response to interleukin-2 (IL-2) in cancer patients. J Clin Immunol. 1988;8(6):426–32.
80. Lange T, et al. Systemic immune parameters and sleep after ultra-low dose administration of IL-2 in healthy men. Brain Behav Immun. 2002;16(6):663–74.
81. De Stefani A, et al. Improved survival with perilymphatic interleukin 2 in patients with resectable squamous cell carcinoma of the oral cavity and oropharynx. Cancer. 2002;95(1):90–7.
82. Yang JC, et al. Randomized comparison of high-dose and low-dose intravenous interleukin-2 for the therapy of metastatic renal cell carcinoma: an interim report. J Clin Oncol. 1994;12(8):1572–6.
83. Kim N, et al. Therapeutic potential of low-dose IL-2 in a chronic GVHD patient by in vivo expansion of regulatory T cells. Cytokine. 2016;78:22–6.
84. Burgess AW, Metcalf D. The nature and action of granulocyte-macrophage colony stimulating factors. Blood. 1980;56(6):947–58.
85. Stanley E, et al. Granulocyte/macrophage colony-stimulating factor-deficient mice show no major perturbation of hematopoiesis but develop a characteristic pulmonary pathology. Proc Natl Acad Sci U S A. 1994;91(12):5592–6.
86. Hamilton JA, et al. Stimulation of macrophage plasminogen activator activity by colony-stimulating factors. J Cell Physiol. 1980;103(3):435–45.
87. Kay AB, et al. Messenger RNA expression of the cytokine gene cluster, interleukin 3 (IL-3), IL-4, IL-5, and granulocyte/macrophage colony-stimulating factor, in allergen-induced late-phase cutaneous reactions in atopic subjects. J Exp Med. 1991;173(3):775–8.
88. Williamson DJ, et al. The detection and initial characterization of colony-stimulating factors in synovial fluid. Clin Exp Immunol. 1988;72(1):67–73.

89. Inaba K, et al. Generation of large numbers of dendritic cells from mouse bone marrow cultures supplemented with granulocyte/macrophage colony-stimulating factor. J Exp Med. 1992;176(6):1693–702.
90. Daro E, et al. Polyethylene glycol-modified GM-CSF expands CD11b(high)CD11c(high) but not CD11b(low)CD11c(high) murine dendritic cells in vivo: a comparative analysis with Flt3 ligand. J Immunol. 2000;165(1):49–58.
91. Kiertscher SM, et al. Granulocyte/macrophage-colony stimulating factor and interleukin-4 expand and activate type-1 dendritic cells (DC1) when administered in vivo to cancer patients. Int J Cancer. 2003;107(2):256–61.
92. Nemunaitis J, et al. Use of recombinant human granulocyte-macrophage colony-stimulating factor in graft failure after bone marrow transplantation. Blood. 1990;76(1):245–53.
93. Brandt SJ, et al. Effect of recombinant human granulocyte-macrophage colony-stimulating factor on hematopoietic reconstitution after high-dose chemotherapy and autologous bone marrow transplantation. N Engl J Med. 1988;318(14):869–76.
94. Ringden O, Leblanc K, Remberger M. Granulocyte and granulocyte-macrophage colony-stimulating factors in allografts: uses, misuses, misconceptions, and future applications. Exp Hematol. 2005;33(5):505–12.
95. Rutella S. Granulocyte colony-stimulating factor for the induction of T-cell tolerance. Transplantation. 2007;84(1 Suppl):S26–30.
96. Luyckx A, et al. G-CSF stem cell mobilization in human donors induces polymorphonuclear and mononuclear myeloid-derived suppressor cells. Clin Immunol. 2012;143(1):83–7.
97. Rowe JM, et al. A randomized placebo-controlled phase III study of granulocyte-macrophage colony-stimulating factor in adult patients (>55 to 70 years of age) with acute myelogenous leukemia: a study of the Eastern Cooperative Oncology Group (E1490). Blood. 1995;86(2):457–62.
98. Ou-Yang P, et al. Co-delivery of GM-CSF gene enhances the immune responses of hepatitis C viral core protein-expressing DNA vaccine: role of dendritic cells. J Med Virol. 2002;66(3):320–8.
99. Dang Y, et al. Dendritic cell-activating vaccine adjuvants differ in the ability to elicit antitumor immunity due to an adjuvant-specific induction of immunosuppressive cells. Clin Cancer Res. 2012;18(11):3122–31.
100. Beyer J, et al. Hematopoietic rescue after high-dose chemotherapy using autologous peripheral-blood progenitor cells or bone marrow: a randomized comparison. J Clin Oncol. 1995;13(6):1328–35.
101. Dranoff G. GM-CSF-based cancer vaccines. Immunol Rev. 2002;188:147–54.
102. Ohm JE, et al. Effect of vascular endothelial growth factor and FLT3 ligand on dendritic cell generation in vivo. J Immunol. 1999;163(6):3260–8.
103. Wing EJ, et al. Recombinant human granulocyte/macrophage colony-stimulating factor enhances monocyte cytotoxicity and secretion of tumor necrosis factor alpha and interferon in cancer patients. Blood. 1989;73(3):643–6.
104. Spitler LE, et al. Adjuvant therapy of stage III and IV malignant melanoma using granulocyte-macrophage colony-stimulating factor. J Clin Oncol. 2000;18(8):1614–21.
105. Grotz TE, et al. Adjuvant GM-CSF improves survival in high-risk stage IIIC melanoma: a single-center study. Am J Clin Oncol. 2014;37(5):467–72.
106. Cheever MA, et al. Translational Research Working Group developmental pathway for immune response modifiers. Clin Cancer Res. 2008;14(18):5692–9.
107. Cheever MA. Twelve immunotherapy drugs that could cure cancers. Immunol Rev. 2008;222:357–68.
108. NCI. Immunotherapy Agent Workshop Proceedings. Immunotherapy Agent Workshop, 2007. Bethesda, MD: National Cancer Institute; 2007.
109. Hashmi MH, Van Veldhuizen PJ. Interleukin-21: updated review of Phase I and II clinical trials in metastatic renal cell carcinoma, metastatic melanoma and relapsed/refractory indolent non-Hodgkin's lymphoma. Expert Opin Biol Ther. 2010;10(5):807–17.
110. Mackall CL, Fry TJ, Gress RE. Harnessing the biology of IL-7 for therapeutic application. Nat Rev Immunol. 2011;11(5):330–42.
111. Perales MA, et al. Recombinant human interleukin-7 (CYT107) promotes T-cell recovery after allogeneic stem cell transplantation. Blood. 2012;120(24):4882–91.
112. Milano-Bausset E, et al. Retrospective comparison of neutropenia in children with Ewing sarcoma treated with chemotherapy and granulocyte colony-stimulating factor (G-CSF) or pegylated G-CSF. Clin Ther. 2009;31(Pt 2):2388–95.
113. Kroschinsky F, Holig K, Ehninger G. The role of pegfilgrastim in mobilization of hematopoietic stem cells. Transfus Apher Sci. 2008;38(3):237–44.
114. Haaff EO, et al. Two courses of intravesical Bacillus Calmette-Guerin for transitional cell carcinoma of the bladder. J Urol. 1986b;136(4):820.
115. Pinsky CM, et al. Intravesical administration of Bacillus Calmette-Guerin in patients with recurrent superficial carcinoma of the urinary bladder: report of a prospective, randomized trial. Cancer Treat Rep. 1985;69:47.
116. Herr HW, et al. Intravesical Bacillus Calmette-Guerin therapy prevents tumor progression and death from superficial bladder cancer: ten-year follow-up of a prospective randomized trial. J Clin Oncol. 1995;13(6):1404–8.
117. Sarosdy MF, Lamm DL. Long-term results of intravesical Bacillus Calmette-Guerin therapy for superficial bladder cancer. J Urol. 1989;142(3):719–22.
118. De Jager RL, et al. Long-term complete remission in bladder carcinoma in situ with intravesical TICE bacillus Calmette Guerin. Overview analysis of six phase II clinical trials. Urology. 1991;38(6):507–13.

119. Herr HW, et al. Bacillus Calmette-Guerin therapy for superficial bladder cancer: a 10-year followup. J Urol. 1992;147(4):1020–3.
120. Lamm DL, et al. Maintenance Bacillus Calmette-Guerin immunotherapy for recurrent TA, T1 and carcinoma in situ transitional cell carcinoma of the bladder: a randomized Southwest Oncology Group Study. J Urol. 2000;163(4):1124–9.
121. Ayres BE, Griffiths TR, Persad RA. Is the role of intravesical Bacillus Calmette-Guerin in non-muscle-invasive bladder cancer changing? BJUInt. 2010;105(Suppl 2):8–13.
122. Jimenez-Cruz JF, et al. Intravesical immunoprophylaxis in recurrent superficial bladder cancer (Stage T1): multicenter trial comparing Bacille Calmette-Guerin and interferon-alpha. Urology. 1997;50(4):529–35.
123. Nepple KG, et al. Combination of BCG and interferon intravesical immunotherapy: an update. World J Urol. 2009;27(3):343–6.
124. Lage JM, et al. Histological parameters and pitfalls in the interpretation of bladder biopsies in Bacillus Calmette-Guerin treatment of superficial bladder cancer. J Urol. 1986;135:916–9.
125. Haaff EO, Caralona WJ, Ratliff TL. Detection of interleukin-2 in the urine of patients with superficial bladder tumors after treatment with intravesical BCG. J Urol. 1986a;136(4):970.
126. Taniguchi K, et al. Systemic immune response after intravesical instillation of Bacille Calmette-Guerin (BCG) for superficial bladder cancer. Clin Exp Immunol. 1999;115(1):131–5.
127. Kaempfer R, et al. Prediction of response to treatment in superficial bladder carcinoma through pattern of interleukin-2 gene expression. J Clin Oncol. 1996;14(6):1778–86.
128. Rigaud J, et al. Development of predictive value of urinary cytokine profile induced during intravesical Bacillus Calmette-Guerin instillations for bladder cancer. Clin Genitourin Cancer. 2015;13(4):e209–15.
129. Amery WK, Bruynseels JP. Levamisole, the story and the lessons. Int J Immunopharmacol. 1992;14(3):481–6.
130. Holcombe RF, et al. Investigating the role of immunomodulation for colon cancer prevention: results of an in vivo dose escalation trial of levamisole with immunologic endpoints. Cancer Detect Prev. 2001;25(2):183–91.
131. Porschen R, et al. Fluorouracil plus leucovorin as effective adjuvant chemotherapy in curatively resected stage III colon cancer: results of the trial adj CCA-01. J Clin Oncol. 2001;19(6):1787–94.
132. Dahl O, et al. Final results of a randomised phase III study on adjuvant chemotherapy with 5 FU and levamisol in colon and rectum cancer stage II and III by the Norwegian Gastrointestinal Cancer Group. Acta Oncol. 2009;48(3):368–76.
133. Paterson AH, et al. Double-blind controlled trial of oral clodronate in patients with bone metastases from breast cancer. J Clin Oncol. 1993;11(1):59–65.
134. Shibuya K, et al. Global and regional estimates of cancer mortality and incidence by site: II. Results for the global burden of disease 2000. BMCCancer. 2002;2:37.
135. Winter MC, Holen I, Coleman RE. Exploring the anti-tumour activity of bisphosphonates in early breast cancer. Cancer Treat Rev. 2008;34(5):453–75.
136. Trinkaus M, et al. Examination of the mechanisms of osteolysis in patients with metastatic breast cancer. Oncol Rep. 2009;21(5):1153–9.
137. Santini D, et al. Pamidronate induces modifications of circulating angiogenetic factors in cancer patients. Clin Cancer Res. 2002;8(5):1080–4.
138. Melani C, et al. Amino-biphosphonate-mediated MMP-9 inhibition breaks the tumor-bone marrow axis responsible for myeloid-derived suppressor cell expansion and macrophage infiltration in tumor stroma. Cancer Res. 2007;67:11438–46.
139. Badger AM, et al. Inhibition of animal models of autoimmune disease and the induction of non-specific suppressor cells by SK&F 105685 and related azaspiranes. Int J Immunopharmacol. 1989;11(7):839–46.
140. Badger AM, et al. Induction of non-specific suppressor cells in normal Lewis rats by a novel azaspirane SK&F 105685. J Autoimmun. 1990;3(4):485–500.
141. King AG, et al. Induction of non-specific suppressor cells and myeloregulatory effects of an immunomodulatory azaspirane, SK&F 105685. Int J Immunopharmacol. 1991;13(1):91–100.
142. Nefedova Y, et al. Hyperactivation of STAT3 is involved in abnormal differentiation of dendritic cells in cancer. J Immunol. 2004;172(1):464–74.
143. Wang T, et al. Regulation of the innate and adaptive immune responses by Stat-3 signaling in tumor cells. Nat Med. 2004;10(1):48–54.
144. Welte T, et al. STAT3 deletion during hematopoiesis causes Crohn's disease-like pathogenesis and lethality: a critical role of STAT3 in innate immunity. Proc Natl Acad Sci U S A. 2003;100(4):1879–84.
145. Hock H, et al. Intrinsic requirement for zinc finger transcription factor Gfi-1 in neutrophil differentiation. Immunity. 2003;18(1):109–20.
146. Passegue E, et al. Chronic myeloid leukemia with increased granulocyte progenitors in mice lacking junB expression in the myeloid lineage. Cell. 2001;104(1):21–32.
147. Ghansah T, et al. Expansion of myeloid suppressor cells in SHIP-deficient mice represses allogeneic T cell responses. J Immunol. 2004;173(12):7324–30.
148. Serafini P, Borrello I, Bronte V. Myeloid suppressor cells in cancer: recruitment, phenotype, properties, and mechanisms of immune suppression. Semin Cancer Biol. 2006a;16(1):53–65.
149. Ochoa AC, et al. Arginase, prostaglandins, and myeloid-derived suppressor cells in renal cell carcinoma. Clin Cancer Res. 2007;13(2):721s–6s.
150. Talmadge JE. Pathways mediating the expansion and immunosuppressive activity of myeloid-derived suppressor cells and their relevance to cancer therapy. Clin Cancer Res. 2007;13(18):5243–8.

151. Almand B, et al. Increased production of immature myeloid cells in cancer patients: a mechanism of immunosuppression in cancer. J Immunol. 2001;166:678–89.
152. Mirza N. All-trans-retinoic acid improves differentiation of myeloid cells and immune response in cancer patients. Cancer Res. 2006;66(18):9299–307.
153. Pak AS, et al. Mechanisms of immune suppression in patients with head and neck cancer: presence of CD34(+) cells which suppress immune functions within cancers that secrete granulocyte-macrophage colony-stimulating factor. Clin Cancer Res. 1995;1(1):95–103.
154. Zea AH, et al. Arginase-producing myeloid suppressor cells in renal cell carcinoma patients: a mechanism of tumor evasion. Cancer Res. 2005;65(8):3044–8.
155. Ko HJ, et al. A combination of chemoimmunotherapies can efficiently break self-tolerance and induce antitumor immunity in a tolerogenic murine tumor model. Cancer Res. 2007;67:7477–86.
156. Seung LP, et al. Radiation can inhibit tumor growth indirectly while depleting circulating leukocytes. Radiat Res. 1996;146(6):612–8.
157. Young MR, Wright MA. Myelopoiesis-associated immune suppressor cells in mice bearing metastatic Lewis lung carcinoma tumors: gamma-interferon plus tumor necrosis factor-alpha synergistically reduces immune suppressor and tumor growth-promoting activities of bone marrow cells and diminishes tumor recurrence and metastasis. Cancer Res. 1992;52(22):6335–40.
158. Mendel DB, et al. In vivo antitumor activity of SU11248, a novel tyrosine kinase inhibitor targeting vascular endothelial growth factor and platelet-derived growth factor receptors: determination of a pharmacokinetic/pharmacodynamic relationship. Clin Cancer Res. 2003;9(1):327–37.
159. Ko JS, Bukowski RM, Fincke JH. Myeloid-derived suppressor cells: a novel therapeutic target. Curr Oncol Rep. 2009a;11(2):87–93.
160. Ko JS, et al. Sunitinib mediates reversal of myeloid-derived suppressor cell accumulation in renal cell carcinoma patients. Clin Cancer Res. 2009b;15:2148–57.
161. Ozao-Choy J, et al. The novel role of tyrosine kinase inhibitor in the reversal of immune suppression and modulation of tumor microenvironment for immune-based cancer therapies. Cancer Res. 2009;69(6):2514–22.
162. van Cruijsen H, et al. Sunitinib-induced myeloid lineage redistribution in renal cell cancer patients: CD1c+ dendritic cell frequency predicts progression-free survival. Clin Cancer Res. 2008;14(18):5884–92.
163. Abrams TJ, et al. SU11248 inhibits KIT and platelet-derived growth factor receptor beta in preclinical models of human small cell lung cancer. MolCancer Ther. 2003a;2(5):471–8.
164. Abrams TJ, et al. Preclinical evaluation of the tyrosine kinase inhibitor SU11248 as a single agent and in combination with "standard of care" therapeutic agents for the treatment of breast cancer. Mol Cancer Ther. 2003b;2(10):1011–21.
165. Shankar G, Pendley C, Stein KE. A risk-based bioanalytical strategy for the assessment of antibody immune responses against biological drugs. Nat Biotechnol. 2007;25(5):555–61.
166. O'Farrell AM, et al. SU11248 is a novel FLT3 tyrosine kinase inhibitor with potent activity in vitro and in vivo. Blood. 2003;101(9):3597–605.
167. Chow LQ, Eckhardt SG. Sunitinib: from rational design to clinical efficacy. J Clin Oncol. 2007;25(7):884–96.
168. Faivre S, et al. Molecular basis for sunitinib efficacy and future clinical development. Nat Rev Drug Discov. 2007;6(9):734–45.
169. Motzer RJ, et al. Overall survival and updated results for sunitinib compared with interferon alfa in patients with metastatic renal cell carcinoma. J Clin Oncol. 2009a;27(22):3573–4.
170. Hutson TE, Figlin RA. Evolving role of novel targeted agents in renal cell carcinoma. Oncology (WillistonPark). 2007;21(10):1175–80.
171. Motzer RJ, et al. Overall survival and updated results for sunitinib compared with interferon alfa in patients with metastatic renal cell carcinoma. J Clin Oncol. 2009b;27(22):3584–90.
172. Figlin RA, Hutson TE, Tomczak P, Michaelson MD, Bukowski RM, Negrier S, Huang X, Kim ST, Chen I, Motzer RJ. Overall survival with sunitinib versus interferon (IFN)-alfa as first-line treatment of metastatic renal cell carcinoma (mRCC). In: 2008 ASCO Annual Meeting, vol. 26. Alexandria, VA: American Society of Clinical Oncology; 2008.
173. Abe F, et al. Therapeutic activity of sunitinib for Her2/neu induced mammary cancer in FVB mice. Int Immunopharmacol. 2010;10(1):140–5.
174. Walsh G. Biopharmaceutical benchmarks--2003. Nat Biotechnol. 2003;21(8):865–70.
175. Xin H, et al. Sunitinib inhibition of Stat3 induces renal cell carcinoma tumor cell apoptosis and reduces immunosuppressive cells. Cancer Res. 2009;69(6):2506–13.
176. Chou AJ, et al. Addition of muramyl tripeptide to chemotherapy for patients with newly diagnosed metastatic osteosarcoma: a report from the Children's Oncology Group. Cancer. 2009;115(22):5339–48.
177. Meyers PA. Muramyl tripeptide (mifamurtide) for the treatment of osteosarcoma. Expert Rev Anticancer Ther. 2009;9(8):1035–49.
178. Wood DD, et al. Role of interleukin-1 in the adjuvanticity of muramyl dipeptide in vivo. In: Oppenheim JJ, Cohen S, editors. Interleukins, lymphokines and cytokines. New York: Raven Press; 1983. p. 691.
179. Ellouz F, et al. Minimal structural requirements for adjuvant activity of bacterial peptidoglycan derivatives. Biochem Biophys Res Commun. 1974;59(4):1317–25.

180. Fedorocko P, et al. Administration of liposomal muramyl tripeptide phosphatidylethanolamine (MTP-PE) and diclofenac in the combination attenuates their anti-tumor activities. Neoplasma. 2003;50(3):176–84.
181. Killion JJ, et al. Maintenance of intestinal epithelium structural integrity and mucosal leukocytes during chemotherapy by oral administration of muramyl tripeptide phosphatidylethanolamine. Cancer Biother Radiopharm. 1996;11(6):363–71.
182. Aoyagi T, et al. Aminopeptidase activities on the surface of mammalian cells. Biochim Biophys Acta. 1976;452(1):131–43.
183. Morahan PS, Edelson PJ, Gass K. Changes in macrophage ectoenzymes associated with anti-tumor activity. J Immunol. 1980;125(3):1312–7.
184. Urabe A, et al. Ubenimex in the treatment of acute nonlymphocytic leukemia in adults. Ann Hematol. 1993;67(2):63–6.
185. Yasumitsu T, et al. Bestatin in resected lung cancer. A randomized clinical trial. Acta Oncol. 1990;29(6):827–31.
186. Hiraoka A, Shibata H, Masaoka T. Immunopotentiation with Ubenimex for prevention of leukemia relapse after allogeneic BMT. The Study Group of Ubenimex for BMT. Transplant Proc. 1992;24(6):3047–8.
187. Goldstein AL. Thymic hormones and lymphokines. New York: Plenum Press; 1984.
188. Ichinose Y, et al. Randomized double-blind placebo-controlled trial of bestatin in patients with resected stage I squamous-cell lung carcinoma. J Natl Cancer Inst. 2003;95(8):605–10.
189. Ito S, et al. Stromal aminopeptidase N expression: correlation with angiogenesis in non-small-cell lung cancer. Gen Thorac Cardiovasc Surg. 2009;57(11):591–8.
190. Kadowaki N, et al. Subsets of human dendritic cell precursors express different toll-like receptors and respond to different microbial antigens. J Exp Med. 2001;194(6):863–9.
191. Ahmad-Nejad P, et al. Bacterial CpG-DNA and lipopolysaccharides activate Toll-like receptors at distinct cellular compartments. Eur J Immunol. 2002;32(7):1958–68.
192. Hemmi H, et al. A Toll-like receptor recognizes bacterial DNA. Nature. 2000;408(6813):740–5.
193. Krieg AM. CpG motifs in bacterial DNA and their immune effects. Annu Rev Immunol. 2002; 20:709–60.
194. Krug A, et al. Identification of CpG oligonucleotide sequences with high induction of IFN-alpha/beta in plasmacytoid dendritic cells. Eur J Immunol. 2001;31(7):2154–63.
195. Krieg AM, et al. CpG motifs in bacterial DNA trigger direct B-cell activation. Nature. 1995;374(6522):546–9.
196. Hartmann G, Krieg AM. Mechanism and function of a newly identified CpG DNA motif in human primary B cells. J Immunol. 2000;164(2):944–53.
197. Marshall JD. Identification of a novel CpG DNA class and motif that optimally stimulate B cell and plasmacytoid dendritic cell functions. J Leukoc Biol. 2003;73(6):781–92.
198. Ballas ZK, Rasmussen WL, Krieg AM. Induction of NK activity in murine and human cells by CpG motifs in oligodeoxynucleotides and bacterial DNA. J Immunol. 1996;157(5):1840–5.
199. Roman M, et al. Immunostimulatory DNA sequences function as T helper-1-promoting adjuvants. Nat Med. 1997;3(8):849–54.
200. Shirota H, et al. Regulation of murine airway eosinophilia and Th2 cells by antigen-conjugated CpG oligodeoxynucleotides as a novel antigen-specific immunomodulator. J Immunol. 2000;164(11):5575–82.
201. Cho HJ, et al. Immunostimulatory DNA-based vaccines induce cytotoxic lymphocyte activity by a T-helper cell-independent mechanism. Nat Biotechnol. 2000b;18(5):509–14.
202. Chu RS, et al. CpG oligodeoxynucleotides act as adjuvants that switch on T helper 1 (Th1) immunity. J Exp Med. 1997;186(10):1623–31.
203. Davis HL. Use of CpG DNA for enhancing specific immune responses. Curr Top Microbiol Immunol. 2000;247:171–83.
204. Zwaveling S, et al. Established human papillomavirus type 16-expressing tumors are effectively eradicated following vaccination with long peptides. J Immunol. 2002;169(1):350–8.
205. Heckelsmiller K, et al. Combined dendritic cell- and CpG oligonucleotide-based immune therapy cures large murine tumors that resist chemotherapy. Eur J Immunol. 2002a;32(11):3235–45.
206. Sandler AD, et al. CpG oligonucleotides enhance the tumor antigen-specific immune response of a granulocyte macrophage colony-stimulating factor-based vaccine strategy in neuroblastoma. Cancer Res. 2003;63(2):394–9.
207. Heckelsmiller K, et al. Peritumoral CpG DNA elicits a coordinated response of CD8 T cells and innate effectors to cure established tumors in a murine colon carcinoma model. J Immunol. 2002b;169(7):3892–9.
208. Ballas ZK, et al. Divergent therapeutic and immunologic effects of oligodeoxynucleotides with distinct CpG motifs. J Immunol. 2001;167(9):4878–86.
209. Krieg AM. Antitumor applications of stimulating toll-like receptor 9 with CpG oligodeoxynucleotides. Curr Oncol Rep. 2004;6(2):88–95.
210. Vollmer J, Krieg AM. Immunotherapeutic applications of CpG oligodeoxynucleotide TLR9 agonists. Adv Drug Deliv Rev. 2009;61(3):195–204.
211. Molenkamp BG, et al. Local administration of PF-3512676 CpG-B instigates tumor-specific CD8+ T-cell reactivity in melanoma patients. Clin Cancer Res. 2008;14(14):4532–42.
212. Manegold C, et al. Randomized phase II trial of a toll-like receptor 9 agonist oligodeoxynucleotide, PF-3512676, in combination with first-line taxane plus platinum chemotherapy

for advanced-stage non-small-cell lung cancer. J Clin Oncol. 2008;26(24):3979–86.
213. Krieg AM. Toll-like receptor 9 (TLR9) agonists in the treatment of cancer. Oncogene. 2008;27(2): 161–7.
214. Francisco LM, Sage PT, Sharpe AH. The PD-1 pathway in tolerance and autoimmunity. Immunol Rev. 2010;236:219–42.
215. Sharpe AH, et al. The function of programmed cell death 1 and its ligands in regulating autoimmunity and infection. Nat Immunol. 2007;8(3):239–45.
216. Barber DL, et al. Restoring function in exhausted CD8 T cells during chronic viral infection. Nature. 2006;439(7077):682–7.
217. Schumacher TN, Schreiber RD. Neoantigens in cancer immunotherapy. Science. 2015;348(6230): 69–74.
218. Chen L, Flies DB. Molecular mechanisms of T cell co-stimulation and co-inhibition. Nat Rev Immunol. 2013;13(4):227–42.
219. Jung K, Choi I. Emerging co-signaling networks in T cell immune regulation. Immune Netw. 2013;13(5):184–93.
220. Maj T, et al. T cells and costimulation in cancer. Cancer J. 2013;19(6):473–82.
221. Bracarda S, et al. Immunologic checkpoints blockade in renal cell, prostate, and urothelial malignancies. Semin Oncol. 2015;42(3):495–505.
222. Hodi FS, et al. Improved survival with ipilimumab in patients with metastatic melanoma. N Engl J Med. 2010;363(8):711–23.
223. Pardoll DM. The blockade of immune checkpoints in cancer immunotherapy. Nat Rev Cancer. 2012;12(4):252–64.
224. Raedler LA. Opdivo (nivolumab): second PD-1 inhibitor receives FDA approval for unresectable or metastatic melanoma. Am Health Drug Benefits. 2015;8(Spec Feature):180–3.
225. Kazandjian D, et al. FDA approval summary: nivolumab for the treatment of metastatic non-small cell lung cancer with progression on or after platinum-based chemotherapy. Oncologist. 2016;21(5):634–42.
226. Sul J, et al. FDA approval summary: Pembrolizumab for the treatment of patients with metastatic non-small cell lung cancer whose tumors express programmed death-ligand 1. Oncologist. 2016;21(5):643–50.
227. Rosenberg JE, et al. Atezolizumab in patients with locally advanced and metastatic urothelial carcinoma who have progressed following treatment with platinum-based chemotherapy: a single-arm, multicentre, phase 2 trial. Lancet. 2016;387(10031):1909–20.
228. Orloff M, et al. Immune check point inhibitors combination in melanoma: worth the toxicity? Rev Recent Clin Trials. 2016;11(2):81–6.
229. Eigentler TK, et al. Effectiveness and tolerability of ipilimumab: experiences from 198 patients included in a named-patient program in various daily-practice settings and multiple institutions. J Immunother. 2014;37(7):374–81.
230. Spranger S, Gajewski T. Rational combinations of immunotherapeutics that target discrete pathways. J Immunother Cancer. 2013;1:16.
231. Wolchok JD, et al. Nivolumab plus ipilimumab in advanced melanoma. N Engl J Med. 2013;369(2):122–33.
232. Postow MA, et al. Nivolumab and ipilimumab versus ipilimumab in untreated melanoma. N Engl J Med. 2015;372(21):2006–17.
233. Larkin J, et al. Combined nivolumab and ipilimumab or monotherapy in untreated melanoma. N Engl J Med. 2015;373(1):23–34.
234. Garon EB, et al. Pembrolizumab for the treatment of non-small-cell lung cancer. N Engl J Med. 2015;372(21):2018–28.
235. Mahoney KM, Atkins MB. Prognostic and predictive markers for the new immunotherapies. Oncology (Williston Park). 2014;28(Suppl 3):39–48.
236. Shin DS, Ribas A. The evolution of checkpoint blockade as a cancer therapy: what's here, what's next? Curr Opin Immunol. 2015;33:23–35.
237. Morrissey KM, et al. Immunotherapy and novel combinations in oncology: current landscape, challenges, and opportunities. Clin Transl Sci. 2016;9(2):89–104.
238. Petrovsky N, Aguilar JC. Vaccine adjuvants: current state and future trends. Immunol Cell Biol. 2004;82(5):488–96.
239. Leroux-Roels G. Unmet needs in modern vaccinology: adjuvants to improve the immune response. Vaccine. 2010;28(Suppl 3):C25–36.
240. Glenny AT, et al. Immunological notes. XVII–XXIV. J Pathol Bacteriol. 1926;29(1):31–40.
241. Grun JL, Maurer PH. Different T helper cell subsets elicited in mice utilizing two different adjuvant vehicles: the role of endogenous interleukin 1 in proliferative responses. Cell Immunol. 1989;121(1):134–45.
242. Wang J, et al. Lenalidomide and cyclophosphamide immunoregulation in patients with metastatic, castration-resistant prostate cancer. Clin Exp Metastasis. 2015;32(2):111–24.
243. Barden A, et al. Short-term n-3 fatty acid supplementation but not aspirin increases plasma proresolving mediators of inflammation. J Lipid Res. 2014;55(11):2401–7.
244. Weintraub HS. Overview of prescription omega-3 fatty acid products for hypertriglyceridemia. Postgrad Med. 2014;126(7):7–18.
245. Calder PC. n-3 fatty acids, inflammation and immunity: new mechanisms to explain old actions. Proc Nutr Soc. 2013;72(3):326–36.
246. Gandini S, et al. Metformin and cancer risk and mortality: a systematic review and meta-analysis taking into account biases and confounders. Cancer Prev Res (Phila). 2014;7(9):867–85.
247. Greenwood J, Steinman L, Zamvil SS. Statin therapy and autoimmune disease: from protein prenylation to immunomodulation. Nat Rev Immunol. 2006;6(5):358–70.

248. Serafini P, et al. Phosphodiesterase-5 inhibition augments endogenous antitumor immunity by reducing myeloid-derived suppressor cell function. J Exp Med. 2006b;203(12):2691–702.
249. Noonan KA, et al. Targeting immune suppression with PDE5 inhibition in end-stage multiple myeloma. Cancer Immunol Res. 2014;2(8):725–31.
250. Kusmartsev S, et al. Reversal of myeloid cell-mediated immunosuppression in patients with metastatic renal cell carcinoma. Clin Cancer Res. 2008;14(24):8270–8.
251. Young MR, Day TA. Immune regulatory activity of vitamin d3 in head and neck cancer. Cancers (Basel). 2013;5(3):1072–85.
252. Storek J, Storb R. T-cell reconstitution after stem-cell transplantation--by which organ? Lancet. 2000;355(9218):1843–4.
253. Talmadge JE, et al. Rapid immunologic reconstitution following transplantation with mobilized peripheral blood stem cells as compared to bone marrow. Bone Marrow Transplant. 1997;19(2):161–72.
254. Horowitz MM, et al. Graft-versus-leukemia reactions after bone marrow transplantation. Blood. 1990;75(3):555–62.
255. Champlin R, et al. Selective depletion of CD8+ T lymphocytes for prevention of graft-versus-host disease after allogeneic bone marrow transplantation. Blood. 1990;76(2):418–23.
256. Walter EA, et al. Reconstitution of cellular immunity against cytomegalovirus in recipients of allogeneic bone marrow by transfer of T-cell clones from the donor. N Engl J Med. 1995;333(16):1038–44.
257. Higuchi CM, et al. Toxicity and immunomodulatory effects of interleukin-2 after autologous bone marrow transplantation for hematologic malignancies. Blood. 1991;77(12):2561–8.
258. Soiffer RJ, et al. Clinical and immunologic effects of prolonged infusion of low-dose recombinant interleukin-2 after autologous and T-cell-depleted allogeneic bone marrow transplantation. Blood. 1992;79(2):517–26.
259. Negrier S, et al. Intravenous interleukin-2 just after high dose BCNU and autologous bone marrow transplantation. Report of a multicentric French pilot study. Bone Marrow Transplant. 1991;8(4): 259–64.
260. Sosman JA, et al. Pilot trial of interleukin-2 with granulocyte colony-stimulating factor for the mobilization of progenitor cells in advanced breast cancer patients undergoing high-dose chemotherapy: expansion of immune effectors within the stem-cell graft and post-stem-cell infusion. J Clin Oncol. 2001;19(3):634–44.
261. Toh HC, et al. High-dose cyclophosphamide + carboplatin and interleukin-2 (IL-2) activated autologous stem cell transplantation followed by maintenance IL-2 therapy in metastatic breast carcinoma - a phase II study. Bone Marrow Transplant. 2000;25(1):19–24.
262. Kennedy-Nasser AA, et al. Ultra low-dose IL-2 for GVHD prophylaxis after allogeneic hematopoietic stem cell transplantation mediates expansion of regulatory T cells without diminishing antiviral and antileukemic activity. Clin Cancer Res. 2014;20(8):2215–25.
263. Meyers JD, et al. Prophylactic use of human leukocyte interferon after allogeneic marrow transplantation. Ann Intern Med. 1987;107(6):809–16.
264. Lizee G. Immunosuppression in melanoma immunotherapy: potential opportunities for intervention. Clin Cancer Res. 2006;12(7):2359s–65s.
265. Finke J, et al. Persistence of circulating t(14;18)-positive cells in long-term remission after radiation therapy for localized-stage follicular lymphoma. J Clin Oncol. 1993;11(9):1668–73.
266. Levitsky HI. Augmentation of host immune responses to cancer: overcoming the barrier of tumor antigen-specific T-cell tolerance. Cancer J. 2000;6(Suppl 3):S281–90.
267. Smyth MJ, Godfrey DI, Trapani JA. A fresh look at tumor immunosurveillance and immunotherapy. Nat Immunol. 2001;2(4):293–9.
268. Dudley ME, et al. A phase I study of nonmyeloablative chemotherapy and adoptive transfer of autologous tumor antigen-specific T lymphocytes in patients with metastatic melanoma. J Immunother. 2002b;25(3):243–51.
269. Gattinoni L, et al. Removal of homeostatic cytokine sinks by lymphodepletion enhances the efficacy of adoptively transferred tumor-specific CD8+ T cells. J Exp Med. 2005;202(7):907–12.
270. Mule JJ, et al. Selective localization of radiolabeled immune lymphocytes into syngeneic tumors. J Immunol. 1979;123(2):600–6.
271. Wang LX, Shu S, Plautz GE. Host lymphodepletion augments T cell adoptive immunotherapy through enhanced intratumoral proliferation of effector cells. Cancer Res. 2005;65(20):9547–54.
272. Cho BK, et al. Homeostasis-stimulated proliferation drives naive T cells to differentiate directly into memory T cells. J Exp Med. 2000a;192(4): 549–56.
273. Dummer W, et al. T cell homeostatic proliferation elicits effective antitumor autoimmunity. J Clin Invest. 2002;110(2):185–92.
274. Dudley ME, et al. Cancer regression and autoimmunity in patients after clonal repopulation with antitumor lymphocytes. Science. 2002a;298(5594): 850–4.
275. Rüttinger D, et al. Adjuvant therapeutic vaccination in patients with non-small cell lung cancer made lymphopenic and reconstituted with autologous PBMC: first clinical experience and evidence of an immune response. J Transl Med. 2007;5(1):43.
276. Gosse ME, Nelson TF. Approval times for supplemental indications for recombinant proteins. Nat Biotechnol. 1977;15(2):130–4.

26 Immunomodulation in Sepsis and Infection

Evangelos J. Giamarellos-Bourboulis

26.1 Introduction

Sepsis is a life-threatening organ dysfunction induced after the dysregulated host response to an infectious microbial insult. Despite the existing progress of our understanding of the mechanisms leading to this dysregulated host response, therapeutic modalities have not been changed. The early administration of antimicrobials remains the mainstay of treatment. Antimicrobials should be given to their highest tolerated dose within the first hour from start of organ failures since every hour of delay increases by 7.6% the risk for unfavorable outcome [1]. The high mortality of severe sepsis, despite antimicrobial therapy, originally led to the development of several treatment strategies that interfere with the immune system. These strategies aimed to modulate the dysregulated immune response of the host in order to improve the final outcome. Although the clinical development of these strategies was initiated already 30 years ago, none has received broad acceptance, and there is considerable debate among experts on the appropriateness of each treatment. As a typical example, drotrecogin alfa, a recombinant analogue of activated protein C which was the only drug ever licensed for severe sepsis and septic shock, was withdrawn from the market in 2010 due to lack of efficacy as proved by the PROWESS-SHOCK study. The debatable success of treatments that are available is mostly due to the fact that we have only recently started to recognize what constitutes the immune status of the host when clinical signs of sepsis appear and what the ideal treatment should be. We currently know that not all patients have a pro-inflammatory response when sepsis appears; some already show signs of an anti-inflammatory process which is probably the cascade that ultimately leads to multiple organ dysfunction (MODS) and death [2]. The pro-inflammatory response is characterized by a huge cytokine storm, whereas during the anti-inflammatory stage, the excessive pro-inflammatory reaction is suppressed. At this stage, blood MONOCYTES are no longer able to generate adequate amounts of CYTOKINES, NEUTROPHIL capacity for phagocytosis is downregulated, and Th2 lymphocyte responses predominate.

Available immunomodulatory therapies can be divided into two categories: (a) therapies aimed toward attenuation of exaggerated immune responses and (b) therapies stimulating host responses. In the present chapter, this classification is used to present available therapies. Therapeutic agents developed during the last decade are listed in Tables 26.1 and 26.2. Those that still remain promising for use in clinical practice are discussed further in this chapter.

Final manuscript submitted on November 28, 2016.

E. J. Giamarellos-Bourboulis (✉)
Medical School, National and Kapodistrian University of Athens, Athens, Greece

Integrated Research and Treatment Center, Center of Sepsis Control and Care, Jena University Hospital, Jena, Germany

4th Department of Internal Medicine, ATTIKON University Hospital, Athens, Greece
e-mail: egiamarel@med.uoa.gr

M. J. Parnham et al. (eds.), *Nijkamp and Parnham's Principles of Immunopharmacology*,
https://doi.org/10.1007/978-3-030-10811-3_26

Table 26.1 Molecules developed during the last decade for the attenuation of the host response in severe sepsis and septic shock (The mode of action and the stage of development are indicated)

Reference	Drug	Chemical structure	Mode of action	Stage of development
[3]	Ulinastatin	Protein	Universal serine protease inhibitor	Decreased 28-day mortality after a small-scale phase II RCT
[4]	AZD9773 (CytoFab™)	Polyclonal preparation of Fab fragments against TNF-α	Binding to and neutralization of TNF-α	Failure of phase II RCT
[5]	Eritoran	Synthetic lipid A structure of LPS	Blockade of LPS binding to the MD2:TLR4 complex	Failure of large-scale phase III RCT
[6]	Drotrecogin alfa	Recombinant human activated protein C	Exerting anticoagulant properties of protein C leading to attenuation of the pro-inflammatory effects of coagulation products	Retraction of the market after failure of PROWESS-SHOCK phase III RCT
[7]	Rosuvastatin	Statin	HMG-CoA reductase inhibitors showing anti-inflammatory effects	Failure of large-scale phase III RCT
[8]	Low-dose hydrocortisone	Corticosteroid	Pleiotropic actions, including attenuation of TNF-α production by blood monocytes	Survival benefit from early start of treatment
[9]	Filter with polymyxin B	Polymyxin B-embedded fiber device for blood filtering	Blockade of LPS in abdominal sepsis	Decreased 28-day mortality after a small-scale phase II RCT
[10]	ART-123	Recombinant human thrombomodulin	Activation of protein C; blockade of LPS and HMGB1	Promising results from a screening phase II RCT

Abbreviations: *HMG* 3-hydroxy-methylglutaryl, *HMGB1* high-mobility group box-1, *RCT* randomized clinical trial

Table 26.2 Molecules developed during the last decade for the stimulation of the host response in severe sepsis and septic shock (The mode of action and the stage of development are indicated)

Reference	Drug	Chemical structure	Mode of action	Stage of development
[15]	Talactoferrin	Analogue of lactoferrin	Stimulation of GALT	Failure of phase III RCT (stopped for futility)
[16]	rhIFN-γ	Recombinant human IFN-γ	Pleiotropic, including stimulation of monocyte cytokine production and neutrophil phagocytosis	Reversal of immunoparalysis after endotoxemia of healthy volunteers
[17]	GM-CSF	Recombinant human GM-CSF	Stimulation of granulocyte and monocyte formation and activities, including an increase of HLD-DR expression on blood monocytes	Earlier improvement, decrease of hospital stay in one phase II RCT
[18, 19]	IV clarithromycin	Macrolide	Various actions, including stimulation of monocyte cytokine production	Survival benefit in septic shock and ARDS
[20]	Pentaglobin	IgM-enriched Ig preparation	Binding of bacteria to form immune complexes that are more effectively phagocytosed	Available in the market

Abbreviations: *ARDS* acute respiratory distress syndrome, *RCT* randomized clinical trial

26.2 Therapies Attenuating Exaggerated Host Responses

Therapies in this section are aimed toward neutralization of pro-inflammatory molecules, like LPS and pro-inflammatory CYTOKINES (Table 26.1) [3–10]. Therapies that remain promising are ulinastatin, statins, intravenous hydrocortisone replacement, polymyxin B-embedded fiber adsorption device, and recombinant human thrombomodulin.

26.2.1 Ulinastatin

Ulinastatin is a natural serine protease inhibitor that behaves as an acute phase reactant, normally produced endogenously to counter-regulate the over-activation of natural serine proteases of the coagulation pathway. Concentrations in plasma and urine are decreased in sepsis, and this generated the concept of exogenous replacement of the inhibitor. Ulinastatin was administered intravenously (IV), in a recent small-scale phase II randomized clinical trial (RCT) in 55 patients at a dose of 200,000 IU within 1 h, repeating the infusion every 12 h for 5 consecutive days; another 59 patients were treated with placebo [3]. This trial was successful for the primary endpoint, i.e., 28-day mortality; 7.3% patients died in the ulinastatin arm versus 20.3% in the placebo arm.

26.2.2 Statins

Considerable enthusiasm has been aroused over the last few years by the retrospective analysis of data showing that mortality from sepsis is lower among patients already on long-term treatment with statins. Statins are inhibitors of 3-hydroxymethylglutaryl coenzyme A reductase, and by this they inhibit the rate-limiting step of cholesterol biosynthesis. This has generated the concept that statins may inhibit the chronic endothelial inflammation and, as a consequence, their role may be broadened for the attenuation of an over-exaggerated inflammatory process like sepsis. Analysis of a prospective noninterventional study revealed that previous statin therapy improves considerably the outcome of severe acute respiratory distress syndrome (ARDS). After propensity score analysis, to match 26 patients on previous statin therapy with 26 patients without previous statin therapy, 28-day mortality was 11.5% for patients on statins compared to 37.5% for patients without statins [10]. Based on these favorable findings, a prospective RCT was conducted to compare the efficacy of rosuvastatin on 379 patients with sepsis-associated ARDS.

Rosuvastatin was administered as a 20 mg daily dose after an initial loading dose of 40 mg; 366 patients were treated with placebo. The RCT failed to show any benefit from rosuvastatin treatment over placebo in the primary endpoint, i.e., death in the healthcare facility by day 60 (28.5% versus 24.9%). The only benefit from treatment was earlier resolution of renal and hepatic failures (10.1 days versus 11.0 days for renal failure; 10.8 days versus 11.8 days for hepatic failure) [7]. These negative results corroborate the findings of a recent meta-analysis of six RCTs that failed to demonstrate any survival benefit from overall statin treatment in patients with severe sepsis [11]. As a consequence, their utility for sepsis remains doubtful.

26.2.3 Low-Dose Hydrocortisone

Although GLUCOCORTICOSTEROIDS were traditionally conceived as molecules with anti-inflammatory properties, the principle of administration in sepsis is quite different. It is based on the understanding that almost 60% of patients with septic shock also suffer from relative adrenal insufficiency so that replacement is mandatory for the patient to overcome this deficiency. Following an initial successful study in France, showing considerably improved survival using this strategy [12], the large-scale, multicenter, international CORTICUS study in 499 patients with septic shock failed to demonstrate any benefit in terms of survival [13]. It seems that the contradictory findings of the two studies are related to the time between the start of vasopressors and the start of hydrocortisone. In the French study, this was set at 8 h, whereas in the CORTICUS RCT, it ranged between 8 and 72 h. An analysis of a prospective cohort of 170 patients with septic shock by the Hellenic Sepsis Study Group [8] showed 30.6% 28-day mortality of patients in whom hydrocortisone was started early and 52.2% 28-day mortality of those in whom hydrocortisone was started later. An early start was taken as initiation of treatment during the first 9 h from the start of vasopressors. However, in the Greek study, it was shown that part of the effect of hydrocortisone was related to its anti-inflammatory properties. More precisely, early start of hydrocortisone treatment was accompanied by considerable attenuation of TNF-α production by circulating MONOCYTES after ex vivo stimulation with LPS and heat-killed Gram-negative bacteria. Overexaggerated TNF-α production was found when the start of hydrocortisone was delayed. The recommended dose is 50 mg bolus infusion four times daily for 7 days followed by gradual tapering as suggested by the 2012 Surviving Sepsis Guidelines [14].

26.2.4 Polymyxin B-Embedded Fiber Device

Polymyxins are an old family of antimicrobials mainly active against Gram-negative bacteria. One mechanism of action is their ability to bind and neutralize LPS. With this as a background, a device was developed composed of fibers into which polymyxin B (PMX cartridge) is embedded for blood hemoperfusion in patients with septic shock. This device was tested in the EUPHAS RCT; 34 patients with septic shock due to intra-abdominal infections were treated with two sessions of hemofiltration and compared with 30 patients treated with placebo. Although 28-day mortality was a secondary endpoint, the study was prematurely stopped due to a large survival benefit from treatment (32% versus 52%) [9]. Three RCTs are currently being performed to confirm in septic shock the efficacy of different filters (PMX and tobramycin) embedded with polymyxin B (ClinicalTrial.gov identifiers NCT02413541, NCT01948778, and NCT01046669). If these trials are positive, this approach will become promising for the management of patients.

26.2.5 Recombinant Thrombomodulin

Thrombomodulin (TM) is normally produced by endothelial cells and activates the secretion of the anticoagulant protein C. In parallel, it attenuates the inflammatory host response by blocking the stimulation through LPS of the production of HMGB1 (high-mobility group box-1). Recombinant TM is licensed in Japan for the treatment of disseminated intravascular coagulation. To this end, its efficacy was tested in a phase IIB RCT showing benefit in terms of survival in patients with acute coagulopathy and mainly those with an international normalized index greater than 1.4 [10]. These favorable results prompted an ongoing phase III trial (ClinicalTrials.gov identifier NCT01598831). The therapeutic relevance of recombinant TM remains to be established.

26.3 Therapies Stimulating Immune Host Responses

These therapies are intended to stimulate the immune response and to reverse sepsis-induced immunosuppression (Table 26.2) [15–20]. Promising therapies include recombinant human IFN-γ and GM-CSF, macrolide antibiotics, and IgM-enriched Ig preparations.

26.3.1 Recombinant Human IFN-γ and GM-CSF

Administration of these molecules is proposed to help the innate immune response evade sepsis-induced immunoparalysis (see also Chaps. 6 and 25). IFN-γ skews the differentiation of T lymphocytes toward the Th1 pro-inflammatory phenotype, whereas GM-CSF primes the differentiation of myeloid cells toward mature NEUTROPHILS to escape from the dysfunctional NEUTROPHIL state of sepsis. This would be expected to offer effective containment of the infection and survival benefit. Evidence from a lethal experimental pyelonephritis model induced by *Escherichia coli* suggests that treatment with rhIFN-γ prolongs survival through an increase in the oxidative burst of NEUTROPHILS and in TNF-α production by peripheral blood mononuclear cells (PBMCs) [21].

Available RCTs on the efficacy of rhIFN-γ utilized the improvement of innate immune responses as their primary study endpoint; this does not allow discrimination of whether these treatments are promising or not for the final outcome of severe sepsis. The efficacy of rhIFNγ was tested in a setting of experimental endotoxemia in healthy human volunteers. After a first injection of LPS, three subcutaneous doses of placebo or rhIFNγ were administered at 48-h intervals. Then a second dose of LPS was given at a time when tolerance to the first dose of LPS had started to appear. In contrast to placebo-treated control volunteers, rhIFNγ treatment of the immunosuppressed subjects increased the low circulating TNF-α and low expression of HLA-DR on MONOCYTES [16]. Similar enhanced ex vivo production of TNF-α, IL-1β, IL-17, and IL-22 from circulating PBMCs was found with daily rhIFN-γ treatment, in contrast to untreated controls, as an adjuvant to standard antifungal therapy of five patients with candidemia [22]. rhIFNγ is currently being tested in a phase II trial of patients with severe sepsis (ClinicalTrials.gov identifier NCT01649921).

The use of GM-CSF in sepsis appears to be very promising since RCTs that have been conducted reported clinical benefit. GM-CSF was administered for 4 consecutive days at a dose of 3 μg/kg, starting before an operation in 28 patients with generalized abdominal peritonitis; 30 patients received identical placebo. Treatment ended with earlier improvement (median 2 days versus median 4 days), a shorter hospital stay (median 9 days versus median 13 days), and a considerable decrease in hospital costs [17]. GM-CSF was also administered for 8 consecutive days to 19 patients with severe sepsis/shock and sepsis-induced immunoparalysis, as defined by decreased expression of HLA-DR on MONOCYTES; another 19 patients received saline comparator treatment. GM-CSF treatment rapidly increased expression of HLA-DR on MONOCYTES and ex vivo production of TNF-α and of IL-6 compared to the comparator group; however, no effect on mortality was found. The above results indicate that the use of GM-CSF as a therapeutic strategy in sepsis has not yet been clarified and further clinical studies are warranted [23].

26.3.2 Macrolides and Intravenous Clarithromycin

Macrolide antibiotics, such as azithromycin and clarithromycin, have been widely reported to exert immunomodulatory effects, reducing NEUTROPHIL infiltration at inflamed sites and promoting the generation of an inflammation-resolving macrophage phenotype [24] (see also Chap. 29). There is robust evidence that systemic administration of macrolides in the event of infection is associated with earlier infection resolution, reduced mortality, and attenuation of the systemic inflammation.

26.3.2.1 Airway Diseases

The investigation of the beneficial effect of macrolides on acute inflammatory states was based on randomized clinical trials showing benefit of long-term macrolide treatment in patients with chronic diseases of the airways (see Chap. 29). Four prospective randomized trials (reviewed in Ref. [24]) have shown that oral administration of azithromycin for 6 months improved lung function and reduced the frequency of infectious exacerbations in patients with CF. More recent trials have shown that long-term oral azithromycin reduced the frequency of exacerbations of COPD [25] and improved lung function in patients with bronchiolitis obliterans developing after lung transplantation [26].

Sixteen studies, all retrospective in design, were published from 1999 until 2013 comparing the efficacy of macrolide-containing regimes over non-macrolide-containing regimens in patients with community-acquired pneumonia (CAP). A meta-analysis of these studies showed considerable survival benefit from macrolide treatment. The cumulative odds ratio (OR) of all 16 studies for death on a macrolide-containing regimen compared to a non-macrolide-containing regimen was 0.66 ($p < 0.0001$). This was 0.66 ($p < 0.0001$) for the seven studies involving patients with severe CAP and 0.59 (p, 0.030) for the six studies involving patients with pneumococcal CAP [27] indicating that the addition of a macrolide to the therapeutic regimen protected patients from an unfavorable outcome. However, the largest published retrospective analysis in patients with CAP was not included in this meta-analysis. In this single large analysis, 38,727 elderly patients were treated with azithromycin for CAP, and they were compared with 34,838 elderly patients treated with guideline-compatible regimens not containing azithromycin. Azithromycin treatment was associated with significantly lower 90-day mortality (17.4% versus 22.2%, $p < 0.0001$) [28].

However, it should be borne in mind that all these are retrospective data. The only prospective RCT on the efficacy of a macrolide to the treatment regimen of CAP was recently published. In this RCT, patients were randomized to either β-lactam monotherapy (IV cefuroxime or amoxicillin/clavulanate) ($n = 291$) or combination of one β-lactam and IV or oral clarithromycin 500 mg two times daily ($n = 289$). The study primary endpoint was clinical instability after 7 days, and the secondary endpoint was hospital readmission during the first 30 days. Although the study was powered for non-inferiority, a considerable trend towards benefit as a result of the addition of clarithromycin was found. More precisely, 41.2% of patients in the monotherapy arm were clinically unstable on day 7 versus 33.6% in the combination arm. Readmission rates were 7.9% and 3.1%, respectively (p, 0.010) [29].

26.3.2.2 Gram-Negative Sepsis

Our study group has conducted a 12-year project on the development of IV clarithromycin as immunomodulatory treatment for sepsis caused by Gram-negative bacteria. The background of clinical development was a series of six publications in an experimental model of acute lethal pyelonephritis induced by multidrug-resistant (MDR) nosocomial isolates against which clarithromycin lacks any antimicrobial activity. All studies (reviewed in Ref. [24]) showed considerable survival benefit from clarithromycin treatment.

As a next step, we conducted an RCT in patients with ventilator-associated pneumonia (VAP) (NCT00297674, ClinicalTrials.gov) from June 2004 to November 2005. In this study [18], 100 patients were allocated to treatment with placebo and guideline-concordant antimicrobials, and another 100 patients were allocated to IV clarithromycin and guideline-concordant antimicrobials. The dose regimen of clarithromycin was 1 g once daily for 3 consecutive days within 1-h infusion via a central catheter. Isolated pathogens from quantitative cultures of tracheobronchial secretions were MDR species of *Acinetobacter baumannii*, *Pseudomonas aeruginosa*, and *Klebsiella pneumoniae*. The appropriateness of co-administered antimicrobials was similar in both groups. Addition of clarithromycin decreased mortality of the most severe patients with septic shock and MODS (28-day mortality 38.9% versus 60%; p, 0.020). This was accompanied by earlier resolution of VAP in the entire study population (10 days versus 15 days, Fig. 26.1) and a reciprocal earlier weaning from mechanical ventilation (16 days versus 22 days; p, 0.049). Although 90-day mortality was not a pre-defined study endpoint, we retrospectively assessed this in the overall study population; the mortality was 60% among the 100 patients allocated to placebo treatment and 43% among the 100 patients allocated to clarithromycin treatment (p, 0.023). In parallel to this retrospective 90-day survival analysis, we measured the total real cost of survivors; this was €26,249.50/patient of the placebo group staying alive after the first 45 days and €19,303.20/patient of the clarithromycin group staying alive after the first 45 days (p, 0.011) [30].

In order to confirm these findings, another prospective RCT was conducted from July 2007 to

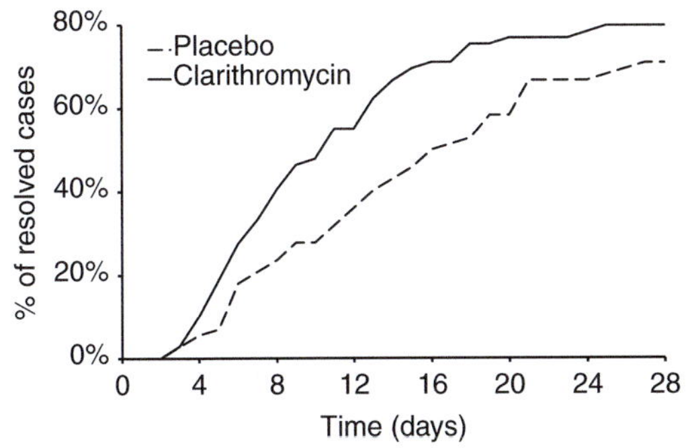

Fig. 26.1 Cumulative incidence of the resolution of ventilator-associated pneumonia (VAP) within the follow-up period of 28 days. Analysis comprised patients who survived until the 28th day ($n = 141$). p of comparisons; Mantel-Cox's log rank test, 0.036; Tarone-Ware's test, 0.017; Breslow's generalized Wilcoxon test, 0.011. (From Ref. [18] with permission © 2008 by The Infectious Diseases Society of America)

April 2011 (NCT01223690, ClinicalTrials.gov) [19]. A total of 600 patients with SIRS developing after acute pyelonephritis or acute intra-abdominal infection or primary Gram-negative bacteremia were allocated to either placebo and antimicrobials ($n = 298$) or IV clarithromycin and antimicrobials ($n = 302$). The appropriateness of co-administered antimicrobials was similar in both groups. Isolated pathogens were species of Gram-negative bacteria mainly enterobacteria against which clarithromycin is non-active. Results largely confirmed the findings of the VAP study. More precisely, addition of IV clarithromycin dramatically decreased 28-day mortality by septic shock and MODS (53.1% versus 73.1% of placebo-treated patients; p, 0.020) and decreased considerably the time until resolution of infection among patients with severe sepsis and septic shock (6 days versus 10 days; p, 0.037). The median cost of hospitalization was 3383.5 Euros in the placebo arm and 2269.3 Euros in the clarithromycin arm (p, 0.044).

26.3.2.3 Mechanism of Action

A major question concerns the mechanism underlying the benefit of clarithromycin. Efficacy in patients with chronic diseases of the airways and beneficiary pharmacokinetics in the lung parenchyma suggest a local effect at the level of alveolar macrophages and/or bronchial epithelia [24]. However, reduction of mortality in septic shock and MODS is indicative of a more systemic mechanism of action. To this end, circulating MONOCYTES were isolated at baseline and on another 6 consecutive days from patients enrolled in our first trial in patients with VAP and assayed by both functional and flow cytometry assays. Findings suggested that survival benefit in the most severe population with septic shock and MODS was associated with improvement of the function of MONOCYTES with regard to the production of IL-6 and increased membrane expression of CD86 and TREM-1. In support of this, the ratio of circulating IL-10 to TNF-α was decreased. Similar immune changes were found in the less severe patient population treated with IV clarithromycin, but they were less profound [31]. These findings suggest that improved survival resulting from clarithromycin treatment was associated with reversal of sepsis-induced immunosuppression. A summary of the immune changes associated with clarithromycin treatment is shown in Table 26.3.

Table 26.3 Measured effect of intravenous clarithromycin treatment in patients with ventilator-associated pneumonia and sepsis

Immune variable	Effect	Patient population
CRP	↓ days 1 and 2	Septic shock and MODS
IL-10/TNF-α	↓ day 4	Septic shock and MODS
TREM-1 on blood monocytes	↑ days 2, 4, and 6	All patients
Apoptosis of blood monocytes	↓ day 4	Septic shock and MODS
CD86 on monocytes	↑ days 2 and 4	All patients
TNF-α production by monocytes	↑ days 2, 4, and 6	Less severe patients
IL-6 production by monocytes	↑ days 4 and 6	Septic shock and MODS
TREM-1 on blood neutrophils	↑ days 2 and 6	Septic shock and MODS

Changes refer to placebo treatment. Treatment lasted for 3 days (day 0 to day 2) and consecutive follow-up was done until day 6

Abbreviations: *MODS* multiple organ dysfunction syndrome, *TREM* triggering receptor expressed on myeloid cells, ↓ decrease, ↑ increase

26.3.3 IgM Immunoglobulins

There are accumulating data indicating that immunosuppression in severe sepsis and septic shock involves B LYMPHOCYTES associated with impairment of the production of Igs. Although there is a consistent decrease in circulating IgG1, IgA, and IgM in severe sepsis [32], a study of the course of septic shock, starting from the time when vasopressor treatment is needed, indicates that failed body distribution of IgM is of paramount importance and this is linked with unfavorable outcome [33]. Lack of IgM is a major component of immunoparalysis since IgM is an Ig pentamer that binds LPS and whole microorganisms and primes NEUTROPHILS for phagocytosis. This provides the rationale for IgM supplementation in severe sepsis and septic shock. Only one IgM-enriched Ig preparation is available, namely, Pentaglobin™.

Many RCTs have been conducted in the past in an attempt to demonstrate clinical benefit from the administration of Ig preparations in severely infected patients. Current evidence suggests that administration of IgG preparations does not offer any clinical benefit [34]. However, as part of a meta-analysis of these RCTs, trials testing the efficacy of IgG preparations were separated from trials testing the efficacy of Pentaglobin. The results of this analysis showed a considerable decrease in the relative risk for death by severe sepsis in both pediatric and adult populations as a result of the administration of Pentaglobin [35]. In a retrospective analysis of patients with septic shock using a propensity score matching, 57 patients treated with Pentaglobin within the first 24 h of shock onset were compared with 57 patients not treated with Pentaglobin. Treatment was associated with a considerable decrease in 28-day mortality (25.5% versus 45.8%; *p*, 0.021) [20]. A retrospective analysis has recently been published by the Hellenic Sepsis Study Group. One hundred patients with septic shock caused by infections from MDR Gram-negative pathogens and treated for 5 consecutive days with Pentaglobin, starting the first 24 h from the start of vasopressor treatment, were compared with another 100 comparators well-matched for severity, type of infection, multidrug resistance of pathogens, appropriateness of empirically administered antimicrobials, and comorbidities. The 28-day mortality was 58% in the comparator group and 39% in the Pentaglobin group (*p*, 0.011) [36]. In the light of the results of the meta-analysis [35], these findings indicate that the administration of the IgM-enriched immunoglobulin preparation is a promising strategy that requires further confirmation with a large prospective trial.

A novel Ig preparation, BT0086, containing 23% of IgM has recently been introduced. The results of an RCT to determine its efficacy when administered on ventilator-free days in patients with severe CAP remain to be presented (ClinicalTrials.gov identifier NCT01420744).

26.4 T Cell Exhaustion: A Promising Field of Immunotherapy

Recent evidence shows that following the phase of sepsis-induced immunosuppression, the host enters a state of T CELL exhaustion. This is characterized by the inability of T cells to proliferate and to produce CYTOKINES and of $CD8^+$ cells to function as cytotoxic cells. The prime effector molecule for this phenomenon is the negative costimulatory receptor, programmed cell death (PD-1), and its ligand PD-L1. In a model of acute lung injury (ALI) after cecal ligation and puncture (CLP) in mice, knockout of PD-1 was accompanied by greater survival rate. This was associated with a lower $CD4^+/CD8^+$ ratio and infiltration by NEUTROPHILS into the lungs of PD-1 knockout mice compared to wild-type mice. These findings were further confirmed in patients with ALI; expression of PD-1 on $CD3^+$ blood cells was lower in survivors than in non-survivors [37].

PD-1 is highly expressed on $CD4^+$ and $CD8^+$ lymphocytes and on $CD14^+$ MONOCYTES in severe sepsis compared to systemic inflammatory response syndrome (SIRS). This high expression remains stable during the course of the illness. To the contrary, expression of PD-L1 on $CD8^+$ cells decreases with time. It is considered that inhibition of the PD-1/PD-L1 pathway may represent a promising target of immunotherapy. Ex vivo treatment with anti-PD-1 and anti-PD-L1 antibodies of peripheral blood mononuclear cells (PBMCs) isolated over the course of the illness in 43 septic patients resulted in (a) a decrease in the rate of apoptosis of $CD4^+$ and of $CD8^+$ lymphocytes, (b) an increase in the capacity of $CD3^+$ cells and of NK cells for IL-2 production, and (c) an increase in the capacity of NK cells to produce IFN-γ [38]. To further investigate the potential for inhibition of the PD-1/PD-L1 pathway in sepsis therapy, the same group used a two-hit experimental model simulating clinical practice; mice survivors from CLP were injected with *Candida albicans* and treated with fluconazole with/without anti-PD-1 and anti-PD-L1. Treatment with antibodies targeting PD-1 and PD-L1 prolonged survival and improved the function of $CD4^+$ cells in the spleen for the production of IFN-γ. The effect was due, at least in part, to better antigen presentation as evidenced by the upregulation of MHC-II [39] (see also Chaps. 12 and 25).

26.5 Conclusions

The current state of knowledge indicates that there is no "golden" immunotherapy for severe sepsis. The main reason for this is the controversial nature of the immune response in the septic host. Available RCTs assume that all patients behave in the same way. However, the Hellenic Sepsis Study Group ran flow cytometry analysis in 505 patients and grouped them according to the type of underlying infection. The results indicated that the immune change upon transition from sepsis to severe sepsis/septic shock is highly individualized and depends on the underlying infection. As such, (1) a decrease in HLA-DR expression on MONOCYTES is found only in acute pyelonephritis and intra-abdominal infections and is not a universal finding; (2) a decrease in $CD4^+$ and $CD8^+$ cells is found only in CAP and intra-abdominal infections; and (3) a decrease in $CD19^+$ cells is characteristic of CAP [40]. VAP complicated by severe sepsis is characterized by profound immunosuppression [41], whereas severe sepsis developing in young survivors from multiple injuries hospitalized in an intensive care unit is associated with high production of TNF-α and IFN-γ by PBMCs [42]. These factors indicate that in the future the immune status of each patient will need to be assessed before individualized therapeutic decisions can be taken. This is of crucial importance in view of the fact that the immune status of a patient can change as the course of the disease is followed.

A schematic representation of the immune changes taking place in sepsis and the available interventions is shown in Fig. 26.2. Some of these interventions have led to marketed therapies. Dose regimes and indications for the marketed compounds are shown in Table 26.4.

26.6 Summary

Despite the great efforts and the vast number of randomized clinical trials (RCT) that have been conducted over the last 30 years, no universally accepted immunotherapy exists for severe sepsis and septic shock. This is due to the wide diversity of immune phenomena that are detectable from one host to the other. Current management relies on the early administration of intravenous fluids and antimicrobials; when possible this management should start within the very first hour from the advent of organ failure(s). The immune response in sepsis is roughly represented by two phases: a pro-inflammatory stage, characterized by overproduction of CYTOKINES from MONOCYTES and tissue infiltration by NEUTROPHILS, and, secondly, a stage of immunosuppression, characterized by failure of MONOCYTES and Th1 cells to generate CYTOKINES, predominance of Th2 responses, and defective phagocytosis by

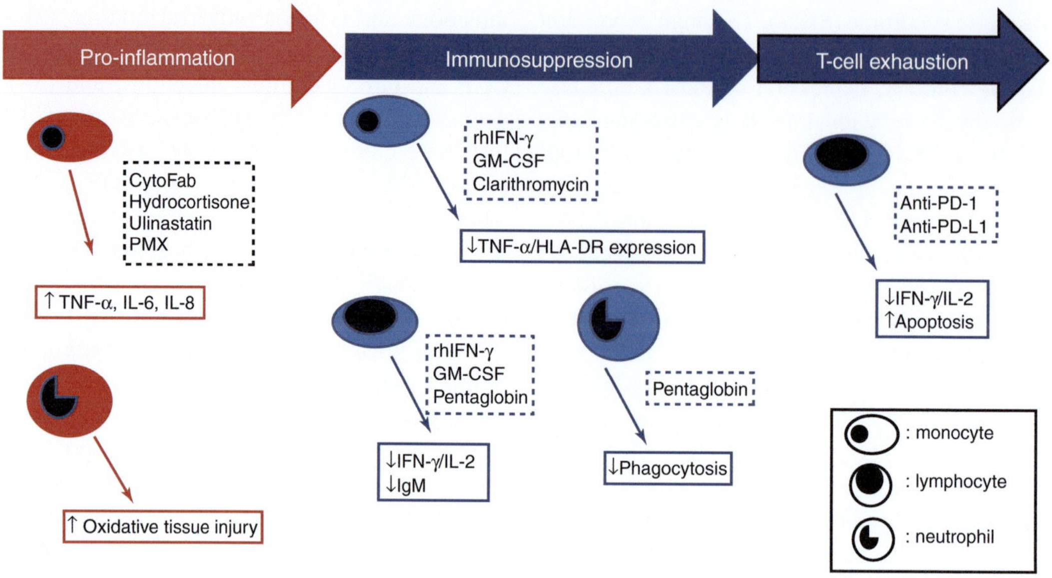

Fig. 26.2 Changes of immune status over the course of sepsis and available therapies. Red color indicates activated cells and blue color indicates suppressed cells. Available therapies to reverse the indicated activation or suppression are shown in dashed boxes. ↑, increase; ↓, decrease

Table 26.4 Chemical names, formulations, doses, and indications for marketed compounds that can be used as immunotherapy in sepsis

Name	Formulation	Dose	Indication
Hydrocortisone	IV	50 mg 6 h for 7 days	Septic shock
rhIFNγ	SC	50 μg/m² body surface	Not yet defined
GM-CSF	SC	3 μg/kg qd for 5 days	Abdominal sepsis and ↓ expression of HLA-DR on monocytes
Clarithromycin	IV	1 g qd 1 h infusion for 4 days	Septic shock and MODS
Clarithromycin	IV	500 mg bid for 7 days	CAP
Azithromycin	PO	500 mg qd for 7 days	CAP
IgM preparations	IV	30 g 24-h infusion for 5 days	Septic shock

Abbreviations: *bid* two times daily, *CAP* community-acquired pneumonia, *IV* intravenous, *PO* per oral, *qd* once daily, *q6h* every 6 h, *SC* subcutaneous

NEUTROPHILS. The immunosuppression stage is followed by T CELL exhaustion. Therapeutic agents under investigation that interfere with the stage of pro-inflammation and that attenuate the exaggerated inflammatory response include ulinastatin, statins, low-dose hydrocortisone, and polymyxin B-embedded fiber device. Those agents that interfere with the stage of immunosuppression and stimulate immune responses include rhIFN-γ, GM-CSF, and IgM-enriched Ig preparations. Of all molecules studied, the greatest clinical benefit has been obtained with macrolides; on co-administration with β-lactams, they significantly decrease mortality of community-acquired pneumonia. Three RCTs have reported benefit with intravenous clarithromycin treatment in two acute situations: severe CAP, where the drug offers early clinical stability, and septic shock aggravated by multiple organ dysfunction where it decreases mortality. The mode of action of clarithromycin is associated with reversal of sepsis-induced immunosuppression.

Selected Readings

Annane D, Sébille V, Charpentier C, et al. Effect of treatment with low doses of hydrocortisone and fludrocortisone on mortality in patients with septic shock. JAMA. 2002;288:862–71.

Boomer JS, Green JM, Hotchkiss RS. The changing immune system in sepsis. Virulence. 2014;5:45–56.

Gogos C, Kotsaki A, Pelekanou A, et al. Early alterations of the innate and adaptive immune statuses in sepsis according to the type of underlying infection. Crit Care. 2010;14:R96.

Kumar A, Roberts D, Wood KE, et al. Duration of hypotension before initiation of effective antimicrobial therapy is the critical determinant of survival in human septic shock. Crit Care Med. 2006;34:1589–96.

Tsaganos T, Raftogiannis M, Pratikaki M, et al. Clarithromycin leads to long-term survival and cost benefit in ventilator-associated pneumonia and sepsis. Antimicrob Agents Chemother. 2016;60:3640–6.

References

1. Kumar A, Roberts D, Wood KE, et al. Duration of hypotension before initiation of effective antimicrobial therapy is the critical determinant of survival in human septic shock. Crit Care Med. 2006;34:1589–96.
2. Boomer JS, Green JM, Hotchkiss RS. The changing immune system in sepsis. Virulence. 2014;5:45–56.
3. Karnad DR, Bhadade R, Verma PK, et al. Intravenous administration of ulinastatin (human urinary trypsin inhibitor) in severe sepsis: a multicenter randomized controlled study. Intensive Care Med. 2014;40:830–8.
4. Bernard GR, Francois B, Mira JP, et al. Evaluating the efficacy and safety of two doses of the polyclonal anti-tumor necrosis factor-α fragment antibody AZD9773 in adult patients with severe sepsis and/or septic shock: randomized, double-blind, placebo-controlled phase IIb study. Crit Care Med. 2014;42:504–11.
5. Opal SM, Laterre PF, Francois B, et al. Effect of eritoran, an antagonist of MD2-TLR4, on mortality in patients with severe sepsis: the ACCESS randomized trial. JAMA. 2013;309:1154–62.
6. Ranieri VM, Thompson BT, Barie PS, et al. Drotrecogin alfa (activated) in adults with septic shock. N Engl J Med. 2012;366:2055–64.
7. Truwit JD, Bernard GR, Steingrub J, et al. Rosuvastatin for sepsis-associated acute respiratory distress syndrome. N Engl J Med. 2014;370:2191–200.
8. Katsenos CS, Antonopoulou AN, Apostolidou EN, et al. Early administration of hydrocortisone replacement after the advent of septic shock: impact on survival and immune response. Crit Care Med. 2014;42:1651–7.
9. Cruz DN, Antonelli M, Fumagalli R, et al. Early use of polymyxin B hemoperfusion in abdominal septic shock: the EUPHAS randomized controlled trial. JAMA. 2009;301:2445–52.
10. Mansur A, Steinau M, Popov AF, et al. Impact of statin therapy on mortality in patients with sepsis-associated acute respiratory distress syndrome (ARDS) depends on ARDS severity: a prospective observational cohort study. BMC Med. 2015;13:128.
11. Deshpande A, Pasuleti V, Rothberg MB. Statin therapy and mortality from sepsis: a meta-analysis of randomized trials. Am J Med. 2015;128:410–7.
12. Annane D, Sébille V, Charpentier C, et al. Effect of treatment with low doses of hydrocortisone and fludrocortisone on mortality in patients with septic shock. JAMA. 2002;288:862–71.
13. Sprung CL, Annane D, Keh D, et al. Hydrocortisone therapy for patients with septic shock. N Engl J Med. 2008;358:111–24.
14. Dellinger RP, Levy MM, Rhodes A, et al. Surviving Sepsis Campaign: international guidelines for management of severe sepsis and septic shock: 2012. Crit Care Med. 2013;41:580–637.
15. Vincent JL, Marshall JC, Dellinger RP, et al. Talactoferrin in severe sepsis: results from the phase II/III oral talactoferrin in severe sepsis trial. Crit Care Med. 2015;43:1832–8.
16. Leentjens J, Kox M, Koch RM, et al. Reversal of immunoparalysis in humans in vivo. Am J Respir Crit Care Med. 2012;186:838–45.
17. Orozco H, Arch J, Medina-Franco H, Pantoja JP, et al. Molgramostim (GM-CSF) associated with antibiotic treatment in nontraumatic abdominal sepsis: a randomized, double-blind, placebo-controlled clinical trial. Arch Surg. 2006;141:150–3.
18. Giamarellos-Bourboulis EJ, Pechère JC, Routsi C, et al. Effect of clarithromycin in patients with sepsis and ventilator-associated pneumonia. Clin Infect Dis. 2008;46:1157–64.
19. Giamarellos-Bourboulis EJ, Mylona V, Antonopoulou A, et al. Effect of clarithromycin in patients with suspected Gram-negative sepsis: results of a randomized controlled trial. J Antimicrob Chemother. 2014;69:1111–8.
20. Cavazzuti I, Serafini G, Busani S, et al. Early therapy with IgM-enriched polyclonal immunoglobulin in patients with septic shock. Intensive Care Med. 2014;40:1888–96.
21. Katsaris MP, Adamis T, Pistiki A, et al. Immunomodulatory intervention with interferon-gamma in *Escherichia coli* pyelonephritis. J Urol. 2014;192:600–6.
22. Delsing CE, Gresnigt MS, Leentjens J, et al. Interferon-gamma as adjunctive immunotherapy for invasive fungal infections: a case series. BMC Infect Dis. 2014;14:166.
23. Meisel C, Schefold JC, Pschowski R, et al. Granulocyte-macrophage colony-stimulating factor to reverse sepsis-associated immunosuppression: a double-blind, randomized, placebo-controlled multicenter trial. Am J Respir Crit Care Med. 2009;180:640–8.

24. Giamarellos-Bourboulis EJ. Macrolides beyond the conventional antimicrobials: a class of potent immunomodulators. Int J Antimicrob Agents. 2008;31:12–20.
25. Han MK, Tayob N, Murray S, et al. Predictors of chronic obstructive pulmonary disease exacerbation reduction in response to daily azithromycin therapy. Am J Respir Crit Care Med. 2014;189:1503–8.
26. Corris PA, Ryan VA, Small T, et al. A randomized controls trial of azithromycin therapy in bronchiolitis obliterans syndrome (BOS) post lung transplantation. Thorax. 2015;70:442–50.
27. Nie W, Li B, Xiu Q. β-Lactam/macrolide dual therapy versus β-lactam monotherapy for the treatment of community-acquired pneumonia in adults: a systematic review and meta-analysis. J Antimicrob Chemother. 2014;69:1441–6.
28. Mortensen EM, Halm EA, Pugh MJ, et al. Association of azithromycin with mortality and cardiovascular events among older patients hospitalized with pneumonia. JAMA. 2014;311:2199–208.
29. Garin N, Genné D, Carballo S, et al. β-Lactam monotherapy vs β-lactam-macrolide combination treatment in moderately severe community-acquired pneumonia: a randomized noninferiority trial. JAMA Intern Med. 2014;174:1894–901.
30. Tsaganos T, Raftogiannis M, Pratikaki M, et al. Clarithromycin leads to long-term survival and cost benefit in ventilator-associated pneumonia and sepsis. Antimicrob Agents Chemother. 2016;60:3640–6.
31. Spyridaki A, Raftogiannis M, Antonopoulou A, et al. Effect of clarithromycin in inflammatory markers of patients with ventilator-associated pneumonia and sepsis caused by Gram-negative bacteria: results from a randomized clinical study. Antimicrob Agents Chemother. 2012;56:3819–25.
32. Bermejo-Martin JF, Rodriguez-Fernandez A, Herran-Monge R, et al. Immunoglobulins IgG1, IgM and IgA: a synergistic team influencing survival in sepsis. J Intern Med. 2014;276:404–12.
33. Giamarellos-Bourboulis EJ, Apostolidou E, Lada M, et al. Kinetics of circulating immunoglobulin M in sepsis: relationship with final outcome. Crit Care. 2013;17:R247.
34. Brocklehurst P, Farrell B, King A, et al. Treatment of neonatal sepsis with intravenous immune globulin. N Engl J Med. 2011;365:1201–11.
35. Kreymann KG, de Heer G, Nierhaus A, Kluge S. Use of polyclonal immunoglobulins as adjunctive therapy for sepsis or septic shock. Crit Care Med. 2007;35:2677–85.
36. Giamarellos-Bourboulis EJ, Tziolos N, Routsi C, et al. Improving outcomes of severe infections by multidrug-resistant pathogens with polyclonal IgM-enriched immunoglobulins. Clin Microbiol Infect. 2016;22:499–506.
37. Monaghan SF, Thakkar RK, Hefferman DS, et al. Mechanisms of indirect acute lung injury. A novel role for the co-inhibitory receptor, programmed death-1. Ann Surg. 2012;255:158–64.
38. Chang K, Svabek C, Vazquez-Guillamet C, et al. Targeting the programmed cell death 1: programmed cell death ligand 1 pathway reverses T cell exhaustion in patients with sepsis. Crit Care. 2014;18:R3.
39. Chang KC, Burnham CA, Compton SM, et al. Blockade of the negative co-stimulatory molecules PD-1 and CTLA-4 improves survival in primary and secondary fungal sepsis. Crit Care. 2013;17:R85.
40. Gogos C, Kotsaki A, Pelekanou A, et al. Early alterations of the innate and adaptive immune statuses in sepsis according to the type of underlying infection. Crit Care. 2010;14:R96.
41. Pelekanou A, Tsangaris I, Kotsaki A, et al. Decrease of CD4-lymphocytes and apoptosis of CD14-monocytes are characteristic alterations in sepsis caused by ventilator-associated pneumonia: results from an observational study. Crit Care. 2009;13:R172.
42. Paraschos MD, Patrani M, Pistiki A, et al. Defective cytokine production early after multiple traumas: modulation in severe sepsis. Cytokine. 2015;76:222–6.

Immunopharmacology of Prebiotics and Probiotics

27

Prescilla V. Jeurink, Negisa Seyed Toutounchi, Hamed Janbaz Acyabar, Gert Folkerts, and Johan Garssen

27.1 Introduction

The medicinal properties of various nutritional components have been appreciated since ancient times. Hippocrates (460–377 B.C.), for example, stated: "Let medicine be thy food and food be thy medicine." Tea brewed from various fruits, shrubs, and trees containing natural salicylates has been consumed for pain relief since the Stone Age. Also the origin of the most well-known painkiller found in almost every home—aspirin—is a willow bark tree extract (see Chap. 33). These examples illustrate how nature can provide the chemical structure for a pharmaceutical. It is now appreciated that 70% of current drugs have their origin in chemical compounds found in plants, fruits, and vegetables. Modern medicinal chemists are capable of isolating and identifying these active chemical compounds and then modifying them to yield compounds with increased activity and less side effects.

There is also a large body of evidence that products from nature modulate human health through interactions with microbes present in the intestine. The beneficial effect of certain microbes was already postulated more than 100 years ago by the Nobel Prize winner Elie Metchnikoff, from the Pasteur Institute in Paris, when he stated that certain lactic acid bacteria present in yoghurt were beneficial to health and had life-prolonging properties. This notion has been validated by a broad array of products that aim to improve human health through direct administration of PROBIOTICS or PREBIOTICS. A sequence of papers has been written about the definition of PREBIOTICS and PROBIOTICS [1–8]. In the recent expert consensus document of the International Scientific Association for Probiotics and Prebiotics (ISAPP), the definitions are as follows: PREBIOTICS are a substrate that is selectively utilized by host microorganisms conferring a health benefit [9], whereas PROBIOTICS, according to the FAO/WHO, are "live micro-organisms which, when administered in adequate amounts, confer a health benefit on the host" [10].

Final manuscript submitted on January 09, 2018.

P. V. Jeurink · J. Garssen (✉)
Department of Pharmacology and Pathophysiology, Utrecht University, Utrecht, The Netherlands

GCoE Immunology, Danone Nutricia Research, Utrecht, The Netherlands
e-mail: johan.garssen@danone.com; j.garssen@uu.nl

N. S. Toutounchi · G. Folkerts
Department of Pharmacology and Pathophysiology, Utrecht University, Utrecht, The Netherlands

H. J. Acyabar
Department of Pharmacology and Pathophysiology, Utrecht University, Utrecht, The Netherlands

Department of Pharmacology and Toxicology, Faculty of Veterinary Medicine, University of Urmia, Urmia, Iran

M. J. Parnham et al. (eds.), *Nijkamp and Parnham's Principles of Immunopharmacology*, https://doi.org/10.1007/978-3-030-10811-3_27

27.1.1 The Intestinal Microbiota

The intestine is inhabited by a vast number of microbes that form an intimate partnership with the human host, contributing to and influencing many aspects of human health [3]. This holds for the functioning of the immune system, as the gastrointestinal (GI) tract is the largest immune organ in the body. It contains approximately 65% of the overall immunological tissues and up to 80% of the immunoglobulin-producing tissues of the body. These immunological areas in the GI tract are involved in suppressing or moderating the potent antimicrobial arsenal that manage explicit pathogenic threats, as well as prevent the induction of overt inflammation by damage-associated molecular patterns. The colonization and establishment of the gut MICROBIOTA during the neonatal period are crucial for maturation of the intestinal immune system [11], development of intestinal morphology, and maintenance of an immunologically balanced inflammatory response [12]. The composition and development of the intestinal MICROBIOTA in babies are influenced by genetic factors and mode of delivery but also largely by the composition of human breast milk (see below) which in turn contributes to the maturation of the baby's immune system [12]. In addition, we have a mutual relationship with our MICROBIOTA, which requires the maintenance of recognizing these organisms as affable and the need to restrain inflammatory responses to these organisms when encountered in hostile settings. Knoop et al. have identified a specific preweaning interval in which gut microbial antigens are encountered by the immune system to induce antigen-specific tolerance to gut bacteria that is more crucial for some bacterial taxa than for others [13]. From the moment that solid foods are introduced and the dietary intake starts to resemble that of adults, both the MICROBIOTA and the intestinal immune system form important parts of our line of defense against ingested (pathogenic) microorganisms. In the gut, there is continuous cross-talk between microbes and the intestinal immune system, and this requires a delicate integration of pro- and anti-inflammatory signals to regulate innate and adaptive immune responses and control inflammation. Throughout life, the dominant species that make up the adult MICROBIOTA remain relatively unchanged. However, in elderly people, there are significant alterations in the proportion and composition of the different taxa, leading to reduced microbial diversity. In addition, the increasing numbers of enteropathogens may lead to chronic inflammation, and together with the aging, mucosal immune system may contribute to these diversity changes [14].

27.1.2 The Development of Gut Microbiota During Infancy

Initially there is still a little oxygen present in the gastrointestinal tract just after birth; therefore, the first colonizers of the infant gut are facultative anaerobic bacteria, such as *Staphylococcus*, *Streptococcus*, *Enterococcus*, and *Enterobacter* spp. Thereafter, oxygen levels increase, and anaerobes such as *Bifidobacterium*, *Bacteroides*, and *Clostridium* spp. can start colonizing the gut [12]. *Bifidobacterium* represents a genus within the phylum *Actinobacteria* which is one of the major phyla in the healthy intestinal tract of humans. *Bifidobacterium* is one of the most abundant genera in adults, but its predominance is even more pronounced in infants, especially during lactation, when they constitute most of the total bacterial population [15]. The type of bacteria that initially colonize the infant gut is highly dependent on the mode of delivery: strains originating from the maternal gut and vagina are transferred to the infant's gut in case of a vaginal delivery, whereas infants born by cesarean section are initially colonized by bacteria from the environment such as from maternal skin, hospital staff, or other neonates [12]. Clinical data show that breast-fed children have high numbers of the health-promoting lactobacilli and bifidobacteria, whereas children that are bottle-fed with a cow's milk formula without additional oligosaccharides have significantly lower numbers of these intestinal bacteria [16]. Addition of specific prebiotics to infant formula brings the microbial composition closer to the ones that are found in breast-fed infants. The large impact of breast milk on the infant's MICROBIOTA is important to achieve an immunological balance which is disturbed during pregnancy

[17]. To prevent rejection of the fetus, a Th2-skewed environment is created inside the womb. This phenomenon is probably the reason why babies are born with an immature immune system. This immaturity results in deficiencies of both the innate and adaptive immune responses in early life.

27.1.3 Microbiota Dysbalance and Disease

Many gut-related disorders are correlated with a dysbalance in the MICROBIOTA and undesired activity of the immune system. Well-known examples include inflammatory bowel diseases (IBD), such as Crohn's disease and ulcerative colitis [18, 19]. Impaired immune functioning is frequently associated with irritable bowel syndrome which is characterized by the frequent occurrence of gut discomfort symptoms such as diarrhea, constipation, or bloating [20–22]. MICROBIOTA and general gut integrity are important in other disease pathologies as well like chronic heart failure [23], HIV [24], autism [25], and allergy [26–29]. For example, there is a potential link between microbial activity and the onset of eczema, which might reflect a suboptimal implementation of gut microbiota at specific developmental stages in infants at high risk for allergy [26].

Highly dysbalanced in vivo disease models are associated with severe clinical symptoms and are frequently used as research models. One of the best studied disease models in this field is represented by IBD. There is overwhelming evidence from animal and human studies that, due to genetic factors in the host, the intestinal MICROBIOTA drives a chronic and often severe inflammation. Specific dysbalances occur in the MICROBIOTA, and by comparing the microbiotic composition in Crohn's disease patients, it was established that a reduction of a usually abundant member of the *Firmicutes*, *Faecalibacterium prausnitzii*, is associated with a higher risk of postoperative recurrence of Crohn's disease in the ileum [30]. In addition, bifidobacterial abundance in patients suffering from gastrointestinal diseases, such as diarrhea, colic, allergy, necrotizing enterocolitis, and obesity, is significantly reduced compared to healthy controls [31]. Besides differences in the MICROBIOTA, cytokine and chemokine levels were strongly elevated in inflamed intestinal regions of IBD patients [32]. In a series of elegant studies in vitro, using Caco-2 cells and peripheral blood mononuclear cells (PBMCs), it was shown that *F. prausnitzii* strongly reduces inflammatory responses. Administration in a murine TNBS-induced colitis model resulted in a strong reduction of inflammation and that *F. prausnitzii* reduces inflammation by blocking NF-κB activation and IL-8 production. These results indicate that counterbalancing this hyper-immune reaction using *F. prausnitzii* or microbes with similar anti-inflammatory activities as a probiotic may be a promising strategy in Crohn's disease treatment.

Autism spectrum disorder (ASD) is a severely neurodevelopmental disorder that impairs a child's ability to communicate and interact with others. Children with such neurodevelopmental disorders are also regularly affected by a dysbiosis of their gut microbiota. Preclinical evidence indicates a bidirectional communication between the gut and central nervous system, called the gut-brain axis [33]. In addition, besides the gut microbes themselves, also their metabolites may be linked to ASD behavior symptoms. When comparing 21 children with ASD and 23 neurotypical children, Kang et al. showed that children with ASD have an altered fecal metabolite profile compared to neurotypical children [34].

Directly studying the intestinal microbes and gut immune responses in infants is difficult. To investigate how different types of early-life gut microbiota affect immune development, Petursdottir et al. collected fecal samples from children with different allergic heredities and inoculated germ-free mice with these fecal samples. They showed that exposure to microbiota associated with an established allergy in later life in children resulted in a T helper 17 signature, both systemically and in the gut mucosa in the mouse offspring, potentially resulting in an increased risk of allergy [35]. Van der Leek et al. summarized how the indoleamine 2,3 dioxygenase (IDO) subset of the kynurenine (KYN) pathway of tryptophan (TRP) metabolism has a relation to allergy and the gut microbiome [36]. This pathway influences T-cell anergy and apoptosis, proliferation of

Treg and Th17 cells, and deviation of the Th1/Th2 response, and the gut microbiota and breast milk are key in determining the functioning of the KYN-IDO pathway. There is a critical window in early life during which microbial dysbiosis can be detected in strong association with a subsequent allergic phenotype predictive of asthma. Arrieta et al. stress not only to focus on the bacterial composition but to also include fungal analyses in all microbial surveys. They showed that a microbial dysbiosis in children from rural Ecuador at 3 months of age was associated with later development of atopic wheeze, similarly as previously seen in Canadian infants [37].

27.2 Modulation of the Immune System

The gut microbiota can be manipulated through passive or active strategies. Passive strategies include diet, lifestyle, and environment, while active strategies comprise antibiotics or pre- and probiotics. Historically, conventional probiotic strategies included a phylogenetically limited diversity of bacteria and some yeast strains. However, biotherapeutic strategies evolved in the last years with the advent of fecal microbiota transplant, successfully applied for treating *Clostridium difficile* infection (CDI) and IBD [38]. Especially in the case of recurrent infections (like seen in CDI), a multifaceted approach is required: an optimized infection control to minimize reinfection, *C. difficile*-targeted antibiotics to minimize microbial dysbiosis, and a gut microbiota restoration to promote colonization resistance [39]. The notion that gut microbes actively modulate the immune system has reinforced interest in PROBIOTICS to bring health benefits. Two major mechanisms of action can be distinguished for immune modulation by PROBIOTICS: direct interaction with the intestinal immune system or modification of the gut MICROBIOTA composition and activity. A number of benefits are achieved through modulation of the immune system, including relief of diarrhea, increased resistance to infections [40, 41], reduction of inflammation and allergies [42, 43], and an improved response to vaccination [44].

There is a huge progress in the field of microbial interactions with the immune system. Probiotics and commensals can modulate intestinal epithelial cell (IEC) function in a variety of ways, including indirect effects on microbial biofilms and direct effects on IECs via enhancement of barrier function by enhancing tight junctions and mucin production, induction of antimicrobial peptides (AMPs) and heat shock protein production, modulation of pro-inflammatory and immunoregulatory cytokines, and interference with pathogenesis. In addition, IECs express the epithelial growth factor receptor (EGFR) which mediates various biological functions, including cellular proliferation, differentiation, and cell survival. EGFR phosphorylates tyrosine and hence activates signaling pathways such as protein kinase C (PKC), PI3K, and MAPK that are involved in the production of tight junction proteins and enhance epithelial barrier integrity. Interestingly, probiotic strains regulate some of these signaling cascades, augmenting tight junction integrity, survival of IECs, and modulation of mucosal barrier immunity through differential activation of MAPK pathways in reparative versus inflammatory responses in epithelial barriers. Some probiotics regulate IEC apoptosis, which is a useful strategy for the prevention of reduced membrane integrity caused by enteric infections and inflammatory disorders [45].

Pathogens are recognized by the innate immune systems pattern recognition receptors (PRR). These receptors bind pathogen-associated molecular patterns (PAMPs), which are common conserved structures shared by most pathogens. PRRs include Toll-like receptors (TLR), NOD-like receptors (NLR), Rig-1-like receptors (RLRs), and C-type lectins (CLRs). TLRs and NLRs provide pathogen surveillance through the activation of NF-κB signaling, leading to the production of pro-inflammatory cytokines, chemokines, and antimicrobial peptides. Interestingly, specific probiotics inhibit pathogen-induced inflammation and thereby contribute to the tolerance of the intestinal barrier to commensal bacteria through the modulation of the TLR-negative regulators in IECs. In addition, probiotic strains can suppress the pro-inflammatory cytokine production from IECs, therefore reducing the detrimental inflammatory damage to the intestinal epithelial barrier caused by pathogen-induced inflammation [45].

Alongside the IECs, multiple other intestinal cells are involved in maintenance of the gut functioning. For example, goblet and Paneth cells secrete mucins and antimicrobial peptides that fortify the barrier against potentially pathogenic microbes. Microfold (M) and goblet cells assist in the transferring of luminal antigens across the epithelial barrier for sampling by mucosal DCs [46]. Members of the commensal MICROBIOTA are continuously sampled by the M cells that lie over Peyer's patches for processing by local dendritic cells and subsequent education of regulatory CD4+ T-cell populations (Fig. 27.1). Tuft cells are important for sensing and responding to protozoa and helminths. Together with intestinal resident immune cells including innate lymphoid cells (ILCs), intraepithelial lymphocytes (IELs), helper T cells, and B cells, a balancing act between barrier protection and microbial tolerance with surveillance and inflammation is maintained [46]. Tolerance results from induction of regulatory T cells that prevent immune responses toward the

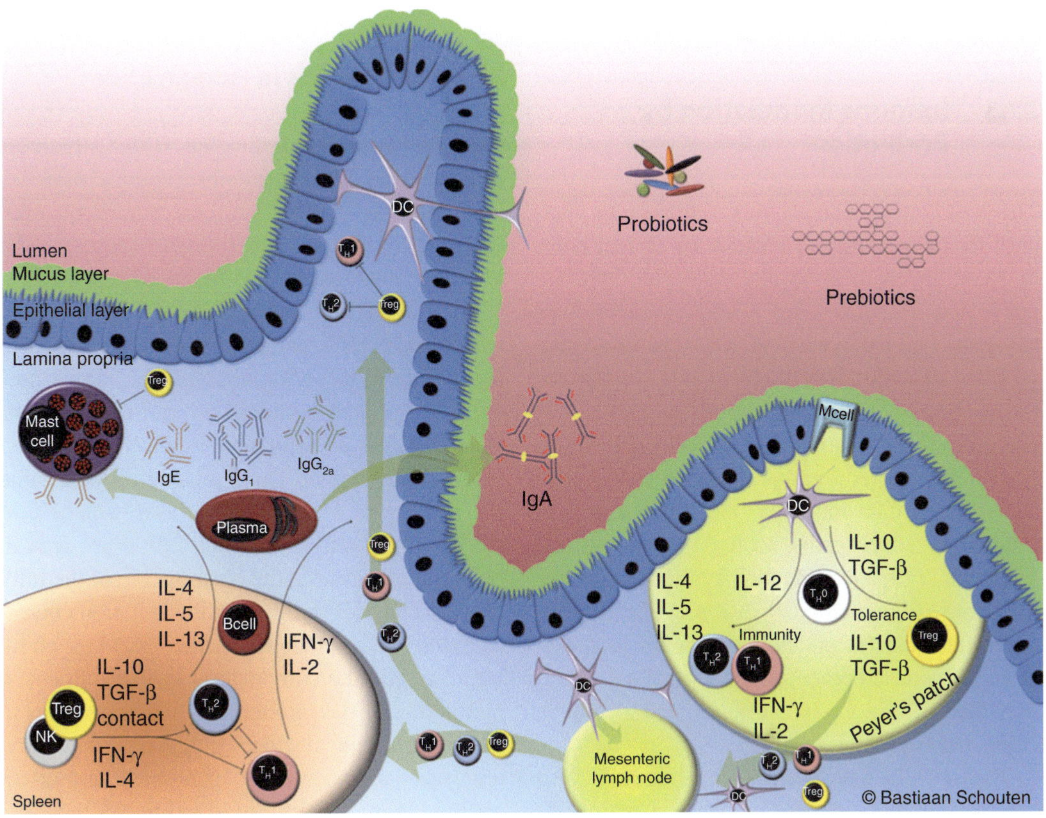

Fig. 27.1 Building tolerance in the intestinal mucosa. Food components and other substances are taken up by M cells, after which the underlying dendritic cells (DCs) in Peyer's patch present these substances to naïve T cells (Th0), which differentiate, dependent on the surrounding cytokine milieu, into Th1 or Th2 effector cells (immunity) or into regulatory T cells (Treg) when tolerance develops. DCs can traffic from Peyer's patch to the mesenteric lymph nodes or sample antigen from the lumen and effector sites and then traffic to the mesenteric lymph nodes where antigens are presented to naïve T cells. Generated Treg, Th1, or Th2 cells enter the blood stream and come back to the intestinal mucosa where they will become resident in the lamina propria. In the mesenteric lymph node, generated Treg, Th1, or Th2 cells can also traffic to the peripheral immune system and transfer tolerance or immunity. B cells expand and mature in the spleen and traffic back to the effector site where they produce specific IgE, IgG1, IgG2a, or IgA. These Igs can be transported into the lumen (IgA) or bind to mast cells (IgE, IgG1, and IgG2a) and wait for another substance exposure. This immunological cascade can be modulated by food components like prebiotics, probiotics, and the combination of these two (synbiotics). Adapted from Schouten et al. [99]

tolerizing antigen through secretion of inhibitory cytokines [e.g., interleukin (IL)-10 and transforming growth factor-β (TGF-β)]. This allows the systemic adaptive immune system to remain ignorant of the ongoing interactions with the normal MICROBIOTA and prevents autoimmune responses. Selected PROBIOTICS exhibit anti-inflammatory capacities both in vitro and in vivo and can prime dendritic cells to confer a protective anti-inflammatory effect [47]. Another potent mechanism by which the mucosal adaptive immune system can mediate inflammatory and immune tolerance toward the MICROBIOTA is humoral immunity via secretory immunoglobulin IgA [48].

27.3 Immune Modulation by Prebiotics

The WHO refers to breast milk as the golden standard for baby nutrition. From a nutritional point of view, breast milk compounds are therefore interesting targets for immunomodulatory research. Human breast milk has anti-infective properties as it reduces the incidence of gastrointestinal and non-enteric infections in infants [17] and is due to its antimicrobial activity against several viruses, bacteria, and protozoa. Breastfeeding can reduce infant mortality and provide protection, for instance, against neonatal meningitis and septicemia [49] and protection against respiratory infections, immunological diseases like insulin-dependent diabetes, and tumors in infancy, as well as reduced development of inflammatory conditions like allergy, Crohn's disease, and ulcerative colitis has been reported.

Breast milk has, besides anti-infective, also anti-inflammatory properties. This is mainly of importance during the bacterial colonization of the newborn's mucosal surfaces, including the skin and gut [50]. A huge number of microbial components are brought in acute contact with the neonate, and coordination of the inflammatory response, after this first contact, is of vital importance. The epithelial layer, together with the intraepithelial and lamina proprial immune-competent cells, is crucial in regulating the recognition of microorganisms and maintenance of gut homeostasis. Taken together, this emphasizes the diverse activity of human breast milk compounds [17]. Breast milk also contains several immunomodulatory compounds including immunoglobulin (Ig)G, IgM and isoforms of immunoglobulins (sIgA), nucleotides, specific amino acids (taurine, polyamines), polyunsaturated fatty acids (PUFAs, eicosapentaenoic acid, docosahexaenoic acid), monoglycerides, lauric acid, linoleic acid, cytokines and chemokines, soluble receptors (CD14, TLR2), antibacterial proteins/peptides (lactoferrin, lysozyme, β-lactoglobulin, casein), carbohydrates, and intact immune cells.

27.3.1 Prebiotics in Human Milk

Over 200 different individual HMO structures ranging from 3 to 14 monosaccharide moieties have been detected by mass spectrometry so far. Of these, >130 have determined structures. The ten most abundant HMOs make up around 75% of the total HMO mass. It is also important to realize that each woman synthesizes her own unique oligosaccharide composition, as its production is related to various physiological parameters like the secretor and Lewis blood group status of the mother. These parameters determine the activity of fucosyl-transferases which form the basis of the four different milk types seen in mothers. In addition, the composition of the milk oligosaccharides varies over the lactation course, adapting to the needs of the growing child. Interestingly, the composition is also influenced by the diet of the mother [51]. Carbohydrates in breast milk, like lactose, glycoconjugates, and oligosaccharides, mainly function as important sources for energy production. Since the nondigestible oligosaccharides present in human breast milk showed a clear bifidogenic effect on the gut MICROBIOTA, these oligosaccharides can be considered as prebiotic. Besides the indirect effects of human oligosaccharides on the immune system through their beneficial influence on the gut MICROBIOTA, the direct effects of these oligosaccharides on immune cells are extensively investigated. In human milk, the proportion of prebiotic carbohydrates is substantial, whereas the prebiotic oligosaccharides in cow's milk are present only in trace amounts.

Thongaram et al. evaluated the ability of 12 lactobacilli and 12 bifidobacteria strains to ferment HMO and their constituent monomers. Each probiotic strain can ferment one or more HMO structures [52]. Bifidobacteria, which dominate during early life, are among the best described gut bacteria with the ability to utilize HMOs. Several species possess glycosyl hydrolases that cleave specific linkages within the HMO molecules. HMOs are preferentially fermented by *B. bifidum* and *B. longum* species which, together with *B. breve*, are the most abundant in breast-fed infant gut microbiota. Their ability to utilize these otherwise nondigestible carbohydrates explains their abundance in breast-fed neonates [15]. Besides the prebiotic effect of the HMOs, these carbohydrate structures directly influence the host. In children below the age of 5, diarrheal disease is the second leading cause of death worldwide. Diarrhea can be initiated by infections with a.o. rotavirus, *Campylobacter*, and calicivirus. The severe dehydrating gastroenteritis that is caused by diarrhea is annually responsible for more than 215,000 deaths. A population study on mother-infant pairs showed that higher concentrations of 2′-linked fucosyl oligosaccharides (2′FL) in human milk were associated with protection against diarrhea caused by *Campylobacter* and calicivirus. Clinical trials with 2′FL-containing milk formula showed growth and 2′FL uptake like the observations in breast-fed infants. This implies that the reduction of (rota)virus infectivity by milk oligosaccharides provides an added incentive for the addition of specific oligosaccharides to infant formula and could confer additional benefits to formula-fed infants [53]. Limited knowledge is available on the involvement of the intestinal immune system, including the altered microbial composition, leaky gut, and altered mucosal immune response, in the etiology of type 1 diabetes and the potential benefits of HMOS and their subsequent short-chain fatty acid profiles [54].

Among the bifidobacterial communities that reside in the gut of infants or adults, certain bifidobacterial species, such as *B. breve*, possess assimilation capabilities for a wide range of carbohydrates, which include both dietary and host-derived glycans [31]. This is probably why these bifidobacteria can ferment oligosaccharide structures that can mimic the prebiotic effect of breast milk, including galacto- (GOS) and fructo-oligosaccharides (FOS).

27.3.1.1 Galacto-Oligosaccharides (GOS)

Several types of GOS with different chemical characteristics are used in research with a focus on immunomodulatory effects. GOS produced by glycosylation of lactose, using β-galactosidase enzymes, is mostly referred to as trans-GOS, β-linked GOS, or TOS in the literature. GOS produced by elongation of galactose by α-galactosidases is often referred to as α-GOS [55]. The α-galactosyl derivatives of sucrose occur widely in nature; the trisaccharide raffinose and the tetrasaccharide stachyose are present in soybeans and many other plants. Examples of the chemical structure of α- and β-galacto-oligosaccharides are depicted in Fig. 27.2. Recently, Akbari et al. compared the direct, microbiota-independent effects of defined oligosaccharides and their related DPs. Monolayers of the human intestinal epithelial cell line, Caco-2, served as a model system for intestinal barrier function, while the fungal toxin deoxynivalenol was used as a model compound to impair the intestinal integrity. The protective effects of GOS in this Caco-2 cell model were compared to the effects of FOS, inulin, and the individual DP fractions of GOS (ranging from DP2 to DP6). As the different DPs all affected different assessed parameters (TEER, paracellular flux of LY, or CXCL8), it remains to be elucidated what the exact structure-function relation is for oligosaccharides and their interaction with the IECs [56].

GOS has not been studied widely with respect to immunomodulating effects as a single dietary agent. Although anti-allergic and anti-inflammatory effects were described for α-GOS [57, 58] and raffinose [58, 59], more work is needed to clarify the effects of GOS and compare these with similar effects of other oligosaccharides and NDC.

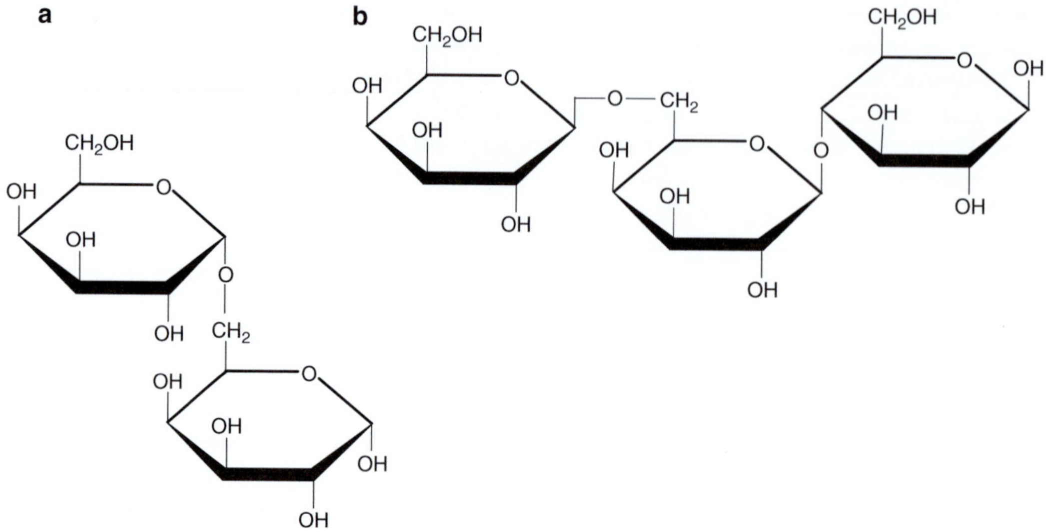

Fig. 27.2 Examples of the chemical structure of α- and β-galacto-oligosaccharides (GOS). Enzymatic elongation of galactose or lactose is used to produce α- or β-GOS, respectively. A disaccharide with α-(1,6)-linkage (**a**) is an abundantly produced component in the production of α-GOS, but various other linkages and trisaccharides and tetrasaccharides occur as well [55]. β-GOS may contain a variety of chain lengths (majority is dp <6) and linkages. An example is shown with a β-(1,6)-linkage (**b**). Adapted from Vos et al. [100]

27.3.1.2 Fructo-Oligosaccharides (FOS)

Definitions of fructans, such as various types of FOS and inulin, vary widely in literature. Most experiments were performed with unprocessed chicory inulin or fructans derived from chicory inulin. The basic structures of carbohydrate chains in fructans are depicted in Fig. 27.3. Unprocessed chicory inulin is mainly composed of fructans with a degree of polymerization (dp) ranging from 2 to 60, ending with a terminal glucose monomer. Partially hydrolyzed inulin (scFOS) has a typical dp range of 2–8, and there are more molecules that end without a terminal glucose monomer compared to inulin. Physical removal of short-chain fructans from chicory inulin (lcFOS) [60, 61] leads to a mixture of fructans with terminal glucose monomers and an approximate average dp of 22. FOS can also be produced by enzymatic elongation of sucrose (eeFOS) resulting in very short fructan molecules of dp 2–4 [62]. Besides the soluble forms, a specific insoluble form of inulin (γ-inulin: a crystallized form of dahlia tuber-derived inulin of dp >50) was shown already decades ago to activate the alternative complement pathway and exhibit adjuvant activity when injected intraperitoneally [63–65].

Fructan supplementation in animal models results in anti-inflammatory effects in IBD models [66], increased survival in infection models [67], anticarcinogenic effects, enhanced mucosal antibody responses, and modulation of mucosal cytokine patterns and cell populations [68]. Similar results were obtained in IBD patients, but other effects are not as clear. The data on infectious diarrhea and immunomodulating effects in infants and young children are interesting, but more data from well-controlled studies are needed to corroborate these findings.

27.3.2 Examples of Immunomodulation by Prebiotics in Infants

The metabolic activity of the intestinal MICROBIOTA observed in breast-fed children is characterized by relatively high acetate and lactate levels and a slightly acidic pH. These physiological conditions restrict the growth of potential

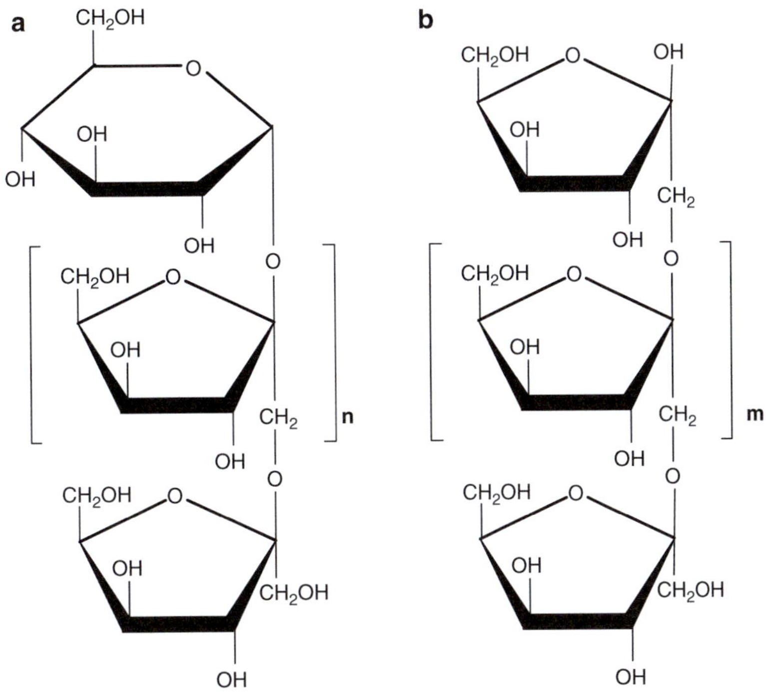

Fig. 27.3 The basic structures of carbohydrate chains in fructans. Most fructose chains in fructan preparations end with a glucose residue at the reducing end (**a**). Chains that do not end with a glucose residue (**b**) are found in small amounts in unprocessed chicory inulin and in larger amounts in short-chain fructo-oligosaccharides (scFOS) derived from enzymatic digestion of inulin. Unprocessed chicory inulin mainly contains chains of $n = 1–59$. scFOS preparations are mainly composed of very small chains (n or $m < 6$). Adapted from Vos et al. [100]

pathogens like *Escherichia coli* and *Clostridium perfringens*. Addition of short-chain GOS/long-chain FOS to infant formula stimulated the acetate and lactate production by the MICROBIOTA and leads to a slightly acidic pH [69]. The scGOS/lcFOS formula increased the secretion of soluble IgA, which is a key factor in GI defense against dietary and microbial antigens [70]. Healthy infants receiving a formula containing scGOS and lcFOS stimulated the growth of bifidobacteria and changed the metabolic activity of the total intestinal MICROBIOTA, resulting in a fermentation profile that is similar to the profile found in breast-fed infants [71]. This increase in bifidobacteria is not only observed in children that start the prebiotic ingestion at 4–6 weeks but also if infants begin the ingestion of scGOS/lcFOS when they are about to start consuming solid foods [72]. Besides an effect of scGOS/lcFOS in healthy infants, the potential of the prebiotic mixture was also investigated in infants that have a high risk for allergy. scGOS/lcFOS supplementation reduced the cumulative incidence of atopic dermatitis in the high-risk infants [59], and in the follow-up to this study, scGOS/lcFOS reduced the total immunoglobulin response, modulating the immune response toward cow's milk proteins, whereas the response to vaccination with DTP (diphtheria, tetanus, and polio) remained unaltered [73]. Moreover, mixtures of scGOS with lcFOS reduced the incidence of infections and atopic dermatitis [74–76].

27.4 Immune Modulation by Probiotics

There are multiple ways to study the interactions of gut microbes with the immune system. In an in vitro study, isolated human cells were co-cultured with microbial cells, and subsequently the immune response can be characterized by analyzing cytokine production profiles (see Chap. 6). Microbial cells are powerful stimulants, as immune cells express specific receptors, including the TLRs that are dedicated to recognize microbial molecules and translate these signals into immune responses. The response is highly specific to the species, and even strain, that is tested. Such in vitro methods provide valuable screening tools and are broadly applied for mechanistic studies, but the translation of findings to effects and clinical observations in vivo is often challenging. The number of studies describing the effect of gut microbes and PROBIOTICS on the mammalian and human immune system using animal models and human subjects is growing. Recent technological advances allow for the first time to integrate molecular analysis in such studies providing further evidence for the mechanisms of cross-talk between microbes and host [77].

As mentioned before, epithelial cells can dampen TLR signaling or expression. In addition, IECs are crucial for maintaining intestinal homeostasis [78], and failure to control inflammatory processes at the epithelial cell level may critically contribute to the disease pathogenesis. IECs react to bacterial as well as immune-derived pro-inflammatory signals by secreting cytokines and chemokines, such as IL-6 and interferon c-induced protein 10 (IP-10, CXCL10), to attract and activate Th1 cells and phagocytic cells for defense at the site of infection. IP-10 is key in uncontrolled disease development, since the blockade of IP-10 by an anti-IP-10 antibody was sufficient to decrease disease severity in IL-10 gene knockout (IL-$10^{-/-}$) mice with an IBD-like condition. This effect was due to reduced Th1 cell generation in lymph nodes and reduced recruitment of Th1 effector cells to the colon [79]. *Lactobacillus casei* impaired IP-10 secretion by a mechanism that did not involve impairment of initial IP-10 production, but a decrease in intracellular IP-10 protein stability. *L. casei* impairs vesicular pathways important for the secretion of IP-10, followed by subsequent degradation of the pro-inflammatory chemokine [80].

Macrophages are another source of various cytokines and are well populated in the normal intestinal lamina propria [81]. Direct interaction between commensal bacteria and the large number of infiltrating peripheral monocytes at the onset of or during active intestinal inflammation is involved in the pathogenesis of inflammatory diseases [82, 83]. Therefore, cytokines released by macrophages in response to PROBIOTICS are particularly crucial for understanding the mechanism(s) of their immunomodulatory effects on the host. Two *L. rhamnosus* strains, GG and GR-1, potently induce production of granulocyte-colony-stimulating factor (G-CSF), which is a crucial mediator for suppressing tumor necrosis factor (TNF) production through an activating signal transducer and activator of transcription (STAT) 3. Subsequently activation of c-Jun-N-terminal kinases (JNKs) in macrophages is inhibited. These results define G-CSF as a key mediator through which PROBIOTICS, particularly *L. rhamnosus*, elicit immunomodulatory effects on the host [84]. *H. hepaticus* produces an immunomodulatory polysaccharide that conditions the macrophage response via a TLR2/CREB-driven anti-inflammatory pathway. The identification of immunomodulatory ligands produced by commensals will require an integrated approach combining immunology, microbiology, bioinformatics, and biochemistry but could ultimately provide new preventive and therapeutic strategies for infectious and inflammatory diseases [85].

Although the role of PROBIOTICS is established in some diseases, evidence of probiotic immune modulation is more difficult to obtain in the healthy population, whereas this is crucial when targeting immune benefits. In this context, the impact of daily consumption of a probiotic dairy drink containing the probiotic strain *L. casei* DN-114 001 was studied on the immune response to influenza vaccination in an elderly population in whom immune responses are known to be weaker compared to younger adults. Vaccination occurred after 4 weeks of product consumption and influenza-specific IgG1 titers for the three viral strains composing the vaccine (H1N1, H3N2, and B) increased after vaccination, being consistently higher in the probiotic product group compared to the control group. Similarly, antibody titers against the B strain increased significantly more in the probiotic group than in the control group at 3, 6, and 9 weeks postvaccination under product consumption. Therefore, consumption of these PROBIOTICS increased antibody responses to influenza vaccination which may provide a health benefit in this population. Recently, Belkacem et al. have investigated the mechanisms of protection conferred by *L. paracasei* CNCM I-1518 strain in a mouse model of influenza infection. Oral consumption of *L. paracasei*-modulated lung immunity was associated with an improved control of the influenza infection. These results further extend the beneficial role for certain lactobacilli to alleviate the burden of respiratory tract infections [86]. Oral administration is effective in treating lactational infectious mastitis, but *Lactobacillus salivarius* PS2 is also able to prevent mastitis when orally administered during late pregnancy to women who had experienced infectious mastitis after previous pregnancies [87]. Besides the direct benefits of probiotics by mothers, it is also possible to reach a more favorable health outcome for the offspring. In a randomized, controlled study (ProPACT trial) with a fermented milk supplemented with *Lactobacillus rhamnosus* GG, *Lactobacillus acidophilus* La-5, and *Bifidobacterium animalis* ssp. *lactis* Bb-12, a reduced incidence of atopic dermatitis in the offspring is observed. This was accompanied by a reduced proportion of Th22 cells in the peripheral blood of their 3-month-old children which may partially explain the preventive effect of probiotics on AD [88]. To exclude that the offspring was exposed to the supplemented probiotics themselves, the presence of the *Lactobacillus rhamnosus* GG, *Lactobacillus acidophilus* La-5, and *Bifidobacterium animalis* ssp. *lactis* Bb-12 in the human milk samples was assessed. The bacteria were only present in breast milk samples for a small subgroup of women. Thus, breast milk was not a meaningful source of LGG, La-5, or Bb-12 for the infants in the ProPACT trial, and the observed preventative effect of the ProPACT regimen on atopic dermatitis cannot be attributed to the transfer of these bacteria through breastfeeding [89].

Further investigation of the mode of action is now possible because of the breakthrough developments in the field of genomics, providing a sensitive tool to assess the global responses of molecular cross-talk between microbes and the host. This potential was exploited in a double-blind randomized control study with healthy adults who ingested preparations of living and heat-killed *L. plantarum* bacteria activating specific mucosal immune responses [90]. Biopsies from the intestinal duodenal mucosa were taken, and altered gene expression profiles were analyzed using whole-genome microarrays and by biological pathway reconstructions. Expression profiles of human mucosa displayed striking differences in the modulation of NF-κB-dependent pathways, notably after consumption of living *L. plantarum* bacteria in different growth phases. This is most likely due to changes in the composition of the cell wall, or physiological state of the bacterium. These results provide a unique and high-resolution view of the molecular interactions between microbe and host and open new avenues toward understanding the mode of action of health ingredients and PROBIOTICS.

27.5 Conclusions and Prospects for Probiotics and Prebiotics

The human gut microbiota is the collection of bacteria, archaea, viruses, and fungi. The focus in this review has been on the bacteria, although more and more attention is given to the virome and the fungome. There are currently two major approaches to study the human gut microbiota: studying phylogenetic markers using 16S rRNA gene amplification by pyrosequencing (called 16S barcode profiling) or studying all DNA using metagenomics shotgun sequencing. The 16S rRNA gene is mainly used to answer the question "who's there?" With shotgun metagenomics, DNA is totally sequenced, and then reads are assembled and annotated to access the functional repertoire. Metagenomics finally answer these two questions: who's there and what are they doing? The next step is microbiomics: making the association of "omics" technique like 16S rRNA analysis, metagenomics, metatranscriptomics, and metaproteomics [91]. When applying these technologies in double-blind placebo-controlled trials with continuously increasing knowledge about the selection criteria of pre- and probiotics and synbiotics, the future promises new products that will provide a more resilient GI MICROBIOTA as well as a more resilient immune system. Sound preclinical testing, including a full toxicological profile required for a novel food status, should precede the use of new ingredients when considering their use in any disease or infant setting. The regulatory constraints that are becoming stricter, not only in Europe but all over the world, demand a translational research approach in which clinical results are backed up by mechanistic studies in preclinical models. This will be crucial to build and sustain the credibility needed to fully harvest the potential of PROBIOTICS and PREBIOTICS in the years to come.

As indicated above, the maternal diet is influencing the composition of her milk. The complexity and variability in human milk composition and the known infant's response to many of the human milk constituents may also explain some of the conflicting results of studies evaluating the effect of prolonged exclusive breastfeeding and the prevention of allergic disease development [92]. Similar findings are observed in the adult population: life style factors (diet, medication use, stress, etc.) modulate the microbiota and the resilience to immune triggers [93]. Future research needs to account for different environmental exposures and use systematic methodologies to characterize variations in the dietary composition in relation to well-defined clinical and immune outcomes. Statistical approaches using cluster analysis should be implemented more frequently to better start understanding the relationship between diet composition and the development of non-communicable diseases. This may allow us to establish a new paradigm in disease prevention not only through direct intervention in the offspring but even through modulation during pregnancy and/or lactation to promote a healthy infant immune development.

Even though the most well-known function of the gut MICROBIOTA is to support gut health by metabolizing dietary nutrients like oligosaccharides [3], commensals additionally actively participate in the metabolism of xenobiotics [94]. Variations in drug metabolism can be attributed to intersubject MICROBIOTA diversity as shown by the different profiles of liver and kidney metabolites in rats [95]. The concept of the gut MICROBIOTA being a potentially new pharmaceutical target stems from the potential of commensals to differentially metabolize nutrients and drugs and thus influence metabolite outcomes [96]. As indicated before, more and more emphasis is given to the relation between systems. Although here the main relation between the gut microbiome and the immune system is depicted, there is a growing body of evidence on the bidirectional interactions of the gut-brain axis. Metabolites of the gut-brain axis are potential targets for treatment and drug design since the interaction or biochemical interplay results in net metabolite production or end products with either positive or negative effects on human health [97, 98].

Selected Readings and Important Websites

Gibson GR, et al. Expert consensus document: The International Scientific Association for Probiotics and Prebiotics (ISAPP) consensus statement on the definition and scope of prebiotics. Nat Rev Gastroenterol Hepatol. 2017;14(8):491–502.

Llewellyn, A. and A. Foey, Probiotic modulation of innate cell pathogen sensing and signaling events. Nutrients. 2017;9(10). https://doi.org/10.3390/nu9101156

Martin R, et al. Early-life events, including mode of delivery and type of feeding, siblings and gender, shape the developing gut microbiota. PLoS One. 2016;11(6):e0158498.

Turroni F, et al. Glycan utilization and cross-feeding activities by bifidobacteria. Trends Microbiol. 2018;26(4):339–50.

Van der Leek AP, Yanishevsky Y, Kozyrskyj AL. The kynurenine pathway as a novel link between allergy and the gut microbiome. Front Immunol. 2017;8:1374.

References

1. Bindels LB, et al. Towards a more comprehensive concept for prebiotics. Nat Rev Gastroenterol Hepatol. 2015;12(5):303–10.
2. Gibson GR, et al. Dietary modulation of the human colonic microbiota: updating the concept of prebiotics. Nutr Res Rev. 2004;17(2):259–75.
3. Gibson GR, Roberfroid MB. Dietary modulation of the human colonic microbiota: introducing the concept of prebiotics. J Nutr. 1995;125(6):1401–12.

4. Hutkins RW, et al. Prebiotics: why definitions matter. Curr Opin Biotechnol. 2016;37:1–7.
5. Louis P, Flint HJ, Michel C. How to manipulate the microbiota: prebiotics. Adv Exp Med Biol. 2016;902:119–42.
6. Pineiro M, et al. FAO Technical meeting on prebiotics. J Clin Gastroenterol. 2008;42(Suppl 3 Pt 2):S156–9.
7. Schrezenmeir J, de Vrese M. Probiotics, prebiotics, and synbiotics--approaching a definition. Am J Clin Nutr. 2001;73(2 Suppl):361S–4S.
8. Verspreet J, et al. A critical look at prebiotics within the dietary fiber concept. Annu Rev Food Sci Technol. 2016;7:167–90.
9. Gibson GR, et al. Expert consensus document: the International Scientific Association for Probiotics and Prebiotics (ISAPP) consensus statement on the definition and scope of prebiotics. Nat Rev Gastroenterol Hepatol. 2017;14(8):491–502.
10. Joint FAO/WHO Expert Consultation. Health and nutritional properties of probiotics in food including powder milk with live lactic acid bacteria. 2001. Available from: http://www.who.int/foodsafety/publications/fs_management/en/probiotics.pdf.
11. Bouskra D, et al. Lymphoid tissue genesis induced by commensals through NOD1 regulates intestinal homeostasis. Nature. 2008;456(7221):507–10.
12. Martin R, et al. Early-life events, including mode of delivery and type of feeding, siblings and gender, shape the developing gut microbiota. PLoS One. 2016;11(6):e0158498.
13. Knoop, K.A., et al., Microbial antigen encounter during a preweaning interval is critical for tolerance to gut bacteria. Sci Immunol. 2017;2(18). https://doi.org/10.1126/sciimmunol.aao1314.
14. Garcia-Pena C, et al. Microbiota and aging. A review and commentary. Arch Med Res. 2017;48(8):681–9.
15. Ruiz L, et al. Bifidobacteria and their molecular communication with the immune system. Front Microbiol. 2017;8:2345.
16. Moro G, et al. Dosage-related bifidogenic effects of galacto- and fructooligosaccharides in formula-fed term infants. J Pediatr Gastroenterol Nutr. 2002;34(3):291–5.
17. van't Land, B., G. Boehm, and J. Garssen, Breast milk: components with immune modulating potential and their possible role in immune mediated disease resistance, in Dietary components and immune function, R.R. Watson, S. Zibadi, and V.R. Preedy, Editors. 2010. p. 685. Humana Press, Totowa, NJ
18. Sartor RB. Microbial influences in inflammatory bowel diseases. Gastroenterology. 2008;134(2):577–94.
19. Strober W, Fuss I, Mannon P. The fundamental basis of inflammatory bowel disease. J Clin Invest. 2007;117(3):514–21.
20. Clavel T, Haller D. Molecular interactions between bacteria, the epithelium, and the mucosal immune system in the intestinal tract: implications for chronic inflammation. Curr Issues Intest Microbiol. 2007;8(2):25–43.
21. Kassinen A, et al. The fecal microbiota of irritable bowel syndrome patients differs significantly from that of healthy subjects. Gastroenterology. 2007;133(1):24–33.
22. Ohman L, et al. B-cell activation in patients with irritable bowel syndrome (IBS). Neurogastroenterol Motil. 2009;21(6):644–50, e27.
23. Sandek A, Anker SD, von Haehling S. The gut and intestinal bacteria in chronic heart failure. Curr Drug Metab. 2009;10(1):22–8.
24. Hofer U, Speck RF. Disturbance of the gut-associated lymphoid tissue is associated with disease progression in chronic HIV infection. Semin Immunopathol. 2009;31(2):257–66.
25. White JF. Intestinal pathophysiology in autism. Exp Biol Med (Maywood). 2003;228(6):639–49.
26. Wopereis H, et al. Intestinal microbiota in infants at high risk for allergy: effects of prebiotics and role in eczema development. J Allergy Clin Immunol. 2018;141(4):1334–42.e5.
27. Hougee S, et al. Oral treatment with probiotics reduces allergic symptoms in ovalbumin-sensitized mice: a bacterial strain comparative study. Int Arch Allergy Immunol. 2010;151(2):107–17.
28. Scholtens PA, et al. Fecal secretory immunoglobulin A is increased in healthy infants who receive a formula with short-chain galacto-oligosaccharides and long-chain fructo-oligosaccharides. J Nutr. 2008;138(6):1141–7.
29. Schouten B, et al. Cow milk allergy symptoms are reduced in mice fed dietary synbiotics during oral sensitization with whey. J Nutr. 2009;139(7):1398–403.
30. Sokol H, et al. Faecalibacterium prausnitzii is an anti-inflammatory commensal bacterium identified by gut microbiota analysis of Crohn disease patients. Proc Natl Acad Sci U S A. 2008;105(43):16731–6.
31. Turroni F, et al. Glycan utilization and cross-feeding activities by bifidobacteria. Trends Microbiol. 2018;26(4):339–50.
32. Banks E, et al. Is psychological distress in people living with cancer related to the fact of diagnosis, current treatment or level of disability? Findings from a large Australian study. Med J Aust. 2010;193(5 Suppl):S62–7.
33. Yang Y, Tian J, Yang B. Targeting gut microbiome: a novel and potential therapy for autism. Life Sci. 2018;194:111–9.
34. Kang DW, et al. Differences in fecal microbial metabolites and microbiota of children with autism spectrum disorders. Anaerobe. 2018;49:121–31.
35. Petursdottir DH, et al. Early-life human microbiota associated with childhood allergy promotes the T helper 17 axis in mice. Front Immunol. 2017;8:1699.
36. Van der Leek AP, Yanishevsky Y, Kozyrskyj AL. The kynurenine pathway as a novel link between allergy and the gut microbiome. Front Immunol. 2017;8:1374.

37. Arrieta MC, et al. Associations between infant fungal and bacterial dysbiosis and childhood atopic wheeze in a nonindustrialized setting. J Allergy Clin Immunol. 2018;142(2):424–434.e10.
38. El Hage R, Hernandez-Sanabria E, Van de Wiele T. Emerging trends in "smart probiotics": functional consideration for the development of novel health and industrial applications. Front Microbiol. 2017;8:1889.
39. Chilton CH, Pickering DS, Freeman J. Microbiological factors affecting *Clostridium difficile* recurrence. Clin Microbiol Infect. 2018;24(5):476–82.
40. Corr SC, et al. Bacteriocin production as a mechanism for the antiinfective activity of Lactobacillus salivarius UCC118. Proc Natl Acad Sci U S A. 2007;104(18):7617–21.
41. Ryan KA, et al. Strain-specific inhibition of Helicobacter pylori by Lactobacillus salivarius and other lactobacilli. J Antimicrob Chemother. 2008;61(4):831–4.
42. Mazmanian SK, Round JL, Kasper DL. A microbial symbiosis factor prevents intestinal inflammatory disease. Nature. 2008;453(7195):620–5.
43. Lee J, Seto D, Bielory L. Meta-analysis of clinical trials of probiotics for prevention and treatment of pediatric atopic dermatitis. J Allergy Clin Immunol. 2008;121(1):116–121.e11.
44. Boge T, et al. A probiotic fermented dairy drink improves antibody response to influenza vaccination in the elderly in two randomised controlled trials. Vaccine. 2009;27(41):5677–84.
45. Llewellyn, A. and A. Foey, Probiotic modulation of innate cell pathogen sensing and signaling events. Nutrients. 2017;9(10). https://doi.org/10.3390/nu9101156.
46. Buck MD, et al. Metabolic instruction of immunity. Cell. 2017;169(4):570–86.
47. Foligne B, et al. A key role of dendritic cells in probiotic functionality. PLoS One. 2007;2(3):e313.
48. Macpherson AJ, Geuking MB, McCoy KD. Immune responses that adapt the intestinal mucosa to commensal intestinal bacteria. Immunology. 2005;115(2):153–62.
49. Jeurink PV, et al. Human milk: a source of more life than we imagine. Benef Microbes. 2013;4(1):17–30.
50. Carvalho R II, et al. Breastfeeding increases microbial community resilience. J Pediatr (Rio J). 2018;94(3):258–67.
51. Grabarics M, et al. Analytical characterization of human milk oligosaccharides - potential applications in pharmaceutical analysis. J Pharm Biomed Anal. 2017;146:168–78.
52. Thongaram T, et al. Human milk oligosaccharide consumption by probiotic and human-associated bifidobacteria and lactobacilli. J Dairy Sci. 2017;100(10):7825–33.
53. Laucirica DR, et al. Milk oligosaccharides inhibit human rotavirus infectivity in MA104 cells. J Nutr. 2017;147(9):1709–14.
54. Xiao L, et al. Early-life nutritional factors and mucosal immunity in the development of autoimmune diabetes. Front Immunol. 2017;8:1219.
55. Yamashita A, et al. Reverse reaction of Aspergillus niger APC-9319 alpha-galactosidase in a supersaturated substrate solution: production of alpha-linked galactooligosaccharide (alpha-GOS). Biosci Biotechnol Biochem. 2005;69(7):1381–8.
56. Akbari P, et al. Characterizing microbiota-independent effects of oligosaccharides on intestinal epithelial cells: insight into the role of structure and size: structure-activity relationships of non-digestible oligosaccharides. Eur J Nutr. 2017;56(5):1919–30.
57. Abe C, et al. Effects of alpha-linked galactooligosaccharide on adjuvant-induced arthritis in Wistar rats and type II collagen-induced arthritis in DBA/1J mice. Int J Tissue React. 2004;26(3–4):65–73.
58. Sonoyama K, et al. Allergic airway eosinophilia is suppressed in ovalbumin-sensitized Brown Norway rats fed raffinose and alpha-linked galactooligosaccharide. J Nutr. 2005;135(3):538–43.
59. Watanabe H, et al. Reduction of allergic airway eosinophilia by dietary raffinose in Brown Norway rats. Br J Nutr. 2004;92(2):247–55.
60. Coussement PA. Inulin and oligofructose: safe intakes and legal status. J Nutr. 1999;129(7 Suppl):1412S–7S.
61. van Loo J, et al. On the presence of inulin and oligofructose as natural ingredients in the western diet. Crit Rev Food Sci Nutr. 1995;35(6):525–52.
62. Mitsuoka T, Hidaka H, Eida T. Effect of fructo-oligosaccharides on intestinal microflora. Nahrung. 1987;31(5–6):427–36.
63. Cooper PD. Solid phase activators of the alternative pathway of complement and their use in vivo. In: Sim RB, editor. Activators and inhibitors of complement. Dordrecht: Kluwer Academic; 1993. p. 99–106.
64. Cooper PD, Carter M. Anti-complementary action of polymorphic "solubility forms" of particulate inulin. Mol Immunol. 1986;23(8):895–901.
65. Silva DG, Cooper PD, Petrovsky N. Inulin-derived adjuvants efficiently promote both Th1 and Th2 immune responses. Immunol Cell Biol. 2004;82(6):611–6.
66. Lara-Villoslada F, et al. Short-chain fructooligosaccharides, in spite of being fermented in the upper part of the large intestine, have anti-inflammatory activity in the TNBS model of colitis. Eur J Nutr. 2006;45(7):418–25.
67. Buddington KK, Donahoo JB, Buddington RK. Dietary oligofructose and inulin protect mice from enteric and systemic pathogens and tumor inducers. J Nutr. 2002;132(3):472–7.
68. Hosono A, et al. Dietary fructooligosaccharides induce immunoregulation of intestinal IgA secretion by murine Peyer's patch cells. Biosci Biotechnol Biochem. 2003;67(4):758–64.
69. Bakker-Zierikzee AM, et al. Effects of infant formula containing a mixture of galacto- and fructo-oligosaccharides or viable Bifidobacterium animalis

on the intestinal microflora during the first 4 months of life. Br J Nutr. 2005;94(5):783–90.
70. Bakker-Zierikzee AM, et al. Faecal SIgA secretion in infants fed on pre- or probiotic infant formula. Pediatr Allergy Immunol. 2006;17(2):134–40.
71. Knol J, et al. Colon microflora in infants fed formula with galacto- and fructo-oligosaccharides: more like breast-fed infants. J Pediatr Gastroenterol Nutr. 2005;40(1):36–42.
72. Scholtens PA, et al. Bifidogenic effects of solid weaning foods with added prebiotic oligosaccharides: a randomised controlled clinical trial. J Pediatr Gastroenterol Nutr. 2006;42(5):553–9.
73. van Hoffen E, et al. A specific mixture of short-chain galacto-oligosaccharides and long-chain fructo-oligosaccharides induces a beneficial immunoglobulin profile in infants at high risk for allergy. Allergy. 2009;64(3):484–7.
74. Boehm G, et al. Prebiotics and immune responses. J Pediatr Gastroenterol Nutr. 2004;39(Suppl 3):S772–3.
75. Coppa GV, et al. Prebiotics in human milk: a review. Dig Liver Dis. 2006;38(Suppl 2):S291–4.
76. Morrow AL, Rangel JM. Human milk protection against infectious diarrhea: implications for prevention and clinical care. Semin Pediatr Infect Dis. 2004;15(4):221–8.
77. Costea PI, et al. Enterotypes in the landscape of gut microbial community composition. Nat Microbiol. 2018;3(1):8–16.
78. Artis D. Epithelial-cell recognition of commensal bacteria and maintenance of immune homeostasis in the gut. Nat Rev Immunol. 2008;8(6):411–20.
79. Hyun JG, et al. Anti-interferon-inducible chemokine, CXCL10, reduces colitis by impairing T helper-1 induction and recruitment in mice. Inflamm Bowel Dis. 2005;11(9):799–805.
80. Hormannsperger G, et al. Post-translational inhibition of IP-10 secretion in IEC by probiotic bacteria: impact on chronic inflammation. PLoS One. 2009;4(2):e4365.
81. Sansonetti PJ. War and peace at mucosal surfaces. Nat Rev Immunol. 2004;4(12):953–64.
82. Mahida YR. The key role of macrophages in the immunopathogenesis of inflammatory bowel disease. Inflamm Bowel Dis. 2000;6(1):21–33.
83. Welte T, et al. STAT3 deletion during hematopoiesis causes Crohn's disease-like pathogenesis and lethality: a critical role of STAT3 in innate immunity. Proc Natl Acad Sci U S A. 2003;100(4):1879–84.
84. Kim SO, et al. G-CSF-mediated inhibition of JNK is a key mechanism for Lactobacillus rhamnosus-induced suppression of TNF production in macrophages. Cell Microbiol. 2006;8(12):1958–71.
85. Danne C, et al. A large polysaccharide produced by Helicobacter hepaticus induces an anti-inflammatory gene signature in macrophages. Cell Host Microbe. 2017;22(6):733–745.e5.
86. Belkacem N, et al. Lactobacillus paracasei feeding improves immune control of influenza infection in mice. PLoS One. 2017;12(9):e0184976.
87. Fernandez L, et al. Prevention of infectious mastitis by oral administration of Lactobacillus salivarius PS2 during late pregnancy. Clin Infect Dis. 2016;62(5):568–73.
88. Ro ADB, et al. Reduced Th22 cell proportion and prevention of atopic dermatitis in infants following maternal probiotic supplementation. Clin Exp Allergy. 2017;47(8):1014–21.
89. Simpson MR, et al. Breastfeeding-associated microbiota in human milk following supplementation with Lactobacillus rhamnosus GG, Lactobacillus acidophilus La-5, and Bifidobacterium animalis subspecies lactis Bb-12. J Dairy Sci. 2018;101(2):889–99.
90. van Baarlen P, et al. Differential NF-kappaB pathways induction by Lactobacillus plantarum in the duodenum of healthy humans correlating with immune tolerance. Proc Natl Acad Sci U S A. 2009;106(7):2371–6.
91. O'Toole PW, Flemer B. From culture to high-throughput sequencing and beyond: a layperson's guide to the "omics" and diagnostic potential of the microbiome. Gastroenterol Clin N Am. 2017;46(1):9–17.
92. Munblit D, et al. Human milk and allergic diseases: an unsolved puzzle. Nutrients. 2017;9(8):E894.
93. Conlon MA, Bird AR. The impact of diet and lifestyle on gut microbiota and human health. Nutrients. 2014;7(1):17–44.
94. Wilson ID, Nicholson JK. The role of gut microbiota in drug response. Curr Pharm Des. 2009;15(13):1519–23.
95. Ebbels TM, et al. Prediction and classification of drug toxicity using probabilistic modeling of temporal metabolic data: the consortium on metabonomic toxicology screening approach. J Proteome Res. 2007;6(11):4407–22.
96. Nicholson JK, Wilson ID. Opinion: understanding 'global' systems biology: metabonomics and the continuum of metabolism. Nat Rev Drug Discov. 2003;2(8):668–76.
97. Obrenovich M, et al. The co-metabolism within the gut-brain metabolic interaction: potential targets for drug treatment and design. CNS Neurol Disord Drug Targets. 2016;15(2):127–34.
98. Slyepchenko A, et al. Gut microbiota, bacterial translocation, and interactions with diet: pathophysiological links between major depressive disorder and non-communicable medical comorbidities. Psychother Psychosom. 2017;86(1): 31–46.
99. Schouten B. Cow's milk allerg - immune modulation by dietary intervention, Pharmacology and pathophysiology. Utrecht: University of Utrecht; 2009. p. 217.
100. Vos P. Preclinical studies on the immune-modulatory effects of dietary oligosaccharides, Medical microbiology. Maastricht: Maastricht University; 2008. p. 227.

28 Mild Plant and Dietary Immunomodulators

Michael J. Parnham, Višnja Stepanić, Nadja Tafferner, Marina Panek, and Donatella Verbanac

28.1 Introduction

Plants and minerals have been used since ancient times for the treatment of many ailments and diseases. Most were used for mystical reasons and others relied on the "doctrine of signatures", which stated that the shape of the plant reflected its potential medicinal use. The root of the mandrake or ginseng, for instance, is shaped like that of the human body and has been used as a general tonic for a variety of illnesses [1]. It is claimed by herbalists to have immunostimulant properties. Siberian ginseng or Taiga root (*Eleutherococcus senticosus*) is also used as a tonic and has been reported to exhibit immunostimulatory properties. The pharmacological bases of these actions are unclear, so these plant medicines cannot be considered unequivocally as immunostimulants.

In recent years, many folklore remedies have been subjected to intensive pharmacological study and some have been shown to exhibit therapeutic immunomodulatory properties in experimental and clinical studies.

The static concept, in which a drug/biological active compound is designed to act on only one target, has changed in recent years, in relation to nutrient compounds, to a dynamic concept of health in which a bioactive (e.g. antioxidant) has a multitude of subtle effects via different targets. In this respect health, as a state of complete physical, mental and social well-being, is defined as a dynamic equilibrium evidencing the ability to adapt. A distinct in vivo pharmacological response often cannot be expected from antioxidants, and the real challenge is how the correct antioxidant balance can be maintained [2].

Dietary antioxidant constituents have also been shown to exert immunoprotective and/or immunostimulant properties and are widely sold

Final manuscript submitted on December 09, 2016.

M. J. Parnham (✉)
Translational Drug Validation, Fraunhofer IME-TMP and Goethe University Frankfurt, Frankfurt am Main, Germany
e-mail: parnham@med.uni-frankfurt.de; michael.parnham@ime.fraunhofer.de

V. Stepanić
Laboratory for Epigenomics, Division of Molecular Medicine, Ruđer Bošković Institute, Zagreb, Croatia
e-mail: Visnja.Stepanic@irb.hr

N. Tafferner
Governmental Affairs, Mundipharma Deutschland GmbH & Co. KG, Frankfurt am Main, Germany
e-mail: nadja.tafferner@mundipharma.de

M. Panek
Alphachrom d.o.o., Zagreb, Croatia
e-mail: marina.panek@alphachrom.hr

D. Verbanac
Department of Medical Biochemistry and Hematology, Faculty of Pharmacy and Biochemistry, University of Zagreb, Zagreb, Croatia
e-mail: dverbanac@pharma.hr

M. J. Parnham et al. (eds.), *Nijkamp and Parnham's Principles of Immunopharmacology*,
https://doi.org/10.1007/978-3-030-10811-3_28

as prophylactic nutritional supplements. Some of the compounds for which clear immunomodulatory actions have been described so far are discussed in this chapter. Combination products are not considered, since there is currently little scientific basis for their efficacy in vivo [3].

28.2 Plant Immunostimulants

The use of herbal remedies has a very long history, but only in recent years have attempts been made to standardize these preparations to bring them in line with other pharmaceutical products. Many reputable manufacturers ensure controlled growing conditions for the plants, with analytical controls and defined harvesting conditions, as well as analytical controls on the content of active ingredients in the final medicinal extract. Regulatory authorities, both in Europe and the United States, have introduced clear guidelines on the quality standards to be applied to these products. Nevertheless, the quality and clinical efficacy of phytopharmaceuticals must be regarded critically.

Fig. 28.1 Purple coneflower (*Echinacea purpurea*)

28.2.1 Purple Coneflower (*Echinacea*)

History. The PURPLE CONEFLOWER (Fig. 28.1) is indigenous to North America and was used by the American Indians of the Great Plains as a universal remedy, particularly for colds, sore throats and pain [4]. Extracts of *Echinacea angustifolia* (narrow-leaved purple coneflower) were introduced into medical practice in the United States at the end of the nineteenth century, becoming the most widely used medicinal plants by the 1930s. With the introduction of antibiotics, *Echinacea* fell into disuse. In Europe, *E. angustifolia* was introduced into homeopathic practice in response to publications from the United States. In 1937, a general lack of supplies subsequently led to the introduction of the common PURPLE CONEFLOWER (*E. purpurea*) to Germany, of which the squeezed sap of the aerial parts of the plant was marketed. Many of the pharmacological studies on *Echinacea* have been performed on this preparation in Germany. Unfortunately, lack of standardization means that information on products—particularly those bought over the counter (OTC)—that claim to contain *Echinacea* requires careful examination. An analysis of 59 products bought in the Denver, Colorado, area in the United States revealed that 6 (10%) contained no *Echinacea* and only 9 (15%) met quality standards described on the label [5].

Chemical constituents. Compounds isolated from *Echinacea* species include caffeic acid derivatives, FLAVONOIDS, ingredients present within the fraction of ethereal oils, polyacetylenes, *N*-ALKYLAMIDES, alkaloids and polysaccharides [6]. Ingredients thought to contribute to the immunostimulatory properties of *Echinacea* include cichoric acid, polysaccharides and *N*-ALKYLAMIDES, the main lipophilic constituents (Fig. 28.2). The *N*-ALKYLAMIDES, which exhibit structural similarities to anandamide, the endogenous ligand of cannabinoid receptors, are the most potent stimulators of innate immunity [7]. Commercial preparations of *Echinacea* contain 60–80 g squeezed sap per 100 g, but the relative proportions of the various

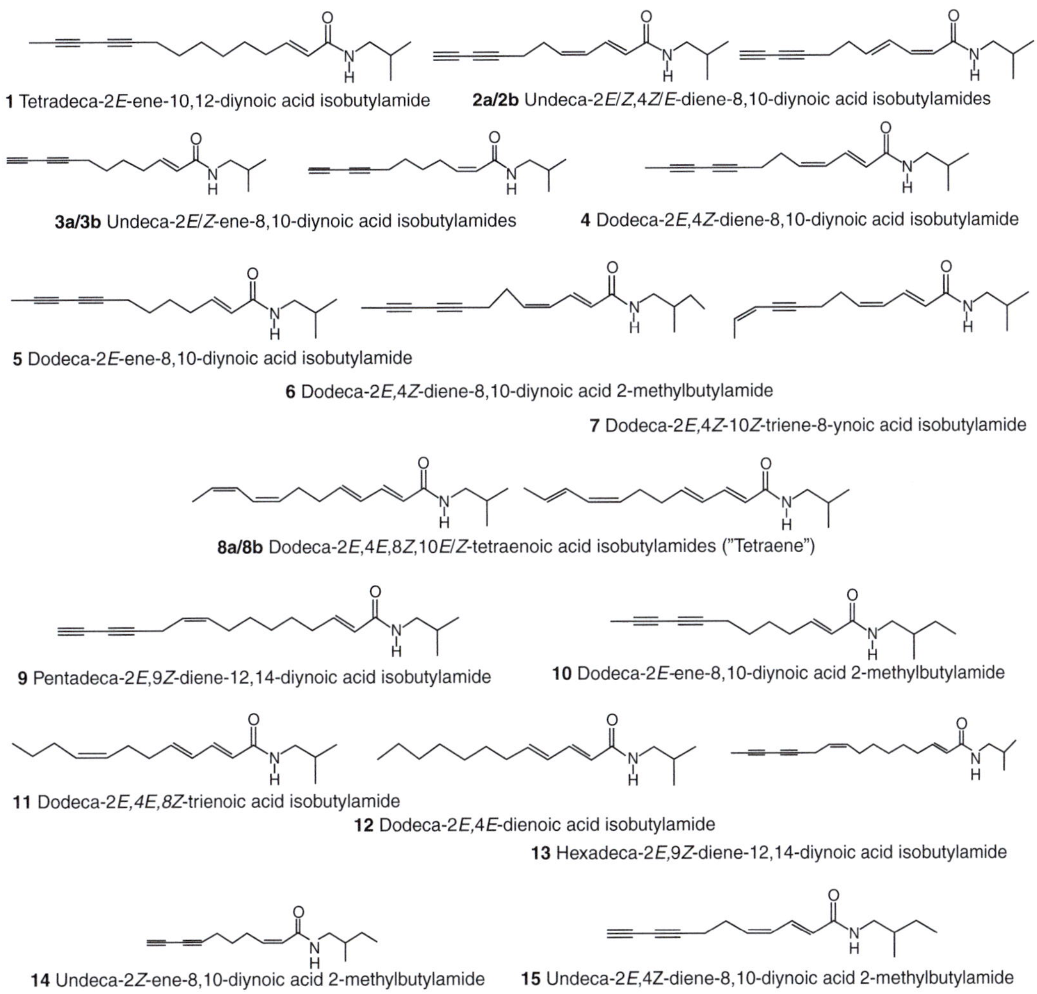

Fig. 28.2 Structures of the main *Echinacea* alkylamides (from [4] with permission)

constituents vary markedly between different products. Because of this lack of uniformity, standardization on the basis of agreed active ingredients is clearly needed.

Modes of action and pharmacological effects. The squeezed sap of *E. purpurea* stimulates the phagocytic activity of neutrophils and macrophages in vitro and in vivo. The response is moderate, but a significant increase in neutrophil phagocytosis has been observed following repeated oral administration to healthy volunteers [6]. With many phytopharmaceuticals, pharmacological effects are thought to be due to a combination of constituents. Stimulation of macrophage phagocytosis appears to be most pronounced with the *N*-ALKYLAMIDES (particularly dodeca-2E,4E,8Z,10E/Z-tetraenoic acid isobutylamides) present in the *E. purpurea* preparation [4]. In contrast, when tested on bacterial lipopolysaccharide (LPS)-stimulated macrophages, these *N*-ALKYLAMIDES weakly reduce cyclooxygenase activity and inhibit the expression of TNF-α, the latter effect (together with reduced IL-8) also being observed after oral administration of *Echinacea*. The release of TNF-α is blocked by antagonists of cannabinoid CB2 receptors, for which the *N*-ALKYLAMIDES have affinity fivefold less than that of anandamide [7]. Consequently, both stimulation of phagocytosis and CB2 receptor-mediated inhibition of inflammatory cytokines appear to contribute to the therapeutic response to *Echinacea*. Other actions, such as stimulation of NK cell activity, modification of circulating leukocyte populations, possible effects on T cell-derived cytokines and antibody formation, as well as direct antiviral actions of *N*-ALKYLAMIDES, also probably contribute to the oral activity of the preparation [7, 8]. In addition, cynarin (1,3-dicaffeoylquinic acid), a minor component of *E. purpurea* extract, is a weak but selective inhibitor of CD28-mediated signal transduction in activated T cells [9]. Administered topically to the skin, the squeezed sap of *E. purpurea* enhances wound healing, probably by inhibiting hyaluronidase, leading to increased hyaluronic acid secretion.

Pharmacokinetics. The 2,4-diene *N*-ALKYLAMIDES are transported, without significant metabolism, across monolayers of Caco-2 colonic epithelial cells, and dodeca-2E,4E,8Z,10E/Z-tetraenoic acid isobutylamides are detectable in human blood after oral administration of *E. purpurea* extract [7, 10, 11]. The t_{max} of the *N*-ALKYLAMIDES is around 30 min after oral *Echinacea* administration, but this is delayed by 1–2 h at higher concentrations of extract, possibly due to micelle formation of these lipophilic compounds. *N*-ALKYLAMIDES accumulate in fat tissues, thereby prolonging their potential duration of action. Formulation in soft gel capsules also prolongs t_{max} to around 2 h [12]. Degradation of *N*-ALKYLAMIDES occurs by hepatic oxidative (CYP) metabolism.

Clinical indications. Non-homeopathic preparations of *E. purpurea* are used mainly for the oral adjuvant treatment of respiratory and urinary tract infections and also topically for wound healing. Several double-blind, controlled clinical trials have confirmed moderate efficacy in the treatment of mild respiratory infections [13]. When administered within a few hours of symptoms arising, *E. purpurea* has been reported to shorten the course of the common cold, but without reducing symptom intensity [14]. However, while *Echinacea* use for respiratory indications remains extensive, negative outcomes in some studies, possibly due to insufficiently standardized product, mean that evidence for efficacy remains inconclusive [15]. A recent Cochrane Review of 24 trials of preparations containing extracts of different parts of the plant, 7 of which involved extracts of the aerial part, failed to detect a significant effect of either prophylaxis or treatment on the symptoms of the common cold [16]. The wide variety in type and quality of the products, however, makes comparison between the studies difficult. Efficacy in the treatment of vaginal candidiasis has only been reported in open studies.

Side effects. Common adverse reactions reported with *Echinacea* include headache, dizziness, nausea, constipation and mild epigastric pain. No adverse effects that are specific to *E. purpurea* have been observed, but, as with all plant extracts, hypersensitivity responses (dermatitis, anaphylaxis) have been reported which in rare cases can be severe. In the United States, because of the occasional occurrence of skin rash, *Echinacea* preparations are not recommended for use in children. In Germany, on the basis of theoretical considerations, *Echinacea* products are to be avoided in patients with autoimmune disorders. Since several components of *Echinacea* interact with cytochrome P450 (CYP)-mediated oxidative metabolism, a variety of interactions are likely with other drugs, particularly those metabolized by CYP3A4.

28.2.2 African Geranium (*Pelargonium sidoides*)

History. Pelargonium sidoides (Geraniaceae) is a native plant of the coastal South Africa, and because of the root's astringent nature, it is traditionally used to treat physical discomfort of the gastrointestinal tract and gynaecological complaints. In the late nineteenth century, Charles Henry Stevens, a resident of Birmingham, United Kingdom, was diagnosed with pulmonary tuberculosis and sent by his physician to South Africa to cure his symptoms. There, a traditional African healer treated him with a root decoction of *Pelargonium sidoides* and Stevens' symptoms completely disappeared within 3 months. After he brought the plant back to England, it became a popular medication to cure tuberculosis but disappeared again when antibiotics were more effective. Still now, alcoholic extracts of *P. sidoides* are indicated for patients with acute upper respiratory tract infections, its therapeutic efficacy being thought due, besides antibacterial and antiviral effects, to immunomodulatory properties. Several patents for the use of *P. sidoides* preparations exist in various countries. In Germany, a liquid OTC product is available based on clinical study data for the indication of acute bronchitis.

Chemical constituents. The root extract of *P. sidoides* contains gallic acid, shikimic acid 3-*O*-gallate, many COUMARIN derivates like umckalin, fraxinol and artelin coumarin glycosides, coumarin sulphates and also PROANTHOCYANIDINS based on (epi)catechin [17, 18]. In contrast to the root, products prepared from the upper parts of *P. sidoides* also contain several FLAVONOIDS.

Modes of action and pharmacological effects. Extracts of *P. sidoides* are well-known for their antiviral and antibacterial effects to which immunomodulatory actions make a significant contribution. Activated macrophages release reactive nitrogen intermediates (RNI) as antimicrobial effector molecules and to induce apoptosis in infected cells. Constituents of the *P. sidoides* extract, namely, 7-hydroxy-5,6-dimethoxycoumarin and 6,8-dihydroxy-5,7-dimethoxycoumarin, have been reported to be highly potent NO-inducers in vitro. In addition, a patented extract EPs® 7630 is also capable of enhancing the expression of mRNA for TNF-α in *Leishmania*-infected RAW macrophages, probably caused by the PHENOLIC constituents [19]. Additionally, mRNA expression of the cytokines IL-1, IL-2 and IL-18 was also increased [20, 21]. A flow cytometric experiment with human phagocytes showed that EPs® 7630 has the capacity to induce the production of reactive oxygen species, improve phagocytosis and promote intracellular killing of yeast in human whole blood phagocyte in vitro [22]. The detailed immunomodulating mechanism of *P. sidoides* extracts, though, is still unknown.

Pharmacokinetics. Extracts of *Pelargonium* are a complex of many active substances; the pharmacokinetics of the single constituents is not known.

Clinical indications. The main indication for orally dosed *Pelargonium* extracts is acute infection of the respiratory tract. Several double-blind, randomized controlled studies for acute bronchitis showed a symptom-relieving effect of EPs® 7630 when taking 30 drops three times daily for 1 week compared to placebo. Three studies in children also indicated an overall positive effect, although the data were not consistent. Interestingly, a lack of efficacy of the tablet form was observed in comparison to the alcoholic extracts. A potential advantage of *Pelargonium* extracts in the treatment of rhinosinusitis or the common cold has been suggested, but the quality of these studies was poor. Too many of the studies reported were poorly designed and mostly financed by the pharmaceutical manufacturer, so that their clinical relevance remains questionable [23].

Side effects. The side effects of *Pelargonium* extracts are very similar to those of *Echinacea purpurea*, cases of mild epigastric pain and nausea having been reported, but also severe cases of hypersensitivity responses can occur. Due to the COUMARIN constituents of *Pelargonium* extracts, they may, like the vitamin K-dependent anticoagulant drugs such as warfarin or phenprocoumon, carry a risk of causing bleeding. Studies in rats, however, failed to show any increased anticoagulant effect or alteration in the pharmacokinetics in the case of warfarin [24]. Possible association between the intake of *Pelargonium* extracts and the occurrence of liver failure was considered unfounded by the European Medicines Agency (EMA).

28.2.3 Mistletoe (*Viscum album*)

History. Extract of the leaves (not berries) of mistletoe (Fig. 28.3) has been used for centuries in Europe as a traditional herbal treatment for infections. In the last century, Rudolf Steiner, the originator of ANTHROPOSOPHY, suggested its use as a remedy for cancer. Biochemical analysis of mistletoe constituents led in the 1980s to the isolation and characterization of specific cytotoxic

Fig. 28.3 Mistletoe (*Viscum album*)

LECTINS that are responsible for the proposed antitumour activity of the extract [25]. The herb is not approved for use in the United States.

Chemical constituents. The main immunostimulatory constituents of mistletoe are the glycosylated LECTINS: ML-I, ML-II and ML-III. The major component is ML-I (viscumin), a member of the type II ribosome-inactivating proteins, which is used to standardize mistletoe extracts. It consists of two polypeptide chains linked by a disulphide bridge. The A-chain has enzymatic rRNA-cleaving activity, and the B-chain binds to the target cell. Other constituents of mistletoe include FLAVONOIDS, viscotoxins, terpenoids and polysaccharides. Amines, such as acetylcholine, histamine and tyramine, are also present and may contribute to hypotensive effects of mistletoe preparations.

Modes of action and pharmacological effects. ML-I has a broad range of affinities for α/β--linked galactopyranosyl residues. High nanogram concentrations of all three mistletoe LECTINS are cytotoxic. This action is due to ribosome inactivation by the rRNA *N*-glycosidase A-chain [26], leading to induction of apoptosis, possibly through activation of cation channels. At lower concentrations, ML-I and ML-I-standardized mistletoe extracts stimulate release of IL-1, IL-6 and TNF-α from peripheral blood mononuclear and skin cells [27]. Repeated doses of mistletoe extract at an ML-I-equivalent of 1 ng/kg s.c. in cancer patients cause an increase in body temperature, increases in circulating Th cells and NK cells and enhanced expression of IL-2 receptors (CD25) on lymphocytes [28]. A direct stimulatory action on T cells is likely. In mice, mistletoe extract reduces formation of melanoma metastases. Direct cytotoxic effects of the extract, on injection into gynaecological tumours in mice, have been observed, but on systemic administration, inactivation by serum glycoproteins and anti-ML antibodies occurs [29]. Non-glycosylated recombinant ML-I (rViscumin, aviscumine) has been shown to exert cytotoxicity on a variety of tumour cells in vitro. It stimulates the generation of pro-inflammatory cytokines from monocytic and lymphocytic T cell lines and enhanced NK and T cell counts in tumour-bearing mice. Inhibition of the growth of a wide variety of tumours was observed in mice in vivo [30].

Mistletoe extract is inactive on oral administration and only exhibits immunostimulatory activity on parenteral injection, possibly through activation of macrophages. In several clinical studies, increases in circulating neutrophil counts have been observed, but it is unclear whether this is related to potential therapeutic or adverse effects.

Pharmacokinetics. Using gold-labelled ML-I, the lectin has been shown to be taken up into L1210 leukaemia cells in vitro via coated pits and via plasma membrane endocytosis. These results correlate, in regard to both time and concentration, with the cytotoxicity of ML-I [31]. Triterpene acids present in mistletoe extracts have also been shown recently to be taken up readily into tumour cell lines in vitro and should be considered as potentially active compounds [32]. In a phase I trial of aviscumine (rML-1), the half-life on i.v. administration was only 13 min, and s.c. administration was selected for further studies [33].

Clinical indication. Mistletoe is one of the most widely used alternative therapies for the treatment of cancer. In German-speaking countries, ML-I-standardized mistletoe extract is administered intracutaneously at increasing dosages from 0.5 to 1.0 ng/kg twice weekly for at least 3 months in cancer patients. It has been used in patients with breast, uterine, cervical, vulvular and ovarian tumours [29]. A recent Cochrane Review evaluated the results of 21 prospective, randomized clinical trials of mistletoe treatment [34]. In the six trials, improvement in survival was reported, but none of the studies were considered to be of high methodological quality. Weaknesses included failures to randomize, small sample sizes and the presence of large numbers of patients who were excluded from analysis or otherwise not assessed. No incontrovertible evidence for tumour remission has been obtained as yet. Two studies in breast cancer patients showed clear benefit of mistletoe treatment on quality of life (QOL), but further high-quality studies are required to confirm this and other potential benefits of the therapy. Case reports of improved QOL in patients with hepatitis C have also appeared but also require confirmation in controlled, randomized trials. More recently, an overview published by the US National Cancer Institute (NCI), under independent editorship, concluded that improvements in survival, quality of life and/or stimulation of the immune system have been frequently reported, but no robust clinical benefit has been shown. In addition, stimulation of the immune system by mistletoe has not been shown unequivocally to enhance ability to fight cancer [35]. Non-glycosylated recombinant ML-I (rViscumin, aviscumine) has been shown to exhibit cytotoxicity in animal tumour models and is under investigation in cancer patients by intravenous infusion [36].

Side effects. In all clinical studies, mistletoe extract has been shown to be well-tolerated at therapeutic doses. The extract can cause fever, headache, leukocytosis, orthostatic hypotension, bradycardia, diarrhoea and hypersensitivity reactions. Toxic doses may cause coma, seizures and death.

28.3 Zinc

History. Zinc (Zn) is an essential mineral that is found in almost every cell. Its importance for human health was first documented in 1963. Zinc is required for many biological functions [37]. It stimulates the activity of approximately 100 enzymes, promoting various biochemical reactions in the human body. Zinc is needed for DNA synthesis, proper immune response and wound healing and helps maintain sense of taste and smell. Furthermore, zinc supports normal growth and development during pregnancy, childhood and adolescence. The element is found in almost all food, but the majority of zinc in the diet is provided by sea food, red meat and poultry. Other good food sources are beans, nuts, whole grains, fortified breakfast cereals and dairy products [38]. Zinc absorption is greater from a diet high in animal protein than a diet rich in plant proteins, because of the presence of PHYTATES, which are found in whole grain cereals and legumes and can interfere with zinc absorption.

Recommendations for adequate dietary zinc intake by humans in the United States and Canada have been revised recently, and a summary is given in Table 28.1.

Pharmacology. Deficiency of zinc in humans due to nutritional factors and several disease states has been recognized gradually, over the

Table 28.1 Daily recommended intakes for zinc

The recommended dietary allowance (RDA) for zinc			
Life stage	Age	Males (mg/day)	Females (mg/day)
Infants	0–6 months	2	2
Infants	7–12 months	3	3
Children	1–3 years	3	3
Children	4–8 years	5	5
Children	9–13 years	8	8
Adolescents	14–18 years	11	9
Adults	19 years and older	11	8
Pregnancy	18 years and younger	–	12
Pregnancy	19 years and older	–	11
Breast-feeding	18 years and younger	–	13
Breast-feeding	19 years and older	–	12

Source: Reference [21]; Updated in February 2008 by Victoria J. Drake, Ph.D. Linus Pauling Institute Oregon State University; Reviewed in February 2008 by: Emily Ho, Ph.D. Associate Professor of Nutrition and Exercise Sciences Principal Investigator, Linus Pauling Institute Oregon State University

past 25 years. Alcoholism, malabsorption, sickle-cell anaemia, chronic renal disease and chronically debilitating diseases are known to be predisposing factors for zinc deficiency. Vegetarians may need as much as 50% more zinc than non-vegetarians because of the lower absorption of zinc from plant foods [38]. Individuals who have had gastrointestinal surgery or who have digestive disorders that result in malabsorption, including sprue, Crohn's disease and short bowel syndrome are also at greater risk of a zinc deficiency.

The immune system is adversely affected by even moderate degrees of zinc deficiency, while severe zinc deficiency depresses immune function [39]. Zinc is unequivocally important for innate, as well as for the adaptive immune response. Decreased zinc concentrations impair NK (natural killer) cell activity, neutrophil and macrophage phagocytosis, chemotaxis and oxidative burst generation [40]. The element is required for recognition by killer cell inhibitory receptors (KIR), expressed on NK cells, of MHC class I molecules (predominantly HLA-C) on target cells. In this way, zinc can influence NK cell-mediated killing of virus-infected and tumour cells and modulate the action of cytolytic T lymphocytes. In addition, zinc is required for the development and activation of T lymphocytes [38].

The element is an essential cofactor in stabilizing thymulin, the thymic hormone that is a key factor for differentiation and maturation of immature T lymphocytes in the thymus and in the periphery. Thymulin acts on cytokine secretion by peripheral blood mononuclear cells and, together with IL-2, induces proliferation of CD8-positive T cells. Zinc also influences mature T cells, inducing the expression of the high-affinity receptor (CD25) for IL-2. This could be the reason why decreased T cell proliferation and anergy are observed after mitogen stimulation during zinc deficiency. Another possible reason could be the fact that zinc participates in binding of the protein tyrosine kinase p56 (Lck) to the alpha chains of the T cell co-receptors, CD4 and CD8, a signalling step necessary for T lymphocyte development and activation. Association of p56 (Lck) with CD4 requires conserved cysteine residues, two in the cytosolic domain of CD4 and two in the amino terminus of p56(Lck), and zinc is essential for this complex formation [41].

It has also been known for more than three decades that zinc ions can induce blast transformation in human lymphocytes. Therefore, zinc can be considered the simplest mitogen known. When added to peripheral blood mononuclear cells at stimulatory concentrations, it induces the release of IL-1, IL-6, TNF-α and IFN-γ, an effect independent of the presence of lymphocytes. In contrast to this direct stimulation of monocytes, the stimulatory effect on T cells is indirect and strictly dependent on the presence of monocytes in the cell culture.

When zinc supplements are given to individuals with low zinc levels, the numbers of circulating T cells increase, and the ability of lymphocytes to fight viral and bacterial infections improves. For instance, malnourished children in India, Africa, South America and Southeast Asia have been shown to experience shorter courses of infectious diarrhoea after taking zinc supplements [42]. The importance of zinc supplementation during aging has also been well documented during the last decade, and some important findings are reported in Box 28.1.

Box 28.1: Interrelationship Between Zinc and Immune Functions During Aging

In addition to other nutritional factors reported to enhance innate immunity in the elderly, the trace element zinc plays a pivotal role in sustaining NK cell cytotoxicity. Zinc (10 μM) improves development in vitro of CD34+ cell progenitors to NK cells from both young and older donors. Since zinc turnover is mediated by metallothioneins (MTs), these proteins are key to understanding the role of NK cells in aging and longevity. The involvement of zinc and MTs in immune function during the regeneration process was first reported in the late 1990s in the model of liver regeneration in young and old mice [43]. Partial hepatectomy and aging both caused a significant increase in MTs, which was associated with low availability of free zinc for uptake into thymocytes and extrathymic T cells. In order to avoid this effect of MTs on zinc availability in the elderly, supplementation with zinc may be useful to improve immune responses in old age. However, this supplementation should be performed with caution in order to avoid possible competition with copper [44]. Moreover, data on the beneficial effects of the zinc supplementation in the elderly are contradictory. The discrepancies are largely due to the different doses of zinc used, the duration of the zinc supplementation and the dietary habits of the elderly subjects with subsequent adverse or toxic effects on the immune response [45] and genomic stability [46].

MT homeostasis and intracellular zinc bioavailability is crucial in attenuating the link between increasing age and chronic inflammation which is becoming increasingly apparent [47]. Deterioration of innate immune function and the worsening of the chronic inflammatory status can lead to the development of common age-related pathologies such as type 2 diabetes or cardiovascular diseases. In contrast, maintenance of the balance between zinc, MT and innate immunity may help elderly subjects to stay healthy to a very old age [48].

Sophisticated mechanisms are involved in the regulation of the activity of the complex intracellular zinc–gene network. In this context, MTs are essential regulators of intracellular zinc homeostasis through the sequestration and release of the metal and thereby control availability of free zinc ions and the zinc–gene network. Zinc–gene interactions in ageing occur both directly (zinc–MT gene) and indirectly (albumin–gene) and should be taken into account when considering zinc supplementation in the deficient elderly [49].

Pharmacokinetics. Zinc supplements are available in oral and parenteral formulations. Available oral formulations include zinc sulphate, zinc gluconate, zinc picolinate and the newest form of supplementary zinc, zinc monomethionine, while zinc chloride and zinc sulphate are available as injections. Zinc monomethionine is the most bioavailable form of zinc, because the molecule is transported through small intestinal epithelial cells using the endogenous transport system for methionine. Foods containing high amounts of phosphorus, calcium or PHYTATES (found in bran, brown bread) can reduce the amount of zinc absorbed. The same effect has been observed with caffeine-rich beverages and food. Commonly used sweeteners, such as sorbitol, mannitol and citric acid, make zinc lozenges ineffective. Drug–zinc interactions have been observed with quinolone and tetracycline antibiotics as well as penicillamine. It is recommended, therefore, to take zinc 6 h before or 2 h after antibiotics.

Clinical indications. Zinc is an essential trace element used to treat zinc deficiency and delayed wound healing associated with zinc deficiency. It is also used for herpes simplex, Hansen's disease, diabetes, dental plaque, Alzheimer's disease, Wilson's disease, colds, acne and other skin problems and to stimulate the immune system to fight infection. With WHO support, inexpensive zinc tablets are being widely distributed in poor areas of Asia and Africa in a successful drive to combat fatal diarrhoea in young children, resulting from inadequate sanitation.

Side effects. Rare side effects with large (<150 mg/day) doses of zinc include chills, sustained ulcers or sores in the mouth or throat, fever, heartburn, indigestion, nausea, sore throat and unusual tiredness or weakness. The symptoms of overdose (>150 mg/day) include chest pain, dizziness, fainting, shortness of breath, vomiting and yellow eyes or skin. Cases of zinc toxicity have been seen in both acute and chronic forms. Intakes of 150–450 mg of zinc per day were accompanied by low copper status, altered iron function, reduced immune function and reduced levels of high-density lipoproteins [50].

28.4 Dietary Antioxidants

Phagocytic cells, including granulocytes and macrophages, generate large quantities of reactive oxygen species (ROS) following activation. This makes the lipid membranes of cells of the immune system particularly susceptible to oxidative damage with subsequent immunosuppression [2]. Several constituents of the normal diet help to protect against the damaging effects on lipid membranes of ROS, such as superoxide radical anion, H_2O_2 and hydroxyl radical, which are formed during a variety of physiological oxidation processes. Vitamin E is a lipid-soluble antioxidant which breaks the chain reaction of lipid peroxidation by scavenging peroxyl radicals. Vitamin C promotes reduction of the vitamin E radical, thereby restoring its reductive potential. Selenium is an essential dietary trace element which is incorporated into the active site of the GLUTATHIONE PEROXIDASE (GPx) enzymes. GPx catalyse the reduction of hydroperoxides by oxidizing glutathione (GSH) to glutathione disulphide (GSSG) and thereby complement the action of vitamin E (Fig. 28.4). Among these nutrients, vitamins C and E and selenium have been shown to have clear immunostimulant/immunoprotective properties and play a role in disease prophylaxis.

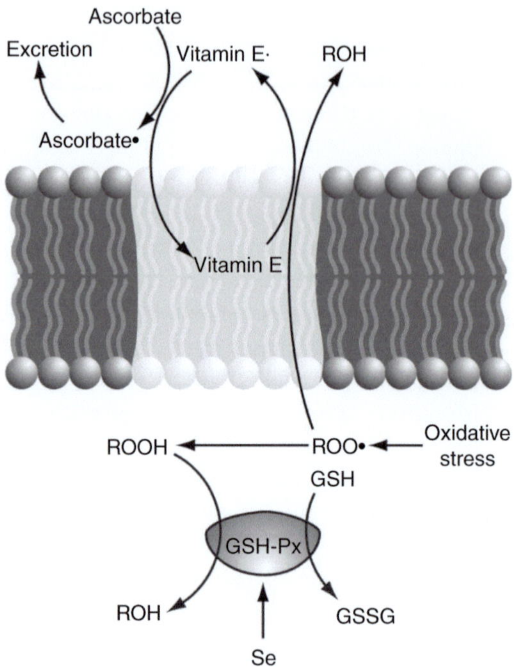

Fig. 28.4 Interplay between selenium, vitamin E and vitamin C in protection of membranes from oxidative damage. (ROOH = lipid hydroperoxide, ROO$^•$ = hydroperoxy radical, ascorbate$^•$ = semidehydroascorbate radical, vitamin E$^•$ = vitamin E radical)

28.4.1 Selenium

History. The element selenium (Se) was discovered in 1818 by the Swedish chemist Berzelius, who named it after Selene, the Greek goddess of the moon. A biological role for selenium was first demonstrated in 1957 by Klaus Schwarz who found that selenium protected against dietary liver degeneration in rats. Subsequently, in 1973, Flohé in Germany and Rotruck in the United States showed that selenium is present at the active site of the enzymes GPx, where it is incorporated as selenocysteine. Since then, 24 new SELENOPROTEINS have been identified in humans and characterized.

Pharmacology. It is well recognized that dietary selenium is important for a healthy immune response. There is also evidence that selenium has a protective effect against some forms of cancer and that it may enhance male fertility, decrease cardiovascular disease mortality and regulate inflammatory mediators in asthma. Selenium status is generally perceived as an important factor for maintaining general health, also in the elderly, even though the mechanisms by which selenium supplementation may improve immune and antioxidant responses in elderly individuals have not yet been fully elucidated. The selenium–gene interactions which are currently under investigation as part of nutrigenomics investigations may clarify this mechanism [49].

In humans, 25 genes encode for selenoproteins and mutations in several of these have been associated with inborn errors of metabolism, including bone and growth defects, glucocorticoid deficiency, neurodevelopmental defects and immune disorders. Incorporated as selenocysteine, now considered an essential amino acid, the pharmacological activity of selenium is mainly expressed through physiologically important enzymes [51]. GLUTATHIONE PEROXIDASE enzymes (cellular GPx-1, gastrointestinal GPx-2, plasma GPx-3, phospholipid hydroperoxide GPx-4) represent a major class of functionally important SELENOPROTEINS, GPx-1 being the most sensitive to changes in dietary selenium levels. GPx-2 appears to provide mucosal defence against dietary pro-oxidants and is much more resistant to dietary selenium deficiency. GPx-3 regulates extracellular oxidant levels, including that of endogenously generated NO. GPx-4 provides protection against membrane lipid peroxidation and is involved in lipid metabolism, including that of arachidonic acid. It is also an important structural protein in the mitochondrial sheath of spermatozoa.

Thioredoxin reductases (Txnrd) are widely distributed selenocysteine-containing enzymes that catalyse the NADPH-dependent reduction of thioredoxin to its oxidized form (Trx). Three mammalian forms have been described: cytoplasmic/nuclear Txnrd1 (also called TR1 or TrxR1) that reduces Trx1, mitochondrial Txnrd2 (also called TR3 or TxnR2) that reduces Trx2 and testes-specific thioredoxin–glutathione reductase (also called Txnrd3, TR2, TxnR3 or TGR). Txnrd1 and Txnrd2 are housekeeping enzymes involved in a variety of cell-signalling processes. Selenoprotein M and Sep15 may be additional members of the Txnrd family.

A third major class of SELENOPROTEINS is represented by the three types of iodothyronine deiodinase enzymes (D1, D2, D3) which catalyse the 5,5′-mono-deiodination of the prohormone thyroxine (T4) into the active thyroid hormone 3,5,3′-triiodothyronine (T3). D2 is believed to generate T3 from T4 locally in specific tissues including the pituitary gland, brown fat and brain, whereas D1 generates T3 from T4 in the thyroid gland and peripheral tissues primarily for export to plasma.

About 60% of selenium in plasma is incorporated into selenoprotein P (SEPP1) which serves as a transport protein for this mineral. Uptake of SEPP1 into tissues is mediated by binding to lipoprotein receptors, including apolipoprotein E receptor-2. Since it is expressed in many tissues, functions other than just transport are considered likely [51]. Selenium intake also affects tissue concentrations of selenoprotein W (Sel W, SEPW1) which is reported to be necessary for muscle metabolism [52].

Diet-induced selenium deficiency is associated with a variety of defects in neutrophil and lymphocyte functions in experimental and domestic animals that are reversed by selenium supplementation [53]. These defects are considered to be due to a decline in the activity of protective GPx in association with increased production of ROS, such as that occurring during the oxidative burst of phagocytes. As a result, cells in the vicinity of actively phagocytosing cells are damaged. This process also occurs to some extent in selenium-adequate animals, in which GPx-1 activity (which is very sensitive to local changes in oxidative stress) decreases in cells at local sites of acute inflammation. Circulating levels of SEPP1 are reduced in several chronic inflammatory diseases, and it appears to act like GPx-1 in macrophages to protect them from auto-oxidative damage during parasite clearance [54]. Studies on selective SEPP1 deletion in mice indicate that it may also regulate polarization of macrophages [55]. In neutrophils from humans with a low selenium status, sodium selenite (Na_2SeO_3) added in vitro is able to enhance the phagocytic and bactericidal activities of the cells [56], probably by protecting them from autolytic damage.

In addition to protecting phagocytes from autolytic damage, inorganic selenium administered to animals in nutritional excess has been shown to enhance antibody titres in response to vaccines or sensitization to erythrocytes. Studies on human lymphocytes in vitro suggest that sodium selenite selectively enhances the synthesis of IgG antibodies [57]. Enhancement of cytotoxic lymphocyte activity is a consistent response to selenium supplementation of animals and humans in nutritional excess, with increased expression of IL-2 receptors on peripheral T cells [53]. Administered to patients on haemodialysis, selenium supplementation (200–500 μg, three times weekly) enhanced T cell responses to mitogens as well as delayed hypersensitivity responses [58]. A similar enhancement of TH1 cell responses (increased IFNγ and IL-10 production and circulating CD4 counts, more rapid T cell proliferation) to oral live attenuated poliomyelitis vaccine has been reported in human subjects receiving selenium supplementation. A more rapid clearance of the poliovirus was observed in the supplemented subjects [59]. However, the effective dose range for selenium supplementation above nutritional requirements is relatively narrow, since increasing the dose leads to immunosuppression.

Pharmacokinetics. Cellular uptake of selenium, GPx activities and cytoprotection have been compared in human hepatoma cells (HepG2). Selenite and selenocysteine serve as selenium donors with high bioavailability. In contrast, selenium from selenomethionine is usually incorporated into cellular proteins but has no effect on GPx activities or cytoprotection. Consequently, not all donor forms of selenium provide selenium in a bioactive form to act as an antioxidant. Cellular selenium content, in general, does not correlate with the cytoprotective activity of this trace element, in contrast to cellular GPx activities, which always correlate, irrespective of the Se donor, with protection against lipid hydroperoxide formation and accumulation. Thus, cellular level of GPx represents a more reliable marker of adequate Se supply [60].

After injection of radiolabelled selenium, the metabolic turnover of SEPP1 in plasma peaks at 6–9 h, whereas that of extracellular GLUTATHIONE PEROXIDASE (eGPx) is sustained for at least 24 h. Selenium is rapidly incorporated into hepatic SEPP1 in the liver, followed by slow and steady incorporation into renal eGPx [61]. Oral selenium (as ^{77}Se in seleno-yeast) is also rapidly absorbed with a plasma peak at 9 h and is retained in the body for several weeks [62].

Clinical indications. Sodium selenite or seleno-yeast is widely available as a nutritional supplement, providing 50–100 µg selenium/day. This is of benefit immunologically in subjects with inadequate selenium intake, including patients on total parenteral nutrition. Keshan's disease is a cardiomyopathy associated with Coxsackie virus (CVB3) in areas of China with endemic soil selenium deficiency. Selenium supplementation completely prevents the disease, at least partially by promoting the development of immunity to the virus. Several chronic viral infections, including that to HIV-1, are associated with low selenium levels, and selenium supplementation has been shown to enhance NK activity and CD4 T counts in HIV-1-infected patients, though the clinical benefit is unclear [63].

Although serum selenium status is low in various inflammatory skin diseases, including psoriasis and rheumatoid arthritis, clear therapeutic benefit of nutritional supplementation with selenium has yet to be demonstrated. There is even some suggestion that the tendency for selenium supplementation to promote preferentially TH1 cell-mediated responses may lead to exacerbation of TH2-mediated responses in asthma [63]. Like vitamin E, selenium supplementation enhances lymphocyte proliferation responses in the elderly, but there is still no clear consensus on the potentially beneficial effects of selenium supplementation on immune function in the aged population [63, 64]. There is growing evidence that prolonged selenium intake in nutritional excess is associated with a reduced incidence of a variety of cancers and both a beneficial association with cancer prevention and effects in the elderly may relate to maintenance of immune responses [49]. Because many geographical areas—particularly Finland, parts of China, New Zealand and the United Kingdom—have low soil selenium content, nutritional supplementation with sodium selenite or selenium-enriched yeast is widespread.

Side effects. Selenium as sodium selenite or selenomethionine is considered to be non-toxic on repeated ingestion up to approximately 1 mg Se per day. Above this dose, hair and nail loss and skin lesions can arise. At higher intakes, nervous system abnormalities, including numbness, convulsions and paralysis, occur.

28.4.2 Vitamin C (Ascorbic Acid)

History. The great seafaring voyages of the Middle Ages meant that sailors were at sea for many months on very poor food rations. Many suffered exhaustion and depression, bleeding gums, haemorrhaging and bruising with fatal diarrhoea and lung and kidney damage—the symptoms of SCURVY. In 1747, the British physician, J. Lind, found that two oranges and one lemon a day could relieve the symptoms of SCURVY, but it wasn't until 1795 that the Royal Navy decreed that all sailors should be given regular lime juice. The "scorbutic principle" was only identified after 1928, the year in which Albert Szent-Gyorgi isolated hexuronic acid as the factor that prevented browning of decaying fruit. The name was changed to vitamin C following structural identification and to ascorbic acid in recognition of its ability to prevent SCURVY. Szent-Gyorgi received the Nobel Prize for Physiology and Medicine in 1937. Vitamin C is now known to be a cofactor for a variety of physiological hydroxylation reactions, including those involved in catecholamine and carnitine synthesis and that of proline during collagen synthesis. It is also a cofactor for the biosynthesis of drug-oxidizing cytochrome P450 enzymes.

Pharmacology. Vitamin C is present at high concentrations in neutrophils and is required for optimal phagocytosis [65]. During vitamin C deficiency (SCURVY) almost every component of the immune system is compromised [66]. This is mainly due to the fact that vitamin C, being a water soluble antioxidant, is able to scavenge free radicals in the extracellular compartments, participating in the prime antioxidant defence in plasma. Lack of vitamin C opens circulating white blood cells to radical attack with subsequent membrane damage and suppression of cell function. Vitamin C in vitro inhibits activation of NFκB, the transcription factor for cytokine expression, and inhibits T cell apoptosis [67, 68]. The extent to which vitamin C, at doses above the dietary requirement, is able to further enhance immune responses is still unclear. Many studies have been confounded by the administration of additional antioxidants which are required for the full protective antioxidant effect (Fig. 28.3).

Pharmacokinetics. Human leukocytes (neutrophils, monocytes and lymphocytes) take up the oxidized form of ascorbic acid, dehydroascorbic acid, actively via glucose transporters, resulting in intracellular concentrations 10- to 100-fold higher than those in plasma. This uptake is facilitated by stimulation of the cells. Oxidation of ascorbic acid by superoxide radical anion, generated by HL-60 neutrophils undergoing an oxidative burst, leads to enhanced dehydroascorbic acid uptake by all cells in the vicinity and its immediate reconversion to ascorbic acid intracellularly [69]. This provides a feedback mechanism to enhance intracellular levels of protective vitamin C in activated leukocytes that are generating large amounts of oxygen radicals.

Clinical indications. Supplementation of subjects deficient in vitamin C (e.g. some poorly nourished elderly persons) clearly restores deficient immune responses, and adequate dietary intake of vitamin C is required to sustain immune responses to infections. However, despite data from a number of clinical trials, there is little unequivocal evidence that mega-doses (>1 g/day) of vitamin C alone are able to stimulate immune responses or increase resistance to the common cold in healthy individuals [70]. Neutrophil responses, though, do appear to be enhanced in healthy subjects and children supplemented with high doses of vitamin C [70].

Side effects. In allergic persons even small amounts of vitamin C (50 mg) may cause breathing problems, tightness in the throat or chest, chest pain, skin hives, rash and itchy or swollen skin. Taking large amounts (in grams) may cause diarrhoea.

28.4.3 Vitamin E

History. Vitamin E was discovered in 1922 by H. Evans and K. Bishop as a dietary factor required for normal rat reproduction. It was officially recognized only in 1968. Vitamin E is a generic description for all tocol and tocotrienol derivatives exhibiting the biological activity of α-tocopherol. Tocopherols, including α-tocopherol, are particularly abundant in olives and olive oil. The ratio of vitamin E and polyunsaturated fatty acids in olive oil is considered optimal for a healthy diet [71].

Pharmacology. Lymphocytes and mononuclear cells have the highest vitamin E content of any cells in the body. Exposure of these cells to oxidative stress, such as that which occurs during inflammation or infection, leads to a loss of vitamin E, damage to cell membranes and cellular dysfunction. Addition of vitamin E in vitro to lymphocytes, which have been subjected to lipid peroxidation, reverses immunosuppression, measured in terms of cell proliferation and antibody formation. This protective action of vitamin E is seen most clearly in experimental vitamin E deficiency in animals. Under these conditions, antibody titres and antibody-forming cells are severely depressed; T cell responses, including proliferation and IL-2 production, are decreased; and mortality to various infections is enhanced [70]. In all cases, supplementation with vitamin E reverses the immunosuppression. Prolonged vitamin E supplementation of mice also partially reverses immunosuppression caused by retrovirus infection [72].

Vitamin E intake is associated with improvement in many processes known to contribute to atherosclerosis. This spectrum of activity includes increase in the resistance of LDL to oxidation, decrease in the cytotoxic effects of oxidized LDL on endothelial cells, inhibition of monocyte release of ROS and pro-inflammatory cytokines, inhibition of smooth muscle cell proliferation and increase in the bioavailability of endothelial NO, to name but a few [73]. Together with vitamin C and selenium, vitamin E contributes to the protection of cell membranes against oxidative damage (Fig. 28.4).

Pharmacokinetics. Vitamin E is highly lipid soluble and therefore rapidly absorbed after oral ingestion and incorporated into cell membranes.

Clinical indications. Vitamin E supplements are available for the treatment of deficiency symptoms and for the protection of muscles, blood vessels and the immune system from the effects of oxidation. Vitamin E deficiency in humans is rare but can arise in preterm infants, in association with impaired neutrophil phagocytic capacity. Phagocytic activity can be restored by administration of vitamin E to newborn children including those with glutathione deficiency [74].

The most convincing indication for clinical supplementation with vitamin E to achieve immunostimulation is the aging subject. The activities of antioxidant enzymes decrease with age, leading to a general increase in lipid peroxide tone in the body. In elderly people, nutrition can be suboptimal, and in such subjects supplementation with vitamin E has been shown to increase DTH skin test responses and enhance antibody responses to vaccines as well as mitogen-induced lymphocyte proliferation and IL-2 production [70, 75, 76]. Whether this enhancement of immune responses by prophylactic vitamin E supplementation leads to an increase in resistance to infectious diseases in humans remains to be demonstrated. Epidemiological studies that suggest a protective effect of vitamin E against cancer cannot be interpreted solely on the basis of possible effects on the immune system, because protection against cell damage in general is also involved in the response to vitamin E. However, studies show that vitamin E can inhibit angiogenesis, a crucial requirement for tumour growth, via suppression of IL-8 and modulation of adhesion molecules [76].

The known anti-inflammatory properties of vitamin E have been linked in observational studies to reduced cardiovascular risk; however, a convincing etiological connection remains to be confirmed [73].

Side effects. The evidence is compelling that intake of vitamin E above the recommended daily allowance (RDA) is of benefit to health. Vitamin E, given at recommended doses orally, has a very low toxicity. The RDA in the United States is 10 mg or 15 IU. At daily doses up to 3 g vitamin E is without any significant side effects. However, there may be an upper limit for immunostimulation, since 300 mg/day of vitamin E depressed bactericidal activity and proliferation of peripheral leucocytes in humans [77].

28.5 Plant-Derived Polyphenols and Biophenols

Polyphenols are abundant in vegetables, fruits, cereals, wine and tea, particularly green and black tea. Taken orally alone or as food constituents, polyphenols are generally poorly bioavailable, achieving nM and low μM concentrations in the blood [78]. The most important reasons for this low polyphenol bioavailability are (1) autooxidation of redox-active polyphenols such as quercetin and (−)-epigallocatechin-3-gallate (EGCG) and (2) extensive metabolic inactivation of polyphenols in general, by tissue and bacterial (microbiota-derived) enzymes. To prevent metabolism and promote efficient delivery to the site of action, significant effort has been made to develop novel targeted formulations of polyphenols [79].

However, polyphenols also have an unexpected action on microbiota [80]. In food, polyphenols are present in the form of glycosides which are cleaved by the gut microbiota. The carbohydrate moieties (usually mono-, di- and trisaccharides) thus released promote microbiota survival and modify the gut composition of bacterial strains. They increase particularly the relative proportion of *Bacteroidetes*, a species known to be reduced in obese people. Consuming polyphenols, therefore, may contribute to maintenance of a healthy body weight as well as to body weight reduction in obese people [81].

Obesity is associated with inflammation and oxidative stress in adipose tissue [82], in which adipocytokines like leptin, IL-6 and TNF-α induce production of ROS. Polyphenols generally show antioxidant and anti-inflammatory effects against these processes at low concentrations, while at high concentrations (>50 μM) they primarily act as pro-oxidants and are cytotoxic in accordance with their generally recognized biphasic dose–response curves [83]. Quercetin and EGCG are very efficient direct free radical scavengers, while polyphenols like apigenin, a weak radical scavenger, contribute to the maintenance of homeostatic ROS levels indirectly by inhibiting ROS-generating enzymes and/or inducing expression of antioxidant and detoxifying enzymes [84]. In the context of obesity, polyphenols have been found to suppress fat absorption from the gut and uptake of glucose by skeletal muscles, inhibit various uptake transporters and CYP450 isoenzymes and also inhibit anabolic pathways and stimulate catabolic pathways in adipose tissues [81].

They modulate activities of various transcription factors, particularly redox-sensitive ones such as NF-κB (the main inflammatory mediator), Nrf2 (the main regulator of antioxidant and detoxifying molecules) and PPAR-γ (the nuclear hormone receptor acting as adipogenic transcription factor). At low doses, polyphenols suppress adipocyte differentiation and proliferation [85].

Some polyphenols have been found to directly target epigenetic enzymes [78]. By regulating genome transcription, polyphenols, even in small doses, impact various signalling pathways and biological processes. For example, quercetin and kaempferol have an inhibitory effect on histone demethylase LSD1 which inversely regulates expression of selenoproteins deiodinases D2 and D3, crucial for thyroid hormone activation [86].

Olea europaea, the olive tree, is an ancient tree that originates from the Mediterranean coast of Asia Minor. The edible olive fruit is also used for its oil, gained by the process of pressing, a nutrient with considerable beneficial effects. Virgin olive oil is the natural juice of the olive fruit, which plays a major role in the healthy Mediterranean diet. The major sources of its health effects, in addition to vitamin E, are oleic acid, BIOPHENOLS (oleocanthal, tyrosol, hydroxytyrosol, oleuropein) and squalenes. The BIOPHENOLS have significant anti-inflammatory activity, and they can remove harmful compounds from the body [87].

28.5.1 Phenolic Compounds as Immunomodulators

Many epidemiological studies have shown that a diet rich in fruits and vegetables can protect against the development of cardiovascular disease [88]. Researchers have examined the composition of these foods and identified the physiologically active components as specific phytochemicals. Plant phytochemicals can be divided into plant sterols, FLAVONOIDS and plant sulphur compounds. FLAVONOIDS are a group of naturally occurring compounds that are widely distributed in nature and are ubiquitous in vegetables and fruits, particularly berries, and provide much of the flavour and colour to fruits and vegetables. At present, more than 500

FLAVONOIDS are known and described, while probably more than 4000 are present in various plants and extracts. A large number of studies have demonstrated the beneficial effects of flavonoid consumption on the development of cancer and cardiovascular disease. The general opinion is that these compounds can serve to help prevent diseases, but are not effective enough to be used as specific therapies.

History. The FLAVONOIDS were first isolated in the 1930s by Albert Szent-Gyorgyi, who also discovered vitamin C. Szent-Gyorgyi found that FLAVONOIDS strengthened capillary walls in ways in which vitamin C could not and, at first, they were referred to as vitamin P. But the chemical diversity of FLAVONOIDS precludes their classification as a single vitamin.

Chemical constituents. According to the central scaffold, the FLAVONOIDS are divided into seven major groups:

- Anthocyanidins
- Flavans including catechins or flavanols
- Flavones
- Flavonols
- Flavanones
- Flavanonols
- Isoflavones

Flavanones and flavanonols contain a 2,3-dihydro skeleton at the 2-phenylchromen-4-one core. The biosynthesis of FLAVONOIDS in plant tissues has been extensively studied in many plants, and several biosynthetic steps have been elucidated. The general metabolism includes shikimic acid, L-phenylalanine and *p*-coumaric acid. The scheme is given in Fig. 28.5. Here, we consider the well-investigated FLAVONOIDS from tea and single nutraceutical compounds that have been subject to the most extensive biological studies: coumarins, quercetin, rosmarinic acid and resveratrol.

28.5.2 Flavonoids from Tea

History. Tea (*Camellia sinensis*) (Fig. 28.6) has been consumed as an infusion for millennia. The first documented use of this particular drink is dated 2700 BC, while the first report of its beneficial effects on human health was written in 1211 by the Japanese monk Eisai [89]. In the sixteenth century, European explorers used tea extracts to fight fever, headache, stomach ache and joint pain. Today, tea is for the most part simply considered a tasty drink, but the scientific community has rediscovered the therapeutic potential of this beverage.

Tea is known as a rich source of antioxidant polyphenols (catechins, flavones, theaflavins and thearubigins). Many of them confer a cardioprotective effect by decreasing LDL oxidative susceptibility, inhibiting LDL lipid peroxidation. Common green and black tea leaves consist of about 25–30% FLAVONOIDS, while the primary sources of polyphenols in green tea are catechins epicatechin, epicatechin-3-gallate, epigallocatechin and mostly EGCG (50% or more) [90, 91].

Pharmacology. Tea catechins have been extensively tested in vitro, and EGCG especially shows inhibitory activity on metallo- and serine proteases that are involved in matrix degradation and act as crucial factors in tumour invasion. In animals, green tea significantly increases the activity of antioxidants and detoxifying enzymes, such as glutathione *S*-transferase, catalase and quinone reductase, in the lungs, liver and small intestine [89–91].

EGCG has been shown to modify (at mRNA and protein levels) the expression [78] and/or activities of various enzymes [e.g. DNA-methyltransferases (DNMT) and histone acetyltransferases (HAT)] involved in transcriptional regulation. EGCG suppresses TNF-α-induced activation of NF-κB by inhibition of p300-induced RelA (p65) acetylation, thereby increasing the level of inhibitory cytosolic IκBα and preventing p65 NF-κB translocation into the nucleus, thus interrupting the TNF-α-induced cascade of events.

Although EGCG is a well-known powerful antioxidant, able to directly reduce more than two free radicals per EGCG molecule, as well as indirectly suppressing the activity of free radical-producing enzymes such as xanthine oxidase, it is also an efficient pro-oxidant [84]. Its

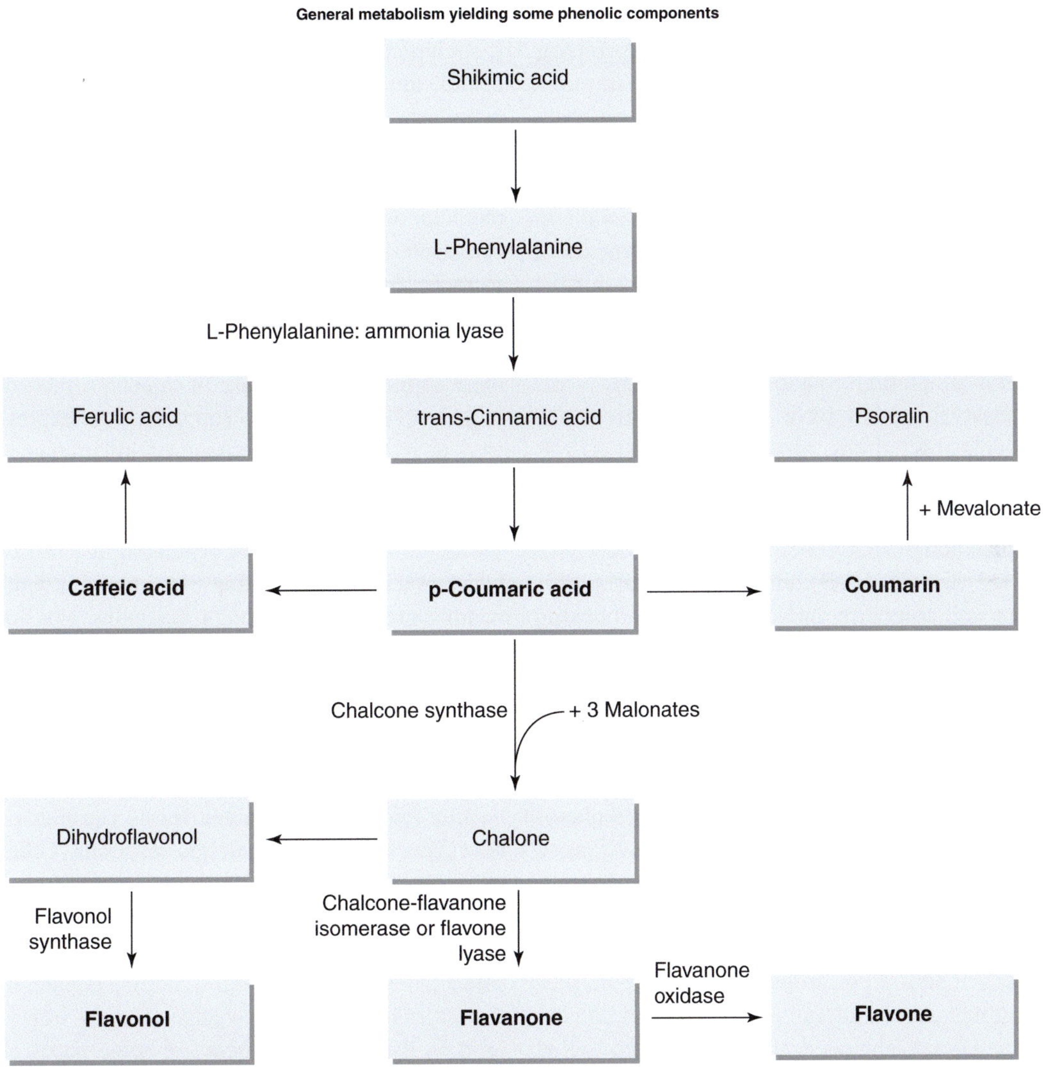

Fig. 28.5 Metabolism of shikimic acid in plants to produce various flavonoids

Fig. 28.6 Tea (*Camellia sinensis*)

manifested redox activity depends upon its concentration as well as the environment (e.g. exposure to air or the cell type). Its pro-oxidative activity is mediated predominantly by H_2O_2 generated by EGCG auto-oxidation.

EGCG, as well as products of its chemical modification by auto-oxidation, oxidative addition and/or condensation, clearly inhibit the expression and activity of telomerase, an enzyme involved in cancer which prolongs cell life by adding protective telomeres to the end of chromosomes. In nude mouse models bearing both telomerase-dependent and telomerase-independent xenograft tumours cloned from a single human cancer progeny, only the telomerase-dependent tumours responded to prolonged oral administration of EGCG [92].

Immunomodulatory effects also have been observed with aqueous extracts of tea. Neopterin production (a sensitive marker of cell-mediated immunity) was slightly enhanced in unstimulated human peripheral mononuclear cells in vitro, whereas, on stimulation with mitogens, a reduction in neopterin formation was seen [93].

Several studies have shown that green tea polyphenols afford protection against UVB-induced inflammatory responses and photocarcinogenesis in murine models. Topical application of EGCG before UVB exposure protected against UVB-induced immunosuppression and tolerance induction by (1) blocking UVB-induced infiltration of CD11b+ cells into the skin, (2) reducing IL-10 production in skin as well as in draining lymph nodes (DLN) and (3) markedly increasing IL-12 production in DLN [94]. These modulatory effects of EGCG could potentially protect against UVB-induced photocarcinogenesis. However, clinical trials of green tea in cancer have not demonstrated clear beneficial effects. It is likely that the alleged antitumour actions require prolonged exposure to the agent. Inhibition of proteolytic enzymes to prevent metastases, alterations in cell communication and anti-angiogenesis all have been touted as explanations for antineoplastic effects of green tea in vitro and in animals. But many of these mechanisms lead to tumour regression only after prolonged exposure to an antineoplastic agent [95].

Black tea also contains theaflavins that inhibit growth of virally transformed human lung cells and cell proliferation of a colon cancer cell line, showing little or no effect on normal human cells [96]. This inhibitory effect appears to be the result of induction of apoptosis and inhibition of the expression of the COX-2 gene.

Pharmacokinetics. EGCG is well absorbed into the plasma after repeated oral administration of EGCG or green tea extract to human volunteers [97]. In plasma, its chemical form may be stabilized by the antioxidant and binding capacity of human serum albumin [98]. EGCG and catechins are detectable in cellular membranes, cytoplasm, mitochondria as well as the nucleus, depending on the cell type [78].

Clinical indications. Green tea is an herbal medicine used in the adjuvant treatment of cancer, dental plaque and heart disease. Green tea extracts are recommended for general use at 300–400 mg daily, which, when "translated" into consumption of tea as a beverage, means about three cups daily. The intake of catechins, present in green tea, significantly suppresses the expression in ischaemic heart tissue of inflammatory factors, including adhesion molecules, cytokines and matrix metalloproteinases (MMPs) [99]. These are all known to be regulated by NF-κB, which is a central signalling molecule in inflammatory processes. Although catechins are not specific inhibitors of NF-κB, they have similar effects to selective inhibitors of adhesion molecules and other inflammatory agents. Therefore, catechins have the potential to suppress clinical inflammatory reactions.

Side effects. Taking green tea as capsules or beverages can provoke allergic reactions. Other more severe side effects include muscle spasms or twitches, nervousness, insomnia (sleeplessness), rapid heart rate, high levels of stomach acid and heartburn (due to the caffeine content in the tea). Green tea may change the way iron is used in the body, and there are some cases of anaemia in children drinking an average of 250 ml of green tea per day.

28.5.3 Single Nutraceutical Compounds

Coumarins. Quite apart from their well-known therapeutic uses as anticoagulants, plant-derived coumarins have received particular interest as potential sources of anti-inflammatory and immunomodulatory drugs. The coumarin scaffold (Fig. 28.7) has served as a starting point for the development of new chemical entities by medicinal chemists. The whole class of compounds possesses a range of

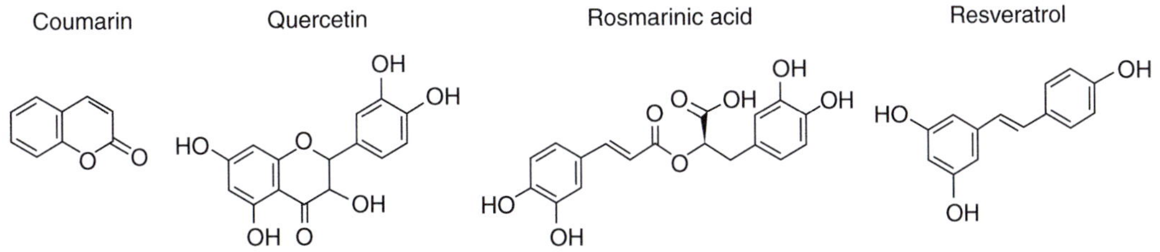

Fig. 28.7 Structures of selected antioxidant nutraceutical agents

different pharmacological activities including anticancer, antioxidant, anti-inflammatory, antiviral, anticoagulant, antibacterial, analgesic and immunomodulatory effects [100]. They exert many of their effects by inhibiting cell proliferation, interfering with mitotic spindle microtubule function, decreasing matrix metalloproteinase (MMP) activity and blocking the cell cycle in the S or G2/M phases to interfere with processes of cell division [101]. They also suppress ROS generation in leukocytes, inhibit different protein kinases and induce carcinogen-detoxifying enzymes. Attenuation of the phosphorylation of Akt/PKB and inhibition of cyclooxygenases (COXs) have been proposed as potential mechanisms of inhibition of inflammation [100]. Coumarins possessing antifungal and antibacterial activities have also been described [102, 103]. The generally poor oral bioavailability of these compounds can be overcome by novel formulations (e.g. nanoparticles, cyclodextrin inclusions) or by linking the molecules to carriers designed for targeted delivery.

Side effects reported with this class of compounds include some non-haemorrhagic responses (i.e. with indanedione derivatives), as well as haemorrhagic events. Therefore, their clinical use should be carefully monitored.

Quercetin. Present in many foods (vegetables, fruit, tea, honey and wine), the flavonol quercetin (Fig. 28.7) is one of the most prominent dietary antioxidants. It is claimed to exert various beneficial health effects. These include protection against cardiovascular diseases, sarcoidosis, certain forms of cancer (see Table 28.2), pulmonary disease and osteoporosis. In addition to being antioxidant, it also exerts anti-inflamma-

Table 28.2 Effect of quercetin on different cancer cell lines (modified from [79])

Cell line [ref]	Quercetin effects
HL-60 [80]	Inhibition of cytosolic and membrane kinases (PKC and TPK) Complete repression of phosphoinositide (PI, PIP and PIP2) activity at sub-microgram concentrations (10^{-7} M)
A549 [81]	At low concentrations promotes cell proliferation, and at higher concentrations, it is cytotoxic At lower concentrations increases total antioxidant capacity (TAC) of cells, leading to a progressive decrease in TAC at higher concentrations
K562 [82]	Suppresses expression of oncogenes (c-myc and Ki-ras) Causes depletion of inositol-1,4,5-triphosphate (IPs)
Glioma cell [83]	Causes arrest at the G2 checkpoint of the cell cycle Decreases the mitotic index
MCF7 [84]	Cytotoxic at low μM concentrations Causes cell cycle arrest in G_2/M phase Inhibits tumour growth in vivo (mice grafted with mammary carcinoma)

tory, antiproliferative and gene expression-regulating effects (Fig. 28.8). Quercetin modulates T helper cell function, inhibits lymphocyte activation and proliferation and induces apoptosis in leukaemia cells, an action to which healthy peripheral blood mononuclear cells are much less sensitive [104, 105]. It is thought that the majority of its beneficial effects are mediated by modulation of radical-induced cell membrane damage.

However, most of these studies have been performed with immortalized or cultured cell lines in vitro. Only its antioxidative and anti-inflammatory effects have been confirmed in vivo [106]. One of the main human metabolites of

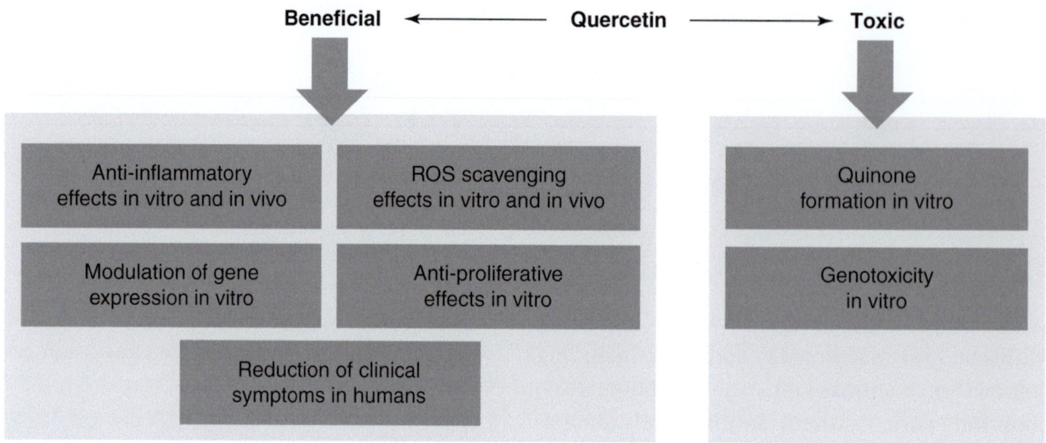

Fig. 28.8 Quercetin—beneficial and toxic effects exerted in vivo and in vitro (from [85] with permission from Elsevier)

quercetin, quercetin-3-*O*-glucuronide, has been found to accumulate selectively in macrophage-derived foam cells in human atherosclerotic lesions. At the inflamed site, extracellular β-glucuronidase secreted from macrophages transforms the glucuronide back into the quercetin aglycone which then enters macrophages and exerts anti-inflammatory and anti-atherosclerotic effects in the damaged aorta [107].

Absorption of quercetin was thought to be restricted to passive diffusion of the aglycone (without a sugar substituent) across the gastrointestinal tract. It is now known that conjugation with a glycoside considerably enhances absorption [108]. Thus, after repeated supplementation, human quercetin plasma concentrations can attain the high nM or low μM range. Limited clinical studies in rheumatoid arthritis, osteoarthritis and asthma have been carried out with quercetin in combination with other antioxidants, but without any clear beneficial effects [109]. So far, toxicity of quercetin has only been reported in vitro (Fig. 28.7). The effects observed are thought to be due to toxic oxidation products which are likely to be metabolized in vivo [106]. Quercetin-quinone, the most important oxidation product of quercetin, is highly thiol reactive, reacting almost immediately with glutathione and may result in tissue injury on chronic administration.

Rosmarinic acid. Rosmarinic acid is an ester of caffeic acid and 3,4-dihydroxyphenyllactic acid (Fig. 28.7). It is commonly found in species of the Boraginaceae and the subfamily Nepetoideae of the Lamiaceae but also in other higher plant families and in some fern and hornwort species. Its phytobiosynthesis starts with the amino acids, L-phenylalanine and L-tyrosine [110]. Rosmarinic acid exerts antiviral, antibacterial, anti-inflammatory and antioxidant effects. Its anti-inflammatory properties are thought to be due to inhibition of (1) lipoxygenases and COXs, (2) interference with the complement cascade and (3) inhibition of Lck, p56, a tyrosine kinase involved in T lymphocyte signalling [111, 112]. Like other phenolic compounds, it may protect cells from pathological transformation.

Rosmarinic acid has poor oral bioavailability and is rapidly eliminated from the blood circulation after intravenous administration, showing very low toxicity [111]. Formulation to increase oral bioavailability would considerably enhance its therapeutic potential. So far, no randomized clinical trials with this product have been reported.

Resveratrol and red wine. While excessive alcohol consumption is well-known to be harmful to health, mild-to-moderate red wine drinking has been linked with reduced cardiovascular, cerebrovascular and peripheral vascular risk, giving rise to the popular concept of the French paradox [113]. Moderate red wine consumption can also prevent undesired cell transformation that could potentially lead to cancer. A large number of reports in the literature indicate that resveratrol present in red wine is primarily responsible for the cardioprotection. Resveratrol (*trans*-3,5,4′-trihydroxystilbene; Fig. 28.7) is a member of a family of polyphenols called stilbenoids. It is biosynthesized from one molecule of *p*-coumaroyl CoA and three molecules of malonyl CoA by stilbene synthase present in higher order plants, such as eucalyptus, spruce, lily, mulberries and peanuts. Several studies indicate that resveratrol may provide protection against a wide variety of mechanisms, such as premature aging, impaired immune function, persistent chronic inflammation and cardiovascular dysfunction, involved in the development of degenerative diseases [114, 115]. This broad spectrum of potential efficacy is based mainly on the fact that resveratrol (like calorie restriction) indirectly activates SIRT1, the prototype of the sirtuins. These are an NAD(+)-dependent group of deacetylases which regulate longevity in lower organisms and, in mammals, control cellular stress resistance, genomic stability, tumorigenesis and energy metabolism. In yeast, resveratrol prolongs lifespan, and in mammals it promotes glucose and lipid metabolism [115]. Resveratrol also modifies apoptosis, inhibiting at very low concentrations and facilitating at higher doses.

In contrast to many other polyphenols, the intestinal absorption of resveratrol in humans is substantial (around 75%) and is thought to occur mainly by transepithelial diffusion. However, due to its extensive metabolism in the intestine and liver, the overall oral bioavailability is low (less than 1%) [79].

The compound, SRT-501, comprising a formulation of resveratrol with fivefold increased bioavailability, has been assessed in clinical trials (type 2 diabetes, cancer, multiple myeloma). Some of the results obtained were promising. For instance, significant lowering of fasting plasma glucose and insulin was noticed in patients affected by type 2 diabetes, and beneficial effects were observed in cancer patients. However, development of SRT-501 was halted in late 2010, due to low efficacy and drug-related adverse effects on renal functions in multiple myeloma study participants. Although some studies question whether SIRT1 was directly activated by SRT-501 [116], subsequent research showed that SIRT1 is directly activated by binding of resveratrol to the N terminus of SIRT1. A mutation in this N-terminal SIRT1 domain (aspartic acid 230 to lysine) was shown to block activation by resveratrol [117]. Resveratrol analogues, such as methylated derivatives with improved bioavailability, may be important for future research [79].

Olive oil and biophenols. Olive oil has been extensively researched both with regard to the activities of its individual components and to their potential synergy when combined in the natural oil. Studies on the primary component responsible for the health benefits indicate that the combined effects of the high monounsaturated fatty acid (MUFA) content of the oil along with several classes of bioactive compounds including vitamin E are the main active components [118].

Oleocanthal is the ester responsible for olive oil's pungent flavour. Inspired by the classic medicinal association of bitterness in plants with potential pharmacologic properties, a research team in Barcelona correlated the intensity of bitterness in olive oil with its anti-inflammatory effects. They confirmed that the mechanism of action of oleocanthal involves inhibition of COX-1 and COX-2 enzymes [119]. The burning sensation in the throat caused by olive oil is due to stimulation of TRPA1 ion channels by oleocanthal. It is questionable to what extent oleocanthal is bioavailable in vivo and BIOPHENOLS generally are more likely to be the source of the anti-inflammatory properties of olive oil. In fact, refined olive oil is devoid of many of these BIOPHENOLS, and therefore, only virgin olive oil contains all the components likely to contribute to its anti-inflammatory properties [120]. BIOPHENOLS in olive oil increase its oxidative stability and enhance the flavour of olive oil, and some authors maintain that they are highly bioavailable [121]. However, such bioavailability has not been unanimously confirmed by all authors yet. Biophenol compounds include hydroxytyrosol, tyrosol and an oleuropein mixture, all of which are biologically active agents. They have also been shown to contribute to the cardioprotective effects of olive oil [122]. Although a minor component of olive oil, hydroxytyrosol is generated as a metabolite of other biophenols. It has direct and Nrf2-mediated antioxidant activity and has been reported to exert neuroprotective effects in vivo [123]. Clinical studies with BIOPHENOLS have not yet been reported.

Altogether, the FLAVONOIDS and BIOPHENOLS are important reminders that the nutritional benefits of wholesome foods go beyond familiar vitamins and minerals. Although it may be convenient to reach for a high-potency flavonoid tablet for a particular disease, the best way to obtain a broad, healthy selection of FLAVONOIDS and BIOPHENOLS is by eating fresh fruits and vegetables, tea, olive oil, honey and soya.

28.6 Emerging Therapies and Summary

Throughout history people have used plants and naturally derived products to cure and prevent diseases. Although the healing properties of plants have been known for a long time, the ability to better exploit the uniqueness of plant therapeutics has been acquired only recently, as a result of the dramatic developments in biochemical engineering, molecular genomics, analytical chemistry separation techniques, molecular characterizations and screening for new pharmaceuticals. The discovery, development and manufacturing of botanical therapeutics, either isolated from plants and different organisms or delivered as food constituents, are becoming a major area of expansion in plant biotechnology.

In addition to the search for the active immunostimulatory agents in extracts of *Echinacea purpurea*, *Viscum album* and *Pelargonium sidoides*, a wide variety of plants are under investigation worldwide for immunostimu1ants and antibacterial and anticancer constituents. It is likely that with increasing emphasis on self-medication to reduce health budgets, the commercial importance of plant and dietary immunostimulants for the therapy and continuous prophylaxis of mild infectious and immune disorders will increase.

Based upon the role of selenium in GPx, an anti-inflammatory benzisoselenazolone, ebselen (harmokisane), with GPx-like and peroxynitrite scavenging activity was developed clinically for cerebral ischemia, though it was not marketed. The discovery of this compound stimulated a search for other selenium-based anti-inflammatory or immunomodulatory agents [124]. Ebselen and a variety of other seleno-organic compounds have been found to be cytokine inducers in vitro and in vivo and have been proposed as potential antiviral agents [125]. Ebselen is currently under clinical investigation for bipolar syndrome.

The immune system is subjected to a wide variety of stress factors in western society. These include overwork, lack of exercise, air pollution and processed foods. Although it is often financially impracticable to perform extensive clinical studies on mild immunostimulants, it is widely agreed that in view of these stress factors, the benefit of dietary antioxidants in nutritional excess is probably greater than was previously thought. Extensive clinical studies on vitamin C and vitamin E suggest that there are indeed some prophylactic health benefits. Therapeutic prospects for the therapy of chronic degenerative or inflammatory diseases are also apparent from zinc supplementation. Even greater benefits are expected from the chemical modification of specific plant phenolic constituents such as quercetin. The poor oral bioavailability of many plant-derived compounds may be circumvented by suitable formulation, and this task remains the biggest challenge for the future. Although the gut and liver are major sites of metabolism of these compounds, breakdown by colonic bacteria and the interactions of the compounds with microbiota are likely to prove crucial in explaining their effects [79].

Acknowledgements Co-authors D.V. and M.P. fully acknowledge the Croatian Science Foundation for the support of this work under the project MINUTE for IBD (HRZZ grant no IP-11-2013-5467).

Selected Readings

Barrett B. Medicinal properties of Echinacea: a critical review. Phytomedicine. 2003;10:66–86.

Firuzi O, Miri R, Tavakkoli M, Saso L. Antioxidant therapy: current status and future prospects. Curr Med Chem. 2011;18(25):3871–88.

Garcıa-Lafuente A, Guillamon E, Villares A, Rostagno MA, Martınez JA. Flavonoids as anti-inflammatory agents: implications in cancer and cardiovascular disease. Inflamm Res. 2009;58:537–52.

Kotsirilos V, Vitetta L, Sali A. A guide to evidence-based integrative and complementary medicine. Australia: Churchill Livingstone - Elsevier; 2011.

Maggini S, Wintergerst ES, Beveridge S, Hornig DH. Selected vitamins and trace elements support immune function by strengthening epithelial barriers and cellular and humoral immune responses. Br J Nutr. 2007;98(Suppl 1):S29–35.

Roxas M, Jurenka J. Colds and influenza: a review of diagnosis and conventional, botanical, and nutritional considerations. Altern Med Rev. 2007;12:25–48.

Saso L, Firuzi O. Pharmacological applications of antioxidants: lights and shadows. Curr Drug Targets. 2014;15(13):1177–99.

Important Websites

American Dietetic Association. www.eatright.org.

Memorial Sloan-Kettering Cancer Center. About herbs, botanicals and other products. www.mskcc.org/aboutherbs.

Natural Standard. US database containing evidence-based information about complementary and alternative therapies. www.naturalstandard.com.

Positive Health. Complementary Medicine Magazine. www.positivehealth.com/permit/Articles/Nutrition.

World Health Organization. Vitamin and mineral nutrition information system. www.who.int/vmnis/en/.

References

1. Leake CD. An historical account of pharmacology to the twentieth century. Springfield, IL: Charles C. Thomas; 1975.
2. Bast A. Ten misconceptions about antioxidants. Trends Pharmacol Sci. 2013;34(8):430–6.
3. Samadi AK, Bilsland A, Georgakilas AG, Amedei A, Amin A, Bishayee A, et al. A multi-targeted approach to suppress tumor-promoting inflammation. Semin Cancer Biol. 2015;35(Suppl):S151–84.
4. Foster S. Echinacea. Nature's immune enhancer. Rochester: Healing Arts Press; 1991.
5. Gilroy CM, Steiner JF, Byers T, Shapiro H, Georgian W. Echinacea and truth in labeling. Arch Intern Med. 2003;163(6):699.
6. Bauer R, Wagner H. Echinacea. Handbuch fur Ärzte, Apotheker und andere Naturwissenschaftler. Stuttgart: Wissenschaftliche; 1990.
7. Woelkart K, Bauer R. The role of alkamides as an active principle of echinacea. Planta Med. 2007;73(7):615–23.
8. Zhai Z, Liu Y, Wu L, Senchina DS, Wurtele ES, Murphy PA, et al. Enhancement of innate and adaptive immune functions by multiple Echinacea species. J Med Food. 2007;10(3):423–34.
9. Dong G-C, Chuang P-H, Chang K, Jan P, Hwang P-I, Wu H-B, et al. Blocking effect of an immunosuppressive agent, cynarin, on CD28 of T-cell receptor. Pharm Res. 2009;26(2):375–81.
10. Jager H, Meinel L, Dietz B, Lapke C, Bauer R, Merkle HP, et al. Transport of alkamides from Echinacea species through Caco-2 monolayers. Planta Med. 2002;68(5):469–71.
11. Guiotto P, Woelkart K, Grabnar I, Voinovich D, Perissutti B, Invernizzi S, et al. Pharmacokinetics and immunomodulatory effects of phytotherapeutic lozenges (bonbons) with *Echinacea purpurea* extract. Phytomedicine. 2008;15(8):547–54.
12. Dall'Acqua S, Perissutti B, Grabnar I, Farra R, Comar M, Agostinis C, et al. Pharmacokinetics and immunomodulatory effect of lipophilic Echinacea extract formulated in softgel capsules. Eur J Pharm Biopharm. 2015;97(Pt A):8–14.
13. Parnham MJ. Benefit-risk assessment of the squeezed sap of the purple coneflower (*Echinacea purpurea*) for long-term oral immunostimulation. Phytomedicine. 1996;3(1):95–102.
14. Hoheisel O, Sandberg M, Bertram S, Bulitta M, Schafer M. Echinagard® treatment shortens the course of the common cold: a double-blind, placebo-controlled clinical trial. Eur J Clin Res. 1997;9:261–8.
15. Roxas M, Jurenka J. Colds and influenza: a review of diagnosis and conventional, botanical, and nutritional considerations. Altern Med Rev. 2007;12(1):25–48.
16. Karsch-Völk M, Barrett B, Kiefer D, Bauer R, Ardjomand-Woelkart K, Linde K. Echinacea for preventing and treating the common cold. Cochrane Database Syst Rev. 2014;(2):CD000530.
17. Brendler T, van Wyk B-E. A historical, scientific and commercial perspective on the medicinal use of Pelargonium sidoides (Geraniaceae). J Ethnopharmacol. 2008;119(3):420–33.
18. Kolodziej H. Fascinating metabolic pools of Pelargonium sidoides and Pelargonium reniforme, traditional and phytomedicinal sources of the herbal medicine Umckaloabo. Phytomedicine. 2007;14(Suppl 6):9–17.
19. Kolodziej H, Kayser O, Radtke OA, Kiderlen AF, Koch E. Pharmacological profile of extracts of Pelargonium sidoides and their constituents. Phytomedicine. 2003;10(Suppl 4):18–24.
20. Kolodziej H, Kiderlen AF. In vitro evaluation of antibacterial and immunomodulatory activities of Pelargonium reniforme, Pelargonium sidoides and the related herbal drug preparation EPs 7630. Phytomedicine. 2007;14(Suppl 6):18–26.
21. Thäle C, Kiderlen A, Kolodziej H. Anti-infective mode of action of EPs 7630 at the molecular level. Planta Med. 2008;74(6):675–81.
22. Conrad A, Hansmann C, Engels I, Daschner FD, Frank U. Extract of Pelargonium sidoides (EPs 7630) improves phagocytosis, oxidative burst, and intracellular killing of human peripheral blood

phagocytes in vitro. Phytomedicine. 2007;14 (Suppl 6):46–51.

23. Dickersin K, Li T. Surgery for nonarteritic anterior ischemic optic neuropathy. Cochrane Database Syst Rev. 2015;(3):CD001538.
24. Koch E, Biber A. Treatment of rats with the Pelargonium sidoides extract EPs 7630 has no effect on blood coagulation parameters or on the pharmacokinetics of warfarin. Phytomedicine. 2007;14(Suppl 6):40–5.
25. Holtskog R, Sandvig K, Olsnes S. Characterization of a toxic lectin in Iscador, a mistletoe preparation with alleged cancerostatic properties. Oncology. 1988;45(3):172–9.
26. Eck J, Langer M, Möckel B, Witthohn K, Zinke H, Lentzen H. Characterization of recombinant and plant-derived mistletoe lectin and their B-chains. Eur J Biochem. 1999;265(2):788–97.
27. Joller PW, Menrad JM, Schwarz T, Pfüller U, Parnham MJ, Weyhenmeyer R, et al. Stimulation of cytokine production via a special standardized mistletoe preparation in an in vitro human skin bioassay. Arzneimittelforschung. 1996;46(6):649–53.
28. Beuth J, Ko HL, Gabius HJ, Burrichter H, Oette K, Pulverer G. Behavior of lymphocyte subsets and expression of activation markers in response to immunotherapy with galactoside-specific lectin from mistletoe in breast cancer patients. Clin Investig. 1992;70(8):658–61.
29. Kienle GS, Glockmann A, Schink M, Kiene H. *Viscum album* L. extracts in breast and gynaecological cancers: a systematic review of clinical and preclinical research. J Exp Clin Cancer Res. 2009;28:79.
30. Zwierzina H, Bergmann L, Fiebig H, Aamdal S, Schöffski P, Witthohn K, et al. The preclinical and clinical activity of aviscumine: a potential anticancer drug. Eur J Cancer. 2011;47(10):1450–7.
31. Walzel H, Jonas L, Rosin T, Brock J. Relationship between internalization kinetics and cytotoxicity of mistletoe lectin I to L1210 leukaemia cells. Folia Biol (Praha). 1990;36(3–4):181–8.
32. Mulsow K, Enzlein T, Delebinski C, Jaeger S, Seifert G, Melzig MF. Impact of mistletoe triterpene acids on the uptake of mistletoe lectin by cultured tumor cells. PLoS One. 2016;11(4):e0153825.
33. Bergmann L, Aamdal S, Marreaud S, Lacombe D, Herold M, Yamaguchi T, et al. Phase I trial of r viscumin (INN: aviscumine) given subcutaneously in patients with advanced cancer: a study of the European Organisation for Research and Treatment of Cancer (EORTC protocol number 13001). Eur J Cancer. 2008;44(12):1657–62.
34. Horneber MA, Bueschel G, Huber R, Linde K, Rostock M. Mistletoe therapy in oncology. Cochrane Database Syst Rev. 2008;(2):CD003297.
35. Mistletoe Extracts (PDQ®): Complementary and alternative medicine - Patient Information [NCI]-Questions and Answers About Mistletoe. 2016. http://www.webmd.com/cancer/tc/mistletoe-extracts-pdq-complementary-and-alternative-medicine%2D%2D-patient-information-nci-questions-and-answers-about-mistletoe.
36. Habeck M. Mistletoe compound enters clinical trials. Drug Discov Today. 2003;8(2):52–3.
37. Sandstead HH. Understanding zinc: recent observations and interpretations. J Lab Clin Med. 1994;124(3):322–7.
38. Institute of Medicine, Food and Nutrition Board. Dietary reference intakes for vitamin A, vitamin K, arsenic, boron, chromium, copper, iodine, iron, manganese, molybdenum, nickel, silicon, vanadium, and zinc. Washington, DC: National Academies Press; 2001.
39. Shankar AH, Prasad AS. Zinc and immune function: the biological basis of altered resistance to infection. Am J Clin Nutr. 1998;68(2 Suppl):447S–63S.
40. Reinhold D, Ansorge S, Grüngreiff K. Immunobiology of zinc and zinc therapy. Immunol Today. 1999;20(2):102–3.
41. Lin RS, Rodriguez C, Veillette A, Lodish HF. Zinc is essential for binding of p56(lck) to CD4 and CD8alpha. J Biol Chem. 1998;273(49):32878–82.
42. Black RE. Therapeutic and preventive effects of zinc on serious childhood infectious diseases in developing countries. Am J Clin Nutr. 1998;68(2 Suppl):476S–9S.
43. Mocchegiani E, Verbanac D, Santarelli L, Tibaldi A, Muzzioli M, Radosevic-Stasic B, et al. Zinc and metallothioneins on cellular immune effectiveness during liver regeneration in young and old mice. Life Sci. 1997;61(12):1125–45.
44. Mocchegiani E, Muzzioli M, Cipriano C, Giacconi R. Zinc, T-cell pathways, aging: role of metallothioneins. Mech Ageing Dev. 1998;106(1–2):183–204.
45. Sandstead HH. Requirements and toxicity of essential trace elements, illustrated by zinc and copper. Am J Clin Nutr. 1995;61(3 Suppl):621S–4S.
46. Sharif R, Thomas P, Zalewski P, Fenech M. Zinc deficiency or excess within the physiological range increases genome instability and cytotoxicity, respectively, in human oral keratinocyte cells. Genes Nutr. 2012;7(2):139–54.
47. Haase H, Rink L. The immune system and the impact of zinc during aging. Immun Ageing. 2009;6:9.
48. Mocchegiani E, Giacconi R, Cipriano C, Malavolta M. NK and NKT cells in aging and longevity: role of zinc and metallothioneins. J Clin Immunol. 2009;29(4):416–25.
49. Mocchegiani E, Costarelli L, Giacconi R, Malavolta M, Basso A, Piacenza F, et al. Micronutrient-gene interactions related to inflammatory/immune response and antioxidant activity in ageing and inflammation. A systematic review. Mech Ageing Dev. 2014;136–137:29–49.
50. Hooper PL, Visconti L, Garry PJ, Johnson GE. Zinc lowers high-density lipoprotein-cholesterol levels. JAMA. 1980;244(17):1960–1.
51. Reeves MA, Hoffmann PR. The human selenoproteome: recent insights into functions and regulation. Cell Mol Life Sci. 2009;66(15):2457–78.

52. Brown KM, Arthur JR. Selenium, selenoproteins and human health: a review. Public Health Nutr. 2001;4(2B):593–9.
53. Finch JM, Turner RJ. Effects of selenium and vitamin E on the immune responses of domestic animals. Res Vet Sci. 1996;60(2):97–106.
54. Bosschaerts T, Guilliams M, Noel W, Hérin M, Burk RF, Hill KE, et al. Alternatively activated myeloid cells limit pathogenicity associated with African trypanosomiasis through the IL-10 inducible gene selenoprotein P. J Immunol. 2008;180(9): 6168–75.
55. Barrett CW, Short SP, Williams CS. Selenoproteins and oxidative stress-induced inflammatory tumorigenesis in the gut. Cell Mol Life Sci. 2017;74(4):607–16. http://www.ncbi.nlm.nih.gov/pubmed/27563706
56. Urban T, Jarstrand C. Selenium effects on human neutrophilic granulocyte function in vitro. Immunopharmacology. 1986;12(2):167–72.
57. Reinhold U, Pawelec G, Enczmann J, Wernet P. Class-specific effects of selenium on PWM-driven human antibody synthesis in vitro. Biol Trace Elem Res. 1989;20(1–2):45–58.
58. Bonomini M, Forster S, De Risio F, Rychly J, Nebe B, Manfrini V, et al. Effects of selenium supplementation on immune parameters in chronic uraemic patients on haemodialysis. Nephrol Dial Transplant. 1995;10(9):1654–61.
59. Broome CS, McArdle F, Kyle JAM, Andrews F, Lowe NM, Hart CA, et al. An increase in selenium intake improves immune function and poliovirus handling in adults with marginal selenium status. Am J Clin Nutr. 2004;80(1):154–62.
60. Leist M, Maurer S, Schultz M, Elsner A, Gawlik D, Brigelius-Flohé R. Cytoprotection against lipid hydroperoxides correlates with increased glutathione peroxidase activities, but not selenium uptake from different selenocompounds. Biol Trace Elem Res. 1999;68(2):159–74.
61. Suzuki KT, Ishiwata K, Ogra Y. Incorporation of selenium into selenoprotein P and extracellular glutathione peroxidase: HPLC-ICPMS data with enriched selenite. Analyst. 1999;124(12): 1749–53.
62. Bügel S, Larsen EH, Sloth JJ, Flytlie K, Overvad K, Steenberg LC, et al. Absorption, excretion, and retention of selenium from a high selenium yeast in men with a high intake of selenium. Food Nutr Res 2008;52. https://doi.org/10.3402/fnr.v52i0.1642
63. Hoffmann PR, Berry MJ. The influence of selenium on immune responses. Mol Nutr Food Res. 2008;52(11):1273–80.
64. Peretz A, Nève J, Desmedt J, Duchateau J, Dramaix M, Famaey JP. Lymphocyte response is enhanced by supplementation of elderly subjects with selenium-enriched yeast. Am J Clin Nutr. 1991;53(5):1323–8.
65. Bendich A. Antioxidant vitamins and their functions in immune responses. Adv Exp Med Biol. 1990;262:35–55.
66. Basu TK. Vitamins in human health and disease. Wallingford: CAB International; 1996.
67. Schwager J, Schulze J. Modulation of interleukin production by ascorbic acid. Vet Immunol Immunopathol. 1998;64(1):45–57.
68. Campbell JD, Cole M, Bunditrutavorn B, Vella AT. Ascorbic acid is a potent inhibitor of various forms of T cell apoptosis. Cell Immunol. 1999;194(1):1–5.
69. Nualart FJ, Rivas CI, Montecinos VP, Godoy AS, Guaiquil VH, Golde DW, et al. Recycling of vitamin C by a bystander effect. J Biol Chem. 2003;278(12):10128–33.
70. Maggini S, Wintergerst ES, Beveridge S, Hornig DH. Selected vitamins and trace elements support immune function by strengthening epithelial barriers and cellular and humoral immune responses. Br J Nutr. 2007;98(Suppl 1):S29–35.
71. Servili M, Esposto S, Fabiani R, Urbani S, Taticchi A, Mariucci F, et al. Phenolic compounds in olive oil: antioxidant, health and organoleptic activities according to their chemical structure. Inflammopharmacology. 2009;17(2):76–84.
72. Wang Y, Huang DS, Eskelson CD, Watson RR. Long-term dietary vitamin E retards development of retrovirus-induced disregulation in cytokine production. Clin Immunol Immunopathol. 1994;72(1):70–5.
73. Kirmizis D, Chatzidimitriou D. Antiatherogenic effects of vitamin E: the search for the Holy Grail. Vasc Health Risk Manag. 2009;5:767–74.
74. Boxer LA, Oliver JM, Spielberg SP, Allen JM, Schulman JD. Protection of granulocytes by vitamin E in glutathione synthetase deficiency. N Engl J Med. 1979;301(17):901–5.
75. Meydani SN, Meydani M, Blumberg JB, Leka LS, Siber G, Loszewski R, et al. Vitamin E supplementation and in vivo immune response in healthy elderly subjects. A randomized controlled trial. JAMA. 1997;277(17):1380–6.
76. Meydani M. Nutrition interventions in aging and age-associated disease. Ann N Y Acad Sci. 2001;928:226–35.
77. Prasad JS. Effect of vitamin E supplementation on leukocyte function. Am J Clin Nutr. 1980;33(3):606–8.
78. Stepanić V, Kujundžić RN, Trošelj KG. Epigenetics and epigenomics. In: Payne CJ, editor. InTech; 2014.
79. Walle T. Bioavailability of resveratrol. Ann N Y Acad Sci. 2011;1215:9–15.
80. Rastmanesh R. High polyphenol, low probiotic diet for weight loss because of intestinal microbiota interaction. Chem Biol Interact. 2011;189(1–2):1–8.
81. Wang S, Moustaid-Moussa N, Chen L, Mo H, Shastri A, Su R, et al. Novel insights of dietary polyphenols and obesity. J Nutr Biochem. 2014;25(1):1–18.
82. Sears B, Ricordi C. Anti-inflammatory nutrition as a pharmacological approach to treat obesity. J Obes. 2011;2011:431985.
83. Portillo P. Beneficial effects of quercetin on obesity and diabetes. Open Nutraceuticals J. 2011;4(1):189–98.
84. Stepanic V, Gasparovic AC, Troselj KG, Amic D, Zarkovic N. Selected attributes of polyphenols in targeting oxidative stress in cancer. Curr Top Med Chem. 2015;15(5):496–509.

85. Kim H-S, Quon MJ, Kim J-A. New insights into the mechanisms of polyphenols beyond antioxidant properties; lessons from the green tea polyphenol, epigallocatechin 3-gallate. Redox Biol. 2014;2:187–95.
86. da-Silva WS, Harney JW, Kim BW, Li J, Bianco SDC, Crescenzi A, et al. The small polyphenolic molecule kaempferol increases cellular energy expenditure and thyroid hormone activation. Diabetes. 2007;56(3):767–76.
87. Parish J, Perić M, Čipčić Paljetak H, Matijašić M, Verbanac D. Translating the Mediterranean diet: from chemistry to kitchen. Period Biol. 2011;113(3):303–10.
88. Bazzano LA, Serdula MK, Liu S. Dietary intake of fruits and vegetables and risk of cardiovascular disease. Curr Atheroscler Rep. 2003;5(6):492–9.
89. Benelli R, Venè R, Bisacchi D, Garbisa S, Albini A. Anti-invasive effects of green tea polyphenol epigallocatechin-3-gallate (EGCG), a natural inhibitor of metallo and serine proteases. Biol Chem. 2002;383(1):101–5.
90. Weisburger JH. Tea and health: the underlying mechanisms. Proc Soc Exp Biol Med. 1999;220(4):271–5.
91. Bushman JL. Green tea and cancer in humans: a review of the literature. Nutr Cancer. 1998;31(3):151–9.
92. Naasani I, Oh-Hashi F, Oh-Hara T, Feng WY, Johnston J, Chan K, et al. Blocking telomerase by dietary polyphenols is a major mechanism for limiting the growth of human cancer cells in vitro and in vivo. Cancer Res. 2003;63(4):824–30.
93. Zvetkova E, Wirleitner B, Tram NT, Schennach H, Fuchs D. Aqueous extracts of Crinum latifolium (L.) and *Camellia sinensis* show immunomodulatory properties in human peripheral blood mononuclear cells. Int Immunopharmacol. 2001;1(12):2143–50.
94. Lu Y-P, Lou Y-R, Xie J-G, Peng Q-Y, Liao J, Yang CS, et al. Topical applications of caffeine or (−)-epigallocatechin gallate (EGCG) inhibit carcinogenesis and selectively increase apoptosis in UVB-induced skin tumors in mice. Proc Natl Acad Sci U S A. 2002;99(19):12455–60.
95. Jatoi A, Ellison N, Burch PA, Sloan JA, Dakhil SR, Novotny P, et al. A phase II trial of green tea in the treatment of patients with androgen independent metastatic prostate carcinoma. Cancer. 2003;97(6):1442–6.
96. Senior K. Tea: a rich brew of anti-cancer magic bullets? Drug Discov Today. 2001;6(21):1079–80.
97. Lamy S, Gingras D, Béliveau R. Green tea catechins inhibit vascular endothelial growth factor receptor phosphorylation. Cancer Res. 2002;62(2):381–5.
98. Bae M-J, Ishii T, Minoda K, Kawada Y, Ichikawa T, Mori T, et al. Albumin stabilizes (−)-epigallocatechin gallate in human serum: binding capacity and antioxidant property. Mol Nutr Food Res. 2009;53(6):709–15.
99. Suzuki J, Isobe M, Morishita R, Nagai R. Tea polyphenols regulate key mediators on inflammatory cardiovascular diseases. Mediat Inflamm. 2009;2009:494928.
100. Wu L, Wang X, Xu W, Farzaneh F, Xu R. The structure and pharmacological functions of coumarins and their derivatives. Curr Med Chem. 2009;16(32):4236–60.
101. Huang W-Y, Cai Y-Z, Zhang Y. Natural phenolic compounds from medicinal herbs and dietary plants: potential use for cancer prevention. Nutr Cancer. 2010;62(1):1–20.
102. Sardari S, Mori Y, Horita K, Micetich RG, Nishibe S, Daneshtalab M. Synthesis and antifungal activity of coumarins and angular furanocoumarins. Bioorg Med Chem. 1999;7(9):1933–40.
103. Kayser O, Kolodziej H. Antibacterial activity of simple coumarins: structural requirements for biological activity. Zeitschrift für Naturforschung C J Biosci. 1999;54(3–4):169–74.
104. Park H, Lee C-M, Jung ID, Lee JS, Jeong Y, Chang JH, et al. Quercetin regulates Th1/Th2 balance in a murine model of asthma. Int Immunopharmacol. 2009;9(3):261–7.
105. Lugli E, Ferraresi R, Roat E, Troiano L, Pinti M, Nasi M, et al. Quercetin inhibits lymphocyte activation and proliferation without inducing apoptosis in peripheral mononuclear cells. Leuk Res. 2009;33(1):140–50.
106. Boots AW, Haenen GRMM, Bast A. Health effects of quercetin: from antioxidant to nutraceutical. Eur J Pharmacol. 2008;585(2–3):325–37.
107. Kawai Y. β-Glucuronidase activity and mitochondrial dysfunction: the sites where flavonoid glucuronides act as anti-inflammatory agents. J Clin Biochem Nutr. 2014;54(3):145–50.
108. Erlund I, Kosonen T, Alfthan G, Mäenpää J, Perttunen K, Kenraali J, et al. Pharmacokinetics of quercetin from quercetin aglycone and rutin in healthy volunteers. Eur J Clin Pharmacol. 2000;56(8):545–53.
109. D'Andrea G. Quercetin: a flavonol with multifaceted therapeutic applications? Fitoterapia. 2015;106:256–71.
110. Zinsmeister HD, Becker H, Eicher T. Moose, eine Quelle biologisch aktiver Naturstoffe? Angew Chemie. 1991;103(2):134–51.
111. Parnham MJ, Kesselring K. Rosmarinic acid. Drugs Future. 1985;10:756–7.
112. Jelić D, Mildner B, Kostrun S, Nujić K, Verbanac D, Culić O, et al. Homology modeling of human Fyn kinase structure: discovery of rosmarinic acid as a new Fyn kinase inhibitor and in silico study of its possible binding modes. J Med Chem. 2007;50(6):1090–100.
113. Halpern GM. A celebration of wine: wine IS medicine. Inflammopharmacology. 2008;16(5):240–4.

114. Bertelli AAA, Das DK. Grapes, wines, resveratrol, and heart health. J Cardiovasc Pharmacol. 2009;54(6):468–76.
115. Marques FZ, Markus MA, Morris BJ. Resveratrol: cellular actions of a potent natural chemical that confers a diversity of health benefits. Int J Biochem Cell Biol. 2009;41(11):2125–8.
116. Beher D, Wu J, Cumine S, Kim KW, Lu S-C, Atangan L, et al. Resveratrol is not a direct activator of SIRT1 enzyme activity. Chem Biol Drug Des. 2009;74(6):619–24.
117. Hubbard BP, Gomes AP, Dai H, Li J, Case AW, Considine T, et al. Evidence for a common mechanism of SIRT1 regulation by allosteric activators. Science. 2013;339(6124):1216–9.
118. Fitó M, Cladellas M, de la Torre R, Martí J, Muñoz D, Schröder H, et al. Anti-inflammatory effect of virgin olive oil in stable coronary disease patients: a randomized, crossover, controlled trial. Eur J Clin Nutr. 2008;62(4):570–4.
119. Beauchamp GK, Keast RSJ, Morel D, Lin J, Pika J, Han Q, et al. Phytochemistry: ibuprofen-like activity in extra-virgin olive oil. Nature. 2005;437(7055):45–6.
120. Visioli F, Bernardini E. Extra virgin olive oil's polyphenols: biological activities. Curr Pharm Des. 2011;17(8):786–804.
121. Cicerale S, Lucas L, Keast R. Biological activities of phenolic compounds present in virgin olive oil. Int J Mol Sci. 2010;11(2):458–79.
122. De la Torre-Carbot K, Chávez-Servín JL, Jaúregui O, Castellote AI, Lamuela-Raventós RM, Nurmi T, et al. Elevated circulating LDL phenol levels in men who consumed virgin rather than refined olive oil are associated with less oxidation of plasma LDL. J Nutr. 2010;140(3):501–8.
123. Rodríguez-Morató J, Xicota L, Fitó M, Farré M, Dierssen M, de la Torre R. Potential role of olive oil phenolic compounds in the prevention of neurodegenerative diseases. Molecules. 2015;20(3):4655–80.
124. Parnham MJ. The pharmaceutical potential of seleno-organic compounds. Expert Opin Invest Drugs. 1996;5:861–70.
125. Inglot AD, Młochowski J, Zielińska-Jenczylik J, Piasecki E, Ledwoń TK, Kloc K. Seleno-organic compounds as immunostimulants: an approach to the structure-activity relationship. Arch Immunol Ther Exp. 1996;44(1):67–75.

Influence of Antibacterial Drugs on Immune and Inflammatory Systems

29

Martina Bosnar, Vesna Eraković Haber, and Garry G. Graham

29.1 Introduction

Interference of ANTIBACTERIAL AGENTS with immune and inflammatory systems, and the possible clinical implications, has long been a focus of attention worldwide. In particular, toxic effects with immunological implications (neutropenia, ALLERGY, INFLAMMATION, etc.) influence the development and clinical use of these drugs. However, favourable effects are also very important. Possible effects of ANTIBACTERIAL AGENTS on immune and inflammatory systems are shown in Table 29.1 and in sections on individual ANTIBACTERIAL AGENTS.

ANTIBACTERIAL AGENTS may interact favourably with immune and inflammatory systems in three ways:

- ANTIBACTERIAL AGENTS stop the growth of, or kill, microorganisms which have initiated an acute excessive and/or chronic immunological or inflammatory state.
- ANTIBACTERIAL AGENTS may modify the actions of antibacterial host cells (MONOCYTES, NEUTROPHILS, etc.), which, together with the ANTIBACTERIAL AGENTS, combine to stop the growth of, or kill, the invading bacteria.
- ANTIBACTERIAL AGENTS may directly affect immune and inflammatory systems and modulate the inflammatory response or correct an immune dysfunction without direct effects on bacteria.

This chapter is concerned mainly with the third possibility with the particular aim of describing ANTIBACTERIAL AGENTS which are or may possibly be useful in the treatment of non-bacteriological immune or inflammatory diseases.

Several classes of ANTIBACTERIAL AGENTS may have activities within the third possibility, most importantly sulphones (dapsone), macrolides, rifampicin, tetracyclines and their analogues. This chapter places emphasis on the basic pharmacology and clinical uses of these four drug groups in isolation from their mechanisms in causing bacteriostatic or bactericidal actions. Actions of other ANTIBACTERIAL AGENTS with less clear non-antibacterial effects are discussed in lesser detail.

The pharmacological effects of the ANTIBACTERIAL AGENTS have been studied widely both in in vitro and in vivo experimental systems. The results of in vitro studies depend, however,

Final manuscript submitted on October 09, 2018.

M. Bosnar
In Vitro Pharmacology, Fidelta, Zagreb, Croatia
e-mail: martina.bosnar@glpg.com

V. Eraković Haber (✉)
Translational Research and Alliances, Fidelta, Zagreb, Croatia
e-mail: Vesna.ErakovicHaber@glpg.com

G. G. Graham (✉)
Department of Clinical Pharmacology, St Vincent's Hospital, Darlinghurst, NSW, Australia
e-mail: g.graham@unsw.edu.au

M. J. Parnham et al. (eds.), *Nijkamp and Parnham's Principles of Immunopharmacology*,
https://doi.org/10.1007/978-3-030-10811-3_29

on the cell type, the technique, the cell activation status, the composition of the media and the concentration of the drug. Table 29.1 shows possible pathophysiological functions and mediators which may be inhibited to produce IMMUNOMODULATION or ANTI-INFLAMMATORY effects of ANTIBACTERIAL AGENTS.

In analysing the effects of any drug in vitro, it is important to compare the effective concentrations with the unbound plasma concentrations produced by therapeutic dosage. There are, however, several reasons why these correlations are often imprecise.

- The effects of the drug may be competitive and therefore depend upon the concentration of the agonist or stimulant of the system.
- The drug may be taken up avidly, even covalently, by the cells used in in vitro incubations. Consequently, the inhibitory activity may decrease with increasing cell densities in the in vitro incubations, as was demonstrated with the gold drug, auranofin, which is bound strongly to NEUTROPHILS [1].

In vivo experimental studies are the gold standard, but are subject to multiple pitfalls, such as

Table 29.1 Possible modulation of immune and inflammatory processes by antibacterial agents

Inhibited function or mediator	Potential consequences	Main antibacterial agents
Respiratory burst/reactive oxygen species (ROS) (including hypochlorous acid)	Cell/tissue protection from ROS but oxidation to reactive metabolites to produce toxicity is possible	Aminoglycosides, dapsone, sulfapyridne, fluoroquinolones, isoniazid, macrolides, tetracyclines
Phagocytosis	Inhibited uptake of necrotic cell material but possible progress of inflammatory diseases (e.g. systemic lupus erythematosus, rheumatoid arthritis) (see Chap. 8)	Ciprofloxacin[a], dapsone, rifampicin
Expression and activity of nitric oxide synthase	Decreased inflammation	Moxifloxacin[a], rifampicin, tetracyclines, tigecycline
Matrix metalloproteinases	Decreased production of active forms of TGF-β and VEGF	Tetracyclines, CMT-3
Expression of cyclooxygenase (COX)/synthesis of prostaglandins	Antinociception and suppression of inflammation	Dapsone, fosfomycin, rifampicin, tetracyclines
Chemotaxis	Decreased access of neutrophils and other inflammatory cells to areas of inflammation	Aminoglycosides, dapsone, rifampicin
TNF	Suppressed inflammatory responses including lesser induction of COX and lipoxygenase enzymes	Dapsone, fluoroquinolones, fosfomycin, fusidic acid, rifampicin, tetracyclines
NF-κB and AP-1	Reduced induction of production of inflammatory cytokines	Ciprofloxacin, macrolides, tigecycline
IL-1	Inflammation decreased by multiple mechanisms	Beta-lactams, fluoroquinolones, fosfomycin, macrolides, tigecycline
IL-2	Decreased immune reactivity (see Chap. 25)	Fusidic acid, fluoroquinolones[a]
IL-6	Decreased acute phase reactants and decreased mobilisation of neutrophils into circulation and synovial fluid (see Chap. 34)	Ciprofloxacin, fosfomycin
IL-8	Decreased angiogenesis and chemotaxis, phagocytosis and respiratory burst of neutrophils	Fluoroquinolones[a], clarithromycin, tetracyclines
IL-10	Lessened downregulation of multiple mechanisms of inflammation including suppression of pro-inflammatory cytokines such as TNF and IL-1	Fluoroquinolones[a], fusidic acid

Note that inhibition of one factor typically leads to more general decreased activity of inflammatory pathways. Several possible immune and anti-inflammatory effects are also involved in their antibacterial actions
[a]Increased activity in some cases

species differences in the composition and functions of the IMMUNE SYSTEM [2], as well as ethical problems and interindividual variability in absorption, plasma concentrations, tissue distribution, metabolism and excretion. Furthermore, it is important to note that, within a particular group of ANTIBACTERIAL AGENTS, variable effects may be obtained. Thus, drugs within the one group of ANTIBACTERIAL AGENTS may have inconsistent pharmacological effects on mammalian systems. Despite these difficulties, major progress has been made in understanding the immune and ANTI-INFLAMMATORY effects of ANTIBACTERIAL AGENTS. Further, novel uses of these drugs and development of their analogues are potentially clinically important.

An early study suggested that the IMMUNOMODULATORY properties of ANTIBACTERIAL AGENTS could be predicted from their modes of action on microbial cells [3]. However, this hypothesis, in general, has not been confirmed.

There are two notable concerns with all ANTIBACTERIAL AGENTS, which may be used clinically for their non-antibacterial activities, possibly for long periods. The concerns are the possibilities of bacterial resistance and changes in gastrointestinal flora. In particular, many ANTIBACTERIAL AGENTS cause pseudomembranous colitis due to an overgrowth of *Clostridium difficile* in the colon. Symptoms range from mild diarrhoea to potentially life-threatening.

Some ANTIBACTERIAL AGENTS have been investigated in attempts to develop analogues which are not antibacterial but have retained their IMMUNOMODULATORY or ANTI-INFLAMMATORY actions. In particular, efforts have been made to synthesise macrolides and tetracyclines with ANTI-INFLAMMATORY actions without significant antibacterial actions (see sections on individual drug groups).

29.2 Aminoglycosides

Aminoglycosides interfere with bacterial protein synthesis by acting on the 30S ribosomal subunit. Although they are considered to be extracellular ANTIBACTERIAL AGENTS, they accumulate slowly in host cells (over days in MACROPHAGES) by fluid-phase pinocytosis.

The aminoglycosides are excreted unchanged to a very large extent, and, consequently, their dosage is reduced in renal impairment in line with their plasma concentrations.

Conflicting data exist on the in vitro inhibitory effect of aminoglycosides (at therapeutic concentrations) on the function of NEUTROPHILS. Inhibition of the respiratory burst is the most common observed effect [4, 5]. Suggested underlying mechanisms include binding to negatively charged membrane phospholipids (leading to membrane disturbances), specific binding to inositol biphosphate (resulting in phospholipase C inhibition) and protein kinase C (PKC) inhibition. Interstingly, amikacin at low concentrations (contrary to other aminoglycosides) enhances the respiratory burst of NEUTROPHILS *in vitro*,whereas supratherapeutic concentrations (> 1 g/l are inhibitory [5].

Most of the non-bacterial clinical interest in the aminoglycosides has been associated with their effects on congenital muscular dystrophic disease. Several trials on the use of aminoglycosides in cystic fibrosis have given encouraging results [6]. However, the need for pharmacokinetic monitoring to avoid toxicity may limit the possible use of aminoglycosides for long-term treatment of congenital muscular dystrophic disease or inflammatory diseases unless they are shown to be extremely beneficial.

29.3 Beta-Lactams

Five groups comprise the beta (β)-lactam, namely, penams (penicillins and β-lactamase inhibitors), penems (faropenem), carbapenems (imipenem, meropenem), cephems (cephalosporins, cephamycins, oxa- and carbacephems) and monobactams (aztreonam), all of which bind to various enzymes (penicillin-binding proteins) and the transpeptidase enzyme involved in the synthesis of PEPTIDOGLYCAN backbone which is normally responsible for the strength of bacterial cell walls.

Cefodizime, a cephalosporin, was the subject of worldwide interest in the 1990s and was referred to as an immune response modifier (IRM) antibiotic. Cefodizime showed a greater effect on Enterobacteriaceae infections than that expected from its in vitro antibacterial effects, even more so in immunocompromised animals [7]. Cefodizime demonstrates pleiotropic effects on immune and inflammatory parameters such as enhanced phagocyte function, B LYMPHOCYTE repressiveness and delayed HYPERSENSITIVITY and it may restore phagocytic activity and activity of NATURAL KILLER (NK) cells, as well as IL-1 and IFN-Y production, in immunocompromised patients and animals. Cefodizime also stimulates the proliferative response of LYMPHOCYTES, increases the phagocytic and bactericidal activity of NEUTROPHILS and downregulates the production of INFLAMMATORY CYTOKINES by stimulated MONOCYTES [7–9]. Another cephalosporin, cefaclor, is considered to "normalise" the immune and inflammatory systems during bacterial infections and may be useful in the clinical treatment of patients with immune disorders leading to chronic INFLAMMATION [10].

There is now little interest in the IRM activity of cefodizime, but further work on the non-antibacterial effects of cephalosporins and other beta-lactams on host cells is still required.

29.4 Fluoroquinolones

The fluoroquinolones are synthetic antibacterial drugs whose activity is due to their inhibition of bacterial TOPOISOMERASE II and thus on DNA replication. Mammalian cells also contain an TOPOISOMERASE II, but it is unclear if an interaction with the enzyme is responsible for any of the therapeutic or adverse effects of fluoroquinolones.

The best-known fluoroquinolone is ciprofloxacin (Fig. 29.1) which is eliminated by both excretion of the unchanged drug and metabolism. Its TERMINAL HALF-LIFE is approximately 4 h, and it is usually administered twice daily. Peak concentrations of unbound ciprofloxacin in

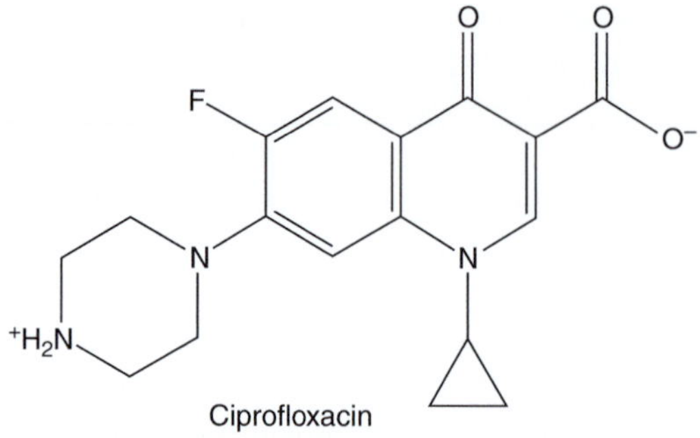

Fig. 29.1 Structure of ciprofloxacin in the unionised form. Like other fluoroquinolones, ciprofloxacin has both acidic (-COOH) and basic (amino group in the piperazine ring system). Consequently, the major species at physiological pH is the zwitterion with no net charge

plasma are approximately 1800 μg/L (5–6 μmol/L).

To date, fluoroquinolones have not proven beneficial in inflammatory diseases. Interest in the potential immunostimulating properties of some fluoroquinolones is, however, growing. Ciprofloxacin, in combination with other drugs, such as pentoxifylline (a vasodilator), accelerates NEUTROPHIL recovery in breast cancer during or following chemotherapy [11].

29.4.1 Basic Immunomodulatory and Anti-inflammatory Pharmacology of Fluoroquinolones

It is difficult to categorise the in vitro pharmacological activities of the fluoroquinolones as they have a variety of effects (increase, decrease, no effect) on PHAGOCYTOSIS, adhesion, and respiratory burst of MONOCYTES and NEUTROPHILS. Their effects on the respiratory burst appear to depend on the animal species and the fluoroquinolone structure. High concentrations decrease levels of the INFLAMMATORY CYTOKINES which have been induced by TNF-α in human cell lines (Table 29.2).

Stimulation of some immune or inflammatory systems in vitro has been reported (Table 29.2). These effects are inconsistent with suppression of immune or inflammatory systems. However,

Table 29.2 Experimental pharmacological actions showing modulation of immune and inflammatory processes by fluoroquinolones

Pharmacological actions	References
In vitro	
Ciprofloxacin increases phagocytosis, intracellular killing, myeloperoxidase activity and malondialdehyde (a marker of lipid peroxidation) levels in neutrophils of healthy subjects and patients with allergic asthma	[52]
Ciprofloxacin increases the production of PGE_2 by monocytes but inhibits the production of IFN-γ and TNF-α after induction by advanced glycation end products (AGE-2 and AGE-3). Proliferation of induced monocytes is also inhibited	[53]
Ciprofloxacin increases the synthesis of IL-2 by PHA-stimulated human lymphocytes. By contrast, IL-1 and IFN-γ are inhibited but to a small degree	[54]
In a lung epithelial cell line, moxifloxacin inhibits the production of nitric oxide (NO) due to reduced expression of inducible NO synthase (iNOS). It also inhibits the expression of NF-κB	[55]
Moxifloxacin decreases TNF-α-induced levels of IL-6, IL-8, p65 factor-κB and phosphorylated ERK in a cystic fibrosis cell line to a much greater extent than ciprofloxacin	[56]
Ex vivo	
After 7 days treatment with ciprofloxacin, LPS-stimulated human monocytes in vitro produce more IL-1, IL-6 and TNF-α	[57]
In vivo	
Ciprofloxacin, trovafloxacin and tosufloxacin (100 mg/kg) diminish serum levels of tumour necrosis factor-α (TNF-α). Levofloxacin (100 mg/kg) does not affect the TNF-α level, whereas a lower dose (10 mg/kg) increases TNF-α level	[58]
Ciprofloxacin treatment decreases colonic inflammation in a model of colitis in mice. Also levels of IL-1β, IL-8, and TNF-α in colon homogenate are decreased	[59]
Ciprofloxacin treatment ameliorates changes in body weight, diarrhoea, colon length and histology in a colitis model. Also NF-κB and TNF-α expression in colon tissue is decreased	[60]

the potential value of fluoroquinolones as immune and ANTI-INFLAMMATORY agents is shown by increases in CYTOKINE responses in rats and mice in vivo (Table 29.2). The fluoroquinolones also decrease the severity of models of colitis in mice and rats where they probably have an ANTI-INFLAMMATORY effect in addition to their antibacterial actions. Further work on the IMMUNOMODULATORY and ANTI-INFLAMMATORY effects of this drug group is required.

29.4.2 Adverse Reactions of Fluoroquinolones

The adverse effects of the fluoroquinolones present a problem for their potential clinical use as non-ANTIBACTERIAL AGENTS. Adverse effects include low incidences of colitis, liver failure, tendon rupture and cardiac arrhythmias, particularly in combination with other drugs which may prolong QT INTERVAL (e.g. some anti-arrhythmics, tricyclic antidepressants and antipsychotic drugs). The Food and Drug Administration of the USA (FDA) recommends that patients should be checked for low blood sugar and adverse mental effects such as disorientation, memory impairment and delirium.

29.5 Fosfomycin

Fosfomycin (1-*cis*-1,2-epoxypropylphosphoric acid) (Fig. 29.2) is a broad-spectrum bactericidal antibiotic which interferes with bacterial cell wall biosynthesis by inhibiting pyruvate-uridine-diphosphate-*N*-acetylglucosamine transferase.

Fosfomycin is a highly ionised compound at physiological pH values which, not surprisingly, enters bacterial cells by a transporter. The glycerol-3-phosphate transporter in bacteria but its transporter into mammalian cells has not been recorded. It is accumulated twofold by NEUTROPHILS. Its availability is about 50% and has a TERMINAL HALF-LIFE of approximately 10 h which is

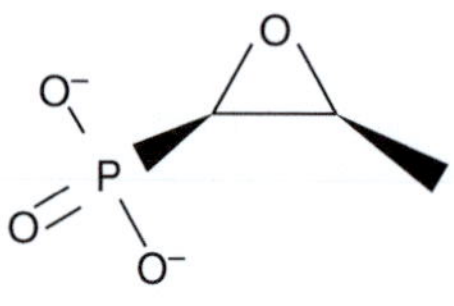

Fig. 29.2 Structure of fosfomycin in the ionised form. Fosfomycin is a strong acid which is present as the anion at physiological pH values. It is available as its sodium and calcium salts and also a salt with an organic base

Table 29.3 Experimental pharmacological actions showing modulation of immune and inflammatory processes by fosfomycin

Pharmacological actions	References
In vitro	
Fosfomycin suppresses NF-κB activation in two human cell lines	[61]
Fosfomycin decreases the levels of IL-6 and TNF-α after the addition of LPS to human blood	[62]
In vivo	
In mice injected with LPS, fosfomycin significantly lowers peak serum levels of TNF-α and IL-1β	[63]
In a rat air pouch model, after carrageenan challenge, fosfomycin decreases the inflammation in the pouch tissues. Also the amounts of PGE_2, TNF-α as well as mRNA encoding COX-2 are reduced	[64]

extended in impairment of kidney function as it is excreted unchanged in urine.

Fosfomycin shows several potential immune and ANTI-INFLAMMATORY effects in in vitro and in vivo systems (Table 29.3). However, following the administration of bacterial LPS to human subjects, fosfomycin did not alter the levels of TNF-α, IL-1β or IL-6 proteins of mRNA [12]. Further work on the immune and ANTI-INFLAMMATORY effects of fosfomycin is required.

29.6 Fusidic Acid

Fusidic acid is a steroid derivative (Fig. 29.3) which is used mainly as an antistaphylococcal agent which interferes with bacterial biosynthesis of proteins. It has an oral BIOAVAILABILITY of about 50% and a TERMINAL HALF-LIFE of elimination of about 10 h after oral dosage.

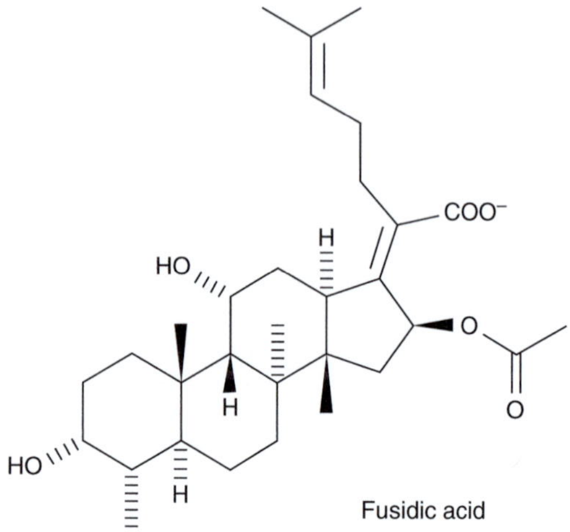

Fig. 29.3 Structure of fusidic acid in the ionised form. Fusidic acid is carboxylic acid which is available as its sodium salt

Table 29.4 Experimental pharmacological actions showing modulation of immune and inflammatory processes by fusidic acid

Pharmacological actions	References
Fusidic acid reduces peak plasma values of TNF-α and increases the survival of neonatal mice and reduces plasma TNF-α in endotoxic shock	[65]
Fusidic acid protected mice from concanavalin-A (ConA)-induced hepatitis. This was accompanied by markedly diminished plasma levels of IL-2, IFN-γ and TNF-α but increased levels of IL-6	[66]
Fusidic acid is beneficial in the treatment of experimental autoimmune neuritis in rats (a model of Guillain-Barré syndrome). Serum levels of interferon-γ, IL-10 and TNF-α decreased	[67]
Fusidic acid reduces tissue oedema in rats after the local injection of formalin	[68]

Fusidic acid has several ANTI-INFLAMMATORY effects in mice and rats in vivo, particularly by decreasing the release of TNF (Table 29.4). These effects require further study.

29.7 Isoniazid

Isoniazid is an antituberculous agent whose antimycobacterial activity has been attributed to its oxidative metabolism by mycobacterial peroxidases. Isoniazid is an irreversible inhibitor of the mammalian enzyme, MYELOPEROXIDASE [13, 14]. Isoniazid therefore decreases the formation of major products, hypochlorous, hypobromous and hypothiocyanous acids of the respiratory burst of NEUTROPHILS. These oxidants are antibacterial but may also cause tissue damage. Isoniazid has, however, not been tested for activity against purely inflammatory diseases in vivo.

29.8 Lincosamides

Clindamycin (Fig. 29.4) is a semisynthetic antibiotic derived from lincomycin, but clindamycin is now the more widely used drug of this group in modern medical practice. Both antibiotics interact with bacterial protein synthesis at the level of the 50S ribosomal subunit. The nucleoside transport system has been suggested to explain the cellular accumulation of clindamycin (12- to 20-fold). Clindamycin is eliminated by hepatic metabolism with a TERMINAL HALF-LIFE of 2–3 h. Orally, it is generally administered twice daily.

Clindamycin was presented as a possible IMMUNOMODULATORY DRUG in infection in the early 1980s. However, controversial effects on phagocyte functions (stimulation, inhibition or no action) have been reported with various techniques and drug concentrations. Interest in the ANTI-INFLAMMATORY effects of clindamycin was stimulated by its potential prophylactic effect in LPS-induced septic shock (see Chap. 26), through inhibition of PRO-INFLAMMATORY CYTOKINE release in vitro and in vivo [15]. Interestingly, modulation of CYTOKINE release in vitro is not accompanied by a parallel change in mRNA expression [16]. In dogs infected with *Babesia gibsoni*, a protozoal parasite which is rare in man, clindamycin does not eliminate the parasite but may stimulate humoral and immune responses to decrease the numbers of the infecting protozoa [17].

The relationship between the therapeutic plasma concentrations of clindamycin and its modulating effect on immunological processes is required for a better understanding of its potential non-antibacterial effects in clinical medicine.

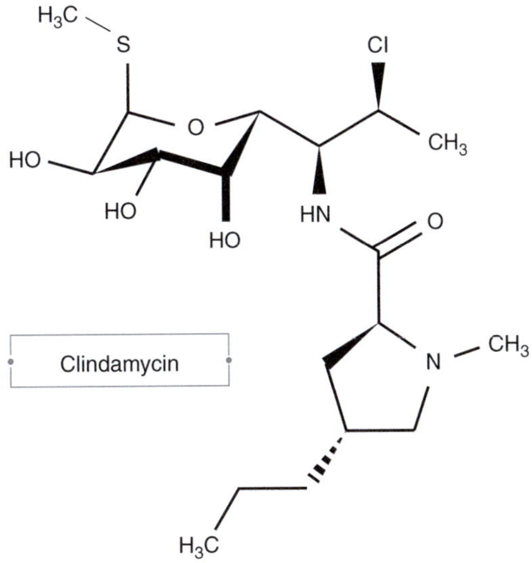

Fig. 29.4 Structure of unionised clindamycin. It is available mainly as the hydrochloride for oral and topical use, while the phosphate ester is included in solution for intravenous use

29.9 Macrolides

Macrolides are widely used antibacterial drugs, which impair bacterial protein synthesis by acting on the 50S bacterial ribosomal subunit. Erythromycin is the original antibiotic in this class.

29.9.1 Chemistry of Macrolides

Macrolide antibiotics have a 12- to 16-membered ring structure containing a lactone group (internal ester) with substitutions by 2 amino groups and/or neutral sugars (Fig. 29.5). Modern semi-synthetic derivatives of erythromycin, azithromycin, roxithromycin and clarithromycin have been obtained by adding new substituents or by introducing a nitrogen atom into the ring structure. These have antibacterial activity. Extensive derivatisation of the scaffold has also resulted in a number of non-ANTIBACTERIAL macrolides with IMMUNOMODULATORY/ANTI-INFLAMMATORY activities [18].

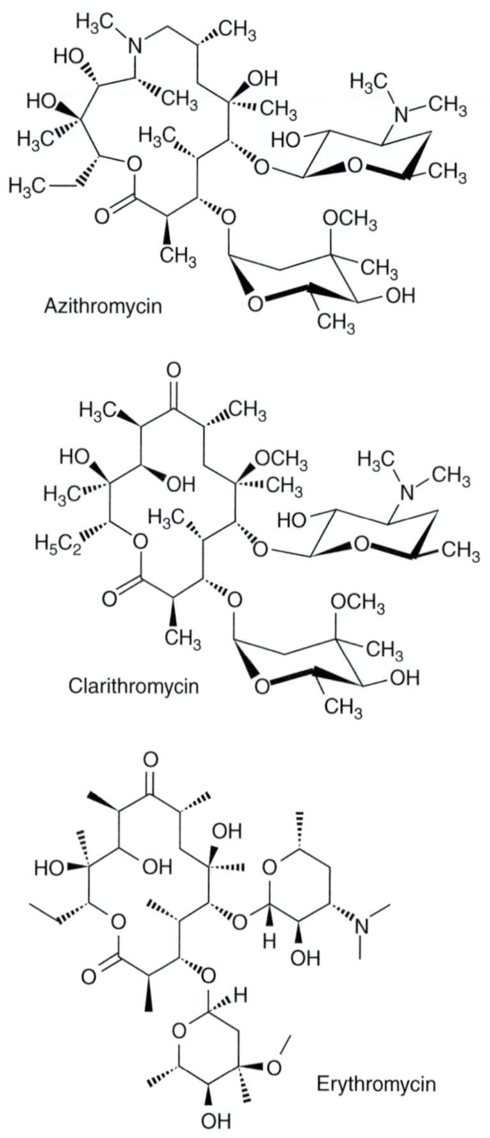

Fig. 29.5 Structures of macrolide antibiotics in the unionised forms. Erythromycin is available as the unionised form, a salt with stearic acid and an ester with ethyl succinic acid for oral dosage. It is also available as the lactobionate salt for intravenous infusion

29.9.2 Metabolism and Pharmacokinetics of Macrolides

The macrolides are eliminated primarily by metabolism. The TERMINAL HALF-LIVES are very much dependent upon the macrolide, ranging from approximately 1.6 h (erythromycin), 3.5 h (clarithromycin), 12 h (roxithromycin) to 40 h (azithromycin). The long TERMINAL HALF-LIFE of azithromycin makes it suitable for long-term dosage where it is often administered every second day. The macrolides are taken up by NEUTROPHILS and MACROPHAGES and concentrate, like other cationic drugs, in the lysosomes, making it difficult to compare concentrations in plasma with those in experiments in vitro (see Sect. 29.1). However, this leucocyte accumulation may act to deliver macrolides to sites of INFLAMMATION and infection.

Erythromycin is unstable in the stomach and is administered orally in two forms: either the base form in ENTERIC-COATED tablets or as the ester form with ethyl succinic acid. The latter form is hydrolysed after absorption from the gastrointestinal tract.

29.9.3 Basic Immune and Anti-inflammatory Pharmacology of Macrolides

The macrolides have several cellular effects both in vitro and in vivo (Table 29.5). These pleiotropic effects indicate that there are several cellular targets of their actions. Among the cellular mechanisms of action, inhibition of MAP kinase/ERK pathway and consequent suppression of transcription factor activation have been suggested. Furthermore, the pharmacological effects of the individual macrolides often vary between the various members of drug class. A non-antibacterial analogue of erythromycin (EM703) has potential ANTI-INFLAMMATORY actions which are indicated by its inhibition of the respiratory burst of NEUTROPHILS (Table 29.5).

29.9.4 Immune and Anti-inflammatory Clinical Effects of Macrolides

Macrolides display IMMUNOMODULATORY properties that may confer beneficial effects on patients with respiratory diseases associated with

Table 29.5 Experimental pharmacological actions showing modulation of immune and inflammatory processes by macrolides

Pharmacological actions	References
In vitro	
Clarithromycin suppresses LPS-induced IL-8 production by human monocytes and human epithelial cells through inhibition of AP-1 and NF-κB transcription factors	[69, 70]
Clarithromycin inhibits NF-κB activation in human mononuclear cells and pulmonary epithelial cells	[71]
Macrolides inhibit the release of a variety of inflammatory cytokines and chemokines in sputum cells isolated from steroid-naïve patients with chronic obstructive pulmonary disease (COPD)	[72]
Azithromycin attenuates LPS-induced production and expression of pro-inflammatory cytokines in alveolar macrophages through inhibition of AP-1	[73]
Azithromycin inhibits specifically the production of pro-inflammatory cytokines IL-1α and IL-1β by human monocytes after stimulation by LPS. There is no inhibition of several other cytokines. Inhibition of the inflammasome/IL-1β axis indicates potential activity in ASTHMA (see Chap. 9)	[74]
The non-antibacterial analogue of erythromycin (EM703) suppresses the production of superoxide by human neutrophils after stimulation by LPS or fMLP	[75]

chronic INFLAMMATION. The macrolides attenuate inflammatory responses in the lung, regulate mucus production and decrease bronchial responsiveness. Panbronchiolitis and cystic fibrosis are the two main clinical indications for macrolide action, but their use in other respiratory diseases is being investigated. Most clinical research has been centred on azithromycin.

29.9.4.1 Diffuse Panbronchiolitis

Diffuse panbronchiolitis is a potentially fatal disease which is an inflammatory disease of bronchioles. It is found most commonly in patients of Asian descent. Erythromycin or azithromycin is recommended widely for the treatment of diffuse panbronchiolitis, with azithromycin favoured because of its long TERMINAL HALF-LIFE. Dosage is once daily to once every 2 or 3 days [19, 20].

29.9.4.2 Cystic Fibrosis (CF)

There is fair evidence for the long-term use of azithromycin in the treatment of cystic fibrosis in patients with and without infection with *Pseudomonas aeruginosa* [19–22]. However, the EFFICACY often decreases markedly after treatment for 1 year with no significant effect after treatment for 3 years. Nevertheless, azithromycin is recommended by the Cystic Fibrosis Foundation for patients who are more than 6 years old [23]. Non-cystic fibrosis bronchiectasis is also treated usefully by azithromycin [23].

29.9.4.3 Asthma

A recent conclusion is that macrolides do not have significant activity in the treatment of acute or chronic ASTHMA [24, 25], although, possible benefits of macrolides in patients with non-eosinophilic ASTHMA was noted in two studies [26]. This finding requires further study.

29.9.4.4 Chronic Obstructive Pulmonary Disease (COPD)

COPD is a major chronic disease. Azithromycin decreases the frequency of acute exacerbations even in patients with optimal treatment with bronchodilators [23] (see also Chap. 23).

29.9.4.5 Bronchiolitis Obliterans Syndrome (BOS)

BOS is major problem after lung transplantation. Some, but not all patients, respond with an improvement in FEV1 during treatment with azithromycin (250 mg every second day) [27, 28].

29.9.5 Adverse Effects of Macrolides

The macrolides are generally well tolerated although gastrointestinal adverse effects, such as discomfort, nausea, etc., are well known and often lead to cessation of their use. The cardiovascular effects of the macrolides are of considerable interest. Prolongation of the QT

INTERVAL is of concern, but patients' underlying cardiac pathophysiology makes it difficult to make definite conclusions about cardiac adverse effects [23]. Nevertheless, they should be taken with care by patients with a history of arrhythmias or taking antiarrhythmic drugs. Because of their antibiotic activities, overgrowth with *Clostridium difficile* is possible, and patients with severe diarrhoea should be examined carefully.

29.10 Rifampicin and Related Drugs

Rifampicin (rifampin) is a member of the ansamycin group of macrocyclic antibiotics. Rifampicin is a major antibiotic for the treatment of tuberculosis, while rifaximin is used for INFLAMMATORY BOWEL DISEASES and travellers' diarrhoea which is not due to species of *Campylobacter*, *Salmonella* or *Shigella*. Rifampicin and the related antibiotics bind to the DNA-dependent RNA polymerase and, thereby, inhibit RNA synthesis and bacterial proliferation.

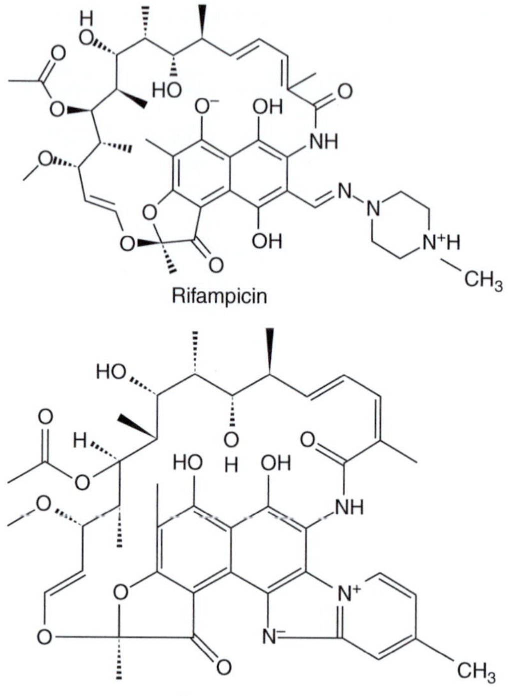

Fig. 29.6 Structures of rifampicin (upper) and rifaximin (lower) in the ZWITTERION forms at physiological pH values. Several ZWITTERION forms of rifaximin may exist and one such form is shown

29.10.1 Chemistry of Rifampicin and Analogues

Rifampicin and rifaximin are ZWITTERIONS at physiological pH values (Fig. 29.6). Several crystal forms of rifaximin exist. These have different absorption characteristics after oral administration (see below).

29.10.2 Metabolism and Pharmacokinetics of Rifampicin and Analogues

Rifampicin and rifaximin are incompletely absorbed after oral dosage, in part, due to their ionised ZWITTERION structures (Fig. 29.6). The absorption of rifaximin is reduced further due its slow dissolution. The alpha-crystalline form is less soluble than that of other crystalline forms, and, in order to achieve maximal availability in the colon, the alpha form is probably the optimal form for the treatment of ulcerative colitis [29]. The TERMINAL HALF-LIFE of rifampicin is approximately 4 h.

29.10.3 Basic Immune and Anti-inflammatory Pharmacology of Rifampicin and Analogues

Rifampicin has potential ANTI-INFLAMMATORY and immunological effects because of its activity in several cellular systems in vitro (Table 29.6). Some effects, such as inhibition of chemotaxis and PHAGOCYTOSIS, are produced in vitro at therapeutic unbound concentrations in plasma. However, several other effects are produced well above therapeutic concentrations in plasma (Table 29.6), and, consequently, some of their clinical relevance is often discounted but cannot be rejected totally at this stage.

Table 29.6 Experimental pharmacological actions showing modulation of immune and inflammatory processes by rifampicin and analogues

Pharmacological effects	References
Effects on inflammatory cells—in vitro	
Therapeutic concentrations of rifampicin decrease chemotaxis of neutrophils	[76]
Supratherapeutic concentrations of rifampicin inhibit production of TNF-α and IL-1β while enhancing IL-6 and IL-10 secretion	[77]
Rifampicin inhibits zymosan phagocytosis and TNF-α production with partial effects at therapeutic concentrations	[78]
Rifampicin and related antibiotics inhibit proliferation of mononuclear cells induced by phytohaemagglutinin, concanavalin A and bacterial superantigen toxic shock syndrome toxin 1	[79]
Supratherapeutic concentrations of rifampicin inhibit both TNF-α and PMA-induced NF-κB activation in Jurkat T cells	[80]
Effects on non-inflammatory cells—in vitro	
Supratherapeutic concentrations of rifampicin inhibit IL-1β-stimulated arachidonic acid release and prostaglandin E2 (PGE_2) production in an alveolar epithelial cell line	[81]
Supratherapeutic concentrations of rifampicin augment cytokine-induced iNOS and consequently NO production in an alveolar cell line	[82]
Rifampicin enhances iNOS expression and production of NO and IL-8 as well as synthesis of IL-1β, IL-8 and IFN-γ-induced protein-10 in HepG2 liver epithelial cells	[83]

The ability of rifampicin to increase the serum levels of IL-10 (Table 29.6) indicates a potential mechanism increased immunosuppressant or ANTI-INFLAMMATORY effects as this CYTOKINE is considered to have ANTI-INFLAMMATORY actions by several mechanisms including the suppression of LPS-induced secretion of PRO-INFLAMMATORY CYTOKINES.

Apart from its direct antibacterial activity, several mechanisms have been suggested for the ANTI-INFLAMMATORY or immunosuppressive actions of rifaximin on the gastrointestinal tract (Table 29.6) [30, 31]. An important mode of action of rifampicin and rifaximin is the activation of pregnane X receptor, which is a nuclear receptor and transcription factor that regulates INFLAMMATION (Table 29.6) [31].

29.10.4 Immune and Anti-inflammatory Clinical Effects of Rifampicin and Analogues

29.10.4.1 Rheumatoid Arthritis

The first trials of rifampicin in early disease showed a possible improvement in RHEUMATOID ARTHRITIS, but later studies showed no improvement [32–34]. Intraarticular injection of rifampicin has been tested but did not improve response to intraarticular CORTICOSTEROID alone [35]. Further studies have not been considered worthwhile because of adverse effects, doubtful clinical effects and weak effects in vitro. The improved modern treatments of RHEUMATOID ARTHRITIS with METHOTREXATE and BIOLOGICALS also have decreased interest in the possible anti-antirheumatic actions of rifampicin (see Chap. 34).

29.10.4.2 Crohn's Disease and Ulcerative Colitis

Rifaximin is beneficial in the treatment of CROHN'S DISEASE and ulcerative colitis although possibly not as effective as other drugs [30, 31, 36] (see Chap. 35). Rifaximin may, however, be useful in combination with other drugs.

29.10.5 Adverse Effects of Rifampicin and Analogues

Liver dysfunction may be produced by rifampicin and should be checked routinely. The severe syndrome drug reaction with eosinophilia and systemic symptom (DRESS) has also been reported.

29.10.6 Drug Interactions of Rifampicin and Analogues

Rifampicin and rifaximin induce the metabolism of drugs by cytochrome P450 3A4 and several cytochrome P450 Cs and transport by P-glycoprotein [37]. The result is that the plasma concentrations of many other drugs are greatly reduced by concurrent dosage with rifampicin and rifaximin. It is strongly advised to check for interactions with any drugs that are taken with rifampicin or rifaximin. Induction of the metabolism of dapsone has been reported (see Sect. 29.11 below).

29.11 Sulphones/Sulphonamides

The major sulphone drug is dapsone (4,4′ diaminophenyl sulphone) (Fig. 29.7) which was initially developed as an antitubercular drug. It was tested in leprosy in the early 1950s and is still a part of drug combinations used in this disease. Apart from its use in leprosy, dapsone is administered orally for several cutaneous diseases, particularly dermatitis herpetiformis [38]. Dapsone is also applied locally for the treatment of acne vulgaris because of its antibacterial and ANTI-INFLAMMATORY actions.

The immunosuppressive and ANTI-INFLAMMATORY activities of two groups, sulphones and sulphonamides, are discussed together in this section because of their similar basic and clinical pharmacological properties. SULFAPYRIDNE is no longer used as an antibacterial agent, but it is still of research and clinical interest as it is the major metabolite of SULFASALAZINE, a widely used drug for RHEUMATOID ARTHRITIS and ulcerative colitis (Fig. 29.7) (see Chap. 34).

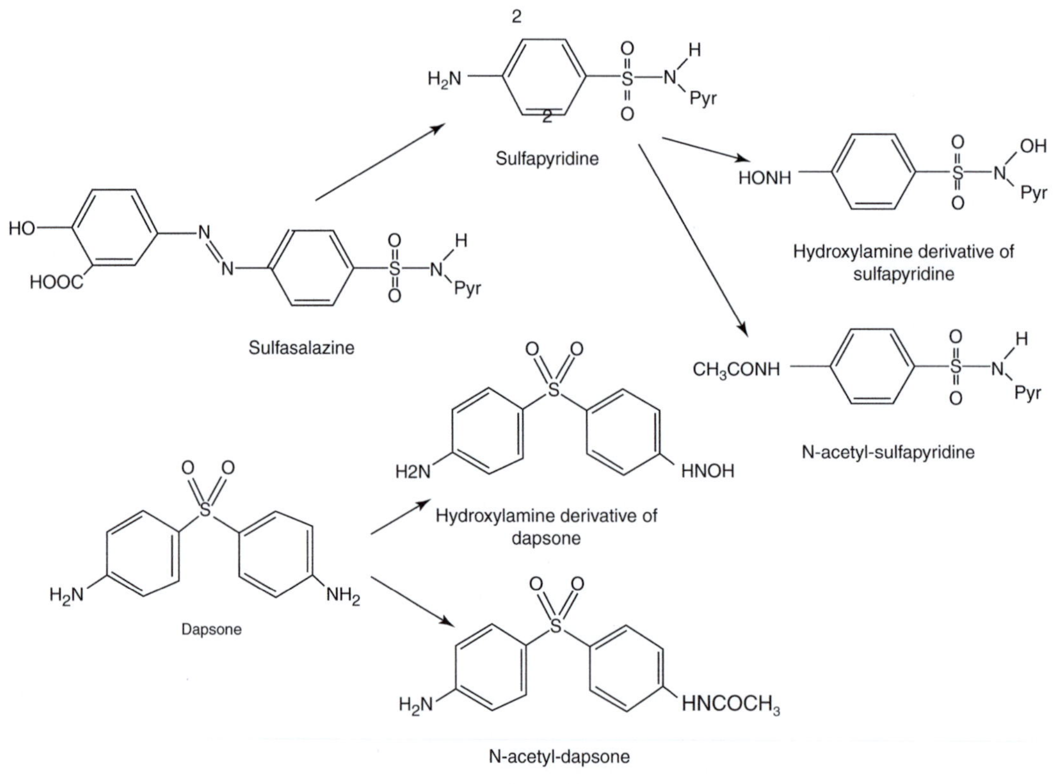

Fig. 29.7 Structures and metabolism of dapsone and sulfapyridine. Sulfapyridine is a metabolite of sulfasalazine. Both dapsone and sulfapyridine are metabolised to reactive hydroxylamine derivatives and also by acetylation, as shown

29.11.1 Metabolism and Pharmacokinetics of Dapsone and Sulfapyridine

The TERMINAL HALF-LIFE OF ELIMINATION of dapsone is about 30 h but is very variable. As a result, dapsone is usually administered once daily, sometimes twice daily. Dapsone is an amino compound which is both acetylated and also oxidised by cytochromc P450 enzymes. Resulting from its two major modes of elimination, the half-life is longer in genetic slow acetylators, while its oxidative metabolism is induced by rifampicin with which it may be administered. SULFAPYRIDNE has a half-life of approximately 14 h in slow acetylators but about 6 h in fast acetylators.

29.11.2 Basic Immune and Anti-inflammatory Pharmacology of Dapsone and Sulphonamides

The antibacterial activities of dapsone and the sulphonamides are due to inhibition of dihydropteroate synthase.

The ANTI-INFLAMMATORY activity of dapsone is indicated by several in vitro and in vivo findings (Table 29.7). The activity of dapsone in rats with implants of carrageenan and cotton pellets is notable because inhibition of INFLAMMATION caused by these treatments is characteristic of the (aspirin-like) non-steroidal ANTI-INFLAMMATORY drugs (see Chap. 33).

There is considerable interest in the ANTI-INFLAMMATORY pharmacology of sulphonamides because the well-known drug, SULFASALAZINE, is metabolised to aminosalicylate and SULFAPYRIDNE (see Chap. 34). Both metabolites are active in the treatment of RHEUMATOID ARTHRITIS, but only the parent, SULFASALAZINE, is used clinically [39]. Like dapsone, SULFAPYRIDNE is active in the treatment of neutrophilic diseases [38]. In particular, SULFAPYRIDNE and dapsone are active in many patients with ocular cicatricial pemphigoid [40].

The antirheumatic activity of SULFAPYRIDNE and some other sulphonamides is backed up by limited experimental pharmacology (Table 29.8). Like dapsone, sulfamethizole decreases the INFLAMMATION caused by carrageenan and cotton pellets (Table 29.8). SULFAPYRIDNE reacts readily with (scavenges) hydroxyl radicals which can be derived from superoxide which is formed in turn from superoxide from the oxidative burst of NEUTROPHILS and MONOCYTES (Table 29.8). Hydroxyl radicals react readily with

Table 29.7 Experimental pharmacological actions showing modulation of immune and inflammatory processes by dapsone

Pharmacological actions	References
In vitro	
In neutrophils, dapsone inhibits myeloperoxidase	[84, 85]
Dapsone inhibits the production of prostaglandin D_2 by rat mast cells	[86]
In vivo	
Dapsone shows anti-inflammatory activity in carrageenan and cotton pellet tests in rats	[87]
Normalisation of mucociliary transport by dapsone after intratracheal administration of LPS to ferrets	[88]
Improved neurological function by dapsone together with increases in the amount of spared tissue after spinal cord injury in rats by inhibiting apoptosis	[89]

Table 29.8 Experimental pharmacological actions showing modulation of immune and inflammatory processess by sulphonamides

Pharmacological actions	References
In vitro	
Sulfapyridne inhibits myeloperoxidase[a]	[85]
Sulfapyridne reacts readily with hydroxyl radical and other free radicals	[90, 91]
Sulfapyridne inhibits ROS of neutrophils after stimulation by receptor, fMLP and calcium ionophore	[92]
40% reduction in mRNA for TNF-α and 50% increase in mRNA for tissue inhibitor of metalloproteinase in rheumatoid synovial fibroblasts	[93]
In vivo	
Sulfamethizole decreases the inflammation caused by carrageenan and cotton pellets.[a] Other sulphonamides, sulphadiazine and sulphanilamide inactive	[87]

[a]Activity also shown by dapsone

biological molecules, and it is doubtful that scavenging could lead to a selective action of SULFAPYRIDNE.

29.11.3 Adverse Effects of Dapsone and Sulfapyridine

The major adverse effects of dapsone and SULFAPYRIDNE are haemolysis and methemoglobinuria. AGRANULOYTOSIS and skin rashes also occur but are uncommon. The adverse effects of both drugs are probably derived from reactive metabolites, particularly the hydroxylamine derivatives which may be cytotoxic [41] (Fig. 29.7). These are formed by oxidation of the amino groups of both dapsone and SULFAPYRIDNE, by MYELOPEROXIDASE and cytochrome P450 enzymes. Alternatively, the reactive metabolites could induce the production of antibodies that could cause the destruction of NEUTROPHIL precursors. In skin, the target cells may be keratinocytes which subsequently activate DENDRITIC CELLS and initiate an immune response within the skin [42].

29.12 Tetracyclines and Related Drugs

The tetracyclines are a group of widely used broad-spectrum antibiotics including minocycline, doxycycline and an individual antibiotic which is termed tetracycline. The tetracyclines interfere with bacterial protein synthesis, by acting on the 30S ribosomal subunit. The tetracyclines have multiple ANTI-INFLAMMATORY and IMMUNOMODULATORY effects which may make them useful in several chronic non-antibacterial diseases.

The tetracycline molecule has been chemically modified in multiple ways, generating second series of semisynthetic tetracyclines (e.g. doxycycline and minocycline) which have both antibiotic and ANTI-INFLAMMATORY properties. A new family of compounds called chemically modified tetracyclines (CMTs) lack antimicrobial activity but have retained some ANTI-INFLAMMATORY actions.

29.12.1 Chemistry of Tetracyclines

The tetracyclines are organic bases with p*K*a values of about 8 and are, therefore, largely present as the cationic (ionised forms) in blood. They are usually administered as their hydrochloride salts (Fig. 29.8). Many chemical analogues of the tetracyclines have been synthesised. The best known is CMT-3 which is a neutral unionised compound (Fig. 29.8).

29.12.2 Metabolism and Pharmacokinetics of Tetracyclines

The tetracyclines are well absorbed and have similar ranges of TERMINAL HALF-LIVES but different modes of elimination (Table 29.9). The dosage of tetracycline and doxycycline should be reduced in renal impairment because of their significant renal excretion. CMT-3 has a long TERMINAL HALF-LIFE (about 57 h) which allows dosage every day or every second day [43].

The tetracyclines are taken up by NEUTROPHILS by saturable, sodium-dependent transport [44]. The observed cellular/extracellular concentration ratios are greater than 60 for minocycline and >7 for doxycycline.

29.12.3 Basic Immune and Anti-inflammatory Pharmacology of Tetracyclines

There are two aspects of the activity of tetracyclines in pharmacological systems in vitro: inhibition of the respiratory burst and suppression of the release of INFLAMMATORY MEDIATORS by NEUTROPHILS. Many reports show the inhibitory action of tetracyclines on various phagocyte functions in vitro (Table 29.10). The production of REACTIVE OXYGEN SPECIES is apparently reduced, but the underlying mechanisms may include chelation of Ca^{2+} or Mg^{2+} or Zn^{2+} and scavenging of reactive products of the respiratory burst of NEUTROPHILS. These chelating actions make

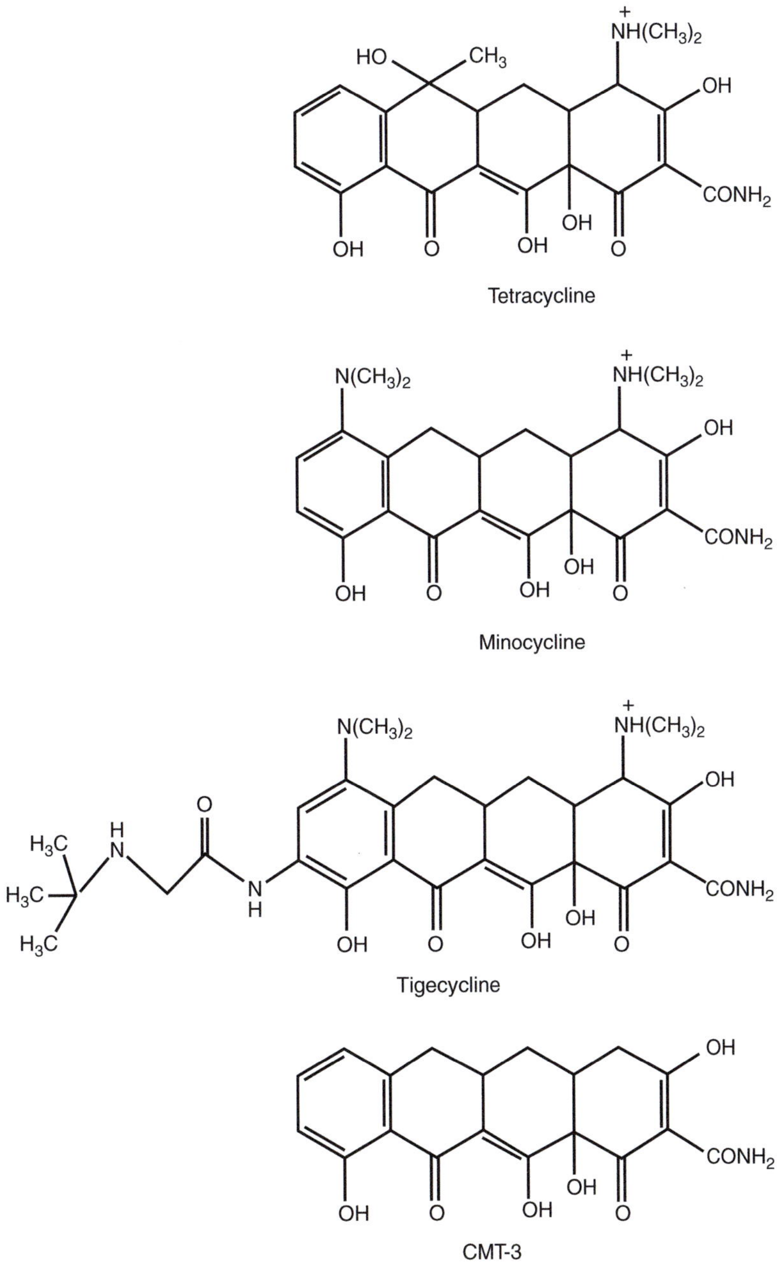

Fig. 29.8 Structures of tetracycline, minocycline and tigecycline in their ionised forms. CMT-3 is a non-antibacterial analogue which is not ionisable

Table 29.9 Pharmacokinetic properties and urinary excretion of tetracyclines

	Tetracycline	Doxycycline	Minocycline
Terminal half-life (h)	6–12	18–22	11–22
% Excretion unchanged in urine	60	40	10

Table 29.10 Experimental pharmacological actions showing modulation of immune and inflammatory processes by tetracyclines

Pharmacological actions	References
In vitro	
Tetracyclines inhibit the respiratory burst (including the production of hypochlorous acid) by neutrophils	[48, 49, 94]
Tetracyclines inhibit matrix metalloproteinases, particularly metalloproteinase-8 through binding to zinc or calcium	[48, 49, 94]
Doxycycline reduces the production of IL-8 and TNF-α by a human mast cell line	[50]
Tetracyclines decrease expression of iNOS by J774 macrophage cell line	[95]
Several proteins (including vimentin and heat shock protein 60) are downregulated by LPS in J774 cells	[95]
Minocycline inhibits collagenase derived from rat gingiva	[94]
Minocycline inhibits the synthesis of prostaglandin E2 and NO due to downregulation of COX-2 and iNOS	[96]
IgE production by PMBCs from with asthma patients is reduced when the cells were co-cultured with IL-4 and CD40 in the presence of minocycline or doxycycline	[97]
CMT-3 inhibits metalloproteinases and human leukocyte elastase by direct inhibition of the enzymes and by decreasing the breakdown of endogenous inhibitors of these enzymes (α_1-proteinase inhibitor and tissue inhibitors of matrix metalloproteinases)	[43]
In vivo	
Doxycycline and minocycline produce antinociception resulting from pain produced by the injection of formalin into the hind paw of the mouse, the carrageenan-induced oedema of the rat paw and leukocyte migration into the mice peritoneal cavity	[98]
CMT-3 prevents respiratory distress syndrome in pigs after the induction of sepsis and ischaemia/reperfusion injury	[99]
Oxytetracycline treatment inhibits influx of inflammatory cells, goblet-cell hyperplasia and concentrations of soluble inflammatory mediators in ovalbumin-induced asthma model in mice	[100]
Minocycline prevents nigrostriatal dopaminergic neurodegeneration in the 1-methyl-4-phenyl-1,2,3,6-tetrahydropyridine (MPTP) mouse model of Parkinson's disease	[101]
CMT-3 decreases brain concentration of TNF-α and activation of microglia after administration of LPS to mice	[102]

it very difficult to correlate therapeutic plasma concentrations with the concentrations in the in vitro experiments. Furthermore, tetracyclines are oxidisable by hypochlorous acid, potentially making the inhibition dependent upon the molecular ratio of concentrations of tetracycline to the yield of hypochlorous acid.

The tetracyclines inhibit metalloproteinases, particularly the matrix metalloproteinases which breakdown several INFLAMMATORY CYTOKINES and many proteins in the intracellular matrix. Matrix metalloproteinases also convert several factors, such as vascular endothelial growth factor (VEGF) and transforming growth factor-β (TGF-β), into their active forms. Overall, inhibition of matrix metalloproteinases may be responsible for much of the ANTI-INFLAMMATORY activity of tetracyclines.

The anti-oxidative actions of the tetracyclines indicate their potential use in the treatment of inflammatory states. Tetracyclines also inhibit several inflammatory diseases in mice and other experimental animals [10]. Treatment of these experimental conditions is consistent with their value in several chronic diseases in people.

29.12.4 Immune and Anti-inflammatory Clinical Effects of Tetracyclines

The tetracyclines have been tested in several inflammatory disease states, while CMT-3 may have activity in Kaposi's sarcoma but not in other sarcomas.

29.12.4.1 Rheumatoid Arthritis

Minocycline and doxycycline relieve the symptoms of RHEUMATOID ARTHRITIS with a better response seen in early-onset seropositive disease. A meta-analysis of ten studies indicates the activity of tetracyclines [45]. Two studies lasted 48 weeks but showed no significant reduction in erosions or joint space narrowing. Treatment of this disease by these tetracyclines is uncommon. Furthermore, alternative antirheumatic drugs are available (see Chap. 34), and it is unlikely that the tetracyclines will achieve widespread use in RHEUMATOID ARTHRITIS.

29.12.4.2 Osteoarthritis

Osteoarthritis may have an inflammatory component. Doxycycline slows the progression of osteoarthritis to a small degree, but there is no reduction in pain or disability [46, 47]. However, some patients may respond to a greater degree than others.

29.12.4.3 Dermatological Diseases

The tetracyclines are useful in treating inflammatory lesions of rosacea and acne such as erythema, papules, pustules and blepharitis but not sebaceous changes that do not appear to be inflammatory [48, 49]. Doses of doxycycline that are low (typically 20 mg twice daily) and considered insufficient for antibacterial activity appear active in the treatment of rosacea and acne. Several other uncommon dermatological diseases have also responded to tetracyclines. These include blistering disorders, chronic wounds (inability of wounds to heal) and GRANULOCYTE disorders (including sarcoidosis) [48, 49].

29.12.4.4 Periodontal Disease

Low-dose doxycycline is approved for treatment of periodontitis (pyorrhoea), a disease in which there is excessive INFLAMMATION due to bacteria adhering to the teeth [48].

29.12.4.5 Asthma

Minocycline lessens symptoms of ASTHMA and allows reduction in the dose of CORTICOSTEROIDS [50] (see also Chap. 23). Serum concentrations of IMMUNOGLOBULIN E decrease during treatment with minocycline without effects on immunoglobulins A, G and M.

29.12.5 Adverse Effects of Tetracyclines

The use of tetracyclines has been limited because of concerns about the development of resistant organisms resulting from their antibiotic activity. This is not a problem with the nonantibiotic CMTs which appear to have lesser adverse effects although a small proportion of patients have developed a lupus-like syndrome.

A variety of adverse effects have been associated with treatment by tetracyclines:

- Continued therapy for more than 2 years is associated with hyperpigmentation of skin and nails in 10–20% of patients and may take a year to resolve.
- Up to age of about 8 years, the tetracyclines may cause discolouration of teeth and hypoplasia of dental enamel, and their use should be avoided.
- Tetracyclines cause discolouration of babies' teeth following administration in the second half of pregnancy, and their use should be avoided at this time.
- Tetracyclines may depress prothrombin activity, and the dosage of warfarin should be decreased accordingly.
- Capsules of tetracyclines have caused ulceration of the oesophagus due to their retention in the oesophagus. They should be taken with milk or food to prevent this.
- The tetracyclines may cause supersensitivity to ultraviolet light (excessive sunburn). This adverse effect is also seen with some CMTs.
- Minocycline is associated with autoimmune reactions, a lupus-like syndrome and hepatitis, which are not shown by other tetracyclines. This tetracycline is metabolised by hepatic cytochrome P450 systems and the NEUTROPHIL enzyme, MYELOPEROXIDASE, to reactive products which may cause adverse effects specific to minocycline [51]. Appropriate monitoring is therefore indicated in patients early in polyarthritis when diagnostic uncertainty may still exist.

29.12.6 Drug Interactions of Tetracyclines

- The absorption of tetracyclines is reduced by complexation with iron salts and by both calcium and bismuth antacids. The combinations should be avoided. Alternatively, a 3 h delay between the administration of the tetracycline and the interacting compounds should decrease the extent of the interactions.
- The EFFICACY of oral anticoagulants (warfarin) may be increased, and oral contraceptives may become ineffective, both interactions occurring because of changes in the gut flora due to the antibiotic activities of the tetracyclines.

29.13 Tigecycline

Tigecycline is the first clinically available drug in a new class of antibiotics termed the glycylcyclines. Structurally, it is tetracycline with a central four-ring carbocyclic skeleton and is closely related to minocycline (Fig. 29.8).

Tigecycline has actions both in vitro and in vivo which indicate modulation of immunological and inflammatory processes. Both inhibition and activation have been reported (Table 29.11). As yet, tigecycline has not been tested for immunological or inflammatory diseases in people.

29.14 Conclusions and Future Research

It is now widely acknowledged that in addition to their antibacterial activity, several ANTIBACTERIAL AGENTS display IMMUNOMODULATORY and ANTI-INFLAMMATORY properties with potential therapeutic importance. Modulation of immune functions is presently a major focus of attention, particularly in inflammatory diseases and cancer. In particular, sulphones (dapsone), tetracyclines, macrolides and rifampicin and their analogues show inhibitory activity towards several initiators of the inflammatory cascade, as well as to mediators of tissue damage. However, lengthy administration and absence of selectivity of these antimicrobial immunomodulators can lead to the induction of microbial resistance. As a result, intensive research is ongoing, to identify IMMUNOMODULATORY antibiotic derivatives which are devoid of antibacterial activity, most notably with tetracycline and macrolide derivatives.

Finally, recent discoveries related to the role that MICROBIOTA plays, both in health and disease, may shed new light on possible ways in which ANTIBACTERIAL AGENTS can influence numerous pathological conditions.

Acknowledgement This is an update of the chapter from the third edition, written by Marie-Therese Labro, who died in 2016.

Table 29.11 Experimental pharmacological actions showing modulation of immune and inflammatory processes by tigecycline

Pharmacological actions	References
In vitro	
In a murine model of *Mycoplasma pneumoniae* pneumonia, tigecycline treatment has a modest microbiological effect, but it significantly decreases histological evidence of lung inflammation and reduces pulmonary cytokines and chemokines	[103]
Tigecycline has no influence on cytokine production by LPS—triggered human blood	[104]
Tigecycline prevents LPS-induced release of pro-inflammatory and mediators of apoptosis, NF-κB, TNF-α, IL-1β and NO, in neuronal cells	[105]
In the therapeutic concentration range, tigecycline potentiates the pro-inflammatory functions of human neutrophils in vitro by acting as a calcium ionophore	[106]
In vivo	
Tigecycline decreases hypotension but not inflammatory cytokines in pigs after infusion of endotoxin. This is a sterile model of sepsis	[107]

Selected Readings

Garrido-Mesa N, Zarzuelo A, Galvez J. Minocycline: far beyond an antibiotic. Br J Pharmacol. 2013;169(2):337–52.

Parnham MJ, Haber Erakovic V, Giamarellos-Bourboulis EJ, Perletti G, Verleden GM, Vos R. Azithromycin: mechanisms of action and their relevance for clinical applications. Pharmacol Ther. 2014;143(2):225–45.

Rothstein DM. Rifamycins, alone and in combination. Cold Spring Harbor Perspect Med. 2016;6(7):a027011.

References

1. Rudkowski R, Ziegler JB, Graham GG, Champion GD. Auranofin inhibits the activation pathways of polymorphonuclear leukocytes at multiple sites. Biochem Pharmacol. 1991;41:1921–9.
2. Haley P. Species differences in the structure and function of the immune system. Toxicology. 2003;188:49–71.
3. Thong YH. Immunomodulation by antimicrobial drugs. Med Hypotheses. 1982;8:361–70.
4. Umeki S. Action of gentamycin through inhibitory effect on neutrophil NADPH oxidase activity. Comp Biochem Physiol B Biochem Mol Biol. 1995;110(4):817–21.
5. Gressier B, Brunet C, Dine T, Luyckx M, Ballester L, Cazin M, et al. In vitro activity of aminoglycosides on the respiratory burst response in human polymorphonuclear neutrophils. Methods Find Exp Clin Pharmacol. 1998;20(10):19–24.
6. Sermet-Gaudelus I, Renouil M, Fajac A, Bidou L, Parbaille B, Pierrot S, et al. In vitro prediction of stop-codon suppression by intravenous gentamicin in patients with cystic fibrosis: a pilot study. BMC Med. 2007;5:5.
7. Labro MT. Cefodizime as a biological response modifier: a review of its in-vivo, ex-vivo and in-vitro immunomodulatory properties. J Antimicrob Chemother. 1990;26(Suppl C):37–47.
8. Meloni F, Ballabio P, Bianchi L, Grassi FA, Gialdroni Grassi GG. Cefodizime modulates in vitro tumor necrosis factor-alpha, interleukin-6 and interleukin-8 release from human peripheral monocytes. Chemotherapy. 1995;41(4):289–95.
9. Fietta A, Bersani C, Bertoletti R, Grassi F, Grassi G. In vitro and ex vivo enhancement of nonspecific phagocytosis by cefodizime. Chemotherapy. 1988;34(5):430–6.
10. Dabrowski MP, Stankiewicz W. Desirable and undesirable immunotropic effects of antibiotics: immunomodulating properties of cefaclor. J Chemother. 2001;13(6):615–20.
11. Bensinger WI, Buckner CD, Lilleby K, Holmberg L, Storb R, Slattery JT. Dose escalation of busulfan with pentoxifylline and ciprofloxacin in patients with breast cancer undergoing autologous transplants. Oncology. 2004;67(5–6):368–75.
12. Sauermann R, Marsik C, Steiner I, Seir K, Cvitko T, Zeitlinger M, et al. Immunomodulatory effects of fosfomycin in experimental human endotoxemia. Antimicrob Agents Chemother. 2007;51(5):1879–81.
13. Hassan HM, Guo H, Yousef BA, Guerram M, Hamdi AM, Zhang L, et al. Role of inflammatory and oxidative Stress, cytochrome P450 2E1, and bile acid disturbance in rat liver Injury induced by Isoniazid and lipopolysaccharide cotreatment. Antimicrob Agents Chemother. 2016;60(9):5265–93.
14. Maiocchi SL, Morris JC, Rees MD, Thomas SR. Regulation of the nitric oxide oxidase activity of myeloperoxidase by pharmacological agents. Biochem Pharmacol. 2017;135(1):90–115.
15. Hirata N, Hiramatsu K, Kishi K, Yamasaki T, Ichimiya T, Nasu M. Pretreatment of mice with clindamycin improves survival of endotoxic shock by modulating the release of inflammatory cytokines. Antimicrob Agents Chemother. 2001;45(9):2638–42.
16. Nakano T, Hiramatsu K, Kishi K, Hirata N, Kadota J, Nasu M. Clindamycin modulates inflammatory-cytokine induction in lipopolysaccharide-stimulated mouse peritoneal macrophages. Antimicrob Agents Chemother. 2003;47(1):363–7.
17. Wulansari R, Wijaya A, Ano H, Horii Y, Makimura S. Lymphocyte subsets and specific IgG antibody levels in clindamycin-treated and untreated dogs experimentally infected with Babesia gibsoni. J Vet Med Sci. 2003;65:579–84.
18. Haber VE, Bosnar M, Kragol G. The design of novel classes of macrolides for neutrophil-dominated inflammatory diseases. Future Med Chem. 2014;6:657–74.
19. Schultz MJ. Macrolide activities beyond their antimicrobial effects: macrolides in diffuse panbronchiolitis and cystic fibrosis. J Antimicrob Chemother. 2004;54:21–8.
20. Li H, Zhou Y, Fan F, Zhang Y, Li X, Yu H, et al. Effect of azithromycin on patients with diffuse panbronchiolitis: retrospective study of 51 cases. Intern Med. 2011;50(16):1663–9.
21. Southern KW, Barker PM. Azithromycin for cystic fibrosis. Eur Respir J. 2004;24:834–8.
22. Saiman L. The use of macrolide antibiotics in patients with cystic fibrosis. Curr Opin Pulmon Med. 2004;10:515–23.
23. Cramer CL, Patterson A, Alchakaki A, Soubani AO. Immunomodulatory indications of azithromycin in respiratory disease: a concise review for the clinician. Postgrad Med. 2017;129(5):493–9.
24. Johnston SL, Szigeti M, Cross M, Brightling C, Chaudhuri R, Harrison T, et al. Azithromycin for acute exacerbations of asthma: the AZALEA randomized clinical trial. JAMA Intern Med. 2016;176(11):1630–7.
25. Richeldi L, Ferrara G, Fabbri LM, Lasserson TJ, Gibson PG. Macrolides for chronic asthma. Cochrane Data Sys Rev. 2005;(4):CD002997.

26. Kew KM, Undela K, Kotortsi I, Ferrara G. Macrolides for chronic asthma. Cochrane Data Sys Rev. 2015;(9):CD002997.
27. Hayes D. A review of bronchiolitis obliterans syndrome and therapeutic strategies. J Cardiothorac Surg. 2011;6:92.
28. Corris PA, Ryan VA, Small T, Lordan J, Fisher AJ, Meachery G, et al. A randomised controlled trial of azithromycin therapy in bronchiolitis obliterans syndrome (BOS) post lung transplantation. Thorax. 2015;70(5):442–50.
29. Blandizzi C, Viscomi G, Marzo A, Scarpignato C. Is generic rifaximin still a poorly absorbed antibiotic? A comparison of branded and generic formulations in healthy volunteers. Pharmacol Res. 2014;85(1): 39–44.
30. DuPont HL. Therapeutic effects and mechanisms of action of rifaximin in gastrointestinal diseases. Mayo Clin Proc. 2015;90(8):1116–24.
31. Sartor RB. Review article: The potential mechanisms of action of rifaximin in the management of inflammatory bowel diseases. Aliment Pharmacol Ther. 2016;43(Suppl 1):27–36.
32. McConkey B, Situnayake RD. Effects of rifampicin with and without isoniazid in rheumatoid arthritis. J Rheumatol. 1988;15(1):46–50.
33. Cox NL, Prowse MV, Maddison MC, Maddison PJ. Treatment of early rheumatoid arthritis with rifampicin. Ann Rheum Dis. 1992;51(1):32–4.
34. Borg AA, Davis MJ, Fowler PD, Shadforth MF, Dawes PT. Rifampicin in early rheumatoid arthritis. Scand J Rheumatol. 1993;22(1):39–42.
35. Blyth T, Stirling A, Coote J, Land D, Hunter JA. Injection of the rheumatoid knee: does intra-articular methotrexate or rifampicin add to the benefits of triamcinolone hexacetonide? Br J Rheumatol. 1998;37(7):770–2.
36. Guslandi M, Giollo P, Testoni PA. Corticosteroid-sparing effect of rifaximin, a nonabsorbable oral antibiotic, in active ulcerative colitis: preliminary clinical experience. Curr Ther Res. 2004;65(3):292–6.
37. Niemi M, Backman JT, Fromm MF, Neuvonen PJ, Kivisto KT. Pharmacokinetic interactions with rifampicin: clinical relevance. Clin Pharmacokinet. 2003;42(9):819–50.
38. Dallegri F, Ottonello L, Bevilacqua M. Possible modes of action of nimesulide in controlling neutrophilic inflammation. Arzneimittelforschung. 1995;45(10):1114–7.
39. Neumann VC, Taggart AJ, Le Gallez P, Astbury C, Hill J, Bird HA. A study to determine the active moiety of sulphasalazine in rheumatoid arthritis. J Rheumatol. 1986;13(2):285–7.
40. Elder MJ, Leonard J, Dart JKG. Sulphapyridine—a new agent for the treatment of ocular cicatricial pemphigoid. Br J Ophthalmol. 1996;80:549–52.
41. Uetrecht J. Drug metabolism by leukocytes and its role in drug-induced lupus and other idiosyncratic drug reactions. Crit Rev Toxicol. 1990;20(4): 213–35.
42. Khan FD, Vyas PM, Gaspari AA, Svensson CK. Effect of arylhydroxylamine metabolites of sulfamethoxazole and dapsone on stress signal expression in human keratinocytes. J Pharmacol Exp Ther. 2007;323:771–7.
43. Gu Y, Lee HM, Simon SR, Golub LM. Chemically modified tetracycline-3 (CMT-3): a novel inhibitor of the serine proteinase, elastase. Pharmacol Res. 2011;64(6):595–601.
44. Walters JD. Characterization of minocycline transport by human neutrophils. J Periodontol. 2006;77:1964–8.
45. Stone M, Fortin PR, Pacheco-Tena C, Inman RD. Should tetracycline treatment be usedmore extensively for rheumatoid arthritis? Metaanalysis demonstrates clinical benefit with reduction in disease activity. J Rheumatol. 2003;30(10):2112–22.
46. Snijders GF, van den Ende CH, van Riel PL, van den Hoogen FH, den Broeder AA. The effects of doxycycline on reducing symptoms in knee osteoarthritis: results from a triple-blinded randomised controlled trial. Ann Rheum Dis. 2011;70(7):1191.
47. da Costa BR, Nësch E, Reichenbach S, Juni P, Rutjes AWS. Doxycycline for osteoarthritis of the knee or hip. Cochrane Database Syst Rev 2012;(11):CD007323.
48. Korting HC, Schöllmann C. Tetracycline actions relevant to rosacea treatment. Skin Pharmacol Physiol. 2009;22(6):287–94.
49. Monk E, Shalita A, Siegel DM. Clinical applications of non-antimicrobial tetracyclines in dermatology. Pharmacol Res. 2011;63(2):130–45.
50. Joks R, Durkin HG. Non-antibiotic properties of tetracyclines as anti-allergy and asthma drugs. Pharmacol Res. 2011;64(6):602–9.
51. Mannargudi B, McNally D, Reynolds W, Uetrecht J. Bioactivation of minocycline to reactive intermediates by myeloperoxidase, horse peroxidase, and hepatic microsomes: implications for minocycline-induced lupus and hepatitis. Drug Metab Dispos. 2009;37(9):1806–18.
52. Rayaman P, Rayaman E, Cevikbas A, Demirtunc R, Sehirli AO, Alagoz SG, et al. Effect of antibiotics on polymorphonuclear leukocyte functions and myeloperoxidase activity, glutathione and malondialdehyde levels in allergic asthma. Polish J Microbiol. 2015;64(1):69–72.
53. Mori S, Takahashi HK, Liu K, Wake H, Zhang J, Liu T, et al. Ciprofloxacin inhibits advanced glycation end products-induced adhesion molecule expression on human monocytes. Br J Pharmacol. 2010;161:229–40.
54. Riesbeck K, Forsgren A. Selective enhancement of synthesis of interleukin-2 in lymphocytes in the presence of ciprofloxacin. Eur J Clin Microbiol Infect Dis. 1990;9:409–13.
55. Werber S, Shalit I, Fabian I, Steuer G, Weiss T, Blau H. Moxifloxacin inhibits cytokine-induced MAP kinase and NF-kB activation as well as nitric oxide synthesis in a human respiratory epithelial cell line. J Antimicrob Chemother. 2005;55:293–300.

56. Blau H, Klein K, Shalit I, Halperin D, Fabian I. Moxifloxacin but not ciprofloxacin or azithromycin selectively inhibits IL-8, IL-6, ERK1/2, JNK, and NF-kappaB activation in a cystic fibrosis epithelial cell line. Am J Physiol Lung Cell Mol Physiol. 2007;292(1):L343–52.
57. Bailly S, Fay M, Ferrua B, Gougerot-Pocidalo MA. Ciprofloxacin treatment in vivo increases the ex vivo capacity of lipopolysaccharide-stimulated human monocytes to produce IL-1, IL-6 and tumour necrosis factor-alpha. Clin Exp Immunol. 1991;85:331–4.
58. Ogino H, Fujii M, Ono M, Maezawa K, Hori S, Kizu J. In vivo and in vitro effects of fluoroquinolones on lipopolysaccharide-induced pro-inflammatory cytokine production. J Infect Chemother. 2009;15(3):166–73.
59. Lahat G, Halperin D, Barazovsky E, Shalit I, Rabau M, Klausner J, et al. Immunomodulatory effects of ciprofloxacin in TNBS-induced colitis in mice. Inflamm Bowel Dis. 2007;13(5):557–65.
60. Toklu HZ, Kabasakal L, Imeryuz N, Kan B, Celikel C, Cetinel S, et al. A study comparing the efficacy of antimicrobial agents versus enzyme (P-gp) inducers in the treatment of 2,4,6 trinitrobenzenesulfonic acid-induced colitis in rats. J Physiol Pharmacol. 2013;64:439–51.
61. Yoneshima Y, Ichiyama T, Ayukawa H, Matsubara T, Furukawa S. Fosfomycin inhibits NF-kappaB activation in U-937 and Jurkat cells. Int J Antimicrob Agents. 2003;21(6):589–92.
62. Zeitlinger M, Marsik C, Steiner I, Sauermann R, Seir K, Jilma B, et al. Immunomodulatory effects of fosfomycin in an endotoxin model in human blood. J Antimicrob Chemother. 2007;59(2):219–23.
63. Matsumoto T, Tateda K, Miyazaki S, Furuya N, Ohno A, Ishii Y, et al. Fosfomycin alters lipopolysaccharide-induced inflammatory cytokine production in mice. Antimicrob Agents Chemother. 1999;43(3):697–8.
64. Morikawa K, Nonaka M, Torii I, Morikawa S. Modulatory effect of fosfomycin on acute inflammation in the rat air pouch model. Int J Antimicrob Agents. 2003;21(4):334–9.
65. Genovese F, Mancuso G, Cuzzola M, Cusumano V, Nicoletti F, Bendtzen K, et al. Improved survival and antagonistic effect of sodium fusidate on tumor necrosis factor alpha in a neonatal mouse model of endotoxin shock. Antimicrob Agents Chemother. 1996;40:1733–5.
66. Nicoletti F, Beltrami B, Raschi E, Di Marco R, Magro G, Grasso S, et al. Protection from concanavalin A (Con A)-induced T cell-dependent hepatic lesions and modulation of cytokine release in mice by sodium fusidate. Clin Exp Immunol. 1997;110(3):479–84.
67. Di Marco R, Khademi M, Wallstrom E, Muhallab S, Nicoletti F, Olsson T. Amelioration of experimental allergic neuritis by sodium fusidate (fusidin): suppression of IFN-gamma and TNF-alpha and enhancement of IL-10. J Autoimmun. 1999;13(2):187–95.
68. Kilic FS, Erol K, Batu O, Yildirim E, Usluer G. The effects of fusidic acid on the inflammatory response in rats. Pharmacol Res. 2002;45(4):265–7.
69. Kikuchi T, Hagiwara K, Honda Y, Gomi K, Kobayashi T, Takahashi H, et al. Clarithromycin suppresses lipopolysaccharide-induced interleukin-8 production by human monocytes through AP-1 and NF-kappa B transcription factors. J Antimicrob Chemother. 2002;45(5):745–55.
70. Abe S, Nakamura H, Inoue S, Takeda H, Saito H, Kato S, et al. Interleukin-8 gene repression by clarithromycin is mediated by the activator protein-1 binding site in human bronchial epithelial cells. Am J Respir Cell Mol Biol. 2000;22(1):51–60.
71. Ichiyama T, Nishikawa M, Yoshitomi T, Hasegawa S, Matsubara T, Hayashi T, et al. Clarithromycin inhibits NF-kB activation in human peripheral blood mononuclear cells and pulmonary epithelial cells. Antimicrob Agents Chemother. 2001;45(1):44–7.
72. Marjanović N, Bosnar M, Michielin F, Willé DR, Anić-Milić T, OgnjenČulić O, et al. Macrolide antibiotics broadly and distinctively inhibit cytokine and chemokine production by COPD sputum cells in vitro. Pharmacol Res. 2011;63(5):389–97.
73. Bosnar M, Snježana Čužić S, Bošnjak B, Nujić K, Ergović G, Marjanović N, et al. Azithromycin inhibits macrophage interleukin-1β production through inhibition of activator protein-1 in lipopolysaccharide-induced murine pulmonary neutrophilia. Int Immunopharmacol. 2011;11:424–34.
74. Gualdoni GA, Lingscheid T, Schmetterer KG, Hennig A, Steinberger P, Zlabinger GJ. Azithromycin inhibits IL-1 secretion and non-canonical inflammasome activation. Sci Rep. 2015;5:12016.
75. Nozoe K, Aida Y, Fukuda T, Sanui T, Nishimura F. Mechanisms of the macrolide-induced inhibition of superoxide generation by neutrophils. Inflammation. 2016;39(3):1039–48.
76. Gray GD, Smith CW, Hollers JC, Chenoweth DE, Fiegel VD, Nelson RD. Rifampin affects polymorphonuclear leukocyte interactions with bacterial and synthetic chemotaxins but not interactions with serum-derived chemotaxins. Antimicrob Agents Chemother. 1983;24:777–83.
77. Ziglam HM, Daniels I, Finch RG. Immunomodulating activity of rifampicin. J Chemother. 2004;16(4):357–61.
78. Mlambo G, Sigola LB. Rifampicin and dexamethasone have similar effects on macrophage phagocytosis of zymosan, but differ in their effects on nitrite and TNF-α production. Int Immunopharmacol. 2003;3:513–22.
79. Maeda M, Ishii H, Tanaka S, Onda K, Hirano T. Suppressive efficacies of antimicrobial agents against human peripheral-blood mononuclear cells stimulated with T cell mitogen and bacterial superantigen. Arzneimittelforschung. 2011;60:760–8.

80. Pahlevan AA, Wright DJM, Bradley L, Smith C, Foxwell BMJ. Potential of rifamides to inhibit TNF-induced NF-kappaB activation. J Antimicrob Chemother. 2002;49:531–4.
81. Yuhas Y, Azoulay-Alfaguter I, Berent E, Ashkenazi S. Rifampin inhibits prostaglandin E2 production and arachidonic acid release in human alveolar epithelial cells. Antimicrob Agents Chemother. 2007;51(12):4225–30.
82. Yuhas Y, Berent E, Ovadiah H, Azoulay I, Ashkenazi S. Rifampin augments cytokine-induced nitric oxide production in human alveolar epithelial cells. Antimicrob Agents Chemother. 2006;50(1):396–8.
83. Yuhas Y, Berent E, Ashkenazi S. Effect of rifampin on production of inflammatory mediators in HepG2 liver epithelial cells. Antimicrob Agents Chemother. 2011;55(12):5541–6.
84. Stendahl O, Molin L, Dahlgren C. The inhibition of polymorphonuclear leukocyte cytotoxicity by dapsone. A possible mechanism in the treatment of dermatitis herpetiformis. J Clin Invest. 1978;62(1):214–20.
85. Ottonello L, Dapino P, Scirocco MC, Balbi A, Bevilacqua M, Dallegri F. Sulphonamides as anti-inflammatory agents: old drugs for new therapeutic strategies in neutrophilic inflammation? Clin Sci. 1995;88(3):331–6.
86. Ruzicka T, Wasserman SI, Soter NA, Printz MP. Inhibition of rat mast cell arachidonic acid cyclooxygenase by dapsone. J Allergy Clin Immunol. 1983;72:365–70.
87. Hiremath SV, Gouripur VV, Patil PA. The anti-inflammatory activity of some sulphonamides in albino rats. Ind J Med Res. 1996;103:120–5.
88. Kanoh S, Tanabe T, Rubin BK. Dapsone inhibits IL-8 secretion from human bronchial epithelial cells stimulated with lipopolysaccharide and resolves airway inflammation in the ferret. Chest. 2011;140:980–90.
89. Ríos C, Orozco-Suarez S, Salgado-Ceballos H, Mendez-Armenta M, Nava-Ruiz C, Santander I, et al. Anti-apoptotic effects of dapsone after spinal cord injury in rats. Neurochem Res. 2015;40:1243–51.
90. Aruoma OI, Wasil M, Halliwell B, Hoey BM, Butler J. The scavenging of oxidants by sulphasalazine and its metabolites. A possible contribution to their anti-inflammatory effects? Biochem Pharmacol. 1987;36(21):3739–42.
91. Joshi R, Kumar S, Unnikrishnan M, Mukherjee T. Free radical scavenging reactions of sulfasalazine, 5-aminosalicylic acid and sulfapyridine: mechanistic aspects and antioxidant activity. Free Rad Res. 2009;39(11):1163–72.
92. Kanerud L, Hafstrom I, Ringertz B. Effect of sulphasalazine and sulphapyridine on neutrophil superoxide production: role of cytosolic free calcium. Ann Rheum Dis. 1990;49(5):296–300.
93. Minghetti PP, Blackburn WDJ. Effects of sulfasalazine and its metabolites on steady state messenger RNA concentrations for inflammatory cytokines, matrix metalloproteinases, and tissue inhibitors of metalloproteinase in rheumatoid synovial fibroblasts. J Rheumatol. 2000;27(3):653–60.
94. Griffin MO, Fricovsky E, Ceballos G, Villarreal F. Tetracyclines: a pleitropic family of compounds with promising therapeutic properties. Review of the literature. Am J Physiol Renal Physiol. 2010;299:C539–48.
95. Dunston CR, Griffiths HR, Lambert PA, Staddon S, Vernallis AB. Proteomic analysis of the anti-inflammatory action of minocycline. Proteomics. 2011;11:42–51.
96. Kim SS, Kong PJ, Kim BS, Sheen DH, Nam SY, Chun W. Inhibitory action of minocycline on lipopolysaccharide-induced release of nitric oxide and prostaglandin E2 in BV2 microglial cells. Arch Pharm Res. 2004;27(3):314–8.
97. Smith-Norowitz TA, Bluth MH, Drew H, Norowitz KB, Chice S, Shah VN, et al. Effect of minocycline and doxycycline on IgE responses. Ann Allergy Asthma Immunol. 2002;89:172–9.
98. Leite LM, Carvalho AGG, Ferreira PL, Pessoa IX, Goncalves DO, Lopes ADA, et al. Anti-inflammatory properties of doxycycline and minocycline in experimental models: an in vivo and in vitro comparative study. Inflammopharmacology. 2011;19(2):99–110.
99. Roy SK, Kubiak BD, Albert SP, Vieau CJ, Gatto L, Golub L. Chemically modified tetracycline 3 prevents acute respiratory distress syndrome in a porcine model of sepsis + ischemia/reperfusion-induced lung injury. Shock. 2012;37(4):424–32.
100. Ci X, Chu X, Chen C, Li X, Yan S, Wang X, et al. Oxytetracycline attenuates allergic airway inflammation in mice via inhibition of the NF-κB pathway. J Clin Immunol. 2011;31:216–27.
101. Du Y, Ma Z, Lin S, Dodel RC, Gao F, Bales KR, et al. Minocycline prevents nigrostriatal dopaminergic neurodegeneration in the MPTP model of Parkinson's disease. Proc Natl Acad Sci U S A. 2001;98:14669–74.
102. Edan RA, Luqmani YA, Masocha W. COL-3, a chemically modified tetracycline, inhibits lipopolysaccharide-induced microglia activation and cytokine expression in the brain. PLoS One. 2013;8(2):e57827.
103. Salvatore CM, Techasaensiri C, Tagliabue C, Katz K, Leos N, Gomez AM, et al. Tigecycline therapy significantly reduces the concentrations of inflammatory pulmonary cytokines and chemokines in a murine model of Mycoplasma pneumoniae pneumonia. Antimicrob Agents Chemother. 2009;53:1546–51.
104. Traunmüller F, Thallinger C, Hausdorfer J, Lambers C, Tzaneva S, Kampitsch T, et al. Tigecycline has no effect on cytokine release in an ex vivo endotoxin model of human whole blood. Int J Antimicrob Agents. 2009;33:583–6.

105. Yagnik RM, Benzeroual KE. Tigecycline prevents LPS-induced release of pro-inflammatory and apoptotic mediators in neuronal cells. Toxicol In Vitro. 2013;27:686–93.
106. Cockeran R, Mutepe ND, Theron AJ, Tintinger GR, Steel HC, Stivaktas PI, et al. Calcium-dependent potentiation of the pro-inflammatory functions of human neutrophils by tigecycline in vitro. J Antimicrob Chemother. 2012;67:130–7.
107. von Seth M, Sjolin J, Larsson A, Eriksson M, Hillered L, Lipcsey M. Effects of tigecycline and doxycycline on inflammation and hemodynamics in porcine endotoxemia: a prospective, randomized, and placebo-controlled trial. Shock. 2015;43(6):604–11.

30 Cytotoxic Drugs

Sabine Grösch, Guido Bocci, Antonello Di Paolo, and Romano Danesi

30.1 Background

Cytotoxic immunosuppressive agents have a long-standing important role in pharmacological IMMUNOSUPPRESSION. Azathioprine was among the first immunosuppressive drugs used in organ transplantation. A further development in this field was landmarked by the introduction of ALKYLATING AGENTS (i.e. cyclophosphamide) and ANTIMETABOLITES (i.e. methotrexate and mycophenolic acid) in therapeutic regimens for the prevention of graft rejection (see Chap. 31) and the treatment of AUTOIMMUNE DISEASES (see Chaps. 34 and 35) because of their well-documented lymphocytolytic effect.

The role of cytotoxic drugs is being challenged by steroids, non-steroidal anti-inflammatory agents (NSAIDs), poly- and MONOCLONAL ANTIBODIES, calcineurin and mTOR inhibitors (discussed in more detail in other chapters). However, recent studies indicate that mycophenolate mofetil is superior in treatment of acute rejection in comparison to azathioprine [1] which makes it the antimetabolite of choice in immunosuppressive protocols.

Final manuscript submitted on February 26, 2018.

S. Grösch (✉)
Institute of Clinical Pharmacology, Faculty of Medicine, Goethe-University Frankfurt, Frankfurt am Main, Germany

G. Bocci · A. Di Paolo · R. Danesi
Clinical Pharmacology and Pharmacogenetics Unit, Department of Clinical and Experimental Medicine, University of Pisa, Pisa, Italy
e-mail: guido.bocci@med.unipi.it; antonello.dipaolo@med.unipi.it; romano.danesi@unipi.it

30.2 Azathioprine

30.2.1 Introduction

Azathioprine, an imidazolyl derivative of 6-mercaptopurine, was developed in the 1950s to improve the BIOAVAILABILITY of its parent drug, mercaptopurine [2]. Animal studies demonstrated that azathioprine had a higher THERAPEUTIC INDEX and was a better immunosuppressant than mercaptopurine. Azathioprine is still used in combination regimens with steroids and calcineurin inhibitors in patients receiving solid-organ transplants, as well as in AUTOIMMUNE DISEASE therapy in rheumatology, dermatology and gastroenterology [2]. Its use as an immunosuppressive and a corticosteroid-sparing agent is being replaced by mycophenolate mofetil, which is considered a safer and more effective agent. However, hepatitis C virus (HCV) load and HCV recurrence seem to be slightly better inhibited by azathioprine than by mycophenolate mofetil [3].

30.2.2 Chemical Structure

Azathioprine is an antimetabolite prodrug for 6-mercaptopurine, with an imidazolyl group

M. J. Parnham et al. (eds.), *Nijkamp and Parnham's Principles of Immunopharmacology*,
https://doi.org/10.1007/978-3-030-10811-3_30

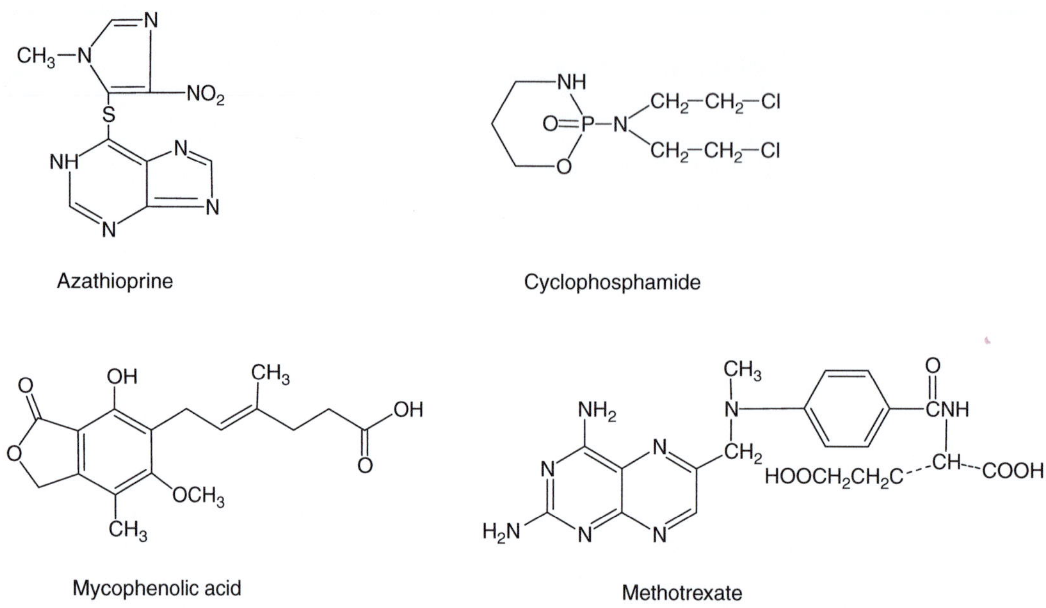

Fig. 30.1 Chemical structures of azathioprine, cyclophosphamide, methotrexate and mycophenolic acid

attached to the SH group of 6-mercaptopurine (Fig. 30.1) to protect it from in vivo oxidation. In tissues, azathioprine is non-enzymatically converted to mercaptopurine.

30.2.3 Mechanism of Action and Pharmacological Effect

Azathioprine is a cytotoxic antimetabolite inhibitor of nucleic acid synthesis. In particular, its metabolite 6-mercaptopurine is further activated intracellularly by ANABOLIC BIOTRANSFORMATION to 6-thioinosinic and 6-thioguanine acid which interfere with the metabolism of inosine monophosphate (IMP) to adenosine monophosphate (AMP) and triphosphate (ATP), thereby impairing the purine de novo biosynthetic pathway. Moreover, active metabolites are incorporated into RNA as well as DNA, and its replication is inhibited. The drug suppresses the proliferation of T and B LYMPHOCYTES and reduces the number of cytotoxic T CELLS and plasma cells in circulation and peripheral organs, thereby decreasing the immunological reactivity of the host [4]. For these reasons, azathioprine may exert a modest anti-inflammatory effect.

30.2.4 Pharmacokinetics

Azathioprine is rapidly absorbed within 1–2 h after administration and evenly distributed in all tissues, although the drug does not cross the blood-brain barrier. Azathioprine is rapidly converted in the liver and ERYTHROCYTES to 6-mercaptopurine and *S*-methyl-4-nitro-5-thioimidazole by sulfhydryl-containing compounds. Although it is generally recognized that 6-mercaptopurine is the active drug, a previous study has suggested that the imidazolyl moiety might have immunosuppressive activity on its own [5]. After standard oral doses, the terminal half-lives (t1/2) of azathioprine and 6-mercaptopurine are 50 and 74 min, respectively. Azathioprine is mainly metabolized by xanthine oxidase, followed by thiopurine methyltransferase (TPMT) and hypoxanthine-guanine phosphoribosyltransferase (HGPRT). 6-Mercaptopurine is inactivated by xanthine oxidase to 6-thiouric acid and by the widely distributed enzyme TPMT to 6-methylmercaptopurine, with *S*-adenosylmethionine as the methyl donor. The anabolic pathway is dependent on the enzyme HGPRT of the purine salvage metabolic pathway, with subsequent multienzymatic steps leading to the formation of cytotoxic 6-thioguanine nucleotides. These active metabolites can accumulate in tissues, where they are catabolized or incorporated into nucleic acids. In addition to its action on 6-mercaptopurine, TPMT can also methylate metabolites of the HGPRT pathway, including thioinosine monophosphate, which is in turn a potent inhibitor of de novo purine synthesis. While xanthine oxidase is not believed to play a significant role in 6-mercaptopurine inactivation at the level of haematopoietic tissues, TPMT activity is the principal detoxification pathway for the cytotoxic thioguanine nucleotides in the BONE MARROW. Haematopoietic toxicity is thus largely dependent on the activity of TPMT, a polymorphic enzyme with genetic variants characterized by low activity and increased risk of severe toxicity in patients [6]. Optimization of azathioprine treatment may thus be performed by TPMT genotyping as well as determination of erythrocyte 6-TGN levels.

30.2.5 Clinical Indications

Azathioprine is an approved drug for renal transplantation and severe rheumatoid arthritis. Its EFFICACY has also been demonstrated in the management of severe ulcerative colitis, and other autoimmune disorders, including bullous diseases in dermatology [7]. In myasthenia gravis, azathioprine, usually in combination with corticosteroids, is a basic immunosuppressive therapy option [8]. Azathioprine is also used as a steroid-sparing agent and has also been administered in cardiac transplantation. The initial dose of azathioprine is 3–5 mg/kg/day; the intravenous formulation may be used postoperatively followed by drug administration by oral route. In combination with cyclosporine and steroids, the dose may be lowered to 1–3 mg/kg/day as a maintenance level. During a 6-month treatment with azathioprine along with cyclosporine microemulsion and steroids in recipients of cadaveric kidney transplants, 35% had clinical rejections, compared with 12% over 15 additional months without steroids [3]. EFFICACY of azathioprine depends on specific genotypes as rheumatoid arthritis patients with the 94C>A mutation in the inosine triphosphatase (ITPA) gene are more responsive to this drug [9].

30.2.6 Adverse Reactions

The dose of azathioprine should be reduced in patients administered allopurinol, because the inhibition of xanthine oxidase reduces the metabolic inactivation of azathioprine and increases the drug's adverse events. Overall, up to 20% of patients may discontinue treatment due to toxicity [10]. Haematopoietic toxicity, including neutropenia, anaemia and thrombocytopenia, is the most common dose-limiting adverse effect, and it is frequently associated with low TPMT activity. Hepatic toxicity is the second most common adverse event and occurs independently of TPMT activity which might be correlated with the production of 6-methylmercaptopurine; therefore, therapeutic monitoring of 6-methylmercaptopurine levels may be useful in identifying patients at risk for hepatotoxicity. Gastrointestinal adverse events (mucositis, nausea, vomiting, abdominal pain, diarrhoea, pancreatitis), neurotoxicity and photosensitive eruptions may occur. Severe hypersensitivity reactions with multi-organ involvement are uncommon. Finally, the incidence of squamous cell carcinoma of the skin and lymphoproliferative malignancies is increased with azathioprine [11].

30.3 Cyclophosphamide

30.3.1 Introduction

Cyclophosphamide is an antineoplastic and immunosuppressive agent used for the treatment of solid and haematological malignancies as well as severe AUTOIMMUNE DISEASES, including systemic lupus erythematosus, sclerodermia and vasculitis. Although cyclophosphamide has been used in clinical practice since the 1950s, its therapeutic use is still widespread and only partially challenged by the introduction of newer drugs.

30.3.2 Chemical Structure

Cyclophosphamide (Fig. 30.1) is a first-generation oxazaphosphorine alkylating agent. It belongs to the group of nitrogen mustards and, as do the other members of the family, has the property of becoming strongly electrophilic in body fluids and forming stable, covalent linkages by alkylation of various nucleophilic moieties, particularly the N7 of guanine residues of DNA.

30.3.3 Mechanism of Action and Pharmacological Effect

Cyclophosphamide acts via its principal active metabolite, phosphoramide mustard, through several mechanisms. At the molecular level, phosphoramide mustard is able to bind DNA [12] reacting with purine bases to form double-strand adducts, but at higher doses, cyclophosphamide may induce strand nicks by destabilizing purine-sugar bonds with the following loss of purine bases. At the cellular level, the drug is able to trigger APOPTOSIS and to induce a pronounced cytotoxic effect on mature lymphocytes with relative sparing of the respective precursor cells [13, 14]. The generation of reactive oxygen free radicals may be considered another mechanism of action, leading to cell death by damaging DNA and inducing lipid peroxidation [15]. Moreover, acrolein—a cyclophosphamide metabolite—seems to be able to inhibit cell proliferation, to induce cell death by APOPTOSIS and to modulate expression of genes and transcription factors, because it reduces the activation of NUCLEAR FACTOR kB (NF-kB) and ACTIVATOR PROTEIN 1 (AP-1). These effects could be increased by the depletion of cellular glutathione, which acts as a detoxifying molecule.

30.3.4 Pharmacokinetics

Cyclophosphamide is well absorbed after oral administration, with a BIOAVAILABILITY greater than 75%. The parent compound is widely distributed throughout the body with low plasma protein binding (20%). The half-life of cyclophosphamide is between 6 and 9 h, and it is eliminated mainly in the urine as metabolites, even if 5–25% of an intravenous dose is excreted unchanged. Cyclophosphamide is quickly metabolized to active alkylating species by the mixed-function oxidase system of the smooth endoplasmic reticulum of hepatocytes, and maximal concentrations of metabolites in plasma may be observed 2–3 h after an intravenous dose. Several cytochrome P450 (CYP) isoforms (CYP2A6, CYP2B6, CYP2C8, CYP2C9 and CYP3A4) are involved in the hydroxylation of the oxazaphosphorine ring of cyclophosphamide [16], leading to 4-OH-cyclophosphamide, which exists in equilibrium with the acyclic tautomer aldophosphamide. Aldophosphamide spontaneously releases acrolein and phosphoramide mustard, the former being a toxic by-product. The involvement of cytochrome P-450 in cyclophosphamide metabolism explains why enzyme induction (mainly of CYP2B, CYP3A4, CYP2C8 and CYP2C9 isoforms), which consists of increased cellular RNA and protein contents and associated catalytic activities, occurs following exposure to cyclophosphamide itself. This phenomenon is responsible for increased CLEARANCE and the shortened half-life of the parent drug, because it influences the rate of 4-hydroxylation.

30.3.5 Clinical Indications

Cyclophosphamide is used for the treatment of systemic lupus erythematosus, vasculitis and other AUTOIMMUNE DISEASES. In multiple sclerosis, it is most effective in young patients and can be used as induction therapy or as second-line therapy in nonresponders to IFN-β [17] (see Chap. 35).

In systemic lupus erythematosus, pulse cyclophosphamide at a dose of 1 g/m^2 administered on a monthly schedule ensures a significant advantage in terms of survival and end-stage renal disease with respect to corticosteroids [18]. A remission rate of 75% was observed in patients affected by severe systemic lupus erythematosus and treated with high-dose cyclophosphamide (10–15 mg/kg) on a monthly schedule for 6 months, followed by quarterly pulses for 18 months. The same schedule has been adopted in lupus nephritis in children and adults, lowering the relapse rate to less than 10%. Positive results have been observed also in the treatment of optic neuritis associated with systemic lupus erythematosus. However, due to toxicity induced by the treatment, it has been proposed to use weekly low-dose pulses of 0.5 g until disease control is achieved, when it is switched to the monthly schedule and subsequently discontinued [19].

Churg-Strauss syndrome, a granulomatous, necrotizing vasculitis affecting small blood vessels, may be treated with daily doses of cyclophosphamide (2 mg) administered orally in combination with steroids, and the same schedule has been adopted for sclerodermia. Remission of Wegener's granulomatosis, a systemic necrotizing vasculitis affecting small- and medium-sized vessels, may be obtained with the use of cyclophosphamide, and daily low doses in combination with steroids are effective against the active disease [20]. Cyclophosphamide plus steroids is effective in the treatment of microscopic polyangiitis, a vasculitis that may be associated with severe pulmonary vasculitis and rapidly progressive glomerulonephritis, and polyarteritis nodosa.

It is noteworthy that the EFFICACY or tolerability of cyclophosphamide depends on the amount of phosphoramide mustard within cells which is controlled by two pathways: (1) aldehyde dehydrogenase which transforms aldophosphamide to carboxyphosphamide, the major urinary inactive metabolite, and (2) the isozymes CYP2B6 and CYP3A4, which catalyse the dechloroethylation of cyclophosphamide [21]. These mechanisms have been extensively investigated in the experimental and clinical setting. CYP2B6 POLYMORPHISM could be responsible for severe toxicities [22], because this CYP isoform catalyses the dechloroethylation of cyclophosphamide to 2- and 3-dechloroethyl-metabolite and chloroacetaldehyde, the latter being a toxic by-product. Furthermore, the concentration of aldophosphamide is associated with the expression of aldehyde dehydrogenase isozymes [23]. Thus far, several polymorphic sites have been identified along the gene sequences of CYP and aldehyde dehydrogenase isoforms. Recently, a single nucleotide POLYMORPHISM (SNP) in the ATP-binding cassette, subfamily C4 (ABCC4) efflux pump, a multidrug resistance (MDR) protein which pumps cytotoxic drugs out of cells, was shown to be associated with both adverse gastrointestinal and leukopenic/neutropenic effects of cyclophosphamide [24]. Such pharmacogenetic studies are helping to identify patients at risk of toxicity who should receive another drug than cyclophosphamide. Finally, *O*6-alkylguanine-DNA alkyltransferases are enzymes capable of protecting cells from the mutagenic effect of DNA alkylation, and therefore a low expression of these genes in target cells may be associated with a better response to cyclophosphamide therapy.

30.3.6 Adverse Reactions

The use of cyclophosphamide is limited by the occurrence of moderate to severe side effects, including gastrointestinal toxicity, alopecia, myelotoxicity, infertility, haemorrhagic cystitis and cardiotoxicity [17, 20]. Nausea and vomiting (10% of treated patients) require adequate prophylactic treatment, with steroids and 5-HT3 antagonists. Bone marrow toxicity (50–100% of cases) is commonly represented by leukopenia 7–14 days after the drug dose, whereas more severe side effects are agranulocytosis and aplastic anaemia. Herpes zoster and other OPPORTUNISTIC INFECTIONS occur in 37% of patients, multiple organ involvement and lower trough leukocyte counts being additional risk factors of severity of infection. Ovarian failure and decreased sperm counts are associated with infertility (up to 100% of subjects), and their severity correlates with the duration of cyclophosphamide treatment and the patient's age. Adverse events may be mitigated by oral administration of the drug instead of using the intravenous route. The severity of haemorrhagic cystitis may be reduced when the drug is co-administered with the thiol-containing agent mesna which inactivates acrolein. Because of its MECHANISMS OF ACTION, cyclophosphamide is teratogenic and carcinogenic, the latter effect being more frequent for BONE MARROW (myeloproliferative disorders, 2% of patients) and bladder (transitional cell carcinoma, 2% of patients).

30.4 Methotrexate

30.4.1 Introduction

Methotrexate is a folate analogue that was introduced into clinical practice more than 50 years ago. It is currently one of the most widely used disease-modifying antirheumatic drugs (DMARDs); its EFFICACY in rheumatoid arthritis has been confirmed in patients refractory or intolerant to other DMARDs or NSAIDs (see Chaps. 33 and 34).

30.4.2 Chemical Structure

Methotrexate (MTX, amethopterin, *N*-[4-[[(2,4-diamino-6-pteridinyl)methyl]methyl-amino] benzoyl]-L-glutamic acid) (Fig. 30.1) remains the only antifolate agent used for clinical IMMUNOSUPPRESSION to date. It is a weak dicarboxylic organic acid, and is negatively charged at neutral pH, resulting in limited lipid solubility.

30.4.3 Mechanism of Action and Pharmacological Effect

Methotrexate enters cells by the reduced folate carrier [25], and its long-chain polyglutamates inhibit the activity of the enzyme dihydrofolate reductase, involved in fundamental metabolic pathways such as de novo synthesis of purines, pyrimidines and polyamines. Methotrexate also inhibits thymidylate synthase indirectly by diminishing levels of the enzyme cosubstrate 5,10-methylenetetrahydrofolate, while polyglutamated metabolites of methotrexate also directly bind and inhibit thymidylate synthase (Fig. 30.2). The use of high-dose methotrexate depletes tumour cells of the purine and pyrimidine precursors required for DNA and RNA synthesis, proliferation and division. Low-dose methotrexate has both immunosuppressive and anti-inflammatory properties resulting in

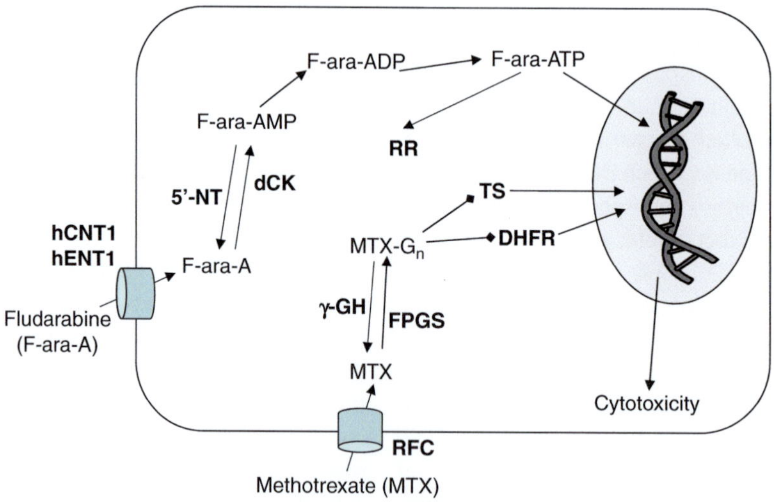

Fig. 30.2 Membrane transport, metabolism and intracellular targets of fludarabine and methotrexate. *hCNT1* human concentrative nucleoside transporter-1, *hENT1* human equilibrative nucleoside transporter-1, *dCK* deoxycytidine kinase, *5′-NT* 5′-nucleotidase, *CdA* cytidine deaminase, *RR* ribonucleotide reductase, *RFC* reduced folate carrier, *FPGS* folylpolyglutamate synthase, *g-GH* g-glutamyl hydrolase, *TS* thymidylate synthase, *DHFR* dihydrofolate reductase

inhibition of proliferation of CD3 and CD4 lymphocytes, MONOCYTES-macrophages and neutrophils. Indeed, at low concentrations, methotrexate induces APOPTOSIS of activated T cells from human peripheral blood in vitro. However, low-dose methotrexate not only acts as a cytotoxic drug against immunocompetent cells but also modulates cytokine secretion from T-helper lymphocytes by increasing IL-4 and IL-10 and decreasing IFN-γ and IL-2. Intracellular methotrexate polyglutamates inhibit the function of 5-amino-imidazole-4-carboxamide ribosyl-5-phosphate formyltransferase (AICAR formyltransferase). The resulting high concentrations of AICAR lead to enhanced release of adenosine into the blood, activating A2a, A2b and A3 extracellular receptors on MONOCYTES-macrophages. In this way, adenosine seems to promote the transcription of mRNA for an IL-1 receptor antagonist and increases the secretion of the potent anti-inflammatory cytokine IL-10. Moreover, adenosine inhibits the production of TNF-α, IL-6 and IL-8 and the expression of E-selectin on the cell surface [26]. Further details are given in Chap. 34.

30.4.4 Pharmacokinetics

After oral administration, active absorption of the drug occurs in the proximal jejunum with a saturable process, and it decreases nonproportionally at increasing oral doses. The extent of absorption is highly variable between patients for doses higher than 10–15 mg/m^2, whereas only a moderate intra-individual pharmacokinetic variability has been described during long-term treatments with low methotrexate doses in patients with psoriasis and rheumatoid arthritis. Low doses of methotrexate are also administered parenterally to ensure compliance and uniform BIOAVAILABILITY. Indeed, the drug is absorbed more rapidly and reaches higher serum concentrations after intramuscular or subcutaneous administration compared with the oral route. Methotrexate may also be injected intra-articularly [27].

The volume of distribution of methotrexate corresponds to the intracellular distribution of the drug and, in blood, 30–70% of the drug is bound to albumin. Four hours after oral or intramuscular administration, the concentrations of methotrexate in the synovial fluid are equivalent to plasma concentrations. Methotrexate transport into cells occurs mainly by a carrier-mediated active transport system that methotrexate shares with folates. Once inside the cell, glutamate residues are progressively added to the drug by the folylpolyglutamate synthetase enzyme. This intracellular accumulation of methotrexate polyglutamates allows a weekly bolus of the drug or the administration of the same dose equally divided in three doses, and relatively high concentrations of the drug are reached in the synovial membrane, cortical and trabecular bone. Methotrexate elimination has been described as biphasic or triphasic with a mean TERMINAL HALF-LIFE of 6–15 h. A longer sampling interval is associated with longer TERMINAL HALF-LIFE estimates of the drug, because of intracellular methotrexate polyglutamylation and the slow release of the drug from cell to plasma.

Methotrexate can be metabolized by three different pathways. (1) In the gastrointestinal tract, intestinal bacteria can degrade the drug to 4-amino-deoxy-*N*10-methylpteroic acid, a metabolite that usually accounts for less than 5% of the administered dose. (2) In the liver, methotrexate is converted to 7-OH-methotrexate, tenfold less potent at inhibiting dihydrofolate reductase but more nephrotoxic than the parent compound because of its precipitation in acidic urine. Despite its extensive binding to serum albumin (>90%), 7-OH-methotrexate does not alter the protein binding of methotrexate. (3) Inside the cells, the drug is converted to pharmacologically active long-chain methotrexate polyglutamates by folylpolyglutamate synthetase and inactivated by gamma-glutamyl hydrolase, potentially contributing to drug resistance [28].

The main elimination route of methotrexate is by renal excretion. The drug is subjected to glomerular filtration and secretion/reabsorption by an active transport system across the renal tubules [29]. In addition, a variable amount of methotrexate is eliminated by active biliary excretion (10–30%) and undergoes enterohepatic recirculation.

30.4.5 Clinical Indications

Methotrexate is a first-line systemic agent for the symptomatic control of moderate to severe psoriasis, though its use is being increasingly challenged by newer biologicals [30]. It has become the gold standard, first-line therapy of patients diagnosed with adult rheumatoid arthritis or juvenile arthritis [31, 32].

The influence of pharmacogenetics on both the immunosuppressive EFFICACY and toxicity of methotrexate in rheumatoid arthritis has attracted the interest of researchers. By affecting the intracellular folate pool, the drug influences the activity of the enzyme methylenetetrahydrofolate reductase (MTHFR), an important step in the generation of 5-methyl-tetrahydrofolate, which is the methyl donor for the conversion of homocysteine to methionine. Numerous POLYMORPHISMs have been described in the MTHFR gene, and among them, the C677T POLYMORPHISM has been associated with altered phenotypes and higher rates of adverse drug events, at least in oncology studies. The C677T variant of the MTHFR gene leads to alanine to valine substitution and a thermolabile MTHFR with decreased enzyme activity and increased plasma homocysteine levels.

Patients with rheumatoid arthritis receiving methotrexate have been assessed for toxicity, disease activity and the presence of the C677T POLYMORPHISM. Homozygous or heterozygous patients have suggested an increased risk of methotrexate discontinuation because of adverse events such as gastrointestinal symptoms (e.g. stomatitis, nausea, vomiting), hair loss, rash and hepatotoxicity (increase in transaminases). Thus, the C677T POLYMORPHISM seems to make patients with rheumatoid arthritis more sensitive to methotrexate toxicity [33, 34]. However, recent studies challenge this proposed association between the C677T POLYMORPHISM and both the EFFICACY and safety of methotrexate in this indication [35]. Single nucleotide POLYMORPHISMs in other enzymes involved in the metabolic pathway of methotrexate, including dihydrofolate reductase and folylpolyglutamate synthase, may be better predictors of methotrexate immunosuppressive EFFICACY and toxicity [36]. Hence, further studies are needed to study POLYMORPHISMs in other enzymes of the folate pathway and their correlations with drug EFFICACY and toxicity in rheumatoid arthritis.

30.4.6 Adverse Reactions

Methotrexate has the potential for severe toxicity, particularly to the liver, mostly related to dose or frequency of administration [37]. For this reason, strict monitoring of drug treatment is recommended as most adverse reactions are reversible if detected early. Severe toxicities are managed with leucovorin rescue and haemodialysis with a high-flux dialyzer. Abnormal liver function tests, nausea/vomiting, stomatitis, diarrhoea, leukopenia, thrombocytopenia, dermatitis, alopecia and interstitial pneumonitis are adverse reactions observed in patients with rheumatoid arthritis treated with low-dose methotrexate (7.5–15 mg/week). With the exception of a higher incidence of alopecia, photosensitivity and "burning of skin lesions", the adverse reaction rates in patients suffering from psoriasis are very similar to those with rheumatoid arthritis. In paediatric patients with juvenile arthritis treated with oral, weekly doses of methotrexate (5–20 mg/m^2/week or 0.1–0.65 mg/kg/week), the most common adverse drug reactions are abnormal liver function tests, nausea, vomiting, diarrhoea, stomatitis and leukopenia [38].

30.5 Mycophenolic Acid

30.5.1 Introduction

The development of mycophenolic acid as an immunosuppressive agent was based on the observation that the proliferation of antigen-responsive T and B LYMPHOCYTES preferentially relies on de novo purine synthesis, with a negligible contribution of the salvage pathway, which is in turn of primary importance for most cells [39]. The antiproliferative, anticancer activity of the drug is thus of secondary importance with respect to the lymphocytolytic effect which led to further clinical characterization, particularly for transplant rejection (see Chap. 31) and more recently for primary glomerulonephritis, such as that during lupus erythematosus [40].

30.5.2 Chemical Structure

Mycophenolic acid (Fig. 30.1) is a FERMENTATION product of several *Penicillium* species. It is the active moiety released by the prodrugs mycophenolate mofetil, a semisynthetic morpholinoethyl ester of mycophenolic acid, and mycophenolate sodium, which is administered as an enteric-coated formulation designed to prevent upper gastrointestinal tract absorption and reduce the gastrointestinal adverse events seen with mycophenolate mofetil [41].

30.5.3 Mechanism of Action and Pharmacological Effect

Mycophenolic acid is a reversible, noncompetitive inhibitor of inosine monophosphate dehydrogenase (IMPDH) and blocks de novo purine synthesis in T and B LYMPHOCYTES (Fig. 30.3), resulting in (1) inhibition of proliferation in response to antigenic stimuli and immunoglobulin production as well as (2) initiation of the apoptotic cascade. Two distinct isoforms of IMPDH have been identified, i.e. type I and type II. IMPDH type I is constitutively expressed mostly in non-replicating cells, and IMPDH type II is the inducible, predominant enzyme in activated lymphocytes. IMPDH type II is approximately five times more susceptible to inhibition by MPA than type I, and this difference explains the unique susceptibility of proliferating lymphocytes to depletion of purine bases by mycophenolic acid. At variance with calcineurin inhibitors, mycophenolic acid has no effect on the production or release of CYTOKINES [42]. Additional pharmacological effects include suppression of antibody production, as a consequence of failure of B-cell activation, suppression of dendritic cell maturation, with a resulting decrease in antigen presentation to T lymphocytes, as well as reduced recruitment of MONOCYTES into sites of graft rejection and inflammation [43]. The effect of MPA on monocyte recruitment is mediated by inhibition of monocyte chemoattractant protein-1 (MCP-1) expression in epithelial cells. MCP-1 is the major chemotactic factor for monocytes, and MPA reduces its expression via the p38 mitogen-activated protein kinase (MAPK) [44]. Anti-inflammatory effects may also result from decreased expression and activity of the inducible form of nitric oxide synthase [45].

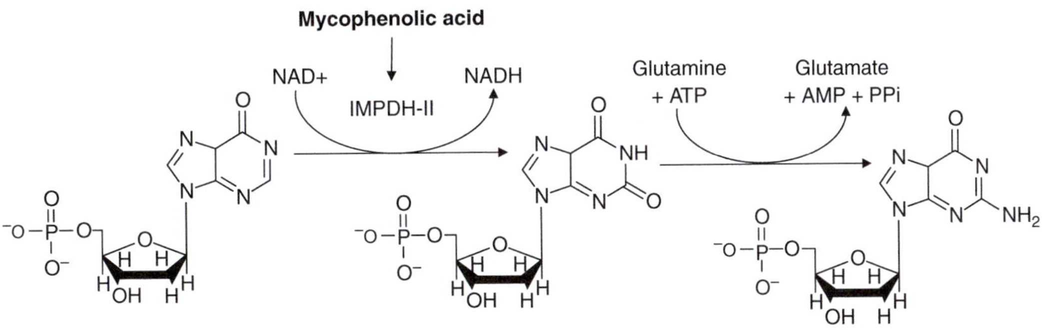

Fig. 30.3 Inhibition of inosine monophosphate dehydrogenase (IMPDH) type II by mycophenolic acid. *PPi* pyrophosphate (inorganic)

30.5.4 Pharmacokinetics

Mycophenolate mofetil (Fig. 30.4) is absorbed in the stomach, and the ester linkage is rapidly hydrolysed by ubiquitous esterases to yield mycophenolic acid, the active immunosuppressive moiety [46]. Therefore, mycophenolate mofetil is undetectable in circulation, even after an intravenous administration. In contrast, enteric-coated mycophenolate sodium (Fig. 30.4) is mainly absorbed in the small intestine. The oral BIOAVAILABILITY of mycophenolic acid from mycophenolate mofetil and mycophenolate sodium is 94% and 71%, respectively, and the peak plasma concentration (C_{max}) occurs 1–2 h after oral administration [41, 47]. After a dose of 1 g of mycophenolate mofetil, the maximum plasma concentration ranges from 10 to 30 mg/L in patients with stable renal function, and the elimination half-life averages 17 h, while pre-dose levels are approximately 1 mg/L [48]. The area under the curve (AUC) of mycophenolic acid is the most significant pharmacokinetic parameter for therapeutic drug monitoring; indeed, low mycophenolic acid AUC is significantly associated with an increased risk of acute rejection of kidney graft [46]. In patients given doses of mycophenolate mofetil tailored to achieve low (16.1 mg/h/L), intermediate (32.2 mg/h/L) or high (60.6 mg/h/L) total AUC of mycophenolic acid, a highly significant relationship was found between the AUC of the active metabolite and the incidence of rejection. In the low, intermediate and high AUC groups, the incidence was 27.5%, 14.9% and 11.5%, respectively [49]. There was also a higher rate of premature withdrawal from the study as the AUC of mycophenolic acid increased: 7.8%, 23.4% and 44.2% for the three groups, respectively [49]. Pharmacokinetic monitoring may be performed

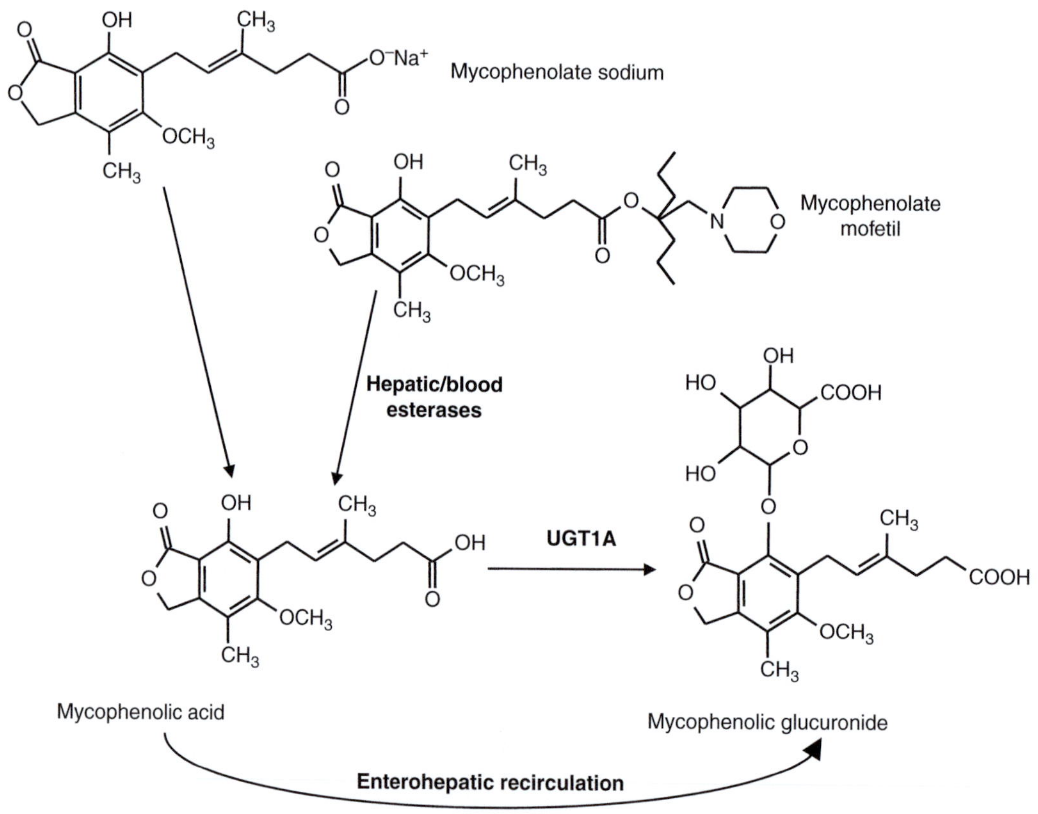

Fig. 30.4 Metabolism of mycophenolate mofetil and mycophenolate sodium to the active moiety mycophenolic acid; UGT1-dependent metabolism yields mycophenolate glucuronide which undergoes enterohepatic recirculation

on the basis of AUC_{0-2h} of mycophenolic acid, this parameter being in good agreement with the AUC_{0-12h} measured across the complete dosing interval period.

Mycophenolic acid undergoes extensive hepatic glucuronidation by glucuronosyltransferase (UGT) isoforms 1A8, 1A9 and 1A10 [50] to form the inactive metabolite mycophenolic acid glucuronide which is excreted into the bile (Fig. 30.4). The glucuronyl moiety of the metabolite is cleaved by enteric β-glucuronidases of intestinal bacteria to release mycophenolic acid which undergoes enterohepatic recirculation producing a secondary peak plasma concentration about 6–12 h after administration. More than 90% of a dose of mycophenolate mofetil is excreted in the urine as mycophenolic acid glucuronide [51]. The comparison of pharmacokinetic parameters of mycophenolate sodium 720 mg and mycophenolate mofetil 1000 mg revealed that C_{max} and AUC_{0-24h} in the former group were consistently higher than in the latter group. Overall, there was a mean increase of 32% in systemic mycophenolic acid exposure over the length of the study in the mycophenolate sodium-treated patients [41].

30.5.5 Clinical Indications

Mycophenolate mofetil is administered at 1–3 g/day orally or, less frequently, by the intravenous route, and it is indicated for the prophylaxis of organ rejection in patients receiving allogeneic renal, cardiac or hepatic transplants mainly in combination with a calcineurin inhibitor (tacrolimus or cyclosporine) and corticosteroids. Clinical studies indicate that mycophenolate mofetil, administered in combination with immunosuppressive antibodies, can be used in calcineurin- and steroid-sparing regimens, in order to reduce the toxicity burden of IMMUNOSUPPRESSION [52]. Mycophenolate sodium 720 mg twice daily has been compared with mycophenolate mofetil 1000 mg twice daily with respect to safety and EFFICACY. Both formulations demonstrated equivalent EFFICACY and safety in renal transplant recipients; in particular, the incidence of gastrointestinal adverse events was similar between the formulations despite the higher serum levels achieved with mycophenolate sodium [41]. Based on its antifibrotic and antiproteinuric effects, mycophenolate mofetil is now being used increasingly in patients with primary glomerulonephritis, particularly those with idiopathic nephritic syndrome or drug-induced deterioration of renal function [53, 54].

Clinically important drug interactions of mycophenolate mofetil and mycophenolate sodium involve acyclovir/valacyclovir (increased haematological toxicity), cholestyramine, antacids containing aluminium hydroxide/magnesium hydroxide and cyclosporine (reduced BIOAVAILABILITY of mycophenolic acid) and levonorgestrel (reduced AUC of the hormone) [41].

30.5.6 Adverse Reactions

Overall, mycophenolate mofetil is better tolerated than azathioprine. The drug produces a lower incidence of leukopenia and fewer immunosuppressive-related malignancies compared with azathioprine and lacks the neurotoxicity and nephrotoxicity associated with calcineurin inhibitors. The most commonly observed adverse events associated with mycophenolate mofetil and mycophenolate sodium are gastrointestinal (i.e. nausea, vomiting, diarrhoea, constipation, dyspepsia, flatulence, anorexia), haematological (leukopenia, thrombocytopenia, anaemia) and OPPORTUNISTIC INFECTIONS (see [55]). Patients experiencing gastrointestinal or haematological adverse events usually respond to dose fractionation or reduction. However, frequent dose changes have been associated with poorer outcomes, including a higher incidence of graft loss.

30.6 Conclusions

The increasing understanding of the pathophysiology of autoimmune disease and graft rejection has revealed a number of potential targets that have been exploited for the design of potent immunosuppressive drugs, including MONOCLONAL ANTIBODIES and calcineurin-binding agents.

However, the occurrence of multi-organ toxicity associated with the use of calcineurin inhibitors, the severe metabolic adverse events induced by corticosteroids, the potential for long-term cancer risk associated with the use of selected MONOCLONAL ANTIBODIES and the occurrence of infectious diseases due to over-IMMUNOSUPPRESSION soon became key management issues that urged a re-evaluation of immunosuppressive treatment schedules. In this context, cytotoxic drugs, particularly ANTIMETABOLITES, still play a crucial role in the control of immune system activation, particularly in steroid- and calcineurin inhibitor-resistant diseases, and are expected to be instrumental in the long-term IMMUNOSUPPRESSION maintenance in steroid- and calcineurin-binding-sparing schedules.

30.7 Summary

Cytotoxic immunosuppressive drugs are a group of heterogeneous compounds characterized by their ability to damage immune cells by non-specific mechanisms, including nucleotide pool depletion, incorporation into DNA and nucleic acid alkylation. Their successful clinical use in IMMUNOSUPPRESSION for organ transplantation and AUTOIMMUNE DISEASES has been proven in a large number of clinical trials.

Azathioprine is an antimetabolite prodrug of 6-mercaptopurine with an imidazolyl group attached to the SH group of 6-mercaptopurine. Active metabolites of azathioprine are inhibitors of nucleic acid synthesis through impairment of the purine de novo biosynthetic pathway and incorporation into RNA and DNA. The drug suppresses the proliferation of T and B LYMPHOCYTES and reduces the number of cytotoxic T CELLS and plasma cells in circulation and peripheral organs. Azathioprine is mainly metabolized by xanthine oxidase; however, thiopurine methyltransferase (TPMT) is the principal detoxification pathway for the cytotoxic thioguanine nucleotides in the BONE MARROW. Haematopoietic toxicity is dependent at least in part on the activity of TPMT, a polymorphic enzyme with genetic variants characterized by low activity and increased risk of severe toxicity in patients. Azathioprine is used in renal transplantation, rheumatoid arthritis, ulcerative colitis and other skin autoimmune disorders. Haematopoietic toxicity, including neutropenia, anaemia and thrombocytopenia, and hepatic toxicity are the most common adverse events and may be predicted by TPMT genotyping.

Cyclophosphamide is a first-generation oxazaphosphorine alkylating agent belonging to the group of nitrogen mustards, characterized by the ability to form stable, covalent linkages by alkylation of the N7 of guanine residues of DNA via its active metabolite phosphoramide mustard. At the cellular level, the drug is able to trigger APOPTOSIS and to induce a pronounced cytotoxic effect on mature lymphocytes. Cyclophosphamide has a good oral BIOAVAILABILITY, and it is eliminated mainly in the urine as metabolites. Cyclophosphamide is transformed to active alkylating species by the mixed-function oxidase system of the smooth endoplasmic reticulum of hepatocytes. Several cytochrome P450 (CYP) isoforms (CYP2A6, CYP2B6, CYP2C8, CYP2C9 and CYP3A4) are involved in the hydroxylation of the oxazaphosphorine ring of cyclophosphamide, leading to 4-OH-cyclophosphamide, which exists in equilibrium with aldophosphamide. Aldophosphamide spontaneously releases acrolein and phosphoramide mustard, the former being a toxic by-product. The EFFICACY or tolerability of cyclophosphamide depends on the amount of phosphoramide mustard within cells. In this respect, key enzymes are aldehyde dehydrogenase, which transforms aldophosphamide to carboxyphosphamide, the major urinary inactive metabolite, and the isozymes CYP2B6 and CYP3A4, which catalyse the dechloroethylation of cyclophosphamide to 2- and 3-dechloroethyl-metabolite and chloroacetaldehyde, the latter being a neurotoxic by-product. The drug is used for the treatment of systemic lupus erythematosus, vasculitis and other AUTOIMMUNE DISEASES. The use of cyclophosphamide is limited by the occurrence of moderate to severe side effects, including gastrointestinal toxicity, alopecia, myelotoxicity, infertility, haemorrhagic cystitis and cardiotoxicity.

Methotrexate, a folate analogue, is currently the gold standard, first-line disease-modifying antirheumatic drug (DMARDs) in adult rheumatoid arthritis and juvenile arthritis. The drug enters cells by the reduced folate carrier, and its long-chain polyglutamates inhibit the enzyme dihydrofolate reductase, thereby blocking de novo synthesis of purines and pyrimidines; methotrexate also inhibits thymidylate synthase. Low-dose methotrexate has both immunosuppressive and anti-inflammatory properties resulting in inhibition of proliferation of CD3 and CD4 lymphocytes, MONOCYTES-macrophages and neutrophils. Oral BIOAVAILABILITY of methotrexate is predictable after administration of low doses, while higher doses should be administered parenterally. The main elimination route of methotrexate is by renal excretion. In addition, a variable amount is eliminated by active biliary excretion and undergoes enterohepatic recirculation. Methotrexate is indicated for the management of rheumatoid arthritis, juvenile arthritis and psoriasis. The drug has the potential for severe toxicity, including hepatotoxicity, nausea/vomiting, stomatitis, diarrhoea, leukopenia, thrombocytopenia, dermatitis, alopecia and interstitial pneumonitis. The homozygous or heterozygous condition of C677T POLYMORPHISM of the enzyme methylenetetrahydrofolate reductase (MTHFR) may be associated with an increased risk of adverse events, but findings are inconclusive.

Finally, mycophenolic acid is an immunosuppressive agent that exerts a reversible, noncompetitive inhibition of inosine monophosphate dehydrogenase (IMPDH) and blocks de novo purine synthesis in T and B LYMPHOCYTES. At variance with calcineurin inhibitors, mycophenolic acid has no effect on the production or release of CYTOKINES but inhibits antigen presentation by dendritic cells to T CELLS. Additional pharmacological effects include suppression of antibody production, as a consequence of failure of B-cell activation, as well as anti-inflammatory effects resulting from reduced monocyte recruitment and decreased expression and activity of the inducible nitric oxide synthase. The oral BIOAVAILABILITY of mycophenolic acid from mycophenolate mofetil and mycophenolate sodium is high, and the drug undergoes extensive hepatic glucuronidation by glucuronosyltransferase (UGT) isoforms 1A8, 1A9 and 1A10 to form the inactive metabolite mycophenolic acid glucuronide which is excreted into the bile. The drug is indicated for the prophylaxis of organ rejection in patients receiving allogeneic renal, cardiac or hepatic transplants mainly in combination with a calcineurin inhibitor and corticosteroids. Overall, mycophenolate mofetil is better tolerated than azathioprine. The drug produces a lower incidence of leukopenia and fewer IMMUNOSUPPRESSION-related malignancies compared with azathioprine and lacks the neurotoxicity and nephrotoxicity associated with calcineurin inhibitors. The most commonly observed adverse events associated with mycophenolate mofetil and mycophenolate sodium are gastrointestinal (i.e. nausea, vomiting, diarrhoea, constipation, dyspepsia, flatulence, anorexia) and haematological (leukopenia, thrombocytopenia, anaemia) as well as OPPORTUNISTIC INFECTIONS.

Selected Readings

Cattaneo D, Perico N, Remuzzi G. From pharmacokinetics to pharmacogenomics: a new approach to tailor immunosuppressive therapy. Am J Transplant. 2004;4:299–310.

Chinen J, Buckley RH. Transplantation immunology: solid organ and bone marrow. J Allergy Clin Immunol. 2010;125(2 Suppl 2):S324–35.

Emadi A, Jones RJ, Brodsky RA. Cyclophosphamide and cancer: golden anniversary. Nat Rev Clin Oncol. 2009;6:638–47.

Fischereder M, Kretzler M. New immunosuppressive strategies in renal transplant recipients. J Nephrol. 2004;17:9–18.

Tedesco Silva H Jr, Pinheiro Machado P, Rosso Felipe C, Medina Pestana JO. Immunotherapy for De Novo renal transplantation: what's in the pipeline? Drugs. 2006;66:1665–84.

Wolff D, Steiner B, Hildebrandt G, Edinger M, Holler E. Pharmaceutical and cellular strategies in prophylaxis and treatment of graft-versus-host disease. Curr Pharm Design. 2009;15:1974–97.

References

1. Knight SR, Russell NK, Barcena L, Morris PJ. Mycophenolate mofetil decreases acute rejection and may improve graft survival in renal transplant recipients when compared with azathioprine: a systematic review. Transplantation. 2009;87:785–94.
2. Barshes NR, Goodpastor SE, Goss JA. Pharmacologic immunosuppression. Front Biosci. 2004;9:411–20.
3. Germani G, Pleguezuelo M, Villamil F, Vaghjiani S, Tsochatzis E, Andreana L, Burroughs AK. Azathioprine in liver transplantation: a reevaluation of its use and a comparison with mycophenolate mofetil. Am J Transplant. 2009;9:1725–31.
4. Mueller XM. Drug immunosuppression therapy for adult heart transplantation. Part 1: Immune response to allograft and mechanism of action of immunosuppressants. Ann Thorac Surg. 2004;77:354–62.
5. Sauer H, Hantke U, Wilmanns W. Azathioprine lymphocytotoxicity. Potentially lethal damage by its imidazole derivatives. Arzneimittelforschung. 1988;38:820–84.
6. Adam L, Phulukdaree A, Soma P. Effective long-term solution to therapeutic remission in Inflammatory Bowel Disease: role of azathioprine. Biomed Pharmacother. 2018;100:8–14.
7. El-Azhary RA. Azathioprine: current status and future considerations. Int J Dermatol. 2003;42:335–41.
8. Gold R, Schneider-Gold C. Current and future standards in treatment of myasthenia gravis. Neurotherapeutics. 2008;5:535–41.
9. Tsuchiya A, Aomori T, Sakamoto M, Takeuchi A, Suzuki S, Jibiki A, Otsuka N, Ishioka E, Kaneko Y, Takeuchi T, Nakamura T. Effect of genetic polymorphisms of azathioprine-metabolizing enzymes on response to rheumatoid arthritis treatment. Pharmazie. 2017;72:22–8.
10. de Jong DJ, Goullet M, Naber TH. Side effects of azathioprine in patients with Crohn's disease. Eur J Gastroenterol Hepatol. 2004;16:207–12.
11. Marcen R, Pascual J, Tato AM, Teruel JL, Villafruela JJ, Fernandez M, Tenorio M, Burgos FJ, Ortuno J. Influence of immunosuppression on the prevalence of cancer after kidney transplantation. Transplant Proc. 2003;35:1714–6.
12. Hengstler JG, Hengst A, Fuchs J, Tanner B, Pohl J, Oesch F. Induction of DNA crosslinks and DNA strand lesions by cyclophosphamide after activation by cytochrome P450 2B1. Mutat Res. 1997;373:215–23.
13. Allison AC. Immunosuppressive drugs: the first 50 years and a glance forward. Immunopharmacology. 2000;47:63–83.
14. Pette M, Gold R, Pette DF, Hartung HP, Toyka KV. Mafosfamide induces DNA fragmentation and APOPTOSIS in human T-lymphocytes. A possible mechanism of its immunosuppressive action. Immunopharmacology. 1995;30:59–69.
15. Sulkowska M, Sulkowski S, Skrzydlewska E, Farbiszewski R. Cyclophosphamide-induced generation of reactive oxygen species. Comparison with morphological changes in type II alveolar epithelial cells and lung capillaries. Exp Toxicol Pathol. 1998;50:209–20.
16. Alan V, Boddy AV, Yule SM. Metabolism and pharmacokinetics of oxazaphosphorines. Clin Pharmacokinet. 2000;38:291–304.
17. Rinaldi L, Perini P, Calabrese M, Gallo P. Cyclophosphamide as second-line therapy in multiple sclerosis: benefits and risks. Neurol Sci. 2009;30(Suppl 2):S171–3.
18. Esdaile JM. How to manage patients with lupus nephritis. Best Pract Res Clin Rheumatol. 2002;16:195–210.
19. Mosca M, Ruiz-Irastorza G, Khamashta MA, Hughes GRV. Treatment of systemic lupus erythematosus. Int Immunopharmacol. 2001;1:1065–75.
20. Langford CA. Management of systemic vasculitis. Best Pract Res Clin Rheumatol. 2001;15:281–97.
21. Huang Z, Roy P, Waxman DJ. Role of human liver microsomal CYP3A4 and CYP2B6 in catalyzing N-dechloroethylation of cyclophosphamide and ifosfamide. Biochem Pharmacol. 2000;59:961–72.
22. Jinno H, Tanaka-Kagawa T, Ohno A, Makino Y, Matsushima E, Hanioka N, Ando M. Functional characterization of cytochrome P450 2B6 allelic variants. Drug Metab Dispos. 2003;31:398–403.
23. Giorgianni F, Bridson PK, Sorrentino BP, Pohl J, Blakley RL. Inactivation of aldophosphamide by human aldehyde dehydrogenase isozyme 3. Biochem Pharmacol. 2000;60:325–38.
24. Low SK, Kiyotani K, Mushiroda T, Daigo Y, Nakamura Y, Zembutsu H. Association study of genetic polymorphism in ABCC4 with cyclophosphamide-induced adverse drug reactions in breast cancer patients. J Hum Genet. 2009;54:564–71.
25. Serra M, Reverter-Branchat G, Maurici D, Benini S, Shen JN, Chano T, Hattinger CM, Manara MC, Pasello M, Scotlandi K, Picci P. Analysis of dihydrofolate reductase and reduced folate carrier gene status in relation to methotrexate resistance in osteosarcoma cells. Ann Oncol. 2004;15:151–60.
26. Fraser AG. Methotrexate: first-line or second-line immunomodulator? Eur J Gastroenterol Hepatol. 2003;15:225–31.
27. Grim J, Chládek J, Martínková J. Pharmacokinetics and pharmacodynamics of methotrexate in non-neoplastic diseases. Clin Pharmacokinet. 2003;42:139–51.
28. Zhao R, Goldman ID. Resistance to antifolates. Oncogene. 2003;22:7431–57.
29. van Aubel RA, Smeets PH, Peters JG, Bindels RJ, Russel FG. The MRP4/ABCC4 gene encodes a novel apical organic anion transporter in human kidney proximal tubules: putative efflux pump

for urinary cAMP and cGMP. J Am Soc Nephrol. 2002;13:595–603.

30. Warren RB, Chalmers RJ, Griffiths CE, Menter A. Methotrexate for psoriasis in the era of biological therapy. Clin Exp Dermatol. 2008;33:551–4.
31. Smolen JS, Landewé R, Breedveld FC, Dougados M, Emery P, Gaujoux-Viala C, Gorter S, Knevel R, Nam J, Schoels M, Aletaha D, Buch M, Gossec L, Huizinga T, Bijlsma JW, Burmester G, Combe B, Cutolo M, Gabay C, Gomez-Reino J, Kouloumas M, Kvien TK, Martin-Mola E, McInnes I, Pavelka K, van Riel P, Scholte M, Scott DL, Sokka T, Valesini G, van Vollenhoven R, Winthrop KL, Wong J, Zink A, van der Heijde D. EULAR recommendations for the management of rheumatoid arthritis with synthetic and biological disease-modifying antirheumatic drugs. Ann Rheum Dis. 2010;69:964–75.
32. Kahn P. Juvenile idiopathic arthritis--current and future therapies. Bull NYU Hosp Jt Dis. 2009;67:291–302.
33. Berkani LM, Rahal F, Allam I, Mouaki Benani S, Laadjouz A, Djidjik R. Association of MTHFR C677T and A1298C gene polymorphisms with methotrexate efficiency and toxicity in Algerian rheumatoid arthritis patients. Heliyon. 2017;3(11):e00467.
34. Shao W, Yuan Y, Li Y. Association between MTHFR C677T polymorphism and methotrexate treatment outcome in rheumatoid arthritis patients: a systematic review and meta-analysis. Genet Test Mol Biomarkers. 2017;21:275–85.
35. Lee YH, Song GG. Associations between the C677T and A1298C polymorphisms of MTHFR and the efficacy and toxicity of methotrexate in rheumatoid arthritis: a meta-analysis. Clin Drug Investig. 2010;30:101–8.
36. Chen Y, Zou K, Sun J, Yang Y, Liu G. Are gene polymorphisms related to treatment outcomes of methotrexate in patients with rheumatoid arthritis? A systematic review and meta-analysis. Pharmacogenomics. 2017;18:175–95.
37. Borchers AT, Keen CL, Cheema GS, Gershwin ME. The use of methotrexate in rheumatoid arthritis. Semin Arthritis Rheum. 2004;34:465–83.
38. Ruperto N, Murray KJ, Gerloni V, Wulffraat N, de Oliveira SK, Falcini F, Dolezalova P, Alessio M, Burgos-Vargas R, Corona F, Vesely R, Foster H, Davidson J, Zulian F, Asplin L, Baildam E, Consuegra JG, Ozdogan H, Saurenmann R, Joos R, Pistorio A, Woo P, Martini A. A randomized trial of parenteral methotrexate comparing an intermediate dose with a higher dose in children with juvenile idiopathic arthritis who failed to respond to standard doses of methotrexate. Arthritis Rheum. 2004;50:2191–201.
39. Smak Gregoor PJ, van Gelder T, Weimar W. Mycophenolate mofetil, Cellcept, a new immunosuppressive drug with great potential in internal medicine. Neth J Med. 2000;57:233–46.
40. Kaul A, Gordon C, Crow MK, Touma Z, Urowitz MB, van Vollenhoven R, Ruiz-Irastorza G, Hughes G. Systemic lupus erythematosus. Nat Rev Dis Primers. 2016;2:16039. https://doi.org/10.1038/nrdp.2016.39.
41. Gabardi S, Tran JL, Clarkson MR. Enteric-coated mycophenolate sodium. Ann Pharmacother. 2003;37:1685–93.
42. Srinivas TR, Kaplan B, Meier-Kriesche HU. Mycophenolate mofetil in solid-organ transplantation. Expert Opin Pharmacother. 2003;4:2325–45.
43. Villarroel MC, Hidalgo M, Jimeno A. Mycophenolate mofetil: an update. Drugs Today. 2009;45:521–32.
44. Shui H, Gao P, Si X, Ding G. Mycophenolic acid inhibits albumin-induced MCP-1 expression in renal tubular epithelial cells through the p38 MAPK pathway. Mol Biol Rep. 2010;37:1749–54.
45. Lui SL, Chan LY, Zhang XH, Zhu W, Chan TM, Fung PC, Lai KN. Effect of mycophenolate mofetil on nitric oxide production and inducible nitric oxide synthase gene expression during renal ischaemia-reperfusion injury. Nephrol Dial Transplant. 2001;16:1577–82.
46. Kelly P, Kahan BD. Review: metabolism of immunosuppressant drugs. Curr Drug Metab. 2002;3:275–87.
47. Del Tacca M. Prospects for personalized immunosuppression: pharmacologic tools – a review. Transplant Proc. 2004;36:687–9.
48. Holt DW. Monitoring mycophenolic acid. Ann Clin Biochem. 2002;39:173–83.
49. Hale MD, Nicholls AJ, Bullingham RE, Hene R, Hoitsma A, Squifflet JP, Weimar W, Vanrenterghem Y, Van de Woude FJ, Verpooten GA. The pharmacokinetic-pharmacodynamic relationship for mycophenolate mofetil in renal transplantation. Clin Pharmacol Ther. 1998;64:672–83.
50. Mackenzie PI. Identification of uridine diphosphate glucuronosyl-transferases involved in the metabolism and clearance of mycophenolic acid. Ther Drug Monit. 2000;22:10–3.
51. Mele TS, Halloran PF. The use of mycophenolate mofetil in transplant recipients. Immunopharmacology. 2000;47:215–45.
52. Giessing M, Fuller TF, Tuellmann M, Slowinski T, Budde K, Liefeldt L. Steroid- and calcineurin inhibitor free immunosuppression in kidney transplantation: state of the art and future developments. World J Urol. 2007;25:325–32.
53. Koukoulaki M, Goumenos DS. The accumulated experience with the use of mycophenolate mofetil in primary glomerulonephritis. Expert Opin Investig Drugs. 2010;19:673–87.
54. Groot N, de Graeff N, Marks SD, Brogan P, Avcin T, Bader-Meunier B, Dolezalova P, Feldman BM, Kone-Paut I, Lahdenne P, McCann L, Özen S, Pilkington CA, Ravelli A, Royen-Kerkhof AV, Uziel Y, Vastert BJ, Wulffraat NM, Beresford MW, Kamphuis S. European evidence-based recommendations for the diagnosis and treatment of childhood-onset lupus nephritis: the SHARE initiative. Ann Rheum Dis. 2017;76:1965–73.
55. Staatz CE, Tett SE. Pharmacology and toxicology of mycophenolate in organ transplant recipients: an update. Arch Toxicol. 2014;88:1351–89.

31 Immunosuppressives in Transplant Rejection

Henk-Jan Schuurman

31.1 Introduction

Suppression of immune reactivity can either be an undesirable effect or a situation which is specifically induced to the benefit of a patient. Examples of the first come from immunotoxicology, e.g., XENOBIOTICS or environmental factors causing immunosuppression. Virus infections, as exemplified by human immunodeficiency virus (HIV), can cause severe immunodeficiency. Under clinical disease conditions, suppression of the immune system is specially indicated in two indications: AUTOIMMUNITY and organ transplantation. In the conventional approach, autoimmune diseases like rheumatoid arthritis (RA) are mainly treated by inhibition of the effector phase with anti-inflammatory drugs like CORTICOSTEROIDS and antimetabolites like methotrexate (MTX). During the last decades, new treatments have been introduced including biologicals (monoclonal antibodies (mAbs) and fusion proteins) with anti-inflammatory and immunosuppressive activity (see Chaps. 33 and 34). Also, immunosuppressants like CYCLOSPORINE A (CsA), which were at first developed for transplantation, are increasingly used in autoimmune diseases, and some like LEFLUNOMIDE have been developed for RA as the first indication.

In contrast, in organ transplantation, there is a principal need for interference with the initiation of an immune response which is induced by the grafted organ. Generally, high-dose immunosuppression is needed in the first period after transplantation (induction treatment) or in the treatment of rejection episodes [1–3]. To keep graft function stable, the so-called maintenance treatment is given. Originally, when transplantation was introduced as a treatment of end-stage organ failure (the first kidney transplant was performed in the early 1950s of the last century), there were few possibilities to prevent or treat allograft rejection. In the 1960s and early 1970s, this was mainly restricted to combinations of AZATHIOPRINE (AZA), CORTICOSTEROIDS, and CYCLOPHOSPHAMIDE (CY). Combinations of these drugs were effective but associated with severe side effects, mainly related to bone marrow depression (myelosuppression, leukopenia, anemia) and gastrointestinal symptoms. A more specific reagent, antilymphocyte globulin (ALG), became available in 1966 and was used in induction treatment immediately after transplantation. Based on the complications working with these drugs, kidney transplantation developed slowly, and heart transplantation did not develop after it was first performed in 1963 because of a lack of efficacious immunosuppressive regimens. The history of transplantation during the second half of the twentieth century has witnessed a major search for innovative approaches to suppress or modulate the immune response toward a graft [4].

Final manuscript submitted on July 21, 2016.

H.-J. Schuurman (✉)
SchuBiomed Consultancy, Utrecht, The Netherlands

M. J. Parnham et al. (eds.), *Nijkamp and Parnham's Principles of Immunopharmacology*,
https://doi.org/10.1007/978-3-030-10811-3_31

The most widely used immunosuppressives at present are XENOBIOTICS, i.e., orally active drugs produced by microorganisms or chemically synthesized molecules (structural formulas of some examples are given in Fig. 31.1). A landmark in immunosuppression for transplantation was the introduction of CsA, an inhibitor of intracellular calcineurin, in 1983. Using CsA as a baseline immunosuppressant in combination with AZA and CORTICOSTEROIDS, 1-year graft survival in kidney transplantation increased to 80–90%, and also heart transplantation reached a 1-year patient survival exceeding 80%. These survival rates evidently are not 100% because of the occurrence of technical (e.g., surgical) complications. The high survival rates make it difficult to develop new immunosuppressants with higher efficacy. Therefore, in pharmaceutical industry, the search for xenobiotic immunosuppressants has almost come to a standstill. The present search mainly addresses the therapeutic window between efficacy and adverse side effects, which are a well-known complication for most xenobiotic immunosuppressives. The present search has therefore shifted in pharma and biotech industry from XENOBIOTICS to biologicals and in academia to cell therapy products.

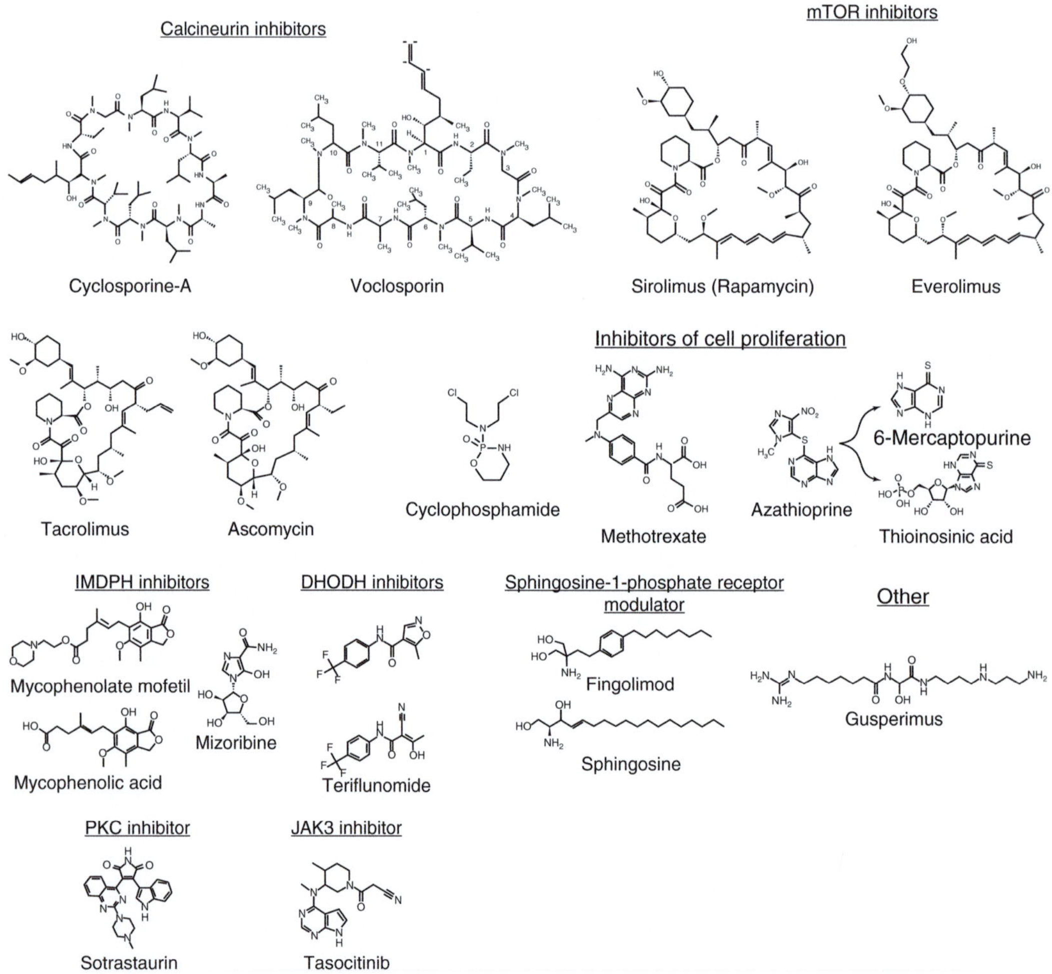

Fig. 31.1 Structures of xenobiotic immunosuppressives: calcineurin inhibitors, inhibitors of the mammalian target of rapamycin (mTOR), cell proliferation, inosine monophosphate dehydrogenase (IMPDH), dihydroorotate dehydrogenase (DHODH), protein kinase C (PKC) and Janus kinase 3 (JAK3), sphingosine-1-phosphate receptor modulator, and gusperimus

Currently, induction immunosuppressive treatment, during the first 2–4 weeks after transplantation, includes either triple therapy, i.e., a calcineurin inhibitor, an inhibitor of cell proliferation like AZA or mycophenolate acid (MPA), and CORTICOSTEROIDS as an anti-inflammatory agent, or quadruple therapy in which an Ab is added to this regimen. Examples are a polyclonal Ab, antithymocyte globulin (ATG) or ALG, or a more specific monoclonal Ab such as anti-CD3 Abs directed to the T-CELL RECEPTOR (TCR) on T LYMPHOCYTES or anti-CD25 Ab directed to the interleukin (IL)-2 receptor on activated T LYMPHOCYTES (see Chaps. 20 and 21). Many alternatives are currently possible in combination drug treatment, in which one of the drugs is replaced by another, such as an inhibitor of the MAMMALIAN TARGET OF RAPAMYCIN (mTOR), which inhibits growth factor-induced cell proliferation.

Most drugs, in particular those with a narrow therapeutic window between efficacy and side effects, require regular monitoring of exposure on the basis of blood concentrations [5–8]. For others, hematological parameters serve as a surrogate marker (like blood leukocyte counts in the case of AZA). Side effects of calcineurin inhibitors primarily concern the kidney, and therefore patients are given such drugs first after good kidney function is achieved. When patients show stable graft function, the induction treatment is gradually converted into maintenance treatment, in which lower doses are used or one component is gradually tapered down and eliminated from the regimen. CORTICOSTEROIDS are an example of this approach, because of the endocrinological side effects (most visible are the Cushingoid features, but others include osteoporosis, diabetes, hyperlipidemia, hypertension, hirsutism, and cataracts). In many transplant centers, steroid-sparing regimens are the first goal in maintenance treatment. Rejection crises (documented by histopathology of a graft biopsy or biochemical markers in blood) are normally treated first with high-dose intravenous CORTICOSTEROIDS (bolus injections on 3–5 successive days); so-called steroid-resistant rejections are to be treated by Ab, either anti-CD3 Ab, ATG, or, in severe forms, ALG.

Biologicals (i.e., poly- or monoclonal Abs and fusion proteins generated by rDNA technology) have long been considered as promising innovative immunosuppressants. This started with the introduction of the polyclonal antilymphocyte Ab ALG and the mouse anti-CD3 mAb muromonab in the 1960s and 1980s of the last century, respectively, and was followed by the introduction of humanized or chimeric anti-IL-2 receptor mAbs for the primary indication transplantation. The main advantage of biologicals is their higher specificity, resulting in a broadened therapeutic window. The main disadvantage is their administration route (parenteral instead of oral for most low-molecular-weight XENOBIOTICS) and their potential immunogenicity, for instance, the formation of anti-mouse Ab in the case of a mouse reagent. In contrast to low-molecular-weight XENOBIOTICS, most biologicals work extracellularly, i.e., at target cell surface molecules. Their potential side effects are therefore mediated by cross-reactivity with other cell populations than the primary target cell. In general, therapeutic Abs to cell surface molecules can affect the target cell by two mechanisms: either by temporary blockade or downregulation of surface molecules, resulting in dysfunction or anergy, or by lysis of the cell, for instance, by the induction of apoptosis, complement-mediated lysis, or Ab-dependent cellular cytotoxicity (ADCC, depleting Ab).

The field of biologicals is progressing, and a number of new reagents have been launched during the last decade or are in advanced development (see Chap. 21). In order to avoid the formation of anti-mouse Ab, mAbs for clinical application are currently generated by genetic engineering. Three major approaches are followed: (1) chimeric Ab, in which the constant part of the immunoglobulin (Ig) heavy and light chains in the mouse Ab molecule is replaced by human Ig sequences; (2) humanized Ab, in which the sequence encoding the complementarity-determining region (CDR) of the variable part of the mouse Ab is inserted into the sequence encoding human Ig; and (3) fully human Ab in which the Ab is produced by a human cell line and not a mouse cell line after immunization. The second approach may lead to a loss of binding affinity, as has been shown for the humanized anti-CD25 Ab daclizumab. Murine antigenic determinants are still present in the variable part of light and heavy chains, albeit less so in humanized Ab, and hence the molecules are still immunogenic, although the immunogenicity is strongly reduced. To reduce this further, current technology allows for the generation of completely human-antihuman Abs. The selection of a relevant Ab is performed using the product of cell lines created by the relevant Ab-encoding sequences inserted into a high-performance human Ab-producing cell line. Engineered Abs have the advantage that the half-life in the circulation can be substantially longer, e.g., from 24 to 48 h for a mouse Ab to 2–4 weeks for an engineered human(ized) Ab.

After the introduction of immunosuppressive XENOBIOTICS and biologicals, cell therapy is gradually entering the transplantation field. Mesenchymal STEM CELLS have received increasing interest because of their immunomodulatory, mainly immunosuppressive, function (see Chap. 36). This is linked to the upregulation of regulatory T cells (Treg) that could be studied after it became possible to isolate this small subset of T LYMPHOCYTES and expand the cells in vitro. Cell therapy in transplantation is not available in the form of products on the market, one reason being the fact that the cells are used in an autologous condition, i.e., the patient self is the donor of the cells. Cell therapy products are further discussed below, in the last subchapter on TOLERANCE induction.

During the last decade, many new immunosuppressants have been introduced to the market or have entered advanced clinical development, both XENOBIOTICS and biologicals (see Chaps. 21 and 30). The main targets in transplantation are solid organs, including the kidney, heart, liver, and lung, which will be the main focus of this chapter. Structural formulas for major XENOBIOTICS are presented in Fig. 31.1, and a summary of immunosuppressants is given in Table 31.1. A selection of general overview papers is given in the Selected Readings list.

The immunosuppressive armamentarium clearly enables the clinician to use a variety of reagents that target various parts of the immune response, i.e., depletion of specific lymphocyte subpopulations, cell trafficking and adhesion, costimulatory blockade in ANTIGEN presentation, TCR signalling, B-lymphocyte signalling, cytokine signalling, and cell proliferation. This is illustrated in Fig. 31.2 for a number of drugs targeting cell surface molecules and intracellular pathways and in Fig. 31.3 for cell surface molecules with a role in ANTIGEN presentation and costimulation. Almost all xenobiotic drugs work intracellularly, by inhibition of early or late events in intracellular signalling after lymphocyte activation or by inhibition of cell proliferation (direct interference in DNA/RNA synthesis) following activation. As these intracellular pathways are not truly selective for lymphocytes, most drugs manifest a quite narrow therapeutic window. Therefore, use of drugs in combination protocols is common practice to optimally exploit synergy in immunosuppression while avoiding adverse side effects. Besides immunosuppressive activity, a number of reagents, in particular those targeting CYTOKINES, manifest anti-inflammatory activity. For a number of reagents, i.e., those targeting costimulation in ANTIGEN presentation, immunomodulation ("tolerance" induction) has been claimed. Calcineurin inhibitors such as CsA and TACROLIMUS are widely used in transplant recipients together with AZA but in a number of centers replaced by MPA derivatives, mTOR inhibitors, and blockers of costimulation. Remarkably, most xenobiotic immunosuppressives originated from anti-infection or cancer drug development programs. Because of their failure in these indications or because immunosuppression was observed as a "side effect," the drugs were subsequently developed as immunosuppressants. One exception is TACROLIMUS which was specifically developed as an immunosuppressant and turned out to have the same mechanism of action as CsA. Other exceptions are the protein kinase C (PKC) inhibitor sotrastaurin and the JANUS-ACTIVATED KINASE 3 (JAK3) inhibitor tasocitinib, which originated from programs on immunosuppression. Some new immunosuppressants were first developed for other indications and then used off-label in transplant patients and subsequently evaluated in clinical trials in transplantation. In general, dose levels in these non-transplant indications are lower than those used in the transplantation setting. Reverse, drugs that were

Table 31.1 Lymphocyte-targeting pharmaceuticals currently approved or in advanced stage of development

Reagent[a]	Trade name	Mechanism of action	Status[b]
Cell proliferation, antimetabolites			
Cyclophosphamide (CY)	Endoxan, Cytoxan, Neosar, Procytox, Revimmune	Alkylating agent	
Azathioprine (AZA)	Azasan, Imuran	Purine synthesis inhibitor; prodrug of 6-mercaptopurine, 6-thioinosinic acid	Transplantation, autoimmune disease
Mizoribine	Bredinin	Inhibitor of IMPDH	Transplantation, RA, autoimmune disease
Mycophenolate mofetil (MMF)	CellCept	Inhibitor of IMPDH	Transplantation, autoimmune disease
Mycophenolic sodium (MPA)	Myfortic	Inhibitor of IMPDH	Transplantation, autoimmune disease
Leflunomide	Arava	Inhibitor of DHODH	RA, psoriatic arthritis
Teriflunomide	Aubagio	Active metabolite of leflunomide	MS
Gusperimus (15-deoxyspergualin)	Spanidin		Transplantation
Methotrexate (MTX)	Rheumatrex	Folate antagonist, inhibits dihydrofolate reductase	RA, cancer, autoimmune disease
Cell surface molecules, lymphocyte-depleting antibodies			
Antilymphocyte globulin (ALG)	Atgam	Depleting polyclonal Ab, raised in horses or rabbits against thymocytes	Aplastic anemia, GVHD, transplantation (rejection treatment)
Antithymocyte globulin (ATG)	Thymoglobulin	Depleting polyclonal anti-T-cell Ab, raised in rabbits	Transplantation (induction or treatment of rejection)
Muromonab (OKT3)	Orthoclone (OKT3)	Mouse antihuman CD3 mAb (to the CD3ε chain)	Acute rejection, GVHD
Teplizumab (hOKT3γ1(Ala-Ala), MGA031)		Humanized OKT3 mAb	Type 1 diabetes (phase 3)
Visilizumab	Nuvion	Humanized anti-CD3 mAb, non-FcR binding	Ulcerative colitis, Crohn's disease, GVHD (all in clinical trials)
Otelixizumab (ChAglyCD3, TRTX4)		Humanized anti-CD3 mAb (to the CD3ε chain)	Type 1 diabetes, autoimmune disease (orphan drug status)
A-dmDT390-bisFv(UCHT1)	Resimmune	Anti-CD3 diphtheria immunotoxin	Cutaneous T-cell lymphoma (phase 2), in clinical trials for metastatic melanoma
Zanolimumab (HuMax-CD4)		Human anti-CD4 mAb	RA, psoriasis, melanoma, cutaneous T-lymphoid malignancies (phase 2)
Siplizumab (MEDI-507)		Humanized anti-CD2 mAb	Psoriasis (phase 2): T-lymphoid malignancies (phase 1)
Alemtuzumab	Campath	Humanized anti-CD52 mAb	B-Lymphocyte chronic lymphocytic leukemia, (cutaneous) T-cell lymphoma, MS, and transplantation
Ofatumumab (HuMax-CD20)	Arzerra	Human anti-CD20 mAb	B-Cell chronic lymphocytic leukemia and other B-lymphoid malignancies: in clinical trials for RA and MS
Rituximab	Rituxan, MabThera	Chimeric anti-CD20 mAb	B-Lymphoid malignancies, RA, autoimmune diseases, antibody rejection in transplantation, induction treatment

(continued)

Table 31.1 (continued)

Reagent[a]	Trade name	Mechanism of action	Status[b]
Ocrelizumab		Humanized anti-CD20 mAb (modified from rituximab)	MS (in registration)
Obinutuzumab (GA101)		Humanized anti-CD20 mAb	B-Lymphoid malignancies (phase 3)
Ocaratuzumab (AME-133v)		Humanized anti-CD20 mAb	B-Lymphoid malignancies (phase 2–3), RA (phase 1)
Epratuzumab	LymphoCide	Humanized anti-CD22 mAb	B-Lymphoid malignancies, autoimmune disease (SLE) (phase 2–3)
CAT-8015		Fusion protein between anti-CD22 mAb and pseudomonas exotoxin A	B-Lymphoid malignancies (phase 2)
Galiximab		Chimeric anti-CD80 mAb (human-monkey)	B-Lymphoid malignancies (phase 3)
Denileukin diftitox	Ontak	Fusion protein between IL-2 and diphtheria toxin	T-Lymphoid malignancies, malignant melanoma
Cell trafficking and adhesion			
Fingolimod	Gilenia	Sphingosine-1-phosphate receptor modulator	MS
Vercirnon (CCX282-B)	Traficet-EN	CCR9 inhibitor	Crohn's disease (phase 3)
Natalizumab	Tysabri	Humanized anti-α_4 integrin mAb	MS, Crohn's disease
Alefacept	Amevive	Fc fusion protein of extracellular portion of LFA-3: blocks CD2	Psoriasis, T-lymphoid malignancies
T-Cell receptor signalling			
Cyclosporine A (CsA)	Gengraf, Neoral, Sandimmune	Inhibitor of calcineurin	Transplantation, RA, autoimmune disease
Voclosporin		Inhibitor of calcineurin	Lupus nephritis, SLE
Tacrolimus (FK506)	Prograf (topical application: Protopic)	Inhibitor of calcineurin	Transplantation, atopic dermatitis, RA, MG, GVHD
Sotrastaurin (AEB071)		Inhibitor of PKC	In clinical trials for diffuse large B-cell lymphoma, uveal melanoma (orphan drug status)
Costimulatory blockade			
Abatacept (CTLA4-Ig)	Orencia	Fc fusion protein of extracellular domain of CTLA-4, blocks CD28-CD80/86	RA, in clinical trials for autoimmune disease and MS
Belatacept (LEA29Y)	Nulojix	Fc fusion protein of extracellular domain of CTLA-4, blocks CD28-CD80/86	Transplantation
ASKP1240		Human antagonistic anti-CD40 mAb	In development for transplantation
B-Lymphocyte signalling			
Belimumab	Benlysta, LymphoStat-B	Human anti-B-lymphocyte stimulator (BLyS) mAb; blocks binding of BLys to its receptor(s)	SLE; Sjögren's syndrome
Atacicept (TACI-Ig)		Fusion protein containing the extracellular ligand-binding portion of the transmembrane activator and calcium modulator and cyclophilin ligand (CAML) interactor (TACI): binds BLys (CD257) and a proliferation-inducing ligand (APRIL; CD256)	SLE (phase 3)

Table 31.1 (continued)

Reagent[a]	Trade name	Mechanism of action	Status[b]
Eculizumab	Soliris	Humanized antibody to complement factor C5, inhibits cleavage to C5a	Paroxysmal nocturnal hemoglobinuria, atypical hemolytic uremic syndrome (orphan drug status); in trials for antibody-mediated rejection
Bortezomib	Velcade, Neomib, Nortecad	Proteasome inhibitor	Multiple myeloma, mantle cell lymphoma; in clinical trials for antibody-mediated rejection
Cytokines/cytokine signalling			
Daclizumab	Zinbryta (formerly Zenapax)	Humanized anti-CD25 mAb (CD25α chain)	MS
Basiliximab	Simulect	Chimeric anti-CD25 mAb (CD25α chain)	Transplantation
Tasocitinib	Xeljanz, Jakvinus	JAK3 inhibitor	RA; psoriasis, inflammatory bowel disease
Sirolimus	Rapamune	mTOR inhibitor	Transplantation, coronary stent coating
Everolimus	Certican or Zortress (transplantation), Afinitor (oncology)	mTOR inhibitor	Transplantation, renal cell carcinoma: (pancreatic) neuroendocrine tumors, other tumors
Ridaforolimus (formerly deforolimus)		mTOR inhibitor	Metastatic soft tissue and bone sarcomas (phase 3)
Ustekinumab	Stelara	Human anti-IL12/IL23 mAb	Psoriasis, psoriatic arthritis
Tocilizumab	Actemra, RoActemra	Humanized anti-IL6 receptor mAb	RA, systemic juvenile idiopathic arthritis (orphan drug status)
Infliximab	Remicade	Chimeric anti-TNF-α mAb	Psoriasis, Crohn's disease, ankylosing spondylitis, psoriatic arthritis, RA, ulcerative colitis
Adalimumab	Humira, Exemptia	Human anti-TNF-α mAb	RA, psoriatic arthritis, ankylosing spondylitis, Crohn's disease, ulcerative colitis, psoriasis, juvenile idiopathic arthritis
Golimumab	Simponi	Human anti-TNF-α mAb	RA, psoriatic arthritis, ankylosing spondylitis, ulcerative colitis
Certolizumab pegol	Cimzia	Humanized anti-TNF-α mAb	Crohn's disease, RA, psoriatic arthritis
Etanercept	Enbrel	Fc fusion protein of extracellular domain of TNF receptor 2: binds TNF-α	RA, juvenile idiopathic arthritis, psoriatic arthritis, ankylosing spondylitis, psoriasis
Pegsunercept		Pegylated TNF receptor 1: binds TNF-α	RA
Anakinra	Kineret	IL-1 receptor antagonist	RA

[a]The inclusion of "tuzu" in the name of the reagent generally indicates an application in oncology and that of "lizu" an application as an immune modulator. The inclusion of "cept" indicates that the product is a fusion protein
[b]Pharmaceuticals in clinical development are presented with the stage of clinical trials; if not otherwise specified, the agents are marketed for the indication shown or used off-label

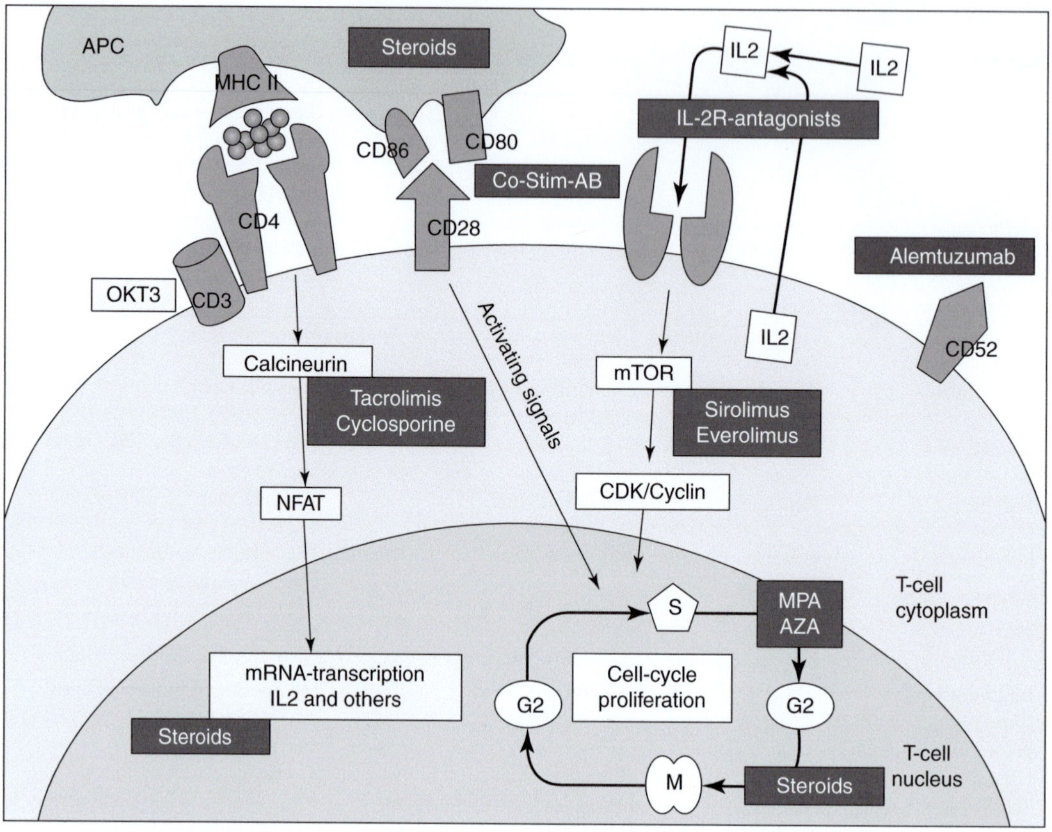

Fig. 31.2 Mechanisms of action of various immunosuppressive agents in T lymphocytes. Alemtuzumab and anti-CD3 mAb (OKT3) target specific lymphocytic surface structures to induce cytolysis or receptor downregulation; a similar principle is involved with antilymphocyte and antithymocyte globulins (not illustrated), which target multiple molecules on the cell surface. Signal transduction from the TCR/CD3 complex (so-called signal 1) is blocked by calcineurin inhibitors (tacrolimus and CsA) which affect nuclear factor of activated T cells (NFAT). Costimulatory signalling (signal 2), via the interaction between CD28 and B7 (CD80, CD86), is blocked by the fusion proteins abatacept and belatacept (Co-stim). Not shown is the costimulatory pathway via the CD40-CD154 interaction (see Fig. 31.3). Protein kinase C is involved in processes downstream of signal 1 and 2, and this step in intracellular signal transduction is inhibited by sotrastaurin (not illustrated). Anti-CD25 mAbs basiliximab and daclizumab (anti-IL-2 mAb) target the receptor for IL-2, required for cell activation mediated by cytokines (signal 3). This pathway involves the mammalian target of rapamycin (mTOR), which is targeted by sirolimus and everolimus and the cyclin-dependent kinase, and the JAK/STAT pathway which is targeted by tasocitinib (not shown). Mycophenolic acid (MPA) derivatives and azathioprine (AZA) interfere with the cell cycle, preventing T- and B-lymphocyte proliferation. Steroids target multiple sites in the process. Modified from Urschel et al. [Selected reading]

originally developed as immunosuppressants were subsequently developed for other indications, mainly in oncology: well-known examples are mTOR inhibitors. Some compounds failed in clinical trials as immunosuppressant and subsequently entered development in an oncologic indication: an example is sotrastaurin.

With the present spectrum of immunosuppressants and extensive clinical experience with various drug combinations, the management of immunosuppression in patients with transplanted organs has markedly improved. However, major complications that are directly associated with immunosuppression still occur, such as drug toxicity, increased susceptibility to infection, and development of tumors, e.g., the so-called post-transplant lymphoproliferative disease [9–11]. But, as a beneficial effect, immunosuppression in the transplant setting can prevent virus infections [12]. On the other hand, long-term graft

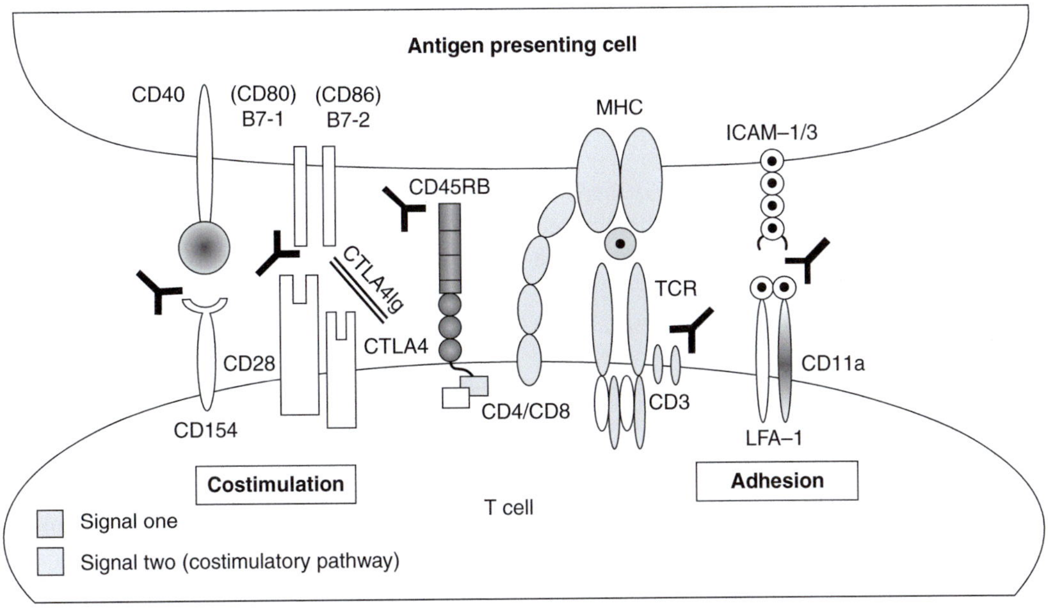

Fig. 31.3 Target molecules on the surface of T lymphocytes and APC involved in the potential induction of T-cell activation and immunological tolerance. T-lymphocyte stimulation requires interaction between the TCR and the antigenic epitope in combination with costimulation via a number of cell surface interactions (depicted are those between CD40 and its ligand CD154, and between CD80/CD86 and CD28/CTLA-4). Inhibition of costimulation can result in an anergic state and subsequent unresponsiveness or tolerance. Inhibition of molecular interactions in cell adhesion promotes inhibition of T-cell activation (the interaction between LFA-1 and ICAM1/3 is depicted; others include that between VLA-4 and ICAM/VCAM). Also depicted are mAbs used to interfere with these interactions. Modified from Vincenti (2002) [Vincenti F. What's in the pipeline? New immunosuppressive drugs in transplantation. Am J Transplant. 2002;2:898–903]

survival is hampered by graft dysfunction due to chronic rejection. This phenomenon not only relates to the ongoing (smoldering) immune response of the recipient to the graft but also to intrinsic changes in the graft itself, mainly regarding the vasculature, the so-called graft vessel disease or accelerated graft arteriosclerosis. This involves thickening of the intima of blood vessels in the graft and is ascribed to migration of smooth muscle cells to this site followed by cellular proliferation and extracellular matrix formation. In lung transplantation, chronic rejection becomes manifest as an obstruction called bronchiolitis obliterans, in which the bronchioles are compressed and narrowed by fibrosis or inflammation. With prolonged graft survival, chronic rejection is now a major cause of graft loss in long-term surviving patients [13, 14]. There are no specific drugs available for the prevention or treatment of chronic rejection. Also, some immunosuppressants, such as mTOR inhibitors and MPA derivatives, are claimed to be effective, as these not only contribute to better immunosuppression (diminished host attack) but also inhibit the vascular (proliferative) response due to their inherent mechanism of action.

Since allograft rejection is a T-cell-mediated process, the main target for immunosuppression in transplantation is the T lymphocyte (see Chap. 3). But other cell types of the immune system, like B LYMPHOCYTES and MACROPHAGES, are involved as well. The role of NK cells in graft rejection is less unclear. This might change in the future when xenotransplantation (transplantation of nonhuman organs into humans) is likely to become available. The rejection of xenografts cannot be prevented by T-cell immunosuppressants since it does not solely involve T cells but also B LYMPHOCYTES which, in addition to T-cell-dependent Ab generation, can be triggered in a T-cell-independent way. Ab-mediated rejection has been identified as a major hurdle in preclinical modelling of solid organ xenografts [15] but appears to be less relevant for tissue or cell xenografts. Ab-mediated rejection is also relevant in allotransplantation because B-cell reactivity appears to be involved in chronic rejection as well; this is further illustrated by the involvement of the complement system [16] (see Chap. 8). Some specific anti-B-cell drugs have been developed for autoimmune diseases or B-lymphoid malignancies, and a number of immunosuppressants discussed below show both T- and B-cell inhibitory activity.

31.2 Calcineurin Inhibitors

31.2.1 Cyclosporine

CsA is a cyclic undecapeptide isolated from the fungus *Tolypocladium inflatum* Gams. Its biological activity in vivo was first discovered in 1973 in a large microbiology screening program, which included Ab formation to sheep red blood cells in mice. Subsequently, it showed efficacy in kidney allotransplantation in rats and pigs. This was followed by first clinical trials in human kidney transplant patients, and its introduction to clinical transplantation in 1983 [17]. Since CsA is a highly lipophilic molecule, it is poorly soluble in water. For oral administration, the first commercial formulation was an oil-based formulation manifesting quite variable absorption. Besides a huge intraindividual variation, there is a large interindividual variability in response to CsA, i.e., the existence of "absorbers" and "non-absorbers" necessitated a drug exposure test in each patient before transplantation. Since the drug shows a relatively narrow therapeutic window, the kidney being the first target for adverse side effects, drug-level monitoring proved necessary to control drug exposure in the therapeutic range. Instead of 16- or 24-h trough levels, the so-called C2 monitoring (levels 2 h after administration) has been introduced as an improved estimate of total exposure. A microemulsion formulation (Neoral) has been marketed since 1995. This formulation shows improved absorption and far less inter- and intra-individual variation.

CsA shows immunosuppressive activity in a large spectrum of animal models of human immune-mediated diseases [18]. Its immunosuppressive effect is mainly restricted to T LYMPHOCYTES. T-cell-independent B-cell responses are not affected. At the time of introduction on the market, the mechanism of action of CsA was largely unknown. A first insight into its mode of action came from the observation that the compound inhibits the production of IL-2, one of the first CYTOKINES produced after T-cell activation. Subsequently, it was demonstrated that CsA inhibits IL-2 gene transcription by interfering with a Ca-dependent intracellular signalling mechanism [19]. A family of cytoplasmic proteins called cyclophilins (CYPs) has been identified which strongly binds to CsA. CYPs are enzymes catalyzing *cis*-trans isomerization of peptidyl-prolyl bonds, i.e., the so-called proline isomerase or rotamase activity, which is important for proper folding of newly synthesized proteins. This inhibition of rotamase activity, however, does not cause the immunosuppressive effect which is actually mediated by the binding of the CsA-CYP complex to the serine/threonine phosphatase calcineurin, which plays a pivotal role in Ca-dependent intracellular signalling.

Activation of T cells via the TCR results in a cascade of events that, among others, involves the activation of the protein tyrosine kinases (PTKs) $p56^{lck}$, $p59^{fyn}$, and ZAP, followed by phosphorylation of phospholipase Cγ1, resulting in generation of second messengers PIP3 and DAGl. These, in turn, result in an increase in cytoplasmic free Ca^{2+} and activation of PKC. Free Ca^{2+} upon complexing with calmodulin activates the phosphatase calcineurin. The CsA-CYP complex, upon binding to calcineurin, inhibits its phosphatase activity. Calcineurin dephosphorylates the NUCLEAR FACTOR OF ACTIVATED T CELLS (NFAT) which is then translocated to the nucleus where it initiates, together with other transcription factors (e.g., NF-κB and AP-1), expression of early T-cell activation genes, especially the gene encoding IL-2 (Fig. 31.2).

CYPs and calcineurin are abundantly expressed in different cell types. The apparent T-cell selectivity of CsA has therefore been related to the fact that Ca-dependent T-cell activation via the TCR uniquely involves the calmodulin-calcineurin pathway. There is as yet no unequivocal proof that potential side effects, like damage to kidney tubules, result from the same intracellular mechanism.

In keeping with its mechanism of action, Ca-independent cell triggering is not affected by CsA. For instance, T cells can be activated via the costimulatory CD28 molecule on the cell surface (Figs. 31.2 and 31.3) which, in combination with activated PKC, causes lymphokine gene transcription and T-cell activation in the absence of calcineurin activation. It has therefore been hypothesized that this pathway is involved in T-cell activation and allograft rejection that is resistant to CsA treatment. Also, inhibition of T-cell activation and growth factor-induced cell proliferation by mTOR inhibitors targets a pathway that differs from calcineurin inhibition, a phenomenon that underlies the synergy in immunosuppression between calcineurin inhibitors and mTOR inhibitors.

There have been many attempts to identify CsA analogues with an improved therapeutic window, but such attempts were largely unsuccessful. Presently, one compound structurally similar to CsA, voclosporin (ISA247; Fig. 31.1), is on the market as treatment for uveitis and in clinical trials for the treatment of lupus nephritis. The compound differs from CsA in the first amino acid residue of the molecule and has been claimed to be more potent than CsA while having fewer adverse side effects, in particular renal side effects. The compound was also evaluated in phase 2 transplantation trials in which a similar efficacy and reduced incidence of posttransplant diabetes was noted.

31.2.2 Tacrolimus (FK506)

TACROLIMUS [20–22] is a macrocyclic lactone isolated from the actinomycete *Streptomyces tsukubaensis*. It was discovered in 1984 in an immunological screening program that was specifically established to identify immunosuppressive compounds. Subsequently, its immunosuppressive activity was demonstrated in various animal models of transplantation (rat, dog). The spectrum of immunosuppressive activity appeared to be identical to that of CsA, but, remarkably, the drug is efficacious at much lower doses than CsA, both in vitro and in vivo. Also therapeutic drug levels in the circulation appear to be much lower than those for CsA.

TACROLIMUS is very lipophilic and poorly soluble in water, resulting in variable absorption and necessitating regular drug monitoring. For oral administration, a solid dispersion formulation in hydroxypropylmethyl cellulose is used. Major side effects are strikingly similar to those of CsA but involve in addition the central nervous system as a target organ. The difference from CsA may at least in part be explained by differences in pharmacokinetics and tissue distribution, as the mechanism of action appears to be the same, i.e., inhibition of calcineurin (Fig. 31.2). This does not hold for other side effects, like the lack of gingiva hyperplasia and hirsutism (adverse side effects of CsA) and the more pronounced diabetogenic effect of TACROLIMUS.

The similarity between TACROLIMUS and CsA regarding immunosuppressive activity is based on the fact that both drugs inhibit calcineurin phosphatase activity and subsequent intranuclear events in T-cell activation. However, in accord with the different molecular structure (Fig. 31.1), TACROLIMUS does not bind to CYP as CsA does. A family of CYPs with rotamase activity has been identified, the FK506-binding proteins (FKBPs). Of these, FKBP12 appears to be most relevant; upon binding to TACROLIMUS, the complex binds to calcineurin at the same site as the CsA-CYP complex and thereby inhibits T-cell activation.

The efficacy of TACROLIMUS in humans was first shown in liver transplantation [22]. In liver disease, e.g., liver allograft dysfunction, the absorption of TACROLIMUS is increased, and its metabolism decreased resulting in higher exposure, which actually is required in cases of rejection. Nowadays, the drug is widely used in patients that poorly tolerate CsA; in addition to this conversion in immunosuppressive regimen, the drug is also used in patients directly after transplantation (the so-called de novo treatment). Interestingly, despite its adverse side effect on pancreas function, TACROLIMUS (together with the mTOR inhibitor SIROLIMUS and the anti-CD25 mAb daclizumab in induction treatment) is part of the immunosuppressive regimen used in clinical islet cell transplantation for patients with type 1 diabetes [23]. This steroid-free immunosuppressive regimen was proposed to bypass potential pancreatoxicity of CORTICOSTEROIDS and was introduced in 2000 as an innovative regimen and boosted islet cell transplantation toward clinical application.

Ascomycin is an immunosuppressant with a similar structure as TACROLIMUS. Ascomycins were first identified as fermentation products of *Streptomyces hygroscopicus* almost 50 years ago. Ascomycin derivatives such as pimecrolimus are currently available for topical administration in skin diseases [24].

31.3 Protein Kinase C Inhibitor

The PKC family of serine/threonine kinases has a central function downstream of T-cell stimulation via the TCR and CD28 costimulatory pathway and affects the NFAT and NF-κB nuclear transactivation pathways. Three PKC categories exist based on their need for cofactors, described as conventional or classical isoforms, novel isoforms, and atypical isoforms. Three isoforms, namely, PKC α, β, and θ, are relevant for intracellular signalling in T and B LYMPHOCYTES. PKC θ activity is mainly restricted to T LYMPHOCYTES. Sotrastaurin (AEB071) selectively inhibits the novel isoforms of PKC α and β and the classical isoforms PKC δ, ε, η, and θ. Its immunosuppressive activity has been demonstrated in animal models of transplantation [25, 26]. Since PKC inhibition is independent of the calcineurin pathway and the mTOR pathway in T-cell activation, the compound shows synergism in immunosuppression with calcineurin inhibitors (CsA) and mTOR inhibitors (everolimus). Synergy with the sphingosine-1-phosphate receptor modulator fingolimod has also been demonstrated in rodent transplant models. The clinical development of the compound for kidney transplantation was halted because of lower efficacy than the comparator CsA: the drug is now in clinical trials for treatment of diffuse large B-cell lymphoma and uveal melanoma.

31.4 Inhibitors of the Mammalian Target of Rapamycin

31.4.1 Sirolimus and Everolimus

SIROLIMUS is a macrocyclic lactone, isolated from the actinomycete *Streptomyces hygroscopicus*, with a long history dating from the mid-1970s [27–30]. It was first discovered as an antifungal compound in a soil sample from Easter Island, a Polynesian island in the southeastern Pacific Ocean also known as "Rapa Nui," which resulted in the name "rapamycin." The compound was not further developed because side effects were encountered, including involution of lymphoid tissue. Subsequently, antitumor effects were documented, as well as immunosuppressive activity in rat models of autoimmune disease. After structural similarities between SIROLIMUS and TACROLIMUS were identified, studies on the efficacy of the compound in rat and mouse organ allograft models were initiated and first reported in 1989. Since then, the compound has been developed for clinical use in transplantation and was launched in 1999. A major complication was the development of a proper oral formulation with acceptable stability, bioavailability, and predictability in absorption characteristics. The compound is very lipophilic and poorly soluble in water. In oily solution, as well as in microemulsion, the drug appears to be readily absorbed after oral administration. Another mTOR inhibitor with improved physicochemical characteristics, EVEROLIMUS [31, 32], was identified in a chemical derivatization program, followed by SIROLIMUS with a launch in Europe in 2003 and in the USA in 2010. The relatively narrow therapeutic window requires regular drug exposure monitoring. Major side effects are hyperlipidemia, followed by impaired wound healing, thrombocytopenia, mouth ulcers, and cholesterolemia.

mTOR inhibitors proved to be extremely potent immunosuppressants, affecting both T LYMPHOCYTES and Ab production by cells in the B-lymphocyte lineage. This was demonstrated in a wide spectrum of experimental animal models and in clinical trials in renal transplantation. Immunosuppression is achieved at relatively low blood levels. In animal models, the drugs show synergistic action in combination with CsA, which points to a difference in mechanism of action. Indeed, whereas CsA and TACROLIMUS inhibit early events in T-cell activation, e.g., the expression of growth factors such as IL-2 in G0–G1 stage of the cell cycle, mTOR inhibitors affect the progression of G1 to S phase in the cell cycle. At the molecular level, it has been demonstrated that these compounds bind, as does TACROLIMUS, to CYPs of the FKBP family, in particular FKBP12. However, the complex with FKBP does not bind to calcineurin but to mTOR (alternative names in literature are FRAP, FKBP-rapamycin-associated protein, and RAFT, rapamycin and FKBP target), a protein with kinase activity. This resultant complex inhibits intracellular cytokine-driven cell proliferation, presumably via p70 S6 kinase, which is involved in translational control. Apparently, this pathway is particularly relevant in lymphoid cells, underlying the peculiar immunosuppressive characteristics of the drug.

Since growth factor-driven cell proliferation applies to other cell types as well, the therapeutic window of compounds in the class of mTOR inhibitors is expected to be narrow. On the other hand, the antiproliferative action of the drug could be beneficial in chronic rejection of solid organ allografts. Indeed, it has been shown that rapamycin, in contrast to CsA and TACROLIMUS, inhibits the proliferation of smooth muscle cells in vitro and in animal transplantation models inhibits intima proliferation of blood vessels as observed in chronic rejection. Such a beneficial effect has also been documented in the clinical situation of patients with an organ allograft [33]. Recently, this beneficial effect on vasculopathies has gained a new application, namely, in relation to stents placed in the vasculature of patients after balloon coronary angioplasty. Such stents can show restenosis, resulting in recurrence of the vessel occlusion. Either treatment of the patient with SIROLIMUS or bathing the stent before implantation in a rapamycin-containing medium reduces this complication [34]. Interestingly, clinical trials in organ transplantation revealed that the incidence of posttransplant malignancies was lower for patients receiving mTOR inhibitors [35], which is essentially in accordance with the original research and discovery of rapamycin as an anticancer drug in the 1980s. The regained interest in mTOR inhibitors as potential anticancer drugs [36] resulted in the launch of EVEROLIMUS for the treatment of advanced renal carcinoma in 2009. EVEROLIMUS is also approved for subependymal giant cell astrocytoma and neuroendocrine tumors.

31.5 Janus Kinase 3 Inhibitor

JAKs are tyrosine kinases involved in intracellular signalling after stimulation of a broad variety of cell surface receptors, in particular those of the cytokine receptor common γ chain family. Cell activation by CYTOKINES via their cell surface receptor results in clustering of receptor chains and recruitment of the inactive cytosolic form of JAK to the γ chain, which is followed by phosphorylation of tyrosine-based docking sites of transcription factors in the SIGNAL TRANSDUCTION AND ACTIVATION OF TRANSCRIPTION (STAT) pathway. Phosphorylated STATs in turn dimerize and migrate to the nuclei where they stimulate gene transcription. In this respect, JAKs play a major role in the so-called signal 3 pathway of T-cell activation, similar to but distinct from mTOR-mediated intracellular signalling (calcineurin/CYP representing signal 1 and costimulation representing signal 2). This signal 3 pathway involves CYTOKINES like IL-2, IL-4, IL-9, IL-15, and IL-21. There are four JAKs in mammalian cells (JAK1–JAK3 and tyrosine kinase 2), of which JAK3 is restricted to hematopoietic cells and mediates cytokine signalling only via receptors that contain a γ chain. Within the hematopoietic cell lineage, JAK3 shows high expression in NK cells and thymocytes, but not in resting T cells. It is inducible in T cells, B cells, and myeloid cells. Therefore, inhibition of JAK3 could be associated with immunosuppression without the adverse side effects observed on inhibition of the signal 1 pathway.

Tasocitinib (CP690, 550) is a specific inhibitor of JAK3, discovered in a screen for inhibitors of catalytic activity using a glutathione S-transferase fusion protein with the kinase domain of human JAK3. In rodent and nonhuman primate transplantation models, the compound showed immunosuppressive efficacy. This was confirmed in clinical trials in kidney transplantation, in which the drug proved to be well tolerated [37]. In a high-dose group, a higher incidence of infections, cytomegalovirus disease, and BK virus nephropathy was noted, and in accordance with the documented presence of JAK3 in NK cells, NK cells were strongly reduced. The drug is since 2012 on the market for the treatment of psoriasis and RA.

31.6 Cytotoxic Drugs

The designation "cytotoxic drugs" is used here to describe drugs that directly interfere with DNA/RNA synthesis and as such affect cell proliferation (see Chap. 30).

31.6.1 Cyclophosphamide

CY is one of the oldest drugs used as an immunosuppressant. It is an alkylating agent that was originally used as an anticancer drug. The compound inhibits cells from entering the S phase of the cell cycle, which is subsequently blocked at the G2 phase. The drug was used in initial trials in clinical transplantation around 1970, in particular in patients with AZA toxicity. Severe side effects were encountered, mainly bone marrow depression with severe leukopenia and anemia. Since the introduction of more selective T-cell immunosuppressants, the drug has barely been used because of these side effects. Interestingly, CY was part of a conditioning regimen used in TOLERANCE induction by mixed hematopoietic chimerism (see below). However, CY is among the most powerful inhibitors of B LYMPHOCYTES and received renewed interest in experimental animal research in xenotransplantation.

31.6.2 Methotrexate

MTX is a folate antagonist, which nowadays is mainly used in a low-dose treatment regimen (weekly administration) in subsets of patients with RA. Its mechanism of action, under these conditions, has not been completely resolved. It has been suggested that MTX, at a low dose, is converted into a polyglutamate that inhibits transmethylation reactions, resulting in an increased release of adenosine and decreased synthesis of guanine. Although this condition affects purine metabolism, an anti-inflammatory signal is delivered by the binding of adenosine to specific adenosine (A2) receptors. Thus, MTX, at a low dose, does not appear to be an immunosuppressant, but rather an anti-inflammatory drug. Presently, MTX is mainly used in the treatment of RA and psoriasis and inflammatory bowel diseases [38].

31.6.3 Azathioprine

AZA has been used since the early days of clinical immunosuppression, being introduced as an immunosuppressant for transplantation in 1961. Its development followed the pioneering work on 6-mercaptopurine as an antileukemic agent in the 1940s. Nowadays, the drug is still in use in conjunction with baseline immunosuppression (e.g., CsA) in transplantation. Its dosing and dose adaptations are based on adverse side effects, i.e., blood leukocyte counts. Despite the long period of clinical use, its exact mechanism of action is still not completely clear. The drug is converted by red blood cell glutathione to 6-mercaptopurine, which in turn is converted into a series of mercaptopurine-containing nucleotides which interfere with the synthesis of DNA and polyadenylate-containing RNA. One of the nucleotides formed is thioguanylic acid which can form thioguanosine triphosphate. This can be incorporated into nucleic acids and induce chromosome breaks and also affects the synthesis of coenzymes. As a general inhibitor of cell proliferation, AZA affects both T- and B-lymphocyte reactivity. Presently, there is an intention to minimize the use of AZA or restrict its use to treatment of rejection episodes, with a proposed replacement by MPA derivatives [39].

31.6.4 Mizoribine

Mizoribine is an imidazole nucleoside, originally isolated as a potential antibiotic from the culture filtrate of the soil fungus *Eupenicillium brefeldianum*. Its immunosuppressive activity was demonstrated first by the inhibition of mouse lymphoma cell lines and subsequently by the inhibition of an Ab response in mice immunized with sheep red blood cells. It was subsequently shown that the drug is phosphorylated intracellularly to the active form, mizoribine 5′-monophosphate, under the influence of adenosine kinase. This compound is a competitive inhibitor of the enzyme inosine monophosphate dehydrogenase (IMPDH), which is a rate-limiting enzyme in purine biosynthesis in lymphoid cells (Fig. 31.4). The drug has been in use in clinical transplantation since 1984, only in Japan, mainly as a replacement for AZA [40]. It is also proposed for the treatment of RA and various renal diseases, including IgA nephropathy, lupus nephritis, and nephritic syndrome [41].

31.6.5 Mycophenolic Acid

Mycophenolate mofetil (MMF, RS-61443) is the morpholinoester of MPA, a fermentation product of various *Penicillium* species, originally isolated and purified in the early 1910s. MPA was originally studied for its antibacterial and antifungal activity and subsequently for its antitumor activity in the late 1960s, but these activities were not further followed in clinical development. The compound was specifically selected as a drug that inhibited IMPDH in the mid-1980s. This selection was based on the fact that MPA was not a nucleoside, that it failed to require phosphorylation to become active, and that it did not show the unwanted side effects of nucleosides, such as induction of chromosome breaks and inhibition of DNA repair enzymes. MPA is a noncompetitive reversible inhibitor of IMPDH (Fig. 31.4). Mycophenolate mofetil was developed as an immunosuppressant. It is rapidly hydrolyzed by esterases to yield MPA. The drug was introduced to the market in 1995 for the transplantation indication, mainly as a replacement for AZA, i.e., in combination treatment with calcineurin inhibitors. Major side effects are gastrointestinal intolerability and bone marrow depression, documented for a marketed formulation (CellCept). An enteric-coated formulation of mycophenolate sodium (Myfortic), with fewer gastrointestinal side effects, was approved for transplantation in 2004. At present the drug is increasingly used in the treatment of autoimmune diseases such as idiopathic thrombocytopenic purpura, systemic lupus erythematosus (SLE), scleroderma (systemic sclerosis), and pemphigus vulgaris [42, 43].

The fact that inhibition of IMPDH causes quite selective immunosuppression is due to the relevance of different pathways of purine metabolism in different cell types (Fig. 31.4). Two pathways exist, the salvage pathway (conversion of guanine into GMP by hypoxanthine-guanine phosphoribosyltransferase (HGPRT)) and the de novo pathway (conversion of IMP into GMP mediated by IMPDH). Lymphocytes highly depend on the de novo pathway and do not use the salvage pathway. At the other end of the spectrum, cells of the central nervous system highly depend on the salvage pathway. Cell types like smooth muscle cells, fibroblasts, endothelial cells, and epithelial cells can use both pathways for purine synthesis. Hence, inhibition of IMPDH results in a quite selective inhibition of purine biosynthesis in lymphocytes. For MPA, an additional selectivity has been documented for the two isoforms of IMPDH; the type I isoform is predominantly expressed in resting lymphocytes, and the type II is strongly expressed in lymphocytes after activation. This type II isoform

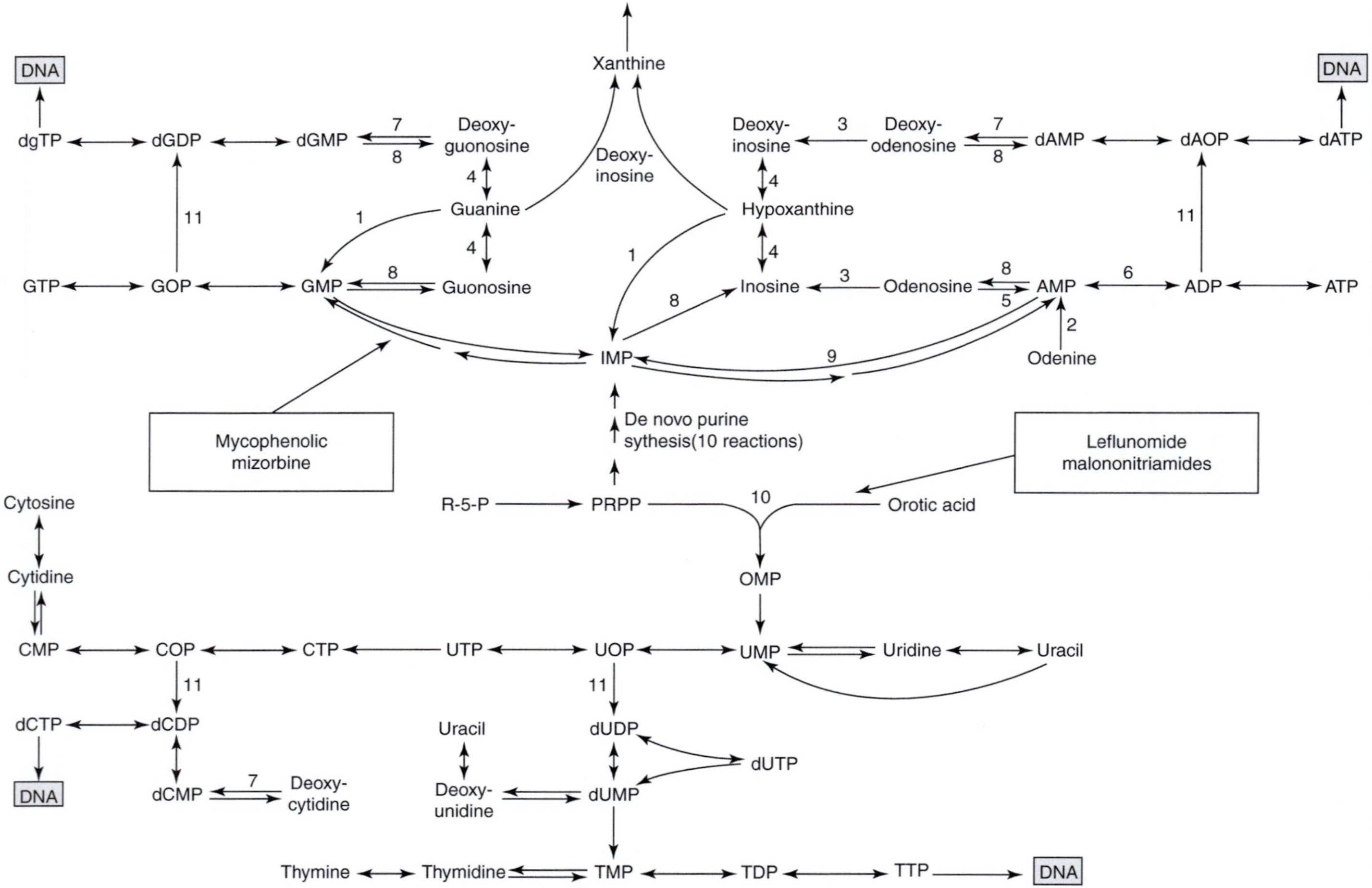

Fig. 31.4 The purine/pyrimidine pathway and sites of inhibition by MPA, mizoribine, leflunomide, and teriflunomide. For purine metabolism, two pathways are shown, the salvage pathway (conversion of guanine to GMP, mediated by HGPRT)) and the de novo pathway (conversion of IMP to GMP, mediated by inosine monophosphate dehydrogenase (IMPDH)). For pyrimidine metabolism, the de novo pathway (conversion of orotic acid to UMP) and the salvage pathway (conversion of uridine to UMP and cytidine to CMP) are shown. Major enzymes involved are hypoxanthine-guanine phosphoribosyltransferase (HGPRT) (1), adenine phosphoribosyltransferase (2), adenosine deaminase (3), purine nucleoside phosphorylase (4), adenosine kinase (5), adenylate kinase (6), deoxycytidine kinase (7), 5′-nucleotidase (8), adenosine monophosphate dehydrogenase (9), dihydroorotate dehydrogenase (DHODH) (10), and ribonucleotide reductase (11)

is 4–5 times more sensitive to inhibition by MPA than the type I isoform. Hence, MPA is a more potent inhibitor of activated lymphocytes. As both T and B LYMPHOCYTES are affected by IMPDH inhibition, MPA is an effective inhibitor of both T and B LYMPHOCYTES, like mizoribine mentioned above. This has not only been demonstrated in rodent models but also in pig-to-nonhuman primate xenotransplantation models, where the suppression of xeno-Ab formation is a critical issue.

Apart from affecting RNA synthesis, IMPDH inhibition has other effects as well. IMPDH inhibition results in depletion of GTP in lymphocytes (Fig. 31.4), which affects the transfer of fucose and mannose to glycoproteins. This can affect adhesion molecules on the cell surface. Examples of adhesion molecules whose expression is inhibited are VLA-4 and ligands for selectins. On the basis of this mechanism, IMPDH inhibitors could affect the recruitment of inflammatory cells into tissue and effector-target cell interactions within tissues. This potential effect has been demonstrated by the inhibition of the adhesion of T LYMPHOCYTES to endothelial cells in vitro when either T cells or endothelial cells are pretreated with MPA.

Finally, because IMPDH inhibition can affect cell growth in other cell types besides lymphocytes, it might have an effect in chronic rejection. MPA inhibits proliferation of human smooth muscle cells in vitro and proved effective in a rat vessel transplantation model which mimics chronic rejection in solid organ allografts. At present, MPA derivatives are mainly used in the transplantation setting; besides being a potential replacement for AZA, the drug has also been studied as a replacement for calcineurin inhibitors [44].

31.6.6 Leflunomide

LEFLUNOMIDE is an isoxazole derivative, originally synthesized as part of an agriculture herbicide program in the mid-1970s. Its anti-inflammatory activity was demonstrated in animal models of adjuvant arthritis and experimental allergic encephalomyelitis. First studies on its immunosuppressive action in models of autoimmune disease were documented in 1990. Since then, the compound has been extensively investigated in animal models of solid organ allo- and xenotransplantation (rodents, dog, nonhuman primates) and proved to be a potent immunosuppressant both for T and B LYMPHOCYTES. LEFLUNOMIDE was subsequently developed as a so-called disease-modifying antirheumatic drug (DMARD) for RA and was launched in 1998 [45, 46]. Following this launch, a number of clinical trials have shown its efficacy in transplantation. A major drawback for its development for the transplantation indication is the long half-life in man (15–18 days). Therefore, analogues with a shorter half-life (malononitrilamides) have been explored as potential immunosuppressants, but these analogues failed in clinical trials. The active metabolite of LEFLUNOMIDE, teriflunomide (A77 1726), is on the market as treatment for multiple sclerosis (MS) [47].

LEFLUNOMIDE is a produg, which under the influence of the intestinal mucosa or liver is metabolized to the active isoxazole open-ring form (compound A77 1726). The antiproliferative action (entry into S phase of the cell cycle), although still not completely understood, is based on its inhibition of two different intracellular pathways. The first involves the enzyme dihydroorotate dehydrogenase (DHODH), which is the fourth rate-limiting sequential enzyme in the de novo pyrimidine biosynthetic pathway (Fig. 31.4). This pathway is particularly relevant for the proliferative response of lymphocytes, as limited intracellular pools of substrates restrict the use of the salvage pathway by these cells. Also, the intracellular concentration of DHODH is relatively low in lymphoid cells, so that this pathway is easily inhibited. Inhibition of DHODH results in depletion of intracellular pyrimidine nucleotides, which have several vital cellular functions, including synthesis of DNA, RNA, glycoproteins (adhesion proteins), and phospholipids. Another mechanism of action is inhibition of protein phosphorylation by inhibition of tyrosine kinase activity. The relevance of this mechanism for the immunosuppressive action of LEFLUNOMIDE is questionable because kinase inhibition generally requires higher concentrations in vitro than does the inhibition of DHODH. Also, there are claims that LEFLUNOMIDE inhibits the activation of NF-κB.

Like inhibition of IMPDH, inhibition of DHODH affects not only T cells but also B cells. Hence, LEFLUNOMIDE has been shown to be a potent drug affecting T-dependent and T-independent Ab synthesis. There are reports that LEFLUNOMIDE affects B cells even more potently than T LYMPHOCYTES. There are also claims that the drug might be effective in chronic rejection, as demonstrated by the in vitro inhibition of rat smooth muscle cell proliferation and its in vivo efficacy in rodent vessel transplantation, a model mimicking vascular pathology in chronic rejection of solid organ allografts. Apart from its immunosuppressive activity, the anti-inflammatory action of the drug (inhibition of the production of inflammatory CYTOKINES) might be relevant for this indication.

31.7 Gusperimus (15-Deoxyspergualin)

Gusperimus [48] is a synthetic derivative of spergualin, an antibiotic isolated from the soil bacterium *Bacillus laterosporus* in a screening program for anticancer drugs in the early 1980s. Its immunosuppressive activity in a mouse skin transplantation model was demonstrated in 1985, and the drug was subsequently developed for use in transplantation. It has been commercially available in Japan since 1994. A major drawback in its clinical application is the low oral bioavailability of the drug, which means that it has to be administered parenterally, and also its instability in aqueous solution. In Europe, gusperimus received orphan drug status in 2001 for the treatment of Wegener's granulomatosis, a serious form of vasculitis frequently associated with permanent disability and/or fatal outcome; a clinical trial showing benefits in these patients has been published [49].

Gusperimus is a potent immunosuppressive compound in animal models of transplantation and AUTOIMMUNITY. B cells are equally well inhibited as T cells, and the drug also shows anti-inflammatory activity. However, the mechanism of action is not very well understood. The compound affects the differentiation of stimulated lymphocytes into effector cells, e.g., cytotoxic T LYMPHOCYTES or Ab-producing plasma cells, and the entry of cells from G0 or G1 into the S phase of the cell cycle. At the molecular level, the drug blocks the nuclear transcription factor NF-κB, which appears to underlie its inhibitory activity in production of proinflammatory CYTOKINES and maturation of dendritic cells.

Also, binding of gusperimus to a cytosolic member of the heat shock protein (HSP) family, HSP70, has been described. HSPs participate in the folding and unfolding of proteins and play a role in protein transport to intracellular organelles (the so-called chaperone function of these molecules, for instance, in ANTIGEN presentation). This capacity of the drug has received attention for a different area, namely, parasite infection. HSP mediates a number of processes that are crucial to parasite survival such as thermoprotection and, through their chaperone function, export of parasite proteins to the erythrocyte, parasite interorganelle protein trafficking, and regulation of parasite infectivity. Gusperimus, by binding HSP70, has therefore been proposed as a new innovative antimalarial drug [50].

31.8 Cell Trafficking and Adhesion

31.8.1 Fingolimod

Fingolimod is a novel immunomodulatory agent which affects lymphocyte recirculation. It is a chemical derivative of myriocin, a metabolite of the ascomycete *Isaria sinclairii*. The compound is an analogue of sphingosine, a major constituent of sphingolipids in the cell membrane which function as lipid signalling molecules. In preclinical studies, the compound prolonged allograft survival with high potency and efficacy in mice, rats, dogs, and nonhuman primates and proved to be effective in other preclinical models of immune-mediated diseases as well [51]. In these models, a high level of synergy with other immunosuppressants like CsA was documented. Also, in contrast to other low-molecular-weight xenobiotic immunosuppressants, the compound has a rather large therapeutic window. Remarkably, it does not affect protective immunity, for example, to virus infections. Fingolimod was tested in phase 3 clinical trials in kidney transplantation but failed to show an additional benefit in the prevention of rejection. The drug was found to be safe; the main adverse side effect was transient and asymptomatic bradycardia, which proved to be a pharmacodynamic effect directly associated with circulating drug concentrations [52]. The drug was subsequently evaluated for the treatment of MS [47, 53], and received market approval in 2010.

The mechanism of action is associated with a physiological response of lymphocytes in immune responses, including the recirculation of T LYMPHOCYTES between blood and lymphoid tissue, contact with ANTIGEN, and entrance into lymph nodes from the blood via high endothelial venules or from other locations via afferent lymphatics. Signalling of the lymphocytic sphingosine-1-phosphate receptor by the endogenous ligand, sphingosine-1-phosphate, is required for lymphocytes to exit from lymph nodes. Upon phosphorylation of the drug by sphingosine kinase 2, the resulting phosphate complex binds to four of the five G protein-coupled sphingosine-1-phosphate receptor subtypes. This causes internalization and degradation of the cell membrane receptor, which affects the capacity of the cells for recirculation, i.e., exit from lymph nodes. As a result, circulating lymphocytes, both T and B cells, accumulate in lymph nodes and no longer recirculate to the periphery so that they cannot participate in immune responses [51]. The same mechanism applies to the beneficial action of fingolimod in MS, namely, inhibition of migration of T LYMPHOCYTES from lymph nodes to the brain and thereby prevention of T-cell-mediated local injury (see Chap. 35). In addition, it has been proposed that the compound inhibits stimulation of neural cells by excess sphingosine-1-phosphate. This stimulation is associated with astrogliosis and perturbed gap junctional communication between cells in the central nervous system.

31.9 Antilymphocyte/ Antithymocyte Globulin/ Anti-CD3 Antibodies

Horse ALG and rabbit ATG [54, 55] were originally introduced as immunosuppressants for induction treatment or for treatment of rejection episodes. These reagents induce a severe but temporary depletion or inactivation of T cells (in the case of ATG) or lymphocytes (ALG) from the circulation. ATG appears more effective and is more widely used than ALG and is more stringent than other induction agents such as anti-CD25 mAb mentioned below [56]. Side effects include fever, leukopenia, and thrombocytopenia: depending on the patient population, a higher incidence of cytomegalovirus disease is commonly observed.

Muromonab is a mouse IgG2a mAb specific for the CD3ε chain of the TCR complex. Upon muromonab binding to CD3, the entire receptor complex is modulated from the cell surface. This modulation results in depression of T-cell activity. Muromonab, in addition, is known to induce apoptosis of the cells and hence has T-cell-depleting activity as well. Muromonab was the first mAb to be approved for clinical application in the mid-1980s. Its use in either induction treatment or treatment of steroid-resistant rejection has revealed the potential side effects of this class of biologicals in general. A major side effect is the so-called cytokine-release syndrome that is related to the potent stimulatory activity of the mAb (which occurs besides its depressing activity). This cytokine-release syndrome can emerge quickly upon first dosing and results in malaise, fever, myalgia, rigors, headache, diarrhea, and in more severe cases hypotension, wheezing, and/or pulmonary edema. Also a temporary rise in serum creatinine is part of the cytokine-release syndrome. A second side effect is related to the fact that muromonab is a mouse Ig and thus can induce anti-mouse Ab formation. The presence of such Ab in the circulation reduces the efficacy of muromonab in subsequent courses of treatment. A number of engineered anti-CD3 Ab have been developed to circumvent this adverse side effect [57] (Table 31.1).

Besides the transplant indication, anti-CD3 antibodies are currently being profiled for application in autoimmune diseases [58], and there are indications that anti-CD3 antibodies might induce immune TOLERANCE as well [59]. A promising application has been documented in patients with recent-onset type 1 diabetes, at a time when the autoimmune destruction of pancreatic islets of Langerhans is not complete. In such patients, the mAb teplizumab reduced the loss of islet cell function [60]: this mAb has not been developed further.

31.10 Other Lymphocyte-Depleting Antibodies

The humanized anti-CD52 mAb alemtuzumab was originally developed for treatment of graft-versus-host disease (GVHD) and approved for treatment of chronic lymphocytic leukemia and T-cell lymphoma. The Ab recognizes a peptide linked to a membrane glycoprotein present on all lymphocytes, monocytes, and MACROPHAGES. It causes a marked and persistent depletion of lymphocytes. Within the T-lymphocyte lineage, this particularly affects the $CD4^+$ population [61]. Alemtuzumab was subsequently evaluated as an induction agent in kidney transplantation, at much lower-dose levels than those used in oncology. Remarkably, it appeared that patients after this induction needed much lower doses of CsA during maintenance treatment, based on which it was suggested that alemtuzumab induction treatment induced a status of "almost" TOLERANCE ("prope" TOLERANCE) [62]. This phenomenon, in conjunction with CsA in maintenance immunosuppression, was not apparent in subsequent trials in which CsA was replaced by mTOR inhibitors. Although the mAb caused long-lasting severe lymphopenia, lymphopenia-associated adverse side effects, such as an increased incidence of infection or posttransplant lymphoproliferative disease, were not apparent [63]. However, the long-lasting lymphopenia is one main reason why the mAb has not gained a strong position in induction immunosuppression after transplantation. The present application is mainly in oncology, i.e., chronic lymphocytic leukemia, cutaneous T-cell lymphoma, and T-cell lymphoma, and also as treatment for MS.

Various mAbs to T-cell subsets have been developed. The humanized anti-CD2 mAb siplizumab was originally tested as a treatment for GVHD and was in clinical trials for psoriasis and in early clinical testing for application in solid organ transplantation and T-lymphoid malignancies: but it has not been developed further. Anti-CD4 mAbs have been tested in transplant models, but this has not been followed by clinical development: the fully human anti-CD4 mAb zanolimumab is in phase 3 trials for treatment of cutaneous T-lymphocyte lymphoma.

A number of mAb to B LYMPHOCYTES, in particular anti-CD20 mAb, have been developed. The humanized mAb rituximab was the first to enter the market in 1997 as treatment for B-cell non-Hodgkin's lymphoma. The mAb is in clinical trials for many other indications, including RA and autoimmune disease in which autoantibodies play a role. In the transplantation setting, rituximab is used in the treatment of antibody-mediated rejection and management of ABO blood group incompatibility, but there are no published trials [64]. Other humanized or human anti-CD20 mAbs have been profiled for an indication in oncology or autoimmune disease. Of more recent date are the anti-CD22 antibodies which are similarly in clinical development for autoimmune disease or B-lymphoid malignancies (Table 31.1).

31.11 Anti-IL-2 Receptor Abs

The development of mAbs to the α chain of the IL-2R (anti-CD25 Abs) is based on the fact that this chain of the receptor is expressed on the surface of T cells only after activation. In peripheral blood, $CD25^+$ cells are present in quite low numbers. Thus, CD25 Abs are presumed to bind only activated T cells. Two mAbs have been approved for the transplantation indication; one is a humanized mAb (daclizumab [65]) and the other a chimeric molecule (basiliximab [66]). Both mAbs reduce the incidence of acute rejection after kidney transplantation. Daclizumab is no longer used in transplantation because of its adverse side effects: the mAb is presently used in the treatment of MS patients [67]. The mAbs permit dose levels of immunosuppressants to be reduced in the posttransplant period, while maintaining adequate immunosuppression. In most patients induction treatment with anti-CD25 mAb is sufficient, but in high-risk patients, a more stringent induction regimen including ATG shows a better outcome after transplantation. Interestingly, this appears not to apply for type 1 diabetes patients receiving an islet transplant. In this condition, a more severe induction treatment using ATG provided a better long-term outcome than induction treatment using basiliximab. There are no relevant side effects for basiliximab, and $CD25^+$ T cells are absent in the circulation as long as receptor-saturating antibody levels are maintained.

31.12 Antibodies Targeting Cytokines or Cytokine Signalling

While mTOR inhibitors and anti-CD25 mAbs have been marketed for use in transplantation, a number of biologicals have been developed to address CYTOKINES or cytokine signalling in inflammatory conditions (see Chap. 21). The application of these antibodies initially was in autoimmune indications, mostly psoriasis and RA (see Chap. 34). This seems logical, as in these conditions an inflammatory condition is a main contributor to the disease manifestations, while this is less the case in a transplant recipient. These biologicals are included in Table 31.1. Their use in the transplant setting may be anticipated to reduce inflammatory reactions, such as those observed in the first posttransplant period, in particular in the case of islet cell transplantation. For instance, the use of etanercept, anakinra, and more recently ustekinumab has been proposed for the induction phase after islet transplantation to reduce inflammation and also for the target CYTOKINES (IL-1, IL-12) to target their role in ANTIGEN presentation. Tocilizumab targeting the IL-6 receptor has been proposed for the same use in experimental pig-to-nonhuman primate transplantation.

31.13 Antibodies to Adhesion Molecules

Therapeutic mAb to adhesion molecules has been generated either for use in transplantation or for immune diseases like psoriasis. Anti-CD11a (anti-LFA-1) mAbs received attention because of their efficacy in experimental transplantation models, including nonhuman primates. Two anti-CD11a mAbs were launched, either for the transplantation indication (odulinomab) or for psoriasis (efalizumab). However, these mAbs were withdrawn after launch because of severe adverse side effects in a number of patients, specifically progressive multifocal leukoencephalopathy associated with JC virus infection.

The humanized anti-VLA-4 mAb natalizumab is on the market for MS and Crohn's disease. This mAb has not been evaluated in transplantation, although molecules like VLA-4 or its ligands (VCAM, ICAM) are interesting targets, based on experimental animal data that the blockade of such interactions results in allograft survival.

Alefacept is a dimeric fusion protein consisting of the CD2-binding portion of human LFA-3 linked to the Fc portion of human IgG1, which blocks the interaction between CD2 and LFA-3. This interaction supports costimulatory signals in ANTIGEN presentation. The fusion protein was approved in 2003 for treatment of psoriasis and is in advanced clinical development for GVHD (based on the results with the anti-CD2 mAb siplizumab mentioned above), T-lymphoid lymphomas, and kidney transplantation. A potential application in the transplant setting is evident from data in a nonhuman primate transplant model, in which alefacept was efficacious in combination with costimulatory blockade [68].

31.14 Costimulatory Blockade

Since the recognition of costimulation (signal 2) in ANTIGEN presentation (Fig. 31.3), interest has been intense in the therapeutic application of mAbs which block costimulatory signals (see Chaps. 12, 21 and 25). Anti-CD80 and anti-CD86 mAbs showed efficacy in a nonhuman primate transplantation model but have not been developed further: another CD80 mAb (galiximab) is in development for B-cell lymphoma. A number of anti-CD154 mAbs have been successfully applied in nonhuman primate transplantation models but failed in subsequent clinical trials. This was due to the emergence of thromboembolic adverse side effects; these side effects were less evident in animal testing because of the different cellular distribution of the corresponding ANTIGEN, being present on human platelets but not on platelets of other species. An anti-CD40 mAb showed efficacy in nonhuman primates, and a derivative of this mAb is currently in early development for psoriatic arthritis. In combination with rituximab induction treatment, continuous treatment with an anti-CD40 mAb that blocks B-cell activation was effective in achieving long-term survival of porcine hearts transplanted in nonhuman primates (the longest survival reported at the time of writing) [69].

Blocking the CD28-B7 (CD80, CD86) integration with fusion proteins has proven to be very successful, owing to the fact that this pathway is most important in costimulation. The first product, abatacept (CTLA4-Ig), is a fusion protein between the extracellular domain of the CD28 antagonist CD152 (cytotoxic T-lymphocyte ANTIGEN, CTLA4, a member of the Ig superfamily expressed on the surface of helper T cells) and an Ig Fc fragment: the Ig Fc fragment is included to increase the half-life of the protein in the circulation. In costimulatory signalling, CTLA4 transmits an inhibitory signal to T cells, whereas CD28 transmits a stimulatory signal. CTLA4-Ig has a 200-fold higher affinity for CD80 than for CD86 and is 100 times more potent in blocking CD80-dependent costimulation than CD86-dependent costimulation. The fusion protein showed efficacy in a nonhuman primate transplantation model, either given alone or in combination with an anti-CD154 mAb, but TOLERANCE, observed in rodent transplant models using costimulatory blockade, was not observed. Abatacept has subsequently been developed and is since 2008 marketed for treatment of RA [70]. The choice of RA and not transplantation as a target indication was based, among other reasons, on the anticipation that a product with a higher binding avidity might be needed in the more stringent setting of a primary immune response to a transplant.

In search of a second generation product with a higher binding avidity, a mutagenesis and screening strategy was used to identify high-avidity mutants with slower dissociation rates. Two amino acid substitutions were identified as potentially useful, and this resulted in the development of belatacept (LEA29Y) for transplantation [71–74]. The drug is now approved for kidney transplantation. Postlaunch trials are addressing, among others, the effectiveness of the combination between belatacept and an mTOR inhibitor in inducing TOLERANCE. In a large clinical trial with 7-year follow-up, belatacept at two different doses was compared with CsA, in all three groups in combination with MMF and steroids: patient and graft survival were significantly better in the belatacept groups [75]. Remarkably, very few adverse side effects have been noted in this and other clinical trials, which gives belatacept a substantial advantage over conventional xenobiotic immunosuppressants.

31.15 Limitations of Presently Available Immunosuppressants

Surgical techniques, including the creation of anastomoses between blood vessels, made it possible to introduce solid organ transplantation into the clinic to replace dysfunctional organs in end-stage organ disease. The first kidney transplant was conducted in 1954 [4]. Since the donor and recipient in this pioneering trial were identical, organ rejection was not an issue, but the management of graft rejection has been a major focus in making transplantation an acceptable clinical procedure. Initial exploratory investigations were performed using antiproliferative drugs, which are well-known for their severe adverse side effects, and CORTICOSTEROIDS. The introduction of the calcineurin inhibitor CsA has revolutionized transplantation. The main problem of current baseline immunosuppressants is still toxicity (direct drug toxicity, increased susceptibility to infection, development of tumors) (see Chaps. 30 and 37). In this regard, the introduction of the antiviral drug ganciclovir (Cytovene) in 1989 for the prevention or treatment of cytomegalovirus infection or disease is often considered as a major hallmark in clinical transplantation [76].

One way to cope with the systemic side effects of immunosuppressants is to deliver drugs specifically to the organ involved. An example of this approach comes from lung transplantation, which generally requires strong immunosuppression and hence has a rather small therapeutic window for immunosuppressives. Local delivery by inhalation has been evaluated for a number of drugs, including anti-inflammatory agents and immunosuppressants [77]. Administration of glucocorticosteroids by inhalation did not improve the outcome of lung transplantation, but for CsA, delivered by aerosol, beneficial results have been reported, not only in acute rejection but also in chronic rejection (bronchiolitis obliterans) [78].

A broad armamentarium of immunosuppressants is nowadays available to increase the therapeutic window of immunosuppressive cocktails in induction and maintenance treatment after transplantation. With regard to induction treatment, the present experience in renal transplantation indicates that the rather mild induction treatment with an anti-CD25 mAb is sufficient to achieve long-term graft function, but in high-risk patients, a more stringent induction regimen using ATG is apparently needed [66]. In heart transplantation [1] and also in islet cell transplantation, the outcome in general appears to be better for patients after ATG induction treatment, despite the fact that ATG induction treatment is associated with an increased risk for cytomegalovirus disease. Also, a more stringent induction regimen appears to be associated with a lower-dose requirement for immunosuppressants during maintenance, as has been most clearly shown for induction treatment with the mAb alemtuzumab. However, this mAb can give long-lasting lymphopenia with the associated concerns for infectious disease [62, 63]. Improvements in maintenance immunosuppression focus on reducing the need of CORTICOSTEROIDS and also focus on a reduction of the use of calcineurin inhibitors. Targeting costimulation offers the promise to fulfill this need. A long-term clinical trial with belatacept showed better efficacy outcomes than CsA, with a much lowered incidence of adverse side effects [75]: it needs to be established whether belatacept manifests also a better outcome than TACROLIMUS in maintenance treatment. The introduction of a steroid-free immunosuppressive regimen in 2000 (the Edmonton protocol) boosted the development of islet cell transplantation to become a clinical procedure [23]. Interestingly, the Edmonton protocol included TACROLIMUS which is well-known for its diabetogenic adverse side effects.

The risk of graft dysfunction after transplantation is related to the quality of the donor organ and the potential of the recipient to reject the graft. With regard to the intrinsic potential for graft rejection, patients that have been sensitized to major histocompatibility complex (MHC) antigens and have circulating antihuman leukocyte ANTIGEN (HLA) Abs represent a high-risk population. Therefore, patients that are on the waiting list for a transplant are tested for such Abs using a cytotoxicity assay with a panel of different target cells. Often, patients with a high PRA (panel-reactive antibody) score remain longer on the waiting list than patients with a low score. Kidney transplantation in high-risk patients can be managed, on the one hand, by avoiding nephrotoxic drugs, such as CsA and TACROLIMUS, and on the other hand by increasing the strength of induction immunosuppression [79]. Most xenobiotic immunosuppressants target intracellular pathways which are more or less shared between different cell types; this phenomenon underlies most of their adverse side effects. It was assumed that biological immunosuppressants, because of their higher level of specificity and their targeting of cell surface structures, should have a better therapeutic window. This is not always the case, as exemplified by the severe adverse side effects noted with a number of biologicals with immunosuppressive function, e.g., anti-CD154 mAbs causing thromboembolic manifestations, the humanized anti-CD20 mAb ocrelizumab showing infections and death in some patients during phase 3 trials in RA, and the anti-CD11a mAb causing progressive multifocal leukoencephalopathy. In the cases mentioned, the biological was halted in clinical development or withdrawn from the market.

These considerations point to the requirement for a more individually customized immunosuppressive regimen. The broad armamentarium of available immunosuppressives allows the physician together with the patient to determine the best immunosuppressive protocol [80]. However, the costs of immunosuppressive drugs can be quite high, and not in all countries there is lifelong reimbursement of medication: e.g., in the USA there is reimbursement by Medicare for only a limited number of years [81]. This has an impact on the compliance in taking medicines and hence the outcomes in terms of graft survival [82, 83]. This aside, keeping an optimal balance between adverse side effects and sufficient immunosuppression is an important item in the long term. There is the general intention to reduce immunosuppression with time after transplantation: this can be associated without rejection, i.e., a state of operational TOLERANCE is created; however, it can also be related with chronic rejection [84]. Operational TOLERANCE is well-known for liver transplantation but has also been documented for kidney transplantation [84, 85].

The often narrow window between efficacy and adverse side effects makes it necessary to monitor patients for drug-level assessment. These analyses address drug exposure and not the intrinsic susceptibility of the patient's immune system to immunosuppression nor the extent to which the individual needs to be immunosuppressed to avoid rejection. Presently, initiatives are in progress to change this empirical approach of patient management to a more logical strategy. For instance, this includes the measurement of pharmacodynamic parameters of immune reactivity before and after transplantation [86–88] and the introduction of pharmacogenomics. There are now a number of examples that the determination of the patient for a particular genotype, i.e., that of cytochrome P450, is useful in the determination of the optimal immunosuppressant dose level [89–94]. This approach is more relevant for XENOBIOTICS than for biologicals.

With the improved outcome of solid organ transplantation, providing better management of acute rejection episodes, chronic rejection is nowadays a major cause of graft failure in the long term after transplantation. Therefore, there is a clinical need for drugs which interfere with this process. Since chronic rejection can include Ab-mediated rejection, a beneficial effect of immunosuppressants which target B cells and Ab formation following indirect ANTIGEN presentation has been proposed. Indirect ANTIGEN presentation includes costimulatory signalling, and it is anticipated that maintenance treatment with effective blockers of costimulation will have a beneficial effect on chronic rejection. Also, prevention or treatment of chronic rejection may require non-immunological approaches (e.g., targeting smooth muscle cell proliferation) in addition to improved novel immunosuppressants, which will not be considered here.

A special category of high-risk patients are those who receive a transplant across the ABO red blood cell barrier. Isohemagglutinins are feared for their potential to cause immediate Ab-mediated rejection of the graft. Therefore, there is a need for Ab management in both the patient with a high PRA score and the patient receiving an ABO-incompatible graft. Current strategies include the removal of antibody, e.g., by plasmapheresis and specific extracorporeal immunoadsorption in an acute phase, treatment with intravenous Ig (IVIG), and treatment with B-lymphocyte immunosuppressants [64, 95–97]. IVIG is used in multiple conditions, not only to give protection against infection in patients with hypogammaglobulinemia but also as treatment of patients with inflammatory disorders or autoimmune disease. The mechanism of action of IVIG in modulation of immune reactivity and inflammation is not completely clear. At the molecular level, neutralization of anti-HLA Ab; inhibition of CYTOKINES IL-1β, IFN-γ, IL-2, and IL-6; and inhibition of terminal complement components C5b-9 have been proposed and at the cellular level regulation of the B-lymphocyte repertoire and activities through binding to FcγRI (CD64). The latter includes induction of B-lymphocyte apoptosis, maturation of dendritic cells, and induction of Treg lymphocytes. There are also claims that IVIG affects NK cell function and the activity of polymorphonuclear granulocytes in inflammation [98, 99].

A major issue in suppressing synthesis of Ab is the fact that Ab-producing plasma cells have low expression of cell surface receptors and are not susceptible to conventional immunosuppressants (see Chap. 4). In addition, there are no immunosuppressants that specifically target T-cell-independent Ab synthesis, i.e., the process in synthesis of isohemagglutinins. The anti-CD20 mAb rituximab, which is marketed for B-lymphoid malignancies, has been evaluated in the transplant setting. Other anti-B-cell mAbs, including anti-CD22 mAb and biologicals affecting intracellular B-lymphocyte signalling (belimumab [100] and atacicept [101]), are on the market or in advanced clinical development, respectively, for autoimmune diseases (Table 31.1).

31.16 New Targets

From the preceding section, it is evident that there is a continued need to develop novel innovative immunosuppressants with fewer adverse side effects. This is most likely to be achieved by selecting targets for pharmacological intervention that are specifically relevant for the immune response and not for physiological processes in other organ systems. Besides selectivity, a critical aspect in the selection of new drug targets is the redundancy of potential targets. Non-redundancy of targets is demonstrated most convincingly by the phenotype of human primary immunodeficiency and knockout mice. Different activation mechanisms are specific for the immune system and might be relevant for direct and indirect presentation of allograft antigens.

ANTIGEN presentation to T cells is still an attractive target (Fig. 31.3) (see Chaps. 3 and 12), considering the promising results with abatacept and belatacept and despite the failure of CD154 mAb. Not only are processes in costimulation unique to T-cell activation, but also it is well-established that inhibition of signal 2 in T-cell activation has the potential to induce T-cell anergy or unresponsiveness which could lead to antigen-specific TOLERANCE. In particular, the combination between costimulatory blockade and mTOR inhibition is of interest as this combination was effective in TOLERANCE induction [102, 103]. This might also apply to interference in adhesion processes, such as the interaction between LFA-1 and ICAM-1, and that between CD2 and LFA-3. Unfortunately, anti-LFA-1 mAbs were withdrawn from the market because of severe adverse side effects. With regard to the interaction between VLA-4 and VCAM-1, the humanized mAb natalizumab, on the market since 2004 for MS and Crohn's disease, is of interest. The application of this mAb is limited because of the incidence of JC virus associated progressive multifocal leukoencephalopathy [104] (see Chap. 35).

At the side of the ANTIGEN-PRESENTING CELL (APC), mAbs to ligands of T-cell costimulatory molecules, like CD80 and CD86, have been used successfully in transplantation models, but the respective mAbs did not enter clinical development (see Chap. 21). The primatized anti-CD80 mAb galiximab is presently in development for B-lymphoid malignancies [105]. Blocking CD40 is especially attractive since CD40 is expressed not only on dendritic APC but also on MACROPHAGES and B LYMPHOCYTES. A number of anti-CD40 mAbs are currently in development for treatment of B-lymphoid malignancies [106].

Intracellular targets other than those described above are found in the signalling pathway emanating from the TCR and include, among others, T-cell-selective molecules such as protein tyrosine kinases of the src family ($p56^{lck}$ and $p59^{fyn}$) and ZAP-70 [107] (see Chap. 9). The pivotal and selective role of ZAP-70 for T-lymphocyte activation is documented by the severe combined immunodeficiency phenotype of humans who lack functional ZAP-70. However, a clinical development program for a specific ZAP-70 inhibitor has not been established thus far. All these kinases are potential targets for the development of new immunosuppressants, but a major challenge is to identify inhibitors with high selectivity that do not inhibit other kinases of critical importance in other cell types. In addition, there are a number of molecules that do not exhibit catalytic function but act as specific adaptor molecules by mediating the interaction between different components of signal transduction pathways. Generally, these proteins contain domains that are important for protein-protein interactions such as SH2/SH3 domains. These interactions are being studied as potential drug targets.

CYTOKINES and cytokine receptors are suitable targets for therapeutic intervention in immune responses and inflammatory processes. Anti-TNF-α mAbs (Table 31.1) were the first biologicals to become available for patients with RA, infliximab being the first in 1988, and these reagents are currently broadly used [108]. Another anti-inflammatory product on the market for RA is the IL1-R antagonist anakinra, which is a so-called biological response modifier and not a DMARD [109]. Concerning CYTOKINES involved in immune responses, the paradigm of CYTOKINES associated with Th1/Th2 cells is relevant, because transplant rejection represents primarily a Th1 response, whereas transplantation TOLERANCE may be favored by Th2 cells. CYTOKINES promoting Th2 responses (e.g., IL-4) or blockade of CYTOKINES promoting Th1 responses (e.g., IL-12) might induce a beneficial shift of the immune system toward a Th2 response. CYTOKINES such as TGF-β, IL-10, and IL-35 might be attractive targets as these CYTOKINES inhibit the Th1 response [110].

31.17 Tolerance

The final goal of immunosuppression in transplantation and AUTOIMMUNITY is the induction of TOLERANCE, often referred to as the “Holy Grail.” Immune TOLERANCE, or unresponsiveness to transplant rejection, which translates to a status of allograft function and life support without the necessity of chronic immunosuppression, can be relatively easily induced in rodent transplantation models, for instance, in rats by using a short 2- to 3-week course of immunosuppression. This is not the case in humans or nonhuman primate models. In the clinic, blood transfusions (either from unrelated donors or donor-specific transfusions) have given a first indication that the immune system can be “modulated” in order to reduce reactivity toward the grafted organ. At the Ig/Ab level, the mechanism of action of IVIG in modulating immune reactions and inflammatory processes has been outlined above.

At the cellular level, much attention is currently given to so-called Treg lymphocytes, T cells expressing CD4 and CD25 and the intracellular transcription factor FoxP3. These cells and their function have been studied extensively in mouse models and have been associated with disturbed immune regulation in autoimmune diseases in humans. In animal models of allogeneic hematopoietic stem cell transplantation as treatment for leukemia, a potent suppressive effect of Treg cells on immune effector cells reactive to host antigens has been demonstrated, resulting in prevention of GVHD while preserving the graft-versus-leukemia effect. This was followed by clinical trials in patients with hematological cancers subjected to allogeneic hematopoietic stem cell transplantation. In the transplant setting, Treg lymphocytes could provide a potential cell therapy to induce TOLERANCE [111–114]. Among others, this suggestion is based on observations of increased numbers of Treg lymphocytes in the circulation of patients with “operational” TOLERANCE.

Also, it has been documented that lymphocyte subpopulations differ in sensitivity to immunosuppressants. For instance, Treg lymphocytes appear to be more resistant to mTOR inhibition than other T-lymphocyte subpopulations, such as effector T LYMPHOCYTES. This phenomenon could contribute to a beneficial role of mTOR inhibitors, in combination with costimulatory blockade, to induce “operational tolerance” [102].

As mentioned above, in clinical transplantation, it is now a well-known fact that patients can reduce immunosuppressive medication late after transplantation. Complete withdrawal of immunosuppression is possible, in particular in patients after liver transplantation. About 20–25% of liver transplant patients develop a state of "operational tolerance" which appears to be unrelated to the immune status of donor and recipient, age, pretransplant liver disease status, length of time for which immune suppression is given, and whether or not the patient shows chimerism with donor-derived cells in the circulation [115, 116]. It has been proposed that this type of operational TOLERANCE occurs especially for liver grafts because of the size of the transplant ("high-dose tolerance"). Other factors related to the size and composition of the graft include the production of MHC antigens by the graft, induction of microchimerism by STEM CELLS transplanted together with the liver, and an overload with donor-derived passenger leukocytes.

Immunosuppressive protocols and other strategies to promote TOLERANCE induction in transplant patients are currently under investigation [117–119]. Blocking costimulatory pathways is one of these strategies. Also, the balance between strength of induction treatment and maintenance of immunosuppression is a focus of investigation, as exemplified by the anti-CD52 mAb alemtuzumab mentioned earlier. The administration of this mAb can result in a substantially lower dose of immunosuppressants during maintenance while sustaining an adequate immunosuppressed state [62, 63]. An even more stringent approach was followed using an anti-CD3 mAb conjugated to a toxin. A chemical conjugate between an anti-CD3 mAb and a mutated diphtheria toxin yielded long-term allograft survival of monkey kidney or islet allografts when given in the peritransplant period. This result is ascribed to the severe T-cell-depleting effect of the immunotoxin, which not only includes T-cell depletion from the blood circulation but also depletion from lymphoid organs. These promising results have not been pursued in patients, and the immunotoxin (Resimmune) has entered clinical development for T-cell lymphoma [120] and metastatic melanoma.

During the last years, mesenchymal STEM CELLS (MSCs) have received much interest because of their immunoregulatory properties. Not only serve such cells in stromal support (and are then called mesenchymal stromal cells), but MSC can also exert anti-inflammatory properties and immunosuppressive activities. This function is ascribed mainly to the secretion of relevant CYTOKINES; with respect to immunomodulation, their activity is associated with upregulation of Treg cells. Transplantation studies in animal models have shown TOLERANCE induction by MSCs, and based on these studies, clinical studies have been initiated to assess the potential of TOLERANCE induction in transplantation [121–124].

Operational TOLERANCE, as described above, is also called peripheral TOLERANCE to differentiate it from the so-called central (or deletional) TOLERANCE, i.e., TOLERANCE at the level of lymphocyte precursor cells. One procedure to achieve central TOLERANCE comprises the elimination of the functioning immune system, including STEM CELLS in the bone marrow, by high-dose chemotherapy and/or whole-body and thymic irradiation, followed by allogeneic hematopoietic stem cell transplantation, treatment with ATG, and a short course of CsA. In this situation, a condition of mixed hematopoietic chimerism is created, in which the individual accepts solid organ transplants from the bone marrow donor without the need of additional immunosuppression [125, 126]. This procedure has been shown to be effective in large animal transplantation models, including pigs and nonhuman primates. Subsequently, success was achieved in exploratory trials in patients with multiple myeloma and end-stage kidney disease requiring a kidney graft [127]. A major drawback of this approach is the very stringent conditioning regimen which limits the treatment to patients with multiple myeloma that are eligible for allogeneic hematopoietic cell transplantation and in addition suffer from end-stage kidney disease. In search for a more subtle approach, a low-intensity conditioning regimen comprising total lymphoid irradiation and ATG given the day after transplantation proved to be effective in mediating TOLERANCE in a recipient of a HLA-matched kidney who had no malignancy [128]. Apparently, the matching of major HLA types, i.e., transplantation across minor MHC antigens, was necessary to achieve TOLERANCE using this protocol. Success has also been noted for a non-myeloablative conditioning regimen in living-donor kidney transplantation in which there was a HLA mismatch for only one haplotype. This regimen comprised CY, the humanized anti-CD2 mAb siplizumab, CsA, and thymic irradiation: subsequent maintenance immunosuppression with CsA was discontinued at about 1 year after transplantation, and in four out of five patients, long-lasting TOLERANCE was obtained. Besides stable kidney graft function, no donor-reactive T-cell response was detectable in laboratory assays [129].

31.18 Summary

A broad armamentarium of immunosuppressives is currently on the market or in advanced clinical development, which are efficacious in prevention or treatment of rejection of a transplant in patients with end-stage organ failure. These drugs can roughly be divided into low-molecular-weight XENOBIOTICS, orally active drugs produced by microorganisms or chemical synthesis, and biologicals (monoclonal), antibodies or rDNA fusion proteins. In addition, cell therapy products are investigated in exploratory research. Most XENOBIOTICS work intracellularly and affect different pathways in lymphocyte activation and/or proliferation. Since such pathways are not truly selective for lymphocytes, most of these drugs show inherent adverse side effects and generally have a low therapeutic window. However, the availability of novel agents with a broader therapeutic window and refinements in combination treatment have greatly added to improved tolerability (synergy in immunosuppression but not adverse side effects). Broadly acting cytotoxic drugs with severe side effects are gradually being replaced by compounds with a more selective action toward lymphoid cells, and corticosteroid-sparing regimens are applied to reduce the adverse side effects of CORTICOSTEROIDS. Most biologicals work by binding to cell surface molecules, resulting in inactivation or depletion of the target cells. Progress in this area has not only resulted in improved immunosuppression but also in potential approaches to induce a status of unresponsiveness, i.e., TOLERANCE to the transplant. This is achieved either by efficient depletion of reactive cells or by inhibition of second signals in costimulation blockade of T-cell activation.

Selected Readings

Cooper JE, Wiseman AC. Novel immunosuppressive agents in kidney transplantation. Clin Nephrol. 2010;73:333–43.

Hartono C, Muthukumar T, Suthanthiran M. Immunosuppressive drug therapy. Cold Spring Harb Perspect Med. 2013;3:a015487.

van den Hoogen MW, Hilbrands LB. Use of monoclonal antibodies in renal transplantation. Immunotherapy. 2011;3:871–80.

Söderlund C, Rådegran G. Immunosuppressive therapies after heart transplantation—the balance between under- and over-immunosuppression. Transplant Rev (Orlando). 2015;29:181–9.

Urschel S, Altamirano-Diaz LA, West LJ. Immunosuppression armamentarium in 2010: mechanistic and clinical considerations. Pediatr Clin N Am. 2010;57:433–57.

Webber A, Hirose R, Vincenti F. Novel strategies in immunosuppression: issues in perspective. Transplantation. 2011;91:1057–64.

Relevant Websites

This is a registry of federally and privately supported clinical trials conducted in the United States and around the world. http://clinicaltrials.gov/

The Immune Tolerance Network is an international clinical research consortium founded by the National Institutes of Health, whose mission is to accelerate the clinical development of immune tolerance therapies through a unique development model. http://www.immunetolerance.org/

The Transplantation Society will provide the focus for global leadership in transplantation, by development of the science and clinical practice, scientific communication, continuing education, and guidance on the ethical practice. http://www.tts.org/

References

1. Ensor CR, Cahoon WD Jr, Hess ML, Kasirajan V, Cooke RH. Induction immunosuppression for orthotopic heart transplantation: a review. Prog Transplant. 2009;19:333–41.

2. Hawksworth JS, Leeser D, Jindal RM, Falta E, Tadaki D, Elster EA. New directions for induction immunosuppression strategy in solid organ transplantation. Am J Surg. 2009;197:515–24.
3. Hardinger KL, Brennan DC, Klein CL. Selection of induction therapy in kidney transplantation. Transpl Int. 2013;26:662–72.
4. Terasaki PI, editor. History of transplantation: thirty-five recollections. Los Angeles: UCLA Tissue Typing Laboratory; 1991.
5. Aucella F, Lauriola V, Vecchione G, Tiscia GL, Grandone E. Liquid chromatography-tandem mass spectrometry method as the golden standard for therapeutic drug monitoring in renal transplant. J Pharm Biomed Anal. 2013;86:123–6.
6. Lemaitre F, Antignac M, Verdier MC, Bellissant E, Fernandez C. Opportunity to monitor immunosuppressive drugs in peripheral blood mononuclear cells: where are we and where are we going? Pharmacol Res. 2013;74:109–12.
7. Christians U, Klawitter J, Klawitter J, Brunner N, Schmitz V. Biomarkers of immunosuppressant organ toxicity after transplantation: status, concepts and misconceptions. Expert Opin Drug Metab Toxicol. 2011;7:175–200.
8. Wieland E, Olbricht CJ, Süsal C, Gurragchaa P, Böhler T, Israeli M, et al. Biomarkers as a tool for management of immunosuppression in transplant patients. Ther Drug Monit. 2010;32:560–72.
9. Gutierrez-Dalmau A, Campistol JM. Immunosuppressive therapy and malignancy in organ transplant recipients: a systematic review. Drugs. 2007;67:1167–98.
10. Marcén R. Immunosuppressive drugs in kidney transplantation: impact on patient survival, and incidence of cardiovascular disease, malignancy and infection. Drugs. 2009;69:2227–43.
11. Carbone J, del Pozo N, Gallego A, Sarmiento E. Immunological risk factors for infection after immunosuppressive and biologic therapies. Expert Rev Anti-Infect Ther. 2011;9:405–13.
12. Brennan DC, Aguado JM, Potena L, Jardine AG, Legendre C, Säemann MD. Effect of maintenance immunosuppressive drugs on virus pathobiology: evidence and potential mechanisms. Rev Med Virol. 2013;23:97–125.
13. Kwun J, Knechtle SJ. Overcoming chronic rejection-can it B? Transplantation. 2009;88:955–61.
14. Suzuki J, Isobe M, Morishita R, Nagai R. Characteristics of chronic rejection in heart transplantation: important elements of pathogenesis and future treatments. Circ J. 2010;74:233–9.
15. Pierson RN III, Dorling A, Ayares D, Rees MA, Seebach JD, Fishman JA, et al. Current status of xenotransplantation and prospects for clinical application. Xenotransplantation. 2009;16:263–80.
16. Cravedi P, Heeger PS. Complement as a multifaceted modulator of kidney transplant injury. J Clin Invest. 2014;124:2348–54.
17. Kahan BD. Forty years of publication of transplantation proceedings—the second decade: the cyclosporine revolution. Transplant Proc. 2009;41:1423–37.
18. Borel JF. Pharmacology of cyclosporine (Sandimmune). IV. Pharmacological properties in vivo. Pharmacol Rev. 1990;41:259–371.
19. Schreier MH. Mechanism of action of cyclosporin. In: Anke T, editor. Fungal biotechnology. New York, NY: Chapmann and Hall; 1997. p. 137–46.
20. Dumont FJ. FK506, an immunosuppressant targeting calcineurin function. Curr Med Chem. 2000;7:731–48.
21. Vicari-Christensen M, Repper S, Basile S, Young D. Tacrolimus: review of pharmacokinetics, pharmacodynamics, and pharmacogenetics to facilitate practitioners' understanding and offer strategies for educating patients and promoting adherence. Prog Transplant. 2009;19:277–84.
22. De Mare-Bredemeijer EL, Metselaar HJ. Optimization of the use of calcineurin inhibitors in liver transplantation. Best Pract Res Clin Gastroenterol. 2012;26:85–95.
23. Barton FB, Rickels MR, Alejandro R, Hering BJ, Wease S, Naziruddin B, et al. Improvement in outcomes of clinical islet transplantation: 1999–2010. Diabetes Care. 2012;35:1436–45.
24. Hebert AA. Review of pimecrolimus cream 1% for the treatment of mild to moderate atopic dermatitis. Clin Ther. 2006;28:1972–82.
25. Evenou JP, Wagner J, Zenke G, Brinkmann V, Wagner K, Kovarik J, et al. The potent protein kinase C-selective inhibitor AEB071 (sotrastaurin) represents a new class of immunosuppressive agents affecting early T-cell activation. J Pharmacol Exp Ther. 2009;330:792–801.
26. Sommerer C, Zeier M. AEB071 – a promising immunosuppressive agent. Clin Transpl. 2009;23(Suppl 21):315–8.
27. Halleck F, Duerr M, Waiser J, Huber L, Matz M, Brakemeier S, et al. An evaluation of sirolimus in renal transplantation. Expert Opin Drug Metab Toxicol. 2012;8:1337–56.
28. Zaza G, Granata S, Tomei P, Masola V, Gambaro G, Lupo A. mTOR inhibitors and renal allograft: Yin and Yang. J Nephrol. 2014;27:495–506.
29. Fantus D, Thomson AW. Evolving perspectives of mTOR complexes in immunity and transplantation. Am J Transplant. 2015;15:891–902.
30. Waldner M, Fantus D, Solari M, Thomson AW. New perspectives on mTOR inhibitors (rapamycin, rapalogs and TORKinibs) in transplantation. Br J Clin Pharmacol. 2016;82(5):1158–70.
31. Sánchez-Fructuoso AI. Everolimus: an update on the mechanism of action, pharmacokinetics and recent clinical trials. Expert Opin Drug Metab Toxicol. 2008;4:807–19.
32. Koch M. Everolimus in liver and lung transplantation. Drugs Today (Barc). 2009;45:11–20.

33. Delgado JF, Manito N, Segovia J, Almenar L, Arizón JM, Campreciós M, et al. The use of proliferation signal inhibitors in the prevention and treatment of allograft vasculopathy in heart transplantation. Transplant Rev. 2009;23:69–79.
34. Byrne RA, Kufner S, Tiroch K, Massberg S, Laugwitz KL, Birkmeier A, et al. Randomised trial of three rapamycin-eluting stents with different coating strategies for the reduction of coronary restenosis: 2-year follow-up results. Heart. 2009;95:1489–94.
35. Campistol JM, Albanell J, Arns W, Boletis I, Dantal J, de Fijter JW, et al. Use of proliferation signal inhibitors in the management of post-transplant malignancies—clinical guidance. Nephrol Dial Transplant. 2007;22(Suppl 1):i36–41.
36. Huang S, Houghton PJ. Inhibitors of mammalian target of rapamycin as novel antitumor agents: from bench to clinic. Curr Opin Investig Drugs. 2002;3:295–304.
37. West K. CP-690550, a JAK3 inhibitor as an immunosuppressant for the treatment of rheumatoid arthritis, transplant rejection, psoriasis and other immune-mediated disorders. Curr Opin Investig Drugs. 2009;10:491–504.
38. Miheller P, Kiss LS, Mandel M, Lakatos PL. Methotrexate: should we start using it in clinical practice? Curr Drug Targets. 2013;14:1480–9.
39. Germani G, Pleguezuelo M, Villamil F, Vaghjiani S, Tsochatzis E, Andreana L, et al. Azathioprine in liver transplantation: a reevaluation of its use and a comparison with mycophenolate mofetil. Am J Transplant. 2009;9:1725–31.
40. Ishikawa H. Mizoribine and mycophenolate mofetil. Curr Med Chem. 1999;6:575–97.
41. Kawasaki Y. Mizoribine: a new approach in the treatment of renal disease. Clin Dev Immunol. 2009;2009:Article ID 681482.
42. Staatz CE, Tett SE. Pharmacology and toxicology of mycophenolate in organ transplant recipients: an update. Arch Toxicol. 2014;88:1351–89.
43. Van Gelder T, Hesselink DA. Mycophenolate revisited. Transpl Int. 2015;28:508–15.
44. Grinyó JM, Cruzado JM. Mycophenolate mofetil and calcineurin-inhibitor reduction: recent progress. Am J Transplant. 2009;9:2447–52.
45. Breedveld FC, Dayer JM. Leflunomide: mode of action in the treatment of rheumatoid arthritis. Ann Rheum Dis. 2000;59:841–9.
46. Alcorn N, Saunders S, Madhok R. Benefit-risk assessment of leflunomide: an appraisal of leflunomide in rheumatoid arthritis 10 years after licensing. Drug Saf. 2009;32:1123–34.
47. Rammohan KW, Shoemaker J. Emerging multiple sclerosis oral therapies. Neurology. 2010;74(Suppl 1):S47–53.
48. Perenyei M, Jayne DR, Flossmann O. Gusperimus: immunological mechanism and clinical applications. Rheumatology (Oxford). 2014;53:1732–41.
49. Flossmann O, Jayne DR. Long-term treatment of relapsing Wegener's granulomatosis with 15-deoxyspergualin. Rheumatology (Oxford). 2010;49:556–62.
50. Shonhai A. Plasmodial heat shock proteins: targets for chemotherapy. FEMS Immunol Med Microbiol. 2010;58:61–74.
51. Brinkmann V, Cyster JG, Hla T. FTY720: sphingosine 1-phosphate receptor-1 in the control of lymphocyte egress and endothelial barrier function. Am J Transplant. 2004;4:1019–25.
52. Martini S, Peters H, Böhler T, Budde K. Current perspectives on FTY720. Expert Opin Investig Drugs. 2007;16:505–18.
53. Brinkmann V. FTY720 (fingolimod) in multiple sclerosis: therapeutic effects in the immune and the central nervous system. Br J Pharmacol. 2009;158:1173–82.
54. Deeks ED, Keating GM. Rabbit antithymocyte globulin (thymoglobulin): a review of its use in the prevention and treatment of acute renal allograft rejection. Drugs. 2009;69:1483–512.
55. Ballen K. New trends in transplantation: the use of thymoglobulin. Expert Opin Drug Metab Toxicol. 2009;5:351–5.
56. Brennan DC, Daller JA, Lake KD, Cibrik D, Del Castillo D. Rabbit antithymocyte globulin versus basiliximab in renal transplantation. N Engl J Med. 2006;355:1967–77.
57. You S, Chatenoud L. New generation CD3 monoclonal antibodies: are we ready to have them back in clinical transplantation? Curr Opin Organ Transplant. 2010;15:720–4.
58. Chatenoud L. Progress towards the clinical use of CD3 monoclonal antibodies in the treatment of autoimmunity. Curr Opin Organ Transplant. 2009;14:351–6.
59. Chatenoud L. CD3-specific antibodies as promising tools to aim at immune tolerance in the clinic. Int Rev Immunol. 2006;25:215–33.
60. Masharani UB, Becker J. Teplizumab therapy for type 1 diabetes. Expert Opin Biol Ther. 2010;10:459–65.
61. Gribben JG, Hallek M. Rediscovering alemtuzumab: current and emerging therapeutic roles. Br J Haematol. 2009;144:818–31.
62. Calne R, Moffat SD, Friend PJ, Jamieson NV, Bradley JA, Hale G, et al. Campath 1H allows low-dose cyclosporine monotherapy in 31 cadaveric renal allograft recipients. Transplantation. 1999;68:1613–6.
63. Gundroo A, Zachariah M, Singh N, Sharma R. Alemtuzumab (Campath-1H) experience in kidney transplantation what we have learned; current practices; and scope for the future? Curr Opin Organ Transplant. 2015;20:638–42.
64. Stegall MD, Gloor JM. Deciphering antibody-mediated rejection: new insights into mechanisms and treatment. Curr Opin Organ Transplant. 2010;15:8–10.
65. Vincenti F, Kirkman R, Light S, Bumgardner G, Pescovitz M, Halloran P, et al. Interleukin-2-

receptor blockade with daclizumab to prevent acute rejection in renal transplantation. N Engl J Med. 1998;338:161–5.
66. McKeage K, McCormack PL. Basiliximab: a review of its use as induction therapy in renal transplantation. BioDrugs. 2010;24:55–76.
67. Bielekova B, Becker BL. Monoclonal antibodies in MS: mechanisms of action. Neurology. 2010;74(Suppl 1):S31–40.
68. Weaver TA, Charafeddine AH, Agarwal A, Turner AP, Russell M, Leopardi FV, et al. Alefacept promotes co-stimulation blockade based allograft survival in nonhuman primates. Nat Med. 2009;15:746–9.
69. Mohiuddin MM, Singh AK, Sorcoran PC, Thomas ML III, Clark T, Lewis BG, et al. Chimeric 2C10R4 anti-CD40 antibody therapy is critical for long-term survival of GTKO.hCD46.hTBM pig-to-primate cardiac xenograft. Nat Commun. 2016;7:11138.
70. Goëb V, Buch MH, Vital EM, Emery P. Costimulation blockade in rheumatic diseases: where we are? Curr Opin Rheumatol. 2009;21:244–50.
71. Emamaullee J, Toso C, Merani S, Shapiro AM. Costimulatory blockade with belatacept in clinical and experimental transplantation – a review. Expert Opin Biol Ther. 2009;9:789–96.
72. Gupta G, Womer KL. Profile of belatacept and its potential role in prevention of graft rejection following renal transplantation. Drug Des Devel Ther. 2010;4:375–82.
73. Su VC, Harrison J, Rogers C, Ensom MH. Belatacept: a new biologic and its role in kidney transplantation. Ann Pharmacother. 2012;46:57–67.
74. Heemann U, Viklicky O. The role of belatacept in transplantation: results and implications of clinical trials in the context of other new biological immunosuppressant agents. Clin Transpl. 2013;27:E3–11.
75. Vincenti F, Rostaing L, Grinyo J, Rice K, Steinberg S, Gaite L, et al. Belatacept and long-term outcomes in kidney transplantation. N Engl J Med. 2016;374:333–43.
76. Mwintshi K, Brennan DC. Prevention and management of cytomegalovirus infection in solid-organ transplantation. Expert Rev Anti-Infect Ther. 2007;5:295–304.
77. Corcoran TE. Aerosol drug delivery in lung transplant recipients. Expert Opin Drug Deliv. 2009;6:139–48.
78. Iacono AT, Johnson BA, Grgurich WF, Youssef JG, Corcoran TE, Seiler DA, et al. A randomized trial of inhaled cyclosporine in lung-transplant recipients. N Engl J Med. 2006;354:141–50.
79. Rhee J, Al-Mana N, Freeman R. Immunosuppression in high-risk transplantation. Curr Opin Organ Transplant. 2009;14:636–42.
80. Popoola J, Greene H, Kyegombe M, MacPhee IA. Patient involvement in selection of immunosuppressive regimen following transplantation. Patient Prefer Adherence. 2014;8:1705–12.
81. Tanriover B, Stone PW, Mohan S, Cohen DJ, Gaston RS. Future of medicare immunosuppressive drug coverage for kidney transplant recipients in the United States. Clin J Am Soc Nephrol. 2013;8:1258–66.
82. Williams AF, Manias E, Gaskin CJ, Crawford KJ. Medicine non-adherence in kidney transplantation. J Ren Care. 2014;40:107–16.
83. Prendergast MB, Gaston RS. Optimizing medication adherence: an ongoing opportunity to improve outcomes after kidney transplantation. Clin J Am Soc Nephrol. 2010;5:1305–11.
84. Orlando G. Finding the right time for weaning off immunosuppression in solid organ transplant recipients. Expert Rev Clin Immunol. 2010;6:879–92.
85. Bamoulid J, Staeck O, Halleck F, Khadzhynov D, Brakemeier S, Dürr M, et al. The need for minimization strategies: current problems of immunosuppression. Transpl Int. 2015;28:891–900.
86. Fleming JN, Weimert NA. Novel strategies for immune monitoring in kidney transplant recipients. Adv Chronic Kidney Dis. 2010;17:e63–77.
87. Barraclough KA, Staatz CE, Isbel NM, McTaggart SJ. Review: pharmacodynamic monitoring of immunosuppression in kidney transplantation. Nephrology (Carlton). 2010;15:522–32.
88. Dieterlen MT, Eberhardt K, Tarnok A, Bittner HB, Barten MJ. Flow cytometry-based pharmacodynamic monitoring after organ transplantation. Methods Cell Biol. 2011;103:267–84.
89. Wavamunno MD, Chapman JR. Individualization of immunosuppression: concepts and rationale. Curr Opin Organ Transplant. 2008;13:604–8.
90. Girnita DM, Webber SA, Zeevi A. Clinical impact of cytokine and growth factor genetic polymorphisms in thoracic organ transplantation. Clin Lab Med. 2008;28:423–40.
91. Barraclough KA, Lee KJ, Staatz CE. Pharmacogenetic influences on mycophenolate therapy. Pharmacogenomics. 2010;11:369–90.
92. Kurzawski M, Droździk M. Pharmacogenetics in solid organ transplantation: genes involved in mechanism of action and pharmacokinetics of immunosuppressive drugs. Pharmacogenomics. 2013;14:1099–118.
93. Elens L, Hesselink DA, van Schaik RH, van Gelder T. Pharmacogenetics in kidney transplantation: recent updates and potential clinical applications. Mol Diagn Ther. 2012;16:331–45.
94. Ware N, MacPhee IA. Current progress in pharmacogenetics and individualized immunosuppressive drug dosing in organ transplantation. Curr Opin Mol Ther. 2010;12:270–83.
95. Singh N, Pirsch J, Samaniego M. Antibody-mediated rejection: treatment alternatives and outcomes. Transplant Rev (Orlando). 2009;23:34–46.
96. Geyer M, Fischer KG, Drognitz O, Walz G, Pisarski P, Wilpert J. ABO-incompatible kidney transplantation with antigen-specific immunoadsorption and rituximab - insights and uncertainties. Contrib Nephrol. 2009;162:47–60.

97. Jordan SC, Peng A, Vo AA. Therapeutic strategies in management of the highly HLA-sensitized and ABO-incompatible transplant recipients. Contrib Nephrol. 2009;162:13–26.
98. Durandy A, Kaveri SV, Kuijpers TW, Basta M, Miescher S, Ravetch JV, et al. Intravenous immunoglobulins—understanding properties and mechanisms. Clin Exp Immunol. 2009;158(Suppl 1):2–13.
99. Jordan SC, Toyoda M, Vo AA. Intravenous immunoglobulin a natural regulator of immunity and inflammation. Transplantation. 2009;88:1–6.
100. Wallace DJ, Stohl W, Furie RA, Lisse JR, McKay JD, Merrill JT, et al. A phase II, randomized, double-blind, placebo-controlled, dose-ranging study of belimumab in patients with active systemic lupus erythematosus. Arthritis Rheum. 2009;61:1168–78.
101. Nestorov I, Papasouliotis O, Pena Rossi C, Munafo A. Pharmacokinetics and immunoglobulin response of subcutaneous and intravenous atacicept in patients with systemic lupus erythematosus. J Pharm Sci. 2010;99:524–38.
102. Ford ML. T Cell cosignaling molecules in transplantation. Immunity. 2016;44:1020–33.
103. Stallone G, Infante B, Di Lorenzo A, Rascio F, Zaza G, Grandaliano G. mTOR inhibitors effects on regulatory T cells and on dendritic cells. J Transl Med. 2016;14:152.
104. Clifford DB, De Luca A, Simpson DM, Arendt G, Giovannoni G, Nath A. Natalizumab-associated progressive multifocal leukoencephalopathy in patients with multiple sclerosis: lessons from 28 cases. Lancet Neurol. 2010;9:438–46.
105. Bhat S, Czuczman MS. Galiximab: a review. Expert Opin Biol Ther. 2010;10:451–8.
106. Hassan SB, Sørensen JF, Olsen BN, Pedersen AE. Anti-CD40-mediated cancer immunotherapy: an update of recent and ongoing clinical trials. Immunopharmacol Immunotoxicol. 2014;36:96–104.
107. Wang H, Kadlecek TA, Au-Yeung BB, Goodfellow HE, Hsu LY, Freedman TS, et al. ZAP-70: an essential kinase in T-cell signaling. Cold Spring Harb Perspect Biol. 2010;2:a002279.
108. Sfikakis PP. The first decade of biologic TNF antagonists in clinical practice: lessons learned, unresolved issues and future directions. Curr Dir Autoimmun. 2010;11:180–210.
109. Mertens M, Singh JA. Anakinra for rheumatoid arthritis: a systematic review. J Rheumatol. 2009;36:1118–25.
110. Bettini M, Vignali DA. Regulatory T cells and inhibitory cytokines in autoimmunity. Curr Opin Immunol. 2009;21:612–8.
111. Hu M, Wang YM, Wang Y, Zhang GY, Zheng G, Yi S, O'Connell PJ, Harris DC, Alexander SI. Regulatory T cells in kidney disease and transplantation. Kidney Int. 2016;90(3):502–14.
112. Safinia N, Scotta C, Vaikunthanathan T, Lechler RI, Lombardi G. Regulatory T cells: serious contenders in the promise for immunological tolerance in transplantation. Front Immunol. 2015;6:438.
113. Rothstein DM, Camirand G. New insights into the mechanisms of Treg function. Curr Opin Organ Transplant. 2015;20:376–84.
114. Ferrer IR, Hester J, Bushell A, Wood KJ. Induction of transplantation tolerance through regulatory cells: from mice to men. Immunol Rev. 2014;258:102–16.
115. Orlando G, Soker S, Wood K. Operational tolerance after liver transplantation. J Hepatol. 2009;50:1247–57.
116. Karimi MH, Geramizadeh B, Malek-Hosseini SA. Tolerance induction in liver. Int J Organ Transpl Med. 2015;6:45–54.
117. Goldman M, Wood K. Translating transplantation tolerance in the clinic: where are we, where do we go? Clin Exp Immunol. 2009;156:185–8.
118. Elias N, Cosimi AB, Kawai T. Clinical trials for induction of renal allograft tolerance. Curr Opin Organ Transplant. 2015;20:406–11.
119. Van der Net JB, Bushell A, Wood KJ, Harden PN. Regulatory T cells: first steps of clinical application in solid organ transplantation. Transpl Int. 2016;29:3–11.
120. Frankel AE, Zuckero SL, Mankin AA, Grable M, Mitchell K, Lee YJ, et al. Anti-CD3 recombinant diphtheria immunotoxin therapy of cutaneous T cell lymphoma. Curr Drug Targets. 2009;10:104–9.
121. Casiraghi F, Perico N, Cortinovis M, Remuzzi G. Mesenchymal stromal cells in renal transplantation: opportunities and challenges. Nat Rev Nephrol. 2016;12:241–53.
122. Vanikar AV, Trivedi HL, Kumar A, Gopal SC, Kute VB. Mesenchymal stem cells and transplant tolerance. Nephrology (Carlton). 2014;19:369–74.
123. Guo K, Ikehara S, Meng X. Mesenchymal stem cells for inducing tolerance in organ transplantation. Front Cell Dev Biol. 2014;2:8.
124. Naji A, Rouas-Freiss N, Durrbach A, Carosella ED, Sensébé L, Deschaseaux F. Concise review: combining human leukocyte antigen G and mesenchymal stem cells for immunosuppressant biotherapy. Stem Cells. 2013;31:2296–303.
125. Sykes M. Hematopoietic cell transplantation for tolerance induction: animal models to clinical trials. Transplantation. 2009;87:309–16.
126. Pilat N, Hock K, Wekerle T. Mixed chimerism through donor bone marrow transplantation: a tolerogenic cell therapy for application in organ transplantation. Curr Opin Organ Transplant. 2012;17:63–70.
127. Chen YB, Kawai T, Spitzer TR. Combined bone marrow and kidney transplantation for the induction of specific tolerance. Adv Hematol. 2016;2016:6471901.
128. Scandling JD, Busque S, Dejbakhsh-Jones S, Benike C, Millan MT, Shizuru JA, et al. Tolerance and chimerism after renal and hematopoietic-cell transplantation. N Engl J Med. 2008;358:362–8.
129. Kawai T, Cosimi AB, Spitzer TR, Tolkoff-Rubin N, Suthanthiran M, Saidman SL, et al. HLA-mismatched renal transplantation without maintenance immunosuppression. N Engl J Med. 2008;358:353–61.

32 Corticosteroids

Gaetano Caramori, Sharon Mumby, Giuseppe Girbino, Kian Fan Chung, and Ian M. Adcock

32.1 Introduction

CORTISOL secretion by the cortex of the adrenal glands (Fig. 32.1) increases in response to any stress in the body, whether physical (such as illness, trauma, surgery or temperature extremes) or psychological. However, this hormone is more than a simple marker of stress levels—it is necessary for the correct functioning of almost every part of the body. Excesses or deficiencies of this crucial hormone also lead to various physical symptoms and disease states [1]. Although cortisol is not essential for life per se, it helps an organism to cope more efficiently with its environment with particular metabolic actions on glucose production and protein and fat catabolism. Nevertheless, loss or profound diminishment of cortisol secretion leads to a state of abnormal metabolism and an inability to deal with stressors, which, if untreated, may be fatal [1, 2].

The body's level of cortisol in the bloodstream displays a DIURNAL VARIATION, that is, normal concentrations of cortisol vary throughout a 24-h period (Fig. 32.1). Cortisol levels in normal individuals are highest in the early morning at around 8 a.m. and are lowest just after midnight. This early morning dip in cortisol level often corresponds to increased symptoms of inflammatory diseases or other pathologies in man [3]. Overlaid upon this diurnal variation is the pulsatile nature of cortisol release under the control of local and central 'clocks' [4]. By mimicking this pulsatile cortisol release, it is hoped to reduce the detrimental side effects of exogenous corticosteroids

Final manuscript submitted on December 17, 2018.

G. Caramori · K. F. Chung · I. M. Adcock (✉)
Unità Operativa Complessa di Pneumologia, Dipartimento di Scienze Biomediche, Odontoiatriche e delle Immagini Morfologiche e Funzionali (BIOMORF), University of Messina, Messina, Italy

Airways Disease Section, National Heart and Lung Institute, Imperial College London, London, UK
e-mail: ian.adcock@imperial.ac.uk

S. Mumby
Unità Operativa Complessa di Pneumologia, Dipartimento di Scienze Biomediche, Odontoiatriche e delle Immagini Morfologiche e Funzionali (BIOMORF), University of Messina, Messina, Italy

Airways Disease Section, National Heart and Lung Institute, Imperial College London, London, UK

Faculty of Medicine, National Heart and Lung Institute, Imperial College London, London, UK
e-mail: s.mumby@imperial.ac.uk

G. Girbino
Unità Operativa Complessa di Pneumologia, Dipartimento di Scienze Biomediche, Odontoiatriche e delle Immagini Morfologiche e Funzionali (BIOMORF), University of Messina, Messina, Italy

Airways Disease Section, National Heart and Lung Institute, Imperial College London, London, UK

Università degli Studi di Messina, Messina, Italy
e-mail: giuseppe.girbino@unime.it

M. J. Parnham et al. (eds.), *Nijkamp and Parnham's Principles of Immunopharmacology*,
https://doi.org/10.1007/978-3-030-10811-3_32

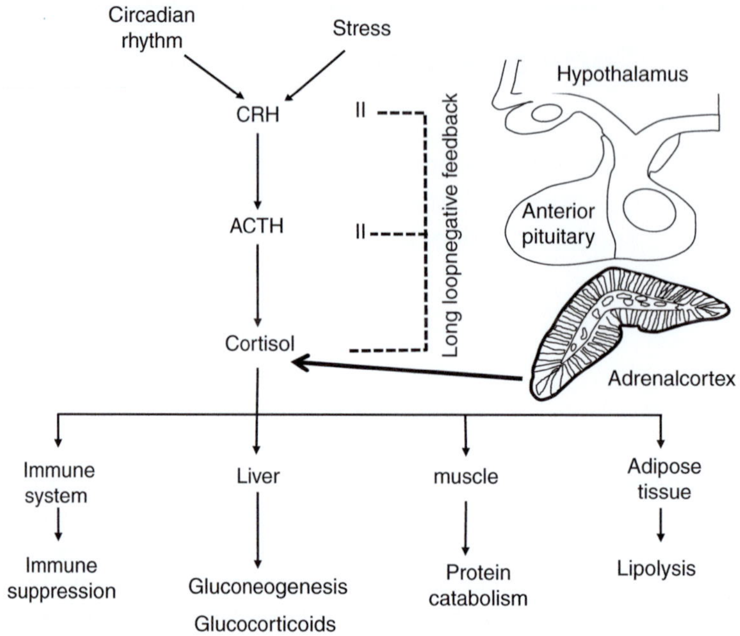

Fig. 32.1 The secretion of cortisol by the adrenal cortex is under the control of many feedback loops. In response to many external and internal stimuli (such as circadian rhythm and stress responses), neurons in the paraventricular nucleus of the hypothalamus release the corticotropin-releasing hormone (CRH) that travels to the anterior pituitary, where it stimulates the corticotroph cells of the anterior pituitary to release the adrenocorticotropic hormone (ACTH) that by binding to cell surface ACTH receptors, located primarily on the adrenocortical cells of the adrenal gland, stimulates the production of both glucocorticoids (cortisol) and mineralocorticoids (aldosterone), which are termed for this reason corticosteroids. Cortisol has many functions in different cells and tissues, including hepatic gluconeogenesis (for this reason the molecules mimicking its effect are also termed glucocorticoids). Cortisol also inhibits the secretion of both CRH and ACTH

whilst enhancing their anti-inflammatory properties [4, 5].

Increased levels of corticosteroids serve as potent suppressors of the IMMUNE AND INFLAMMATORY SYSTEMS. This is particularly evident when they are administered at pharmacological doses but is also important in controlling normal immune responses. As a consequence, corticosteroids are widely used as drugs to treat many different inflammatory and autoimmune diseases such as rheumatic diseases [e.g., rheumatoid arthritis (RA) and systemic lupus erythematosus (SLE)], inflammatory diseases of the upper airways (rhinitis, chronic rhinosinusitis), pulmonary inflammatory diseases [bronchial asthma, chronic obstructive pulmonary disease (COPD), bronchiectasis, interstitial lung diseases (such as sarcoidosis, hypersensitivity pneumonias, idiopathic eosinophilic pneumonias, idiopathic fibrosing interstitial pneumonias)], inflammatory bowel disease (IBD, Crohn's disease and ulcerative colitis), infections (including tuberculosis), inflammatory skin diseases (e.g., psoriasis, atopic dermatitis) and kidney diseases (e.g., glomerulonephritis). Corticosteroids may also be used in organ transplantation to reduce the chance of rejection (see Chap. 32). Thus, although the early effect of cortisol is to stimulate the immune system, cortisol and synthetic corticosteroids predominantly repress the inflammatory response by decreasing the activity and production of immunomodulatory and inflammatory cells.

The usefulness of corticosteroids in treating inflammatory diseases was exemplified by the early work of Kendall and Hench [6]. In a classic experiment, 100 mg of cortisone was injected into the muscle of a patient (Mrs. G.) suffering from chronic rheumatoid arthritis on Sept 21, 1948. Seven days later the patient was able to walk to the shops for the first time in years. Kendall and Hench were awarded the Nobel prize for this work in 1950, and it represented a new approach to therapy with natural hormones by utilising pharmacological, rather than physiological, doses.

There are five main aspects of inflammation: (1) the release of inflammatory mediators such as histamine, lipid mediators, complement factors, CYTOKINES AND CHEMOKINES, other growth factors and neuropeptides and gaseous mediators; (2) increased blood flow in the inflamed area (erythema) caused by some of these inflammatory mediators; (3) leakage of plasma from the vasculature into the damaged area (oedema) due to increased capillary permeability; (4) cellular infiltration signalled by chemoattractants; and (5) repair processes such as tissue fibrosis. Corticosteroids can modify all of these processes.

Inflammation is a central feature of many chronic diseases (please see above). The specific characteristics of the INFLAMMATORY SYSTEM and the inflammatory response in each disease and the site of inflammation differ, but both involve the recruitment and activation of inflammatory cells and changes in the structural cells of the target organ. All are characterised by an increased expression of many INFLAMMATORY MEDIATORS including cytokines, chemokines, growth factors, enzymes, receptors and ADHESION MOLECULES. The increased expression of these proteins is the result of enhanced gene transcription since many of the genes are not expressed in normal cells but are induced in a cell-specific manner during the inflammatory process [7].

32.2 Chemical Structures

Corticosteroids are 21-carbon steroid hormones (Fig. 32.2) composed of four rings [8, 9]. The basic structure of the A ring is a 1α, 2β-half-chair, whatever the substitutions. Rings B and C are semi-rigid chairs with minimal structural influence by substituent groups. In contrast, the shape of the D-ring depends on the nature and environment of the substituent groups. Modern corticosteroids such as prednisone, prednisolone, fluticasone, budesonide, dexamethasone and deflazacort are based on the cortisol (hydrocortisone) structure with modification to enhance the anti-inflammatory effects such as insertion of a C=C double bond at C1,C2 or by the introduction of 6α-fluoro, 6α-methyl and 9α-fluoro and/or further substitutions with α-hydroxyl, α-methyl or β-methyl at the 16 position, for example, in dexamethasone (Fig. 32.2) [8, 9]. Reduced binding to the mineralocorticoid receptor is achieved by insertion of the C=C double bond at C1,C2, and lipophilic substituents such as 21α-esters attached to the D-ring increase glucocorticoid receptor (nuclear receptor subfamily 3, group C, member 1; NR3C1; GR) binding and enhance topical deposition and hepatic metabolism. These substitutions are seen with budesonide and fluticasone two of the most commonly used inhaled corticosteroids [8, 9]. Beclomethasone dipropionate (BDP) is a prodrug of the active form, beclomethasone (beclomethasone-17-monopropionate, BMP). Ciclesonide is a prodrug that is enzymatically hydrolysed, by esterase enzymes of the airways, to a pharmacologically active metabolite, C21-desisobutyryl-ciclesonide (also termed desciclesonide or RM1). The ligand-binding domain (LBD) of GR has a pocket on the floor of the binding cleft that lies beneath the C17 residue of the steroid backbone. The degree of occupancy of this pocket affects the affinity, duration of action and side effect profile of ligands, and computational chemistry can design drugs with improved clinical characteristics including those without a steroid backbone to improve safety [10].

Using this structural knowledge has allowed the production of non-steroidal GR agonists (SEGRAs) or selective glucocorticoid receptor modulators (SEGRMs) which fill the GR ligand-binding domain spatially and have many classical GR activities but can avoid the side effects associated with the steroid backbone such as association with other steroid receptors [11]. The latter class is able to modulate the activity of a GR agonist and/or may not classically bind

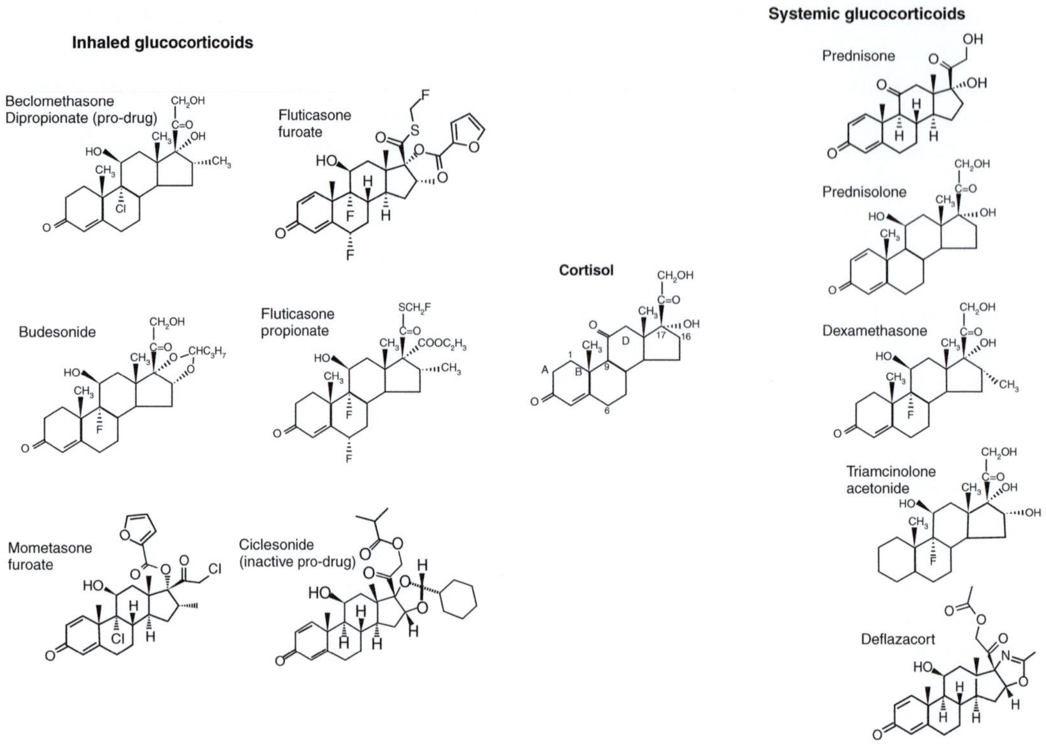

Fig. 32.2 Structural modifications of cortisol exhibited by the clinically used corticosteroids prednisone, prednisolone, deflazacort, dexamethasone and triamcinolone, beclomethasone dipropionate, budesonide, ciclesonide, fluticasone (propionate and furoate) and mometasone. Most of the images have been obtained from https://commons.wikimedia.org/wiki/Main_Page

the GR LBD. SEGRMs were expected to present the same or better efficacy compared to classical corticosteroids but cause minimal side effects. SEGRAs and SEGRMs are collectively denominated SEGRAMs (selective glucocorticoid receptor agonists and modulators) [11]. Although this transrepression vs transactivation concept has been proved to be too simplistic, the SEGRAMs have been helpful in elucidating various molecular actions of the glucocorticoid receptor [11]. The exact structural and lipophilic requirements to optimise corticosteroid pharmacokinetics and pharmacodynamics to separate their anti-inflammatory efficacy from their side effects are still unclear, but corticosteroids with improved clinical characteristics are likely to be synthesised as our knowledge in this area increases.

32.3 Mechanisms of Corticosteroid Action

32.3.1 Glucocorticoid Receptors

Classically, corticosteroids exert their effects by binding to a single receptor but with many isoforms. The glucocorticoid receptors [GRs or nuclear receptor subfamily 3 group C member 1 (NR3C1)] are transcription factors belonging to the superfamily of nuclear receptors that are localised to the cytoplasm of target cells. GRs are expressed in almost all cell types, and their density varies from 200 to 30,000 per cell [9, 12] with an affinity for cortisol of ~30 nM, which falls within the normal range for plasma concentrations of free hormone. The human GR gene is located on the long arm of the chromosome 5 and expresses two major mRNAs variants, termed GRa (previously GRα) and GRb (GRβ). The human GRa mRNA further expresses multiple isoforms [named GRa-A (classic form), GRa-B, GRa-C1, GRa-C2, GRa-C3, GRa-D1, GRa-D2 and GRa-D3]. All human GRa isoforms translocate into the nucleus in response to ligand, while they are differentially distributed in the cytoplasm and/or the nucleus in the absence of ligand. GRa-B and GRa-C1 possess transcriptional activities similar to that of GRa-A, whereas GRa-C2 and GRa-C3 isoforms have stronger transcriptional activities, while GRa-D1, GRa-D2 and GRa-D3 demonstrate weaker activities [9, 12–15].

GR exists in all cells within the airways as the predominant GRa-A isoform explaining the pronounced effect that corticosteroids have on airway resident and inflammatory cells and their clinical efficacy in most subjects with asthma [9, 12–15].

The GRb isoform has been implicated in corticosteroid insensitivity in some patients by acting as a dominant negative regulator of GRa-A [9, 12–16].

The GR has several functional domains (Fig. 32.3). The corticosteroid ligand-binding domain (LBD) is sited at the carboxyl terminus of the molecule and is separated from the DNA binding domain (DBD) by a hinge region. There is an N-terminal transactivation domain which is involved in gene activation following DNA binding. This region may also be involved in binding to other transcription factors. The inactive GR is part of a large protein complex (~300 kDa) that includes two subunits of the heat shock protein hsp90, which blocks the nuclear localisation of GR and one molecule of the immunophilin p59, termed FK506-binding protein 2 (FKBP2 also known as PPIase or FKBP-13), based on its ability to bind the immunosuppressive drug FK506 [9, 12].

Corticosteroids are lipophilic and diffuse freely from the circulation into cells across the cell membrane and bind the ligand-binding domain of their GR to induce activation (Fig. 32.4) [9, 12–15]. Once activated, GR translocates into the nucleus where it interacts with transcriptional coactivators or repressors to modulate GENE TRANSCRIPTION repressing inflammatory genes (TRANSREPRESSION) or enhancing the expression of anti-inflammatory genes (TRANSACTIVATION). This because once the corticosteroid binds to GR, hsp90 dissociates revealing nuclear localisation signals (NLS) allowing the nuclear translocation of the activated GR-corticosteroid complex and its binding to DNA.

Shuttling of GR between the nucleus and cytoplasm is regulated by nuclear import and export receptors in a dynamic manner. GR possess two NLS, NLS1 and NLS2. GR interacts with several importins including importins 7, 8 and 13 and the α/β heterodimer. Defects in nuclear translocation observed in patients with relative steroid-resistant severe bronchial asthma may result from abnormal levels of importin 7 or its ability to interact with GR under the influence of oxidative stress [9, 14, 17–20]. GR translocates into mitochondria and lysosomes as well [14].

The function of GR is affected by many post-translational modifications, particularly phosphorylation, acetylation, sumoylation and nitration, and these can have major effects on all aspects of GR function, from ligand binding and nuclear translocation to cofactor association and control of gene transcription with the effect of phosphorylation being the most studied. Correct GR phosphorylation is essential for optimal GR function with phosphorylation at both Ser226 and Ser221 being seen with GR activation. Ser211 phosphorylation has been linked to alterations in ligand binding, nuclear translocation and transactivation and cofactor association. GR Ser226 phosphorylation, in contrast, is associated with greater transcription efficacy [9, 12, 14, 15, 20–22].

GR can also be acetylated on lysines K494 and K495 following activation. Acetylation of GR affects the ability of GR to interact with p65 (see below), and removal of these tags is important for the suppression of subsets of inflammatory genes [9, 12, 14, 15, 23]. Small ubiquitin-like modifier (SUMO) proteins can also modify GR and affect its function. Sumoylation affects GR transactivation potential particularly at promoters with multiple GREs, whilst K293 GR

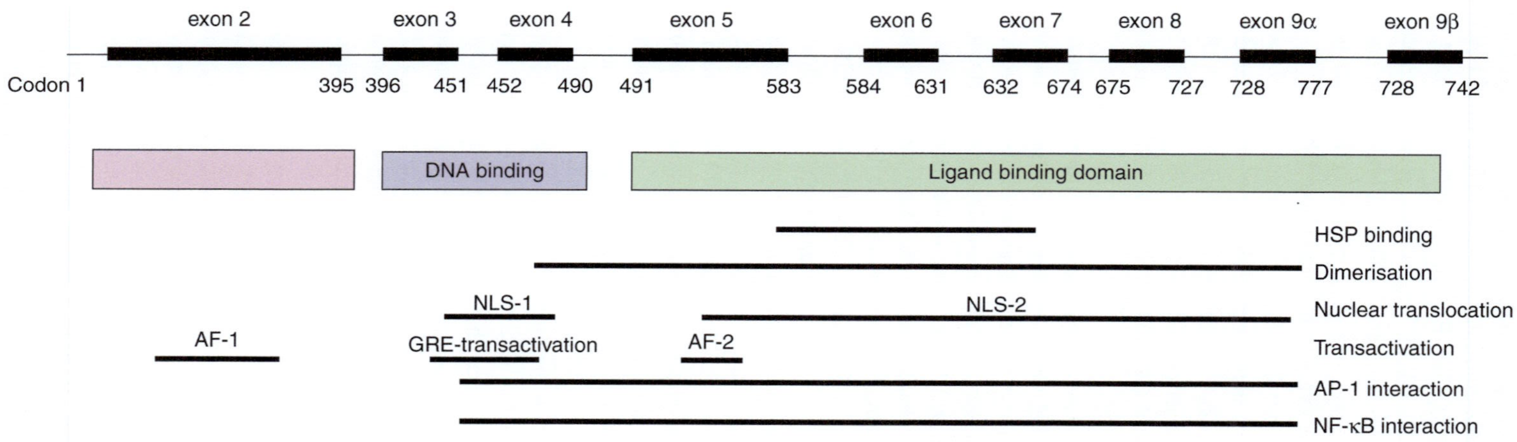

Fig. 32.3 Modular structure of the glucocorticoid receptor (GR). The coding region of GR results from splicing together of exons 2–9 of the GR gene. The GRb isoform of GR results from the use of the short 9β exon which removes the ligand-binding domain seen in GRa. The modular design of GR enables distinct regions of the protein to function in isolation as ligand-binding domains, dimerisation domains, nuclear localisation domains, transactivation and transrepression (AP-1 and NF-κB interacting) domains. *NLS* nuclear localisation signal, *AF-1/2* activating factor 1/2, *GRE* glucocorticoid response element (composed of two palindromic half sites (AGAACA) separated by three nucleotides)

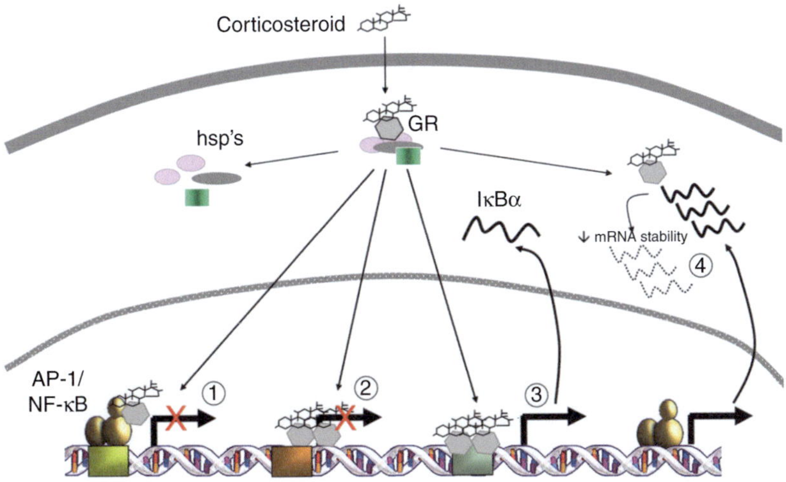

Fig. 32.4 Mechanisms of gene repression by the glucocorticoid receptor (GR). The glucocorticoid can freely migrate across the plasma membrane where it associates with the cytoplasmic GR. This results in activation of the GR and dissociation from the heat shock protein (hsp90) chaperone complex. Firstly, activated GR translocates to the nucleus where it can bind as a monomer either directly or indirectly with the transcription factors, activator protein-1 (AP-1) and nuclear factor kappa B (NF-κB), preventing their ability to switch on inflammatory gene expression ①. Secondly, the GR dimer can bind to a glucocorticoid response element (GRE) that overlaps the DNA binding site for a pro-inflammatory transcription factor or the start site of transcription thereby preventing inflammatory gene expression ②. Thirdly, the GR dimer can induce the expression of the NF-κB inhibitor IκBα ③, and fourthly glucorticoids can increase the levels of cell ribonucleases and mRNA destabilising proteins, thereby reducing the levels of mRNA ④

SUMOylation is essential for GC-induced inverted repeated negative GC response element (IR nGRE)-mediated direct transrepression and for NF-κB/AP1-mediated GC-induced tethered indirect transrepression. In addition, cells with sumoylation-deficient FK506-binding protein 51 (FKBP51) fail to interact with Hsp90 and GR, thus facilitating the recruitment of the closely related protein, FKBP52, which enhances GR transcriptional activity [9, 24–26]. Finally, nitration of the GR results in an enhancement of GR-mediated transcriptional activity.

Nuclear GR can induce or repress gene expression following DNA binding at specific glucocorticoid response elements (GREs) or acting as a monomer interact with DNA-bound pro-inflammatory factors and thereby enable transcriptional regulator proteins to be positioned such that they repress activated gene expression. Pro-inflammatory transcription factors such as ACTIVATOR PROTEIN-1 (AP-1) and NUCLEAR FACTOR KAPPA B (NF-KB), which are upregulated during inflammation, are the major targets for this tethering process although tethering between GR and p65 (the major subunit of NF-κB), for example, is not essential for repression in airway epithelial cells [9, 20, 27, 28].

It is likely that the altered transcription of many different genes is involved in the anti-inflammatory action of corticosteroids, but the most important action of these drugs is likely to be inhibition of transcription of cytokine and chemokine genes implicated in inflammation. Evidence for this has been presented in a series of elegant experiments using mice expressing mutated GRs unable to dimerise and subsequently bind to DNA [9, 15, 20].

32.3.2 Gene Induction by Corticosteroids

The GRE is the imperfect palindrome AGAACA*nnn*TGTTCT with GR able to interact with each hexamer independently. Even small changes in the GRE sequence can have a profound effect on transcriptional activity. Indeed, the GRE may be considered as a different type of GR ligand which is able to modify GR function by altering the association with transcriptional cofactors, changing the local chromatin configuration and thereby affecting downstream functional actions of GR [9, 28–30].

The activated GR only remains associated with the GRE for a few seconds before being replaced by a different GR in a process called assisted loading. Binding of the first GR to a GRE initiates an ATP-dependent chromatin remodelling process that provides site more amenable for GR-GRE interaction highlighting the importance of co-ordinated GRE interactions to obtain the full glucocorticoid (GC) response in a cell- and tissue-dependent manner [9, 31]. Combining DNase I accessibility assays with chromatin immunoprecipitation and high-throughput sequencing, the transcription factor activator protein 1 (AP-1) was identified as a major partner for productive GR-chromatin interactions. This highlighted the critical role for AP-1 in regulating GR-mediated transcription and recruitment to co-regulatory elements. Indeed, the baseline chromatin accessibility of GR recruitment to GREs is dependent on AP-1 binding. This may account for the importance of AP-1 and its components such as c-Jun in gene regulatory networks that distinguish asthmatic subjects who respond poorly to corticosteroids [32].

In airway epithelial cells, there are >10,000 GR binding sites (GBS), of which only 13% are able to induce transcriptional activation in response to GC exposure. The GBS lacking activation potential clustered around the inducible GBS, and interactions between these direct and tethered GBS are necessary for the full gene activation response to GC [9, 33].

Several genes are upregulated by glucocorticoids, including the β_2 adrenergic receptor (β_2-AR), MAPK phosphatase (MKP-1/DUSP1) and serum leukoprotease inhibitor (SLPI). Interestingly, corticosteroids can also induce the expression of the NF-κB inhibitor IκBα in certain cell types.

Several other mechanisms of GR function have been reported including effects on mRNA stability (Fig. 32.4). GCs affect the expression of pro-inflammatory gene mRNAs which contain adenylate-uridylate-rich elements (AREs) within their 3′ untranslated regions through targeting the RNA-binding proteins tristetraprolin (TTP) and Hu antigen R (HuR) family members which control mRNA decay and stability, respectively. This mechanism is used by dexamethasone, for example, to downregulate COX-2 and CCL11 expression acting via the p38 MAPK-MKP-1 axis [9].

Also non-coding RNAs (ncRNAs), such as microRNAs (miRNAs), modify GR expression and function. The expression of certain key miRNAs is regulated by GR, and GR is itself the target of other miRNAs. Induction of GILZ expression by GR is reduced by miR18 and miR124a in human cells, and aberrant expression of these miRNAs may be involved in the relative corticosteroid insensitivity in some patients with severe asthma. Hydrocortisone increases miR124 expression in sepsis patients which causes GRa downregulation and corticosteroid insensitivity [9]. Furthermore, inhibition of miR145 prevents eosinophilia, mucous secretion and airway hyperresponsiveness to the same extent as dexamethasone in an animal model of asthma.

Long ncRNAs (lncRNAs) are defined as being >200 nucleotides in length, and two specific lncRNAs have opposite effects on GR function. Steroid receptor RNA activator (SRA) is a constituent of the steroid receptor coactivator (SRC)-1/SRC-2 complex, and it increases GR transcriptional. In contrast, growth arrest-specific 5 (Gas5) is a GRE decoy by binding to the DNA binding site of active GR [9].

32.3.3 Gene Repression by Corticosteroids

GR plays a critical role in suppressing inflammatory gene expression. The mechanisms involved generally evoke tethering of activated GR to an activated transcriptional complex driven by DNA-bound NF-κB, for example. The interaction between GR and NF-κB is mutually antagonistic with GR repression seen with NF-κB activation. Importantly, increased NF-κB activation at the nuclear localisation and expression level is associated with severe asthma.

This process is driven in part by HISTONE DEACETYLASE (HDAC)2-mediated alterations in GR acetylation status. HDAC2 expression and/or activity linked to enhanced HAT activity is reduced in severe asthma patients, particularly children. Interestingly, GRb has been reported to reduce HDAC2 expression in human BAL macrophages. A lack of HDAC activity may also evoke local changes in histone acetylation at inflammatory gene promoters, thereby modulating gene expression [9, 13, 15, 23, 30, 34–36]. Alterations in the phosphorylation status of RNA polymerase 2 C-terminal domain (CTD) have also been implicated in the mechanism of dexamethasone-induced suppression of TNF-α-/NF-κB-induced CXCL8 activation by preventing phosphorylated CTD from interacting with the basal transcription factor P-TEFb. A reduction in HDAC2 expression results in a failure to remove the CTD phosphorylation tag and stalling of RNA polymerase 2 on the promoters of steroid-responsive genes.

In addition to interactions with AP-1 and NF-κB, GR can also associate with, and repress, the function of many other transcription factors including the signal transducer and activator of transcription (STAT) family of transcription factors. Many inflammatory (including some of the acute-phase response) genes are under STAT regulation induced by mediators such as interferons (IFNs), interleukin (IL)-5 and IL-6, for example. Interestingly, inflammatory mediators induced by IFNγ-stimulated airway epithelial cells can be inhibited by JAK-STAT inhibitors but not by corticosteroids [9, 37–39].

32.3.4 Non-genomic Rapid Actions of Corticosteroids

The traditional genomic theory of steroid action, whether through direct interaction with DNA or involving cross-talk with other transcription factors, does not fully explain the rapid effects of hormonal steroids, and it is thought that the non-genomic actions are mediated by a distinct membrane receptor [40]. These receptors have distinctive hormone-binding properties, compared to the well-characterised cytoplasmic receptor, and are probably linked to a number of intracellular signalling pathways, acting through G-protein coupled receptors and a number of kinase pathways [40]. In addition, the classical receptor is associated with a variety of kinases and phosphatases within the inactive GR/hsp90 complex. These enzymes are released upon hormone binding and may also account for the rapid induction of tyrosine kinase activity seen in some cell types by glucocorticoids [41]. Evidence of immediate responses is also seen clinically since systemic doses of corticosteroid can lead to very rapid clinical improvement and inhibition of allergic/anaphylactic responses [42].

32.4 Pharmacological Effects of Corticosteroids

32.4.1 Effects of Corticosteroids on Inflammatory and Structural Cells

Corticosteroids are the only therapeutic agents that clearly reverse the inflammation present in many chronic inflammatory diseases of different organs and causes (see above) (see also Chaps. 23, 31 and 33). Topical and systemic corticosteroids have similar pharmacological effects, with differences related to the dose delivered to the target organ and to the enhanced effect of systemic corticosteroids on the mobilisation and recruitment of inflammatory cells from the blood and bone marrow.

In general, in all chronic inflammatory and immune diseases, corticosteroids cause a marked reduction in the number and activation of infiltrating cells, including mast cells, macrophages, T lymphocytes and eosinophils, in the inflamed tissue [9, 15]. Furthermore, topical and oral corticosteroids can have effects on structural cells and in asthma, for example, can reverse the bronchial epithelial damage, mucus hyperplasia and basement-membrane thickening that is characteristically seen in the bronchial biopsies from these patients [9, 15].

Corticosteroids may have direct inhibitory effects on many of the cells involved in inflammation, including macrophages, T and B lymphocytes, eosinophils, smooth muscle and endothelial and epithelial cells, resulting in their reduced inflammatory mediator synthesis and release [9, 15, 43–45] (Fig. 32.5). In general, corticosteroids substantially reduce mast cell-/eosinophil-/lymphocyte-driven processes while leaving unaltered, or even augmenting, neutrophil-mediated processes [9, 15, 43–45]. For example, corticosteroids may enhance neutrophil function as a result of increased leukotriene B4 and superoxide anion production, in addition to inhibiting their apoptosis [44]. GCs modulate inflammatory cell survival, inducing apoptosis in T and B lymphocytes and eosinophils while delaying constitutive neutrophil apoptosis and promoting non-inflammatory phagocytosis of apoptotic cell targets, a process important for the successful resolution of inflammation [45, 46]. Corticosteroids in autoimmune diseases decrease the cell and tissue damage mediated by T cells and autoantibodies and immunocomplexes. Interestingly in allergic diseases, corticosteroids reduce the number of mast cells within the inflamed tissue; however, they do not appear to inhibit mediator release from these cells [9, 15, 43, 44]. Treatment with topical corticosteroids also reduces the number of activated T lymphocytes (CD25+ and HLA-DR+) in bronchoalveolar lavage (BAL) fluid and peripheral blood from asthmatic patients [47].

Corticosteroids are particularly effective against eosinophilic inflammation, possibly as a result of decreasing eosinophil survival by stimulating apoptosis [46, 48]. In addition to their suppressive effects on inflammatory cells, corticosteroids may also decrease endothelial permeability and inhibit plasma exudation and/or leucocyte transendothelial migration in most tissues [9, 15]. Inflammation drives angiogenesis by direct and indirect mechanisms, promoting endothelial proliferation, migration and vessel

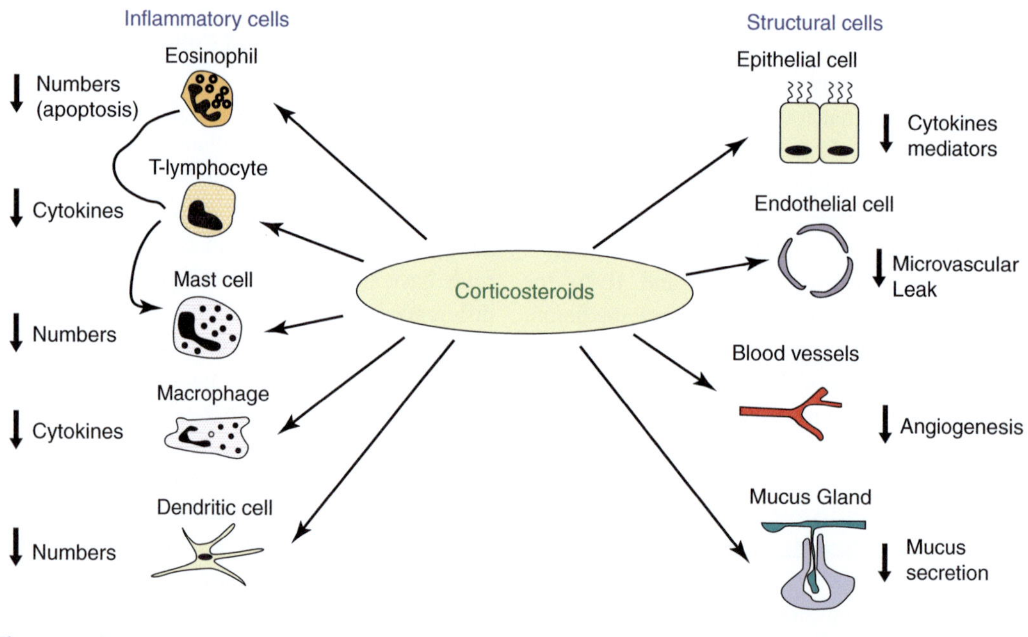

Fig. 32.5 Glucocorticoids act on most inflammatory and structural cells of the tissues to suppress inflammation. The activity (T and B lymphocytes and macrophages) and number of infiltrating inflammatory cells (eosinophils, T and B lymphocytes, macrophages, mast cells and dendritic cells) are decreased by glucocorticoids. Glucocorticoids also have a suppressive effect on structural cells of the tissues and can reduce inflammatory mediator release and adhesion molecule expression on epithelial and endothelial cells, microvascular leakage from blood vessels, angiogenesis and both the numbers of mucus cells and release of mucus from these cells

sprouting, but also by mediating extracellular matrix remodelling and release of sequestered growth factors and recruitment of proangiogenic leucocyte subsets. By facilitating greater infiltration of leucocytes and plasma proteins into inflamed tissues, angiogenesis can also propagate chronic inflammation [49], and high doses of topical systemic corticosteroids may reduce both neo-angiogenesis and the increased blood flow present at sites of inflammation [9, 15]. At sites of acute and/or chronic mucosal inflammation, there is often an increased secretion of the major secretory mucins (MUC5AC and MUC5B). The effect of the glucocorticoids on the expression of these mucins is still quite controversial and may be cell- and tissue-type dependent [50].

32.4.2 Effects of Corticosteroids on Inflammatory Mediator Production and/or Secretion and/or Activation

Corticosteroids block the generation of most pro-inflammatory cytokines and chemokines [9, 15, 51, 52]. Despite the wide pleiotropy (multiple actions) and redundancy that exists in the cytokine and chemokine families, and the subsequent inability to ascribe precise roles to most of these molecules in inflammatory disease pathogenesis, it is clear that these proteins are important mediators in chronic inflammation (see Chap. 5). The development of TUMOUR NECROSIS FACTOR-α (TNF-α) and INTERLEUKIN (IL)-1β, IL-6, IL-12 and IL-23 antagonists has provided evidence that, in many inflammatory bowel, rheumatic and skin diseases, these mediators plays a key driving role in inflammation, despite different clinical relevance of each inflammatory mediator in different tissues and diseases [53, 54] (see Chap. 33). This does not appear to be the case with all inflammatory diseases, however [9, 15, 51, 52]. Interestingly, corticosteroids can also enhance the expression of key anti-inflammatory molecules such as IL-10 and IL-1ra in some inflammatory diseases but again not all [9, 15]. For example, corticosteroids increase the production of IL-10, but not IL-1ra, at sites of inflammation in asthma [9, 15].

Arachidonic acid metabolism via 5-lipoxygenase gives rise to a group of biologically active lipids known as LEUKOTRIENES: leukotriene B_4, which is a potent activator of leucocyte chemotaxis, and cysteinyl leukotrienes (leukotriene C_4, D_4 and E_4) which account for the spasmogenic activity previously described as slow-reacting substance of anaphylaxis (see Chap. 7). Leukotrienes, particularly cysteinyl leukotrienes, are thought to play a key role in some chronic and acute inflammatory diseases but do not appear to be major targets for corticosteroids that are unable to block leukotriene biosynthesis and their release [55]. However, corticosteroids in vitro accelerate LTC_4 catabolism by inducing the activity of a LTC_4-degrading enzyme, gamma-glutamyl transpeptidase-related enzyme (γ-GTPRE) [56]. Analysis of serum from patients with increasing severity of asthma identified 15 metabolites that were significantly altered in asthma although some such as dehydroepiandrosterone sulphate, cortisone, cortisol, prolylhydroxyproline, pipecolate and N-palmitoyltaurine correlated significantly with ICS and oral corticosteroid use. In contrast, oleoylethanolamide increased with asthma severity independently of steroid treatment. Overall, the data indicated that asthma was characterised by a systemic metabolic shift according to disease severity and that corticosteroid treatment significantly affects metabolism [57].

32.4.3 Oxidative Stress and Reduced Response to the Corticosteroid Effects

The inflammation in COPD, in severe asthma and in a high number of patients with IBD is scarcely suppressed by topical or oral corticosteroids, even at very high doses. Potential reasons for the failure of corticosteroids to function effectively in reducing inflammation in these diseases include the fact that all of them have a high oxidative stress and oxidative stress may reduce corticosteroid receptor (GR) nuclear translocation with reduced GRα expression or altered GR-cofactor interactions within the nucleus [9, 15, 58, 59].

Interestingly, cigarette smoke contains 10^{17} oxidant particles per puff, and asthmatic subjects who smoke have a reduced responsiveness to both topical and oral corticosteroids [60–62].

32.4.4 Mechanisms of Reduced Corticosteroid Responsiveness in COPD

Oxidative stress reduces HDAC2 expression and activity, thus potentially limiting glucocorticoid effectiveness in suppressing inflammation in vitro studies and in patients with COPD. Overexpression of HDAC2, but not HDAC1, improves corticosteroid sensitivity in bronchoalveolar lavage (BAL) macrophages from stable COPD patients [23] through a mechanism that involves the phosphoinositide-3-kinase (PI3K)-δ pathway [9, 15]. Sub-bronchodilator low doses of theophylline, at concentrations that do not inhibit phosphodiesterase (PDE)4 activity, can enhance HDAC2 activity in vitro, and functionally this enhances glucocorticoid effects. Combined theophylline and ICS treatment improves lung function and sputum neutrophilia in stable COPD patients and lung function in smoking asthmatics [9, 15]. This effect may be via phosphoinositide-3-kinase (PI3K)δ-induced hyperphosphorylation of HDAC2 particularly since PI3Kδ is upregulated in peripheral lung tissue of patients with COPD. Use of inhaled PI3Kδ-selective inhibitors may even prove more efficacious in improving patient responses to ICS [63]. The results of an ongoing phase IIa large controlled study [https://clinicaltrials.gov/ct2/show/NCT03345407] on the efficacy of a highly selective inhaled PI3Kδ inhibitor (GSK2269557) in patients with stable COPD are awaited with interest.

32.4.5 Mechanisms of Reduced Corticosteroid Responsiveness in Severe Bronchial Asthma

Some patients with severe asthma are unable to suppress asthmatic inflammation with high-dose ICS or even oral glucocorticoids. These patients are distinct from those who are non-compliant with their treatment or subjects without access to the correct therapies. The reduced GC function in refractory asthma may be multifactorial, and each stage of GR activation, namely, GR expression, ligand binding, nuclear translocation and/or binding to the GRE and other transcription factors, has been proposed as a mechanism [9, 15, 64, 65].

It is possible that redox-sensitive activation of the AP pathway may drive relative steroid refractoriness in peripheral blood mononuclear cells (PBMCs) from severe steroid-refractory asthmatics [23]. Indeed, the expression of AP-1 components and its upstream activators is greater in PBMCs and bronchial biopsies from patients with corticosteroid-resistant asthma [35], and their expression is not altered by high doses of oral glucocorticoids [61]. In immortalised peripheral blood B cells, there is a distinct network connectivity and gene ontology pattern between good and poor corticosteroid responders which is linked to AP-1 components and a differential response to apoptosis [32].

Nitrosative stress may also impact upon corticosteroid responsiveness, and peroxynitrite formation causes nitration of specific tyrosine (Y) residues which results in the loss of enzymatic activity (Y146) and degradation (Y253) of HDAC2 [62]. However, reduced HDAC2 expression and/or activity is not seen in all patients with therapy-refractory asthma possibly reflecting the heterogeneity of severe asthma phenotype [64, 65].

The transcriptome of bronchial epithelial cells of mild/moderate asthmatics has identified a gene profile that predicts ICS responsiveness—namely, an IL-13-induced gene signature. The expression of this signature is variable in asthma and is inversely correlated with Th17 cells which are linked with steroid insensitivity and with IL-6 a marker of neutrophilic asthma which is also associated with more severe disease [9, 15, 64, 65].

In vitro high levels of IL-2, IL-4 and IL-13 reduce corticosteroid responses in T cells by reducing the affinity of GR for its ligand. This may reflect differences in GR phosphorylation status under the control of the p38 MAPK pathway. Increased p38 MAPK activity is also seen in peripheral blood monocytes, and BAL macrophages from patients with severe asthma and p38 MAPK inhibitors restored GC responsiveness in these cells. Similar results are seen in cells from COPD patients. There is some evidence that this may be linked to changes in HDAC and HAT activities. p38 MAPK may also modulate GR responses by changing GR phosphorylation status. Phosphorylation of GR on Ser134 is p38 MAPK-dependent and significantly downregulates dexamethasone-dependent genome-wide transcriptional responses and cell functions. MKP-1/DUSP1 is a GC-inducible gene which dephosphorylates and inactivates p38 MAP kinases, and its expression and induction are impaired in severe asthma [9, 15, 64, 65].

The exact stimulus given to a cell modifies the intracellular pathway(s) activated, and other signalling pathways such as the MEK/ERK pathway have been implicated in controlling relative GC refractoriness. Furthermore, cyclin-dependent kinases (CDK), glycogen synthase kinase-3 and JNKs can also target GR phosphorylation or phosphorylation of GR-associated cofactors. Neutrophilic asthma is associated with GC refractoriness and increased IL-17 expression and Th17 cells. An animal model of asthma demonstrated that Th17 cell transfer causes a dexamethasone-insensitive neutrophilic inflammatory response and bronchial hyperresponsiveness to methacholine. More importantly, IL-17 inhibits budesonide sensitivity in primary human bronchial epithelial cells through modulating PI3K and HDAC2 expression [9, 15, 64, 65].

32.4.6 Mechanisms of Reduced Corticosteroid Responsiveness During Viral-Induced Asthmatic Exacerbations

Viral infections cause most of the asthmatic exacerbations in children and adults, and these exacerbations are not readily resolved by glucocorticoids, even when administered systemically at high doses. In human experimental models of virally induced asthmatic exacerbations, neither ICS nor oral prednisolone prevents the worsening of airway inflammation or improves clinical symptoms. IFNγ may be one of the dominant mediators responsible for chronic persistent airflow obstruction in severe asthma, and primary bronchial epithelial cells stimulated with IFNγ do not respond to GC. These inflammatory responses are, however, completely ablated by treatment of cells with a JAK-STAT inhibitor. Exposure of primary human airway epithelial cells to RV-16 causes a relative GC resistance by preventing GR nuclear import. This is reversed by suppression of the RV-16-induced JNK and NF-κB pathways [9, 15, 66].

32.5 Pharmacokinetics of Corticosteroids

The pharmacokinetics of many corticosteroids has been well described. In general, plasma concentrations of corticosteroids vary considerably (up to tenfold) after oral administration of the same dose by normal volunteers and asthmatic patients although the reasons for this are unclear [67]. The plasma half-life of currently used ICS varies from <2 h (budesonide) to >5 h [BDP/BMP, fluticasone (propionate and furoate) and mometasone]. This is in contrast to their biological effects which last for 18–36 h [10, 63, 68]. The pharmacokinetic properties of topical drugs depend upon a combination of tissue deposition/targeting, receptor binding, volume of distribution, tissue retention and lipid conjugation. In addition, in order to achieve a good therapeutic index, drugs need to possess a low oral bioavailability and small particle size, rapid metabolism, high clearance, high plasma protein binding and a low systemic half-life. Furthermore, an ideal compound would be inactive at sites distal to the target organ/tissues [69, 70].

There are two main methods of reducing the systemic activity of topical corticosteroids: (1) reducing gastrointestinal BIOAVAILABILITY and (2) prolonging TISSUE RESIDENCY. For example, oral administration of ileal release budesonide capsules for the treatment of Crohn's disease gives similar levels of systemic exposure to active drug, bioavailability and cortisol suppression in adults and children as seen with prednisolone, but importantly, no clinically relevant adverse side effects were reported [71]. Alternatively, for IBD the corticosteroid can be altered to reduce gastrointestinal absorption and/or enhance first-pass hepatic metabolism. Prolonged retention in the tissue can be achieved by increasing lipophilicity, as with fluticasone propionate (FP) and fluticasone furoate (FF) and mometasone furoate, or by forming soluble intracellular fatty acid esters, as with budesonide and ciclesonide [9, 15, 63, 72, 73].

The lipophilic nature of synthetic GCs enables their rapid absorption after topical administration and helps prolong their retention in the airways [74]. When corticosteroids are delivered by inhalation, changing the inhaler device can also decrease oral delivery and subsequently gastrointestinal availability and enhance deposition in the lower airways by altering the particle size [75]. Metered dose inhalers (MDI) and dry powder inhalers (DPI) deliver 10–20% of the inhaled dose to the lungs, but >50% is deposited in the oropharynx and mouth. The drug may then be swallowed and taken up from the gut and become systemically available.

ICSs as a group all have a good therapeutic index resulting from a small particle size enabling low oral bioavailability and rapid metabolism/clearance combined with high plasma protein binding to give a short systemic half-life [10, 74]. Lipophilicity generally correlates well with absorption characteristics. For example, fluticasone (both propionate and furoate) has high lipophilicity and binding affinity for the GR, resulting in a high volume of distribution and long plasma half-life. However, the systemic side effects of fluticasone that arise from systemic absorption are limited due to its almost complete first-pass metabolism in the liver and, in the case of gastrointestinal delivery, low absorption from the gastrointestinal tract [10, 63, 74]. In general, for topical corticosteroids, treatment efficacy and side effects are directly related to tissue dose. The pharmacokinetic profile of topical corticosteroids, therefore, varies with the individual drug, the delivery mechanism and patient profile [10, 74].

32.6 Clinical Indications for Corticosteroids

Pharmacological control of inflammation may be obtained in most patients with varying doses of oral or TOPICAL corticosteroids, depending upon the disease severity [9, 15]. In most patients with asthma, COPD, IBD, allergic rhinitis and inflammatory skin diseases (such as contact dermatitis and psoriasis), adequate doses of topical corticosteroids allow systemic administration to be reduced or withdrawn completely [9, 15]. As such, corticosteroids are standard therapy for these disorders (see also Chap. 23).

Whereas in inflammatory rheumatic diseases (see Chap. 33) and in many other autoimmune diseases (e.g. systemic vasculitis) the gold standard is still represented by systemic corticosteroids, usually taken orally or parenterally during life-threatening exacerbations (flares) of these diseases.

32.6.1 Use of Corticosteroids in the Treatment of Bronchial Asthma

Asthma has long been known as a chronic inflammatory disease of the lower airways, and the beneficial effect of the potent anti-inflammatory prednisolone in asthmatic patients further emphasised this point. Treatment with prednisolone was associated with adverse side effects however. Dramatic improvements in asthma symptoms were also seen with the introduction of ICS which had few systemic side effects and systemic CORTICOSTEROID-SPARING effects. In these initial studies, only 40% of asthmatics responded well to ICS with respect to improvements in lung function—it was not investigated whether this is related to a lack of PATIENT ADHERENCE, poor inhaler technique or a true relative insensitivity to ICS [9, 15].

The routine use of ICS to prevent airway inflammation in combination with relievers such as β_2 agonists, which help the airway smooth muscle to relax after contraction, is effective in treating symptoms, reducing exacerbations and improving lung function in most asthmatics and has resulted in great improvements in asthma control and the quality of life of most asthmatics [9, 13, 15, 61, 64]. Unfortunately a minority of asthmatics show refractoriness to GC treatment (see above and Chap. 23). In asthma, corticosteroids consistently lessen airway hyperresponsiveness (AHR) and the maximal response to a number of spasmogens and irritants. Interestingly, the reduction in airway hyperresponsiveness (AHR) may not be maximal until treatment has been given for several months. The magnitude of the reduction varies, and airway responsiveness can remain abnormal [9, 13, 15, 42, 61, 64].

As with other chronic inflammatory diseases, ICSs reduce the inflammatory markers seen in the asthmatic airways, and this results in the improvement in FEV_1 and the reversal of AHR back to levels seen in healthy non-asthmatic subjects in most subjects with mild-moderate disease. However, since discontinuation of ICS leads to a return of the symptoms of asthma and both AHR and airway inflammation to pretreatment levels, they are not a cure for asthma [9, 13, 15, 43, 61, 64].

GCs are the most successful anti-inflammatory treatment used in asthma as they target all the cells implicated in asthmatic inflammation. GCs have profound effects on infiltrating immune cells as well as on the function of airway structural cells. ICS prevents eosinophil recruitment from the bone marrow as well as their migration into the airways, and this probably explains the greater beneficial effect of oral GCs. GCs also suppress the expression of eosinophil survival factors and induce eosinophil apoptosis [9, 13, 15, 43, 61, 64].

Total blood lymphocyte numbers are reduced in asthmatic subjects who receive oral GCs. GCs can also affect $CD4^{+}CD25^{+}$ $Foxp3^{+}$ regulatory T cell (Treg) expression and function. In comparison to the marked effects on T-cell function, ICSs have little effect on B-cell IgE production in vivo in asthma although higher doses may be effective in COPD and in vitro [9, 13, 15, 43, 61, 63, 64]. In addition, ICSs have profound effects on the function, terminal differentiation and activation status of macrophages and monocytes in asthma. In particular, they reduce the expression of macrophage-derived pro-inflammatory cytokines and chemokines. ICS treatment reduces peripheral blood levels of monocytes and also their low affinity IgE receptor expression. ICS, by regulating DC CCR7 expression, can modulate DC migration to local lymphoid collections. Furthermore, the release of Th1- and Th2-polarising cytokines is suppressed by GCs, whilst that of IL-10 is increased [9, 13, 15, 43, 61, 64].

Overall, although most inflammatory responses in the airway are suppressed by GCs, some innate immune responses including neutrophil production and survival (GCs enhance peripheral blood neutrophilia and prevent neutrophil apoptosis), macrophage phagocytosis and airway epithelial cell survival are either unaffected or even increased. Furthermore, GCs often increase rather than suppress the expression of Toll-like receptors, complement, pentraxins, collectins, serum amyloid A (SAA) and other host defence genes [9, 13, 15, 43, 61, 64]. Despite this, a short course (at least 7 days) of systemic glucocorticoids (50 mg/daily of prednisone equivalent) is mandatory in patients with moderate-to-severe exacerbations of asthma (https://ginasthma.org; see also Chap. 22).

32.6.2 Use of Corticosteroids in the Treatment of Chronic Obstructive Pulmonary Disease

In contrast to asthma, glucocorticoid treatment of stable COPD is rather ineffective in reducing airway inflammation and the decline of lung function. Current national and international guidelines for the management of stable COPD patients recommend the use of inhaled long-acting bronchodilators, ICS and their combination for maintenance treatment of moderate-to-severe stable COPD (https://goldcopd.org) [63]. Several large controlled clinical trials of inhaled combination therapy with ICS and LABAs/ULABAs in stable COPD have shown that this combination therapy is well tolerated and produces a modest but statistically significant reduction in the number of severe exacerbations and improvement in FEV_1, quality of life and respiratory symptoms in stable COPD patients (see also Chap. 23). In addition, the Towards a Revolution in COPD Health (TORCH) study showed a 17% relative reduction in mortality over 3 years for patients receiving salmeterol/fluticasone propionate, although this just failed to reach significance. Some patients with COPD do respond to ICS however. The Groningen and Leiden Universities Corticosteroids in Obstructive Lung Disease (GLUCOLD) study suggested that particular phenotypes of COPD benefit from fluticasone propionate ± salmeterol by reducing the rate of lung function decline. In ICS-responsive patients, the expression of genes associated with cell cycle, oxidative phosphorylation, epithelial cell signalling, p53 signalling and T-cell signalling was decreased. Overall, the long-term benefits of ICS on FEV1 decline in moderate-to-severe COPD were most pronounced in patients with fewer pack-years and less severe emphysema and inflammation [76].

Blood eosinophil counts are a promising biomarker of the response to ICS in COPD and could potentially be used to stratify patients for different exacerbation risk and reduction strategies despite we need more long-term controlled clinical trials in this area to better understand the precise cut-off of the blood eosinophil count to use for this purpose in daily clinical practice. [9, 15, 63].

A short course (at least 3 days) of systemic glucocorticoids (50 mg/daily of prednisone equivalent) is mandatory in patients with moderate-to-severe exacerbations of COPD (goldcopd.org; see also Chap. 23).

32.6.3 Use of Corticosteroids in Inflammatory Bowel Disease (IBD)

The clinical management of IBD aims to induce and maintain remission in patients with active disease. Treatment strategies are complex, consisting of pharmacological treatment and surgery depending on disease location, severity and patients' treatment history. The traditional step-up approach consists of first-line therapy with 'conventional' or standard-of-care treatments such as aminosalicylates, corticosteroids and immunomodulators. Systemic corticosteroids are highly efficacious but burdened by side effects. Thiopurines or methotrexate can be utilised as systemic corticosteroid-sparing agents. Biologic agents targeting TNF-α remain important for systemic corticosteroid-sparing therapy in moderate-to-severe IBD [77, 78].

Topical budesonide provides an alternative to aminosalicylates when targeted to appropriate sites in the distal small bowel and colon, as do conventional corticosteroids when applied rectally [78]. At present, oral budesonide is the first-line therapy to induce remission in microscopic colitis and mild-to-moderate ileocaecal Crohn's disease patients, and oral beclomethasone is effective in treating mild-to-moderate ulcerative colitis patients with left-sided or extensive disease [79].

32.6.4 Use of Corticosteroids in Rhinology

Corticosteroids are widely used today in the field of rhinology. Allergic and non-allergic rhinitis, acute rhinosinusitis and chronic rhinosinusitis with and without nasal polyps are common diseases treated effectively with intranasal glucocorticoids, sometimes associated with intranasal histamine 1 receptor antagonists. Several studies have demonstrated the symptomatic efficacy of intranasal corticosteroids in these diseases and the prevention of nasal polyposis. Oral corticosteroids are usually reserved only to nasal polyps unresponsive to maximal pharmacological treatment with intranasal glucocorticoids associated with oral antileukotrienes [80].

There might be an improvement in symptom severity, polyp size and condition of the sinuses when assessed using CT scans in patients taking oral corticosteroids when these are used as an adjunct therapy to antibiotics or intranasal corticosteroids, but the quality of the evidence supporting this is low or very low. It is also unclear whether the benefits of oral corticosteroids as an adjunct therapy are sustained beyond the short follow-up period reported (up to 30 days), as no longer-term data are available. More research in this area, particularly research evaluating longer-term outcomes, is required [81].

32.6.5 Use of Corticosteroids in Inflammatory Skin Diseases

Topical corticosteroids are the pillars of dermatotherapeutics with hydrocortisone being the first compound used successfully. Modifications in the basic structure generated greater in vivo activity, and thus a very long list of different topically active compounds is now available on the market. Apart from the classic Stoughton skin vasoconstrictor assay, various other methods are used for potency assessment of topical corticosteroids. Topical corticosteroids are classified based upon potency and action of these molecules. There is a significant evidence favouring the use of these drugs in atopic dermatitis, localised vitiligo, psoriasis, contact dermatitis and localised bullous pemphigoid. However, high-level scientific evidence in support of their use in cutaneous lichen planus, cutaneous sarcoidosis and seborrheic dermatitis is not available [82–84].

32.6.6 Use of Corticosteroids in Autoimmune Diseases

The treatment of the autoimmune inflammatory rheumatic diseases and other autoimmune conditions is covered in the chapters on disease-modifying antirheumatic drugs and on drugs for soft tissue autoimmune disorders.

Immunosuppressive therapy in combination with glucocorticoids should be used for the management of antineutrophil cytoplasmic antibody (ANCA)-associated vasculitis according to the European League Against Rheumatism (EULAR) recommendations. For the induction of remission in life-threatening or organ-threatening disease, cyclophosphamide and rituximab are considered to have similar efficacy; in addition, plasma exchange is recommended, where licenced, in the setting of rapidly progressive renal failure or severe diffuse pulmonary haemorrhage [85].

Immune thrombocytopenia (ITP) is characterised by immune-mediated platelet destruction and impaired production, resulting in a platelet count of less than 100,000 mm^{-3} and varying degrees of bleeding risk. ITP is classified as newly diagnosed or transient (less of 3 months of thrombocytopenia), persistent (3–12 months) or chronic (>12 months) on the basis of the time since diagnosis. In primary (or idiopathic) ITP, the thrombocytopenia is isolated, whereas in secondary ITP, it is associated with other disorders (e.g., human immunodeficiency virus infection or systemic lupus erythematosus) [86]. Treatment is rarely indicated in patients with platelet counts above 50,000 mm^{-3} in the absence of the following, bleeding due to platelet dysfunction or another haemostatic defect, trauma, surgery, clearly identified comorbidities for bleeding or mandated anticoagulation therapy, or in persons whose profession or lifestyle predisposes them to trauma [86].

Initial treatment for ITP is generally a course of systemic glucocorticoids, intravenous immune globulin or both. The recommended second-line treatment is splenectomy. Oral, non-peptide thrombopoietin receptor agonists (such as romiplostim and eltrombopag) are used in patients with chronic immune thrombocytopenia who have an insufficient response to glucocorticoids, intravenous immune globulin or splenectomy [86].

Oral prednisone is usually given at 0.5–2 mg/kg/day until the platelet count increases ($\geq$50,000 mm^{-3}), which may require several days to several weeks. To avoid systemic corticosteroid-related complications, prednisone should be rapidly tapered and usually stopped in responders and especially in nonresponders after 4 weeks [86, 87]. Good response rates to systemic corticosteroid therapy range from 65% to 78% for ITP [87].

Autoimmune liver diseases coexist with rheumatic disorders in approximately 30% of cases and may also share pathogenic mechanisms. Autoimmune liver diseases result from an immune-mediated injury of different tissues, with autoimmune hepatitis targeting hepatocytes and primary biliary cholangitis and primary sclerosing cholangitis targeting cholangiocyte. The American Association for the Study of Liver Diseases (AASLD) recommends the use of combination therapy with prednisone (or prednisolone) and azathioprine as first choice treatment. Monotherapy with prednisone is preferred in cases of pregnancy, however. For combination therapy, the induction dose of prednisone is 30 mg daily for 1 week, followed by 20 mg daily for 1 week, followed by 15 mg daily for 2 weeks. The maintenance dose is 10 mg daily until treatment is stopped. For monotherapy, a typical induction dose of prednisone is 60 mg daily for 1 week followed by 40 mg in the second week and 30 mg daily in the third and fourth week. The maintenance dose of prednisone is 20 mg daily until the endpoint or deep clinical remission. Prednisone should be tapered over time and eventually discontinued. AASLD recommends at least 3 years of treatment. Upon completion of prednisone, patients are classified as in remission, relapsed or treatment failure based on their histopathological and laboratory response to corticosteroids and the presence or absence of clinical symptoms. Mycophenolate mofetil, calcineurin inhibitors, mTOR inhibitors and biological agents are reserved for selected nonresponsive patients and administered only in experienced centres. Liver transplantation is a life-saving option for those patients who progress to end-stage liver disease [88–90].

32.6.7 Use of Corticosteroids in Inflammatory Eye Disease

Uncomplicated allergic eye disease can be managed in primary care with one or more cold compresses, lubricants, topical and/or oral antihistamines and topical mast cell stabilisers. Topical and oral corticosteroids and immunomodulatory agents should be prescribed only under the care of an ophthalmologist in refractory cases [91]. Oral corticosteroids and immunosuppression may be a preferred initial therapy for many non-infectious, intermediate, posterior and panuveitides cases [92]. Corticosteroids are also used to prevent and reduce inflammation and the risk of complications following cataract surgery [93].

32.6.8 Use of Corticosteroids in Infections

Despite their potential to decrease the immune responses against most infectious agents (see below), corticosteroids have proven to be efficacious in the management of selected infections in controlled clinical trials.

Topical antibiotics are the best treatment for bacterial keratitis. However, outcomes remain poor with secondary corneal melting, scarring and perforation remaining. Adjuvant therapies aimed at reducing the immune response associated with keratitis include topical corticosteroids. The large, randomised, controlled Steroids for Corneal Ulcers Trial found that although corticosteroids provided no significant improvement overall, they did seem beneficial for ulcers that were central, deep or large, non-*Nocardia* or classically invasive *Pseudomonas aeruginosa*; for patients with low baseline vision; and when started early after the initiation of antibiotics. The Herpetic Eye Disease Study has also shown a significant benefit of topical corticosteroids and oral acyclovir for stromal keratitis [94].

Overall, low-quality evidence indicates that systemic corticosteroids reduce mortality among patients with sepsis. Moderate-quality evidence suggests that a long course of low-dose systemic corticosteroids reduced 28-day mortality without inducing major complications but leading to an increase in metabolic disorders [95]. There is no clear evidence that any one corticosteroid drug or treatment regimen is more likely to be effective in reducing mortality or reducing the incidence of gastrointestinal bleeding or superinfection in septic shock. Hydrocortisone delivered as a bolus or as an infusion is more likely than placebo and methylprednisolone to result in septic shock reversal [96].

In addition to antituberculous chemotherapy, systemic corticosteroids reduce mortality from tuberculous meningitis but may have no effect on the number of people who survive tuberculous meningitis with disabling neurological deficits, and the data on HIV-positive subjects are still insufficient [97]. The treatment of tuberculous pericarditis includes systemic corticosteroids, drainage and surgery. For HIV-negative patients, corticosteroids may reduce death. For HIV-positive patients not on antiretroviral drugs, corticosteroids may reduce constriction [98].

In cases of urinary tuberculosis, ureteral stenosis can deteriorate and result in ureteral obstruction during antituberculosis treatment. Pre-emptive administration of systemic corticosteroids may be beneficial for preventing such stenosis in patients with a pre-existing ureteral lesion [99].

The evidence for a benefit from systemic corticosteroid treatment of pleural tuberculosis is still inconclusive. In addition, the information on the impact of pleural tuberculosis on long-term respiratory function (potentially the most important outcome) is unknown and needs to be quantified to help decide whether further trials of corticosteroids for pleural tuberculosis are worthwhile [100].

The adjunct of systemic corticosteroids to standard care for patients hospitalised with community-acquired pneumonia (CAP) reduces time to clinical stability and length of hospital stay by approximately 1 day without a significant effect on overall mortality but with an increased risk for CAP-related rehospitalisation and hyperglycaemia [101].

The number and size of trials investigating adjunctive systemic corticosteroids for HIV-infected patients with *Pneumocystis jiroveci* pneumonia (PCP) is small, but the evidence suggests a beneficial effect for adult patients with acute respiratory failure, whereas adjunctive corticosteroids did not improve the outcome of *P. jiroveci* pneumonia in non-HIV patients [102, 103]. Aerosolized ribavirin is the first choice treatment for immunocompromised adults with respiratory syncytial virus infections, but the addition of an immunomodulator (intravenous immunoglobulins, corticosteroids and palivizumab) may provide a survival benefit over ribavirin alone [104]. Currently there is not sufficient evidence to determine the effectiveness of corticosteroids for patients with influenza [105].

For patients with bacterial arthritis, corticosteroids are also beneficial and reduce long-term disability. Dexamethasone did not show a clear decrease in cough episodes and length of hospital stay in patients with pertussis [106]. However, corticosteroids are harmful in viral hepatitis and cerebral malaria [107].

32.6.9 Use of Corticosteroids in Other Inflammatory and Neoplastic Diseases

Low-dose hydrocortisone or a corresponding low-dose corticosteroid therapy may improve morbidity in specific target groups of critically ill patients [such as patients with acute respiratory distress syndrome (ARDS) or burns]. However beneficial effects on mortality remain to be demonstrated in large-scale randomised controlled trials [108]. There is moderate-quality evidence that suggests there is no effect of corticosteroids on critical illness polyneuropathy or myopathy (a frequent complication in the intensive care units) and high-quality evidence that corticosteroids do not affect secondary outcomes, except for fewer new shock episodes [109].

Improvements in lung function have been reported for long-term treatment with ICSs of adult patients with non-CF bronchiectasis though this study was small and not clinically relevant. Improvements in dyspnoea, wheeze and cough-free days are reported in small trials of ICSs and ICS/LABA combinations for the same patients [110].

Primary therapy of allergic bronchopulmonary aspergillosis (ABPA) consists of oral corticosteroids to control exacerbations, itraconazole as a systemic corticosteroid-sparing agent and optimised asthma therapy [111].

Oral corticosteroid treatment of patients with sarcoidosis improves the chest X-ray symptoms and spirometry over 3–24 months. However, there is little evidence of an improvement in lung function. There are limited data beyond 2 years to indicate whether oral corticosteroids have any modifying effect on long-term disease progression. For these reasons oral corticosteroids may be of benefit for patients with stage 2 and 3 sarcoidosis with moderate-to-severe or progressive symptoms or chest X-ray changes [112].

Most patients with sarcoidosis are not disabled by their illness, so the decision to provide treatment should reflect a weighing of the risks of using corticosteroids, the most common therapy, against the potential benefits. A general rule is to consider instituting treatment when organ function is threatened. Detection of granulomatous disease on physical examination, biopsy, imaging studies or serologic testing is not a mandate to provide treatment. Systemic corticosteroids remain the first-line treatment. Other immunosuppressive agents may be considered for the patients who respond poorly to corticosteroids or who experience significant adverse effects. An international expert panel has suggested initiating treatment with oral prednisone at a dose of 20–40 mg/day. The panel recommends evaluating the response to treatment after 1–3 months. If there has been a response, the prednisone dose should be tapered to 5–15 mg/day, with treatment planned for an additional 9–12 months. Lack of a response after 3 months suggests the presence of irreversible fibrotic disease, nonadherence to therapy or an inadequate dose of prednisone [113, 114].

Neurological and cardiac involvement in sarcoidosis are rare but two of the major causes of death in these patients. One-third of patients with neurosarcoidosis do not respond to treatment with systemic corticosteroids or other therapies [115]. There is a clear need for large multicentre prospective registries and trials in patients with cardiac sarcoidosis [116].

At present, there is no evidence for an effect of corticosteroid treatment in patients with idiopathic pulmonary fibrosis(IPF)/usual interstitial pneumonia (UIP) [117]. In contrast, the prognosis of nonspecific interstitial pneumonia (NSIP) is good compared with IPF/UIP, because usually these patients respond to systemic corticosteroids and/or immunosuppressive agents [118].

Because corticosteroids act through a variety of mechanisms to inhibit eosinophil function and induce apoptosis, they are first-line therapy for primary (or idiopathic) eosinophilic lung diseases, including primary acute eosinophilic pneumonia, primary chronic eosinophilic pneumonia (Carrington's disease), eosinophilic granulomatosis with polyangiitis [(EGPA), also known as Churg-Strauss syndrome] and primary hypereosinophilic syndrome (IHES). In these disorders elevated levels of activated eosinophil in the lung tissue lead to inflammation and tissue damage [119].

Hypersensitivity pneumonitis, also known as extrinsic allergic alveolitis, is a complex immunopathological pulmonary syndrome caused by inhalation of a wide variety of antigens to which the individual has been previously sensitised. The treatment of first choice is represented by the avoidance of inciting antigens. Systemic corticosteroids may be useful in acute episodes for symptomatic relief or in chronic and progressive disease, but their long-term efficacy has never been validated in prospective clinical trials [120].

Langerhans cell histiocytosis (LCH) limited to the lungs is low-grade histiocytic neoplasm and usually responds well to complete smoking cessation. The current standard of care for multisystem LCH is empirically derived chemotherapy with vinblastine and prednisone but cures fewer than 50% of patients, and optimal therapies for relapse and neurodegenerative disease remain uncertain [121].

In patients with lung cancer, drug-induced interstitial lung disease has a more unfavourable outcome but requires higher-dose systemic corticosteroid therapy as compared with those with radiation-induced pneumonitis [122].

There are short-term benefits in using systemic corticosteroids for the symptomatic treatment of cancer-related fatigue and anorexia cachexia in advanced incurable cancer. Future studies are needed to determine the optimal dose, the type and the role of corticosteroid rotation so as to optimise long-term efficacy and minimise side effects [123].

32.7 Side Effects of Corticosteroid Therapy

Overall, the duration, dosage and dosing regime, the particular corticosteroid used and its mode of administration along with the patient's individual susceptibility appear to determine the incidence of adverse events. Not surprisingly, side effects are much more severe with systemic corticosteroid use, although even topical application can induce both local and systemic side effects (Table 32.1). When corticosteroids are administered by systemic routes over a long period of time at reasonably high concentrations, their beneficial effects are often overshadowed by a number of side effects (iatrogenic Cushing syndrome) [124].

Side effects of oral corticosteroids include skin and muscle atrophy, delayed wound healing and increased risk of infections, OSTEOPOROSIS and bone necrosis, glaucoma and cataracts, behavioural changes, hypertension, peptic ulcers, gastrointestinal bleeding and diabetes mellitus. Interestingly, it appears that early skin atrophy induced by corticosteroid therapy is reversible, whereas major atrophy leading to striae formation is not [124]. These side effects often occur together, and this is exemplified by Cushing's syndrome (hypercortisolism), the signs and symptoms of which include elevated systemic arterial blood pressure, development of diabetes mellitus, pink-to-purple striae (stretch marks) on the abdominal skin, fatigue, depression, moodiness and accentuated accumulation of fatty tissue on the face and upper back (Buffalo hump). Women with Cushing's syndrome often have irregular menstrual periods and develop new facial hair growth. Men may show a decrease in sex drive [9].

Commonly cited side effects associated with long-term systemic corticosteroid exposure included systemic arterial hypertension (prevalence >30%); bone fractures (21–30%); cataract (1–3%); nausea, vomiting, and other gastrointestinal conditions (1–5%); and metabolic issues (e.g. weight gain, hyperglycaemia and type 2 diabetes mellitus where cases had fourfold risk of controls). However, the association of dose and duration with increased adverse effect risk is not well-quantified [125]. Systemic glucocorticoid long-term use is also associated with an increased risk of bacterial, fungal and viral infections, including tuberculosis, disseminated strongyloidiasis and hepatitis B virus and hepatitis C virus

Table 32.1 Potential tissue-/organ-specific side effects of topical and/or systemic corticosteroids

Cardiovascular system and metabolism:
Systemic arterial hypertension
Dyslipidaemia (increased VLDL and/or LDL serum levels)
Hypercoagulability with thrombosis
Central nervous system:
Disturbances in mood, behaviour, memory and cognition, appetite increase
Decreased fatigue, "steroid psychosis", steroid dependence
Endocrine and renal systems:
Iatrogenic Cushing's syndrome
Iatrogenic diabetes mellitus
Iatrogenic Addison's disease
Growth retardation
Hypogonadism, delayed puberty
Increased sodium retention and potassium excretion
Eye:
Glaucoma
Cataract
Gastrointestinal:
Peptic ulcer
Increased risk of gastrointestinal bleeding
Hepatic steatosis
Acute pancreatitis
Immune system:
Increased risk of infections
Re-activation of latent viral (varicella-zoster virus), bacterial (*M. tuberculosis*) or helminthic (*S. stercoralis*) infections
Skeleton and muscle:
Muscle atrophy/myopathy
Osteoporosis and bone fractures
Bone necrosis
Skin and subcutaneous:
Weight gain, skin thinning, striae rubrae
Delayed wound healing
Steroid acne, perioral dermatitis
Erythema, telangiectasia, purpura, hypertrichosis
Adipose tissue:
Adipose tissue expansion and redistribution [visceral obesity, subcutaneous localised adiposity (buffalo hump)]

VLDL very-low-density lipoproteins, *LDL* low-density lipoproteins

reactivation. Patients with undiagnosed and untreated HIV infection may be at increased risk of developing infectious complications with the initiation of chronic glucocorticoid therapy [126].

Patients with conditions necessitating moderate or high doses of systemic corticosteroids (≥20 mg/day of prednisone) for >2 weeks should be asked about vaccination history to ensure that they are up to date on the following vaccinations: *Haemophilus influenzae* B; hepatitis A virus (HAV) and HBV; human papillomavirus; influenza; *Neisseria meningitidis*; measles, mumps and rubella (MMR); *Streptococcus pneumoniae*; and tetanus. Patients >50–60 years of age (with or without a history of varicella-zoster virus infection) who have not received the varicella-zoster virus vaccine should receive it, if possible, at least 2–4 weeks before the initiation of moderate- or high-dose systemic glucocorticoids. Patients taking ≤20 mg prednisone per day (or the equivalent) can safely receive the zoster vaccine at any time [126]. In addition, invasive aspergillosis is increasingly found in COPD patients during exacerbations treated with systemic corticosteroids [127].

All currently available topical corticosteroids are absorbed into the systemic circulation and, therefore, inevitably have some systemic effect, although this is considerably less than those seen with systemic corticosteroids (Table 32.1). The occurrence and severity of the side effects seen depend upon the duration of use, dosage, dosing regime and specific drug used, along with individual patient variability. However, the highest risk factor appears to be prolonged use. Side effects of topical corticosteroids are tissue-dependent and include glaucoma, cataracts, tissue atrophy and delayed wound healing, whilst at high doses, there is an increased risk of local and systemic infections, adrenal suppression (iatrogenic Addison's disease) and osteoporosis (and bone fractures). The growth retardation seen with oral corticosteroids does not appear to be a problem with modern topical corticosteroids, although there may be an initial reduction in growth velocity on starting therapy [125]. Currently, most patients with asthma and COPD are treated with ICSs, whereas the systemic preparations are being limited to patients with severe asthma or during exacerbations of these diseases which reduces the incidence of side effects (see above and Chap. 23).

Modern inhaled GCs (ICS) have high receptor affinity, are retained in the airways and are rapidly metabolised after absorption from the gastrointestinal tract which accounts for their good safety profile even when used in more severe asthmatics and at high nebulised doses during exacerbations of asthma or COPD. The side effects seen with ICSs are usually limited to oropharyngeal candidiasis and/or dysphonia. In general, ICS treatment efficacy and side effects are directly related to tissue dose although there is some evidence that this may vary with the drug and patient profile [9, 15].

The incidence of pneumonia is slightly increased in COPD patients treated with inhaled glucocorticoids compared to placebo in most studies, regardless of the type of glucocorticoid inhalation used, suggesting a class effect. Older age, low body mass index, low FEV_1 and being a current smoker are all factors variously associated with increased risk of pneumonia [128]. Compared with non-ICS treatment, ICS treatment also seems associated with a significantly higher risk of tuberculosis (OR, 2.29) but not influenza. The number needed to harm to cause one additional tuberculosis event is lower for patients with COPD treated with ICSs in endemic areas than for those in nonendemic areas (909 vs. 1667, respectively) [129]. Furthermore, ICS use may increase the risk of upper respiratory tract infection (URTI) in patients with COPD, but this should be further investigated [130].

Second-generation intranasal corticosteroids have pharmacokinetic characteristics that minimise their systemic bioavailability, resulting in minimum risk for systemic adverse events [80].

Guidelines for the management of asthma recommend the continued use of ICSs in pregnancy, with budesonide having a particularly good safety profile. Recent data suggest small effects of asthma and/or asthma medication use on congenital malformations; however, there is less data available on the safety of ICS/LABA combinations, which are increasingly used for maintenance treatment [131]. The risk-benefit ratio should always be considered before prescribing any intranasal corticosteroid sprays during pregnancy. Lacking sufficient clinical trials on the use of intranasal corticosteroid sprays in pregnancy, it has been suggested that the intranasal use of fluticasone furoate, mometasone and budesonide is safe if they are used at the recommended therapeutic dose after a proper medical evaluation. Intranasal fluticasone propionate might also be a safe option in the absence of other intranasal corticosteroids [132].

Similarly, no association between maternal use of topical corticosteroids for dermatological use of any potency and an increase in adverse pregnancy outcomes, including mode of delivery, congenital abnormality, preterm delivery, foetal death and low Apgar score, have been noted. However a probable association between low birth weight and maternal use of potent to very potent skin topical corticosteroids, especially when the cumulative dosage of topical corticosteroids throughout the pregnancy is very large, warrants further investigation [133]. Besides, evidence clearly advocates judicious use of mild-to-moderate topical corticosteroids for dermatological use (if required) in pregnancy and lactation, and there is no risk of any foetal abnormality [133].

In dermatology, 'steroid phobia' is still a considerable concern in connection with topical corticosteroids, particularly with regard to skin thinning (the more potent drugs) and systemic absorption with effects on growth and development. Tachyphylaxis and allergic contact dermatitis (due to a delayed type allergic reaction) are the real potential problems in clinical practice when using topical corticosteroids for dermatological diseases [83, 84, 134]. It should also be noted that rare allergic hypersensitivity to topical corticosteroids applied to other sites than skin and even to systemic corticosteroids has been described [134].

Taken together, the side effects seriously limit the value of long-term treatment with systemic corticosteroids in severe inflammation where the risk/benefit ratio is compromised. This has driven the need to develop novel agents with the anti-inflammatory capacity of corticosteroids but with reduced side effects.

While the major anti-inflammatory effects of corticosteroids are almost certainly due to transrepression, the underlying molecular mechanisms for the side effects of corticosteroids are complex and not fully understood. Certain side effects such as diabetes mellitus and glaucoma are due to transactivation events, whilst others are due to transrepression (hypothalamic-pituitary axis, HPA, suppression). In addition, the precise molecular events underlying corticosteroid induction of osteoporosis is unclear but probably requires both gene induction and gene repression.

Despite this uncertainty, there has been a search for 'dissociated' corticosteroids that selectively transrepress without significant transactivation, thus potentially reducing the risk of systemic side effects. Several SEGRAMs have been reported which show dissociated properties in human cells and are now in clinical development, showing good separation between transrepression and transactivation actions in the skin. This suggests that the development of SEGRAMs with a greater margin of safety is possible and may even lead to the development of oral compounds that have reduced adverse effects. Furthermore, the newer inhaled corticosteroids used today, such as fluticasone (both propionate and furoate), mometasone and budesonide, appear to have more potent transrepressing than transactivating effects, which may account, at least in part, for their selection as potent anti-inflammatory agents [9, 15]. These new potent corticosteroids are particularly effective as topical agents, and their use has overtaken that of oral/systemic corticosteroids for many diseases. An alternative approach to obtain safer drugs is the use of soft drugs, such as ciclesonide, which are only activated at the site of inflammation. Ciclesonide is a novel inhaled corticosteroid for the treatment of asthma. Ciclesonide itself is inactive and needs to be cleaved by lung-specific esterases in order to bind to the GR. According to data from healthy volunteers and asthma patients, ciclesonide affects serum cortisol levels significantly less compared to beclomethasone dipropionate or fluticasone propionate, indicating that ciclesonide might have less systemic effects and hence a superior safety profile [9, 15].

32.8 Summary

Corticosteroids are the most effective therapy for chronic immune and inflammatory diseases in current use. Despite their success over the past 70 years, and especially since the advent of new potent halogenated compounds, worries about the detrimental side effects of systemic corticosteroids have limited their long-term use. This has resulted in the increasing use of topical corticosteroids targeted to the site of inflammation rather than systemic administration. Improvements in risk/benefit ratios are likely to occur, as greater understanding of the role of chemical substitution of the synthetic corticosteroids becomes clear and more potent tissue selective drugs are developed. Drugs that target distinct aspects of corticosteroid function, switching on or off genes, are also under development and, along with non-steroidal agents that target different aspects of the inflammatory response, are likely to lead to safer drugs with a much reduced side effect profile. However, until these become widely available, current systemic and topical corticosteroids are likely to remain the major treatment for most acute and chronic inflammatory diseases.

Recommended Websites

https://www.ncbi.nlm.nih.gov/pubmed
http://www.cochranelibrary.com/

References

1. Cutright A, Ducey S, Barthold CL, Kim J. Recognizing and managing adrenal disorders in the emergency department. Emerg Med Pract. 2017;9:1–24.
2. Pazderska A, Pearce SH. Adrenal insufficiency-recognition and management. Clin Med (Lond). 2017;17:258–62.
3. Collomp K, Baillot A, Forget H, Coquerel A, Rieth N, Vibarel-Rebot N. Altered diurnal pattern of steroid hormones in relation to various behaviors, external factors and pathologies: a review. Physiol Behav. 2016;164:68–85.
4. Gamble KL, Berry R, Frank SJ, Young ME. Circadian clock control of endocrine factors. Nat Rev Endocrinol. 2014;10:466–75.
5. Bedrosian TA, Fonken LK, Nelson RJ. Endocrine effects of circadian disruption. Annu Rev Physiol. 2016;78:109–31.
6. Raju TN. The Nobel chronicles. 1950: Edward Calvin Kendall (1886–1972); Philip Showalter Hench (1896–1965); and Tadeus Reichstein (1897–1996). Lancet. 1999;353:1370.
7. Caramori G, Casolari P, Adcock I. Role of transcription factors in the pathogenesis of asthma and COPD. Cell Commun Adhes. 2013;20:21–40.
8. Johnson M. Pharmacodynamics and pharmacokinetics of inhaled glucocorticoids. J Allergy Clin Immunol. 1996;97:169–76.
9. Adcock IM, Mumby S. Glucocorticoids. Handb Exp Pharmacol. 2017;237:171–96.
10. Daley-Yates PT. Inhaled corticosteroids: potency, dose equivalence and therapeutic index. Br J Clin Pharmacol. 2015;80:372–80.
11. Sundahl N, Bridelance J, Libert C, De Bosscher K, Beck IM. Selective glucocorticoid receptor modulation: new directions with non-steroidal scaffolds. Pharmacol Ther. 2015;152:28–41.
12. Ramamoorthy S, Cidlowski JA. Corticosteroids: mechanisms of action in health and disease. Rheum Dis Clin North Am. 2016;42:15–31.
13. Adcock IM, Marwick J, Casolari P, Contoli M, Chung KF, Kirkham P, Papi A, Caramori G. Mechanisms of corticosteroid resistance in severe asthma and chronic obstructive pulmonary disease (COPD). Curr Pharm Des. 2010;16:3554–73.
14. Kino T. Glucocorticoid receptor. In: De Groot LJ, Chrousos G, Dungan K, Feingold KR, Grossman A, Hershman JM, Koch C, Korbonits M, McLachlan R, New M, Purnell J, Rebar R, Singer F, Vinik A, editors. Endotext. South Dartmouth, MA: MDText.com, Inc.; 2017. p. 2000.
15. Barnes PJ. Glucocorticosteroids. Handb Exp Pharmacol. 2017;237:93–115.
16. Marwick JA, Caramori G, Stevenson CS, Casolari P, Jazrawi E, Barnes PJ, Ito K, Adcock IM, Kirkham PA, Papi A. Inhibition of PI3Kdelta restores glucocorticoid function in smoking-induced airway inflammation in mice. Am J Respir Crit Care Med. 2009;179:542–8.
17. Maneechotesuwan K, Yao X, Ito K, Jazrawi E, Usmani OS, Adcock IM, Barnes PJ. Suppression of GATA-3 nuclear import and phosphorylation: a novel mechanism of corticosteroid action in allergic disease. PLoS Med. 2009;6:e1000076.
18. Hakim A, Barnes PJ, Adcock IM, Usmani OS. Importin-7 mediates glucocorticoid receptor nuclear import and is impaired by oxidative stress, leading to glucocorticoid insensitivity. FASEB J. 2013;27:4510–9.
19. Chang PJ, Michaeloudes C, Zhu J, Shaikh N, Baker J, Chung KF, Bhavsar PK. Impaired nuclear translocation of the glucocorticoid receptor in corticosteroid-insensitive airway smooth muscle in severe asthma. Am J Respir Crit Care Med. 2015;191:54–62.

20. Scheschowitsch K, Leite JA, Assreuy J. New insights in glucocorticoid receptor signaling-more than just a ligand-binding receptor. Front Endocrinol (Lausanne). 2017;8:16.
21. Anbalagan M, Huderson B, Murphy L, Rowan BG. Post-translational modifications of nuclear receptors and human disease. Nucl Recept Signal. 2012;10:e001.
22. Beck IM, Vanden Berghe W, Vermeulen L, Yamamoto KR, Haegeman G, De BK. Crosstalk in inflammation: the interplay of glucocorticoid receptor-based mechanisms and kinases and phosphatases. Endocr Rev. 2009;30:830–82.
23. Ito K, Yamamura S, Essilfie-Quaye S, Cosio B, Ito M, Barnes PJ, Adcock IM. Histone deacetylase 2-mediated deacetylation of the glucocorticoid receptor enables NF-kappaB suppression. J Exp Med. 2006;203:7–13.
24. Antunica-Noguerol M, Budziñski ML, Druker J, Gassen NC, Sokn MC, Senin S, Aprile-Garcia F, Holsboer F, Rein T, Liberman AC, Arzt E. The activity of the glucocorticoid receptor is regulated by SUMO conjugation to FKBP51. Cell Death Differ. 2016;23:1579–91.
25. Hua G, Ganti KP, Chambon P. Glucocorticoid-induced tethered transrepression requires SUMOylation of GR and formation of a SUMO-SMRT/NCoR1-HDAC3 repressing complex. Proc Natl Acad Sci U S A. 2016;113:E635–43.
26. Hua G, Paulen L, Chambon P. GR SUMOylation and formation of an SUMO-SMRT/NCoR1-HDAC3 repressing complex is mandatory for GC-induced IR nGRE-mediated transrepression. Proc Natl Acad Sci U S A. 2016;113:E626–34.
27. Kadiyala V, Sasse SK, Altonsy MO, Berman R, Chu HW, Phang TL, Gerber AN. Cistrome-based cooperation between airway epithelial glucocorticoid receptor and NF-κB orchestrates anti-inflammatory effects. J Biol Chem. 2016;291:12673–87.
28. Adcock IM, Caramori G. Cross-talk between pro-inflammatory transcription factors and glucocorticoids. Immunol Cell Biol. 2001;79:376–84.
29. Meijsing SH, Pufall MA, So AY, Bates DL, Chen L, Yamamoto KR. DNA binding site sequence directs glucocorticoid receptor structure and activity. Science. 2009;324:407–10.
30. Xavier AM, Anunciato AK, Rosenstock TR, Glezer I. Gene expression control by glucocorticoid receptors during innate immune responses. Front Endocrinol (Lausanne). 2016;7:31.
31. Biddie SC, Conway-Campbell BL, Lightman SL. Dynamic regulation of glucocorticoid signalling in health and disease. Rheumatology (Oxford). 2012;51:403–12.
32. Qiu W, Guo F, Glass K, Yuan GC, Quackenbush J, Zhou X, Tantisira KG. Differential connectivity of gene regulatory networks distinguishes corticosteroid response in asthma. J Allergy Clin Immunol. 2017;pii:S0091-6749(17)31109-0. https://doi.org/10.1016/j.jaci.2017.05.052.
33. Vockley CM, D'Ippolito AM, McDowell IC, Majoros WH, Safi A, Song L, Crawford GE, Reddy TE. Direct GR binding sites potentiate clusters of TF binding across the human genome. Cell. 2016;166:1269–81.
34. Li LB, Leung DY, Martin RJ, Goleva E. Inhibition of histone deacetylase 2 expression by elevated glucocorticoid receptor beta in steroid resistant asthma. Am J Respir Crit Care Med. 2010;182:877–83.
35. Ito K, Chung K, Adcock IM. Update on glucocorticoid action and resistance. J Allergy Clin Immunol. 2006;117:522–43.
36. Barnes PJ, Adcock IM. Glucocorticoid resistance in inflammatory diseases. Lancet. 2009;373:1905–17.
37. Langlais D, Couture C, Balsalobre A, Drouin J. The Stat3/GR interaction code: predictive value of direct/indirect DNA recruitment for transcription outcome. Mol Cell. 2012;47:38–49.
38. O'Shea JJ, Plenge R. JAK and STAT signaling molecules in immunoregulation and immune-mediated disease. Immunity. 2012;36:542–50.
39. Fenwick PS, Macedo P, Kilty IC, Barnes PJ, Donnelly LE. Effect of JAK inhibitors on release of CXCL9, CXCL10 and CXCL11 from human airway epithelial cells. PLoS One. 2015;10:e0128757.
40. Norman AW, Mizwicki MT, Norman DP. Steroid-hormone rapid actions, membrane receptors and a conformational ensemble model. Nat Rev Drug Discov. 2004;3:27–41.
41. Levin ER. Rapid signaling by steroid receptors. Am J Physiol Regul Integr Comp Physiol. 2008;295:R1425–30.
42. Liyanage CK, Galappatthy P, Seneviratne SL. Corticosteroids in management of anaphylaxis; a systematic review of evidence. Eur Ann Allergy Clin Immunol. 2017;49:196–207.
43. Caramori G, Adcock I. Pharmacology of airway inflammation in asthma and COPD. Pulm Pharmacol Ther. 2003;16:247–77.
44. Adcock IM, Chung KF. Overview: why are corticosteroids ineffective in COPD? Curr Opin Investig Drugs. 2002;3:58–60.
45. Zen M, Canova M, Campana C, Bettio S, Nalotto L, Rampudda M, Ramonda R, Iaccarino L, Doria A. The kaleidoscope of glucorticoid effects on immune system. Autoimmun Rev. 2011;10:305–10.
46. McColl A, Michlewska S, Dransfield I, Rossi AG. Effects of glucocorticoids on apoptosis and clearance of apoptotic cells. ScientificWorldJournal. 2007;7:1165–81.
47. Wilson JW, Djukanovic R, Howarth PH, Holgate ST. Inhaled beclomethasone dipropionate downregulates airway lymphocyte activation in atopic asthma. Am J Respir Crit Care Med. 1994;149:86–90.
48. Schleimer RP, Bochner BS. The effects of glucocorticoids on human eosinophils. J Allergy Clin Immunol. 1994;94:1202–13.
49. Whiteford JR, De Rossi G, Woodfin A. Mutually supportive mechanisms of inflammation and vascular remodeling. Int Rev Cell Mol Biol. 2016;326:201–78.

50. Poletti D, Iannini V, Casolari P, Contoli M, Papi A, Kirkham P, Hansel TT, Chung KF, Barnes PJ, Pastore A, Pelucchi S, Adcock IM, Caramori G. Nasal inflammation and its response to local glucocorticoid regular treatment in patients with persistent non-allergic rhinitis: a pilot study. J Inflamm (Lond). 2016;13:26.
51. Caramori G, Adcock IM, Di Stefano A, Chung KF. Cytokine inhibition in the treatment of COPD. Int J Chron Obstruct Pulmon Dis. 2014;9:397–412.
52. Caramori G, Di Stefano A, Casolari P, Kirkham PA, Padovani A, Chung KF, Papi A, Adcock IM. Chemokines and chemokine receptors blockers as new drugs for the treatment of chronic obstructive pulmonary disease. Curr Med Chem. 2013;20:4317–49.
53. Cantini F, Niccoli L, Nannini C, Cassarà E, Kaloudi O, Giulio Favalli E, Becciolini A, Biggioggero M, Benucci M, Li Gobbi F, Grossi V, Infantino M, Meacci F, Manfredi M, Guiducci S, Bellando-Randone S, Matucci-Cerinic M, Foti R, Di Gangi M, Mosca M, Tani C, Palmieri F, Goletti D, Italian board for the TAilored BIOlogic therapy (ITABIO). Tailored first-line biologic therapy in patients with rheumatoid arthritis, spondyloarthritis, and psoriatic arthritis. Semin Arthritis Rheum. 2016;45:519–32.
54. Park SH, Aniwan S, Loftus EV Jr. Advances in the use of biologics and other novel drugs for managing inflammatory bowel disease. Curr Opin Pharmacol. 2017;37:65–71.
55. Peters-Golden M, Sampson AP. Cysteinyl leukotriene interactions with other mediators and with glucocorticosteroids during airway inflammation. J Allergy Clin Immunol. 2003;111(Suppl 1):S37–42; discussion S43–8
56. Zaitsu M, Hamasaki Y, Aoki Y, Miyazaki S. A novel pharmacologic action of glucocorticosteroids on leukotriene C4 catabolism. J Allergy Clin Immunol. 2001;108:122–4.
57. Reinke SN, Gallart-Ayala H, Gómez C, Checa A, Fauland A, Naz S, Kamleh MA, Djukanović R, Hinks TS, Wheelock CE. Metabolomics analysis identifies different metabotypes of asthma severity. Eur Respir J. 2017;49(3):1601740. https://doi.org/10.1183/13993003.01740-2016.
58. Kirkham PA, Barnes PJ. Oxidative stress in COPD. Chest. 2013;144:266–73.
59. Heijink I, van Oosterhout A, Kliphuis N, Jonker M, Hoffmann R, Telenga E, Klooster K, Slebos DJ, ten Hacken N, Postma D, van den Berge M. Oxidant-induced corticosteroid unresponsiveness in human bronchial epithelial cells. Thorax. 2014;69:5–13.
60. Thomson NC, Spears M. The influence of smoking on the treatment response in patients with asthma. Curr Opin Allergy Clin Immunol. 2005;5:57–63.
61. Adcock IM, Caramori G, Chung KF. New targets for drug development in asthma. Lancet. 2008;372:1073–87.
62. Osoata GO, Yamamura S, Ito M, Vuppusetty C, Adcock IM, Barnes PJ, Ito K. Nitration of distinct tyrosine residues causes inactivation of histone deacetylase 2. Biochem Biophys Res Commun. 2009;384:366–71.
63. Caramori G, Ruggeri P, Casolari P, Chung KF, Girbino G, Adcock IM. Fluticasone furoate and vilanterol for the treatment of chronic obstructive pulmonary disease. Expert Rev Respir Med. 2017;11:955–67.
64. Chung KF, Wenzel SE, Brozek JL, Bush A, Castro M, Sterk PJ, Adcock IM, Bateman ED, Bel EH, Bleecker ER, Boulet LP, Brightling C, Chanez P, Dahlen SE, Djukanovic R, Frey U, Gaga M, Gibson P, Hamid Q, Jajour NN, Mauad T, Sorkness RL, Teague WG. International ERS/ATS guidelines on definition, evaluation and treatment of severe asthma. Eur Respir J. 2014;43:343–73.
65. Chung KF. Clinical management of severe therapy-resistant asthma. Expert Rev Respir Med. 2017;11:395–402.
66. Papi A, Contoli M, Adcock IM, Bellettato C, Padovani A, Casolari P, Stanciu LA, Barnes PJ, Johnston SL, Ito K, Caramori G. Rhinovirus infection causes steroid resistance in airway epithelium through nuclear factor κB and c-Jun N-terminal kinase activation. J Allergy Clin Immunol. 2013;132:1075–1085.e6.
67. Pickup ME. Clinical pharmacokinetics of prednisone and prednisolone. Clin Pharmacokinet. 1979;4:111–28.
68. Swartz SL, Dluhy RG. Corticosteroids: clinical pharmacology and therapeutic use. Drugs. 1978;16:238–55.
69. Colice GL. Comparing inhaled corticosteroids. Respir Care. 2000;45:846–53.
70. Edsbacker S. Pharmacological factors that influence the choice of inhaled corticosteroids. Drugs. 1999;58(suppl 4):7–16.
71. Kwapisz L, Jairath V, Khanna R, Feagan B. Pharmacokinetic drug evaluation of budesonide in the treatment of Crohn's disease. Expert Opin Drug Metab Toxicol. 2017;13:793–801.
72. Mukker JK, Singh RS, Derendorf H. Ciclesonide: a pro-soft drug approach for mitigation of side effects of inhaled corticosteroids. J Pharm Sci. 2016;105:2509–14.
73. Edsbäcker S, Brattsand R. Budesonide fatty-acid esterification: a novel mechanism prolonging binding to airway tissue. Review of available data. Ann Allergy Asthma Immunol. 2002;88:609–16.
74. O'Connor D, Adams WP, Chen ML, Daley-Yates P, Davis J, Derendorf H, Ducharme MP, Fuglsang A, Herrle M, Hochhaus G, Holmes SM, Lee SL, Li BV, Lyapustina S, Newman S, Oliver M, Patterson B, Peart J, Poochikian G, Roy P, Shah T, Singh GJ, Sharp SS. Role of pharmacokinetics in establishing bioequivalence for orally inhaled drug products: workshop summary report. J Aerosol Med Pulm Drug Deliv. 2011;24:119–35.
75. Dolovich MB, Dhand R. Aerosol drug delivery: developments in device design and clinical use. Lancet. 2011;377:1032–45.

76. Snoeck-Stroband JB, Lapperre TS, Sterk PJ, Hiemstra PS, Thiadens HA, Boezen HM, Ten Hacken NH, Kerstjens HA, Postma DS, Timens W, Sont JK, GLUCOLD Study Group. Prediction of long-term benefits of inhaled steroids by phenotypic markers in moderate-to-severe COPD: a randomized controlled trial. PLoS One. 2015;10:e0143793.
77. Pillai N, Dusheiko M, Burnand B, Pittet V. A systematic review of cost-effectiveness studies comparing conventional, biological and surgical interventions for inflammatory bowel disease. PLoS One. 2017;12:e0185500.
78. Chang S, Hanauer S. Optimizing pharmacologic management of inflammatory bowel disease. Expert Rev Clin Pharmacol. 2017;10:595–607.
79. Nunes T, Barreiro-de Acosta M, Marin-Jiménez I, Nos P, Sans M. Oral locally active steroids in inflammatory bowel disease. J Crohns Colitis. 2013;7:183–91.
80. Karatzanis A, Chatzidakis A, Milioni A, Vlaminck S, Kawauchi H, Velegrakis S, Prokopakis E. Contemporary use of corticosteroids in rhinology. Curr Allergy Asthma Rep. 2017;17:11.
81. Chong LY, Head K, Hopkins C, Philpott C, Schilder AG, Burton MJ. Intranasal steroids versus placebo or no intervention for chronic rhinosinusitis. Cochrane Database Syst Rev. 2016;4:CD011996.
82. Humbert P, Guichard A. The topical corticosteroid classification called into question: towards a new approach. Exp Dermatol. 2015;24:393–5.
83. Mehta AB, Nadkarni NJ, Patil SP, Godse KV, Gautam M, Agarwal S. Topical corticosteroids in dermatology. Indian J Dermatol Venereol Leprol. 2016;82:371–8.
84. Das A, Panda S. Use of topical corticosteroids in dermatology: an evidence-based approach. Indian J Dermatol. 2017;62:237–50.
85. Yates M, Watts RA, Bajema IM, Cid MC, Crestani B, Hauser T, Hellmich B, Holle JU, Laudien M, Little MA, Luqmani RA, Mahr A, Merkel PA, Mills J, Mooney J, Segelmark M, Tesar V, Westman K, Vaglio A, Yalçındağ N, Jayne DR, Mukhtyar C. EULAR/ERA-EDTA recommendations for the management of ANCA-associated vasculitis. Ann Rheum Dis. 2016;75:1583–94.
86. Imbach P, Crowther M. Thrombopoietin-receptor agonists for primary immune thrombocytopenia. N Engl J Med. 2011;365:734–41.
87. Provan D, Stasi R, Newland AC, Blanchette VS, Bolton-Maggs P, Bussel JB, Chong BH, Cines DB, Gernsheimer TB, Godeau B, Grainger J, Greer I, Hunt BJ, Imbach PA, Lyons G, McMillan R, Rodeghiero F, Sanz MA, Tarantino M, Watson S, Young J, Kuter DJ. International consensus report on the investigation and management of primary immune thrombocytopenia. Blood. 2010;115:168–86.
88. Linzay CD, Pandit S. Hepatitis, autoimmune. Treasure Island, FL: StatPearls Publishing; 2017.
89. Liberal R, Krawitt EL, Vierling JM, Manns MP, Mieli-Vergani G, Vergani D. Cutting edge issues in autoimmune hepatitis. J Autoimmun. 2016;75:6–19.
90. Terziroli Beretta-Piccoli B, Mieli-Vergani G, Vergani D. Autoimmune hepatitis: standard treatment and systematic review of alternative treatments. World J Gastroenterol. 2017;23:6030–48.
91. Patel DS, Arunakirinathan M, Stuart A, Angunawela R. Allergic eye disease. BMJ. 2017;359:j4706.
92. Jabs DA. Immunosuppression for the uveitides. Ophthalmology. 2017;pii:S0161-6420(17)31557-9. https://doi.org/10.1016/j.ophtha.2017.08.007.
93. Juthani VV, Clearfield E, Chuck RS. Non-steroidal anti-inflammatory drugs versus corticosteroids for controlling inflammation after uncomplicated cataract surgery. Cochrane Database Syst Rev. 2017;7:CD010516.
94. Austin A, Lietman T, Rose-Nussbaumer J. Update on the management of infectious keratitis. Ophthalmology. 2017;124:1678–89.
95. Annane D, Bellissant E, Bollaert PE, Briegel J, Keh D, Kupfer Y. Corticosteroids for treating sepsis. Cochrane Database Syst Rev. 2015;12:CD002243.
96. Gibbison B, López-López JA, Higgins JP, Miller T, Angelini GD, Lightman SL, Annane D. Corticosteroids in septic shock: a systematic review and network meta-analysis. Crit Care. 2017;21:78.
97. Prasad K, Singh MB, Ryan H. Corticosteroids for managing tuberculous meningitis. Cochrane Database Syst Rev. 2016;4:CD002244.
98. Wiysonge CS, Ntsekhe M, Thabane L, Volmink J, Majombozi D, Gumedze F, Pandie S, Mayosi BM. Interventions for treating tuberculous pericarditis. Cochrane Database Syst Rev. 2017;9:CD000526.
99. Matsui K, Furumoto A, Ohba K, Mochizuki K, Tanaka T, Takaki M, Morimoto K, Ariyoshi K. Use of corticosteroids for urinary tuberculosis patients at risk of developing ureteral obstruction. Intern Med. 2016;55:3539–42.
100. Ryan H, Yoo J, Darsini P. Corticosteroids for tuberculous pleurisy. Cochrane Database Syst Rev. 2017;3:CD001876.
101. Briel M, Spoorenberg SMC, Snijders D, Torres A, Fernandez-Serrano S, Meduri GU, Gabarrús A, Blum CA, Confalonieri M, Kasenda B, Siemieniuk RAC, Boersma W, Bos WJW, Christ-Crain M, Ovidius Study Group; Capisce Study Group; STEP Study Group. Corticosteroids in patients hospitalized with community-acquired pneumonia: systematic review and individual patient data meta-analysis. Clin Infect Dis. 2017; https://doi.org/10.1093/cid/cix801.
102. Ewald H, Raatz H, Boscacci R, Furrer H, Bucher HC, Briel M. Adjunctive corticosteroids for Pneumocystis jiroveci pneumonia in patients with HIV infection. Cochrane Database Syst Rev. 2015;4:CD006150.
103. Fujikura Y, Manabe T, Kawana A, Kohno S. Adjunctive corticosteroids for Pneumocystis jirovecii pneumonia in non-HIV-infected patients: a systematic review and meta-analysis of observational studies. Arch Bronconeumol. 2017;53:55–61.
104. Hynicka LM, Ensor CR. Prophylaxis and treatment of respiratory syncytial virus in adult immunocompromised patients. Ann Pharmacother. 2012;46:558–66.

105. Rodrigo C, Leonardi-Bee J, Nguyen-Van-Tam J, Lim WS. Corticosteroids as adjunctive therapy in the treatment of influenza. Cochrane Database Syst Rev. 2016;3:CD010406.
106. Wang K, Bettiol S, Thompson MJ, Roberts NW, Perera R, Heneghan CJ, Harnden A. Symptomatic treatment of the cough in whooping cough. Cochrane Database Syst Rev. 2014;9:CD003257.
107. McGee S, Hirschmann J. Use of corticosteroids in treating infectious diseases. Arch Intern Med. 2008;168:1034–46.
108. Briegel J, Bein T, Möhnle P. Update on low-dose corticosteroids. Curr Opin Anaesthesiol. 2017;30:186–91.
109. Hermans G, De Jonghe B, Bruyninckx F, Van den Berghe G. Interventions for preventing critical illness polyneuropathy and critical illness myopathy. Cochrane Database Syst Rev. 2014;1:CD006832.
110. Welsh EJ, Evans DJ, Fowler SJ, Spencer S. Interventions for bronchiectasis: an overview of Cochrane systematic reviews. Cochrane Database Syst Rev. 2015;7:CD010337.
111. Agarwal R, Chakrabarti A, Shah A, Gupta D, Meis JF, Guleria R, Moss R, Denning DW, ABPA Complicating Asthma ISHAM Working Group. Allergic bronchopulmonary aspergillosis: review of literature and proposal of new diagnostic and classification criteria. Clin Exp Allergy. 2013;43:850–73.
112. Paramothayan NS, Lasserson TJ, Jones PW. Corticosteroids for pulmonary sarcoidosis. Cochrane Database Syst Rev. 2005;2:CD001114.
113. Sah BP, Goyal S, Iannuzzi MC. Novel pharmacotherapy of sarcoidosis. Pharmacol Ther. 2016;157:1–9.
114. Iannuzzi MC, Rybicki BA, Teirstein AS. Sarcoidosis. N Engl J Med. 2007;357:2153–65.
115. Fritz D, van de Beek D, Brouwer MC. Clinical features, treatment and outcome in neurosarcoidosis: systematic review and meta-analysis. BMC Neurol. 2016;16:220.
116. Sadek MM, Yung D, Birnie DH, Beanlands RS, Nery PB. Corticosteroid therapy for cardiac sarcoidosis: a systematic review. Can J Cardiol. 2013 Sep;29(9):1034–41.
117. Richeldi L, Davies HR, Ferrara G, Franco F. Corticosteroids for idiopathic pulmonary fibrosis. Cochrane Database Syst Rev. 2003;3:CD002880.
118. Tomassetti S, Ryu JH, Piciucchi S, Chilosi M, Poletti V. Nonspecific interstitial pneumonia: what is the optimal approach to management? Semin Respir Crit Care Med. 2016;37:378–94.
119. Rose DM, Hrncir DE. Primary eosinophilic lung diseases. Allergy Asthma Proc. 2013;34:19–25.
120. Spagnolo P, Rossi G, Cavazza A, Bonifazi M, Paladini I, Bonella F, Sverzellati N, Costabel U. Hypersensitivity pneumonitis: a comprehensive review. J Investig Allergol Clin Immunol. 2015;25:237–50; quiz follow 250.
121. Zinn DJ, Chakraborty R, Allen CE. Langerhans cell histiocytosis: emerging Insights and clinical implications. Oncology (Williston Park). 2016;30:122–32,139.
122. Kim S, Oh IJ, Park SY, Song JH, Seon HJ, Kim YH, Yoon SH, Yu JY, Lee BR, Kim KS, Kim YC. Corticosteroid therapy against treatment-related pulmonary toxicities in patients with lung cancer. J Thorac Dis. 2014;6:1209–17.
123. Yennurajalingam S, Bruera E. Role of corticosteroids for fatigue in advanced incurable cancer: is it a 'wonder drug' or 'deal with the devil'. Curr Opin Support Palliat Care. 2014;8:346–51.
124. Hopkins RL, Leinung MC. Exogenous Cushing's syndrome and glucocorticoid withdrawal. Endocrinol Metab Clin North Am. 2005;34: 371–84.
125. Rice JB, White AG, Scarpati LM, Wan G, Nelson WW. Long-term systemic corticosteroid exposure: a systematic literature review. Clin Ther. 2017;39:2216–29.
126. Caplan A, Fett N, Rosenbach M, Werth VP, Micheletti RG. Prevention and management of glucocorticoid-induced side effects: a comprehensive review: infectious complications and vaccination recommendations. J Am Acad Dermatol. 2017;76:191–8.
127. Schmiedel Y, Zimmerli S. Common invasive fungal diseases: an overview of invasive candidiasis, aspergillosis, cryptococcosis, and Pneumocystis pneumonia. Swiss Med Wkly. 2016;146:w14281.
128. Arpinelli F, Caramori G, Magnoni MS. Risk of pneumonia during long term regular treatment of stable COPD with inhaled glucocorticoids: a systematic review. Recenti Prog Med. 2017;108: 175–82.
129. Dong YH, Chang CH, Wu FL, Shen LJ, Calverley PMA, Löfdahl CG, Lai MS, Mahler DA. Use of inhaled corticosteroids in patients with COPD and the risk of TB and influenza: a systematic review and meta-analysis of randomized controlled trials a systematic review and meta-analysis of randomized controlled trials. Chest. 2014;145:1286–97.
130. Yang M, Chen H, Zhang Y, Du Y, Xu Y, Jiang P, Xu Z. Long-term use of inhaled corticosteroids and risk of upper respiratory tract infection in chronic obstructive pulmonary disease: a meta-analysis. Inhal Toxicol. 2017;29:219–26.
131. Murphy VE, Jensen ME, Gibson PG. Asthma during pregnancy: exacerbations, management, and health outcomes for mother and infant. Semin Respir Crit Care Med. 2017;38:160–73.
132. Alhussien AH, Alhedaithy RA, Alsaleh SA. Safety of intranasal corticosteroid sprays during pregnancy: an updated review. Eur Arch Otorhinolaryngol. 2017; https://doi.org/10.1007/s00405-017-4785-3.
133. Chi CC, Wang SH, Wojnarowska F, Kirtschig G, Davies E, Bennett C. Safety of topical corticosteroids in pregnancy. Cochrane Database Syst Rev. 2015;10:CD007346.
134. Baeck M, Marot L, Nicolas JF, Pilette C, Tennstedt D, Goossens A. Allergic hypersensitivity to topical and systemic corticosteroids: a review. Allergy. 2009;64:978–94.

33 Nonsteroidal Anti-inflammatory Drugs

Eeva Moilanen and Katriina Vuolteenaho

33.1 Introduction

NONSTEROIDAL ANTI-INFLAMMATORY DRUGS (NSAIDs) are the most widely used medicines globally. NSAIDs relieve pain (analgesic effect) and fever (antipyretic effect). They have also some anti-inflammatory properties, but as anti-inflammatories, they are noticeably less potent than GLUCOCORTICOIDS or DISEASE-MODIFYING ANTI-RHEUMATIC DRUGS (DMARDs). Non-selective NSAIDs also inhibit PLATELET aggregation, and acetylsalicylic acid (aspirin, ASA) is used in the (secondary) prevention of myocardial infarction and stroke.

NSAIDs reduce the production of PROSTAGLANDINS, PROSTACYCLIN, and THROMBOXANE (collectively known as PROSTANOIDS) by inhibiting the enzyme CYCLOOXYGENASE (also known as prostaglandin synthase). PROSTANOIDS are lipid mediators, which regulate various physiological functions in the human body, but they also mediate pain, fever and inflammatory responses (see Chap. 9). Therefore, inhibition of PROSTANOID synthesis serves as the mechanism of action for the pharmacological effects of NSAIDs. However, the same property results in certain adverse effects such as those targeting the gastrointestinal tract, cardiovascular system, and kidneys. Although the overall safety profile of NSAIDs is favorable, patients at risk to develop serious adverse events should be recognized, and special precaution should be followed in their treatment.

33.2 History

The use of bark and leaves of willow tree to relieve fever and pain has been known for more than 2000 years in herbal folklore and was first clearly documented by Edmund Stone in his 1763 letter to the Royal Society. The effective compound salicin was crystallized and its active metabolite, salicylic acid, isolated in the first half of the nineteen century. A couple of decades later, salicylic acid was being produced industrially and used for fever, rheumatism, and gout. Hoffman, a chemist at Bayer Laboratories, sought to improve the adverse effect profile of salicylic acid by investigating acetylated salicylic acid, and acetylsalicylic acid (aspirin, ASA) was brought to market in 1899.

Paracetamol (acetaminophen) was also first used in medicine at the end of the nineteenth century. However, it gained popularity only 50 years later when it was identified as the active metabolite of the earlier used analgesics acetanilide and phenacetin.

Final manuscript submitted on April 21, 2018.

E. Moilanen (✉) · K. Vuolteenaho
The Immunopharmacology Research Group,
Tampere University, Tampere, Finland
e-mail: eeva.moilanen@tuni.fi

M. J. Parnham et al. (eds.), *Nijkamp and Parnham's Principles of Immunopharmacology*,
https://doi.org/10.1007/978-3-030-10811-3_33

In 1971 three research groups, Vane, Moncada, and Ferreira, Smith and Willis, and Collier and Flower, discovered that aspirin-like drugs inhibit PROSTAGLANDIN synthesis [1–4]. Soon it became clear that this effect presents the mechanism of action of NSAIDs. Thereafter, several novel drugs with the same mechanism of action were developed. Later in the 1970s, PROSTACYCLIN and THROMBOXANE were found as new members of the PROSTAGLANDIN/PROSTANOID family, and their contribution in the therapeutic and adverse effects of NSAIDs started to be revealed.

At the turn of the 1980s and the 1990s, inducible form of PROSTAGLANDIN synthase was discovered[5–8].ItwasnamedasCYCLOOXYGENASE-2 (COX-2), and its expression was found to be enhanced by bacterial products and inflammatory CYTOKINES and suppressed by GLUCOCORTICOIDs. In an inflammatory focus, high amounts of PROSTANOIDS, especially PROSTAGLANDIN E_2, were found to be produced following the significantly enhanced expression of COX-2 and the inducible form of PROSTAGLANDIN E synthase (MICROSOMAL PROSTAGLANDIN E SYNTHASE-1, mPGES-1) that was discovered in 1999 [9].

At that time, it was assumed that the therapeutic effects of NSAIDs were caused by inhibition of COX-2, while most of their adverse effects were due to inhibition of the physiological prostaglandin synthase, CYCLOOXYGENASE-1 (COX-1) [10]. Accordingly, several COX-2 selective NSAIDs were developed. As expected, they had therapeutic efficacy similar to that of non-selective NSAIDs but less gastrointestinal adverse effects. Unfortunately, highly selective COX-2 inhibitors were associated with more frequent and more severe cardiovascular adverse effects than non-selective NSAIDs, especially when used at higher doses in patients with cardiovascular diseases. Therefore, some of the highly COX-2 selective NSAIDs were withdrawn from the market during the first decade of the twenty-first century. Based on those lessons, inhibitors of mPGES-1 as well as different NSAID derivatives (such as NO-releasing NSAIDs) have been and still are under development to improve the safety profile of NSAIDs.

33.3 Molecules

NSAIDs are a chemically heterogeneous group of compounds that shares a common mechanism of action and similar pharmacological and adverse effects. Most widely used NSAIDs are presented in Table 33.1 and their structures are shown in Appendix 3.

Some NSAIDs, especially propionic acid derivatives ibuprofen, ketoprofen, and flurbiprofen, exist as (*R*) and (*S*) enantiomers, and their clinically used formulations normally consist of a racemic mixture of these two. The (*S*) enantiomers inhibit PROSTANOID production, while the (*R*) enantiomers do not, and the latter have been considered pharmacologically rather inactive. Interestingly, the (*R*) enantiomers may have some PROSTANOID-independent analgesic effects through inhibition of the endocannabinoid metabolism by COX-2 [11].

NSAIDs are weak acids with pKa between 3 and 6. They are rather lipid soluble, which augments their transport to tissues, and their protein binding is typically >90%. Their half-life varies from 2 to 70 hours, and some of them are used as prodrugs. Most NSAIDs are metabolized in the liver through the microsomal cytochrome P450 oxidase system (most often by CYP3A, CYP2C9, or both), thereafter glucuronidated or otherwise conjugated and excreted in the urine. Varying degrees of biliary excretion and enterohepatic circulation may also be involved. Generally, use of NSAIDs is not recommended for patients with advanced hepatic or renal disease.

33.4 Mechanism of Action

33.4.1 Inhibition of Cyclooxygenase

NONSTEROIDAL ANTI-INFLAMMATORY DRUGS reduce PROSTAGLANDIN, PROSTACYCLIN, and THROMBOXANE synthesis by inhibiting the enzyme CYCLOOXYGENASE (COX), also known as prostaglandin synthase (Fig. 33.1). Two isoenzymes of COX (COX-1 and COX-2) are known. COX-1 is constitutively expressed in practically all cell types, and it is responsible for low-level PROSTANOID

Table 33.1 Chemical classification and characteristics of nonsteroidal anti-inflammatory drugs

Chemical classification	Drug	T½ (h)	Protein binding (%)	Notes
Propionic acid derivatives	Flurbiprofen	3–4	>99	– Non-selective NSAIDs – Over-the-counter (OTC) preparations available in many countries – Patients taking high doses of ibuprofen (>2400 mg/day) have an increased risk of arterial thromboembolic complications (myocardial infarction or stroke) compared to other NSAIDs
	Ibuprofen	2–3	>90	
	Ketoprofen	2	>90	
	Naproxen	12–15	>99	
Acetic acid derivatives	Diclofenac	1–2	>99	– Non-selective NSAID – Transient elevations of serum aminotransferases may occur, but only rare cases of clinically important liver failures have been reported – Increased risk of arterial thromboembolic complications (myocardial infarction or stroke) compared to other NSAIDs
	Etodolac	7	>99	– Preferential COX-2 inhibitor – Reduced risk of GI adverse events compared to other NSAIDs
	Indomethacin	2–3	90	– Non-selective NSAID – High risk of GI adverse events – Central nervous system adverse effects reported by up to 10% of patients taking indomethacin (headache, dizziness, confusion and depression)
	Ketorolac	5	>99	– Non-selective NSAID – Analgesic used mainly to treat postoperative pain – Weak anti-inflammatory effect
	Sulindac	8 (sulindac) 16 (sulindac sulphide)	>90	– Non-selective NSAID – Prodrug metabolized to sulindac sulphide – Extensive enterohepatic cycling
Oxicam derivatives	Meloxicam	20	>99	– Preferential COX-2 inhibitor – Reduced risk of GI adverse events compared to other NSAIDs
	Piroxicam	50	>99	– Non-selective NSAID – Increased risk of GI adverse events
	Tenoxicam	60–70	>99	– Non-selective NSAID
Fenamate derivatives	Meclofenamic acid	1–5	>99	– Non-selective NSAIDs – Fenamates cause more commonly diarrhea than other NSAIDs – Tolfenamic acid has reduced risk of upper GI adverse events compared to other non-selective NSAIDs, but it may cause dysuria
	Mefenamic acid	2–4	>90	
	Tolfenamic acid	2–3	>99	
Salicylates	Acetylsalicylic acid (ASA)	15–20 min (ASA) 2–3 (salicylic acid) up to 15 h at high doses	50 (ASA) >70 (salicylic acid)	– Non-selective NSAID – ASA metabolizes rapidly to salicylate – Increased risk of GI adverse events when used at analgesic/anti-inflammatory doses – Low-dose aspirin is used for the prevention of myocardial infarction and stroke; ASA inhibits COX in an irreversible manner through acetylation

(continued)

Table 33.1 (continued)

Chemical classification	Drug	T½ (h)	Protein binding (%)	Notes
Coxibs	Celecoxib	8–12	97	– COX-2 selective NSAIDs – Reduced risk of GI adverse events compared to other NSAIDs – Increased risk of arterial thromboembolic complications (myocardial infarction and stroke) compared to other NSAIDs – Etoricoxib may increase blood pressure, and it should not be prescribed to patients with blood pressure above 140/90 mmHg – Parecoxib is an injectable prodrug of valdecoxib
	Etoricoxib	22	92	
	Parecoxib	8[a]	98[a]	
Other NSAIDs	Nabumetone	24[a]	99[a]	– Preferential COX-2 inhibitor – Prodrug metabolized to 6-methoxy-2-naphthylacetic acid (6-MNA) – Reduced risk of GI adverse events compared to other NSAIDs
	Nimesulide	3–6	>97.5	– Preferential COX-2 inhibitor – Reduced risk of GI adverse events campared to other NSAIDs – Increased risk of hepatotoxicity compared to other NSAIDs – Should only be used as a second choice and only in the treatment for acute pain or dysmenorrhea
Aniline derivatives	Paracetamol (acetaminophen)	2–4	20–50	– Different mechanism of action than actual NSAIDs – Antipyretic and analgesic, but no anti-inflammatory action – Safer than NSAIDs – Toxic doses cause potentially fatal liver injury

[a]Refers to the pharmacokinetic features of the active metabolite

production needed for physiological regulation of various processes in the human body. In contrast, COX-2 is an inducible enzyme, the expression of which is highly increased in INFLAMMATION in response to bacterial products and inflammatory CYTOKINES. COX-2 is constitutively expressed only in a very few tissues like macula densa in the kidneys, and its expression is enhanced by some physiological factors, cyclic expression in ovaries in response to hormonal factors as an example.

The most important therapeutic effects of NSAIDs are linked to their inhibitory effect of COX-2, except inhibition of PLATELET aggregation which is caused by inhibition of COX-1. While the adverse effects of NSAIDs are mediated by inhibition of both COX-1 and COX-2.

33.4.1.1 Non-selective and COX-2 Selective NSAIDs

NSAIDs vary in their selectivity against the two COX isoenzymes [12, 13], and they are divided into groups based on that property. Most of the currently used NSAIDs are non-selective and inhibit both COX-1 and COX-2 at therapeutic doses/concentrations. Among the non-selective NSAIDs, acetylsalicylic acid is a relatively COX-1 selective NSAID that inhibits COX-1 at lower concentrations than COX-2. Etodolac, meloxicam, nabumetone, and nimesulide are examples of the so-called preferential COX-2 inhibitors. When used at therapeutic doses/concentrations, they show some selectivity toward COX-2, inhibiting COX-2 at lower concentrations than COX-1. Coxibs (for instance, etoricoxib, celecoxib, and parecoxib/valdecoxib) belong to COX-2 selective

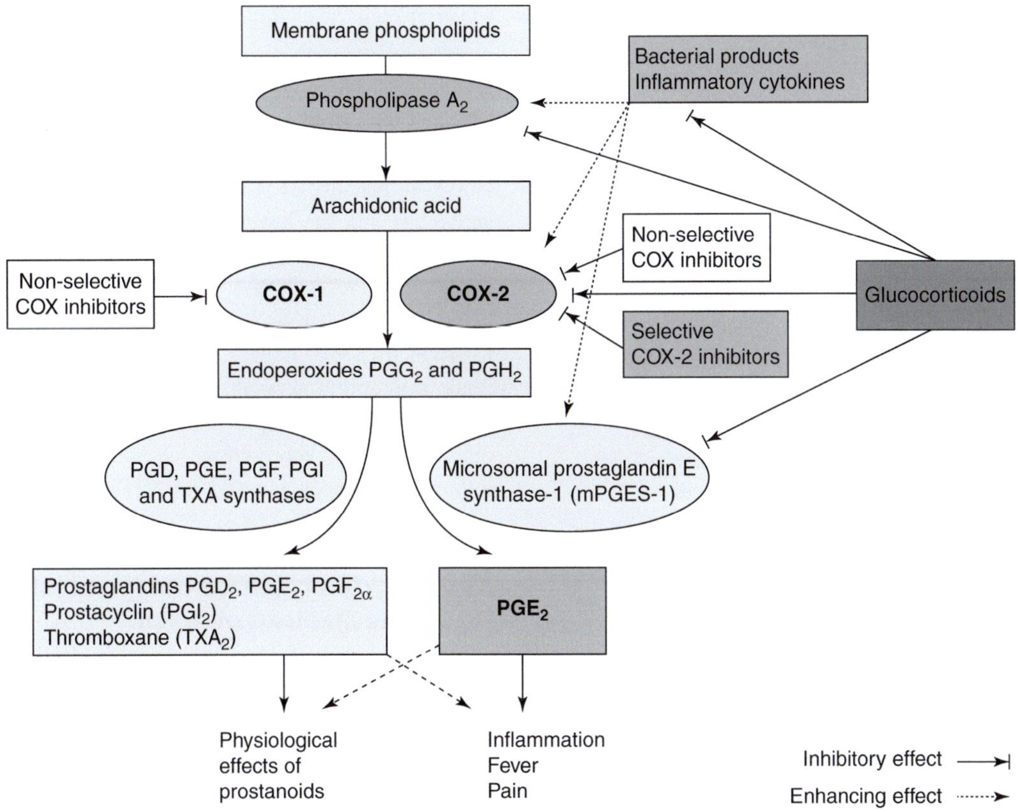

Fig. 33.1 Prostanoid synthesis and the effects of non-selective and COX-2 selective inhibitors and glucocorticoids. Non-selective NSAIDs inhibit both COX-1 and COX-2 activity, while COX-2 selective NSAIDs inhibit only COX-2. Glucocorticoids inhibit / down-regulate the expression of COX-2 and mPGES-1, and induce a phosholipase A_2 inhibiting factor. Bacterial products and inflammatory cytokines enhance the expression of COX-2 and mPGES-1, and activate phospholipase A_2 with various mechanisms, while glucocorticoids inhihit the expression of many proinflammatory cytokines. *COX* cyclooxygenase, *PG* prostaglandin

NSAIDs that inhibit COX-2 at noticeably lower concentrations than COX-1, and their effect on COX-1 at therapeutic doses is negligible. The classification of NSAIDs to non-selective COX inhibitors, preferential COX-2 inhibitors, and COX-2 selective inhibitors is given in Table 33.2.

33.4.1.2 Reversible Inhibition of COX

Most NSAIDs are reversible competitive inhibitors of COX enzymes. They interact with or bind to the active site of the COX enzymes, thus competing with the substrate (arachidonic acid) for binding. NSAIDs can be subdivided further on the basis of whether they bind to the COX active site in a time-dependent or time-independent manner.

COX-1 and COX-2 are bifunctional enzymes which mediate (1) the cyclooxygenase reaction whereby arachidonic acid and oxygen are converted to the cyclic endoperoxide PGG_2 followed by (2) the hydroperoxidase reaction in which PGG_2 undergoes reduction to PGH_2. Structurally COX enzymes are homodimer membrane proteins with a long and narrow hydrophobic channel in the center of the catalytic domain that contains the cyclooxygenase and peroxidase activities. NSAIDs function by blocking the access of the substrate arachidonic acid to the active site of the COX enzyme within the hydrophobic channel. The structural differences in this substrate-binding channel in the catalytic domain between COX-1 and COX-2 formed the basis for the development of COX-2 selective inhibitors [13, 14].

NSAIDs differ in the rates how long it takes to inhibit the COX activity and leave the active site when the circulating drug concentration decreases. Ibuprofen is an example of a drug

Table 33.2 Classification of NSAIDs based on their COX-1 versus COX-2 selectivity

Non-selective COX inhibitors	Acetylsalicylic acid
	Diclofenac
	Flurbiprofen
	Ibuprofen
	Indomethacin
	Ketoprofen
	Ketorolac
	Meclofenamic acid
	Mefenamic acid
	Naproxen
	Piroxicam
	Sulindac
	Tenoxicam
	Tolfenamic acid
Preferential COX-2 inhibitors	Etodolac
	Nabumetone
	Nimesulide
	Meloxicam
COX-2 selective inhibitors	Celecoxib
	Etoricoxib
	Parecoxib[a]

[a]Prodrug of valdecoxib

which binds and inhibits COX enzyme rapidly, but it is also removed from the active site quickly when drug levels in the body decrease. Most NSAIDs possess time-dependent activity and initially form a loose complex with the active site in COX before a stronger binding is established. This time is needed for proper orientation of the drug in the substrate-binding channel in the catalytic domain and conformational changes. Diclofenac, which binds to serine 530 in the catalytic site of COX [15], serves as an example of a time-dependent inhibitor of the enzyme. It is slower than ibuprofen to block COX activity and needs hours to exit the enzyme active site.

33.4.1.3 Irreversible Inhibition of COX by Acetylsalicylic Acid

In contrast to other NSAIDs, acetylsalicylic acid (ASA) inhibits COX-1 in a noncompetitive manner. It acetylates serine 530 in the active site of the enzyme, leading to complete and irreversible inhibition of COX-1 [16]. Most cells recover their PROSTANOID synthesis through COX-1 by *de novo* synthesis of the enzyme. PLATELETS are an exception here, because as nonnuclear cells, they are unable to synthesize new enzyme and ASA blocks THROMBOXANE synthesis in PLATELETS for their lifetime. PLATELET population recovers from the effect of ASA in a couple of days through release of novel PLATELETS from the bone marrow. ASA acetylates the amino acid serine 530 also in COX-2. However, because of structural differences in the active site and the lower acetylation efficacy for COX-2, the inhibitory effect on COX-2 is much lower than that on COX-1. Interestingly, ASA has been shown to alter the catalytic activity of COX-2 to produce 15-HETE and LIPOXINS from arachidonic acid and to generate anti-inflammatory and resolution-inducing lipid mediators also from other fatty acids, representing an additional anti-inflammatory mechanism for ASA [17].

33.4.2 Cyclooxygenase-Independent Mechanisms of Action

Some NSAIDs have been reported to have also COX-independent effects, which may contribute to their therapeutic action. However, most of these effects are only seen at noticeably higher drug concentrations than their inhibitory effect on COX enzymes. Examples of the COX-independent effects are inhibition of LEUKOTRIENE production and some other LEUKOCYTE functions by fenamates, piroxicam, ketoprofen, and nimesulide and radical scavenging effects of piroxicam. ASA and salicylates have been reported to inhibit the inflammatory transcription factor NUCLEAR FACTOR-κB (NF-κB) and some NSAIDs to downregulate inducible NITRIC OXIDE synthesis or to inhibit PHOSPHODIESTERASES to enhance intracellular CYCLIC AMP (cAMP) levels. The clinical significance of these mainly *in vitro* detected effects remain, however, unclear.

33.4.3 Mechanism of Action of Paracetamol

The mechanism of action of paracetamol (acetaminophen) is not known in detail but it seems to be different from that of other NSAIDs [18–21]. In addition, its pharmacological profile differs from that of classical NSAIDs, and therefore it is sometimes excluded from the group of NSAIDs and classified in its own group of "centrally acting analgesics." Paracetamol has analgesic and antipyretic effects most probably through a central mechanism but does not have anti-inflammatory properties. Accordingly, the inhibitory effect of paracetamol on PROSTANOID synthesis in peripheral tissues is very low or non-existing. Paracetamol does not interact with the active site of COX enzymes like NSAIDs, but it has been proposed to serve as a reducing co-substrate for the peroxidase activity of COX; it has been reported to inhibit also other peroxidases such as myeloperoxidase through the same mechanism [18]. In INFLAMMATION, peroxide concentrations in tissues are increased and that has been suggested as a reason for the negligible effect of paracetamol on PROSTANOID synthesis in INFLAMMATION. As paracetamol does not significantly inhibit PROSTANOID synthesis in periphery, it is also practically free of the PROSTANOID-dependent adverse effects of NSAIDs, such as gastrointestinal, cardiovascular, and renal side effects.

Preclinical experiments have suggested an additional mechanism of action for paracetamol [19–21]. Paracetamol is metabolized in the liver to aminophenol, which can cross the blood-brain barrier and conjugate in the brain with arachidonic acid. In the brain, this conjugate (AM404) has been found to augment the activity of the endocannabinoid system and to modify the pain-modulating transient receptor potential (TRP) ion channels. In the CENTRAL NERVOUS SYSTEM, paracetamol has been proposed to inhibit PROSTANOID synthesis by different mechanisms [18–20]. In addition, paracetamol (or the conjugate AM404) has been reported to activate analgesic serotonin and opioid pathways in the CENTRAL NERVOUS SYSTEM [18, 19, 21]. It remains to be studied if these mechanisms seen mostly in experimental studies contribute to the analgesic and antipyretic effects of paracetamol also in humans.

33.5 Pharmacological Effects and Therapeutic Uses

NONSTEROIDAL ANTI-INFLAMMATORY DRUGS have three main therapeutic effects: they are analgesic and antipyretic and possess mild anti-inflammatory properties. All these effects are linked to their inhibitory effect on COX-2, and in clinical studies the efficacy of non-selective and COX-2 selective NSAIDs is comparable. In addition, ASA has a long-lasting antiplatelet activity, which is mediated by inhibition of COX-1 and utilized in the (secondary) prevention of thrombotic disorders, especially cardiac infarction and stroke.

33.5.1 Analgesic Effect

NSAIDs have both central and peripheral analgesic effects through inhibition of PROSTANOID synthesis [22]. In periphery, PROSTAGLANDIN E_2 (PGE_2) and PROSTACYCLIN (PGI_2) do not directly cause pain, but they induce hyperalgesia, i.e., sensitize the nociceptors to pain caused by chemical or physiological factors. The sensitization is believed to be mediated through PROSTANOID receptor-activated increase in the intracellular cAMP, which in turn activates protein kinase A (PKA). PKA is one of the mechanisms, which augment, through phosphorylation, the sensitivity of the pain triggering ion channels in nociceptors. By inhibiting PGE_2 and PROSTACYCLIN production, NSAIDs reduce the hyperalgesia typical for conditions related to INFLAMMATION and tissue injury where PROSTANOID production is enhanced.

PROSTANOIDS are also produced in the CENTRAL NERVOUS SYSTEM where they modulate pain signals and participate in central sensitization. COX-2 is constitutively expressed in the dorsal horn of the spinal cord (which is a significant site to modulate afferent pain pathways), and its expression is enhanced during peripheral INFLAMMATION [23]. NSAIDs have been reported to have also PROSTANOID-independent analgesic effects in the spinal cord, for instance, through their effects on ion channels or endocannabinoid system [22].

In addition to inflammatory pain, NSAIDs are also used in the treatment of migraine [24]. PROSTANOIDS seem to be involved in the vascular dysfunction associated with the development of migraine-type headache. NSAIDs are also effective in relieving dysmenorrhea [25]. Locally formed PROSTANOIDS (especially PGE_2 and $PGF_{2\alpha}$) are involved in causing painful muscle constructions in the uterus in patients with dysmenorrhea, and when PROSTANOID production is inhibited, the symptoms are alleviated.

NSAIDs seem to have both peripheral and central analgesic effects, and therefore it is likely that compounds, which easily cross the blood-brain barrier, have a better and faster action. Based on experiments with COX-1 or COX-2 deficient animals, both enzymes seem to have a role in pain processes. However, non-selective and COX-2 selective NSAIDs are equally effective analgesics in large clinical trials.

33.5.2 Antipyretic Effect

Thermoregulatory center in the anterior hypothalamus maintains normal body temperature by regulating the balance between heat production and loss. In infection or INFLAMMATION, bacterial products and/or inflammatory CYTOKINES (exogenous and endogenous pyrogens) enhance the expression of COX-2 and mPGES-1 in the hypothalamus. This in turn leads to increased production of PROSTANOIDS, primarily PGE_2, which generate neuronal signals to activate the thermoregulatory center to reset the body temperature to fever. NSAIDs reduce fever by inhibiting PROSTANOID production in the anterior hypothalamus, and the body temperature then decreases through normal regulatory mechanisms such as perspiration and vasodilatation in cutaneous vessels. NSAIDs do not alter body temperature in physiological conditions. In addition to NSAIDs, paracetamol also has a rapid-onset antipyretic effect [26].

33.5.3 Anti-inflammatory Effect

PROSTANOID production is increased in INFLAMMATION through increased expression of COX-2 and mPGES-1 and enhanced release of their substrate arachidonic acid. PGE_2 and PROSTACYCLIN cause vasodilatation and augment the increased vascular permeability induced by other inflammatory mediators, and they are also involved in the mechanisms of inflammatory pain [22]. PROSTANOIDS contribute to the development of the signs and symptoms of acute INFLAMMATION: redness, swelling, heat, and pain. The role of PROSTANOIDS in chronic INFLAMMATION is, however, less clear. PROSTANOIDS have regulatory effects on LEUKOCYTES and CYTOKINE production, and in experimental models anti-inflammatory effects have been seen when PROSTANOIDS have been administered at pharmacological doses.

In addition to PROSTANOIDS, the inflammatory response is mediated and modulated by dozens or even hundreds of CYTOKINES and other inflammatory mediators (see Chaps. 6 and 9). Therefore it is understandable that by blocking of the synthesis of just a few of those mediators (like PROSTANOIDS) does not cause such a major anti-inflammatory effect as GLUCOCORTICOIDS or some anti-rheumatic drugs, which reduce the synthesis of a wide array of inflammatory factors. The anti-inflammatory effect of NSAIDs is seen in acute inflammatory responses, such as sunburns or sore throat associated with the flu, but in chronic inflammatory diseases NSAIDs have mainly symptomatic, i.e., pain-relieving effect.

33.5.4 Antiplatelet Effect

Activated PLATELETS metabolize arachidonic acid by COX-1 to THROMBOXANE A_2 (TXA_2), which is a powerful vasoconstrictor and PLATELET-aggregating and PLATELET-activating agent. Another PROSTANOID, PROSTACYCLIN, has opposite effects: it causes vasodilatation and inhibits PLATELET adhesion, aggregation, and activation. PROSTACYCLIN is produced in endothelial cells, in normal conditions by COX-1 but in hypoxia also by COX-2. If the balance between THROMBOXANE and PROSTACYCLIN production is altered to favor relatively increased THROMBOXANE levels, the risk for thrombus formation is increased, whereas the opposite balance (i.e., relatively increased PROSTACYCLIN levels) reduces the risk of thrombi. [27]

Non-selective NSAIDs inhibit both THROMBOXANE and PROSTACYCLIN production. COX-2 selective NSAIDs do not affect THROMBOXANE production in PLATELETS but may reduce PROSTACYCLIN synthesis in the endothelium especially in hypoxic conditions. Therefore, COX-2 selective NSAIDs have an unfavorable effect on the balance between THROMBOXANE and PROSTACYCLIN production and increase the risk of myocardial infarction and stroke in patients with cardiovascular diseases (see below the chapter on adverse effects of NSAIDs).

In contrast, ASA has a favorable effect on THROMBOXANE and PROSTACYCLIN balance in patients with cardiovascular diseases. ASA acetylates and thus inactivates COX enzyme in an irreversible manner, and it is relatively selective toward COX-1. In PLATELETS, ASA blocks THROMBOXANE production for the lifetime of the cell by acetylating COX-1, which cannot be resynthesized. The antiaggregatory effect of a single low dose of ASA lasts up to 4–6 days until novel PLATELETS with intact COX-1 enzyme have been released from the BONE MARROW. The endothelium recovers from the effect of ASA in a much shorter time through *de novo* synthesis of COX capable to normal PROSTACYCLIN production.

In clinical trials, treatment with a low dose of ASA (75–375 mg every or every other day) has been proved effective in (secondary) prevention of myocardial infarction and stroke [27]. It is of note that concomitant use of other NSAIDs may reduce the beneficial effect of low-dose ASA therapy because of blocking the access of ASA to its binding site in the COX enzyme (especially if NSAIDs are taken prior to ASA) or because of longer-term reduction of PROSTACYCLIN production [28, 29].

In addition to the beneficial effect described above, the antiplatelet effect of ASA (and to a lower extent also other NSAIDs) may also increase bleeding time. This may predispose to gastrointestinal and other bleeds and can also cause harm in surgical operations. Special precaution should be followed in concomitant use of NSAIDs with warfarin and other antithrombotic agents and anticoagulants because of pharmacodynamic interactions.

33.5.5 Therapeutic Uses

NSAIDs are used widely in the symptomatic treatment of osteoarthritis and other rheumatic diseases, as well as other musculoskeletal diseases and trauma. They are also used in migraine, dysmenorrhea, respiratory infections, and other occasional conditions with pain and fever, headache, and toothache. NSAIDs have an established role in the treatment of pain following surgical and dental operations where they can reduce the opioid doses. NSAIDs are used in cancer pain, especially in bone metastases, and they may even have chemopreventive effects in cancer. As described above, low-dose ASA treatment is used in (secondary) prevention of myocardial infarction and stroke.

33.6 Adverse Effects

PROSTANOIDS regulate various physiological functions including vascular tone, PLATELET function, renal excretion, reproduction, and protection of gastrointestinal MUCOSA. NSAIDs inhibit PROSTANOID synthesis, and that is the basis of their therapeutic effects, but they share also a common spectrum of adverse effects related to the same mechanism. In general, severe adverse reactions are rare and NSAIDs can be regarded as safe drugs. However, because they are the most commonly used drugs worldwide, their adverse effects should be considered. Prescribers should be aware of the adverse effects of NSAIDs and of patient groups at increased risk to develop severe adverse reactions to be able to treat them appropriately.

The most common adverse effects of NSAIDs are gastrointestinal mucosal injury and its consequences. Much less common are renal and cardiovascular adverse effects as well as bronchoconstriction in patients with aspirin-exacerbated ASTHMA and liver, skin, and BONE MARROW toxicity. In addition, NSAIDs may retard fracture healing. They are not recommended at the last trimester of pregnancy as they may have an effect on ductus arteriosus in the fetal heart and retard parturition.

33.6.1 Gastrointestinal Tract Effects

33.6.1.1 Mechanisms of Gastrointestinal Adverse Effects

PROSTANOIDS, especially PGE_2 and PROSTACYCLIN, are a part of the physiological mechanisms protecting gastrointestinal MUCOSA against the acidic and irritating contents of the stomach and intestine. The two PROSTANOIDS maintain the normal mucosal blood flow and integrity and promote amelioration of mucosal erosions. PGE_2 stimulates the secretion of protective mucus and acid-neutralizing bicarbonate in the mucosal membrane. PGE_2 also inhibits gastric acid secretion triggered by HISTAMINE, gastrin, and stimulation of the parasympathetic nervous system. Therefore, inhibition of PGE_2 production by NSAIDs paralyzes one of the major mechanisms protecting the gastrointestinal MUCOSA (see Fig. 33.2).

NSAIDs have also been reported to increase the accumulation of NEUTROPHIL GRANULOCYTES into the gastrointestinal MUCOSA. NEUTROPHILS produce REACTIVE OXYGEN SPECIES (ROS) and other factors, which further aggravate mucosal INFLAMMATION and injury. NSAIDs seem to differ from each other in their effects on NEUTROPHILS.

NSAID-induced gastrointestinal mucosal injury is also effected by the phenomenon called ion trapping, which results in increased intracellular concentrations of the drug. The

PGE$_2$ production protects stomach mucosal lining

- Mucus secretion ↑
- Bicarbonate secretion ↑
- Gastric acid / HCl secretion ↓
- Mucosal blood flow ↑
- Healing of mucosal erosions ↑

NSAIDs inhibit PGE$_2$ production, which results in

- Mucus secretion ↓
- Bicarbonate secretion ↓
- Gastric acid / HCl secretion ↑
- Mucosal blood flow ↓
- Healing of mucosal erosions ↓

Risk factors for NSAID-induced gastrointestinal adverse effects

- Previous history of peptic ulcer
- Use of high-dose NSAIDs and/or multiple NSAIDs
- Concurrent use of
  - Low-dose aspirin
  - Glucocorticoids
  - Anticoagulants
  - SSRI-antidepressants
- Age > 65 years
- Chronic disease
- Helicobacter pylori infection

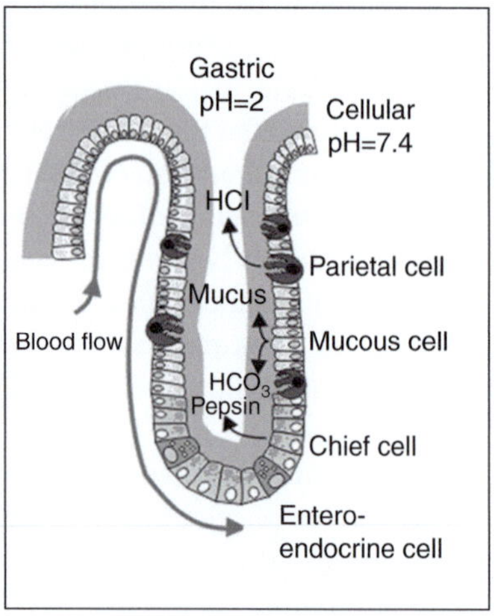

Fig. 33.2 Mechanisms and risk factors for NSAID-induced gastrointestinal adverse effects. *HCl* hydrochloric acid, *NSAID* non-steroidal anti-inflammatory drug, *SSRI* serotonin selective reuptake inhibitor (group of antidepressive drugs) *HCO_3-* bicarbonate *PGE_2* prostaglandin E_2

acidic content of the stomach has a low pH of about 2, and in that pH NSAIDs (which are weak acids) remain mostly in their uncharged form. The cell membrane is permeable to lipid-soluble (uncharged or non-ionized) drug molecules, but water-soluble (charged or ionized) molecules cannot cross it easily. Therefore, NSAID molecules (which are uncharged in the acidic stomach) diffuse easily into the mucosal cells where they arc ionized because of the physiological intracellular pH of 7.4. The ionized drug molecules are not easily diffused back across the cell membrane but are "trapped" in the intracellular space. As a consequence, the intracellular drug concentration increases and may have cytotoxic effects. In addition, when the drug concentration increases, the primarily relatively COX-2 selective compounds start to inhibit COX-1 as well.

In healthy gastrointestinal MUCOSA, the physiological PGE_2 production is catalyzed by COX-1, and the gastrointestinal mucosal lesions of NSAIDs are mainly associated with their inhibitory effect on COX-1. Therefore, it was hypothesized that COX-2 selective NSAIDs have less gastrointestinal adverse effects than non-selective NSAIDs. This hypothesis has been approved in large clinical trials, which show that preferential COX-2 inhibitors and COX-2 selective inhibitors cause less gastric ulcers and its complications (bleeding, perforation) than non-selective NSAIDs, especially in high-risk patients [30–32].

In addition, COX-2 may have a significant role in producing protective PROSTANOIDS in certain conditions, for instance, in *Helicobacter pylori* infection and associated gastritis and in healing mucosal lesions. In those conditions, also COX-2 selective NSAIDs inhibit the production of protective PROSTANOIDS and may aggravate the mucosal lesions and retard their recovery.

Based on clinical trials, when low-dose ASA treatment is used concomitantly with COX-2 selective NSAIDs, it reverses, at least partly, the gastrointestinal benefit of the COX-2 selective NSAIDs over non-selective NSAIDs. That may be explained by the fact that when the combination is used, it leads to inhibition of both COX-1 and COX-2 in the gastric MUCOSA and thus equals the situation when a non-selective NSAID is administered [33]. An additional explanation may lie on the recent finding that acetylation of COX-2 by ASA alters the catalytic activity of COX-2 to produce gastroprotective lipids instead of PROSTANOIDS. When COX-2 selective NSAIDs are used together with low-dose ASA, they inhibit this ASA-induced production of gastroprotective mediators, which may increase the ulcerogenic activity of that drug combination [17, 34].

33.6.1.2 Clinical Presentation of Gastrointestinal Adverse Effects

Based on endoscopic examinations, up to one third of patients develop gastrointestinal mucosal erosions during treatment with NSAIDs. These are, however, normally harmless, symptomless, and rapidly healing lesions also under continuing NSAID treatment. About 10–20% of NSAID users report gastrointestinal symptoms including dyspepsia, gastric pain, and diarrhea. Dyspepsia may be treated with proton pump inhibitors or other acid-lowering or acid-neutralizing drugs. Alternatively, non-selective NSAID may be replaced with a COX-2 preferential or selective drug. Only a minority of patients with serious gastrointestinal events report antecedent dyspepsia.

Very few patients develop a serious gastrointestinal adverse event: ulcer, bleeding, perforation, or stricture. These may, at worst, lead to peritonitis, bowel obstruction, or death. Antecedent subjective symptoms correlate poorly with the serious gastrointestinal adverse events, which may develop acutely to a previously symptom-free patient. Risk factors for serious NSAID-induced gastrointestinal adverse events include increased age (>65 years); previous history of peptic ulcer disease; use of high-dose or multiple NSAIDs; concomitant use of low-dose ASA, GLUCOCORTICOIDS, anticoagulants, or serotonin-selective antidepressants; *Helicobacter pylori* infection; infirmity or chronic illness as listed in Fig. 33.2. Special care should be taken on patients who have several of those risk factors.

NSAIDs may also cause mucosal injury in the esophagus, especially in patients with esophageal reflux, and in the small bowel. The latter may cause intestinal (occult) bleeding and development of anemia. There is also a concern that NSAIDs may exacerbate inflammatory bowel disease, and therefore NSAIDs should be prescribed with special precaution to those patients.

33.6.1.3 Prevention of Gastrointestinal Adverse Effects

To prevent the serious gastrointestinal adverse effects of NSAIDs, it is crucial to identify the individuals at increased risk to develop those reactions and treat them with safe regimens: (1) use paracetamol or "mild" opioids (e.g., tramadol or codeine) instead of NSAIDs, (2) use preferential COX-2 inhibitors or COX-2 selective inhibitors instead of non-selective COX inhibitors, or (3) combine gastroprotective drug (proton pump inhibitor or possibly misoprostol) to NSAID treatment [35, 36]. In high-risk patients, it may be appropriate to combine a proton pump inhibitor and COX-2 selective NSAID [37, 38], but large clinical trials and meta-analyses are still needed to evaluate the clinical outcome and cost-benefit ratio of such combination treatment.

33.6.2 Cardiovascular Adverse Effects

Treatment with NSAIDs is associated with a slightly increased risk of cardiovascular adverse events especially in patients with a history of cardiovascular diseases. Based on wide clinical trials and meta-analyses, it has been estimated that treatment with NSAIDs increases the thrombosis risk (myocardial infarction or stroke) to 1.5-fold, which is a relatively minor increase as compared to the increase caused by smoking, hypertension, or mental stress.

33.6.2.1 Mechanisms of Cardiovascular Adverse Effects

The balance between the COX-1-derived THROMBOXANE A_2 production in PLATELETS and COX-1 and COX-2-derived PROSTACYCLIN production in vascular endothelium, and consequences of the changes in this balance on PLATELET function and thrombus formation, has been described above in the paragraph 33.5.4 [13, 39, 40]. The accumulated literature supports the idea that the NSAID-coupled increased risk of thrombosis (myocardial infarction or stroke) is associated with the degree of COX-2 inhibition with the particular dose and compound and not on COX-2 selectivity *per se* (as previously assumed). In ischemic conditions, endothelium produces antiaggregatory PROSTACYCLIN through COX-1 and COX-2 pathways, and inhibition of the latter in conditions where THROMBOXANE A_2 in PLATELETS is not fully inhibited increases the risk of thrombus formation in patients with increased risk (i.e., in patients with cardiovascular diseases). Therefore, the risk of cardiovascular adverse effects is associated with both non-selective and COX-2 selective NSAIDs, especially when used at high doses in high-risk patients. Unlike the other NSAIDs, naproxen has a long half-life and at higher doses inhibits COX-1 mediated THROMBOXANE production by more than 95%; in some (but not all) meta-analyses, it has shown less cardiovascular adverse effects than many other NSAIDs, but its safety profile is hampered by increased risk of gastrointestinal toxicity as compared to many other NSAIDs [41–45].

There are also other mechanisms, which may contribute to the cardiovascular adverse effects of NSAIDs. NSAIDs may increase blood pressure (see below) and cause heart failure following reduced renal excretion of sodium and water. In addition, NSAIDs have been reported to have direct harmful effects on the blood vessels and the heart, e.g., through inhibition of NITRIC OXIDE synthesis [40].

33.6.2.2 Official Guidance to Reduce Cardiovascular Adverse Effects

European Medicines Agency (EMA) has repeatedly assessed the cardiovascular safety of NSAIDs. They have estimated that the overall benefit-risk balance is favorable, but NSAIDs are advised to be used at the lowest effective dose for the shortest possible duration of treatment. Exceptions are COX-2 inhibitors, diclofenac, and ibuprofen, the latter at high doses (i.e., at doses above 2400 mg daily), which must not be used in patients with ischemic heart disease, congestive heart failure (NYHA class II–IV), cerebrovascular disease (stroke), or peripheral arterial disease. In addition, caution should be exercised when prescribing those drugs to patients with risk factors for heart disease such as hypertension, hyperlipidemia, diabetes, or smoking [46–49].

33.6.2.3 Hypertension

NSAIDs may increase blood pressure with an average increase of mean arterial pressure between 5 and 10 mmHg. In healthy subjects, the effect is normally slight and transient. However, special attention should be paid to patients with prehypertension or established disease. The mechanisms on how NSAIDs increase blood pressure may be associated with their renal effects (see below, paragraph 33.6.3) as well as with their direct effects on vessel wall and/or their inhibitory effect on endothelial NITRIC OXIDE production. NSAIDs may also attenuate the effects of antihypertensive drugs including diuretics, angiotensin-converting enzyme inhibitors, angiotensin receptor blockers, and beta-blockers [50–52].

Etoricoxib may have the most obvious effect on blood pressure. European Medicines Agency has advised that patients whose blood pressure is persistently above 140/90 mmHg and has not been adequately controlled should not take etoricoxib [47]. In addition, high blood pressure should be controlled before etoricoxib treatment is begun and should be monitored for 2 weeks after the start of treatment and regularly thereafter. Although not officially advised, it is recommendable to monitor blood pressure in patients with hypertension who are treated with NSAIDs, especially at high doses.

33.6.3 Renal Adverse Effects

Therapeutic doses of NSAIDs in healthy adults pose little threat to kidney function because of the redundancy of the systems regulating renal function. In situations that challenge the normal regulation of kidney function, such as dehydration, hypovolemia, congestive heart failure, hepatic cirrhosis, or chronic kidney disease, the regulatory effects of PROSTANOIDS become crucial, and inhibition of PROSTANOID synthesis causes clinically significant adverse effects. These are typical for both non-selective and COX-2 selective NSAIDs.

In the kidneys, PROSTANOIDS regulate vascular tone, sodium and water excretion, and renin release. Both COX-1 and COX-2 are normally expressed in the kidneys. COX-1 is expressed in renal vasculature, collecting ducts, and glomerular mesangial cells, while COX-2 is found in macula densa, medullar interstitial cells, and renal blood vessels. States such as dehydration, salt restriction, and renovascular hypertension regulate renal COX-2 expression, and inhibition of COX-2 decreases plasma renin levels and renal renin activity [53, 54].

33.6.3.1 Effects on Sodium Excretion

COX-1 and COX-2-derived PROSTANOIDS promote natriuresis by regulating sodium reabsorption in various parts of the renal tubule. NSAIDs inhibit sodium excretion, and this can lead to weight gain, peripheral edema, increased blood pressure, and exacerbations of congestive heart failure in susceptible patients [55]. Due to their effects on renin-aldosterone system through inhibition of PROSTANOID synthesis, NSAIDs may increase potassium reabsorption and cause hyperkalemia. Concomitant use of NSAIDs may also attenuate the therapeutic effects of diuretics [53, 54].

33.6.3.2 Acute Renal Failure

Acute renal failure is a rare reversible complication of NSAID treatment. It is caused by vasoconstrictive effects of NSAIDs and manifests in patients with reduced intravascular volume, e.g., dehydration, congestive heart failure, cirrhosis of the liver, or renal insufficiency. A major reduction of medullary blood flow may cause injury of medullary interstitial cells and papillary necrosis [53, 54, 56, 57].

33.6.3.3 Interstitial Nephritis and Chronic Kidney Disease

Another uncommon renal adverse effects of NSAIDs is chronic renal failure, which is associated with long-term use of high doses of NSAIDs, especially aspirin and paracetamol, in patients with pre-existing renal disease [53, 54, 58]. In addition, infrequent idiosyncratic reactions with massive proteinuria and interstitial nephritis, fever, rash, and eosinophilia have been observed with the use of most NSAIDs.

33.6.4 Aspirin-Exacerbated Respiratory Disease

NSAIDs induce bronchoconstriction in a small group of patients with ASTHMA, and the reaction is often associated with naso-ocular symptoms. This ASTHMA phenotype is called aspirin-induced ASTHMA (AIA) or aspirin-exacerbated respiratory disease (AERD), and it is characterized by the triad of vasomotor rhinitis, nasal polyposis, and ASTHMA [59–61]. AIA is not an allergic reaction to aspirin or any other NSAID but a severe adverse reaction caused by all NSAIDs in susceptible patients (sometimes referred to as PSEUDO-ALLERGIC reaction). All NSAIDs are contraindicated in patients who have a history of an AIA/AERD attack.

The detailed ethio-pathogenesis of AIA is not known, but it is believed to be mediated by cysteinyl LEUKOTRIENES. Cysteinyl LEUKOTRIENES (LTC_4, LTD_4, and LTE_4) are potent bronchoconstrictors. They are formed from arachidonic acid by 5-LIPOXYGENASE enzymes, while CYCLOOXYGENASEs metabolize arachidonic acid to PROSTANOIDS. It is assumed that when CYCLOOXYGENASE pathway is blocked with NSAIDs, a larger proportion of arachidonic acid is available for metabolism by 5-LIPOXYGENASE resulting in increased production of cysteinyl LEUKOTRIENES probably especially in patients with an altered 5-LIPOXYGENASE variant. There are data showing that patients with AIA/AERD-type ASTHMA excrete in the urine increased amounts of metabolites of cysteinyl LEUKOTRIENES following exposure to NSAIDs as compared to healthy controls or ASTHMA patients without NSAID-induced bronchoconstriction [60, 61].

33.6.5 Hypersensitivity and Skin Reactions

NSAIDs may cause different types of skin reactions, varying from mild rash to urticaria, solar erythema, and severe skin reactions (e.g., erythema multiforme or Stevens-Johnson syndrome). Severe HYPERSENSITIVITY reactions are very rare, but single cases of angioedema, eosinophilic lung reactions, and anaphylactic reactions (ANAPHYLAXIS) have been reported. BONE MARROW toxicity (aplastic anemia, agranulocytosis, and/or thrombocytopenia) is also an extremely infrequent adverse event. It should be considered that celecoxib and valdecoxib contain a sulfonamide group and should not be given to patients who report ALLERGY to sulfa-containing drugs.

33.6.6 Liver Toxicity

Among NSAIDs, liver toxicity is a relatively rare adverse event. Some compounds, especially diclofenac, sulindac, and nimesulide, have been reported to increase circulating levels of liver enzymes already at therapeutic doses, and the effect is reversible following discontinuation of the drug [62, 63].

ASA should be avoided in the treatment of fever in children with infectious diseases.

ASA has been associated with the development of hepatocellular failure and encephalopathy (Reye's syndrome) with high mortality rates when used in some viral infections [64, 65].

Paracetamol intoxication is associated with severe liver toxicity, which is discussed below, paragraph 33.6.9.

33.6.7 Effects on Fertility and Pregnancy

PROSTANOIDS, especially those derived from COX-2, have been proposed to regulate multiple stages in the female reproductive cycle as well as in parturition [66, 67]. In experimental models, COX-2-deficient mice have reduced fertility because of failures in ovulation, fertilization, implantation, and decidualization [68]. It is believed that COX-2-derived PROSTANOIDS may signal the time of ovulation in mammals. PROSTANOIDS may also mediate the generation of the proteolytic enzymes that rupture the follicle and seem to have a role in the implantation of the fertilized embryo into the myometrium. In addition, PROSTANOID concentrations in the semen are high, and PROSTANOIDS are supposed to participate in the augmenting of the sperm movement in the uterus and fallopian tubes. However, the possible contribution of NSAIDs use in human infertility problems remains unknown.

PROSTANOIDS belong to the endogenous mediators that induce uterine contractions during labor. NSAIDs may retard the onset and cause of parturition. In addition, experimental studies have shown that NSAIDs may induce a premature closure of ductus arteriosus in the offspring. For these reasons, NSAIDs are recommended not to be used at the end of pregnancy or as analgesics during the labor.

33.6.8 Effects on Bone Healing

PROSTANOIDS and NSAIDs may have clinically meaningful effects on bone. PROSTANOIDS regulate osteoblast and osteoclast functions, and in experimental models NSAIDs have been shown to retard fracture healing. The same phenomenon can be regarded as a beneficial and adverse effect of NSAID treatment: NSAIDs are occasionally used to prevent ectopic ossification of soft tissues, which is a known complication of certain types of major trauma or surgery. On the other hand, NSAIDs may retard fracture union, especially during the first days of the healing process. The clinical significance of that adverse effect in healthy adults is not clear but should be considered when treating patients with increased risk to fracture non-union [69–72].

33.6.9 Adverse Effects of Paracetamol: Liver Toxicity

The mechanism of action of paracetamol is different from that of NSAIDs as discussed previously in the paragraph 33.4.3: paracetamol does not significantly inhibit PROSTANOID synthesis in the periphery. Therefore, paracetamol is not anti-inflammatory although it has analgesic and antipyretic properties comparable to those of NSAIDs. Furthermore, paracetamol is free of the common adverse effects of NSAIDs that are mediated through their inhibitory effect on PROSTANOID synthesis. The gastrointestinal, cardiovascular, and renal adverse effects of paracetamol are comparable to placebo. In addition, paracetamol is safe for most asthmatics with AIA/AERD, and it does not affect fertility, labor, or fracture healing [18, 20, 21].

The most noteworthy adverse effect of paracetamol is liver toxicity. Therapeutic doses of paracetamol (up to 4 g daily) are not associated with liver toxicity in healthy adults, but liver toxicity is a severe problem associated with paracetamol overdose. A single dose of 10–15 g causes a severe liver failure and 20–25 g can be fatal. Normally, liver toxicity is seen in connection with intentional overdose (attempted suicide) and is not a consequence of therapeutic use of paracetamol [20, 21].

When used at therapeutic doses, paracetamol is metabolized in the liver to glucuronide and sulfate conjugates. In overdose, those metabolic pathways are saturated, and an increased amount of a toxic and highly reactive metabolite, *N*-acetyl-*p*-benzoquinoneimine (NAPQI), is formed in reactions catalyzed by cytochrome P450 enzymes. Glutathione protects hepatocytes from the toxic effects of *N*-acetyl-*p*-benzoquinoneimine, but when glutathione reservoir is depleted, severe liver injury starts to develop [18, 20, 21].

Paracetamol intoxication is clinically insidious because during the first 2 days, there are only mild if any symptoms, possibly nausea and abdominal pain. The signs and symptoms of liver toxicity develop in 2–3 days. Plasma aminotransferases increase (in severe intoxication to very high values), and prothrombin time prolongs. *N*-Acetylcysteine is used as an antidote to treat paracetamol intoxication. It replenishes glutathione and thus increases the capacity of the liver to inactivate the toxic reactive metabolite of paracetamol. *N*-Acetylcysteine is effective only when given early in the course of the developing liver toxicity, preferably during the first 10–12 h (i.e., during the asymptomatic phase). In most severe cases, liver transplantation is needed to save the life of the patient [73–75].

33.7 Conclusions and Future Prospects

The concept to inhibit PROSTANOID synthesis has proved to be an effective means to treat pain and fever, and it also has some anti-inflammatory efficacy. However, it is accompanied with some potentially severe adverse effects. Therefore, attempts have been made to develop NSAIDs with a more favorable safety profile. COX-2 selective NSAIDs solved the problem partly as they have less gastrointestinal adverse effects, but they unfortunately appeared to be less safe than non-selective NSAIDs in patients with cardiovascular diseases.

Compounds to block specific PROSTANOID receptors are under investigation as well as inhibitors of the inducible form of PROSTAGLANDIN E synthase, namely, MICROSOMAL PROSTAGLANDIN E SYNTHASE 1 (mPGES-1). The former could specifically inhibit certain effects of a single PROSTANOID and refrain the other PROSTANOID functions intact. While the inhibitors of mPGES-1 are supposed to inhibit the increased synthesis of the most important PROSTANOID in connection to INFLAMMATION and inflammatory pain, namely PGE_2, in those conditions, but leave unaltered the synthesis of other PROSTANOIDS as well as physiological PGE_2 synthesis through other PGE synthases.

Dual inhibitors of COX and 5-LIPOXYGENASE enzymes were believed to have increased therapeutic effect because they inhibit also LEUKOTRIENE synthesis, and there were also early hopes of improved gastrointestinal safety. Unfortunately, they have not fulfilled the promise so far. Another development has been NITRIC OXIDE-releasing NSAIDs, which were promising in preclinical testing in their improved efficacy and safety, but the clinical studies did not show superiority over NSAIDs. In addition, hydrogen sulphide-releasing NSAIDs and lipoxin co-treatment has been investigated.

Paracetamol has an improved safety profile as compared to NSAIDs because it does not significantly inhibit PROSTANOID production in the periphery. Further understanding of its mechanism of action may open new avenues for the development of second-generation paracetamol-type analgesics.

Paracetamol and NSAIDs remain effective first-line drugs in several common indications through their analgesic, antipyretic, and, in the case of NSAIDs, anti-inflammatory properties. Their overall safety profile is favorable, but care should be taken to recognize patients with increased risk to develop serious adverse effects and treat them accordingly.

References

1. Vane JR. Inhibition of prostaglandin synthesis as a mechanism of action for aspirin-like drugs. Nat New Biol. 1971;231:232–5.
2. Ferreira SH, Moncada S, Vane JR. Indomethacin and aspirin abolish prostaglandin release from the spleen. Nat New Biol. 1971;231:237–9.

3. Smith JB, Willis AL. Aspirin selectively inhibits prostaglandin production in human platelets. Nat New Biol. 1971;231:235–7.
4. Collier JG, Flower RJ. Effect of aspirin on human seminal prostaglandins. Lancet. 1971;2:852–3.
5. Fu J-Y, Masferrer JL, Seibert K, Needleman P. The induction and suppression of prostaglandin H_2 synthase (cyclooxygenase) in human monocytes. J Biol Chem. 1990;265:16737–40.
6. Masferrer JL, Zweifel BS, Seibert K, Needleman P. Selective regulation of cellular cyclooxygenase by dexamethasone and endotoxin in mice. J Clin Invest. 1990;86:1375–9.
7. Xie W, Chipman JG, Robertson DL, Erikson RL, Simmons DL. Expression of a mitogen-responsive gene encoding prostaglandin synthase is regulated by mRNA splicing. Proc Natl Acad Sci U S A. 1991;88:2692–6.
8. Hla T, Neilson K. Human cyclooxygenase-2 cDNA. Proc Natl Acad Sci U S A. 1992;89:7384–8.
9. Jakobsson PJ, Thoren S, Morgenstern R, Samuelsson B. Identification of human prostaglandin E synthase: a microsomal, glutathione-dependent, inducible enzyme, constituting a potential novel drug target. Proc Natl Acad Sci U S A. 1999;96:7220–5.
10. Simmons DL, Botting RM, Hla T. Cyclooxygenase isozymes: the biology of prostaglandin synthesis and inhibition. Pharmacol Rev. 2004;56:387–437.
11. Duggan KC, Hermanson DJ, Musee J, Prusakiewicz JJ, Scheib JL, Carter BD, Banerjee S, Oates JA, Marnett LJ. (R)-profens are substrate-selective inhibitors of endocannabinoid oxygenation by COX-2. Nat Chem Biol. 2011;7:803–9.
12. Warner TD, Giuliano F, Vojnovic I, Bukasa A, Mitchell JA, Vane JR. Nonsteroid drug selectivities for cyclo-oxygenase-1 rather than cyclo-oxygenase-2 are associated with human gastrointestinal toxicity: a full in vitro analysis. Proc Natl Acad Sci U S A. 1999;96:7563–8.
13. Grosser T, Fries S, FitzGerald GA. Biological basis for the cardiovascular consequences of COX-2 inhibition: therapeutic challenges and opportunities. J Clin Invest. 2006;116:4–15.
14. Marnett LJ. Cyclooxygenase mechanisms. Curr Opin Chem Biol. 2000;4:545–52.
15. Rowlinson SW, Kiefer JR, Prusakiewicz JJ, Pawlitz JL, Kozak KR, Kalgutkar AS, Stallings WC, Kurumbail RG, Marnett LJ. A novel mechanism of cyclooxygenase-2 inhibition involving interactions with ser-530 and tyr-385. J Biol Chem. 2003;278:45763–9.
16. Loll PJ, Picot D, Garavito RM. The structural basis of aspirin activity inferred from the crystal structure of inactivated prostaglandin H2 synthase. Nat Struct Biol. 1995;2:637–43.
17. Spite M, Serhan C. Novel lipid mediators promote resolution of acute inflammation: impact of aspirin and statins. Circ Res. 2010;107:1170–84.
18. Graham GG, Davies MJ, Day RO, Mohamudally A, Scott KF. The modern pharmacology of paracetamol: therapeutic actions, mechanisms of action, metabolism, toxicity and recent pharmacological findings. Inflammopharmacology. 2013;21:201–32.
19. Högestätt ED, Jönsson BA, Ermund A, Andersson DA, Björk H, Alexander JP, Cravatt BF, Basbaum AI, Zygmunt PM. Conversion of acetaminophen to the bioactive N-acylphenolamine AM404 via fatty acid amide hydrolase-dependent arachidonic acid conjugation in the nervous system. J Biol Chem. 2005;280:31405–12.
20. Bertolini A, Ferrari A, Ottani A, Guerzoni S, Tacchi R, Leone S. Paracetamol: new vistas of an old drug. CNS Drug Rev. 2006;12:250–75.
21. Ghanem CI, Pérez MJ, Manautou JE, Mottino AD. Acetaminophen from liver to brain: new insights into drug pharmacological action and toxicity. Pharmacol Res. 2016;109:119–31.
22. Grosser T, Theken KN, FitzGerald GA. Cyclooxygenase inhibition: pain, inflammation, and the cardiovascular system. Clin Pharmacol Ther. 2017;102:611–22.
23. Samad TA, Moore KA, Sapirstein A, Billet S, Allchorne A, Poole S, Bonventre JV, Woolf CJ. Interleukin-1beta-mediated induction of cox-2 in the CNS contributes to inflammatory pain hypersensitivity. Nature. 2001;410:471–5.
24. Becker WJ. Acute migraine treatment in adults. Headache. 2015;55:778–93.
25. Marjoribanks J, Ayeleke RO, Farquhar C, Proctor M. Nonsteroidal anti-inflammatory drugs for dysmenorrhoea. Cochrane Database Syst Rev. 2015;7:CD001751.
26. Bartfai T, Conti B. Fever. SciWorldJ. 2010;10:490–503.
27. Patrono C. The multifaceted clinical readouts of platelet inhibition by low-dose aspirin. J Am Coll Cardiol. 2015;66:74–85.
28. Catella-Lawson F, Reilly MP, Kapoor SC, Cucchiara AJ, DeMarco S, Tournier B, Vyas SN, FitzGerald GA. Cyclooxygenase inhibitors and the antiplatelet effects of aspirin. N Engl J Med. 2001;345:1809–17.
29. Schuijt MP, Huntjens-Fleuren H, de Metz M, Vollaard EJ. The interaction of ibuprofen and diclofenac with aspirin in healthy volunteers. Br J Pharmacol. 2009;157:931–4.
30. Chen YF, Jobanputra P, Barton P, Bryan S, Fry-Smith A, Harris G, Taylor RS. Cyclooxygenase-2 selective non-steroidal anti-inflammatory drugs (etodolac, meloxicam, celecoxib, rofecoxib, etoricoxib, valdecoxib and lumiracoxib) for osteoarthritis and rheumatoid arthritis: a systematic review and economic evaluation. Health Technol Assess. 2008;12:1–278.
31. Laine L, Curtis SP, Cryer B, Kaur A, Cannon CP, MEDAL Steering Committee. Assessment of upper gastrointestinal safety of etoricoxib and diclofenac in patients with osteoarthritis and rheumatoid arthritis in the multinational etoricoxib and diclofenac arthritis long-term (MEDAL) programme: a randomised comparison. Lancet. 2007;369:465–73.
32. Mallen SR, Essex MN, Zhang R. Gastrointestinal tolerability of NSAIDs in elderly patients: a pooled analysis of 21 randomized clinical trials with celecoxib and nonselective NSAIDs. Curr Med Res Opin. 2011;27:1359–66.

33. Yuan JQ, Yang M, Threapleton DE, Qi XS, Ye DQ, Mao C, Tang JL, Higgins JP. Systematic review with meta-analysis: the gastrointestinal benefits of COX-2 selective inhibitors with concomitant use of low-dose aspirin. Aliment Pharmacol Ther. 2016;44:785–95.
34. Pajdo R, Brzozowski T, Szlachcic A, Konturek PC, Ptak-Belowska A, Drozdowicz D, Targosz A, Konturek SJ, Pawlik WW. Lipoxins, the novel mediators of gastroprotection and gastric adaptation to ulcerogenic action of aspirin. Curr Pharm Des. 2011;17:1541–51.
35. Rostom A, Dube C, Wells GA, Tugwell P, Welch V, Jolicoeur E, McGowan J, Lanas A. Prevention of NSAID-induced gastroduodenal ulcers. Cochrane Database Syst Rev. 2011;6:CD002296.
36. Lanza FL, Chan FK, Quigley EM, Practice Parameters Committee of the American College of Gastroenterology. Guidelines for prevention of NSAID-related ulcer complications. Am J Gastroenterol. 2009;104:728–38.
37. Chan FK, Wong VW, Suen BY, Wu JC, Ching JY, Hung LC, Hui AJ, Leung VK, Lee VW, Lai LH, Wong GL, Chow DK, To KF, Leung WK, Chiu PW, Lee YT, Lau JY, Chan HL, Ng EK, Sung JJ. Combination of a cyclo-oxygenase-2 inhibitor and a proton-pump inhibitor for prevention of recurrent ulcer bleeding in patients at very high risk: a double-blind, randomised trial. Lancet. 2007;369:1621–6.
38. Bakhriansyah M, Souverein PC, de Boer A, Klungel OH. Gastrointestinal toxicity among patients taking selective COX-2 inhibitors or conventional NSAIDs, alone or combined with proton pump inhibitors: a case-control study. Pharmacoepidemiol Drug Saf. 2017;26:1141–8.
39. Patrono C. Cardiovascular effects of cyclooxygenase-2 inhibitors: a mechanistic and clinical perspective. Br J Clin Pharmacol. 2016;82:957–64.
40. Mitchell JA, Kirkby NS. Eicosanoids, prostacyclin and cyclooxygenase in the cardiovascular system. Br J Pharmacol. 2019;176:1038-50.
41. Coxib and traditional NSAID Trialists' (CNT) Collaboration. Vascular and upper gastrointestinal effects of non-steroidal anti-inflammatory drugs: meta-analyses of individual participant data from randomised trials. Lancet. 2013;382:769–79.
42. Bally M, Dendukuri N, Rich B, Nadeau L, Helin-Salmivaara A, Garbe E, Brophy JM. Risk of acute myocardial infarction with NSAIDs in real world use: Bayesian meta-analysis of individual patient data. BMJ. 2017;357:j1909.
43. García Rodríguez LA, Tacconelli S, Patrignani P. Role of dose potency in the prediction of risk of myocardial infarction associated with nonsteroidal anti-inflammatory drugs in the general population. J Am Coll Cardiol. 2008;52:1628–36.
44. Bresalier RS, Sandler RS, Quan H, Bolognese JA, Oxenius B, Horgan K, Lines C, Riddell R, Morton D, Lanas A, Konstam MA, Baron JA, Adenomatous Polyp Prevention on Vioxx (APPROVe) Trial Investigators. Cardiovascular events associated with rofecoxib in a colorectal adenoma chemoprevention trial. N Engl J Med. 2005;352:1092–102.
45. Solomon SD, Wittes J, Finn PV, Fowler R, Viner J, Bertagnolli MM, Arber N, Levin B, Meinert CL, Martin B, Pater JL, Goss PE, Lance P, Obara S, Chew EY, Kim J, Arndt G, Hawk E, Cross Trial Safety Assessment Group. Cardiovascular risk of celecoxib in 6 randomized placebo-controlled trials: the cross trial safety analysis. Circulation. 2008;117:2104–13.
46. https://www.ema.europa.eu/en/medicines/human/referrals/celecoxib
47. https://www.ema.europa.eu/en/medicines/human/referrals/etoricoxib
48. https://www.ema.europa.eu/en/medicines/human/referrals/diclofenac-containing-medicines
49. https://www.ema.europa.eu/en/medicines/human/referrals/ibuprofen-dexibuprofen-containing-medicines
50. Chan C, Reid C, Aw T, Liew D, Haas S, Krum H. Do COX-2 inhibitors raise blood pressure more than nonselective NSAIDs and placebo? an updated meta-analysis. J Hypertens. 2009;27:2332–41.
51. Ruschitzka F, Borer JS, Krum H, Flammer AJ, Yeomans ND, Libby P, Luscher TF, Solomon DH, Husni ME, Graham DY, Davey DA, Wisniewski LM, Menon V, Fayyad R, Beckerman B, Iorga D, Lincoff AM, Nissen SE. Differential blood pressure effects of ibuprofen, naproxen, and celecoxib in patients with arthritis: the PRECISION-ABPM (prospective randomized evaluation of celecoxib integrated safety versus ibuprofen or naproxen ambulatory blood pressure measurement) trial. Eur Heart J. 2017;38:3282–92.
52. Kalafutova S, Juraskova B, Vlcek J. The impact of combinations of non-steroidal anti-inflammatory drugs and anti-hypertensive agents on blood pressure. Adv Clin Exp Med. 2014;23:993–1000.
53. Rahman S, Malcoun A. Nonsteroidal anti-inflammatory drugs, cyclooxygenase-2, and the kidneys. Prim Care Clin Office Pract. 2014;41:803–21.
54. Vučković S, Vujovic KS, Srebro D, Medic B, Ilic-Mostic T. Prevention of renal complications induced by non-steroidal anti-inflammatory drugs. Curr Med Chem. 2016;23:1953–64.
55. Arfè A, Scotti L, Varas-Lorenzo C, Nicotra F, Zambon A, Kollhorst B, Schink T, Garbe E, Herings R, Straatman H, Schade R, Villa M, Lucchi S, Valkhoff V, Romio S, Thiessard F, Schuemie M, Pariente A, Sturkenboom M, Corrao G. Non-steroidal anti-inflammatory drugs and risk of heart failure in four european countries: nested case-control study. BMJ. 2016;354:i4857.
56. Ungprasert P, Cheungpasitporn W, Crowson CS, Matteson EL. Individual non-steroidal anti-inflammatory drugs and risk of acute kidney injury: a systematic review and meta-analysis of observational studies. Eur J intern Med. 2015;26:285–91.
57. Zhang X, Donnan PT, Bell S, Guthrie B. Non-steroidal anti-inflammatory drug induced acute kidney injury

in the community dwelling general population and people with chronic kidney disease: systematic review and meta-analysis. BCM Nephrol. 2017;18:256.
58. Fored CM, Ejerblad E, Lindblad P, Fryzek JP, Dickman PW, Signorello LB, Lipworth L, Elinder CG, Blot WJ, McLaughlin JK, Zack MM, Nyren O. Acetaminophen, aspirin, and chronic renal failure. N Engl J Med. 2001;345:1801–8.
59. Jenkins C, Costello J, Hodge L. Systematic review of prevalence of aspirin induced asthma and its implications for clinical practice. BMJ. 2004;328:434.
60. Rodriguez-Jiminez JC, Moreno-Paz FJ, Teran LM, Guani-Guerra E. Aspirin exacerbated respiratory disease: current topics and trends. Resp Med. 2018;135:62–75.
61. Walgama ES, Hwang PH. Aspirin-exacerbated respiratory disease. Otolaryngol Clin North Am. 2017;50:83–94.
62. Aithal GP, Day CP. Nonsteroidal anti-inflammatory drug-induced hepatotoxicity. Clin Liver Dis. 2007;11:563–75.
63. Unzueta A, Vargas HE. Nonsteroidal anti-inflammatory drug–induced hepatoxicity. Clin Liver Dis. 2013;17:643–56.
64. Belay ED, Bresee JS, Holman RC, Khan AS, Shahriari A, Schonberger LB. Reye's syndrome in the united states from 1981 through 1997. N Engl J Med. 1999;340:1377–82.
65. Schrör K. Aspirin and reye syndrome: a review of the evidence. Pediatr Drugs. 2007;9:195–204.
66. Stone S, Khamashta MA, Nelson-Piercy C. Nonsteroidal anti-inflammatory drugs and reversible female infertility: is there a link? Drug Saf. 2002;25:545–51.
67. Bloor M, Paech M. Nonsteroidal anti-inflammatory drugs during pregnancy and the initiation of lactation. Anesth Analg. 2013;116:1063–75.
68. Lim H, Paria BC, Das SK, Dinchuk JE, Langenbach R, Trzaskos JM, Dey SK. Multiple female reproductive failures in cyclooxygenase 2–Deficient mice. Cell. 1997;91:197–208.
69. Vuolteenaho K, Moilanen T, Moilanen E. Nonsteroidal anti-inflammatory drugs, cyclooxygenase-2 and the bone healing process. Basic Clin Pharmacol Toxicol. 2008;102:10–4.
70. Ranganathan K, Loder S, Agarwal S, Wong VW, Forsberg J, Davis TA, Wang S, James AW, Levi B. Heterotopic ossification: basic-science principles and clinical correlates. J Bone Joint Surg Am. 2015;97:1101–11.
71. Dodwell E, Latorre J, Parisini E, Zwettler E, Chandra D, Mulpuri K, Snyder B. NSAID exposure and risk of nonunion: a meta-analysis of case–control and cohort studies. Calcif Tissue Int. 2010;87:193–202.
72. Kurmis AP, Kurmis TP, O'Brien JX, Dalén T. The effect of nonsteroidal anti-inflammatory drug administration on acute phase fracture-healing: a review. J Bone Joint Surg Am. 2012;94:815–23.
73. Heard KJ. Acetylcysteine for acetaminophen poisoning. N Engl J Med. 2008;359:285–92.
74. Ferner RE, Dear JW, Bateman DN. Management of paracetamol poisoning. BMJ. 2011;342:d2218.
75. Chiew AL, Gluud C, Brok J, Buckley NA. Interventions for paracetamol (acetaminophen) overdose. Cochrane Database Syst Rev. 2018;2:CD003328.

34 Disease-Modifying Anti-rheumatic Drugs

Kevin D. Pile, Garry G. Graham, Stephen M. Mahler, and Richard O. Day

34.1 Introduction

Rheumatoid arthritis (RA) is the most common INFLAMMATORY rheumatic disease, with a worldwide prevalence of about 1%. Targeting the synovial membrane, cartilage and bone, untreated RA leads to joint destruction, disability and increased mortality. Although the total incidence of this disease is low, the level of ill health and economic burden is significant with the patients often partially or totally unemployed. Generally, the patients require long-term drug treatment and non-pharmacological approaches such as physiotherapy and psychosocial support. There are no reliably curative or disease-remitting therapies, although considerable gains have been made utilizing biologic therapies and novel small molecules to target specific CYTOKINES (TUMOUR NECROSIS FACTOR (TNF) and INTERLEUKINS such as IL-1, IL-6, IL-12/IL-23 and IL-17), cellular subsets (B CELLS, T_H17 CELLS) and immune regulatory steps in RA (JANUS KINASE (JAK) pathway).

All RA patients require treatment with a medication that, in the absence of a cure, suppresses RA, i.e. a DISEASE-MODIFYING ANTI-RHEUMATIC DRUG (DMARD). Most recent additions to the DMARD armamentarium are biologically engineered products, primarily IMMUNOGLOBULIN constructs which are referred to as biological DMARDs (bDMARDs). CONVENTIONAL SYNTHETIC DMARD (csDMARD) refers to the older small molecule agents (SAARDs). TARGETED SYNTHETIC DMARDs (tsDMARDs) refer to the developing area of small molecule inhibitors of signalling pathway proteins. These terms are discussed in more detail below.

Long-term deleterious effects such as erosions of bone begin in the first 1–2 years of disease, and, consequently, initiation of therapy early in RA is important. This creates a window of opportunity whereby early disease suppression prevents lasting joint damage [1]. Early treatment with DMARDs, in addition to reducing erosive progression, improves symptoms and limits disability. For all INFLAMMATORY arthritic diseases, NON-STEROIDAL ANTI-INFLAMMATORY DRUGS (NSAIDs) are useful symptomatically but have no clear effect on the progression of the diseases (see Chap. 33). Many patients with RA

Final manuscript submitted on May 31, 2017.

K. D. Pile
School of Medicine, Western Sydney University, Sydney, NSW, Australia
e-mail: kevin.pile@sswahs.nsw.gov.au

G. G. Graham (✉) · R. O. Day
School of Medical Sciences, University of New South Wales, Sydney, NSW, Australia

Department of Clinical Pharmacology, St Vincent's Hospital, Sydney, NSW, Australia
e-mail: g.graham@unsw.edu.au

S. M. Mahler
Australian Institute for Bioengineering and Nanotechnology, University of Queensland, Brisbane, QLD, Australia
e-mail: s.mahler@eng.uq.edu.au

M. J. Parnham et al. (eds.), *Nijkamp and Parnham's Principles of Immunopharmacology*,
https://doi.org/10.1007/978-3-030-10811-3_34

or other INFLAMMATORY arthritic diseases are also administered intra-articular and/or low-dose oral CORTICOSTEROID. These are frequently administered as bridging therapy until the effect of a DMARD is fully established. In some patients, CORTICOSTEROIDS are used longer term to satisfactorily suppress the disease. Overall, the major challenge in therapeutics of RA is optimizing the timing, combination and dosage of available therapies to preserve joint function, quality of life and life expectancy.

It should be noted that the principles of the use of DMARDs in RA are paralleled in their treatment of other INFLAMMATORY diseases such as PSORIATIC ARTHRITIS, ANKYLOSING SPONDYLITIS, juvenile chronic arthritis and INFLAMMATORY BOWEL DISEASE. However, NSAIDs may be used in the long-term treatment of ANKYLOSING SPONDYLITIS for some further improvement in the patient's condition.

34.2 Pathophysiology of RA

The concept of a genetic predisposition and an environmental trigger has been applied to nearly all AUTOIMMUNE DISEASES, including RA.

Genetic factors contributing a substantial 50–65% to the risk of developing RA include:

1. The MAJOR HISTOCOMPATIBILITY COMPLEX genes, HLA-DR1 and HLA-DR4 (containing the shared EPITOPE) which associate strongly with disease severity.

   Other RA-associated loci include:
2. CYTOTOXIC T-LYMPHOCYTE ANTIGEN 4 (CTLA4) affecting the second co-stimulatory signal between ANTIGEN-presenting cells and T CELLS.
3. Protein tyrosine phosphatase, non-RECEPTOR type 22 (PTPN22) which reduces T CELL signalling allowing autoreactive T CELLS to escape deletion.
4. CYTOKINE and CYTOKINE RECEPTOR loci (TNF, IL-1, IL-4, IL-10, IL-18 and IL-12/IL-23).
5. PEPTIDYL ARGININE DEIMINASE TYPE 4 (PADI 4), the enzyme catalysing the conversion of arginine residues to citrulline residues in some peptides and proteins which are ANTIGENS to which ANTIBODIES develop, e.g. anti-cyclic citrullinated peptide (anti-CP). The time of appearance of such ANTIBODIES may precede the development of clinical synovial inflammation and therefore act as markers of RA.

Environmental triggers for RA are postulated, although despite an intensive search for transmissible agents, no infectious cause has been proven. Smoking is the strongest environmental risk factor for RA, with a significant multiplicative interaction between heavy smoking (>10 pack years) and any HLA shared EPITOPE resulting in an increased risk of seropositive RA with an odds ratio of 7.5 if double copy of the shared EPITOPE is present [2]. Smoking reduces NATURAL KILLER CELLS and depresses hormonal systems and cell-mediated immunity, leading to dysfunction of T CELLS [3], and may enhance the citrullination of self-peptides within the lung and initiate the formation of anti-cyclic citrullinated peptide ANTIBODIES. RA is associated with periodontitis, and its causative organism *Porphyromonas gingivalis* may facilitate local citrullinated peptide production via its endogenous PADI. Epidemiological studies confirm the association of RA with pulmonary silica exposure and have broadened this to include textile dust, thus maintaining a focus on the immune-environmental interaction of the respiratory tract mucosa.

Despite the uncertainly relating to the initiating events, the ultimate pathology of RA, namely, synovitis, is well established. The normal synovium is a thin and delicate layer that reflects off the cartilage-periosteal border on to the underlying fibrous joint capsule. Synovium is composed of cells of FIBROBLAST and MACROPHAGE origin, and it has two major functions: the provision of oxygen and nutrients to cartilage via the synovial fluid and the production of lubricants, notably hyaluronic acid, that allow the articular surfaces to glide smoothly across each other.

In RA, the synovium is transformed into a chronically inflamed tissue (Fig. 34.1). The normally thin synovial layer thickens dramatically due to accumulation of MACROPHAGE-like and

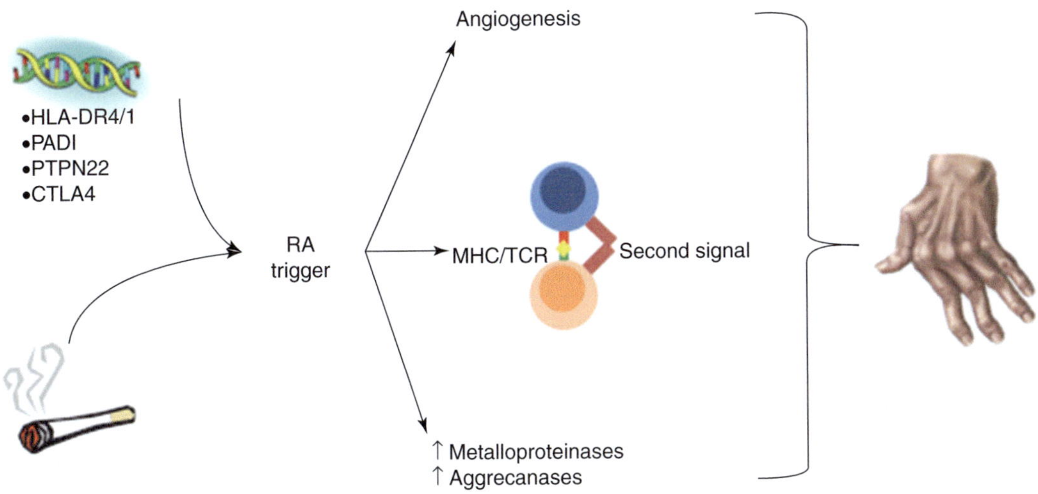

Fig. 34.1 Schematic diagram of events in development of RA

FIBROBLAST-like synoviocytes, and the subsynovial layer becomes oedematous, hypervascular and hypercellular with the accumulation of MACROPHAGES, MAST CELLS, CD4+ T CELLS (T_H1, T_H17, and Treg), CD8+ T CELLS, NATURAL KILLER (NK) CELLS, B CELLS and plasma cells. The increased number of cells in the synovium in RA is believed to result from recruitment of blood-derived LEUKOCYTES, as well as increased proliferation and reduced APOPTOSIS. NEUTROPHILS are abundant in rheumatoid synovial fluid but are sparse within the synovium.

In concert with this INFLAMMATORY process, the inflamed synovium invades adjacent cartilage and bone. Cartilage injury is caused by the generation of degradative enzymes, including MATRIX METALLOPROTEINASE (MMP), AGGRECANASES and MYELOPEROXIDASE which may be a source of neoANTIGEN and perpetuation of the inflammation (Fig. 34.1). Bone injury is separately mediated by a process of osteoclast activation, regulated by the RANKL (RECEPTOR ACTIVATOR OF NF-κB-RANK-LIGAND) system which is induced by MACROPHAGES, LYMPHOCYTES and/or FIBROBLAST production of CYTOKINES. The cells, CYTOKINES, receptors and signalling pathways operative in these processes constitute therapeutic targets for biological therapies in RA, and many are influenced by existing DMARDs.

Cellular expansion in the synovium is accompanied by significant overproduction of CYTOKINES which function as crucial effectors in the pathogenesis of RA and are potential therapeutic targets [4]. Compelling evidence from in vitro and animal studies, as well as the effects of human treatment with CYTOKINE antagonists, indicates that various CYTOKINES drive continued cell recruitment and expansion, ANGIOGENESIS and the production of pro-INFLAMMATORY mediators such as PROSTAGLANDINS and reactive oxygen and nitrogen species (Table 34.1). As CYTOKINES such as TNF, IL-1 and MACROPHAGE migration inhibitory factor (MIF) can induce their own synthesis and secretion, a self-perpetuating INFLAMMATORY cycle of amplification ensues, leading to the chronic INFLAMMATION of RA.

Advances in our understanding of the pathophysiology of RA have been mirrored in the development of targeting therapies, with the largest group to date being anti-CYTOKINE therapy (Table 34.1). Early CYTOKINE research identified TNF and IL-1 as important in joint inflammation and destruction. Subsequently a range of TNF inhibitors such as MONOCLONAL ANTIBODIES (INFLIXIMAB, ADALIMUMAB, GOLIMUMAB), PEGYLATED MONOCLONAL part ANTIBODIES, (CERTOLIZUMAB PEGOL) and soluble TNF receptors (ETANERCEPT) have been found to be active in the treatment of RA. However, an IL-1 receptor antagonist (ANAKINRA) has little clinical use in RA.

TNF can be produced by a variety of cells including monocytes, T CELLS, B CELLS, NK cells, GRANULOCYTES, keratinocytes and smooth muscle cells, from an equally diverse range of stimuli: bacteria, viruses, trauma, irradiation, hypoxia and CYTOKINES (e.g. IL-1 and IL-17). TNF in particular has key roles within the INFLAMMATORY network as well as heading an INFLAMMATORY cascade, leading to MONOCYTE and NEUTROPHIL activation, PROSTAGLANDIN release, APOPTOSIS of NEUTROPHILS and T CELLS, increased expression of adhesion molecules and increased MMPs (Table 34.1).

T CELLS are implicated in RA on the basis of the genetic association with the HLA shared EPITOPES and the protein tyrosine phosphatase, non-receptor type 22 (PTPN22) receptor loci as well as their presence in the synovium and their requirement in animal models of disease. The functionality of T CELLS is dependent on their activation that, in contrast to many other signalling pathways, requires a dual signalling mechanism (see Chap. 3). The first is the presentation by an ANTIGEN-presenting cell of an HLA-peptide complex to the ANTIGEN-specific T CELL receptor. The second or co-stimulatory signal is the linking of the ANTIGEN-presenting cell surface ligand, CD80/CD86, with its cognate receptor CD28 on the T CELL. In the absence of co-stimulation, the HLA-peptide complex-T CELL receptor interaction induces a state of T CELL anergy in which the T CELLS remain in a functionally inactivated, hyporesponsive state. A naturally occurring attenuator of T CELL activation is CTLA4, which is homologous to CD28 and binds to CD80/CD86 with tenfold greater affinity. Current concepts suggest that the inhibition of co-stimulation of T CELLS is the key mode of action of CTLA4. As well as impacting on ANTIGEN presentation, CTLA4 appears to inhibit the RANKL system and TNF-mediated differentiation of MONOCYTES to osteoclasts and may explain part of the anti-erosive effect of ABATACEPT in RA [5]. ABATACEPT is a fusion molecule of the extracellular domain of CTLA4 and IgG1 Fc domain, which binds to CD80/CD86, thus preventing co-stimulatory binding on naive T CELLS resulting in anergy. Because ABATACEPT predominately targets naive T CELLS, other pathways are less likely to

Table 34.1 Cytokines involved in the pathogenesis of RA, potential therapeutic targets and targeting bDMARD

Cytokine	Source cells	Cytokine function	Targeting bDMARD
TNF	Monocytes, T and B cells, NK cells, mast cells, synoviocytes and osteoblasts	Proliferation and differentiation of T and B cells and NK cells. Activation of monocytes and neutrophils, prostaglandin synthesis, apoptosis of neutrophils and T cells, increased expression of leukocyte and endothelial adhesion molecules, increased MMP production and increased free fatty acid release from adipocytes	Infliximab Adalimumab Golimumab Certolizumab pegol Etanercept
IL-6	Monocytes, macrophages, T and B cells	B cell maturation (and hence antibody production), stimulation of hepatocyte CRP production, haematopoiesis and thrombopoiesis	Tocilizumab Sirukumab
IL-17	T_H17 cells	Osteoclastogenesis, MMP and inhibition of chondrocyte glycosaminoglycan synthesis, angiogenesis, enhance synovial infiltration by immune cells	Secukinumab
IL-12/IL-23	Macrophages and dendritic cells	IL-12 is a potent inducer of interferon-gamma and promotes differentiation of T cells to T_H1 cells IL-23 promotes the production of the pro-inflammatory cytokines IL-17 and IL-22 from T_H17 cells in turn promoting increased levels of TNF and IL-1β	Ustekinumab

be affected, preserving innate immunity and reactivation of memory T CELLS [6].

The RA paradigm has broadened from a predominantly T-helper 1 (T_H1) cell-mediated disorder to a model in which T_H17 (IL-17-producing) T CELLS are also crucial. T_H17 cells derive from naive CD4+ cells under the influence of IL-21 and IL-23, producing IL-17, IL-21 and IL-22 (Table 34.1). IL-17 drives NEUTROPHIL differentiation and maturation; activation of NEUTROPHILS, monocytes and synovial FIBROBLASTS; and increased synthesis of PROSTAGLANDIN and MMP synthesis. The cumulative effects of IL-17 suggest a potent role for this CYTOKINE in joint damage. T_H17 cells are induced to produce IL-17 by IL-6, IL-23 and TGF-β. The latter in combination with IL-6 promotes differentiation into T_H17 cells but alone favours the induction of regulatory T cells. IL-6 and IL-23 can induce T_H17 cells to produce IL-22, with the latter shown to promote INFLAMMATION in the skin, but its production may be modulated in the synovium by the opposing effect of TGF-β. IL-6 additionally is important in B CELL maturation and therefore production of AUTOANTIBODIES (rheumatoid factor, anti-CP ANTIBODIES) It is also important inducing the hepatic production of C-REACTIVE PROTEIN (CRP) and HEPCIDIN (responsible for the anaemia of RA) and affect the HYPOTHALAMIC-PITUITARY-ADRENAL AXIS to cause fatigue. IL-6 is therapeutically targeted by the anti-IL-6R ANTIBODY, TOCILIZUMAB, and anti-IL-6, sirukumab, which has shown activity in clinical trials. The growing importance of the IL-17/IL-12/IL-23 axis is demonstrated by the successful introduction of the specific anti-IL-17A monoclonal ANTIBODY USTEKINUMAB, and SECUKINUMAB, which target both IL-12 and IL-23 (Table 34.1).

CYTOKINES, CHEMOKINES, ANTIBODIES/ANTIGENS and cell-cell interaction via signalling pathways leads to altered gene expression and stabilization of mRNA or permits the cell to change its activation status for a particular *response*. These coordinated intracellular signal transduction cascades utilize the NF-κB pathway, MITOGEN-activated protein (MAP) kinases, JAK/signal transducers and activators of transcription (STAT) and SPLEEN TYROSINE KINASE (SYK) (see Chap. 9). Because of the cross-signalling and amplification loops involved, targeting a component of one pathway may also lead to inhibition of another pathway. These kinase (phosphorylating) enzymes are amenable to inhibition by new small molecule tsDMARDs (see definitions below), which can be produced using lower-cost techniques compared to bDMARD [7].

34.3 Disease-Modifying Anti-rheumatic Drugs (DMARDs): Definitions

DMARDs are drugs which decrease the inflammation, slow the damage to joints and decrease the systemic effects of RA and are used or investigated for other INFLAMMATORY diseases. There are three major groups of DMARDs:

- CONVENTIONAL SYNTHETIC DMARDs (csDMARDs). These are the older smaller molecular mass drugs (e.g. ANTIMALARIALS, METHOTREXATE, SULFASALAZINE and LEFLUNOMIDE). Members of this group are often termed slow-acting anti-rheumatic drugs (SAARDs) because their actions develop slowly.
- TARGETED SYNTHETIC DMARDs (tsDMARDs). This is a new group which including inhibitors of JAK enzymes and phosphodiesterase 4 inhibitors. These drugs have low molecular masses and have been synthesized to target specific processes.
- BIOLOGICAL DMARDs (bDMARDs).The bDMARDs contain protein (polypeptide) structures (mostly monoclonal ANTIBODIES) which are targeted to various INFLAMMATORY processes. Their effects are produced more quickly than csDMARDs, and they are not termed SAARDs.

34.4 csDMARDs

The most commonly used traditional csDMARD is METHOTREXATE, and, consequently, particular attention has been paid to this agent in this chapter. Other widely used csDMARDs are LEFLUNOMIDE, SULFASALAZINE and, the antimalarial, HYDROXYCHLOROQUINE. The majority of RA patients are initially treated with a single or combination of older csDMARDs. However, as efficacy data mounts and costs fall, earlier introduction of bDMARDs and the newer csDMARDs is occurring. Their profound effects on the immune system result in an increased risk of acquired and reactivated infections. Pragmatically, all RA patients should be screened, before treatment, for tuberculosis, hepatitis B and C virus, with updated screening if an exposure risk has occurred.

METHOTREXATE monotherapy has become the standard first-line treatment for RA. Subsequent treatment is more variable and depends upon a variety of factors including concomitant diseases and the severity of the RA. If the response to METHOTREXATE is insufficient, then another csDMARD is often added quickly, and, if the efficacy is still inadequate, a bDMARD is frequently used with METHOTREXATE. Combinations, particularly with METHOTREXATE, are recommended on the basis of systematic randomized clinical trials (RCT) showing that combination treatment is more effective in both early and late RA (see Sect. 34.11.7).

The British guidelines [8] are unchanged since 2009 and exclude a bDMARD for first-line therapy as not cost-effective. US guidelines similarly recommend METHOTREXATE alone for early and active RA, but appropriate treatment for moderate or severe RA can be very variable. Combinations are the most common recommendations [9].

Therapy with a csDMARD results in up to 15% of RA patients achieving a sustained disease-free remission, which is predicted by acute onset of symptoms, shorter duration before treatment, non-smoker status, undetectable IgM RHEUMATOID FACTOR, absence of HLA shared EPITOPE alleles and minimal radiographic damage of joints at baseline [10].

Several active csDMARDs have little current use and are grouped together in this chapter as minor DMARDs. These include AZATHIOPRINE, CYCLOSPORIN, GOLD COMPLEXES, PENICILLAMINE and TETRACYCLINES. The TETRACYCLINES are major antibiotics, while AZATHIOPRINE and CYCLOSPORIN have important uses in diseases other than RA, and the orally administered gold complex, AURANOFIN, is being evaluated for several diseases other than RA. The bDMARD, ANAKINRA, is an inhibitor of IL-1 but is a minor DMARD for the treatment of RA and is discussed under Minor bDMARDs (see below) (Table 34.2).

34.4.1 Clinical Measures Used in the Evaluation of Anti-rheumatic Drugs

ACR20/50/70: Clinical trials in RA commonly use the ACR percentage response developed by the American College of Rheumatology. The ACR20 score is a composite improvement in the symptoms of RA and indicates a 20% improvement in tender and swollen joint count plus 20% improvement in at least three of five of the following criteria: patient pain assessment, physician global assessment, patient global assessment, patient self-assessed disability and acute-phase reactant concentrations in plasma. Although a reliable discriminator between active and placebo in RCT, the ACR20 represents only a small clinical improvement, with ACR50 and ACR70 scores being more clinically meaningful. As percentage changes, the ACR measures require comparisons in clinical status before and during treatment.

Disease activity score (DAS) was developed by the European League Against Rheumatism. Like the ACR scores, the DAS is a composite score calculated from a count of active and swollen joints, INFLAMMATORY marker levels (CRP, erythrocyte sedimentation rate) and patients' general health rating from a visual analogue scale. The DAS score is a complex calculation requiring computer assistance.

The simplified and clinical disease activity indexes (SDAI, CDAI) are the numerical sum of their components, being the 28 tender joint count, 28 swollen joint count, patient global assessment

Table 34.2 Anti-TNF peptides used as bDMARD

Substem in name	Example	Structure
-xi-	Infliximab	Mouse-human chimeric antibody The variable (V) domains of the heavy (V_H) and light (V_L) chains of a mouse antibody to human TNF are cloned and fused to the constant domains of a human antibody, thus generating a chimeric antibody. Thus, the constant regions of the mouse antibody are removed and replaced by the constant human regions. This reduces the amount of mouse antibody sequence by approximately two thirds, but the remaining framework and complementarity determining regions (CDRs) still originate from the mouse
-zu-	Certolizumab pegol	Humanized The human CDR sequences of an irrelevant, fully human antibody are mutated to those of the sequences of the CDR regions of a mouse antibody to human TNF. The only remaining sequence of mouse origin is the CDRs, i.e. the peptide loops located in the V domain and supported by the framework scaffold that are responsible for contacting and binding antigen. All the structure is human apart from mouse CDRs
-u-	Adalimumab	Fully human antibody Adalimumab was isolated through humanizing an existing anti-human TNF mouse antibody using the technique termed guided selection and phage display technology. The V_H and V_L genes of a mouse antibody to human TNF are separated and sequentially paired with separate human genes for V_H and V_L domains from human antibody gene libraries; i.e. the V_H of the mouse antibody are paired with a diverse range of human V_L domains. antibody proteins are generated and screened for binding to human TNF by phage display. Subsequent pairing of the isolated human TNF-specific V_L with a library of human V_H regions, followed by screening for optimal binding to human TNF, results in the isolation of a fully human antibody V_H and V_L domains. In vitro affinity maturation to increase affinity is performed by mutating CDR regions and selecting for optimal TNF binders. Selected V_H and V_L genes are subsequently fused to a human IgG1 Fc domain. The final protein is a variant of a natural fully human antibody with variation occurring in the CDR regions
-u-	Golimumab	Golimumab was isolated using genetically engineered mice with human humoral immune systems; i.e. human immunoglobulin gene loci are incorporated into the genomes of the mice, so that a humoral immune response generates antibodies with human sequence. Immunization with human TNF allows the subsequent generation of mouse hybridomas secreting human TNF-specific human antibodies
	Etanercept	Etanercept is a soluble TNF receptor fusion protein dimer. The TNF type 2 high affinity p75 receptor is fused to the Fc portion of human IgG1. Similarly to antibodies, the Fc regions form a dimer

and evaluator global assessment for both indices, while the SDAI includes CRP in mg/dL. They all have the advantage of providing individualized values that give a result at any point in time reflective of disease activity (Table 34.3). Earlier therapeutic intervention has been combined with strategies to achieve a predetermined target of minimum disease activity utilizing these

Table 34.3 Clinical assessment and measures of severity of symptoms of rheumatoid arthritis[a]

Activity index	Cutoffs			
	Remission	Low disease activity	Moderate disease activity	High disease activity
DAS28-ESR or CRP	<2.5	2.6–3.2	3.3 to <5.1	5.2+
SDAI	<3.3	3.4–11	12–26	27+
CDAI	<2.8	2.9–10	11–22	23+

Nomenclature and structures (Fig. 34.2)
[a]ACR20, ACR50 and ACR70 (see text) are indicators of improvement, whereas these measures are indicators of disease activity at any time

composite outcome measures or a simpler pragmatic approach of aiming at normalization of INFLAMMATORY markers and lowest possible count of swollen and tender joints.

Radiographic progression commonly utilizes the modified Sharp score, which scores both joint space narrowing and erosions in the hands and feet. Results are presented as total scores, joint space scores and erosion scores.

34.4.2 Difficulties in Evaluating the Therapeutic Efficacy of Anti-rheumatic Drugs

The RCT is the standard method of evaluating the efficacy of drugs, and the results of such trials on anti-rheumatic drugs are quoted widely in this chapter. As examples, comparisons of treatments including active drug versus placebo or between two or more active drugs are discussed. However, the results of RCTs often do not indicate the overall benefit of anti-rheumatic drugs in clinical practice. Many patients treated with anti-rheumatic drugs in normal clinical practice have less active disease than patients in RCTs [11]. Other problems of RCTs in RA include the often constant dosage regimens, the lack of published data on individual patients, the use of single drugs and, because of the limited number of patients in a clinical trial, the inability to detect rare adverse reactions. RCTs are, of course, still of great value, but they do have limitations.

34.5 bDMARDs

As outlined above (see Sect. 34.2), several CYTOKINES have major pro-INFLAMMATORY actions and are also involved with the systemic effects of RA. Thus, the neutralization of these agents leads to profound inhibition of INFLAMMATION in joints. bDMARDs target CYTOKINES in the extracellular environment using MONOCLONAL ANTIBODIES against the CYTOKINE or its receptor, or soluble receptor constructs to act as a decoy to the CYTOKINE. A return to traditional pharmacological small molecule inhibition is seen with the development of intracellular CYTOKINE modulators targeting the pathways between cell surface receptor and CYTOKINE translation. These are the tsDMARDs.

Several bDMARDs are now available or in clinical trial for the treatment of RA, the major bDMARDs being the TNF inhibitors (Table 34.2). The TNF inhibitors and other anti-CYTOKINES have revolutionized the treatment of RA and other INFLAMMATORY diseases. Indeed, their success is leading to an unprecedented phase of therapeutic target development. A problem with the bDMARDs is that they require parenteral administration either by subcutaneous injection or intravenous infusion, plus attention to temperature controlled storage, which are not factors with most csDMARDs or tsDMARDs.

The available anti-CYTOKINE biological agents are effective in the treatment of RA either when used alone or in combination with csDMARDs, particularly METHOTREXATE. Overall, about two thirds of patients respond with a clinically significant degree of disease control. The bDMARDs targeting TNF, IL-6, B CELLS and CTLA4 generally have similar clinical efficacy with ANAKINRA less effective. Compared to traditional csDMARDs, bDMARDs generally show a more rapid response with slowing of the radiographic progression of RA. Combinations of TNF antagonists with low-dose METHOTREXATE have generally been more effective than either drug alone in the treatment of RA (see Table 34.6 below).

The bDMARDs typically have long half-lives of elimination with considerable interpatient variation. Typically, the bDMARDs have an initial faster elimination phase which can be seen after their intravenous dosage [12]. Subcutaneous dosage generally leads to slow and incomplete absorption, again with interpatient variation. An example is shown in Fig. 34.2.

The inter-patient variation may have considerable clinical significance. The bDMARD inhibitors typically have poor effect in 20–40% of patients, while some patients may respond to particular bDMARD and not others. It has been suggested that a higher than average clearance and the consequent lower plasma concentrations may, in some patients, lead to impaired anti-rheumatic

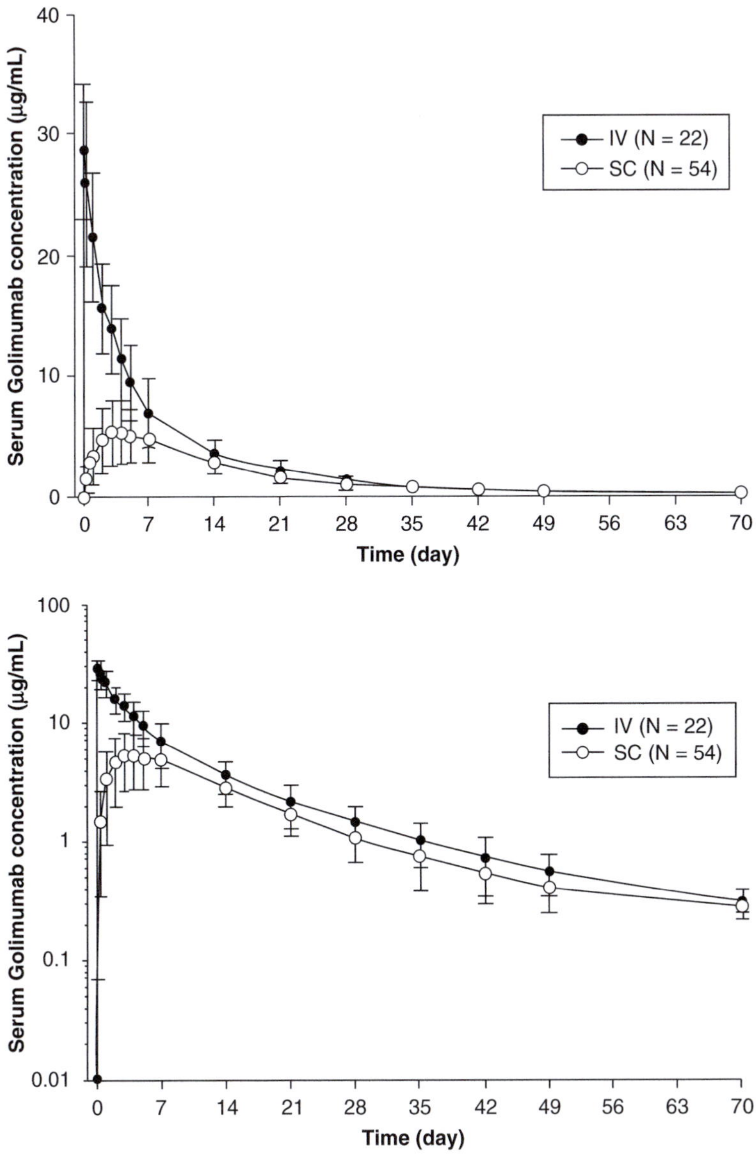

Fig. 34.2 Time courses of serum concentrations of golimumab. Upper. Linear concentration scale. The relative areas under the time courses (subcutaneous/intravenous) indicate the fractional absorption of golimumab after subcutaneous dosage (about 50%). Concentrations = mean ± SD. Lower. Logarithmic concentration scale. The nonlinearity indicates multiphasic elimination of golimumab. From Xu et al. [12]

response. This is a principle of the clinical pharmacology of conventional drugs, but, as yet, there is only limited evidence for this relationship with bDMARDs. However, INFLIXIMAB has shown an improved response if trough plasma concentrations are over approximately 1 μg/mL [13–15].

34.6 Chemistry and Targets

34.6.1 Tumour Necrosis Factor (TNF)

34.6.1.1 Rationale

TNF has key roles within the INFLAMMATORY network as well as heading an INFLAMMATORY cascade (Table 34.1). Anti-TNF therapy quickly suppresses LEUKOCYTE migration, deactivates endothelial cells and promotes recovery of regulatory T CELL function and phenotype.

There are three different types of anti-TNF agents (Table 34.2; Fig. 34.3) [16]. INFLIXIMAB targets TNF and was the first CHIMERIC (murine-human) MONOCLONAL ANTIBODY to be approved for the treatment of AUTOIMMUNE DISEASES. Subsequent MONOCLONAL ANTIBODIES targeting TNF include ADALIMUMAB and GOLIMUMAB (Table 34.2; Fig. 34.3). CERTOLIZUMAB PEGOL and ETANERCEPT are not complete IMMUNOGLOBULINS. CERTOLIZUMAB PEGOL is a humanized Fab fragment combined to polyethylene glycol. Thus, it does not have an Fc domain. As a consequence of its structure, CERTOLIZUMAB PEGOL does not bind to transmembrane TNF and is not actively transported across the placenta. ETANERCEPT is a soluble RECEPTOR construct containing an Fc IMMUNOGLOBULIN backbone and a dimer of the extracellular portion of human TNF

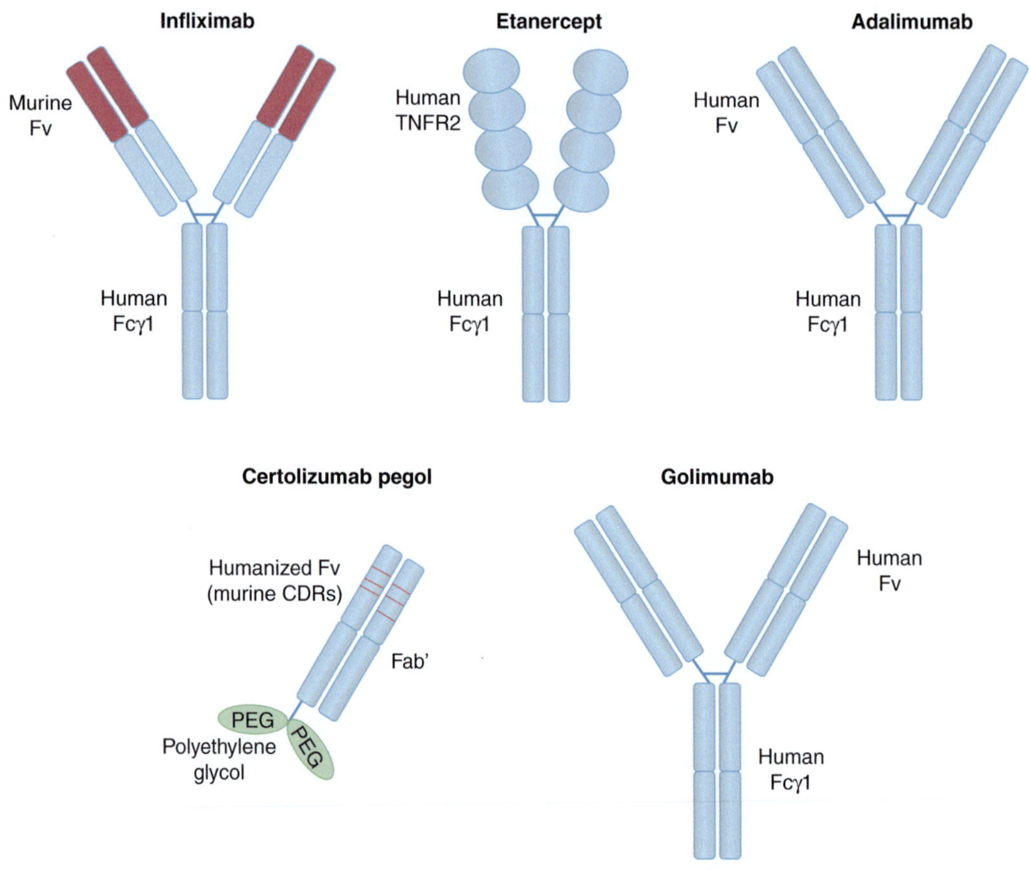

Fig. 34.3 Diagrammatic structures of TNF inhibitors

RECEPTORS (Fig. 34.3). It is thus a recombinant soluble TNF receptor fusion protein which binds both TNF and lymphotoxin although lymphotoxin blockade may not be clinically significant.

34.6.1.2 Clinical Indications and Efficacy

Therapeutic blockade of TNF yields a clinical response in ≈70% of RA patients, with the 60:40:20 rule indicating approximately 60% of patients achieve at least an ACR20 response, 40% at least an ACR50 or response and 20% an ACR70 response. Clinical response and remission are decreased in older patients, as well as patients with low functional status [17]. Introductory data indicates that the anti-rheumatic response to ADALIMUMAB increases with increasing expression of the mRNA of CD11c [18]. CD11c is a gene leading to the synthesis of integrins, proteins which are involved in the adherence of neutrophils and monocytes to endothelial cells. This clinical correlation, if confirmed, may indicate a further mode of action of the anti-TNF ANTIBODIES.

INFLIXIMAB is used in combination with METHOTREXATE which limits the development of neutralizing antidrug ANTIBODIES to the murine (mouse)-derived protein material in INFLIXIMAB. These ANTIBODIES may be responsible for the tachyphylaxis observed in some patients treated with this agent alone. Although other anti-CYTOKINES, such as ABATACEPT, and RITUXIMAB (see below) are not required to be administered with METHOTREXATE, they often are used with METHOTREXATE because the combinations are more effective than anti-cytokine treatment alone.

While anti-TNF ANTIBODIES are most often in the first line of bDMARD therapy, TOCILIZUMAB (see below) may be considered as first line, although background therapy with METHOTREXATE is often used concomitantly. There are patients who either do not respond, have adverse events, or due to co-morbidity are excluded from therapy with anti-TNF agents. Many of these apparent failures may be due to low plasma concentrations of the anti-TNF ANTIBODIES (see MONOCLONAL ANTIBODIES—bDMARD, PHARMACOKINETICS below), but research is still required to determine if a better response can be obtained by increasing the dose and hence the plasma concentrations of the anti-TNF ANTIBODIES.

In addition to their use in RA, most anti-TNF agents are active in the treatment of PSORIASIS, PSORIATIC ARTHRITIS and ANKYLOSING SPONDYLITIS. Although ETANERCEPT has approximately equal activity to the other anti-TNF agents in the treatment of RA, it may be less efficacious in the treatment of PSORIASIS [16]. Except for ETANERCEPT, the anti-TNF agents are also useful for GRANULOMATOUS diseases, such as Crohn's disease, Wegener's granulomatosis and sarcoidosis.

34.6.1.3 Pharmacokinetics

The anti-TNF ANTIBODIES have long half-lives of the order of 1–4 weeks [19, 20] (Fig. 34.2) which allow relatively infrequent dosage. The newer agents GOLIMUMAB and CERTOLIZUMAB PEGOL also have long $t_{1/2}s$ [21, 22], with GOLIMUMAB able to be administered subcutaneously monthly or intravenously three-monthly. The conjugation of peptides with polyethylene glycol (pegylation) is well known to slow the elimination of peptides. In the case of CERTOLIZUMAB PEGOL, pegylation extends the half-life of this partial ANTIBODY to the range seen with the full ANTIBODIES, i.e. approximately 20 days.

Interestingly, the initial half-life of INFLIXIMAB decreases with increasing severity of the rheumatic disease, possibly due to higher concentrations of TNF in plasma and tissues and the formation of ANTIBODY complexes with TNF.

The $t_{1/2}$ of ETANERCEPT is shorter than the other anti-TNF products with a half-life of 3–5 days [23], and, with absorption occurring over 2–3 days, once or twice a week dosage produces relatively constant plasma concentrations.

34.6.1.4 Adverse Effects

An increased incidence of infections is reported during trials and in post-marketing surveillance, presumably related to the IMMUNOSUPPRESSIVE effects of these agents, with particular risk on the reactivation of chronic infections such as tuberculosis. Meta-analysis shows that INFLIXIMAB and ADALIMUMAB double the risk of serious infection with a tendency towards greater likelihood of infections with increasing doses [24]. Anti-TNF agents can reactivate latent tuberculosis, and screening for tuberculosis is therefore required before treatment. There have also been reported increases in intracellular or opportunistic pathogens including *Legionella*, *Listeria* and *Salmonella*, with caution subsequently recommended on the consumption of raw eggs and unpasteurized milk. In general, anti-TNF therapy should be avoided in patients with active or recurrent infections. In particular, concomitant therapy with CORTICOSTEROIDS may increase the susceptibility to infections. Combined treatment with ANAKINRA and a TNF inhibitor causes an increased risk of infection, as has also been recently shown with ABATACEPT and ETANERCEPT. These combinations are contraindicated.

It has been postulated that inhibitors of TNF could reduce immune surveillance and lead to the development of tumours. A population-based study linking three RA cohorts to the Swedish cancer registry found the RA cohort was only at a marginally elevated overall risk of solid cancers. Not surprisingly, given smoking is associated with RA, smoking-related cancers were increased by 20–50%, and the risk for non-melanoma skin cancer increased by 70%. The good side was a 25% decreased risk for breast and colorectal cancer, with the cancer pattern in patients treated with anti-TNF agents mirroring historic RA cohorts [25]. In the same studies, there was two-fold increased risk of lymphoma and leukaemia, but not for myeloma. Among RA patients treated with anti-TNF agents, the risk of lymphoma was tripled, but was not higher than in other RA cohorts [26].

In children using TNF blockers, post-marketing surveillance found an increased risk of malignancy, with INFLIXIMAB having a higher reporting rate for lymphoma and all malignancies and ETANERCEPT having a higher reported rate for lymphoma only [27]. However, a clear causal link could not be established because of confounding due to the underlying illness and concomitant IMMUNOSUPPRESSIVES.

34.6.2 Inhibitors of B Cells, Tituximab

34.6.2.1 Rationale

Despite the presence of B CELLS in inflamed RA synovium and the presence of RHEUMATOID FACTOR being a poor prognostic factor, B CELLS were not specifically targeted in the treatment of RA until recent years. Originally developed for the treatment of B CELL lymphoma, the use of anti-CD20 ANTIBODY, RITUXIMAB, in RA did not have a strong rationale, but was noted to improve the synovitis of an RA patient with B CELL lymphoma. B CELLS are now realized to act as efficient ANTIGEN-presenting cells, promote T CELL accumulation and provide co-stimulatory signals and CYTOKINES to sustain T CELL activation in the RA synovium. CD20 is expressed after maturation B CELLS from stem cells and is lost when they mature to plasma cells. The CD20 ligand is an attractive target for IMMUNOTHERAPY on the basis that it does not internalize upon MONOCLONAL ANTIBODY binding and because it does not shed from the cell surface. Depletion of all CD20-bearing B CELLS, leaving only stem cells and plasma cells, reduces AUTOANTIBODY production and subsequent immune-complex formation and also reduces B CELL-derived CYTOKINES such as IL-6, TNF and lymphotoxin. B CELL depletion also prevents the formation of ectopic germinal centres and the optimal activation of B CELLS.

RITUXIMAB is a chimeric MONOCLONAL ANTIBODY made of mouse variable light and heavy variable domains of the anti-CD20 ANTIBODY together with the kappa-light chain domain and heavy chain constant domain of human IgG1 [28]. RITUXIMAB leads to rapid and prolonged depletion of CD20+ B CELLS via ANTIBODY-dependent cell-mediated CYTOTOXICITY, complement-dependent CYTOTOXICITY and induction of B CELL APOPTOSIS [28, 29].

Ocrelizumab and ofatumumab are the humanized and fully human anti-CD20 MONOCLONAL

ANTIBODIES, respectively, which have completed trials in RA, but have not advanced to market for RA. Ocrelizumab is likely to gain FDA approval for primary progressive and relapsing-remitting multiple sclerosis. Other strategies to modify B CELL activation and survival involve targeting of the B CELL proliferation-inducing ligand (APRIL) and B CELL-activating factor (BAFF), although the latter has not been successful in RA patients to date.

34.6.2.2 Clinical Indications and Efficacy

Controlled trials have demonstrated the efficacy of RITUXIMAB in RA, with the European Medicines Agency recommending it be considered in patients with clinically active RA with an inadequate response or intolerance to at least one other DMARD, including one or more anti-TNF agents. The recommended dose is 1 g, is repeated 14 days later and is different to the lymphoma schedule of 375 mg/m^2 given weekly for 4 weeks [30]. RITUXIMAB is given in combination with weekly METHOTREXATE 10–25 mg which improves the clinical response. This recommendation obviously limits the use of RITUXIMAB to patients who tolerate METHOTREXATE.

Peripheral B CELL count can be monitored during treatment with RITUXIMAB, with only partial depletion predicting a poor clinical response. The repopulation with B CELLS can be detected by the highly sensitive technique of fluorescence-activated cell sorting (FACS) of blood, but the dilemma is whether RITUXIMAB should be administered again on the return of B CELLS or active RA or what to do if disease returns without easily measurable B CELL repopulation. Repeat dosing with RITUXIMAB at approximately 6-month intervals may be an effective option with the efficacy observed after subsequent RITUXIMAB courses comparable to that seen during the initial course, irrespective of prior anti-TNF exposure. Further improvement has been noted with subsequent courses, even in patients who did not respond well to the first course of treatment. These patients often have higher numbers of preplasma cells before treatment and may respond to an earlier retreatment with RITUXIMAB [31].

34.6.2.3 Adverse Effects

In the clinical trials of RITUXIMAB, the overall frequency of adverse events and serious adverse events was similar in both RITUXIMAB and placebo-treated groups. A slightly higher proportion of patients receiving RITUXIMAB experienced an infusion reaction (23%) compared with placebo (18%), and this was less at the second infusion (see below). The frequency of adverse events remained stable after the first, second, third and fourth courses. Reduction in serum IgM concentration and to a lesser extent IgG was observed in some patients following repeated courses of therapy; however there was no link to an enhanced infection risk.

The risk of developing bacterial and viral infections with RITUXIMAB therapy was not significantly different to those with other bDMARDs and was predominantly recorded as respiratory tract bacterial infection. Serious infections occurred in 2.3% of RITUXIMAB-treated patients and 1.5% of controls, which is similar to the placebo rate but less than the 3.7% seen in anti-TNF-treated RA patients. Viral load increases in patients with hepatitis C-related mixed cryoglobulinaemia, reinforcing the need for screening, and dose modification and/or concomitant antiviral therapy is required.

34.6.3 Inhibitors of Interleukin-6 (IL-6), Tocilizumab

34.6.3.1 Rationale

The major inhibitor of IL-6 is TOCILIZUMAB which is a humanized MONOCLONAL ANTIBODY that binds both soluble and membrane-expressed IL-6R. TOCILIZUMAB was developed by grafting the complementarity determining regions of a nude mouse anti-human IL-6R ANTIBODY onto human IgG1. It is produced in suspensions of Chinese hamster ovary cells. Further anti-IL-6 ANTIBODIES are being developed: sarilumab which binds to IL-6R (like TOCILIZUMAB) and ANTIBODIES which bind directly to IL-6 (clazakizumab and olokizumab) [32, 33].

IL-6 is a MONOCYTE-derived CYTOKINE important in B CELL maturation (and hence

ANTIBODY production), osteoclastogenesis, T_H17 differentiation, stimulation of hepatocyte CRP production, haematopoiesis and thrombopoiesis. IL-6 signals primarily through a membrane-bound complex of the non-signalling IL-6 receptor (IL-6R) and two signal-transducing gp130 subunits [32, 34]. IL-6R is predominantly expressed on NEUTROPHILS, MONOCYTE/MACROPHAGESS, hepatocytes and some LYMPHOCYTES. However soluble IL-6R can be generated by proteolysis of the membrane-bound IL-6R or by alternative mRNA splicing. Released sIL-6R is able to complex with IL-6 and be transported in bodily fluids, and the IL-6: IL-6R complex is able to bind and activate the ubiquitously expressed signalling gp130 subunit which is found on a range of cells such as endothelial cells and synoviocytes. Thus, IL-6 has predictable actions via IL-6R-bearing cells but has a broader potential effect on any cell expressing gp130.

IL-6 mobilizes marginated NEUTROPHILS in the circulation, with large numbers of NEUTROPHILS found in synovial fluid. Binding of IL-6 to NEUTROPHIL membrane-bound IL-6R induces secretion of proteolytic enzymes and reactive oxygen intermediates causing cartilage degradation. During acute inflammation in RA, MONOCYTES, MACROPHAGES and endothelial cells release IL-6 accompanied by an increase in NEUTROPHILS in synovial fluid. Subsequently, IL-6 is thought to influence the shift from acute to chronic INFLAMMATION, marked by an increase in the recruitment of MONOCYTES.

IL-6 is the principal stimulator of acute-phase protein synthesis through hepatocyte stimulation, with levels correlating with CRP levels in RA patients. Hepatocyte production of HEPCIDIN is stimulated by IL-6 preventing iron transport and the release of iron from MACROPHAGES causing anaemia of chronic inflammation.

34.6.3.2 Clinical Indications and Efficacy

TOCILIZUMAB has a licensed dose of 8mg/kg infused over 1 h every 4 weeks. Several studies have shown its benefit in RA:

- SAMURAI study [35]. TOCILIZUMAB monotherapy was significantly superior to csDMARDs with ACR20/50/70 responses of 78%, 64% and 44%, with significantly less worsening of joint damage. The majority of the control patients in the SAMURAI study received METHOTREXATE either alone or in combination with other DMARDs, particularly SULFASALAZINE.
- LITHE study [36, 37]. The combination of METHOTREXATE and TOCILIZUMAB inhibited joint damage and improved function greater than METHOTREXATE alone. In RA patients who had failed to respond or been intolerant to an anti-TNF agent within the previous year, TOCILIZUMAB (8 mg/kg) plus METHOTREXATE was superior to METHOTREXATE alone.
- ADACTA trial [38]. In RA patients who had failed treatment with METHOTREXATE, TOCILIZUMAB monotherapy was superior to ADALIMUMAB monotherapy.

In giant cell arteritis, the addition of TOCILIZUMAB to a protocol-driven reduction of CORTICOSTEROID yielded higher rates of clinical remission at 12 weeks and greater relapse-free survival at 1 year [39].

34.6.3.3 Adverse Effects

Events particularly related to TOCILIZUMAB are neutropaenia, an increase in hepatic enzymes and cholesterol elevation. Caution should be taken in patients with a previous history of intestinal ulceration or diverticulitis.

Reduction in NEUTROPHIL count occurs frequently in patients with high baseline values and very frequently in those with normal baseline values. Few counts go below 1000 cells/mm^3, and there is no association between low NEUTROPHIL counts and infection-related serious adverse reactions. In trials, both TOCILIZUMAB interruption and DMARD dose modification were used, with a recommendation not to treat with NEUTROPHIL counts $<0.5 \times 10^{9\ \mathrm{L}^{-1}}$ or platelet counts $<50 \times 10^9\ \mathrm{L}^{-1}$. Elevation of transaminases (AST and ALT) up to three times the upper limit of normal occurs three times more often than that seen with DMARD only, with highest mean increases 2 weeks after each infusion. More than threefold elevation of enzymes is much less prevalent (2–4%) but remains 3–5 times higher than with DMARD alone. Total bilirubin increase within three times the upper limit of normal is also seen. It is recommended to be cautious in the use of TOCILIZUMAB in patients with active liver disease or elevated hepatic enzymes (50% above the upper limit of normal) and not to treat patients with ALT or AST more than five times the upper normal levels). Transaminases should be monitored every 1–2 months during the first 6 months of therapy and three monthly thereafter.

Upper gastrointestinal events suggestive of INFLAMMATION, gastritis or ulcer are also more common. Increases in fasting plasma lipids occur early after treatment and remain elevated during therapy with approximate mean changes: total cholesterol 0.8 mmol/L, HDL cholesterol 0.1 mmol/L and LDL cholesterol 0.5 mmol/L. Approximately 25% of patients receiving TOCILIZUMAB experienced sustained elevations in total cholesterol to at least 6.2 mmol/L. Lipid parameters should be measured 4–8 weeks after commencing therapy and managed according to local guidelines and taking into account the individual risk factors.

The rates of tuberculosis do not seem to increase, but all RA patients should be screened for latent disease.

34.6.4 Inhibitors of Second Co-stimulatory Signals, Abatacept

34.6.4.1 Rationale

ABATACEPT is a fusion protein produced from CTLA4 with the Fc portion of IgG1 [40]. As outlined above (see CYTOKINES in RA, Table 34.1), T CELLS are implicated in the pathogenesis of RA. ABATACEPT binds to CD80/CD86 on the surfaces of B CELLS and MONOCYTES leading to decreased activation of T CELLS and preventing the release of INFLAMMATORY mediators [40]. It may also inhibit myeloid cell function.

34.6.4.2 Clinical Indications and Efficacy

ABATACEPT is administered intravenously at 0, 2 and 4 weeks and then every 4 weeks (<60 kg, 500 mg dose; 60–100 kg, 750 mg dose; and >100 kg, 1000 mg dose). Background therapy with METHOTREXATE is recommended.

Monotherapy with ABATACEPT is effective in refractory RA previously treated with one or more DMARDS [41]. After 2-year treatment, 50% of patients had no progression of structural damage, with a suggestion that the disease-modifying effect may increase over time [42].

34.6.4.3 Adverse Effects

ABATACEPT is well tolerated but should not be administered in combination with ETANERCEPT due to increased serious adverse events, including serious infections, without any significant additional efficacy [43].

34.6.5 Inhibitors of Interleukins 12 and 23, Ustekinumab

34.6.5.1 Rationale

The involvement of IL-12 and IL-23 is described above (see CYTOKINES in RA above, Table 34.1). USTEKINUMAB is a fully human IgG1κ MONOCLONAL ANTIBODY, anti-IL-12p40, against the p40 subunit common to both IL-12 and IL-23 which prevents interaction with the IL-12 Rβ1 RECEPTOR. IL-12 is pivotal in the differentiation of naive Th cells to Th1 cells, and IL-23 is involved in the differentiation of naive Th cells to Th17. IL-23 has been implicated in the enthesitis and aortic root disease seen within the seronegative spondyloarthritis cluster of PSORIATIC ARTHRITIS and ANKYLOSING SPONDYLITIS. Th17-derived IL-12 facilitates INFLAMMATION and osteoproliferation as seen in the spine of patients with ANKYLOSING SPONDYLITIS, and the Th17-derived IL-17 is implicated in the inflammation and bone erosion seen with the sacroiliac joints and peripheral joints in PSORIATIC ARTHRITIS. Guselkumab, a human MONOCLONAL ANTIBODY, and tildrakizumab, a humanized monoclonal ANTIBODY, are being developed against the p19 subunit of IL-23.

34.6.5.2 Clinical Indications and Efficacy

On the basis of IL-12 and IL-23 expression in lesions, USTEKINUMAB has been trialled in PSORIASIS, PSORIATIC ARTHRITIS, CROHN'S DISEASE and multiple sclerosis. Significant efficacy has been shown in chronic plaque PSORIASIS, but the drug is of no value in multiple sclerosis. In PSORIATIC ARTHRITIS, USTEKINUMAB 45–90 mg is administered subcutaneously at 0 and week 4 and then every 12 weeks (Appendix 4). Significant gains over placebo were demonstrated with ACR joint improvement, in skin scores, dactylitis and enthesitis [44]. Radiographic score of joint appearance also improved. Significant activity has also been seen in Crohn's disease [45]. There have been insufficient studies to date to make any definite conclusion about its activity in RA.

Two of the newer ANTIBODIES, guselkumab and tildrakizumab, appear active in early trials in plaque PSORIASIS, PSORIATIC ARTHRITIS and ANKYLOSING SPONDYLITIS [46].

34.6.5.3 Adverse Effects

USTEKINUMAB is an immunosuppressant agent, and standard precautions for pre-existing serious infections, chronic viral disease and latent tuberculosis apply. HYPERSENSITIVITY occurs in <1%, and infections are similar to placebo-treated patients.

34.6.6 Inhibitors of Interleukin-17 (IL-17), Secukinumab, Ixekizumab and Brodalumab

34.6.6.1 Rationale

IL-17 is a key effector cytokine of the Th17 subset of Th cells and constitutes a family of six CYTOKINES of which IL-17A and IL-17F are the most important members (A more potent than F). These cytokines increase bone destruction via osteoclastogenesis and the production of MMPs. Enhanced IL-17-mediated inflammation occurs via the pro-INFLAMMATORY CYTOKINES TNF, IL-1, IL-6 and IL-8. The Th17 cellular pathway has been implicated as being particularly important in PSORIASIS, PSORIATIC ARTHRITIS and ANKYLOSING SPONDYLITIS. IL-17 is highly expressed in the synovium of PSORIATIC ARTHRITIS and skin plaques of PSORIASIS. Increased levels of IL-17 have also been reported in serum and synovial fluid of patients with RA.

SECUKINUMAB and the newer ANTIBODY, IXEKIZUMAB, both bind IL-17A, thus decreasing its actions. The mechanism of action of BRODALUMAB is different. It blocks the actions of IL-17 by binding to the IL-17 receptor.

34.6.6.2 Clinical Indications and Efficacy

SECUKINUMAB is the most important IL-17 inhibitor. It is a human IgG1 MONOCLONAL ANTIBODY against IL-17A, with a half-life of 22–31 days, and is approved by the FDA for moderate-to-severe plaque PSORIASIS, PSORIATIC ARTHRITIS and ANKYLOSING SPONDYLITIS. Dramatic responses to PSORIASIS have been noted. In two clinical trials [47], 150 or 300 mg SECUKINUMAB was administered subcutaneously weekly for 4 weeks and then every 4 weeks for 48 weeks in patients with PSORIASIS. A reduction of at least 75% from baseline in the Area and Severity Index of PSORASIS was seen in 77–82% of patients treated with 300 mg and 67–72% of those treated with 150 mg compared to 5% placebo rate. ETANERCEPT was also more active than placebo but at a lower rate than seen with either dose of SECUKINUMAB [47]. Over a quarter of those receiving 300 mg achieved a 100% reduction in the Area and Severity Index versus 0.8% in placebo.

In the FUTURE 2 trial, PSORIATIC ARTHRITIS patients received SECUKINUMAB 300, 150, 75 mg or placebo subcutaneously weekly for 4 weeks then every 4 weeks.

Significant benefit using ACR20 outcome was seen at weeks 24 and 52 in patients. Many patients were also receiving anti-TNF and/or METHOTREXATE. Although responses were generally higher in the anti-TNF-naive population, clinical benefits were also seen in patients receiving anti-TNF treatment. It was concluded that secukinumab is a suitable additional treatment option for both populations.

In the MEASURE 2 trial [48], ANKYLOSING SPONDYLITIS patients received SECUKINUMAB 300, 150, 75 mg or placebo. The ASAS20 responses were 54%, 51%, 29% and 15%, respectively. A Phase II study of SECUKINUMAB in RA did not meet the primary ACR20 endpoint, but a systematic literature review and meta-analysis of RCTs using anti-IL17 therapy showed IXEKIZUMAB was more effective than placebo. SECUKINUMAB trended towards a positive response in RA but did not achieve an ACR20 response, while BRODALUMAB was not effective [49].

34.6.6.3 Adverse Effects

The main adverse effect reported in trials on SECUKINUMAB is nasopharyngitis and upper respiratory symptoms, with an increased incidence of diarrhoea (Appendix 4). Approximately 1% develops neutropaenia. Inhibition of Th17 cell-mediated activity by SECUKINUMAB increases the incidence of mild to moderate *Candida* infection [48]. BRODALUMAB and IXEKIZUMAB also caused an increased rate of fungal infections. BRODALUMAB has been associated with an increased rate of suicide, but confirmation is required from further clinical trials.

34.7 Biosimilars

The development and production of bDMARDs and other biological pharmaceuticals (also known as BIOPHARMACEUTICALS, biologicals or biologics) make them highly individual products. The original biologics developed by the innovator (the innovator product) has market exclusivity for the lifetime of the patent. BIOPHARMACEUTICALS which are similar but not identical to the existing innovator biologics are being developed upon expiry of the innovator patent and are termed BIOSIMILARS. The marketing of BIOSIMILARS introduces competition with the innovator product in the market and is expected to lower costs of the bDMARDs and thus create savings for healthcare systems [50]. However, since bDMARDs are made through complex cellular processes by living cells in bioreactors, exact copies are not possible, and their beneficial clinical effects and toxicity may not be identical to the innovator product. Consequently, a comparability exercise is undertaken to determine quality, safety and efficacy of the candidate BIOSIMILAR to the innovator product; this includes extensive physicochemical characterization, PHARMACOKINETICS, cellular actions in vitro and clinical testing to show close similarity to the innovator product, before BIOSIMILARS are approved and marketed.

BIOSIMILARS differ from the classical low molecular pharmaceuticals, which have singular, defined structures, whereby exact copies can be made. Low molecular mass pharmaceuticals may, however, contain very low amounts of impurities from their chemical manufacture. Different batches may, therefore, contain differing amounts of the impurities although they must always be below the very stringent official limit tests.

Some BIOSIMILARS of have been approved, and there are several at various stages of development. A BIOSIMILAR of INFLIXIMAB is marketed. Extensive physicochemical characterization of INFLIXIMAB BIOSIMILAR was conducted utilizing several batches of manufactured BIOSIMILAR, to gain insight into the consistency of physicochemical properties compared to the originator's infliximab. These data confirmed that the secondary and tertiary structures of BIOSIMILAR samples tested were highly similar to the innovator product. The comparability exercise showed that the majority of the physicochemical properties of the INFLIXIMAB BIOSIMILAR are similar to those of the innovator product of INFLIXIMAB, even though some differences were noted in material presented to the European Medicines

Agency. For example, there were differences in the amount of afucosylated INFLIXIMAB (Fucose is a C5 sugar), there is also a lower binding AFFINITY towards specific Fc receptors and a lower ex vivo ANTIBODY-dependent cellular CYTOTOXICITY (ADCC) activity in the most sensitive ADCC assay. However, there was no meaningful clinical difference in biological activity [51, 52].

The innovator product of ADALIMUMAB was approved in 2002, with patent expiry in the USA and the EU in December 2016 and April 2018, respectively. The innovator product is the largest selling drug globally, with sales of $US 12.5 B in 2014. The approval of ADALIMUMAB BIOSIMILARS would be expected to make the biological available to a broader spectrum of the population for a variety of INFLAMMATORY disease indications at a lower cost than the innovator product. An Indian ADALIMUMAB BIOSIMILAR was approved in 2014 with several others in clinical trials including the advanced Phase III level.

The innovator product of ETANERCEPT was first marketed in 1998 with two BIOSIMILARS approved in 2016. The BIOSIMILARS are useful in all disease indications treated using the innovator product.

34.8 Minor bDMARDs

34.8.1 Interleukin-1 Inhibitors

34.8.1.1 Rationale

IL-1α and IL-1β are expressed in RA synovium related to MONOCYTES, B CELLS, FIBROBLASTS and CHONDROCYTES. Mice deficient in the naturally occurring receptor antagonist (IL-1ra) develop a spontaneous erosive arthritis. ANAKINRA is very similar to IL-1ra, differing in that ANAKINRA is not glycosylated and it also contains a terminal methionine residue which is necessary for its biological production. ANAKINRA binds to IL-1 receptors and thus, prevents IL-1 interaction with its native RECEPTOR. ANAKINRA is effective in disorders that are driven by IL-1 dysregulation such as the cold auto-inflammatory disorders, but it has only been of modest efficacy in treating RA in adults, although it may be effective in the treatment of juvenile RA.

ANAKINRA has the greatest clearance of all the present bDMARDs with an apparent half-life of elimination of about 3 h after the usual daily subcutaneous injection. ANAKINRA is eliminated predominately by renal excretion, and its dosage must, therefore, be reduced in patients with severe renal impairment [53].

34.8.2 General Adverse Effects of bDMARDs

34.8.2.1 Monitoring

Detection of potential adverse effects is an integral part of each medical consultation, reviewing recent and current events and anticipating potential effects related to new prescribing, overseas travel, or surgery. Any suspected sepsis or exposure risks should be thoroughly characterized as bDMARDs may significantly ameliorate both clinical and laboratory indicators of severity. RA patients have increased risks of cardiovascular events, smoking-related malignancy and skin cancers and lymphomas. Monitoring should include 6–12 monthly skin and lymph node examination; 6–12 monthly blood pressure, glucose and fasting lipid profiles; and monthly renal and liver function as indicated by concomitant medication. Patients having received B CELL-depleting therapy may require measurement of B CELL repopulation and immunoglobulin levels.

34.8.2.2 Progressive Multifocal Leukoencephalopathy (PML)

PML is a very rare, but usually fatal, CNS infection caused by the JC polyoma virus which leads to demyelination. JC polyoma virus is a common and usually clinically silent human infection, which reactivates with IMMUNOSUPPRESSION leading to mental, motor and visual decline with a final rapid decline to coma and death. T CELL lymphopaenia, HIV infection and immunosuppressive therapy are primary risk factors, with cases reported in SLE and RA patients receiving conventional IMMUNOSUPPRESSIVE therapy. PML has been reported in

RA patients treated with INFLIXIMAB, ETANERCEPT and TOCILIZUMAB [54]. In some cases reactivation of virus has been proven, but it has been suggested that the treatment itself may be associated with noninfectious demyelination, with a recommendation that the polymerase chain reaction test (PCR) be performed on the cerebrospinal fluid of all RA patients treated with newer immunosuppressive agents who present with a demyelination syndrome. It is also recommended that bDMARDs should not be used in patients with known pre-existing demyelinating conditions such as multiple sclerosis.

In theory B CELL depletion should have no effect on activation of this virus, but sufficient confirmed or suspected cases have been reported, primarily in oncology patients, for the manufacturer in Europe to recommend all RITUXIMAB-treated patients carry a "patient alert card". The cumulative reporting rate of 2.2 cases per 100,000 RA patients treated with rituximab is approximately double the expected cumulative incidence rate in RA of 1 per 100,000 [55], and it has been proposed that post-marketing surveillance of patients taking RITUXIMAB or other bDMARD is essential. Patients should be made aware of the risk of this infection, and informed consent is strongly recommended before treatment with this bDMARD is commenced [29].

34.8.2.3 Induced Autoimmunity

Chimeric ANTIBODIES are recognized as foreign by the human immune system and may provoke an ALLERGIC reaction and the development of human anti-chimeric ANTIBODIES (HACA) which may decrease the activity or increase the clearance of the ANTIBODY. Notwithstanding the increased "humanness" of human and humanized ANTIBODIES, they remain immunogenic, and a small proportion of users develop human anti-human ANTIBODIES (HAHA). ANTIBODY production against bDMARDs is reduced by concomitant use of METHOTREXATE, presumably due to its immunosuppressive activity [56].

csDMARDs and bDMARDs have been associated with the development of other autoimmune DISEASES, such as lupus syndromes reported with LEFLUNOMIDE and TNF blockers and PSORIASIS in patients treated with RITUXIMAB. The use of anti-TNF agents has been associated with formation of antinuclear antibodies and anti-double-stranded DNA antibodies. However, lupus reactions without end organ damage are very rare, and there is presently no case for testing for these antibodies [57].

34.8.2.4 Infusion Reactions

Treatment for anaphylactic reactions should be available for all infusion therapies, and closer monitoring is required during the initial infusions. Immunogenicity induced by biological therapies decreases with increasing humanness of the product but still remains. Immunogenicity and antibody formation is associated with acute or delayed ALLERGIC reactions. Acute infusion reactions occurred in 23% of patients within the ACCENT study (placebo around 10%), with 4% classified as severe and 2.3% experiencing delayed HYPERSENSITIVITY [57]. Use of concomitant IMMUNOSUPPRESSIVES reduces the rate of infusion reactions, and it is important not to exceed the recommended rate of infusion. If a reaction has occurred, pretreatment with antihistamine and CORTICOSTEROID should be considered before subsequent doses.

B CELL lysis with RITUXIMAB can be associated with fever, chills, nausea and in severe cases hypotension and chest pain. To reduce both the incidence and severity of reactions, it is recommended to premedicate with an antipyretic, antihistamine and 100 mg intravenous methylprednisolone. Infusion reactions from ABATACEPT and TOCILIZUMAB occur only in a small proportion of persons and are usually mild requiring no premedication.

34.9 Conventional Synthetic Disease-Modifying Anti-rheumatic Drugs (csDMARDs)

34.9.1 Methotrexate

METHOTREXATE is a folate analogue (Fig. 34.4) originally developed in the 1940s as a CYTOTOXIC drug for the treatment of various

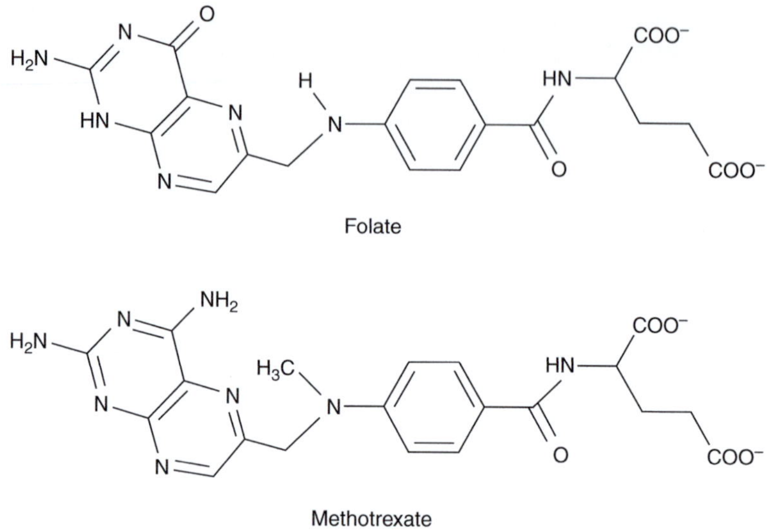

Fig. 34.4 Comparative structures of folic acid and methotrexate. Folic acid is only active as a cofactor for one-carbon transfers after reduction to tetrahydrofolic acid. Both folic acid and methotrexate form polyglutamates in cells

tumours. An older folate analogue, aminopterin, was shown to be useful in treatment of RA in 1951 [58] and its replacement with METHOTREXATE as a cytotoxic agent led ultimately to the testing of METHOTREXATE in the treatment of RA. METHOTREXATE, like its close analogue, folic acid (Fig. 34.4), is a hydrophilic ionized compound at physiological pH, indicating that it should not diffuse passively through cell membranes. Several transporters have been identified recently (see below).

34.9.1.1 Pharmacokinetics and Metabolism

METHOTREXATE is usually administered orally but may also be administered by subcutaneous or intramuscular injection if excessive nausea occurs when the drug is taken orally and the nausea is not controlled by folic acid treatment.

Urinary recovery of METHOTREXATE indicates that the bioavailability is about 80%. Following oral, subcutaneous or intramuscular dosage, the initial half-life of METHOTREXATE is about 7 h. This is followed by a very slow phase with a half-life about 5 days [59]. This slow elimination phase, possibly caused by the slow intracellular accumulation and loss of the active polyglutamate metabolites (see below), probably serves to prolong the effects of METHOTREXATE and to allow for its once a week dosage in RA.

METHOTREXATE is primarily eliminated via the kidneys. Therefore, a lower dose should be used in patients with chronic renal impairment, and temporary cessation of METHOTREXATE treatment may be required at times of volume depletion (such as perioperatively). Dosage should also be decreased in older age because of decreasing renal function. Co-prescription of agents known to impair renal function, such as aminoglycosides and CYCLOSPORIN, should be undertaken with caution. It has also been reported that prolonged use of METHOTREXATE itself may reduce renal function and hence its own clearance [60], a possible mechanism being increased plasma adenosine concentrations as a consequence of methotrexate activating adenosine A_1 receptors in the renal parenchyma, thereby diminishing renal blood flow and salt and water excretion [61].

Many patients taking low-dose METHOTREXATE are also treated with a NSAID in order to suppress the symptoms of inflammation although toxicity from METHOTREXATE in occasional patients has been attributed to this combination of drugs. Although renal blood flow and renal function can be decreased by NSAIDs, prospective studies do not indicate any NSAID-induced decrease in the renal clearance of METHOTREXATE, except during treatment with high doses of aspirin [62].

Probenecid decreases the renal excretion of METHOTREXATE and should be avoided [63]. Additionally, bone marrow suppression has occasionally been seen with the combination of cotrimoxazole and METHOTREXATE, probably because cotrimoxazole has weak anti-folate activity in humans.

METHOTREXATE contains a glutamate moiety, and, after entering the cell, up to six glutamates are added by the action of folylpolyglutamyl synthase. This polyglutamation maintains a low intracellular concentration of METHOTREXATE as the polyglutamates cannot be transported extracellularly, unless hydrolysed to the monoglutamate (i.e. METHOTREXATE itself), by polyglutamate hydrolase. Thus, the polyglutamation of METHOTREXATE effectively increases its intracellular life and enhances its enzyme inhibitory potency because the polyglutamates are active inhibitors of dihydrofolate reductase.

Given the variable anti-rheumatic activity and adverse reactions of METHOTREXATE, correlations have also been sought between the clinical effects and the concentrations of METHOTREXATE or its glutamates in plasma or red blood cells. However, no consistent correlates have been found [64, 65]. Possibly, monitoring of plasma or blood concentrations early in treatment with METHOTREXATE may indicate probability of success or failure to treatment with this drug, i.e. before the dosage of METHOTREXATE has been increased in non-responding patients in attempts to improve the clinical response.

Genetic variants of enzymes or transporters have been sought as predictors of the response to METHOTREXATE, but no single variant has been associated with response or failure to treatment [66, 67]. Interpatient differences in the expression of enzymes or transporters (i.e. the level of enzyme or transporter protein) could be major causes of clinical variation in the response to METHOTREXATE, but data are lacking.

34.9.1.2 Mechanism of Action

In the treatment of tumours, the major action of METHOTREXATE is inhibition of dihydrofolate reductase, the result being the blockade of the intracellular production of reduced tetrahydrofolate which is important in the transfer of one-carbon units. These are necessary for the synthesis of some amino acids and nucleic acid bases. An action on dihydrofolate reductase is also indicated in RA because the trough concentration of unbound METHOTREXATE (see above) exceeds the approximate dissociation constant of METHOTREXATE from dihydrofolate reductase (0.01–0.2 nmol/L) [68].

Based on its actions on tumours, the mechanism of action of METHOTREXATE in RA was postulated initially as the inhibition of the proliferation of activated LYMPHOCYTES. There is, however, no convincing evidence that LYMPHOCYTE proliferation is inhibited in RA patients. More recently, it been suggested that low-dose METHOTREXATE may inhibit the recruitment of immature and INFLAMMATORY MONOCYTES into INFLAMMATORY sites and reduce their survival in the inflamed synovium but with little or no effect on tissue-infiltrating MONOCYTES and resident MACROPHAGES [69].

Current hypotheses suggest beneficial effects of low-dose METHOTREXATE are secondary to an anti-inflammatory action rather than its anti-proliferative action. In general, low-dose METHOTREXATE alters the CYTOKINE balance by inhibiting the production of pro-inflammatory cytokines (TNF, IL-6) and enhancing anti-inflammatory CYTOKINES (IL-1 receptor antagonist, IL-1ra).

The major anti-inflammatory effect of low-dose METHOTREXATE may be inhibition of the enzyme AICAR transformylase, the result being the intracellular accumulation of adenosine monophosphate (AMP) and its conversion to adenosine in the extracellular space (Fig. 34.5). It is suggested that the higher levels of extracellular adenosine then cause it to bind to the transmembrane G protein-coupled adenosine cell surface receptors (A_1, $A_{2\alpha}$, $A_{2\beta}$, A_3) [61, 70]. According to this hypothesis, METHOTREXATE predominantly acts via ligation of the $A_{2\alpha}$ receptors that are present on NEUTROPHILS, MACROPHAGES, MONOCYTES, LYMPHOCYTES and BASOPHILS. Binding increases intracellular cAMP leading to immunosuppression by inhibition of phagocytosis; inhibition of secretion of TNF, IFNγ,

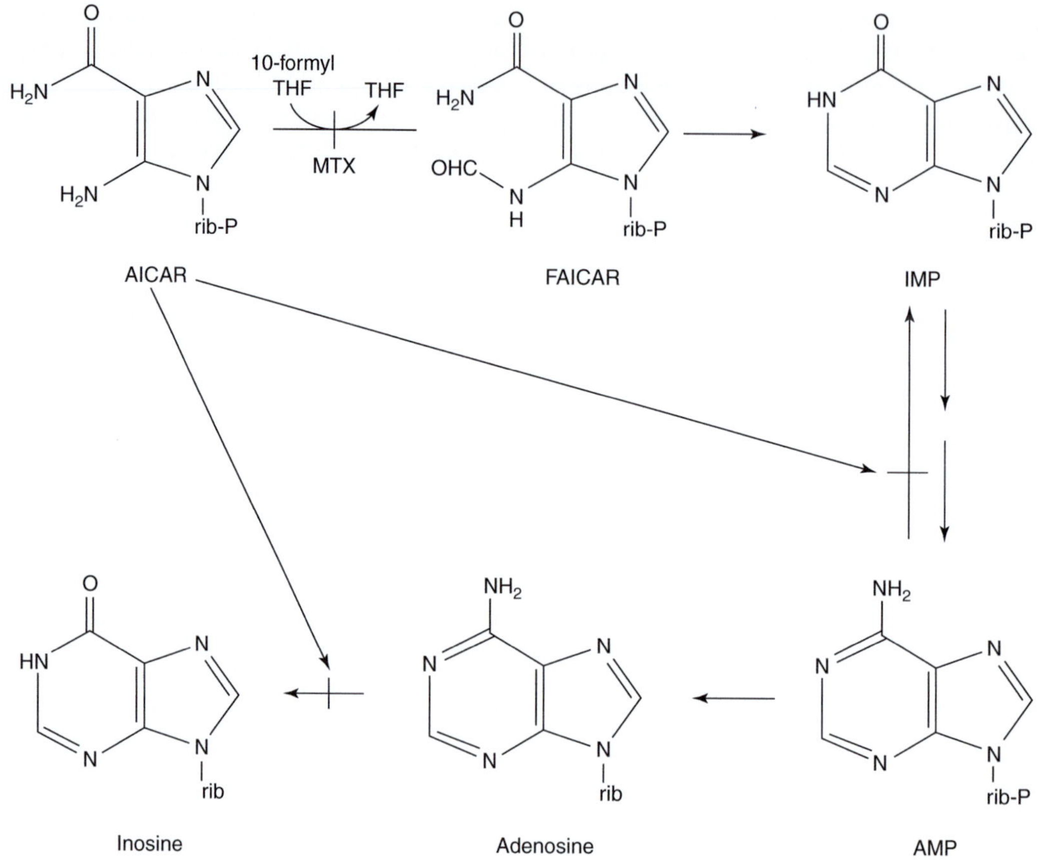

Fig. 34.5 Proposed mode of action of methotrexate. Methotrexate (MTX) inhibits the 5-amino-4-carboxamide ribonucleotide (AICAR) transformylase which catalyzes the conversion of AICAR to N-formylAICAR (FAICAR). This leads to the accumulation of AICAR which decreases the activity of intracellular enzymes which deaminate adenosine monophosphate (AMP) and adenosine. The consequent accumulation of adenosine interacts with adenosine receptors causing suppression of inflammation

IL-2, IL-6, IL-8 and HLA expression; and increased secretion of IL-10, an anti-INFLAMMATORY CYTOKINE. Binding of adenosine to A_3 receptors on macrophage-monocytes leads to inhibition of secretion of TNF, IL-12, IFNγ and IL-1ra [61, 70]. Results in A_{2A} and A_3 knockout mice are consistent with adenosine mediating the anti-INFLAMMATORY effects of METHOTREXATE because it does not have anti-INFLAMMATORY activity in mice lacking either receptor [71]. Further, METHOTREXATE increases adenosine concentrations in air pouch exudates, a model of inflammation. By contrast, METHOTREXATE does not increase the blood concentration of adenosine in patients, although changes at peripheral sites cannot be excluded [72].

34.9.1.3 Clinical Indications and Efficacy

METHOTREXATE is an established DMARD for RA, PSORIATIC ARTHRITIS and other INFLAMMATORY joint diseases (Table 34.4). METHOTREXATE is used in the treatment of RA at lower doses than when used as a cytotoxic drug. A recent review of dosing strategies found a starting dose of 15 mg/week orally, escalating at 5 mg/month to 25–30 mg/week or the highest tolerable dose, yields the best results [73]. Overall, the aim is to increase the dosage up to a level which produces satisfactory suppression of the activity of the disease with limited adverse effects.

An outline of recent evaluations of methotrexate and other DMARD (Table 34.4) indicates the

Table 34.4 Comparative results of clinical trials on methotrexate and other dmards

Treatments	Duration (months)	Methotrexate dose (mg/week)	Folate Supplement	ACR[a] 20% 50% 70%	Contrasts (ACR and radiographic progression)	Health assessment questionnaire (HAQ)[a]	Reference
Methotrexate vs leflunomide vs. placebo	12	7.5–15	Yes	46% 23% 9%	Methotrexate = leflunomide > placebo by ACR and X-ray	−0.26 methotrexate 0 placebo	[74]
Continuation of trial above	24	7.5–15	Yes	48% 28% 12%	Methotrexate = leflunomide by ACR and X-ray	−0.6 leflunomide greater improvement than with methotrexate −0.37	[75]
Methotrexate vs. leflunomide	12	10–15	10%[b]	65% 44% 10%	Methotrexate > leflunomide by ACR but not by X ray	−0.7 all groups	[76]
Methotrexate vs. sulfasalazine vs. combination	12	7.5–15	No	59%	No differences		[77]
Methotrexate vs. etanercept	12	19 (mean)	Yes	65% 42% 22%	Etanercept > methotrexate but only in first 6 months		[78]

[a]A decrease in HAQ of 0.22 is considered to be the minimum clinically meaningful decrease
[b]Folate usually started after an adverse reaction

required complexity of modern clinical trials in RA. Measures of symptoms, clinical signs and X-ray examination are all required. In terms of the results, these trials indicate little difference between METHOTREXATE and other DMARDs [74–78]. The initial response is greater for both LEFLUNOMIDE and ETANERCEPT than during treatment with METHOTREXATE, but it is unclear whether this translates into a longer-term benefit. An important indication of the long-term therapeutic benefit of DMARDs is their effects on the degree of damage to joints. Present evidence is that METHOTREXATE retards, but does not entirely block, joint damage in many patients (Table 34.4).

The utility of METHOTREXATE is seen from the high maintenance on treatment with about 60–70% of patients still taking METHOTREXATE 6 years after initiation of the treatment [79, 80]. This retention rate is generally greater than seen with the older csDMARDs [80].

The present view on METHOTREXATE is that it should be considered for all patients at the time of diagnosis of RA. Individual factors such as pregnancy and alcohol intake (see Sects. 34.9.1.4 and 34.9.1.5) may impact on that decision, but methotrexate needs to be considered. METHOTREXATE is frequently used with other DMARDs depending upon the progress of the treatment of RA (see Sect. 34.11.7).

34.9.1.4 Adverse Effects

Low-dose METHOTREXATE produces a large number of adverse reactions (Appendix 4). Ten to 35% of patients treated with METHOTREXATE cease therapy due to toxicity, which is less than for SULFASALAZINE and GOLD COMPLEXES but higher than for HYDROXYCHLOROQUINE. There does not appear to be an increased risk of herpes zoster or other serious infections. Low-dose methotrexate could reduce cardiovascular mortality which is increased in RA. The prevalence of raised liver enzymes to above twice the upper limit of normal is around 13–15%. The prevalence of liver fibrosis/cirrhosis is much lower. There is no strong evidence of increased lymphoma or malignancy risk, but the data is suboptimal to draw strong conclusions.

Oral ulceration, nausea and fatigue occur very frequently and are probably related to intracellular depletion of folates, resulting in increased levels of adenosine and homocysteine. This provides the rationale for the usefulness of supplementation with oral folic acid. Various doses of folic acid have been recommended, but present advice is that a single dose of 5 mg folic acid should be administered to all patients on the morning following the dose of METHOTREXATE [81]. Supplementation at this level does not reduce the anti-rheumatic efficacy of low-dose methotrexate. Folinic acid contains the fully reduced form of folic acid and is used to treat methotrexate-induced haematotoxicity and overdose with the drug.

An unexpected side effect of METHOTREXATE is the accelerated formation of rheumatoid nodules, particularly around the fingers. This may be due to activation of adenosine A1 receptors leading to the development of multinucleated giant cells and the nodules [82]. Colchicine may prevent their formation [83]. Interstitial pneumonitis is a serious adverse effect of METHOTREXATE and occurs in 2–7% of patients. It is potentially fatal. Treatment consists of cessation of METHOTREXATE, general supportive measures and high doses of CORTICOSTEROIDS. Although most patients with METHOTREXATE-induced lung disease have a complete recovery, some have permanent lung damage. The strongest predictors for lung injury are age above 60 years, diabetes mellitus, rheumatoid pulmonary involvement, previous use of DMARDs and hypoalbuminemia [84]. METHOTREXATE should not be reintroduced after recovery from pneumonitis. Although pneumonitis is clearly an adverse effect of METHOTREXATE, many reported cases of pneumonitis were the result of pulmonary infections which were not differentiated from METHOTREXATE-induced pneumonitis.

Pregnancy should be avoided during treatment with low-dose METHOTREXATE because of the high risk of teratogenic effects. Consequently, treatment with METHOTREXATE should be stopped 3 months before conception and not restarted until after delivery. Fortunately, the disease activity generally decreases during pregnancy, but exacerbations can be treated with low-dose CORTICOSTEROIDS.

34.9.1.5 Monitoring

METHOTREXATE should be avoided in patients with significant pre-existing liver or lung disease. Therefore, at the start of therapy, a complete blood count, a chest radiograph and liver function tests including measurements of aspartate aminotransferase, alanine aminotransferase, albumin and alkaline phosphatase should be undertaken [85]. Tests for hepatitis B and C should be conducted in patients who are at risk of these diseases. Serum creatinine should be assayed as a measure of renal function. The complete blood count and tests of liver and renal function should be monitored every month for 6 months and then every 1–2 months subsequently. Treatment with METHOTREXATE should be stopped in patients with transaminase concentrations persistently at twice the upper limit of normal or at three times the upper level of normal at any time. At this stage, measurement of plasma concentrations of methotrexate does not appear to be useful in predicting significant hepatotoxicity [86]. Liver biopsy is required only for those patients who need to continue METHOTREXATE and who have sustained enzyme abnormalities.

34.9.2 Antimalarials (Chloroquine and Hydroxychloroquine)

HYDROXYCHLOROQUINE and CHLOROQUINE are antimalarial drugs. Their introduction for the treatment of RA followed the chance discovery of the value of an older antimalarial, mepacrine, in the treatment of RA. Both HYDROXYCHLOROQUINE and CHLOROQUINE are now used in the treatment of RA, but HYDROXYCHLOROQUINE is the more widely used. Both are also used for discoid lupus erythematosus. HYDROXYCHLOROQUINE and CHLOROQUINE are relatively small molecular mass bases which are mainly present as the ionized species (dications) at physiological pH values (Fig. 34.6). High intracellular concentrations of these ionized species may mediate their cellular effects.

34.9.2.1 Pharmacokinetics

The BIOAVAILABILITY of HYDROXYCHLOROQUINE is very variable, ranging from below 20% up to 100%, but the bioavailability remains fairly constant within an individual [87]. The variable BIOAVAILABILITY may be responsible for much of the interpatient variation in the response to HYDROXYCHLOROQUINE. An important feature of the PHARMACOKINETICS of HYDROXYCHLOROQUINE and CHLOROQUINE is their extremely long terminal half-lives of about 40 days. This means that steady-state concentrations may not be achieved until after 3–6 months of daily dosing. The use of loading regimens may decrease the time until the onset of effect of the antimalarials, but this is not standard clinical practice.

34.9.2.2 Mechanism of Action

The basic character of the ANTIMALARIALS indicates that they should accumulate in acidic organelles in cells, particularly lysosomes, where

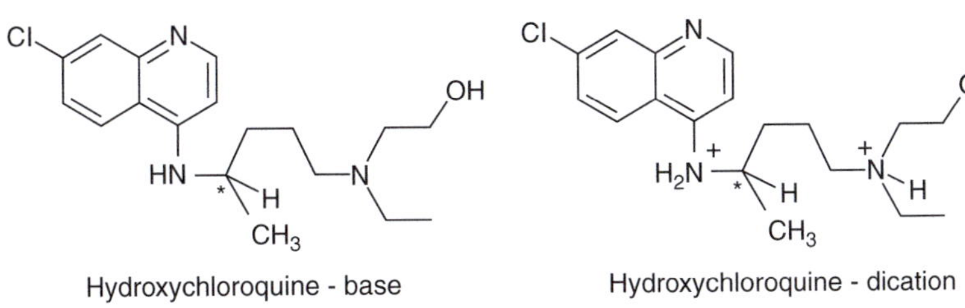

Fig. 34.6 Structure of hydroxychloroquine. Structures of the unionized base form and the dication form (+ charges) which is largely present at physiological pH values. The structure of chloroquine is the same but does not have the hydroxyl group. The available drug is available as the racemate. The asterisk indicates the chiral centre

they may raise the pH. This may affect the function of the acidic organelles, particularly lysosomal enzymes. One such enzyme is acidic sphingomyelinase which is located within the lipid membrane of lysosomes and is an important mediator in the signal transduction pathway between the TNF receptor on the cell surface and activation of transcription factor NF-κB in the nucleus [88]. Raising the pH of lysosomes inhibits acidic sphingomyelinase activity, consequently inhibiting NF-κB activity and pro-INFLAMMATORY gene expression. ANTIMALARIALS have also been reported to inhibit the activity of many other enzymes, including phospholipase A_2 and the production of IL-1.

34.9.2.3 Clinical Indications and Efficacy

ANTIMALARIALS are mild anti-rheumatic agents with low toxicity. This makes them attractive for use in the early stages of RA or in combination with other DMARDs. However, in contrast to methotrexate and several other DMARDs, the ANTIMALARIALS do not appear to retard the damage to joints produced by RA [89].

34.9.2.4 Adverse Effects

HYDROXYCHLOROQUINE and CHLOROQUINE have a very similar range of adverse effects which are mostly mild and transient not requiring cessation of the drug (Appendix 4). HYDROXYCHLOROQUINE is associated with less toxicity than CHLOROQUINE. Of most concern is a rare, irreversible retinopathy, resulting in permanent visual loss. The most important risk factors are daily dosage of HYDROXYCHLOROQUINE and duration of therapy with most reports of toxicity in those taking more than 6.5 mg/kg (lean weight estimation in those overweight) or taking it for over 5 years. Other risk factors are coexisting renal or liver disease, obesity, age, and pre-existing retinal disease.

34.9.2.5 Monitoring

It is recommended that there should be a baseline eye evaluation and examinations of the fundi and visual fields every 6–12 months. A relationship between the plasma concentrations of HYDROXYCHLOROQUINE and efficacy has been shown, indicating that there may be some value in measuring blood HYDROXYCHLOROQUINE concentrations to optimize dosing regimens [90]. However, adjusting the dosage of HYDROXYCHLOROQUINE after monitoring its plasma concentrations is still most uncommon.

34.9.3 Leflunomide

34.9.3.1 Mechanism of Action

LEFLUNOMIDE is a csDMARD which is active after oral dosage. Through its active metabolite, TERIFLUNOMIDE, the primary mode of action of LEFLUNOMIDE is the selective and reversible inhibition of dihydroorotate dehydrogenase (Fig. 34.7) [91]. This enzyme provides the rate-limiting step in the pyrimidine synthesis that is accelerated in the activated CD4+ T CELLS that proliferate rapidly during the progression of RA. This anti-proliferative effect on activated LYMPHOCYTES is likely the key effect of LEFLUNOMIDE on the pathophysiology of RA. Additionally, LEFLUNOMIDE interferes with T CELL signalling and also has broad anti-INFLAMMATORY effects. For example, the active metabolite is a potent inhibitor of NF-κB activation and causes a dose-dependent inhibition of CYTOKINE production (including TNF). LEFLUNOMIDE also decreases the local production of synovial MMP that suggests a mechanism by which it acts to prevent joint destruction.

34.9.3.2 Pharmacokinetics and Metabolism

LEFLUNOMIDE is rapidly and almost completely converted to the active open chain metabolite, TERIFLUNOMIDE, by first-pass metabolism in the gut wall and liver (Fig. 34.7). TERIFLUNOMIDE has a long half-life of between 15 and 18 days because of its enterohepatic recirculation [92]. About 90% of a single dose of LEFLUNOMIDE is eliminated; about

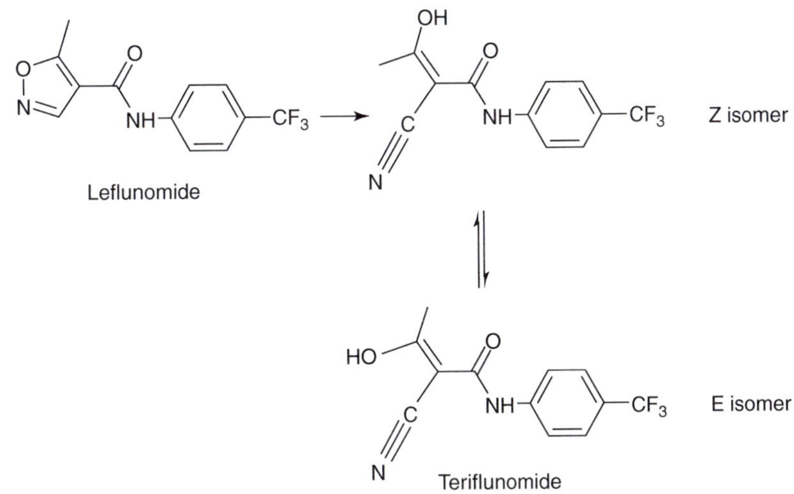

Fig. 34.7 Structures of leflunomide and its active metabolite, teriflunomide. Teriflunomide is the form which circulates in plasma. Teriflunomide exists as two isomers which are in rapid equilibrium

half is present in urine primarily as metabolites, while about 50% is secreted in bile as the active metabolite and is ultimately excreted in faeces. Because TERIFLUNOMIDE relies heavily on biliary excretion for its clearance, and also given its risk of hepatotoxicity, LEFLUNOMIDE is contraindicated in patients with hepatic impairment.

TERIFLUNOMIDE may take 15–20 weeks to reach steady-state plasma concentrations. In order to achieve therapeutic concentrations rapidly, it is common to administer loading doses (Appendix 4). In practice, the loading dose is often decreased or not used to avoid "nuisance" problems with diarrhoea or nausea, both of which may influence early patient adherence. Anecdotally, clinical efficacy in this way is maintained but delayed by a few weeks.

TERIFLUNOMIDE binds strongly to cholestyramine within the gastrointestinal tract. The result is that its plasma half-life is reduced to approximately 1 day and cholestyramine is used when rapid elimination of TERIFLUNOMIDE is required (see Sect. 34.9.3.4).

34.9.3.3 Clinical Indications and Efficacy

The efficacy of LEFLUNOMIDE has been well examined in RA. Its efficacy is similar to that of METHOTREXATE or SULFASALAZINE (Table 34.4). It not only decreases symptoms and increases function and quality of life in RA but also retards radiographic joint damage [93]. Clinical improvement has been sustained for up to 5 years. Combination therapy with LEFLUNOMIDE and METHOTREXATE is also effective and well tolerated in patients responding inadequately to METHOTREXATE alone [94].

Apart from its value in the treatment of RA, LEFLUNOMIDE is useful in the treatment of PSORIATIC ARTHRITIS and has been used in a small cohort of patients with systemic lupus erythematosus with the drug appearing to be efficacious and safe [95].

34.9.3.4 Adverse Effects

The most common adverse events associated with LEFLUNOMIDE treatment are gastrointestinal. They usually decrease with time and/or dose reduction. Rash and reversible alopecia are also common.

The cytostatic effect of TERIFLUNOMIDE may explain some of the side effect profile of LEFLUNOMIDE, such as reversible alopecia and, conversely, the lack of opportunistic infections. Most memory T CELLS circulate in the G0 phase and therefore do not require dihydroorotate for any de novo pyrimidine synthesis and are not susceptible to the anti-proliferative effect of LEFLUNOMIDE. In addition, because of the sparing of the salvage pathway, the replicating cells in the gastrointestinal tract and haemopoietic system are relatively unaffected, thus explaining the lack of mucositis or marrow toxicity.

When LEFLUNOMIDE has been used as monotherapy in clinical trials, abnormal transaminase levels are noted in 5–15% of patients, but these effects were generally mild (less than twofold elevations) and reversible and usually resolved while continuing treatment. Furthermore, post-marketing surveillance shows that almost all cases of hepatic dysfunction were associated with other confounding factors [96]. LEFLUNOMIDE may increase plasma levels of cholesterol and low-density lipoproteins in a progressive manner, but long-term effects of this are unknown. Any suspected toxicity may be further evaluated by use of a short course (1–2 days) of cholestyramine at lower dose (4 g three times daily). This will often reverse the side effect, be it rash or diarrhoea or other, quite quickly.

LEFLUNOMIDE is absolutely contraindicated in women who are or may become pregnant, because of its teratogenic effects (lymphomas) in animals. Because of the prolonged half-life of TERIFLUNOMIDE, any woman taking LEFLUNOMIDE who is contemplating pregnancy should allow the plasma concentrations of TERIFLUNOMIDE to fall below 0.02 mg/L. This may take several months because of its long half-life. Alternatively, the elimination of the TERIFLUNOMIDE can be accelerated by cholestyramine. TERIFLUNOMIDE diffuses into breast milk although it is not known if the concentrations are sufficient to cause toxicity. At this stage, however, it is contraindicated in nursing mothers.

The rate of birth defects in women undergoing cholestyramine washout in three months prior to conception or upon detection of pregnancy was only marginally higher than background, and management of such pregnancies requires specialist obstetric medicine guidance, with a number of reported cases of delivery of full-term healthy infants [97].

34.9.3.5 Monitoring

Baseline investigation should include hepatitis B and C serology and any persistent hepatic dysfunction further investigated. Patients should be advised to reduce alcohol consumption because of possible greater liver impairment. Liver function tests (including transaminases) should be monitored every 4–6 weeks for at least 6 months and longer if patients are taking combination therapy with METHOTREXATE or other hepatotoxic drugs. Thereafter, liver function tests should be repeated at least every 3 months, more frequently if transaminases have increased. Usually complete blood counts are performed at the same frequency. Alcohol should be avoided if patients are taking the combination of METHOTREXATE and LEFLUNOMIDE. All women of childbearing age must have a negative pregnancy test before beginning the drug and should be counselled to use effective forms of contraception.

34.9.4 Sulfasalazine

SULFASALAZINE consists of MESALAZINE (5-aminosalicylate) and a sulphonamide, SULFAPYRIDINE, linked by an azo bond (Fig. 34.8). This drug was synthesized in the late 1930s and originally developed and used on the basis of a belief in an infectious cause of RA [98]. The mode of action of SULFASALAZINE is unclear although it has been reported to inhibit AICAR transformylase and therefore may act in a similar fashion to METHOTREXATE through the accumulation of adenosine [99]. Oddly, it is not known whether SULFASALAZINE or its metabolites are the major clinically active species. The apparent anti-rheumatic activity of SULFAPYRIDINE indicates that it is an active metabolite [100]. On the other hand, both SULFASALAZINE and OLSALAZINE appear active in the treatment of ANKYLOSING SPONDYLITIS [101], yet OLSALAZINE is a dimer of MESALAZINE only. OLSALAZINE, like SULFASALAZINE, is metabolized in the large bowel, but, as is evident from its structure, OLSALAZINE can yield only MESALAZINE. It is of note that SULFASALAZINE is used in both the treatment of RA and INFLAMMATORY BOWEL DISEASES. Furthermore, many patients with INFLAMMATORY BOWEL DISEASES also develop arthritic states of varying

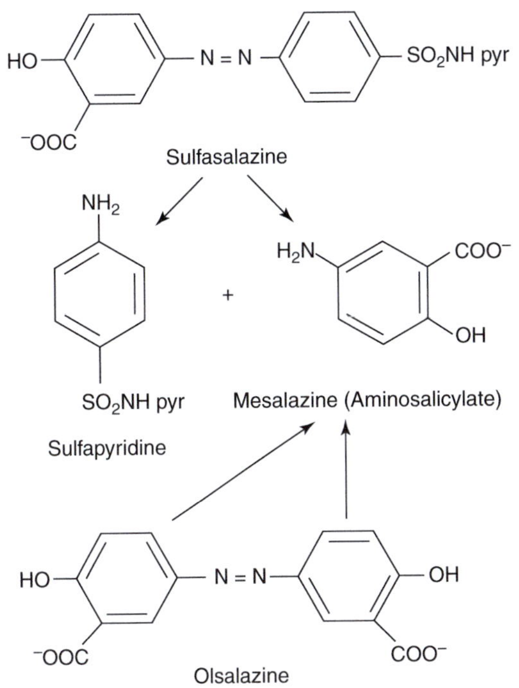

Fig. 34.8 Structure and initial metabolism of sulfasalazine and olsalazine. Reduction of the azo bond of sulfasalazine within the large intestine yields sulphapyridine and mesalazine (aminosalicylate). Sulfapyridine is subsequently absorbed, while the absorption of aminosalicylate is incomplete. Both metabolites are further converted to other inactive compounds. *Pyr* pyridine residue. Reduction of the azo bond of olsalazine yields only mesalazine (aminosalicylate)

severity. These similarities indicate a commonality in the cause of the diseases and a common mode of action of SULFASALAZINE, although the reason for these parallel diseases and treatments is unknown.

34.9.4.1 Metabolism and Pharmacokinetics

SULFASALAZINE is administered orally, and upon reaching the large bowel, the azo bond is reduced by colonic bacteria to yield SULFAPYRIDINE and MESALAZINE (Fig. 34.8). SULFAPYRIDINE is absorbed, whereas MESALAZINE is poorly absorbed, although low concentrations of its acetylated metabolite are found in plasma. Although metabolized in the large intestine, unchanged SULFASALAZINE also achieves substantial concentrations in plasma. During daily treatment with 2 g SULFASALAZINE, the average concentrations of the unchanged drug are about 4 mg/L (10 μM) [102].

34.9.4.2 Clinical Indications and Efficacy

SULFASALAZINE is widely used and effective in the treatment of RA as well as ANKYLOSING SPONDYLITIS and HLA B27-related arthropathies. In particular, SULFASALAZINE slows the joint destruction of RA [103].

34.9.4.3 Adverse Effects

Although SULFASALAZINE can produce a wide range of side effects, it is among the best tolerated csDMARD (Appendix 4), along with HYDROXYCHLOROQUINE and METHOTREXATE. Toxicity is most frequent in the first 2–3 months of usage, but its likelihood can be reduced by gradually increasing the dosage and the use of ENTERIC COATED formulations. Serious side effects are rare, and most adverse effects are eliminated if the dose is reduced from the usual 1 g twice a day. Nausea and upper abdominal discomfort are the most frequent side effects at the start of the therapy. Leukopaenia is very uncommon but can develop rapidly. Its occurrence is most likely in the first 6 months of therapy but can develop later. The metabolite, SULFAPYRIDINE, can produce haemolysis in patients with a deficiency of glucose-6-phosphate dehydrogenase. SULFASALAZINE should be avoided at around the time of conception and pregnancy, although no teratogenicity has been reported for this drug.

34.9.4.4 Monitoring

For the rapid detection of haematological side effects, it is recommended that complete blood cell count should be performed before commencement of treatment with SULFASALAZINE. For the follow-up visits, complete blood cell counts every 2–4 weeks for the first 3 months are suggested and at greater intervals subsequently. The haemolytic anaemia in patients with a deficiency of glucose-6-phosphate dehydrogenase will be detected by low haematocrit levels.

Baseline measurement of hepatic transaminases is advised in patients with known or suspected liver disease.

34.10 Targeted Synthetic Disease-Modifying Anti-rheumatic Drugs (csDMARD)

34.10.1 Inhibitors of Janus Kinase (JAK) Enzymes, Tofacitinib

34.10.1.1 Rationale

JAK inhibitors are small molecular mass compounds (Fig. 34.9) that block the phosphorylation activity of one or more isoforms of the four JAK enzymes: JAK1, JAK2, JAK3 and Tyk2. The JAK enzymes are large proteins that mediate the intracellular signalling of a range of pro-INFLAMMATORY CYTOKINES, INTERFERONS and hormones. JAKs are bound to the receptor cytoplasmic tail as homo- or heterodimers (Table 34.5). JAK1, JAK2, JAK4 and Tyk2 bind to many different CYTOKINE receptors, whereas JAK3 binds only to the common gamma chain.

As outlined in Fig. 34.10, the binding of the CYTOKINE or other ligand to their receptors causes a conformational change in the RECEPTOR and subsequent activation and phosphorylation of the JAK enzymes. In turn, this leads to phosphorylation of both ligand RECEPTORS and its related signal transducers and activators of transcription (STAT) protein. Phosphorylated STATs form dimers which translocate into the nucleus where they upregulate gene transcription and expression [104]. Thus, a particular pro-INFLAMMATORY CYTOKINE cluster can be modulated through inhibition of a single or multiple JAKs.

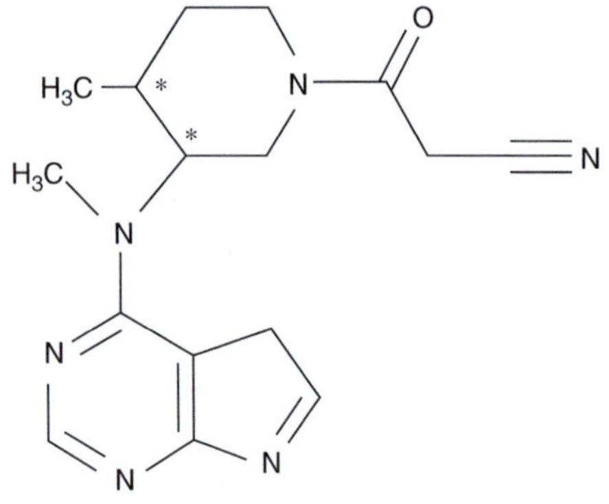

Fig. 34.9 Structure of tofacitinib. The drug is available as a single active isomer. The asterisks indicate the two chiral centres

34.10.1.2 Pharmacokinetics

TOFACITINIB is well absorbed orally and has a half-life of about 3 h [105]. It is therefore, normally administered twice daily (Appendix 4) although there is still considerable fluctuation in its plasma concentrations. A sustained release formulation has been prepared and is administered once daily.

34.10.1.3 Clinical Indications and Efficacy

TOFACITINIB inhibits JAK1 and JAK3 and is a broad-acting IMMUNOMODULATORY drug, impacting on a range of CYTOKINE signals (Table 34.5) [104]. At 5 mg twice daily, it is approved for the treatment of patients with moderate to severe RA who have responded inadequately to or have not tolerated METHOTREXATE or other DMARDs. Phase II and Phase III clinical trials have shown that TOFACITINIB is efficacious compared to placebo, when administered either as monotherapy or in combination with METHOTREXATE. In monotherapy, TOFACITINIB is superior to METHOTREXATE alone to reduce signs and symptoms, to improve physical function and inhibit the progression of structural damage in RA [106].

Table 34.5 Major inflammatory interleukins and other hormones which are potentially blocked by inhibitors of janus kinase (JAK) enzymes

JAK dimer	Receptor inhibition
JAK1, JAK3	IL-2, IL-4, IL-7, IL-9, IL-15, IL-21—the common γ-chain cytokines
JAK2, JAK2	Erythropoietin, thrombopoietin, GM-CSF, growth hormone, IL-3, IL-5
JAK1, Tyk2; JAL2, Tyk2	Il-6
JAK2, Tyk2	IL-12, IL-23
JAK1, Tyk2	Type 1 interferon α, β
JAK1, JAK2	Type 2 interferon γ

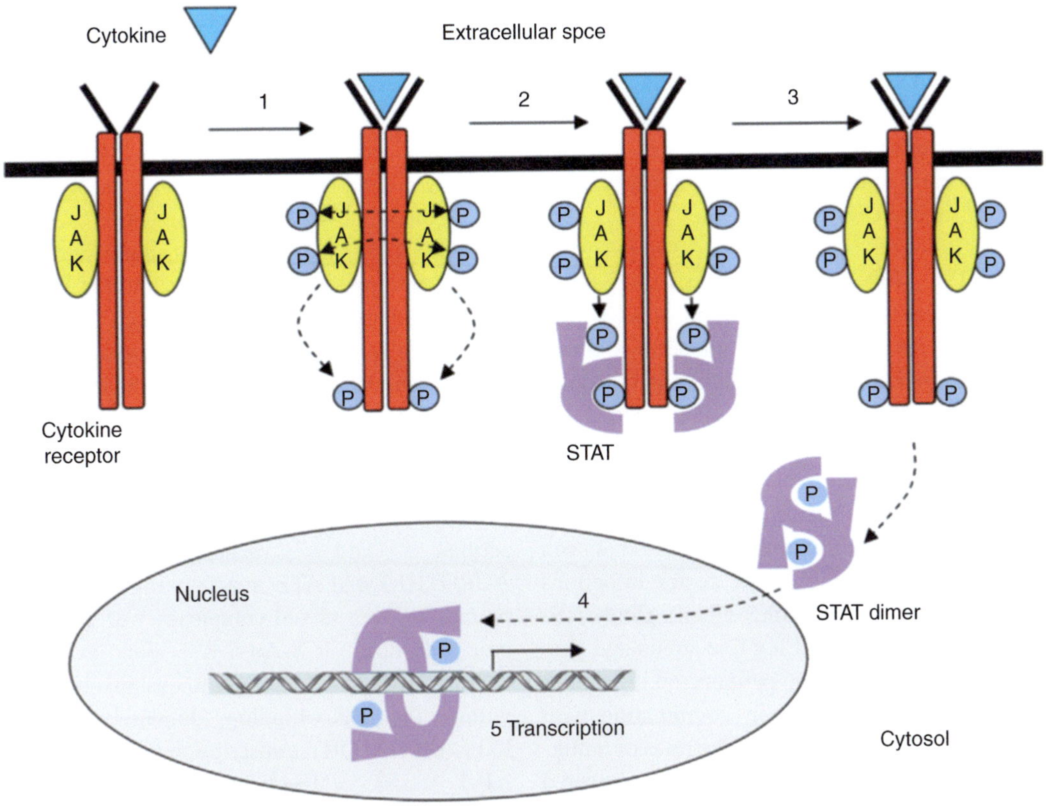

Fig. 34.10 Janus kinases (JAK) mediate the intracellular signalling of type 1 and type 2 cytokines. JAK homo- or heterodimers bind to the intracytoplasmic receptor chain. Cytokine binding initiates intermolecular *trans*-phosphorylation of the opposite JAK and subsequent intramolecular *cis*-phosphorylation of the receptor tail to which the JAK is bound. The signal transducer and activator of transcription (STAT) appropriate for the JAK pairing binds to the phosphorylated receptor tail and in turn is *cis*-phosphorylated. The phosphorylated STATs are released from the receptor tail, dimerize and translocate into the nucleus where they mediate the transcription of target genes. JAK inhibitors are ATP analogues that bind to the ATP-binding site of the JAK thus blocking the critical phosphorylation steps depicted. Inhibition of either of the JAK molecules completely blocks the intracellular cytokine signalling. From Roskoski Jr. [104]

Significant efficacy has been demonstrated in the treatment of moderate-to-severe plaque PSORIASIS, and a small Phase III study of Japanese patients with PSORIATIC ARTHRITIS receiving TOFACITINIB 5 or 10 mg twice daily demonstrated 100% ACR20 responses at week 16, with ACR50 in 75–87% and ACR70 in 50–62% of patients [107, 108]. TOFACITINIB is also being tested in the treatment of ANKYLOSING SPONDYLITIS, but clear-cut results are not available at this stage.

BARICITINIB, an inhibitor of JAK1 and JAK2, has been trialled in patients with moderate to severe RA which has responded inadequately to treatment with METHOTREXATE. For example, significantly more patients treated with BARICITINIB achieved ACR 20 response (76% BARICITINIB vs. 41% placebo) at week 12 [109]. Similarly in RA patients with an inadequate response to bDMARDs, BARICITINIB 4 mg daily was efficacious [110].

Decernotinib, a putative highly selective JAK3 inhibitor, was given to RA patients not responding to methotrexate alone. Significant improvement in ACR 20 and DAS-CRP response at week 12 was seen in doses of 100, 150 and 200 mg daily or 100 mg twice daily. Improvements were maintained at week 24 [111].

The related and well-known drug, RUXOLITINIB, is a potent inhibitor of both JAK1 and JAK2 and is available for the treatment of the neoplastic diseases, polycythaemia vera and myelofibrosis. RUXOLITINIB appears active in early trials on RA, but it has not reached more advanced clinical trials.

34.10.1.4 Adverse Effects

The most common adverse effects of JAK inhibition are infections (including serious infections), hyperlipidemia and cytopaenias (including anaemia). The anaemia is attributed to the inhibition of JAK2 which mediates erythropoietin receptor signalling. A notable increase in herpes zoster infections is seen in patients on TOFACITINIB, approximately double that seen on bDMARDS, with serious zoster infections occurring in those on concomitant glucocorticoids [112]. It is unclear if a live attenuated vaccine should be used in patients taking either TOFACITINAB alone or in combination with CORTICOSTEROIDS [112]. The rates of malignancy for TOFACITINIB are within the expected range for patients with moderate to severe RA [106, 113].

34.11 Minor DMARDs

34.11.1 Azathioprine

Azathioprine is a synthetic purine and antimetabolite drug which is commonly used as an immunosuppressant to prevent transplant rejection (see Chap. 31). It has limited use in the treatment of RA.

34.11.2 Cyclosporin

CYCLOSPORIN is a fungal anti-metabolite which is also widely used to prevent the rejection of transplanted tissues (see Chap. 31). CYCLOSPORIN is efficacious in the treatment of RA but has a low therapeutic index and is rarely used alone in the treatment of RA. It has been most recently tested in combination with METHOTREXATE (see Sect. 34.11.7).

34.11.3 Gold Complexes, Aurothiomalate and Auranofin

Two gold complexes are active csDMARDs but have little current therapeutic use. These are the polymeric complex, sodium AUROTHIOMALATE, which is administered by weekly intramuscular injections, and the monomeric complex, AURANOFIN, which is administered orally. Both have modest anti-rheumatic activity, particularly during long-term treatment. Sodium AUROTHIOMALATE is the more efficacious [114].

Two chemical reactions of AUROTHIOMALATE are important in its cellular actions and metabolism. Firstly, AUROTHIOMALATE reacts with thiol compounds to form mixed complexes with the general formula, R-S-Au-S-R′. For example, AUROTHIOMALATE binds to cysteine residues within the DNA-binding domains of pro-INFLAMMATORY transcription factors, such as AP-1 and NF-kB [115]. Secondly, AUROTHIOMALATE may be activated by its conversion to aurocyanide ($Au(CN)_2^-$) by small amounts of cyanide produced from thiocyanate (SCN^-) by NEUTROPHILS and MACROPHAGES [116]. Aurocyanide potently inhibits the oxidative burst of NEUTROPHILS and the proliferation of LYMPHOCYTES [116].

AURANOFIN is a reactive compound with effects on several cell types and may be suitable treatment for a variety of diseases including leukaemia, carcinomas, parasitic, bacterial and viral infections, if selectivity over normal cells is achieved [117].

Much of sodium AUROTHIOMALATE is eliminated quickly but some accumulates, particularly as deposits (aurosomes) in MACROPHAGES in which gold may be present for many years following the last dose of gold.

Sodium AUROTHIOMALATE has several adverse effects, including blood dyscrasias, mouth ulcers, rashes and proteinuria. Adverse effects are a common cause of withdrawal of treatment over 2–4 years of therapy [114]. Diarrhoea is a common adverse effect of AURANOFIN but not AUROTHIOMALATE.

34.11.4 Penicillamine

PENICILLAMINE is dimethyl cysteine but has the D configuration unlike the naturally occurring cysteine and all other amino acids which have the L configuration and are the forms incorporated into proteins. PENICILLAMINE has anti-rheumatic actions equivalent to those of the gold complexes. However, like the gold complexes, PENICILLAMINE is not used in modern rheumatological practice because of both the introduction of more effective DMARDs and the high incidence of adverse effects such as rashes, proteinuria, neutropaenia and thrombocytopenia, which lead to cessation of treatment in many patients.

Like gold complexes, PENICILLAMINE (Pen-SH) reacts oxidatively with thiol (sulfhydryl) compounds, an activity which may be involved in its DMARD action. Not surprisingly, the major species in plasma is albumin bound through the thiol of group of albumin (albumin-S-H) forming the oxidized species, albumin-S-S-pen [118]. PENICILLAMINE also reacts with reactive aldehydes which may be a cause of the cellular damage of RA [119].

Oxidative metabolism with small molecular mass compounds also occurs. The species formed include the dimer, Pen-S-S-Pen, and a mixed disulphide with cysteine, Cys-S-S-Pen. The mixed disulphide is more water-soluble than cystine (Cys-S-S-Cys). Consequently, PENICILLAMINE is used in the treatment of cystinuria, a disease in which deposits of cystine occur in kidney tubules. D-PENICILLAMINE also chelates copper and is used for Wilson's disease, a rare disease of excessive copper accumulation in the liver and other tissues.

34.11.5 Tetracyclines

The TETRACYCLINES are well-known antibiotics which decrease the activity of several diseases which have an INFLAMMATORY component. These include periodontitis (pyorrhoea), rosacea, acne and asthma. The activity of TETRACYCLINES in rheumatoid arthritis is modest with no significant reduction in erosions or joint space narrowing during long-term treatment [120]. In INFLAMMATORY tests in experimental animals, doxycycline and minocycline decrease the nociception produced by the injection of formalin into the hind paw of the mouse, carrageenan-induced oedema of the rat paw and LEUKOCYTE migration into the mouse peritoneal cavity [121].

The TETRACYCLINES inhibit several enzymes, particularly MMPs which may cleave some cytokincs, such as vascular endothelial growth factor (VEGF) and transforming growth factor-β (TGF-β), into their active forms [122].

The use of the TETRACYCLINES as anti-INFLAMMATORY drugs will always be limited by their antibiotic activities as their widespread use may produce resistance. Non-antibiotic analogues of the TETRACYCLINES have been examined in detail in a variety of experimental systems, but none have been approved for clinical use.

34.11.6 Corticosteroids

The general chemistry, physiology and pharmacology of the CORTICOSTEROIDS are discussed in Chap. 32 of this volume. Aspects of the clinical effects of the CORTICOSTEROIDS which are relevant to the treatment of RA are discussed below.

34.11.6.1 Rationale

In relation to the rheumatic diseases, the most profound effects of CORTICOSTEROIDS are mediated through inhibition of the expression of pro-INFLAMMATORY genes, including many of the mediators involved in the pathogenesis of RA. This broad spectrum of effects is mediated by the interaction between the occupied glucocorticoid receptor and the NF-κB complex, resulting in inactivation of the transcriptional effects of NF-κB [123]. Although these effects of CORTICOSTEROIDS are very significant, CORTICOSTEROIDS also induce the expression of a number of anti-INFLAMMATORY genes, including annexin I (formerly known as lipocortin 1) and MAP kinase phosphatase 1, an endogenous inhibitor of MAP kinase activation and hence of cell activity in INFLAMMATION.

34.11.6.2 Metabolism and Pharmacokinetics

Prednisolone, the active metabolite of prednisone, is the major CORTICOSTEROID used in the treatment of RA. The plasma half-life of prednisolone is short, of the order of 2 h, but the pharmacological effects of prednisolone, as is the case with all CORTICOSTEROIDS, are prolonged because of strong binding to nuclear receptors. The changes induced in protein synthesis also develop slowly and are lost slowly. As a result, prednisolone is usually only administered once a day. Other CORTICOSTEROIDS, such as betamethasone, have also been used for the treatment of rheumatic diseases, but their duration of action is generally longer with greater suppression of the HYPOTHALAMIC-PITUITARY-ADRENAL AXIS.

CORTICOSTEROIDS are also administered by intra-articular injection. This provides a depot in the joint from which the CORTICOSTEROIDS dissolve slowly, thereby producing a prolonged anti-INFLAMMATORY effect in the joint. Systemic effects are seen due to absorption from the joint and may also be prolonged.

34.11.6.3 Clinical Indications and Efficacy

In addition to their wide use in the treatment of patients with RA, CORTICOSTEROIDS are used for all INFLAMMATORY arthritic states and conditions such as vasculitis and polymyalgia rheumatica. Oral or intra-articular CORTICOSTEROIDS significantly improve the clinical symptoms of RA, and they are superior to the NSAIDs in this respect. The effect is rapid and most pronounced in the first weeks of administration. Patients with severe RA, vasculitis or active systemic lupus erythematosus may require high doses of oral CORTICOSTEROIDS, or if the patients are acutely ill, high doses of intravenous CORTICOSTEROIDS. Large single doses of CORTICOSTEROIDS (pulse therapy) have a beneficial effect for at least several weeks and may be useful, while the effects of DMARDs are developing [124].

The long-term use of CORTICOSTEROIDS in RA is controversial. A reduced rate of joint destruction has been reported, but, in one study, no additional effect of low-dose prednisolone (7 mg daily) was found in patients treated with SULFASALAZINE [125]. Overall, the aims of treatment with the CORTICOSTEROIDS should always be to keep the dose and duration of therapy to the minimum which is compatible with good disease control.

Intra-articular injections of CORTICOSTEROIDS are an important treatment for monoarticular inflammatory synovitis or for single joints that are difficult to control in the polyarthritic patient. The long-term effects of intra-articular therapy are unclear, with animal studies suggesting both beneficial and detrimental effects on articular cartilage. Generally, an individual joint may be injected up to 3–4 times per year but no more in weight-bearing joints because of the risk of degeneration of cartilage.

34.11.6.4 Adverse Effects

The adverse effects of the CORTICOSTEROIDS are outlined in Chap. 32 of this volume. In contrast to their anti-INFLAMMATORY effects, the side effects of CORTICOSTEROIDS are mostly due to the induction of genes outside the immune system. Partly for this reason, CORTICOSTEROID side effects are generally dose-related. The catabolic effect of CORTICOSTEROIDS on bone is strongest in the first 6–12 months after initiation of the drug. It is recommended that a dose equivalent to the physiological replacement dosage of cortisol (7.5 mg prednisolone per day) should not be exceeded, as increased osteoporosis occurs with increasing dose [126]. Larger doses are often needed in other INFLAMMATORY rheumatic conditions such as polymyalgia rheumatica, but this limitation of dosage is often possible in the treatment of RA. In order to minimize the dose of CORTICOSTEROIDS, it is common to administer CORTICOSTEROIDS in conjunction with a DMARD.

A major problem with the long-term use of CORTICOSTEROIDS is suppression of the HYPOTHALAMIC-PITUITARY-ADRENAL AXIS. This problem is reduced, but not eliminated, by administration in the morning. Alternate day administration reduces the suppressive effect further, although it is often not practicable because of inadequate clinical response. In order for the hypothalamic-pituitary axis to recover, systemic CORTICOSTEROIDS must be withdrawn slowly, particularly when daily dosage above the equivalent of 10–15 mg prednisolone has been used for more than about 3 weeks. Dosage of the systemic CORTICOSTEROIDS must be increased temporarily if patients become acutely unwell or if there is evidence of adrenal insufficiency.

34.11.7 Combination DMARD Therapies

Combination therapy for the treatment of RA was developed because:

1. There may be additive or synergistic effect from combinations. Despite the lack of knowledge on mechanisms of action of csDMARDs, it is likely that these drugs work in different ways. Thus, modifying or inhibiting the disease processes at multiple sites may gain a greater degree of suppression of RA with less treatment resistance.
2. Combinations of DMARD may allow a decreased dose of one or both agents leading to a concomitant decrease in adverse effects without diminished efficacy.

Combination therapy can be delivered in at least three different styles. Multiple medications can be initiated together and combined dosage of DMARD maintained. Alternatively, one or more of the DMARDs is gradually withdrawn. This is the "step-down" approach. Alternatively the medications can be "stepped up", commencing with one agent and adding another DMARD if the desired outcome is not achieved. A csDMARD can be combined with a bDMARD and a widely advocated combination is METHOTREXATE and an anti-TNF agent.

Table 34.6 shows the results from several recent trials on combinations of DMARD, including bDMARD, in early RA. As is the case with many modern clinical trials, these trials have been very expensive with complex designs. To date, however, it is not possible to identify or advocate one combination as being the most effective. As is the case with many large-scale clinical trials, the trials of anti-rheumatic combinations are often referenced by acronyms of the trial designs (Table 34.6).

The COMET trial demonstrates the complex design of modern clinical trials on DMARD (Table 34.6). This trial compared the treatment with a variable dose of METHOTREXATE alone (starting at 7.5 and increasing as necessary to 20 mg/week over 8 weeks) against concurrent treatment with thc variable dose of METHOTREXATE and a fixed dose of ETANERCEPT (50 mg/week) (Table 34.6) [127]. The apparent response to ETANERCEPT and METHOTREXATE may be confounded by changing doses of other anti-rheumatic drugs. In order to reduce this problem, yet make the trial clinically relevant, concurrent dosage with CORTICOSTEROIDS (≤10 mg prednisolone) or a single NSAID was allowed, provided that these drugs were started at least 4 weeks before commencing the trial. After 24 weeks of constant dosage of these additional drugs, reductions in the dose of prednisolone by 1 mg/week were permitted provided that prednisolone was tapered to 3 mg daily before the dosage of the NSAID was decreased. After 1 year, the combination treatment with ETANERCEPT and METHOTREXATE led to a less severe disease than was seen with METHOTREXATE alone (Table 34.6).

This trial was continued for a further year in which the groups were split (Table 34.6) [128]. Overall, the patients who continued treatment with the combination had less severe disease than the other groups. The number of patients whose joint damage had not progressed during continued treatment with the combination was remarkable (Table 34.6).

A step-up trial on CYCLOSPORIN and METHOTREXATE shows the necessity for pharmacokinetic examination in the study. Thus, the addition of CYCLOSPORIN (2.5–5 mg/kg/day) to the maximally tolerated dose of METHOTREXATE increased the response to treatment in patients with active RA [129]. There was no difference in the reported toxicities. The basis for this improvement, however, may be more PHARMACOKINETIC than an additive or synergistic response. Thus, cyclosporine

Table 34.6 Combination therapies in early RA. Adapted from Emery et al. [127, 128]

Trial name	Design[a]	Therapy	Results
COMET	1-year early RA	Combination methotrexate (titrated doses) + etanercept (50 mg/week) vs. methotrexate (titrated doses)	Patients in remission Combination 50% Methotrexate only 28% Lack of progression of X-ray of joints Combination 80% Methotrexate only 59%
COMET continuation	Groups split after 1 year, and trial continued for further 1 year	Combination continuation vs. methotrexate step up to combination vs. methotrexate only for 1 year and then step up to combination vs. Combination step down to methotrexate only vs. methotrexate only continuation	Patients in remission Combination continuation 57% Methotrexate only step up to combination 58% Combination step down to etanercept only 50% Methotrexate only continuation 35% Lack of progression of X-ray of joints Combination continuation 90% Methotrexate only step up to combination 75% Combination step down to etanercept only 75% Methotrexate continuation 67%
TEMPO 3	3 years, 3-year follow-up	Etanercept + methotrexate vs. etanercept vs. methotrexate	Combination clinically better than either monotherapy, with less X-ray progression at follow-up if received etanercept
COBRA 1	28 week, step down 28 week, observed 24 week, 11-year follow-up	Prednisolone + methotrexate + sulfasalazine vs. sulfasalazine, followed by step-down of prednisolone and then step-down of methotrexate	Rapid benefit from combination, reducing after withdrawal of prednisone and disappearing after methotrexate withdrawal at 5-year minimal clinical difference. Reduced X-ray progression with combination to 5 years, at 11 years no difference
Dougados et al.	1 year	Sulfasalazine + methotrexate vs. sulfasalazine vs. methotrexate	Combination not clinically superior to individual drugs
PREMIER 4	2 years	Adalimumab + methotrexate vs. adalimumab vs. methotrexate	Combination better than monotherapy, with less X-ray progression
BeST 5	2 year, 4-year follow-up	Sequential DMARD monotherapy vs. step-up combination therapy vs. initial combination therapy with tapered high-dose prednisone vs. initial combination therapy with infliximab	Combination therapy with either prednisone or infliximab resulted in early functional benefit but no clinical difference at 2 years. Initial combinations had less X-ray progression than sequential monotherapy or step-up combination therapy at 4 years

FIN-RACo 6,7	2-year trial, 11-year follow-up	Monotherapy with sulfasalazine (later methotrexate) ± prednisolone vs. sulfasalazine + methotrexate + hydroxychloroquine and prednisolone. Treatment restrictions removed at 2 years, targeting remission continued	Initial combination therapy better with more remissions during trial. 11-year significantly greater minimal disease activity and remission in combination group but HAQ score similar
ASPIRE 9	1 year	Placebo + methotrexate vs. methotrexate + infliximab (3 mg/kg and 6 mg/kg)	Combined therapy better
SWEFOT 10	1 year	Run in methotrexate, and if inadequate control addition of either sulfasalazine + hydroxychloroquine or infliximab	When methotrexate failed, addition of infliximab superior to conventional triple therapy, methotrexate + sulfasalazine + hydroxychloroquine
CARDERA 11	2 years	Methotrexate vs. methotrexate + cyclosporin vs. methotrexate and step-down prednisolone vs. triple therapy	Triple therapy initially better clinically but at 2 years no difference, lower HAQ scores with triple therapy and methotrexate/prednisolone. Triple therapy had least erosions, followed by methotrexate with either cyclosporine or prednisolone
Methotrexate vs. methotrexate + cyclosporine with intra-articular betamethasone as needed hydroxychloroquine added week 68 to both arms, cyclosporin tapered from week 76 to zero at week 104	2 years		Methotrexate monotherapy with intra-articular steroids gave excellent control, addition of cyclosporine did not effect DAS remission or X-ray progression, At 2 years ACR20/50 but not 70 better with combination. Higher creatinine in combination, hypertension prevalence not increased

[a]All trials were double blind except where noted

(3 mg/kg/day) produced a 26% increase in mean peak plasma concentration of METHOTREXATE and a 80% reduction in the metabolite 7-hydroxymethotrexate [130]. The metabolite is less efficacious than METHOTREXATE in rat adjuvant arthritis and 4- to 17-fold less cytotoxic in human cell culture. By altering the PHARMACOKINETIC balance in favour of METHOTREXATE, the increased efficacy of the combination of methotrexate and cyclosporine can be explained.

Combining METHOTREXATE and HYDROXYCHLOROQUINE may produce less acute liver damage than METHOTREXATE alone. Triple therapy of METHOTREXATE, SULFASALAZINE and HYDROXYCHLOROQUINE also appears useful although it is difficult to make good comparisons of the various treatments (Table 34.6).

References

1. Nell VP, Machold KP, Eberl G, et al. Benefit of very early referral and very early therapy with disease-modifying anti-rheumatic drugs in patients with early rheumatoid arthritis. Rheumatology. 2004;43:906–14.
2. Karlson EW, Chang SC, Cui J, et al. Gene-environment interaction between HLA-DRB1 shared epitope and heavy cigarette smoking in predicting incident rheumatoid arthritis. Ann Rheum Dis. 2010;69(1):54–60.
3. Sugiyama D, Nishimura K, Tamaki K, et al. Impact of smoking as a risk factor for developing rheumatoid arthritis: a meta-analysis of observational studies. Ann Rheum Dis. 2010;69(1):70–81.
4. McInnes IB, Schett G. Cytokines in the pathogenesis of rheumatoid arthritis. Nat Rev Immunol. 2007;7(6):429–42.
5. Axmann R, Herman S, Zaiss M, et al. CTLA-4 directly inhibits osteoclast formation. Ann Rheum Dis. 2008;67(11):1603–9.
6. Kremer JM. Selective costimulation modulators: a novel approach for the treatment of rheumatoid arthritis. J Clin Rheumatol. 2005;11(3 Suppl):S55–62.
7. Mavers M, Ruderman EM, Perlman H. Intracellular signal pathways: potential for therapies. Curr Rheumatol Rep. 2009;11(5):378–85.
8. National Collaborating Centre for Chronic Conditions. Rheumatoid arthritis: national clinical guideline for management and treatment in adults. London: Royal College of Physicians; 2009.
9. Singh JA, Saag KG, Bridges SL, et al. 2015 American College of Rheumatology guideline for the treatment of rheumatoid arthritis. Arthritis Care Res. 2016;68(1):1–25.
10. van der Woude D, Young A, Jayakumar K, et al. Prevalence of and predictive factors for sustained disease-modifying antirheumatic drug-free remission in rheumatoid arthritis: results from two large early arthritis cohorts. Arthritis Rheum. 2009;60(8):2262–71.
11. Pincus T, Stein CM. Why randomised controlled trials do not depict accurately long-term ourcomes in rheumatoid arthritis: some explanations and suggestions for future studies. Clin Exp Rheumatol. 1997;15(Suppl 17):S27–38.
12. Xu Z, Wang Q, Zhuang Y, et al. Subcutaneous bioavailability of golimumab at 3 different injection sites in healthy subjects. J Clin Pharmacol. 2010;50:276–84.
13. St Clair EW, Wagner CL, Fasanmade AA, et al. The relationship of serum infliximab concentrations to clinical improvement in rheumatoid arthritis: results from ATTRACT, a multicenter, randomized, double-blind, placebo-controlled trial. Arthritis Rheum. 2002;46(6):1451–9.
14. Mulleman D, Chu Miow Lin D, Ducourau E, et al. Trough infliximab concentrations predict efficacy and sustained control of disease activity in rheumatoid arthritis. Ther Drug Monit. 2010;32(2):232–6.
15. Mori S. A relationship between pharmacokinetics (PK) and the efficacy of infliximab for patients with rheumatoid arthritis: characterization of infliximab-resistant cases and PK-based modified therapy. Mod Rheumatol. 2007;17(2):83–91.
16. Tracey D, Klareskog L, Sasso EH, et al. Tumor necrosis factor antagonist mechanisms of action: a comprehensive review. Pharmacol Ther. 2008;117(2):244–79.
17. Hetland ML, Christensen IJ, Tarp U, et al. Direct comparison of treatment responses, remission rates, and drug adherence in patients with rheumatoid arthritis treated with adalimumab, etanercept, or infliximab: results from eight years of surveillance of clinical practice in the nationwide Danish DANBIO registry. Arthritis Rheum. 2010;62(1):22–32.
18. Stuhlmuller B, Haupl T, Hernandez MM, et al. CD11c as a transcriptional biomarker to predict response to anti-TNF monotherapy with adalimumab in patients with rheumatoid arthritis. Clin Pharmacol Ther. 2010;87(3):311–21.
19. Zhou H, Jang H, Fleischmann RM, et al. Pharmacokinetics and safety of golimumab, a fully human anti-TNF-alpha monoclonal antibody, in subjects with rheumatoid arthritis. J Clin Pharmacol. 2007;47(3):383–96.
20. Weisman MH, Moreland LW, Furst DE, et al. Efficacy, pharmacokinetic, and safety assessment of adalimumab, a fully human anti-tumor necrosis factor-alpha monoclonal antibody, in adults with rheumatoid arthritis receiving concomitant methotrexate: a pilot study. Clin Ther. 2003;25(6):1700–21.

21. Breedveld F, Agarwal S, Yin M, et al. Rituximab pharmacokinetics in patients with rheumatoid arthritis: B-cell levels do not correlate with clinical response. J Clin Pharmacol. 2007;47(9):1119–28.
22. Ma Y, Lin BR, Lin B, et al. Pharmacokinetics of CTLA4Ig fusion protein in healthy volunteers and patients with rheumatoid arthritis. Acta Pharmacol Sinica. 2009;30:363–71.
23. Korth-Bradley JM, Rubin AS, Hanna RK, et al. The pharmacokinetics of etanercept in healthy volunteers. Ann Pharmacother. 2000;34(2):161–4.
24. Askling J, Dixon W. The safety of anti-tumour necrosis factor therapy in rheumatoid arthritis. Curr Opin Rheumatol. 2008;20(2):138–44.
25. Askling J, Fored CM, Brandt L, et al. Risks of solid cancers in patients with rheumatoid arthritis and after treatment with tumour necrosis factor antagonists. Ann Rheum Dis. 2005;64(10):1421–6.
26. Askling J, Fored CM, Baecklund E, et al. Haematopoietic malignancies in rheumatoid arthritis: lymphoma risk and characteristics after exposure to tumour necrosis factor antagonists. Ann Rheum Dis. 2005;64(10):1414–20.
27. Diak P, Siegel J, La Grenade L, et al. Malignancy in children and tumor necrosis factor- blockers: forty-eight cases reported to the food and drug administration. Arthritis Rheum. 2010;62(8):2517–24.
28. Korhonen R, Moilanen E. Anti-CD20 antibody rituximab in the treatment of rheumatoid arthritis. Basic Clin Pharmacol Toxicol. 2010;106(1):13–21.
29. Perosa F, Prete M, Racanelli V, et al. CD20-depleting therapy in autoimmune diseases: from basic research to the clinic. J Intern Med. 2010;267(3):260–77.
30. Smolen JS, Keystone EC, Emery P, et al. Consensus statement on the use of rituximab in patients with rheumatoid arthritis. Ann Rheum Dis. 2007;66(2):143–50.
31. Vital EM, Dass S, Rawstron AC, et al. Management of nonresponse to rituximab in rheumatoid arthritis: predictors and outcome of re-treatment. Arthritis Rheum. 2010;62(5):1273–9.
32. Kim GW, Lee NR, Pi RH, et al. Il-6 inhibitors for treatment of rheumatoid arthritis: past, present, and future. Arch Pharm Res. 2015;38:575–84.
33. Semerano L, Thiolat A, Minichiello E, et al. Targeting IL-6 for the treatment of rheumatoid arthritis: phase II investigational drugs. Expert Opin Investig Drugs. 2014;23(7):979–99.
34. Dayer JM, Choy E. Therapeutic targets in rheumatoid arthritis: the interleukin-6 receptor. Rheumatology (Oxford). 2010;49(1):15–24.
35. Nishimoto N, Hashimoto J, Miyasaka N, et al. Study of active controlled monotherapy used for rheumatoid arthritis, an IL-6 inhibitor (SAMURAI): evidence of clinical and radiographic benefit from an x ray reader-blinded randomised controlled trial of tocilizumab. Ann Rheum Dis. 2007;66(9):1162–7.
36. Fleischmann RM, Halland AM, Brzosko M, et al. Tocilizumab inhibits structural joint damage and improves physical function in patients with rheumatoid arthritis and inadequate responses to methotrexate: LITHE study 2-year results. J Rheumatol. 2013;40(2):113–26.
37. Maini RN, Taylor PC, Szechinski J, et al. Double-blind randomized controlled clinical trial of the interleukin-6 receptor antagonist, tocilizumab, in European patients with rheumatoid arthritis who had an incomplete response to methotrexate. Arthritis Rheum. 2006;54(9):2817–29.
38. Gabay C, Emery P, van Vollenhoven R, et al. Tocilizumab monotherapy versus adalimumab monotherapy for treatment of rheumatoid arthritis (ADACTA): a randomised, double-blind, controlled phase 4 trial. Lancet. 2013;381:1541–50.
39. Villiger PM, Adler S, Kuchen S, et al. Tocilizumab for induction and maintenance of remission in giant cell arteritis: a phase 2, randomised, double-blind, placebo-controlled trial. Lancet. 2016;387(10031):1921–7.
40. Bruce SP, Boyce EG. Update on abatacept: a selective costimulation modulator for rheumatoid arthritis. Ann Pharmacother. 2007;41(7):1153–62.
41. Goëb V, Buch MH, Vital EM, et al. Costimulation blockade in rheumatic diseases: where we are? Curr Opin Rheumatol. 2009;21(3):244–50.
42. Genant HK, Peterfy CG, Westhovens R, et al. Abatacept inhibits progression of structural damage in rheumatoid arthritis: results from the long-term extension of the AIM trial. Ann Rheum Dis. 2008;67(8):1084–9.
43. Weinblatt M, Schiff M, Goldman A, et al. Selective costimulation modulation using abatacept in patients with active rheumatoid arthritis while receiving etanercept: a randomised clinical trial. Ann Rheum Dis. 2007;66(2):228–34.
44. Kavanaugh A, Puig L, Gottlieb AB, et al. Maintenance of clinical efficacy and radiographic benefit through two years of ustekinumab therapy in patients with active psoriatic arthritis: Results from a randomized, placebo-controlled phase III trial. Arthritis Care Res. 2015;67(12):1739–49.
45. Feagan BG, Sandborn WJ, Gasink C, et al. Ustekinumab as induction and maintenance therapy for Crohn's disease. N Engl J Med. 2016;375(20):1946–60.
46. Fragoulis GE, Siebert S, McInnes IB. Therapeutic targeting of IL-17 and IL-23 cytokines in immune-mediated diseases. Annu Rev Med. 2016;67:337–53.
47. Langley RG, Elewski BE, Lebwohl M, et al. Secukinumab in plaque psoriasis—results of two phase 3 trials. N Engl J Med. 2014;371(4):326–38.
48. McInnes IB, Mease PJ, Kirkham B, et al. Secukinumab, a human anti-interleukin-17A monoclonal antibody, in patients with psoriatic arthritis (FUTURE 2): a randomised, double-blind, placebo-controlled, phase 3 trial. Lancet. 2015;386(9999):1137–46.
49. Kunwar S, Dahal K, Sharma S. Anti-IL-17 therapy in treatment of rheumatoid arthritis: a systematic literature review and meta-analysis of randomized controlled trials. Rheumatol Int. 2016;36:1065–75.

50. Aronson JK, Furner RE. How similar are biosimilars. BMJ. 2016;353:i2721.
51. Jung SK, Lee KH, Jeon JW, et al. Physicochemical characterization of Remsima. MAbs. 2014;6(5):1163–77.
52. Committee for Medicinal Products for Human Use EMA. Inflectra: assessment report. EMA/CHMP/589422/2013. 2013. http://www.ema.europa.eu/docs/en_GB/document_library/EPAR_-_Public_assessment. Accessed 1 Nov2016.
53. Yang BB, Baughman S, Sullivan JT. Pharmacokinetics of anakinra in subjects with different levels of renal function. Clin Pharmacol Ther. 2003;74(1):85–94.
54. Roos JC, Ostor AJ. Anti-tumor necrosis factor å and the risk of JC virus infection. Arthritis Rheum. 2006;54:381–2.
55. Fleischmann RM. Progressive multifocal leukoencephalopathy following rituximab treatment in a patient with rheumatoid arthritis. Arthritis Rheum. 2009;60(11):3225–8.
56. Kremer J, Ritchlin C, Mendelsohn A, et al. Golimumab, a new human anti-tumor necrosis factor alpha antibody, administered intravenously in patients with active rheumatoid arthritis: forty-eight-week efficacy and safety results of a phase III randomized, double-blind, placebo-controlled study. Arthritis Rheum. 2010;62(4):917–28.
57. Vermeire S, Van Assche G, Rutgeerts P. Serum sickness, encephalitis and other complications of anticytokine therapy. Best Pract Res Clin Gastroenterol. 2009;23(1):101–12.
58. Gubner R, August S, Ginsberg V. Therapeutic suppression of tissue reactivity. II. Effect of aminopterin in rheumatoid arthritis and psoriasis. Am J Med Sci. 1951;221(2):176–82.
59. Seideman P, Beck O, Eksborg S, et al. The pharmacokinetics of methotrexate and its 7-hydroxy metabolite in patients with rheumatoid arthritis. Brit J Clin Pharmacol. 1993;35(4):409–12.
60. Kremer JM, Petrillo GF, Hamilton RA. Pharmacokinetics and renal function in patients with rheumatoid arthritis receiving a standard dose of oral weekly methotrexate: association with significant decreases in creatinine clearance and renal clearance of the drug after 6 months of therapy. J Rheumatol. 1995;22:38–40.
61. Cronstein BN. Molecular therapeutics. Methotrexate and its mechanism of action. Arthritis Rheum. 1996;39:1951–60.
62. Stewart CF, Fleming RA, Germain BF, et al. Aspirin alters methotrexate disposition in rheumatoid arthritis. Arthritis Rheum. 1991;34:1514–20.
63. Bannwarth B, Pehourcq F, Schaeverbeke T, et al. Clinical pharmacokinetics of low-dose pulse methotrexate in rheumatoid arthritis. Clin Pharmacokinet. 1996;30:194–210.
64. Stamp LK, O'Donnell JL, Chapman PT, et al. Methotrexate polyglutamate concentrations are not associated with disease control in rheumatoid arthritis patients receiving long-term methotrexate therapy. Arthritis Rheum. 2010;62(2):359–68.
65. Dervieux T, Kremer J. Methotrexate polyglutamate concentrations are not associated with disease control in rheumatoid arthritis patients receiving long-term methotrexate therapy: comments on the article by Stamp et al. Arthritis Rheum. 2010;62(8):2559–60.
66. Ranganathan P, Culverhouse R, Marsh S, et al. Methotrexate (MTX) pathway gene polymorphisms and their effects on MTX toxicity in Caucasian and African American patients with rheumatoid arthritis. J Rheumatol. 2008;35(4):572–9.
67. Wessels JA, van der Kooij SM, le Cessie S, et al. A clinical pharmacogenetic model to predict the efficacy of methotrexate monotherapy in recent-onset rheumatoid arthritis. Arthritis Rheum. 2007;56(6):1765–75.
68. Seideman P, Beck O, Eksborg S, et al. The pharmacokinetics of methotrexate and its 7-hydroxy metabolite in patients with rheumatoid arthritis. Br J Clin Pharmacol. 1993;35:409–12.
69. Cutolo M, Bisso A, Sulli A, et al. Antiproliferative and antiinflammatory effects of methotrexate on cultured differentiating myeloid monocytic cells (THP-1) but not on synovial macrophages from patients with rheumatoid arthritis. J Rheumatol. 2000;27:2551–1557.
70. Cutolo M, Sulli A, Pizzorni C, et al. Anti-inflammatory mechanisms of methotrexate in rheumatoid arthritis. Ann Rheum Dis. 2001;60:729–35.
71. Montesinos MC, Desai A, Delano D, et al. Adenosine A2A or A3 receptors are required for inhibition of inflammation by methotrexate and its analog MX-68. Arthritis Rheum. 2003;48:240–7.
72. Smolenska Z, Kaznowska Z, Zarowny D, et al. Effect of methotrexate on blood purine and pyrimidine levels in patients with rheumatoid arthritis. Rheumatology (Oxford). 1999;38:997–1002.
73. Visser K, van der Heijde D. Optimal dosage and route of administration of methotrexate in rheumatoid arthritis: a systematic review of the literature. Ann Rheum Dis. 2009;68(7):1094–9. https://doi.org/10.1136/ard.2008.092668.
74. Strand V, Cohen S, Schiff M, et al. Treatment of active rheumatoid arthritis with leflunomide compared with placebo and methotrexate. Arch Intern Med. 1999;159:2542–50.
75. Cohen S, Cannon GW, Schiff M, et al. Two-year, blinded, randomized, controlled trial of treatment of active rheumatoid arthritis with leflunomide compared with methotrexate. Arthritis Rheum. 2001;44:1984–92.
76. Emery P, Breedveld F, Lemmel E, et al. A comparison of the efficacy and safety of leflunomide and methotrexate for the treatment of rheumatoid arthritis. Rheumatology. 2000;39:655–65.
77. Dougados M, Combe B, Cantagrel A, et al. Combination therapy in early rheumatoid arthritis: a randomised, controlled, double blind 52 week clinical trial of sulphasalazine and methotrexate compared with the single components. Ann Rheum Dis. 1999;58:220–5.

78. Bathon JM, Martin RW, Fleischmann RM, et al. A comparison of etanercept and methotrexate in patients with early rheumatoid arthritis. N Engl J Med. 2000;343:1586–93.
79. Buchbinder R, Hall S, Sambrook PN, et al. Methotrexate therapy in rheumatoid arthritis: a life table review of 587 patients treated in community practice. J Rheumatol. 1993;20:639–44.
80. Hoekstra M, van de Laar M, Bernelot Moens H, et al. Longterm observational study of methotrexate use in a Dutch cohort of 1022 patients with rheumatoid arthritis. J Rheumatol. 2003;30:2325–9.
81. Whittle SL, Hughes RA. Folate supplementation and methotrexate treatment in rheumatoid arthritis: a review. Rheumatology (Oxford). 2004;43:267–71.
82. Merrill JT, Shen C, Schreibman D, et al. Adenosine A1 receptor promotion of multinucleated giant cell formation by human monocytes: a mechanism for methotrexate-induced nodulosis in rheumatoid arthritis. Arthritis Rheum. 1997;40(7):1308–15.
83. Abraham Z, Rozenbaum M, Rosner I. Colchicine therapy for low-dose-methotrexate-induced accelerated nodulosis in a rheumatoid arthritis patient. J Dermatol. 1999;26(10):691–4.
84. Alarcón GS, Kremer JM, Macaluso M, et al. Risk factors for methotrexate-induced lung injury in patients with rheumatoid arthritis: A multicentre, case-control study. Methotrexate study group. Ann Intern Med. 1997;127:356–64.
85. Saag KG, Teng GG, Patkar NM, et al. American College of Rheumatology 2008 recommendations for the use of nonbiologic and biologic disease-modifying antirheumatic drugs in rheumatoid arthritis. Arthritis Rheum. 2008;59(6):762–84.
86. Fathi N, Mitros F, Hoffman J, et al. Longitudinal measurement of methotrexate liver concentrations does not correlate with liver damage, clinical efficacy, or toxicity during a 3.5 year double blind study in rheumatoid arthritis. J Rheumatol. 2002;29:2092–8.
87. McLachlan A, Tett S, Cutler D, et al. Bioavailability of hydroxychloroquine tablets in patients with rheumatoid arthritis. Br J Clin Rheumatol. 1994;33:235–9.
88. Wiegmann K, Schutze S, Machleidt T, et al. Functional dichotomy of neutral and acidic sphingomyelinases in tumour necrosis factor signaling. Cell. 1994;78:1005–15.
89. Sanders M. A review of controlled clinical trials examining the effects of antimalarial compounds and gold compounds on radiographic progression in rheumatoid arthritis. J Rheumatol. 2000;27:523–9.
90. Tett S, Day R, Cutler D. Concentration-effect relationship of hydroxychloroquine in rheumatoid arthritis-a cross sectional study. J Rheumatol. 1993;20:1874–9.
91. Davis JP, Cain GA, Pitts WJ, et al. The immunosuppressive metabolite of leflunomide is a potent inhibitor of human dihydroorotate dehydrogenase. Biochemistry. 1996;35(4):1270–3.
92. Rozman B. Clinical pharmacokinetics of leflunomide. Clin Pharmacokinet. 2002;41(6):421–30.
93. Kalden JR, Schattenkirchner M, Sorensen H, et al. The efficacy and safety of leflunomide in patients with active rheumatoid arthritis: a five-year followup study. Arthritis Rheum. 2003;48(6):1513–20.
94. Weinblatt ME, Kremer JM, Coblyn JS, et al. Pharmacokinetics, safety, and efficacy of combination treatment with methotrexate and leflunomide in patients with active rheumatoid arthritis. Arthritis Rheum. 1999;42(7):1322–8.
95. Remer CF, Weisman MH, Wallace DJ. Benefits of leflunomide in systemic lupus erythematosus: a pilot observational study. Lupus. 2001;10(7):480–3.
96. Kaplan MJ. Leflunomide aventis pharma. Curr Opin Investig Drugs. 2001;2(2):222–30.
97. Chakravarty EF, Sanchez-Yamamoto D, Bush TM. The use of disease modifying antirheumatic drugs in women with rheumatoid arthritis of childbearing age: a survey of practice patterns and pregnancy outcomes. J Rheumatol. 2003;30(2):241–6.
98. Svartz N. Salazopyrin, a new sulfanilamide preparation: a. therapeutic results in rheumatic polyarthritis; b. therapeutic results in ulcerative colitis; c. toxic manifestations in treatment with sulfanilamide preparations. Acta Med Scand. 1942;110:577–98.
99. Gadangi P, Longaker M, Naime D, et al. The antiinflammatory mechanism of sulfasalazine is related to adenosine release at inflamed sites. J Immunol. 1996;156(5):1937–41.
100. Neumann VC, Taggart AJ, Le Gallez P, et al. A study to determine the active moiety of sulphasalazine in rheumatoid arthritis. J Rheumatol. 1986;13(2):285–7.
101. Ferraz MB, Tugwell P, Goldsmith CH, et al. Metaanalysis of sulfasalazine in ankylosing spondylitis. J Rheumatol. 1990;17(11):1482–6.
102. Taggart AJ, McDermott BJ, Roberts SD. The effect of age and acetylator phenotype on the pharmacokinetics of sulfasalazine in patients with rheumatoid arthritis. Clin Pharmacokinet. 1992;23(4):311–20.
103. van der Heijde DM, van Riel PL, Nuver-Zwart IH, et al. Sulphasalazine versus hydroxychloroquine in rheumatoid arthritis: 3-year follow-up. Lancet. 1990;335(8688):539.
104. Roskoski R. Janus kinase inhibitors in the treatment of inflammatory and neoplastic diseases. Pharm Res. 2016;111(1):784–803.
105. Dowty ME, Lin J, Ryder TF, et al. The pharmacokinetics, metabolism, and clearance mechanisms of tofacitinib, a janus kinase inhibitor, in humans. Drug Metab Dispos. 2014;42(4):759–73.
106. Charles-Schoeman C, Burmester G, Nash P, et al. Efficacy and safety of tofacitinib following inadequate response to conventional synthetic or biological disease-modifying antirheumatic drugs. Ann Rheum Dis. 2015;75:1293–301.
107. Ritchlin CT, Krueger JG. New therapies for psoriasis and psoriatic arthritis. Curr Opin Rheumatol. 2016;28:204–10.

108. Asahina A, Etoh T, Igarashi A, et al. Oral tofacitinib efficacy, safety and tolerability in Japanese patients with moderate to severe plaque psoriasis and psoriatic arthritis: a randomized, double-blind, phase 3 study. J Dermatol. 2016;43:869–80.
109. Keystone EC, Taylor PC, Drescher E, et al. Safety and efficacy of baricitinib at 24 weeks in patients with rheumatoid arthritis who have had an inadequate response to methotrexate. Ann Rheum Dis. 2015;74:333–40.
110. Genovese MC, Kremer J, Zamani O, et al. Baricitinib in patients with refractory rheumatoid arthritis. N Engl J Med. 2016;374:1243–52.
111. Genovese MC, van Vollenhoven RF, Pacheco-Tena C, et al. VX-509 (decernotinib), an oral selective JAK-3 inhibitor, in combination with methotrexate in patients with rheumatoid arthritis. Arthritis Rheumatol. 2016;68(1):46–55.
112. Winthrop KL, Yamanaka H, Valdez H, et al. Herpes zoster and tfacitinib therapy in patients with rheumatoid arthritis. Arthritis Rheum. 2014;66(10):2675–84.
113. Curtis JR, Lee EB, Kaplan IV, et al. Tofacitinib, an oral Janus kinase inhibitor: analysis of malignancies across the rheumatoid arthritis clinical development programme. Ann Rheum Dis. 2016;75:831–41.
114. Champion GD, Graham GG, Ziegler JB. The gold complexes. In: Brooks P, editor. Bailliere's clinical rheumatology, slow acting anti-rheumatic drugs and iminunosuppressives, vol. 3. London: Bailliere; 1990. p. 491–534.
115. Handel ML, Watts CKW, deFazio A, et al. Inhibition of AP-1 binding and transcription by gold and selenium involving conserved cysteine residues in Jun and Fos. Proc Natl Acad Sci U S A. 1995;92:4497–501.
116. Graham GG, Champion GD, Ziegler JB. The cellular metabolism and effects of gold complexes. Metal Based Drugs. 1994;1:395–404.
117. Madeira JM, Gibson DL, Kean WF, et al. The biological activity of auranofin; implications fo novel treatment of dieases. Inflammopharmacology. 2012;20:297–306.
118. Williams KM, Day RO, Breit SN. Biochemical actions and clinical pharmacology of anti-inflammatory drugs. Adv Drug Res. 1993;24:121–98.
119. Wood PL, Khan MA, Moskal JR. Mechanism of action of the disease-modifying anti-arthritic thiol agents D-penicillamine and sodium aurothiomalate: Restoration of cellular free thiols and sequestration of reactive aldehydes. Eur J Pharmacol. 2008;580(1):48–54.
120. Stone M, Fortin PR, Pacheco-Tena C, et al. Should tetracycline treatment be used more extensively for rheumatoid arthritis? Metaanalysis demonstrates clinical benefit with reduction in disease activity. J Rheumatol. 2003;30(10):2112–22.
121. Leite LM, Carvalho AGG, Ferreira PL, et al. Anti-inflammatory properties of doxycycline amd minocycline in experimental models: an in vivo and in vitro comparative study. Inflammopharmacology. 2011;19(2):99–110.
122. Monk E, Shalita A, Siegel DM. Clinical applications of non-antimicrobial tetracyclines in dermatology. Pharmacol Res. 2011;63(2):130–45.
123. Barnes PJ. Anti-inflammatory actions of glucocorticoids: molecular mechanisms. Clin Sci. 1998;94:557–72.
124. Smith MD, Ahern MJ, Roberts-Thompson PJ. Pulse methylprednisolone therapy in rheumatoid arthritis: unproved therapy, unjustified therapy, or effective adjunctive treatment? Ann Rheum Dis. 1990;49:265–7.
125. Capell HA, Madhok R, Hunter JA, et al. Lack of radiological and clinical benefit over two years of low dose prednisolone for rheumatoid arthritis: results of a randomised controlled trial. Ann Rheum Dis. 2004;63:797–803.
126. Sambrook PN, Jones G. Corticosteroid osteoporosis. Br J Rheumatol. 1995;34:8–12.
127. Emery P, Breedveld FC, Hall S, et al. Comparison of methotrexate monotherapy with a combination of methotrexate and etanercept in active, early, moderate to severe rheumatoid arthritis (COMET): a randomised, double-blind, parallel treatment trial. Lancet. 2008;372(9636):375–82.
128. Emery P, Breedveld F, van der Heijde D, et al. Two-year clinical and radiographic results with combination etanercept-methotrexate therapy versus monotherapy in early rheumatoid arthritis: a two-year, double-blind, randomized study. Arthritis Rheum. 2010;62(3):674–82.
129. Tugwell P, Pincus T, Yocum D, et al. Combination therapy with cyclosporine and methotrexate in severe rheumatoid arthritis. The Methotrexate-Cyclosporine Combination Study Group. N Engl J Med. 1995;333(3):137–41.
130. Fox RI, Morgan SL, Smith HT, et al. Combined oral cyclosporin and methotrexate therapy in patients with rheumatoid arthritis elevates methotrexate levels and reduces 7-hydroxymethotrexate levels when compared with methotrexate alone. Rheumatology (Oxford). 2003;42(8):989–94.

35 Drugs for Soft Tissue Autoimmune Disorders

Carolin F. Manthey, Marcial Sebode, Helen Hayward-Koennecke, Andreas Lutterotti, and Samuel Huber

35.1 Inflammatory Bowel Diseases

35.1.1 Introduction

CROHN'S DISEASE (CD) and ulcerative colitis (UC) are the two major forms of INFLAMMATORY BOWEL DISEASES (IBD). The exact cause for IBD is unknown; however, a genetic predisposition seems to lead to a disturbed immune response in the gut towards the commensal flora which results in intestinal inflammation [1]. Both types of IBD show relapsing courses and mainly affect children and young adults, although a second peak of disease onset can be observed in ages 50–70 years. The first description of ulcerative colitis dates back to the nineteenth century when Sir Samuel Wilks described an idiopathic colitis distinct from epidemic dysentery [2]. However, the most recognized paper was published in 1932 by Crohn, Ginzburg and Oppenheimer on "Regional Enteritis, a pathological and clinical entity". The description was based on the examination of 14 cases and sums up the presentation of patients with CD in a way that is still current knowledge [3].

Patients suffer mainly from chronic inflammation of the gastrointestinal tract; symptoms include abdominal pain, cramping and diarrhoea with the passage of blood and/or mucus. In patients with CROHN'S DISEASE, perianal disease involving fistulas occurs frequently. Additionally, other organs can be affected as extra-intestinal manifestations of disease. In male patients with ulcerative colitis, the risk for primary sclerosing cholangitis (PSC) is elevated, and other affected organs include the joints (arthritis), eyes (uveitis), skin (pyoderma gangraenosum, erythema nodosum) and spine (spondylarthritis) [4]. In CROHN'S DISEASE, all parts of the gastrointestinal tract can be involved with a predilection for the terminal ileum, and inflammation usually is discontinuous, whereas in ulcerative colitis, inflammation starts in the rectum and extends proximally in a continuous way which can involve the complete colon. The diagnosis is made based on the clinical history, family history, physical examination, laboratory results like elevated inflammation markers and endoscopic evaluation (ileocolonoscopy, gastroscopy). Furthermore, radiologic imaging (MRI, CT scan) is performed for determining the extent of disease. There is no single test to confirm the diagnosis which is rather assessed on the combined findings of the above-mentioned tests.

Final manuscript submitted on September 25, 2018.

C. F. Manthey · M. Sebode · S. Huber (✉)
I. Medizinische Klinik und Poliklinik, Universitätsklinikum Hamburg-Eppendorf, Hamburg, Germany
e-mail: shuber@uke.de

H. Hayward-Koennecke · A. Lutterotti
Klinik für Neurologie, UniversitätsSpital Zürich, Zürich, Switzerland
e-mail: andreas.lutterotti@usz.ch

M. J. Parnham et al. (eds.), *Nijkamp and Parnham's Principles of Immunopharmacology*, https://doi.org/10.1007/978-3-030-10811-3_35

In terms of underlying disease causes, familial aggregation has been observed for a substantial amount of time [5]. Several concordance studies in twins support the role for genetic predisposition for IBD where high concordance was found in monozygotic twins [6, 7]. Also, the risk to develop disease is up to 15.5% in first-degree relatives of a person with ulcerative colitis [8, 9].

Genome-wide association studies (GWAS) and meta-analyses have found 71 susceptibility loci in patients for CROHN'S DISEASE [10], 20 of them overlapping with the 47 loci found to be associated with ulcerative colitis [11]. Furthermore, environmental factors play an important role in disease development. The incidence for UC and CD is higher in developed countries than in third-world countries and in cities compared to rural areas. Increased hygiene, living in smaller families and change of diet can lead to a different and diminished exposure towards environmental MICROBIOTA at younger age, which might predispose to a pathological immune response to commensals and pathogenic microbes at later stages [12, 13]. The disease-promoting role of the environment is supported by the fact that incidence is rising in immigrants coming to developed countries from areas with reported low incidence for IBD [14]. Other factors involved in disease development are lifestyle related, and here, the early exposure to tobacco smoke is associated with higher risk for CROHN'S DISEASE. Interestingly, the opposite was shown for ulcerative colitis where smoking is protective and disease activity is often increased in patients who stop smoking.

Complications in IBD patients include the development of bowel stenoses, strictures and intraenteric and enterovesical or enterovaginal fistulas. The long-term risk for colon cancer is significantly increased in patients with ulcerative colitis, especially in patients with pancolitis, with a cumulative risk of 2% after 10 years of diagnosis, 8% after 20 years and 18% after 30 years [15]. In patients with concomitant diagnosis of PSC, the risk of colorectal cancer is up to four times greater than in those without PSC [16]. Also in patients with CROHN'S DISEASE, the risk for early small bowel and colon cancer is increased [17]. In general, 50% of all patients with CROHN'S DISEASE require surgery within 10 years after diagnosis, and 50% of adult patients suffer from an intestinal complication within 20 years of diagnosis [18].

35.1.2 Microbiota

The intestinal MICROBIOTA has been shown to have a principal role in maintaining homeostasis by inducing tolerance in immune cells as well as by fighting obligatory pathogens [19]. The four dominant bacterial phyla in the human gut are *Firmicutes*, *Bacteroidetes*, *Actinobacteria* and *Proteobacteria*. Dysbiosis has been linked to intestinal disease, and it has been demonstrated that patients with IBD show a reduced diversity in the composition of their intestinal MICROBIOTA. A reduction in *Faecalibacterium prausnitzii*, one of the most abundant species in the gut, has been associated with increased risk of postoperative recurrence of ileal CROHN'S DISEASE [20]. Until now, dysbiosis is not considered the driving cause for IBD but can facilitate disease progression and is representing a susceptibility trait as confirmed by mouse models with human-relevant susceptibility mutations [21].

35.1.3 Intestinal Barrier

The epithelial barrier represents the first line of defence against intraluminal pathogens and separates luminal content from the underlying tissue. Defects in the epithelial barrier, possibly through defective tight junctions, are linked to increased permeability and invasion of luminal microbes into the lamina propria. In UC and CD, the expressions of mucins have been found to be reduced [22, 23], indicating loss of the protective mucin layer and leading to a vulnerable epithelium. Furthermore, specialized cells within the epithelium, Paneth cells, produce antimicrobial peptides that support the intestinal immune system. Paneth cells of patients with CROHN'S DISEASE and certain genetic susceptibility in autophagy genes show diminished and functionally altered granules [24].

35.1.4 Microbial Processing and Adaptive Immunity

It has been shown that the dysregulation of innate and adaptive immune cells is an important factor for disease progression in IBD. MACROPHAGES and DENDRITIC CELLS populate the lamina propria and sample bacteria to present antigens to B cells and T CELLS which initiate an adaptive immune response [25] (see Chap. 11). The number of DENDRITIC CELLS is greatly increased in patients with ulcerative colitis, and their circulating number correlates with disease activity [26] indicating a significant role for them to start inflammation. Microbial products are recognized by the so-called TLRs (TOLL-LIKE RECEPTORS), and activation of TLRs induces signalling cascades that cause inflammatory or tolerant immune responses. It was shown that the distribution and phenotype of different TLRs is altered in IBD patients, thus leading to an exaggerated lipopolysaccharide response [27–29].

Furthermore, an imbalance between regulatory and effector T CELLS has been observed in IBD patients (see Chap. 3). Effector T CELLS, mainly T helper (Th) 1 or Th17 cells, seem to be differentially activated in patients with CROHN'S DISEASE, whereas in ulcerative colitis, an inappropriate Th2 cell regulation occurs [30, 31] involving IL-5 and IL-13, the latter one representing an aggressive cytotoxic phenotype. On the other hand, REGULATORY T CELLS and inducible regulatory cells (Tr1, Th3) that secrete IL-10 and TGF-β are dysregulated as well, since mutations in crucial regulators of T cell fate have been linked to IBD [10]. Loss of IL-10 receptor-1 or IL-10 receptor leads to severe ulcerative colitis probably due to missing IL-10 signalling [32].

35.1.5 General Remarks on IBD Therapy

There are currently no reliable biomarkers available which would allow to predict the response to one or the other drug. Therefore, therapy is chosen individually for each patient based on clinical examination and completion of diagnostic workup. The choice of drug is based on phenotype (extent of disease), severity of symptoms, endoscopic findings, comorbidities and other patient-specific characteristics. Furthermore, the treatment is guided by the definition of disease extent and activity according to diagnostic tools that score clinical, laboratory and endoscopic findings (Harvey-Bradshaw index, CROHN'S DISEASE activity index (CDAI) and Mayo score for ulcerative colitis): mild to moderate or severe disease, localized or extensive disease, refractory disease and presence of (perianal) fistulas in CROHN'S DISEASE. Medications are defined as being suited for induction and/or maintenance therapy. In general, IBD therapy is usually consisting of two steps: induction of remission and maintenance of remission.

Usually, a fast-acting therapy is initiated to induce remission. Initially, treatment is commenced using steroids in a tapered regimen (see Chap. 32). In patients who show severe, steroid-refractory disease (persisting symptoms despite initiation of steroids), therapy with AZATHIOPRINE, biologics, cyclosporine or TACROLIMUS can be employed (as further outlined in the next sections; see also Chaps. 30 and 31). Also, in patients with isolated terminal ileitis or in patients with steroid-refractory ulcerative colitis, surgery should also be considered. Generally, the patient with a severe disease course should be hospitalized, and diagnostics should include ultrasound or computerized tomography to exclude an abscess in the presence of an abdominal mass. Furthermore, in the evidence of systemic infection such as sepsis or severe perianal disease, antibiotic therapy is added to the regimen. The overall therapy goal is to achieve clinical and endoscopic remission which is defined as "mucosal healing" to reduce late complications [33].

Upon induction of remission, a therapy to maintain remission is needed in patients with steroid-dependent disease (recurring symptoms during steroid tapering) or in patients with multiple disease exacerbations or with a high risk for a complicative disease course. Here, purine analogues (AZATHIOPRINE, 6-MERCAPTOPURINE, thioguanine), biologics and methotrexate are recommended (as outlined in the next sections).

The management of the individual patient includes the consideration of two different approaches: top-down or step-up strategy. For example, in a patient with multiple risk factors for severe disease course (e.g. young age at disease onset, perianal disease, structuring phenotype), initial treatment should be aggressive and includes early use of anti-TNF (top-down), often in combination with another immunosuppressive agent, i.e. AZATHIOPRINE. On the contrary, patients who present with mild disease and lack risk factors can be treated in the conventional step-up manner, i.e. initial use of MESALAZINE followed by steroids if needed (see also Chap. 34).

35.2 Autoimmune Hepatitis

35.2.1 Introduction

Autoimmune hepatitis (AIH) is a chronic inflammatory liver disease with unknown aetiology [34]. It mainly affects women at a median age between 40 and 70 years and is mostly asymptomatic. AIH has a wide clinical spectrum: it can manifest as fulminant acute liver failure, it can present with mild elevation of liver enzymes, or it becomes apparent at the stage of liver cirrhosis. Its diagnosis is based on elevated transaminases and IgG/gamma-globulin levels; on the presence of autoantibodies like ANA (anti-nuclear antibodies), anti-SMA (antibodies directed against smooth muscle antigen), anti-LKM (antibodies directed against liver-kidney microsomes) or anti-SLA/LP (antibodies directed against soluble liver antigen/liver-pancreas antigen); also on a typical histological pattern; and on the exclusion of other causes of acute or chronic hepatitis [35–37].

In general, the local microenvironment turns the liver into a tolerant organ [38]. This prevents an immunological overreaction against the continuous exposition of foreign antigens by portal venous blood flow: the liver is one of the first organs that come into contact with food and drugs after oral intake. For AIH, the main antigen-driving chronic inflammation of the liver is unknown. Regulatory mechanisms fail to control the autoimmune reaction that is directed against hepatocytes. This results in chronic inflammation leading to liver fibrosis and cirrhosis. To control inflammation and to prevent hepatic damage, long-term, mostly life-long immunosuppressive treatment is the gold standard for AIH [39]. Recently, more selective, immunomodulatory treatment has been applied as second-line therapy for AIH in order to avoid side effects of classical immunosuppression or for difficult-to-treat cases.

35.2.2 Standard Treatment of AIH

The prognostic benefit of immunosuppressive treatment for AIH has been shown since the 1970s [40–42]. The usage of predniso(lo)ne for the induction of remission is well established since these early days of AIH treatment. However, there is still no international consensus about dosage and schedules of steroid treatment because no comparative studies have been performed yet. In the USA, a starting dose of about 40–60 mg prednisolone is recommended, whereas in several European centres, starting with 1 mg per kilogram body weight is preferred [43–45]. Steroids are then tapered by a weekly scheme to reach a minimal dosage (about 5–10 mg/day) that still controls hepatic inflammation. Typical side effects of steroids are metabolic changes (gain of weight, hyperglycaemia), water retention and swellings, acne, cataract and glaucoma, osteoporosis and psychiatric disorders (see Chap. 32).

AIH therapy aims at eliminating hepatic inflammation. Transaminases (alanine aminotransferase (ALT) and aspartate aminotransferase (AST)) and IgG/gamma-globulin levels are good surrogates for histological inflammation in AIH, and they both have to be regularly checked under treatment. Therefore, complete biochemical remission, defined as total normalization of both transaminases and IgG/gamma-globulins, is the goal of AIH treatment. A prolonged elevation of these biochemical markers is associated with a worse clinical outcome of AIH [46].

Budesonide is an alternative to systemic predniso(lo)ne in AIH for minimizing steroid side effects and thereby increasing patients' adherence to treatment [47]. However, the efficacy of budesonide to induce remission in AIH in comparison to predniso(lo)ne is controversial. It is started at a dose of 3 × 3 mg/day. No recommendations can be made for further dose reduction since these have not been studied in AIH yet. There are only limited data on the long-term outcome of AIH patients on budesonide treatment [48]. Budesonide is contraindicated in cirrhotic patients.

One to two weeks after the beginning of steroid treatment, AZATHIOPRINE is introduced for the maintenance of remission of AIH. AZATHIOPRINE is a purine analogue with antimetabolite effects (see also Chap. 31). In AIH, an AZATHIOPRINE dosage of 1–1.5 mg/kg body weight is usually sufficient to control hepatic inflammation. Typical side effects are allergic reactions, bone marrow suppression, cholestatic hepatitis, pancreatitis and, in the long run, an elevated risk for non-melanoma skin cancer and lymphoma. Regular testing for liver enzymes, lipase and blood counts is recommended. Due to its low predictive value, screening for thiopurine *S*-methyltransferase (TPMT) activity is not recommended to predict the risk of side effects for AIH patients.

35.2.3 Second-Line Immunosuppressive Treatment for AIH

Two clinical scenarios are in need of second-line therapy for AIH: intolerance of side effects or an aggressive form of the disease that cannot be controlled sufficiently with standard therapy.

For patients with intolerance to AZATHIOPRINE, alternatives are 6-MERCAPTOPURINE (6-MP) and MYCOPHENOLATE MOFETIL (MMF) [49] (see Chap. 30). Despite a very close pharmacological relationship between 6-MP and AZATHIOPRINE, adverse reactions decline after a switch from AZATHIOPRINE to 6-MP. MMF inhibits inosine monophosphatase and thereby interferes with purine synthesis. Due to its teratogenic potential, contraception is recommended under MMF therapy [50].

For AIH patients, showing pronounced inflammatory activity and not reaching complete biochemical remission under standard therapy, an increase in AZATHIOPRINE dosage is one of the first steps (up to 1.5–2.0 mg/kg body weight) to adapt treatment. Prior to this, incompliance of the patient and concurrent or alternative liver diseases have to be excluded. TACROLIMUS at a dose targeting a serum level of 2–5 ng/ml is an effective option for AIH patients with an aggressive form of the disease [51]. Another calcineurin inhibitor (CNI), CYCLOSPORINE A (see Chap. 31), has also been used in small case series as second-line therapy for AIH [52]. Side effects of CNIs are nephro- and neurotoxicity and metabolic changes. The mTOR (mammalian target of rapamycin) inhibitor EVEROLIMUS has been applied as another rescue therapy for AIH; however, data are very limited due to small patient groups [53]. The inhibition of mTOR suppresses effector T CELLS and relatively spares REGULATORY T CELLS (Treg) as cellular representatives of immune regulation.

35.2.4 Immunomodulatory Treatment of AIH

To avoid side effects of unselective immunosuppression, a more selective and effective treatment of AIH is desirable. The mechanistic insights into AIH's pathogenesis made progress in the last 20–30 years. However, there are still conflicting data as to which effector cell type and which inflammatory cytokines are predominating in AIH [54]. Nevertheless, immunomodulatory treatment has been applied for AIH patients. Antibody blockage of TNF-α by INFLIXIMAB (see also Chap. 34) has been successfully used in a small case series of difficult-to-treat AIH patients [55]. Of note, treatment of AIH with biologicals is controversial since several case reports delineated the induction of immune-mediated drug-induced liver injuries by its usage, especially for anti-TNF-α inhibitors [56]. Also, selective immunotherapy in AIH aims at B cells, though their exact pathogenic role in its pathogenesis is not clear. RITUXIMAB, a monoclonal antibody to the B cell surface marker CD20 (see also Chap. 34), has already been used in a small case series of AIH patients [57].

35.3 Multiple Sclerosis

35.3.1 Introduction

Multiple sclerosis (MS) is an autoimmune disease of the central nervous system (CNS), affecting young adults with a strong female preponderance [58]. With worldwide more than two million patients, MS is one of the leading causes of neurological disability in early adulthood. Like many other chronic autoimmune diseases, MS is a complex disease, in which both aetiology and disease course are strongly influenced by an interplay of genetic, environmental and lifestyle factors [59]. Like other complex diseases, MS is characterized by a high degree of heterogeneity in clinical presentation, imaging and main pathomechanisms of disease leading to variability in disease courses and response to different treatments.

Neurologic symptoms in MS patients are caused by inflammatory lesions occurring at different sites in the brain and spinal cord, leading to demyelination and axonal damage. Depending on the strategic localization of the lesion, MS can lead to symptoms in most functional systems of the CNS including sensory symptoms, weakness, coordination problems, visual disturbances, bladder and bowel dysfunction, fatigue, cognitive symptoms, mood disorders and others. The combination and extent of these symptoms vary between patients. Heterogeneity of the clinical presentation is also driven by insidious progression of disabilities caused by an additional neurodegenerative process in MS patients, which can occur independently of relapses. The majority of patients show a relapsing-remitting course (RRMS) with transition into a progressive clinical course after 10–20 years (secondary progressive MS, SPMS). A progressive course from onset can be found in approximately 15% of MS patients (primary progressive MS, PPMS).

Early pathological studies described an inflammatory disease with typical multifocal lesions of demyelination occurring in the white matter of the brain and spinal cord [60]. Later studies recognized the importance of axonal damage and axonal loss within the lesions as a hallmark of the disease [61, 62]. Driven by high-resolution imaging studies, the disease is now recognized as affecting both the white and the grey matter with cortical lesions dominating in patients with high disease progression [63, 64].

Although the aetiology of MS is as yet not fully understood, all current therapeutic approaches are based on the concept that the disease is driven by an immune response directed against self-antigens in the central nervous system (CNS) [65]. The autoimmune response in MS involves several components of the immune system including T CELLS, B cells/antibodies, DENDRITIC CELLS, MACROPHAGES, microglia and cytokines/chemokines and a close interplay with microglia and neurons. However, T CELLS and B cells are the main target cells of currently approved therapies. The impressive efficacy of B cell-depleting therapies in MS has further corroborated the importance of the adaptive immune system and the interaction of B and T CELLS in the pathogenesis of MS [66].

Over the last 15 years, there have been enormous therapeutic advances with several new drugs being approved for MS. This has led to more than a dozen approved disease-modifying therapies (DMT) in Europe and the USA at the beginning of 2018, and there are more treatments to come in the near future. Overall, treatment efficacy has increased considerably; however, this has been paralleled by a significant increase in risks for patients. A concept that has become clear in recent years is the importance of early initiation of DMT [67–70]. There is good evidence that long term the burden of disease can be reduced through early initiation of DMT. However, the individual courses of the disease vary considerably between patients, and there is a lack of clear prognostic parameters to base treatment decision in the early stage. Current treatment guidelines in MS recommend the selection of a DMT from the spectrum of available therapies considering disease activity as "mild to moderately active" and "highly active" disease. Disease activity in MS is quantified either by a clinical measure, i.e. the number of relapses or a radiological parameter, i.e. the number of evolving lesions on MRI of the brain and spinal cord. There is no general consensus on the definition of "highly active" disease, but it has been proposed to categorize patients as "highly active" if they have either suffered breakthrough disease, despite an adequate DMT, or if untreated, they have experienced two or more disabling relapses in 1 year and one or more contrast-enhancing lesions on brain MRI or a significant increase in T2 lesion load as compared to a previous recent MRI. Thus, the field has moved from a classical treatment escalation approach, using first- and second-line therapies, towards a targeted approach where treatment selection is established on the basis of individual disease activity. The success of more effective therapies has also modified the treatment goals to a treat-to-target concept as already used in rheumatologic diseases. "No evidence of disease activity" (NEDA) measured by clinical and imaging parameters has evolved as important outcome measure in clinical trials [71, 72]. Of note, unlike rheumatologic diseases (see Chap. 34), immunomodulating therapies are not combined in MS treatment regimes, and only monotherapies are currently used.

35.4 Drugs and Therapeutics/ Drug Classes

35.4.1 5-ASA (Mesalazine) and Derivatives

SULFASALAZINE was the first drug of aminosalicylates to be introduced for treatment of IBD as early as 1942 [73], and 5-aminosalicylic acid (syn. 5-ASA, MESALAZINE, mesalamine) drugs have been employed for decades in the therapy of IBD. MESALAZINE is the first-line treatment for mild to moderately active ulcerative colitis. Basic therapy for patients with ulcerative colitis regularly includes treatment with MESALAZINE for an indefinite amount of time due to its protective effect on the colonic mucosa. Despite its proven efficacy, patients often require intensified treatment with additional regimens, e.g. steroids, to induce remission.

MESALAZINE drugs interfere with inflammation, proliferation and apoptosis. It has been shown that MESALAZINE reversibly inhibits COX-1 and COX-2 in different cell systems including lymphocytes and polymorphonuclear leukocytes (PMNLs) [74]. Furthermore, the activation of NfκB is inhibited, mostly by inhibiting the degradation of IκB, the inhibitory unit of the NfκB complex. NfκB can activate the inflammatory cascade. Patients with UC who were treated with MESALAZINE for 8 weeks showed significant inhibition of NfκB activation in colon specimens [75]. Further research demonstrated that MESALAZINE was able to reduce the level of REACTIVE OXYGEN METABOLITES (ROMs) by scavenging superoxide and hydroxyl radicals. ROMs are thought to increase inflammation and tissue stress in IBD patients as well as in carcinogenesis [76]. These mechanisms of action of MESALAZINE also led to the presumed protective effect of reducing the risk for colon cancer which was shown in a meta-analysis [77].

SULFASALAZINE is composed of MESALAZINE linked to sulfapyridine via a diazo bond. This bond is cleaved in the colon by bacterial azoreductases to release the active drug, MESALAZINE [78]. Unbound or uncoated MESALAZINE is unable to reach the colon since it is already absorbed in the upper jejunum [79], whereas SULFASALAZINE is able to serve as a delivery system to transport 5-ASA to the lower gastrointestinal tract. However, up to 30% of patients report side effects of the medication, mostly due to the sulfapyridine moiety.

Side effects include nausea, indigestion, headache, vomiting and abdominal pain. Other side effects have been described as myelosuppression, sulfasalazine-induced lupus erythematosus or immune thrombocytopenia [80]. To improve tolerance and achieve better bioavailability, 5-ASA is now covered in resins (Eudragit S or L) that dissolve at pH greater than 7.

Different formulations of oral MESALAZINE are available, in addition to topical MESALAZINE (suppositories, enemas, foams). Depending on the extent of disease, either topical formulations for proctitis or left-sided colitis are chosen opposed to oral MESALAZINE for extended colitis. Combined treatment is often chosen to increase efficacy [81, 82]. To induce remission, 2–4 g/day of MESALAZINE is given for 2–4 weeks. In maintenance treatment dosage is reduced depending on the formulation.

35.4.2 Azathioprine, 6-Mercaptopurine, and Thioguanine

AZATHIOPRINE (6-(1-methyl-4-nitro-5-imidazolythio)purine) and 6-MERCAPTOPURINE (6-MP, beta-thiopurine) represent antimetabolite drugs (see also Chap. 30). 6-MP was originally developed for the treatment of leukaemia [83] but is nowadays applied as an immunosuppressive for immune-mediated diseases such as inflammatory bowel disease (IBD) and autoimmune hepatitis and after solid organ transplantation, while it is not used anymore for the treatment of MS. AZATHIOPRINE and 6-MP have immunosuppressive and cytotoxic properties as they inhibit purine nucleotide biosynthesis [84]. Further molecular mechanisms of azathioprine and 6-MP consist in modulation of pro-inflammatory genes and of induction of T cell apoptosis [85, 86]. For patients with IBD, weight-based dosage of AZATHIOPRINE requires 2–2.5 mg/kg body weight per day. For autoimmune hepatitis (AIH), on the other hand, lower doses of 1.5 mg/kg body weight can in most cases sufficiently control hepatic inflammation. Dosage of 6-MP is lower than that of AZATHIOPRINE and ranges at about 1–1.5 mg/kg body weight in IBD.

Both AZATHIOPRINE and 6-MP constitute prodrugs that are activated through a complex metabolic pathway. After oral intake, AZATHIOPRINE is directly converted to 6-MP in the liver. Thus, both drugs share further metabolic steps and biological effects. 6-MP is then metabolized by three alternative routes, one for activation and two for inactivation. Variations of these pathways lead to individual differences of AZATHIOPRINE and 6-MP metabolism.

6-MP is activated by the enzyme hypoxanthine phosphoribosyltransferase (HPRT) and by further enzymatic steps. This finally leads to the formation of the main active metabolite, 6-thioguanine nucleotide (6-TGN), which has immunosuppressive and cytotoxic effects exerted by its insertion into the DNA of leukocytes [2]. 6-TGN levels can be measured intracellularly in peripheral red blood cells and correlate with those in leukocyte DNA [87]. 6-TGN levels of 235–450 pmol/8 × 10^8 erythrocytes correlate with a clinical response in IBD [88, 89].

Apart from this route of 6-MP activation, two other pathways for its inactivation exist. In one route, the enzyme thiopurine *S*-methyltransferase (TPMT) causes 6-MP methylation resulting in the formation of the inactive metabolite 6-methyl-MP (6-MMP). Elevated levels of 6-MMP in red blood cells have been associated with adverse effects like hepatotoxicity and myelosuppression [90, 91]. TPMT activity depends on genetic polymorphisms [92]. Heterozygosity for a non-functional TPMT allele results in intermediate enzymatic activity, while homozygosity for two non-functional TMPT alleles will result in low or absent enzymatic activity of TPMT. Patients with low or absent enzymatic TPMT activity are at increased risk for myelotoxicity after the application of conventional doses of AZATHIOPRINE or 6-MP [93]. TPMT genotyping or phenotyping (analysis of TPMT activity in erythrocytes) can help to identify patients at risk for this adverse reaction. TPMT activity in erythrocytes correlates with enzymatic activity in lymphocytes or organs like the liver [94, 95].

The other route of 6-MP inactivation comprises oxidation of 6-MP by xanthine oxidase (XO), resulting in 6-thiouric acid formation. This metabolic step is responsible for thiopurine accumulation and thereby for increased toxicity in patients under AZATHIOPRINE or 6-MP who are simultaneously receiving XO inhibitors like allopurinol. In addition to accumulating thiopurine levels, allopurinol also shifts metabolites favouring a pronounced increase of therapeutic 6-TGN as well as a relative decrease of toxic 6-MMP [96]. This interference of allopurinol with 6-MP metabolism has been taken advantage of by combining therapy of low-dose AZATHIOPRINE or 6-MP with allopurinol to booster therapeutic levels of

6-TGN and to avoid an accumulation of toxic 6-MMP. For instance, combination therapy of AZATHIOPRINE or 6-MP with allopurinol has been used for IBD or AIH patients with insufficient response to AZATHIOPRINE or 6-MP monotherapy [90, 97].

Thioguanine (6-TG) treatment is an alternative for patients intolerant to AZATHIOPRINE or 6-MP. 6-TG is directly converted to active 6-TGN and is a poor substrate for TPMT, and as a result, production of the toxic metabolite 6-MMP remains minimal. However, clinical experience with 6-TG in immune-mediated diseases is limited, mostly due to its restricted use because of hepatic side effects (see below).

Patients under AZATHIOPRINE and 6-MP are at increased risk of lymphoma and non-melanoma skin cancer [98–100]. Severe bone marrow suppression can occur, and complete blood counts have to be performed regularly, especially in the first month of treatment [101]. Gastrointestinal hypersensitivity with nausea, vomiting, diarrhoea, rash and fever can occur, often during the first weeks of therapy. These symptoms of hypersensitivity will cease after discontinuation of the drug. Acute pancreatitis is a severe but rare form of gastrointestinal hypersensitivity under thiopurine drugs. Hepatotoxicity often becomes apparent by an elevation of cholestatic liver enzymes. Nodular regenerative hyperplasia (NRH) has been reported primarily as a side effect of 6-TG [102] but can also occur under 6-MP and AZATHIOPRINE. There is a general warning for the usage of AZATHIOPRINE during pregnancy due to potential teratogenic effects. However, its usage seems to be safe, and, due to current recommendations for INFLAMMATORY BOWEL DISEASES and autoimmune hepatitis, intake can be continued during pregnancy and nursing.

35.4.3 Anti-TNF-α Inhibitors (Adalimumab, Certolizumab Pegol, Etanercept, Golimumab, Infliximab)

Tumour necrosis factor-α (TNF-α) acts as a pro-inflammatory cytokine (see Chap. 6) and plays a role in several immune-mediated diseases like IBD or rheumatoid arthritis (RA) but also has a protective role in host defence mechanisms [103, 104]. TNF-α can be membrane-bound or, after proteolytic cleavage, soluble [105]. Both forms are capable of binding two different TNF receptors (TNFR1 and TNFR2) that promote pro-inflammatory gene transcription. Membrane-bound TNF-α can act both as a ligand and as a receptor. That is why some anti-TNF-α inhibitors exhibit agonistic effects. Anti-TNF-α inhibitors have been approved for the treatment of rheumatoid arthritis (see Chap. 34), ankylosing spondylitis, psoriatic arthritis and IBD. Five anti-TNF-α inhibitors with different molecular structures and half-lives are available [106]. ADALIMUMAB and GOLIMUMAB constitute fully human, recombinant antibodies, whereas INFLIXIMAB is a chimeric antibody with a murine variable region and a human constant region. CERTOLIZUMAB is a humanized Fab fragment conjugated to polyethylene glycol. ETANERCEPT is a fusion protein that competitively inhibits binding of TNF-α to its receptor. It consists of two TNFR2 extracellular domains and the Fc fragment of IgG1. ETANERCEPT has the shortest half-life with above 3 days, whereas the half-life of the other anti-TNF-α inhibitors varies between 10 and 15 days.

INFLIXIMAB was the first biological to reach clinical use. It is the only anti-TNF-α inhibitor which is administered intravenously, whereas the others are applied subcutaneously. INFLIXIMAB is usually administered at a dose of 5 mg/kg bodyweight IN week 0, 2, 6 and 8, followed by an infusion every 8 weeks. In case of a partial response, INFLIXIMAB can be administered every 4 weeks. ADALIMUMAB is used every 2 weeks. The initial doses are 160 mg, then 80 mg and then 40 mg. In case of a partial response, ADALIMUMAB can be administered weekly. ETANERCEPT is administered at a dose of 25–50 mg once or twice per week. GOLIMUMAB is applied at a dose of 50–100 mg every 2–4 weeks. For ulcerative colitis, a high induction dose of GOLIMUMAB of 200 mg is initially used. The dosing regimen of CERTOLIZUMAB usually requires 200–400 mg every 2–4 weeks.

Efficacy and clinical response to anti-TNF-α inhibitors can be limited by their immunogenicity:

the development of anti-drug antibodies leads to inactivation or increased clearance of anti-TNF-α inhibitors [107]. Measurement of anti-drug antibodies or anti-TNF levels helps to identify patients with established immunogenicity. An increase of dosage and shortening of application intervals are attempts to regain adequate drug levels. Combination therapy with other immunosuppressants seems to limit immunogenicity [108, 109].

Severe side effects under anti-TNF-α inhibitors are opportunistic infections, particularly the reactivation of latent tuberculosis, and malignancies [110]. The risk of lymphoma under anti-TNF-α inhibitors is controversial: it is not clear whether the risk is due to the disease itself or owed to immunosuppression in general. Nevertheless, a potential risk of lymphoma should be discussed with the patient. Infusion reactions can occur after administration of INFLIXIMAB, ranging from mild forms with headache, urticaria or nausea to severe forms like allergic shock. The other anti-TNF-α inhibitors can lead to injection site reactions such as local erythema or swelling of the skin. Anti-TNF-α inhibitors are contraindicated in patients with moderate to severe congestive heart failure and should be avoided in patients with demyelinating diseases [111, 112]. Their use seems to be safe during pregnancy and breastfeeding [113].

Since the patent for most of the anti-TNF-α inhibitors expired recently, follow-up products of the originals, so-called BIOSIMILARS, are on the market today. Anti-TNF-α inhibitors are biological, meaning that they are manufactured from living cells. The exact conditions of this manufacturing process of the original biologicals are unknown. Therefore, BIOSIMILARS are not generic equivalents of biologicals and can differ from the original with regard to efficacy, side effects and immunogenicity.

35.4.4 Integrin Antibodies (Natalizumab, Vedolizumab, Etrolizumab)

The migration of innate and adaptive leukocytes is mediated by different adhesion receptors, among them selectins, integrins and their ligands as well as chemokine receptors. In IBD, the migration of leukocytes is disturbed, and leukocytes show an abnormal retention in the inflamed gut. MAdCAM-1 (mucosal addressin cell adhesion molecule 1) is an adhesion molecule which is expressed on the vascular endothelium of mucosal blood vessels; it enables leukocyte adhesion and extravasation into the tissue, thereby promoting intestinal inflammation [114] by binding to α4β7 INTEGRIN. Another INTEGRIN, αEβ7 INTEGRIN, binds to E-cadherin and retains intraepithelial lymphocytes near or inside the epithelial monolayer that lines the intestine.

The first successful inhibitor of leukocyte migration was natalizumab, a α4 blocker, which binds to all α4-containing integrins. It is successfully employed in the treatment of multiple sclerosis as it inhibits the interaction between VCAM1 and α4β1 which reduces T cell infiltration in the central nervous system. Natalizumab (Tysabri®) was the first highly active treatment approved for relapsing-remitting MS and clearly laid the foundation for development of new treatment goals involving the "No evidence of disease activity (NEDA)" concept [115]. In the pivotal phase III trials (AFFIRM, SENTINEL), treatment with natalizumab showed a profound effect in reducing both the annualized relapse rate and the risk of disability progression compared to placebo or active comparator (interferon-ß-1a i.m.) [116, 117]. The high efficacy of natalizumab on clinical outcome measures was also confirmed in the robust reduction of new lesion formation on brain MRI, with more than half of treated patients remaining free of new or enlarging T2 lesions on brain MRI compared to only 15% in the placebo-treated group.

Natalizumab was also employed for patients with IBD and showed good efficacy [118]. However, a serious side effect has been observed: progressive multifocal leukoencephalopathy (PML). It is a result of the reactivation of the John Cunningham virus (JCV), a polyoma virus. The risk has been described as approximately 2 in 1000 patients but is dependent on positive testing for JC virus antibodies (STRATIFY test), duration of treatment and previous immunosuppressive therapy. The infection usually leads to death or severe disability, and until now, there is no cure. The pathogenesis is uncertain; it might be associated with the blockade of JCV-specific leukocytes across the blood-brain barrier [119].

Due to good efficacy of natalizumab in treating IBD, research aimed at finding gut-specific antibodies targeting α4-containing integrins. Vedolizumab (Entyvio®) is a humanized IgG1 monoclonal antibody that prevents α4β7 INTEGRIN from binding to MAdCAM-1; it is manufactured in Chinese hamster ovary cells. Vedolizumab was approved by the European Medicines Agency (EMA) in June 2014 for use in patients with ulcerative colitis and CROHN'S DISEASE. It is administered intravenously in a constant dose of 300 mg, starting in week 0, continuing in week 2 and week 6, and then it is given every 8 weeks. The reported prevalence on the development of anti-drug antibodies is low [120]. Post hoc analyses from phase III trials showed that the efficacy of vedolizumab in patients with ulcerative colitis who have not been treated with anti-TNF before is enhanced. After 6 weeks of therapy, the response rates were 53.1% (vedolizumab) and 26.3% (placebo) for anti-TNF-naïve patients opposed to 39.0% and 20.6% in patients with anti-TNF failure [121].

The most common side effects include nasopharyngitis, headache, arthralgia and nausea. Infusion-related reactions are uncommon. A recent report on the integrated safety of vedolizumab, analysing data from six clinical trials, demonstrated that there is no increased risk for serious infections [122]. In <0.6% of patients, serious clostridial infections and tuberculosis were reported in <1% of patients. Overall, the safety profile of vedolizumab is beneficial compared to other drugs employed in IBD, although observation was usually <1 year. There have been no cases of PML to date after treatment with vedolizumab.

Currently, the treatment with vedolizumab is an alternative option for patients who do not respond to anti-TNF treatment or who show secondary treatment failure to anti-TNF. On the other hand, due to its favourable safety profile, it is also considered as a first-line treatment to maintain remission in patients with increased risk for infectious complications such as elderly patients.

A new antibody which is still under investigation in phase III trials, etrolizumab, targets not only α4β7 INTEGRIN but also αEβ7 INTEGRIN and thereby also blocks the interaction with E-cadherin. It is thought to inhibit the pathological retention of leukocytes as well as T cell homing to the gut [123].

35.4.5 IL-12/IL-23 Antibodies (Ustekinumab)

USTEKINUMAB is a fully human immunoglobulin G1 (IgG1) kappa monoclonal antibody that blocks the activity of IL-12 and IL-23 by inhibiting their receptors on T CELLS, natural killer cells and antigen-presenting cells (APCs). It targets the p40 subunit that is shared by both cytokines. It was originally designed to block IL-12 activity and thereby block Th1 function. Only later on, it was discovered that IL-23 also contains the p40 subunit as well [124]. IL-12 and IL-23 are pro-inflammatory cytokines that are involved in the pathophysiology of several autoimmune disorders as well as in CROHN'S DISEASE. With the assistance of other factors such as IL-6 and transforming growth factor (TGF)-β, IL-23 helps to induce the differentiation of CD4 T CELLS into Th17 cells. In addition, other innate immune cells expressing the retinoic acid receptor-related orphan receptor γt (RORγt) transcription factor are responsive to IL-23 and are named type 17 cells. Th17 and type 17 cells can stimulate the production of inflammatory cytokines like IL-17, IL-22, tumour necrosis factor-α (TNF-α) and granulocyte-macrophage colony-stimulating factor (GM-CSF), thereby inducing local inflammation [125]. IL-12 is an important activator of T cell proliferation and drives Th1 differentiation. Therefore, these two cytokines play an important role in disease development in CROHN'S DISEASE.

USTEKINUMAB was initially approved for use in patients with moderate to severe psoriasis in 2009 and has consequently been used in patients with psoriatic arthritis as well [126]. Only recently in 2016, it was also approved by the European Medicines Agency (EMA), the US Food and Drug Administration (FDA) and Health Canada, for the treatment of adult patients with moderately to severely active CROHN'S DISEASE, who have had an inadequate response with, lost response to or were intolerant to either conventional therapy or a TNF antagonist or have medical contraindications to such therapies.

In patients with CD, USTEKINUMAB is administered intravenously for induction therapy, and in case of clinical response, the following dosages are applied subcutaneously every 8 weeks. The initial dose is adapted to the patient's weight (55 kg or less, 260 mg; 55–85 kg, 390 mg; >85 kg, 520 mg), and the following dosage is universal (90 mg).

The pharmacokinetic properties of USTEKINUMAB are similar to human endogenous IgG1. The approximate half-life of the drug after i.v. or s.c. injection is 3 weeks, probably due to the neonatal Fc (FcRn) receptor which prohibits lysosomal degradation of antibodies. The peak serum concentration of USTEKINUMAB was reached after 7–14 days in patients with psoriasis [124].

The safety profile of USTEKINUMAB that was assessed in clinical studies until now is favourable. Analyses in patients with psoriasis did not indicate a shift from a TH1 to a TH2 bias in the systemic immune system after USTEKINUMAB treatment. Furthermore, the reported side effects including infections, serious infections and administration site reactions are similar between treatment groups, pointing to USTEKINUMAB as a rather safe drug.

USTEKINUMAB was shown to be efficient in inducing remission in patients with CROHN'S DISEASE in three phase III trials, corresponding to significant reduction in disease activity after 6 weeks of the initial dose in 34.3% of patients (ustekinumab) vs. 21.5% (placebo). In further trials patients who had not been treated with anti-TNF before showed even greater response rates with an overall response of 55.5% vs. 28.7% (placebo) after 6 weeks. Also, data are promising for maintenance of remission under USTEKINUMAB.

Until now, the position of USTEKINUMAB in the clinical management of patients with CD is not clearly defined. In certain patients, e.g. elderly patients, predominant small bowel disease, it might be reasonable to use USTEKINUMAB as a first-line therapy. Still, further trials have to establish the role and usage of USTEKINUMAB in this cohort of patients. For now, it remains a treatment for patients who did not respond well to anti-TNF and/or show contraindications for the treatment with other immunomodulators.

35.4.5.1 Interferon-Beta

Interferon-ß (IFN-ß) is a type I INTERFERON involved in the regulation of the immune system with predominant antiviral and antitumour activity (see also Chap. 6). Several mechanisms have been reported to account for the immunomodulatory activity of IFN-ß in the treatment of MS patients including modulation of phenotypic T cell activity, i.e. Th1, Th2 and Th17 type T cell responses, regulation of the blood-brain barrier permeability and regulatory effects on antigen-presenting cells [127]. Currently, two formulations of recombinantly produced IFN-ß, known as IFN-ß-1a and IFN-ß-1b, are available, which differ in their glycosylation, secondary to the manufacturing processes, whereby IFN-ß-1a is produced in Chinese hamster ovary (CHO) cells and IFN-ß-1b in *Escherichia coli*. IFN-ß-1b (Betaferon®/Betaseron®, Bayer) was the first compound approved for the treatment of RRMS. Another formulation of IFN-ß (30 μg, IFN-ß-1a, Avonex®, Biogen) given intramuscularly once weekly was soon approved for the treatment of MS [128]. Consequently, further formulations of IFN-ß-1a (Rebif®, EMD Serono) and pegylated IFN-ß-1a (Plegridy®, Biogen) were approved for the treatment of RRMS, after demonstrating their efficacy in reducing relapse rate and new contrast-enhancing lesions on MRI compared to placebo in several pivotal trials [129, 130]. Later, IFN-ß-1b and subcutaneous IFN-ß-1a were approved for the treatment of SPMS [131, 132]. Meanwhile, IFN-ß-1a and IFN-ß-1b are approved for treatment of patients with clinically isolated syndrome (CIS), which represents the first manifestation of MS, to reduce the risk of a second relapse and consequently conversion to clinically definite MS [67, 133]. Nowadays, IFN-ß is mainly used as a first-line therapy in CIS and early MS patients. The advantages over other approved therapies are their proven long-term safety. The frequency of the application and the overall dose vary between the different drugs from every other day (Betaferon®/Betaseron®, Bayer; Extavia®, Novartis), three times weekly (Rebif®, EMD Serono), once weekly (Avonex®, Biogen) to every 2 weeks (Plegridy®, Biogen). Treatment with IFN-ß is associated with the risk of developing of neutralizing antibodies, which can reduce treatment efficacy [134, 135]. Other side effects include flu-like symptoms with fatigue, muscle aching, sometimes transient elevations of liver enzymes, injection site reactions and mood disorders. Overall, all IFN-ß preparations have a favourable long-term safety profile and are thus frequently used in early MS patients and those with low disease activity.

Family planning is an important theme in MS care, since most MS patients are young women. IFN-ß might be associated with a higher risk of abortion, nevertheless continuation of the therapy throughout pregnancy can be considered in active patients.

35.4.5.2 Glatiramer Acetate

Glatiramer acetate (GA) is a synthetic copolymer consisting of L-glutamic acid, L-lysine, L-alanine and L-tyrosine in a molar ratio of 4.2:3.4:1.4:1.0, which was developed at the Weizmann Institute (Israel) to mimic the antigenicity of myelin basic protein (MBP) to induce experimental autoimmune encephalomyelitis (EAE), an animal model of multiple sclerosis. However, GA was highly efficient in both preventing and treating EAE independent of the species or the disease-inducing antigen [136]. The mechanisms of action involve several immunomodulatory effects including MHC-binding competition; T cell receptor antagonism; induction of an anti-inflammatory phenotype of T CELLS, B cells and antigen-presenting cells; and also the induction of neurotrophic factors in the brain [137]. Consequently, GA was further explored as immunotherapy for patients with MS. The efficacy of 20 mg GA (Copaxone®, Teva) given subcutaneously every day in reducing the number of relapses compared to placebo was demonstrated in a pivotal clinical trial in RRMS patients and led to the approval of the drug in 1996 [138]. A further study confirmed the efficacy of GA to reduce the number of new contrast-enhancing lesions on magnetic resonance imaging (MRI) at month 9 compared to placebo [139]. GA was well tolerated with local injection side reactions as the main side effects. In 2011 GA (20 mg daily s.c.) was approved for patients with CIS following result of the PRECISE trial where it significantly reduced the risk for conversion to clinically definite MS (CDMS) by 45% and the time to convert to CDMS by 115% compared to placebo [68]. The GALA trial established the efficacy of a new treatment regimen of 40 mg GA given s.c. three times a week, and the new dosing scheme was approved in 2014 [140]. Meanwhile, several generic formulations of GA have accessed the market both in the USA and Europe.

The long-term safety and efficacy of GA was established in long-term extension of the pivotal clinical trials and post-marketing observations [141, 142]. The efficacy of GA was compared to interferon-ß-1a and interferon-ß-1b in head-to-head trials and proved similar in efficacy on clinical and MRI parameters of inflammatory disease activity [143, 144]. Overall, GA has a favourable safety profile, with local injection side reactions, hypersensitivity reactions and rarely elevation of liver enzymes as the main side effects. GA can be used throughout pregnancy, and no foetal toxicity, malformation or low birth weight was observed [145–147].

35.4.5.3 Fingolimod

Fingolimod is an orally available, immunomodulating agent acting as an agonist of the lysophospholipid sphingosine-1-phosphate (S1P) receptors. After binding to its ligand, fingolimod downregulates S1P receptors 1, 3, 4 and 5 (not S1P2) on target cells and renders them unresponsive to the natural ligand S1P, a lipid mediator that regulates various biological processes including cell proliferation, migration, survival and differentiation [148]. S1P receptors are found on a variety of cells and exert different physiological effects on the vascular system, cardiac and endothelial function and tumour angiogenesis. The S1P1 receptor is the predominant S1P receptor expressed on lymphocytes. Signalling of the S1P1 receptor is required for the recirculation of lymphocytes from lymph nodes, and its down-regulation entraps autoreactive lymphocytes within lymph nodes [149]. The latter is thought to be the main mode of action of fingolimod in the treatment of MS.

Fingolimod was the first oral drug approved for treatment of relapsing-remitting MS (EMA and FDA) after demonstrating efficacy in reducing clinical disease activity and development of new lesion on MRI in two pivotal phase III trials. In the FREEDOMS trial, a daily dose of 0.5 mg fingolimod reduced the annualized relapse rate by 54% compared to placebo, and in the TRANSFORMS trial, annualized relapse rate was reduced by 50% compared to intramuscular INTERFERON beta-1a [150, 151]. Fingolimod also reduces the number of contrast-enhancing T1 lesions or new or enlarged T2 MRI lesions. In the placebo-controlled FREEDOMS study, fingolimod reduced the risk of disability progression over time compared with placebo for the entire study period, but progression rates were comparable between fingolimod and INTERFERON beta-1a in the TRANSFORMS trial. In a recent phase III trial in primary progressive MS, fingolimod (0.5 mg daily) did not slow disease progression compared to placebo [152]. Thus, there is no clear neuroprotective effect of fingolimod as had been proposed by studies in animal models [153, 154].

Although it is generally well tolerated, the drug carries the risk of potential side effects, because of the widespread expression of S1P receptors in different tissues. A strict monitoring programme for all patients is recommended including differential blood count, liver enzymes and bilirubin, varicella zoster virus (VZV) serology, ECG, ophthalmologic examination and dermatological examination in patients with history of basalioma or melanoma before commencing therapy and 3 monthly differential blood count and liver function test under treatment. VZV seronegative patients should be vaccinated. Possible bradycardia after the first dose has led to the recommendation for a 6 h observation period with ECG, pulse and blood pressure measurements on the day of first intake. It should not be prescribed in patients with antiarrhythmic drugs and only with caution in combination with beta-blockers or calcium channel blockers. PML cases in patients treated with fingolimod have been described.

Because of potential teratogenic effects, women of childbearing potential should use adequate contraception, and a washout period of 3 months is recommended before conception.

Several novel S1P1 receptor blockers are in late-stage clinical development for the treatment of relapsing and progressive MS.

35.4.5.4 Dimethyl Fumarate

Fumaric acid is an immunomodulating compound for which a relatively large body of medical experience exists in the topical treatment of psoriasis. The oral formulation, Tecfidera®, contains only the dimethyl fumarate (DMF) as active substance. The exact mechanism of its action in MS is not yet fully understood but is thought to involve not only leukocyte-depleting and anti-inflammatory effects but also neuroprotective effects through induction of antioxidative and stress response genes. The primary metabolite monomethyl fumarate is able to activate the nuclear factor E2-related factor 2 (Nrf2) transcriptional pathway [155]. Via this pathway, fumaric acid leads to activation of phase II detoxifying enzymes in astroglial and microglial cells. Fumaric acids alter the immune response through reducing the number of leukocytes, mainly T CELLS, and causing a shift in cytokine response from Th1 to Th2. Fumaric acids have a well-known favourable long-term safety profile in patients with psoriasis, with more than 30,000 patient-years of fumaric acid treatment [156].

Both pivotal phase III studies, DEFINE and CONFIRM, showed a reduction in the annualized relapse rate with dimethyl fumarate (with the now approved dose of 240 mg twice daily) vs. placebo (53% and 44%) [157, 158]. Primary endpoints were the numbers of patients experiencing a relapse (DEFINE) or annual relapse rate (CONFIRM) after 96 weeks. Twenty-seven percent of patients with DMF relapsed vs. 46.1% in the placebo group (DEFINE). Compared to placebo, DMF reduced sustained disability progression by 38% in the DEFINE study. Furthermore, 45% of the patients had no new T2 lesions or Gd-enhancing lesion (vs. 27% in the placebo group). The CONFIRM study additionally tested DMF against GA as an active comparator, although the study did not address non-inferiority or superiority of DMF over GA but was focused on testing superiority over placebo. GA reduced the annualized relapse rate (AAR) by 29%, while DMF showed a 51% reduction.

In MS, most common side effects are gastrointestinal side effects and flush occurring in up to 30% of patients in the first month of treatment, usually disappearing thereafter. Lymphocyte counts should be monitored closely. Long-term lymphopenia was linked to the development of PML [159], but the exact pathomechanisms still need to be determined [160]. An interim analyses of the follow-up study, ENDORSE, supported the favourable safety and tolerability profile over a period of 8 years [161].

No adverse pregnancy outcomes or foetal abnormalities after exposure to DMF have been reported [162]. The treatment with DMF should be discontinued after a positive pregnancy test.

35.4.5.5 Teriflunomide

TERIFLUNOMIDE is the active metabolite of LEFLUNOMIDE, which is licensed for the treatment of rheumatoid arthritis (see Chap. 34). Pharmacologically, TERIFLUNOMIDE inhibits the mitochondrial enzyme dihydroorotate dehydrogenase, which is crucial for pyrimidine synthesis. Activation and proliferation of lymphocytes largely depend on de novo pyrimidine synthesis, and TERIFLUNOMIDE, therefore, interferes with effective immune responses. Additionally, TERIFLUNOMIDE inhibits protein tyrosine kinase activity in vitro, which leads to reduced T cell proliferation and secretion of the inflammatory cytokines IFN-γ, granzyme B and interleukin-2 (IL-2) as well as reduced B cell immunoglobulin (Ig)G1 production [163, 164]. Several other immunomodulating properties of TERIFLUNOMIDE have also been described in vitro through blockade of nuclear factor-κB signalling and a T helper 2 (Th2) shift [165–167].

Two phase III studies (TEMSO, TOWER) showed a reduction of relapse rate as compared to placebo (31.5% and 36.3%) and a significant effect on the confirmed disability progression over 3 months [168, 169]. The comparative study, TENERE, did not show superiority of TERIFLUNOMIDE 14 mg vs. the active comparator interferon-β-1a (44 μg 3×/week) in regard to annual relapse rate [170]. TERIFLUNOMIDE is approved for daily administration of 14 mg/day orally. Long-term use of LEFLUNOMIDE in rheumatology patients provided a rather well-described safety profile (see Chap. 34). Liver enzyme elevation and reversible hair loss/thinning are common side effects. TERIFLUNOMIDE causes a reduction in lymphocyte counts (median <15% from baseline value). Periodic laboratory controls are required (hematogram and liver values every 4 weeks, in case of stable values over 12 weeks, regular controls every 3 months). Due to a strong enterohepatic circulation, TERIFLUNOMIDE displays a long half-life of 19 days. After discontinuation, TERIFLUNOMIDE remains in the organism for months. A quick and safe elimination of TERIFLUNOMIDE is achieved by the use of cholestyramine or activated carbon.

TERIFLUNOMIDE is contraindicated in pregnancy, and due to its long half-life, it should not be used in women who plan to become pregnant.

35.4.5.6 Alemtuzumab

ALEMTUZUMAB is a humanized monoclonal antibody against CD52, a cell surface molecule expressed on lymphocytes, in particular B and T CELLS, monocytes, mature NK cells, plasma cells and eosinophils [171]. The anti-CD52 mAb was developed as a therapy for lymphocytic malignancies with cytotoxic effects on lymphocytes through complement-dependent cytotoxicity (CDC) and ADCC, thereby rapidly producing a profound lymphopenia [172]. ALEMTUZUMAB selectively depletes CD4 T CELLS for a median of 60 months, while total B cells and monocytes return to their pretreatment levels within 3–6 months [173]. The first anti-CD52 monoclonal antibody was called Cambridge Pathology 1

(Campath-1) in reference to the inventing institution. Later it was humanized on an IgG1 framework creating Campath-1H or ALEMTUZUMAB. Hematopoietic stem cells do not express CD52.

The first studies of Campath-1 in MS patients were done in secondary progressive disease cohort. In this patient group, treatment with Campath-1 failed to show clinical improvement despite significantly decreased radiological findings of disease activity and a prolonged depletion of circulating lymphocytes [174, 175]. The strong effect on inflammatory of Campath-1H could be confirmed in a cohort of 17 drug-naïve and five breakthrough RRMS patients, where treatment led to a significant reduction in relapse rate and improvement of disability [176]. The multicentre, rater-blinded, phase II trial (CAMMS223) compared two dose levels of ALEMTUZUMAB (24 and 12 mg/day) with high-dose interferon-ß1a (44 µg three times per week) in 334 treatment-naïve RRMS patients. The study met its primary endpoints, and treatment with ALEMTUZUMAB reduced clinical and MRI disease activity and the risk for accumulation of disability significantly [177].

In the pivotal phase III, CARE-MS I and CARE-MS II studies, treatment with ALEMTUZUMAB led to a significant reduction in inflammatory disease activity and the risk for disability progression compared to interferon-ß-1a [178, 179]. The study cohort consisted of both naïve patients (CARE-MS I) and patients with previous breakthrough disease under conventional DMT (CARE-MS II). Study patients were treated with ALEMTUZUMAB 12 mg intravenously over a total of 8 days (first application on 5 consecutive days, second administration 12 months later over 3 days). Co-primary endpoints were reduction of relapse rate and sustained accumulation of disability. Secondary endpoints included percentage of relapse-free patients after 2 years, change in EDSS and changes in MRI lesion load. The CARE-MS I study showed a reduced annual relapse rate of 55% as compared to IFN-β-1a treatment. Over an observation period of 2 years in the CARE-MS II study, relapse rate in the ALEMTUZUMAB group was 49% lower than in the IFN-β-1a group (0.26 per year vs. 0.52 relapses per year in IFN-β-1a group) [178]. Sustained accumulation of disability over 6 months was significantly lower in alemtuzumab-treated patients (12.7% vs. 21.1%) in the CARE-MS II study. The percentage of patients with Gd-enhancing lesions was also significantly lower as compared to patients treated with IFN-β-1a (19% vs. 34%). After 2 years, no evidence of disease activity (NEDA) was reached in 32% of ALEMTUZUMAB patients vs. 14% in the control group. After treatment with ALEMTUZUMAB, there is a considerable fraction of patients with long-term reduction of disease activity following a single cycle, only 22% of patients in the CARE-MS I received a third application. More than 90% of the patients from both studies participated in a 5-year follow-up study; 69% (CARE-MS I) and 60% (CARE-MS II) of the patients did not receive another ALEMTUZUMAB application [180, 181]. Ninety-eight percent of the patients did not require further DMT treatment. Annualized relapse rate remained low without any further treatment (CARE-MS I, 0.15 in year 5 vs. 0.19 in year 3; CARE-MS II, 0.18 in year 5 vs. 0.2 in year 3). The percentage of patients with NEDA in year 5 was 62.4% resp. 58.2%. Brain atrophy rate was normalized over 5 years [182].

ALEMTUZUMAB treatment is associated with a high risk of acute and long-term side effects. Acute cytokine release syndrome with pyrexia, headache malaise and urticarial rash occurs in the majority of patients. Pretreatment with corticosteroids ameliorates these symptoms. The risk of infections is increased in patients in particular early after treatment for *listeria monocytogenes* infections and reactivation of herpes virus. Secondary, antibody-mediated autoimmunity is a major concern, most importantly autoimmune thyroiditis in approximately one third of treated patients (maximum in the third year), but also idiopathic thrombocytopenia (ITP) and Goodpasture's disease [183, 184] in 10% of patients show a severe course [181]. The mechanisms underlying the emergence of antibody-mediated autoimmune disease are not clear [183]. A potential explanation is the differential repopulation of B cells together with regulatory cell elements. The frequency of unwanted side effects and infusion-related reactions decreases with treatment duration [181]. With regard to the potential severity of side effects and the latency with which some of them occur, consequent sustained clinical vigilance is mandatory [185].

ALEMTUZUMAB has a half-life of 4–5 days; approximately 30 days after infusion, the antibody is no detectable in plasma. ALEMTUZUMAB can cross the placenta; in animal studies, reproductive toxicity was reported. Women should use effective contraception during and 4 months after ALEMTUZUMAB treatment. The observation in 193 women that were exposed to ALEMTUZUMAB during pregnancy showed no increased risk for congenital malformations [186].

35.4.5.7 Rituximab/Ocrelizumab

An important role of B cells and antibodies has been reported in the pathogenesis of many autoimmune disease [187] including MS [188]. RITUXIMAB has been approved for the treatment of RA patients, in whom anti-TNF-α therapy has failed. Particularly in MS, intrathecal Ig synthesis and oligoclonal Ig production are strong evidence for localized B cell expansion and the involvement of B cells in the pathogenesis of the disease. Further, a subset of B cells accumulates in the CSF of MS patients, and the number correlates strongly with the intrathecal IgG production and CNS inflammation determined by MRI [189]. The importance of B cells is clearly established in a subgroup of MS by pathological studies [190]. The two monoclonal antibodies, RITUXIMAB and ocrelizumab, target the CD20 protein present on the surface of immature and mature B cells, which are consequently depleted, but neither monoclonal antibodies target CD20-negative plasma cells. The effector mechanisms of RITUXIMAB and ocrelizumab leading to depletion of B cells involve complement-dependent cytotoxicity and antibody-dependent cellular cytotoxicity. However, there are differences in the magnitude of the different mechanisms and ocrelizumab, to a larger extent, exerts antibody-dependent, cell-mediated cytotoxic effects compared to complement-dependent cytotoxicity [191]. The depleting effect of both antibodies is complete within 2 weeks after treatment. In a phase II study of ocrelizumab in MS patients, the median time to B cell repopulation was 72 weeks.

Ocrelizumab has been approved for the treatment of RRMS following two pivotal multicentre phase III trials (OPERA I and OPERA II) with identical double-blind, double-dummy study design [192]. Intravenous infusions of ocrelizumab (600 mg) every 24 weeks demonstrated higher efficacy in reducing the annualized relapse rate (primary endpoint: OPERA I, 46%; OPERA II, 47%) compared to subcutaneous injection of INTERFERON beta-1a three times per week. Treatment with ocrelizumab led to a 40% reduction in 12- and 24-week confirmed disability progression (CDP) and a reduction in the number of contrast-enhancing T1 lesions (OPERA I, 94%; OPERA II, 95%) and a reduction in the number of new/enlarging T2 hyperintense lesions on MRI scans (OPERA I, 77%; OPERA II, 83%). RITUXIMAB is not approved for the treatment of RRMS; however it is frequently used as off-label treatment in different countries [193].

Ocrelizumab is the first disease-modifying therapy approved for primary progressive MS. In a randomized placebo-controlled phase III trial (ORATORIO), 6 monthly infusions of 600 mg ocrelizumab reduced the percentage of patients meeting 12-week confirmed disability progression (CDP) compared to placebo-treated patients (32.9% vs. 39.3%) [194]. Several secondary endpoints with imaging parameters corroborated the beneficial effects of ocrelizumab in PPMS, among these a reduction in the mean number of new or enlarging hyperintense T2 lesions and the total volume of T2 lesions on brain MRI, which decreased in the ocrelizumab-treated group, while there was an increase in the total volume of T2 lesions in the placebo group. The percentage of brain volume loss, as a marker of neurodegeneration in MS patients, was significantly lower in ocrelizumab-treated patients compared to the placebo group (0.9% vs. 1.09%).

The most common side effects of ocrelizumab are infusion-related reactions, which can occur up to 24 h following administration. It is recommended that all patients receive premedication with antihistamines and corticosteroid to prevent hypersensitivity reactions and be monitored during the infusion.

In the pivotal clinical trials, there was only a slight increase in the rate of infections, including nasopharyngitis, upper respiratory tract and herpes in patients treated with ocrelizumab. Active hepatitis B infection should be excluded prior to starting ocrelizumab, because of the risk of hepatitis B reactivation. Patients should be assessed for vaccination status and brought up to date on all vaccines, specifically live or live-attenuated vaccines, at least 6 weeks prior to initiating ocrelizumab. Live vaccines should be avoided during ocrelizumab therapy. Contraception is recommended until 6 months after the last infusion of ocrelizumab. There was a safety signal in the clinical development programme so far, with an increased frequency of different tumours (2.3%, 11 from 486 patients), most common of the breast (n = 4), in patients treated with ocrelizumab [194]. Overall, the incidence of neoplasms in ocrelizumab-treated patients remained within the expected background rate of MS patients [195].

35.4.5.8 Cladribine

Cladribine is a cytotoxic drug that predominantly affects lymphocytes and monocytes through the drug's action as an adenosine deaminase (ADA)-resistant deoxyadenosine. Intracellularly, successive phosphorylation by deoxycytidine kinase (DCK) leads to the accumulation of triphosphorylated cladribine (see also Chap. 31). Because of low levels of 5′-NTase, a dephosphorylating enzyme involved in removing the phosphate functional group from deoxyadenosine, lymphocytes rely on ADA to degrade deoxyadenosine phosphate. This results in the accumulation of deoxyadenosine nucleotides in lymphocytes and subsequent cell death [196]. Cladribine rapidly depletes both dividing and quiescent lymphocytes, with the strongest effects seen within CD4+ and CD8+ T CELLS and only transient effect on B cells and natural killer (NK) cells. The effect of cladribine is long-lasting and irreversible for the affected cell populations.

Treatment with cladribine (3.5 mg/kg) has been approved for the treatment of patients with RRMS in Europe following a phase III trial (CLARITY, Cladribine Tablets Treating MS Orally), which compared 3.5 or 5.25 mg/kg cladribine with placebo [197]. Patients received two or four short courses during the first 48 weeks and then two short courses starting at week 48 and week 52 for a total study period of 96 weeks. In each course, cladribine tablets (or placebo) were given once daily for the first 4–5 days of a 28-day period.

Treatment with cladribine met the primary endpoint with a significant reduction of the relapse rate by 58% in the low-dose group and 55% in the high-dose group compared with placebo. Significantly more patients remained relapse-free in the treated groups (79.7% and 78.9% in the low- and high-dose groups, respectively, vs. 60.9% placebo). Patients in the cladribine group had a profound reduction of new contrast-enhancing T1 lesions or new non-enhancing T2 or enlarging T2 lesions. Oral cladribine also led to a reduction of the risk of 3-month sustained progression of disability in both treatment groups compared with placebo.

Adverse events include lymphocytopenia, neutropenia and thrombocytopenia, herpes zoster infections and malignancy. The occurrence of neoplasms was reported in 1.4% of patients with cladribine 3.5 mg/kg and in 0.9% of the 5.25 mg/kg group, but none occurred in the placebo-treated patients. The neoplasms included five benign uterine leiomyomas and three cases of cancer (melanoma, pancreas carcinoma and ovarian carcinoma); additionally, a case of cervical carcinoma in situ was reported, and one choriocarcinoma occurred in a patient 9 months after completion of the study. Thus, adverse event surveillance of cladribine, especially for neoplasms and the effects of long-term immunosuppression, will be essential.

The use of cladribine is contraindicated in pregnant women. Contraception is mandatory during and 6 months after each treatment phase. Interference of cladribine with DNA synthesis in gametogenesis is possible; therefore men should not father children during and 6 months after the last cladribine dose.

35.5 Summary and Concluding Remarks

The incidence and prevalence of immune-mediated inflammatory diseases in the liver, intestine and CNS are steadily increasing. During the last 10 years, major improvements in the medical therapies, most notably antibodies against several inflammatory pathways, have been achieved. These therapies have significantly improved the welfare of the patients. Unfortunately, most of these therapies used against these diseases have as of yet palliative character and mostly do not offer a cure. Thus life-long treatment using immune-suppressive agents is required in the majority of cases. As a consequence, patients with immune-mediated diseases must live with the side effects of these treatments, such as increased risk of opportunistic infections and of relapsing flares of the disease itself. Furthermore, chronic inflammation, another possible side effect, can promote the development of certain forms of cancer. For example, it has been observed that inflammatory bowel disease (IBD) is associated with an increased risk of developing colorectal cancer, especially in patients suffering from chronic uncontrolled intestinal inflammation. Therefore, there is major need for new therapies, which can modulate the immune response more specifically. Accordingly, key aim of current research is to establish therapies, which would be able to re-establish the immune homeostasis and thus would be able to induce long-term remission with less side effects. An overview of drugs used in soft tissue autoimmunity is given in Appendix 5.

References

1. Abraham C, Cho JH. Inflammatory bowel disease. N Engl J Med. 2009;361:2066–78.
2. Banerjee AK. Sir Samuel Wilks: a founding father of clinical science. J R Soc Med. 1991;84:44–5.
3. Crohn BB, Ginzburg L, Oppenheimer GD. Landmark article Oct 15, 1932. Regional ileitis. A pathological and clinical entity. JAMA. 1984;251:73–9.
4. Baumgart DC, Sandborn WJ. Crohn's disease. Lancet. 2012;380:1590–605.
5. Bengtson MB, Solberg C, Aamodt G, et al. Familial aggregation in Crohn's disease and ulcerative colitis in a Norwegian population-based cohort followed for ten years. J Crohn's Colitis. 2009;3:92–9.
6. Ng SC, Woodrow S, Patel N, Subhani J, Harbord M. Role of genetic and environmental factors in British twins with inflammatory bowel disease. Inflamm Bowel Dis. 2012;18:725–36.
7. Spehlmann ME, Begun AZ, Burghardt J, Lepage P, Raedler A, Schreiber S. Epidemiology of inflammatory bowel disease in a German twin cohort: results of a nationwide study. Inflamm Bowel Dis. 2008;14:968–76.
8. Farmer RG, Michener WM, Mortimer EA. Studies of family history among patients with inflammatory bowel disease. Clin Gastroenterol. 1980;9:271–7.
9. Monsen U, Brostrom O, Nordenvall B, Sorstad J, Hellers G. Prevalence of inflammatory bowel disease among relatives of patients with ulcerative colitis. Scand J Gastroenterol. 1987;22:214–8.
10. Franke A, McGovern DP, Barrett JC, et al. Genome-wide meta-analysis increases to 71 the number of confirmed Crohn's disease susceptibility loci. Nat Genet. 2010;42:1118–25.
11. Anderson CA, Boucher G, Lees CW, et al. Meta-analysis identifies 29 additional ulcerative colitis risk loci, increasing the number of confirmed associations to 47. Nat Genet. 2011;43:246–52.
12. Bernstein CN, Rawsthorne P, Cheang M, Blanchard JF. A population-based case control study of potential risk factors for IBD. Am J Gastroenterol. 2006;101:993–1002.
13. Lopez-Serrano P, Perez-Calle JL, Perez-Fernandez MT, Fernandez-Font JM, Boixeda de Miguel D, Fernandez-Rodriguez CM. Environmental risk factors in inflammatory bowel diseases. Investigating the hygiene hypothesis: a Spanish case-control study. Scand J Gastroenterol. 2010;45:1464–71.
14. Damas OM, Avalos DJ, Palacio AM, et al. Inflammatory bowel disease is presenting sooner after immigration in more recent US immigrants from Cuba. Aliment Pharmacol Ther. 2017;46:303–9.
15. Eaden JA, Abrams KR, Mayberry JF. The risk of colorectal cancer in ulcerative colitis: a meta-analysis. Gut. 2001;48:526–35.
16. Kornfeld D, Ekbom A, Ihre T. Survival and risk of cholangiocarcinoma in patients with primary sclerosing cholangitis. A population-based study. Scand J Gastroenterol. 1997;32:1042–5.
17. Canavan C, Abrams KR, Mayberry J. Meta-analysis: colorectal and small bowel cancer risk in patients with Crohn's disease. Aliment Pharmacol Ther. 2006;23:1097–104.
18. Peyrin-Biroulet L, Loftus EV Jr, Colombel JF, Sandborn WJ. Long-term complications, extraintestinal manifestations, and mortality in adult Crohn's disease in population-based cohorts. Inflamm Bowel Dis. 2011;17:471–8.
19. Nagao-Kitamoto H, Kamada N. Host-microbial cross-talk in inflammatory Bowel disease. Immune Netw. 2017;17(1):12.
20. Sokol H, Pigneur B, Watterlot L, et al. Faecalibacterium prausnitzii is an anti-inflammatory commensal bacterium identified by gut microbiota analysis of Crohn disease patients. Proc Natl Acad Sci U S A. 2008;105:16731–6.
21. Bloom SM, Bijanki VN, Nava GM, et al. Commensal bacteroides species induce colitis in host-genotype-specific fashion in a mouse model of inflammatory bowel disease. Cell Host Microbe. 2011;9:390–403.
22. Buisine MP, Desreumaux P, Debailleul V, et al. Abnormalities in mucin gene expression in Crohn's disease. Inflamm Bowel Dis. 1999;5:24–32.

23. Van Klinken BJ, Van der Wal JW, Einerhand AW, Buller HA, Dekker J. Sulphation and secretion of the predominant secretory human colonic mucin MUC2 in ulcerative colitis. Gut. 1999;44:387–93.
24. Cadwell K, Liu JY, Brown SL, et al. A key role for autophagy and the autophagy gene Atg16l1 in mouse and human intestinal Paneth cells. Nature. 2008;456:259–63.
25. Niess JH, Brand S, Gu X, et al. CX3CR1-mediated dendritic cell access to the intestinal lumen and bacterial clearance. Science. 2005;307:254–8.
26. Aad G, Abbott B, Abdallah J, et al. Measurement of the W+ W− cross section in sqrt(s) = 7 TeV pp collisions with ATLAS. Phys Rev Lett. 2011;107: 041802.
27. Baumgart DC, Metzke D, Guckelberger O, et al. Aberrant plasmacytoid dendritic cell distribution and function in patients with Crohn's disease and ulcerative colitis. Clin Exp Immunol. 2011;166:46–54.
28. Baumgart DC, Metzke D, Schmitz J, et al. Patients with active inflammatory bowel disease lack immature peripheral blood plasmacytoid and myeloid dendritic cells. Gut. 2005;54:228–36.
29. Baumgart DC, Thomas S, Przesdzing I, et al. Exaggerated inflammatory response of primary human myeloid dendritic cells to lipopolysaccharide in patients with inflammatory bowel disease. Clin Exp Immunol. 2009;157:423–36.
30. Heller F, Florian P, Bojarski C, et al. Interleukin-13 is the key effector Th2 cytokine in ulcerative colitis that affects epithelial tight junctions, apoptosis, and cell restitution. Gastroenterology. 2005;129: 550–64.
31. Heller F, Fromm A, Gitter AH, Mankertz J, Schulzke JD. Epithelial apoptosis is a prominent feature of the epithelial barrier disturbance in intestinal inflammation: effect of pro-inflammatory interleukin-13 on epithelial cell function. Mucosal Immunol. 2008;1(Suppl 1):S58–61.
32. Glocker EO, Kotlarz D, Boztug K, et al. Inflammatory bowel disease and mutations affecting the interleukin-10 receptor. N Engl J Med. 2009;361:2033–45.
33. Baert F, Moortgat L, Van Assche G, et al. Mucosal healing predicts sustained clinical remission in patients with early-stage Crohn's disease. Gastroenterology. 2010;138:463–8; quiz e10–1.
34. Manns MP, Lohse AW, Vergani D. Autoimmune hepatitis—update 2015. J Hepatol. 2015;62:S100–11.
35. Johnson PJ, McFarlane IG. Meeting report: International Autoimmune Hepatitis Group. Hepatology. 1993;18:998–1005.
36. Alvarez F, Berg PA, Bianchi FB, et al. International Autoimmune Hepatitis Group report: review of criteria for diagnosis of autoimmune hepatitis. J Hepatol. 1999;31:929–38.
37. Hennes EM, Zeniya M, Czaja AJ, et al. Simplified criteria for the diagnosis of autoimmune hepatitis. Hepatology. 2008;48:169–76.
38. Tiegs G, Lohse AW. Immune tolerance: what is unique about the liver. J Autoimmun. 2010;34:1–6.
39. van Gerven NM, Verwer BJ, Witte BI, et al. Relapse is almost universal after withdrawal of immunosuppressive medication in patients with autoimmune hepatitis in remission. J Hepatol. 2013;58:141–7.
40. Cook GC, Mulligan R, Sherlock S. Controlled prospective trial of corticosteroid therapy in active chronic hepatitis. Q J Med. 1971;40:159–85.
41. Soloway RD, Summerskill WH, Baggenstoss AH, et al. Clinical, biochemical, and histological remission of severe chronic active liver disease: a controlled study of treatments and early prognosis. Gastroenterology. 1972;63:820–33.
42. Murray-Lyon IM, Stern RB, Williams R. Controlled trial of prednisone and azathioprine in active chronic hepatitis. Lancet. 1973;1:735–7.
43. Manns MP, Czaja AJ, Gorham JD, et al. Diagnosis and management of autoimmune hepatitis. Hepatology. 2010;51:2193–213.
44. Gleeson D, Heneghan MA. British Society of Gastroenterology (BSG) guidelines for management of autoimmune hepatitis. Gut. 2011;60:1611–29.
45. Schramm C, Weiler-Normann C, Wiegard C, Hellweg S, Muller S, Lohse AW. Treatment response in patients with autoimmune hepatitis. Hepatology. 2010;52:2247–8.
46. Muratori L, Muratori P, Lanzoni G, Ferri S, Lenzi M. Application of the 2010 American Association for the study of liver diseases criteria of remission to a cohort of Italian patients with autoimmune hepatitis. Hepatology. 2010;52:1857; author reply –8.
47. Manns MP, Woynarowski M, Kreisel W, et al. Budesonide induces remission more effectively than prednisone in a controlled trial of patients with autoimmune hepatitis. Gastroenterology. 2010;139:1198–206.
48. Peiseler M, Liebscher T, Sebode M, et al. Efficacy and limitations of budesonide as a second-line treatment for patients with autoimmune hepatitis. Clin Gastroenterol Hepatol. 2018;16(2):260–267.e1.
49. Hubener S, Oo YH, Than NN, et al. Efficacy of 6-mercaptopurine as second-line treatment for patients with autoimmune hepatitis and azathioprine intolerance. Clin Gastroenterol Hepatol. 2016;14:445–53.
50. Hoeltzenbein M, Elefant E, Vial T, et al. Teratogenicity of mycophenolate confirmed in a prospective study of the European Network of Teratology Information Services. Am J Med Genet A. 2012;158a:588–96.
51. Than NN, Wiegard C, Weiler-Normann C, et al. Long-term follow-up of patients with difficult to treat type 1 autoimmune hepatitis on Tacrolimus therapy. Scand J Gastroenterol. 2016;51:329–36.
52. Fernandes NF, Redeker AG, Vierling JM, Villamil FG, Fong TL. Cyclosporine therapy in patients with steroid resistant autoimmune hepatitis. Am J Gastroenterol. 1999;94:241–8.
53. Ytting H, Larsen FS. Everolimus treatment for patients with autoimmune hepatitis and poor response to standard therapy and drug alternatives in use. Scand J Gastroenterol. 2015;50:1025–31.

54. Sebode MHJ, Vergani D, et al. Autoimmune hepatitis: from current knowledge and clinical practice to future research agenda. Liver Int. 2018;38(1):15–22.
55. Weiler-Normann C, Schramm C, Quaas A, et al. Infliximab as a rescue treatment in difficult-to-treat autoimmune hepatitis. J Hepatol. 2013;58:529–34.
56. Bjornsson ES, Gunnarsson BI, Grondal G, et al. Risk of drug-induced liver injury from tumor necrosis factor antagonists. Clin Gastroenterol Hepatol. 2015;13:602–8.
57. Burak KW, Swain MG, Santodomingo-Garzon T, et al. Rituximab for the treatment of patients with autoimmune hepatitis who are refractory or intolerant to standard therapy. Can J Gastroenterol. 2013;27:273–80.
58. Reich DS, Lucchinetti CF, Calabresi PA. Multiple sclerosis. New Engl J Med. 2018;378:169–80.
59. Olsson T, Barcellos LF, Alfredsson L. Interactions between genetic, lifestyle and environmental risk factors for multiple sclerosis. Nat Rev Neurol. 2017;13:25–36.
60. Charcot. Lecons sur les maladies du systeme nerveux faites a la Salpetrière. 1880.
61. Trapp BD, Peterson J, Ransohoff RM, Rudick R, Mork S, Bo L. Axonal transection in the lesions of multiple sclerosis. New Engl J Med. 1998;338:278–85.
62. Ferguson B, Matyszak MK, Esiri MM, Perry VH. Axonal damage in acute multiple sclerosis lesions. Brain. 1997;120:393–9.
63. Bo L, Geurts JJ, Ravid R, Barkhof F. Magnetic resonance imaging as a tool to examine the neuropathology of multiple sclerosis. Neuropathol Appl Neurobiol. 2004;30:106–17.
64. Kutzelnigg A, Lucchinetti CF, Stadelmann C, et al. Cortical demyelination and diffuse white matter injury in multiple sclerosis. Brain. 2005;128:2705–12.
65. Sospedra M, Martin R. Immunology of multiple sclerosis. Annu Rev Immunol. 2005;23:683–747.
66. Greenfield AL, Hauser SL. B-Cell therapy for multiple sclerosis: entering an era. Ann Neurol. 2018;83:13–26.
67. Comi G, Filippi M, Barkhof F, et al. Effect of early interferon treatment on conversion to definite multiple sclerosis: a randomised study. Lancet. 2001;357:1576–82.
68. Comi G, Martinelli V, Rodegher M, et al. Effect of glatiramer acetate on conversion to clinically definite multiple sclerosis in patients with clinically isolated syndrome (PreCISe study): a randomised, double-blind, placebo-controlled trial. Lancet. 2009;374:1503–11.
69. Jacobs LD, Beck RW, Simon JH, et al. Intramuscular interferon beta-1a therapy initiated during a first demyelinating event in multiple sclerosis. New Engl J Med. 2000;343:898–904.
70. Kappos L, Polman CH, Freedman MS, et al. Treatment with interferon beta-1b delays conversion to clinically definite and McDonald MS in patients with clinically isolated syndromes. Neurology. 2006;67:1242–9.
71. Giovannoni G, Turner B, Gnanapavan S, Offiah C, Schmierer K, Marta M. Is it time to target no evident disease activity (NEDA) in multiple sclerosis? Hand Clin. 2015;4:329–33.
72. Kaunzner UW, Al-Kawaz M, Gauthier SA. Defining disease activity and response to therapy in MS. Curr Treat Options Neurol. 2017;19:20.
73. Salazopyrin NS. a new sulfanilamide preparation: A. Therapeutic results in rheumatic polyarthritis. B. Therapeutic results in ulcerative colitis. C. Toxic manifestations in treatment with sulfanilamide preparation. Acta Med Scand. 1942;110:557–90.
74. Egan LJ, Mays DC, Huntoon CJ, et al. Inhibition of interleukin-1-stimulated NF-kappaB RelA/p65 phosphorylation by mesalamine is accompanied by decreased transcriptional activity. J Biol Chem. 1999;274:26448–53.
75. Bantel H, Berg C, Vieth M, Stolte M, Kruis W, Schulze-Osthoff K. Mesalazine inhibits activation of transcription factor NF-kappaB in inflamed mucosa of patients with ulcerative colitis. Am J Gastroenterol. 2000;95:3452–7.
76. Allgayer H. Review article: mechanisms of action of mesalazine in preventing colorectal carcinoma in inflammatory bowel disease. Aliment Pharmacol Ther. 2003;18(Suppl 2):10–4.
77. Velayos FS, Terdiman JP, Walsh JM. Effect of 5-aminosalicylate use on colorectal cancer and dysplasia risk: a systematic review and metaanalysis of observational studies. Am J Gastroenterol. 2005;100:1345–53.
78. Peppercorn MA, Goldman P. The role of intestinal bacteria in the metabolism of salicylazosulfapyridine. J Pharmacol Exp Ther. 1972;181:555–62.
79. Myers B, Evans DN, Rhodes J, et al. Metabolism and urinary excretion of 5-amino salicylic acid in healthy volunteers when given intravenously or released for absorption at different sites in the gastrointestinal tract. Gut. 1987;28:196–200.
80. Narayan N, Rigby S, Carlucci F. Sulfasalazine induced immune thrombocytopenia in a patient with rheumatoid arthritis. Clin Rheumatol. 2017;36:477–9.
81. Marteau P, Probert CS, Lindgren S, et al. Combined oral and enema treatment with Pentasa (mesalazine) is superior to oral therapy alone in patients with extensive mild/moderate active ulcerative colitis: a randomised, double blind, placebo controlled study. Gut. 2005;54:960–5.
82. Safdi M, DeMicco M, Sninsky C, et al. A double-blind comparison of oral versus rectal mesalamine versus combination therapy in the treatment of distal ulcerative colitis. Am J Gastroenterol. 1997;92:1867–71.
83. Burchenal JH, Murphy ML, Ellison RR, et al. Clinical evaluation of a new antimetabolite, 6-mercaptopurine, in the treatment of leukemia and allied diseases. Blood. 1953;8:965–99.

84. Lennard L. The clinical pharmacology of 6-mercaptopurine. Eur J Clin Pharmacol. 1992;43:329–39.
85. Thomas CW, Myhre GM, Tschumper R, et al. Selective inhibition of inflammatory gene expression in activated T lymphocytes: a mechanism of immune suppression by thiopurines. J Pharmacol Exp Ther. 2005;312:537–45.
86. Tiede I, Fritz G, Strand S, et al. CD28-dependent Rac1 activation is the molecular target of azathioprine in primary human CD4+ T lymphocytes. J Clin Invest. 2003;111:1133–45.
87. Cuffari C, Seidman EG, Latour S, Theoret Y. Quantitation of 6-thioguanine in peripheral blood leukocyte DNA in Crohn's disease patients on maintenance 6-mercaptopurine therapy. Can J Physiol Pharmacol. 1996;74:580–5.
88. Cuffari C, Hunt S, Bayless T. Utilisation of erythrocyte 6-thioguanine metabolite levels to optimise azathioprine therapy in patients with inflammatory bowel disease. Gut. 2001;48:642–6.
89. Roblin X, Serre-Debeauvais F, Phelip JM, et al. 6-Tioguanine monitoring in steroid-dependent patients with inflammatory bowel diseases receiving azathioprine. Aliment Pharmacol Ther. 2005;21:829–39.
90. Dubinsky MC, Lamothe S, Yang HY, et al. Pharmacogenomics and metabolite measurement for 6-mercaptopurine therapy in inflammatory bowel disease. Gastroenterology. 2000;118:705–13.
91. Hindorf U, Lindqvist M, Peterson C, et al. Pharmacogenetics during standardised initiation of thiopurine treatment in inflammatory bowel disease. Gut. 2006;55:1423–31.
92. Weinshilboum RM, Sladek SL. Mercaptopurine pharmacogenetics: monogenic inheritance of erythrocyte thiopurine methyltransferase activity. Am J Hum Genet. 1980;32:651–62.
93. Lennard L, Van Loon JA, Weinshilboum RM. Pharmacogenetics of acute azathioprine toxicity: relationship to thiopurine methyltransferase genetic polymorphism. Clin Pharmacol Ther. 1989;46:149–54.
94. Van Loon JA, Weinshilboum RM. Thiopurine methyltransferase biochemical genetics: human lymphocyte activity. Biochem Genet. 1982;20:637–58.
95. Szumlanski CL, Honchel R, Scott MC, Weinshilboum RM. Human liver thiopurine methyltransferase pharmacogenetics: biochemical properties, liver-erythrocyte correlation and presence of isozymes. Pharmacogenetics. 1992;2:148–59.
96. Ansari A, Patel N, Sanderson J, O'Donohue J, Duley JA, Florin TH. Low-dose azathioprine or mercaptopurine in combination with allopurinol can bypass many adverse drug reactions in patients with inflammatory bowel disease. Aliment Pharmacol Ther. 2010;31:640–7.
97. Dhaliwal HK, Anderson R, Thornhill EL, et al. Clinical significance of azathioprine metabolites for the maintenance of remission in autoimmune hepatitis. Hepatology. 2012;56:1401–8.
98. Khan N, Abbas AM, Lichtenstein GR, Loftus EV Jr, Bazzano LA. Risk of lymphoma in patients with ulcerative colitis treated with thiopurines: a nationwide retrospective cohort study. Gastroenterology. 2013;145:1007–15.e3.
99. Beaugerie L, Brousse N, Bouvier AM, et al. Lymphoproliferative disorders in patients receiving thiopurines for inflammatory bowel disease: a prospective observational cohort study. Lancet. 2009;374:1617–25.
100. Lennard L, Thomas S, Harrington CI, Maddocks JL. Skin cancer in renal transplant recipients is associated with increased concentrations of 6-thioguanine nucleotide in red blood cells. Br J Dermatol. 1985;113:723–9.
101. Gisbert JP, Gomollon F. Thiopurine-induced myelotoxicity in patients with inflammatory bowel disease: a review. Am J Gastroenterol. 2008;103:1783–800.
102. Seiderer J, Zech CJ, Reinisch W, et al. A multicenter assessment of liver toxicity by MRI and biopsy in IBD patients on 6-thioguanine. J Hepatol. 2005;43:303–9.
103. Kodama S, Davis M, Faustman DL. The therapeutic potential of tumor necrosis factor for autoimmune disease: a mechanistically based hypothesis. Cell Mol Life Sci. 2005;62:1850–62.
104. Peschon JJ, Torrance DS, Stocking KL, et al. TNF receptor-deficient mice reveal divergent roles for p55 and p75 in several models of inflammation. J Immunol. 1998;160:943–52.
105. Hehlgans T, Pfeffer K. The intriguing biology of the tumour necrosis factor/tumour necrosis factor receptor superfamily: players, rules and the games. Immunology. 2005;115:1–20.
106. Thalayasingam N, Isaacs JD. Anti-TNF therapy. Best Pract Res Clin Rheumatol. 2011;25:549–67.
107. Maneiro JR, Salgado E, Gomez-Reino JJ. Immunogenicity of monoclonal antibodies against tumor necrosis factor used in chronic immune-mediated Inflammatory conditions: systematic review and meta-analysis. JAMA Intern Med. 2013;173:1416–28.
108. Baert F, Noman M, Vermeire S, et al. Influence of immunogenicity on the long-term efficacy of infliximab in Crohn's disease. N Engl J Med. 2003;348:601–8.
109. Farrell RJ, Alsahli M, Jeen YT, Falchuk KR, Peppercorn MA, Michetti P. Intravenous hydrocortisone premedication reduces antibodies to infliximab in Crohn's disease: a randomized controlled trial. Gastroenterology. 2003;124:917–24.
110. Targownik LE, Bernstein CN. Infectious and malignant complications of TNF inhibitor therapy in IBD. Am J Gastroenterol. 2013;108:1835–42; quiz 43.
111. Chung ES, Packer M, Lo KH, Fasanmade AA, Willerson JT. Randomized, double-blind, placebo-controlled, pilot trial of infliximab, a chimeric monoclonal antibody to tumor necrosis factor-alpha, in patients with moderate-to-severe heart failure: results of the anti-TNF Therapy Against Congestive Heart Failure (ATTACH) trial. Circulation. 2003;107:3133–40.

112. Nanavati SA, Ergun GA, Schwartz JT. Avoiding infliximab in the treatment of Crohn's disease in patients with multiple sclerosis. Am J Gastroenterol. 2003;98:2333–4.
113. Gisbert JP, Chaparro M. Safety of anti-TNF agents during pregnancy and breastfeeding in women with inflammatory bowel disease. Am J Gastroenterol. 2013;108:1426–38.
114. Briskin M, Winsor-Hines D, Shyjan A, et al. Human mucosal addressin cell adhesion molecule-1 is preferentially expressed in intestinal tract and associated lymphoid tissue. Am J Pathol. 1997;151:97–110.
115. Havrdova E, Galetta S, Hutchinson M, et al. Effect of natalizumab on clinical and radiological disease activity in multiple sclerosis: a retrospective analysis of the Natalizumab Safety and Efficacy in Relapsing-Remitting Multiple Sclerosis (AFFIRM) study. Lancet Neurol. 2009;8:254–60.
116. Polman CH, O'Connor PW, Havrdova E, et al. A randomized, placebo-controlled trial of natalizumab for relapsing multiple sclerosis. N Engl J Med. 2006;354:899–910.
117. Rudick RA, Stuart WH, Calabresi PA, et al. Natalizumab plus interferon beta-1a for relapsing multiple sclerosis. N Engl J Med. 2006;354:911–23.
118. Sandborn WJ, Colombel JF, Enns R, et al. Natalizumab induction and maintenance therapy for Crohn's disease. N Engl J Med. 2005;353:1912–25.
119. Du Pasquier RA, Autissier P, Zheng Y, Jean-Jacques J, Koralnik IJ. Presence of JC virus-specific CTL in the cerebrospinal fluid of PML patients: rationale for immune-based therapeutic strategies. AIDS. 2005;19:2069–76.
120. Parikh A, Leach T, Wyant T, et al. Vedolizumab for the treatment of active ulcerative colitis: a randomized controlled phase 2 dose-ranging study. Inflamm Bowel Dis. 2012;18:1470–9.
121. Feagan BG, Rubin DT, Danese S, et al. Efficacy of vedolizumab induction and maintenance therapy in patients with ulcerative colitis, regardless of prior exposure to tumor necrosis factor antagonists. Clin Gastroenterol Hepatol. 2017;15:229–39 e5.
122. Colombel JF, Sands BE, Rutgeerts P, et al. The safety of vedolizumab for ulcerative colitis and Crohn's disease. Gut. 2017;66:839–51.
123. Rutgeerts PJ, Fedorak RN, Hommes DW, et al. A randomised phase I study of etrolizumab (rhuMAb beta7) in moderate to severe ulcerative colitis. Gut. 2013;62:1122–30.
124. Benson JM, Sachs CW, Treacy G, et al. Therapeutic targeting of the IL-12/23 pathways: generation and characterization of ustekinumab. Nat Biotechnol. 2011;29:615–24.
125. Frieder J, Kivelevitch D, Haugh I, Watson I, Menter A. Anti-IL-23 and anti-IL-17 biologic agents for the treatment of immune-mediated inflammatory conditions. Clin Pharmacol Ther. 2018;103(1):88–101.
126. D'Angelo S, Tramontano G, Gilio M, Leccese P, Olivieri I. Review of the treatment of psoriatic arthritis with biological agents: choice of drug for initial therapy and switch therapy for non-responders. Open Access Rheumatol. 2017;9:21–8.
127. Yong VW, Chabot S, Stuve O, Williams G. Interferon beta in the treatment of multiple sclerosis: mechanisms of action. Neurology. 1998;51:682–9.
128. Jacobs LD. Intramuscular interferon beta-1a for disease progression in relapsing multiple sclerosis (vol. 39, p. 285, 1996). Ann Neurol. 1996;40:480.
129. PRISMS (Prevention of Relapses and Disability by Interferon beta-1a Subcutaneously in Multiple Sclerosis) Study Group. Randomised double-blind placebo-controlled study of interferon beta-1a in relapsing/remitting multiple sclerosis. Lancet. 1998;352:1498–504.
130. Kappos L, Polman C, Freedman MS, et al. Betaferon (R) in newly emerging multiple sclerosis for initial treatment (BENEFIT): clinical results. Mult Scler. 2005;11:S10.
131. Secondary Progressive Efficacy Clinical Trial of Recombinant Interferon-Beta-1a in MS (SPECTRIMS) Study Group. Randomized controlled trial of interferon- beta-1a in secondary progressive MS: clinical results. Neurology. 2001;56:1496–504.
132. European Study Group. Placebo-controlled multicentre randomised trial of interferon beta-1b in treatment of secondary progressive multiple sclerosis. European Study Group on interferon beta-1b in secondary progressive MS. Lancet. 1998;352:1491–7.
133. Jacobs LD, Beck R, Simon JH, et al. Results of a phase III trial of interferon beta-1a treatment in patients at high risk of developing multiple sclerosis. Neurology. 2000;54:2351.
134. Ross C, Clemmesen KM, Svenson M, et al. Immunogenicity of interferon-beta in multiple sclerosis patients: influence of preparation, dosage, dose frequency, and route of administration. Ann Neurol. 2000;48:706–12.
135. Bertolotto A, Sala A, Malucchi S, et al. Biological activity of interferon betas in patients with multiple sclerosis is affected by treatment regimen and neutralising antibodies. J Neurol Neurosurg Psychiatry. 2004;75:1294–9.
136. Teitelbaum D, Meshorer A, Hirshfeld T, Arnon R, Sela M. Suppression of experimental allergic encephalomyelitis by a synthetic polypeptide. Eur J Immunol. 1971;1:242–8.
137. Aharoni R. The mechanism of action of glatiramer acetate in multiple sclerosis and beyond. Autoimmun Rev. 2013;12:543–53.
138. Johnson KP, Brooks BR, Cohen JA, et al. Copolymer-1 reduces relapse rate and improves disability in relapsing-remitting multiple-sclerosis – results of a phase-III multicenter, double-blind, placebo-controlled trial. Neurology. 1995;45:1268–76.
139. Comi G, Filippi M, Wolinsky JS, Aceta ECG. European/Canadian multicenter, double-blind, randomized, placebo-controlled study of the effects of glatiramer acetate on magnetic resonance

imaging-measured disease activity and burden in patients with relapsing multiple sclerosis. Ann Neurol. 2001;49:290–7.

140. Khan O, Rieckmann P, Boyko A, Selmaj K, Zivadinov R, Grp GS. Three times weekly glatiramer acetate in relapsing-remitting multiple sclerosis. Ann Neurol. 2013;73:705–13.
141. Ford C, Goodman AD, Johnson K, et al. Continuous long-term immunomodulatory therapy in relapsing multiple sclerosis: results from the 15-year analysis of the US prospective open-label study of glatiramer acetate. Mult Scler. 2010;16:342–50.
142. Miller A, Spada V, Beerkircher D, Kreitman RR. Long-term (up to 22 years), open-label, compassionate-use study of glatiramer acetate in relapsing-remitting multiple sclerosis. Mult Scler. 2008;14:494–9.
143. Mikol DD, Barkhof F, Chang P, et al. Comparison of subcutaneous interferon beta-1a with glatiramer acetate in patients with relapsing multiple sclerosis (the REbif vs glatiramer acetate in relapsing MS disease [REGARD] study): a multicentre, randomised, parallel, open-label trial. Lancet Neurol. 2008;7:903–14.
144. O'Connor P, Filippi M, Arnason B. 250 μg or 500 μg interferon beta-1b versus 20 mg glatiramer acetate in relapsing-remitting multiple sclerosis: a prospective, randomised, multicentre study. Lancet Neurol. 2009;8:889–981.
145. Herbstritt S, Langer-Gould A, Rockhoff M, et al. Glatiramer acetate during early pregnancy: a prospective cohort study. Mult Scler J. 2016;22:810–6.
146. Sandberg-Wollheim M, Neudorfer O, Grinspan A, et al. Pregnancy outcomes from the branded glatiramer acetate pregnancy database. Int J MS Care. 2018;20:9–14.
147. Giannini M, Portaccio E, Ghezzi A, et al. Pregnancy and fetal outcomes after glatiramer acetate exposure in patients with multiple sclerosis: a prospective observational multicentric study. BMC Neurol. 2012;12:124.
148. Spiegel S, Milstien S. Sphingosine-1-phosphate: an enigmatic signalling lipid. Nat Rev. 2003;4:397–407.
149. Pelletier D, Hafler DA. Fingolimod for multiple sclerosis. N Engl J Med. 2012;366:339–47.
150. Kappos L, Radue EW, O'Connor P, et al. A placebo-controlled trial of oral fingolimod in relapsing multiple sclerosis. New Engl J Med. 2010;362:387–401.
151. Cohen JA, Barkhof F, Comi G, et al. Oral fingolimod or intramuscular interferon for relapsing multiple sclerosis. N Engl J Med. 2010;362:402–15.
152. Lublin F, Miller DH, Freedman MS, et al. Oral fingolimod in primary progressive multiple sclerosis (INFORMS): a phase 3, randomised, double-blind, placebo-controlled trial. Lancet. 2016;387:1075–84.
153. Miron VE, Jung CG, Kim HJ, Kennedy TE, Soliven B, Antel JP. FTY720 modulates human oligodendrocyte progenitor process extension and survival. Ann Neurol. 2008;63:61–71.
154. Colombo E, Di Dario M, Capitolo E, et al. Fingolimod may support neuroprotection via blockade of astrocyte nitric oxide. Ann Neurol. 2014;76:325–37.
155. Linker RA, Lee DH, Ryan S, et al. Fumaric acid esters exert neuroprotective effects in neuroinflammation via activation of the Nrf2 antioxidant pathway. Brain. 2011;134:678–92.
156. Lee DH, Linker RA, Gold R. Spotlight on fumarates. Int MS J. 2008;15:12–8.
157. Fox RJ. Placebo-controlled phase 3 study of oral BG-12 or glatiramer in multiple sclerosis. New Engl J Med. 2012;367:1087–97.
158. Gold R. Placebo-controlled phase 3 study of oral BG-12 for relapsing multiple sclerosis. New Engl J Med. 2012;367:1098–107.
159. Rosenkranz T, Novas M, Terborg C. PML in a patient with lymphocytopenia treated with dimethyl fumarate. New Engl J Med. 2015;372:1476–8.
160. Nieuwkamp DJ, Murk JL, van Oosten BW. Consortium PDMP. PML in a patient without severe lymphocytopenia receiving dimethyl fumarate. New Engl J Med. 2015;372:1474–6.
161. Gold R, Arnold DL, Bar-Or A, et al. Long-term effects of delayed-release dimethyl fumarate in multiple sclerosis: interim analysis of ENDORSE, a randomized extension study. Mult Scler. 2017;23(2):253–65.
162. Gold R, Phillips JT, Havrdova E, et al. Delayed-release dimethyl fumarate and pregnancy: preclinical studies and pregnancy outcomes from clinical trials and postmarketing experience. Neurol Ther. 2015;4:93–104.
163. Siemasko K, Chong AS, Jack HM, Gong H, Williams JW, Finnegan A. Inhibition of JAK3 and STAT6 tyrosine phosphorylation by the immunosuppressive drug leflunomide leads to a block in IgG1 production. J Immunol. 1998;160:1581–8.
164. Xu X, Williams JW, Bremer EG, Finnegan A, Chong AS. Inhibition of protein tyrosine phosphorylation in T cells by a novel immunosuppressive agent, leflunomide. J Biol Chem. 1995;270:12398–403.
165. Dimitrova P, Skapenko A, Herrmann ML, Schleyerbach R, Kalden JR, Schulze-Koops H. Restriction of de novo pyrimidine biosynthesis inhibits Th1 cell activation and promotes Th2 cell differentiation. J Immunol. 2002;169:3392–9.
166. Korn T, Magnus T, Toyka K, Jung S. Modulation of effector cell functions in experimental autoimmune encephalomyelitis by leflunomide—mechanisms independent of pyrimidine depletion. J Leukoc Biol. 2004;76:950–60.
167. Manna SK, Aggarwal BB. Immunosuppressive leflunomide metabolite (A77 1726) blocks TNF-dependent nuclear factor-kappa B activation and gene expression. J Immunol. 1999;162:2095–102.
168. O'Connor P, Wolinsky JS, Confavreux C, et al. Randomized trial of oral teriflunomide for relapsing multiple sclerosis. New Engl J Med. 2011;365:1293–303.

169. Confavreux C, O'Connor P, Comi G, et al. Oral teriflunomide for patients with relapsing multiple sclerosis (TOWER): a randomised, double-blind, placebo-controlled, phase 3 trial. Lancet Neurol. 2014;13:247–56.
170. Vermersch P, Czlonkowska A, Grimaldi LME, et al. Teriflunomide versus subcutaneous interferon beta-1a in patients with relapsing multiple sclerosis: a randomised, controlled phase 3 trial. Mult Scler J. 2014;20:705–16.
171. Lutterotti A, Martin R. Getting specific: monoclonal antibodies in multiple sclerosis. Lancet Neurol. 2008;7:538–47.
172. Waldmann H, Hale G. CAMPATH: from concept to clinic. Philos Trans R Soc Lond Ser B Biol Sci. 2005;360:1707–11.
173. Cox AL, Thompson SA, Jones JL, et al. Lymphocyte homeostasis following therapeutic lymphocyte depletion in multiple sclerosis. Eur J Immunol. 2005;35:3332–42.
174. Coles AJ, Wing MG, Molyneux P, et al. Monoclonal antibody treatment exposes three mechanisms underlying the clinical course of multiple sclerosis. Ann Neurol. 1999;46:296–304.
175. Moreau T, Thorpe J, Miller D, et al. Preliminary evidence from magnetic resonance imaging for reduction in disease activity after lymphocyte depletion in multiple sclerosis. Lancet. 1994;344:298–301.
176. Coles AJ, Cox A, Le Page E, et al. The window of therapeutic opportunity in multiple sclerosis: evidence from monoclonal antibody therapy. J Neurol. 2006;253:98–108.
177. Coles A, Group oboTCIS. Alemtuzumab improved multiple sclerosis functional composite scores and delayed time to first relapse at 2-year interim analysis compared to subcutaneous interferon beta-1a. Mult Scler. 2007;13:S166.
178. Coles AJ, Twyman CL, Arnold DL, et al. Alemtuzumab for patients with relapsing multiple sclerosis after disease-modifying therapy: a randomised controlled phase 3 trial. Lancet. 2012;380:1829–39.
179. Cohen JA, Coles AJ, Arnold DL, et al. Alemtuzumab versus interferon beta 1a as first-line treatment for patients with relapsing-remitting multiple sclerosis: a randomised controlled phase 3 trial. Lancet. 2012;380:1819–28.
180. Havrdova E, Arnold DL, Cohen JA, et al. Alemtuzumab CARE-MS I 5-year follow-up durable efficacy in the absence of continuous MS therapy. Neurology. 2017;89:1107–16.
181. Coles AJ, Cohen JA, Fox EJ, et al. Alemtuzumab CARE-MS II 5-year follow-up efficacy and safety findings. Neurology. 2017;89:1117–26.
182. Schippling S, Cohen J, Coles A, et al. Alemtuzumab-treated patients with RRMS demonstrate durable slowing of brain volume loss over 5 years despite most being treatment-free for 4 years: CARE-MS I and II extension study. Neurology. 2016;86:S51.001.
183. Simpson BS, Coles AJ. Rationale for cytotoxic monoclonal antibodies in MS. Int MS J. 2007;14:48–56.
184. Coles AJ, Wing M, Smith S, et al. Pulsed monoclonal antibody treatment and autoimmune thyroid disease in multiple sclerosis. Lancet. 1999;354:1691–5.
185. Senior P, Arnold D, Cohen J, et al. Incidence and timing of thyroid adverse events in patients with RRMS treated with alemtuzumab through 5 years of the CARE-MS studies. Neurology. 2016;86:P2.086.
186. Achiron A, Chambers C, Fox EJ, et al. Pregnancy outcomes in patients with active RRMS who received alemtuzumab in the clinical development program. Mult Scler J. 2015;21:581–2.
187. Shlomchik MJ. Sites and stages of autoreactive B cell activation and regulation. Immunity. 2008;28:18–28.
188. Meinl E, Krumbholz M, Hohlfeld R. B Lineage cells in the inflammatory central nervous system environment: migration, maintenance, local antibody production, and therapeutic modulation. Ann Neurol. 2006;59:880–92.
189. Cepok S, Rosche B, Grummel V, et al. Short-lived plasma blasts are the main B cell effector subset during the course of multiple sclerosis. Brain. 2005;128:1667–76.
190. Lucchinetti C, Bruck W, Parisi J, Scheithauer B, Rodriguez M, Lassmann H. Heterogeneity of multiple sclerosis lesions: implications for the pathogenesis of demyelination. Ann Neurol. 2000;47:707–17.
191. Krumbholz M, Derfuss T, Hohlfeld R, Meinl E. B Cells and antibodies in multiple sclerosis pathogenesis and therapy. Nat Rev Neurol. 2012;8:613–23.
192. Hauser SL, Bar-Or A, Comi G, et al. Ocrelizumab versus interferon beta-1a in relapsing multiple sclerosis. N Engl J Med. 2017;376:221–34.
193. Hauser SL, Waubant E, Arnold DL, et al. B-Cell depletion with Rituximab in relapsing-remitting multiple sclerosis. New Engl J Med. 2008;358:676–88.
194. Montalban X, Hauser SL, Kappos L, et al. Ocrelizumab versus placebo in primary progressive multiple sclerosis. N Engl J Med. 2017;376:209–20.
195. Hauser SL, Kappos L, Montalban X, et al. Safety of ocrelizumab in multiple sclerosis: updated analysis in patients with relapsing and primary progressive multiple sclerosis. Mult Scler J. 2017;23:324–5.
196. Beutler E. Cladribine (2-chlorodeoxyadenosine). Lancet. 1992;340:952–6.
197. Giovannoni G, Comi G, Cook S, et al. A placebo-controlled trial of oral cladribine for relapsing multiple sclerosis. N Engl J Med. 2010;362:416–26.

The Pharmacology of Stem Cell Transplant for Autoimmune Disease

36

Christina Wiedl and Kamar Godder

36.1 Introduction

Hematopoietic and mesenchymal stem cells are multipotent cells which have the ability to re-establish the bone marrow and immune systems. They can be derived from a multitude of sources, and the choice of using a specific stem cell source for a specific patient depends on the patient's underlying disease state and comorbidities. Stem cells can be harvested from the patient's own hemato-immune system, i.e., AUTOLOGOUS collection, or from another person (a donor), i.e., ALLOGENIC sources. The cells can be harvested through an APHERESIS (peripheral stem cell collection), from the placenta and CORD BLOOD or from the donor's bone marrow. Unlike hematopoietic stem cells, mesenchymal stem cells are not specific for a patient (see also Chap. 2). They are undifferentiated cells with the capacity of self-renewal and the ability to undergo multilineage differentiation which can be obtained from a multitude of sources.

The potential role of stem cell transplant in autoimmune disease was first suggested in the 1960s and 1970s based on xenograft studies in immunodeficient mice. These studies indicated that lupus could be induced in an irradiated mouse following transplant of spleen or marrow cells from a mouse with murine lupus [1]. Subsequent studies have documented the reverse: stem cell transplant from healthy mice is capable of curing murine lupus. The clinical benefit of stem cell transplant in autoimmune disease was first demonstrated in patients who were cured of their autoimmune disease after receiving an ALLOGENEIC stem cell transplant for a coinciding malignancy or aplastic anemia [2]. A review of these case reports revealed a reduction in the risk of relapse of the autoimmune disease post-discontinuation of immunosuppression in patients who developed GRAFT VS. HOST DISEASE, suggesting a graft vs. autoimmune effect. Likewise, patients who received an ALLOGENEIC vs. AUTOLOGOUS transplant for malignancy or aplastic anemia had marked improvement in relapse-free survival (of their autoimmune disease), 89% at 18 years vs. 38% at 5 years [3]. However, due to the significant transplant-related mortality associated with ALLOGENEIC transplant related to intensive CONDITIONING REGIMENS and GRAFT VS. HOST DISEASE, recent clinical trials have focused on AUTOLOGOUS transplantation. Most studies have relied on hematopoietic stem cells; however, in recent years there have been several case reports using mesenchymal stem

Final manuscript submitted on November 11, 2017.

C. Wiedl (✉) · K. Godder (✉)
Division of Hematology/Oncology, Department of Pediatrics, Virginia Commonwealth University, Richmond, VA, USA
e-mail: Christina.wiedl@vcuhealth.org; kammar.godder@mch.com

M. J. Parnham et al. (eds.), *Nijkamp and Parnham's Principles of Immunopharmacology*, https://doi.org/10.1007/978-3-030-10811-3_36

cells, thereby eliminating the need for pretransplant CONDITIONING regimens and reducing immunogenicity and potentially fatal GRAFT VS. HOST DISEASE.

36.2 Allogenic vs. Autologous Stem Cell Transplant vs. Mesenchymal Stem Cell Therapy

ALLOGENEIC transplant involves the ENGRAFTMENT of either a related or unrelated donors' hematopoietic cells. Typically, this must be preceded by a CONDITIONING REGIMEN to suppress the patient's immune system to reduce the risk of graft rejection. AUTOLOGOUS transplant involves the engraftment of the patient's own stem cells. AUTOLOGOUS transplant is often used as "rescue therapy" following high doses of chemotherapy or immunotherapy. AUTOLOGOUS transplant can also be used following graft engineering/gene therapy. Mesenchymal stem cells involve the use of undifferentiated stem cells which have the ability to regenerate most of the normal body tissues; these stem cells do not require HLA (human leukocyte antigen) matching (see also Chap. 5).

36.3 Mechanism of Action of Stem Cell Transplant in Autoimmune Disease

The role of transplant in replacing the dysfunctional immune system has been documented both in animal studies and in the literature reports for patients transplanted for malignant conditions. Clearance or marked reduction in autoimmune antibodies following ENGRAFTMENT coincided with resolution of autoimmune symptoms [4]. The potential role for graft vs. autoimmune effect suggested by the reduced risk of relapse in patients who developed GRAFT VS. HOST DISEASE, as well as the response to donor lymphocyte infusions, raises the possibility of ALLOGENEIC transplant as a potential curative option if treatment-related mortality and the occurrence of severe GRAFT VS. HOST DISEASE could be reduced. In AUTOLOGOUS transplant, the mechanism of action is not completely clear; however, it is thought that the initial reduction of autoreactive cells by the cytoreduction chemotherapy allows for a "reeducation" of the immune system. Typically, the harvested infused stem cells are mainly used for "rescue" rather than transplant of a new population of cells. Subsequently, the re-boosting of the immune system leading to the clinical response is attributed to (1) lymphopenia secondary to the induction therapy, reducing production of antibodies, and (2) increased immune tolerance by increased number of regulatory Foxp3 T cells (Roord 2008), shifts in T-cell populations toward naive populations of $CD31^{+}CD45RA^{+}CD4^{+}$ cells, and later establishment of a wide T-cell repertoire with TREC and CD31 expression, as shown by Muraro et al. [5] in multiple sclerosis (see Chap. 35).

36.3.1 Transplant Complications

HEMATOPOIETIC STEM CELL TRANSPLANTATION (HSCT) in malignant disorders has been associated with severe, and at times lifelong, morbidity and mortality. Whereas transplant for malignant disorders is used for life-threatening illnesses, the immediate goal in autoimmune disease is improvement in quality of life (QOL). Daikeler et al. [6] reviewed transplant complications of AID and reported the factors most associated with increased complications were pre-existing reduced organ function and lymphopenia which increased the risk of opportunistic infection. In addition, secondary autoimmunity, endocrine late effects, and malignancy may develop post-transplant. All of these factors must be considered when designing protocols for stem cell transplantation.

36.4 Allogeneic Bone Marrow Transplant

The majority of literature regarding the use of ALLOGENEIC transplant for autoimmune disease is case reports. Although in theory

ALLOGENEIC HSCT would provide the means for truly curative therapy by replacing the diseased immune system, the repeatedly documented relapses of autoimmune disease despite maintaining full donor CHIMERISM would appear to contradict the theory of graft vs. autoimmune effect. Of note, these case reports are based on sibling HLA identical transplants, and in some cases weakly positive autoantibodies were noted in the donor, raising the possibility of autoimmune predisposition. It has been hypothesized that either residual recipient cells could sensitize the HLA identical donor T cells to reactivate the autoimmune disease or the survival of long-lived recipient plasma cells could perpetuate humoral autoimmunity. Given the combination of limited patient numbers, higher TRM, and the failure of ALLOGENEIC transplant to clearly provide a survival benefit over AUTOLOGOUS transplant, the majority of ongoing studies are focusing on AUTOLOGOUS or mesenchymal stem cells [4].

36.5 Autologous Bone Marrow Transplant

In the mid-1990s the first case reports of patients with severe autoimmune disease, including several patients with refractory SLE/Lupus, treated with AUTOLOGOUS stem cell transplant were published. Based on these preliminary successes, a multicenter prospective phase I/II trial was initiated by the European Group for Blood and Marrow Transplantation (EBMT) to demonstrate the feasibility and safety of AUTOLOGOUS stem cell transplant for severe autoimmune disease. In 1999, the preliminary results of 74 consecutive patients from 22 centers were published. Please see Table 36.1 for patient characteristics, transplant methods, and outcomes. Patients who had severe refractory autoimmune disease including rheumatologic disorders, neurological disorders, immune CYTOPENIAS, and vasculitides were included. Patients were between 7 and 64 years old (median 39) with a female to male predominance of 47:27. Patients with severe organ damage were excluded. Two patients died before transplant following mobilization. Seventy-three transplants were performed (one patient was transplanted twice); mobilized peripheral blood stem cell sources were predominately used. Forty-three grafts were manipulated with CD34 selection and/or T-cell depletion. The majority of the conditioning regimens were cyclophosphamide based. Median follow-up time at the time of publication was 308 days, range 25–1026 days. Seven patients died, corresponding to a survival probability at 1 year of 87%. Five deaths were directly attributed to transplant complications, three to infections, and two to bleeding. The preliminary results showed a response rate of 65%, although the median follow-up time was short, and 23% of patients were noted to have progressive disease or relapse. Overall, the preliminary results indicated AUTOLOGOUS transplant in autoimmune disease had transplant-related mortality (TRM) of 9%, similar to that of transplant in malignant conditions, and it had at least some efficacy in the majority of patients [7].

In 2010, the EBMT published the results of an observational study of 12 years of experience with AUTOLOGOUS transplant between 1996 and 2007. This paper expanded on the preliminary results of the 1999 publication and clarified the difference in TRM associated with various underlying autoimmune diseases. Please see Table 36.1 for patient characteristics, transplant methods, and outcomes. Nine hundred patients were included, with a median age 35 years and female predominance of 64%. The disease types included: multiple sclerosis, systemic sclerosis, SLE, rheumatoid arthritis, juvenile arthritis, and immune CYTOPENIA. Peripheral blood stem cell use remained predominant at 93% of cases, and most were mobilized with cyclophosphamide 1.5–4 g/m^2 and GCSF. Forty-four percentage of grafts were manipulated. The majority of patients received cyclophosphamide-based CONDITIONING regimens. The 5-year survival was 85% with progression-free survival of 43%. 38.8% of the deaths were from the original disease and 53.1% from transplant-related causes with infection (45.7%) being the leading cause of TRM. The overall 100-day TRM was 5%; however, there were significant differences in TRM,

Table 36.1 Results of the first two European Working Group studies of stem cell transplant in autoimmune disorders

Study group, author (year)	EBMT, Tindell (1999)	EBMT, Farge (2010)
Study duration	1995–1999	1996–2007
Numbers of patients	74	900
Median age of patients	39 (7–64)	35 (2.7–76)
Disorders no (%)		
Rheumatologic	35 (47)	414 (175 SSC/85 SLE/89 RA/65 JRA) (46)
Neurologic	31 (42)	345 (38)
Hematologic	5 (7)	37 (6)
Vasculitides	3 (4)	0
Stem cell source		
BM/PBSC	7 BM/66 PBSC	63 BM/837 PBSC
PBSC priming		
GCSF alone	15	
Cyclophosphamide + GCSF + other	38	
Cyclophosphamide + GCSF	9	
Cyclophosphamide alone	7	
Conditioning regimen (%)		
Cy + ATG	36	52
BEAM + ATG	39	34
BuCy	8	4
TBICy + ATG	9	7
Other	4	3
Survival	1 year 87%	5 year 85%
RR/PFS	RR 65% at <1 year	PFS 43% at 5 years

progression-free survival, and overall survival related to disease type. Please see Table 36.2 for an illustration of TRM, progression-free, and overall survival based on disease type. TRM was as high as 11% for SLE patients and progression-free survival as low as 18% for rheumatoid arthritis patients. The 5-year overall survival varied with disease type, with multiple sclerosis and rheumatoid arthritis patients faring the best. In a multivariate analysis, the 100-day TRM varied according to diagnosis ($P = 0.003$), and was lower in experienced centers ($P = 0.003$), age less than 35 years ($P = 0.004$), and transplant performed after December 2000 ($P = 0.0015$) [8]. Please see Table 36.3 for guidelines on the use of stem cell therapy in patients with autoimmune disease.

Table 36.2 Results of the European Working Group studies on autoimmune disease: transplant-associated mortality, 5-year progression-free, and overall survival based on disease type

Disease	100-Day TRM (%)	5-Year PFS (%)	5-Year OS (%)
SLE	11	44	76
JRA	11	52	82
Immune cytopenias	8	34	80
Systemic sclerosis	6	55	76
Multiple sclerosis	2	45	92
Rheumatoid arthritis	1	18	94

The results of the European Bone Marrow Transplant Group AUTOLOGOUS Stem Cell Transplant International Scleroderma (ASTIS) trial were presented at the 2012 American Society of Hematology meeting. This phase III trial was a multicenter, randomized active-controlled, parallel-group trial comparing MYELOABLATIVE AUTOLOGOUS stem cell transplant vs. “conventional” therapy with pulse dose of monthly cyclophosphamide. MYELOABLATION was performed with cyclophosphamide 200 mg/kg and rabbit ATG 7.5 mg/kg followed by the infusion of LEUKAPHERESIS CD34-selected AUTOLOGOUS stem cells mobilized with cyclophosphamide 2×2 g/m^2 and filgrastim 10 mg/kg/

Table 36.3 Guidelines for the use of stem cell transplant in autoimmune diseases

Factor		Guideline
Patient		Minimal pre-existing organ dysfunction
	Cardiac	Left ventricle ejection fraction (LVEF) of 50% for SSC, 40% for other diseases. No evidence of uncontrolled arrhythmia, pericardial effusion >1 cm
	Renal	Creatinine clearance 40 mL/min in SSC, 30 mL/min in other diseases
	Pulmonary	DLCO <40% of predicted, mean pulmonary artery pressure 50 mmHg in SSC
	Infection	No evidence of uncontrolled infection
	Disease	Severe progressive disease despite treatment with standard therapy including biologicals
Center		Performed on a clinical trial in an experienced whenever possible
Graft source		Autologous rather than allogeneic stem cell source
Graft manipulation (CD34+ selection)		Benefit not documented
Mobilization		Cyclophosphamide/GCSF
Conditioning		Cyclophosphamide +/ATG

day. Patients on the cyclophosphamide pulse arm received monthly doses of 750 mg/m^2 for 1 year. The trial was open to patients 16–65 years of age with early progressive systemic sclerosis of 4 years or less duration and evidence of organ involvement and a modified Rodnan skin score of 15 or more, or 2 years of duration or less with evidence of systemic inflammation and a skin score of 20 or more. Patients with severe disease were excluded based on the criteria of evidence of severe pulmonary hypertension demonstrated by mean pulmonary arterial pressure of >50 mmHg, DLCO <40% of predicted, severe renal disease with a creatinine clearance of <40 mL/min, cardiac dysfunction with LVEF <45%, uncontrolled arrhythmia, infection, liver failure, or prior cyclophosphamide administration of >5 g IV or >3 months oral treatment. One hundred fifty-six patients were included in this study between March 2001 and October 2009 from 27 transplant centers in ten countries. Fifty-nine percent of the patients were female with a mean age of 44, mean disease duration of 1.4 years, mean BMI of 24, and mean Rodnan skin score of 25 with no significant differences reported between the two arms. Seventy-nine patients were randomized to the transplant arm and 77 to the pulse cyclophosphamide control arm. Seventy patients in the transplant arm and 58 in the control arm completed treatment with median follow-up at 33 months in the transplant arm and 27 months in the control arm. There were 18 events, 16 deaths, and 2 irreversible renal failures in the transplant arm. Of these, eight deaths were deemed treatment related including three due to heart failure, two ARDS, two multi-organ system failures, and one episode of pulmonary edema. There were no deaths deemed treatment related in the control arm; however, there were 24 events related to disease progression, all of which resulted in death. Eight patients in the control arm were transferred to the HSCT arm due to progressive disease, one of whom later died from secondary acute myeloid leukemia. Although there was significant transplant-related mortality (10%), the event-free survival and overall survival were better in the transplant compared with the control arm [9]. The US counterpart to the ASTIS trial, the SCOT trial, is ongoing. The SCOT trial compares high-dose cyclophosphamide vs. AUTOLOGOUS stem cell transplant following cyclophosphamide 4 g/m^2 plus GCSF priming. The CONDITIONING regimen consists of TBI, cyclophosphamide, and ATG. The stem cell product is CD34 selected prior to infusion. Between 2006 and 2011, the SCOT enrolled 205 patients and randomized 75. The last subject was enrolled in September 2011; based on a 54-month follow-up endpoint, there was marked improvement in global right composite scores (GRCS, which includes event-free survival, forced vital capacity, health assessment questionnaire, disability index, and modified Rodnan score) for patients enrolled on the transplant arm. The ASTIC trial evaluated the role of autologous stem cell transplant following high intensity conditioning in patients with refractory Crohn's disease. Unfortunately, there was a high rate of toxicity and adverse events and few patients achieved the primary endpoint of three month remission off all immunosuppression. However, many patients did have evidence of disease improvement and QOL scores. The ASTIClite trial currently evaluating the efficacy and toxicity of utilizing a reduced intensity conditioning regimen. The primary endpoint is mucosal healing at week 48 [10].

36.6 Mesenchymal Stem Cell Therapy

In their review of potential future therapies in IBD, Plevy and Targan [11] discussed cell-based therapy in addition to therapy that targets mucosal immunity and therapy which targets the site of inflammation. In regard to the role of stem cell transplant, they expressed concern regarding the morbidity and mortality of the transplant in the face of the relatively high risk of disease relapse. They discuss promising preclinical studies in which isolated Treg (Regulatory T cells) and dendritic cells, both shown to play an important role in the recovering immune system, were expanded in vitro and then reinfused without MYELOABLATION.

Mesenchymal stem cells (MSCs) have been proposed as a potential alternate option for treatment of IBD. These cells are undifferentiated cells with the capacity of self-renewal and the ability to undergo multilineage differentiation. They reside in the bone marrow, adipose tissue, gastrointestinal tract, and liver. In preclinical studies MSCs differentiate into cells that can reduce the effect of inflammation, promote repair in the gut, and downregulate the immune response. MSCs have been used successfully in treatment of posttransplant GRAFT VS. HOST DISEASE of the gut. Studies with mesenchymal cells in IBD are underway. MSCs may be particularly useful in perianal fistulizing disease, a form of IBD which tends to be refractory despite aggressive surgical and medical management (Lopez et al. 2011). There are several early-phase clinical trials for MSCs in IBD currently recruiting, and two phase I studies were recently completed. These studies revealed MSC therapy was well tolerated with no significant toxicities noted. Improvement in Crohn's disease activity index scores (Duijvestein et al. 2010) and increased mucosal and circulatory Treg cells were identified in some patients (Ciccocioppo et al. 2011).

Mouse model studies of MRL/lpr mice found reduction in anti-dsDNA levels and proteinuria as well as decreased CD4 T cells with increased T helper cell populations following ALLOGENEIC MSC injection [12]. Mouse model studies by Ji et al. found high-dose MSCs inhibited the abnormal activation of Akt/GSK3β signaling pathway of T cells. Based on preclinical results, pilot studies of MSC therapy in SLE patients were undertaken. Liang et al. [12] report the early results of 15 patients treated with ALLOGENEIC MSCs for refractory SLE. At a follow-up of 1 year, all patients were found to have clinical improvement with decrease in SLE disease activity index, reduction in proteinuria, and decreasing anti-dsDNA antibody levels. Sun et al. [13] reported the results of 16 patients with refractory SLE transplanted with umbilical cord MSCs. At a median follow-up of 8 months, all patients had a significant reduction in disease activity scores, levels of ANA, anti-dsDNA, and improvement in renal function. In patients who achieved clinical remission, increased peripheral blood Treg cells, increased TGF-β levels, and rebalanced Th1 and Th2 cytokine levels. Yuan et al. [14] have documented CD1c+ dendritic cells and serum levels of FLT3L are upregulated following MSC treatment in SLE patients, possibly leading to inflammatory suppression.

36.7 Conclusion

Stem cell therapy may play a powerful role in treating patients with life-threatening refractory autoimmune disorders. However, transplant does carry a significant risk of morbidity and mortality particularly in patients with pre-existing comorbidities. There are several ongoing trials regarding the use of stem cell therapy for this patient population and the use of reduced intensity conditioning may reduce toxicity and improve outcomes. As more data emerges, medical teams may be better able to predict which patients are best served using stem cell therapy.

References

1. Denman AS, Russell AS, et al. Adoptive transfer of the diseases of New Zealand black mice to normal mouse strains. Clin Exp Immunol. 1969;5(6):567–95.
2. Marmont A. Coincidental autoimmune disease in patients transplanted for conventional indications. Best Pract Res Clin Haematol. 2004;17(2):233–2.
3. Hinterberger W, Hinterberger-Fischer M, Marmont A. Clinically demonstrable anti-autoimmunity mediated by allogeneic immune cells favourably affects outcome after stem cell transplant in human autoimmune diseases. Bone Marrow Transplant. 2002;30:753–9.
4. Marmont A. Allogeneic haematopoietic stem cell transplant for severe autoimmune diseases: great expectations but controversial evidence. Bone Marrow Transplant. 2006;38:1–4.
5. Muraro PA, et al. Thymic output generates a new and diverse TCR repertoire after autologous stem cell transplantation in multiple sclerosis patients. J Exp Med. 2005;201(5):805–16.
6. Daikeler T, et al. Complications of autologous stem cell transplantation for patients with autoimmune diseases. Pediatr Res. 2012;71:439–44.
7. Tyndall A, Fassas A, et al. Autologous haematopoietic stem cell transplants for autoimmune disease-feasibility and transplant-related mortality. Bone Marrow Transplant. 1999;24:729–34.
8. Farge D, Labopin M, et al. Autologous hematopoietic stem cell transplantation for autoimmune disease: an observational study on 12 years' experience from the European Group for Blood and Marrow Transplantation working party on autoimmune disease. Haematologica. 2010;95:284–92.
9. Farge D, Van Laar J, et al. Autologous hematopoietic stem cell transplant versus intravenous cyclophosphamide for severe or rapidly progressive systemic sclerosis, the ASTIS trial. Blood. 2012;120:964.
10. Snowden JA, et al. Autlogous stem cell transplantation in refractory Crohn's disease- low intensity therapy evaluation. BMC Gastroenterol. 2019;19(1):82.
11. Plevy SC, Targan SR. Future therapeutic approaches for inflammatory Bowel diseases. Gastroenterology. 2011;140:1838–2846.
12. Liang J, Zhang H, et al. Allogenic mesenchymal stem cells transplantation in refractory SLE: a pilot clinical study. Ann Rheum Dis. 2010;69:1423–9.
13. Sun L, Wang D, et al. Umbilical cord mesenchymal stem cell transplantation in severe and refractory SLE. Arthritis Rheum. 2010;62:2467–75.
14. Yuan X, et al. Mesenchymal stem cell therapy induces FLT3L and CD1c+ dendritic cells in systemic lupus erythematosus patients. Nat Commun. 2019;10(1):2498.

Part IV

Immunotoxicology

Immunotoxicology

37

Raymond Pieters, Jan Willem van der Laan,
and Henk van Loveren

37.1 Introduction

Pharmaceuticals for human use comprise a very wide variety of product types. These include traditional products (i.e. chemically synthesized or derived from natural resources) as well as biological products (such as vaccines and blood products isolated from biological sources) and biotechnology-derived pharmaceuticals (such as peptide/protein products manufactured by recombinant DNA techniques, monoclonal antibodies and gene therapy products). In the interest of the public, these medicinal products are subject to worldwide regulatory control by government authorities. The major objective of this regulation is to ensure that the benefit of the products to the patients is not outweighed by their adverse effects. To achieve this goal, the authorities carefully assess the balance between efficacy and safety. If this balance is positive, they allow marketing. To support applications for marketing authorization, the pharmaceutical industry, therefore, has to submit scientific data which prove that their products are efficacious and acceptably safe in the proposed therapeutic indication. Furthermore, the pharmaceutical quality of the products applied for has to meet high standards.

Chemicals used for a variety of purposes can have adverse effects on the immune system of both animals and humans. In the case of pharmaceuticals, this can be the result of pharmacological interference with the immune system or an undesired reaction. One form of immunotoxicity is the direct toxicity of the compound to components of the immune system, which often leads to suppressed function. This may result in decreased resistance to infection, the development of certain types

Final manuscript submitted on May 07, 2018.

R. Pieters
Immunotoxicology, Institute for Risk Assessment Sciences (IRAS), Utrecht University, Utrecht, The Netherlands

University of Applied Sciences Utrecht, Utrecht, The Netherlands
e-mail: r.h.h.pieters@uu.nl

J. W. van der Laan (✉)
Section on Pharmacology, Toxicology and Kinetics, Medicines Evaluation Board, Utrecht, The Netherlands

Division on Toxicology, Leiden Academic Center for Drug Research, Leiden, The Netherlands
e-mail: jw.vd.laan@cbg-meb.nl

H. van Loveren
Department of Toxicogenomics, Maastricht University, Maastricht, The Netherlands
e-mail: h.vanloveren@maastrichtuniversity.nl

M. J. Parnham et al. (eds.), *Nijkamp and Parnham's Principles of Immunopharmacology*,
https://doi.org/10.1007/978-3-030-10811-3_37

of tumours or immune dysregulation and stimulation, thereby promoting ALLERGY or AUTOIMMUNITY. Other types or manifestations of immunotoxicity include allergy or autoimmunity in which the compound causes the immune system to respond as if the compound were an antigen or to respond to self-antigens that have been altered or released by the chemical or its toxic effect. A more recently recognized manifestation of immunotoxicity, called cytokine release syndrome, is due to overt release of cytokines, which may cause acute clinical phenomena.

Except for cancer patients on chemotherapy and organ transplant patients on long-term immunosuppressive therapy, there is little evidence that drugs are associated with undesired, clinically significant, direct IMMUNOSUPPRESSION. Only a few valid epidemiological studies of immunologically based diseases have been carried out [1], probably due to the complication of such studies by confounding factors such as (disease-associated) stress, nutritional status, lifestyle, (co)medication and genetics. Few conventional drugs have been shown to induce unexpected enhancement of immune competence. Unwanted immunostimulation has gained attention primarily through the introduction of new biotechnologically manufactured drugs such as cytokines. Drug-induced hypersensitivity reactions and autoimmune disorders are a major concern and often the reason for withdrawing drugs from the market or restricting their use. For instance, recent epidemiological research based upon large patient data have revealed that long-term use of HMG-CoA reductase inhibitors (statins) is associated with an increase in incidence of autoimmune diseases such as rheumatoid arthritis [2] and systemic lupus erythematosus [3]. A causal mechanism is difficult to establish but is likely to be related to direct effects of statins on lymphocyte function [4, 5]. Another particular cause of drug withdrawal is drug-induced liver injury (DILI), which is in at least a substantial number of cases also related to immunostimulation [6, 7].

For the detection of chemically induced direct immunotoxicity, animal models have been developed, and a number of these methods have been validated. Several compounds, including certain drugs, have been shown in this way to cause immunosuppression. Methods are also available for the detection of skin allergic responses, whereas no validated general test is available to predict potential induction of AUTOIMMUNITY or DILI.

In this chapter, the various mechanisms of immunotoxicity are introduced and discussed, by which pharmaceuticals affect different cell types and interfere with immune responses, ultimately leading to immunotoxicity. Further, procedures for preclinical testing of drugs are covered, comprising direct immunotoxicity as well as sensitizing capacity. This section is followed by consideration of procedures for clinical and epidemiological testing of drugs. Finally, regulatory aspects of immunotoxicity are discussed, including current guidelines and new developments in immunotoxicity assessment.

37.2 Mechanisms of Immunotoxicity by Pharmaceuticals

37.2.1 Effects on Precursor Stem Cells

Precursor stem cells that are responsible for replenishing peripheral leukocytes reside in the bone marrow, making it an organ that harbours many highly proliferating cells. All leukocyte lineages originate from these stem cells, but once distinct subsets of leukocytes are established, their dependence on replenishment from the bone marrow differs vastly. The short-lived neutrophils rely heavily on proliferation and new formation in the bone marrow, as each day more than 108 neutrophils enter and leave the circulation in a normal adult. In contrast, tissue macrophages are usually long-lived and have limited dependence on new formation of precursor cells. The adaptive immune system, comprising antigen-specific T- and B-lymphocytes, is almost completely established around puberty and therefore is essentially bone marrow-independent in the adult.

As a consequence of their high proliferation rate, stem cells in the bone marrow are extremely vulnerable to antiproliferative cytostatic drugs such as the antineoplastic drugs, cyclophosphamide and methotrexate, and the antirheumatic azathioprine [1–8] (Fig. 37.1). This is particularly the case at high doses of these drugs, and conceivably lineages like neutrophils that are extremely bone marrow-dependent will be most vulnerable and are affected first by treatment with these drugs. Notably, since neutrophils are crucial in first-line defence to opportunistic infections (e.g. *Staphylococcus aureus*, *Pseudomonas aeruginosa*), anti-proliferative drugs are often linked to such infections. After prolonged exposure, macrophages and T or B cells of the adaptive immune system are also suppressed.

37.2.2 Effects on Maturation of Lymphocytes

After leaving the bone marrow, cells of both the T-cell and the B-cell lineages mature into antigen-specific lymphocytes. T-lymphocytes mature in the thymus during a process referred to

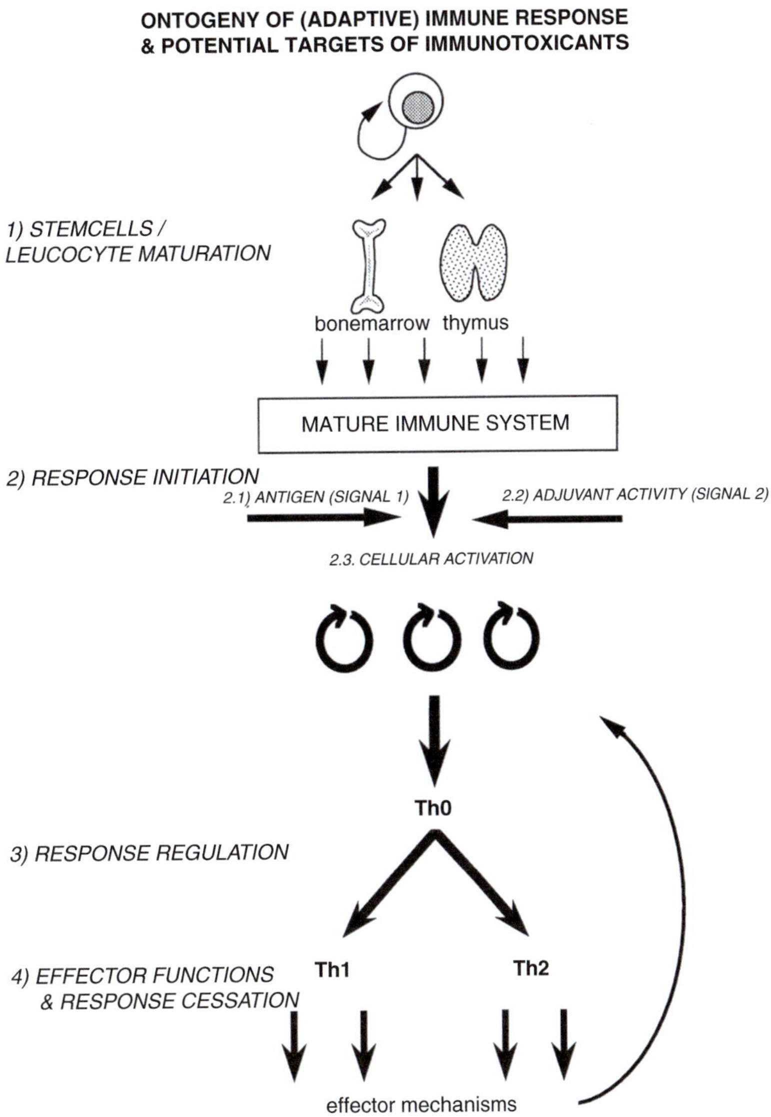

Fig. 37.1 Ontogeny of the immune response and targets of immunotoxic pharmaceuticals. This figure represents the different steps in the ontogeny of adaptive immune responses from stem cell to response cessation. It forms our conception framework to identify potential mechanisms of immunotoxicity. Effects of pharmaceuticals on stem cells and leucocyte maturation are indicated, on the two signals that are essential for lymphocyte activation and on the resulting cellular activation. Also indicated are the regulation and cessation of immune responses as potential targets for immunotoxic pharmaceuticals

as thymocyte differentiation, which is a very complex selection process that takes place under the influence of the thymic microenvironment and ultimately generates an antigen-specific, host-tolerant population of mature T cells (see Chaps. 2 and 3). Because this process involves cellular proliferation, gene rearrangement, apoptotic cell death, receptor up- and downregulation and antigen presentation processes, it is very vulnerable to a number of chemicals, including pharmaceuticals (Fig. 37.1). Drugs may target different stages of T-cell differentiation: bone marrow precursors (azathioprine), proliferating and differentiating thymocytes (azathioprine), antigen-presenting thymic epithelial cells and dendritic cells (cyclosporin A) [9] and cell death processes (corticosteroids) [10] (Fig. 37.2).

In general, immunosuppressive drugs that affect the thymus cause a depletion of peripheral T cells, particularly after prolonged treatment and during early stages of life when thymus activity is high and important in establishing a mature T-cell population, e.g. in children up to adulthood.

After the bone marrow stage, B cells mature in the spleen. With the exception of monoclonal antibodies such as rituximab, there are no identified drugs as yet that specifically affect B-cell development, although some studies claim a more or less B-cell-specific effect of immunosuppressive pharmaceuticals [11]. In general, suppression of the adaptive immune system at the antibody level is the result of an effect on T cells or their development.

37.2.3 Effects on Initiation of Immune Responses

Once a mature immune system has been established, the innate and adaptive arms of the immune system co-operate to eliminate invading pathogens. Ideally, T cells tailor the responses to neutralize invaders with minimal damage to the host. After elimination of T cells with high affinity for self-antigens in the thymus, tolerance for autoantigens is further maintained in the periphery by the two distinct signals that govern lymphocyte activation. Signal 1 is the specific recognition of antigen via clonally distributed antigen receptors. Signal 2 consists of antigen non-specific costimulation or "help" and involves interactions of various adhesive and signalling molecules [12]. It is

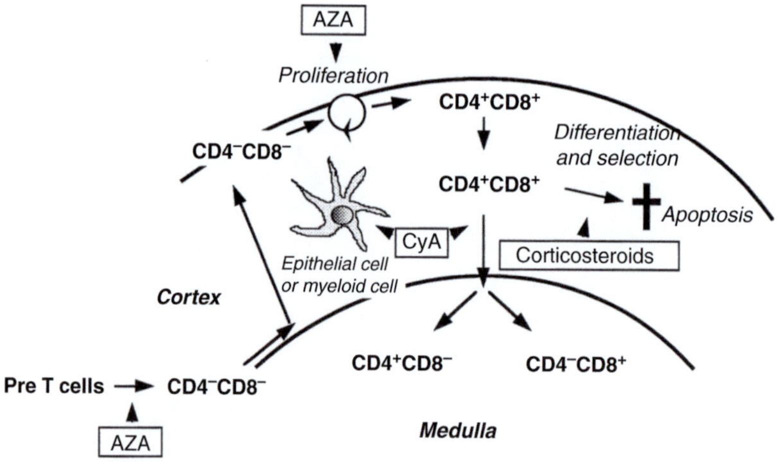

Fig. 37.2 Schematic view of the thymus showing the cortical and medullary region. Immature pre-T cells ($CD4^-CD^-$) enter the thymus at the cortico-medullary region and migrate to the subcapsular region where they show high proliferative activity and differentiate into $CD4^+CD8^+$ thymocytes. In the cortex, most thymocytes are $CD4^+CD8^+$, and at this stage thymocytes are selected under the influence of thymic epithelial cells and are prone to apoptotic cell death. After the $CD4^+CD8^+$ stage, cells differentiate either to $CD4^+$ or $CD8^+$ cells. Stages sensitive to pharmaceutical attack are indicated: azathioprine inhibits formation of pre-T cells and may inhibit immature thymocyte proliferation; cyclosporin A interferes with thymocyte selection, possibly through an effect on thymic dendritic cells; and corticosteroids stimulate apoptosis

imperative that lymphocytes receive both signals, as antigen recognition without costimulation induces tolerance, and lymphocytes are unresponsive to costimulation without an antigen-specific signal. The molecules transmitting signal 2 are thought to be expressed mainly in response to tissue damage, linking initiation of immune responses to situations of acute "danger" for the host [13]. This helps to aim immune responses at potentially dangerous microorganisms (non-self) while minimizing deleterious reactions to innocuous (non-self, e.g. food proteins) antigens and to the host (self) (Chap. 3).

Pharmaceuticals and other XENOBIOTICs may interfere with the initiation of immune responses by forming complexes with self-proteins (e.g. HAPTEN-CARRIER complexes) or by releasing previously hidden self-antigens. By doing so, these chemicals may provide signal 1 to antigen-specific T cells that were previously not addressed. Since such neo-antigen-specific T cells still need costimulatory signals, chemicals may, directly or after becoming reactive after bioactivation, also induce conditions (e.g. cellular damage, proinflammatory conditions) that favour upregulation of signal 2. However, signal 2 may also arise from other, e.g. microbial, substances. Finally, pharmaceuticals (e.g. certain Toll-like receptor (TLR)-binding substances, such as imiquimod) may directly affect cellular activation following occupation of the receptors involved in the two activation signals (Fig. 37.1).

37.2.3.1 Interference with Antigen Recognition (Signal 1)

High Molecular Weight Pharmaceuticals Can Be Antigens

High molecular weight pharmaceuticals (>4000 Da) can function as antigens and become targets of specific immune responses themselves (reviewed in [14]). This is particularly relevant for foreign biopharmaceuticals of protein origin, as these may be recognizable by both T- and B-lymphocytes. The resulting specific immune responses may lead to formation of antibodies and induce specific memory, which can lead to allergic responses to the substance. For example, passive immunization to tetanus toxin or snake venoms with serum from immunized horses causes the temporary formation of immune complexes with symptoms of fever, joint tenderness and proteinuria (serum sickness). Because serum proteins are given in large amounts and have a long half-life, SENSITIZATION and allergic reactions take place after a single dose. Similar adverse effects due to immunogenicity may occur after repeated treatment with pharmaceuticals like porcine insulin, murine antibodies and biotechnologically engineered "novel proteins". In patients developing neutralizing antibodies, absence of response or reversal of clinical efficacy have been described [15]. The risk of immunotoxic effects due to immunogenicity is much lower when homologous recombinant human or "humanized" proteins are used as pharmaceuticals. Other factors that determine whether a certain biopharmaceutical raises an immune response are route of exposure, contaminants and formulations [14].

Reactive Pharmaceuticals Can Form Haptens

Low molecular weight pharmaceuticals cannot function as antigens, because they are not proteins, and hence cannot as such be presented by MHC molecules to T cells. Reactive drugs that bind to proteins and thus form HAPTEN-protein complexes (also called drug-protein adducts) may, however, become detectable by the adaptive immune system. In other words, if haptenated EPITOPEs derived from the protein adducts are presented by MHC, T cells may become primed T cells to the low molecular weight pharmaceutical. Once primed these drug-specific T cells may in turn provide costimulation to specific B cells and activate effector mechanisms, e.g. macrophages and neutrophils. This effect is responsible for allergic responses to many new (NEO)EPITOPEs formed by chemical haptens, including pharmaceuticals, occupational contact sensitizers and respiratory sensitizers (Fig. 37.3). Other compounds (called pro-haptens) require metabolic activation to form reactive metabolites that bind to proteins. The anaesthetic halothane, for instance, is metabolized to alkyl halides by cytochrome P-450 in the liver. The alkyl halides bind to microsomal proteins including P-450, and the bound haptens induce an immune response that causes so-called halothane hepatitis. Other compounds can be activated by extrahepatic metabolism, in particular by the myeloperoxidase system in phagocytic cells. For instance, activated macrophages and granulocytes can metabolize the antiarrhythmic procainamide to reactive metabolites that can bind to proteins, and immune responses to these haptens are considered to be responsible for the initiation of procainamide-related agranulocytosis and lupus [16].

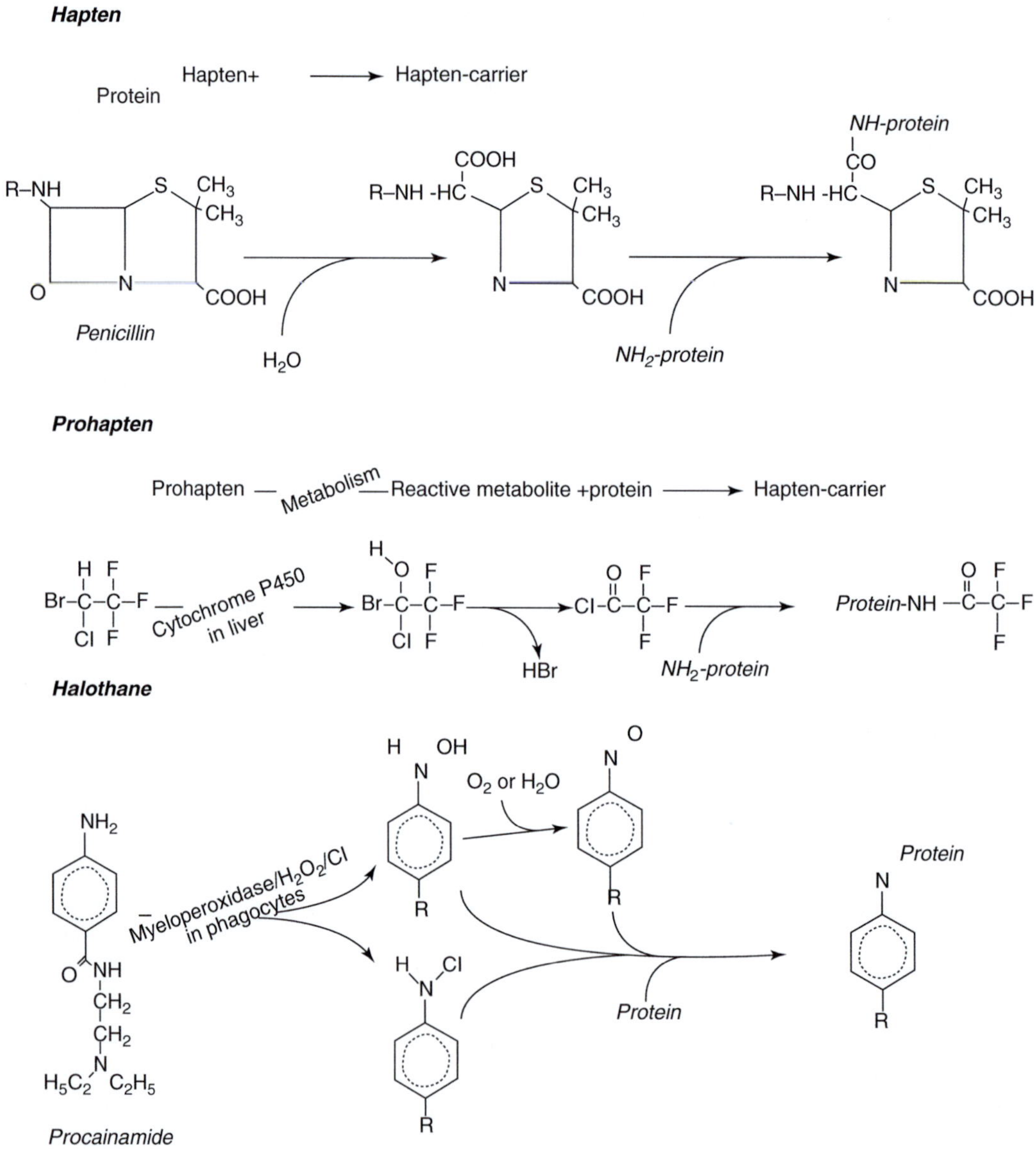

Fig. 37.3 Haptens and prohaptens. Pharmaceuticals that are too small to attract a T cell response can become antigenic when they bind as hapten to a protein carrier. In this case, T cells responding to chemically induced hypersensitivities on the carrier provide costimulation for B cells responding to the hapten. Prohaptens require metabolic activation to a reactive metabolite that can function as a hapten. Penicillin is a well-known example of a pharmaceutical that can form haptens by direct binding to proteins. In contrast, halothane itself does not form haptens, but cytochrome P450-mediated metabolism in the liver results in reactive metabolites that do bind to proteins. Procainamide can be metabolized by the myeloperoxidase/H_2O_2/Cl^- system of phagocytes. These metabolites are very reactive and can bind covalently to nucleophilic thiol and amino groups of proteins

Induction of Cross-Reactivity by Pharmaceuticals

Formation of neo-antigenic structures, in particular covalent or non-covalent HAPTEN-CARRIER complexes, may cause stimulation of cross-reactive T cells. Cross-reactivity implies that T cells recognize not only the best fitting MHC-(neo-)peptide complex but also other structurally less related (neo-)peptides, even in the groove of unrelated MHC molecules (also termed allo-reactivity) (reviewed in [17]).

It has been found that certain drug-reactive T cell clones from patients were MHC allele-unrestricted and at the same time highly drug-specific, i.e. they did not respond to drug derivatives with small chemical alterations. Other drug-induced T cell clones appeared less stringent with respect to the structure of the drug they recognized, but they were highly MHC allele-restricted. Still, other drug-induced T cells responded to MHC-peptide complexes in the absence of the initiating drug and in a MHC allele-unrestricted manner. So the specificity of drug-reactive T cells may range from highly drug-specific and non-MHC-restricted to highly MHC-restricted and nondrug-specific. From these findings it can be inferred that drug-induced T cells may also react with autoantigens through cross-reactivity.

Responses to Haptens Can Spread to Autoreactive Responses

Chemical modification of AUTOANTIGENS can also lead to autoreactive responses to unmodified self-epitopes by a mechanism unrelated to cross-reactivity. This may particularly concern antigens composed of neo- and self-epitopes (i.e. haptenated autoantigens), which conceivably would occur after exposure to a haptenating drug (the neo-part of the conjugated self-antigen), which will mostly encounter self-proteins in the body.

Haptenated AUTOANTIGENS can be recognized and internalized by B cells (which do not rely on MHC restriction to recognize antigens) specific either to the HAPTEN or to the unmodified B-cell EPITOPES on the haptenated autoantigen. These cells subsequently present a mixture of neo- and self-epitopes complexed to distinct class II major histocompatibility (MHC-II) molecules on their surface (Fig. 37.4). Since T-cell tolerance is obviously

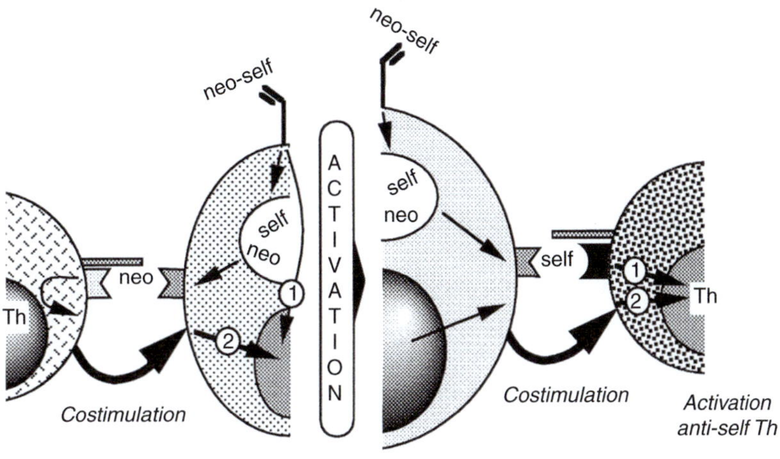

Fig. 37.4 Determinant spreading. Immune responses to haptens can spread to autoreactive responses to the carrier protein. Haptenated autoantigens are recognized and internalized by specific B cells. After uptake and processing, the B cells present a mixture of the neo- and self-epitopes complexed to distinct MHC-II molecules. Naive T cells do not respond to the self-epitopes, but neo-specific T cells provide costimulation for such B cells. This leads to activation of the B cell and production of antibodies. Moreover, when the B cells are activated, they can provide costimulation for naive T cells that recognize unmodified self-epitopes, leading to their activation and breaking of T cell tolerance. *1* antigen-specific or first signal, *2* costimulatory or second signal, *neo* hypersensitivity

not established for the NEO-EPITOPES (e.g. the hapten), neo-specific Th cells provide signal 2 to the B cell. This leads to production of either anti-hapten or anti-self antibodies depending on the exact specificity of the B cell. Moreover, once these B cells are activated, they can stimulate autoreactive Th cells that recognize unmodified self-epitopes. The underlying process is called EPITOPE (determinant) SPREADING and causes the diversification of adaptive immune responses. Responses induced by injection of mercury salts, for instance, are initially directed only to unidentified chemically created NEO-EPITOPES but after 3–4 weeks include reactivity to unmodified self-epitopes [18]. The distinction between allergic and autoimmune responses induced by haptens may therefore only be gradual, reflecting the relative antigenicity of the neo- and self-epitopes involved [19].

Pharmaceuticals Can Expose (Epitopes of) Autoantigens

Induction of self-tolerance involves specific recognition of autoantigen leading to selective inactivation of autoreactive lymphocytes, and tolerance is therefore not established for (EPITOPES of) AUTOANTIGENS that are normally not available for immune recognition. Pharmaceuticals can expose such sequestered (epitopes of) autoantigens by disrupting barriers between the antigen and the immune system (i.e. blood-brain barrier, blood-testis barrier, cell membranes). Tissue damage, cell death and protein denaturation induced or enhanced by pharmaceuticals can largely increase the availability of such (epitopes of) autoantigens for immune recognition (Fig. 37.5). Moreover, altered antigen processing, augmenting the presentation of previously undisclosed epitopes, increases the availability of these so-called subdominant or CRYPTIC EPITOPES for recognition by T cells [20, 21]. It has been shown, for instance, in mouse studies that antigen pre-incubated with Au(III), the oxidized metabolite of the anti-rheumatic auranofin, elicits additional T-cell responses to cryptic epitopes [19].

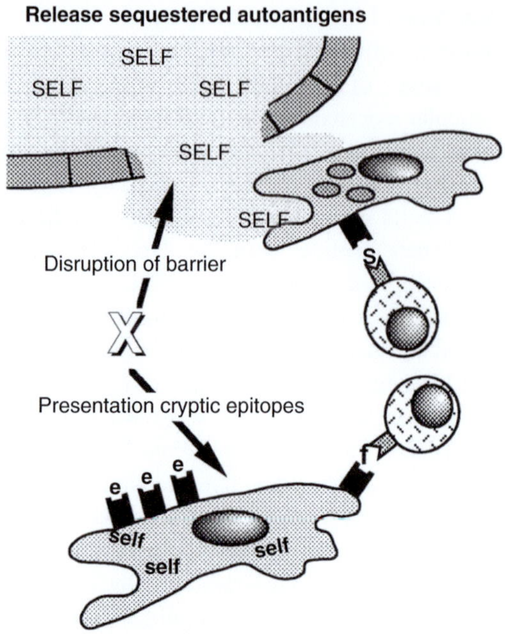

Fig. 37.5 Release of sequestered self-epitopes. Pharmaceuticals can expose previously sequestered (epitopes of) self-antigens by disrupting barriers between the antigen and the immune system (i.e. blood-brain barrier, blood-testis barrier, cell membranes). Similarly, augmented presentation of cryptic epitopes, as a result of altered antigen processing, increases the availability of these epitopes for recognition by T cells

37.2.3.2 Interference with Costimulation (Signal 2)

It is important to stress that antigen recognition by itself does not lead to activation of lymphocytes (above and Chap. 3) but that costimulation (i.e. signal 2) is required for the initiation of immune responses [22–24]. Many XENOBIOTICs have the inherent capacity to induce this costimulation; they have intrinsic adjuvant activity. For instance, immunostimulatory responses in mice induced by D-penicillamine and phenytoin could be inhibited by blocking costimulatory interactions (i.e. mediated by CD40–CD154) with a specific monoclonal antibody to CD154 [25]. The underlying mechanisms are not always understood, but several mutually non-exclusive possibilities have been described.

Induction of Inflammation

Cytotoxic pharmaceuticals or their reactive metabolites can induce tissue damage, which leads to accumulation of tissue debris and release of proinflammatory cytokines like tumour necrosis factor-α (TNF-α), interleukin-1 (IL-1) and IL-6, and attract inflammatory cells like granulocytes and macrophages. Cytokines produced during this inflammatory response activate antigen-presenting cells. These present selected EPITOPES of antigens from the debris and provide costimulation for Th cells, which leads to the initiation of an adaptive immune response [22]. Conceivably, all reactive and cytotoxic pharmaceuticals can have this effect to some extent. Side effects reported after the therapeutic use of cytokines have provided evidence that activation of the immune response may sometimes have deleterious consequences, such as flulike reactions, vascular leak syndrome and cytokine release syndrome. Cytokine-induced exacerbations of underlying autoimmune or inflammatory diseases may be other complications of concern [15]. The occurrence of cytokine release syndrome has also been reported as a serious consequence of the administration of liposomal CARRIERS or certain monoclonal antibodies (see below) [26].

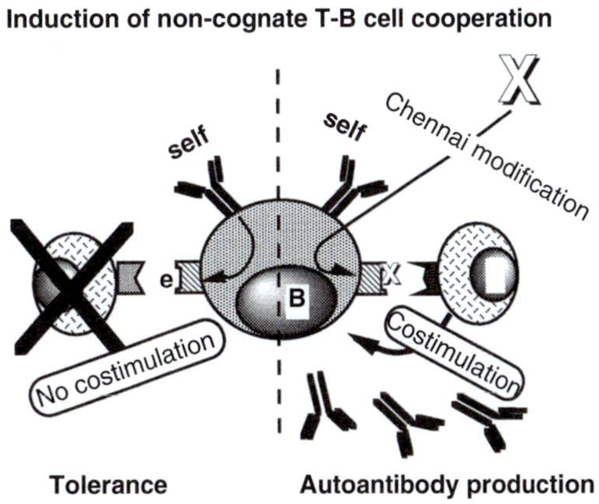

Fig. 37.6 Non-cognate T-B cell cooperation. Normally, activation of B cells requires costimulation from activated T cells that cognately recognize epitopes of the same antigen. B Cells with specificity for autoantigens do not receive this signal because T cells have learned to ignore autoantigens. However, certain pharmaceuticals can bind to B-cell proteins and Th cells activated by the neo-epitopes thus created can non-cognately provide costimulation for the (e-specific) B cell (e = a part of self for which tolerance exists). This can bypass tolerance in the T cell compartment and can lead to activation of autoreactive B cells and production of autoantibodies. This resembles stimulation of host B cells by graft T cells responding to the MHC molecules on the B cells during graft-versus-host reactions and leads to a similar spectrum of autoantibodies

Non-cognate T-B Cooperation

Reactive XENOBIOTICs may also stimulate adaptive immune responses by disturbing the normal cooperation between Th and B cells. Normally, B cells receive costimulation from Th cells that cognately recognize (epitopes of) the same antigen. As such, B-cell tolerance for AUTOANTIGENS is a corollary of the T-cell tolerance for such antigens. However, when Th cells respond to non-self epitopes on B cells, such B cells may be non-cognately stimulated by the Th cell. This occurs during graft-versus-host responses following bone marrow transplantation, when Th cells of the host recognize non-self epitopes on B cells of the graft and vice versa. This leads to T- and B-cell activation and results in production of autoantibodies to distinct autoantigens like DNA, nucleoli, nuclear proteins, erythrocytes and basal membranes. Drug-related lupus is characterized by a similar spectrum of autoantibodies, and it has therefore been suggested that non-cognate—graft-versus-host-like—T-B cooperation caused by T-cell reactivity to HAPTENs on (autoreactive) B cells is one of the underlying mechanisms (Fig. 37.6) [27]. Chlorpromazine, hydralazine, phenytoin, isoniazid, α-methyldopa and procainamide are just a few of the pharmaceuticals that are associated with drug-related lupus [28].

37.2.3.3 Interference with Cellular Activation

Occupation of various lymphocyte receptors results in a cascade of molecular processes that eventually lead to production of growth factors and cellular proliferation and/or activation. Its complexity makes this cascade vulnerable to pharmaceuticals at numerous stages, although most of these chemicals target very crucial processes like purine metabolism, as in the case of azathioprine. Most drugs that interfere with cellular activation are not cell type-selective at higher doses, as all living cells depend on the same basic molecular processes. At low doses, however, drugs like azathioprine, cyclophosphamide and methotrexate appear to have a more selective effect. Azathioprine, for instance, is claimed to selectively suppress T-lymphocyte function at low doses, whereas cyclophosphamide or methotrexate preferentially affects B cells [11].

Suppressive effects of cyclosporin A and tacrolimus, both interfering with the activation of the T-cell-specific transcription factor NF-ATc, and rapamycin, preventing IL-2 receptor activation, are obviously more specific to the T-lymphocyte.

37.2.4 Cytokine Release Syndrome

In 2006, eight healthy young male volunteers were enrolled in a first-inhuman clinical trial with TGN1412, a new anti-CD28 monoclonal antibody that was expected to specifically stimulate regulatory T cells but appeared to directly and polyclonally stimulate T cells. The serious life-threatening symptoms that six of them (two placebos) experienced were summarized as a cytokine storm [29]. Earlier cases of cytokine release syndrome were observed with muromonab (anti-CD3/Orthoclone OKT3; [30]) and alemtuzumab, CAMPATH 1H [31]. The mechanism behind this induction of cytokine release might be different among the various monoclonal antibodies. It is a challenge for the usually small companies developing new concepts in antibody-based therapeutic research to predict this type of adverse effect. Stebbings et al. [32] have developed an in vitro screening method based on cytokine responses to immobilize TGN1412 of human PBMC. Assays should be used rather as a hazard identification tool, followed by adequate risk mitigation strategies (e.g. use of corticosteroids). A positive outcome should not be a reason to discontinue product development [33]. Further research with a range of monoclonal antibodies has revealed that antibodies, which are entirely devoid of the potential to induce cytokine release, are rare. But TGN1412 is thus far the most potent without any doubt. It is important to note that some assays available in this respect are not sensitive enough. In 2015, TGN1412 has received renewed attention in the treatment of graft-versus-host disease [34, 35].

37.2.5 Regulation of the Immune Response

Ongoing immune responses have to be carefully regulated in order to mount the most suited defence (Fig. 37.1). Elimination of (intra)cellular targets, like virally infected or neoplastic cells, is most efficient by Th1-driven cellular responses using cytotoxic T cells and macrophages as effector mechanisms. Soluble targets, like extracellular bacteria and proteins, on the other hand, are most effectively eliminated by Th2-driven humoral responses, which rely on the formation of specific antibodies. The regulation of the type of immune response elicited and of the effector mechanisms activated is the result of a complex interplay of cytokines produced by macrophages, dendritic cells, mast cells, granulocytes and lymphocytes (Chap. 5) and is influenced by a number of endo- and exogenous factors. Genetic make-up, in particular genes encoding for MHC molecules, but also gender (oestrogens) are among the endogenous factors, whereas the type and dose of antigen, the route of exposure but also the type of (ongoing) costimulatory adjuvant activity are among the exogenous factors [34]. The role of the genetic make-up is illustrated by the strain-dependent effects of $HgCl_2$ in small laboratory animals. This chemical is capable of inducing a Th1-dependent *immunosuppressive* state in an H2d strain of mice or an RT1l strain of rat, whereas it induces an *autoreactive* Th2-dependent response in an H2s strain of mice or RT1n strain of rat [36, 37]. However, in some cases, the outcome of the response may also depend on the chemical, and not so much on the strain. For instance, $HgCl_2$ induces a Th2-like response in BALB/c mice, whereas the diabetogenic antitumour compound streptozotocin (STZ) induces a Th1 response in the same mouse strain [38]. Other examples are the adjuvants, complete Freund's adjuvant (CFA) and alum, which stimulate the formation of immunoglobulin (Ig)G2a and IgG1/IgG2a isotypes of antibodies, respectively [34].

How chemicals exactly modulate the immune response is largely unknown, but modulation of EPITOPE selection by MHC molecules, selective activation of the innate immune system (e.g. mast cells in the case of $HgCl_2$) and chemical-specific factors (e.g. macrophage activation by STZ) as well as interaction with other environmental factors (microbial influences) may all contribute to the ultimate immunotoxicological effect.

The discovery of Toll-like receptors as mediators of regulation of immunologic response has led to the rational development of immunostimulatory compounds, with the goal to enhance efficacy of vaccination, protective responses against infectious antigens. On the other hand, vaccine adjuvants are associated with a reputation of adverse effects in the lay press because of individual reports of autoimmune-like adverse reactions (see [39]).

Apart from regulation of the type of immune response, the occurrence of an immune response per se is tightly controlled. Recently, a number of regulatory T-cell subsets have been identified that suppress autoaggressive responses and ALLERGY [40]. Studies indicate that regulatory cells are also involved in tolerance to orally encountered XENOBIOTICs. For instance, it has been demonstrated that regulatory $CD4^+CD25^+FOXP3^+$ T cells prevent autoantibody production by procainamide, gold sodium thiomalate or $HgCl_2$ in mice [41]. Similarly, induction of tolerance to adverse immune effects by D-penicillamine appears at least partly mediated by other T cells as well, possibly interferon-γ-producing $CD8^+$ T cells [42].

37.2.6 Effector Functions and Response Cessation

To avoid unnecessary damage, the immune system has several feedback mechanisms to stop ineffective and obsolete responses (Fig. 37.1). The simplest feedback is the antigen itself, as complete degradation of the response-inducing antigen usually leads to response cessation. Pharmaceuticals that impair the activity of effector mechanisms delay antigen degradation and lead to accumulation of debris. It has been demonstrated, for instance, that several drugs, including D-penicillamine and procainamide, inhibit complement factor C4. This hampers the clearance of immune complexes and may therefore lead to their deposition and excessive tissue damage (reviewed in [43]).

Some drugs may also directly stimulate effector mechanisms, such as the complement system (contrast media) or effector cells, such as mast cells and basophils. For instance, the antibiotic vancomycin may induce histamine release, and some non-steroidal anti-inflammatory drugs (NSAIDs) such as aspirin directly modulate the arachidonic acid pathway. In any case, anaphylactoid clinical effects may occur without involvement of specific immune recognition. For those cases, in which effector mechanisms are directly altered by drugs and clinically apparent allergic symptoms are seen, the term pseudoallergy is used (reviewed in [44]).

37.2.7 Clinical Consequences of Immunotoxicity of Pharmaceuticals

In general, pharmaceuticals that inhibit cellular replication or activation induce IMMUNOSUPPRESSION, which is dose-dependent. Particularly, impaired activity of the first line of defence formed by the natural immune system can have disastrous consequences. These are generally not influenced by the genetic predisposition of the exposed individual, but an actual outbreak of (opportunistic) infections or increased frequency of neoplasms may depend on the general immune status prior to exposition. This explains why immunosuppressive pharmaceuticals are most likely to have clinical consequences in immunocompromised individuals such as young children, the elderly and transplant recipients.

Immunotoxic pharmaceuticals that somehow activate the immune system can lead to autoimmune or allergic diseases. Although on a drug-by-drug basis immune-mediated adverse effects are mostly rare, overall, it is estimated that 6–10% of all adverse drug reactions are immune-mediated and these immune-mediated adverse effects are the most frequent cause of failure during clinical development [45]. Drug-induced ALLERGY and autoimmune phenomena may have very serious clinical outcomes, such as acute liver diseases (also designated, drug-induced liver injury, DILI), BLOOD

DYSCRASIAS and skin diseases. But in fact, all organs can be subject to clinical consequences induced by drugs, in which case a multi-organ syndrome can be seen. In these cases, autoimmune phenomena may be initiated by drug-specific reactions [46].

ADJUVANT system-induced autoimmunity (ASIA) is a group of autoimmune-related adverse responses said to be associated with adjuvant-like approaches. There is a huge societal power criticizing more vaccination in children because of deteriorating effects. There is, however, no compelling evidence thus far that specific adjuvants are causally related to AUTOIMMUNITY phenomena [39].

DILI may occur in various pathological types (e.g. cholestasis, steatosis or hepatocellular damage) and involve various mechanisms. DILI is often associated with other adverse effects, e.g. skin rashes of cytopenias, adding to the suggestion that the immune system is involved in DILI, in particular the idiosyncratic type. Other well-known examples of drugs causing DILI include halothane, diclofenac and a number of fluoroquinolones. Although the incidence of idiosyncratic DILI per drug is often low, DILI as a whole appears responsible for 12% of all liver diseases and is a huge concern for the pharma industry for every new drug [47]. Preclinical prediction of DILI, in particular the idiosyncratic type of DILI, is subject to many cofounding factors. Nevertheless, various clear examples of animal studies, in rat and mouse, with a range of drugs show that idiosyncratic DILI can be detected using specific protocols, in which animals are co-exposed to microbial substances such as LPS [7] or in which immunoregulatory mechanisms are inhibited [48, 49]. Although none of these models is standardized for testing series of drugs, they have demonstrated an important contribution of the immune system in the development of the disease.

Actual development of clinical symptoms in most of the above cases is influenced by the route and duration of exposure, gender, the dosage of the pharmaceutical, co-exposures (e.g. LPS) and immunogenetic (MHC haplotype, Th1-type versus Th2-type responders) and pharmacogenetic (acetylator phenotype, sulphoxidizer, Ah receptor, etc.) predisposition of the exposed individual. From an immunological point of view, it is clear that the polymorphic MHC molecules select the EPITOPES that are presented to T cells and therefore influence all immune responses, including allergic and autoimmune responses induced by pharmaceuticals. Moreover, atopic individuals that tend to mount Th2 immune responses are more susceptible to anaphylaxis triggered by an IgE response to chemical HAPTENS than typical Th1 responders. Genetic variation in metabolism of pharmaceuticals is important as it determines the formation and clearance of immunotoxic metabolites. The slow acetylating phenotype, for instance, predisposes to drug-related lupus because reactive intermediates of phase I metabolism have an increased opportunity to bind proteins as they are only slowly conjugated.

Drugs known to interfere with the immune system are normally prescribed by well-trained physicians and are taken under more or less controlled conditions. As a result, adverse effects should be recognized as soon as they become apparent and measures can be taken before permanent harm is done. However, in the case of allergenic drugs, the immune system is sensitized, which may hamper future treatments with the same or a structurally related chemical (penicillin). In the case of strong immunosuppressant drugs, used, for instance, to prevent allograft transplant rejection or as anti-inflammatory treatment, an immunocompromised state is taken for granted. However, prolonged and severe IMMUNOSUPPRESSION may cause permanent detrimental effects when (other) neoplasms are formed or when (opportunistic) infections occur. In the latter case, precautions (e.g. antimicrobial compounds) have to be taken. In case of mild immunosuppressant drugs, more subtle changes in resistance to infections, such as influenza or common cold, may occur.

As the clinical consequences of exposure to immunotoxic pharmaceuticals range from immunodepressed conditions, on the one hand, to allergic and autoimmune diseases, on the other hand, preclinical testing of pharmaceuticals in laboratory animals requires different approaches. In the following sections, procedures are covered that comprise direct immunotoxicity as well as sensitizing capacity.

37.3 Procedures for Preclinical Testing of Direct Immunotoxicity

37.3.1 Testing in Rodents by Tiered Approach

Several laboratories have developed and validated a variety of methods to determine the effects of chemicals on the immune system of rats and mice [50]. Most employ a tier-testing system, whereas some investigators have advocated multiple testing in a single animal. The tier-testing approaches are similar in design, in that the first tier is a screen for immunotoxicity with the second tier consisting of more specific or confirmatory studies, host resistance studies or in-depth mechanistic studies. At present, most information regarding these models comes from the model developed at the National Institute for Public Health and the Environment (RIVM) in Bilthoven, the Netherlands, and the model developed at the US National Institute of Environmental Health Sciences National Toxicology Program (NIEHS-NTP). The RIVM tiered system [51, 52] is based on the guideline 407 of the Organization for Economic Co-operation and Development (OECD) and performed in the rat using at least three dose levels, i.e. one resulting in overt toxicity, one aimed at producing no toxicity and one intermediate level. There is no immunization or challenge with an infectious agent. The first tier comprises general parameters, including conventional haematology, serum immunoglobulin concentrations, bone marrow cellularity, weight and histology of LYMPHOID ORGANS [thymus, spleen, lymph nodes, MUCOSA-ASSOCIATED LYMPHOID TISSUE (MALT)], flow cytometric analysis of spleen cells and possibly immunophenotyping of tissue sections (Table 37.1). This approach has been used for the immunotoxic evaluation of pesticides [53] and pharmaceuticals [54].

Table 37.1 Panel of the Dutch National Institute for Public Health and the Environment for detecting immunotoxic alterations in the rat

Parameters	Procedures
Tier 1	
Non-functional	– Routine haematology, including differential cell counting
	– Serum IgM, IgG, IgA and IgE determination
	– Lymphoid organ weights (thymus, spleen, local and distant lymph nodes)
	– Histopathology of the thymus, spleen, lymph nodes and mucosa-associated lymphoid tissue
	– Bone marrow cellularity
	– Analysis of lymphocyte subpopulations in the spleen by flow cytometry
Tier 2 panel	
Cell-mediated	– Sensitization to T-cell-dependent antigens (e.g. oval-immunity bumin, tuberculin, *Listeria*) and skin test challenge
	– Lymphoproliferative responses to specific antigens (*Listeria*), mitogen responses (Con-A, PHA)
Humoral immunity	– Serum titration of IgM, IgG, IgA and IgE responses to T-dependent antigens (ovalbumin, tetanus toxoid, *Trichinella spiralis, sheep red blood cells*) with ELISA
	– Serum titration of T-cell-independent IgM response to LPS with ELISA
	– Mitogen response to LPS
Macrophage function	– In vitro phagocytosis and killing of *Listeria monocytogenes* by adherent spleen and peritoneal cells
	– Cytolysis of YAC-1 lymphoma cells by adherent spleen and peritoneal cells
NK cell function	– Cytolysis of YAC-1 lymphoma cells by non-adherent spleen and peritoneal cells
Host resistance	– *Trichinella spiralis* challenge (muscle larvae counts and worm expulsion)
	– *Listeria monocytogenes* challenge (spleen clearance)
	– Rat cytomegalovirus challenge (clearance from salivary gland)
	– Endotoxin hypersensitivity
	– Autoimmune models (adjuvant arthritis, experimental allergic encephalomyelitis)

The OECD guideline 407 includes the weight of the spleen and thymus and histopathology of these organs, in addition to the lymph nodes, PEYER'S PATCHes and bone marrow [55]. But it should be borne in mind that in this guideline 407, some immunotoxic compounds may not be identified as such [56]. For instance, the opiate analgesic buprenorphine [57] and the long-acting b2-adrenoreceptor agonist salmeterol [58] affect serum immunoglobulins in rats. Results of an interlaboratory validation study in the rat, using cyclosporin A and hexachlorobenzene, showed the importance of updating the 407 protocol with LYMPHOID ORGAN weights and serum immunoglobulin levels. The study concluded that histomorphological examination of lymphoid tissues resulted in the most reliable and sensitive data to be considered in regulatory toxicology and risk assessment [59].

In this OECD guideline 407 for testing toxicants, the immune system is not evaluated functionally. The inclusion of an in vivo antigen challenge test, e.g. with sheep red blood cells (SRBC), is currently considered to improve the sensitivity of the toxicity test. Experimental results [60] indicate that intravenous injection with SRBC during a 30- and 90-day toxicity study did not alter haematological and clinical chemistry parameters. With the expected exception of the spleen, administration of SRBC did not significantly alter the weights or morphology of routinely analysed tissues.

It should be noted that the array of tests currently included in the updated OECD guideline 407 is aimed at detecting potential immunotoxicity. Once immunotoxicity has been identified, further testing is required to confirm and extend the earlier findings. Further testing should include immune function testing (Table 37.1). In addition to confirming functional implications of the immunotoxicity identified, functional tests will likely provide information on no-adverse-effect levels and are therefore valuable for the process of risk assessment. Caution is needed in determining the relevance of slight effects on immune parameters in view of the functional reserve capacity of the immune system. In those cases, infection models can be very helpful for risk assessment, as they are tools to elucidate the actual consequences of disturbances of immune function; effects observed using such infection models have surpassed the reserve capacity of the immune system. The fate of the pathogen and the associated host pathology may serve as indicators of the health implications of the immunotoxicity of the test chemical. Pathogens used in these host resistance models are chosen so that they are good models for human disease [42]. With some compounds, induction of immunotoxicity occurs especially during prenatal exposure. Yet, so far, there are no immune parameters included in current OECD Guidelines for developmental or reproductive toxicity testing [61].

The US NTP has developed a tiered approach in mice that is linked closely to the standard protocol for chronic oral toxicity and carcinogenicity studies [62]. Routinely, exposure periods of 14–30 days have been used at dose levels that have no effect on body weight or other toxicological endpoints. In this way, compounds are identified for which the immune system represents the most sensitive target organ system. Tier 1 includes conventional haematology; LYMPHOID ORGAN weight; cellularity and histology of the spleen, thymus and lymph nodes; ex vivo splenic IgM-antibody plaque-forming cell assay following SRBC immunization; in vitro lymphocyte proliferation after stimulation with MITOGENs and allogeneic cells; and an in vitro assay for natural killer (NK) cell activity. In an adapted form of this approach, 51 different chemicals were evaluated, selected on the basis of structural relationships with previously identified immunotoxic chemicals [63]. The splenic SRBC IgM plaque-forming cell response and cell surface marker analyses showed the highest accuracy for identification of potential immunotoxicity.

In 2004, a study on the interlaboratory reproducibility of extended histopathology was published, evaluating the thymus, spleen and mesenteric lymph node of these past NTP studies, which were performed in the mouse using ten chemicals and three positive controls [64]. The consistency was examined between four experienced toxicological pathologists with varied expertise in immunohistopathology. Agreement between pathologists was highest in the thymus, in particular when evaluating thymus cortical cellularity, good in spleen follicular cellularity and in spleen and lymph node germinal centre development and poorest in spleen red pulp changes. The ability to identify histopathological changes in lymphoid tissues was dependent upon the experience/training that the individual possessed in examining lymphoid tissue and the apparent severity of the specific lesion. In a further study, the accuracy of extended histopathology to detect immunotoxic chemicals was investigated [65]. While overall there was good agreement between histopathology and functional tests, the antibody-forming cell (AFC) assay detected immune suppression in two instances where no changes in pathology were indicated. In contrast, the AFC assay failed to detect oxymetholone as an IMMUNOTOXICANT, although extended histopathology indicated immunological changes. These data suggest that, while not as sensitive as functional tests, extended histopathology may provide a reasonable level of accuracy to identify immunotoxic chemicals. In the NTP protocol, in contrast to the OECD 407 guideline, the high dose was selected so as not to produce overt toxicity, thus limiting the likelihood of producing severe histopathological effects.

37.3.2 Immunotoxicity Testing in Non-rodent Species

Various non-human primates, including *Macaca mulatta* (rhesus macaque), *Macaca nemestrina* (pig-tailed macaque), *Macaca fascicularis* (cynomolgus monkey) and the marmoset, have been used in immunotoxicological studies. Virtually all of the immunotoxicology assays which are carried out in the mouse or rat can be and have been adapted for use with the non-human primates [66, 67]. Phenotypic markers and functional assays in three different species of non-human primates were evaluated [68]. Functional assays included NK-cell activity, lymphocyte transformation and antigen presentation. The extensive phenotypic marker studies included the evaluation of over 20 markers or combination of markers for each of the three monkey species. Otherwise, strategies and methods applied in studies in humans have been introduced in studies on non-human primates (see Table 37.2).

Also, other mammalian species have been used. While dogs are not the species of choice for immunotoxicological studies, they are one of the species predominantly used in toxicological safety assessments. Virtually all of the assays used for assessing immunotoxic potential have been adapted for use in the dog [50]. Among these are assays to evaluate basal immunoglobulin levels for IgA, IgG and IgM, ALLERGEN-specific serum IgE, mononuclear phagocyte function, NK-cell activity, cytotoxic T-cell activity and MITOGEN and cell-mediated immune responses.

Table 37.2 Assays recommended for immunotoxicity assessment in humans

1. Complete blood count with differential count
2. Antibody-mediated immunity (one or more of the following):
• Primary antibody response to protein antigen (e.g. epitope-labelled influenza vaccine)
• Immunoglobulin concentrations in serum (e.g. IgM, IgG, IgA, IgE)
• Secondary antibody response to protein antigen (e.g. diphtheria, tetanus or polio)
• Natural immunity to blood group antigens (e.g. anti-A, anti-B)
3. Phenotypic analysis of lymphocytes by flow cytometry:
• Surface analysis of CD3, CD4, CD8 and CD20
4. Cellular immunity:
• Delayed-type hypersensitivity (DTH) skin testing
• Primary DTH reaction to protein (KLH)
• Proliferation to recall antigens
5. Autoantibodies and inflammation:
• C-reactive protein
• Autoantibody titres to nuclei (ANA), DNA, mitochondria and IgE (rheumatoid factor)
• IgE to allergens
6. Measure of non-specific immunity:
• NK cell enumerations (CD56 or CD60) or cytolytic activity against K562 tumour cell line
• Phagocytosis (NBT or chemiluminesce)
7. Clinical chemistry screen
• Proposal for all persons exposed to immunotoxicants

From [30]

In 2014, experience gained from studies in Göttingen-Ellegaard Minipigs was published [69]. Using KLH as a T-cell-dependent antigen, the authors characterized the effects of cyclosporin A and dexamethasone, revealing a similar response of the minipig immune system to that observed earlier in other non-rodent species. The minipig is a non-clinical species that is receiving growing attention as an additional choice for toxicological testing [70]. Due to the commercial interest in pigs, there is a great amount of available knowledge on the pig immune system, which is applicable also to the minipig [71].

37.4 Procedures for Preclinical Testing of Sensitizing Capacity

37.4.1 Structure-Activity Relationships

The intrinsic capacity of chemicals to exert adverse effects is linked to the structure of the compound. Structure-activity relationships with respect to direct toxicity of compounds to components of the immune system have received little attention. More attention has been given to structure-activity relationships with respect to the induction of ALLERGY. Here, structure-activity relationship models are directed towards a fuller understanding of the relationship between chemical structure and physicochemical properties and skin-sensitizing activity, in order ideally to derive quantitative structure-activity relationships (QSAR), linked perhaps to the development of expert, rule-based systems. In this context, parameters that appear to be of particular importance are protein reactivity and lipophilicity associated with the capacity to penetrate into the viable epidermis [72]. The correlation of the protein reactivity of chemicals with their skin SENSITIZATION potential is well established [73], so that it is now accepted that if a chemical is capable of reacting with a protein, either directly or after appropriate (bio)chemical transformation, it has the potential to be a contact ALLERGEN, assuming of course that it can accumulate in the appropriate epidermal compartment. Each of the existing structure-activity relationship (SAR) models proposes structural alerts, i.e. moieties associated with sensitizing activity. In all cases, the structural alerts comprise electrophilic moieties, or moieties which can be metabolized into electrophilic fragments (proelectrophiles).

37.4.2 Testing for Skin Allergy

37.4.2.1 Guinea Pig Models

The guinea pig was for many years the animal of choice for experimental studies of contact SENSITIZATION, and several test methods were developed in this species (reviewed in [74]). The best-known and most widely applied test methods are the *Buehler test*, the *guinea pig maximization test*, and the *guinea pig optimization test*, which have formed the basis of hazard assessment for many years. Both the guinea pig maximization test and the Buehler test are now recommended according to an OECD guideline, accepted in 1992. While these tests differ with respect to procedural details, they are similar in principle. Guinea pigs are exposed to the test material or to the relevant vehicle. In the Buehler test, both induction and challenge exposures are done topically. The Buehler test is sensitive, but false negatives are frequently observed. The test was improved by occluded application of the test compound. In the guinea pig maximization test, induction involves intradermal and occluded epidermal exposure, and in the optimization test, induction is done by intradermal, challenged by intradermal and occluded epidermal exposure. ADJUVANT is employed in the guinea pig maximization test and the optimization test to augment induced immune responses. Challenge-induced inflammatory reactions, measured as a function of erythema and/or oedema, are recorded 24 and 48 h later. Classification of sensitizing activity in the guinea pig tests is qualitative and not quantitative. It is based usually on the percentage of test animals that display macroscopically detectable challenge reactions. Any compound resulting in positive induction in at least 30% of the animals in an adjuvant test is labelled as a sensitizer; in the case of a non-adjuvant test, 15% positivity is sufficient to classify the compound as a sensitizer.

37.4.2.2 Mouse Models

In the recent years, increased understanding of the cellular and molecular mechanisms associated with contact ALLERGY has been derived largely from experimental investigations in the mouse [75, 76]. The most important test to be developed in mice is the *local lymph node assay* (LLNA) [77]. In contrast to the guinea pig assays described above, activity in the LLNA is measured by the primary T-cell response in the draining lymph node following topical application to the mouse ear. Mice are treated daily, for 3 consecutive days, on the dorsum of both ears, with the test material or with an equal volume of vehicle alone. Proliferative activity in draining lymph nodes (measured by the incorporation in situ of radiolabelled thymidine) is evaluated 5 days following the initiation of exposure. Currently, chemicals are classified as possessing sensitizing potential if, with one test concentration, a stimulation index of 3 or greater, relative to vehicle-treated controls, is induced. The method has the advantages of short duration and objective measurement of proliferation and minimal animal treatment. In contrast to guinea pig assays, activity is measured as a function of events occurring during the induction, rather than elicitation phase of contact SENSITIZATION.

Risk assessment of sensitizing chemicals requires, besides hazard identification, the assessment of potency. By using dose-response modelling (employment of a regression method that includes determination of the uncertainty margins) in the LLNA test, the potency of sensitizing chemicals can be determined, thus offering a possibility for classification [78]. The LLNA has been developed further to discriminate skin sensitizers from respiratory sensitizers based on the induction of $CD4^+$ T helper subsets (Th1- versus Th2-mediated responses) by the analysis of cytokine profiles in draining lymph node cells [79]. Chemicals differ with respect to the types of HYPERSENSITIVITY they induce. Compounds that induce Th1 cells and mediate type IV-delayed hypersensitivity are generally skin sensitizers. Such responses are associated with the production by draining lymph node cells of interferon-γ (IFNγ). Compounds that induce Th2 cells and mediate type I-immediate hypersensitivity by the production of IgE and IgG1 are generally respiratory sensitizers and are associated with the production by draining lymph node cells of high levels of IL-4. However, this is not true in all cases, as skin sensitization with some low molecular compounds such as picrylchloride [80] and toluene diisocyanate (TDI) [81] can induce respiratory hypersensitivity with features of type IV hypersensitivity in mice. Also, in humans, specific IgE is only detected in a minority of patients suffering from respiratory allergy induced by TDI. Some while ago, it was proposed that, by direct linkage of proliferation and cytokine production, in a dose-response manner, distinction of contact ALLERGENs from respiratory allergens may be improved [82].

37.4.3 Testing for Respiratory Allergy

Most of the animal models that are used for studying specific respiratory tract HYPERSENSITIVITY were developed using high molecular weight ALLERGENS, notably proteins. Very few animal models have been developed as predictive tests for hazard identification and risk assessment in the area of chemically induced respiratory ALLERGY [83]. The majority of these models are based on antibody-mediated events. The models differ with regard to the following aspects: the animal species utilized, the route of administration of the agent, the protocol for both induction and elicitation of responses, type of response measured and judgment of significant response.

37.4.3.1 Guinea Pig Models

The guinea pig has been used for decades for the study of anaphylactic shock and pulmonary HYPERSENSITIVITY. The guinea pig is similar to humans in that the lung is a major shock organ for anaphylactic responses to antigens. The guinea pig responds to histamine and can experience both immediate-onset and late-onset responses. Airway hyperreactivity and eosinophil influx and inflammation can also be demonstrated in this animal species. Mechanistic studies have been hampered by the lack of reagents needed to identify cells and mediators in respiratory ALLERGY. In addition, the major anaphylactic antibody is IgG1, whereas it is IgE in humans and other rodent species.

The guinea pig model developed by Karol et al. [84] has proven to be valuable for low molecular weight chemical ALLERGENS. Guinea pigs sensitized by inhalation of free or protein-bound chemical allergens, such as TDI, will exhibit symptoms of pulmonary hypersensitivity following subsequent inhalation challenge. Hypersensitivity reactions are measured, usually as a function of challenge-induced changes in respiratory rate or alterations in other breathing parameters such as tidal volume. Changes in breathing patterns can also be provoked in dermally sensitized guinea pigs by inhalation challenge with the free chemical. In this approach, it is not necessary to use HAPTEN-protein conjugates.

A tiered approach to hazard assessment in guinea pigs proposed by Sarlo and Clark [85] comprises sequential analyses of physicochemical similarities with known allergens, the potential to associate covalently with protein, the ability to stimulate antibody responses and finally, activity in a model of respiratory hypersensitivity in which animals sensitized by subcutaneous injection are challenged by intratracheal instillation.

37.4.3.2 Mouse Models

Models to investigate airway responses to sensitizing compounds have been developed in the mouse and comprise responses mediated by IgE [86] and non-IgE-mediated reactions [80, 81]. These models have not been used so far for predictive purposes.

As discussed earlier, analysis of the cytokine profile in the mouse LLNA may provide information on whether a compound is a respiratory ALLERGEN. In the same series of investigations, it was found that topical administration to mice of chemical respiratory allergens stimulated a substantial increase in the serum concentration of total IgE, a response not seen with contact allergens considered to lack the ability to cause SENSITIZATION of the respiratory tract [79]. These observations suggested that it might be possible to identify chemical respiratory sensitizers as a function of induced changes in serum IgE concentration. The advantage of this approach, which forms the basis of the mouse IgE test, is that measurement of a serum protein is required rather than of HAPTEN-specific antibody.

Investigations suggest that the mouse IgE test may provide a useful method for the prospective identification of chemical respiratory allergens [85]. It must be emphasized, however, that to date the assay has been evaluated only with a limited number of chemicals and that most of the analyses have been performed in a single laboratory. Difficulties arise from the assumption of IgE mediation of respiratory HYPERSENSITIVITY response in mice. As mentioned earlier, respiratory allergic responses, associated with increased reactivity of airways, may occur by a delayed type IV immune response-inducing compound [82]. For this reason, actual testing of lung functions in vivo seems prudent for those chemicals that are known to sensitize but are unable to produce IgE responses.

37.4.4 Testing for Autoimmunity and Drug Allergy

Drugs may elicit allergic responses to the drug itself (as HAPTEN-CARRIER complex) or induce autoimmune responses. As the distinction between allergic and autoimmune responses induced by haptens may be gradual, depending on multifactorial aetiology, it is difficult, if possible at all, to discriminate between allergenic and autoimmunogenic potential of drugs. In addition, clinical outcomes may differ from one to the other drug and from one to the other individual. Probably, individual sensibilities and circumstances are decisive here. Predictive models, which mimic the complete development of AUTOIMMUNITY or drug ALLERGY, are not available, and prediction in these cases may depend on a set of well-designed tests.

37.4.4.1 Induced and Genetic Models

For the detection of the potential of compounds to exacerbate induced or genetically predisposed autoimmunity, a range of animal models is available [87]. In induced models, a susceptible animal strain is immunized with a mixture of an adjuvant and an autoantigen isolated from the target organ. Examples are *adjuvant arthritis*, *experimental encephalomyelitis* and *experimental uveitis* in the Lewis strain rat. Examples of spontaneous models of autoimmune disease are the BB rat and the NOD mouse that develop autoimmune *pancreatitis* and subsequently *diabetes* and the (NZBxNZW) F1 mouse or MRL/lpr mouse that develop pathology that resembles human *systemic lupus erythematosus*. These models are mainly used in the study of the pathogenesis of AUTOIMMUNITY and the preclinical evaluation of immunosuppressive drugs. Very few studies have addressed the potential of these models to assess whether a drug (or chemical) exacerbates autoimmune reactions or disease. An animal model that has been under scrutiny for some time now in this respect is the Brown Norway rat model [88]. In this rat strain, penicillamine and nevirapine induce various immunological effects (formation of autoantibodies) and skin rashes as typical clinical effects. Other suspected drugs (captopril and felbamate) appeared to be ineffective in this model.

All these models may eventually help to understand underlying mechanisms, but may not be suitable as stand-alone predictive assays.

37.4.4.2 Popliteal Lymph Node Assay

Although, currently, no predictive assays have been developed and validated to identify the potential of drugs to induce drug ALLERGY or autoimmune responses, it should be noted that available assays to identify contact sensitizers may also be helpful to identify systemic sensitizers. Clinical signs of systemic adverse immune-mediated effects usually become manifest only during advanced clinical development of drugs. The conditions used in routine preclinical toxicological screening are obviously not optimal for the detection of allergic or autoimmunogenic potential of drugs and chemicals (e.g. small animal number, use of outbred animal strains, dynamics of disease development versus snapshot determinations, lack of predictive parameters).

Autoimmunity often results from the interaction of the compound with normal tissue components, thereby rendering them immunogenic. A variety of chemicals and drugs, in particular the latter, have been found to induce autoimmune-like responses [89]. For the detection of chemicals that produce this type of reaction, the popliteal lymph node assay (PLNA) in mice is a possible tool. The PLNA [90] is based on the hyperplasia (increase in weight) of lymph nodes observed in experimental graft-versus-host reactions and has been modified to assess the immunomodulatory potential of drugs. The test substance is injected subcutaneously into one hind footpad, and the contralateral side is either untreated or inoculated with vehicle alone. Comparison of popliteal lymph nodes from both sides allows the effect of the test drug to be measured. Apart from differences in weight, histological evidence of in vivo immunostimulatory activity can be discerned. These pseudo-graft-versus-host reactions with follicular hyperplasia have been documented in mice for drugs such as diphenylhydantoin, D-penicillamine and streptozotocin. The assay appears to be appropriate to recognize sensitizing, i.e. allergenic and autoimmunogenic chemicals, as well as non-sensitizing immunostimulating compounds, and has important advantages as it is a simple model based on local reactions that indicate direct immunostimulation with less interference by immunoregulatory mechanisms. So far, many compounds (mainly pharmaceuticals and structural homologues) have been tested in the PLNA, and outcomes (i.e. PLN-index: ratio of weight and cell numbers of PLN of compound-treated over vehicle-injected animals) appear to correlate well with documented adverse immune effects in humans [91]. Caution must be exercised in the case of autoimmunogenic drugs, such as procainamide, that act as prohaptens. They are false-negative in the PLNA, as such, and require co-injection of metabolizing systems (S9 mix of granulocytes) to become positive.

Results of preliminary interlaboratory validation studies indicate the potential predictive value of the PLNA or variations of it using other sites of injection, in the mouse [92] as well as in the rat [93]. Thus, the direct PLNA seems to be a versatile tool to recognize T-cell-activating drugs and chemicals, including autoimmunogenic chemicals, bearing in mind the possible false-negative results with prohaptens. With the adoptive transfer PLNA, sensitized cells are used as probes to detect the formation in vivo of immunogenic metabolites of low molecular weight chemicals [91].

A modification of the PLNA uses bystander antigens to report the nature and type of immune response that is elicited by a given pharmaceutical [94]. In this assay, so-called reporter antigens (RA), either TNP-Ficoll or TNP-ovalbumin (TNP-OVA), are injected together with the compound of interest, and the IgG-response to the RA is measured. Here, it is important that unlike TNP-OVA, TNP-Ficoll cannot directly induce specific T cell activation. The IgG response to TNP-Ficoll, however, is susceptible to neo-antigen-specific T cell help, implying that an IgG response to TNP-Ficoll indicates that a co-injected compound has immunosensitizing potential. If a compound increases an IgG response to TNP-OVA and not to TNP-Ficoll, it can be concluded that the compound has the capacity to act merely as an adjuvant to an immune response. So, dependent on the type of RA, the IgG measured in this so-called RA-PLNA indicates whether compound-induced immunostimulation involves immunosensitization or proinflammatory adjuvant activity. The RA approach can also be applied in animal tests that use oral exposures to drugs [95].

It is important to note that the PLNA in any of its forms is essentially a hazard identification test that can indicate whether a compound has the potency to induce allergic or autoimmune phenomena in man.

37.5 Procedures for Immunotoxicity Testing in Humans

37.5.1 Epidemiology Design

It is obvious that many of the compounds causing direct immunotoxicity have been identified in rodent studies, as the database in humans is less complete and often inconclusive. The most common design used in immunotoxicity research in humans is the cross-sectional study, in which exposure parameters and effect parameters are assessed at the same time point [96]. The immune function of "exposed" subjects is compared with the immune function of "non-exposed" subjects by the measurement of various immunological parameters. For this reason, proper definition of exposure criteria in the exposed group is necessary. This group should include subjects at the upper end of exposure. Where possible, the study should incorporate individual estimates of exposure or actual measurements of the compound. In the broadest sense, biomarkers are measurements on biological specimens that will elucidate the relationship between environmental exposure and human diseases, so that exposure and diseases can be prevented. In clinical medicine, biomarkers are valued as surrogates for the presence or absence of diseases or the course of the disease during therapeutic intervention. As such indicators are available for *exposure, effect* or *susceptibility* [97].

37.5.1.1 Markers of Exposure

A biological marker of exposure is the presence of a XENOBIOTIC compound or its metabolite or the product of an interaction between the compound and some target cell or biomolecule. The most common markers of exposure are the concentration of the compound in urine, blood or target organ or tissue. Immune-specific biomarkers of exposure are antibodies or positive skin tests to the particular compound.

37.5.1.2 Markers of Effect

A biomarker of effect is a measurable cellular or biochemical parameter that, depending on magnitude, can be used to recognize and identify an established or potential health impairment or disease. These range from markers of slight structural or functional changes to markers that are indicators of a subclinical stage of a disease or the manifestation of the disease itself. Functional changes in cells of the immune system by an immunotoxic chemical may be the first step in the process towards disease. For instance, longitudinal studies on asymptomatic individuals with low NK activity showed that these individuals had an increased risk for upper respiratory infection and morbidity [98]. Immunosuppression may lead to more subtle changes in resistance to infections, such as influenza or common cold, rather than opportunistic infection. Data in experimental animals also indicate that small changes in immune function could increase the likelihood of disease [63].

37.5.1.3 Markers of Susceptibility

Markers of susceptibility, also called effect modifiers, can act at any point along the exposure-disease continuum. Important sources of variability are genetic, endocrine, age-related and environmental factors. Over the last two decades, it has become clear that many immunological disorders are linked to alleles of the major histocompatibility gene complex (MHC). The products of MHC alleles in humans [human lymphocyte antigens (HLA)] have aroused interest at a clinical level as potential biomarkers of disease susceptibility. In some instances, there is a remarkable increase in relative risk of disease in individuals possessing particular alleles. Similar associations have been described in drug-induced immunological disorders. However, it should be noted that other genetic factors as well as environmental factors are also of importance. Stress of various types can also affect the immune system and influence the susceptibility to and recovery from infectious, autoimmune and neoplastic diseases. The developing foetus shows age-related variability and is more susceptible to immunotoxic effects than the adult.

37.5.2 Assays for Assessment of Immune Status

There is a plethora of tests developed to assess immunity in humans [99, 100], as described in laboratory manuals [101–103]. Many of these tests are nowadays commercially available as kits. A systematic approach to the evaluation of immune function, which is based on simple screening procedures followed by appropriate specialized tests of immune function, usually permits the definition of the immune alteration. This should include evaluation of the B-cell system, of the T-cell system and of non-specific resistance (polymorphonuclear leukocytes, monocytes and macrophages, NK cells, the complement system).

Testing schemes for evaluation of individuals exposed to IMMUNOTOXICANTS are proposed, among others, by the Subcommittee on Immunotoxicology of the US National Research Council [88] and by a task group of the World Health Organization (WHO) [50]. The panel proposed by the WHO is listed in Table 37.2 and is composed of assays that cover all major aspects of the immune system. Included are functional assays to test for HUMORAL IMMUNITY, i.e. specific antibodies to tetanus or diphtheria (for which vaccination programs exist), and for cellular immunity using recall antigens. It should be mentioned that these tests were all developed for diagnostic purposes, but in the context of immunotoxicity testing in humans, they are to be used in an epidemiological setting. This means that distinctions found in parameters between an exposed group and a control group may have a different biological significance than an altered value in an individual. Whereas a decrease in a single immune parameter in an individual may not indicate increased susceptibility to disease, a subtle alteration in an immune biomarker in a population may indicate immunotoxicity.

Establishing immune changes in humans is considerably more complex than in animals, considering that non-invasive tests are limited, exposure levels to the agent (i.e. dose) are difficult to establish and responses in the population are extremely heterogeneous. Also, the normal population exhibits a wide range of immunological responses with no apparent health impact. In addition to this underlying population variability, certain host characteristics or common exposures may be associated with significant, predictable alterations in immunological parameters. If not recognized and effectively addressed in the study design or statistical analysis, these confounding factors may severely alter the results of population studies. Examples of factors associated with measurable alterations in immunological parameters include age, race, gender, pregnancy, acute stress and the behavioural ability to cope with stress, coexistent diseases or infections, nutritional status, lifestyle, tobacco smoking and some medications. Paracetamol has been reported to interfere with lymphocyte function, as established by immunotoxicogenomic approaches [102]. Independently, it has been shown that prevaccination administration of paracetamol (acetaminophen) has resulted in lower antibody titres after vaccination with Synflorix (pneumococcal vaccine), whereas administration after vaccination did not. The lower antibody titres might still be high enough to protect the vaccinated children, although the final proof for this is lacking. Later, this prevaccination effect of paracetamol was confirmed with a viral vaccine, i.e. hepatitis B [112].

Besides the variables mentioned, periodical (ranging from daily to seasonal) influences also exist. This should be taken into account when evaluating the influence of immunotoxic agents in humans as some of these effects may be sufficiently large to exceed the expected effect of the agent. These environmental factors are therefore of primary concern in large epidemiological studies. For a review on the influence of endogenous and environmental factors on vaccination responses, see [103]. The importance of genetic factors for the response to a vaccine is shown by the role of cytokine polymorphisms in the susceptibility of humans to ultraviolet B-induced modulation of immune responses after hepatitis B vaccination [104].

37.5.3 Predictive Testing for Allergy in Humans

There are a variety of skin test procedures for the diagnosis of several types of allergic reactions, dealt with above. Basically, predictive tests in humans for skin ALLERGY are similar to diagnostic tests, but the aims are different (see Chap. 14). For diagnostic tests, the aim is to determine SENSITIZATION to chemicals to which there has been a prior exposure, whereas sensitization as a result of the procedure should be avoided. For predictive testing in humans, the aim is to show sensitizing capacity in individuals who have not been exposed previously to the compound.

For obvious reasons, predictive testing for respiratory sensitization is not done in humans. Occasionally, case reports may serve as adequate hazard identifications, but not as a risk estimate, because data on route and extent of exposure and on the "population at risk" are usually missing. In the absence of case reports, it cannot be concluded that no potential for sensitization exists.

37.6 Immunotoxicity Regulations

37.6.1 Regulatory Guidance

There is great variation in the approaches adopted by regulatory agencies throughout the world to the control of human pharmaceuticals. Leading agencies involved in the regulation of pharmaceuticals for human use are the US Food and Drug Administration (FDA), the Committee on Human Medicinal Products (CHMP) of the European Medicines Agency (EMA) in the European Union (EU) and the Ministry of Health, Labour and Welfare (MHLW), with the Pharmaceuticals and Medical Devices Agency (PMDA) in Japan. The requirements the industry has to meet, in order to gain marketing approval for its products, have been laid down in official guidelines. These guidelines inform the industry about the data needed to demonstrate to the authorities the pharmaceutical quality of new pharmaceuticals to be marketed, as well as their benefit and safety for the patient. The guidance given has a major impact on the development programmes adopted by the industry.

The regulations administered by government agencies are greatly influenced by the history, culture and legislations of the countries concerned. This still accounts for many national differences [105, 106]. However, worldwide harmonization of regulatory requirements is ongoing. Since the 1990s, the International Council for Harmonisation of Technical Requirements for Pharmaceuticals for Human Use (ICH) has proved to be a success. In this international forum, government regulators and industry representatives of the three major regions of the world pharmaceutical market participated (i.e. the United States, the EU and Japan). A number of harmonized guidelines have been developed by the ICH. Most of these have already been adopted officially by the regulatory authorities, even in non-ICH countries. Existing national guidelines thus have been or will be replaced by new ones based on the ICH consensus.

37.6.1.1 Regulatory Aspects of Laboratory Animal Immunotoxicology

To identify potential target organs of toxicity in humans, the industry must screen the toxicity of pharmaceuticals in laboratory animals. Among targets such as the liver and kidney, adverse effects on the immune system need to be assessed. As a rule, the regulatory authorities do not dictate how specific tests have to be conducted. The detailed technical requirements defined by the OECD may or may not be followed. This approach allows deviations from routine protocol toxicity testing whenever justified. The study protocols may be adjusted in such a way that they provide the most relevant information, depending on the nature and therapeutic indication of the pharmaceutical to be tested. Guidelines on immunotoxicology were introduced more than a decade ago (2000–2001), which focus particularly on immunotoxicological issues regarding pharmaceuticals. In this regard, it is necessary to differentiate between the various aspects of immunotoxicity, i.e. IMMUNOSUPPRESSION and immunostimulation, ALLERGY and AUTOIMMUNITY. Within the framework of ICH, a harmonized guideline has been written based on a cause-for-concern approach (see below).

37.6.2 Unintended Immunosuppression

Current predictive immunotoxicity testing is mainly done in the context of general toxicity, according to OECD guideline 407. Tests on the immune system comprise haematology, including differential cell counting, and histopathology of lymphoid tissues. Within the OECD, a debate over functional testing, i.e. measurement of antibody responses to sheep red blood cells, is in progress, in addition to discussion regarding the inclusion of measurement of NK activity and inclusion of FACS analysis of lymphocyte subpopulations.

The guidelines from all regions introduced in the past followed similar lines for immunotoxicity testing as those drawn up by the OECD. The major difference between the EU and the guidance in the other regions is that the CHMP had included functional testing for all compounds, whereas the FDA document advocates a cause-for-concern approach on the need for functional assays. A cause-for-concern is generally an adverse effect found in the toxicity endpoints but might also be the intended use or the pharmacological activity of compounds (e.g. anti-HIV drugs or anti-inflammatory activity).

Under the auspices of the International Conference (now Council) on Harmonisation, a process was started in November 2003 to harmonize the approach to screening human pharmaceuticals with regard to unintended IMMUNOSUPPRESSION. The industry provided a small database of pharmaceutical compounds being tested with respect to immune function, which has been gathered in two steps, and analysed [107]. A total of 45 out of 64 compounds were included in a formal evaluation. The main question was: Would additional immunotoxicity testing be more sensitive than the immunological endpoints of standard toxicity testing? Eventually, only 6 of 45 compounds could be called immunotoxic on the basis of additional immunotoxicity testing, without evidence in standard toxicity studies. It was decided that this number was too low to justify a routine requirement for functional testing, as was requested by the EU. Therefore, a cause-for-concern approach was adopted. Factors that might prompt additional immunotoxicity studies were (1) findings from standard toxicity studies (see Table 37.1, tier 1), (2) pharmacological properties of the drug, (3) the intended patient population, (4) structural similarities to known immunomodulators, (5) disposition of the drug and (6) clinical information. A list of additional studies mentioned in the ICH guidance document is given in Table 37.3. A weight of evidence review is needed to justify the absence of additional immunotoxicity studies. In case the outcome of the assessment of the drug-induced immunotoxicity suggests a risk of immunotoxicity, it should be weighed whether this risk is considered acceptable and/or should be addressed clinically in a so-called risk management plan. In the latter case, no further animal studies are needed.

37.6.3 Sensitizing Capacity

Regulatory HYPERSENSITIVITY testing generally focuses on locally applied compounds. Guidance documents focus predominantly on type IV hypersensitivity. The EMA/CHMP Note for Guidance on Local tolerance testing refers to the OECD Guidelines with regard to test methods (EMA 2016). The OECD test 406 guideline on skin SENSITIZATION covers the guinea pig maximization test (GMPT) and the Buehler test (BT); OECD guideline 429 covers the LLNA. The FDA and CHMP regard the LLNA as a suitable stand-alone method for detecting hypersensitivity potential. The guidance document of the MHLW requests that all dermatological preparations should be tested in at least one skin sensitization study. The following tests are regarded as acceptable by the MHLW: ADJUVANT and patch test, Buehler test, Draize test, Freud's complete adjuvant test, maximization test, open cutaneous test, optimization test and the split adjuvant test. The most commonly used test in Japan is the maximization test, followed by the adjuvant patch test or the Buehler test. Table 37.4 provides an overview of the regional requirements.

Table 37.3 List of additional studies mentioned in the ICH S8 document

Additional immunotoxicity studies (AIS)
T-cell-dependent antibody response (TDAR)
Immunophenotyping
Natural killer cell activity assays
Host resistance models
Macrophage/neutrophil function
Assay to measure cell-mediated immunity

Drugs intended for inhalation should be tested for their potential to induce type I hypersensitivity reaction according to the FDA. The Japanese and the European authorities have not issued a non-clinical requirement for inhalation ALLERGY testing.

37.6.4 Autoimmunity

Because of the lack of suitable and validated animal models, the potential risk for inducing AUTOIMMUNITY cannot be predicted. As regulatory guidance follows scientific developments, such guidance for autoimmunity is not available. The FDA guidance document touches the issue of autoimmunity, mentioning the PLNA as a proposed test for autoimmunity.

Table 37.4 Test requirements for topically applied drugs

Assays	EMA	FDA	MHLW
GPMT	All^{A}	All^{A}	All^{A}
BA	All^{A}	All^{A}	All^{A}
LLNA	All^{A}	All^{A}	All^{A}
MEST	All	All	All^{A}
Adjuvant and patch test	–	–	All^{A}
Draize test	–	–	All^{A}
FCA	–	–	All^{A}
Open epicutaneous test	–	–	All^{A}
Optimization test	–	–	All^{A}
Spilt adjuvant test	–	–	All^{A}
MIGET	–	Inhalation drugs	–

All drugs all topically applied drugs, *All drugs*^{A} test regarded as stand-alone assay

It recognizes, however, that an extensive evaluation to support its use currently is lacking. Biomarkers of T-cell activation or markers of Th2 cell induction and autoantibody induction in experimental animals may be helpful, but their predictive value for determining the potential to induce autoimmune disease in humans has not been ascertained. Neither the EMA nor the MHLW has released any guidance on autoimmunity testing.

37.6.5 Immunotoxicology of Biotechnological Products

In 1997, the ICH issued a guideline on "Preclinical Safety Assessment of Biotechnology-Derived Pharmaceuticals", which has received an update by the Addendum finished in 2011 (for the EMA link, see Appendix 6). This document contains guidance on the assessment of immunogenicity as many products of this type are immunogenic in animals. Antibody responses should be characterised, especially when any pharmacological changes have been identified. The finding of antibodies against the product should, however, not be the sole criterion for the early termination of an animal study. Only in case of neutralisation of the response in the majority of the animals should the study be stopped. This reflects the growing insight that immunogenicity of human proteins in animals, as such, is not predictive for the human situation and testing for anti-drug-antibodies is only needed to understand the outcome of the toxicity studies performed.

Little specific information is present on general immunotoxicity. Inflammatory responses might be indicative of a stimulatory immune response but may also be the result of toxic changes. Routine tiered-testing approaches are not recommended for biotechnology-derived pharmaceuticals, as it is supposed that on the one hand the character of most of recombinant proteins is well-known and not directed to the immune system. On the other hand, the effects on the immune system of immunosuppressive monoclonal antibodies will be tested anyway by characterising their pharmacodynamics.

37.6.6 Developmental Immunotoxicity

At the start of the twenty-first century, there has been an increased focus on chemical risk assessment for children as a specific subpopulation, as hazard and risk assessment for adults may not necessarily be sufficiently suitable for children. Developing organs may be more susceptible to chemical-induced toxicity than mature adult organs. Especially, the nervous system, immune system and reproductive system are more vulnerable to toxicants during the developmental stages. The development of the immune system starts in utero and is essentially complete and functioning at birth [108]. However, postnatally, there is continued maturation in early life [109] up to the age of 13 [110] by the acquisition of immune competence.

Immune-based diseases in the developed countries affect around 25% of the children as early-onset diseases related to malfunctioning of the immune system show a high prevalence in children in Western societies, including recurrent otitis media (18–26%), asthma (15–26%), atopic dermatitis (15%) and allergic rhinitis (8–12%). Moreover, juvenile-onset allergic, inflammatory and autoimmune diseases have shown an increasing prevalence in the last decades. The underlying causes are probably multifactorial, and a possible causative role of chemical exposures during pre- and postnatal development of the immune system cannot be excluded. Yet, direct evidence for a relation with chemical exposure is poor. Part of this lack of knowledge is the current virtual absence of developmental immunotoxicity testing in the regulatory requirements of chemical hazard assessment.

The introduction of legislation for paediatric medicines in the United States and Europe in the early years of the twenty-first century has also drawn the attention of the pharmaceutical regulators to developmental immunotoxicity. The ICH S6 document, originally published in 1997, is the first one to mention developmental immunotoxicity, but the guidance is very general. It indicates that the study design of the toxicity studies may be modified, e.g. in the case of monoclonal antibodies with prolonged immunological effects. The FDA guidance on juvenile toxicity (2005) suggests the incorporation of immunotoxicological determinations in the ICH Stage C-F reproductive toxicology study if a drug is expected to be used in pregnant women. Regulatory guidance documents from Europe and Japan do not specifically address developmental immunotoxicity as yet. The European Medicines Agency (EMA-CHMP) released a guideline on juvenile toxicity in 2008, indicating that immunotoxicity studies in juvenile animals are only required if the chemical/pharmacological class of compounds or previous studies in humans or animals give cause for concern for the developing immune system (with a reference to the ICH S8 Guideline) (EMA 2008).

OECD Guidelines for the evaluation of reproductive and developmental toxicity include the prenatal developmental toxicity study, the one- and two-generational reproductive toxicity study and developmental neurotoxicity study. These protocols do not contain parameters for developmental immunotoxicity (DIT). Parameters indicating immune toxicity such as spleen and thymus weight, differential blood cell counts and immunohistopathology parameters could easily be incorporated into the reproductive toxicity studies. Moreover, functional assays may provide additional sensitivity to DIT testing. In 2011, OECD adopted the guideline for the Extended One-Generation Reproduction Toxicity Study (EOGRTS), with dosing from the premating period to offspring adulthood, which now includes structural and functional endpoints for developmental neurotoxicity and DIT. Besides the above-mentioned structural immune parameters, a T-cell-independent antibody response assay, i.e. the accepted primary IgM antibody response to a T-cell-dependent antigen, such as sheep red blood cells (SRBC) or keyhole limpet haemocyanin (KLH), may provide additional functional information.

The acceptance of the OECD TG 443, including specific attention for developmental immunotoxicity parameters, has been recognized as a significant hallmark. It confirms the importance of assessing developmental immune parameters in regulatory toxicology. However, this study is only applied for a subset of chemicals of high tonnage under the European legislation for chemical safety (Registration, Evaluation and Authorisation of Chemicals, REACH). Moreover, it is a logistically complicated and expensive study. Developmental immune toxicity (DIT) testing could perhaps be more practical in other study designs. In the past, several research groups have embarked on defining exposure protocols considering windows of sensitivity of the developing immune system.

A series of developmental immunotoxicity studies in rats have been performed according to various study designs and using exposure windows based on various existing protocols (pre-/perinatal exposure, juvenile exposure). Collectively, these studies demonstrated that the most sensitive immune parameters showed effects at lower-dose levels compared to more general developmental parameters, indicating the relevance of including them by default in protocols for DIT testing. Furthermore, functional immune parameters largely made up to the most sensitive parameter sets, demonstrating the relevance of including them by default in protocols for DIT testing [111].

Differences exist when comparing the different exposure designs and route of exposure (this differs per compound). Comparing the EOGRTS and juvenile model shows that both exposure designs have some advantages. On the one hand, the EOGRTS model studies continuous exposure during the development of the immune system, whereas the juvenile model is limited in that sense but may allow higher sensitivity of endpoints especially vulnerable during this window of development.

Developing organs are considered more susceptible to chemical-induced toxicity than mature adult organs. Especially, the nervous system, immune system and reproductive system are more vulnerable to toxicants during the developmental stages. There is quite some concern about effects of chemicals, including pharmaceutical on the developing immune system, as it appears that in the recent decades, a number of diseases showed increased prevalence, in particular in Western and westernized societies. Although a number of possible reasons have been proposed for this, e.g. increased hygiene, lifestyle factors and environmental pollution, altered or increased drug use in children may be one of these.

The lack of detailed instructions for human pharmaceuticals in the regional Guideline on Juvenile Toxicity is due to the fact that the original three regions within ICH have adopted a flexible, case-by-case, science-based approach to evaluate developmental immunotoxicity. It is to be expected that the increasing knowledge on developmental immunotoxicity will lead to new approaches in this field. This might be stimulated by the way human studies are conducted with vaccines such as with hepatitis B antigens or with pneumococcal polysaccharide conjugate vaccine (Synflorix). Earlier in this chapter it has been described that using the latter vaccine, an immunosuppressive effect of acetaminophen (paracetamol) has been revealed. This immunosuppressive effect was also suggested from animal studies using an immunotoxicogenomic approach [102]. Consequently, the Experts involved in ICH S6 were reluctant to formulate in-depth requirements, which may turn out to be too strict. Such guidance might be valid for some biotechnology-derived products, but not for others.

The National Institute of Environmental Health Sciences (NIEHS) and the National Institute of Occupational Safety and Health (NIOSH) have started a discussion on a potential consensus document on the most appropriate experimental approaches and assays available to assess developmental immunotoxicity [97]. This initiative is also supported by the International Life Sciences Institute (ILSI) in Washington that organized a conference on this topic in 2003. Examples of pharmaceuticals as developmental IMMUNOTOXICANTS in rodents are acyclovir, cyclosporin A, cyclophosphamide, corticosteroids, benzodiazepines, azathioprine and tacrolimus. As mentioned above, the FDA suggests incorporation of immunotoxicological determinations in reproductive toxicology studies.

37.6.7 New Developments

In the future, transgenic animals or adoptive transfer models might contribute to our insight into the immunotoxic mechanisms of compounds in vivo. Further contributions to the screening for immunotoxicity can be expected from immunotoxicogenomics (microarray technology) and proteomics, as illustrated by the case of paracetamol mentioned above [102].

Importantly, as with other toxicological disciplines, results of tests should be translatable to the organism of interest, taking into account the relevance of test animals or in vitro methods [98]. Translational aspects of hazard and risk evaluation will receive more attention in the coming years, in particular in the field of immunotoxicology.

37.7 Conclusions

Preclinical testing of pharmaceuticals in laboratory animals requires different approaches to detect direct immunotoxicity, resulting in unwanted IMMUNOSUPPRESSION or immunostimulation, or to detect drug-induced HYPERSENSITIVITY and AUTOIMMUNITY. Tiered immunotoxicity-testing procedures have been developed and validated in the rat and mouse and are being used successfully to detect drug-induced direct immunotoxicity.

Drug-induced hypersensitivity and autoimmune reactions are of great clinical concern. For contact ALLERGY, routine contact SENSITIZATION testing in guinea pigs has been extended by the LLNA in the mouse as a stand-alone method for detecting hypersensitivity potential. An important issue in contact allergy is the development of quantitative measurements of the potency of ALLERGENs. No validated models are yet available to investigate the ability of drugs to induce respiratory sensitization. The LLNA in mice or skin sensitization testing in guinea pigs should be recommended as a first screen.

Animal models are currently available to detect the potential of compounds to exacerbate induced or genetically predisposed autoimmunity but are seldom used in immunotoxicity studies. Models to investigate the ability of chemicals to induce autoimmunity, as a result of an immune response to self-proteins modified by the chemical, are virtually limited to the PLNA. As human data show that chemical agents, in particular drugs, can cause autoimmune diseases, new models should be developed.

In conclusion, immunotoxicology is a rapidly evolving field in the regulation of pharmaceuticals. This is reflected in the rapid harmonization of the regional approaches on immunotoxicity testing under the auspices of ICH. The standard use of a large immunotoxicological test battery is not recommended. Instead, a flexible approach on a case-by-case basis is accepted, taking into account that some classes of drugs and some indications may be a greater cause of concern than others.

Important Websites

Guidelines on hypersensitivity testing: http://oecdpublications.gfi-nb.com/cgibin/OECDBookShop.storefront/

ILSI-HESI information on immunotoxicology: http://www.hesiglobal.org/i4a/pages/index.cfm?pageid=3471

FDA guidelines: http://www.fda.gov/MedicalDevices/DeviceRegulationandGuidance/GuidanceDocuments/ucm080495.htm, http://www.fda.gov/downloads/Drugs/GuidanceComplianceRegulatoryInformation/Guidances/ucm079239.pdf

EMA guidelines: https://www.ema.europa.eu/en/non-clinical-local-tolerance-testing-medicinal-products, https://www.ema.europa.eu/en/documents/scientific-guideline/guideline-need-non-clinical-testing-juvenile-animals-pharmaceuticals-paediatric-indications_en.pdf

EMEA guidelines: http://www.ema.europa.eu/pdfs/human/swp/214500en.pdf, http://www.ema.europa.eu/docs/en_GB/document_library/Scientific_guideline/2009/09/WC500002828.pdf

ICH guidelines: ICH S6(R1) Preclinical testing of Biotechnology-derived pharmaceuticals. http://www.ema.europa.eu/docs/en_GB/document_library/Scientific_guideline/2009/09/WC500002828.pdf, https://www.ich.org/fileadmin/Public_Web_Site/ICH_Products/Guidelines/Safety/S6_R1/Step4/S6_R1_Guideline.pdf

ICH S8: International conference on harmonization of technical requirements for pharmaceuticals for human use. Note for guidance on immunotoxicity studies for human pharmaceuticals. 2006. www.ich.org, www.ema.europa.eu/humanmedicines, https://www.ich.org/fileadmin/Public_Web_Site/ICH_Products/Guidelines/Safety/S8/Step4/S8_Guideline.pdf

References

1. Descotes JG, Vial T. Cytoreductive drugs. In: Dean JH, Luster MI, Munson AE, Kimber I, editors. Immunotoxicology and immunopharmacology. 2nd ed. New York, NY: Raven Press; 1994. p. 293–301.
2. De Jong HJI, Klungel OH, van Dijk L, Vandebriel RJ, HGM L, Cohen Tervaert JW, Van der Laan JW, Van Loveren H. Use of statins is associated with an increased risk of rheumatoid arthritis. Ann Rheum Dis. 2012;71:648–54.
3. De Jong HJ, Cohen Tervaert JW, Saldi SR, Vandebriel RJ, Souverein PC, Meyboom RH, Van Loveren H, Klungel OH. Association between statin use and lupus-like syndrome using spontaneous reports. Semin Arthritis Rheum. 2011;41:373–38.
4. Vandebriel RJ, De Jong HJI, Gremmer ER, Klungel OH, Cohen Tervaert JW, Slob W, van der Laan JW, van Loveren H. Statins accelerate the onset of collagen type II-induced arthritis in mice. Arthritis Res Ther. 2012;14:R90.
5. Vandebriel RJ, De Jong HJI, Gremmer ER, Klungel OH, Cohen Tervaert JW, Slob W, van der Laan JW, van Loveren H. Reply: statins accelerate the onset of collagen type II-induced arthritis in mice. Arthritis Res Ther. 2013;15:403.
6. Chalasani N, Bonkovsky HL, Fontana R, Lee W, Stolz A, Talwalkar J, Reddy KR, Watkins PB, Navarro V, Barnhart H, Gu J, Serrano J, United States Drug Induced Liver Injury Network. Features and outcomes of 899 patients with drug-induced liver injury: the DILIN Prospective Study. Gastroenterology. 2015;148:1340–52.
7. Shaw PJ, Ganey PE, Roth RA. Idiosyncratic drug-induced liver injury and the role of inflammatory stress with an emphasis on an animal model of trovafloxacin hepatotoxicity. Toxicol Sci. 2010;118:7–18.
8. Ryffel B, Car BD, Eugster H-P, Woerly G. Transplantation agents. In: Dean JH, Luster MI, Munson AE, Kimber I, editors. Immunotoxicology and immunopharmacology. 2nd ed. New York, NY: Raven Press; 1994. p. 267–92.
9. Majoor GD, Wodzig WH, Vriesman PJCVB. Cyclosporin-induced autoimmunity. In: Kendall MD, Ritter MA, editors. Thymus update: thymus in immunotoxicology. London: Harwood Academic Publishers; 1991. p. 179–200.
10. Cohen JJ. Glucocorticoid-induced apoptosis in the thymus. Seminars in immunology. New York, NY: Academic; 1992.
11. Salinas-Carmona MC, Perez LI, Galan K, Vazquez AV. Immunosuppressive drugs have different effect on B lymphocyte subsets and IgM antibody production in immunized BALB/c mice. Autoimmunity. 2009;42:537–44.
12. Clark EA, Ledbetter JA. How B- and T-cells talk to each other. Nature. 1994;367:425–8.
13. Matzinger P. Tolerance, danger and the extended family. Annu Rev Immunol. 1994;12:991–1045.
14. Crommelin DJA, Storm G, Verrijk R, de Leede L, Jiskoot W, Hennink WE. Shifting paradigms: biopharmaceuticals versus low molecular weight drugs. Int J Pharm. 2003;266:3–16.
15. Vial T, Descotes J. Immune-mediated side effects of cytokines in humans. Toxicology. 1995;105:31–57.

16. Uetrecht J. Drug metabolism by leukocytes and its role in drug-induced lupus and other idiosyncratic drug reactions. Crit Rev Toxicol. 1990;20:213–35.
17. Kubicka-Muranyi M, Kremer J, Rottmann N, Lübben B, Albers R, Bloksma N, Lührmann R, Gleichmann E. Murine systemic autoimmune disease induced by mercuric chloride: T helper cells reacting to self proteins. Int Arch Allergy Immunol. 1996;109:11–20.
18. Pichler WJ. Drug-induced autoimmunity. Curr Opin Allergy Clin Immunol. 2003;3:249–53.
19. Griem P, Panthel K, Kalbacher H, Gleichmann E. Alteration of a model antigen by Au (III) leads to T-cell sensitization to cryptic peptides. Eur J Immunol. 1996;26:279–87.
20. Bloksma N, Kubicka-Muranyi M, Schuppe H-C, Gleichmann E, Gleichmann H. Predictive immunotoxicological test systems: suitability of the popliteal lymph node assay in mice and rats. Crit Rev Toxicol. 1995;25:369–96.
21. Griem P, Gleichmann E. Metal ion induced autoimmunity. Curr Opin Immunol. 1995;7:831–8.
22. Janeway CA. The immune system evolved to discriminate infectious nonself from noninfectious self. Immunol Today. 1992;13:11–6.
23. Cohen IR. The cognitive paradigm and the immunological homunculus. Immunol Today. 1992;13:490–4.
24. Ibrahim MAA, Chain BM, Katz DR. The injured cell: the role of the dendritic cell system as a sentinel receptor pathway. Immunol Today. 1995;16:181–6.
25. Nierkens S, Van Helden P, Bol M, Bleumink R, Van Kooten P, Ramdien-Murli S, Boon L, Pieters R. Selective requirement for CD40-CD154 in drug-induced type 1versus type 2 responses to trinitrophenyl-ovalbumin. J Immunol. 2002;168: 3747–54.
26. Sgro C. Side effects of a monoclonal antibody, muromonab CD3/orthoclone OKT3: bibliographic review. Toxicology. 1995;105:23–9.
27. Gleichmann E, Pals ST, Rolink AG, Radaszkiewicz T, Gleichmann H. Graft-versus-host reactions: clues to the etiopathology of a spectrum of immunological diseases. Immunol Today. 1984;5:324–32.
28. Adams LE, Hess EV. Drug-related lupus. Incidence, mechanisms and clinical implications. Drug Saf. 1991;6:431–49.
29. Suntharalingam G, Perry MR, Ward S, Brett SJ, Castello-Cortes A, Brunner MD, Panoskaltsis N. Cytokine storm in a phase I trial of the anti-CD28 Monoclonal antibody TGN1412. N Engl J Med. 2006;355:1018–28.
30. Chatenoud L, Ferran C, Legendre C, Thouard I, Merote S, Reuter A, Gevaert Y, Kreis H, Franchimont P, Bach JF. In vivo cell activation by OKT3 administration. Transplantation. 1990;49:697–702.
31. Moreau T, Coles A, Wing MG, Isaacs J, Hale G, Waldmann H, Compston A. Transient increase in symptoms associated with cytokine release in patients with multiple sclerosis. Brain. 1996;119:225–37.
32. Stebbings R, Findlay L, Edwards C, Eastwood D, Bird C, North D, Mistry Y, Dilger P, Liefooghe E, Cludts I, Fox B, Tarrant G, Robinson J, Meager T, Dolman C, Thorpe SJ, Bristow A, Wadhwa M, Thorpe R, Poole S. "Cytokine storm"in the phase I trial of monoclonal antibody TGN1412: better understanding the causes to improve preclinical testing of immunotherapeutics. J Immunol. 2007;179:3325–31.
33. Vidal JM, Kawabata TT, Thorpe R, Lima BS, Cederbrant K, Poole S, Mueller-Berghaus J, Pallardy M, Van der Laan JW. European Medicines Agency-sponsored workshop: in vitro cytokine release assays for predicting cytokine release syndrome: the current state of the science. Report of a European Medicines Agency Workshop. Cytokine. 2010;51:213–5.
34. Murphy K. Janeway's immunobiology: the immune system in health and disease. Current biology. Oxford: Garland Science; 2011.
35. Beyersdorf N, Werner S, Wolf N, Hünig T, Kerkau T. In vitro polyclonal activation of conventional T cells with a CD28 superagonist protects mice from acute graft versus host disease. Eur J Immunol. 2015;45:1997–2007.
36. Druet P, Pelletier L, Rossert J, Druet E, Hirsch F, Sapin C. Autoimmune reactions induced by metals. In: Kammuller ME, Bloksma N, Seinen W, editors. Autoimmunity and toxicology: immune disregulation induced by drugs and chemicals. Amsterdam: Elsevier; 1989. p. 349–61.
37. Van Vliet E, Uhrberg M, Stein C, Gleichmann E. MHC control of IL-4-dependent enhancement of B-cell Ia expression and Ig class switching in mice treated with mercuric chloride. Int Arch Allergy Immunol. 1993;101:392–401.
38. Albers R. Chemical-induced autoimmunity -Immune disregulation assessed with reporter antigens. Utrecht University, PhD thesis; 1996.
39. van der Laan JW, Gould S, Tanir JY. ILSI HESI Vaccines and Adjuvants Safety Project Committee. Safety of vaccine adjuvants: focus on autoimmunity. Vaccine. 2015;33:1507–14.
40. Jonulit H, Schmitt E. The regulatory T cell family: distinct subsets and their interrelations. J Immunol. 2003;171:6323–7.
41. Layland LE, Wulferink M, Dierkes S, Gleichmann E. Drug-induced autoantibody formation in mice: triggering by primed CD4+CD25− T cells, prevention by primed CD4+CD25+ T cells. Eur J Immunol. 2004;34:36–46.
42. Mason MJ, Uetrecht J. Tolerance induced by low dose D-penicillamine in the Brown Norway rat model of drug-induced autoimmunity is immune mediated. Chem Res Toxicol. 2004;17:82–94.30.
43. Coleman JW, Sim E. Autoallergic responses to drugs: mechanistic aspects. In: Dean JH, Luster MI, Munson AE, Kimber I, editors. Immunotoxicology and immunopharmacology. 2nd ed. New York, NY: Raven Press; 1994. p. 553–72.

44. Descotes J, Choquet-Kastylevsky G. Gell and Coombs's classification: is it still valid? Toxicology. 2001;158:43–9.
45. Adkinson NF Jr, Essayan D, Gruchalla R, Haggerty H, Kawabata T, Sandler JD, Updyke L, Shear NH, Wierda D. Task force report: future research needs for the prevention and management of immune-mediated drug hypersensitivity reactions. J Allergy Clin Immunol. 2002;109:S461–78.
46. Uetrecht J. Immune-mediated adverse drug reactions. Chem Res Toxicol. 2009;22:24–34.
47. Ostapowicz G, Fontana RJ, Schiødt FV, Larson A, Davern TJ, Han SH, TM MC, Shakil AO, Hay JE, Hynan L, Crippin JS, Blei AT, Samuel G, Reisch J, Lee WM, U.S. Acute Liver Failure Study Group. Results of a prospective study of acute liver failure at 17 tertiary care centers in the United States. Ann Intern Med. 2002;137:947–54.
48. Metushi IG, Hayes MA, Uetrecht J. Treatment of PD-1(−/−) mice with amodiaquine and anti-CTLA4 leads to liver injury similar to idiosyncratic liver injury in patients. Hepatology. 2015;61:1332–42.
49. Mak A, Uetrecht J. The combination of anti-CTLA-4 and PD1−/− mice unmasks the potential of isoniazid and nevirapine to cause liver injury. Chem Res Toxicol. 2015;28(12):2287–91.
50. IPCS. International programme for chemical safety, environmental health criteria document 160. Principles and methods for assessing direct immunotoxicity associated with chemical exposure. Geneva: World Health Organization; 1996.
51. Vos JG. Immunotoxicity assessment: screening and function studies. Arch Toxicol Suppl. 1980;4:95–108.
52. Van Loveren H, Vos JG. Immunotoxicological considerations: a practical approach to immunotoxicity testing in the rat. In: Dayan AD, Paine AJ, editors. Advances in applied toxicology. London: Taylor and Francis; 1989. p. 143–65.
53. Vos JG, Krajnc EI. Immunotoxicity of pesticides. In: Hayes AW, Schnell RC, Miya TS, editors. Developments in the science and practice of toxicology. Amsterdam: Elsevier; 1983. p. 229–40.
54. De Waal EJ, Van Loveren H, Vos JG. Practice of tiered testing for immunosuppression in rodents. Drug Inf J. 1997;31:1317–23.
55. Koeter HBWM. International harmonisation of immunotoxicity testing. Hum Exp Toxicol. 1995;14:151–4.
56. Van Loveren H, Vos JG, De Waal EJ. Testing immunotoxicity of chemicals as a guide for testing approaches for pharmaceuticals. Drug Inf J. 1996;30:275–9.
57. Van Loveren H, Gianotten N, Hendriksen CFM, Schuurman H-J, Van Der Laan JW. Assessment of immunotoxicity of buprenorphine. Lab Anim. 1994;28:355–63.
58. De Waal EJ, De Jong WH, Van Der Vliet H, Verlaan B, Van Loveren H. An immunotoxicity screening study on salmeterol in rats. Int J Immunopharmacol. 1996;18:523–8.
59. Schulte A, Althoff J, Ewe S, Richter-Reichhelm HB. Two immunotoxicity ring studies according to OECD TG 407-comparison of data on cyclosporin A and hexachlorobenzene. Regul Toxicol Pharmacol. 2002;36:12–21.
60. Ladics GS, Smith C, Heaps K, Ellioo GS, Slone TW, Loveless SE. Possible incorporation of an immunotoxicological functional assay for assessing humoral immunity for hazard identification purposes in rats on standard toxicology study. Toxicology. 1995;96:225–38.
61. Van Loveren H, Vos J, Putman E, Piersma A. Immunotoxicological consequences of perinatal chemical exposures: a plea for inclusion of immune parameters in reproduction studies. Toxicology. 2003;185:185–91.
62. Luster MI, Munson AE, Thomas PT, Holsapple MP, Fenters JD, White KL, Lauer LD, Germolec DR, Rosenthal GJ, Dean JH. Development of a testing battery to assess chemical-induced immunotoxicity: National Toxicology Program's guidelines for immunotoxicity evaluation in mice. Fund Appl Toxicol. 1988;10:2–19.
63. Luster MI, Portier C, Pait DG, Rosenthal GJ, Germolec DR, Corsini E, Blaylock BL, Pollock P, Kouchi Y, Craig W, et al. Risk assessment in immunotoxicology. II. Relationship between immune and host resistance tests. Fund Appl Toxicol. 1993;21:71–82.
64. Germolec DR, Nyska A, Kashon M, Kuper CF, Portier C, Kommineni C, Johnson KA, Luster MI. Extended histopathology in immunotoxicity testing: interlaboratory validation studies. Toxicol Sci. 2004;78:107–15.
65. Germolec DR, Kashon M, Nyska A, Kuper CF, Portier C, Kommineni C, Johnson KA, Luster MI. The accuracy of extended histopathology to detect immunotoxic chemicals. Toxicol Sci. 2004;82:504–14.
66. Bugelski PJ, Thiem PA, Solleveld HA, Morgan DG. Effects of sensitization to dinitrochlorobenzene (DNCB) on clinical pathology parameters and mitogen-mediated blastogenesis in cynomolgus monkeys (*Macaca fascicularis*). Toxicol Pathol. 1990;18:643–50.
67. Kozlosky JC, Mysore J, Clark SP, Burr HN, Li J, Aranibar N, Vuppugalla R, West RC, Mangipudy RS, Graziano MJ. Comparison of physiologic and pharmacologic parameters in Asian and mauritius cynomolgus macaques. Regul Toxicol Pharmacol. 2015;73:27–42.
68. Ahmed-Ansari A, Brodie AR, Fultz PN, Anderson DC, Sell KW, McClure HM. Flow microfluorometric analysis of peripheral blood mononuclear cells from nonhuman primates: correlation of phenotype with immune function. Am J Primatol. 1989;17:107–31.
69. van Mierlo GJ, Cnubben NH, Wouters D, Wolbink GJ, Hart MH, Rispens T, Ganderup NC, Kuper CF, Aarden L, Penninks AH. The minipig as an alternative non-rodent model for immunogenicity testing using the TNFα blockers adalimumab and infliximab. J Immunotoxicol. 2014;11:62–71.

70. Van der Laan JW, Brightwell J, McAnulty P, Ratky J, Stark C. Regulatory acceptability of the minipig in the development of pharmaceuticals, chemicals and other products. J Pharmacol Toxicol Methods. 2010;62:184–95.
71. Bode G, Clausing P, Gervais F, Loegsted J, Luft J, Nogues V, Sims J, Steering Group of the RETHINK Project. The utility of the minipig as an animal model in regulatory toxicology. J Pharmacol Toxicol Methods. 2010;62:196–220.
72. Barratt MD, Basketter DA, Chamberlain M, Admans GD, Langowski JJ. An expert system rule base for identifying contact allergens. Toxicol In Vitro. 1994;8:1053–60.
73. Dupuis G, Benezra C. Contact dermatitis to simple chemicals: a molecular approach. New York, NY: Marcel Dekker; 1982.
74. Maurer T. Predictive testing for skin allergy. In: Vos JG, Younes M, Smith E, editors. Allergic hypersensitivities induced by chemicals. Recommendations for prevention. Boca Raton, FL: CRC Press; 1996. p. 237–59.
75. Garssen J, Vandebriel RJ, Kimber I, Van Loveren H (1996) Hypersensitivity reactions: definitions, basic mechanisms, and localizations. In: JG Vos, M Younes, E Smith (eds): Allergic hypersensitivities induced by chemicals. Recommendations for prevention. CRC Press, Boca Raton, FL, 19–58.
76. Kimber I, Dearman RJ. Contact hypersensitivity: immunological mechanisms. In: Kimber I, Maurer T, editors. Toxicology of contact hypersensitivity. London: Taylor and Francis; 1996. p. 14–25.
77. Kimber I, Mitchell JA, Griffin AC. Development of a murine local lymph node assay for the determination of sensitizing potential. Food Chem Toxicol. 1986;24:585–6.
78. Van Och FM, Slob W, de Jong WH, Vandebriel RJ, van Loveren H. A quantitative method for assessing the sensitizing potency of low molecular weight chemicals using a local lymph node assay: employment of a regression method that includes determination of the uncertainty margins. Toxicology. 2000;20:49–59.
79. Dearman RJ, Mitchell JA, Basketter DA, Kimber I. Differential ability of occupational chemical contact and respiratory allergens to cause immediate and delayed dermal hypersensitivity reactions in mice. Int Arch Allergy Appl Immunol. 1992;97:315–21.
80. Garssen J, Nijkamp FP, Van Der Vliet H, Van Loveren H. T-cell mediated induction of airway hyperreactivity in mice. Am Rev Respir Dis. 1991;144:931–8.
81. Scheerens H, Buckley TL, Davidse EM, Garssen J, Nijkamp FP, Van Loveren H. Toluene diisocyanate-induced in vitro tracheal hyperreactivity in the mouse. Am J Respir Crit Care Med. 1996;154:858–65.
82. Van Och FM, Van Loveren H, De Jong WH, Vandebriel RJ. Cytokine production induced by low-molecular-weight chemicals as a function of the stimulation index in a modified local lymph node assay: an approach to discriminate contact sensitizers from respiratory sensitizers. Toxicol Appl Pharmacol. 2002;184:46–56.
83. Karol MH. Predictive testing for respiratory allergy. In: Vos JG, Younes M, Smith E, editors. Allergic hypersensitivities induced by chemicals. Recommendations for prevention. Boca Raton, FL: CRC Press; 1996. p. 125–37.
84. Karol MH, Griffiths-Johnson DA, Skoner DP. Chemically induced pulmonary hypersensitivity, airway hyperreactivity and asthma. In: Gardner D, et al., editors. Toxicology of the lung. 2nd ed. New York, NY: Raven Press; 1993. p. 417–33.
85. Sarlo K, Clark ED. A tier approach for evaluating the respiratory allergenicity of low molecular weight chemicals. Fund Appl Toxicol. 1992;85:55–8.
86. Hessel EM, Van Oosterhout AJM, Hofstra CL, De Bie JJ, Garssen J, Van Loveren H, Verheyen AKCP, Savelkoul HFJ, Nijkamp FP. Bronchoconstriction and airway hyperresponsiveness after ovalbumin inhalation in sensitized mice. Eur J Pharmacol. 1995;293:401–12.
87. Kammüller ME, Bloksma N, Seinen W, editors. Autoimmunity and toxicology: immune dysregulation induced by drugs and chemicals. Amsterdam: Elsevier; 1989.
88. Pieters R, Nierkens S. Experimental models of autoimmunity. In: House R, Luebke RV, Kimber I, editors. Immunotoxicology and Immunopharmacology. 4th ed. Boca Raton, FL: CRC press; 2007. p. 469–85.
89. Gleichmann E, Vohr HW, Stringer C, Nuyens J, Gleichmann H. Testing the sensitization of T-cells to chemicals. From murine graft-versus-host (GVH) reactions to chemical-induced GVH-like immunological diseases. In: Kammüller ME, Bloksma N, Seinen W, editors. Autoimmunity and toxicology: immune dysregulation induced by drugs and chemicals. Amsterdam: Elsevier; 1989. p. 364–90.
90. Gleichmann E, Kind P, Schuppe HC, Merk H. Tests for predicting sensitization to chemicals and their metabolites, with special reference to heavy metals. In: Dayan AD, Hertel RF, Heseltine E, Kazantzis G, Smith EM, Van Der Venne MT, editors. Immunotoxicity of metals and immunotoxicology. New York, NY: Plenum; 1990. p. 139–52.
91. Pieters R, Albers R. Screening tests for autoimmune-related immunotoxicity. Environ Health Perspect. 1999;107(Suppl 5):673–7.
92. Weaver J, Chapdelaine J, Descotes J, Germolic D, Holsapple M, House R, Lebrec H, Meade J, Pieters R, Hastings K, Dean J. Evaluation of a lymph node proliferation assay for its ability to detect pharmaceuticals with the potential to cause immune mediated drug reactions. J Immunotoxicol. 2005;2:11–20.
93. Vial T, Carleer J, Legrain B, Verdier F, Descotes J. The popliteal lymph node assay: results of a preliminary interlaboratory validation study. Toxicology. 1997;122:213–8.
94. Albers R, Broeders A, ven der Pijl A, Seinen W, Pieters R. The use of reporter antigens in the popliteal lymph node assay to assess immunomodulation by chemicals. Toxicol Appl Pharmacol. 1997;143:102–9.

95. Gutting BW, Updyke LW, Amacher DE. BALB/c mice orally pretreated with diclofenac have augmented and accelerated PLNA responses to diclofenac. Toxicology. 2002;172:217–30.
96. Selgrade MJ, Cooper KD, Devlin RB, Van Loveren H, Biagini RE, Luster MI. Immunotoxicity-bridging the gap between animal research and human health effects. Fund Appl Toxicol. 1995;24:13–21.
97. National Research Council. Biological markers in immunotoxicology. Washington: National Academy Press; 1992.
98. Levy SM, Herberman RB, Lee J, Whiteside T, Beadle M, Heiden L, Simons A. Persistantly low natural killer activity, age, and environmental stress as predictors of infectious morbidity. Nat Immun Cell Growth Regul. 1991;10:289–307.
99. Bentwich Z, Bianco N, Jager L, Houba V, Lambert PH, Knaap W, Rose N, Seligman M, Thompson R, Torrigiani G, et al. Use and abuse of laboratory tests in clinical immunology: Critical considerations of eight widely used diagnostic procedures. Report of a joint IUIS/WHO meeting on assessment of tests used in clinical immunology. Clin Immunol Immunopathol. 1982;24:122–38.
100. Bentwich Z, Beverley PCL, Hammarstrom L, Kalden JR, Lambert PH, Rose NR, Thompson RA. Laboratory investigations in clinical immunology: methods, pitfalls, and clinical indications. Clin Immunol Immunopathol. 1988;49:478–97.
101. Miller LE, Ludke HR, Peacock JE, Tomar RH, editors. Manual of laboratory immunology. 2nd ed. Lea and Fabiger: Philadelphia, PA; 1991.
102. Baken KA, Pennings JL, Jonker MJ, Schaap MM, de Vries A, van Steeg H, Breit TM, van Loveren H. Overlapping gene expression profiles of model compounds provide opportunities for immunotoxicity screening. Toxicol Appl Pharmacol. 2008;226:46–59.
103. Van Loveren H, Van Amsterdam JG, Vandebriel RJ, Kimman TG, Rumke HC, Steerenberg PS, Vos JG. Vaccine-induced antibody responses as parameters of the influence of endogenous and environmental factors. Environ Health Perspect. 2001;109:757–64.
104. Sleijffers A, Yucesoy B, Kashon M, Garssen J, De Gruijl FR, Boland GJ, Van Hattum J, Luster MI, Van Loveren H. Cytokine polymorphisms play a role in susceptibility to ultraviolet B-induced modulation of immune responses after hepatitis B vaccination. J Immunol. 2003;170:3423–8.
105. Diggle GE. Overview of regulatory agencies. In: Ballantyne B, Marrs T, Turner P, editors. General and applied toxicology. Basingstoke: Macmillan; 1993. p. 1071–90.
106. Putman E, Van der Laan JW, Van Loveren H. Assessing immunotoxicity: guidelines. Fund Clin Pharmacol. 2003;17:615–26.
107. Weaver J, Tsutsui N, Hisada S, Vidal J-M, Spanhaak S, Sawada J-I, Hastings KL, Van der Laan JW, Van Loveren H, Kawabata TT, Sims J, Durham SK, Fueki O, Matula T, Kusunoki H, Ulrich P, Nakamura K. Development of the ICH Guidelines on Immunotoxicology. Evaluation of pharmaceuticals using a survey of industry practices. J Immunotoxicol. 2005;2:171–80.
108. Buse E. Development of the immune system in the cynomolgus monkey: the appropriate model in human targeted toxicology? J Immunotoxicol. 2005;2:211–6.
109. Gans H, DeHovitz R, Forghani B, Beeler J, Maldonado Y, Arvin AM. Measles and mumps vaccination as a model to investigate the developing immune system: passive and active immunity during the first year of life. Vaccine. 2003;21:3398–405.
110. Holsapple MP, West LJ, Landreth KS. Species comparison of anatomical and functional immune system development. Birth Defects Res B Dev Reprod Toxicol. 2003;68:321–34.
111. Hessel EV, Tonk EC, Bos PM, van Loveren H, Piersma AH. Developmental immunotoxicity of chemicals in rodents and its possible regulatory impact. Crit Rev Toxicol. 2015;45:68–82.
112. Doedée AMCM, Boland GJ, Pennings JLA, de Klerk A, Berbers GAM, van der Klis FRM, de Melker HE, Van Loveren H, Janssen R. Effects of prophylactic and therapeutic paracetamol treatment during vaccination on hepatitis B antibody levels in adults: two open-label, randomized controlled trials. PLoS ONE. 2014;9(6):e98175.

Appendix 1: Licensed Cytokines

Drug	Corporation	Indication	Approval
Actimmune® interferon-gamma 1b (IFN-γ-1b)	InterMune Pharm.	Management of chronic granulomatous disease (CGD)	December 1, 1990
Actimmune® IFN-γ-1b	InterMune Pharm.	Osteopetrosis	February 1, 2000
Alferon N Injection® interferon alpha n3 (IFN-α-n3), human leukocyte derived	Interferon Sciences	Genital warts	October 1, 1989
Betaseron® recombinant interferon beta-1b (IFN-β-1b)	Berlex Laboratories	Relapsing, remitting multiple sclerosis (MS)	July 1, 1993
Bone morphogenetic protein 2 (BMP-2)	Medtronic Sofamor Danek	Treatment of spinal degenerative disc disease	July 2, 2014
Denosumab	Genentech	Treatment and prevention of osteopetrosis (EU)	December 9, 2014
Enbrel®, tumor necrosis factor receptor:Fc (TNFR:Fc)	Amgen	Active ankylosing spondylitis	July 3, 2014
Enbrel® TNFR:Fc	Immunex	Moderate to severe active rheumatoid arthritis (RA)	November 1, 1998
Enbrel® TNFR:Fc	Immunex	Moderate to severe active juvenile RA	May 1, 1999
EPOGEN® epoetin alfa (rEPO)	Amgen	Anemia caused by chemotherapy	April 1, 1993
EPOGEN® epoetin alfa (rEPO)	Amgen	Anemia, chronic renal failure, anemia in Retrovir®-treated human immunodeficiency virus (HIV)-infected	June 1, 1989
EPOGEN® epoetin alfa (rEPO)	Amgen	Chronic renal failure, dialysis	November 1, 1999
EPOGEN® epoetin alfa (rEPO)	Amgen	Surgical blood loss	December 1, 1996
Erythropoiesis-stimulating agent (ESA) darbepoetin alfa	Roche	Stimulating agent	June 23, 2005
Infergen® IFN alfacon-1	Amgen	Treatment of chronic hepatitis C viral infection	October 1, 1997
Intron® A IFN-α-2b	Schering-Plough	Acquired immune deficiency syndrome (AIDS)-related Kaposi's sarcoma	November 1, 1988
Intron® A IFN-α-2b	Schering-Plough	Follicular lymphoma	November 1, 1997
Intron® A IFN-α-2b	Schering-Plough	Follicular lymphoma	November 1, 1997
Intron® A IFN-α-2b	Schering-Plough	Genital warts	June 1, 1988
Intron® A IFN-α-2b	Schering-Plough	Hairy cell leukemia	June 1, 1986
Intron® A IFN-α-2b	Schering-Plough	Hepatitis B	July 1, 1992
Intron® A IFN-α-2b	Schering-Plough	Hepatitis C	February 1, 1991
Intron® A IFN-α-2b	Schering-Plough	Malignant melanoma (MM)	December 1, 1995

(continued)

M. J. Parnham et al. (eds.), *Nijkamp and Parnham's Principles of Immunopharmacology*,
https://doi.org/10.1007/978-3-030-10811-3

Drug	Corporation	Indication	Approval
Leukine™ sargramostim granulocyte-macrophage colony-stimulating factor (GM-CSF)	Immunex	Allogeneic bone marrow (BM) transplantation	November 1, 1995
Leukine™ sargramostim (GM-CSF)	Immunex	Post auto or allotransplant with delayed or failed engraftment	December 1991
Leukine™ sargramostim (GM-CSF)	Immunex	Autologous BM transplantation	March 1, 1991
Leukine™ sargramostim (GM-CSF)	Immunex	Neutropenia resulting from chemotherapy	September 1, 1995
Leukine™ sargramostim (GM-CSF)	Immunex	PB progenitor cell mobilization	December 1, 1995
Leukocyte function antigen-1 (LFA-1)/IgG1	Biogen	Moderate to severe chronic plaque psoriasis	January 3, 2014
Mircera®, methoxy polyethylene glycol epoetin beta	Roche	Anemia associated with chronic renal failure	November 7, 2014
Neumega®, oprelvekin, recombinant human interleukin-11 (rHu IL-11)	Genetics Institute	Chemotherapy-induced thrombocytopenia	November 1, 1997
NEUPOGEN® filgrastim, recombinant granulocyte colony-stimulating factor (rG-CSF)	Amgen	Acute myelogenous leukemia (AML)	April 1, 1998
NEUPOGEN® Filgrastim (rG-CSF)	Amgen	Autologous or allogeneic BM transplantation	June 1994
NEUPOGEN® Filgrastim (rG-CSF)	Amgen	Chemotherapy-induced neutropenia	February 1, 1991
NEUPOGEN® Filgrastim (rG-CSF)	Amgen	Chronic severe neutropenia	December 1, 1994
NEUPOGEN® Filgrastim (rG-CSF)	Amgen	PB progenitor cell transplantation	December 1, 1995
NEUPOGEN® Filgrastim (rG-CSF)	Amgen	Reduction of time required for polymorphonuclear cell (PMN) recovery post chemotherapy in adults with AML	April 1998
Nplate®, romiplostim peptide agonist for thrombopoietin receptor (TPO-R)	Amgen	Treatment of thrombocytopenia in patients with idiopathic thrombocytopenia purpura	August 8, 2014
Ontak®, diphtheria toxin—IL-2	Sevagen	Cutaneous T-cell lymphoma (CTCL)	February 1, 1999
Palifermin® recombinant human (rHu) keratinocyte growth factor (KGF)	Amgen	Reduction of incidence and duration of severe osteomucositis	2004
Proleukin®, aldesleukin, IL-2	Chiron	Metastatic melanoma	January 1998
Proleukin®, aldesleukin, IL-2	Chiron	Renal cell carcinoma (RCC)	May 1, 1992
Rebetron™ ribavirin/IFN-α-2b	Schering-Plough	Chronic hepatitis C	June 1, 1998
Rebetron™ ribavirin/IFN-α-2b	Schering-Plough	Chronic hepatitis C, compensated liver disease	December 1, 1999
Rebif®, Avonex® IFN-β-1a	Biogen/Idec/Serono	Relapsing forms of MS	March 2, 1996
Regranex®, becaplermin, recombinant human platelet-derived growth factor BB (rHPDGF-BB)	Ortho-McNeil	Diabetic neuropathy, foot ulcers	December 1, 1997

Drug	Corporation	Indication	Approval
Roferon®, IFN-α-2a	Hoffman-La Roche	AIDS-related Kaposi's sarcoma	November 1, 1988
Roferon®, IFN-α-2a	Hoffman-La Roche	Chronic myelogenous leukemia	November 1, 1995
Roferon®, IFN-α-2a	Hoffman-La Roche	Hairy cell leukemia	June 1, 1986
Roferon®, IFN-α-2a	Hoffman-La Roche	Hepatitis C	November 1, 1996
Stemgen®, stem cell factor (SCF)	Amgen	Mobilization (Australia, New Zealand, Canada)	June 19, 1995
Tbo-filgrastim, Granix®	Teva	Severe neutropenia in patients with nonmyeloid malignances	August 30, 2012
Wellferon®, IFN-n	Glaxo Welcome	Treatment of hepatitis C in patients 18 years-of-age or older without decompensated liver disease	March 1, 1999

Appendix 2: Recombinant Cytokines in Clinical Development

Target	Company	Activity	Indications
Diphtheria toxin (DT) fused to IL-3 (SL-401)	University of Texas Southwestern Medical Center	Blastic plasmacytoid dendritic cell (DC) neoplasm (BPDCN) tumor cells over express IL-3R	BPDCN, acute myeloid leukemia (AML), and myelodysplasia
Fms-like tyrosine kinase 3 (Flt3) ligand	Celldex	Hematopoietic growth factor both MDSC, HSC, and DC	Stem cell mobilization, vaccine adjuvant
IL-1b (OCT-43)	Otsuka	IL-1b affects hematopoiesis and inflammation	Myelosuppression
IL-3	Immunex/ Behringwerke	Stimulates progenitor cell proliferation and maturation, as well as differentiated cell functions including histamine release	Myelosuppression
IL-4	Schering-Plough	Regulation of monocyte differentiation, regulation of T-cell function	Renal cell carcinoma (RCC), melanoma, B-chronic lymphocytic leukemia (B-CLL), non-Hodgkin lymphoma (NHL)
IL-6	Sandoz	Immunoregulatory, hematopoietic, and inflammatory activities	Thrombocytopenia cancer patients
IL-7 (CYT107)	Cytheris	T-cell development, hematopoiesis, and post-developmental immune functions	Acquired immunodeficiency syndrome (AIDS), various solid tumors
IL-10 ilodecakin/ Tenovil®	Schering-Plough	Immunoregulatory cytokine with anti-inflammatory activity, suppression of COX-2, NK-cell activation	Psoriasis, rheumatoid arthritis (RA), graft-versus-host disease (GVHD), irritable bowel disease (IBD)
IL-12	National Cancer Institute (NCI)	Cytokine mediator of hematopoiesis and T-cell immunity, significant toxicity	Head and neck squamous cell carcinoma (HNSCC), normal, melanoma, RCC, ovarian cancer, NHL, HD
IL-15	NCI	Cytokine activator of antitumor CD8 T cells and natural killer (NK) cells	Recurrent or metastatic melanoma, RCC, H&NSCC, NSCLC
IL-18	GlaxoSmithKline (GSK)	Immunostimulatory cytokine that regulates both innate and adaptive immune responses	Lymphoma, melanoma, ovarian cancer
IL-21 (denenicokin)	ZymoGenetics, Novo Nordisk	Angiostatic activity, increases in NK- and T-cell activity	Recurrent or metastatic melanoma, RCC, NHL, ovarian cancer
IL-22, F-652	Generon Corp. (Shanghai)	Regulation of cells of epithelial origin	Hepatic injury
IL-24	Introgen	Adenovirus vector with IL-24 transgene injected intratumorally to induce tumor cell apoptosis	Advanced cancer patients

M. J. Parnham et al. (eds.), *Nijkamp and Parnham's Principles of Immunopharmacology*,
https://doi.org/10.1007/978-3-030-10811-3

Appendix 3: Chemical Structures of the NSAIDS

Propionic acid derivatives

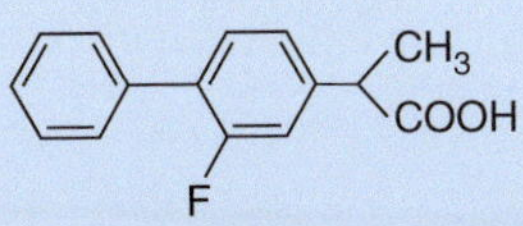

Flurbiprofen

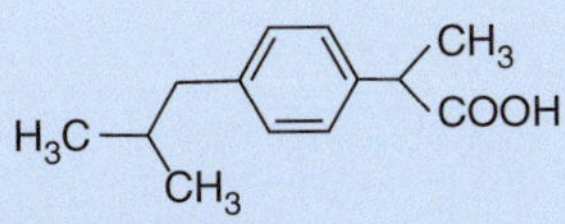

Ibuprofen

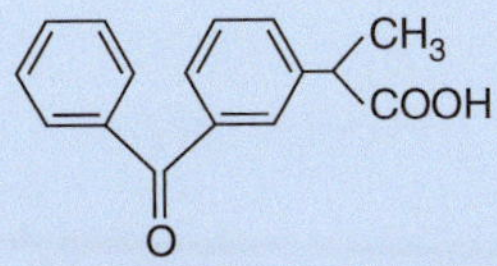

Ketoprofen

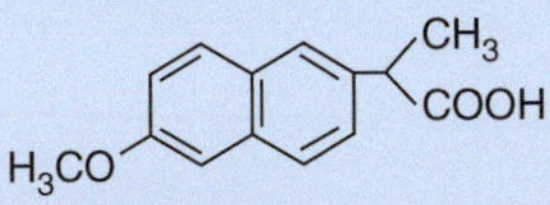

Naproxen

Acetic acid derivatives

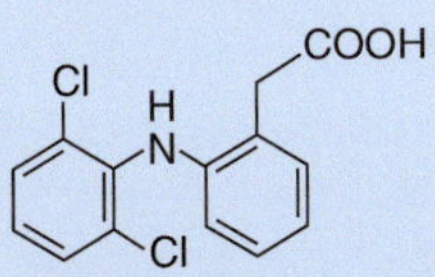

Diclofenac

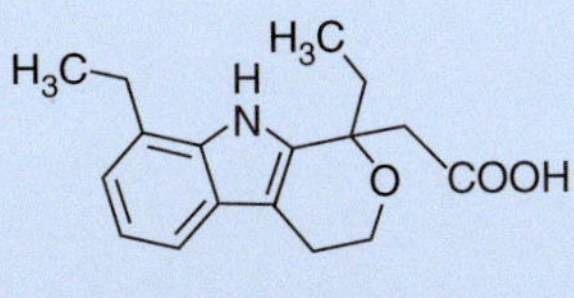

Etodolac

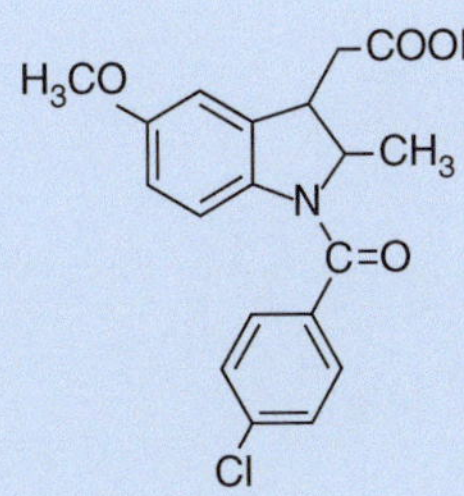

Indomethacin

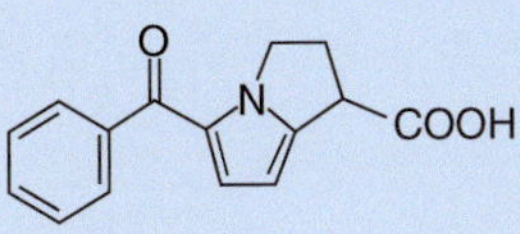

Ketorolac

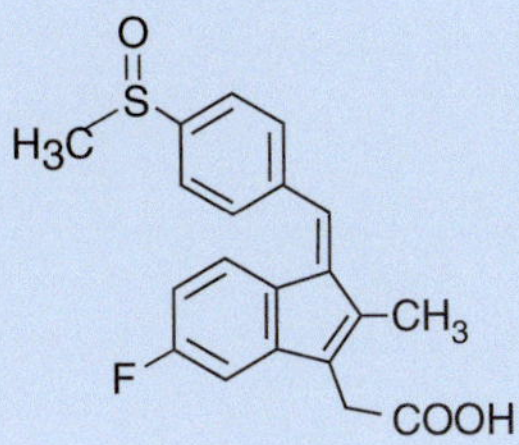

Sulindac

M. J. Parnham et al. (eds.), *Nijkamp and Parnham's Principles of Immunopharmacology*,
https://doi.org/10.1007/978-3-030-10811-3

Oxicam derivatives

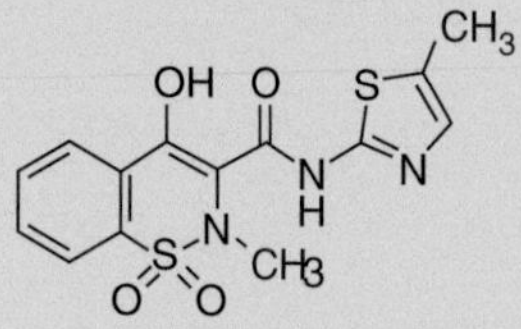
Meloxicam

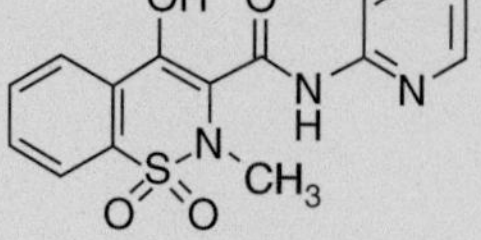
Piroxicam

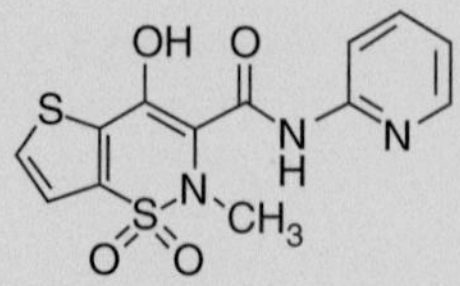
Tenoxicam

Fenamate derivatives

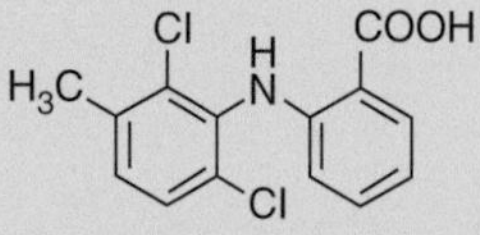
Meclofenamic acid

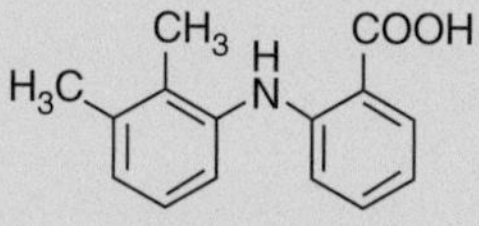
Mefenamic acid

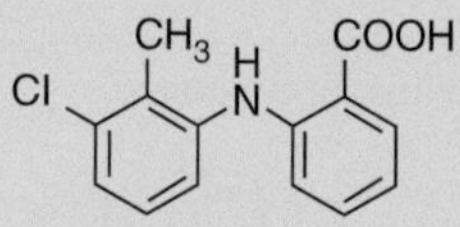
Tolfenamic acid

Salicylates

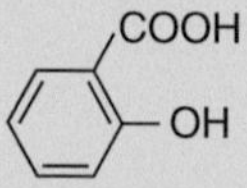
Salicylic acid

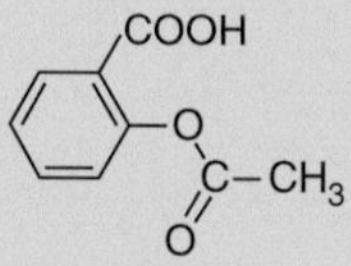
Acetylsalicylic acid (ASA)

Coxibs

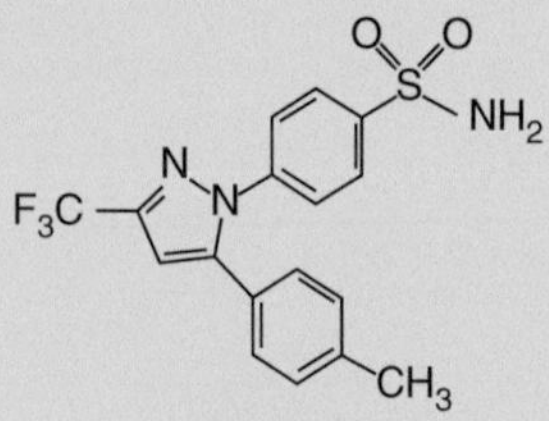
Celecoxib

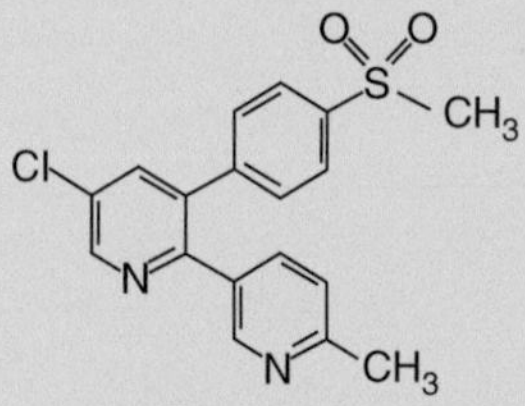
Etoricoxib

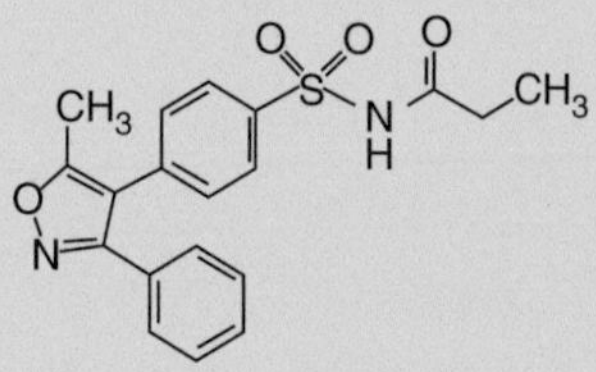
Parecoxib

Other NSAIDs

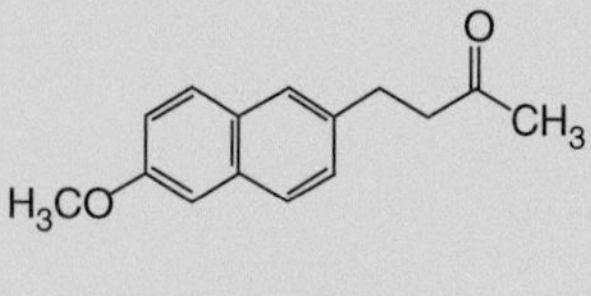
Nabumetone

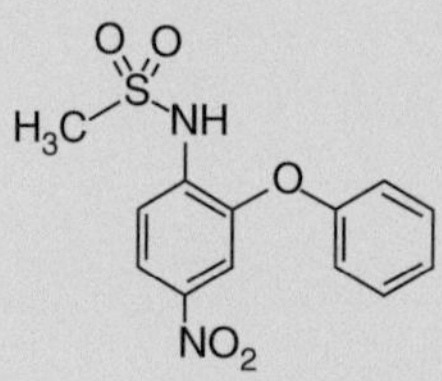
Nimesulide

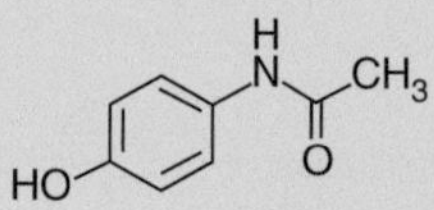
Paracetamol (acetaminophen)

Appendix 4: Overview of Major DMARDs Use in the Treatment of RA

Drug	Approximate time to benefit	Usual maintenance dose for arthritic diseases	Toxicity
Anti-TNF antibodies			Infections (all), demyelinating disorders, worsening cardiac failure, lupus-like reactions, malignancies, TB and hepatitis B/C reactivation, injection/infusions of fever, hypotension, chills, rash, local irritation
Adalimumab	1 month	s.c. 40 mg every 2 weeks. Alone or with other DMARDs	
Infliximab	1 month	i.v. 3 mg/kg every 8 weeks. Used with methotrexate	
Golimumab	1 month	s.c. 50 mg monthly. i.v. 2–4 mg every 3 months. Used with methotrexate	
Certolizumab pegol	1 month	s.c. 400 mg 0, 2 and 4 weeks then 200 mg fortnightly. Used with methotrexate	
TNF receptor construct			
Etanercept	1 month	s.c. 25 mg twice weekly or 50 mg weekly. Used alone or with methotrexate	Infections, injection site reactions
Anti-CD20 antibody			
Rituximab	1 month	i.v. 1 g repeated after 2 weeks. Used with methotrexate	Infusion reactions, increasing hepatitis C viral load
Anti-IL-6 antibody			
Tocilizumab	1 month	i.v. 2–8 mg every 4 weeks. Used with methotrexate	Neutropaenia, elevations of plasma cholesterol and hepatic transaminases
Anti-IL-12/23			
Ustekinumab	4 months	s.c. 45–90 mg at 0 and 4 weeks then once every 12 weeks	Increased risk of infections including tuberculosis
Anti-IL-17			

(*continued*)

M. J. Parnham et al. (eds.), *Nijkamp and Parnham's Principles of Immunopharmacology*,
https://doi.org/10.1007/978-3-030-10811-3

Drug	Approximate time to benefit	Usual maintenance dose for arthritic diseases	Toxicity
Secukinumab	1–2 months	s.c. 150–300 mg weekly for 4 weeks then 150–300 mg every 4 weeks	Nasopharyngitis, diarrhoea, rare neutropaenia, increased fungal infections
CD80/86 binding			
Abatacept	2–3 months	i.v. 500–1000 mg at 0 and 2 weeks then every 4 weeks. Used with methotrexate	Should not be combined with etanercept
csDMARDs			
Methotrexate	1–2 months	Oral initial as a single dose or split into 3 doses over 36 h. Maintenance individualized dose 15–25 mg/week	Gastrointestinal (oral ulcers stomatitis, nausea vomiting, diarrhoea) CNS (headache, fatigue, "fuzziness"), rash alopecia, myelosuppression, hepatotoxicity, rare but serious pneumonitis
Hydroxychloroquine	2–4 months	Oral initial 400–600 mg daily reducing to 200–400 mg daily.	Infrequent rash, diarrhoea, rare retinal toxicity
Leflunomide	1–4 months	Oral 10–20 mg daily. Rarely preceded by loading dose, 100 mg daily for 3 days	Diarrhoea, dyspepsia, mild leukopaenia, rash, alopecia, headache, dizziness, possible severe hepatotoxicity
Sulfasalazine	1–2 months	Oral 1000 mg two or three times daily	Rash, infrequent myelosuppression, gastrointestinal intolerance
tsDMARDs			
Tofacitinib	2 weeks	Oral 5–10 mg twice daily	Increased blood cholesterol, increased incidence of herpes zoster, rare anaemia

Appendix 5: Overview of Drugs Used for Soft Tissue Autoimmunity

Tabular overview of all relevant marketed drugs/therapeutics with the following information: chem. name (INN), trade name, formulation, usual dose (range), pharmacokinetics, indications and contraindications.

M. J. Parnham et al. (eds.), *Nijkamp and Parnham's Principles of Immunopharmacology*,
https://doi.org/10.1007/978-3-030-10811-3

Chemical name (INN)	Trade name[a]	Formulation	Usual dose (range)[b]	Pharmacokinetics	Indications	Contraindications
Mesalazine	Salofalk® Pentasa® Mezavant® Claversal®	Oral, rectal	2–4 g/day (active ulcerative colitis) 1.5–4 g/day (maintenance of remission in UC)	– (5-ASA) Tmax mean 5.5 h, Median 6.5 – Cmax 10.9 nmol/ml, Median 8.5 – AUC Mean 38.3 nmol/ml × h, Median 18.3[c] [1]	Mild to moderate ulcerative colitis	– Hypersensitivity – Severe liver or kidney disease
Sulfasalazine	Azulfidine® Salazopyrin®	Oral	500 mg to maximum 4 g/day	– (5-ASA) Tmax 11.6–14.3, Median 15 h – Cmax 0.7–3.5 nmol/ml – AUC Mean 9.6–27.5 nmol/ml × h [1]	Mild to moderate ulcerative colitis, especially in case of inflammatory arthritis	– Hypersensitivity – Severe liver or kidney disease – Porphyries – Bone marrow deficiency (leucopenia)
Azathioprine	Imurek® Imuran®	Oral	2.5 mg/kg body weight (IBD) 1–1.5 mg/kg body weight (AIH)	– (6-MP) Tmax median 2 h – Cmax 16.9 ng/ml – AUC Mean 41.6 ng/ml × h (after 50 mg Azathioprine) [2]	– Severe rheumatoid arthritis not responding to less toxic DMARD – Moderate to severe ulcerative colitis and Crohn's disease – Lupus erythematodes – Dermatomyositis – Autoimmune hepatitis – Polyarteritis nodosa – Refractory haemolytic anaemia – Chronic refractory idiopathic thrombocytopenic purpura	– Hypersensitivity – Severe infection – Severe liver disease – Bone marrow deficiency – Pancreatitis – Vaccination with live vaccines (pregnancy, breastfeeding)[d]
6-Mercaptopurine	Purinethol®	Oral	1.5 mg/kg body weight	– (6-MP) Tmax 0.5–2 h – Cmax 10–38 ng/ml – AUC Mean 13–45 ng/ml × h (after 75 mg/m^2 6-MP) [3]	See azathioprine	See azathioprine

Infliximab	Remicade® *Biosimilars*: Remsima® Inflectra®[e] Flixabi®	Intravenously (IV)	3–5 mg/kg body weight in week 0, 2, 6, then every 8 weeks	– Cmax 132.3 μg/ml – AUC Mean 32526.76 μg/ml × h – Half-life T½-α 5.1 (men), 4 (women) days – T½-β (days) 19.7 (men), 18.0 (women) days (after 5 mg/kg Infliximab) ([4], [5])	– Severe rheumatoid arthritis not responding to less toxic DMARD – Moderate to severe ulcerative colitis and Crohn's disease – Ankylosing spondylitis – Psoriasis – Psoriatic arthritis	– Hypersensitivity – Severe infection: tuberculosis, sepsis, abscess, opportunistic infections – Congestive heart failure NYHA class III-IV – History of malignancy[f] (pregnancy, breastfeeding)[d]
Adalimumab	Humira®	Subcutaneous injection	Initial dose 160 mg in week 0, followed by 80 mg in week 2, then 40 mg every other week	– Tmax 190.75 h – Cmax 3.6 μg/ml – AUC Mean 2167.38 μg/ml × h (after 40 mg Adalimumab) [6]	– Severe rheumatoid arthritis not responding to less toxic DMARD – Juvenile idiopathic arthritis – Moderate to severe ulcerative colitis and Crohn's disease – Ankylosing spondylitis – Psoriasis – Psoriatic arthritis – Uveitis	– Hypersensitivity – Severe infection: tuberculosis, sepsis, abscess, opportunistic infections – Congestive heart failure NYHA class III-IV – History of malignancy[f] (pregnancy, breastfeeding)[d]
Golimumab	Simponi®	Subcutaneous injection	Initial dose 200 mg in week 0, followed by 100 mg in week 2, then 50/100 mg every other week[g]	– Tmax 2–6 days – Cmax 3.2 ± 1.4 μg/ml (50 mg Golimumab s.c.) Mean steady state trough concentration 1.8 ± 1.1 μg/ml (100 mg every 4 weeks s.c.) [7]	– Severe rheumatoid arthritis not responding to less toxic DMARD – Juvenile idiopathic arthritis – Moderate to severe ulcerative colitis[h] – Ankylosing spondylitis – Psoriasis – Psoriatic arthritis	– Hypersensitivity – Severe infection: tuberculosis, sepsis, abscess, opportunistic infections – Congestive heart failure NYHA class III-IV – History of malignancy[f] (pregnancy, breastfeeding)[d]

(*continued*)

Chemical name (INN)	Trade name[a]	Formulation	Usual dose (range)[b]	Pharmacokinetics	Indications	Contraindications
Etanercept	Enbrel®	Subcutaneous injection	50 mg weekly	– Cmax 1.37 ± 0.72 μg/ml – AUC being 217 ± 85.7 μg h/ml [8]	– Severe rheumatoid arthritis not responding to less toxic DMARD – Juvenile idiopathic arthritis – Ankylosing spondylitis – Psoriasis – Psoriatic arthritis	– Hypersensitivity – Severe infection: tuberculosis, sepsis, abscess, opportunistic infections – Congestive heart failure NYHA class III-IV – History of malignancy[f] (pregnancy, breastfeeding)[d]
Vedolizumab	Entyvio®	Intravenously (IV)	300 mg in week 0, 2, 6, then every 8 weeks	– $T_{1/2}$ 18.3 days – Cmax 115 μg/ml – AUC Mean 1990 μg/ml × day (after 300 mg vedolizumab) [9]	– Moderate to severe ulcerative colitis and Crohn's disease not responding to conventional therapy and/or contraindications for conventional therapy	– Hypersensitivity – Severe infection: tuberculosis, sepsis, abscess, opportunistic infections (pregnancy, breastfeeding)[d]
Ustekinumab	Stelara®	IV and subcutaneous injection	Initial dose 55 kg or less: 260 mg, 55–85 kg: 390 mg, >85 kg: 520 mg IV, then 90 mg s.c. every 8/12 weeks	– $T_{1/2}$ 15–32 days – Tmax 8.5 days – Cmax 3.1–14.1 μg/mL (after 90 mg ustekinumab) [10]	– Psoriasis – Psoriatic arthritis – Moderate to severe Crohn's disease[i]	– Hypersensitivity – Severe infection: tuberculosis, sepsis (pregnancy, breastfeeding)[d]

[a]Due to multiple trade names for mesalazine and sulfasalazine only the most relevant trade names in Europe are listed
[b]Regarding the indications ulcerative colitis, Crohn's disease and autoimmune hepatitis
[c]Exemplary data for Salofalk® 500 mg
[d]Pregnancy is a relative contraindication, in individual cases therapy can be continued during pregnancy, see text for details
[e]Pharmacokinetics comparable to the originator product, see [5]
[f]Individual therapy decision, relative contraindication
[g]Body weight >80 kg: 100 mg every other week, body weight <80 kg: 50 mg every other week
[h]Not approved for treatment of Crohn's disease
[i]Not approved for treatment of ulcerative colitis

References

1. Sandborn WJ, Hanauer SB. Systematic review: the pharmacokinetic profiles of oral mesalazine formulations and mesalazine pro-drugs used in the management of ulcerative colitis. Aliment Pharmacol Ther. 2003;17(1):29–42.
2. Van Os EC, Zins BJ, Sandborn WJ, Mays DC, Tremaine WJ, Mahoney DW, Zinsmeister AR, Lipsky JJ. Azathioprine pharmacokinetics after intravenous, oral, delayed release oral and rectal foam administration. Gut. 1996;39(1):63–8.
3. Endresen L, Lie SO, Storm-Mathisen I, Rugstad HE, Stokke O. Pharmacokinetics of oral 6-mercaptopurine: relationship between plasma levels and urine excretion of parent drug. Ther Drug Monit. 1990;12(3):227–34.
4. Ternant D, Aubourg A, Magdelaine-Beuzelin C, Degenne D, Watier H, Picon L, Paintaud G. Infliximab pharmacokinetics in inflammatory bowel disease patients. Ther Drug Monit. 2008;30(4):523–9.
5. Yoo DH, Oh C, Hong S, Park W. Analysis of clinical trials of biosimilar infliximab (CT-P13) and comparison against historical clinical studies with the infliximab reference medicinal product. Expert Rev Clin Immunol. 2015;11(Suppl 1):S15–24.
6. Hyland E, Mant T, Vlachos P, Attkins N, Ullmann M, Roy S, Wagner V. Comparison of the pharmacokinetics, safety, and immunogenicity of MSB11022, a biosimilar of adalimumab, with Humira((R)) in healthy subjects. Br J Clin Pharmacol. 2016;82(4):983–93.
7. Harzallah I, Rigaill J, Williet N, Paul S, Roblin X. Golimumab pharmacokinetics in ulcerative colitis: a literature review. Therap Adv Gastroenterol. 2017;10(1):89–100.
8. Takeuchi T, Miyasaka N, Kawai S, Sugiyama N, Yuasa H, Yamashita N, Sugiyama N, Wagerle LC, Vlahos B, Wajdula J. Pharmacokinetics, efficacy and safety profiles of etanercept monotherapy in Japanese patients with rheumatoid arthritis: review of seven clinical trials. Mod Rheumatol. 2015;25(2):173–86.
9. Rosario M, Dirks NL, Milch C, Parikh A, Bargfrede M, Wyant T, Fedyk E, Fox I. A review of the clinical pharmacokinetics, pharmacodynamics, and immunogenicity of vedolizumab. Clin Pharmacokinet. 2017;56(11):1287–301.
10. Famenini S, Wu JJ. The safety of ustekinumab in psoriasis. J Drugs Dermatol. 2012;11(8):907–10.

Appendix 6: ICH S8 Note for Guidance on Immunotoxicity Studies for Human Pharmaceuticals

Objectives of the Guideline.

The objectives of this guideline are to provide (1) recommendations on nonclinical testing approaches to identify compounds which have the potential to be immunotoxic and (2) guidance on a weight-of-evidence decision-making approach for immunotoxicity testing. Immunotoxicity is, for the purpose of this guideline, defined as unintended immunosuppression or enhancement. Drug-induced hypersensitivity and autoimmunity are excluded.

Overview.

The general principles that apply to this guideline are:

1. All new human pharmaceuticals should be evaluated for the potential to produce immunotoxicity.
2. Methods include standard toxicity studies (STS) and additional immunotoxicity studies conducted as appropriate. Whether additional immunotoxicity studies are appropriate should be determined by a weight of evidence review of the following factor(s).

Factors to consider:

1. Standard toxicity studies.
2. Pharmacological properties.
3. Intended patient population.
4. Structural similarity.
5. Disposition of the drug.
6. Signs observed in clinical trials or clinical use.

Regarding Standard Toxicity Studies.

Changes in these parameters could reflect immunosuppression or enhanced activation of the immune system. Immunosuppression is usually reflected by reduced values of immune parameters, whereas immunoenhancement is usually reflected by increased values. However, these relationships are not absolute and can be inverted in some cases. Similar to the assessment of risk with toxicities in other organ systems, the assessment of immunotoxicity should include the following:

- Statistical and biological significance of the changes.
- Severity of the effects.
- Dose/exposure relationship.
- Safety factor above the expected clinical dose.
- Treatment duration.
- Number of species and endpoints affected.
- Changes that may occur secondarily to other factors (e.g. stress).
- Possible cellular targets and/or mechanism of action.
- Doses which produce these changes in relation to doses which produce other toxicities.
- Reversibility of effect(s).

Regarding Pharmacological Properties.

If the pharmacological properties of a test compound indicate it has the potential to affect immune function (e.g. anti-inflammatory drugs), additional immunotoxicity testing should be considered. Information obtained from the nonclinical pharmacology studies on the ability

M. J. Parnham et al. (eds.), *Nijkamp and Parnham's Principles of Immunopharmacology*,
https://doi.org/10.1007/978-3-030-10811-3

of the compound to affect the immune system could be used in a weight of evidence approach to decide if additional immunotoxicity studies are needed.

Weight of Evidence Review.

A weight of evidence review should be performed on information from all the factors outlined above to determine whether a cause for concern exists. A finding of sufficient magnitude in a single area should trigger additional immunotoxicity studies. Findings from two or more factors, each one of which would not be sufficient on its own, could trigger additional studies. If additional immunotoxicity studies are not performed, the sponsor should provide justification.

Glossary

Abatacept Humanized monoclonal antibody which is a fusion molecule of the extracellular domain of CTLA4 and IgG1 Fc domain. It is useful in the treatment of RA due to its binding of CD80/CD86.

Accuracy Accuracy describes how close the measured value is to the nominal value. Mostly expressed as % recovery.

Acetylcholine A neurotransmitter that controls several neurobiological functions such as attention, memory, sleep, and muscular activity. It is also a neurotransmitter in the autonomic nervous system, being the final product released by the parasympathetic nervous system.

Activator protein-1 (AP-1) A key DNA binding protein that is required to activate the expression of many inflammatory and immune genes.

ADA (antidrug antibodies) Antibodies produced in the host against the biotherapeutic that can be directed against the backbone structure or against the epitope where the latter leads to neutralization of the biological activity.

Adalimumab Complete human monoclonal antibody directed against human TNFα.

Adaptive (acquired) immunity Antigen-specific, lymphocyte-mediated defense mechanisms that take several days to become protective and are designed to remove the specific antigen.

ADCC (antibody-dependent cell cytotoxicity) Is the killing of an antibody-coated target cell by a cytotoxic effector cell through a nonphagocytic process followed by the release of cytotoxic granules or cell death-inducing molecules.

ADCP (antibody-dependent cell phagocytosis) Phagocytosis of target cells expressing FcγRIIa receptor (primarily) coated with antibody.

Adhesion molecule A protein that enables cells to interact with each other.

Adjuvant A substance co-administered with antigen with the aim of increasing the immunogenicity of the antigen. Usually, the adjuvant itself is not immunogenic.

Adoptive (acquired) immunity Refers to Ag-specific defense mechanisms that take several days to become protective and are designed to remove a specific Ag.

Adoptive immunotherapy Immunotherapeutic approach used to treat cancer or viral infection. Immune cells (usually lymphocytes) are collected from the patient and subjected to selection, proliferation, activation, and/or genetic modification to increase the number of immune cells able to fight cancer or certain infections. These cells are afterward reinfused into the patient.

Affinity column A separation column packed with an affinity resin, which selectively binds a specific substance and passes through all other constituents. (e.g., streptavidin affinity column binds only biotin and passes through all other constituents.)

Affinity maturation The affinity of a particular antibody is increased as a result of somatic hypermutation and selection of the B cell receptor by competition for pathogens.

M. J. Parnham et al. (eds.), *Nijkamp and Parnham's Principles of Immunopharmacology*,
https://doi.org/10.1007/978-3-030-10811-3

Affinity The strength with which an antibody or antigen molecule binds an epitope (= antigenic determinant).

Agammaglobulinemia An immune disorder characterized by very low levels of protective immunoglobulins; affected people develop repeated infections.

Age-related macular degeneration (AMD) Is an eye condition that affects your central vision, but not your peripheral vision, and occurs gradually over time.

Aggrecanases Proteolytic enzymes which break down proteoglycans in cartilage in RA and other arthritic diseases.

Agranulocytosis Loss of granulocytes, neutrophils, eosinophils and basophilic granulocytes. Agranulocytosis increases the risk of infectious diseases.

Allergen An antigen that provokes allergy.

Allergic Causing Allergy

Allergy Hypersensitivity caused by exposure to an exogenous antigen (allergen) resulting in a marked increase in reactivity and responsiveness to that antigen on subsequent exposure, resulting in adverse health effects.

Allogeneic Cell or tissue which is genetically different because of being derived from separate individuals of the same species.

Anakinra Therapeutic form of a recombinant human interleukin-1 receptor antagonist.

Analyte A molecule being quantified or analyzed.

Anaphylaxis This is a type of immune-mediated hypersensitivity reactions. Anaphylactic (type I) or immediate hypersensitivity reactions involve specific IgE antibodies. The response is a serious reaction, which develops fast and may cause death. Typical symptoms are itchy rash, throat and/or tongue swelling, bronchoconstriction and reduced blood pressure.

Angiogenesis The formation of new blood vessels.

Angiostasis Cessation of angiogenesis.

Ankylosing spondylitis An inflammatory arthritis primarily of the sacroiliac joints and spinal ligaments, but other large peripheral joints may be affected. Inflammation of the anterior chamber of the eye is common. Most patients have the antigen HLA-B27.

Anthroposophy A system of beliefs and practices based on the philosophy of Rudolf Steiner which claims to integrate the practical and psychological in child-centred education.

Antibacterial agents Compounds that have one or two actions: bactericidal agents which kill bacteria; bacteriostatic agents which stop the growth of bacteria.

Antibody (Ab; see also Immunoglobulin) Immunoglobulin molecule produced by B lymphocytes in response to immunization/sensitization with a specific antigen that specifically reacts with that antigen.

Antidrug antibodies Antibodies against the drug. They are produced by an organism (e.g., humans) after drug exposure (e.g., injected therapeutic antibody) or therapeutic peptide. Also known as anti-therapeutic antibodies (ATA).

Antigen (Ag) Substance to which a specific immunological reaction mediated by either antibody or lymphocyte is directed.

Antigen-presenting cell (APC) A cell that presents antigen to lymphocytes, enabling its specific recognition by receptors on the cell surface. In a more restricted way, this term is used to describe MHC class II-positive (accessory) cells that are able to present (processed) antigenic peptides complexed with MHC class II molecules to T helper-inducer lymphocytes. These cells include macrophage populations (in particular Langerhans cells and dendritic or interdigitating cells), B lymphocytes, activated T lymphocytes and certain epithelial and endothelial cells (after MHC class II antigen induction by, e.g. interferon-γ).

Anti-IgE Antibody directed against immunoglobulin E; omalizumab is the first therapeutically used monoclonal anti-IgE antibody for the treatment of allergy and asthma.

Anti-inflammatory Term applied to drugs and cytokines which decrease inflammation.

Anxiety disorders A group of psychiatric conditions characterized by a state of chronic and/or recurrent pathological anxiety defined as an excessive uneasiness and apprehension.

Aorta-gonad-mesonephros Early-stage embryonic structure that gives rise to the large blood vessels, gonads, and kidney.

Apoptosis (programmed cell death) A genetically determined process whereby the cell self-destructs after activation, by Ca^{2+}-dependent endonuclease-induced nuclear DNA (chromosomal) fragmentation into sections of about 200 base pairs. It can be spontaneous (as in neutrophils, explaining the short life span of these cells) or induced (as in tumor cells by interaction with cytotoxic lymphocytes or NK cells).

Asthma Respiratory disease due to airway constriction, associated with chronic inflammation and airway remodelling; either due to repeated inhaled exposure to allergen or to other environmental agent.

Attenuated vaccine Vaccine based on live bacteria or viruses that are made non-virulent, usually via serial passage in vitro. The mechanism by which mutations are introduced via this empirical approach is not well understood.

Autoantibodies Immunoglobulins (antibodies) that are directed against endogenous molecules of the host. They circulate in the serum but may be also detectable in other body fluids or bound in target tissue structures. Autoantibodies may occur as a part of the natural immunoglobulin repertoire (natural antibodies) or are induced by different mechanisms (non-natural or pathological autoantibodies). A number of non-natural autoantibodies are diagnostic markers of defined autoimmune diseases regardless of their pathogenetic activity. They may be directed against conserved non-organ-specific autoantigens, organ-specific autoantigens or cell-specific autoantigens.

Autoantigens Self-antigens of the organism, which are also targets for autoimmune responses of autoreactive B cells (see autoantibodies) or T cells, including proteins, glycoproteins, nucleic acids, phospholipids and glycosphingolipids.

Autoimmune disease Disease caused by antibodies or T cells targeting self-antigens.

Autoimmunity A state of immune reactivity towards self-constituents (*see* Autoantigens) that may be either destructive or non-destructive. Destructive autoimmunity is associated with the development of autoimmune diseases.

Autologous Cells or tissues obtained from the same individual.

Azathioprine Cytostatic immunosuppressant used for autoimmune disease which mainly decreases the number of T lymphocytes.

B lymphocyte (B cells) Lymphocytes expressing immunoglobulin (antibody) surface receptors (on naive B cells IgM and IgD) that recognize nominal antigen and, after activation, proliferate and differentiate into antibody-producing plasma cells. During a T cell-dependent antibody-producing process, there is an immunoglobulin class switch (IgM into IgG, lgA, IgD, or IgE) by the cells with maintenance of the antigen-combining structure. For T cell-independent antigens, cells differentiate only into IgM- or IgA-producing plasma cells.

Baricitinib An inhibitor of JAK1/2 which has shown activity in clinical trials on RA.

Basiliximab Chimeric mouse/human monoclonal antibody against the IL-2 receptor.

Basophils White blood cells with granules that stain with basic dyes and which have a function similar to mast cells (*see* Mast cell).

Bell-shaped Symmetrical curve of a normal distribution of data, sometimes used in reference to immune augmentation.

Beta galactosidase A glycoside hydrolase enzyme that catalyzes the hydrolysis of substrates (β-galactosidase) into monosaccharides. In assay systems, beta galactosidase is commonly used to hydrolyze substrates to produce chromogenic, fluorescent, or luminescent molecules, which can be detected.

Bioactivity The effect of a given drug or biologic, such as a vaccine or cytokine, upon a living organism or on living tissue.

Bioavailability The degree to which a drug or other substance becomes available to the target tissue after administration.

Biological response modifiers (BRM) Natural, synthetic, or engineered products that are used to boost, suppress, direct, or restore the body's ability to fight the disease.

Biologicals/Biopharmaceuticals Proteins, poly- or monoclonal antibodies and fusion proteins generated by recombinant DNA technology.

Biophenols Biologically active compounds present in a variety of natural sources containing *a single* benzene ring substituted with one or more OH groups.

Biosimilars Biological DMARDs which are similar to an innovator's product.

Biotherapeutic Therapeutic microorganism that has a beneficial effect because of its antagonistic activities against specific pathogens for the prevention and treatment of diseases.

Bipolar disorder Also known as manic-depressive illness, it is a chronic psychiatric disorder characterized by alternating episodes of depression or mania.

Blood coagulation The protease-controlled cascade, initiated by surface activation of blood, by which fibrin is generated, blood is changed to a gel and the hemostatic process of stopping blood loss from a damaged vessel is initiated.

Blood dyscrasias A general term to describe any abnormality in the blood or bone marrow's cellular components, such as low white blood cell count, low red blood cell count or low platelet count.

Bone marrow Soft tissue in hollow bones, containing hematopoietic stem cells and precursor cells of all blood cell subpopulations (primary lymphoid organ). This is a major site of plasma cell and antibody production (secondary lymphoid organ).

Brodalumab A human antibody which is used and on clinical trial for psoriasis, psoriatic arthritis and ankylosing spondylitis with activity through binding to the IL-17 receptor.

Brownian motion Random movement of particles resulting from collision of molecules or atoms in gas or liquid.

Canakinumab Complete human monoclonal antibody directed against IL-1β.

Cancer immunotherapy Therapeutic strategies use the immune system or its individual components to attack cancer. Different approaches have been explored, ranging from the use of antibodies against tumor-associated antigens, injection of antitumor immune effector, to inhibition of immune-restricting signaling pathways. In principal, cancer immunotherapy may activate antitumor effector mechanisms or block immune inhibitory mechanisms.

Carrier An immunogenic macromolecule (usually protein) to which a hapten is attached, allowing this hapten to be immunogenic. Term also applied to transporter carrying small molecules or ions through cell membranes.

Catecholamines A group of monoamine compounds that act as neurotransmitters/hormones and have sympathomimetic action.

CD Cluster of differentiation, e.g. CD4, CD8, etc. This is a standard naming system for cell surface proteins of the immune system. For example, CD4 and CD8 identify different subsets of T cells, and CD69 is a cell surface protein induced upon short-term activation of T cells.

CD3 Molecule consisting of at least four invariant polypeptide chains, present on the surface of T lymphocytes associated with the T cell receptor which mediates transmembrane signaling (tyrosine phosphorylation) after antigen binding.

CD4 Glycoprotein of 55 kDa on the surface of T lymphocytes and a proportion of monocytes/macrophages. On mature T cells, the presence is restricted to T helper cells; the molecule has an accessory function to antigen binding by the T cell receptor, by binding to a non-polymorphic determinant of the MHC class II molecule.

CD4+ T helper cells recognize antigenic peptide in association with MHC class II molecules. They mediate their effector functions by enhancing the persistence of antigen-stimulated T cells or through secretion of effector cytokines.

CD8 Complex of dimers or higher multimers of 32–34 kDa glycosylated polypeptides linked together by disulfide bridges, on the surface of T lymphocytes. On mature T cells, the presence of CD8 is restricted to cytotoxic T cells; the molecule has an accessory function to antigen binding by the T cell receptor, by binding to a non-polymorphic determinant of the MHC class I molecule.

CD8+ Cytotoxic T cells recognizing antigenic peptide in association with MHC class I molecules. CD8+ T cells mediate their effector functions by killing the cells presenting the relevant antigenic peptide or secreting effector cytokines.

Cell adhesion molecules (CAMS) Group of proteins of the immunoglobulin supergene family involved in intercellular adhesion, including ICAM-1, ICAM-2, VCAM-1 and PECAM-1.

Cell-mediated immunity Immunological reactivity mediated by T lymphocytes.

Cellular therapy Type of treatment involving the isolation of a patient or control mononuclear cells, in vitro manipulation of cells, and infusion of modified cells back to the patient.

Central nervous system (CNS) System composed of the brain and the spinal cord, being responsible for processing and integrating sensory information, planning and coordinating adequate responses, and providing control over the activities of other systems, including motor and autonomic systems.

Certolizumab pegol A humanized monoclonal anti-TNF antibody consisting of a Fab' fragment bound covalently to polyethylene glycol used in the treatment of chronic inflammatory diseases.

Chemiluminescence Phenomenon where light is produced as the result of a chemical reaction.

Chemokines Small-molecular-weight, pro-inflammatory peptide cytokines which attract cells of the immune system (chemotaxis) along a concentration gradient and activate them.

Chimeric protein (antibody) A human-engineered protein (antibody) that is encoded by a nucleotide sequence made by a splicing together of two or more complete or partial genes.

Chimeric antigen receptor T Artificial genetically engineered T-cell receptor, combining the specificity of a monoclonal antibody with the activation capacity of a T-cell receptor. Extracellularly this receptor consists of the antigen-binding domains of an antibody and intracellularly T-cell receptor and coreceptor activation domains. Modified T cells with chimeric antigen receptor are used as adoptive cellular immunotherapy to treat certain types of cancer.

Chimerism A genetic chimerism is the condition of a single organism or tissue that is composed of cells from two different sets of DNA.

Chloroquine An antimalarial drug which is also a csDMARD.

Crohn's disease (CD) Crohn's disease is a long-term chronic disease that involves inflammation of the digestive tract lining.

Chronic obstructive pulmonary disease (COPD) Describes a group of lung conditions that make it difficult to empty air out of the lungs as the airways have been narrowed.

Ciclosporin See Cyclosporin

Class switch The shift of a B cell or its progeny from the secretion of an immunoglobulin of one isotype or class to an immunoglobulin with the same V regions but a different heavy-chain constant region and, hence, a different isotype.

Clearance The volume of plasma or serum freed of a drug during a specific time interval.

Clinical hypothesis Theory(ies) to be tested during the clinical phase, such as dose, frequency, duration, and route of administration usually predetermined based on data from in vitro, animal, human safety, and dose-finding studies.

Colony-stimulating factors (CSFs) Cytokines which predominant inducing the differentiation of bone marrow precursor cells into mature leukocytes and activate non-lymphocytic leukocytes.

Combination therapy Treatment incorporating two or more types of therapy, i.e., surgery, chemotherapy, radiation therapy, hormonal therapy, gene therapy, drug therapy, etc. Investigators have found that one type of therapy does completely treat the disease or if the therapy is taken on its own over a period of time, the disease is no longer affected by the therapy. If two or more therapies are taken together, it can increase response rates and reduce the resistance rate.

Combination vaccine Vaccine containing antigens derived from more than one pathogen. Examples include diphtheria-tetanus, diphtheria-pertussis-tetanus, diphtheria-pertussis-tetanus-polio, and measles-mumps-rubella vaccines.

Compensation The application of a correcting factor to a fluorescence measurement, which deconvolutes the effects of optical spillover from other fluorochromes contributing fluorescence to a given detector.

Complement system Series of proteolytic enzymes in blood, capable of lysing microbes and enhancing the uptake of microbes by phagocytes.

Complementarity determining regions Three regions (CDR1, CDR2, and CDR3) of amino acid sequence in the immunoglobulin variable region that are highly divergent (hypervariable).

Complement-dependent cytotoxicity (CDC) The immune process involving complement by which the antibody-antigen complex activates a cascade of proteolytic enzymes that ultimately results in the formation of a terminal lytic complex that is inserted into a cell membrane, resulting in lysis and cell death.

Concanavalin-A (ConA) A plant mitogen which stimulates the production of four types of T-cells and immune-mediated liver injury but inhibits the development hepatoma in mice.

Conditioning therapy/regimen The treatment given to the recipient prior to the administration of hematopoietic stem cells.

Confocal microscopy Multiple two-dimensional images captured through a pinhole lens that blocks out-of-focus light, with subsequent reconstruction of a three-dimensional image.

Conventional synthetic DMARDs (csDMARDs) Older class of DMARDs which include the antimalarials, methotrexate, sulfasalazine, gold complexes and leflunomide.

Cord blood The blood collected from the placenta and umbilical cord after birth.

Corticosteroid (See also Glucocorticoid) Class of steroid hormones from the cortex of the adrenal gland and synthetic drugs with pronounced anti-inflammatory and immunomodulatory actions and metabolic and cardiovascular effects.

Corticosteroid-sparing Therapeutic agent that produces the same functional response as a higher dose of corticosteroid.

Cortisol The most abundant corticosteroid produced by the adrenal gland cortex with glucocorticoid effects.

C-Reactive protein Protein of hepatic origin whose plasma concentrations increase in inflammation due to stimulation by IL-6.

Cryptic epitope A hidden or sequestered epitope that is processed and presented as a result of an inflammatory immune responses initiated by an infectious agent, tissue damage or chemicals such as pharmaceuticals.

CSAID Cytokine suppressive anti-inflammatory drugs.

Cyclic adenosine monophosphate (cyclic AMP or cAMP) A second messenger important in many biological processes. cAMP is derived from adenosine triphosphate (ATP) in response to G-protein coupled receptor activation and used for intracellular signal transduction in many different organisms.

Cyclooxygenase (COX) Enzyme catalyzing the oxidative metabolism of arachidonic acid to the biologically active prostaglandins, prostacyclin and thromboxane (collectively known as prostanoids). Also known as prostanoid synthase or prostaglandin synthase.

Cyclosporin Also known as ciclosporin and cyclosporine. It is a T cell modulating immunosuppressive drug and minor csDMARD whose use in RA is very limited because of its adverse effects, particularly renal impairment. Its most important use is to produce immunosuppression after tissue transplantation.

Cytokines Proteins secreted by activated immunocompetent cells that act as intercellular mediators regulating cellular differentiation and activation, particularly within the immune system. They are produced by a number of tissue or cell types rather than by specialized glands and generally act locally in a paracrine or autocrine manner often with overlapping or synergistic actions.

Cytokine release syndrome Referring to a pathological reaction to certain immunotherapies caused by an acute release of inflammatory cytokines from activated immune cells. The severity of the syndrome depends on the amount, the velocity, and the type of cytokines released but generally includes symptoms such as fever, rash, chills, headache, and low blood pressure.

Cytopenia Low blood cell count.

Cytotoxic T-lymphocyte-associated protein 4 (CTLA-4) An inhibitory immunoglobulin produced by activated T cells and regulatory T cells. It binds to CD80 and CD86 on antigen presenting cells and blocks the co-stimulatory signal.

Cytotoxicity Induced cell death, either by binding of peptide mediators to specific death receptors, by insertion of membrane-penetrating components (e.g., complement), or by products released by specific lymphocytes (NK cells, cytotoxic T cells) or by granulocytes with cell-destroying properties.

Degranulation Fusion of intracellular granules with the plasma membrane or with the phagosomal membrane, leading to the release of granule contents into the extracellular space or into the phagosome, respectively.

Dendritic cells White blood cell with protrusions of cytoplasm in tissue that are specialized in antigen presentation. Dendritic cells take up antigen and sense for danger signals at the mucosal surfaces. They then migrate toward draining lymph nodes and present the processed antigen to naïve lymphocytes. In this way, dendritic cells can induce and steer the adaptive immune response.

Detergents A substance that is partly hydrophilic and partly hydrophobic. In immunoassay detergents are used to avoid non-specific interactions between sample components, assay reagents, and the solid phase of the assay.

Differentially methylated region (DMR) A large proportion of the cytidines in the genome are either methylated or not methylated, and thus, the methylation pattern is stable. However, there are regions, so-called differently methylated regions, in the genome whose methylation pattern may change over time, for example, due to an external or internal stimulus or as part of the cell's differentiation or aging.

Disease-modifying antirheumatic drugs (DMARDs) Drugs which decrease inflammation, slow damage to joints and decrease the systemic effects of RA. Separate types of DMARDs are known: biological DMARDs (bDMARDs), conventional DMARDs (csDMARDs) and targeted synthetic DMARDs (tsDMARDs). They are often useful in the treatment of other inflammatory and non-inflammatory diseases.

Diurnal variation Fluctuations that occur during each day.

DNA methylation A modification, mostly known as the methylation of the nucleoside cytidine (C). In mammals, this modification may predominantly occur to C nucleosides, which are followed by a guanosine (G) on the DNA strand in regard to the 5′ to 3′ direction. Due to the double helix nature of DNA, the "CG" sequence on one strand is paired with the "CG" sequence on the opposed strand. A family of enzymes, called DNA methyltransferases, performs this modification.

DNA methyltransferase Here, the term DNA methyltransferase (DNMTs) refers to enzymes that catalyze the methylation of cytidine's pyrimidine ring at the 5′-carbon position. Other enzymes catalyzing the methylation of other nucleosides are known, but are not addressed here. In human, three active DNMTs are known (DNMT1, DNMT3a, DNMT3b). DNMT1 is responsible for the maintenance of the DNA methylation pattern copying the methylation from the mother strand to the daughter strand after DNA replication. DNMT3a and DNMT3b are responsible for de novo DNA methylation.

Dopamine (DA) A catecholamine neurotransmitter present in different circuits of the brain that are involved in motor control and reward-motivated behavior, among other activities.

Drift A directional (increase or decrease) trend of sample results. Can be observed in sample replicates in one assay run or during a long-term use of the assay.

Effector T cells T cells which carry out immune functions, i.e. T helper cells and cytotoxic T cells.

Efficacy Capacity or effectiveness of a drug to control or cure an illness. Efficacy should be distinguished from activity.

Electrochemiluminescence Emission of light, as the result of an electrochemical reaction.

ELISA Enzyme-linked immunosorbent assay. An immunoassay where the detection antibody is labeled with an enzyme.

Endocrine system A network of glands distributed throughout the body. The glands produce hormones that are transported through the blood until their targets where they function as chemical messengers to control several body functions.

Endothelium The layer of (vascular or lymphatic endothelial) cells which lines the inner surface of blood and lymphatic vessels, forming a mechanical barrier which also permits active transport of fluid, proteins and cells between the blood circulation and the underlying tissues.

Engraftment The timing and condition when the body of the recipient accepts the transplanted bone marrow or stem cells and begin to produce new blood cells and immune system cells.

Enteric coated Coating of whole tablets or granules in capsules. The enteric coating resists breakdown in the stomach but disintegrates in the small intestine to release the active drug. It is used with drugs which irritate the stomach or are unstable in the acidic conditions of the stomach.

Eosinophils (Eosinocyte, eosinophilic leukocyte) are granular leukocytes stained by eosin that contain a typically bilobed nucleus and large specific granules. The eosinophils reside predominantly in submucosal tissue and normally low in blood. The cells participate in phagocytosis and inflammatory responses.

Epinephrine Also known as adrenaline, it is a catecholamine released by the adrenal medulla and specialized sites in the brain, acting both as a hormone and a neurotransmitter. It is released into the bloodstream during stress, driving the body into a general state of arousal, which enables it to cope with the stressful event.

Epitope spreading Diversification of epitope specificity from the initial focused dominant epitope-specific immune response to subdominant or cryptic epitopes on that protein.

Epitope The recognition site on an antigenic protein to which either a specific antibody or T cell receptor binds.

ERK One of a chain of mitogen activated protein kinases (phosphorylating enzymes) involved in the transduction of signal from cell membrane to nuclear DNA.

Erythrocytes Red blood cell, involved in oxygen transport to tissue. Contains a nucleus in distinct avian species like chickens but does not have a nucleus in mammals.

Etanercept Fusion protein between the extracellular parts of the human type 1 TNF receptor and human IgG used in the treatment of rheumatoid arthritis.

Everolimus Derivative of sirolimus with similar immunosuppressant properties.

Fab region Region of an antibody that contains the antigen-binding site.

Fc receptor Cell surface receptor on phagocytes and other (mostly immune) cells involved in the recognition of Fc regions. Fc receptor triggering in phagocytes can trigger various effector functions, including phagocytosis, degranulation and intracellular (oxidative) killing.

Fc region Region of an antibody responsible for binding to Fc receptors and the C1q component of complement.

FCγRI High-affinity receptor for immunoglobulin E antibodies present on effector cells such as mast cells and basophils.

Fibre-based imaging The use of optical fibres to physically penetrate tissues and allow imaging deeper within tissues.

Fibroblast A cell which synthesizes collagen and other extracellular proteins. It is the activated form of the fibrocyte.

Ficoll A solution of high molecular weight carbohydrate that is used as a density gradient for the isolation of mononuclear cells (lymphocytes and monocytes) from whole blood; also used for separation of viable from dead lymphocytes.

Filgrastim Recombinant human G-CSF with an additional methionine used therapeutically to stimulate granulocyte production.

Flavonoids Secondary metabolites of the phenylpropanoid class with important functions as antioxidants.

Flow cytometry The analysis of fluorescence and light scatter properties of cells as they pass in suspension through a laser beam.

Fluorescence molecular tomography (FMT) The collection of multiple data points from multiple orientations creating a three-dimensional image of a fluorescent probe or protein concentration within a tissue.

Fluorescence When a substance becomes excited by the absorbance of a photon (or photons) and then returns to its resting state through the emission of a photon.

Fluorochrome An organic dye or protein that has fluorescent properties and thus can be conjugated to an antibody for use in flow cytometric staining assays.

Fluorophore The substance which becomes excited by the absorbance of a photon.

Follicles Round to oval structures in lymphoid tissue, where B cells are lodged. Primary follicles only contain small-sized resting B cells: secondary follicles comprise a pale-stained germinal centre, containing centrocytes and centroblasts (B lymphocytes in a state of activation/proliferation), macrophages and the stroma consisting of follicular dendritic cells. This germinal centre is surrounded by a mantle (corona) with small B lymphocytes. The germinal centre is the pale-staining centre in follicles of lymphoid tissue, where B lymphocytes are activated by antigen in a T lymphocyte-dependent manner and subsequently proliferate and differentiate, acquiring the morphology of centroblasts, centrocytes and plasma cells. The germinal centre has a specialized microenvironment made up of follicular dendritic cells.

Förster resonance energy transfer (FRET) The transfer of photon energy from one fluorophore to another fluorophore.

Forward scatter The measurement of light deflected by a cell at narrow angles to the laser beam in a flow cytometer. This measurement correlates with cell size.

Gene therapy Treatment correcting a genetic defect by the introduction of a normal gene into a cell.

Gene transcription Refers to the process where the DNA sequence of a gene is transcribed (copied out) to produce a molecule of messenger ribonucleic acid (mRNA).

Glucocorticoids (See also Corticosteroid) Major steroid hormones synthetized and secreted by the cortex of the adrenal glands in response to stress and used as potent immunosuppressant and anti-inflammatory drugs. Glucocorticoids have potent immunosuppressive and anti-inflammatory effects; however, persistent elevation of glucocorticoids may induce desensitization of their receptors and, hence, lead to systemic pro-inflammatory status.

Glutathione peroxidase Selenium- and non-selenium-containing enzymes that convert lipid peroxides to hydroxyl moieties, thereby reducing their biological reactivity.

Gold complexes Also known as gold salts. Gold complexes are one of the oldest csDMARDs but are minor drugs for the treatment of RA.

Golimumab Fully human monoclonal antibody directed against TNF. Used for the treatment of RA.

GR Glucocorticoid receptor. The proteins that mediate all the glucocorticoid effects.

Graft versus host disease A condition that occurs when donor bone marrow or stem cells attack the recipient.

Granulocyte Granule-containing myeloid cells, comprising neutrophilic granulocytes (neutrophils), eosinophilic granulocytes (eosinophils), and basophilic granulocytes (basophils).

GRE Consensus GR DNA binding sequence found in the upstream regions of many corticosteroid genes.

Growth factor A growth factor is a naturally occurring substance capable of stimulating cellular growth, cellular proliferation, and cellular differentiation. Usually it is a protein or a steroid hormone. Growth factors are important for regulating a variety of cellular processes. Growth factors typically act as signaling molecules between cells. Examples are cytokines and hormones that bind to specific receptors on the surface of their target cells.

Hapten A non-immunogenic compound of low relative molecular mass which becomes immunogenic after conjugation with a carrier protein or cell and in this form induces immune responses. Antibodies, but not T-cells, can bind the hapten alone in the absence of carrier.

Hematopoiesis The formation of blood cells—erythrocytes (red blood cells), thrombocytes (platelets), and leukocytes (white blood cells).

Hematopoietic stem cell Multipotent stem cells that give rise to all blood cell lineages. They reside in the fetal liver and fetal and postnatal bone marrow and can be recruited to peripheral blood.

Hematopoietic stem cell transplantation Transplantation of multipotent hematopoietic stem cells, usually derived from the bone marrow, peripheral blood, or umbilical cord blood. It may be autologous, allogeneic, or syngeneic.

Hepatitis B virus DNA virus specifically infecting the liver, resulting in acute or chronic hepatitis.

Hepcidin A peptide which regulates serum iron. Its plasma concentrations are high in inflammatory diseases and lead to anaemia.

High-dose chemotherapy Type of chemotherapy treatment in which myeloablative doses are given.

Histamine Vasoactive amine released from mast cell and basophil granules. Triggers allergic symptoms, activates eosinophil granulocytes, increases gastric secretion and regulates the release of neurotransmitters.

Histone acetylation Histone acetylation is the process by which lysine residues within the N-terminal tail protruding from the histone core of the nucleosome are acetylated by a class of enzymes called histone acetyltransferases (HATs). Acetylation results in charge neutralization of conserved, often invariant, lysine residues of the histone resulting in more accessible DNA which allows transcription factors to gain access to the DNA.

Histone acetyltransferase (HATs) Histone acetyltransferases are a family of bi-substrate enzymes that catalyzes acetylation not only of histone tails in the nucleosome but also of cellular proteins such as TFs, nuclear receptors, and enzymes. HATs function enzymatically by transferring an acetyl group from acetyl-coenzyme A (acetyl-CoA) to an ε-amino group of certain lysine side chains within a histone's basic N-terminal tail region. HAT proteins act in multiple complexes or locations and thus do not fit easily into a class.

Histone code The term histone code refers to the various posttranslational modifications such as acetylation, methylation, and phosphorylation on histone tails which define all aspects of chromatin biology. Most modifications are regulated by enzymes that add or remove the changes. Proteins that recognize a single or a specific combination of modifications translate these signals in order to activate or repress processes including transcription, DNA repair, or DNA replication.

Histone deacetylases (HDACs) Histone deacetylases are a class of enzymes that remove acetyl groups from an ε-N-acetyl lysine amino acid on a histone, allowing the histones to wrap the DNA together more tightly, thereby preventing transcription. HDACs also interact with non-histone proteins such as TFs, coregulators and enzymes, thereby controlling the activity and stability of these proteins. HDACs are broadly classified into four classes depending on sequence homology to the original enzymes and domain organization.

Histone lysine demethylase Histone lysine demethylases or short KDMs catalyze the removal of specific methylations at lysine residues. Similar to lysine methyltransferases, these enzymes are often part of a larger complex, thereby restricting their activities to specific genome areas. They counterbalance the lysine methyltransferases and thus, are important for regulation of signals mediated by histone methylation.

Histone methylation Histone proteins H2A, H2B, H3, or H4 can be methylated at lysine or arginine residues. The modification has no direct effect on nucleosome structures but is recognized by proteins involved in processes that modify chromatin structures by introducing or removing nucleosomes, thereby regulating transcription, DNA replication or DNA repair.

HLA See Major histocompatibility complex.

Hormones Chemical messengers produced by glands and transported by the circulatory system for the purpose of regulating the activity of target cells.

Horseradish peroxidase Enzyme from horseradish. Generates a chromogenic, fluorescent, or luminescent assay signal by oxidizing a substrate.

Humoral immunity Immunological reactivity mediated by antibodies.

Hydroxychloroquine An antimalarial drug which is a csDMARD.

Hypersensitivity Abnormally increased, immunologically mediated response to a stimulus. Sometimes used loosely for any increased response or to describe allergy. The reaction can be mimicked by non-immunological mechanisms (e.g. chemical stimulation of mast cell degranulation).

Hypothalamic-pituitary-adrenal (HPA) axis A major neuroendocrine pathway involving the hypothalamus, the pituitary gland (also known as hypophysis), and the adrenal (also known as suprarenal) glands. The set of interactions and feedbacks within the HPA axis is responsible for controlling the response to stress and maintain body homeostasis.

ICAM-1 and ICAM-2 Intercellular adhesion molecules on leukocytes and tissue cells that interact with β2-integrins and mediate binding of leukocytes to other cells.

Idiosyncratic The mechanism of a variety of adverse effects is not known. Idiosyncrasy refers to those adverse effects where an unexpected susceptibility of the host is presumably involved. It is recommended to restrict this term either to adverse effects reflecting a genetic predisposition, or whose mechanism is not elucidated.

Immune cells Effector cells of the immune system, also known as white blood cells (WBC) or leukocytes.

Immune system A system encompassing all aspects of host defense mechanisms against xenobiotics and pathogens that are encoded in the genes of the host. It includes barrier mechanisms (e.g. membranes), all organs of immunity, effectors (proteins, bioactive molecules and cells—mainly phagocytes) of the innate (immediate, non-specific) immune response, as well as effectors (T and B lymphocytes and their products) of the adaptive (delayed, specific) immune response. The two responses (specific and non-specific) act synergistically for a fully effective immune response.

Immune thrombocytopenia purpura (ITP) Is a clinical syndrome in which a decreased number of circulating platelets (thrombocytopenia) manifest as a bleeding tendency, easy bruising (purpura) or extravasation of blood from capillaries into skin and mucous membranes (petechiae).

Immune checkpoint inhibitors Monoclonal antibodies designed to block immune checkpoint receptor or ligands, impeding their inhibitory interaction and thereby (re)-activating antitumor immune cells. Immune checkpoint inhibitors are currently the most widely explored immunotherapeutic option to treat cancer.

Immune checkpoints Regulatory molecules of the immune system. These regulatory pathways are mediated by receptors on the surface of lymphocytes, which interact with their respective ligands on, e.g., antigen-presenting cells or tumor cells, thereby restricting (inhibitory immune checkpoints) or activating (activating immune checkpoints) lymphocyte responses. These molecules have a key role in maintaining homeostasis and tolerance.

Immunoallergic This term is used to describe immune-mediated HYPERSENSITIVITY reactions.

Immunoglobulin (See also Antibody) Immunoglobulins (Ig) are synthesized by plasma cells. The basic subunit consists of 2 identical heavy chains (about 500 amino acid residues, organized into 4 homologous domains; for μ-chain in IgM about 600 amino acid residues, organized in 5 homology domains) and 2 identical light chains (about 250 amino acid residues organized into 2 homologous domains). Each chain consists of a variable domain and one constant domain (light chain) or three or four constant domains (heavy chain). The antigen-binding fragment (Fab) consists of variable domains of heavy and light chain (two per basic subunit). Five classes of immunoglobulins exist, which differ according to heavy chain type (constant domains): IgG (major Ig in blood), IgM (pentamer consisting of five basic units), IgA (major Ig in secretions, here present mainly as a dimeric Ig molecule), IgD (major function as receptor on B lymphocytes), and IgE. Effector functions after antigen binding are mediated by constant domains of the heavy chain (Fc part of the molecule) and include complement activation (IgG, IgM), binding to phagocytic cells (IgG), sensitization and antibody-dependent cell-mediated cytotoxicity (IgG), adherence to platelets (IgG), and sensitization and degranulation of mast cells and basophils (IgE). IgA lacks these effector functions and acts mainly in immune exclusion (prevention of entry in the body) at secretory surfaces.

Immunomodulation Immunomodulation is directed towards either enhancement or suppression of host immunological mechanisms, such as phagocytosis and bactericidal activity, cytokine production, lymphocyte proliferation, antibody response and cellular immunity.

Immunophenotyping The use of fluorochrome-conjugated antibodies and flow cytometry to identify subsets of leukocytes in a sample such as peripheral blood. Such assays can be used to identify disease states such as leukemia/lymphoma.

Immunopsychiatry A term to designate the recent interest in the study of peripheral immune mechanisms that may influence cognition, emotion, and/or behavior.

Immunosuppression Defects in one or more components of the non-specific/innate or specific/adaptive immune system, resulting in inability to eliminate or neutralize non-self-antigens. Congenital or primary immunodeficiencies are genetic or due to developmental disorders. Acquired or secondary immunodeficiencies develop as a consequence of immunosuppressive compounds, malnutrition, malignancies, radiation or infection. This may result in decreased resistance to infection, the development of certain types of tumours or immune dysregulation and stimulation, thereby promoting allergy or autoimmunity.

Immunotherapy Treatment of a disease by the artificial stimulation of the body's immune system to induce or suppress an immune response. *See also* Specific immunotherapy.

Immunotoxicant Drug, chemical or other agent that is toxic to cells or other components of the immune system.One form of immunotoxicity is the direct toxicity of the compound to components of the immune system, which often leads to suppressed function. This may result in decreased resistance to infection, the development of certain types of tumours or immune dysregulation and stimulation, thereby promoting allergy or autoimmunity. Other types or manifestations of immunotoxicity include allergy or autoimmunity in which the compound causes the immune system to respond as if the compound were an antigen or to respond to self-antigens that have been altered by the chemical.

Inflammasome Intracellular multiprotein complex consisting of caspase 1, PYCARD, a NALP, and sometimes caspase 5 or caspase 11. The inflammasome promotes the maturation of inflammatory cytokines IL-1β and IL-18.

Inflammation/Inflammatory response A complex biological and biochemical process involving cells of the immune system and a plethora of biological mediators (particularly cytokines); it may be defined as the normal response of living tissue to mechanical injury, chemical toxins, invasion by microorganisms or hypersensitivity reactions. Excessive or chronic inflammation can have disastrous consequences for the host.

Inflammatory bowel disease (IBD) A group of inflammatory conditions of the colon and small intestine, including Crohn's disease and ulcerative colitis.

Inflammatory cytokines Cytokines that primarily contribute to inflammatory reactions, including interferon-γ, interleukin-1, tumour necrosis factor and chemokines.

Inflammatory mediators A group of molecules with varying structures (lipids, proteins), preformed or newly synthesised by both inflammatory and structural cells, which modulate the inflammatory response, encompassing regulation of cell function by intracrine, autocrine paracrine, endocrine and exocrine signalling, leading to pro- or anti-inflammatory effects.

Inflammatory system A group of physiologically or anatomically related organs and cells that control the reaction of living tissues to injury, infection or irritation.

Infliximab Chimeric mouse/human monoclonal antibody (fusion protein between the variable portions of a murine antibody with the constant parts of human IgG) directed against human TNFα.

Innate immune system/innate (non-specific) immunity Nonadaptive, non-antigen-specific host defence system against pathogens and injurious stimuli, present at birth and consisting of phagocytes, natural killer cells, innate lymphoid cells and the complement system.

Innate lymphoid cells A group of innate immune cells that belong to the lymphoid lineage and are defined by the absence of antigen-specific B or T cell receptors. Several subsets can be identified with functions analogous to helper T cells, and the group also includes NK cells.

Integrins Family of heterodimeric cell surface molecules sharing in part a β-chain (β 1, β 2, β 3, about 750 amino acids long), each with a different α-chain (about 1100 amino acids long), which mediate cell adhesion and migration by binding to other cell adhesion molecules (CAMS), complement fragments or extracellular matrix. Based on strong structural and functional similarities, integrins form a protein family rather than a superfamily. Examples: leukocyte function-related antigen LFA-1 (αL/β1, CD11a/CD18; receptor for ICAM-1, ICAM-2 and ICAM-3); Mac-1 (αM/β2, CD11b/CD18; complement C3 receptor CR3); p150,95 (αX/β2, CD11c/CD18); very late antigens (VLA)-1 (α1/β1, CD49a/CD29; laminin, collagen receptor), VLA-2 (α 2/β 1, CD49b/CD29; laminin, collagen receptor), VLA-3 (α 3/β 1, CD49c/CD29; laminin, collagen, fibronectin receptor), VLA-4/LPAM-1 (α 4/β 1, CD49d/CD29; receptor for fibronectin and VCAM-1), VLA5 (α 5/β 1, CD49e/CD29, fibronectin receptor) and VLA-6 (α 6/β 1, CD49f/CD29; laminin receptor, and α V/β 1, CD51/CD29; vitronectin receptor) and LPAM-2 (α 4/β p, D49d/…, or α 4/β 7).

Interferons (IFN) A group of antiviral and immunoregulatory glycoprotein cytokines induced in different cell types by appropriate (mostly viral) stimuli, conferring resistance of neighbouring cells to infection with homologous or heterologous viruses. They are divided into types I, II and III based on their three different target receptors. Interferons α and ß are type I interferons mainly produced by virally stimulated fibroblasts or monocytes.

Interferon alpha 2a or alpha 2b (unconjugated or pegylated) Recombinant human interferons α either unconjugated or conjugated with monomethoxy-polyethylene glycol (mPEG) used for the treatment of viral infections and several cancers.

Interferon beta 1a, 1b Recombinant human interferons β used for the treatment of multiple sclerosis

Interleukins (ILs) Heterogenous group of immunoregulatory protein cytokines, also including lymphokines, monokines and interferons, acting as communication signals between cells. They generally have relatively low molecular masses (<80 kDa) and are frequently glycosylated. They also regulate immune cell function and inflammation at picomolar concentrations by binding to specific cell surfaces. Interleukins are transiently and locally produced; have a wide range of overlapping functions and act in a paracrine (i.e. act locally) or autocrine manner (i.e. act on cells producing the interleukin).

Interleukin-1 (IL-1) Two proteins are grouped together because of their very similar inflammatory actions which are due, in part, to their stimulation of the secretion of IL-6 and TNF. Both induce cell proliferation, apoptosis, fever and central pain. IL-1β is synthesised by activated macrophages but IL-1β can be produced by normal barrier cells, such as endothelial and epithelial cells.

Interleukin-6 (IL-6) A pro-inflammatory cytokine produced by several cell types including macrophages, skeletal muscle after exercise, smooth muscle, osteoblasts and adipocytes. IL-6 has many effects: stimulation of the maturation of B cells (and hence the production of immunoglobulins), induction of the cute phase reaction, margination of neutrophils into the circulation, stimulation of these cells to release proteolytic enzymes and to develop the respiratory burst.

Interleukin-8 (IL-8) A pro-inflammatory cytokine produced by many cells including macrophages, monocytes, neutrophils, epithelial cells, endothelial cells and airway smooth muscle cells. IL-8 induces angiogenesis and chemotaxis of neutrophils to sites of tissue damage and infection where there is induction of the respiratory burst of neutrophils. IL-8 may be involved in the development of colorectal cancer, bronchiolitis and the symptoms of cystic fibrosis.

Interleukin 10 (IL-10) An anti-inflammatory cytokine produced by monocytes and, to a lesser extent, several other cell types. Its main effects are inhibition of the synthesis of several inflammatory cytokines including TNF, IL1-β, IL-12, IFN-γ, IL-10 also stimulate Th2 cells, mast cells and maturation of B cells with consequent increased production of immunoglobulins.

Interleukin-12 (IL-12) A pro-inflammatory cytokine IL-12 which is produced by activated antigen-presenting cells, including dendritic cells and macrophages. IL-12 is also produced by neutrophils. IL-12 assists in the conversion of naïve T cells into Th1 cells. IL-12 is a heterodimeric cytokine composed of IL-12p40 (common to IL-23) and IL-12p35 subunits.

Interleukin-17 (IL-17) A pro-inflammatory group of cytokines which are induced by IL-23 in combination with other cytokines. Together with other cytokines, IL-17 mediates inflammatory diseases including RA, psoriasis and inflammatory bowel disease. IL-17F may be significant cytokine in asthma.

Interleukin-23 (IL-23) A pro-inflammatory cytokine which stimulates the production of the pro-inflammatory cytokines, interleukin-17 (IL-17) and interleukin-22 (IL-22) thereby increasing the synthesis of TNF and IL-1β. IL-23 is involved in several inflammatory diseases including rheumatoid arthritis. IL-23 is a heterodimeric cytokine composed of IL-12p40 (common to IL-12) and IL-23p19 subunits.

Isotype Class of immunoglobulin (IgG, IgM, IgD, IgA, or IgE).

Ixekizumab A humanized IL-17A antibody, undergoing clinical trials for various chronic inflammatory diseases.

Janus kinase (JAK) enzymes Phosphorylating enzymes which are mediators of the intracellular signalling of a range of pro-inflammatory substances including CYTOKINES.

Kinases A large group of different enzymes catalysing the transfer of phosphate groups from phosphate-donating molecules to specific substrates (phosphorylation). The expression and/or functional activity of many of these kinases is regulated by the corticosteroids. Other kinases may modulate the anti-inflammatory effects of the corticosteroids.

Lectins Unique group of proteins found in plants, viruses, microorganisms and animals. Despite their ubiquity, their function in nature is unclear. Some lectins induce mitosis in cells which normally are not dividing.

Leflunomide A conventional DMARD (csDMARD) whose is activity is due to its metabolite, teriflunomide.

Lenograstim Recombinant human G-CSF.

Leukocytes Effector cells of the immune system, also known as white blood cells (WBC) or immune cells. They comprise granulocytes (polymorphonuclear neutrophilic, eosinophilic, and basophilic granulocytes), monocytes, and lymphocytes.

Leukopheresis/apheresis The process of collecting white cells or stem cells from the peripheral blood.

Leukotrienes A family of biologically active compounds derived by enzymatic oxidation from arachidonic acid. They participate in host defence reactions and pathophysiological conditions such as inflammation.

Levamisole Orally active DMARD previously used for eliminating intestinal parasites in animals.

Ligand Molecule that binds to a receptor.

Lipopolysaccharide (LPS) Product of some Gram-negative bacterial cell walls that bind to specific toll-like receptors, to CD14 and to LPS-binding protein (LBP). It activates B lymphocytes, macrophages and neutrophils. Also referred to as endotoxin.

Lipoxins Biologically active lipid mediators synthesized from arachidonic acid by lipoxygenases, have anti-inflammatory properties and resolve inflammatory processes.

Lymph node Lymphoid organ made up of an encapsulated, highly organized structure containing lymphocytes. They usually occur in groups where lymphatic vessels converge (such as in the armpit, groin, neck, and lung hila and along the aorta). The lymphocytes are organized in lymphoid follicles within the lymph node. These lymphoid follicles consist of a T cell-rich lymphocyte corona and a B cell-rich germinal center. Lymph nodes are the primary location for antigen presentation and subsequent lymphocyte proliferation and differentiation.

Lymphocytes Cells belonging to the lymphoid lineage of bone marrow-derived haematopoietic cells. A more restricted designation is that of a small resting or recirculating mononuclear cell in the blood or lymphoid tissue that measures about 7–8 μm and has a round nucleus containing densely aggregated chromatin and little cytoplasm. Lymphocytes play a key role in immune reactions through specific recognition of antigens.

Lymphoid cells Innate immune cells that belong to the lymphoid lineage but lack antigen-specific receptors.

Lymphoid organ Organ in which cells of the immune system, mainly lymphocytes, are lodged in an organized microenvironment, either in a resting stage or in a stage of activation/differentiation/proliferation. Lymphoid organs include the bone marrow, thymus, lymph nodes, spleen and mucosal-associated lymphoid tissue. Central (primary) lymphoid organs are those in which T and B lymphocytes develop and mature (bone marrow, thymus); peripheral (secondary) lymphoid organs are those where immunocompetent lymphocytes recognize antigen and subsequently initiate immunological reactions and produce effector elements of these reactions.

Lysine methyltransferase Abbreviated as KMTs, this is a group of ubiquitously expressed enzymes that catalyze the methylation of a small number of specific lysine residues in histone proteins using *S*-adenosyl methionine (SAM) as a methyl donor. Lysine residues can be mono-, di-, or trimethylated. The enzymes are usually organized in larger protein complexes that coordinate distribution along the DNA and activity of the enzymes to restrict the modification to defined areas in the genome.

Macrophage migration inhibitory factor (MIF) An inflammatory cytokine released by leukocytes, named because it inhibits macrophage chemotaxis, but which exerts multiple pro-inflammatory effects and acts as an important regulator of innate immunity.

Macrophages Large 12–20 μm mononuclear phagocytic and antigen-presenting cells, present in tissue, contributing to the mononuclear phagocytic system that includes monocytes, macrophages, dendritic cells (in lymphoid organs), Langerhans cells (in the skin) and Kupffer cells (in the liver).

Magnetic resonance imaging (MRI) The use of radio waves to deflect magnetically susceptible spinning protons in a specimen to generate an image.

Major depression Mood disorder characterized by low mood and often accompanied by hopelessness, pessimism, loss of interest in life, reduced emotional wellbeing, and abnormalities in sleep, appetite, and energy level.

Major histocompatibility complex (MHC) Set of genes that code for tissue compatibility markers. The MHC complex in man is called HLA (human leukocyte antigens), in mice H2 and in rat RT-1. These markers, which are targets in the rejection of an allograft (matched grafted tissue or organ from a different individual) and hence determine the fate of allografts, also play a central role in control of cellular interactions during immunological reactions. Tissue compatibility is coded by class I and class II gene loci. MHC class I molecules are coded by the A, B or C gene locus in the HLA complex and the K and D locus in the mouse H2 complex, in association with the β2-microglobulin molecule. These two-chain molecules are present on all nucleated cells. MHC class II molecules are coded by the D (DR, DP, DQ) gene locus in the HLA complex and the I-A and I-E locus in the mouse H2 complex and comprise an α- and a β-chain (intracellularly associated with an "invariant" chain). The two-chain molecules are present on B lymphocytes, activated T lymphocytes, monocytes/macrophages/interdigitating dendritic cells and some epithelial and endothelial cells (variable, dependent on species and state of activation), which are also called antigen-presenting cells. Genes within or closely linked to the MHC control certain complement components (MHC class III genes). MHC restriction is the phenomenon whereby immunological reactions can only occur in association with or parallel to recognition of the polymorphic determinant of a given MHC molecule, and not with that of another MHC molecule. This applies to T lymphocytes with an αβ T cell receptor, which recognize antigenic peptides in association with the polymorphic determinant of MHC molecules and a fraction of the T cell population with a γδT cell receptor.

Mast cell Tissue (mainly skin and mucosa)-associated cell activated by antigen/allergen bridging of surface-bound IgE antibodies, releasing enzymes and vasoactive mediators, especially histamine.

Matrix (pl. matrices) Term used in analysis to describe all components of biological fluids other than the analyte.

Matrix metalloproteinase (MMP) Group of calcium-dependent zinc-containing endopeptidases which degrade extracellular proteins and may be important in the pathophysiological processes in rheumatoid arthritis, cancer and cardiovascular diseases.

Mechanism of action Complex process whereby a drug, which itself is not the triggering component of therapy but which manipulates cellular and cytokine elements of the immune system that are the ultimate antitumor effectors.

Membrane attack complex (MAC) Assembled complex of complement components C5b–C9 of the lytic pathway that is inserted into target cell membranes and causes cell lysis.

Mesalazine Also known as 5-aminosalicylic acid. It is an anti-inflammatory drug used in the treatment of inflammatory bowel disease and is a constituent and metabolite of sulfasalazine which is a csDMARD.

Metabolic syndrome The co-occurrence of several biochemical and physiological abnormalities that are well-known risk factors for cardiovascular diseases.

Metalloproteinase (or metalloproteases) Are a group of proteolytic enzymes whose catalytic mechanism involves a metal. Most metalloproteases are zinc dependent, some use cobalt.

Metallothionein A family of cysteine-rich, low molecular weight proteins (MW ranging from 3500 to 14,000 Da). They bind both physiological (e.g. zinc, copper, selenium) and xenobiotic heavy metals (e.g. cadmium, mercury, silver, arsenic) via the thiol group of its cysteine residues, which represent nearly 30% of their amino acidic residues.

Methotrexate Antagonist of folic acid used therapeutically as a cytostatic and as an antirheumatic drug.

Microbiota Refers to the commensal bacteria and other microorganisms in the GI tract.

Microglia Resident phagocytic immune cells inside the central nervous system (CNS). Microglia and monocyte-derived macrophage are different cells, although both cell types share phagocytic properties.

Microsomal prostaglandin E synthase-1 (mPGES-1) Inducible form of prostaglandin E synthase; in inflammation, its expression is increased by bacterial products and pro-inflammatory cytokines and it contributes to the increased formation of PGE_2 in inflammation.

Mitogen Substance that activates resting cells to transform and proliferate.

Molgramostim Recombinant human GM-CSF used therapeutically to stimulate granulocyte and monocyte generation.

Monoclonal antibodies (mAbs) Identical copies of antibody with the same antigen specificity that consist of one heavy-chain class and one light-chain type. Typically, monoclonal antibodies are produced by a hybridoma, which is a transformed cell line grown in vivo or in vitro and is a somatic hybrid of two parent cell lines, one of which is a plasma cell originally producing the single antibody.

Monocytes Large 10–15 μm non-differentiated mononuclear cells, present in the blood and in lymphatics, comprising the circulating component of the mononuclear phagocyte system.

Mononuclear phagocytes Bone-marrow-derived cells with a large rounded nucleus, including monocytes in the blood, macrophages in connective tissues, Kupffer cells in the liver and Langerhans cells in the skin, which have the capacity to engulf particles (phagocytosis), release a variety of enzymes, proteins, lipid and other mediators in order to regulate the inflammatory process and its resolution. The cells have considerable plasticity and can change from one phenotype to another. Most also have the capacity to process antigens and present these to T lymphocytes for the induction of an adaptive immune response.

Mood disorder A psychiatric disorder characterized by altered mood, i.e., depressed or elated (manic).

Mucosa Structural tissue unit that delineates the lumen of an internal organ (such as the gut or airways) and its orifice. It consists of an epithelial layer next to the lumen. Underneath the epithelium a region of connective tissue is present usually containing glands small blood vessels and smooth muscle cells. Cells of the immune system are dispersed throughout the mucosa.

Mucosa-associated lymphoid tissue (MALT) Lymphoid tissue in immediate contact with the mucous-secreting mucosal layer in the nasal cavity and nasopharynx (nasal-associated lymphoid tissue), airways (bronchus-associated lymphoid tissue) and intestinal tract (gut-associated lymphoid tissue). Serves as the immunological defence system at secretory surfaces, to some extent independent of the systemic (internal) response.

Multiphoton microscopy The use of two or more photons to excite the fluorophore with wavelengths longer than the resulting emitted light.

Multiple sclerosis (MS) An autoimmune disease that affects the brain and spinal cord (central nervous system). Symptoms vary depending on the location and severity of each attack. Episodes can last for days, weeks or months. These episodes alternate with periods of reduced or no symptoms (remissions). Because nerves in any part of the brain or spinal cord may be damaged, patients with multiple sclerosis can have symptoms in many parts of the body.

Muramyl dipeptides Peptide that stimulates resistance to bacterial infection by binding to receptors on macrophages causing them to release cytokines.

Muteins Cytokine mutants, generated by gene technology, behaving mainly as receptor antagonists.

Myeloablation High-dose chemotherapy that kills cells in the bone marrow, including cancer cells. It lowers the number of normal blood-forming cells in the bone marrow.

Myeloperoxidase Enzyme which interacts with chloride and hydrogen peroxide to produce hypochlorous acid.

***N*-alkylamides** Biologically active, immunomodulatory compounds, consisting of an alkyl group attached to the nitrogen of an amide, present particularly in the roots of the purple coneflower (*Echinacea*). *N*-alkylamides have been defined also as a class of cannabinomimetics.

Natural killer (NK) cells Lymphocyte-like cells of the innate immune system capable of antigen-independent killing of virus-infected and tumour-transformed cells in an antigen-independent manner.

Natural Product that is extracted from and identical to its natural self.

Neo-angiogenesis The growth of new capillaries from pre-existing vessels.

Neo-epitope A newly formed epitope on an existing antigen that can be a cryptic epitope or a hapten-carrier conjugate.

Nephelometry Method to detect particles in liquid samples. The measurement is made by measuring the scattered light.

Neurodegenerative diseases A heterogeneous group of diseases that are characterized by the progressive loss of neuron structure or function, including Alzheimer's disease, Parkinson's disease, and amyotrophic lateral sclerosis, among others.

Neuroimmunoendocrinology The branch of physiology and medicine concerned with the interactions between the nervous, endocrine, and immune systems. These systems can communicate closely with each other by several mechanisms that work in parallel.

Neuroinflammation A set of inflammatory changes inside the central nervous system.

Neuropeptides Short-chain polypeptides that act as neurotransmitters.

Neurotransmitters Chemical messengers released by neurons at a synapse aiming at transmitting information to other cells (mainly neurons) via receptors.

Neutrophils (Neutrophilic Granulocytes) Also called polymorphonuclear leukocytes, neutrophils are highly specialized white blood cells characterized by a multilobed nucleus (polymorphonuclear) and a granular cytoplasm that is "neutral" to histological staining under the light microscope. Specialized constituents of the neutrophil membrane, cytoplasmic granules and cytosol together mediate ingestion and killing of bacteria. After attachment to the cell membrane and internalization of the microorganism into the phagocytic vacuole (phagosome), its destruction is mediated by the release of an array of antimicrobial polypeptides and reactive oxidant species.

Nitric oxide (NO) Gaseous highly reactive signaling molecule with vasoactive, proinflammatory, bacteriocidal and neurotransmitter activities; NO is synthesized from the amino acid arginine by two constitutive (nNOS and eNOS) and one inducible (iNOS) nitric oxide synthase enzymes.

Non-steroidal anti-inflammatory drugs (NSAIDs) Analgesic, antipyretic and anti-inflammatory drugs with some antithrombotic effects. NSAIDs inhibit the activity of the cyclo-oxygenase enzyme and the production of prostanoid mediators. They are structurally different from steroids.

Norepinephrine (NE) A catecholamine neurotransmitter produced both in the brain and in the peripheral nervous system (sympathomimetic action). Norepinephrine (also called noradrenaline) is involved in arousal, sleep, and blood pressure regulation, among other functions.

Nuclear factor kappa B (NF-κB) Ubiquitous, inducible, transcriptional activator that binds to enhancer elements in most immune and inflammatory cells to stimulate inflammatory genes.

Nucleosome DNA in eukaryotic cells, which includes humans and mice, is organized in nucleosomes. Nucleosomes are composed of four core histone proteins (H2A, H2B, H3, and H4) each wrapped by approximately 147 base pairs of DNA. The N-terminal tail of the histones protrudes from the nucleosome and can possess multiple posttranslational modifications (see Histone code).

Obsessive-compulsive disorder A psychiatric disorder characterized by recurring thoughts (obsessions) and behaviors (compulsions) that the person is unable to control.

Olsalazine An azo compound linking two molecules of mesalazine. It is used for the treatment of ulcerative colitis and Crohn's disease

Opsonization Covering of pathogenic particles and microorganisms with antibody or complement proteins that enhance uptake of these by phagocytes.

Optical imaging Using light to visualize and investigate cellular and molecular biological process.

Optimal immunomodulatory dose Treatment dose that maximally activates the chosen immunomodulatory action with acceptable toxicity.

Osteoarthritis (OA) Inflammatory joint disease causing cartilage degradation, disability and pain.

Osteoporosis Metabolic disorder associated with fractures of the femoral neck, vertebrae and distal forearm.

Patient adherence (compliance) Voluntary cooperation of the patient in following a prescribed regimen.

Pattern recognition receptors (PRRs) Surface molecules ound on most body defense cells, which recognize common pathogen-associated molecular patterns (PAMPs) and produce an immediate response of the cell against the invading microorganism.

Pegylation Process whereby polyethylene glycol (PEG) is attached to a protein in order to extend its circulating half-life and thereby enhance its biological activity.

Penicillamine An amino acid (D-dimethylcysteine) which has the opposite configuration to the L-amino acids which are incorporated into proteins. It is a minor csDMARD which is used in modern medicine in the treatment of Wilson's disease (because its sulfhydryl (thiol) binds copper) and in patients with cystinuria (because of the formation of a disulphide, penicillamine – cysteine).

Peyer's patch Lymphoid tissue in the wall of the small intestine, separated from the gut lumen by a domed area and an epithelial layer; forms part of the mucosa-associated lymphoid tissue (MALT). Its main function is initiation of immunological reactions towards pathogens entering through dome epithelium.

Phagocytosis Process by which phagocytes bind and engulf material >1 μm (e.g. microbes) in an Fc receptor-dependent manner, with accessory help of complement receptors. Phagocytosis occurs via a "zipper" mechanism, whereby the particle, opsonized (coated) with antibody or complement, becomes enclosed by the cell membrane of the phagocyte. The particle is then incorporated into a vacuole (phagosome) where it is degraded by proteases and an NADPH oxidase-mediated oxidative burst with formation of superoxide anion, peroxide anion and hydroxyl radicals.

Pharmacodynamics (PD) The study of the biochemical and physiological effects of the drug on the target molecule or a representative protein related to the target.

Pharmacokinetics (PK) Mathematically describes the rates of absorption, distribution and elimination of drugs and is also referred to as dug disposition. The disposition of a drug is defined by its pharmacokinetic parameters determined from concentration versus time curves. This may include volume of distribution, area under the curve (AUC), clearance and half-life ($t_{1/2}$).

Phosphodiesterase (PDE) Enzymes that metabolize (and inactivate) cyclic nucleotides (cAMP, cGMP).

Photoacoustic imaging Generation of a single image by combining optical excitation and acoustic detection.

Phytates Phosphorus compounds found primarily in cereal grains, legumes and nuts. They bind to minerals such as iron, calcium and zinc and interfere with their absorption. They also have beneficial health effects, help disease prevention and can be considered as antioxidant compounds in food.

Plasmacytoid dendritic cells (pDCs) This name refers to DCs which do not have a dendritic cell morphology but a rounded shape with an eccentric nucleus and a prominent endoplasmic reticulum and Golgi. pDCs resembles plasma cells and is needed for the production of prominent amounts of type I interferon.

Platelet (Thrombocyte) Small cytoplasmic fragment in blood that is responsible for coagulation. Its main role is to block damaged vessel walls and prevent hemorrhage, by clumping and aggregation. Platelets contain heparin and serotonin, which contribute after release to the acute vascular response in hypersensitivity reactions, and produce oxygen radicals.

Platelet-activating factor (PAF) Molecule derived from phosphatidylcholine; released by immune cells and tissue cells; acts as a chemoattractant for and activator of phagocytes as well as an adhesion molecule.

Polyclonal antibodies Antibodies that consist of a mixture of different antibodies clones. Each clone in the polyclonal mixture might bind to a different epitope or with different affinity to the antigen.

Positron emission tomography (PET) The detection of gamma rays released by a radionuclide (tracer) conjugated to a biologically active molecule with three-dimensional images of tracer concentrations subsequently detected. Usually detects cellular metabolic activity.

Prebiotics Non-digestible food ingredients that beneficially affect the host by selectively stimulating the growth and/or activity of one or a limited number of bacteria in the colon that can improve host health.

Probiotics Live microorganisms which, when administered in adequate amounts, confer a health benefit on the host.

Progressive multifocal leukoencephalopathy (PML) This is a rare and usually fatal viral disease characterized by progressive damage (-pathy) or inflammation of the white matter (leuko-) of the brain (-encephalo-) at multiple locations (multifocal). It is caused by the JC virus, which is normally present and kept under control by the immune system. JC virus is harmless except in cases of weakened immune systems. In general, PML has a mortality rate of 30–50% in the first few months, and those who survive can be left with varying degrees of neurological disabilities.

Pro-inflammatory cytokines Cytokines which primarily contribute to inflammatory reactions, including interferon γ, interleukin 1, tumor necrosis factor, or chemokines.

Prostacyclin (PGI_2) Biologically active lipid mediator synthesized from arachidonic acid by cyclo-oxygenases and prostacyclin synthase which has vasodilatatory, antiproliferative and antithrombotic effects.

Prostaglandin(s) (PG) Biologically active lipid mediators synthesized from arachidonic acid by cyclo-oxygenasess and specific prostaglandin synthases. Examples of prostaglandins are PGE_2 and PGD_2, which are involved in inflammation.

Prostanoid(s) Biologically active lipid mediators synthesized from arachidonic acid by cyclo-oxygenass and specific synthases to form prostaglandins, prostacyclins, and thromboxanes, which are collectively known as prostanoids.

Protein isoform A variant of structurally related proteins. Isoforms are generated as products of alternative splicing of mRNA.

Pseudo-allergic This term is used to describe nonimmune-mediated hypersensitivity reactions. Owing to the fact that the same mediators are released as in immune-mediated hypersensitivity reactions, the two types of reaction bear many clinical similarities. The term anaphylactoid is often used to depict pseudo-anaphylactic reactions in which nonimmune-mediated hypersensitivity reactions are involved.

Psoriasis (Ps) A predominantly T cell-mediated autoimmune disease of the skin, characterized by patches (plaques) of abnormal skin, which are typically red, itchy and scaly. They may vary in severity from small and localized to complete body coverage. The most common form is plaque psoriasis in which red patches of skin are covered with white scales. Many patients with psoriasis also develop psoriatic arthritis.

Psoriatic arthritis (PsA) A type of arthritis that is often associated with psoriasis of the skin. The arthritis may be mild and involve only a few joints, particularly those at the end of the fingers or toes. Patients who also have arthritis usually have the skin and nail changes of psoriasis.

Psychiatric disorder An illness that affects someone's emotion, cognition, and behavior, influencing his/her social interactions and functioning in daily life.

Purple coneflower (*E. purpurea*) Indigenous North American plant, an extract of which is used as a mild immunostimulant.

QT interval Prolongation of the QT interval in the electrocardiogram indicates prolongation of the cardiac action potential. While this action may be clinically useful, it may also lead to potentially cardiotoxic arrhythmias, particularly in combination with other drugs that may also lengthen the QT interval, such as tricyclic antidepressants and anti-psychotic drugs including chlorpromazine.

Raman spectroscopy The detection of changes in light scatter to determine molecular characteristics of a tissue or crystal structure of a compound.

Reactive oxygen species (ROS) Unstable, highly reactive form of oxygen or oxygen-containing molecule (e.g. H_2O_2, superoxide anion, hydroxyl radical), which can be formed, for example, by phagocyte membrane NADPH oxidase and is involved in bacterial killing, protein degradation and acute inflammatory reactions.

RANKL (RANKL, receptor activator of nuclear factor kappa-B ligand) binds to its receptor RANK on the osteoclast surface inducing osteoclast differentiation and bone resorption leading to RA erosions, and osteoporosis. Denosumab, an anti-RANKL antibody, is a treatment for osteoporosis.

Receptor Cell surface molecule that binds specifically certain extracellular molecules, present either in solution or bound to the surface of other cells usually to elicit a biological response.

Recombinant Protein produced in vitro from eukaryotic or prokaryotic cells as a result of alteration of the gene for the protein by mutation, addition or deletion in the laboratory.

Repertoire The total spectrum of specific antigen-recognizing capacities (diversity) within the population of T or B lymphocytes.

Resolution The ability of an imaging system to separate two distinct points.

Rheumatoid arthritis (RA) Systemic autoimmune disease mediated by autoantibodies and autoreactive lymphocytes, characterized by loss of bone density and painful inflammatory destruction of joint cartilage and bone, leading to deformation of the joints and limbs.

Rheumatoid factor (RF) An inflammatory autoantibody which reacts to the Fc portion of immunoglobulin G to form immune complexes. Plasma levels of RF are high in many patients with RA and chronic inflammatory and infectious diseases.

Rilonacept (IL1-Trap) Fusion protein between the extracellular parts of the IL-1 receptor, the (IL-1 receptor) accessory protein, and human IgG which neutralizes IL-1 and is used in the treatment of several orphan inflammatory autoimmune diseases (e.g. Muckle-Wells syndrome).

Rituximab Anti-CD20 antibody targeting B cells used for treatment of B cell lymphoma and RA.

Ruxolitinib A potent inhibitor of both JAK1 and JAK2 available for the treatment of the neoplastic diseases, polycythaemia vera and myelofibrosis.

Schizophrenia A chronic and severe psychiatric disorder characterized mainly by psychosis (hallucinations and delusions). Other symptoms, such as apathy, flattened affect, and impaired cognition, can occur.

Scurvy Disease characterized by exhaustion and bleeding due to vitamin C deficiency.

Secukinumab A humanized IL-17A antibody which is used for the treatment of psoriasis, psoriatic arthritis and ankylosing spondylitis.

Selenoproteins Endogenous proteins, with or without known functions, into which selenium is incorporated as selenocysteine.

Sensitization Induction of specialized immunological memory in an individual by exposure to an antigen.

Shock Acute hypotensive response resulting from systemic immunological activation of mast cells, with release of histamine (anaphylactic shock), or from systemic bacterial infection, with stimulation of inflammatory cytokine release by bacterial lipopolysaccharide (LPS; septic shock); the latter is often associated with extensive complement activation and subsequent inflammatory necrosis of essential organs (multiple organ failure).

Sickness behavior A set of cytokine-induced behaviors in which mammals, including humans, exhibit symptoms resembling a depressive episode, including fatigue, psychomotor slowness, and appetite, sexual, and sleep changes.

Side scatter The measurement of light deflected by a cell at wide angles to the laser beam in a flow cytometer. This measurement correlates with cell granularity.

Signal transducer and activator of transcription (STAT) Family (with six members) of transcription factors that modulate the synthesis of many inflammatory genes.

Single-photon emission computed tomography (SPECT) The use of the emission of high-energy photons from a decaying radionuclide conjugated to a biologically relevant ligand to generate three-dimensional images.

Sirolimus Immunosuppressant inhibiting mTOR (mammalian target of rapamycin, the original name of sirolimus) and thereby the IL-2-dependent proliferation of T lymphocytes.

Slow-acting anti-rheumatic drugs (SAARD) Term also used for csDMARDs because of their slow onset to effect.

Solid phase A solid support used to immobilize reagents in an immunoassay (e.g., plastic well surface).

Somatic mutation A change in the genetic structure that is neither inherited nor passed to offspring.

Specialized pro-resolving mediators (SPMs) A growing class of cell signaling molecules generated from polyunsaturated fatty acids by different enzymatic oxidative metabolic pathways. They act on inflammatory and many other cells to limit the inflammatory process and promote its resolution. They also act on connective tissue, neuronal and other cells to regulate tissue damage and pain and to promote tissue repair.

Spleen tyrosine kinase (SYK) A phosphorylating enzyme which is involved in the transmission of signals from B cells to T cells.

Stem cell transplantation Therapeutic process of infusing healthy stem cells into persons who have undergone high-dose chemotherapy for one of many forms of leukemia, immunodeficiency, lymphoma, anemias, or metabolic disorders.

Stimulated excitation depletion (STED) microscopy The selective deactivation of fluorophores to reduce the area of illumination at the focal point, thereby increasing the resolution within an imaging system.

Streptavidin 52 kDa protein from the bacterium *Streptomyces*. It forms an extraordinarily strong non-covalent bond with biotin.

Stress Any external stimulus that threatens body homeostasis.

Stress-related disorder A psychiatric disorder associated with the exposure to a traumatic or stressful event.

Subunit vaccine Vaccine containing (purified) antigen(s) derived from pathogens.

Sulfapyridine An antibacterial sulfonamide which is a constituent of and metabolite of sulfasalazine which is a csDMARD.

Sulfasalazine An azo compound formed by linking sulfapyridine and mesalazine. It is split in the large intestine into its constituents which are clinically active. It is a csDMARD used for rheumatoid arthritis and psoriatic arthritis. It is also used for inflammatory bowel disease.

Super-resolution microscopy (SRM) Images taken using techniques that allow a higher resolution than those limited by the conventional diffraction limit.

Surrogate Biological marker that is considered likely to predict therapeutic benefit and is sufficiently correlated with the primary endpoint.

Sympathetic nervous system A branch of the autonomic nervous system responsible for mobilizing the body's energy and resources during times of stress or arousal.

Synthetic Man-made product/drug made from chemical sources that are approximate, but need not be identical to a related natural product. Synthetic products are often designed by computer and can be derived from natural sources that are modifed or manufactured.

Systemic lupus erythematosus (SLE) Human autoimmune disease, usually mediated by antinuclear antibodies.

T cell immunity Immune responses involving activated T lymphocytes as helper or effector (cytotoxic) cells.

T cell receptor (TCR) Heterodimeric molecule on the surface of the T lymphocyte that recognizes antigen. The polypeptide chains have a variable and a constant part and can be an α-, β- or γ-chain. The αβT cell receptor occurs on most T cells and recognizes antigenic peptides in combination with the polymorphic determinant of MHC molecules (self-MHC restricted). The γδT cell receptor occurs on a small subpopulation, e.g. in mucosal epithelium, and can recognize antigen in a non-MHC-restricted manner. The T cell receptor occurs exclusively in association with the CD3 molecule that mediates transmembrane signalling.

T cell/lymphocyte Thymus-derived lymphocytes that induce, regulate and effect specific immunological reactions stimulated by antigen, mostly in the form of processed antigen complexed with MHC on an antigen-presenting cell. Most T lymphocytes recognize antigen by a heterodimeric αβ surface receptor molecule associated with CD3 molecule mediating transmembrane signalling. Subsets include helper-inducer (Th) and suppressor-cytotoxic (Tc) cells. Th1 and Th2 subpopulations exist. Th1 cells produce interleukin (IL)-2 and IL-3, interferon (IFN)-γ, tumour necrosis factor (TNF)-α and TNF-β and granulocyte/macrophage colony-stimulating factor (GM-CSF) and function in induction of delayed-type hypersensitivity, macrophage activation and IgG2a synthesis. Th2 cells produce IL-3, IL-4 and IL-5, TNF-α and GM-CSF and function in induction of IgG1, IgA and IgE synthesis and induction of eosinophilic granulocytes. Th17 cells produce IL-17. Cytotoxic T lymphocytes differentiate from precursor to effector cytotoxic cells and subsequently kill target cells.

(Terminal) Half-life The time taken for the plasma concentration of a drug to fall by one-half during the elimination phase.

T helper (Th) cells/lymphocytes Functional subset of T lymphocytes that can help to generate cytotoxic T lymphocytes and cooperates with B lymphocytes in the production of antibody. Th1 and Th2 responses are immune responses mediated by particular types of T lymphocytes. T lymphocytes are subdivided into T helper (CD4+) and T effector/cytotoxic (CD8+) lymphocytes. CD4+ T cells assure the regulation of immune responses via the release of different patterns of cytokines, primarily IL-2 and interferon-γ that characterize Th1 responses, and IL-4, IL-5 and IL-10 that characterize Th2 responses.

T lymphocyte subsets CD8+ cytotoxic T cells recognize antigenic peptide in association with MHC class I molecules. CD8+ T cells mediate their effector functions by killing the cells presenting the relevant antigenic peptide or secreting effector cytokines. CD4+ T helper cells recognize antigenic peptide in association with MHC class II molecules. They mediate their effector functions by enhancing the persistence of antigen-stimulated T cells or through secretion of effector cytokines.

T regulatory cells (Tregs) T lymphocytes that regulate the activity and function of other T lymphocytes.

Tacrolimus Immunosuppressive drug mainly inhibiting interleukin-2 synthesis in T lymphocytes mainly used to prevent rejection of transplanted organs.

Target Molecule of interest in an analyte or a molecule at which a drug acts..

Targeted synthetic DMARDs (tsDMARDs) A group of small molecular mass DMARDs which have been designed to inhibit specific inflammatory processes.

T-cell-depleted In patients who receive myeloablative therapy and an allogeneic stem cell transplant, removal of the T cells (T-cell depletion) from the stem cell product has been used to decrease the incidence and severity of GVHD. However, T-cell depletion has also been found to significantly delay immune reconstitution and increase the rate of graft failure and tumor relapse.

Teriflunomide The active metabolite of the csDMARD, leflunomide which is used in the treatment of multiple sclerosis.

Tetracyclines A family of antibiotics which also have anti-inflammatory activity and are minor drugs in the treatment of RA.

Th cells T helper lymphocytes.

T_{H1} and T_{H2} responses Refer to particular types of T lymphocyte-mediated immune responses. T lymphocytes mediate cell-mediated (cellular) immunity. They are further subdivided into T helper ($CD4^+$) and T effector/cytotoxic ($CD8^+$) lymphocytes. $CD4^+$ T cells assure the regulation of immune responses via the release of different patterns of cytokines, primarily IL-2 and IFN-γ which characterize T_{H1} responses (mainly anti-viral and antibacterial, but also in autoimmunity), and IL-4, IL-5 and IL-10 which characterize T_{H2} responses (mainly anti-parasitic and also allergic responses).

Therapeutic drug monitoring Measurement of therapeutic drug plasma concentrations in clinical practice. Useful for drugs with a narrow therapeutic window (generally the ratio between minimum effective concentrations and the minimum toxic concentration).

Therapeutic strategy A pharmacologically based approach to the treatment of a disease based on the drug mechanism of action, pharmacokinetics, and toxicology, generally involving multiple modalities that integrate with clinical pathophysiological considerations.

Thrombocyte See Platelet

Thrombosis The process, following injury to a blood vessel or endothelium (q.v.), by which a blood clot is formed, blood loss is prevented and the injury to the vessel is initially covered. The process is initiated by the blood coagulation process (q.v.) and progresses with activation and aggregation of platelets (thrombocytes) and the formation of thrombin. This protective clot leads to activation of subsequent repair processes but may also become pathological, blocking blood flow and causing tissue ischemia, myocardial infarction or stroke.

Thromboxane (TXA_2) Lipid mediator synthesized from arachidonic acid by cycloxygenase and thromboxane synthetase with platelet activating and vasoconstricting effects.

Thymus A specialized primary lymphoid organ of the immune system located mediastinum of the thoracic cavity anterior and superior to the heart and posterior to the sternum. The thymus is seeded by early progenitor cells from the bone marrow which develop within the thymus into mature T lymphocytes.

Tissue engineering Development and manipulation of laboratory-grown molecules, cells, tissues, or organs to replace or support the function of defective or injured body parts.

Tissue residency The amount of time a drug spends in a specific tissue—generally the target tissue of interest.

Tocilizumab A humanized IL-6 antibody that binds both soluble and membrane-expressed IL-6R used for the treatment of RA.

Tofacitinib An inhibitor of JAK 1 and JAK 3. It is active orally in the treatment of RA, psoriasis and psoriatic arthritis.

Tolerance A state of unresponsiveness to antigenic stimulation, due to the absence of responding elements or the loss of capacity of existing elements to mount a reaction. Synonym for anergy.

Toll-like receptors (TLRs) Leukocyte receptors essential for microbial recognition and immune responses.

Topical Administration of a drug directly to the site of disease/inflammation (e.g., skin, mucosa) as opposed to systemic administration by which the drug reaches the inflamed tissue/site via absorption into the bloodstream. However, topically applied drugs may also reach the bloodstream in variable amounts.

Topoisomerases Enzymes that cut off one phosphate backbone chain of DNA (Type I enzymes) or both phosphate chains (Type II enzymes). These processes are essential for cellular division as they allow DNA to unwind. Anthracycline anti-cancer drugs, such as daunorubicin and doxorubicin, are inhibitors of topisomerase II.

Transactivation Stimulation of transcription by a transcription factor binding to DNA and activating adjacent proteins.

Transforming growth factor beta (TGFβ) A cytokine that is secreted by various cell types, including macrophages, and functions in cell growth, cell differentiation, apoptosis and cellular homeostasis by binding to its receptor, the TGFβ type II receptor. Multiple signalling pathways lie downstream of the receptor, including the SMAD transcription factors and cell cycle proteins.

Transrepression Reduction of transcription by a transcription factor binding to DNA and inhibiting adjacent proteins.

Tumor microenvironment The cellular and molecular composition of a tumor, which can vary greatly even throughout individual tumors. It consists of transformed cells and stromal non-transformed cells including endothelial cells, fibroblasts, adipocytes, mesenchymal stem cells, cells of the peripheral nervous system, and immune cells and is further characterized by an altered extracellular matrix and metabolite and oxygen availability. The tumor microenvironment is believed to impact all aspects of tumorigenesis.

Tumor-associated antigens Antigens that are expressed at low level in normal tissues and whose expression is substantially upregulated in tumor cells as a result of activation of oncogenic signaling cascades. They are usually expressed in a large number of patients harboring the same type of tumor and can therefore be therapeutically exploited more broadly than tumor-specific antigens. However, the risk of targeting normal tissues expressing tumor-associated antigens exists.

Tumor-specific antigens Antigens that are not expressed in normal tissues but arise due to viral infection, chromosomal translocation, or somatic mutations in tumor cells, each resulting in the production of "foreign" peptides. They often vary from patient to patient. Therefore, their therapeutic utility beyond personalized medicine is limited.

Tumour necrosis factor (TNF) General mediator of inflammation and septic shock originally described as a tumour-degrading factor induced by bacterial lipopolysaccharide. It comprises two forms, TNF-α and TNF-β, produced by monocytes/macrophages; TNF-β is also produced by T lymphocytes and natural killer cells. Tumour necrosis factor has activity similar to IL-1 and acts synergistically with IL-1. It promotes an antiviral state and is cytotoxic to tumour cells. It stimulates granulocytes and eosinophils, activates macrophages to IL-1 synthesis, stimulates B cells to proliferate and differentiate and T cells to proliferate and synthesize IL-2 receptor and IFN-γ. It induces fibroblasts to synthesize prostaglandin and to proliferate and induces fever and synthesis of acute-phase proteins. It reduces cytochrome p450 synthesis, activates endothelium, promotes adherence of neutrophilic granulocytes to endothelium and induces cell adhesion molecules like lymphocyte function-associated antigens LFA-1 and LFA-3, ICAM-1, and ELAM-1. It reduces lipoprotein lipase synthesis by adipocytes and activates osteoclasts to bone resorption.

Turbidimetry Measuring the light remaining after passage through a turbid solution.

Ultrasound Sound with a frequency in the range of 1–20 MHz which is projected through tissue with the reflected sound collected to allow image construction.

Ustekinumab A fully human IgG1κ monoclonal antibody which binds the p40 subunit common to IL-12 and IL-23. It is used for the treatment of psoriasis, psoriatic arthritis and ankylosing spondylitis.

V(D)J recombination Mechanism of genetic modification of the genomic DNA encoding the B cell and T cell antigen receptors. Through a cut-and-paste mechanism, genetic elements are pasted together, while the intervening DNA is removed to form circular excision product. This process occurs during early differentiation in the bone marrow (B cells) and thymus (T cells).

Vaccine delivery system Colloidal carrier allowing multimeric antigen presentation (i.e. containing more than one antigen per particle), aiming to increase the immunogenicity of the antigen; it may also contain adjuvants.

Vaccine Substance or group of substances which induces the immune system to respond in an antigen-specific manner to a tumor or to microorganisms, such as bacteria or viruses. A vaccine can help the body recognize and destroy cancer cells or microorganisms. It consists of a preparation of a weakened or killed pathogen, such as a bacterium or virus, or of a portion of the pathogen's structure that upon administration stimulates Ab production or cellular immunity against the pathogen.

Xenobiotic Chemical or substance that is foreign to the biological system.

Zwitterion A molecule with at least one positive and one or more negative charges. Zwitterions bearing only one positive and one negative charge have no net charge. Naturally occurring amino acids are zwitterions at physiological pH values. The simplest amino acid zwitterion is glycine whose structure is $H3N^+$-CH_2-COO^-.

Index

M. J. Parnham et al. (eds.), *Nijkamp and Parnham's Principles of Immunopharmacology*,
https://doi.org/10.1007/978-3-030-10811-3

O

P

Q

R

S

U

V

MIX
Papier aus verantwortungsvollen Quellen
Paper from responsible sources
FSC® C105338

If you have any concerns about our products,
you can contact us on
ProductSafety@springernature.com

In case Publisher is established outside the EU,
the EU authorized representative is:
Springer Nature Customer Service Center GmbH
Europaplatz 3, 69115 Heidelberg, Germany

Printed by Libri Plureos GmbH
in Hamburg, Germany